Handbook of Bioengineering

Other McGraw-Hill Handbooks and Reference Books of Interest

Baumeister, Avallone, and Baumeister MARKS' STANDARD HANDBOOK FOR MECHANICAL ENGINEERS, *8/e*

Beer and Johnston MECHANICS FOR ENGINEERS, *3/e*

Beer and Johnston MECHANICS OF MATERIALS

Flugge HANDBOOK OF ENGINEERING MECHANICS

Oden/Ripperger MECHANICS OF ELASTIC STRUCTURES, *2/e*

Shames MECHANICS OF FLUIDS, *2/e*

Shelley ENGINEERING MECHANICS

Streeter HANDBOOK OF FLUID DYNAMICS

Woodson HUMAN FACTORS DESIGN HANDBOOK

Handbook of Bioengineering

CO-EDITORS:

Richard Skalak, Ph.D.

James Kip Finch Professor of Engineering Mechanics
Department of Civil Engineering and Engineering Mechanics
Director, Bioengineering Institute
Columbia University, New York, N.Y.

Shu Chien, M.D., Ph.D.

Professor of Physiology and Cellular Biophysics
Department of Physiology and Cellular Biophysics
Director, Division of Circulatory Physiology and Biophysics
College of Physicians and Surgeons
Columbia University, New York, N.Y.

McGraw-Hill Book Company

New York St. Louis San Francisco Auckland Bogotá
Hamburg Johannesburg London Madrid Mexico
Milan Montreal New Delhi Panama
Paris São Paulo Singapore
Sydney Tokyo Toronto

Library of Congress Cataloging-in-Publication Data

Handbook of bioengineering.

Includes index.
1. Biomedical engineering. 2. Biomechanics.
I. Skalak, Richard. II. Chien, Shu.
R856.H36 1986 610'.28 86-10313
ISBN 0-07-057783-8

1234567890 DOCDOC 8932109876

ISBN 0-07-057783-8

The editors for this book were Betty Sun, Jim Halston, and Nancy Warren, the designer was Mark E. Safran, and the production supervisor was Teresa F. Leaden. It was set in Times Roman by Bi-Comp.

Printed and bound by R.R. Donnelley

Contents

Contributors

Max Anliker, Ph.D. *Professor of Biomedical Engineering, Institut fuer Biomedizinische Technik der Universitaet und der ETH Zuerich, Moussonstrasse 18, CH-8044 Zuerich, Switzerland*

Richard B. Ashman, Ph.D. *Director of Biomechanics, Department of Orthopaedic Surgery, University of Texas Health Science Center, Dallas, TX 75219*

Gertrude L. Blackshear, M.D., Ph.D. *Assistant Professor, Department of Physiology, 6-255 Millard Hall, University of Minnesota, Minneapolis, MN 55455*

Perry L. Blackshear, Ph.D. *Professor, Department of Mechanical Engineering, 125 Mechanical Engineering Building, University of Minnesota, Minneapolis, MN 55455*

Thomas R. Canfield, Ph.D. *Assistant Professor, Department of Radiotherapy, Hines Veterans Administration Hospital and Loyola University Medical Center, Hines, IL 60141*

William M. Chardack, M.D. *Associate Professor of Surgery, School of Medicine, State University of New York at Buffalo, Buffalo, NY 14226*

John C. Chato, Ph.D. *Professor of Mechanical Engineering and Bioengineering, Department of Mechanical and Industrial Engineering, University of Illinois at Urbana-Champaign, Urbana, IL 61801*

Neil S. Cherniack, M.D. *Professor of Medicine and Physiology, Department of Medicine, Case Western Reserve University, Cleveland, OH 44106*

Giles R. Cokelet, Ph.D. *Professor, Department of Radiation Biology and Biophysics, University of Rochester Medical Center, 601 Elmwood Avenue, Rochester, NY 14642*

Stephen C. Cowin, Ph.D. *Professor, Department of Biomedical Engineering, Tulane University, New Orleans, LA 70118*

Philip B. Dobrin, M.D., Ph.D. *Associate Professor, Department of Surgery, Hines Veterans Administration Hospital and Loyola University Medical Center, Hines, IL 60141*

Gary M. Drzewiecki, Ph.D. *Assistant Professor, Department of Biomedical Engineering, College of Engineering, Rutgers University, Piscataway, NJ 08854*

Janie M. Fouke, Ph.D. *Assistant Professor, Department of Biomedical Engineering, Case Western Reserve University, Cleveland, OH 44106*

Yuan-Cheng B. Fung, Ph.D. *Professor, Departments of Applied Mechanics and Engineering Sciences, and Bioengineering, M-005, University of California, San Diego, La Jolla, CA 92093*

Thomas K. Goldstick, Ph.D. *Professor, Departments of Chemical Engineering, Biomedical Engineering, and Neurobiology & Physiology, Northwestern University, Evanston, IL 60201*

J. Wallace Grant, Ph.D. *Professor, Department of Engineering Science and Mechanics, Virginia Polytechnic Institute and State University, Blacksburg, VA 24061*

Wilson Greatbatch, P.E. *Adjunct Professor of Electrical Engineering, State University of New York at Buffalo, Buffalo, NY 14226*

Alan R. Hargens, Ph.D. *Professor, Department of Surgery, Division of Orthopedics and Rehabilitation (V-151), University of California, San Diego, CA 92161*

Lowell D. Harris, Ph.D. *Director of Display Systems, Bio-Imaging Research Inc., 425 Barclay Blvd., Lincolnshire, IL 60069*

Robert M. Hochmuth, Ph.D. *Professor, Departments of Mechanical Engineering, Material Science, and Biomedical Engineering, Duke University, Durham, NC 27706*

Jay D. Humphrey, Ph.D. *Georgia Institute of Technology, School of Engineering Science and Mechanics, 225 No. Ave. N.W., Atlanta, GA 30332*

Marcos Intaglietta, Ph.D. *Professor, Department of AMES-Bioengineering, University of California, San Diego, La Jolla, CA 92093*

Marvin W. Johnson, Ph.D. *Professor, Department of Physics, California State University, Stanislaus, 801 W. Monte vista Avenue, Turlock, CA 95380*

Roger D. Kamm, Ph.D. *Associate Professor of Mechanical Engineering, Massachusetts Institute of Technology, 77 Massachusetts Ave., Cambridge, MA 02139*

Zvi Karni, Ph.D. *Deceased. Formerly, Associate Professor, Department of Biomedical Engineering, Technion-Israel Institute of Technology, Technion City, Haifa 32000, Israel*

J. Lawrence Katz, Ph.D. *Professor, Department of Biomedical Engineering, Rensselaer Polytechnic Institute, Troy, NY 12181*

W. Michael Lai, Ph.D. *Professor, Departments of Mechanical Engineering, Aeronautical Engineering and Mechanics, Rensselaer Polytechnic Institute, Troy, NY 12181*

David Landowne, Ph.D. *Associate Professor, Departments of Physiology and Biophysics, University of Miami School of Medicine, P.O. Box 016430, Miami, FL 33101*

Yoram Lanir, Ph.D. *Associate Professor, Department of Biomedical Engineering, Technion-Israel Institute of Technology, Technion City, Haifa 32000, Israel*

Murina J. Levesque, Ph.D. *Senior Research Scientist, Physiological Fluid Mechanics Laboratory, Department of Mechanical Engineering, University of Houston, Houston, TX 77004*

William D. Lew *Rehabilitation Engineering Program, Northwestern University, Chicago, IL 60201*

Jack L. Lewis, Ph.D. *Professor, Department of Civil Engineering, Bioengineering, and Orthopaedic Surgery, Northwestern University, Evanston, IL 60201*

Herbert H. Lipowsky, Ph.D. *Associate Professor, Department of Physiology and Cellular Biophysics, College of Physicians and Surgeons, Columbia University, 630 West 168 St., New York, NY 10032*

Guy Longobardo, Ph.D. *IBM Europe/Middle East/Africa Corporation, 360 Hamilton Avenue, White Plains, NY 10601*

Arthur F. Mak, Ph.D. *Assistant Professor, Department of Bioengineering, University of Pennsylvania, Philadelphia, PA 19104*

Odile Mathieu-Costello, Ph.D. *Assistant Professor in Residence, Department of Medicine M-023A, University of California, San Diego, La Jolla, CA 92093*

Thomas A. McMahon, Ph.D. *Gordon McKay Professor of Applied Mechanics and Professor of Biology, Harvard University, Pierce Hall 325, Cambridge, MA 02138*

Julius Melbin, Ph.D. *Professor, Cardiovascular Studies Unit, Departments of Bioengineering and Animal Biology, University of Pennsylvania, Philadelphia, PA 19104*

Joseph Mizrahi, Ph.D. *Associate Professor, Department of Biomedical Engineering, Technion-Israel Institute of Technology, Technion City, Haifa 32000, Israel*

David C. Mountain, Ph.D. *Associate Professor, Department of Biomedical Engineering, Boston University, 110 Cummington St., Boston, MA 02215*

Van C. Mow, Ph.D. *Clark and Crossan Professor of Engineering, Departments of Mechanical Engineering, Aeronautical Engineering and Mechanics, Rensselaer Polytechnic Institute, Troy, NY 12181*

Robert M. Nerem, Ph.D. *Professor, Department of Mechanical Engineering, University of Houston-University Park, Houston, TX 77004*

Peter Niederer, Ph.D. *Associate Professor, Institut fuer Biomedizinische Technik der Universitaet und der ETH Zuerich, Moussonstrasse 18, CH-8044 Zuerich, Switzerland*

Abraham Noordergraaf, Ph.D. *Professor, Cardiovascular Studies Unit, Departments of Bioengineering and Animal Biology, University of Pennsylvania, Philadelphia, PA 19104*

Russell L. Pimmel, Ph.D. *Professor, Department of Electrical and Computer Engineering, University of Missouri-Columbia, Columbia, MO 65211*

Aleksander S. Popel, Ph.D. *Professor, Department of Biomedical Engineering, School of Medicine, The Johns Hopkins University, 720 Rutland Ave., Baltimore, MD 21205*

Peter D. Richardson, Ph.D. *Professor, Division of Engineering and Section of Physiology, Brown University, Providence, RI 02912*

Erik L. Ritman, M.D., Ph.D. *Professor, Biodynamics Research Unit, Mayo Foundation, 200 First Street, S.W. Rochester, MN 55905*

Richard A. Robb, Ph.D. *Professor, Biodynamics Research Unit, Mayo Foundation, 200 First Street, S.W. Rochester, MN 55905*

Geert W. Schmid-Schönbein, Ph.D. *Professor, Department of Applied Mechanics and Engineering Sciences-Bioengineering, University of California, San Diego, M-005, La Jolla, CA 92093*

Albert B. Schultz, Ph.D. *Professor, Department of Mechanical Engineering and Applied Mechanics, University of Michigan, Ann Arbor, MI 48109*

Henry M. Spotnitz, Ph.D. *Professor of Surgery, 17-430 College of Physicians and Surgeons, Columbia University, 630 West 168 St., New York, NY 10032*

Charles R. Steele, Ph.D. *Professor, Department of Mechanical Engineering, Stanford University, Stanford, CA 94305*

Jean-Claude Stettler, Ph.D. *Research Scientist, Institut fuer Biomedizinische Technik der Universitaet und der ETH Zuerich, Moussonstrasse 18, CH-8044 Zuerich, Switzerland*

Kingman P. Strohl, M.D. *Associate Professor, Department of Medicine, Case Western Reserve University, Cleveland, OH 44106*

William C. Van Buskirk, Ph.D. *Professor, Department of Biomedical Engineering, Tulane University, New Orleans, LA 70118*

Donald L. Vawter, Ph.D. *Assistant Professor of Engineering Science and Mechanics, Georgia Institute of Technology, 225 No. Ave. N.W., Atlanta, GA 30332*

Andrus Viidik, M.D., Ph.D. *Professor, Department of Connective Tissue Biology, Institute of Anatomy, University of Aarhus, DK-8000 Aarhus C, Denmark*

Peter D. Wagner, M.D. *Professor, Department of Medicine M-023, University of California, San Diego, La Jolla, CA 92093*

Alvin Wald, M.D., Ph.D. *Research Associate, Technical Director, Department of Anesthesiology, College of Physicians and Surgeons, Columbia University, 630 West 168 St., New York, NY 10032*

Savio L-Y. Woo, Ph.D. *Professor of Surgery and Bioengineering, University of California, San Diego, La Jolla, CA 92093*

Allen Zelman, Ph.D. *Professor, Department of Biomedical Engineering, Rensselaer Polytechnic Institute, Troy, NY 12181*

Preface

The field of bioengineering has been developing rapidly in the last two decades. Although its precise definition is still evolving, there is a growing body of knowledge that forms the basis of this interdisciplinary subject. The purpose of this handbook is to collect, in one place, authoritative summary accounts of the various topics that comprise the field of bioengineering. The handbook is intended to serve as a reference for practicing bioengineers, medical doctors, biological scientists, and students interested in the subject. It will also serve to introduce bioengineering to other physical scientists such as physicists, mathematicians, and engineers who seek to enter the field.

The handbook is intended to be representative of the main areas of research and technology which have become known as bioengineering. The principal areas are orthopedics, cardiovascular mechanics, blood rheology, respiratory mechanics, and properties of hard and soft tissues. There are many additional topics that are not covered in this handbook, but sometimes included in the term bioengineering, such as fermentation processes, genetic engineering, and agricultural or food technology. To include all the possibilities would yield an unwieldy volume, and the present handbook is centered on aspects related to biomedical engineering.

The interdisciplinary nature and the diverse subject matter of bioengineering necessitates a collaborative effort of many individuals to present the state-of-the-art of the field. The authors of each chapter of the handbook have been chosen for their expertise in the particular fields. The technical merit of this work is entirely to the credit of the chapter authors, as the editors have served mainly in a coordinating capacity. The choice of subject content and presentation within each chapter has been the prerogative of the authors. We are most grateful and appreciative of their contributions, because they have given comprehensive accounts derived from their professional experience, condensing and presenting the best information available. Their insights define the developing field of bioengineering.

The handbook is organized to present the basic tissue properties first and proceeds to more applied subjects in the later chapters. Properties of soft tissues, bone, tendon, ligaments, muscle, skin, and blood are covered in the early chapters. Later chapters cover models of the arterial system, the microcirculation, lung mechanics, cochlear mechanics, joint replacement, and other applications. It is the hallmark of bioengineering that quantitative, mathematical, and computational techniques such as those used in other disciplines of engineering are brought to bear on biomedical problems, both in the interpretation of biological phenomena and in the design of medical devices.

While the accounts in the handbook are as mathematical as the subject demands, sufficient descriptive background has been included so that nonbiologists can follow the medical applications and nonengineers can appreciate the physical principles involved.

It is anticipated that this handbook will be useful to students and faculty in the many departments of bioengineering that have been established at engineering and medical schools as a reference in lecture courses and as a starting point in research. It is also intended to be useful to bioengineers in industry, responsible for research and development of medical instrumentation and devices. It should also be helpful to biologists and physicians whose research or clinical practice calls for the acquisition of new instruments and in the quantitive modeling of research results. As medical practice becomes increasingly quantitative, bioengineering aspects become more important to both research and clinical applications.

Many people have contributed to the development of this handbook, particularly the many authors who have given it substance. It has been a pleasure to work with Betty Sun and Jim Halston of McGraw-Hill who have given us unfailing cooperation during the entire process of producing this volume. We also acknowledge with thanks, the very competent and thorough assistance of Louis Soslowsky in proofreading and coordinating manuscripts at Columbia University. His help has facilitated the progress of the project greatly.

For the omissions or imbalance of the material covered, the editors take responsibility and will appreciate comments from readers as to topics or types of coverage that may be desirable in future editions.

Richard Skalak
Shu Chien

CO-EDITORS

Columbia University
New York

CHAPTER 1

Mechanics of Soft Tissues

Yuan-Cheng B. Fung
University of California, San Diego
La Jolla, California

INTRODUCTION

In this chapter the general features of the mechanics of soft tissues are presented. It will be shown that the mechanical properties of soft tissue are related to its structure. Moreover, the mechanical properties of an organ depend not only on its tissue, but also on its geometry and relationship to the neighboring organs. A typical example is the blood vessel. The capillary blood vessels of the mesentery are "rigid," whereas those in the bat's wing, for example, are "distensible." The capillaries of the lung can be both rigid and distensible. They are rigid in the plane of the capillary network, and compliant in the direction perpendicular to this plane. The stress-strain relationship of the systemic arteries is highly nonlinear, showing stiffening exponentially with increasing strains; yet that of the pulmonary arteries in the lung is linear. The systemic veins are easily collapsible; yet the pulmonary veins in the lung are not: they remain patent when the blood pressure falls below the alveolar gas pressure. These differences in mechanical properties arise because of the interaction between the blood vessels and the tissues surrounding the vessels. These differences in the mechanical properties of the blood vessels in different organs produce significant, sometimes dramatic, effects on the blood circulation of different organs.

Similarly, the mechanical properties of each tissue depend not only on its chemical composition but also on its structural or ultrastructural details. Two tissues made of the same proportions of collagen, elastin, and ground substances may behave quite differently depending on how these basic elements are put together.

More details about the properties of muscle, blood vessels, skin, uterus, lung, kidney, cochlea, spine, bone, cartilage, tendons, and ligaments are presented in separate chapters of this handbook.

1.1 PSEUDOELASTICITY

Soft tissues are pseudoelastic; that is, they are not elastic, but under periodic loading and unloading each tissue will have a steady-state stress-strain relationship which is not very sensi-

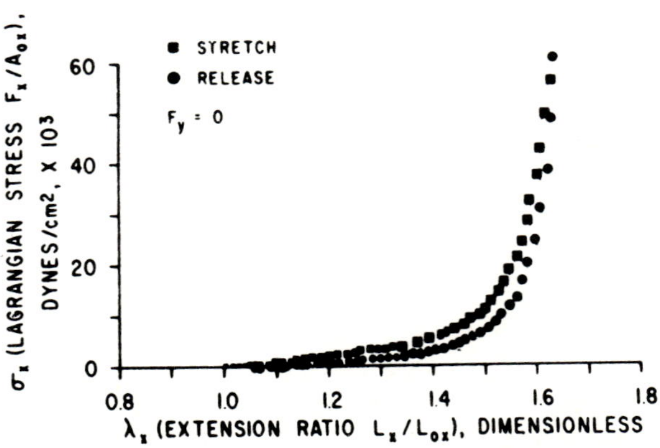

FIG. 1.1 A typical stress-strain curve for uniaxial loading. Every fourth data point is plotted. Note that the unloading curve is different from the loading curve, showing the existence of hysteresis. (*From Vawter, Fung, and West, 1978, by permission.*)

tive to strain rate. For example, Fig. 1.1 shows the stress-strain relationship of the lung tissue of the dog (with the airspace filled with saline so that the surface tension acting on the alveolar walls is very small) subjected to biaxial loading. After a number of cycles of loading and unloading, a repeatable stress-strain loop as shown in Fig. 1.1 was obtained. The existence of the loop shows that the tissue is viscoelastic and not elastic. But since the loop is repeatable, we can treat the loading and unloading curves separately and borrow the method of the theory of elasticity to describe the mechanical properties. Hence the term *pseudoelasticity*. Figure 1.2

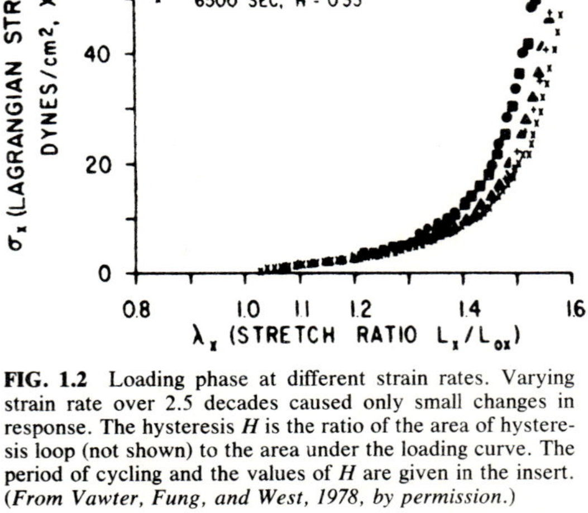

FIG. 1.2 Loading phase at different strain rates. Varying strain rate over 2.5 decades caused only small changes in response. The hysteresis H is the ratio of the area of hysteresis loop (not shown) to the area under the loading curve. The period of cycling and the values of H are given in the insert. (*From Vawter, Fung, and West, 1978, by permission.*)

shows the stress-strain relationship of the same lung tissue in *loading* at different strain rates. Each cycle was done at a constant rate. The period of each cycle is noted in the figure. It is seen that over a 360-fold change in strain rate there was only a minor change in the stress-strain relationship. The *hysteresis H,* defined as the ratio of the area of the hysteresis loop divided by the area under the loading curve, is also noted in Fig. 1.2. H is seen to be variable, but its variation with strain rate is not large. Similar experience is encountered with other tissues. Records of skeletal and cardiac muscles, ureter, teniae coli, arteries, veins, pericardium, mesentery, bile duct, skin, tendon, elastin, cartilage, and other tissues show similar characteristics. Typically, in a 1000-fold change in strain rate, the stress at a given strain in a loading (or unloading) process does not change by more than a factor of 2.

The features shown in Figs. 1.1 and 1.2 may be described by saying that living soft tissues are *nonlinearly pseudoelastic*. The stress-strain relationship is nonlinear, the viscoelasticity is pseudoelastic; hysteresis may be sizable, but it varies only mildly over a wide range of strain rates.

1.2 NONLINEAR STRESS-STRAIN RELATIONSHIP

Treating the loading and unloading curves separately, we can borrow the method of elasticity to describe the stress-strain relationship. The simplest way to represent the pseudoelasticity of a nonlinear material is to introduce a *pseudoelastic potential* (or *strain energy function*) $\rho_0 W$, which is a function of Green's strain components E_{ij}, symmetric with respect to E_{ij} and E_{ji}. The partial derivatives of $\rho_0 W$ with respect to E_{ij} gives the corresponding Kirchhoff stress components S_{ij}. W is defined for a unit mass of the tissue, and ρ_0 is the density of the tissue in the initial state; hence $\rho_0 W$ is the strain energy per unit initial volume. Thus

$$S_{ij} = \frac{\partial \rho_0 W}{\partial E_{ij}} \qquad (i, j = 1, 2, 3) \tag{1.1}$$

If the material is incompressible (volume does not change), then it can take on a *pressure* that is independent of the deformation of the body. In that case a pressure term should be added to the right-hand side of Eq. (1.1). The value of the pressure (as in water) can vary from point to point, and it can be determined only when the equations of motion, continuity, constitutivity, and boundary conditions are all satisfied.

An example of pseudoelastic potential for arteries and veins is the following (Chuong and Fung, 1983):

$$\rho_0 W = Ce^Q \tag{1.2}$$

where

$$Q = a_1E_1^2 + a_2E_2^2 + a_3E_3^2 + 2a_4E_1E_2 + 2a_5E_2E_3 + 2a_6E_3E_1 \tag{1.3}$$

Here the vessel is assumed cylindrical and the coordinate axes x_1, x_2, x_3 are pointing in the circumferential, longitudinal, and radial directions, respectively; and $E_1 = E_{11}$, $E_2 = E_{22}$, $E_3 = E_{33}$, whereas E_{ij} $(i \neq j)$ is assumed to be zero. The constants C, a_1, . . ., a_6 are the material constants that characterize the artery. For simplicity, cylindrical symmetry is assumed in this expression; otherwise one would have to generalize the quadratic form Q to include all six strain components. If the vessel wall can be considered homogeneous, then the constants C, a_1, . . ., a_6 are independent of the location in the blood vessel. If, in addition, the arterial wall material is assumed to be incompressible, then the condition of incompressibility

$$(1 + 2E_1)(1 + 2E_2)(1 + 2E_3) = 1 \tag{1.4}$$

may be introduced through a lagrangian multiplier H. We define a new pseudoelastic function

$$\rho_0 W^* = Ce^Q + H[(1 + 2E_1)(1 + 2E_2)(1 + 2E_3) - 1] \tag{1.5}$$

so that

$$S_i = \frac{\partial(\rho_0 W^*)}{\partial E_i} \qquad (i = 1, 2, 3) \tag{1.6}$$

It is well known that H has the significance of a hydrostatic pressure which can assume an arbitrary value. The true value is determined by the equations of motion (or equilibrium) and boundary conditions.

For a body subjected to small changes in strains, the range of the incremental strains may be chosen to be so small that the relationship between the incremental stresses and strains is linear. This linear relationship is Hooke's law, for which the familiar material constants are the incremental Young's modulus and incremental shear modulus. For soft tissues a general feature implied by Eq. (1.2) is that the incremental moduli increase with increasing stresses.

1.3 VISCOELASTICITY

When a tissue is stretched suddenly to a new dimension which is then held constant, it will show the phenomenon of stress relaxation; i.e., the stress gradually decreases with increasing time.

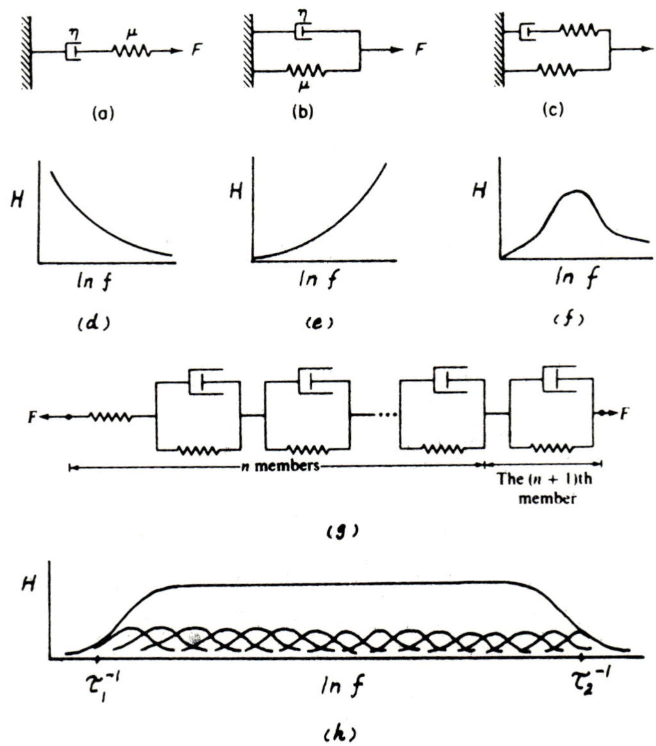

FIG. 1.3 Several models of viscoelasticity and their corresponding hysteresis-frequency relationships: (*a*) Maxwell model, (*b*) Voigt model, (*c*) Kelvin model; (*d*), (*e*), and (*f*) show hysteresis of models (*a*), (*b*), and (*c*), respectively. The hysteresis H is defined as the ratio of the area of hysteresis loop to the area under the loading curve. The abscissa is frequency in log scale. (*g*) is a generalized model for living soft tissues; (*h*) is the corresponding hysteresis versus frequency curve. Each member of the generalized model contributes a small bell-shaped curve. The sum of these small bell-shaped curves yields a curve of H which is almost constant over a wide range of frequencies. (See text for explanation.)

The degree of relaxation depends on the tissue. In time a ureter can relax off almost all the stress, whereas an artery can relax off only 20 to 30 percent, and a ligament nuchae relaxes off only a few percent. When the tissues are stressed suddenly by a step function and then the stress is kept constant, the tissue creeps; i.e., its strain increases with increasing time. The degree of creep also depends on the tissue but is usually fairly small in tendons, ligaments, skin, and arteries.

Under cyclic loading, these tissues show the phenomenon of hysteresis. As shown in Figs. 1.1 and 1.2, the hysteresis loops of soft tissues are relatively insensitive to strain rate.

These features—relaxation, creep, hysteresis, and strain rate insensitivity—can be condensed into a mathematical formulation called *quasi-linear viscoelasticity theory* (Fung, 1972). In this theory, the instantaneous stress response to strain is called the "elastic stress." The relationship between the elastic stress and strain is nonlinear. The stress at any given time is, however, related to the entire past history of the elastic stress. This history dependence is linear. The present stress is obtained by multiplying the past elastic stress increment with a relaxation function and integrating the product convolutionally from the beginning of time to the present. The relaxation function has a *continuous relaxation spectrum,* the meaning of which will be explained presently in the following paragraphs.

One of the two most popular models of viscoelasticity is the Maxwell model of a spring in series with a dashpot. Another is the Voigt model with a spring and a dashpot in parallel. A third is the Kelvin model, which is a combination of a spring in parallel with a Maxwell body. See Fig. 1.3*a* to *c*. None of these models can represent a soft tissue because, when a material represented by any one of these models is subjected to a cyclic strain, the hysteresis will not be insensitive to strain rate. As frequency increases, the dashpot in the Maxwell body will move less and less at the same load so that damping decreases with frequency. On the other hand, as frequency increases, the dashpot in the Voigt body will take up more and more of the load so that the damping increases with frequency. See Fig. 1.3*d* and *e*. For the Kelvin body there exists a characteristic frequency at which the damping is a maximum (see Fig. 1.3*f*). None of these has the feature of a nearly constant damping, independent of frequency, as soft tissues do.

A model suitable for the soft tissue is shown in Fig. 1.3*g*; it has an infinite number of springs and dashpots. In the corresponding hysteresis diagram shown in Fig. 1.3*h*, there are an infinite number of bell-shaped curves which add up to a continuous curve of nearly constant height over a very wide range of frequencies. In this situation, we say that the soft tissue has a *continuous relaxation spectrum*. The two ends of the spectrum, marked by frequencies τ_1^{-1} and τ_2^{-1} in Fig. 1.3*h*, define two characteristic times τ_1 and τ_2 which can be determined from experimental data. (See Fung, 1972, or Fung, 1981, pp. 232 ff., for mathematical details.) Tanaka and Fung (1974) have found τ_1 and τ_2 for various arteries of the dog: τ_1 lies in the range of several hundred to thousands of seconds; τ_2 lies in the range of 0.05 to 0.36 s. Chen and Fung (1973) showed that for the mesentery τ_1, τ_2 are 1.869×10^4 and 1.735×10^{-5} s, respectively. Woo et al. (1979) have found that for the cartilage $\tau_2 = 0.006$ s, $\tau_1 = 8.38$ s.

1.4 WHY DIFFERENT BLOOD VESSELS BEHAVE SO DIFFERENTLY

The aorta and thoracic arteries have nonlinear stress-strain curves, as indicated in Eq. (1.2). The pulmonary arteries and veins, in contrast, have linear pressure-diameter relationships, as shown in Fig. 1.4. The capillary blood vessels of the mesentery appear to be rigid—that is, without measurable change in diameter when blood pressure changes from 0 to 100 mmHg (Baez et al., 1960; Fung et al. 1966). However, the capillary blood vessels in the lung are very distensible. Figure 1.5 shows the variation of the thickness of the capillary sheet (the interalveolar septa) with the transmural pressure ΔP (blood pressure minus alveolar gas pressure). The pulmonary capillaries are closely knit into a sheetlike dense network which occupies about 90 percent of the total space in the interalveolar septa. Each sheet is exposed to gas on both sides. Figure 1.5 shows that the thickness of the pulmonary capillary sheets of the cat increases linearly with increasing transmural pressure at a rate of 0.22 μm/cmH_2O when ΔP is positive. But when ΔP is negative, the thickness quickly drops to zero. There is a sudden change of

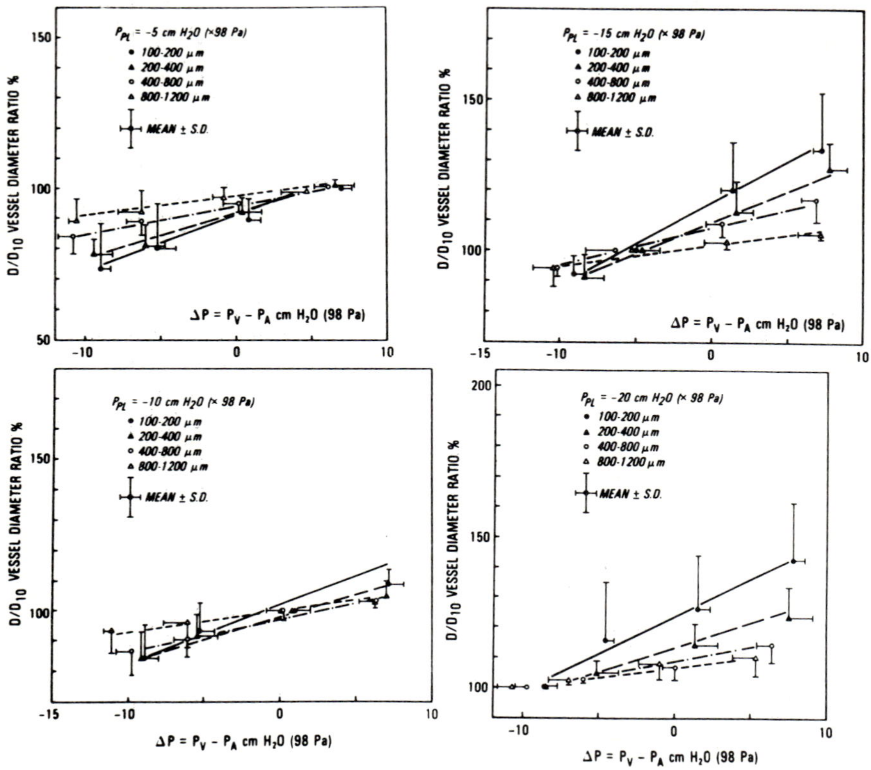

FIG. 1.4 The distensibility of pulmonary veins subjected to positive and negative $p_v - p_A$. The vessel diameter is normalized against its value when $p_v - p_{pL}$ is 10 cmH$_2$0, at which the vessel cross section is circular. The points are classified according to their diameter at $p_v - p_{pL} = 10$. (*From Yen and Foppiano, 1981, by permission.*)

thickness at $\Delta P = 0$. When ΔP tends to 0 from the positive side, the limiting value of the sheet thickness is 4.28 μm for the cat. When $\Delta P < -1$ cmH$_2$O, the capillaries are all collapsed. Why do the capillaries of the lung behave so differently from those of the peripheral circulation?

There is another property of the blood vessels that has an important physiological effect: the stability of the vessel when the external pressure exceeds the internal pressure. We have seen in Fig. 1.5 that the pulmonary capillaries collapse when $\Delta P < 0$. But we know that peripheral capillaries do not collapse when blood pressure falls below tissue pressure (Fung et al., 1966). On the other hand, it is common knowledge that peripheral veins and venae cavae collapse when the blood pressure falls below the pressure in the surrounding media. But the pulmonary veins and venules do not collapse when the airway pressure exceeds the blood pressure (see Fig. 1.5 and Fung et al., 1983). Why do these vessels behave so differently when their composition and histology are very similar?

The answer is that the properties of a blood vessel depend not only on the intrinsic properties of the blood vessel wall but also on the properties of neighboring tissues. The peripheral capillaries in the mesentery are embedded in a gel. Since a gel behaves as a solid, a capillary blood vessel embedded in it behaves as a tunnel in solid earth. That is why the capillary is so rigid with respect to blood pressure and so stable with respect to compression (Fung et al., 1966). On the other hand, a pulmonary capillary is exposed to alveolar gas which provides little

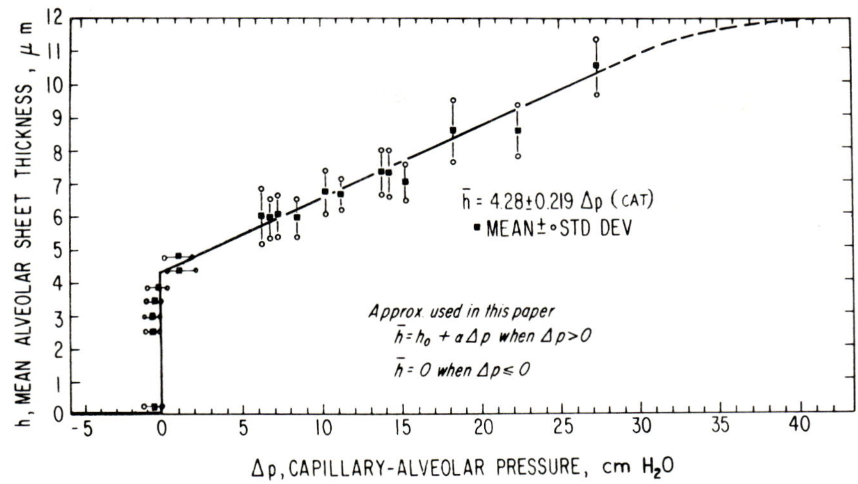

FIG. 1.5 Variation of the thickness of the pulmonary capillary sheet of the cat with the transmural pressure ΔP = capillary blood pressure minus the alveolar gas pressure. (*From Fung and Sobin, 1972, by permission.*)

resistance to deformation, and as a consequence the pulmonary capillaries are distensible and collapsible (Fung and Sobin, 1972).

If we compare a pulmonary artery with a peripheral artery, we see that the latter is an isolated vessel, whereas the pulmonary vessel is embedded in the lung parenchyma. The lung tissue (alveolar walls) pulls on the pulmonary blood vessel wall and keeps it patent when the alveolar gas pressure exceeds the blood pressure. The tissue stress influences the vessel deformation so much that a nearly linear pressure-diameter relationship (Fig. 1.4) is obtained. If the pulmonary artery were isolated, it would most likely behave in the same way as the peripheral vessels.

The same reasoning explains the difference between the incollapsibility of pulmonary veins versus the collapsibility of peripheral veins: the pulmonary veins have the support of the lung parenchyma while the peripheral veins have little support from their neighbors.

1.5 STRUCTURE, RHEOLOGY, AND FUNCTION: AN EXAMPLE OF THE PULMONARY BLOOD FLOW

It is often said that the function of an organ depends on its structure. However, the function of an organ depends on both its structure *and* its rheological properties. One example will suffice to illustrate this point: the blood flow in the lung. First, if the radius of a cylindrical blood vessel, a, varies linearly with the transmural pressure ΔP in such a way that

$$a = a_0 + \alpha\, \Delta P \tag{1.7}$$

where α is the compliance constant, a_0 is the radius when $\Delta P = 0$, and ΔP is the difference of the blood pressure and the pressure outside the blood vessel, then the flow in the vessel is given by Fung (1984) to be

$$\dot{Q} = \frac{\pi}{40\mu\alpha L}\left[(a_0 + \alpha\, \Delta P_{\text{entry}})^5 - (a_0 + \alpha\, \Delta P_{\text{exit}})^5\right] \tag{1.8a}$$

or
$$\dot{Q} = \frac{\pi}{40\mu\alpha L}(a^5_{\text{entry}} - a^5_{\text{exit}}) \tag{1.8b}$$

Here, L is the length of the blood vessel between the entry and exit sections, μ is the viscosity of the blood, and the subscripts *entry* and *exit* refer to the entry and exit sections of the vessel. The power 5 is a powerful factor. If the compliance is large, the effect of vessel elasticity on flow can be very large.

For the lung, the pulmonary capillary blood vessels are arranged in two-dimensional sheets, and the total flow is given by the following formula (Fung and Sobin, 1972):

$$\dot{Q} = \frac{A}{4\mu k f \bar{L}^2 \alpha}(h^4_{\text{entry}} - h^4_{\text{exit}}) \tag{1.9}$$

Here h denotes the alveolar sheet thickness, which, as shown in Fig. 1.5, varies linearly with ΔP when $\Delta P > 0$:

$$h = h_0 + \alpha\,\Delta P \tag{1.10}$$

and is zero when ΔP is negative. α is the compliance constant of the alveolar capillaries, A is the area of the alveolar wall, μ is the viscosity of blood, and k (about 12) and f (about 2 to 3) are numerical factors which depend on the sheet geometry. $\bar{L}$ is the mean path length of blood between the entry and exit sections. In the lung, the entry section is located at a pulmonary arteriole; the exit is located at a pulmonary venule.

Equation (1.9), together with the knowledge that pulmonary veins and venules do not collapse when the pulmonary alveolar gas pressure exceeds the blood pressure, explains an important phenomenon known as *waterfall*, or *sluicing*. To explain this term let us consider a lung perfused under the condition of a fixed airway pressure P_A and a fixed arterial pressure p_a, while the left atrium pressure p_v is varied. Let the flow be measured when p_v is gradually reduced. The result is shown in Fig. 1.6. When p_v is decreased below p_a, flow starts, and it increases with decreasing p_v. But an upper limit is reached when p_v approaches P_A. From there

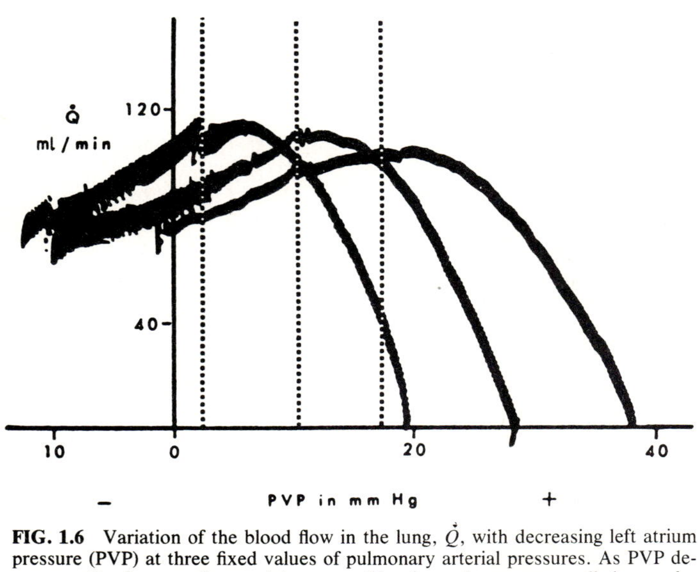

FIG. 1.6 Variation of the blood flow in the lung, $\dot{Q}$, with decreasing left atrium pressure (PVP) at three fixed values of pulmonary arterial pressures. As PVP decreases, the flow reaches a plateau, resulting in a phenomenon called *vascular waterfall*. (*From Permutt et al., 1962, by permission.*)

on further decrease of p_v does not increase the flow; a flow limitation is reached (see Permutt et al., 1962, and Fung, 1984). This is analogous to a waterfall whose volume flow rate does not depend on the height of the fall. The explanation lies in Eq. (1.9): when p_v tends to P_A, the h_{exit} tends to zero, and the flow depends entirely on h_{entry}, i.e., ΔP at the pulmonary arteriole. Thus, when $p_v < P_A$, the last term in Eq. (1.9) is either negligible or is zero, and $\dot{Q}$ remains constant for all values of venous pressure p_v if $p_v < P_A$.

The similarity of Eqs. (1.8) and (1.9) suggests that other flow-limitation phenomena can be similarly explained, including the phenomena of flow limitation in the airway in forced expiration, and flow limitation in micturation due to action of muscle sphincters in the urethra.

If Eqs. (1.8) and (1.9) are used for every generation of the blood vessels in the lung, then a synthesis blood flow in the whole lung can be obtained (Fung, 1984; Zhuang et al., 1983). Such a synthesis includes the viscous losses due to disturbed flow at points of bifurcation of blood vessels, turbulence, the secondary flow due to curvature of the vessel, as well as the effect of dynamic pressure change due to difference in velocity of flow in different generations, and the inertial force due to acceleration or deceleration (Fung, 1984). The same method can be used, of course, for other organs. The point is that rheology and structure together define the function of the organ.

1.6 ULTRASTRUCTURE

All the properties of the organs and tissues have an ultrastructural basis. For soft tissues, the mechanical properties can be analyzed ultimately in terms of the networks of collagen and elastin fibers, the muscle and other cells, the ground substances, and the interstitial fluids. If the ultrastructure is known, then we should be able to deduce theoretically the mechanical properties of the tissue. We have shown above, however, that at a higher level, from tissues to organs, we must consider the geometric structure and interaction of various components. By analogy, at a lower level, from fibers, cells, and ground substances to tissues, we must also consider the geometric structure and interaction of these components.

1.7 NEED FOR A DATA BASE

One of the most crucial steps in the development of biomechanics is the identification of the constitutive equations of the tissues involved, that is, a concise mathematical description of the mechanical properties. That is so especially from the point of view of applying biorheology to biology, physiology, and medicine. Biology and physiology are very extensive fields. Subjects in these fields are usually complex, with many factors interacting. Rheology participates as one of the factors; and in order to analyze its interaction with other factors, one needs to express it concisely and precisely. This is why mathematical expressions are needed. They simplify and condense, with known precision.

Formulas such as Eqs. (1.1) to (1.6) represent many curves, and a few parameters such as the coefficients $c, a_1, \ldots, a_6$ characterize the material properties precisely. Reams of data sheets can be condensed into a table of these numbers. States of health and disease can be identified from these numbers. Effectiveness of drugs or other intervention can be similarly evaluated. Therefore, for practical applications, obtaining these numbers is the objective of biorheological measurements.

Constitutive equations are required so that biomechanical problems can be formulated as mathematical problems and solutions can be definitive. Without constitutive equations, biomechanics would remain qualitative in character. Only when the full power of biomechanics to predict the behavior of an animal when certain parameters are changed is developed can biomechanics be of real service to medicine, surgery, health preservation and improvement, sports, welfare, and quality of life of humans and animals. The need to complete the data base for constitutive equations and material constants may be identified as the most important task for the immediate future.

1.8 RESEARCH TRENDS

At the present time, the constitutive equations for many biological materials are still unknown. The most important unknowns are those of the muscles (skeletal, heart, and smooth muscles). Without a constitutive equation for the active contraction of the heart muscle we cannot build a rigorous theory of cardiology. Without the constitutive equations for the vascular smooth muscle, we cannot develop a mathematical theory of circulation regulation and control. Similarly, for theories of other internal organs we need the constitutive equations of their smooth muscles. However, reliable constitutive equations for these muscles are not yet available.

Soft tissue mechanics has roots in ancient literature, but vigorous development took place only in the last couple of decades. Currently there are several directions in which the study of soft tissue mechanics is developing. One direction seeks to go further and further into the microstructural world for basic understanding of the macroscopic mechanical properties of a tissue. Another direction consists of the collection of data to enrich the field. Further, two other kinds of constitutive equations are being developed: those describing transport of matter in tissue, and those describing growth or resorption of a tissue in response to a change of the state of stress in the tissue. The importance of these topics is widely recognized, because living materials have a major property that has no counterpart in all other materials in engineering: namely, the ability to grow and resorb. Living tissues can change their dimensions, or, sometimes, their mechanical properties. These changes are often regulated by stresses as well as by biochemical processes. The effect is that the strain and stress fields in a living organism can change with time. Because of this effect, we cannot be certain about the residual stresses in organs (i.e., the stresses that remain when all the external loads are removed). Unless a new principle is found which tells us how residual stresses in tissues or an organ can be found, we shall not be able to determine the stress distribution in organs. The search for such a principle is of supreme interest.

The real activity in the field of biomechanics is the search for new biological principles concerning how living organisms work. Biomechanics has a place in the broad field of biology because many principles cannot be clarified without a thorough understanding of the mechanics involved. For example, atherosclerosis is a serious concern to many people. We would like to know what causes atherosclerosis and how to treat it. In spite of a tremendous effort in research and many claims to the contrary, atherosclerosis remains a problem and a mystery. In order to advance, further research in biomechanics is needed; atherogenesis cannot be understood without a thorough understanding of the blood vessels and the blood flow. As another example, in locomotion many muscles act on bone and joints. These muscles form a highly redundant structural system. How can the forces in these muscles be evaluated? We need a biological principle.

It is safe to predict that biomechanics will continue to develop until its potential to medicine and biology is fully realized. Understanding physiology is a big attraction in itself. Predicting the consequences of artificial interventions will be a legitimate objective of this science. Developing methods of preserving and improving health will benefit humanity. To achieve these and other objectives, systematic collection of the needed data and searching for new principles are of equal importance.

1.9 A BRIEF SURVEY OF BOOK LITERATURE

Books about physiology written with soft tissue mechanics held in clear view include Burton (1965), Wetterer and Kenner (1968), Bergel (1972), Fung et al. (1972), Middleman (1972), Lightfoot (1974), Caro et al. (1978), Pedley (1980), Patel and Vaishnav (1980), Poorinya and Kasyanov (1980), Brankov (1981), and Knets et al. (1980).

These books attest to a high state of development in biomechanics, but much of the basic data on morphology and rheology are not covered, and the theory remains nonspecific; no detailed prediction of the function of any specific organ is possible without additional collection of data. A systematic collection of basic data is an essential task for the future.

The term *biorheology* was first proposed by Alfred L. Copley (1960, 1965). Hartmann (1974) edited the proceedings of a conference on the rheology of connective tissues. For biorheology in general, consult Frey-Wyssling (1952), Scott Blair (1974), Syoten Oka (1974, 1981), Evans and Skalak (1980), and Fung (1981).

Mechanics of hard tissues and soft tissues are generally inseparable in the consideration of physiological problems. For example, orthopedics, locomotion, sports, and surgery involve soft tissues as much as bone. For literature on these subjects, see other chapters of this handbook.

Original papers are published in journals. The books referred to above may serve as guides to literature, but one must consult the originals.

ACKNOWLEDGMENT

The support of the National Institutes of Health (NIH) through grants HL-26647 and HL-07089 and of the National Science Foundation (NSF) through grant MEA 79-10560 is gratefully acknowledged.

REFERENCES

Baez, S., Lamport, H., and Baez, A. (1960). Pressure effects in living microscopic vessels. In A. L. Copley and G. Stainsby (eds.), *Flow Properties of Blood and Other Biological Systems*. Pergamon Press, Oxford, pp. 122–136.

Bergel, D. H. (1972). *Cardiovascular Fluid Dynamics*. Academic Press, New York, vols. 1, 2.

Brankov, G. (1981). *Osnovi Biomechaniki* [Basic Biomechanics]. Izdatelstvo "MIR," Moscow.

Burton, A. C. (1965). *Physiology and Biophysics of the Circulation*. Year Book, Chicago.

Caro, C., Pedley, T. J., Schroter, R. C., and Seed, W. A. (1978). *The Mechanics of the Circulation*. Oxford University Press, Oxford.

Chen, Y. L., and Fung, Y. C. (1973). Stress-history relations of rabbit mesentery in simple elongation. In *1973 Biomechanics Symposium*. ASME Pub. No. AMD-2, American Society of Mechanical Engineers, New York, pp. 9–10.

Chuong, C. J., and Fung, Y. C. (1983). Three-dimensional stress distribution in arteries. *J. Biomech. Eng., 105,* 268–274.

Copley, A. L. (ed.). (1965). *Symposium on Biorheology*. Interscience, New York.

Copley, A. L., and Stainsby, G. (1960). *Flow Properties of Blood and Other Biological Systems*. Pergamon Press, New York.

Evans, E. A., and Skalak, R. (1980). *Mechanics and Thermodynamics of Biomembranes*. CRC Press, Boca Raton, Fla.

Frey-Wyssling, A. (1952). *Deformation and Flow of Biological Systems,* North Holland, Amsterdam.

Fung, Y. C. (1972). Stress-strain-history relations of soft tissues in simple elongation. In Y. C. Fung (ed.), *Biomechanics: Its Foundations and Objectives*. Prentice-Hall, Englewood Cliffs, N.J., pp. 181–208.

Fung, Y. C. (1981). *Biomechanics: Mechanical Properties of Living Tissues*. Springer-Verlag, New York.

Fung, Y. C. (1984). *Biodynamics: Circulation*. Springer Verlag, New York.

Fung, Y. C., Fronek, K., and Patitucci, P. (1979). On pseudo-elasticity of arteries and the choice of its mathematical expression. *Am. J. Physiol., 237*(5), H620–H631.

Fung, Y. C., Perrone, N., and Anliker, M. (1972). *Biomechanics: Its Foundations and Objectives*. Prentice-Hall, Englewood Cliffs, N.J.

Fung, Y. C., and Sobin, S. S. (1972). Pulmonary alveolar blood flow. *Circ. Res., 30,* 470–490.

Fung, Y. C., Sobin, S. S., Tremer, H., Yen, M. R. T., and Ho, H. H. (1983). Patency and compliance of pulmonary veins when airway pressure exceeds blood pressure. *J. Appl. Physiol.: Respir., Envir. Exercise Physiol., 54*(6), 1538–1549.

Fung, Y. C., Zweifach, B. W., and Intaglietta, M. (1966). Elastic environment of the capillary bed. *Circ. Res., 19,* 441–461.

Hartmann, F. (ed.). (1974). *Biopolymere und Biomechanik von Bindegewebssystemen.* Springer-Verlag, Heidelberg.

Knets, E. V., Pfafrod, G. O., and Saulgozis, U. J. (1980). *Deformation and Failure of Solid Biological Tissues* [*Deformerovanie e Razryshenie Tverdi Biologichskeh Tkanee*]. Riga, "Zenatne."

Lightfoot, E. N. (1974). *Transport Phenomena and Living Systems.* Wiley, New York.

Middleman, S. (1972). *Transport Phenomenon in the Cardiovascular System.* Wiley, New York.

Oka, Syoten. (1974). *Rheology-Biorheology* (in Japanese). Syokabo, Tokyo.

Oka, Syoten. (1981). *Cardiovascular Hemorheology.* Cambridge University Press, Cambridge.

Patel, D. J., and Vaishnav, R. N. (1980). *Basic Hemodynamics and Its Role in Disease Process.* University Park Press, Baltimore.

Pedley, T. J. (1980). *The Fluid Mechanics of Large Blood Vessels.* Cambridge University Press, London.

Permutt, S., Bromberger-Barnea, B., and Bane, H. N. (1962). Alveolar pressure, pulmonary venous pressure and the vascular waterfall. *Med. Thorac., 19,* 239–260.

Poorinya, B. A., and Kasyanov, V. A. (1980). *Biomechanika Kropnich Krovenosnich Sosoodov Cheloveka* [*Biomechanics of Large Blood Vessels of Man*]. Riga, "Zenatne."

Scott Blair, G. W. (1974). *An Introduction to Biorheology.* Elsevier, Amsterdam.

Sobin, S. S., Fung, Y. C., Tremer, H. M., and Rosenquist, T. H. (1972). Elasticity of the pulmonary alveolar microvascular sheet in the cat. *Circ. Res., 30:* 440–450.

Tanaka, T. T., and Fung, Y. C. (1974). Elastic and inelastic properties of the canine aorta and their variation along the aortic tree. *J. Biomech., 7,* 357–370.

Vawter, D. L., Fung, Y. C., and West, J. B. (1978). Elasticity of excised dog lung parenchyma. *J. Appl. Physiol., 45:* 261–269.

Vawter, D. L., Fung, Y. C., and West, J. B. (1979). Constitutive equation of lung tissue elasticity. *J. Biomech. Eng., Trans. ASME 101*(1), 38–45.

Wall, R. J., Sobin, S. S., Karspeck, M., Lindal, R. G., Tremer, H. M., and Fung, Y. C. (1981). Computer-derived image compositing. *J. Appl. Physiol., 51*(1), 84–89.

Wetterer, E., and Kenner, T. (1968). *Grundlagen der Dynamik des Arterienpulses,* Springer-Verlag, Berlin, Heidelberg.

Woo, S. L-Y., Simon, B. R., Kuei, S. D., and Akeson, W. H. (1979). Quasi-linear viscoelastic properties of normal articular cartilage. *J. Biomech. Eng., Trans. ASME, 102,* 85–90.

Yen, M. R. T., and Foppiano, L. (1981). Elasticity of small pulmonary veins in the cat. *J. Biomech. Eng., Trans. ASME, 103,* 38–42.

Zhuang, F. Y., Fung, Y. C., and Yen, R. T. (1983). Analysis of blood flow in cat's lung with detailed anatomical and elasticity data. *J. Appl. Physiol.: Respir., Envir. Exercise Physiol., 55*(4), 1341–1348.

CHAPTER 2

Properties of Bone

Stephen C. Cowin
William C. Van Buskirk
Richard B. Ashman
Department of Biomedical Engineering
Tulane University
New Orleans, Louisiana

INTRODUCTION

In this chapter we present a selective summary, current to the fall of 1983, of the reported mechanical properties of bone and the methods commonly used to determine these properties. Earlier summaries of certain portions of the material covered here appear in the books by Yamada (1970) and Evans (1973), which summarize the literature to 1969, and more recent articles by Reilly and Burstein (1974), Carter and Spengler (1978), and Van Buskirk and Ashman (1981). We are, and will continue to be, interested in maintaining the currency of the data and information on the mechanical properties of bone. It would be appreciated if any corrections, omissions, additions, or updates would be communicated to the first author.

2.1 BONE STRUCTURE

In biological terms bone is described as a connective tissue. A *tissue* is an aggregation of similarly specialized cells united in the performance of a particular function. *Connective tissue* is the tissue which binds together and is the support of the various structures of the body. In mechanical terms bone is a composite material with several distinct solid and fluid phases. Bone is unique among the connective tissues because it is hard, and it is hard because the extracellular collagenous matrix, the major organic component of all connective tissue, is impregnated with a mineral phase, principally hydroxyapatite [$Ca_{10}(PO_4)_6OH_2$] or very similar crystals. Loosely, the structure is similar to that of fiberglass because fiberglass is constructed by impregnating a resin matrix with a mineral phase, glass fibers. However, bone, as we shall see below, is much more highly organized than fiberglass.

At the macroscopic level there are two major forms of bone tissue: *compact,* or *cortical,* bone and *cancellous,* or *trabecular,* bone. The location of these bone types in a femur is

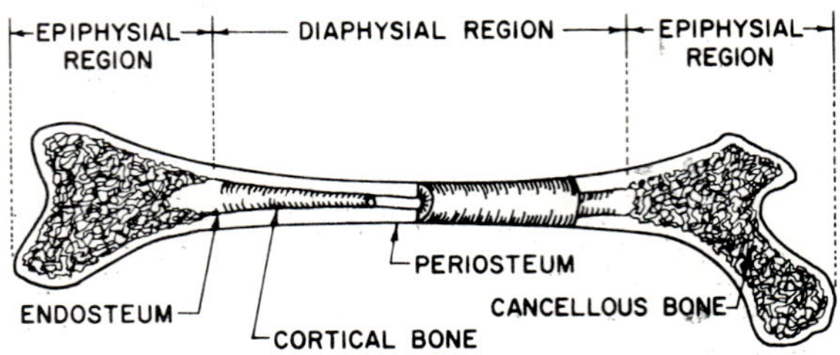

FIG. 2.1 Longitudinal section of the femur, illustrating cancellous and cortical bone.

illustrated in Fig. 2.1. Cortical or compact bone is a dense material with a specific gravity of about 2. The external surface of bone is generally smooth and is called the *periosteal* surface. The interior surface is called the *endosteal* surface. It has a roughened texture which resembles cancellous bone. Cancellous bone also exists in the epiphysial and metaphyseal region of long bones and within the confines of the cortical bone coverings in the smaller flat and short bones. Cancellous bone is also called trabecular bone because it is composed of short struts of bone material called *trabeculae*. The connected trabeculae give cancellous bone a spongy appearance, and it is often called *spongy bone*. There are no blood vessels within the trabeculae, but there are vessels immediately adjacent to the tissue, and they weave in and out of the large spaces between the individual trabeculae. Cancellous bone has a vast surface area as would be suggested by its spongy appearance. This is illustrated by the human pelvis, which has an average volume of 40 cm^3 and an average periosteal surface of 80 cm^2, but the average surface of its trabecular bone is 1600 cm^2.

We return now to discussing the denser cortical bone. From a microscopic viewpoint there are three types of cortical bone: *woven, laminar,* and *haversian* (see Fig. 2.2). Woven bone is found typically in both cortical and cancellous bone of young, growing animals and in adults after some bone injury. However, during normal maturation it is gradually replaced by laminar bone so that, in a human being, for example, there is normally no woven bone present after the age of 14 to 16 years. An additional distinguishing feature of woven bone is the lack of a relationship of mineral to collagen. In laminar and haversian bone these elements are closely related, and it seems virtually impossible for laminar bone to become hypermineralized. However, the mineral density of woven bone varies enormously; the mineral appears to be weakly related to the organic matrix, mineralization occurs in a haphazard fashion, and hypermineralization is often observed. Woven bone appears to bear the same relationship to laminar bone as a temporary scaffolding bears to a completed structure.

When it occurs in the midshaft of a long bone, laminar bone consists of a number of concentrically arranged laminae, as laminae in laminar bone, about 10 to 20 mm long. The thickness is the same because the blood supply for the haversian system is a central lumen containing a blood vessel, and thus every point in the haversian system is no more than 100 μm from the blood supply, as with laminar bone. Haversian bone is organized to accommodate small arteries, arterioles, capillaries, and venules of the microcirculating system. Haversian bone is never formed as a primary event, but forms as the result of the vascular invasion of bone. In young animals woven bone is formed initially, then endosteal capillaries invade the avascular bone, thus forming haversian systems. The biological process might be described in engineering terms as simultaneous tunneling and pipe laying.

The osteons of haversian bone and the laminae of laminar bone are basically just different geometric configurations of the same material (see Fig. 2.3). In both geometric configurations no point in the tissue is more than 100 μm away from the blood supply. The interfaces between the laminae in both haversian and laminar bone contain an array of roughly ellipsoidally shaped cavities called *lacunae* which contain bone cells, and from which extend numerous fine canals called the *canaliculi*. The thin layer between adjacent osteons is called the *cement line,* and the three-dimensional region between osteons is filled with irregular pieces of lamellar bone. The

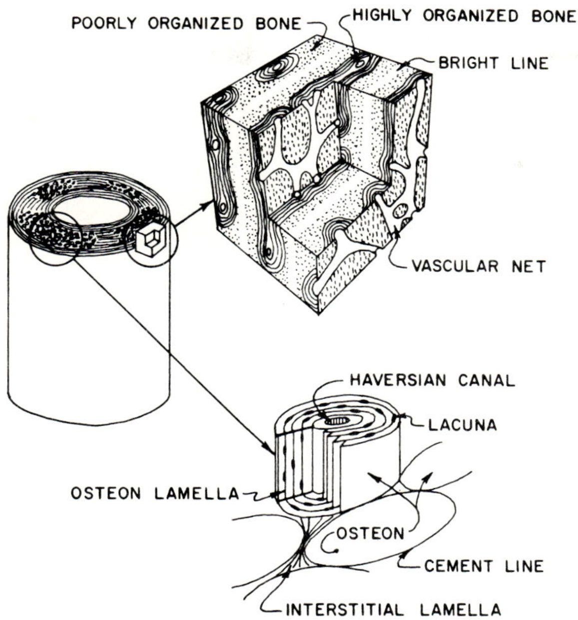

FIG. 2.2 Typical bone structure in the diaphysis of the femur. Two types of cortical bone are illustrated: plexiform and haversian.

canaliculi do not cross the cement lines, nor do they cross the bright lines between laminae in laminar bone.

Both haversian and laminar bone occur simultaneously in the long bones of humans and many animals, including cattle. In the very young the long bones are composed of woven bone with a few osteons, called *primary osteons*. With maturation the woven bone is converted to laminar bone, and, at maturity, there is a partial conversion to haversian bone. The conversion

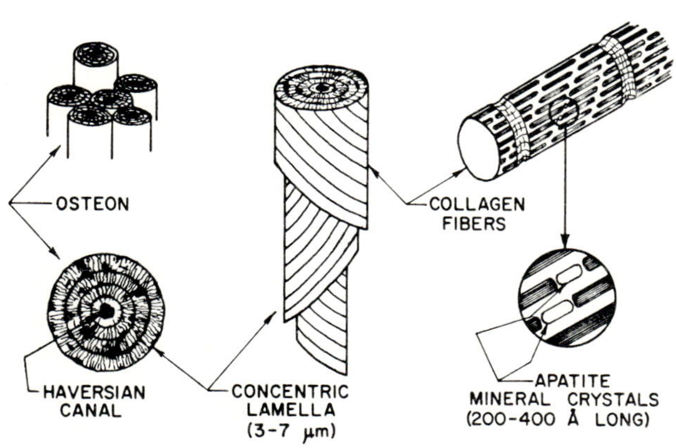

FIG. 2.3 Detailed structure of an osteon.

from laminar to haversian bone is somewhat of a biological enigma. Haversian bone is known to have a less efficient local circulation system (Currey, 1960) and to have less mechanical strength than laminar bone, yet the percentage of haversian bone generally increases with age.

It is possible to proceed further down the microstructural scales in bone tissue and to identify structural elements beyond the molecular scale. An understanding of this finer structure is not necessary for our purposes, and so the discussion is terminated at this level. The relationship between the finer structure and the molecular structure is discussed by Glimcher (1976).

2.2 BONE COMPOSITION

The composition of bone tissue is, very roughly, equal thirds by volume of minerals, water, and the extracellular collagenous matrix. In order to be more precise about bone composition, one must specify species, age, sex, the specific bone in question, the type of bone tissue (cancellous or cortical), and whether or not bone disease is present. Details of some age and species differences are given by Jowsey (1968). Table 2.1 presents data on the specific gravity, water

TABLE 2.1 Results of hydrated bone assays for 16 species using cortical bone from the tibia and the femur

Species*	Specific gravity	Water content, vol %	Mineral ash, vol %	Organic $+CO_2$, vol %
Fish (2)	1.80	39.6	29.5	36.9
Turtle (6)	1.81	37.0	29.2	40.1
Frog (4)	1.93	35.2	34.5	38.5
Polar bear (1)	1.92	33.0	36.2	40.1
Human being (15)	1.94	15.5	39.9	41.8
Elephant (1)	2.00	20.0	41.4	41.5
Monkey (3)	2.09	23.0	42.6	41.1
Cat (1)	2.05	23.6	42.2	40.5
Horse (3)	2.02	25.0	41.0	40.5
Chicken (4)	2.04	24.5	41.7	38.7
Dog (10)	1.94	28.0	38.7	35.5
Goose (2)	2.04	23.0	42.7	37.6
Cow (5)	2.05	26.2	42.6	36.2
Guinea pig (2)	2.10	25.0	43.5	37.0
Rabbit (2)	2.12	24.5	45.0	37.2
Rat (12)	2.24	20.2	49.9	38.3

Source: Data from Blitz and Pellegrino (1969).

* The number of adults of each species sampled are indicated in parentheses after the common species name.

fraction, mineral fraction, and organic fraction of cortical bone for 16 different vertebrates, including humans. These data are taken from a paper by Blitz and Pellegrino (1969), who give a detailed chemical assay of the bone for each of the 16 vertebrates. They assayed the percentages of calcium, phosphorus, magnesium, sodium, carbon dioxide, OH-proline, hexosamine, citrate, and lactate. Results of hydrated bone assays for both cortical and trabecular bone for four species were reported by Gong, Arnold, and Cohn (1964). Some of their data are summarized in Table 2.2. Comparison of the data given in Table 2.2 for trabecular bone with that given for cortical bone shows that the water fraction is greater and the ash fraction is less for trabecular bone. The organic fraction for the two bone types is fairly close.

The role of water in bone remains obscure. Timmins and Wall (1977) concluded from a survey of the published data that very little is understood concerning the role and structure of the water in bone, although variations in water content between species and with age and

TABLE 2.2 Results of hydrated bone assays of cortical and trabecular bone for four species using cortical bone from the tibia or femur and trabecular bone from the vertebrae

Species	Specific gravity*	Water fraction, vol %	Ash fraction, vol %	Organic fraction, vol %	Volatile inorganic fraction, vol %
			Trabecular bone		
Human being	1.92	27	33.9	34.9	4.2
Monkey	1.89	27.1	32.9	36.1	4.0
Cow	1.93	28.1	33.5	34.2	4.2
Dog	1.91	28.8	32.6	34.5	4.2
			Cortical bone		
Human being	1.99	23.9	37.7	33.8	4.6
Monkey	2.04	23.7	38.2	33.7	4.7
Cow	2.00	25.2	36.6	33.6	4.6
Dog	2.00	22.3	36.8	36.3	4.6

Source: Data from Gong, Arnold, and Cohn (1964).

* This is the specific gravity of the trabeculae, not the apparent specific gravity of the cancellous structure, which would be a number less than 1.

pathological conditions are fairly well documented. In the section on experimental methods it will be pointed out that the mechanical properties of bone vary significantly with water content. The work of Marino, Becker, and Bachman (1967) indicates that the distinction between free and bound water in bone occurs at a critical hydration value of 37 to 48 mgH_2O per gram of bone. When water is extracted from bone, the free water is extracted first, then the water bound into the bone structure. The hydration value is the value at which the bound water begins to be extracted.

2.3 EXPERIMENTAL TECHNIQUES

2.3.1 Preservation of Bone for Testing

The in vitro properties of bone can vary significantly from the in vivo properties depending on the freshness of the tissue and the method by which the tissue was preserved. Yamada (1970) gives a great deal of data on the variation of the mechanical properties with the method of tissue preservation. For measurements of density, mineral content, elasticity, and strength that accurately reflect values in the fresh state, the following methods have been shown by Ashman et al. (1982) and Sedlin and Hirsch (1966) to preserve the tissue adequately. The most important of these procedures is to keep the bone moistened with normal saline. The properties of fresh tissue can vary in a matter of minutes if the bone is allowed to dry. For longer storage the moist bone should be frozen. If all the musculature is to be removed before freezing, the bones must be wrapped in gauze soaked in normal saline and placed in plastic bags. The bones must be then frozen to −20°C within 1 h after excision. If freezer space permits, the musculature may be left intact. This eliminates the wrapping with gauze and soaking with saline. A plastic bag should cover the musculature to prevent freezer burn. After removal from the freezer, the bone should be maintained wet in saline during all further stages of tissue preparation. For time periods of up to 3 months, small specimens may be preserved at room temperature in a solution of 50% normal saline and 50% alcohol.

It has been reported by Sedlin and Hirsch (1966) that embalmed bone exhibits different values of strength and elasticity compared with fresh tissue. Whenever possible, the tissue should remain unembalmed. If embalmed tissue is used, it should be treated similarly to unembalmed tissue to prevent embrittlement due to degradation of the organic phase of the bone due to drying or decay.

2.3.2 Measurement of Bone Density

Several methods have been used for the measurement of bone density (Ashman et al., 1983; Yoon and Katz, 1976; Abendschein and Hyatt, 1970). Of the methods available, the simplest and most repeatable is that which is based on Archimedes' principle. This method was employed, for example, by Ashman et al. (1983). It exploits the fact that the buoyant force on a submerged object is equal to the volume of fluid displaced times the density of the fluid. In practice, distilled water is used as the fluid. The bone specimen is weighed suspended in the water from the beam of a balance. The volume of the specimen is equal to the moist specimen's weight in air minus the specimen's weight suspended in water, divided by the density of water.

An alternative method to that of the application of Archimedes' principle is to determine the volume of the specimen directly from a measurement of the specimen's physical dimensions. However, in order to determine the specimen's volume accurately by measurement of the physical dimensions, the specimen must be a perfect right parallelepiped. This requirement makes specimen preparation rather difficult.

There are three other points of interest concerning density measurements of bone tissue. First, the density of both cortical and cancellous bone should be measured with the bone in its natural wet state, that is, the weight in air when the bone is wet. Second, it should be noted that fat and grease trapped in the pores of bone, especially cancellous bone, can invalidate density measurements. In the measurement of the density of cortical bone, care should be taken to ensure that the specimen contains only cortical bone and not portions of cancellous bone with trapped fat. Third, there are two different density measurements performed on cancellous bone, the *density of the trabeculae* themselves and the *apparent density* of the cancellous structure. The density of the trabeculae of cancellous bone can be measured easily by the application of Archimedes' principle as described above. It is important to remove trapped fat by boiling the bone in detergent or cleaning it with a high-pressure water jet before the density measurements are made. The apparent density is defined as the weight of the defatted cancellous structure divided by the overall volume of the specimen. Typical values of the density of the trabeculae are the same as those of cortical bone, namely, 1.85 to 2.00 g/cm^3, whereas the apparent density of cancellous bone ranges from 0.15 to 1.0 g/cm^3.

2.3.3 Measurement of Bone Mineral Content

Bone mineral content is the ratio between the unit weight of the mineral phase and the unit weight of the dry specimen of bone. It is often reported as a percentage. The most common method used to determine the mineral content is that of "ashing" the bone (see Ashman et al., 1983 and Currey, 1969). The resulting mineral content measured by this technique is often referred to as the *ash weight* of bone, although *ash fraction* is a more accurate term since the mineral content is a fraction, not a weight. In this technique, the bone specimen is first dried in desiccant at room temperature or in air at a temperature of 60°C. Generally, the specimen must be dried for at least a week. The specimen is assumed dry when the daily change in weight is less than 1 percent. The specimen is then ashed in a furnace at 700°C. The mineral content (ash fraction) is equal to the weight of the specimen after ashing divided by the dry weight of the specimen. When this technique is used, care must be taken to ensure that the crucibles which hold the specimens during ashing are completely dry. It is suggested that the crucibles be dried at 700°C for at least 1 day prior to the ashing of the bone.

Another technique for determining the mineral content is that of demineralizing the bone in a commercially available decalcifying solution (Scientific Products D1208). This technique is more difficult than ashing the bone since the rate at which the calcium is removed decays exponentially. If this technique is used, it should be noted that small, thin specimens, which expose more surface to the solution, work best. Also, the decalcifying solution should be changed daily. Complete decalcification can be determined by testing with lithium carbonate. As with the ash weight technique, the ratio of dry specimen weights should be reported.

An indirect method for the determination of mineral content involves the measurement of the radiographic density of thin or pulverized undecalcified sections of bone as was used by Vose and Kubala (1959). The radiographic density of whole bones has also been used by Pelker and Saha (1983) to determine mineral content. Though these techniques have the advantage of ease of application, the measurements are hard to quantify accurately.

Several other radiographic and spectrographic methods are available for the measurement of bone mineral content. These methods involve the use of scattering or absorption at various wavelengths and generally yield information on the relative components of the mineral phase of bone. More complete descriptions of these techniques can be found in textbooks of quantitative analysis, e.g., Fritz and Shenk (1974), and articles by Katoh et al. (1981), Leichter et al. (1982), Kerr et al. (1980), and Krolner et al. (1980).

2.3.4 Measurement of the Stress-Strain Properties by Mechanical Testing

The elastic properties of cortical bone have been measured by two different methods, mechanical testing and ultrasonic testing. Mechanical testing methods involve the application of measurable forces to specimens of uniform dimensions for which changes in length are measured. Stress is equal to the applied force divided by the specimen's cross-sectional area, and strain is equal to the change in length divided by the original length. Stress-strain curves are plotted from which the *elastic modulus,* or the slope of the stress-strain curve in the linear region, is obtained. This is illustrated in Fig. 2.4 for the case of a tension test. Also derivable from the stress-strain curve is the yield stress. The *yield stress* is the point of deviation from the initial linear region. In materials in which there is no abrupt deviation from the initial linear region, such as

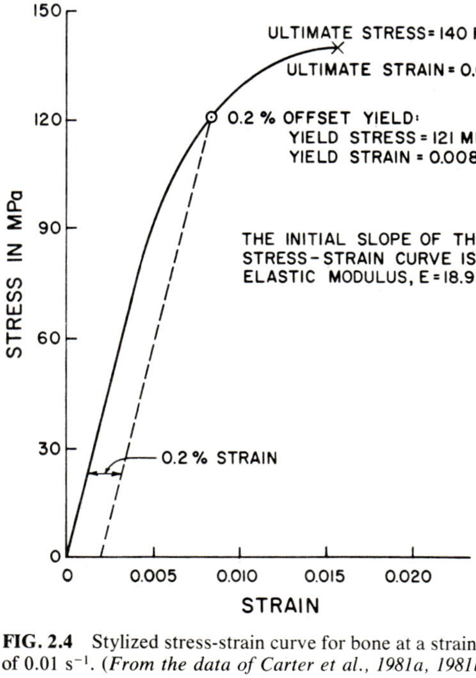

FIG. 2.4 Stylized stress-strain curve for bone at a strain rate of 0.01 s^{-1}. (*From the data of Carter et al., 1981a, 1981b.*)

bone, the yield stress is defined by the intersection of an offset yield line and the stress-strain curve, as illustrated in Fig. 2.4. More detailed descriptions of stress-strain curves may be found in any textbook on the mechanics of materials, e.g., Crandall, Dahl, and Lardner (1978). The loads in mechanical tests can be applied in tension (Burris, 1983), compression (Carter and Hayes, 1976, 1977), torsion (Burris, 1983), or in bending (Currey, 1969) or biaxially. Strains can be measured with dial gauges, with foil strain gauges (Knets et al., 1977), with electrical extensometers (Burris, 1983) or optically. Of the mechanical techniques available, the most common method used is the tensile test. Tensile specimens with dimensions that are approximately those employed by Reilly and Burstein (1974), illustrated in Fig. 2.5, are most common. Strain rates should be approximately physiological. Physiological strain rates are discussed in the section on stress-strain relations for cortical bone.

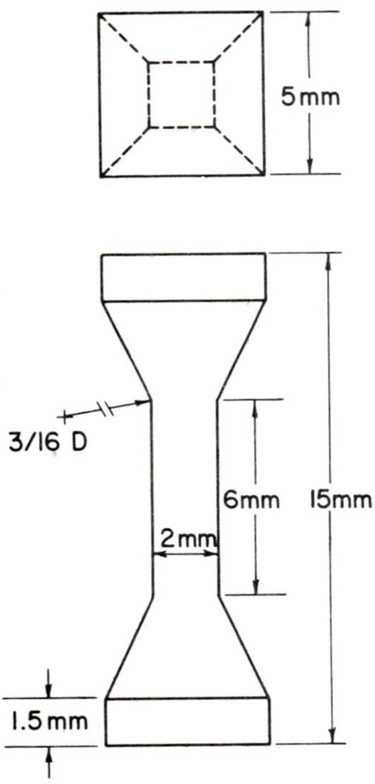

FIG. 2.5 Typical bone specimen. (*Redrawn from Reilly and Burstein, 1974.*)

2.3.5 Measurements of the Elastic Properties by Ultrasonic Methods

Though mechanical testing methods have been used with success on bone, these methods are difficult to apply to bone (Ashman et al., 1983). The overall dimensions of the bone limit the size of the test specimen, and the heterogeneity of bone requires that the specimens be small in order to ensure that the properties are nearly uniform throughout the test specimen. An additional problem arises from the anisotropy of bone. This anisotropy requires that traditional mechanical tests be applied in several different directions in order to obtain enough information for the calculation of all of the independent elastic coefficients. At most, four elastic coefficients can be obtained from one mechanical test specimen. One Young's modulus and two Poisson's ratios can be obtained from a tensile test. If the cross section of the specimen is circular, a torsion test may be applied to obtain the value of one shear modulus. Therefore, three separate specimens would be needed to measure the orthotropic elastic coefficients by mechanical means. Since it is impossible to fabricate three different specimens containing the same portion of bone, the heterogeneity and anisotropy make it impossible to measure the elastic properties at a point by traditional mechanical testing methods.

Ultrasonic techniques offer an advantage in that they allow the use of smaller specimens of less complicated geometries than do mechanical testing methods. These techniques make use of relations between the speed of sound in a material and the elastic properties of that material (Ashman et al., 1983). Another advantage of the use of ultrasound is that more than one elastic coefficient can be measured on each specimen. By measuring ultrasonic velocities in a number of directions on a single specimen, it is possible to measure all the elastic coefficients and hence to characterize the anisotropy of bone. Furthermore, information on the heterogeneity of bone can be obtained with ultrasonic techniques by measuring the elastic properties on specimens taken from many different locations within a single whole bone. Descriptions of various ultrasonic techniques can be found in an article by McSkimin (1961). The techniques used to measure the elastic properties of bone can be found in papers by Lang (1970), Yoon and Katz (1976), Van Buskirk et al. (1981), and Ashman et al. (1983).

2.4 STRESS-STRAIN RELATIONS FOR CORTICAL BONE

2.4.1 The Stress-Strain Curve

Consider a specimen of cortical bone subjected to a tensile test. A stylized version of the stress-strain curve found in such a test is shown in Fig. 2.4. There is an elastic region in which Hooke's law ($\sigma = E\varepsilon$) is valid and a plastic region prior to fracture. There is a smooth transition between the elastic and plastic regions, so a specific yield stress cannot be easily picked off the curve. This is also the case for many engineering materials in which it is typical to use a 0.2 percent offset method as indicated in Fig. 2.4 to define the yield properties. Carter et al. (1981a and b) conducted tensile tests of human cortical bone at a strain rate of 0.01 s^{-1}. Except for the value of the elastic modulus, the means of the results they obtained are shown in Fig. 2.4. The elastic modulus was adjusted so that the other average values could be consistently represented on a "typical" curve.

Effect of Strain Rate on the Stress-Strain Curve. The stress-strain curve exhibited by bone is a function of the strain rate to which the bone is subjected. As an example of this phenomenon consider the data of McElhaney (1966) for embalmed human femur in compression shown in Fig. 2.6. As indicated, bone is stiffer and stronger at higher strain rates. Carter and Hayes (1977) found that both strength and elastic modulus were approximately proportional to strain rate raised to the 0.06 power. These relationships were said to hold for all bone in the skeleton.

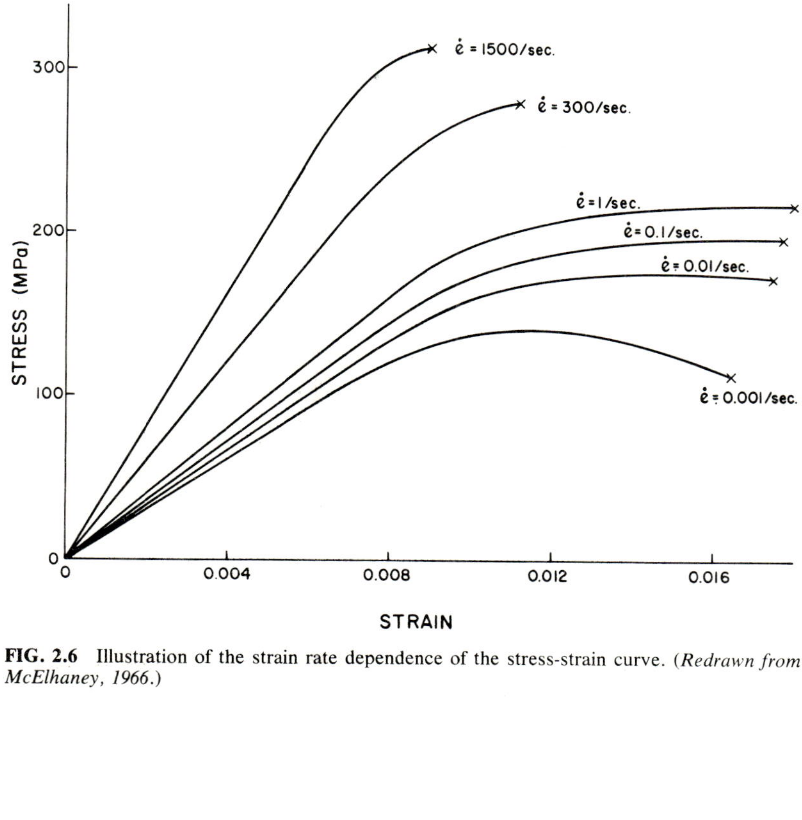

FIG. 2.6 Illustration of the strain rate dependence of the stress-strain curve. (*Redrawn from McElhaney, 1966.*)

Hight and Brandeau (1983) investigated the use of the Ramberg-Osgood equation to model the strain rate–dependent stress-strain relationship of bone. They achieved very reasonable curve fits for three sets of data from the literature: the data of McElhaney (1966), who tested embalmed human femur in compression; the data of Wood (1971), who examined fresh human cranial bone in tension; and the data of Crowninshield and Pope (1974), who tested fresh bovine bone in longitudinal tension. The form of the Ramberg-Osgood equation they used is

$$\varepsilon = \frac{\sigma}{c\dot{\varepsilon}^d} + a\sigma^n\dot{\varepsilon}^b \tag{2.1}$$

The parameters they found for each of the three data sets are given in Table 2.3. For physiological levels of stress and strain the second term in Eq. (2.1) is negligible. It essentially describes

TABLE 2.3 Parameter values* for the Ramberg-Osgood equation as determined by Hight and Brandeau (1983)

Data set†	c	d	a	n	b
A	24.50	0.06707	2.02 (10^{-18})	6.534	−0.3740
B	15.21	0.05670	1.37 (10^{-18})	7.662	−0.4127
C	11.69	0.01798	4.09 (10^{-106})	45.25	−2.336

* In the use of these values the unit for stress in the equation is in gigapascals.
† Data sets A, B, and C are from McElhaney (1966), Wood (1971), and Crowninshield and Pope (1974), respectively.

the plastic portion of the stress-strain curve. Thus for the stresses and strains to which bone is normally exposed, the Young's modulus is given by

$$E = \frac{\sigma}{\varepsilon} = c\dot{\varepsilon}^d \tag{2.2}$$

Note that for the data of McElhaney (1966) and Wood (1971) the coefficient d is approximately 0.06, confirming the results of Carter and Hayes (1977). The strain rate to which bone is normally subjected ranges from 0.001 s^{-1} for slow walking to 0.01 s^{-1} for vigorous activity. Using $d = 0.06$, we find that the elastic modulus can vary as much as 15 percent during normal activity.

2.4.2 Anisotropic Elastic Relations

Bone is a nonhomogeneous, anisotropic, nonlinearly viscoelastic material, and as such its detailed properties are difficult to measure. But for most studies of stress and strain in bone it is adequate to assume that it is linearly elastic with a constitutive equation

$$T_{ij} = C_{ijkl}E_{kl} \tag{2.3}$$

where T_{ij} is the stress tensor, E_{kl} is the infinitesimal strain tensor, and C_{ijkl} is a fourth-order tensor known as the *elasticity tensor*. As is customary in the discussion of linear anisotropic elastic materials, we introduce the single-index notation for stress and strain shown in Table 2.4. With this notation the constitutive equation for bone becomes $\sigma_i = c_{ij}\varepsilon_j$, where the usual summation convention is employed with a range of 6 and c_{ij} is known as the *stiffness matrix*.

TABLE 2.4 Equivalent notations for stress and strain

Stresses	Strains
$\sigma_1 = T_{11}$	$\varepsilon_1 = E_{11}$
$\sigma_2 = T_{22}$	$\varepsilon_2 = E_{22}$
$\sigma_3 = T_{33}$	$\varepsilon_3 = E_{33}$
$\sigma_4 = T_{23} = T_{32}$	$\varepsilon_4 = 2E_{23} = 2E_{32}$
$\sigma_5 = T_{13} = T_{31}$	$\varepsilon_5 = 2E_{13} = 2E_{31}$
$\sigma_6 = T_{12} = T_{21}$	$\varepsilon_6 = 2E_{12} = 2E_{21}$

If we assume bone to be an orthotropic material, then the stiffness matrix is given by

$$[c_{ij}] = \begin{bmatrix} c_{11} & c_{12} & c_{13} & 0 & 0 & 0 \\ c_{12} & c_{22} & c_{23} & 0 & 0 & 0 \\ c_{13} & c_{23} & c_{33} & 0 & 0 & 0 \\ 0 & 0 & 0 & c_{44} & 0 & 0 \\ 0 & 0 & 0 & 0 & c_{55} & 0 \\ 0 & 0 & 0 & 0 & 0 & c_{66} \end{bmatrix} \tag{2.4}$$

There are 12 nonzero components, of which 9 are independent.

Bone has generally been assumed by researchers to be transversely isotropic (cf. Lang, 1970, Reilly and Burstein, 1975, and Yoon and Katz, 1976). For a transversely isotropic material $c_{11} = c_{22}$, $c_{13} = c_{23}$, $c_{44} = c_{55}$, and $c_{66} = \frac{1}{2}(c_{11} - c_{12})$. The 3 axis is assumed to be the axis of symmetry. For bone, the 3 axis is parallel to the long axis of the bone. There are 12 nonzero components, but only 5 are independent.

The strain-stress relationships can be given in terms of a compliance matrix s_{ij} so that $\varepsilon_i = s_{ij}\sigma_j$. The compliance coefficients can be further expressed in terms of certain technical constants, namely, the generalized Young's moduli, Poisson's ratios, and shear moduli. These constants can be measured in simple tests such as uniaxial tension or pure shear tests. These constants, with their obvious physical interpretation, have more direct meaning than the components of the relatively abstract stiffness matrix. The technical constants are defined in the following way:

E_i = Young's modulus in the i direction

ν_{ij} = Poisson's ratio for strain in the j direction when stressed in the i direction $(i \neq j)$ (2.5)

G_{ij} = shear modulus $= T_{ij}/2E_{ij}$ for $i \neq j$

For an orthotropic material the compliance matrix expressed in terms of the technical constants is given by

$$[s_{ij}] = \begin{bmatrix} E_1^{-1} & -\nu_{21}E_2^{-1} & -\nu_{31}E_3^{-1} & 0 & 0 & 0 \\ -\nu_{12}E_1^{-1} & E_2^{-1} & -\nu_{32}E_3^{-1} & 0 & 0 & 0 \\ -\nu_{13}E_1^{-1} & -\nu_{23}E_2^{-1} & E_3^{-1} & 0 & 0 & 0 \\ 0 & 0 & 0 & G_{23}^{-1} & 0 & 0 \\ 0 & 0 & 0 & 0 & G_{31}^{-1} & 0 \\ 0 & 0 & 0 & 0 & 0 & G_{12}^{-1} \end{bmatrix} \tag{2.6}$$

which has 12 different components. Only nine are independent since symmetry of the compliance matrix requires that

$$\frac{\nu_{12}}{E_1} = \frac{\nu_{21}}{E_2} \qquad \frac{\nu_{13}}{E_1} = \frac{\nu_{31}}{E_3} \qquad \text{and} \qquad \frac{\nu_{23}}{E_2} = \frac{\nu_{32}}{E_3} \tag{2.7}$$

For a transversely isotropic material

$$E_1 = E_2 \qquad \nu_{12} = \nu_{21} \qquad \nu_{31} = \nu_{32} \qquad G_{23} = G_{31} \qquad \text{and} \qquad G_{12} = \frac{E_1}{2(1 + \nu_{12})} \tag{2.8}$$

2.4.3 Elastic Constants for Human, Bovine, and Canine Cortical Bone

Table 2.5 presents the technical constants for human bone measured by various investigators. When ultrasound is employed, various velocities are measured from which the elastic coefficients are determined. The technical constants are then found by matrix inversion. The material symmetry generally assumed is that of transverse isotropy (TI) or orthotropy (abbreviated *orth.* in tables).

TABLE 2.5 Technical constants for human bone (values in gigapascals)*

Group	Reilly and Burstein (1975)	Yoon and Katz (1976)	Knets and Malmeisters (1977)	Ashman et al. (1983)
Bone	Femur	Femur	Tibia	Femur
Symmetry	TI	TI	OrTH.	OrTH.
Method†	M	U	M	U
E_1	11.5	18.8	6.91	12.0
E_2	11.5	18.8	8.51	13.4
E_3	17.0	27.4	18.4	20.0
G_{12}	3.6†	7.17	2.41	4.53
G_{13}	3.3	8.71	3.56	5.61
G_{23}	3.3	8.71	4.91	6.23
ν_{12}	0.58	0.312	0.49	0.376
ν_{13}	0.31‡	0.193	0.12	0.222
ν_{23}	0.31‡	0.193	0.14	0.235
ν_{21}	0.58	0.312	0.62	0.422
ν_{31}	0.46	0.281	0.32	0.371
ν_{32}	0.46	0.281	0.31	0.350

* The 3 direction is coincident with the long axis of the bone; the 1 and 2 directions are radial and circumferential, respectively.
† M = standard machine testing; U = ultrasound.
‡ Not measured.

The most readily accessible mammalian bone comes from slaughtered beef cattle. Because of its availability and its large size, the bovine femur has been subjected to numerous studies. The results of those studies are recorded in Table 2.6. Again the material symmetry generally assumed is that of transverse isotropy or orthotropy. In the case of two of the ultrasound studies, it was possible to measure only seven of the nine orthotropic constants. All nine stiffness coefficients must be known in order to perform matrix inversion. The other two constants (c_{12} and c_{13}) were estimated by the values they would possess if the material were transversely isotropic. This situation is referred to in Table 2.6 under "Symmetry" as TI+.

There are questions, however, about the validity of bovine bone as a model for human bone. The structure of bovine bone in young cattle is typically laminar or plexiform, whereas for the adult human, the cortical tissue of the long bones is primarily haversian. As one can see, however, from a comparison of Tables 2.5 and 2.6, the elastic properties of bovine and human bone are quite similar.

TABLE 2.6 Technical constants for bovine and canine bone (values in gigapascals)*

Group	Lang (1970)	Reilly and Burstein (1975)	Ambardar and Ferris (1976)	Van Buskirk et al. (1981)	Burris (1983)	Burris (1983)	Ashman et al. (1983)
Bone	Bovine phalanx	Bovine femur	Bovine femur	Bovine femur	Bovine femur	Bovine femur	Canine femur
Symmetry	TI	TI	TI	OrTH.	TI	TI+	TI+
Method†	U	M	U	U	M	U	U
E_1	11.3	10.2	6.97	11.6	13.3	10.79	12.8
E_2	11.3	10.2	6.97	14.6	13.3	12.24	15.6
E_3	22.0	22.6	20.9	21.9	18.1	18.90	20.1
G_{12}	3.8	‡	2.2	5.29	3.5	3.38	4.68
G_{13}	5.4	3.6	6.9	6.29	5.0	4.47	5.68
G_{23}	5.4	3.6	6.9	6.99	5.0	5.96	6.67
ν_{12}	0.484	0.51	0.55	0.302	‡	0.45	0.282
ν_{13}	0.203	‡	0.15	0.109	‡	0.24	0.289
ν_{23}	0.203	‡	0.15	0.205	‡	0.22	0.265
ν_{21}	0.484	0.51	0.55	0.380	‡	0.51	0.366
ν_{31}	0.396	0.36	0.44	0.206	0.22	0.42	0.454
ν_{32}	0.396	0.36	0.44	0.307	0.22	0.33	0.341

* The 3 direction is coincident with the long axis of bone; the 1 and 2 directions are radial and circumferential, respectively.

† U = ultrasound; M = standard machine testing.

‡ Not measured.

The dog is frequently used as an experimental animal in bone research. To perform analytical work in conjunction with the experimental work, one needs to know the elastic properties of canine bone. Data on canine bone are also given in Table 2.6.

In his study of human tibia, Saulgozis (1976) found the following statistically significant inequalities.

$$E_3 > E_2 > E_1 \qquad G_{23} > G_{13} > G_{12} \tag{2.9}$$

and

$$\nu_{21} > \nu_{12} > \nu_{31} > \nu_{23} > \nu_{13} \tag{2.10}$$

An examination of the results in Tables 2.5 and 2.6 reveals that, in general, these inequalities are borne out. This suggests that the usual assumption of transverse isotropy for bone is not sufficiently general.

The choice of material symmetry for an elastic model of bone depends to a large extent on the intended application of the elastic model by the investigator. As an example, Huiskes (1982) indicates that the stress analysis of a human femur is adequately served with an isotropic elastic model. The data presented above indicate that the percentage difference between the elastic constants obtained using the assumption of orthotropy and those obtained using the assumption of transverse isotropy are quite small and are unlikely to have a significant effect on the stress analysis of a bone. Thus, if the intended use of the elastic constants is the stress analysis of a particular bone, the assumption of the greater symmetry of transverse isotropy can be easily justified.

When the physiological behavior of bone tissue as a composite material is under consideration, the difference in the assumptions of orthotropy or transverse isotropy might be significant. As an example, consider the suggestion of Yanson et al. (1974) that the elastic symmetry of bone may be linked to the circulation of blood in the bone. Although there is little difference in the constants determined by using the assumptions of orthotropy and transverse isotropy, the difference is always consistent in that the stiffness in the circumferential direction is always greater than the stiffness in the radial direction. If bone were transversely isotropic, these two elastic constants would be the same. Blood normally flows from the medullary canal to the periosteum, that is, in an outward radial direction. Yanson et al. (1974) suggest that the lower stiffness in the radial direction is associated with the greater permeability in that direction. There is a significantly smaller need for blood flow in the circumferential direction, so the permeability is less and the stiffness is higher. Thus, there appears to be at least one physiological reason for the cortical tissue of long bones to be orthotropic rather than transversely isotropic.

Temperature Dependence of Elastic Constants. The temperature dependence of the elastic coefficients c_{ij} was investigated by three groups of researchers, each using ultrasonic techniques similar to those explained in the section on experimental methods. The temperature dependence in each study was found to be linear. Slopes of the regression line were −0.24%/°C (Bonfield and Tully, 1982), −0.20%/°C (Yoon and Katz, 1976), and −0.17%/°C (Ashman, 1982). The slope of the regression line can be used to estimate the elastic coefficient at one temperature given its value at another, known temperature.

Relationship of Elastic Constants Measured Using Ultrasound and Standard Mechanical Testing. Ultrasound is being more widely used as a tool for determining elastic constants. The reasons for this have been given above. But, also as shown above, the stress-strain properties of bone are strain rate–dependent. It is legitimate then to ask what relation the properties determined using ultrasound have to properties determined at physiological strain rates using standard mechanical testing.

This question was addressed by Burris (1983) in his doctoral dissertation. He performed quasi-static mechanical tests on one set of bovine haversian bone specimens and ultrasound tests at 2.25 MHz on a set of adjacent specimens. He assumed bone to be transversely isotropic. The mean values he obtained are shown in Table 2.6. Except for the Poisson's ratio ν_{31}, the mean values he found in the two types of tests appear to be in good agreement. The question occurs, however, whether for a particular bone one can adequately estimate its quasi-static properties from ultrasound tests or from other bone properties such as density and ash fraction. To assess this possibility, a single-variable, linear regression analysis was applied to each of the technical constants obtained from machine testing as a function of each of density, ash fraction, and the ultrasonically measured elastic properties of the adjacent specimens. He found that only three of the five independent constants for a transversely isotropic model of bone could be adequately estimated using the results of simple regression analysis involving only one other variable. The regression equations of these three variables against their ultrasonic counterparts are

$$
\begin{aligned}
(E_3)_M &= 12.47 + 0.330(E_3)_U \qquad & r &= 0.531 \\
(\nu_{31})_M &= 0.156 + 0.188(\nu_{31})_U \qquad & r &= 0.336 \\
(E_2)_M &= 10.20 + 0.234(E_2)_U \qquad & r &= 0.246
\end{aligned}
\tag{2.11}
$$

where E_i is in gigapascals. The subscripts M and U denote mechanical testing and ultrasound results, respectively.

Even better regression relations were obtained for the Young's moduli using $v_{3/3}$ and $v_{2/2}$, the longitudinal velocities in the axial and circumferential directions, respectively. These relations are

$$
\begin{aligned}
(E_3)_M &= -6.48 + 0.0065v_{3/3} \qquad & r &= 0.591 \\
(E_2)_M &= -0.337 + 0.004v_{2/2} \qquad & r &= 0.333
\end{aligned}
\tag{2.12}
$$

where the velocities are in meters per second. A better estimate for $(\nu_{31})_M$ is obtained using the density ρ,

$$
(\nu_{31})_M = -0.0363 + 0.00014\rho \qquad r = 0.398 \tag{2.13}
$$

where ρ is in kilograms per cubic meter.

Adequate estimates of the moduli of rigidity could only be obtained using linear regression equations with three variables: an appropriate wave speed, the density, and the ash fraction (m, a dimensionless quantity measured by an ashing technique described in the section on methods). The best regression relations for these moduli were found to be

$$
\begin{aligned}
(G_{12})_M &= -15.37 - 0.01194v_{1/2} + 64.64m - 0.00541\rho \qquad & r &= 0.789 \\
(G_{13})_M &= 0.037 + 0.0078v_{1/3} - 64.867m + 0.0261\rho \qquad & r &= 0.968
\end{aligned}
\tag{2.14}
$$

where $v_{i/j}$ is in meters per second and is the transverse wave propagating in the i direction with particle displacements in the j direction. Multiple regression did not provide significantly improved relations to estimate $(E_3)_M$, $(E_2)_M$, and $(\nu_{31})_M$.

An additional point should be made concerning the relatively low values of the correlation coefficient r in the above relations. The range of possible values of the moduli of normal bone is rather narrow (10 to 30 GPa in the axial direction). With a given amount of experimental scatter (for material inhomogeneity, specimen preparation, and induced bending in mechanical testing) the quality of the correlation will be a function of the range of possible values. Hence, in order to show what the ultrasonic technique is capable of, in terms of agreement with mechanical testing, the range of moduli was increased by including specimens of aluminum and plexiglass with the specimens of bone. Figure 2.7 shows the mechanical moduli plotted against the ultrasonically determined moduli. The line shown has a slope of 1, indicating perfect agreement between the two techniques. The revised slope is 1.05 and the correlation coefficient is 0.99.

2.4.4 Viscoelastic Stress-Strain Relations

Only in the 1970s have quantitative studies of the viscoelastic properties of bone been reported. Earlier studies are observations or simple demonstrations of viscoelastic phenomena in bone

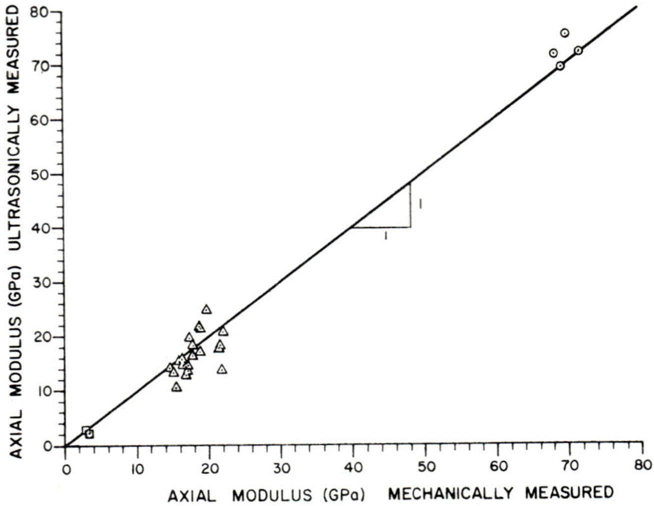

FIG. 2.7 Axial elastic modulus measured by the ultrasonic technique versus axial elastic modulus measured by mechanical testing techniques. Circles are aluminum, triangles are bovine femoral bone, squares are Plexiglas (a total of eight overlapping Plexiglas points). The line shown has a slope of 1, indicating perfect agreement between the ultrasonically measured moduli and the mechanically measured moduli.

tissue. Two of the consistently reported viscoelastic effects are the strain rate dependence of the elastic modulus represented by Eq. (2.2) and the absence of any permanent deformation after the removal of load, although the recovery from the deformed state is a time-dependent process. For example, Currey (1965) studied the long-term creep behavior of wet bovine bone using cantilever beam specimens. He found that the deformation recovery was always asymptotic to zero strain after 100 days, and that the creep deflection was proportional to load.

A number of investigators have tried to identify the particular viscoelastic stress-strain relationship that describes the time-dependent mechanical behavior of bone. Only the fitting of the integral equation form of the viscoelastic stress-strain relations will be considered here because the integral form is more general than the forms with a finite number of parameters associated with spring and dashpot models. An integral form of the constitutive relation for a homogeneous anisotropic linear viscoelastic material is

$$T_{ij}(t) = \int_{-\infty}^{t} C_{ijkl}(t - \tau)\dot{E}_{kl}(\tau)\, d\tau \tag{2.15}$$

where $T_{ij}(t)$ is the stress at time t, $\dot{E}_{kl}(\tau)$ is the strain rate at time τ, and $C_{ijkl}(t)$ is the fourth-rank relaxation modulus tensor which, in general, has 36 independent material functions. The constitutive relations for viscoelastic materials are discussed by Christensen (1971) and Lockett (1972). All the viscoelasticity theory discussed below is described in detail in these and other volumes on viscoelasticity.

Hayes and Morland (1969) proposed a series of tests to determine the possible 36 independent relaxation moduli $C_{ijkl}(t)$. The complete series comprises three simple shear tests and 21 tests of a single type described as pure compression. In the case of orthotropic symmetry, the number of independent relaxation moduli to be determined reduces to 12, and they can be determined from 12 tests: three simple shear tests and nine uniaxial compression tests. Nine of the twelve relaxation moduli of an orthotropic material, namely, $C_{1111}(t)$, $C_{2222}(t)$, $C_{3333}(t)$, $C_{1122}(t)$, $C_{2211}(t)$, $C_{1133}(t)$, $C_{3311}(t)$, $C_{2233}(t)$, and $C_{3322}(t)$, are rather complicated functions of the parameters measured in a uniaxial tension or compression test. This point is discussed, for example, by Lakes and Katz (1974). The other three moduli, $C_{1212}(t)$, $C_{2323}(t)$, and $C_{1313}(t)$, can be determined directly from the data of simple shear tests, but they cannot be determined directly from torsion tests because, for example, a pure twist about the x_3 axis involves both the functions $C_{1212}(t)$ and $C_{1313}(t)$. However, if the material is transversely isotropic in planes perpendicular to the x_3 axis, then $C_{1212}(t)$ and $C_{1313}(t)$ are equal to, say, $2G(t)$,

$$2G(t) = C_{1212}(t) = C_{1313}(t) \tag{2.16}$$

and can be determined from a torsion test about the x_3 axis. The only detailed data on a viscoelastic relaxation function over a wide range of frequency are the data on $G(t)$ reported by Lakes, Katz, and Sternstein (1979). These data will be discussed below.

Before discussing these data, the definitions of storage and loss moduli and loss tangent must be recorded. If the viscoelastic material is transversely isotropic about its x_3 axis, then the viscoelastic stress-strain relation [Eq. (2.15)] specifies the following relationship between the shearing stresses T_{12} and T_{13} and the shearing strain rates E_{12} and E_{13}:

$$\begin{aligned} T_{12} &= 2\int_{-\infty}^{t} G(t-\tau)\dot{E}_{12}(\tau)\,d\tau \\ T_{13} &= 2\int_{-\infty}^{t} G(t-\tau)\dot{E}_{13}(\tau)\,d\tau \end{aligned} \tag{2.17}$$

If a cylindrical specimen with axis along the x_3 direction is subjected to a forced harmonic motion of the form

$$E_{13}(t) = E_{12}(t) = E_0 e^{i\omega t} \tag{2.18}$$

where E_0 is the strain amplitude and ω the frequency, then the stresses T_{12} and T_{13} are given by

$$T_{12}(t) = T_{13}(t) = 2G(i\omega)E_0 e^{i\omega t} \tag{2.19}$$

where $G(i\omega)$ is called the *complex modulus*. The complex modulus $G(i\omega)$ is written as the sum of a *storage modulus* $G'(\omega)$ and a *loss modulus* $G''(\omega)$,

$$G(i\omega) = G'(\omega) + iG''(\omega) \tag{2.20}$$

The storage and loss moduli are related to the relaxation modulus $G(t)$ by

$$G'(\omega) = G_\infty + \omega \int_0^\infty \hat{G}(\tau) \sin \omega\tau\, d\tau \tag{2.21}$$

and

$$G''(\omega) = \omega \int_0^\infty \hat{G}(\tau) \cos \omega\tau\, d\tau \tag{2.22}$$

where

$$G_\infty = \lim_{t\to\infty} G(\tau) \tag{2.23}$$

and

$$\hat{G}(t) = G(t) - G_\infty \tag{2.24}$$

The *loss tangent* $\delta(\omega)$ is defined as

$$\tan \delta(\omega) = \frac{G''(\omega)}{G'(\omega)} \tag{2.25}$$

Lakes, Katz, and Sternstein (1979) reported data on the storage and loss moduli $G'(\omega)$ and $G''(\omega)$ for frequencies between 2×10^{-3} and 100 Hz and on the relaxation modulus $G(t)$ between 1 and 10^5 S for wet human and bovine cortical bone as a function of strain level E_0 and superposed axial load. The data on a typical specimen of human tibial bone at 37°C reported by these authors are illustrated in Fig. 2.8. The relaxation function $G(t)$ was determined from a

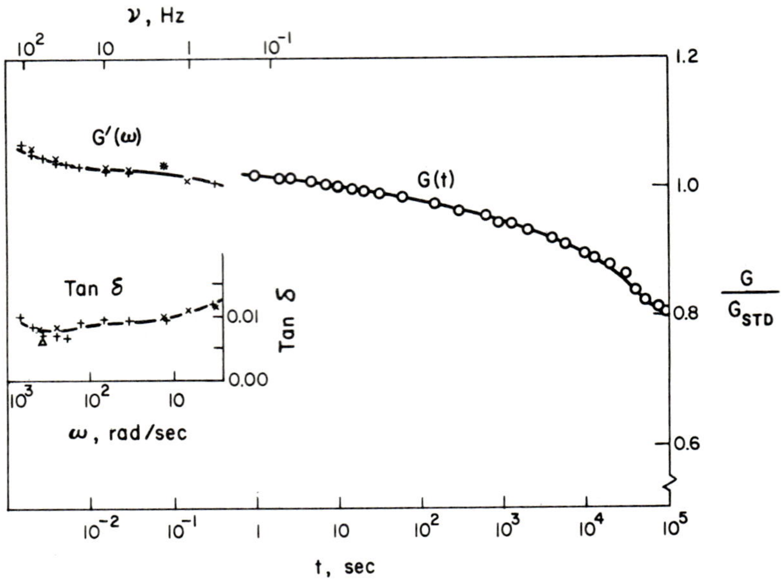

FIG. 2.8 Viscoelastic shear behavior of human tibial bone at 37°C. The ratio of the stress relaxation function $G(t)$ to a G_{STD} = 4.15 GPa is plotted as a function of time. The ratio of the storage modulus $G'(\omega)$ to the same G_{STD} is plotted as a function of frequency. The ratio of the storage modulus $G'(\omega)$ to the loss modulus $G''(\omega)$, called the loss tangent and denoted by tan $\delta(\omega)$, is plotted as a function of frequency. (*Redrawn from Lakes and Katz, 1979.*)

creep test in which a strain of 9×10^{-4} was imposed. According to linear viscoelasticity theory, the relaxation function $G(t)$ should be independent of the imposed strain in a creep test, but some dependence upon the imposed strain was found in this study. The storage modulus $G'(\omega)$ and the loss tangent are also given in the figure. They were determined at shear strains E_0 of magnitudes 3.6×10^{-n}, where n = 4, 5, and 6. They do not appear to have a dependence upon E_0 over this range of strain. According to linear viscoelasticity theory, $G'(\omega)$ and $G''(\omega)$ should be independent of E_0.

Lakes, Katz, and Sternstein (1979) were able to draw some general conclusions from their data. First, they observed that the loss tangent is of the order 0.01 in the domain 0.5 to 100 Hz and generally has a larger value at the lower frequencies. This means that the storage modulus is 100 times greater than the loss modulus in the domain 0.5 to 100 Hz. Below 0.5 Hz the loss tangent increased to about 0.025 at about 0.1 Hz. Second, the authors observed that most of the change in the relaxation shear modulus $G(t)$ occurred at long times of relaxation. Lastly, a nonlinear viscoelastic response of the bone tissue was noted. It was found that the relaxation shear modulus $G(t)$ increased with increasing strain level.

The results reported here suggest that linear viscoelasticity theory is applicable to bone tissue only under a relatively restricted set of conditions. In most situations involving physiological strains and strain rates, linear elasticity will suffice as a stress-strain relation for bone tissue. If viscoelastic effects are observed, then it is possible that they are nonlinear effects and a nonlinear theory of viscoelasticity will be necessary. The observed dependence of the elastic modulus of bone upon strain rate described by Eq. (2.2) cannot be predicted on the basis of a linear viscoelasticity theory; a nonlinear theory is needed. The studies of Lakes and Katz (1974) and Lakes, Katz, and Sternstein (1979) both indicate that the superposition principle of linear viscoelasticity is not satisfied by bone tissue. Lakes and Katz (1979) have proposed a nonlinear viscoelastic constitutive relation for bone tissue.

2.5 CORTICAL BONE STRENGTH

2.5.1 Mechanical Strength

The stress at which a material yields is called its *yield strength,* or *yield stress,* and the maximum stress a material can sustain before failure is called its *ultimate strength,* or *ultimate stress.* The maximum strain the material can sustain before failure is called the *ultimate strain* or, in the case of a torsion test, the *ultimate angle of twist* or *rotation.*

The two principal studies concerned with the determination of the yield stress, the ultimate stress, and the ultimate strain of human and bovine bone tissue are those of Reilly and Burstein (1975) and Cezayirlioglu, Bahniuk, Davy, and Heiple (1985). In both these studies the femur and tibia have been assumed to possess transversely isotropic symmetry with respect to mechanical strength. Because of the scatter in strength data, the assumption of transverse isotropy is quite reasonable. The plane of isotropy is the transverse plane of the long bone, and the unique direction is along the axis of the bone. The yield strengths in tension and compression in the axial direction will be denoted here by σ_0^+ and σ_0^-, respectively. In general, the yield strengths in tension and compression in a direction making an angle of $\theta°$ with respect to the long axis of the bone will be denoted by σ_θ^+ and σ_θ^-. Thus, for example, the yield strengths in tension and compression in the plane perpendicular to the long axis of the bone are denoted by σ_{90}^+ and σ_{90}^-, respectively. The shear strength of a torsion specimen whose axis is coincident with the long axis of the bone is denoted by τ.

Mean values for the ultimate strength of human bone as determined by Reilly and Burstein (1975) are

$$\sigma_0^+ = 133 \text{ MPa}$$

$$\sigma_0^- = 193 \text{ MPa}$$

$$\sigma_{90}^+ = 51 \text{ MPa}$$

$$\sigma_{90}^- = 133 \text{ MPa}$$

$$\tau = 68 \text{ MPa}$$

As with other bone properties, the strength of bone is dependent on the rate of loading. As stated earlier, Carter and Hayes (1977) proposed that both elastic modulus and strength were approximately proportional to the strain rate raised to the 0.06 power. In a more recent paper, Carter and Caler (1983) found that the ultimate tensile strength of bone could be expressed as

$$\sigma_0^+ = 147\dot{\varepsilon}^{0.055} \text{ MPa}$$

where $\dot{\varepsilon}$ is the strain rate expressed in s^{-1}.

The experimentally determined values of σ_0^+, σ_0^-, σ_{30}^+, σ_{30}^-, σ_{60}^+, σ_{60}^-, σ_{90}^+, σ_{90}^-, and τ are recorded in Table 2.7 for bovine femoral bone and in Table 2.8 for human femoral bone. These tables pool the data from Reilly and Burstein (1975) and Cezayirlioglu et al. (1985). These tables list not only the yield strengths but also the associated ultimate strengths and ultimate strains. We do not introduce a special notation for the ultimate strengths and ultimate strains, but the reader can easily deduce whether a particular strength is tensile or compressive or normal or shear, and its spatial orientation from the notation for the associated yield strength. The yield strengths reported by Burris (1983) were determined by the 0.3 percent offset strain method. (The 0.2 percent offset strain method is illustrated in Fig. 2.4.) The yield strengths reported by Reilly and Burstein (1975) were determined by a graphical procedure that the authors suggested was equivalent to the 0.2 percent offset strain method. Other differences in test procedure and reporting of data exist between Reilly and Burstein (1975) and Cezayirlioglu et al. (1985). The strain rates employed by Burris (1983) varied from 0.01 to 0.06 s^{-1}, while those employed by Reilly and Burstein (1975) were 0.025 s^{-1} for tension and 0.005, 0.05, and 0.5 s^{-1} for torsion. Reilly and Burstein (1975) made a distinction between haversian and laminar bovine bone, while Cezayirlioglu et al. (1985) did not make such a distinction. In spite of these differences the data reported in Tables 2.7 and 2.8 are pooled from these two references. In both references there

TABLE 2.7 The yield stress, ultimate stress, and ultimate strain for bovine bone tissue

Notation for yield stress*	Yield stress,† MPa	Ultimate stress,† MPa	Ultimate strain or rotation‡
σ_0^+	141 (17.1) [58]	156 (23.5) [66]	0.0324 (0.0161) [66]
σ_0^-	196 (18.5) [25]	237 (38.6) [49]	0.0253 (0.0096) [49]
σ_{30}^+	....	109 (9.0) [21]	0.0118 (0.0025) [21]
σ_{30}^-	....	190 (12.6) [14]	0.0250 (0.0038) [14]
σ_{60}^+	....	60 (9.8) [12]	0.0075 (0.0019) [12]
σ_{60}^-	....	148 (2.6) [2]	0.0320 (0.0006) [2]
σ_{90}^+	....	50 (12.5) [52]	0.0067 (0.0024) [52]
σ_{90}^-	150 (30.7) [22]	178 (31.1) [26]	0.0517 (0.0199) [26]
τ	57 (8.4) [21]	73 (9.6) [49]	0.3910 (0.0908) [21]

Source: Summarized and condensed from Reilly and Burstein (1975) and Cezayirlioglu et al. (1985).

* See text for explanation.

† The first number is the yield or ultimate stress in megapascals, the number in parentheses is the standard deviation, and the number in brackets is the number of specimens.

‡ The first number is the ultimate strain or the ultimate rotation in radians, the number in parentheses is the standard deviation, and the number in brackets is the number of specimens.

are reports of the age of the human donors. We have also pooled that data in Table 2.8. The strength properties of bone tissue are relatively constant across the third, fourth, fifth, and sixth decades of human life, but drop in the seventh decade.

It was pointed out by Cowin (1979) and Cezayirlioglu et al. (1985) that the strength data reported by Reilly and Burstein (1975) and Cezayirlioglu et al. (1985) can be adequately represented by the quadratic form of the Tsai-Wu criterion (1971) specialized to transversely isotropic materials. A strength criterion is a function of the stress components of $f(T_{11}, T_{22}, T_{33}, T_{12}, T_{13}, T_{23})$ which takes on a particular value, say 1, when yield occurs. The purpose of the criterion is to distinguish between stress states that will cause yielding and those that will not. Thus, if $f(T_{11}, T_{22}, T_{33}, T_{12}, T_{13}, T_{23})$ has a value less than 1, then yielding will not occur for the stress state characterized by T_{11}, T_{22}, T_{33}, T_{12}, T_{13}, and T_{23}. However, yield will occur for the stress state characterized by T_{11}, T_{22}, T_{33}, T_{12}, T_{13}, and T_{23} that satisfies the criterion

$$f(T_{11}, T_{22}, T_{33}, T_{12}, T_{13}, T_{23}) = 1 \tag{2.26}$$

The functional form of the dependence of the function f upon the stress state is restricted by the assumption of transverse isotropy and by neglecting terms of third and higher order in the stress. The form of the restricted equation is given by Cowin (1979) as

$$a_1 T_{33} + a_2(T_{11} + T_{22}) + a_3 T_{33}^2 + a_4(T_{11} + T_{22})^2 + a_5(T_{23}^2 + T_{13}^2) + a_6(T_{12}^2 - T_{22}T_{11}) + a_7 T_{33}(T_{11} + T_{22}) = 1 \tag{2.27}$$

where the x_3 direction lies along the axis of the bone and the coefficients a_1 to a_7 are to be

TABLE 2.8 The yield stress, ultimate stress, and ultimate strain for human bone tissue

Notation for yield stress*	Yield stress,† MPa	Ultimate stress,† MPa	Ultimate strain or rotation‡
σ_0^+	115 (11.2) [202]	133 (14.1) [202]	0.0293 (0.0094) [202]
σ_0^-	182 (14.4) [19]	195 (19.6) [114]	0.0220 (0.0057) [114]
σ_{30}^+	....	100 (10.4) [23]	0.0198 (0.0083) [23]
σ_{30}^-	....	173 (13.8) [5]	0.0280 (0.0052) [5]
σ_{60}^+	....	61 (12.2) [19]	0.0069 (0.0022) [19]
σ_{60}^-	....	133 (15.0) [7]	0.0311 (0.0010) [7]
σ_{90}^+	....	51 (10.1) [31]	0.0072 (0.0016) [31]
σ_{90}^-	121 (9.2) [3]	133 (16.7) [13]	0.0462 (0.0260) [13]
τ	54 (5.2) [19]	69 (4.4) [31]	0.3299 (0.0890) [19]

Source: Summarized and condensed from Reilly and Burstein (1975) and Cezayirlioglu et al. (1985).

* See text for explanation.

† The first number is the yield or ultimate stress in megapascals. The number in parentheses is the standard deviation, and the number in brackets is the number of specimens.

‡ The first number is the ultimate strain or the ultimate rotation in radians, the number in parentheses is the standard deviation, and the number in brackets is the number of specimens.

determined from strength tests. Following Tsai and Wu (1971) the coefficients a_1 to a_7 are associated with the data from strength tests by the following prescriptions:

$$a_1 = \frac{1}{\sigma_0^+} - \frac{1}{\sigma_0^-} \qquad a_2 = \frac{1}{\sigma_{90}^+} - \frac{1}{\sigma_{90}^-} \qquad a_3 = \frac{1}{\sigma_0^+\sigma_0^-}$$
$$a_4 = \frac{1}{\sigma_{90}^+\sigma_{90}^-} \qquad a_5 = \frac{1}{\tau^2} \qquad a_6 = \frac{1}{(\tau')^2} \qquad a_7 = 2F \tag{2.28}$$

where σ_0^+, σ_0^-, σ_{90}^+, σ_{90}^-, and τ are as defined above, τ' is the shear strength in the plane of symmetry, and F is a stress interaction coefficient that represents the effect of a transverse stress on the strength in a particular direction. The experimental data on σ_0^+, σ_0^-, σ_{90}^+, σ_{90}^-, and τ are presented in Tables 2.7 and 2.8. There are no data reported on τ'. Tsai and Wu (1971) argue that F should be determined in a biaxial stress experiment, but this has not been done for bone. Cowin (1979) estimated the value of F from the uniaxial strength data of Reilly and Burstein (1975). For bovine femoral bone, he found that $F = -(145 \text{ Mpa})^{-2}$, and for human femoral bone, $F = -(200 \text{ MPa})^{-2}$.

The advantage of the strength criterion shown in Eq. (2.27) is that the normal or shear strength in any direction can be predicted from the criterion and the knowledge of the values of σ_0^+, σ_0^-, σ_{90}^+, σ_{90}^-, τ, τ', and F. On the basis of data reported to date, two types of predictions can be compared with the experimental results. These predictions are for the off-the-grain angle strength σ_θ^+ or σ_θ^- and the strength in a combined tension-torsion experiment. The strength criterion in Eq. (2.27) predicts that, in the case of uniaxial tension and compression tests on a specimen whose axis makes an angle of $\theta°$ with the long axis of the bone, the product of the tensile and compressive strengths σ_θ^2,

$$\sigma_\theta = \sigma_\theta^+ \sigma_\theta^- \tag{2.29}$$

is related to the products σ_0^2, σ_{90}^2, F, and τ by

$$\sigma_\theta^2 = \sigma_0^2 \sigma_{90}^2 \left[\sigma_0^2 \sin^4 \theta + \sigma_{90}^2 \cos^4 \theta + \sigma_0^2 \sigma_{90}^2 \left(2F + \frac{1}{\tau^2} \right) \sin^2 \theta \cos^2 \theta \right]^{-1} \tag{2.30}$$

This formula may be used to predict σ_{30}^2 and σ_{60}^2, and when the predictions are compared with the data of Reilly and Burstein given in Tables 2.7 and 2.8, the results are quite satisfactory. The details of this comparison are given by Cowin (1979).

The second situation in which a prediction of the strength criterion of Eq. (2.27) has been compared with the experimental data is that of combined torsion and axial loading. This comparison, reported by Cezayirlioglu et al. (1985), showed that the criterion (2.27) was an excellent predictor of yield strength. In the special case of combined torsion and axial loading, the strength criterion (2.27), with the values of a_1 to a_7 given by (2.28), reduces to

$$\left(\frac{1}{\sigma_0^+} - \frac{1}{\sigma_0^-} \right) T_{33} + \frac{T_{33}^2}{\sigma_0^+ \sigma_0^-} + \frac{1}{\tau^2} T_{23}^2 = 1 \tag{2.31}$$

There are several curves showing the excellent comparison of this formula with experimental data in Cezayirlioglu et al. (1985).

2.5.2 Fatigue Strength

The fatigue strength of a material is normally determined by subjecting the material to a cyclic stress and determining the number of cycles to failure. Most tests are conducted with bending loads at high loading frequencies. The results are presented in terms of a graph of the stress amplitude versus the number of cycles to failure. Bone has also been subjected to such testing. The results of several fatigue tests of bone are shown in Fig. 2.9.

Carter et al. (1981a and b) conducted uniaxial fatigue tests of human cortical bone at a physiological strain rate of 0.01 s^{-1}. In plotting their results in terms of stress amplitude, they found a great deal of scatter. Further investigation revealed a strong correlation between fatigue life and the Young's modulus of the specimen. Carter et al. (1981a and b) conducted their fatigue testing using strain ranges from 0.005 to 0.01 with mean strains of −0.002, 0, and +0.002. They found that the fatigue life did not correlate well with maximum stress, maximum strain, mean stress, or mean strain, but was determined primarily by the cyclic strain range. The results of their testing are shown in Fig. 2.10. Their results suggest that the fatigue life of bone is very much less than that indicated by earlier researchers. They attribute this difference primarily to the use of uniaxial testing rather than rotating bending or flexural testing. They point out that this low fatigue life suggests that bone is constantly accumulating fatigue damage during normal activity and that the normal process of bone remodeling is necessary to maintain the long-term structural integrity of the skeletal system.

2.6 THE ELASTIC MODULUS OF CANCELLOUS BONE

The mechanical properties of cancellous bone have not been investigated to the same degree as those for cortical bone. Cancellous bone is known to be anisotropic, but the degree of elastic anisotropy has not been measured. The elastic modulus has been shown to depend on the type of trabecular architecture and, very strongly, on the apparent density or porosity of the tissue. Most of the studies of elastic modulus have measured the modulus as a function of the apparent

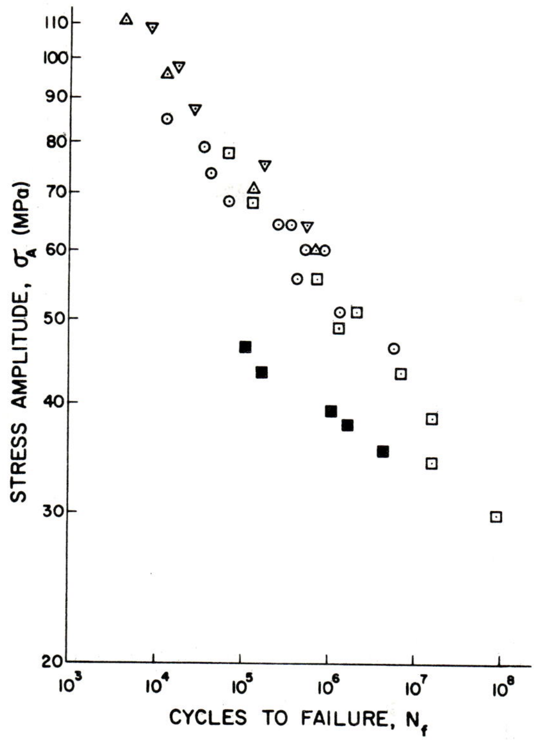

FIG. 2.9 The stress-related fatigue behavior of in vitro bone under completely reversed loading, as reported in several studies: solid squares (Gray and Korbacher, 1974), open circles (Swanson et al., 1971), upward triangles (Lafferty and Raju, 1979), downward triangles (Carter et al., 1976), open squares (King and Evans, 1967). (*Redrawn from Carter et al., 1981b.*)

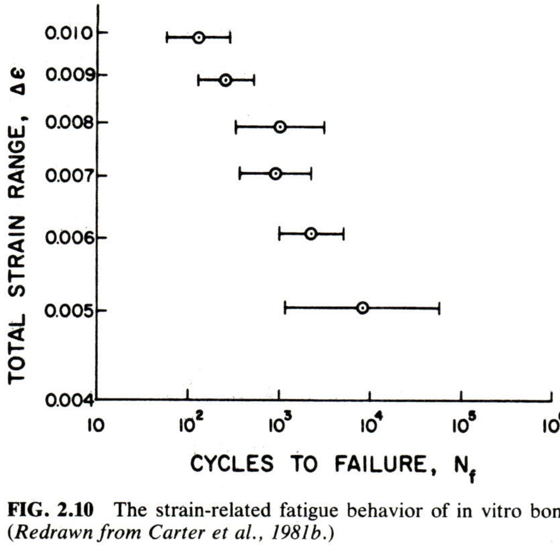

FIG. 2.10 The strain-related fatigue behavior of in vitro bone. (*Redrawn from Carter et al., 1981b.*)

or bulk density. The apparent density of cancellous bone was defined in the section on the measurement of bone density.

Carter and Hayes (1977) determined the elastic modulus of both bovine and human trabecular bone as a function of both strain rate and apparent density. In their analysis of these data they lumped both the fresh bovine and human trabecular bone with embalmed human and fresh bovine cortical bone to obtain a representation which suggests that this combination of bone tissues has a dependence upon the third power of apparent density and the 0.06 power of strain rate:

$$E = E_0 \left(\frac{\dot{\varepsilon}}{\dot{\varepsilon}_0}\right)^{0.06} \left(\frac{\rho}{\rho_0}\right)^3 \tag{2.32}$$

where E is the modulus, $\dot{\varepsilon}$ is the strain rate, ρ is the density, and E_0, ρ, and $\dot{\varepsilon}_0$ are reference values of these three parameters. The reference values E_0, $\dot{\varepsilon}_0$, *and* ρ_0 are taken to be those for human cortical bone tested at a strain rate of 1.0 s^{-1}. The modulus is 22.1 GPa and the density is 1.8 g/cm^3. With these numbers, Eq. (2.32) can be written in the form

$$E = 3790\dot{\varepsilon}^{0.06}\rho^3 \tag{2.33}$$

where E is in GPa, $\dot{\varepsilon}$ is in s^{-1}, and ρ is in g/cm^3. The data of Carter and Hayes (1977) were obtained in compression tests. Carter, Schwab, and Spengler (1980) conducted tensile tests on the trabecular bone from an 84-year-old human, and they suggest that their tensile test data were consistent with the compressive test data given earlier by Carter and Hayes (1977). Bensusan, Davy, Heiple, and Verdin (1983) recently reported data that suggest the modulus of bovine cancellous bone depends on the square of the apparent density.

2.7 THE STRENGTH OF CANCELLOUS BONE

The compressive strength of cancellous bone is generally taken to be the maximum stress achieved prior to pore collapse (Carter and Hayes, 1977). The tensile strength (Carter, Schwab, and Spengler, 1980) and the shear strength (Stone, Beaupre, and Hayes, 1983) are the stress at fracture of the trabeculae.

Carter and Hayes (1977) determined the strength of bovine and human cancellous bone. As was the case with their analysis of the elastic modulus data, they lumped both the fresh bovine and human cancellous bone compressive strength data with embalmed human and fresh bovine cortical bone data to obtain a multi-bone tissue representation for bone strength. This multitissue representation has a dependence on the square of apparent density and the strain rate to the 0.06 power:

$$\sigma^- = \sigma_0^- \left(\frac{\dot{\varepsilon}}{\dot{\varepsilon}_0}\right)^{0.06} \left(\frac{\rho}{\rho_0}\right)^2 \tag{2.34}$$

where σ^- is the compressive strength, $\dot{\varepsilon}$ is the strain rate, ρ is the density, and σ_0^-, $\dot{\varepsilon}_0$, and ρ_0 are reference values of these quantities. When the reference value of strength is taken as that of compact bone tested at a strain rate of 1.0 s^{-1}, σ_0^- is 221 MPa at a ρ_0 of 1.8 g/cm^3. With these numbers, Eq. (2.34) can be written in the form

$$\sigma^- = 68\dot{\varepsilon}^{0.06}\rho^2 \tag{2.35}$$

where σ^- is in MPa, $\dot{\varepsilon}$ is in s^{-1}, and ρ is in g/cm^3. Carter, Schwab, and Spengler (1980) indicated that the tensile strength of cancellous human bone also depends upon the square of the apparent density. Other investigators have noted a linear dependence of strength on apparent density for specific cancellous tissues. For cancellous tissue from the human vertebral bodies, Galante, Rostoker, and Day (1970) obtain a relation of the form

$$\sigma^- = -0.68 + 12.55\rho \tag{2.36}$$

while Robertson and Smith (1978) obtain a relation of the form

$$\sigma^- = -146 + 98\rho \tag{2.37}$$

for porcine mandibular tissue. In these representations the dimensions of σ^- are megapascals and those of ρ are grams per cubic centimeter. However, recent data on bovine cancellous bone reported by Bensusan et al. (1983) indicate that the strength depends upon the square of the apparent density as suggested by the multitissue relation (2.35) of Carter and Hayes (1977).

Stone et al. (1983) report shear strength data on bovine trabecular bone. The shear strength τ was found to be related to the apparent density ρ by

$$\tau = 21.6\rho^{1.65} \tag{2.38}$$

where τ is in megapascals and ρ is in grams per cubic centimeter.

ACKNOWLEDGMENT

This work was supported by an NIH-NIADDKD grant entitled "Bone Remodeling of the Femur" and by the Solid Mechanics Program at the National Science Foundation. We thank Dennis Carter, Dwight Davy, Richard T. Hart, and Roderick Lakes for helpful comments on an earlier draft of this manuscript.

REFERENCES

Abendschein, W., and Hyatt, G. W. (1970). Ultrasonic and selected properties of bone. *Clinical Orthop. Relat. Res.*, *69*, 294–301.

Ambardar, J. D., and Ferris, C. D. (1976). A simple technique for measuring certain elastic moduli in bone. *Biomed. Sci. Instr.*, *12*, 23–27.

Ashman, R. B. (1982). Ultrasonic determination of the elastic properties of cortical bone: Techniques and limitations. Ph.D. dissertation, Tulane University, New Orleans.

Ashman, R. B., Cowin, S. C., Van Buskirk, W. C., and Rice, J. C. (1984). A continuous wave technique for the measurement of the elastic properties of bone. *J. Biomechan.*, *17*, 349–361.

Ashman, R. B., Donofrio, M., Cowin, S. C., and Van Buskirk, W. C. (1982). Postmortem changes in the elastic properties of bone. *Proc. 28th Annu. ORS*. New Orleans, p. 63.

Bensusan, J. S., Davy, D. T., Heiple, K. G., and Verdin, P. J. (1983). Tensile, compressive and torsional testing of cancellous bone. *Trans. Orthop. Res. Soc.*, *8*, 132.

Biltz, R. M., and Pellegrino, E. D. (1969). The chemical anatomy of bone. *J. Bone Jt. Surg.*, *51A*, 456–466.

Bonfield, W., and Tully, A. E. (1982). Ultrasonic analysis of the Young's modulus of cortical bone. *J. Biomed. Eng.*, *4*, 23–27.

Burris, C. L. (May 1983). A correlation of quasistatic and ultrasonic measurements of the elastic properties of cortical bone. Ph.D. dissertation, Tulane University, New Orleans.

Carter, D. R., and Caler, W. E. (1983). Cycle dependent and time dependent bone fracture with repeated loading. *J. Biomech. Eng.*, *105*, 166–170.

Carter, D. R., Caler, W. E., Spengler, D. M., and Frankel, V. H. (1981a). Fatigue behavior of adult cortical bone—The influence of mean strain and strain range. *Acta Orthop. Scand.*, *52*, 481–491.

Carter, D. R., Caler, W. E., Spengler, D. M., and Frankel, V. H. (1981b). Uniaxial fatigue of human cortical bone—The influence of tissue physical characteristics. *J. Biomech.*, *14*, 461–474.

Carter, D. R., and Hayes, W. C. (1976). Bone compressive strength: The influence of density and strain rate. *Science*, *194*, 1174–1176.

Carter, D. R., and Hayes, W. C. (1977). The compressive behavior of bone as a two-phase porous structure. *J. Bone Jt. Surg.*, *59A*, 954–962.

Carter, D. R., Hayes, W. C., and Schurman, D. J. (1976). Fatigue life of compact bone: II. Effects of microstructure and density. *J. Biomech.*, *9*, 211.

Carter, D. R., Schwab, G. H., and Spengler, D. M. (1980). Tensile fracture of cancellous bone. *Acta Orthop. Scand., 51,* 733–741.

Carter, D. R., and Spengler, D. M. (1978). Mechanical properties and composition of cortical bone. *Clin. Orthop., 135,* 192–214.

Cezayirlioglu, H., Bahniuk, E., Davy, D. T., and Heiple, K. G. (1985). Anisotropic yield behavior of bone under combined axial force and torque, *J. Biomech., 18,* 61–70.

Christensen, R. M. (1971). *Theory of Viscoelasticity*. Academic Press, New York.

Cowin, S. C. (1979). On the strength anisotropy of bone and wood. *J. Appl. Mech., 46,* 832–838.

Crandall, S. H., Dahl, N., and Lardner, T. J. (1978). *An Introduction to the Mechanics of Solids* (2d ed.). McGraw-Hill, New York.

Crowninshield, R. D., and Pope, M. H. (1974). The response of compact bone in tension at various strain rates. *Ann. Biomed. Eng., 2,* 217–225.

Currey, J. D. (1960). Differences in the blood supply of bone of different histological types. *Q. J. Microsc. Sci., 101,* 351–370.

Currey, J. D. (1965). Anelasticity of inelastic flow in bone. *J. Exp. Biol., 43,* 279–292.

Currey, J. D. (1969). The mechanical consequences of variation in the mineral content of bone. *J. Biomech., 2,* 1–11.

Currey, J. D. (1969). The relationship between the stiffness and the mineral content of bone. *J. Biomech., 2,* 477–480.

Evans, F. G. (1973). *Mechanical Properties of Bone*. Charles C Thomas, Springfield, Ill.

Fritz, J. S., and Schenk, G. H. (1974). *Quantitative Analytical Chemistry* (3d ed.). Allyn and Bacon, Boston.

Galante, J., Rostoker, W., and Day, R. D. (1970). Physical properties of trabecular bone. *Calif. Tiss. Res, 5,* 236–246.

Glimcher, M. J. (1976). Composition, structure and organization of bone and other mineralized tissues and the mechanisms of calcification. In *Handbook of Physiology: Endocrinology*. American Physiological Society vol. VII, pp. 25–116.

Gong, J. K., Arnold, J. S., and Cohn, S. H. (1964). Composition of trabecular and cortical bone. *Anat. Rec., 149,* 325–331.

Gray, R. J., and Korbacher, G. K. (1974). Compressive fatigue behavior of bovine compact bone. *J. Biomech., 7,* 287.

Hayes, M. A., and Morland, L. W. (1969). The response functions of an anisotropic linear viscoelastic material. *Trans. Soc. Rheol., 13,* 231–240.

Hight, T. K., and Brandeau, J. F. (1983). Mathematical modeling of the stress strain–strain rate behavior of bone using the Ramberg-Osgood Equation. *J. Biomech., 16,* 445–450.

Huiskes, R. (1982). On the modelling of long bones in structural analysis. *J. Biomech., 15,* 65–69.

Jowsey, J. (1968). Age and species differences in bone. *Cornell Vet. 83* (suppl.: *Symposium on Equine Bone and Joint Diseases*), 74–94.

Katoh, T., Sakamaki, K., and Nakamura, T. (1981). A new method for measurement of bone mineral deposition pattern—using a dual energy radiographic densitometry method in a single exposure. *Bull. Tokyo Med. Dent. Univ., 28,* 91–98.

Kerr, S. A., Kouris, K., Webber, C. E., and Kennett, T. J. (1980). Coherent scattering and the assessment of mineral concentration in trabecular bone. *Phys. Med. Biol., 25,* 1037–1047.

King, A. I., and Evans, F. G. (1967). Analysis of the fatigue strength of human compact bone by the weibull method. In *Digest 7th Int. Conf. Med. Biol. Eng.,* p. 514.

Knets, I., and Malmeisters, A. (1977). Deformability and strength of human compact bone tissue. In G. Brankov (ed.), *Mechanics of Biological Solids: Proceedings of the Euromech Colloquium 68,* Varna, Bulgaria, September 1975. Bulgarian Academy of Sciences, Sofia, pp. 133–141.

Krolner, B., Pors Nielsen, S., Lund, B., Sorensen, O. H., and Chrenholdt, A. (1980). Measurement of bone mineral content (BMC) of the lumbar spine: I. Theory and application of a new two-dimensional dual-photon attenuation method. *Scand. J. Clin. Lab. Invest., 40,* 653–663.

Lafferty, J. F., and Raju, P. V. V. (1979). The influence of stress frequency on the fatigue strength of cortical bone. *J. Biomed. Eng., 101,* 112–117.

Lakes, R. S., and Katz, J. L. (1974). Interrelationships among the viscoelastic functions for anisotropic solids: Application to calcified tissues and related systems. *J. Biomech., 7,* 259–270.

Lakes, R. S., and Katz, J. L. (1979). Viscoelastic properties of wet cortical bone: III. A non-linear constitutive equation. *J. Biomech., 12,* 689–698.

Lakes, R. S., Katz, J. L., and Sternstein, S. S. (1979). Viscoelastic properties of wet cortical bone: I. Torsional and biaxial studies. *J. Biomech., 12,* 657–678.

Lang, S. B. (1970). Ultrasonic method for measuring elastic coefficients of bone and results on fresh and dried bovine bones. *IEEE Trans., Biomed. Eng., 17,* 101–105.

Leichter, I., Margulies, S. Y., Weinreb, A., Mizrahi, J., Robin, G. C., Conforty, B., Makin, M., and Bloch, B. (1982). The relationship between bone density, mineral content and mechanical strength in the femoral neck. *Clin. Orthop., 163,* 272–281.

Lockett, F. J. (1972). *Nonlinear Viscoelastic Solids.* Academic Press, London.

Marino, A. A., Becker, R. O., and Bachman, C. H. (1967). Dielectric determination of bound water of bone. *Phys. Med. Biol., 12,* 367–378.

McElhaney, J. H. (1966). Dynamic response of bone and muscle tissue. *J. Appl. Physiol. 21,* 1231–1236.

McSkimin, H. S. (1961). Notes and references for the measurement of elastic moduli by means of ultrasonic waves. *J. Acoust. Soc., 33,* 606–615.

Pelker, R. R., and Saha, S. (1983). Stress wave propagation in bone. *J. Biomech., 16,* 481–489.

Reilly, D. T., and Burstein, A. H. (1974). The mechanical properties of cortical bone. *J. Bone Jt. Surg., 56A,* 1001–1022.

Reilly, D. T., and Burstein, A. H. (1975). The elastic and ultimate properties of compact bone tissue. *J. Biomech., 8,* 393–405.

Reilly, D. T., Burstein, A. H., and Frankel, V. H. (1974). The elastic modulus for bone. *J. Biomech., 7,* 271–275.

Robertson, D. M., and Smith, D. C. (1978). Compressive strength of mandibular bone as a function of microstructure and strain rate. *J. Biomech., 11,* 455–471.

Saulgozis, Y. (1976). Differences and correlations between the elastic characteristics of the compact bone tissue of the human tibia. *Polymer Mechanics* [trans. of *Mekhanika Polimerov*], *11,* 559–563.

Sedlin, E. D., and Hirsch, C. (1966). Factors affecting the determination of the physical properties of femoral cortical bone. *Acta Orthop. Scand., 37,* 29–48.

Stone, J. L., Beaupre, G. S., and Hayes, W. C. (1983). Multiaxial strength characteristics of trabecular bone. *J. Biomech., 9,* 743–752.

Swanson, S. A. V., Freeman, M. A. R., and Day, W. H. (1971). The fatigue properties of human cortical bone. *Med. Biol. Eng., 9,* 23.

Timmins, P. A., and Wall, J. C. (1977). Bone water. *Calif. Tiss. Res., 23,* 1–5.

Torzilli, P. A., Takebe, K., Burstein, A. H., and Heiple, K. G. (1981). Structural properties of immature canine bone. *J. Biomech. Eng., 103,* 232–238.

Torzilli, P. A., Takebe, K., Burstein, A. H., Zika, J. M., and Heiple, K. G. (1982). The material properties of immature bone. *J. Biomech. Eng., 104,* 12–20.

Tsai, S. W., and Wu, E. M. (1971). A general theory of strength for anisotropic materials. *J. Compos. Mater., 5,* 58–80.

Van Buskirk, W. C., and Ashman, R. B. (1981). The elastic moduli of bone. In S. C. Cowin (ed.), *Mechanical Properties of Bone.* American Society of Mechanical Engineers, New York, p. 45.

Van Buskirk, W. C., Cowin, S. C., and Ward, R. N. (1981). Ultrasonic Measurement of orthotropic elastic constants of bovine femoral bone. *J. Biomech. Eng., 103,* 67–72.

Vose, G. P., and Kubala, A. L. (1959). Bone strength—Its relationship to x-ray determined ash content. *Hum. Biol., 31,* 262–270.

Wood, J. L. (1971). Dynamic response of human cranial bone. *J. Biomech., 4,* 1–12.

Yamada, H. (1970). In F. G. Evans (ed.), *Strength of Biological Materials.* Williams & Wilkins, Baltimore.

Yanson, K. A., Knets, I. V., and Saulgozis, Y. Z. (1974). Physiological significance of changes in bone volume associated with deformation. *Polymer Mechanics* [trans. of *Mekhanika Polimerov*], *10,* 695–703.

Yoon, H. S., and Katz, J. L. (1976). Ultrasonic wave propagation in human cortical bone: II. Measurements of elastic properties and micro-hardness. *J. Biomech., 9,* 459–464.

Yoon, H. S., and Katz, J. L. (1979). Temperature dependence of the ultrasonic velocities in bone. *1979 IEEE Ultrasonics Symposium Proceedings,* 395–398.

CHAPTER 3

Electromechanical Effects in Bone

Marvin W. Johnson
Physics Department
California State University, Stanislaus
Turlock, California

J. Lawrence Katz
Biomedical Engineering Department
Rensselaer Polytechnic Institute
Troy, New York

INTRODUCTION

Since the report on the piezoelectric effect in bone by Fukada and Yasuda (1957) and its independent confirmation by Bassett and Becker (1962), this property of bone has been a subject of considerable interest to both the researcher and clinician. Several authors have suggested that the electromechanical effect in bone may be an important factor affecting bone remodeling processes (Bassett, 1971), a hypothesis made plausible by the reports that electrical stimulation produces osteogenesis. However, it is not known how electrical stimulation influences bone formation or healing. Although much of the work on electromechanical effects is justified because of the possible relation to bone remodeling, there is little or no direct evidence that electromechanical effects actually influence the remodeling process.

A number of people have studied the electromechanical properties of bone under different sample conditions: wet, dry, partially hydrated, in vitro, and even in vivo. Recent work indicates that two different mechanisms are responsible for the effects in wet and dry bone. In dry bone a piezoelectric effect produces the observed signals, and in wet bone the signals are due to electrokinetic phenomena, i.e., streaming potentials (Eriksson, 1976; Johnson et al., 1980a; Gross and Williams, 1982; and Pienkowski and Pollack, 1983). Recent reviews by Williams (1982) and Grodzinsky (1983—primarily on cartilage, but it contains some information on bone) have covered some aspects of electromechanical effects in bone.

In this article we will focus on recent work and on our understanding of the microscopic origins and the physical characteristics of electromechanical effects and the influence of sample conditions on those effects.

3.1 PIEZOELECTRIC EFFECTS: DRY BONE

3.1.1 Theory: Homogeneous Materials

This section offers a brief summary of piezoelectric effects; a more complete treatment may be found in Cady (1964). The basic piezoelectric effect is the production of a polarization by a stress (or strain) in the material:

$$P_i = T_j d_{ij} \tag{3.1}$$

where P is the polarization, T is the stress, and d is the piezoelectric constant (sum over repeated indices; note that i runs from 1 to 3 and j from 1 to 6, using the usual convention for elasticity and piezoelectricity). The units of d are coulombs per newton (C/N) and are often quoted in picocoulombs per newton (pC/N). For a thin, flat sample with the polarization along the thickness, the 1 direction (Fig. 3.1), the charge density on the surface is

$$q = P_1 = T_j d_{ij} \tag{3.2}$$

The open-circuit voltage is just the total charge divided by the capacitance of the parallel-plate capacitor formed by the sample, or

$$V = \frac{tP_1}{\varepsilon_{11}} = \frac{tT_j d_{1j}}{\varepsilon_{11}} \tag{3.3}$$

where t is the thickness and ε is the permittivity (a second-rank tensor). The commonly measured quantity is the short-circuited charge (the charge necessary to hold the potential across the sample at zero). It is q times the area A of the plate. Suppressing the subscripts, it is written

$$Q = TAd \tag{3.4}$$

For a homogeneous, single crystal, d does not depend on the dielectric constants, which implies

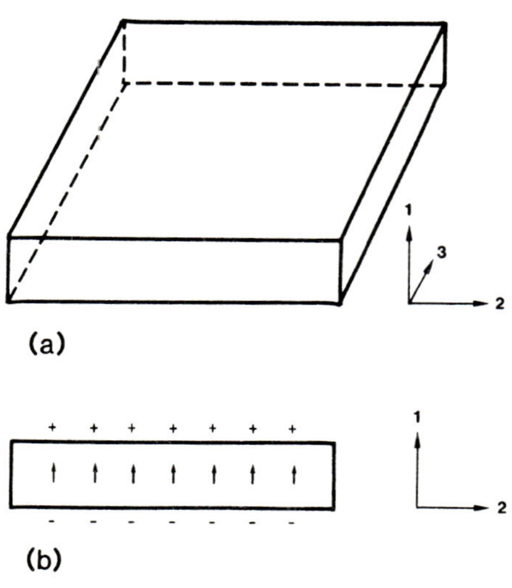

FIG. 3.1 (*a*) Flat slab of a piezoelectric material, showing the coordinate system. (*b*) Side view of the slab, showing the polarization (arrows) produced by the piezoelectric effect. The + and − signs indicate the bound charge due to the polarization.

that the charge does not depend on the dielectric constant. However, the open-circuit voltage does depend on ε. Thus a measurement of the voltage is influenced by the dielectric properties, while a charge measurement is not.

3.1.2 Theory: Composites

For composite materials, the macroscopic piezoelectric constant is a function of the piezoelectric moduli, elastic moduli, dielectric constants, and geometry of the different phases (Date, 1972). Since bone is a composite made up of mineral and protein phases, its piezoelectric properties will depend on these same quantities. It is generally believed that the protein collagen is responsible for the piezoelectric effect in bone. To quantify the macroscopic piezoelectric behavior of bone in terms of the piezoelectric properties of the collagen, one would have to take all these factors into account. One cannot simply add the tensor properties, but must solve the stress distribution and the electric boundary value problems associated with the microscopic texture. For example, one would have to know what stresses are coupled to the collagen fibers when bone is stressed. Although there have been attempts to relate the piezoelectric properties of bone to those of collagen, the authors are not aware of any treatment that comes close to solving these problems.

It is also apparent that the dielectric properties (including the conductivity) have a strong influence on the piezoelectric effect, especially in composites, and the presence of a finite conductivity introduces a frequency (or time) dependence into the measured voltages. A simple system is represented by a dielectric (a capacitor) in parallel with a resistance (Fig. 3.2), where the ratio ε_1/g is just the RC time constant. In this case ε can be a complex quantity; the real part, ε_1, represents the dielectric constant, and the imaginary part, ε_2, represents the conductivity g divided by the frequency ω. Then

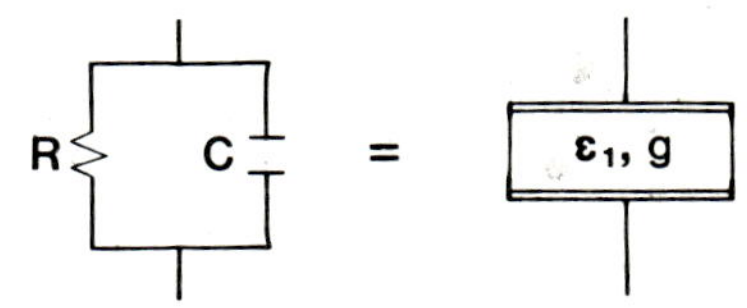

FIG. 3.2 Dielectric represented by a parallel *RC* circuit, where *R* corresponds to $1/g$ and *C* corresponds to ε_1.

$$\varepsilon = \varepsilon_1 - \frac{ig}{\omega} \tag{3.5}$$

For $\varepsilon_1 > g/\omega$, the permittivity, and therefore the voltage in Eq. (3.3), is independent of frequency. For $\varepsilon_1 < g/\omega$, the permittivity is inversely proportional to ω and the voltage is proportional to ω. The expected frequency dependence of the open-circuit voltage is shown in Fig. 3.3. In reality the resistance and the capacitance in Fig. 3.2 depend on frequency, so ε_1 is a function of frequency as well (cf. Reinish and Nowick, 1976; and Lakes et al. 1977). Nevertheless the simple model on Fig. 3.2 provides a good conceptual tool for understanding the behavior of bone.

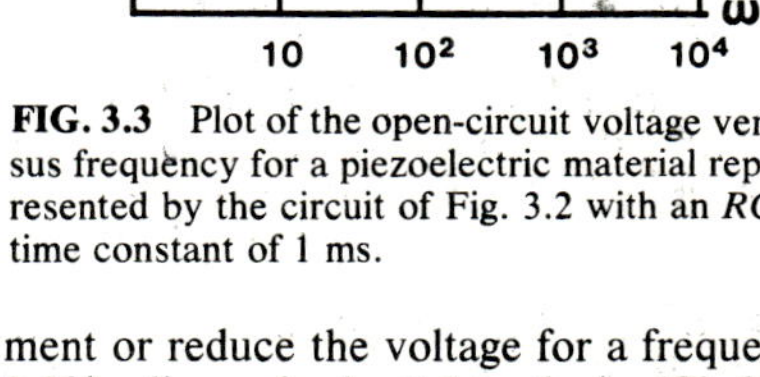

FIG. 3.3 Plot of the open-circuit voltage versus frequency for a piezoelectric material represented by the circuit of Fig. 3.2 with an *RC* time constant of 1 ms.

For very dry bone, $g < 10^{-10}$ mho/m and $\varepsilon_1 = 6$ to 10×10^{-11}. Therefore the voltage in Eq. (3.2) will be independent of frequency for $\omega > 2$ rad/s, 0.3 Hz, or, for a step function stress, the decay time will be about 0.5 s. (Decay times on the order of several seconds have been measured in very dry bone.)

As bone becomes hydrated, g increases and ε_1 increases too, but more slowly. This will shorten the decay time for a step function voltage measurement or reduce the voltage for a frequency measurement. By the time bone is saturated with 0.9% saline, g is about 0.1 mho/m (Chakkalakal et al., 1980; Kosterich et al., 1983), resulting in an expected reduction in voltage (from the dry bone value) of more than 10^6 at $\omega = 10$ (Johnson et al., 1980a). The corresponding decay time for a step function stress would be $< 10\ \mu$s.

The short-circuited charge given by Eq. (3.2) would not be directly affected by the changes in the dielectric properties. The voltage is affected because it is inversely proportional to ε. The only way ε will influence Q is if d depends on ε. For a composite, the ε's of the different phases will affect the macroscopic d. The dependence of d on the ε's will be determined by the geometric relationship of the different phases, and for certain geometries an increase in ε_1 or g will result in an increase in d (cf. the effect of removing an air gap between the electrodes and the sample in Cady, 1964).

Because of the complex geometry involved, it is difficult to predict exactly what will happen to d as bone is taken from the dry state to a wet state. The best one can do is to estimate limits on its behavior (Johnson et al., 1980a), and these suggest that there could be a small increase in d. However, it is hard to perform charge measurements on highly conductive materials.

3.1.3 Experimental Results

It is difficult to compare the results of different experiments because few investigators have reported on the dielectric properties of their samples or have determined the state of hydration. One of the better studies is that of Reinish and Nowick (1975, 1976), but they primarily measured the dielectric properties and measured the piezoelectric coefficients only at higher frequencies.

Reinish and Nowick reported, as did Ghilardinetto and Zimmerman (1975), that d_{14}, the largest piezoelectric constant for bone, decreased with increasing moisture content. Reinish and Nowick made their measurements in the kilohertz region, and Ghilardinetto and Zimmerman made their measurements at 50 Hz.

Bur's work (1976) is also interesting. He measured the low-frequency piezoelectric constants as a function of hydration and frequency. At high hydrations (relative humidity $> 80\%$) and low frequencies (< 10 Hz) he found that some of the piezoelectric constants increased with decreasing frequency. While small increases might be expected, it would be difficult to explain a large increase.

In all these measurements the pores were probably *not* filled with fluid. (Reinish's dc resistivities for bone at 100% relative humidity were > 50 times those measured for fluid saturated bone, implying that the pores were not completely filled with fluid.)

The open-circuit voltages measured for very dry bone are fairly large. For a shear stress of 10^6 Pa (10 atm) the shear term d_{14} (about 0.1 to 0.2 pC/N) produces an electric field of about 1 V/mm. This is the largest piezoelectric term for bone, and other terms (d_{24} and the compressive coefficients) are an order of magnitude smaller (cf. Liboff and Shamos, 1973). For a given stress, the open-circuit voltages for the piezoelectric effect in dry bone are much larger than similar voltages (due to streaming potentials) measured in wet bone.

When bone is saturated with fluid, one would not be able to measure an appreciable voltage due to a piezoelectric effect at low frequencies because the conductivity of the fluid will "short out" the voltage produced by the piezoelectric effect. This seems to be the case. In a cantilever bending experiment Gross and Williams (1982) measured the signal that would be produced by d_{14}, an electric field perpendicular to the plane of bending. The signal was prominent in dry bone but appeared to disappear in wet bone. Innacone et al. (1979) reported that they could not measure a net voltage across a wet sample in simple compression, a result consistent with Gross's measurements. However, at high frequencies, perhaps in the upper audio range, the open-circuit voltage produced by the piezoelectric effect might be appreciable.

One other point should be mentioned with respect to piezoelectricity in bone, and that is the role of the inhomogeneous nature of bone. There were some anomalies in the signals measured in cantilever bending that could not be explained by a piezoelectric effect in a homogeneous material. As a result a polarization proportional to the gradient of the stress was proposed to explain the data. Recently Johnson et al. (1980b) suggested that the results might be due to a piezoelectric effect and the inhomogeneous nature of bone. This is supported by evidence that the signal in cantilever bending is quite variable (Johnson et al., 1980a). This would mean that piezoelectric measurements from different samples (with different orientations) cannot be compared to derive a piezoelectric matrix for bone.

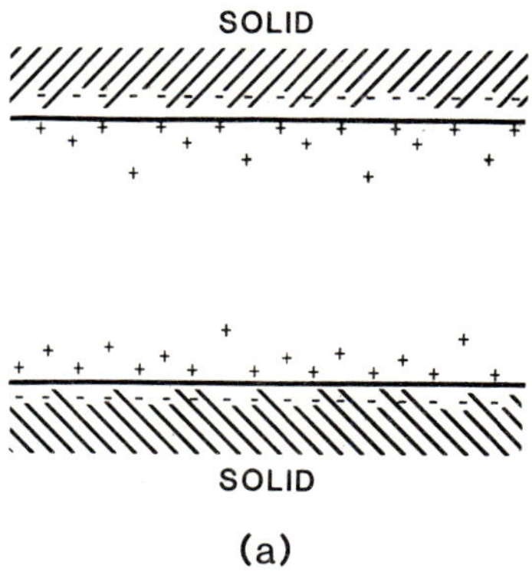

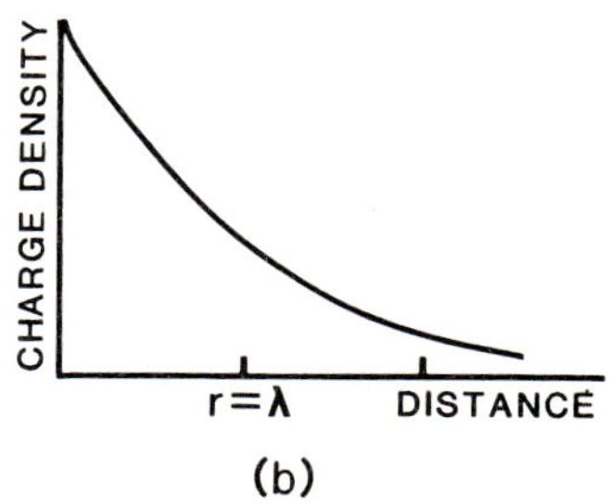

FIG. 3.4 (*a*) Diagram of a tube filled with fluid, showing the electrified interface between the fluid and the solid. (*b*) Plot of the charge density as a function of distance from the solid, representing a Gouy-Chapman model for the interface.

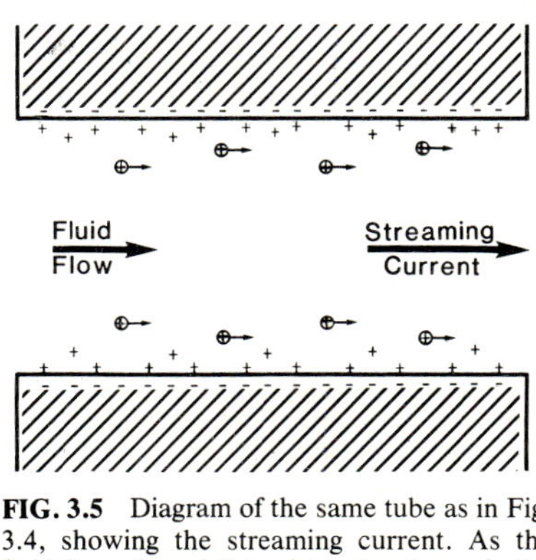

FIG. 3.5 Diagram of the same tube as in Fig. 3.4, showing the streaming current. As the fluid flows from left to right, the charge in the interfacial region is dragged along, producing a current.

3.2 ELECTROKINETIC EFFECTS: WET BONE

3.2.1 Theory

Electrokinetic effects are due to the flow of fluid past an electrified solid-liquid interface (Fig. 3.4). The electrified interface produces a charge imbalance that extends into the fluid approximately one screening length. As fluid flows past the solid, it drags along the charge in the fluid at the interface, producing an electric current called a *streaming current* (Fig. 3.5). The magnitude of the current depends on the magnitude of the charge imbalance and fluid velocity in the interfacial region.

This effect can be measured by forcing fluid through a tube and measuring the current (holding the potential difference between the ends of the tube at zero). Alternatively, one can measure the open-circuit voltage between the two ends of the tube, which is called the *streaming potential.*

The streaming potential is a steady-state phenomenon. As the streaming current carries charge downstream, it will produce a charge imbalance between the two ends of the tube. This produces an electric field in the opposite direction, which in turn produces a back current. In the steady state the back current exactly cancels the streaming current. The voltage resulting from the back electric field is the streaming potential.

For Poiseuille flow through a cylindrical tube, the streaming potential is just (Bockriss and Reddy, 1970)

$$V = \frac{X\varepsilon}{g\eta}\,\Delta P \tag{3.6}$$

where X is the zeta potential characterizing the electrified interface, ε is the dielectric constant of the interfacial region, g is the conductivity of the fluid, η is the viscosity of the fluid, and ΔP is the pressure difference between the ends of the tube. Note that the diameter and length of the tube do not influence the result. The streaming current depends on the geometry of the tube, but the resistance also depends on the geometry, and they cancel. (This result assumes that the resistance between the ends of the tube is entirely due to the fluid in the tube.) The streaming potentials are inversely proportional to the conductivity and the viscosity. X depends on the type of ions in the fluid, the nature of the solid, and also on the conductivity of the fluid.

The charged nature of the interfacial region is also responsible for electrophoresis, and measurements of the electrophoretic mobility of bone particles provides information about X.

The streaming potentials measured in stressed bone will be due to pressure gradients and fluid motion produced by the stress. Only recently have models for fluid flow in porous materials been applied to bone (Nowinski, 1972; Johnson et al., 1982; Johnson and Katz, 1984; and Pollack et al., 1984), and they will be discussed after the experimental results have been presented.

3.2.2 Experimental Results

The early measurements of the electromechanical effect in wet or fluid saturated bone include measurements by Cochran et al. (1968), and Bassett and Becker (1962). Cochran et al. used bone in vivo, and they measured signals similar to those measured for bone in vitro. All these measurements were made by applying a step function force and measuring the result in voltage. Typically the samples were in cantilever bending, and the investigators measured the voltage in the plane of bending (Fig. 3.6). The bandwidth of these measurements appears to have been limited to a few hertz.

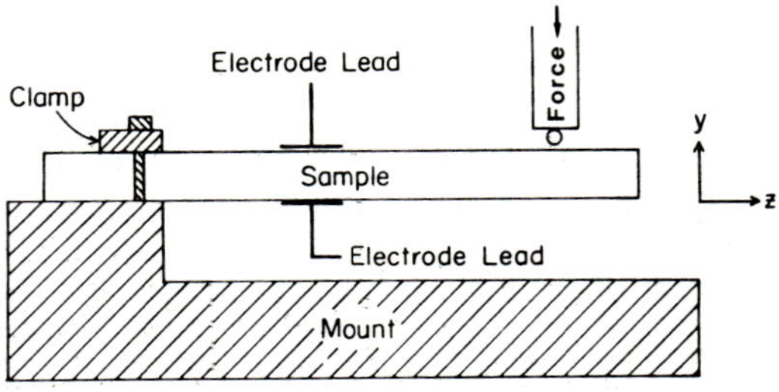

FIG. 3.6 Schematic of a jig for measuring the electromechanical effect during cantilever bending. For wet bone the electrodes would be wick electrodes.

Later measurements (Eriksson, 1976; Gross and Williams, 1982; and Pienkowski and Pollack, 1983) produced similar results. For step function measurements a typical record is shown in Fig. 3.7. Three of these later experimental investigations stand out. The first are those of Starkebaum et al. (1979), who measured the microscopic potentials around an osteon when the bone was in four-point bending.

The other two are those of Gross and Williams (1982) and Pienkowski and Pollack (1983), who investigated the dependence of the streaming potentials on fluid viscosity and conductivity. They confirmed the prediction of Johnson et al. (1980a) that the slow decay of the signal in wet bone is not due to an electrical decay (as for a piezoelectric effect), but is instead due to the relaxation of fluid flow in the stressed bone (as it would be if the effect were due to a streaming potential). In their step function measurements, Gross and Williams and Pienkowski and Pollack found that the decay of the voltage was related to the viscosity of the fluid and not its conductivity, an indication that the signal was due to a streaming potential and not piezoelectricity.

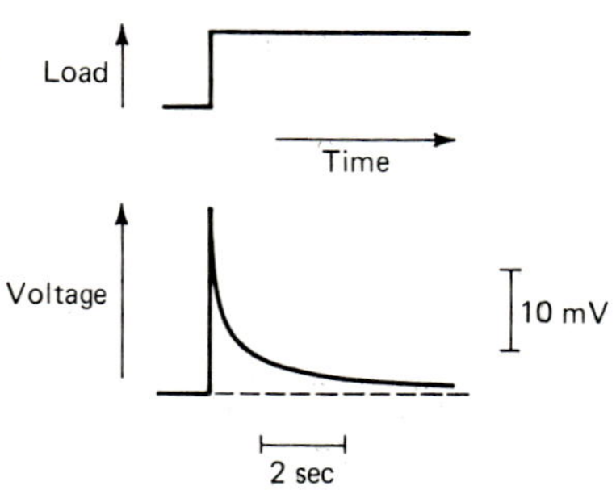

FIG. 3.7 Voltage generated in bone during four-point bending (from Pienkowski and Pollack, 1983). The fluid has a very low conductivity, $g < 0.01$ S/m, and the viscosity is about six times that of normal saline.

The magnitude of the signal was approximately

inversely proportional to the conductivity of the fluid; however, that could be the case for either a streaming potential or a piezoelectric effect. Pienkowski and Pollack also found that the sign of the signal reversed at high NaCl concentrations [> 0.77 *M* (molar)], which they interpreted as a change in the charge distribution in the double-layer region.

The bandwidth of Pollack's measurements was on the order of 10 Hz (τ = 0.015 s), and the measurements of Gross and Williams appeared to have a similar bandwidth, but they do not state what it was. They applied a step function force and measured the voltage, and the question is how rapidly they applied the force. A bandwidth of 10 Hz would be a reasonable estimate.

The amplitude of the macroscopic voltages from bent, wet bone is on the order of 2 to 3 mV/mm for a maximum strain of 0.002 (maximum strain in bending). The microscopic signals observed around osteons by Starkebaum et al. (1979) are almost an order of magnitude larger. The values depend on the type of fluid, and the above results are for normal saline solutions (0.15 g · mol/L NaCl).

Pienkowski and Pollack (1983) reported that at high viscosities the sign of the voltage changed as the signal relaxed. This intriguing result is unexplained at the present time. They also reported two relaxation times in the decay of the signal, which could be due either to relaxation in different pore systems or to the nature of the relaxation of the fluid flow (see Sec. 3.3). Both these observations may be important for testing any microscopic models of the electromechancial effect in wet bone.

3.3 STRESS-INDUCED FLUID FLOW

All the measurements discussed previously were made in bending (cantilever, or four-point, bending), where there is a stress gradient across the sample in the direction of the measured electric field. Furthermore the stress is compressive or tensile, producing a local dilatation (change in volume) in the sample. The dilatation is necessary to produce a change in the local fluid pressure, and the gradient of the stress would result in a gradient in the dilatation or a gradient in the fluid pressure. It is hypothesized that this gradient of the fluid pressure is responsible for the fluid flow and the resulting streaming potentials.

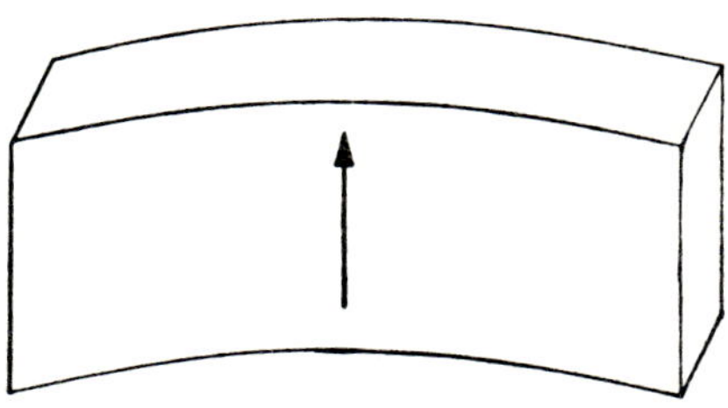

FIG. 3.8 Diagram of the hypothesized fluid flow for a bone under four-point bending. The fluid would flow from the bottom surface (in compression) to the top surface (in tension). The corresponding streaming potential would produce an electric field in the direction of the arrow.

The hypothesized flow pattern for these electromechanical measurements is shown in Fig. 3.8. It requires a flow from the compression side of the sample to the tension side. The relaxation times observed in these measurements are (for saline solutions with viscosities of about 10^{-3} Pa · s) on the order of 0.3 to 1 s.

The major theoretical questions are whether one can model the flow patterns to account for the measured voltages and relaxation times. Therefore the nature of the electromechanical effect in wet bone is intimately related to the nature of the stress-induced fluid flow. Unfortunately we know little about the relationship between the stress in the solid matrix and the resulting fluid behavior.

The simplest models for the fluid flow in bone are similar to Biot's model (1941). They assume that under compression the solid matrix collapses around the pores. This collapse is resisted by a temporary increase in the fluid pressure. The increase in the fluid pressure is temporary because the fluid can flow out of the bone or to other parts of the bone where the fluid pressure is lower. The major problem in the model is to relate the fluid pressure to the stress in the surrounding matrix. One can use simple models to do this (Johnson and Katz, 1984), and the result is Biot's equation:

$$\frac{d^2P}{dx^2} = \frac{C\,dP}{dt} \tag{3.7}$$

where C is a constant related to the geometry of the pores and the properties of the fluid and matrix.

3.3.1 Relaxation Times

A simple application of this model to the vascular systems in bone leads to an estimate of 10 to 100 μs for the relaxation time for fluid flow in the vascular systems in vitro (Johnson et al., 1982; Johnson and Katz, 1984). An extrapolation for flow in the vascular channels in vivo predicts relaxation times on the order of 1 to 5 ms (Johnson, 1984).

A result of the solution of Eq. (3.7) is that the decay of the fluid pressure is a series solution, each term having a different spatial dependence and a different time constant (see Johnson et al., 1982). This could account for the two time constants reported by Pienkowski and Pollack (1983).

The calculated relaxation times are much faster than those observed experimentally. (None of the experiments on wet bone has had the bandwidth to measure relaxation times less than 1 ms.) The question is what determines the relaxation time for the fluid flow observed experimentally. It may be the relaxation for flow in the canalicula and lacunae.

Johnson et al. (1982) and Pollack et al. (1984) have calculated relaxation times for flow in the canalicula and lacunae, both predicting relaxation times of around 1 ms or less if the canalicula are free of debris. However, the relaxation times are very sensitive to the state of the pores, and if they are not clear, the relaxation time will be much longer. Of course, in vivo they are occupied by cells, and one expects a long relaxation time. In vitro they will contain the remnants of the dead cells, and it is not known how much that will influence the impedance of the pores to fluid flow. Still, it is doubtful whether they will have relaxation times of 200 to 500 ms.

Another possibility is that a microporosity in the matrix is responsible for the long relaxation time. Further experiments are needed to clarify the point.

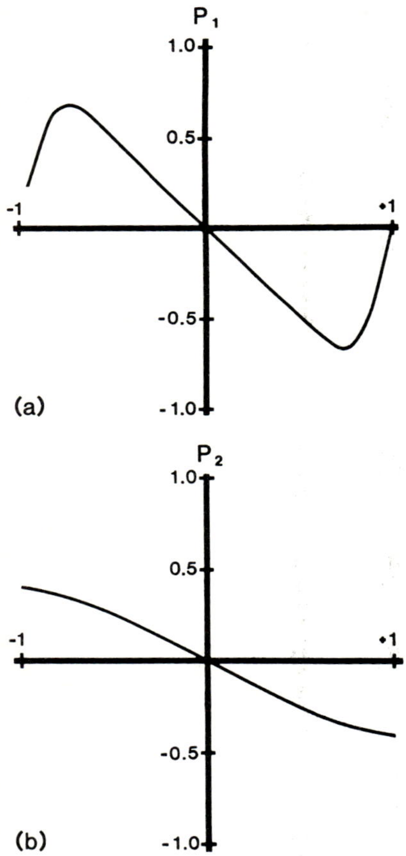

FIG. 3.9 (*a*) Plot of the pressure for a bent bone when the fluid can flow out of the bone. The plot is for a time $t = 0.1\tau$ after the application of a step function load, where τ is the decay time. (*b*) A similar plot for the case when the fluid is constrained to stay inside the bone.

3.3.2 Flow Patterns

Another major problem with our understanding of the electromechanical effect is the question of why we observe any voltages in bent bone. A simple application of Eq. (3.7) to a bone in four-point bending predicts that, if fluid can flow out of the surfaces of the sample, there is no *net* flow from the compression side to the tension side (cf. Johnson, 1984). This can be seen because the fluid pressure at the neutral axis is initially zero (gage pressure), and the pressure at the two surfaces is zero. The pressure is shown in Fig. 3.9*a*. A net flow across the sample occurs only if the fluid is constrained to remain in the sample. Then the pressure is given by Fig. 3.9*b*. Figure 3.10*a* and *b* gives the corresponding flow patterns.

If the streaming potentials are proportional to the macroscopic pressure gradients, then there would not be a net potential across the sample for the flow pattern of Fig. 3.10*a*. However, net potentials are observed. This suggests either that the fluid cannot flow out of the sample or that the macroscopic potentials are not proportional to the macroscopic pressure differences.

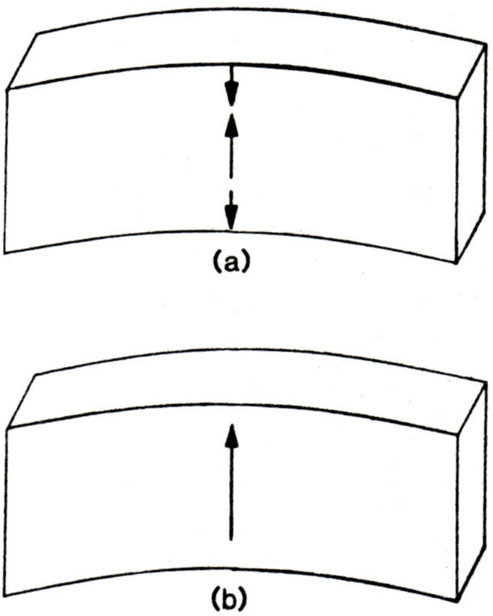

FIG. 3.10 (*a*) Arrows indicate the direction of fluid flow due to the pressure distribution in Fig. 3.9*a*. (*b*) Corresponding flow pattern for the pressure distribution in Fig. 3.9*b*.

The second alternative would mean that the microscopic flow patterns could produce a macroscopic potential even though the macroscopic pressure difference is zero. In this case a detailed analysis of the microscopic flow patterns and the streaming currents generated by those patterns is needed. The work of Pollack et al. (1984), which analyzes the flow patterns and streaming potentials generated in an osteon, is a good start in that direction and provides a good explanation of the observations of Starkebaum et al. (1979).

Johnson (1984) has suggested that capillary pressure may act to hold the fluid inside the bone, so that fluid is not actually free to flow out. This would produce a flow pattern like that in Fig. 3.10*b*, allowing a macroscopic potential proportional to the pressure difference across the sample. An interesting consequence of this model is an amplitude dependence of the signal. Once the fluid pressure exceeded the capillary pressure, the flow pattern would change and the signal would not increase with a further increase in the fluid pressure (i.e., with an increase in the amplitude of vibration). The capillary pressure could be altered with surfactants to change the onset of the nonlinear behavior.

Both these alternatives need to be examined closely and investigated experimentally. For example, the role of capillary pressure could be investigated by looking for an amplitude dependence of the macroscopic potentials.

3.3.3 Sample Conditions

The relaxation of the fluid pressure is sensitive to the permeability of bone, which is strongly dependent on sample conditions (Rouhana et al., 1981). It is very important to know what type of materials occupy the pore spaces, especially in vitro. Furthermore, the material in the pores will probably affect the streaming potentials. The differences between in vivo and in vitro measurements are likely to be large.

In vivo, the vascular channels are lined with cells, and much of the fluid flow would be through blood vessels. In that case the fluid-solid interface would be with the lining cells or the walls of the blood vessels. In vitro, the interface might be between the fluid and the calcium phosphate–collagen matrix. These interfaces might have very different properties.

Measurements of the influence of sample preparation on the electromechanical properties are needed, with respect to both streaming potentials and the relaxation times.

SUMMARY

Although there has been considerable progress in work on the electromechanical properties of bone, especially wet bone, we still do not have a good microscopic understanding of the phenomena that would allow us to estimate what is occurring at a cellular level in vivo. Most of the measurements have been in vitro, and the behavior might be somewhat different in bone in vivo. Furthermore, the measurements on wet bone have been done at low frequencies ($f <$ 10 Hz), and experiments at higher frequencies (100 Hz to 10 kHz) are needed to test the models that have been developed and to resolve the questions raised by those models.

ACKNOWLEDGMENTS

The authors would like to acknowledge the advice and assistance of Hyo Sub Yoon and Surendra Singh. The work was supported in part by NIH grant AM35377.

REFERENCES

Bassett, C. A. L. (1971). Biophysical principles affecting bone structure. *The Biochemistry and Physiology of Bone*. Academic Press, New York, vol. III, pp. 1–76.

Bassett, C. A. L., and Becker, R. O. (1962). Generation of electric potentials in bone in response to mechanical stress. *Science, 137,* 1063.

Biot, M. A. (1941). General theory of three-dimensional consolidation. *J. Appl. Phys., 12,* 155–164.

Bockris, J. O., and Reddy, Amulya K. (1970). *Modern Electrochemistry*. Plenum Press, New York.

Bur, A. J. (1976). Measurements of the dynamic piezoelectric properties of bone as a function of temperature and humidity. *J. Biomech., 9,* 495.

Cady, W. G. (1964). *Piezoelectricity*. Dover, New York.

Chakkalakal, D. A., Johnson, M. W., Harper, R. H., and Katz, J. L. (1980). Dielectric properties of fluid saturated bone. *IEEE Trans. Biomed. Eng., BME-27*(2), 95.

Cochran, G. V. B., Pawluk, R., and Bassett, C. A. L. (1968). Electromechanical characteristics of bone under physiological moisture conditions. *Clin. Orthop. Relat. Res., 58,* 249.

Date, M. (1972). Piezoelectric constant in dispersed systems. *Rep. Progr. Polymer Phys. Jpn., 15,* 385–387.

Eriksson, C. (1976). Electrical properties of bone. *Biochemistry and Physiology of Bone*. Academic Press, New York, vol. IV, pp. 329–384.

Fukada, E., and Yasuda, I. (1957). On the piezoelectric effect of bone. *J. Phys. Soc. Jpn., 12,* 1158.

Fukada, E., Veda, H., and Rinaldi, R. (1976). Piezoelectric and related properties of hydrated collagen. *Biophys. J., 16,* 911.

Ghilardinetto, T., and Zimmerman, R. L. (1975). Effect of water on piezoelectricity in bone and collagen. *Biophys. J., 14,* 573–576.

Grodzinsky, A. J. (1983). Electromechanical and physicochemical properties of connective tissue. *CRC Crit. Rev. Biomed. Eng., 9,* 133.

Gross, D., and Williams, W. S. (1982). Streaming potentials and the electromechanical response of physiologically moist bone. *J. Biomech., 15*(4), 277.

Innacone, W., Korostoff, E., and Pollack, S. R. (1979). Microelectrode study of stress-generated potentials obtained from uniform and nonuniform compression of human bone. *J. Biomed. Mater. Res., 13,* 753.

Johnson, M. W. (1984). Behavior of fluid in stressed bone and cellular stimulation. *Calcif. Tissue Int. 36,* S72–S76.

Johnson, M. W., Chakkalakal, D. A., Harper, R. A., and Katz, J. L. (1980a). Comparison of the electromechanical effects in wet and dry bone. *J. Biomech., 13,* 437.

Johnson M. W., and Katz, J. L. (1984). Some new developments in the rheology of bone. *Biorheology, 198,* Suppl. I, 169–174.

Johnson, M. W., Rouhana, S. W., et al. (1982). Fluid flow in bone in vitro. *J. Biomech., 15*(11), 881–885.

Johnson, M. W., William, W. S., and Gross, D. (1980b). Ceramic model for piezoelectricity in dry bone. *J. Biomech., 13,* 565.

Kosterich, J. D., Foster, K. R., and Pollack, S. R. (1983). Dielectric permittivity and electrical conductivity of fluid saturated bone. *IEEE Trans. Biomed. Eng., 30*(2), 81.

Lakes, R. S., Harper, R. A., and Katz, J. L. (1977). Dielectric relaxation in cortical bone. *J. Appl. Physiol., 48,* 808.

Liboff, A., and Shamos, M. (1973). Solid state physics of bone. In I. Zipkin (ed.), *Biological Mineralization.* John Wiley & Sons, New York, pp. 335–396.

Nowinski, J. L. (1972). Stress concentration around a cylindrical cavity in a bone treated as a poroelastic body. *Acta Mech., 13,* 281–292.

Pienkowaski, D., and Pollack, S. R. (1983). The origin of stress generated potentials in fluid saturated bone. *J. Orthop. Res., 1,* 30.

Pollack, S. R., Petrov, N., Saltzstein, R., Brankov, G., and Blagoeva, R. (1984). An anatomical model for streaming potentials in osteons. *J. Biomech., 17,* 627.

Reinish, G. B., and Nowick, A. S. (1975). Piezoelectric properties of bone as functions of moisture content. *Nature, 253,* 626–627.

Reinish, G. B., and Nowick, A. S. (1976). Effect of moisture on the electrical properties of bone. *J. Electrochem. Soc., 123,* 1451.

Rouhana, S. W., Johnson, M. W., Chakkalakal, D. A., Harper, R. A., and Katz, J. L. (1981). Permeability of compact bone. *Joint ASME-ASCE Conference Biomechanics Symposium, AMD 43,* 169–172.

Starkebaum, W., Pollack, S. R., and Korostoff, E. (1979). Microelectrode studies of stress generated potentials in fourpoint bending of bone. *J. Biomed. Mater. Res., 13,* 751.

Williams, W. S. (1982). Piezoelectric effects in biological materials. *Ferroelectric, 41,* 225–246.

CHAPTER 4

Biomechanical Properties of Articular Cartilage

Savio L-Y. Woo
Division of Orthopaedic Surgery
San Diego Veterans Administration Medical Center
and University of California, San Diego
La Jolla, California

Van C. Mow
W. Michael Lai
Department of Mechanical Engineering
Aeronautical Engineering and Mechanics
Rensselaer Polytechnic Institute
Troy, New York

INTRODUCTION

Mathematical descriptions of the mechanical properties of articular cartilage are requisite in the analysis of its function within diarthrodial joints. Examples of this dependence on the fundamental properties of articular cartilage in the interpretation of the experimental and theoretical results of diarthrodial joint function are numerous, particularly in the literature on lubrication (Charnley, 1959; Dintenfass, 1963; Linn, 1967, 1968; Dowson, 1967; McCutchen, 1962; Walker et al., 1968; Greenwald and Haynes, 1972; Radin et al., 1970; Radin and Paul, 1972; Dowson et al., 1981). For a detailed description of the lubrication of diarthrodial joints, see Chapter 5 of this handbook. Further, an understanding of interstitial fluid movement in the mechanics of joint function and of metabolite transport through the tissue for chondrocyte nutrition in the long-term survivability of this avascular tissue necessitates that laws governing mass transport through this tissue be known (Ekholm and Norback, 1951; McCutchen, 1962; Linn and Sokoloff, 1965; Edwards, 1967; Honner and Thompson, 1971; Mow and Torzilli, 1975; Maroudas, 1975a, b, 1979; Mow and Lai, 1980; Salter et al., 1980; Torzilli et al., 1983).

Unfortunately, this tissue is very difficult to work with, because it is rarely over 2 mm thick and it overlays highly curved bony surfaces. Thus accurate and repeatable measurements of its properties are extremely difficult to achieve (see the review article by Mow, Holmes, and Lai, 1984). The situation is further complicated by the fact that articular cartilage, because of the architecture of its macromolecular components, exhibits nonhomogeneous and anisotropic material behaviors. It can easily be stretched and compressed; thus, finite deformational theories are occasionally required. Its material behavior can be characterized by history-dependent responses as well as environmentally dependent responses (i.e., those dependent on temperature, ionic concentration, and pH). The history dependence can manifest itself when the tissue is loaded at different rates and is probably due to two specific mechanisms: (1) the intrinsic viscoelastic response of the macromolecular solid matrix—made up of collagen (~60% by dry weight) and proteoglycans (~30% by dry weight) and (2) the relative velocity between the solid matrix and the interstitial water (~80% by wet weight). The environmental dependence can manifest itself by changes in sample temperatures and interstitial ion and/or proton concentrations, which produce changes in the intrinsic material properties of the collagen-proteoglycan solid matrix.

In this chapter we will first review the historical background and methods of analysis previously used to study properties of articular cartilage, and then present the procedures and analyses currently used to determine its intrinsic material properties in tension, compression, and shear, as well as methods for assessing its permeability. These intrinsic properties should provide the foundation for understanding cartilage function in diarthrodial joints as well as their correlation with the biochemical composition and molecular structure of the collagen network and proteoglycan conformation (Armstrong and Mow, 1982a, b).

4.1 PREVIOUS WORK ON THE DETERMINATION OF ARTICULAR CARTILAGE PROPERTIES

4.1.1 Measurement of Indentation Properties

The most frequently employed experiment for studying the biomechanical properties of articular cartilage is the indentation experiment. This experiment is used because the loading condition can come close to yielding the compressive response of the tissue under physiological conditions. As early as the 1920s, Bär (1926) and Göcke (1927) utilized this experimental procedure to test human articular cartilage. Hirsch (1944) used the indentation test to study the compressive properties of human patellar cartilage and their correlation with the histological and histochemical appearance of the tissue. From these studies, one comes to accept the fact that when a compressive load is instantaneously applied, an instantaneous deformation occurs, followed by a slow, time-dependent (creep) deformation. Upon unloading, an instantaneous recoil occurs, followed by a slower, time-dependent recovery.

Hirsch also found that cartilage would not completely regain its original thickness upon removal of the indenting load, giving rise to a phenomenon which became known as "imperfect elasticity." This imperfect elasticity was felt to be an artifact associated with the friction of the indenting apparatus since the tissue, Hirsch believed, must always regain its original thickness in vivo when the load is removed. This anomaly was eventually resolved by Elmore and coworkers (1963), who tested the specimens immersed in a solution bath; as a result, complete recovery occurred. They concluded that the incomplete recovery resulted from performing the experiments in air, which caused fluids to be expelled from the tissue and to be evaporated. Hence, an insufficient amount of fluid was available for resorption upon unloading.

The indentation technique has the advantage that it allows for the study of the topographic variation of tissue properties over the joint surface, since the indenting probe is usually no more than a few millimeters in diameter. With this, Hirsch found a difference in the indenting response of normal and degenerate areas of human patellar cartilage. Elmore and coworkers also found large local variations of compressibility of cartilage as a structure (since no thickness measurements were made) on joints obtained from humans, calves, dogs, and guinea pigs. Later, Kempson and coworkers (1970, 1971) examined up to 40 areas within one human femoral head and found large topographic variation of the compressive stiffness of articular cartilage. In general, lower stiffnesses were obtained from mildly fibrillated areas than from visually normal areas. Correlative chemical studies of the tissue taken from the fibrillated areas showed a depletion of glycosaminoglycans (Kempson et al., 1971; Kempson, 1979).

Early on, various attempts to model the elastic and viscoelastic properties of cartilage were based upon these indentation tests. Hirsch employed the Hertz solution for two contacting elastic spherical bodies to describe his results. Later Sokoloff (1966) also used a linear elastic material to model his indentation tests. He used a rigid, plane-ended, circular, cylindrical indenter to indent the tissue. Further, Sokoloff assumed that cartilage is of infinite depth and incompressible; that is, Poisson's ratio was taken to be 0.5. The formula used by Sokoloff to calculate the Young's modulus (E) was

$$E = \frac{P}{2.67 w_0 a} \tag{4.1}$$

where P is the load applied by the indenter, w_0 is the depth of indentation, and a is the radius of the circular indenter. This formula comes from the exact solution of Boussinesq (1885) for an elastic half space:

$$E = \frac{P(1 - \nu^2)}{2 w_0 a} \tag{4.2}$$

Here ν is Poisson's ratio. With the incompressibility assumption, that is, with $\nu = 0.5$, Eq. (4.2) reduces to Eq. (4.1). It is clear that Sokoloff's solution must be regarded as only an early attempt to come to grips with the problem of accurately describing and precisely determining the material properties of cartilage.

The most obvious improvement upon these early attempts was to take into consideration the thinness of the tissue. Thus Kempson and coworkers (1971) modified the elastic half-space assumption by introducing the work of Waters (1965a, b). Waters obtained an empirical factor modifying the indentation solution for a thin rubber sheet resting on a rigid half space. The elastic modulus of articular cartilage was thus calculated from the amount of indentation obtained after 2 s of creep. Both the hemisphere-ended indenter and the plane-ended indenter were used to determine the so-called 2-s creep modulus. However, the experimental work of Simon (1971) on the indentation properties of bovine, canine, and human articular cartilage resulted in values for E that could not be verified by Kempson's approach using Water's empirical solution for E at either the 2-s or the long-time response taken 30 s after loading. The discrepancy may have arisen from using Water's approach for determining the modulus of thin rubber sheets as well as the arbitrariness of picking the 2-s time or the 30-s time to take the displacement data. Further, rubber is naturally incompressible, while cartilage is compressible upon indentation, because of water exudation. Thus, the incompressibility question must be further scrutinized.

Hayes and coworkers (1972) obtained a mathematical solution to the problem of indentation by assuming that cartilage, bounded to the subchondral bone, may be represented as a thin elastic layer bonded to a rigid half space. Using the displacement equations with appropriate mixed boundary conditions, they obtained an analytical solution for the case of a plane-ended, rigid, cylindrical, frictionless indenter. This yielded

$$E = \frac{P(1 - \nu^2)}{2 a w_0 \kappa(a/h, \nu)} \tag{4.3}$$

where κ is numerically determined from an integral equation and depends on the parameters a/h (aspect ratio) and ν (Poisson's ratio); see Fig. 4.1. From this result, with an assumed Poisson's ratio $\nu = 0.5$, and an aspect ratio $a/h = 0.3$, for example, the instantaneous Young's modulus determined by Sokoloff (2.28 MPa) is reduced to 1.63 MPa. Thus, the high reading obtained by Sokoloff was the result of the stiffer bone manifesting its influence upon the indentation.

In Hayes's 1972 analysis, assumptions of small strain, isotropic and homogeneous material properties, and a linear stress-strain relationship were made. Later, Hori and Mockros (1976) measured the short-time shear and bulk moduli of human articular cartilage, using torsion and confined compression tests similar to those of Hayes and Mockros (1971). These material

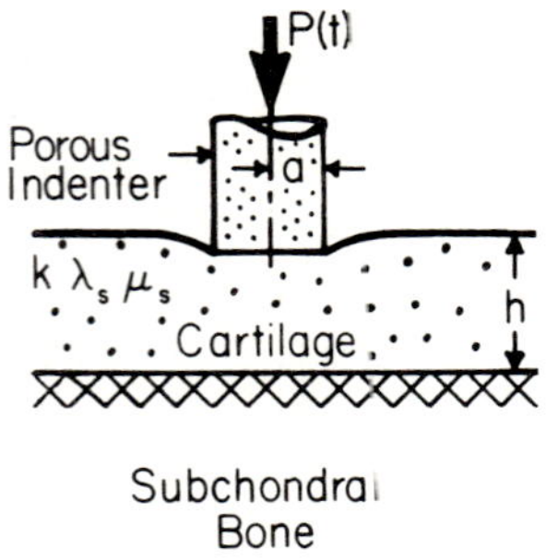

FIG. 4.1 Representation of a plane-ended circular cylindrical indenter applied to a thin layer of articular cartilage. Cartilage has been modeled as a single-phase, semi-infinite, elastic medium (Sokoloff, 1966), a layered, single-phase elastic medium (Hayes and coworkers, 1972; Kempson and coworkers, 1971), a single-phase, viscoelastic medium (Coletti and coworkers, 1972; Parsons and Black, 1977), and a layered, biphasic medium (Mow and coworkers, 1982b).

properties (converted to E and ν by assuming isotropy) did not compare well with the theoretical predictions of Eq. (4.3). Hence, these authors felt that indentation tests are not suitable to determine the material properties of articular cartilage because of its anisotropic, nonhomogeneous, and nonlinear behavior. Even if these material complications did not exist and the material were linearly elastic and isotropic, in all these tests, Poisson's ratio must be known a priori before one can assess the shear modulus μ or E. For articular cartilage, this is a great uncertainty (Mow et al., 1982).

The viscoelastic behavior of articular cartilage during indentation was first modeled by Coletti and coworkers (1972). These authors used a combination of discrete Kelvin and Voigt elements to describe the creep of normal articular cartilage and experimental arthroplastic (regenerated cartilage) surfaces. Apparent elastic and viscous moduli were obtained for a three-element viscoelastic material used to represent the response of these tissues. Later, Parsons and Black (1977, 1979) used the solution of Hayes and coworkers (1972), Eq. (4.3), to obtain the "unrelaxed" and "relaxed" shear moduli, since $E = 2(1 + \nu)\mu$, for rabbit articular cartilage in an indentation (creep) experiment by assuming $\nu = 0.40$. This is an ad hoc model, since it was assumed that while the shear compliance changed with time during the indentation creep response, the Poisson's ratio ν did not. However, this technique did appear to fit the creep data reasonably well. In any case, these studies indicate that in indentation tests, one must consider either load-controlled creep or displacement-controlled stress-relaxation experiments in order to have a well-posed test. In all earlier studies, this was not a primary concern; consequently, there were major discrepancies, along with other considerations mentioned above, in the determined material properties.

It now seems clear that there are serious drawbacks in using the existing theories for indentation tests as means to predict the mechanical properties of articular cartilage. These theories have not accommodated the nonhomogeneous and directional properties of articular cartilage, nor have they taken into consideration possible finite deformational effects or the flow of the interstitial fluid relative to its porous permeable solid matrix (Mow et al., 1982). Most recently, Mak and coworkers (1982) showed theoretically that the exudation and fluid redistribution within articular cartilage could account for the creep and stress-relaxation behavior of the tissue during an indentation test. The exact details of using this approach to determine the relevant material constants have not yet been completed, but such techniques could be of great utility in resolving some of the contradictions in the data obtained from the previous indentation tests.

In addition, when the entire joint surface is used as a test specimen, the present theories give little attention to the geometry of the joint surface. For example, Simon (1971) showed that differences exist in the deformability of articular cartilage from bovine, canine, and human joints, although the applied load and cartilage thickness were the same. One probable explanation for the differences in the experimental results may be that the joint geometry of these species is different, and thus stress and strain in the tissue may be different even for the same applied load. Also, the composition of articular cartilage varies from species to species, joint to joint, and site to site over the articulating surface (Muir, 1980; Hall, 1983); and the material properties of articular cartilage have been shown to depend strongly on its composition, age, and degree of degeneration (Armstrong and Mow, 1982a, b). Therefore, one cannot be certain from earlier studies whether the difference in deformation is a result of the varying intrinsic properties resulting from its molecular characteristics (Muir, 1977) or the varying geometric properties of the joint surface.

4.1.2 Measurement of Tensile Properties

In 1963 Zarek and Edwards analyzed the principal stress trajectories in articular cartilage in their attempt to verify the general rule that the arrangement of the collagen fibrils within the tissue is related to the tensile stress they must withstand. The goal was to investigate the fibril arrangements in articular cartilage as suggested by Hultkrantz (1898), Benninghoff (1925), and MacConaill (1951). Using the hertzian solution for an elastic half space, Zarek and Edwards showed that the largest principal tensile stresses are at and parallel to the articular surface, and tensile stress could exist throughout the depth of the tissue, except in the area directly beneath the contact loading, where the principal stresses are all compressive. Beneath the surface the tensile stress trajectory showed no predominance in either the radial or oblique direction; this refuted both Benninghoff's and MacConaill's hypotheses for the collagen ultrastructural arrangements.

The surface traction is another factor in the determination of the stress trajectory in articular cartilage. Simon felt that the deformation of articular cartilage served to increase the congruence of joints, thereby reducing the surface compressive stresses. Burstein and coworkers (1971) showed that a reduction in the area of load application significantly altered the magnitude and distribution of the radial and tangential stresses in articular cartilage, and yet little or no change in stress distribution was observed when the cartilage thickness was reduced. Later, Askew and Mow (1978), using a three-layer elastic anisotropic model for articular cartilage and an elastic layer for bone, analyzed the role of collagen ultrastructure in influencing the stresses acting within the cartilage. It was found that tensile stresses do exist in such a layered system and that the system is largely dependent upon the stiffness of the superficial tangential zone—the collagen-rich region of the tissue—and the aspect ratio—the extent of loaded surface area relative to the thickness of the tissue. Also, large tensile principal stresses exist at approximately 45° from and at the cartilage-bone interface. More recently, Mow and Lai (1980) determined the principal stress trajectories in the cartilage layer modeled as a mixture of incompressible fluid and an elastic solid. Conclusions similar to those of Askew and Mow were deduced. These studies all imply that it is essential to know the tensile stiffness and strength characteristics of articular cartilage, including its nonlinearity, nonhomogeneity, and anisotropic properties.

In the literature, implications of cartilage failure prior to degenerative joint disease stem from tensile fatigue failure of cartilage strips in aging tissues (Weightman, 1976). The earliest record of tensile studies of cartilage goes back to Rauber (1876). The next experimental study of articular cartilage in tension was performed by Ranu (1967) on full-thickness articular cartilage specimens taken from the femoral heads of dogs. Later a more systematic study of the tensile properties of articular cartilage was performed by Kempson and coworkers (1968, 1973), using human tissue taken from the femoral condyles. Full-thickness, standardized, dumbbell-shaped cartilage specimens were taken from the distal portions of the femur parallel or perpendicular to the surface fiber patterns or cleavage-line (split-line) pattern (Hultkrantz, 1898). Each specimen was then sliced into $\sim$200-μm sections in order to study the variation of tensile properties with depth. These authors found a gradual reduction of stiffness from the surface to the deeper layers for 40- to 70-year-old human specimens. In the surface layer, specimens taken parallel to the cleavage-line direction were significantly stiffer than those taken perpendicular to the cleavage-line direction. The trends of diminishing tensile stiffness were not always observed, however, for healthy immature bovine humeral and femoral cartilage (Woo et al., 1976; Roth and Mow, 1980). Ruth and Mow found that prior to closure of the growth plate, i.e., in skeletally immature animals, the layerwise variation of cartilage tensile properties are very different from those of skeletally matured animals—the latter being very similar to those obtained by Kempson and coworkers. However, the anisotropic characteristics of immature and mature cartilage in tension appear to be similar.

4.2 TENSILE PROPERTIES OF ARTICULAR CARTILAGE

To determine the mechanical properties of articular cartilage in tension, it would be essential to restrict the type of material to be tested to prevent large statistical variations. Woo and coworkers (1976, 1979, 1980) restricted their study to articular cartilage specimens obtained from the bovine humeral joint of 15- to 18-month-old animals. The availability of a large number of bovine samples also allowed the tensile properties to be studied on a statistical basis. Further, because the specimens, small in dimension, were obtained from this relatively large joint, minimal straightening of the initially curved cartilage specimen was necessary during the tensile tests. Using similar criteria, Roth and Mow (1980) used the bovine patellofemoral groove, further refining the age category by noting the existence of a growth plate.

4.2.1 Morphological Characterization

In low-magnification scanning electron microscopy, normal cartilage can be separated into three zones, i.e., superficial, intermediate, and deep zones (McCall, 1969). The superficial zone appears to be comprised of large parallel bundles of collagen fibers. In the intermediate zone, the fine collagen fibrils are randomly oriented, exhibiting copious interfibrillar space (Redler, 1974), and in the deep zone, the collagen fibers are the largest (0.3 to 1.4 μm in diameter). The majority of collagen fibers in the deep zone appear to be closely packed together and oriented radially relative to the cartilage-bone junction. Clarke (1974) further observed that, at higher magnification, the collagen fiber bundles parallel to the articular surface in the superficial zone were really made of smaller, more randomly oriented fibrils (see Fig. 5.6), while in the middle zone, this distinction was not apparent. These detailed scanning electron microscopy studies are at odds with the light microscopy studies of Benninghoff (1925) and Bullough and Goodfellow (1968), and, to date, no explanation of this discrepancy is available. Excellent surveys, up to 1974, describing the details of the collagen architecture of articular cartilage as observed by scanning electron microscopy, may be found in Clarke's 1971 and 1974 reviews. Models for these views of collagen ultrastructure may be found in Mow and coworkers (1974) and Lane and Weiss (1975); see Fig. 4.2. More recent surveys advancing understanding of detailed fine struc-

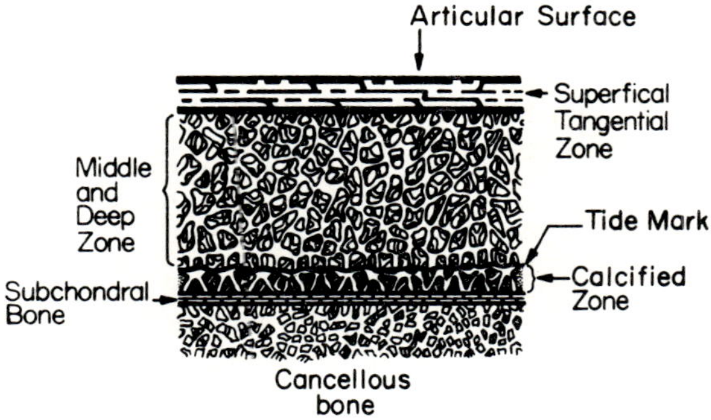

FIG. 4.2 Representation of collagen fibril ultrastructure throughout the depth of the cartilage depicting the distinct idealized zones: (*a*) A superficial tangential zone composed of densely packed collagen fibrils in sheets parallel to the surface; (*b*) a randomly arrayed middle zone with abundant interfibrillar space to accommodate the proteoglycans and water; and (*c*) a radial deep zone where collagen fibrils appear to anastomose into larger bundles as they insert into the calcified zones. The cartilage, i.e., the noncalcified and calcified zones, rests upon a stiff subchondral plate supported by the soft, cancellous marrow composite structure.

ture of the cartilaginous matrix may be found in Ghadially (1978), Speer and Dahners (1979), and Boyde and Jones (1983). Further advances have been made, using x-ray diffraction methods, by Aspden and Hukins (1981) in resolving the fine details of collagen architecture.

On the basis of all these morphological observations, it is safe to assume that cartilage has three major specific structural zones: the surface zone, the middle zone, and the deep zone. Most tensile studies done to date have examined the layered variations of the properties of the tissue, grouped roughly into these three zones.

4.2.2 Specimen Preparation Procedures

In the present description, the cartilage specimens used were taken from fresh bovine humeral joints from animals 15 to 18 months old. First, the specimens should be of specific orientation relative to the local split-line pattern. Most investigators have taken specimens either parallel or perpendicular to the split-line pattern; in addition, Woo and coworkers (1976) took specimens oriented at 45° relative to the split lines. Figure 4.3 shows a general split-line pattern of the

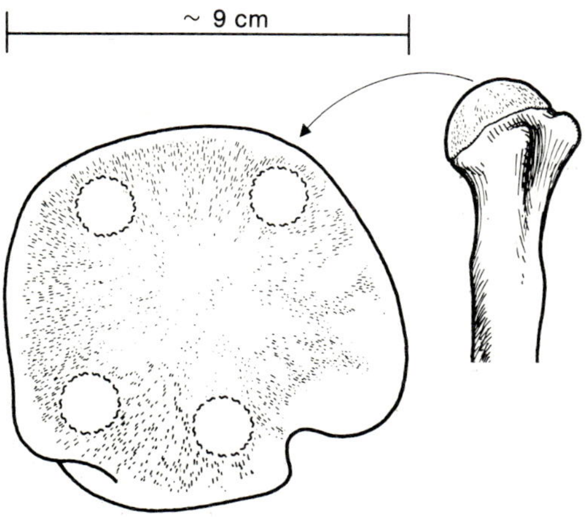

FIG. 4.3 Schematic diagram of the bovine humeral head, showing the general cleavage-line patterns over the entire joint surface. In the preparation of tensile specimens, generally four smooth and flat 12.5-mm-diameter circular areas can be selected, and the cleavage-line patterns outside the circles can be determined.

entire bovine humeral joint. Generally, four very smooth and flat 12.5-mm-diameter plugs can be selected. The split-line pattern at the areas outside of each 12.5-mm-diameter circle can be determined by using the india-ink staining technique (Kempson, 1968, 1973). Areas failing to exhibit uniform split-line patterns should not be used. The split-line direction should then be marked on the plug, and the plug of cartilage with bone can be cored out vertically.

A die cutter should be used to stamp a dumbbell-shaped specimen through the thickness of the articular cartilage to the subchondral bone to obtain standardized specimens at 0, 45, and 90° from the axis of split-line patterns. Slices, 250 to 325 μm each, can be obtained from each sample by using a freezing stage sledge microtome. The first slice is considered to be the surface zone of the tissue, the second and third slices, the middle zone, and the fourth slice, the deep zone. Thus a total of nine different types of articular cartilage specimens, i.e., three zones in three directions, can be studied.

4.3 EXPERIMENTAL APPARATUS FOR TENSILE TESTS

4.3.1 Cross-Sectional Dimensions of the Test Specimens

The width of the middle portion of the dumbbell-shaped cartilage specimen can be easily measured using an apparatus such as a cathetometer accurate to 0.01 mm. The thickness of the specimen can be measured by specially designing a micrometer instrument with a weightless, flat glass foot mounted on a double cantilever beam system (Woo et al., 1976) or a micrometer attached to a dc circuit to indicate contact (Woo et al., 1976; Roth and Mow, 1980). These measurement techniques, for specimens with thicknesses of ~250 μm, are accurate to within ± 2 μm.

4.3.2 Measurement of Tensile Strains—A Video Dimensional Analyzer

To measure the deformation or strain of the specimen, a noncontact, optical dimensional analyzer system should be used. Such a system may consist of a television camera, an electronic dimensional analyzer, and television monitors (Fig. 4.4). These video dimensional analyzers (VDAs) have been used successfully to measure with great accuracy axial tensile strains. The dimensional analyzer of the VDA system has two windows which can be set to follow automatically two contrast phases or gage lengths (Fig. 4.5). The scan time, which corresponds to the gage length, increases during the tensile stress; hence, the strain can be measured continuously. The frequency response of the VDA is 25 Hz. The error in linearity in accuracy of the VDA system is less than 0.5 percent. The VDA system was originally developed and used by Intaglietta and Tompkins (1972).

There are a number of advantages in using this VDA system to measure the tensile strain of soft biological tissues such as articular cartilage: (1) no external devices are attached to the small specimen to measure local strain, (2) only the strain in the area where tensile stress is uniform is measured, and (3) inaccuracies due to the slippage between the specimen and its

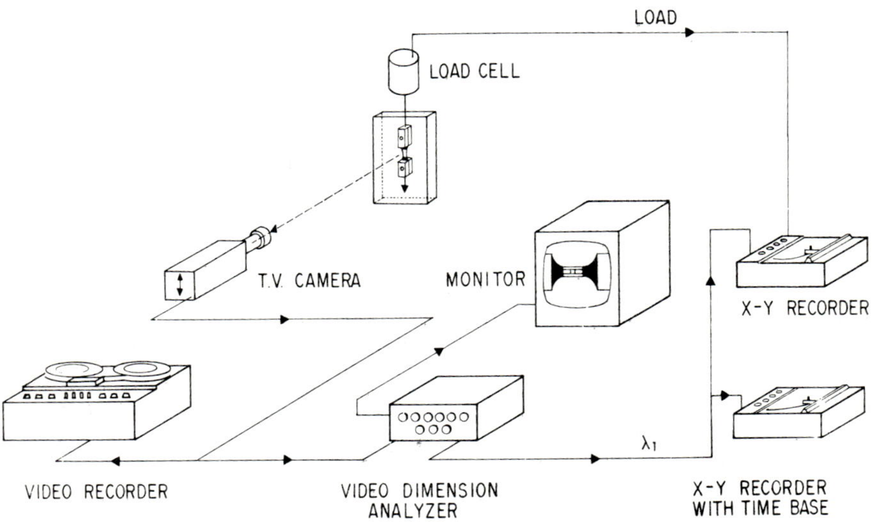

FIG. 4.4 Flow diagram showing the experimental apparatus for measuring the load deformation characteristics of articular cartilage in tension and the video dimensional analyzer system used to measure the tensile strain of the specimen. The entire test can be recorded by a video tape recorder for later analysis of λ_2. The tensile stretch (λ_1) versus time can also be recorded on a second *XY* recorder.

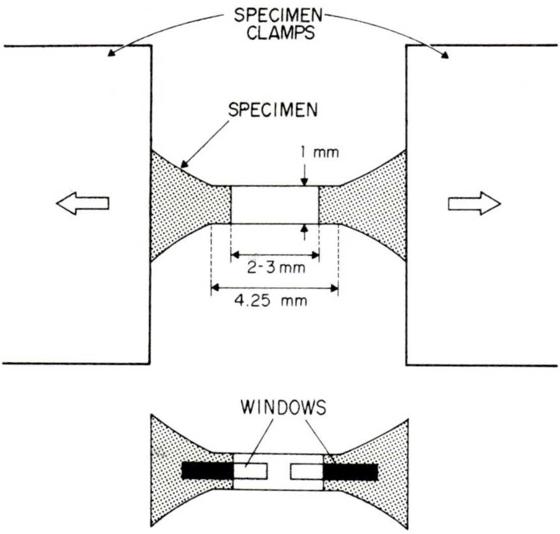

FIG. 4.5 A typical TV monitor display showing the dimension of cartilage specimens used for uniaxial tension tests. The lower sketch shows the image of the two video dimension analyzer windows which automatically follow the gage marks on the specimen during tensile stretch, thus determining the tensile strain.

grips are eliminated. It is interesting to note that in the same test the discrepancy in strain values measured by the VDA and those measured by the clamp-to-clamp deformation could be as much as 30 percent.

4.3.3 Testing Procedures

After width and thickness measurements within the gage length are made, the specimen can be clamped at the ends and mounted onto a tensile testing machine. All tensile tests should be performed with the specimen completely immersed in a physiological saline solution maintained at a constant temperature, e.g., 37°C. The data from a total of 76 articular cartilage specimens tested up to failure, at a strain rate of approximately 1.2 percent per second (Woo et al., 1976), will be discussed in this chapter. The reader is referred to the cited literature for tensile property data on human tissue (Kempson, 1976) and on other bovine joints (Roth and Mow, 1980). In Woo's studies, all tests were recorded using a video tape machine so that the specimen deformation histories were stored for repeated analyses.

4.4 TENSILE STRESS-STRAIN RELATIONSHIPS

4.4.1 One-Dimensional Theory

Since the original cross-sectional area and the initial gage length are known, the nominal stress (σ = load/undeformed area) and engineering strain ($\varepsilon = \Delta l/l_0$) relationships could be calculated from the load deformation recordings. The nonlinear stress-strain relationships obtained indicate that the traditional, single Young's modulus description is inadequate (Viidik, 1980), but an exponential stress-strain relationship, such as the one given below, has been used by several authors (Morgan, 1960; Kenedi et al., 1964; Ridge and Wright, 1964; Fung, 1967).

$$\sigma = A(e^{B\varepsilon} - 1) \tag{4.4}$$

$$\frac{d\sigma}{d\varepsilon} = B\sigma + C \tag{4.5}$$

where $$C = AB$$

In this exponential representation where $\sigma \sim \varepsilon$, $d\sigma/d\varepsilon$ is linearly related to σ. The proportionality constant measures the rate of change of the tangent modulus for a nonlinear material with respect to σ. The other constant represents the initial slope of the $\sigma \sim \varepsilon$ diagram. The constants B and C for all nine categories of the 76 bovine humeral cartilage specimens tested in tension are summarized in Table 4.1*a*. For 215 bovine patellofemoral specimens, these constants are shown in Table 4.1*b* (Roth and Mow, 1980).

From Fig. 4.6, it can be seen that the directional properties of articular cartilage specimens taken from the same zone showed that the 0° specimens were much stiffer than the 45 and 90° specimens. With the exception of the surface zone where the 45 and 90° specimens exhibited similar stiffness characteristics, a gradual reduction of stiffness from 0 to 90° was found for the middle and deep zones. Variations of zonal stiffness also existed in the 0° specimens, where the

TABLE 4.1*a* The coefficients A, B, C in the exponential equation, Eq. (4.4), representing the uniaxial tensile stress-strain relationships of bovine articular cartilage*

Zone	Specimen	A, MN/m²	B	$C = AB$, MN/m²
	0° (‖)	2.0	3.38	6.76
Surface	45°	1.2	3.52	4.22
	90° (⊥)	1.3	3.35	4.36
	0° (‖)	2.6	3.06	7.96
Middle	45°	1.1	3.51	3.86
	90° (⊥)	1.5	1.90	2.85
	0° (‖)	2.3	2.66	6.12
Deep	45°	2.5	2.24	5.60
	90° (⊥)	3.3	1.65	5.45

* Nine anatomical categories of young, 12 to 18 months, bovine humeral articular cartilage were tested in uniaxial tension. Infinitesimal strain and lagrangian stress were used by Woo and coworkers (1976), in Eq. (4.4), for the curve-fitting procedure to determine the coefficients A, B, C.

TABLE 4.1*b* The coefficients A, B, C in the exponential equation, Eq. (4.4), representing the tensile stress-strain relationships for skeletally immature (SI) and skeletally mature (SM) bovine patellofemoral articular cartilage*

	Group ‖ SI			Group ‖ SM			Group ⊥ SI			Group ⊥ SM		
Parameter	1	2	3	1	2	3	1	2	3	1	2	3
A	1.625	1.859	2.757	2.108	1.037	0.563	1.385	0.904	0.939	0.902	0.457	0.248
B	5.089	4.321	5.703	5.044	3.203	1.557	4.373	4.162	4.331	3.594	2.214	1.296
$C = AB$	8.270	8.033	15.723	10.633	3.322	0.877	6.057	3.762	4.067	3.242	1.012	0.321

* ‖ and ⊥ represent parallel, or 0°, and perpendicular, or 90°, specimens, respectively, and 1, 2, and 3 represent the first three layers from the articular surface (Roth and Mow, 1980). Infinitesimal strain and lagrangian stress were used in Eq. (4.4).

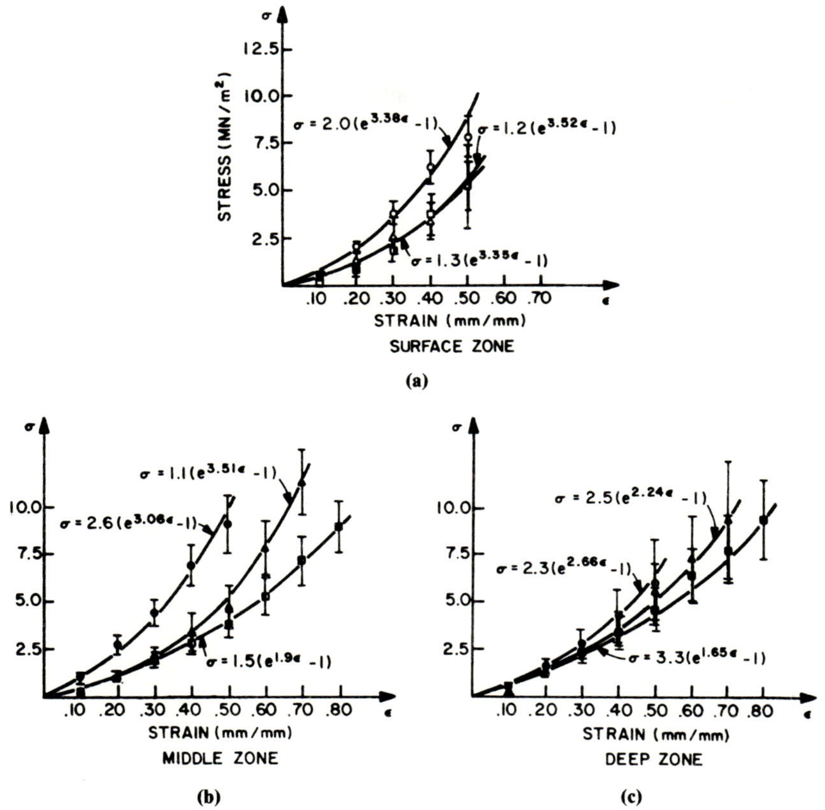

FIG. 4.6 Nominal stress $[F(t)/A_0]$ versus strain ($\varepsilon = \delta l/l_0$) curves of articular cartilage specimens subjected to uniaxial tension. This figure compares the difference in the directional properties of specimens taken from (*a*) surface, (*b*) middle, and (*c*) deep zones of the tissue. *Note:* Circular, triangular, and square symbols represent 0, 45, and 90° specimens, respectively.

middle zone exhibited similar stiffness characteristics to those of the surface zone and both were stiffer than that of the deep zone.

The ultimate tensile stress and strain for all nine categories of specimens are shown in Fig. 4.7. Ultimate tensile strength ranged from 9 to 18 MN/m^2. The ultimate strain values ranged from 60 to 120 percent. Other empirical one-dimensional stress-strain laws have been proposed. However, all these empirical laws, not derived from a strain energy function, are not generally valid for a three-dimensional state of stress.

4.4.2 Finite Deformation Theory

Since the articular cartilage specimen tested in tension undergoes large deformation before failure, i.e., occasionally the specimen could be stretched to double its original length in a material test, a rational description of its mechanical properties would have to be based upon finite deformation theory. One representation of stress and strain used in a finite deformation theory would be to use the eulerian stress, i.e., load/deformed area, and the stretch ratio, λ =

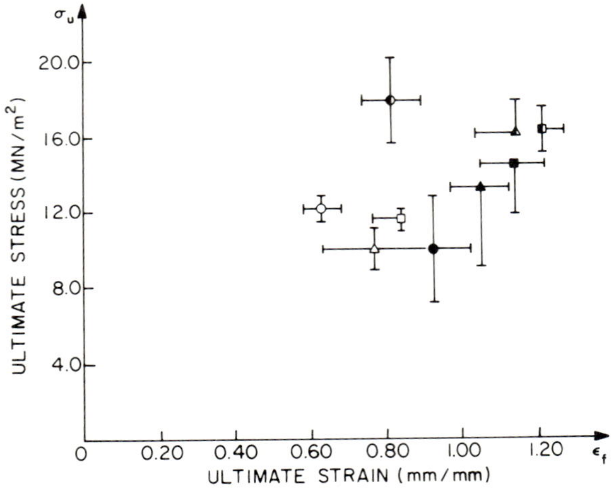

FIG. 4.7 Range of the ultimate tensile stress and ultimate tensile strain of all nine types of articular cartilage specimens tested to failure in tension. *Note:* Open, half-closed, and closed symbols represent surface, middle, and deep zone specimens, respectively. Circular, triangular, and square symbols represent 0, 45, and 90° specimens, respectively. Symbols represent mean values, while bars represent standard deviations in ultimate stress and failure strain.

l/l_0 (where l = deformed length and l_0 = reference, or undeformed, length) (Woo et al., 1979). In finite deformation, a measure of strain is given by (Green and Zerna, 1968):

$$\gamma_{ij} = \frac{1}{2}(G_{ij} - g_{ij}) \tag{4.6}$$

where g_{ij} and G_{ij} are the components of the metric tensors for the unstrained and strained material coordinates embedded in the body, respectively.

In uniaxial tension, the strain components are

$$\begin{aligned} \varepsilon_1 &= \gamma_{11} = \frac{1}{2}(\lambda_1^2 - 1) \\ \varepsilon_2 &= \gamma_{22} = \frac{1}{2}(\lambda_2^2 - 1) \\ \varepsilon_3 &= \gamma_{33} = \frac{1}{2}(\lambda_3^2 - 1) \\ \gamma_{12} &= \gamma_{13} = \gamma_{23} = 0 \end{aligned} \tag{4.7}$$

where λ_1 represents the longitudinal stretch, and λ_2 and λ_3 represent the lateral contraction and the thickness reduction, respectively, during the longitudinal stretch.

***Measurement of Lateral Contraction* (λ_2).** An experimental apparatus for measuring the lateral contraction is shown in Fig. 4.8. The video tape recorder displayed the recorded deformation history of the cartilage specimen during the uniaxial tension test on one of the monitors, e.g., monitor no. 1. The television camera is used to transmit information on monitor no. 1 through the video dimensional analyzer as displayed on the other monitor, i.e., monitor no. 2.

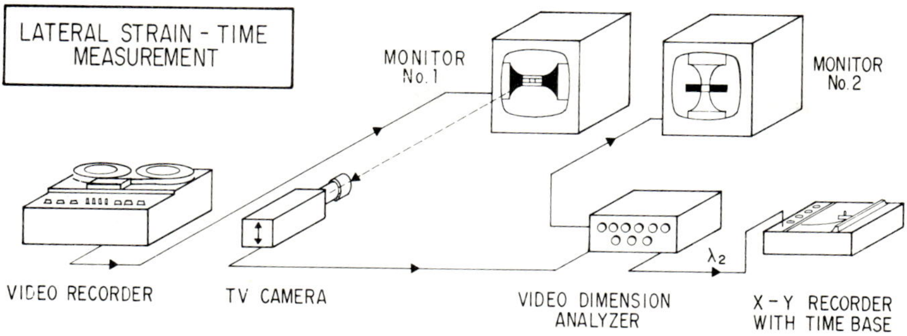

FIG. 4.8 Flow diagram showing the experimental apparatus for measuring the lateral contraction (λ_2) versus time during the longitudinal stretch (λ_1). The video tape recorder displayed the test on real time on monitor no. 1 for the analysis of λ_2.

Notice that both TV monitors displayed the same test specimen subjected to the same test. Thus λ_2 can be measured using the two edges of the articular cartilage specimen (as seen on monitor no. 2) to serve as the contrast phases. The λ_2 versus time can then be plotted as shown in Fig. 4.9*a*, and λ_1 can also be plotted against λ_2, as shown in Fig. 4.9*b*.

***Measurement of Thickness Reduction* (λ_3).** Figure 4.10 shows an experimental device developed by Woo and coworkers (1979) which was used to measure the thickness reduction, λ_3, during the longitudinal stretch, λ_1. For continuous measurement of the thickness of the specimen during tensile stress, an electronic micrometer* which has a range of 0 to 500 μm was used. The sensitivity of this instrument is well within 1.25 μm, and the frequency response is 50 kHz.

During testing, the tip of the micrometer is placed directly beneath and in contact with the central portion of the cartilage specimen (see Fig. 4.10). A 2-mm-wide strip of silver (an excellent conductor) is placed on top of the specimen directly above the electronic micrometer to serve as a target. Hence, a direct and continuous measurement of the thickness reduction of

* Multi-Vit Electronic Measuring System, Kaman Scientific Corporation, P.O. Box 7463, Colorado Springs, Colorado.

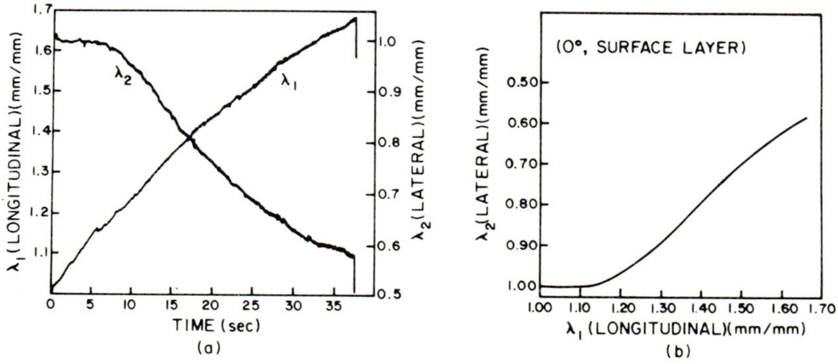

FIG. 4.9 (*a*) A typical recording of λ_1 and λ_2 versus time of the same articular cartilage specimen. (*b*) λ_1 versus λ_2 curve of the same specimen as shown in (*a*). Note that for an anisotropic material subjected to large deformation ($\lambda_1 > 1.20$), the λ_2/λ_1 ratio is higher than 0.5.

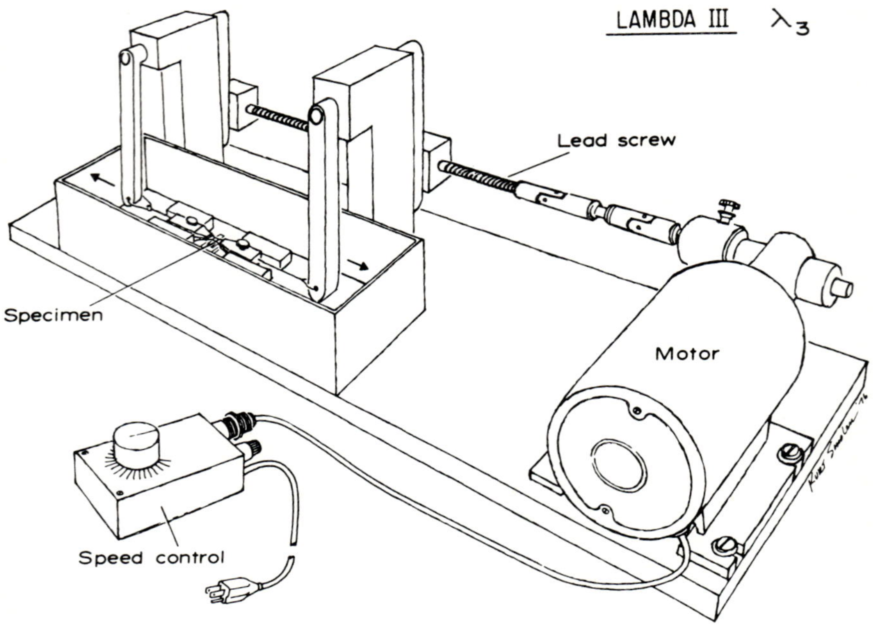

FIG. 4.10 Experimental apparatus designed to measure the thickness reduction of an articular cartilage specimen (λ_3) during the longitudinal stretch (λ_1). The rate of λ_1 used was similar to that of the Instron testing machine. (See text for detailed principles of operation of this device.)

the specimen may be obtained. It is also possible to measure λ_1 simultaneously with λ_2 by mounting the VDA system directly above the specimen. The data from a series of 30 specimens, stretched at 1.2 percent per second, are reported.

It has been tacitly assumed that most of the soft biological materials are incompressible because of their high water content (Fung, 1967; Woo et al., 1972; Crisp, 1972):

$$\lambda_1\lambda_2\lambda_3 = \frac{\text{deformed volume}}{\text{original volume}}$$

$$= 1 \tag{4.8}$$

Therefore, for incompressible materials, the lagrangian stress = σ_L = load/original cross-sectional area (F/A_0) and the eulerian stress = σ_E = load/deformed cross-sectional area [$F/A(t)$] are related by

$$\sigma_E = \lambda_1\sigma_L \tag{4.9}$$

Thus one can readily see the analytical advantage of treating the material as incompressible. Not only can λ_3 be calculated from λ_1 and λ_2, but also the eulerian stress σ_E can be obtained from the lagrangian stress (σ_L) from Eq. (4.9). For cartilage, however, Eq. (4.9) is not true because $\lambda_1\lambda_2\lambda_3 < 1$ (see Fig. 4.11).

4.4.3 Nonlinear Finite Stress-Strain Relationships

Generally, finite deformation stress-strain relationships are obtained through the use of a strain energy function (Fung, 1982). However, in the literature thus far, only the stress-strain relation-

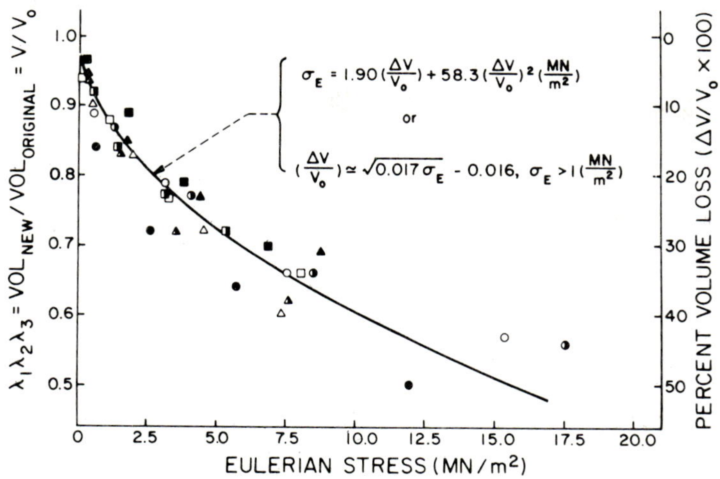

FIG. 4.11 Averaged value of volume versus the eulerian stress. Very little scatter occurs among all nine types of cartilage samples. A least-square quadratic equation was used to approximate the eulerian stress–volume loss relationship. Open, half-closed, and closed symbols represent surface, middle, and deep zone specimens, respectively. Circular, triangular, and square symbols represent 0, 45, and 90° specimens, respectively.

ship in the form of Eq. (4.4), where now σ is the eulerian stress in the extensional direction and ε_1 of Eq. (4.7) is the measure of extensional strain, has been used to describe articular cartilage. Using this form, Woo and coworkers (1979) obtained the coefficients *A*, *B*, and *C* (Table 4.2) by a numerical approximation procedure. The two representations of Eq. (4.4) give rise to significantly different values of the coefficients. Comparison of the directional and zonal properties of articular specimens, as described by the finite stress-strain relationships, showed trends similar to those described earlier (Woo et al., 1979).

TABLE 4.2 The coefficients *A*, *B*, *C* in the exponential equation, Eq. (4.4), for the two representations of stress: lagrangian, $\sigma_L = F/A_0$, and eulerian, $\sigma_E = F/A_0\lambda_2\lambda_3$, used in the uniaxial tensile stress-strain relationship*

		Lagrangian stress: $\sigma_L = \frac{F}{A_0}$			Eulerian stress: $\sigma_E = \frac{F}{A_0\lambda_2\lambda_3}$		
Zone	Specimen	*A*, MN/m²	*B*	*C*, MN/m²	*A*, MN/m²	*B*	*C*, MN/m²
	0° (‖)	4.00	1.85	7.40	2.50	3.74	9.35
Surface	45°	2.49	1.86	4.63	1.51	3.59	5.42
	90° (⊥)	2.41	1.90	4.58	1.26	3.67	4.62
	0° (‖)	5.31	1.65	8.76	3.05	3.50	10.68
Middle	45°	2.06	2.00	4.12	1.44	3.49	5.03
	90° (⊥)	4.44	1.01	4.48	1.58	2.84	4.48
	0° (‖)	4.54	1.39	6.31	1.49	4.08	6.08
Deep	45°	4.90	1.18	5.78	1.54	3.36	5.17
	90° (⊥)	5.18	0.97	5.02	2.49	2.64	6.57

* The tensile strain ε_1 is now calculated from the extension ratio λ_1 on the basis of finite deformation theory, Eq. (4.7).

4.4.4 Water Loss during Finite Deformation in Tension

Edwards (1967) found 33 percent water loss in cartilage after 30 min of compressive stress at 0.31 MPa. In tension, a volume loss for articular cartilage is also observed. This cannot be unless there is a loss of mass from within the tissue. In fact, Woo and coworkers (1979) measured a monotonic decrease of $\lambda_1\lambda_2\lambda_3$ with increasing eulerian stress σ_E for $\sigma_E > 1.0$ MN/m^2 (Fig. 4.11). From measurements described above, and using the relationship

$$\frac{V}{V_0} = \lambda_1\lambda_2\lambda_3 \tag{4.10}$$

the change in volume, $\Delta V = V_0 - V$, may be represented as

$$\frac{\Delta V}{V_0} = 1 - \lambda_1\lambda_2\lambda_3 \tag{4.11}$$

It was possible to use a quadratic equation to obtain a least-square curve fit of the volume loss–eulerian stress data.

$$\sigma_E = 1.90\left(\frac{\Delta V}{V_0}\right) + 58.3\left(\frac{\Delta V}{V_0}\right)^2 \tag{4.12}$$

or

$$\left(\frac{\Delta V}{V_0}\right) = \sqrt{0.017\sigma_E} - 0.016 \qquad \sigma_E > 1 \text{ MN/m}^2 \tag{4.13}$$

with a correlation coefficient of 0.92($p < 0.001$); see Fig. 4.11.

This stress–volume loss relationship suggests that the water exudation from articular cartilage during a tensile experiment may be similar for all zones. In a similar tensile study, Roth and Mow (1980) showed that exudate formed by stretching the specimen collects on the surface and is noticeable to the naked eye. Most recently, Li and coworkers (1983) showed that, for articular cartilage, fluid flow transverse to the specimen during a tensile experiment can dominate the stress-strain behavior of the tissue at the highest strain rate used (0.12 s^{-1}). At lower strain rates, however, the flow of the interstitial fluid does not contribute substantially to the measured tensile stress.

4.5 HISTORY AND TIME-DEPENDENT TENSILE PROPERTIES

4.5.1 Quasi-Linear Viscoelastic Theory

It is known that the stress-strain curves and the size of the hysteresis loops may change dramatically during the first few cycles of loading and unloading, a process called *preconditioning* (Fung, 1972). Cyclic testing of articular cartilage specimens in tension at a constant strain rate and at moderate strain levels will show repeatable stress-strain relationships within 7 to 10 cycles of preconditioning. The hysteresis loop during loading and unloading at a constant rate after preconditioning will also remain essentially unchanged.

For the rate-dependent behavior of articular cartilage, experimental evidence suggests that the stiffness of cartilage will increase with increasing strain rate, and the size of the hysteresis loop will also increase, though moderately, as a result of the higher strain rate (Woo et al., 1979). Both these effects are consistent with the biphasic description (see below) of articular cartilage: (1) at higher strain rates, the frictional resistance due to interstitial fluid flow relative to the solid matrix will increase, thus a higher stiffness, and (2) at higher strain rates, the dissipation due to interstitial fluid flow will also increase, thus a larger hysteresis loop.

The stress-relaxation properties of articular cartilage in tension following a rapid stretch rate of 30 mm/s was studied by Woo and coworkers (1980). For these tests, it is convenient to plot a "reduced relaxation function," $G(t) = \sigma(t)/\sigma(t_0)$ against the logarithmic scale of time (Fig.

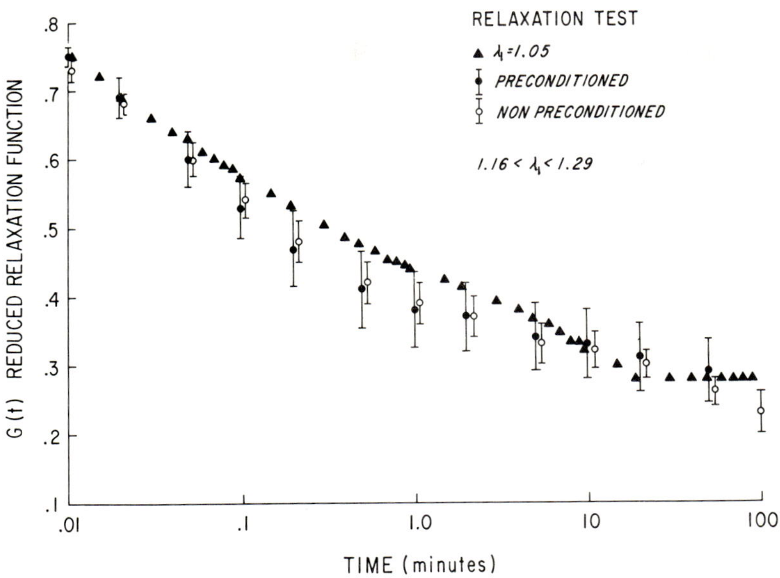

FIG. 4.12 Results of relaxation test up to 100 min. At a low strain level (λ_1 = 1.05) stress leveled out after 15 min. At higher stress levels (λ_1 = 1.16 to 1.29) stresses usually do not level off for over 100 min. Preconditioning and nonpreconditioning of the cartilage samples had no effect on the stress-relaxation characteristics.

4.12). If the initial stretch is small, e.g., $\lambda_1 = 1.05$, the stress will usually level off after 15 min. For higher values of stretch (λ_1 between 1.16 and 1.29), the relaxation curves do not level off until after 100 min. Also, preconditioning the specimens appeared to have no effect on their relaxation behavior. For very high stretch values (λ_1 between 1.50 and 1.55), complete relaxation usually does not occur within 1000 min. However, it is important to note that the slope of the $G(t)$-versus-time was similar for all strain levels, and the major portion of stress relaxation occurred in the first 15 min.

Stress-relaxation data provide a basis for determining the time- and history-dependent properties of any viscoelastic material. Fung (1972) proposed that the stress response of biological tissues to a step change in strain (or relaxation function Φ) could be written in terms of a "quasi-linear" viscoelastic law; that is,

$$\Phi = \Phi[\gamma(t), t] = G(t)*S^e[\gamma(t)] \tag{4.14}$$

where $S^e(\gamma)$ represents the nonlinear "elastic response," $G(t)$, where $G(0) = 1$ and $\gamma = (\lambda_1^2 - 1)/2$. The stress-strain-history integral takes the following form:

$$\begin{aligned} S(t) &= \int_{-\infty}^{t} G(t - \tau)\dot{S}^e(\tau)\, d\tau \\ &= S^e[\gamma(t)] - \int_0^t \frac{\partial G(t - \tau)}{\partial \tau} S^e(\tau)\, d\tau \end{aligned} \tag{4.15}$$

Woo and coworkers (1980) used this representation to describe, phenomenologically, the stress-relaxation behavior of articular cartilage in tension. Once the functionals in this model, $G(t)$ and $S^e[\gamma(t)]$, are determined, the time-dependent stress $S(t)$, for a known history of strain $\gamma(t)$, can be determined by Eq. (4.15). Since the reduced stress-relaxation behavior of articular cartilage appears to satisfy the condition described above, that is, $G(t)$ is independent of λ_1, Woo and coworkers showed that the curve fit for the experimental values of $G(t)$, chosen in the form

$$G(t) = \frac{1 + c[E_1(t/\tau_2) - E_1(t/\tau_1)]}{1 + c \ln (\tau_2/\tau_1)} \tag{4.16}$$

appears to work well. Here, E_1 is the exponential integral function, and c, τ_1, and τ_2 are constants. These three constants were determined by a least-square curve-fitting procedure (Powell, 1975). Good agreement between this phenomenological model and experimental $G(t)$ values (Fig. 4.13) was obtained for this specimen with $\tau_1 = 0.006$ s, $\tau_2 = 8.38$ s, and $c = 2.02$.

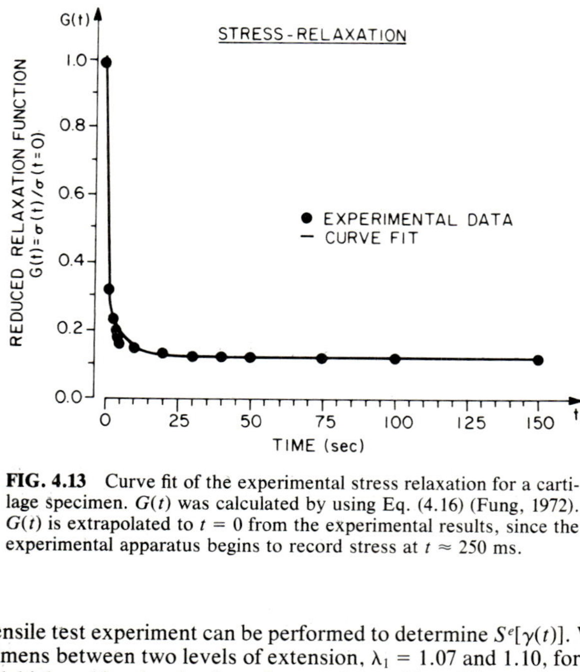

FIG. 4.13 Curve fit of the experimental stress relaxation for a cartilage specimen. $G(t)$ was calculated by using Eq. (4.16) (Fung, 1972). $G(t)$ is extrapolated to $t = 0$ from the experimental results, since the experimental apparatus begins to record stress at $t \approx 250$ ms.

The cyclic tensile test experiment can be performed to determine $S^e[\gamma(t)]$. Woo and coworkers cycled specimens between two levels of extension, $\lambda_1 = 1.07$ and 1.10, for 10 cycles. Stress relaxation at both high and low levels of tensile stretch is evident. The cyclic stress data are used to determine the elastic response S^e, where it can be assumed to be a power series in γ of the form

$$S^e(\gamma) = \sum_{i=1}^{n} a_i \gamma^i \tag{4.17}$$

Substitution of Eq. (4.17) into Eq. (4.15) with $G(t)$ now given by Eq. (4.16) yields a linear set of equations that can be solved for the constants a_i by a least-square procedure for each n, on the basis of measured values of the stress $S(t)$. For $n = 2$, the constants a_1 and a_2 are 30 and 56 MN/m^2, respectively. One finds good agreement between the Kirchhoff stress-time curves as measured experimentally with those computed from the quasi-linear viscoelastic equation (Fig. 4.14). Similar agreement can be obtained for higher orders of n.

It has been shown that the quasi-linear viscoelastic theory, Eqs. (4.14) to (4.16), proposed by Fung (1972), provides a powerful means to describe the observed phenomenological viscoelastic behavior of articular cartilage in tension (Woo and coworkers, 1979; Simon and coworkers,

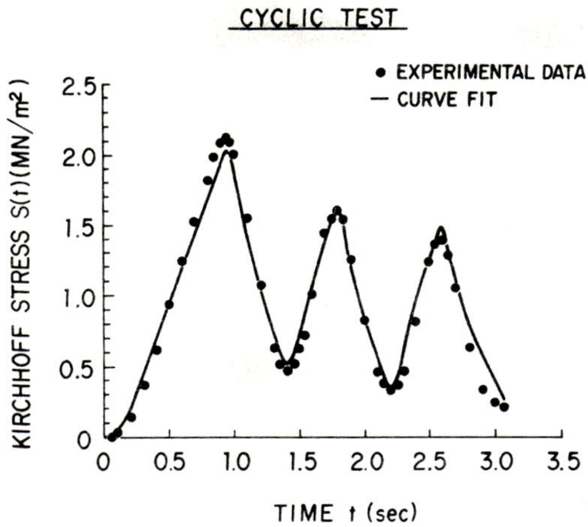

FIG. 4.14 With $G(t)$, $S(t)$, and $\gamma(t)$ known from cyclic testing, Eq. (4.15) was used to calculate the elastic response of cartilage by assuming $S^e(\gamma)$ to be a power series in γ. Good agreement between the experimental stress history and the computed stress versus time was found, suggesting that the quasi-linear viscoelastic model may be used to represent the phenomenological behavior of normal articular cartilage.

1982). To do so, the reduced relaxation function $G(t)$ and the nonlinear elastic response S^e must be determined. A method has been outlined to determine these quantities in this chapter, specifically for articular cartilage, but it is generally applicable to many biological tissues and viscoelastic materials (Woo et al., 1981). Once $G(t)$ and S^e are determined, the phenomenological time and history dependence of stress on any strain history may be calculated.

4.6 BIPHASIC THEORY OF ARTICULAR CARTILAGE

4.6.1 Permeability of Articular Cartilage

The movement of the interstitial fluid within cartilage and across the articular surface are two fundamental phenomena that govern the mechanical response, biomechanical function (e.g., the lubrication of synovial joints, see Chapter 5), and biological integrity of the tissue (e.g., the nutrition required by the chondrocytes of the adult cartilage). There are two primary fluid transport mechanisms through the cartilage matrix:

1. The interstitial fluid is transported through the porous permeable matrix via the application of a direct fluid pressure gradient across the tissue. The transport of the fluid increases with increasing pressure. For one-dimensional permeation flows through a finite thickness h of the tissue, an apparent permeability k_{app} is often defined by Darcy's law in the form

$$Q = \frac{A k_{app} P_A}{h} \tag{4.18}$$

where Q is the volume flow rate across an area A of the cartilage, and P_A is the pressure drop across the thickness h of the specimen. An average value of this apparent permeability coefficient was first determined by McCutchen (1962) to be 0.58×10^{-15} m^4/(N · s). Such a small value of permeability indicates a very high frictional resistance to fluid flow. It was also observed that permeability decreases with depth, with a value of 0.76×10^{-15} $m^4/(N \cdot s)$ near the middle region and a value of 0.43×10^{-15} $m^4/(N \cdot s)$ for the underlying cartilage. The value of permeability also varies with the nature of the permeating fluid. For example, Edwards (1967) determined permeabilities of 0.11×10^{-15} and 0.33×10^{-15} $m^4/(N \cdot s)$ for normal saline and normal Ringer's solutions, respectively.

2. Fluid transport can be achieved by deformation of the cartilage matrix. Exudation occurs by squeezing the tissue between two rigid, porous platens with the fluid in the pores of

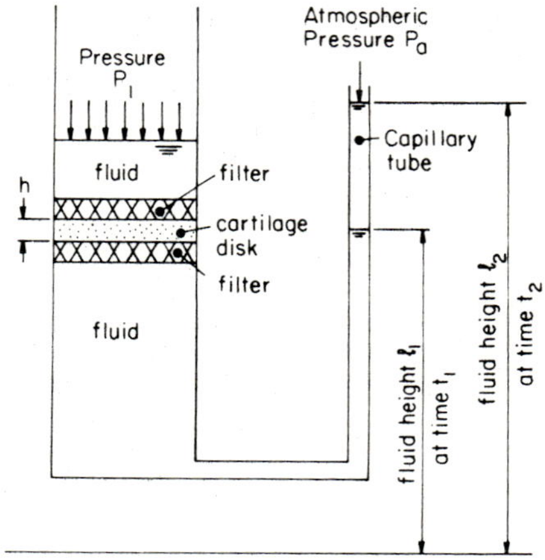

(a)

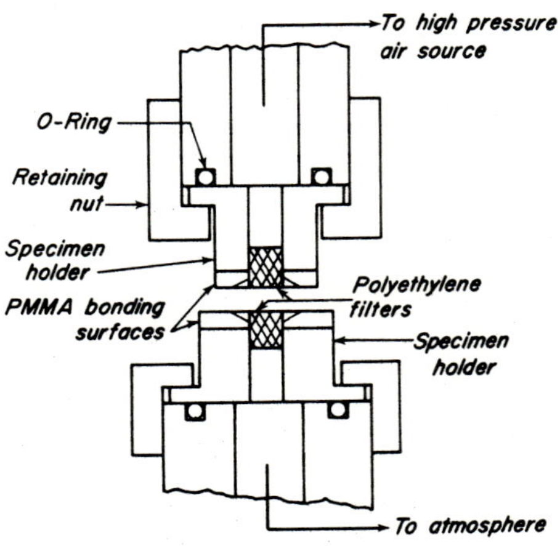

(b)

FIG. 4.15 (*a*) Schematic representation of the permeability experiment method of depicting flow measurement. (*b*) Cross-sectional view of the test chamber showing the manner in which the specimens should be mounted. Upper section may be moved in the vertical direction to compress the specimen up to 40 percent; applied pressure can range up to 2.75 MN/m^2.

the platen at ambient pressure. When this is done, the compressed solid matrix causes a rise of the internal pressure acting on the fluid within the tissue. This creates a pressure gradient across the surface of the specimen, favoring exudation of the interstitial fluid. This mechanism is known as *flow due to consolidation*.

Because articular cartilage is very soft, these two mechanisms are coupled, which gives rise to a nonlinear interaction phenomenon. Compression decreases the permeability of the tissue. An apparatus was designed in 1976 by Mansour and Mow to determine this nonlinear interaction phenomenon by clamping the tissue with a precise amount of compressive strain during the permeation experiment. Figure 4.15*a* and *b* shows the schematic diagram of the apparatus and a section through the specimen holders, respectively. The general features of the apparatus are as follows: the cartilage specimen is bonded between two specimen holders; a precise amount of axial displacement (i.e., the clamping strain) can be applied to and maintained on the specimen (the specimen holders are guided and maintained in alignment by the four linear bearings); the displacement of the surface of the cartilage is measured by a linear variable displacement transducer (LVDT) mounted on the top holder; the permeating fluid, any aqueous solution, is driven through the specimen by air pressure; and the flow rates are measured by the change of height of fluid in the capillary tube by a cathetometer. This apparatus is capable of measuring specimen permeability up to a pressure of 2.75 MN/m^2 and 40 percent compression.

The actual fluid flow through the specimen and the test apparatus is better seen by referring to the schematic diagram shown in Fig. 4.15*b*. Fluid is placed in the upper holder as well as in the base of the apparatus and in the capillary. When a pressure (P_1) is applied to one side of the cartilage, fluid will be forced through the tissue and rise in the capillary tube, where the fluid pressure is maintained at the atmospheric level P_a.

Using bovine cartilage slices approximately 200 μm thick, Mansour and Mow found that the permeability dramatically decreased with increasing clamping strain and applied pressure. More extensive measurements by Gordon (1978), using the same apparatus, provided the experimental data for the empirical deduction of Mow and Lai (1980). They showed that the empirical form of the functional dependence of apparent permeability with clamping strain and applied pressure is

$$\begin{aligned} k_{\text{app}} &= A(P_A) \exp [\alpha(P_A)\varepsilon_{\text{clamp}}] \\ P_A &= P_1 - P_a \end{aligned} \tag{4.19}$$

$A(P_A)$ and $\alpha(P_A)$ depend on the pressure differential P_A and were determined numerically by a nonlinear regression curve-fitting procedure. The solid lines of Fig. 4.16 are plots of this exponential law, and the points are from one set of experimental data. Gordon obtained 10 such sets of data, all of which can be empirically described in the same manner. In Eq. (4.19), the subscript *app* for the permeability *k* emphasizes that for nonzero P_A only "apparent permeability" is directly measured; see explanation below.

The decrease of permeability with increasing compressive strain can be explained on the basis that compaction of the tissue decreases the average pore size of the tissue, thereby making it more difficult for the permeating liquid to flow through it. However, there are two sources of such compaction. The obvious one is the compaction due to the clamping strain $\varepsilon_{\text{clamp}}$ imposed on the tissue; the other is the compaction due to the action of diffusive drag caused by the permeating fluid. The latter source originates from the P_A required for the maintenance of the steady perfusion. Figure 4.16*b* and *c* demonstrates how the empirical coefficients A and α change with P_A.

Intrinsic Permeability. The applied clamping strain $\varepsilon_{\text{clamp}}$ is uniform, whereas the drag-induced strain is nonuniform. This nonuniform strain in the tissue must therefore result in a nonuniform permeability throughout the depth of the specimen. Thus, the permeation experiments which measure only the permeability from flow rates across the entire thickness can measure only the "apparent" permeability. It seems logical then to introduce a concept of "intrinsic" permeability, which is defined as the permeability function of the tissue that depends only on the uniform clamping strain. This condition could be achieved under the limiting condition of $P_A \to 0$, that is, when the nonuniform strain field caused by the pressure differential becomes vanishingly small. Even though such an experimental condition is impossible to

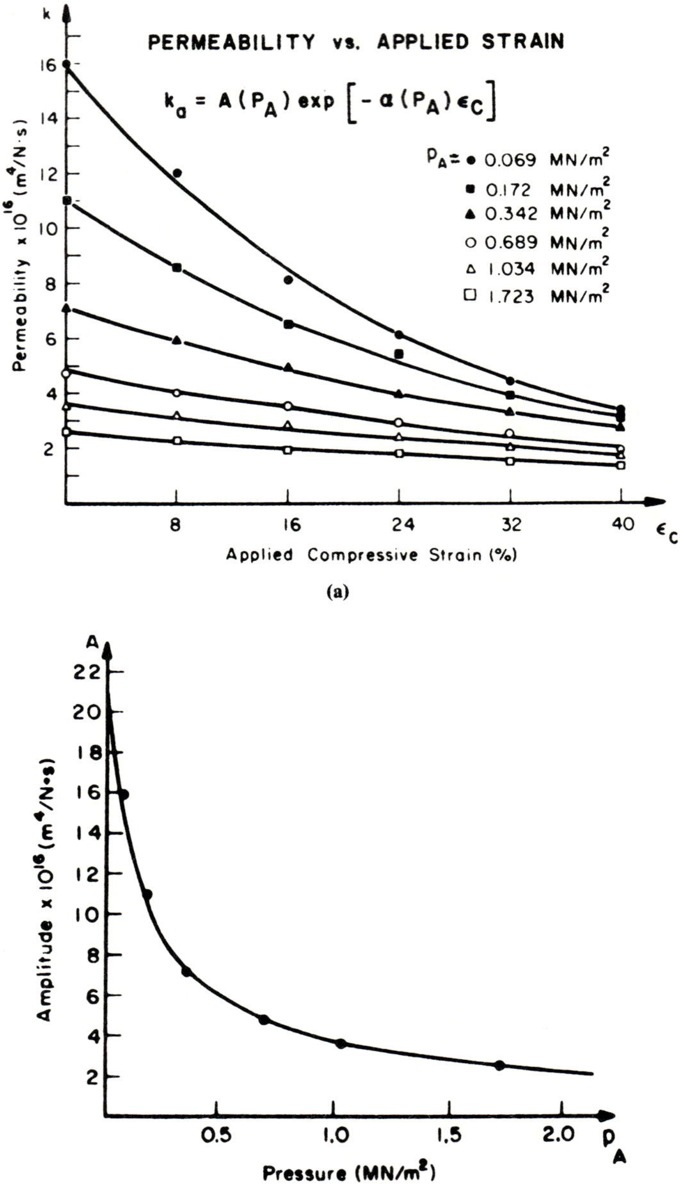

(a)

(b)

FIG. 4.16 (*a*) Experimental curves of permeability versus applied compressive strain at various levels of applied pressure for a sample of bovine articular cartilage. The solid curves are least-square best fits of the indicated exponential decay law, Eq. (4.19). (*b*) Curve of the amplitude function $A(P_A)$ and experimental data points versus P_A for the same bovine specimen.

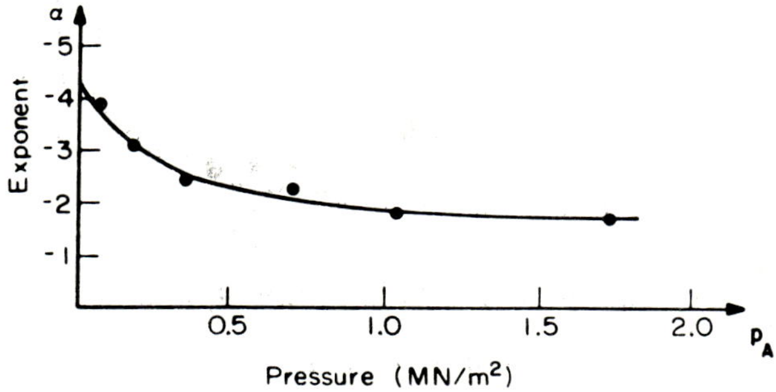

(c)

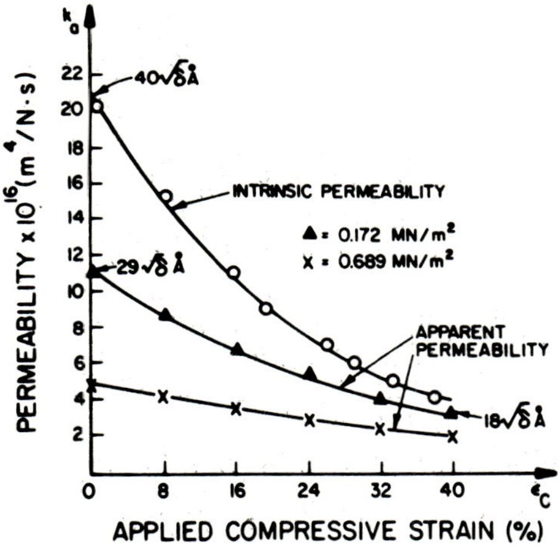

(d)

FIG. 4.16 (*c*) Curve of exponent $\alpha(P_A)$ versus P_A for the same bovine specimen (Mow et al., 1980). (*d*) Intrinsic permeability curve as obtained by the extrapolation procedure (Lai and Mow, 1980); see text.

achieve in reality, the intrinsic permeability function can be obtained by extrapolating to the limit of $P_A \rightarrow 0$. An intrinsic permeability curve, Fig. 4.16*d* was obtained by such an extrapolation procedure. This intrinsic permeability can be described by the expression:

$$k_{\text{intrinsic}} = k_0 \exp M\varepsilon \qquad (4.20)$$

where k_0 and M are *constants* and ε is the true strain. The intrinsic permeability curve corresponding to the data in Fig. 4.16*a* to *c* is adequately described by the values $k_0 = 1.71 \times 10^{-15}$ $m^4/(N \cdot s)$ and $M = 4.3$ in Fig. 4.16*d* (Lai and Mow, 1980). Confirmation of this intrinsic permeability concept has been made by an extensive independent experimental study via rate-controlled compression stress-relaxation phenomenon by McCormack (1983), and the asymptotic analysis of this nonlinear stress-relaxation phenomenon by Holmes and coworkers (1983a). These models are derived from the recent biphasic theory developed by Mow and Lai (1980) to describe the fluid-solid interactions existing within articular cartilage during deformation.

4.6.2 Biphasic Model of Articular Cartilage

A completed biphasic model for articular cartilage, based on the mixture theory of Craine and coworkers (1970) and Bowen (1976), was formulated by Mow and coworkers (1980). In essence, this model depicts the tissues as consisting of a soft, permeable, elastic solid mixed with water. Assuming that both phases are intrinsically incompressible, the continuity equation for this binary mixture is:

$$\text{div}\, \bar{v}^f + \alpha\, \text{div}\, \bar{v}^s + \alpha(\bar{v}^s - \bar{v}^f) \cdot \text{grad}\ln \rho^s = 0 \tag{4.21}$$

where $\bar{v}^f$ and $\bar{v}^s$ are the velocities of the fluid phase and the solid phase, respectively, and ρ^s is the apparent density of the solid. The coefficient α is the solid content defined by the ratio of solidity (V_s/V_T) to porosity (V_f/V_T) of the tissue. Here V_s, V_f, and V_T are the solid, fluid, and total volume of the organic matrix, interstitial fluid, and tissue, respectively. Under most circumstances, the momentum balance equation may be used by omitting inertial forces

$$\text{div}\, \boldsymbol{\sigma}^s - K(\bar{v}^s - \bar{v}^f) = 0 \tag{4.22a}$$

and

$$\text{div}\, \boldsymbol{\sigma}^f + K(\bar{v}^s - \bar{v}^f) = 0 \tag{4.22b}$$

to describe the deformational response of cartilage. The second term in Eqs. (4.22) represents the diffusive drag arising from the relative velocities between the fluid and solid components. Even in the unloaded state, the diffusive coefficient K has values on the order of 10^{15} N · s/m^4, which explains why the inertia terms may be omitted in Eqs. (4.22). Under slow flow conditions, the diffusive coefficient is related to the permeability k by the inverse relation $K = 1/[k(1 + \alpha)^2]$ (Lai and Mow, 1980). The permeability has been directly measured by steady perfusion studies described in the previous section.

Under the condition of infinitesimal deformations and velocities, the isotropic stress-strain relationships of the first-order theory are

$$\boldsymbol{\sigma}^s = -\alpha P\mathbf{I} + \lambda_s e\mathbf{I} + 2\mu_s\mathbf{e} + \lambda_s'\, \text{div}\, \bar{v}^s\mathbf{I} + 2\mu_s'\mathbf{D}^s - 2K_c\boldsymbol{\Gamma} \tag{4.23a}$$

and

$$\boldsymbol{\sigma}^f = -P\mathbf{I} + \lambda_f\, \text{div}\, \bar{v}^f\mathbf{I} + 2\mu_f\mathbf{D}^f + 2K_c\boldsymbol{\Gamma} \tag{4.23b}$$

where P is the fluid-solid interactive pressure term; (λ_s, μ_s) and (λ_s', μ_s') are the elastic and viscoelastic moduli of the solid phase in the mixture; $\mathbf{e}$ and $\mathbf{D}^s$ are the infinitesimal strain and rate of deformation tensors for the solid phase; $\boldsymbol{\Gamma}$ is the spin tensor associated with the relative motion of the solid with respect to the fluid; $\mathbf{D}^f$ is the rate of deformation tensor for the fluid phase; and (λ_f, μ_f) are the viscosity coefficients of the fluid phase in the mixture.

Even though the isotropic case is presented here, formulas for the more general case of biphasic anisotropic stress-strain laws may be found in the paper by Mow and coworkers (1980). The significant anisotropic structure of cartilage, as manifested, for example, in the split-line pattern or anisotropic tensile properties, has not been pursued, in deference to the simpler isotropic biphasic theory aimed at the more pressing historical need for elucidating the role of fluid motion in the observed viscoelastic creep and stress-relaxation behavior of cartilage in compression. Other generalizations, such as finite deformation of the solid matrix, nonlinear permeability effects, physicochemical effects, and interstitial capillary force effects, can be undertaken using the existing general biphasic theory. In this chapter we will describe the effect of nonlinear permeability only as it relates to the compressive creep and stress-relaxation phenomena.

In the constitutive relationships embodied in Eqs. (4.23), most of the coefficients are unknown, and experiments have yet to be devised to determine all of them. Accordingly, in most of the work to date it is assumed that the solid matrix is elastic ($\lambda_s' = \mu_s' = 0$), the viscosity of the interstitial fluid enters only through the diffusive drag ($\lambda_f = \mu_f = 0$), effects of diffusional couples are absent ($K_c = 0$), and the solid content is spatially homogeneous. Even with these restrictive assumptions, the resulting biphasic model of articular cartilage (known as the *linear KLM model*) appears to describe adequately its observed creep and stress-relaxation behavior in compression under isothermal and constant electrolytic conditions. However, each assumption is only approximately correct. For example, Hayes and Bodine (1978) and Mow and coworkers (1984a) found that the tissue is slightly viscoelastic ($0.10 \leq \tan \delta \leq 0.25$), and so a more accurate model for the tissue would have to include terms to describe the viscoelastic behavior of the solid matrix (for example, $\lambda_s' \neq 0$, $\mu_s' \neq 0$). The intrinsic viscoelastic properties of the solid matrix are described in a later section. However, it is generally believed that most of the viscoelastic behavior of cartilage in *compression* can be attributed to the diffusive drag caused by the interstitial water flowing relative to the solid matrix, and not to the viscoelastic properties of the solid matrix itself. With these assumptions, investigators have begun to determine the diffusive coefficient K and elastic constants λ_s, μ_s of the linear *KLM* model for articular cartilage, as well as for other tissues (Favenesi et al., 1983; Mow and Schoonbeck, 1982) to test the accuracy and general applicability of the model, and to see how the constants are affected by changes in the structural integrity and biochemical composition of the tissue.

4.7 APPLICATION OF THE BIPHASIC THEORY

4.7.1 Confined Compression Creep and Stress-Relaxation Tests

Mow and coworkers (1980) and Armstrong and Mow (1982a, b) have conducted tests on bovine and human articular cartilage under the confined compression configuration in creep and stress relaxation. These tests provide measurements of the material properties of the tissue as well as an understanding of the mechanism by which the tissue exhibits its viscoelastic behavior. The confined compression experiments for creep and stress relaxation will be described below.

Confined Compression Creep Test (Constant Permeability). In this test, the cartilage surface of a small cylindrical plug is loaded against a rigid, highly permeable filter. The subchondral bone–cartilage junction is kept intact, so negligible fluid flow across this boundary is assured for specimens from skeletally mature animals. The specimen is mounted in an impermeable, cylindrical confining chamber, schematically shown in Fig. 4.17*a*. Ideally, therefore, all radial bulging of the cartilage and radial fluid leakage are prevented.

A load is suddenly applied to and constantly maintained on the specimen. The lateral confinement of the specimen and the intrinsic incompressibility of the tissue as a whole prevent any instantaneous displacement of the surface. However, exudation of the interstitial fluid begins immediately. The loss of fluid volume from the specimen by this exudation causes the specimen to creep in compression (Fig. 4.17*b*). Thus the permeability must be one of the parameters which controls the rate of creep in this experiment.

During the creep process, the applied load is supported by a nonhomogeneous compaction of the solid matrix and a buildup of the interstitial fluid pressure. The compaction of the matrix is caused by the load and the frictional drag accompanying the exudation. The effect of the frictional drag is initially to restrict the compaction to a surface region. However, with time, the compacted region gradually diffuses into the deeper layer of the specimen; eventually, this creep process ceases, and an equilibrium state of homogeneous compression is reached. At equilibrium, the load is carried by the solid matrix, providing a method to determine the intrinsic, confined compression modulus of the material.

Under these conditions, applying the biphasic constitutive equations discussed above, Mow and coworkers (1980) showed that the matrix displacement u in the axial direction is governed by a simple linear diffusion equation, and obtained the following solution for the surface displacement u of a specimen of thickness h, creeping under a compressive load of intensity F_0:

$$\frac{u(0, t)}{h} = \frac{F_0}{H}\left\{1 - 2\sum_{n=0}^{\infty} \frac{\exp\left[-(n + \frac{1}{2})^2\pi^2 Hkt/h^2\right]}{(n + \frac{1}{2})^2\pi^2}\right\} \tag{4.24}$$

where $H = \lambda_s + 2\mu_s$ is the equilibrium elastic modulus under the confined compression configuration (also known as the aggregate elastic modulus H_A). Note that as $t \to \infty$, $(u/h) \to (F_0/H)$, providing the measure for H. At time $t = 0$, this expression gives $u = 0$; that is, there is no instantaneous displacement of the cartilage surface. From this expression, clearly it can be seen

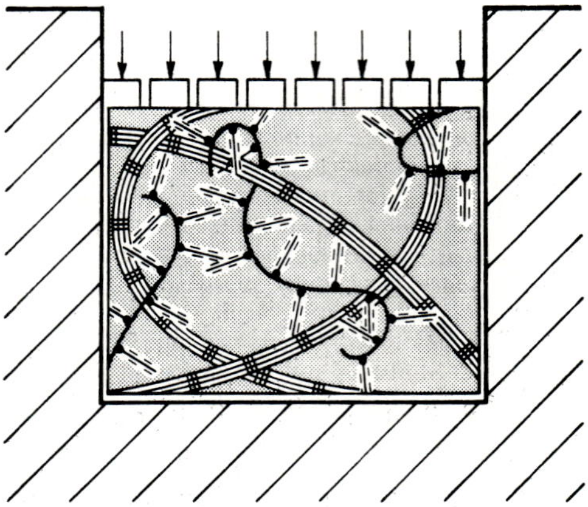

(a)

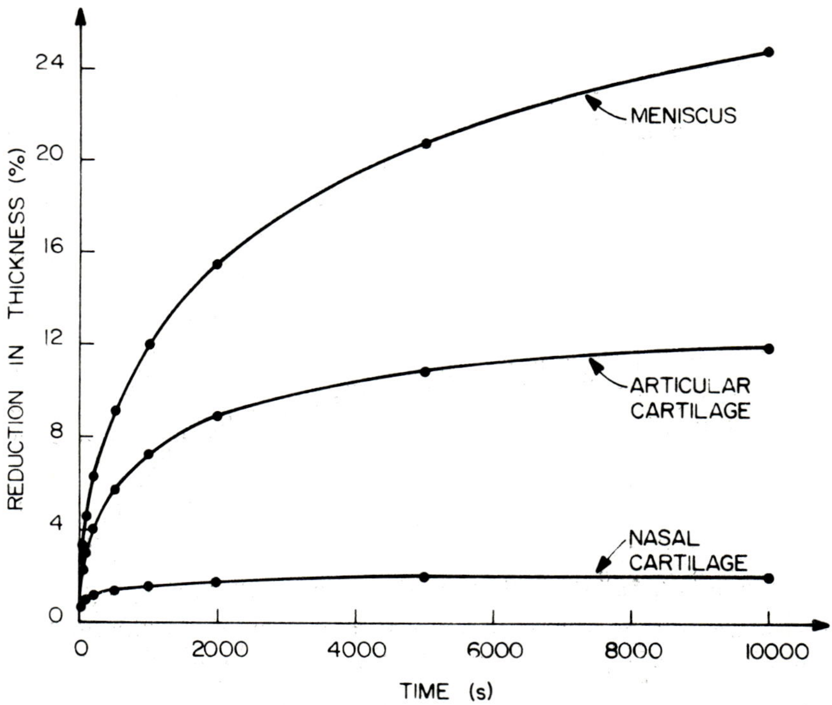

(b)

FIG. 4.17 (*a*) Schematic representation of the confined compression experiment on a plug of biphasic articular cartilage. Load is applied by a rigid porous filter under free-draining conditions. (*b*) Biphasic creep response associated with a suddenly applied load. Creep is controlled by fluid exudation, i.e., the tissue permeability, while equilibrium displacement is determined by the intrinsic modulus of the solid matrix. Creep response of three types of biphasic materials: meniscus, articular cartilage, and nasal cartilage. (See Table 4.3*b* for comparison of material properties.)

that the characteristic time τ of creep is controlled by the combination of physical parameters given by h^2/Hk. In other words, a decrease in modulus or permeability or an increase in specimen thickness will cause the specimen to creep longer before reaching equilibrium. It has been found that τ provides an excellent assessment of the rate of creep as well as other rates of deformation processes. Typical experimental measurements and theoretical predictions (obtained by a least-square curve-fitting procedure) of cartilage surface displacement are shown in Fig. 4.17*b* for various types of cartilages. Table 4.3*a* summarizes the values of permeability k and modulus H from tests of 103 human patellar samples, and Table 4.3*b* and *c* shows values of k and H for bovine nasal and meniscus specimens along with other compositional parameters.

Confined Compression Stress-Relaxation Test (Constant Permeability). In this test, the experimental setup is the same as that for the confined creep test. However, a ramp displacement function (Fig. 4.18) is imposed on the articular surface instead. During the compression phase, the fluid is forced out of the tissue into the porous filter used to load the surface. Because of the large frictional drag associated with fluid flow through the solid matrix, large loads are needed to compress the tissue at seemingly small rates of compression. Recall that the diffusive coefficient

TABLE 4.3*a* The biphasic properties of human patellar articular cartilage at autopsy as determined from the confined compression experiment

	Variable	Number	Minimum	Maximum	Mean	Standard deviation
Permeability × 10^{14}m^4/(N · s)	k	103	0.05	1.95	0.47	0.36
Modulus, MPa	H	103	0.13	1.91	0.79	0.36
Thickness, mm	h	103	1.69	5.17	3.12	0.72
Water content, %	....	58	72.80	88.40	78.63	3.86

Source: Armstrong and Mow, 1982.

TABLE 4.3*b* Comparison of biphasic properties of bovine nasal and bovine and human articular cartilage*

Cartilage	H, MPa	k × 10^{15}m^4/(N · s)	Water, % total wt.	Uronic acid, % wet wt.	Ash content, % dry wt.
Bovine nasal	5.64 (3.19) n = 19	0.49 (0.3) n = 19	75.6 (1.6) n = 36	1.30 (1.18) n = 36	8.00 (1.95) n = 12
Bovine articular	0.85 (0.21) n = 28	4.67 (3.8) n = 28	80.9 (4.0) n = 10	0.97 (0.25) n = 9	0.85 (0.24) n = 12
Human articular	0.79 (0.36) n = 103	4.70 (3.6) n = 103	78.6 (3.9) n = 58	0.65 (0.08) n = 26	

Source: Mow and Schoonbeck, 1982.
* n = number in sample. Numbers in parentheses represent standard deviations for each sample.

TABLE 4.3c Comparison of biphasic properties of bovine menisci and articular cartilage

	Menisci							
	Superficial	Deep	Posterior	Central posterior	Central anterior	Anterior	Averaged	Articular cartilage
%H_2O content	74.4	73.4	73.2	73.1	73.1	76.3	73.9	80.9
k × 10^{16} m^4/(N · s)	7.52	8.73	6.80	9.25	8.46	8.02	8.13	46.70
H, MPa	0.41	0.41	0.40	0.38	0.41	0.45	0.41	0.85

Source: Favenesi and coworkers (1983).

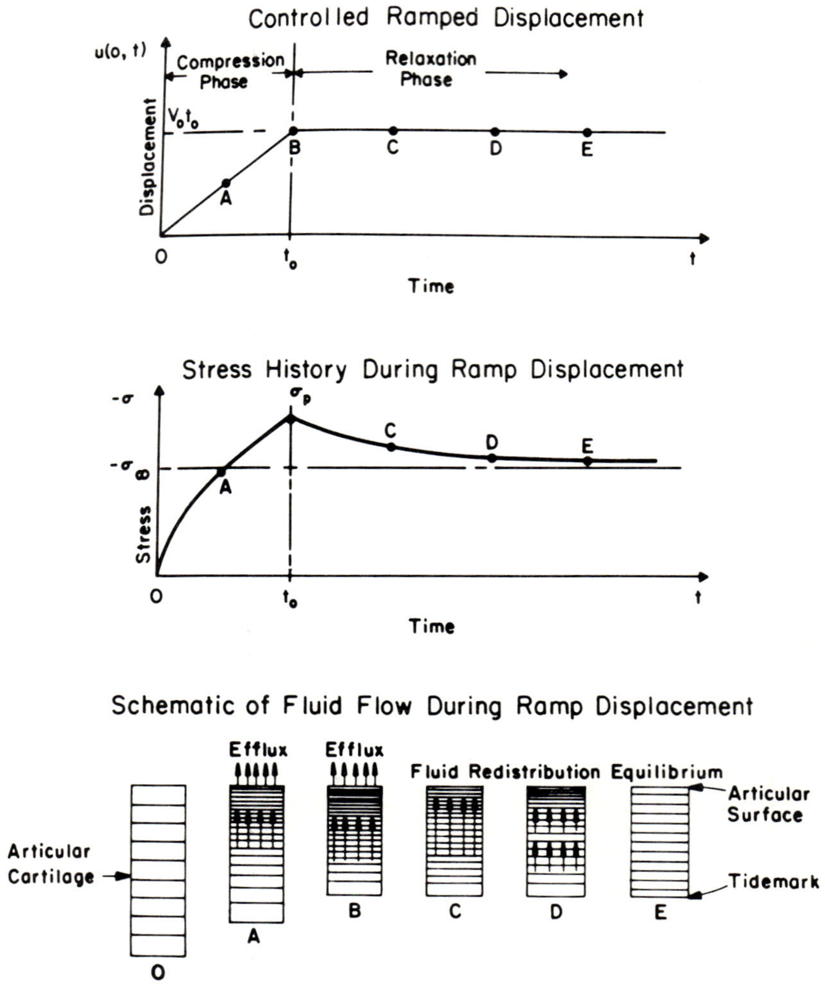

FIG. 4.18 A stress history in confined compression. The rate of compression is controlled by specifying V_0, and the equilibrium strain is set by specifying $V_0 t_0$. The middle curve shows a typical stress rise during the compression phase and a typical stress-relaxation response in the relaxation phase. Lower schematics show how the movement of interstitial fluid predominates the stress-history response. During the compression phase, fluid exudation gives rise to a peak stress σ_p, and during the relaxation phase, fluid redistribution gives rise to the relief of the compacted region at the surface and hence to stress relaxation.

K is on the order of 10^{15} N · s/m^4. In addition, this drag acts to compact the matrix nonuniformly; a boundary layer of very large compaction is experienced near the surface (Fig. 4.18). Exudation occurs only after the fluid has flowed past this compacted region, thereby further amplifying the stress required to compress the tissue. This is a manifestation of the nonlinear permeability effect, Eq. (4.20), and cannot be accounted for in the present constant permeability case. (A brief account of the influence of the nonlinear permeability on the stress-relaxation process will be presented in the next section.)

In the relaxation phase, Fig. 4.18, fluid exudation stops. This is a consequence of the continuity equation, Eq. (4.21). For these tissues, stress relaxation in compression occurs because of fluid redistribution within the tissue. This process continues until all relative fluid motion ceases. At equilibrium, as in the creep case, a uniform state of compression exists in the matrix and all the load is carried by the solid matrix. Figure 4.18 also depicts the complete stress history for this experiment. According to the constant permeability theory (Mow et al., 1980) the complete stress history is:

$$\sigma_{zz}^s(0, t) = -\frac{HV_0}{h}\left[t + \frac{h^2}{3kH} - 2\sum_{n=1}^{\infty}\frac{h^2}{n^2\pi^2kH}\exp\left(-n^2\pi^2kHt/h^2\right)\right] \tag{4.25a}$$

for the compression phase $0 \le t \le t_0$ and

$$\sigma_{zz}^s(0, t) = -\frac{HV_0}{h}\left\{t_0 - 2\sum_{n=1}^{\infty}\frac{h^2}{n^2\pi^2kH}\left[\exp\left(-\frac{n^2\pi^2kHt}{h^2}\right) - 1\right]\right\} \tag{4.25b}$$

for the relaxation phase $t_0 \le t < \infty$. Note that as $t \to \infty$, Eq. (4.25*b*) reduces to $\sigma_{zz}^s = HV_0t_0/h = \sigma_{eq}(= \sigma_\infty)$, which is used to determine H. Stahurski and coworkers (1981) and McCormack (1983) used the stress-relaxation test to determine H and k by using a nonlinear regression analysis which agrees quite well with those values obtained from the creep studies. The stress-relaxation test is preferable because the relaxation process is four times faster [see Eqs. (4.24) and (4.25*b*)] than the creep process; however, the creep test is preferable because the apparatus and the loading situation are far simpler to achieve than the controlled strain rate experiment required for the stress-relaxation test.

Extensive tests have been performed on all types of human materials—normal autopsy patellar cartilage, normal autopsy femoral condyles, and tibial plateau cartilage and specimens removed from total knee joint replacements (TKR)—using confined compression creep analysis. These properties have been correlated with age (Fig. 4.19), water content (Fig. 4.20*a* and *b*), and uronic acid content (Fig. 4.21*a*). Figure 4.21*b* shows the progression of change of the

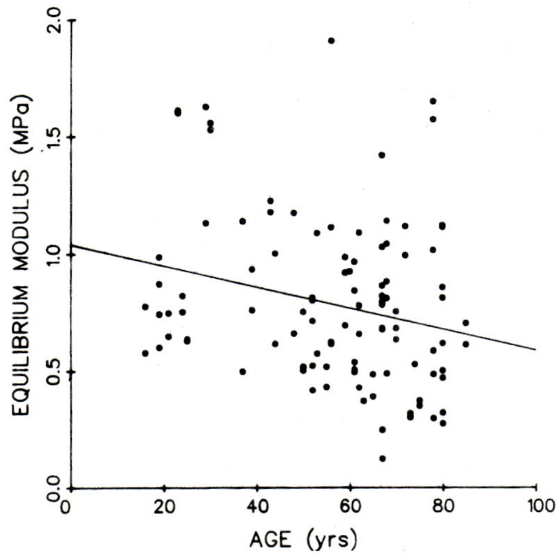

FIG. 4.19 Variation of the equilibrium modulus $H = \lambda_s + 2\mu_s$ obtained from confined compression creep with age. (*From Armstrong and Mow, 1982b.*)

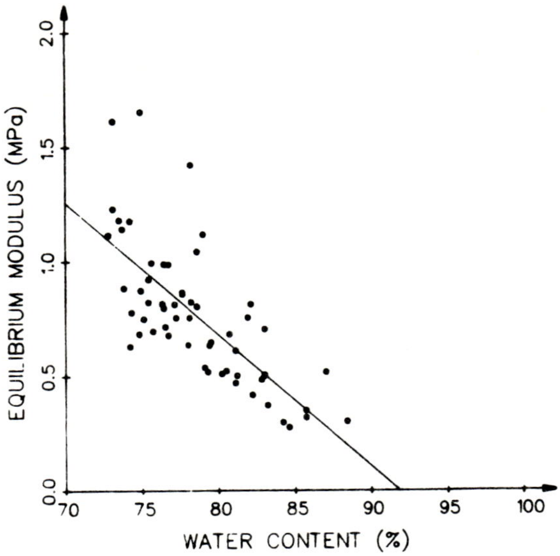

(a)

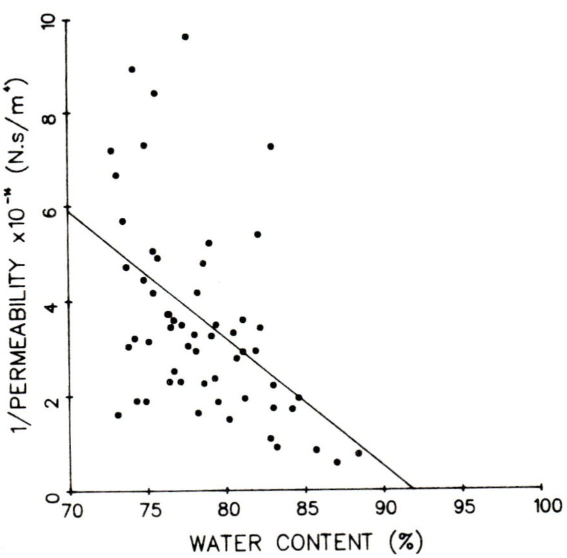

(b)

FIG. 4.20 (*a*) Variation of the equilibrium modulus $H = \lambda_s + 2\mu_s$ with tissue water content. (*b*) Variation of the reciprocal of permeability (which is directly proportional to the diffusive resistance coefficient K) with water content. (*From Armstrong and Mow, 1982b.*)

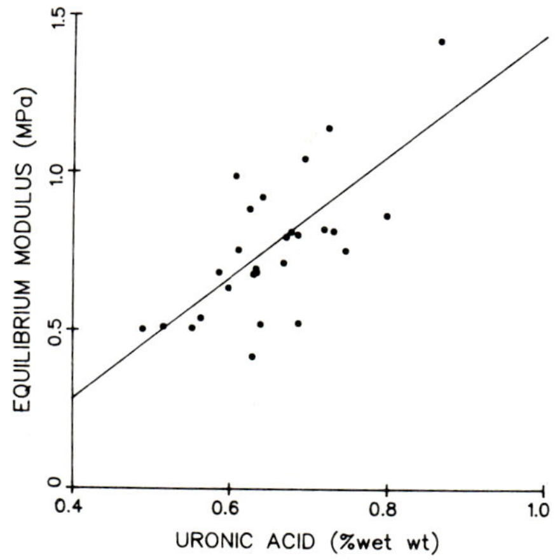

(a)

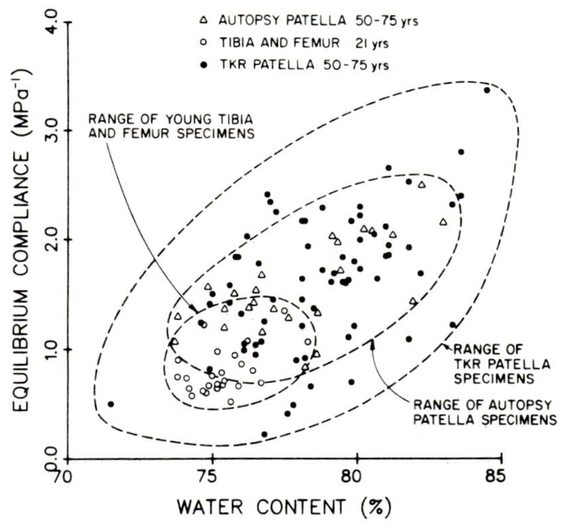

(b)

FIG. 4.21 (*a*) Variation of the equilibrium modulus *H* with uronic acid content. (*From Armstrong and Mow, 1982a.*) (*b*) Scattergram of equilibrium modulus with water content for various human tissue. Normal articular cartilage shows little scatter followed by random autopsy materials. Material obtained from total knee replacement shows large amounts of scatter.

equilibrium elastic compliance ($1/H$) from the various sources of human materials. Note that TKR specimens are stiffer in general than degenerating tissues, which indicates that an active biological repair process is in progress. (Note: Uronic acid is a major component of the glycosaminoglycans making up the proteoglycan content of cartilage.) Changes in proteoglycan macromolecular structure have been implicated as being early initiating factors involved in the development of osteoarthritis (Rosenberg and coworkers, 1982; Armstrong and Mow, 1982a, b; Muir, 1977). These results provide a good measure of tissue properties when either water content or uronic acid content is known. Notice that the variation of the equilibrium elastic modulus H with age is statistically significant only to a low correlation factor of $r = -0.24$. This means that not all aging materials are biomechanically degenerated or, conversely, that degeneration of cartilage may occur at any age leading up to the development of arthritis.

4.7.2 The Effect of Nonlinear Intrinsic Permeability on Creep and Stress Relaxation

Articular cartilage and other soft, highly hydrated connective tissues exhibit a compression-dependent permeability effect. This effect has been analyzed in considerable detail (Lai and Mow, 1980). This intrinsic permeability function gives rise to a nonlinear diffusion equation

$$H_A \frac{\partial^2 u}{\partial z^2} = \frac{1}{k_0} \exp\left(-M \frac{\partial u}{\partial z}\right) \frac{\partial u}{\partial t} \tag{4.26}$$

governing confined compression creep and stress relaxation. This nonlinearity essentially precludes the possibility of any closed-form exact solutions which are analogous to those derived for the constant permeability case, Eqs. (4.24) and (4.25). Using a numerical iteration procedure to obtain the inversion of a finite Fourier cosine transform for the stress-relaxation problem, Lai and coworkers (1981) showed that the stress history depends on two parameters: $R_0 = k_0 H_A / V_0 h$ and M. The parameter R_0 characterizes the *fluid-flow effect*, i.e., the rise of stress above the equilibrium value due to fluid exudation, and M characterizes the *nonlinear flow-limiting effect*, i.e., the rise of stress above the constant permeability fluid-flow effect. In Fig. 4.22*a*, the linear constant permeability case corresponds to $M = 0$. Here, increasing R_0 means either increasing k_0 or decreasing V_0. Both correspond to diminishing fluid-flow effects; obviously the converse is also true. For constant R_0, e.g., constant speed of compression, increasing M increases the stress. Thus the frictional drag associated with the nonlinear flow-limiting effect increases the stress required to compress the tissue. Figure 4.22*b* shows four experimental curves (dashed lines) for such tests. The two top curves correspond to a strain rate of $0.333 \times 10^{-3}\ \text{s}^{-1}$, and the two bottom curves correspond to a lower strain rate of $0.333 \times 10^{-4}\ \text{s}^{-1}$. Reasonable agreements were obtained with values of H_A, k_0, and M taken to be $0.33\ \text{MN/m}^2$, $7.4 \times 10^{-15}\ \text{m}^4/(\text{N} \cdot \text{s})$, 3 and $0.58\ \text{MN/m}^2$, $4.2 \times 10^{-15}\ \text{m}^4/(\text{N} \cdot \text{s})$, 3 for the two specimens, respectively. These values of permeability compare favorably with those obtained from the direct permeability experiment. The equilibrium elastic moduli also agree well with those obtained by other methods.

The difficulty with this numerical iterative approach, as well as any other numerical solution, such as that obtained by the Crank-Nikelson scheme, is that it is not convenient to use in a nonlinear regression method to determine the parameters. Also, it is difficult to gain insights into the physical processes from tedious computations without some simpler analytical forms. Recently, Holmes and coworkers (1983a, b) obtained asymptotic mathematical solutions of Eq. (4.26) to both the creep and stress-relaxation problem. For slow rates of compression, say 5 percent in 3000 s, the stress $\sigma_{zz}^s(0, t)$ required during the compression phase is given by:

$$\sigma_{zz}^s(0, t) \approx \frac{V_0 H_A}{h} \left[t + \frac{h^2}{3k_0 H_A} \exp\left(\frac{M V_0 t}{h}\right) \right] \qquad \frac{h^2}{k_0 H_A} < t \le t_0 \tag{4.27a}$$

while for the stress-relaxation phase, $\sigma_{zz}^s(0, t)$ is given by

$$\sigma_{zz}^s(0, t) \approx \sigma_p - M_r (t - t_0)^{1/2} \qquad (t - t_0) \ll 1 \tag{4.27b}$$

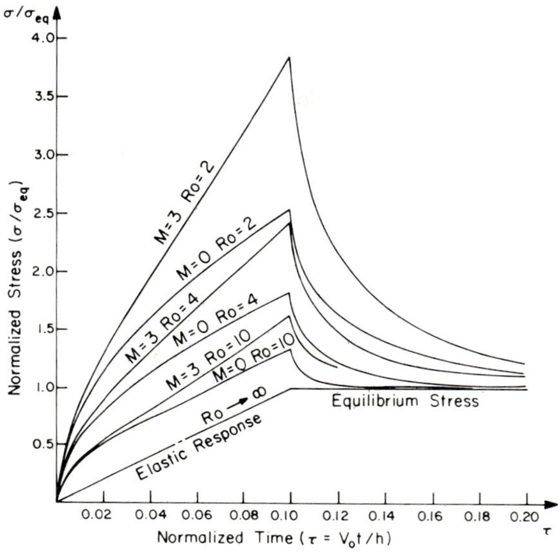

(a)

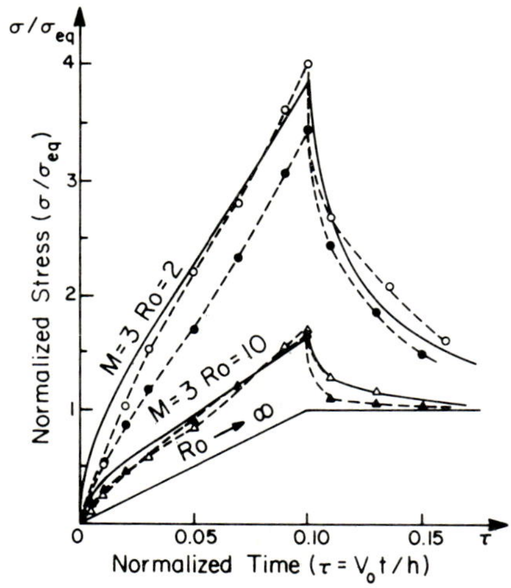

(b)

FIG. 4.22 (*a*) Theoretical predictions of the confined-compression stress-history response. The fluid-flow effect may be seen with the case where $M = 0$. A case of the nonlinear flow-limiting effect, $M = 3$, is shown. The parameter R_0 is the ratio of characteristic "flow speed" of the tissue (k_0H/h) to that of compression speed V_0 on tissue parameters k_0, H, and h. (*b*) Comparison of the predicted stress history with that obtained experimentally (*From Lai et al., 1981.*)

Here the peak stress σ_p is obtained from Eq. (4.27*a*) by substituting $t = t_0$, and

$$M_r = \frac{2h\sigma_{eq}}{t_0[\pi H_A k_0 \exp(-MV_0 t_0/h)]^{1/2}} \tag{4.27c}$$

defines the relaxation constant. Equation (4.27*a*) clearly shows the fluid-flow effect; decreasing V_0 or increasing k_0 will diminish the fluid-flow effect and vice versa, and will diminish the nonlinear flow-limiting effect; increasing M increases the stress required to compress the tissue. Further, because of these two effects, when k_0 is small and/or MV_0t_0 is large, the strain field at peak stress is so nonuniform that the rate of stress relaxation is large. That is, the rate of relaxation increases with decreasing k and/or increasing MV_0t_0 as is evidenced from Eqs. (4.27*b* and *c*). Note that Fig. 4.23*a* summarizes the asymptotic results of Eqs. (4.27), while Fig. 23*b* illustrates one such experimental test using Eqs. (4.27) for analysis. These results also provide a self-consistent check upon the theory. While the values of k_0 and M were determined from the compression phase by a nonlinear regression curve-fitting procedure, with H determined at equilibrium, the stress-relaxation curve is thus a predicted result. The prediction corresponds very well with the experimental data.

The asymptotic expression for the creep response of this nonlinearly permeable material is:

$$u(0, t) = 2\varepsilon_0 \left[k_0 H_A \exp\left(-\frac{\varepsilon_0 M}{\pi}\right)\right]\left(\frac{t}{\pi}\right)^{1/2} \qquad 0 < t < t_M \tag{4.28}$$

To calculate k_0, M, and H, two creep tests are required using different values of applied loads; the equilibrium elastic constant may be determined from one of the equilibrium creep results, while the two creep curves are required to determine k_0 and M. Holmes and coworkers (1983b) have verified that the early creep response, $0 < t < t_M$, is indeed governed by the $t^{1/2}$ relationship.

4.7.3 Biphasic Description of Tensile Properties

Constant Strain Rate Tensile Tests. The earliest tensile studies on articular cartilage were constant strain rate experiments with little or no facilities to accurately measure strain (Ranu, 1967; Kempson and coworkers, 1968, 1973). More recent studies focused on the need to assess the true strains by using various optical methods: the video dimensional analyzer (VDA) (Woo and coworkers, 1976) and the photographic method (Roth and Mow, 1980). The VDA is particularly versatile because it provides a rapid capability to assess finite strains biaxially. Concomitant with a third measure of strain in the thickness direction, Woo and coworkers (1979) developed an appreciation of the tissue volume change associated with the finite strains that occur in a tensile experiment. Large volume loss was measured, and this is associated with the observed exudation of fluid from the cartilage specimen pulled in tension (Roth and Mow, 1980). A single-phase, quasi-linear viscoelastic theory was successfully employed by Woo and coworkers (1979), replacing the previous ad hoc polynomial or exponential time-independent constitutive laws, to describe phenomenologically the tensile behavior of cartilage under creep, stress relaxation, constant strain rate, and cyclic conditions. This approach has been outlined in the earlier sections of this chapter.

An alternative description—though only limited results are at present available in the literature—is to employ the biphasic theory. There are two essential differences between the two alternatives. The first is obvious: the quasi-linear viscoelastic theory is single-phase theory ascribing the viscoelasticity of the tissue to some unknown molecular processes, while the biphasic theory, until now, has ascribed the viscoelasticity of the tissue completely to the flow of the interstitial fluid through the solid matrix. Obviously, both theories provide approximations. The second difference is as follows: the use of the quasi-linear viscoelastic theory tacitly assumes that the strain field in the gage section is homogeneous; for the biphasic theory, interstitial fluid flow necessitates a nonhomogeneous strain field in the gage section and the use of the equations of motion.

Recently, using the assumptions involved with the symmetry of a tensile specimen pertain-

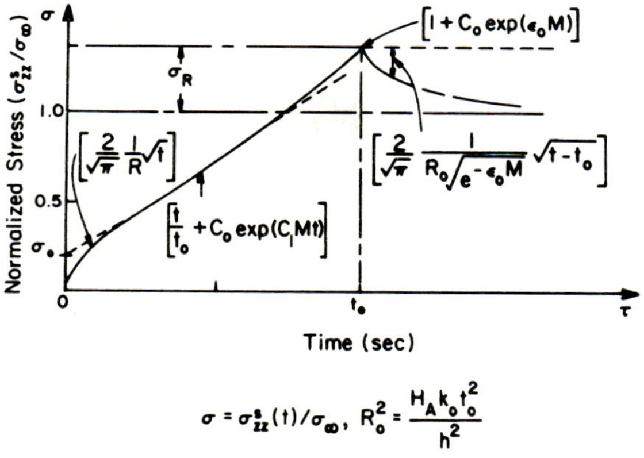

(a)

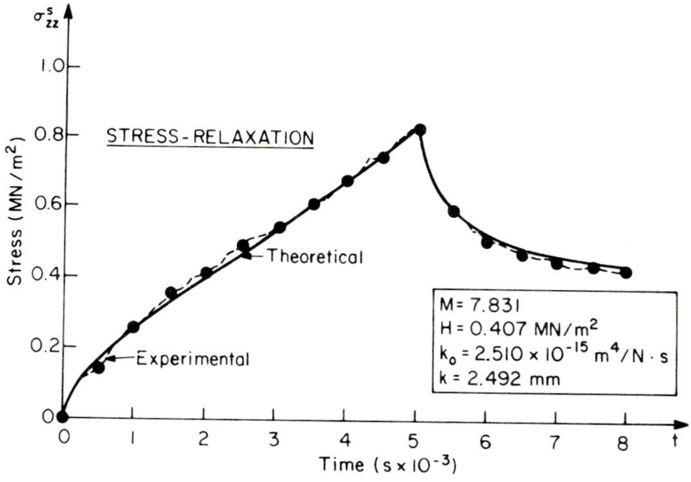

(b)

FIG. 4.23 (*a*) Mathematical structure of the stress-history response deduced from an asymptotic analysis of the nonlinear *KLM* theory for cartilage. (*From Holmes et al., 1983a.*) (*b*) Comparison of experimental data and exact numerical predictions using data obtained from the asymptotic analysis.

ing to the nonhomogeneous nature of the strain field, Li and coworkers (1983) derived an expression for the stress exerted by the tissue during a constant strain rate ($\dot{\varepsilon}_0$) experiment [$\varepsilon(t) = \dot{\varepsilon}_0 t$]:

$$\frac{F(t)}{A} = E_s \varepsilon(t) + \frac{h^2 \dot{\varepsilon}_0}{Hk} \left[A_0 - \sum_{n=1}^{\infty} \frac{A_n \exp(-4\alpha_n^2 Hkt/h^2)}{\alpha_n^2} \right] \tag{4.29}$$

using the linear *KLM* biphasic theory. Here E_s is the Young's modulus of the solid matrix, ν_s is Poisson's ratio, α_n are the roots of the equation $(\tan x)/x = 2(1 - \nu_s)/(1 - 2\nu_s)$, A_n are constants depending on ν_s of the solid matrix, and A is the original cross-sectional area. Further, Li and coworkers obtained an asymptotic result from Eq. (4.29) for large $\dot{\varepsilon}_0$:

$$\frac{F(t)}{A} = 3\mu_s\varepsilon(t) - \frac{8\mu_s\dot{\varepsilon}_0}{h\sqrt{\pi}}\left(\frac{k_0}{H}\right)^{1/2} t^{3/2} + \cdots \tag{4.30}$$

It can be seen that E_s of the tissue in tension may be obtained from Eq. (4.29) for the limiting, slow strain rate experiment, i.e., $\dot{\varepsilon}_0 \rightarrow 0$. At slow strain rates, the fluid-flow effect, characterized by h^2/Hk, is negated. From Eq. (4.30), one can obtain μ_s from large strain rate experiments at small times. Note that under such circumstances, $F(t)/A \rightarrow 3\mu_s\varepsilon(t)$; the tissue is essentially incompressible, exhibiting an *apparent* single-phase Young's modulus $E = 3\mu_s$. This is a consequence of the intrinsic incompressibility of the interstitial fluid and the solid matrix, and because for $t \rightarrow 0$ insufficient exudation has occurred so that the tissue as a whole is incompressible. From 74 normal bovine tensile specimens, the values of E_s and μ_s were determined (Li and coworkers, 1983).

Under low strain rates or high strain rates at small-time conditions, the fluid-flow effects are negated. Under other circumstances, fluid exudation across the surface will influence the tensile stress-strain behavior. Figure 4.24 shows the stress-strain behaviors of the superficial zones of bovine patellofemoral cartilage at three different strain rates. All strains are measured with an Optron electromechanical biaxial extensometer. At low strain rates, $\dot{\varepsilon} = 0.0012\ \text{s}^{-1}$, the stress-strain relationship does not exhibit an inflection point. At high strain rates, $\dot{\varepsilon} = 0.12\ \text{s}^{-1}$, the stress-strain relationship exhibits a definite inflection point. At these rates, the rise of stress early in the stress-strain diagram is due to fluid exudation. At the intermediate strain rate, $\dot{\varepsilon} = 0.012\ \text{s}^{-1}$, one can detect some influence of the fluid-flow effect. The high strain rate result is similar to that observed in the rate-controlled compression stress-relaxation case due to fluid-flow effects.

It would be safe to use the infinitesimal strain, linear *KLM* biphasic theory for strains less than 10 percent. For large tensile strains, not only is the theory not valid, but the extracellular matrix of collagen fibrils and proteoglycans no longer functions as a cohesive solid material. Straightening of the normally coiled collagen fibrils and collagen pullout through the proteoglycan gel occurs. This is evident in Fig. 4.24*b*, where for strains greater than 40 percent, all stress-strain curves are nearly parallel. In this zone, the collagen fibrils are the predominant load-supporting material in the tissue. Thus, the tissue may be viewed to have three distinct zones of tensile stress-strain behavior: region I (biphasic zone)—occurring at low strain levels, say, less than 20 percent strain, where the extracellular matrix of collagen fibrils and proteoglycans behaves like a porous permeable solid matrix; region II (transition zone)—occurring at intermediate strain levels, say, for strains between 20 and 40 percent, where straightening of the collagen fibrils and collagen pullout occur; and region III (collagen zone)—occurring at high strain levels, say, for strain greater than 40 percent, where the straightened collagen network supports the load.

No theoretical work exists to describe the induced anisotropy associated with collagen pullout and alignment. In fact, very little theoretical work exists which can be used to model the experimentally observed pronounced anisotropic and nonhomogeneous properties of articular cartilage in tension. A major challenge for the next phase of research is the development of more advanced models to include the anisotropic, nonhomogeneous, nonlinearly permeable, and viscoelastic nature of the extracellular matrix in a biphasic description of the tissue. More accurate characterization of the tissue is obviously necessary to understand its normal function in diarthrodial joint mechanics and its degeneration leading to osteoarthritis.

4.7.4 Intrinsic Viscoelasticity of the Solid Matrix

Collagen fibrils and proteoglycans are known as structural macromolecules because, together with water, they provide connective tissues such as articular cartilage, tendons, and ligaments with their load-carrying capacities. The architectural arrangement of these two structural macromolecules seems to be optimized to meet the type of high stress levels experienced by each of the connective tissues. For example, tendons, required to carry only tensile loads and sometimes at very high magnitudes (< 5000 N), usually have only unidirectionally oriented fibrils and

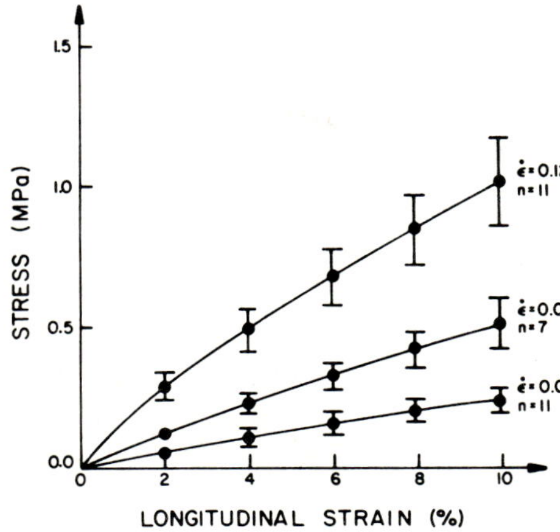

(a)

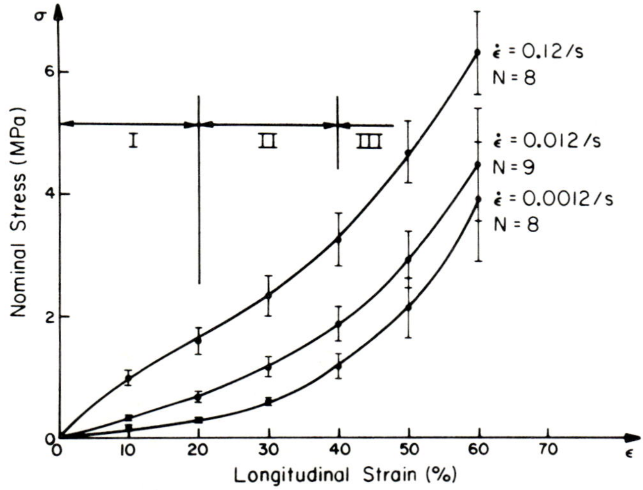

(b)

FIG. 4.24 (*a*) Variations of the nominal tensile stress $[F(t)/A]$ versus longitudinal strain $(\Delta l/l_0)$ of bovine articular cartilage with strain rate. At the highest strain rate, $\dot{\varepsilon} = 0.12\ s^{-1}$, the tissue exhibits a convex stress-versus-strain pattern which is characteristic of a biphasic response. (*b*) Variations of the behavior of nominal tensile stress versus longitudinal strain of bovine articular cartilage for the three strain rates used up to 60 percent tensile strain. See text for description of regions I, II, and III.

a very high collagen content. No strength or stiffness perpendicular to the axis of the structure is needed; thus the individual fiber bundles of the tendon are not associated with each other nor with the low concentration of proteoglycans in the extrafibrillar space. For articular cartilage, because of the complexity of the loading pattern at the joint surfaces, the high stress fields in the tissue are also complex, especially near the surface and the cartilage–subchondral bone interface. High shear stresses have been predicted in these regions (Askew and Mow, 1978; Mow and Lai, 1980). Correspondingly, the ultrastructural arrangement of the collagen fibrils and

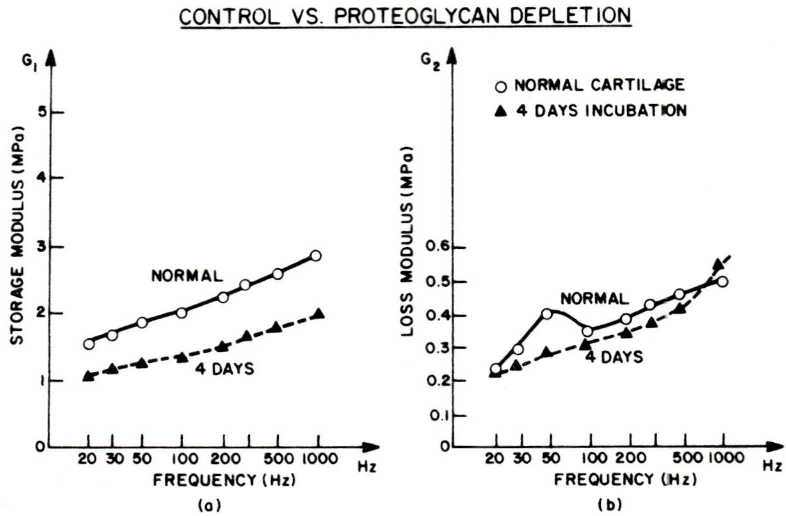

FIG. 4.25 (*a*) Flow-independent viscoelastic shear response of articular cartilage assessed from the simple shear experiment: $G_1 = \text{Re}\{G^*\}$ denotes the storage modulus; shown is G_1 for normal and proteoglycan extracted tissues for frequencies ranging from 20 to 1000 Hz. (*b*) The loss modulus $G_2 = \text{Im}\{G^*\}$ for normal and proteoglycan extracted tissues. (*From Hayes and Bodine, 1978.*)

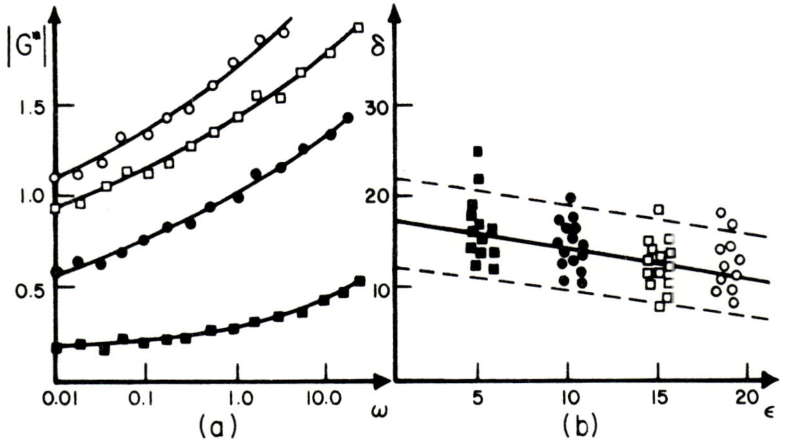

FIG. 4.26 (*a*) Variation of the magnitude of the complex shear modulus $|G^*|$ with frequency ranging from 0.01 to 20 Hz. ■, ●, □, and ○ denote 5, 10, 15, and 20 percent compressive clamping strains, respectively. (*b*) Variation of $\delta = \tan^{-1}(G_2/G_1)$ with frequency and amount of compression. (*From Mow et al., 1982.*)

proteoglycans in the articular cartilage is very complicated (see Fig. 4.2). Thus, functionally, as well as from the standpoint of the need to characterize completely the intrinsic properties of the solid matrix, the shear response of the tissue must be directly assessed.

With the tissue composed of 75 to 80% water, general deformations usually cause the interstitial fluid to flow, which is the major source of the internal friction predominantly responsible for viscoelastic behavior. In order to assess the intrinsic viscoelastic behavior of the collagen-proteoglycan solid matrix, the experiment must be precisely controlled so as to produce a deformational field in the solid matrix which does not induce any interstitial fluid flow. Under conditions of infinitesimal pure shear, no volume change occurs in the solid matrix. Since interstitial fluid flow can only occur under constant ionic conditions (Myers, Armstrong, and Mow, 1983; Myers et al., 1983b), if there is a pressure gradient or a volume change in the matrix, then the infinitesimal pure shear experiment can be used to assess the intrinsic shear properties of the solid matrix.

In 1978 Hayes and Bodine reported their findings on the "flow-independent" viscoelastic shear properties of the solid matrix using a simple shear experiment where a thin strip of the tissue is sinusoidally excited parallel to the two plane surfaces of the specimen for frequencies ranging from 20 to 1000 Hz. Figure 4.25*a* and *b* depicts their major findings: (1) the normal cartilage matrix is slightly viscoelastic, with the phase angle $\delta = \tan^{-1}(G_2/G_1)$ ranging from 7 to 10°, and (2) the proteoglycan-depleted matrix is less rigid and more dissipative; i.e., δ increases with proteoglycan depletion. The magnitude of the shear modulus, $|G^*| = G_1^2 + G_2^2$, of normal tissues ranged between 1.5 to 3.0 MPa over the frequency range of 20 to 1000 Hz. Hayes and Bodine also found that $|G^*|$ increased with clamping strain required to grip the specimen.

Mow and coworkers (1983a) reported their findings on the intrinsic viscoelastic shear properties of the solid matrix, using a torsional pure shear test for infinitesimal sinusoidal excitations with frequencies ranging from 0.01 to 20 Hz. Also, the compression stiffening effect reported by Hayes and Bodine was investigated. Figure 4.26*a* and *b* shows: (1) over the low frequency range, $|G^*|$ is frequency- as well as compression-dependent, and (2) the phase angle δ is not frequency-dependent; however, it decreases with increasing compression. Further, they found a strong negative correlation between $|G^*|$ and water content and a strong positive correlation between $|G^*|$ and collagen content. The variations of δ with these two compositional parameters is less significant.

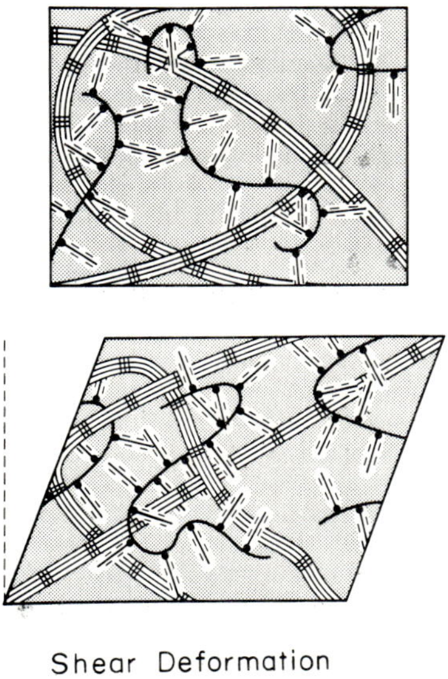

FIG. 4.27 Schematic representation of a unit cell of cartilage depicting the collagen-proteoglycan solid matrix and interstitial water (top) and the manner with which collagen may function during shear deformation (bottom).

It is significant to note that $|G^*|$ and δ are of the order of 1 MN/m^2 and 15°, respectively. For purified proteoglycan solutions at 50 mg/mL, the concentration in situ, $|G^*|$ and δ are of the order of 10 N/m^2 and 70°, respectively (Mow, Holmes, and Lai, 1984). Thus the proteoglycans not only appear to be required to provide the shear rigidity of the collagen-proteoglycan composite matrix, but they are also required to maintain the collagen ultrastructural form as well—via their swelling properties (Myers, Armstrong, and Mow, 1983)—so that the individual fibrils may become tensed during a shear deformation (Fig. 4.27). For randomly dispersed collagen arrangements, it is likely that the fibril will be stressed during shear. The low value of $\delta \approx 15°$ is further evidence that the more elastic collagen fibril is providing the shear strength for the collagen-proteoglycan network. Loss of proteoglycan by enzymatic treatment (Hayes and Bodine, 1978; Mow et al., 1983a) does decrease the shear stiffness, but not by very great amounts. Compression of the enzymatically treated, proteoglycan-reduced cartilage tends to restore its original stiffness, and the strong dependence of $|G^*|$ with collagen content also points to the fact that collagen is the structural macromolecule providing the shear rigidity of the matrix.

4.8 CONCLUDING REMARKS

A collection of collagen fibrils dispersed in solution does not possess properties resembling the extracellular matrix, nor does a purified proteoglycan solution. A mixture of collagen and proteoglycan, hydrated with 75 to 80% water, could. The collagen fibrils must be correctly arranged ultrastructurally and the proteoglycans must be aggregated. They must interact to provide a cohesive solid matrix. The exact nature of these interactions—chemical (collagen cross-links), electrostatic (between charged groups), and mechanical (steric exclusion and entanglement)—are not very well known. However, changes in the characteristics of these constituents and their mutual interactions can and will influence the response of the tissue. The efforts of the past 20 years have only begun to explore the interrelationship between the composition, structure (molecular and macroscopic), and biomechanical properties of articular cartilage. This knowledge and future findings are essential for an understanding of how animal joints function in health and in disease.

SUMMARY

In this chapter, we have tried to present a historical account of the efforts of various investigators to quantify and describe the biomechanical response of articular cartilage. Data are presented which we believe reflect the current state of knowledge in terms of the response of the tissue in tension, compression, and shear. Discussions of the quasi-linear viscoelastic theory and biphasic theories used by many investigators to describe articular cartilage are also presented. The use of the indentation method for determination of cartilage properties is critically analyzed. Finally, an attempt has been made to correlate material property changes with the composition, molecular structure, and ultrastructural arrangement of articular cartilage.

ACKNOWLEDGMENTS

This paper is based on research sponsored by National Institute of Arthritis, Diabetes, Digestive, and Kidney Diseases grants AM 19094, AM 26440, and AM 28467. National Science Foundation grant MEA 82-11968, and Rehabilitation Research & Development of the Veterans Administration. Any opinions, findings, and conclusions or recommendations expressed in this publication are those of the authors and do not necessarily reflect the views of the National Science Foundation. We wish to thank Ms. Lynnette Fleck and Ms. Joyce A. Brock for typing the equations and Ms. Rose A. Boshoff for assisting with the editing.

REFERENCES

Armstrong, C. G., and Mow, V. C. (1982a). Biomechanics of normal and osteoarthrotic articular cartilage. In P. D. Wilson and L. R. Straub (eds.), *Clinical Trends in Orthopaedics*. Thieme-Stratton, New York, pp. 189–197.

Armstrong, C. G., and Mow, V. C. (1982b). Variations in the intrinsic mechanical properties of human articular cartilage with age, degeneration and water content. *J. Bone Jt. Surg., 64-A*, 88–94.

Askew, M. J., and Mow, V. C. (1978). The biomechanical function of the collagen ultrastructure of articular cartilage. *J. Biomech. Eng., 100K*, 105–115.

Aspden, R. M., and Hukins, D. W. L. (1981). Collagen organization in articular cartilage, determined by x-ray diffraction and its relationship to tissue function. *Proc. R. Soc. London, 212-B,* 299–307.

Bär, E. (1926). Elasticitatsprufungen der gelenkknorpel. *Arch. Entwicklungsmech. Organ., 108,* 739–760.

Benninghoff, A. (1925). Form und bau der gelenkknorpel in ihren beziehungen zu funktion: I. Die modellierenden und formerhaltenden faktoren des knorpelrelieps. *Z. Anat. Entwicklungsgesch., 76,* 43–63.

Boussinesq, J. (1885). Applications des potentiels a l'etude de l'equilibre et du mouvement des solides elastique. Paris: Gauthier-Villars.

Bowen, R. M. (1976). Theory of mixtures. In A. E. Erigen (ed.), *Continuum Physics.* Academic Press, New York, vol. III, pp. 1–127.

Boyde, A., and Jones, S. J. (1983). Scanning electron microscopy of cartilage. In B. K. Hall (ed.), *Cartilage,* vol. I: *Structure, Function and Biochemistry.* Academic Press, New York, pp. 105–148.

Bullough, P. G., and Goodfellow, J. (1968). The significance of the fine structure of articular cartilage. *J. Bone Jt. Surg., 50-B,* 852–857.

Burstein, A. H., Shaffer, B. W., and Frankel, V. H. (1971). Elastic analyses of condylar structures. *ASME Pub., 70-WA,* BHF-1.

Charnley, J. (1959). The lubrication of animal joints. *Symposium on Biomechanics.* Institute of Mechanical Engineering, London, pp. 12–22.

Clarke, I. C. (1971). Articular cartilage: A review and scanning electron microscope study: I. The interterritorial fibrillar architecture. *J. Bone Jt. Surg., 53-B,* 732–750.

Clarke, I. C. (1974). Articular cartilage: A review and scanning electron microscope study: II. The territorial fibrillar architecture. *J. Anat., 118,* 261–280.

Coletti, J. M., Akeson, W. H., and Woo, S. L-Y. (1972). A comparison of the physical behavior of normal articular cartilage and the arthoplasty surface. *J. Bone Jt. Surg., 54-A,* 147–160.

Craine, R. E., Green, A. E., and Naghdi, P. M. (1970). A mixture of viscoelastic material with different temperature. *Q. J. Mech. Appl. Math., 23,* 171–184.

Crisp, J. D. C. (1972). Properties of tendon and skin. In Y. C. Fung, N. Perrone, and M. Anliker (eds.), *Bio-Mechanics: Its Foundations and Objectives.* Prentice-Hall, Englewood Cliffs, N.J., pp. 141–180.

Dintenfass, L. (1963). Lubrication in synovial joints: A theoretical analysis. *J. Bone Jt. Surg., 45-A,* 1241–1256.

Dowson, D. (1967). Modes of lubrication in human joints. *Proc. Inst. Mech. Eng., 181,* 45–54.

Dowson, D., Unsworth, A., Cooke, A. F., and Gvozdanovic, D. (1981). Lubrication of joints. In D. Dowson and V. Wright (eds.), *An Introduction to the Biomechanics of Joints and Joints Replacements.* Mechanical Engineering Publ., London. pp. 120–145.

Edwards, J. (1967). Physical characteristics of articular cartilage. *Proc. Inst. Mech. Eng., 181,* 16–24.

Ekholm, R., and Norback, B. (1951). On the relationship between cartilage change and function. *Acta Orthop. Scand., 21,* 81–89.

Elmore, S. M., Sokoloff, L., Norris, G., and Carmeci, P. (1963). Nature of "imperfect" elasticity of articular cartilage. *J. Appl. Physiol., 18,* 393–396.

Eyre, D. R. (1980). Collagen: Molecular diversity in the body's protein scaffold. *Science, 207,* 1315–1322.

Favenesi, J. A., Shaffer, J. C., and Mow, V. C. (1983). Biphasic mechanical properties of knee menicus. *Trans. 29th Annu. Mtg. Orthop. Res. Soc., 8,* 57.

Fung, Y. C. B. (1967). Elasticity of soft tissues in simple elongation. *Am. J. Physiol., 213,* 1532–1544.

Fung, Y. C. B. (1972). Stress-strain-history relations of soft tissues in simple elongation. In Y. C. Fung, N. Perrone, and M. Anliker (eds.), *Biomechanics: Its Foundations and Objectives.* Prentice-Hall, Englewood Cliffs, N.J., pp. 181–208.

Fung, Y. C. B. (1982). *Biomechanics.* Springer-Verlag, New York, Chap. 7.

Ghadially, F. N. (1978). Fine structure of joints. In L. Sokoloff (ed.), *The Joints and Synovial Fluid.* Academic Press, New York, pp. 105–176.

Göcke, E. (1927). Elastizitasstudien an jungen und alten gelenkknorpel, *Verh. Dtsch. Orthop. Ges., 22,* 130–147.

Gordon, S. S. (1978). Permeability and streaming potential of normal and trypsin treated articular cartilage. M.S. thesis, Rensselaer Polytechnic Institute, Troy, New York, p. 79.

Green, A. E., and Zerna, W. (1968). *Theoretical Elasticity* (2d ed.). Oxford University Press, New York, pp. 79–81.

Greenwald, A. S., and Haynes, D. W. (1972). Weight bearing areas in human hip joint. *J. Bone Jt. Surg., 54-B,* 157–163.

Hall, B. K. (1983). *Cartilage:* vol. I. *Structure, Function and Biochemistry.* Academic Press, New York.

Hayes, W. C., and Bodine, A. J. (1978). Flow-independent viscoelastic properties of articular cartilage matrix. *J. Biomech., 11,* 407–420.

Hayes, W. C., and Mockros, L. F. (1971). Viscoelastic properties of human articular cartilage. *J. Appl. Physiol., 31,* 562–568.

Hayes, W. C., Keer, L. M., Herrmann, G., and Mockros, L. F. (1972). A mathematical analysis for indentation tests of articular cartilage. *J. Biomech., 5,* 541–551.

Hirsch, C. (1944). The pathogenesis of chondromalacia of the patella. *Acta Chir. Scand., 90,* suppl. 83.

Holmes, M. H., Lai, W. M., and Mow, V. C. (1983a). Compression effects on cartilage permeability. In A. Hargens (ed.), *Tissue Nutrition and Viability.* Springer-Verlag, New York, pp. 1–51.

Holmes, M. H., Lai, W. M., and Mow, V. C. (1983b). The nonlinear interaction of solic and fluid in the creep response of articular cartilage. *Proc. V. Int. Congr. Biorheol.,* Baden-Baden, FDR 111.

Honner, R., and Thompson, R. C. (1971). The nutritional pathways of articular cartilage. *J. Bone Jt. Surg., 53-A,* 742–748.

Hori, R. Y., and Mockros, L. F. (1976). Indentation tests of human articular cartilage. *J. Biomech., 9,* 259–268.

Hultkrantz, W. (1898). Ueber die spaltrichtungen der gelenkknorpel. *Verh. Anat. Ges. Jena, 12,* 248–256.

Intaglietta, M., and Tompkins, W. R. (1972). On-line measurement of microvascular dimensions by television microscopy. *J. Appl. Physiol., 32,* 546–551.

Kempson, G. E. (1979). Mechanical properties of articular cartilage. In M. A. R. Freeman (ed.), *Adult Articular Cartilage.* Pitman, Kent, England, pp. 333–414.

Kempson, G. E., Freeman, M. A. R., and Swanson, S. A. V. (1968). Tensile properties of articular cartilage. *Nature, 220,* 1127–1128.

Kempson, G. E., Muir, H., Freeman, M. A. R., and Swanson, S. A. V. (1970). Correlations between stiffness and chemical constitutents of human articular cartilage. *Biochim. Biophys Acta, 215,* 70–77.

Kempson, G. E., Muir, H., Polland, C., and Tuke, M. (1973). The tensile properties of the cartilage of human femoral condyles related to the content of collagen and glycosaminoglycans. *Biochim. Biophys. Acta, 297,* 456–472.

Kempson, G. E., Spivey, C. J., Swanson, S. A. V., and Freeman, M. A. R. (1971). Patterns of cartilage stiffness on normal and degenerate human femoral heads. *J. Biomech., 4,* 597–609.

Kenedi, R. M., Gibson, T., and Daly, C. H. (1964). Bioengineering studies of the human skin, the effects of unidirectional tension. In Jackson, Harkness, and Tristrain (eds.), *Structure and Function of Connective and Skeletal Tissue.* Scientific Committee, St. Andrews, Scotland, pp. 388–395.

Lai, W. M., and Mow, V. C. (1980). Drag-induced compression of articular cartilage during a permeation experiment. *Biorheology, 17,* 111–123.

Lai, W. M., Mow, V. C., and Roth, V. (1981). Effects of a nonlinear strain-dependent permeability and rate of compression on the stress behavior of articular cartilage. *J. Biomech. Eng., 103,* 61–66.

Lane, J. M., and Weiss, C. (1975). Reviews of articular cartilage collagen research. *Arthritis Rheum., 18,* 553–563.

Li, J. T., Armstrong, C. G., and Mow, V. C. (1983). The effect of strain rate on mechanical properties of articular cartilage in tension. In S. L-Y. Woo and R. Mates (eds.), *1983 Biomechanics Symposium.* American Society of Mechanical Engineers, New York, pp. 117–120.

Linn, F. C. (1967). Lubrication of animal joints: I. The arthrotripsometer. *J. Bone Jt. Surg., 49-A,* 1079–1098.

Linn, F. C. (1968). Lubrication of animal joints: II. The mechanism. *J. Biomech., 1,* 193–206.

Linn, F. C., and Sokoloff, L. (1965). Movement and composition of interstitial fluid of cartilage. *Arthritis Rheum., 8,* 481–494.

MacConaill, M. A. (1951). The movements of bones and joints: IV. The mechanical structure of articular cartilage. *J. Bone Jt. Surg., 33-B,* 251–257.

Mak, A. F., Lai, W. M., and Mow, V. C. (1982). Indentation of articular cartilage: A biphasic analysis. In L. Thibault (ed.), *1982 Advances in Bioengineering.* American Society of Mechanical Engineers, New York, pp. 71–74.

Mansour, J. M., and Mow, V. C. (1976). The permeability of articular cartilage under compressive strain and high pressures. *J. Bone Jt. Surg., 58-A*, 509–516.

Maroudas, A. (1975a). Biophysical chemistry of cartilaginous tissue with special reference to solute and fluid transport. *Biorheology, 12*, 233–248.

Maroudas, A. (1975b). Fluid transport in cartilage. *Ann. Rheum. Dis., 34*, suppl. 2, 77.

Maroudas, A. (1979). Physico-chemical properties of articular cartilage. In M. A. R. Freeman (ed.), *Adult Articular Cartilage*. Pitman, Kent, England, pp. 215–290.

McCall, J. G. (1969). Load deformation response of the microstructure of articular cartilage. In V. Wright (ed.), *Lubrication and Wear in Joints*. Lippincott, London, pp. 39–48.

McCormack, B. A. O. (1983). Determination of intrinsic permeability coefficients of human articular cartilage from rate-controlled compression stress-relaxation experiments. Rensselaer Polytechnic Institute, Troy, N.Y.

McCutchen, C. W. (1962). The frictional properties of animal joint. *Wear, 5*, 1–17.

Morgan, F. R. (1960). The mechanical properties of collagen fibers: Stress-strain curves. *J. Soc. Leather Trades Chem., 44*, 171–182.

Mow, V. C., and Lai, W. M. (1980). Recent developments in synovial joint biomechanics, *SIAM Rev., 22*, 275–317.

Mow, V. C., and Schoonbeck, J. M. (1982a). Comparison of compressive properties and water, uronic acid and ash contents of bovine nasal bovine articular and human articular cartilage. *Trans. 28th Annu. Mtg. Orthop. Res. Soc., 7*, 209.

Mow, V. C., and Torzilli, P. A. (1975). Fundamental fluid transport mechanism through articular cartilage. *Ann. Rheum. Dis., 34* (suppl.), 82–84.

Mow, V. C., Holmes, M. H., and Lai, W. M. (1984). Fluid transport and mechanical properties of articular cartilage. *J. Biomech. 17(5)*, 377–394.

Mow, V. C., Lai, W. M., and Holmes, M. H. (1982b). Advanced theoretical and experimental techniques in cartilage research. In R. Huiskes, D. Van Campen, and J. DeWijn (eds.), *Biomechanics: Principles and Applications*. Martinus Nijhoff, The Hague, pp. 47–74.

Mow, V. C., Lai, W. M., and Redler, J. (1974). Some surface characteristics of articular cartilage: I. *J. Biomech., 7*, 449–456.

Mow, V. C., Mak, A. F., Lai, W. M., Rosenberg, L. C., and Tan, L-H. (1984a). Viscoelastic properties of proteoglycan subunits and aggregates in varying solution concentrations, *J. Biomech. 17(5)*, 325–338.

Mow, V. C., Schoonbeck, J. M., Koob, T. J., and Eyre, D. R. (1983a). Correlation studies on viscoelastic shear properties and chemical composition of articular cartilage. *Trans. 29th Annu. Mtg. Orthopo. Res. Soc., 8*, 201.

Muir, H. (1977). A molecular approach to the understanding of osteoarthrosis. *Ann. Rheum. Dis., 36*, 199–208.

Muir, I. H. M. (1980). The chemistry of the ground substance of joint cartilage. In L. Sokoloff (ed.), *The Joints and Synovial Fluid*. Academic Press, New York. Vol. II, pp. 27–94.

Myers, E. R., Armstrong, C. G., and Mow, V. C. (1983c). Swelling pressure and collagen tension. In D. W. L. Huskins (ed.), *The Connective Tissue Matrix*. Macmillian, London (to appear).

Myers, E. R., Lai, W. M., and Mow, V. C. (1983a). A continuum theory and experimental verification of the ion-induced swelling behavior of articular cartilage. *J. Biomech. Eng.* (to appear).

Myers, E. R., Lai, W. M., and Mow, V. C. (1983b). Isometric swelling studies of articular cartilage: Experiments and Theory. In *Proc. 1st China-Japan-U.S. Symp. Biomech*, Beijiung, China.

Parsons, J. R., and Black, J. (1977). The viscoelastic shear behavior of normal rabbit articular cartilage. *J. Biomech. 10*, 21–30.

Parsons, J. R., and Black, J. (1979). Mechanical behavior of articular cartilage: Quantitative changes with alteration of ionic environment. *J. Biomech. 12*, 765–773.

Powell, M. J. D. (1975). A method for minimizing a sum of species of nonlinear functions without calculating derivatives. *Computer J., 7*, 303–307.

Radin, E. L., and Paul, I. L. (1972). A consolidated concept of joint lubrication. *J. Bone Jt. Surg., 54-A*, 607–612.

Radin, E. L., Paul, I. L., and Lowy, M. (1970). A comparison of dynamic force transmitting properties of subchondral bone and articular cartilage. *J. Bone Jt. Surg., 52-A*, 444–456.

Ranu, H. S. (1967). Rheological behavior of articular cartilage under tensile loads. M.S. thesis, University of Surrey, Surrey, England.

Rauber, A. A. (1876). Elasticitat und festigkeit der knochen. In *Anatomisch-Physiologische Studie*. E. W. Engelmann, Leipzig.

Redler, I. (1974). A scanning electron microscope study of human normal and osteoarthritic articular cartilage. *Clin. Orthop. Rel. Res., 103*, 262–268.

Ridge, M. D., and Wright, V. (1964). The description of skin stiffness. *Biorheology, 2*, 67–74.

Rosenberg, L. C., Mow, V. C., Buckwalter, J. A., and Poole, A. R. (1982). Structure and function of proteoglycans in developing cartilage. In W. H. Akeson, P. Bornstein, and M. J. Glimcher (eds.), *Symposium on Heritable Disorders of Connective Tissue*. American Academy of Orthopedic Surgeons, Mosby, St. Louis, pp. 285–305.

Roth, V., and Mow, V. C. (1980). The intrinsic tensile behavior of the matrix of bovine articular cartilage and its variation with age. *J. Bone Jt. Surg., 62-A*, 1102–1117.

Salter, R. B., Simmonds, D. F., Malcolm, B. W., Rumble, E. J., MacMichael, D., and Clements, N. D. (1980). The biological effect of continuous passive motion on the healing of full-thickness defects in articular cartilage. *J. Bone Jt. Surg., 62-A*, 1232–1251.

Simon, B. R., Woo, S. L-Y., and Coats, R. S. (1982). Relaxation and creep quasilinear viscoelastic models for normal articular cartilage. In L. Thibault (ed.), *1982 Advances in Bioengineering*. American Society of Mechanical Engineers, New York, pp. 75–78.

Simon, W. H. (1971). Scale effects in animal joints—II. Thickness and elasticity in the deformability of articular cartilage. *Arthritis Rheum., 14*, 493–502.

Sokoloff, L. (1966). Elasticity of aging cartilage. *Fed. Proc., 25*, 1089–1095.

Speer, D. P., and Dahners, L. (1979). The collagenous architecture of articular cartilage. *Clin. Orthop. Rel. Res., 139*, 267–276.

Stahurski, T. M., Armstrong, C. G., and Mow, V. C. (1981). Variation of the intrinsic aggregate modulus and permeability of articular cartilage with trypsin digestion. In Van Buskirk and S. L-Y. Woo (eds.), *Proc. 1981 Biomech. Symp*. American Society of Mechanical Engineers, New York, pp. 137–140.

Timoshenko, S., and Goodier, J. N. (1953). *Theory of Elasticity*. McGraw-Hill, New York.

Torzilli, P. A., Dethmers, D. A., Rose, D. E., and Schryer, H. F. (1983). Movement of interstitial water through loaded articular cartilage. *J. Biomech., 16*, 169–179.

Viidik, A. (1980). Interdependence between structure and function in collagenous tissues. In A. Viidik and J. Vuust (eds.), *Biology of Collagen*. Academic Press, New York, pp. 257–280.

Walker, P. S., Dowson, D., Longfield, M. D., and Wright, V. (1968). Boosted lubrication in synovial joints by fluid entrapment and enrichment. *Ann. Rheum. Dis., 27*, 512–520.

Waters, N. E. (1965a). The indentation of thin rubber sheets by cylindrical indenters. *Br. J. Appl. Phys., 16*, 1387–1392.

Waters, N. E. (1965b). The indentation of thin rubber sheets by spherical indenters. *Br. J. Appl. Phys., 16*, 557–563.

Weightman, B. (1976). Tensile fatigue of human articular cartilage. *J. Biomech., 9*, 193–200.

Woo, S. L-Y., Akeson, W. H., and Jemmott, G. F. (1976). Measurements of nonhomogeneous, directional mechanical properties of articular cartilage in tension. *J. Biomech., 9*, 785–791.

Woo, S. L-Y., Gomez, M. A., and Akeson, W. J. (1981). Time and history-dependent viscoelastic properties of the canine medial collateral ligament. *J. Biomech. Eng., 103*, 293–298.

Woo, L-Y., Kobayashi, A. S., Schlegel, W. A., and Lawrence, C. (1972). Nonlinear material properties of intact cornea and sclera. *Exp. Eye Res., 14*, 29–39.

Woo, S. L-Y., Lubock, P., Gomez, M. A., Jemmott, G. F., Kuei, S. C., and Akeson, W. H. (1979). Large deformation nonhomogeneous and directional properties of articular cartilage in uniaxial tension. *J. Biomech., 12*, 437–446.

Woo, S. L-Y., Simon, B. R., Kuei, S. C., and Akeson, W. H. (1980). Quasi-linear viscoelastic properties of normal articular cartilage. *J. Biomech. Eng., 102*, 85–90.

Zarek, J. M., and Edwards, J. (1963). The stress-structure relationship in articular cartiagle. *Med. Electron. Biol. Eng., 1*, 497–507.

CHAPTER 5

Lubrication of Diarthrodial Joints

Van C. Mow
Department of Mechanical Engineering
Aeronautical Engineering and Mechanics
Rensselaer Polytechnic Institute
Troy, New York

Arthur F. Mak
Department of Bioengineering
University of Pennsylvania
Philadelphia, Pennsylvania

INTRODUCTION

Joints are formed between bones during the growth of the skeleton [1]. Those that allow various degrees of relative motion of the bones are called *diarthrodial joints*, or *synovial joints;* the hip, knee, shoulder, elbow, ankle, and finger joints are familiar examples of this type of joint [2]. Figure 5.1 shows schematic representations of some of the principal features of a human knee joint. The bone ends come together within a fibrous enclosure called the *joint capsule*. The inner lining of this joint capsule is a metabolically active tissue called the *synovium*. The ends of the bones are covered by a thin and highly deformable layer of dense connective tissue known as *articular cartilage*. The joint cavity, formed by the cartilaginous surfaces and the synovium, is filled with a small amount of synovial fluid (*syn,* meaning "like," *-ovial,* meaning "egg"; from its similarity to egg white). Ligaments, tendons, and other soft tissues inside and outside the joint cavity give stability to the joint and maintain the proper alignment of the articulating bone ends during motion. Daily use of the human body's lower extremity synovial joints—hips, knees, and ankles—involves large ranges of relative motion in multiple directions, with loads often as high as six times that of body weight during a normal walking cycle [3, 4]. These loads must be sustained by these biological bearings, the diarthrodial joints, with tribological characteristics (friction, lubrication, and wear) which must be the envy of modern engineering science—i.e., the ability of the bearings to function effectively for seven or eight decades, under normal conditions, in a relatively physically isolated environment. Because of this ability, the mechanics of synovial joint lubrication have been the subject of detailed biomechanical investigation over the past 50 years [5–38].

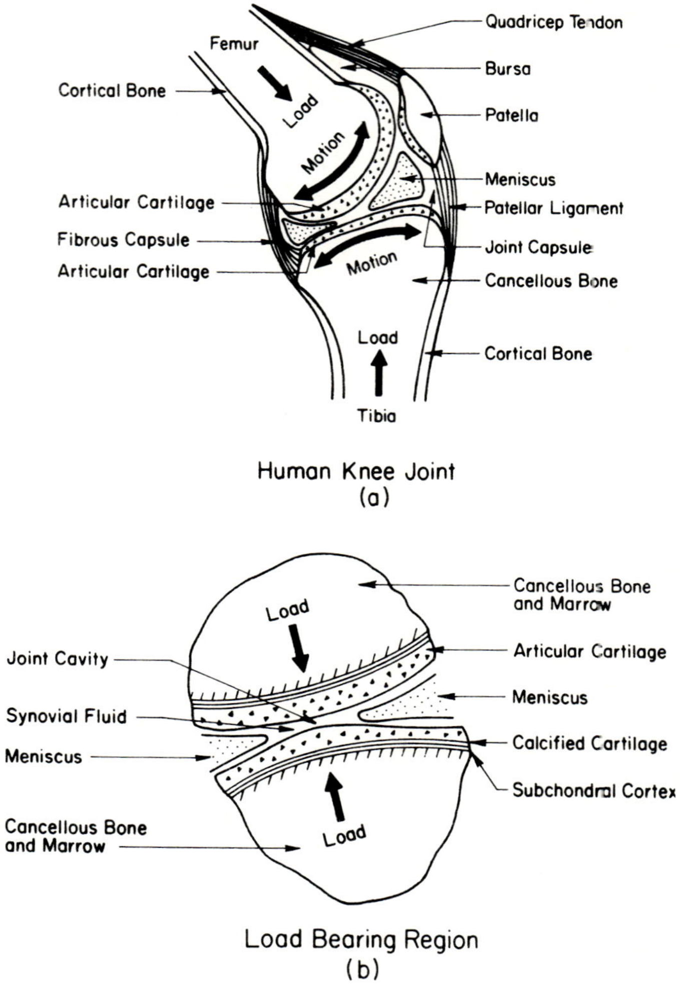

FIG. 5.1 (*a*) Schematic representation of the human knee joint, showing salient features which are important in its mechanical function. (*b*) Enlargement of the region of load bearing, showing the articular cartilage, calcified cartilage, subchondral cortex, and cancellous bone. The calcified cartilage and subchondral cortex function as a shell, providing backing to the articular cartilage, and are supported by the cancellous bone. The cancellous bone–bone marrow system provides the shock-absorbing mechanism for the diarthrodial joint.

Today, our understanding of diarthrodial joint lubrication is rooted in basic knowledge of the structural and deformational characteristics of articular cartilage, the biochemical and biorheological properties of synovial fluid, the topological design and microscopic roughness of the articulating surfaces of the joint, the kinematics of the joints, and the manner in which loads are applied across these joint surfaces. Research in these areas has resulted in the detailed modeling of various components of the joints, and the results of the quantitative analyses of these models have provided basic insight into the biomechanical functions of nature's bearings.

However, it must be stressed that our understanding is far from complete—there are many difficult unresolved problems in the field—since both synovial fluid and articular cartilage are nonlinear in the load deformation response in the range of joint loading and joint motion. A number of comprehensive reviews on the nonlinear deformational characteristics of synovial fluid [39], the nonlinear properties of articular cartilage [40–42], and the mechanics of joint lubrication already exist in the literature (see, for example, Refs. 37, 40, 43, and 44). In this chapter, the present understanding of the lubrication mechanisms and the related phenomenon of articular cartilage wear will be discussed in light of the most recent experimental and theoretical findings in the bioengineering literature (ca. 1984).

5.1 ARTICULAR CARTILAGE

The thickness of the articular cartilage layer varies with the species, the particular joint, and the location within the joint [45]. Generally, it ranges from about 0.5 mm in rabbit knee joints, to 1.0 mm in the patellofemoral groove of the bovine knee joint, to 1.5 mm in the human patella. The tissue is composed of a composite organic solid matrix which is swollen by water and in which a variety of mobile electrolytes (Na^+, Ca^{2+}, etc.) are distributed. The composite organic solid matrix of the tissue consists of collagen fibrils (~65% by dry weight) that are interspersed among proteoglycan aggregates (~25% by dry weight), with the rest consisting of nonspecific glycoproteins, lipids, and chondrocytes [46]. However, the nonspecific protein fraction can increase up to 50% with age [46, 47].

The solid phase constitutes ~22% of the total tissue weight, depending on the physiological state of the tissue [46]. With degeneration, the solid phase may decrease to ~10% before total disaggregation of the tissue occurs [48]. The intrinsic mechanical properties of this organic phase depend not only on the structure and properties of the comprising elements, collagen fibrils (type II) [49] and proteoglycan aggregates [48, 50], but also on the character of their mutual interactions [42, 46, 51]. Further, the matrix appears to be stratified: the most superficial zone (200 μm) is rich in collagen and poor in proteoglycan, and, with increasing depth, the collagen content remains constant while proteoglycan content increases [33]. Degradation of the constituents or of the interactions between the constituents will often lead to considerable reduction of the capacity of the tissue to function as the bearing material of diarthrodial joints [42].

The other phase, ~78% by weight of normal cartilage tissue, is water. Most of this water is free and resides in the intermolecular space between the proteoglycan and collagen solution domains. It is estimated that about 70% of the water is located in this compartment [52]. Most of the remaining water probably resides in the intrafibrillar space of the collagen fibrils and some in the solution domain of the proteoglycans. Recent experimental evidence suggests that most of the water in the intermolecular space can be expelled upon compaction of the solid phase and resorbed after the load is removed, in a manner analogous to compressing and releasing a fully saturated sponge. Even with the large weight fraction of water content, however, "pore" size estimates for the tissue range from 2 to 6.5 nm [53, 54]; thus, articular cartilage can be described as a "molecular sponge." Mechanically, articular cartilage can be considered as a binary continuum mixture of a porous, permeable solid filled with fluid (Fig. 5.2).

In general, depending on the degree of sophistication required of the theory, the solid matrix may be inhomogeneous, anisotropic, and nonlinear. Mow and coworkers have developed a linear biphasic theory describing the mechanical behavior of cartilage [40, 41]. Many compressive deformational properties of the tissue have been successfully modeled using this linear biphasic theory. It has been found that most of the compressive "viscoelastic" behavior of articular cartilage can be ascribed to the frictional resistance to relative interstitial fluid flow offered by the soft, permeable, solid matrix during deformation. For example, the compressive viscoelastic creep behavior of the tissue is a direct manifestation of fluid exudation when a load is applied to the surface of the tissue via a free-draining, rigid, porous platen (Fig. 5.3). Fluid exudation and imbibition are of primary importance to the normal functioning of a diarthrodial joint [40, 42, 53, 55]. For a detailed description of articular cartilage deformation, see Chapter 4, "Biomechanical Properties of Articular Cartilage."

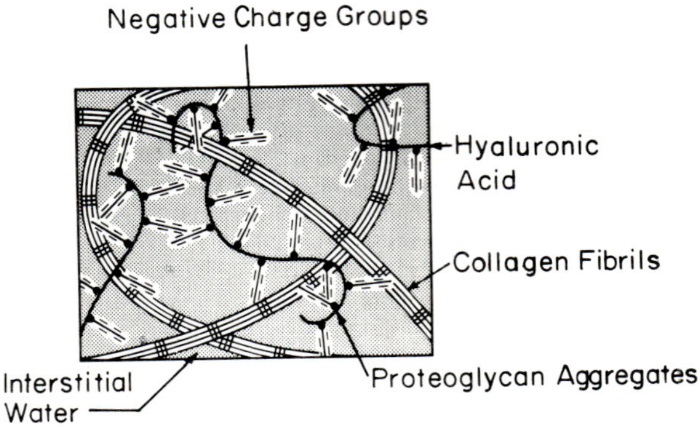

FIG. 5.2 Schematic model of a unit cell of articular cartilage, depicting the collagen-proteoglycan solid matrix and interstitial water. The solid matrix may be viewed as a molecular sponge whose "pores" range in size from 2 to 6.5 nm in diameter. This material has been modeled as a biphasic continuum mixture [41].

Pronounced changes in the mechanical properties of articular cartilage take place with age and degenerative joint disease. Compared to normal cartilage, degenerate cartilage is more permeable and softer, has a lower ultimate tensile strength, and swells more [48, 56–60]. Similar decreases in tensile stiffness, ultimate strength, and fatigue resistance have also been observed in aging cartilage [58, 61]. In situ compression in loaded human hip cartilage has been observed to increase with age [62]. Many of these changes are statistically correlatable in a highly significant manner to water-content increases and biochemical changes, which correspond to very early (< 2 weeks) osteoarthritic lesions on cartilage in experimental animal models [63, 64]; a decrease in solid matrix stiffness and an increase in permeability with increasing water content and decreasing uronic acid content are generally observed [48, 65]. These material property changes can and do adversely affect the lubrication of synovial joints.

5.2 SYNOVIAL FLUID

Synovial fluid is secreted into the joint cavity by the synovium. Normally, it is a clear and colorless or slightly yellowish fluid. Aspiration of the normal human knee joint usually yields about 0.2 mL. Biochemically, synovial fluid is a dialysate of blood plasma and varying amounts of the polysaccharide protein association known as *hyaluronic acid protein* (HAP) *complex* [66–71]. Hyaluronic acid (HA) is an unbranched macromolecule whose basic dimer is a disaccharide composed of glucuronic acid linked with *N*-acetyl glucosamine (Fig. 5.4). A macromolecule of HA, $\sim 10^6$ daltons, is associated in synovial fluid with some glycoproteins. The average molecular weight of the HAP complex is about 2 million. In solution, this macromolecule is a polyelectrolyte that is stiffly extended into a spherical or ellipsoidal shape occupying a solvent domain much larger than the volume of the polymer chain itself [72, 73]. The presence of these macromolecules endows synovial fluid with highly non-Newtonian biorheological (flow) properties that include (1) a shear thinning effect—i.e., the viscosity, or resistance to shearing, of synovial fluid decreases as the rate of shear increases [74, 75]; (2) a normal stress effect—i.e., as synovial fluid is sheared, a force in the direction perpendicular to the flow is generated [20, 76]; and (3) an elastic effect—i.e., as synovial fluid is sheared, energy may be elastically stored by the deforming HAP complex [77–79]. Such flow behaviors may play an important role in synovial joint lubrication. Synovial fluids obtained from degenerative joints show that many of these flow properties may be diminished [39, 73], and fluids from rheumatoid joints may lose all these

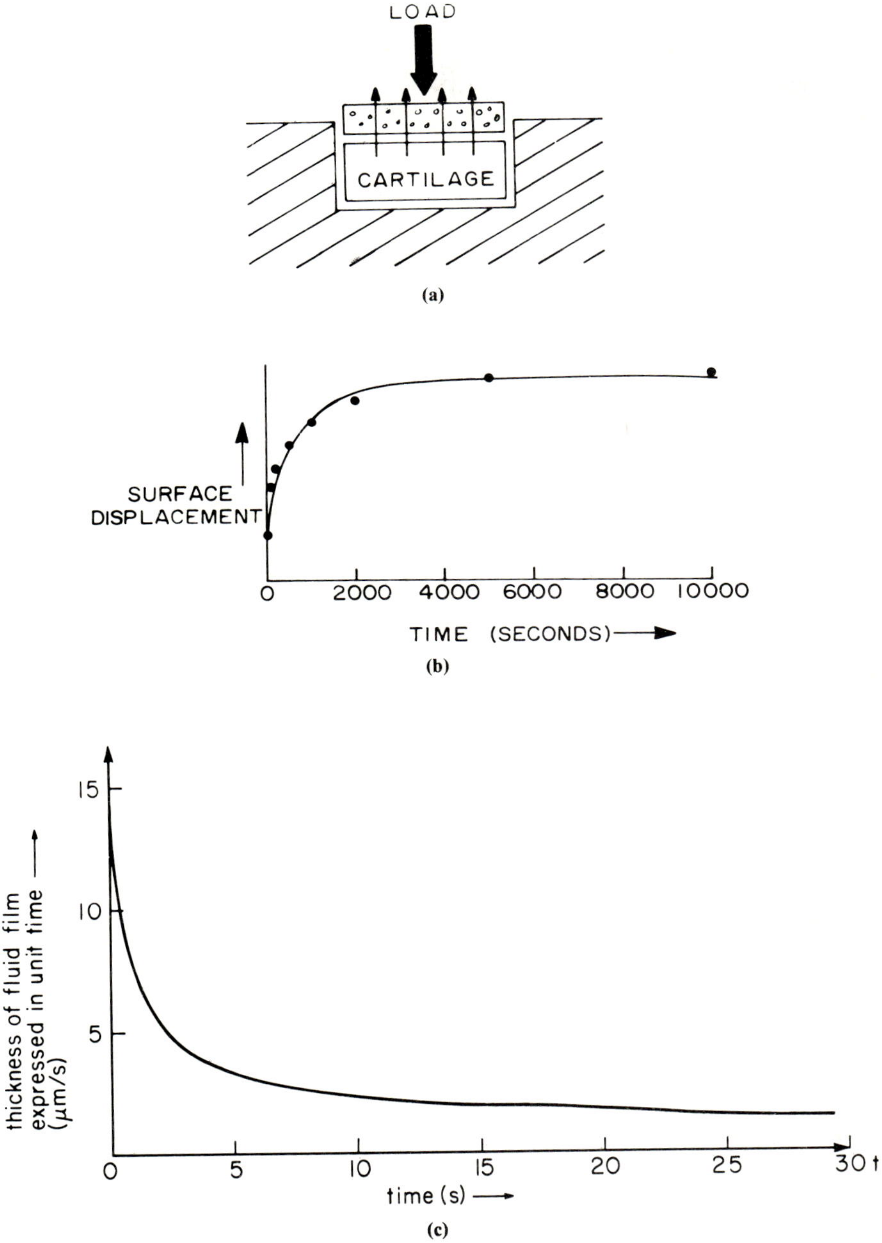

FIG. 5.3 Under uniaxial conditions (*a*), compression creep (*b*) occurs because the interstitial fluid is expressed into the free-draining porous filter. At equilibrium, no fluid movement occurs, and the compressive load is supported by the intrinsic stiffness of the solid matrix [48]. (*c*) Theoretical prediction of the rate of fluid expression as a function of time upon sudden application of the compressive load [41]. Note that within the initial second, ~10 μm of fluid may be expressed into the loading filter under free-draining conditions.

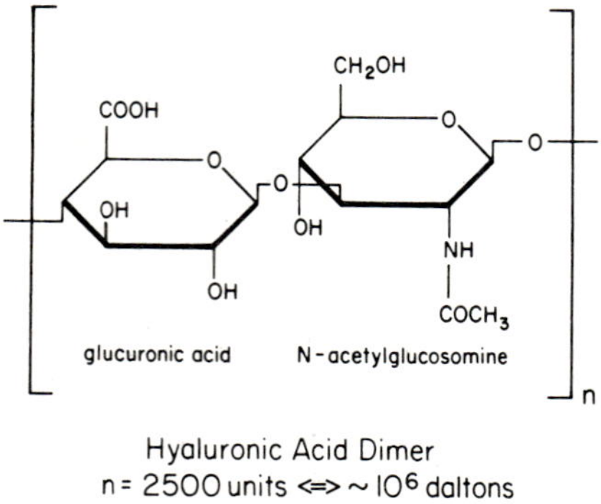

FIG. 5.4 The repeating disaccharide unit of hyaluronic acid. In synovial fluid, this nonsulfated dimer is polymerized into long chains consisting of approximately 2500 of these repeating units. The highest molecular weight reported for bovine synovial fluid is 12×10^6 daltons.

properties [74, 76, 80]. Detailed rheological modeling of synovial fluid-flow properties and numerical constants may be obtained in Lai and coworkers [39].

5.3 ARTICULAR SURFACE TOPOLOGY

Because an important feature determining the frictional characteristics between two bodies sliding over each other is the topographical nature of their surfaces [81], the gross anatomical form and microsurface irregularities of the articular surface must be appreciated. The gross geometric designs of the articulating surfaces differ greatly from joint to joint within the body, depending on the specific kinematics required of them. The ball-and-socket hip joints, for example, have very conforming surfaces compared with, say, the knee joints. Detailed map-

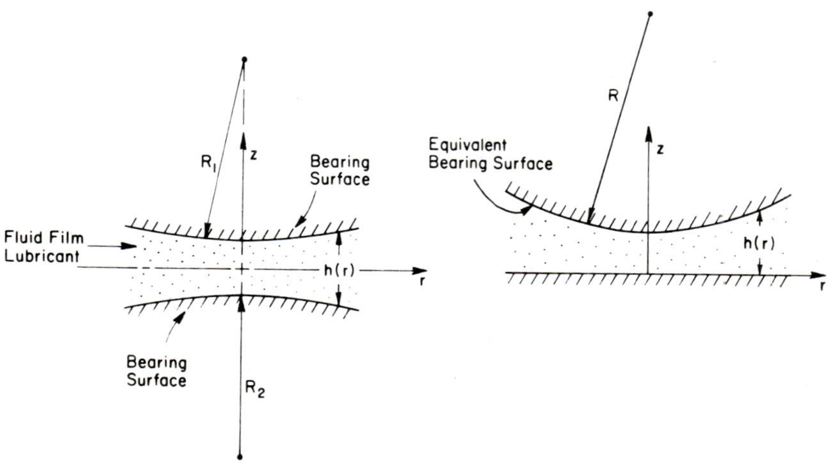

FIG. 5.5 (*a*) Two bearing surfaces separated by a thin layer of fluid lubricant in the load-bearing region. The thin film geometry is described by $h(r)$, and the two radii of curvatures are R_1 and R_2. (*b*) The bearing depicted in (*a*) is completely equivalent to a curved surface of radius R, Eq. (5.1), on a flat plane, since the film geometry is still $h(r)$.

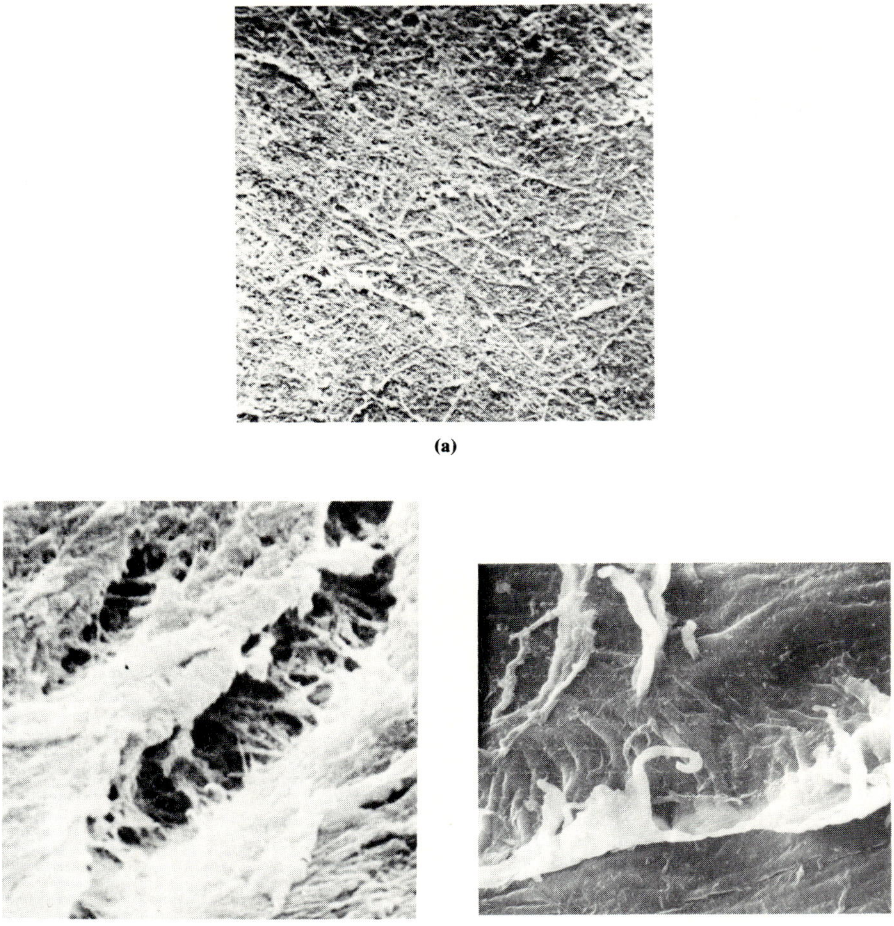

(a) (b) (c)

FIG. 5.6 (*a*) Scanning electron microscopic view of the texture of normal human articular cartilage surface, showing a dense-pack random arrangement of collagen fibrils at the surface; specimen is a 21-year-old male femoral head retrieved from autopsy and magnified 3000× [113]. (*b*) Typical appearance of articular cartilage surface from an osteoarthritic cartilage specimen, 3000×. Note individual collagen fibrils beneath the surface spanning the large fissure [113]. (*c*) Scanning electron micrograph of an aging femoral head surface retrieved from a fracture neck of the femur, 1000×. Note the peeling and ridging effect seen on the most superficial layer of the cartilage.

pings of normal human knee joint articulating surfaces have been reported by Wismans and coworkers [82]. With age and disease, these normal anatomical forms do change. For example, it is known that normal aged hip joints are not as congruous as normal young hip joints [83]. Osteoarthritis and other arthritic diseases can grossly alter the anatomical forms of the afflicted joint [84]. In conjunction with the joint kinematics and the manners in which the forces are transmitted across the joint, such variations in the anatomic contours play an important role in affecting the mode of lubrication for that joint.

At the macroscopic level, to facilitate mathematical analyses of these effects, the two articulating surfaces are often approximated in two dimensions as an equivalent system of a cylindrical surface interacting with a plane, or in three dimensions by a sphere in articulation with a half space. For example, if R_1 and R_2 are the radii of curvature of two osculating cylinders approximating the articulating surfaces at the point of contact, a geometrically equivalent cylinder-plane system would involve a cylinder of effective radius R given by

$$\frac{1}{R} = \frac{1}{R_1} \pm \frac{1}{R_2} \tag{5.1}$$

where the plus or minus sign would be chosen according to whether the two centers of curvature lie on the opposite or the same side of the articulation, respectively (Fig. 5.5). For hip joints, the radius of the equivalent sphere near a plane has been estimated to be as high as 1.0 m, whereas for the knee, it has been suggested to assume values ranging from 0.02 to 0.1 m [14, 17, 22, 43, 85, 86]. We shall see how this effective radius of curvature R can be used to estimate the feasibility of fluid-film lubrication mechanisms for a given joint.

At the microscopic level, quite contrary to visual impressions obtained by observing the smooth and shiny appearance of the normal articular surface, Talysurf tracings of surface replicas reveal that these surfaces are, in fact, relatively rough [87–90]. The *arithmetical mean deviation* R_a, which is defined as the arithmetical average value of the departure of the microscopic surface profile above and below a certain reference center line, has been reported for cartilage by Dowson and coworkers [91] in comparison to other engineering and bioengineering surfaces. For example, the typical R_a value measured for the femoral head of a common artificial hip joint is roughly 0.025 μm compared with the corresponding value of 1 to 6 μm for articular cartilage. The following orders of topographical grade can be defined [83, 87]:

1. Primary anatomical contours.
2. Secondary roughness $<$ 0.5 mm in diameter and up to 50 μm deep (as in old, osteoarthritic tissue).
3. Tertiary hollows 20 to 45 μm deep.
4. Quarternary ridges 1 to 4 μm in diameter and 0.1 to 0.3 μm deep.

Even in the loaded configuration, the higher-order surface roughness has been noted to persist [88]. Such microscopic characteristics of the surface are important factors in determining the friction of the articulation, not only in the case when the two surfaces come into actual contact with their asperities engaged in direct interactions, but also in situations with fluid-film lubrication, since the surface topology defines the minimum thickness of the fluid film required to separate the two surfaces completely. Scanning electron micrographs of osteoarthritic cartilage show an enormous range of surface damage: pronounced ridging, splits, and loosely attached tufts of torn surface materials (Fig. 5.6*a* to *c*). Such changes would undoubtedly affect adversely any lubrication mechanisms, which would in turn accelerate the attrition rate experienced by articular cartilage.

5.4 JOINT KINEMATICS AND LOADING

Because of the tremendous difficulties involved with the experimental measurement of the complex joint motions in vivo, precise data are still lacking on the relative motions between the articulating surfaces of a joint corresponding to daily activities [3, 82]. Descriptions in the literature are often qualitative in nature. Quantitative data, if available, are generally derived from theoretical models with little or no direct experimental verification. Generally, the motions studied are gait-related and thus intermittent (Fig. 5.7*a*) [3, 4, 92–97]. Both sliding and rolling may occur simultaneously between articulating surfaces [98]. Depending on the motion at the instant of interest, the relative translational speed between the two articulating surfaces may range from 0.03 m/s, say, between the femoral head surface and acetabulum surface during normal walking, and up to 0.3 m/s, say, between the humeral head surface and the surface of the

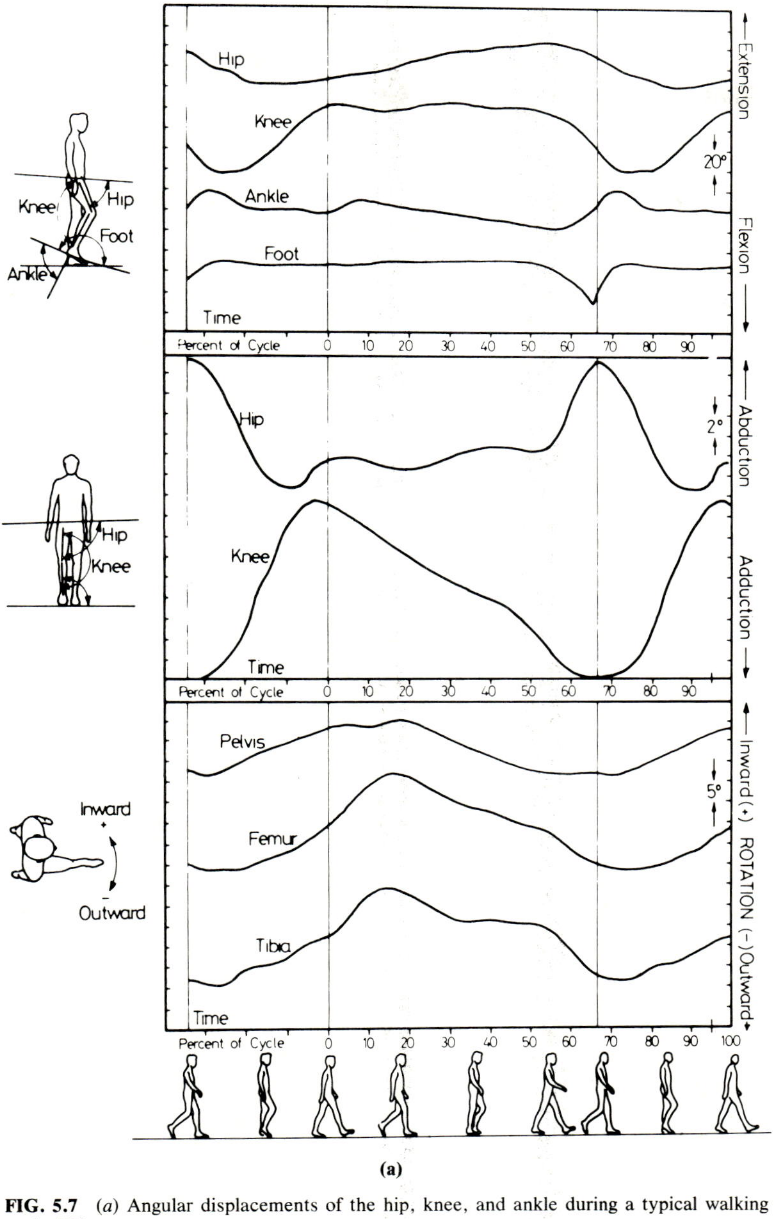

(a)

FIG. 5.7 (*a*) Angular displacements of the hip, knee, and ankle during a typical walking cycle [92].

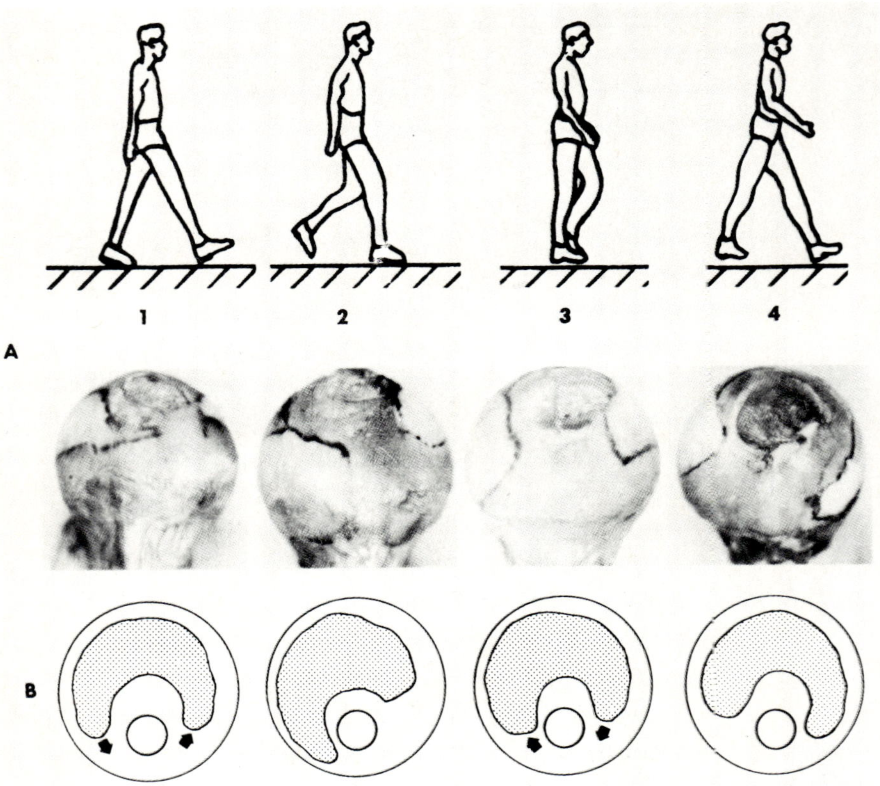

(b)

FIG. 5.7 (*b*) Movement of the contact area across the femoral head during a typical walking cycle [99].

glenoid cavity of the shoulder for a baseball pitcher throwing the ball at 100 mi/h.* However, for animal joints, these motions are never steady, and at best are repetitive, with a characteristic frequency of 1 to 2 Hz.

The load transmitted across the joint is normally carried by the opposing joint surfaces either by direct contact or through a layer of fluid film. Additional stability is also provided by the tensile forces in the ligaments and the surrounding musculature. The magnitude of the load transmitted is again dependent upon the activity, e.g., walking or running, and the loading site can fluctuate rapidly as the articulating surfaces move relative to each other [86, 99–101] (Fig. 5.7*b*). During normal walking on level ground, the human hip and knee joints can be subjected to joint forces as high as six times body weight [3, 4, 95] (Fig. 5.7*c*). The peaks usually occur right after heel strike and right before toe-off. The joint load averages around three times body weight in the course of the stance phase, which normally lasts up to 60 percent of the walking cycle. (This level of loading is similar to those observed during static two-leg standing within the knee joint and the hip joint [93].) Throughout the swing phase of the walking cycle, the joints are only lightly loaded while the articular surfaces move rapidly over each other. The magnitude

* These order-of-magnitude estimates can easily be obtained by using elementary rigid-body mechanics and crude anatomical models for the joint [22].

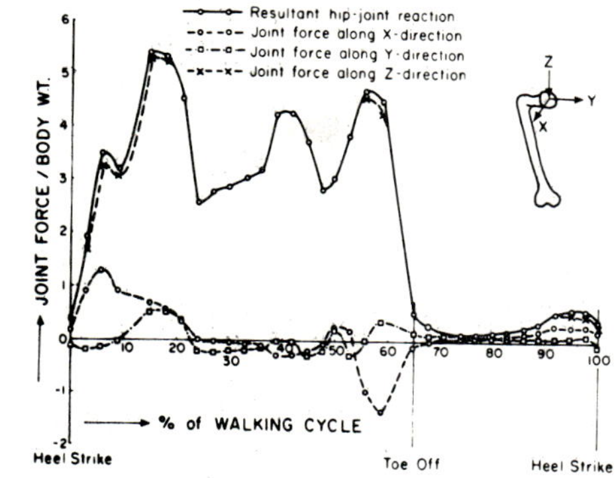

Hip joint reaction and components.

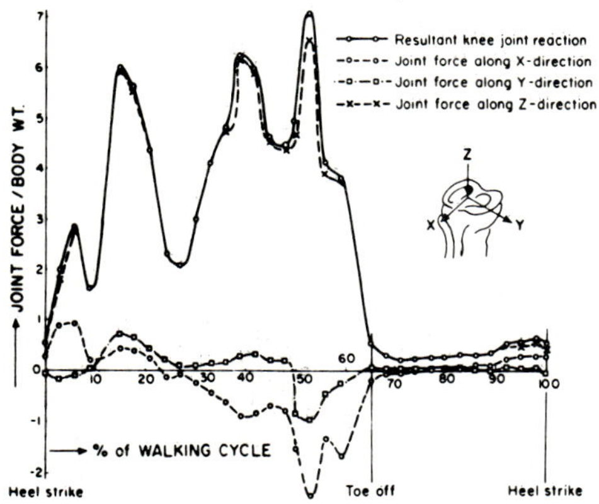

Knee joint reaction and components.

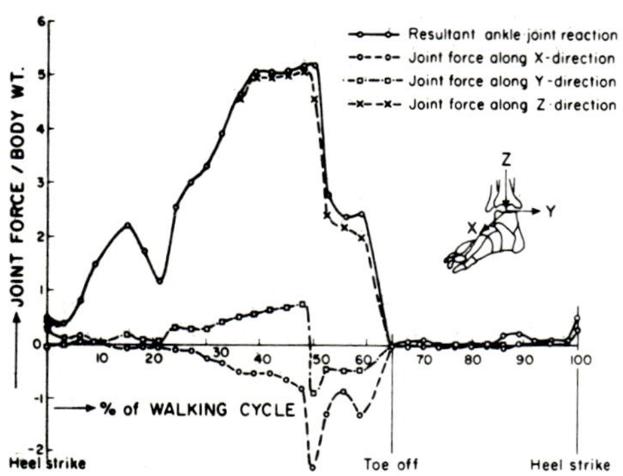

Ankle joint reaction and components.

(c)

FIG. 5.7 (*c*) Typical history of hip, knee, and ankle forces experienced during a walking cycle [95].

and duration of the joint loading and the relative speed of the two articulating surfaces are the critical factors in bearing operation, and discussion of the effectiveness of the mechanisms of lubrication in synovial joints must take into account these factors.

5.5 LUBRICATION THEORIES FOR SYNOVIAL JOINTS

The ability of synovial joints to function with almost negligible frictional resistance, ranging from 0.001 to 0.03 [6, 7, 11, 18, 19, 34], and wear rates [33] suggests that some very remarkable lubricating mechanisms operate in these systems. In 1932, MacConaill, the first to use engineering analysis applied to joint lubrication, ascribed the low frictional characteristics of synovial joints to the fact that the two articulating surfaces might not be touching each other but could be separated by a thin layer of synovial fluid [5]. This thin layer of fluid, he claimed, was continuously dragged into the gap between the wedge-shaped meniscus and the femoral and tibial articulating surfaces of the knee as a result of the high viscosity of synovial fluid and the relative motion between the articulating surfaces. The meniscus was believed to function to provide joint conformity, thus promoting the viscous hydrodynamic action required for full fluid-film lubrication. According to this theory, because of the very nature in which this mechanism operates, the viscosity of synovial fluid is the only important fluid property required for the remarkable lubrication efficacy of diarthrodial joints. If hydrodynamic lubrication is indeed the mode of lubrication, then the primary resistance to relative surface motions would come from the shearing of the viscous synovial fluid layer during joint motion and neither from the viscoelastic deformation of the articular cartilage nor from the flow of interstitial fluid through the highly dissipative organic matrix of the tissue. This mechanism of lubrication is known as *hydrodynamic lubrication*.

However, some severely restrictive assumptions must be satisfied in order for this hydrodynamic lubrication theory to work. Sir John Charnley (1960) [7, 8], as well as many others, pointed out that the classical hydrodynamic lubrication mechanism demands a continuous, high-speed relative motion between the two bearing surfaces; unfortunately, such conditions could never be met under in vivo situations, where joint motions are at best intermittent in nature. Charnley suggested an alternative mechanism whereby the articular surfaces might, in fact, be separated by a monolayer of lubricant molecules. He and others believed that the HAP complex absorbed on each of the articular surfaces is probably the boundary lubricant for the synovial joint [7, 8, 16].

Boundary lubrication phenomena are well-known to tribologists, surface chemists, and surface physicists [102]. More recently, Swann and Radin were able to isolate the molecule in synovial fluid believed responsible for lubricating the articular surfaces [26]. They call this the *lubricating glycoprotein fraction* (LGF) of synovial fluid. This turns out to be not the HAP complex but a single polypeptide chain containing oligosaccharides, or single-sugar molecules, distributed along the length of a protein core. The molecular weight is about 250,000 [71]. This molecule has a tremendous affinity for the articular surface and is not easily rubbed off; see the discussion in Sec. 5.6, "Experimental Evidence from Measurements of Friction." If the joint were indeed in this mode of operation, degradation of the boundary material alone would drastically reduce the ease of relative sliding motion, while merely a change in the rheological properties of synovial fluid would not. This of course is also not true. Thus both fluid-film lubrication and boundary lubrication must coexist in vivo. The operative regime of each mode depends on the kinematics and loading, joint anatomy, and material properties within the joint.

In order to assess how fluid-film lubrication might exist within the joint, it would be instructive to apply the various engineering lubrication theories for those conditions of joint kinematics, loading, and material properties. First, the effectiveness of hydrodynamic lubrication for diarthrodial joints under in vivo conditions can be assessed by idealizing the articulation as either a geometrically equivalent rigid sphere on a half space for the hip or a rigid cylinder on a half plane for the knee (Fig. 5.5). These are extreme geometric simplifications, and thus the analysis can be used to provide only a qualitative assessment. However, these extreme geometric simplifications are all that are available short of a full-blown computer simulation using full-

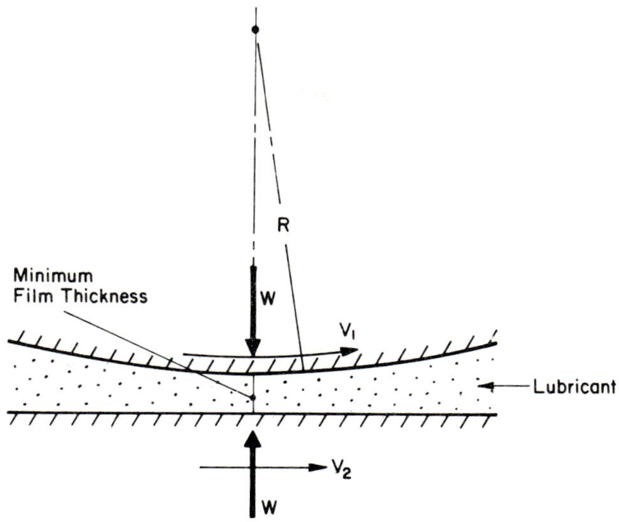

FIG. 5.8 Cylinder on plane bearing configuration; u = entraining velocity = $(V_1 + V_2)/2$, R = effective radius of curvature, w = bearing load supported by the lubricant of viscosity η.

scale anatomical forms.* To assess whether hydrodynamic lubrication is possible, the approach was to assume a priori that a film of lubricant exists and to calculate, using the idealized material model for synovial fluid, the minimal thickness of fluid film being entrained into the wedge-shaped gap between the two articular surfaces. The criteria for acceptance of hydraulic fluid-film lubrication would be if the predicted minimal film thickness is greater than the total distance necessary to clear the microscopic asperities above a centerline average defining the articular surface.

In 1916 Martin studied the hydrodynamic lubrication of a cylinder-plane system and presented an estimate of the minimum film thickness in the form [103]

$$\frac{h}{R} = 4.9 \frac{\eta u l}{W} \tag{5.2}$$

whereas in 1955 Kapitza [104] found the relationship

$$\frac{h}{R} = 113.7 \left(\frac{\eta u R}{W}\right)^2 \tag{5.3}$$

for the sphere-plane system. In the above equations, h is the minimum film thickness; R is the radius of the geometrically equivalent cylinder or sphere; l is the axial length of the cylinder; η is the viscosity of the Newtonian fluid lubricant; u, the entraining velocity, is equal to one-half the sum of the velocities of the interacting surfaces; and W is the load carried by the bearing (see Fig. 5.8).

Using $R = 1.0$ m for the hip joint and $R = 0.1$ m and $l = 0.1$ m for the knee joint, and $\eta = 2 \times$

* In view of the simplifications involved—e.g., the synovial fluid is Newtonian; the articular cartilage is assumed to be rigid—it would indeed be foolish to pursue a detailed computer simulation using these models.

10^{-3} N · s/m^2, one finds from Eqs. (5.2) and (5.3) that the minimum fluid-film thickness is on the order of 10^{-13} and 10^{-9} m, respectively, during the typical peak loading period of the stance phase of level walking [44]. These values are much smaller than the order of roughness on the cartilage surfaces, ~1.0 μm, indicating that hydrodynamic lubrication alone under these loading conditions is not likely to be the main lubrication mechanism for these joints in vivo. However, as in the case of the baseball pitcher throwing a fastball, the articulating surfaces of the shoulder may be slightly distracted, so the joint is operating under a lightly loaded condition with high sliding speeds, and the hydrodynamic fluid-film mechanism may indeed be the mode of lubrication.

Dintenfass (1963) proposed that the pseudoplasticity, or shear thinning non-Newtonian effect, exhibited by synovial fluid might enhance the fluid-film thickness within the joint [11, 12]. At lower speeds, the lubricant, i.e., synovial fluid, would exhibit a higher viscosity and thus a greater fluid-film thickness. Such a mechanism could indeed work at the end of each period during reciprocating motion.

5.5.1 Elastohydrodynamic Lubrication

Dintenfass [11] also brought out the significance of the deformability of articular cartilage in fluid-film lubrication. He postulated that elastic deformation of cartilage could not only spread the joint load to a larger bearing area but also reduce the velocity gradient between the two surfaces. When both the viscous resistance of the lubricant and the elastic deformation of the bearing surfaces play a prominent role in the lubrication process, the term *elastohydrodynamic lubrication* is used to describe the mechanism. Tanner (1966) demonstrated the importance of elastohydrodynamic action in synovial joints by showing that the film thickness could be on the order of 0.1 μm when the deformations of the bearing surfaces are taken into consideration [14]. More recently (1981) Dowson and coworkers [44, 105], following their earlier results [17], showed that the elastohydrodynamic film thickness can be as high as 1.3 and 1 25 μm for the hip joint and the knee joint, respectively.

In all the above estimates, the articular bearing surfaces are assumed to be made up of a homogeneous, linearly elastic material. However, at each end of the bone within the synovial joint is a layered structure. Each end is covered by a soft layer of articular cartilage supported by a much harder calcified cartilage, which is in turn backed by a subchondral cortical shell (Fig. 5.1*b*). This rigid backing stiffens the cartilage layer, which would then result in a thinner fluid film. In fact, Hooke and O'Donoghue studied the effects of such a layered structure and presented the following formula for the cylinder-plane configuration:

$$\frac{h}{R} = \frac{l^{0.2}(\eta u)^{0.6}}{W^{0.2}(E'R)^{0.4}} \tag{5.4}$$

where E' is the effective elastic modulus calculated from

$$\frac{1}{E'} = \frac{1}{2}\left(\frac{1-\nu_1^2}{E_1} + \frac{1-\nu_2^2}{E_2}\right) \tag{5.5}$$

where E_1, ν_1, E_2, and ν_2 are the elastic moduli for the two-layer articulating model [106]. When Eq. (5.4) was applied to the human knee joint, the film thickness was predicted to be roughly 0.5 μm [44].

5.5.2 Weeping Lubrication

Lewis and McCutchen postulated an alternative source of lubricant for fluid-film joint lubrication. According to these investigators, a lubricant fluid film is generated between the bearing surfaces by the natural compression of articular cartilage during joint function, and this fluid exudation occurs over the entire articulating surface [9, 107]. In a simple experiment using fully saturated sponges and excised specimens of cartilage pressing against a glass plate, McCutchen demonstrated that a layer of fluid film can be formed as a result of interstitial water being expressed when a load is applied. Thus, it was believed that when joint cartilages are pressed against each other in vivo, fluid is uniformly exuded upon compaction of the tissue, creating the fluid required for fluid-film lubrication. This mechanism very closely resembles that of a human-made hydrostatic bearing in which a fluid film is constantly maintained between the two bearing surfaces by an external pump. In the synovial joint, however, McCutchen described a hypothetical self-pressurized mechanism which was termed *self-pressurized hydrostatic lubrication*, or *weeping lubrication*.

For this hypothetical mechanism to work, one must consider articular cartilage as a porous, permeable material whose interstitial fluid can be expressed upon compaction of the tissue. During the period from 1959 to 1980, some significant results were published pertaining to the rate of fluid exudation [41, 54, 55], permeability [108, 109], and the effect of electrolytes in the fluid bath surrounding the tissue upon its deformation [110]. See Chap. 4 for a complete discussion of these concepts.

5.5.3 Squeeze-Film Lubrication

For fluid-film lubrication, the thickness of the fluid layer and the load that the bearing must sustain determine its effectiveness. When two bearing surfaces, separated by the film of fluid lubricant, are squeezed together, a large viscous resistance is generated in the lubricant to retard the closing of the gap. Typically, a 20-μm-thick, viscous lubricant film can resist several megapascals of pressure for many minutes before the lubricant film becomes depleted. This mode of lubrication involves two bearing surfaces approaching each other, where no sliding action occurs, and is known as *squeeze-film lubrication*. Furthermore, if the bearing surfaces are covered by a layer of soft material such as cartilage, the pressure in the squeeze film can cause substantial deformation to the mating surfaces. Localized depressions occur under the regions of highest pressures and thus would promote a longer-lasting fluid-film lubrication effect, because fluid would be naturally trapped by the depressions.

Fein (1967) first demonstrated the importance of such a mechanism in synovial joint lubrication and showed that under squeeze-film conditions for a sphere-plane system, the film thickness can be estimated from

$$\frac{h}{R} = 2.86\,\frac{(W/E'R^2)^{1/6}}{(TE'/\eta)^{1/2}} \tag{5.6}$$

where T is the squeeze time, i.e., the time required to reduce an infinitely thick film to thickness h [15]. Later Higginson (1978) estimated that a (physiologically) useful squeeze-film time (~60 s) can be obtained between two compliant bearing surfaces [35]. Dowson and coworkers applied this concept; for example, if a synovial fluid film of 2.5 μm was generated in the human hip joint during the course of the lightly loaded motions of the swing phase of normal level walking, from whatever source, the film thickness would decrease by only 0.06 μm in 0.5 s after heel strike [44].

The effects of the non-Newtonian properties of synovial fluid on the squeeze-film phenomenon have also been discussed in the works of Tanner (1966) [14] and Mow (1969) [22]. In the latter, it was shown that for a viscoelastic liquid such as synovial fluid, a lower peak hydrodynamic pressure would exist within the fluid film, and thus a larger contact area would be required to support the load. Thus squeeze-film time would be prolonged by the viscoelastic nature of the lubricant film.

5.5.4 Boosted Lubrication

In the context of this squeeze-film mechanism, Maroudas (1967) [16] and Walker and coworkers (1968) [19] apparently independently proposed another mechanism of lubrication, which was termed *boosted lubrication* by Walker and coworkers. It was suggested that as the articulating surfaces approach each other, the solvent of the thin layer of synovial fluid (< 5 μm), i.e., water, would be uniformly squeezed out from the entire contact region and into the articular cartilage. These authors believed that this would occur because the closing gap could increase the resistance of sideways efflux of the lubricant from the gap tangent to the articulating surfaces to a point beyond the resistance of flow into the cartilaginous bearing material normal to the articulating surface (Fig. 5.9).

However, the pore size of the normal cartilage surface (< 5 nm) is much too small to allow passage of the major macromolecular component of synovial fluid (the HAP complex has a solution domain with a diameter of ~400 nm) to enter the tissue. Thus, the normal articulating surface would act as an ultrafiltration membrane, permitting only water and small electrolytes to pass into the tissue. As a result, a highly viscous concentrated layer of HAP complex is left in the joint space. Seller and coworkers (1971) presented additional evidence from scanning electron microscopic observations which apparently showed the existence of such a layer of HAP complexes between the normal articular surfaces under squeeze-film conditions [111] (Fig. 5.10). Walker and coworkers postulated that this "gel" layer is capable of supporting greater loads and lasting longer periods of time than synovial fluid with normal concentrations of HAP complexes [19]. They also suggested that the microroughness of the articular surfaces can play an important role in trapping the gel, thus enhancing the formation of many micropockets of gel over the joint surface.

The process of ultrafiltration at the articular surface for understanding boosted lubrication or for understanding the effect of convection and diffusion on macromolecular transport across the articular surface was rigorously analyzed by Lai and Mow [112]. These authors used a one-dimensional convection-diffusion model to solve the ultrafiltration phenomenon of a macromolecular solution and an ultrafiltrating membrane. Of considerable importance in this convection-diffusion problem is the Peclet number Vh/K, which is the ratio of the rate of convection to the rate of diffusion. Here V is the rate at which the two bearing surfaces are forced together, h is the gap thickness, and K is the diffusivity of the solute in the lubricant film. Obviously, the relative rates of these two mass transport processes are important in understanding how the solute would distribute throughout the lubricant film in the gap. Also, the effectiveness of the articulat-

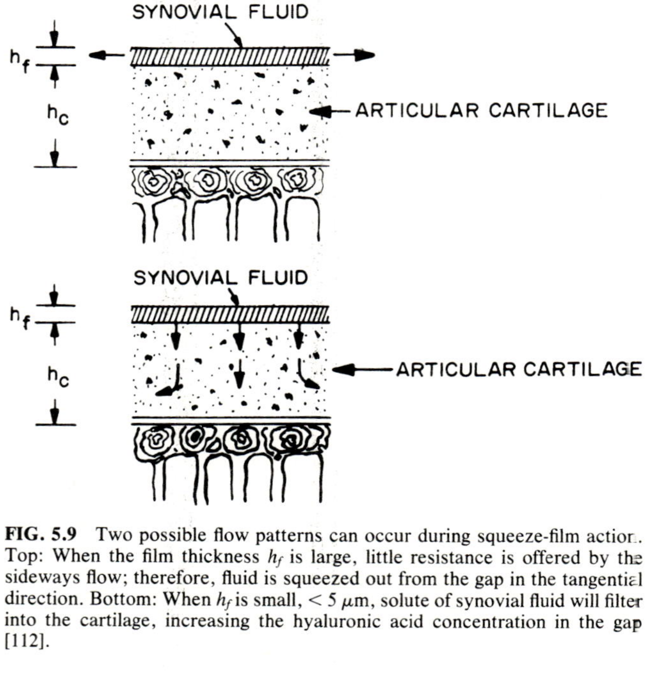

FIG. 5.9 Two possible flow patterns can occur during squeeze-film action. Top: When the film thickness h_f is large, little resistance is offered by the sideways flow; therefore, fluid is squeezed out from the gap in the tangential direction. Bottom: When h_f is small, < 5 μm, solute of synovial fluid will filter into the cartilage, increasing the hyaluronic acid concentration in the gap [112].

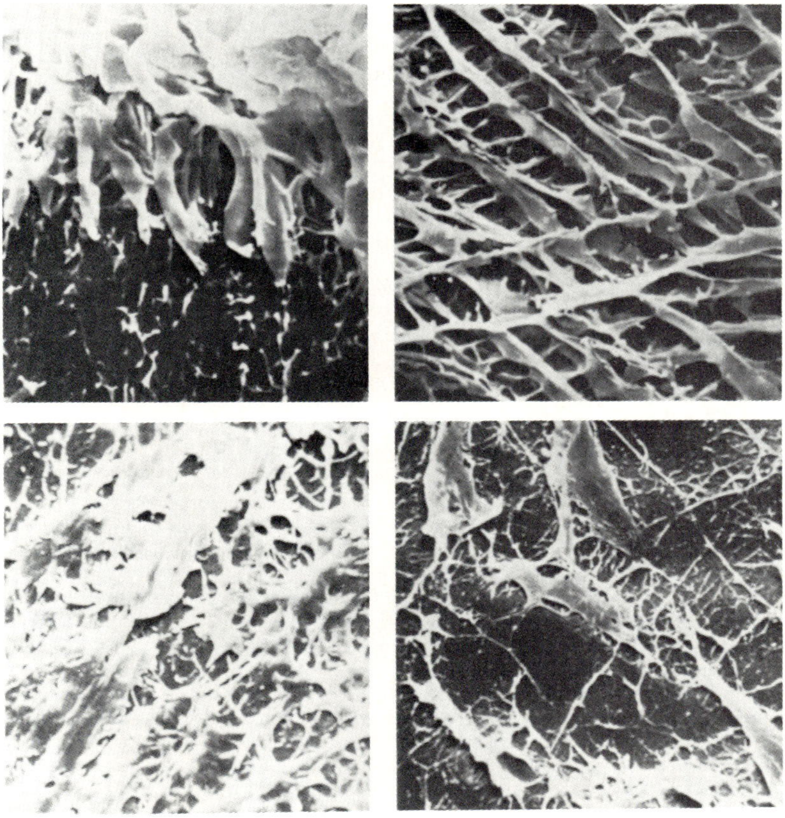

FIG. 5.10 Scanning electron microscopy showing surface aggregation of dried hyaluronic acid protein complexes on cartilage surfaces [111].

ing surface as a filtering membrane—from the total blockage to total transmission for a certain molecular species—is obviously an important parameter. Both these conditions, i.e., the kinematics of the problem and the parameters describing the material surface, must be assessed a priori.*

The diversity of possible lubrication mechanisms has created considerable interest and controversy over the past decade. Evidence cited to support the various hypotheses has often been indirect and conjectural. The apparent contradiction between weeping and boosting lubrication—that the former requires fluid exudation over the entire contacting surface, whereas the latter demands fluid imbibition over the entire contacting surface—has bewildered many researchers in the field. Most recently a rational view of diarthrodial joint mechanics has come forth [40, 55], based on an engineering-science point of view. The first task was to develop accurate constitutive laws for both synovial fluid [39] and articular cartilage [40] and to determine the range of applicability of the laws [39–42]. Then, using known anatomical geometric

* An appreciation of these fundamental aspects of physical modeling would have lessened the intensity of scientific debate on joint lubrication which occurred throughout the early years of the 1970s between the proponents of weeping lubrication and those of boosted lubrication.

factors and kinematic and loading characteristics for a specific joint of interest, with the appropriate interfacial boundary conditions, one can calculate how fluid flow occurs within the joint cavity and within the articular cartilage layers. To date, no fully coupled synovial fluid–articular cartilage (non-Newtonian fluid-biphasic mixture) problems have been solved [42, 55].

5.5.5 Joint Lubrication

The central point of controversy between the two hypothetical lubrication mechanisms, weeping and boosted, concerns the source of the fluid film. In weeping lubrication, cartilage is considered as a reservoir of lubricant for the joint, and, upon compression, the interstitial fluid is exuded uniformly into the gap *regardless* of joint motion, joint loading, or the material properties of cartilage and synovial fluid. In boosted lubrication, the water in cartilage is irrelevant; synovial fluid, and in particular the HAP complex, is the lubricant for the joint. Fluid flows uniformly into the cartilage *regardless* of joint motion, joint loading, and the material properties of cartilage and synovial fluid.*

To assess whether cartilage could yield its 70 to 80% fluid content as the lubricant for diarthrodial joints and to determine the exact manner of interstitial fluid flow within cartilage and how the lubricant film might be formed, Mansour and Mow (1977) [113] and Lai and Mow (1979) [114] solved a series of boundary value problems where articular cartilage is modeled as a layer of biphasic medium [41]. In the earlier work, the material properties for cartilage were obtained from those in the literature [59, 115, 116], whereas in the latter, the true biphasic material properties were used [41, 55]. Most recently, the effect of tissue degeneration in terms of the influence of the rise of its water content, as in early osteoarthritis, on the tissue modulus and permeability were assessed [48]. All these studies indicate that in order to understand joint lubrication or fluid flow within the joint cavity and within articular cartilage, one must know a priori the relevant material properties, joint anatomy, joint kinematics, and joint loading. No uniform concept of joint lubrication should be expected; one must seek specific answers to the specific operating range of a specific joint.

Under the action of a constant-width, parabolically distributed normal load sliding over the articular surface at physiological speeds, normal cartilage could yield its interstitial fluid in a manner suitable for generating a fluid film for lubrication (Fig. 5.11). No synovial fluid was considered in these models proposed by Mow and coworkers [55, 114]. The cartilage is subjected to a prescribed traction—one similar to those predicted in Hertzian contact problems [117]. Since the traction is prescribed on a surface where both a solid phase and a fluid phase coexist, a load partition factor (β) must be assumed a priori [55]. It is generally believed that this load partition factor is bounded by the surface porosity (V_f/V_T). The specific variation of this load partition factor β with surface morphology (Fig. 5.6), joint kinematics and loading (Fig. 5.7), and cartilage deformation and interaction with synovial fluid is not known. Thus β must be used as a parameter [55].

For $0 \leq \beta \leq V_f/V_T$, the theory predicts what Mow and coworkers have termed a *self-generating mechanism* whereby fluid exudation occurs over the "leading-edge region" of the advancing loaded area, and fluid imbibition occurs over the "trailing-edge region" of the passing loaded area (Fig. 5.11). The amount of this fluid transported into the joint space is a function of the intrinsic material properties of the organic matrix of cartilage and the speed and magnitude of the sliding load. As seen from a particular point on the cartilage surface at some distance in front of the load, fluid exudation begins well in advance of the oncoming load, and fluid expression becomes more prominent as the load approaches (Fig. 5.11*b*). After passage of the sliding load, elastic recovery of the organic solid matrix causes the expelled fluid to be resorbed. The extent of the fluid exudation region and the volume of the exudate determine the effectiveness of the cartilaginous interstitial fluid as a lubricant. The quantity and extent depend on how the load is partitioned between the solid and fluid phases at the articular surface, which in turn depends on surface morphology, the intrinsic compressive stiffness of the organic solid matrix,

* To appreciate the intensity of this controversy, refer to the transcription of the debates between Dr. McCutchen and Dr. Walker in *Lubrication and Wear in Joints* (1969) [132].

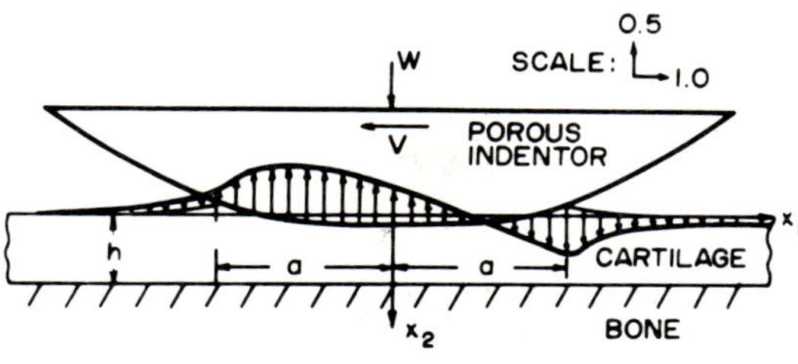

(a)

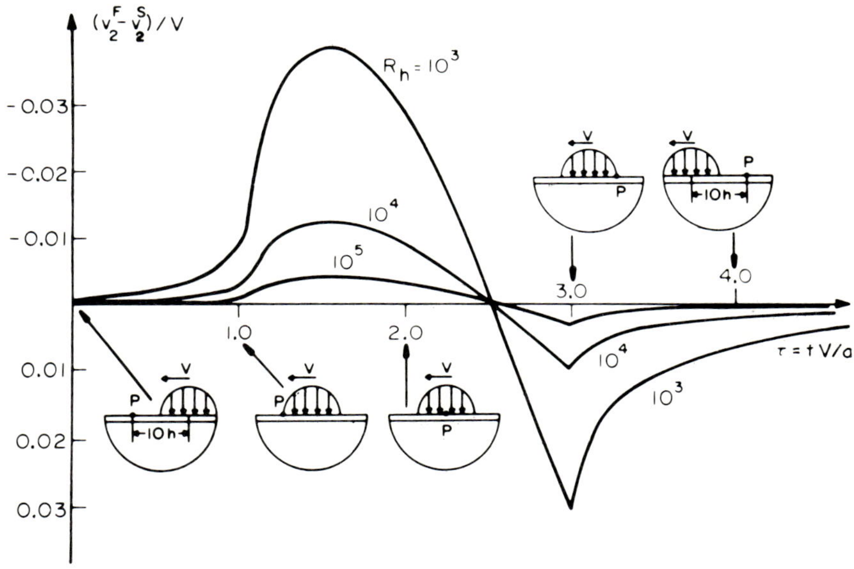

(b)

FIG. 5.11 (*a*) Pattern of predicted fluid exudation and imbibition over the articular surface resulting from a frictionless, free-draining ($\beta = 0$) moving indenter. V is the speed of the horizontal translation, and W is the resulting line load of a surface pressure distribution $p_A(X_1)$ acting over a spread of $|X_1| \leq a$. Flow exudation occurs in front of and beneath the leading portion of the moving indenter, and fluid imbibition occurs after the load passes [55]. (*b*) Time history of the X_2 component of fluid velocity relative to the solid matrix at a material point P on the surface; h is the thickness of the cartilage layer [55].

and its permeability [41, 55]. Surface fibrillation, proteoglycan depletion, and swelling of the tissue could defeat the self-generating mechanism of lubrication [55]. Figure 11*c* and *d* shows the importance of surface structure (β) in controlling the interstitial fluid-flow process near the surface and the effect of changes in material properties as they affect R_h on fluid pressure within the tissue. The importance of these theoretical predictions is that they show a pattern of the simultaneous existence of fluid exudation and imbibition under a moving load. Thus the earlier controversy about joint lubrication and fluid flow is really pointless, and both the hypothetical models of lubrication—weeping and boosted—are only partially correct.

To verify the self-generating and circulatory flow mechanism of fluid-film creation by articular cartilage, Mow and Lai constructed an optical, sliding-contact analytical rheometer to produce a cinematographic record of this flow pattern [118]. Excised strips of normal articular

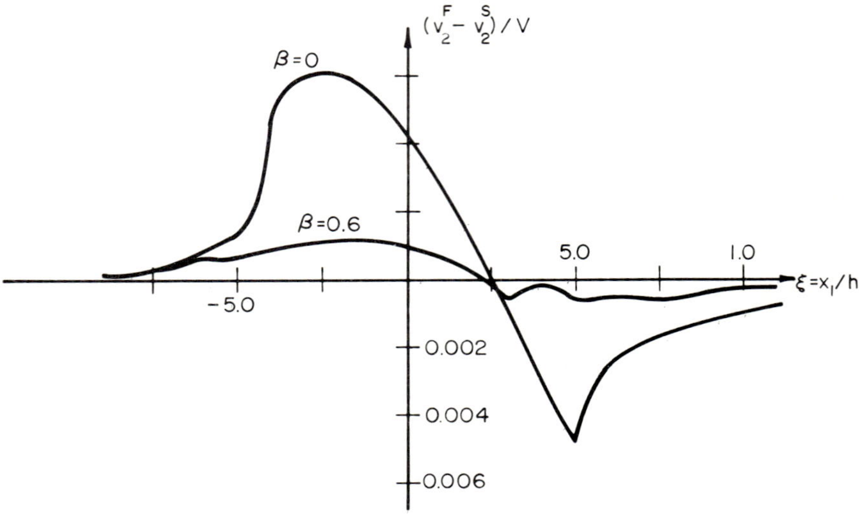

(c)

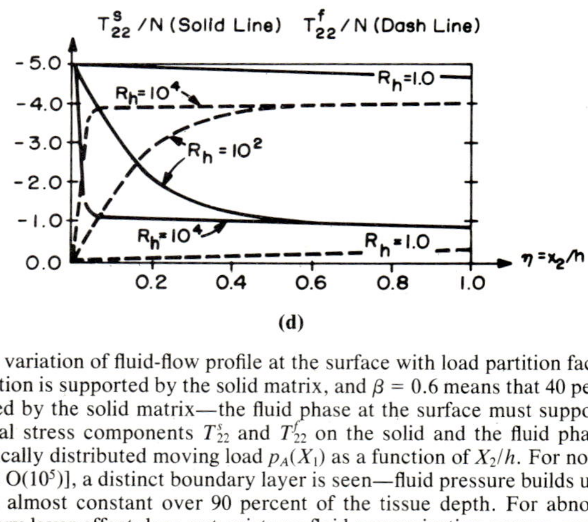

(d)

FIG. 5.11 (*c*) The variation of fluid-flow profile at the surface with load partition factor β. $\beta = 0$ means that all surface traction is supported by the solid matrix, and $\beta = 0.6$ means that 40 percent of the surface traction is supported by the solid matrix—the fluid phase at the surface must support the remaining 60 percent. (*d*) Normal stress components T_{22}^s and T_{22}^f on the solid and the fluid phases underneath the center of a parabolically distributed moving load $p_A(X_1)$ as a function of X_2/h. For normal tissue parameters $[O(10^3) \leq R_h \leq O(10^5)]$, a distinct boundary layer is seen—fluid pressure builds up rapidly within the tissue and remains almost constant over 90 percent of the tissue depth. For abnormal tissues $[R_h \leq O(10^2)]$, the boundary layer effect does not exist; no fluid pressurization occurs, and the solid matrix is highly stressed [55].

cartilage were mounted on a glass slide and loaded normally by an optical glass lens. The load caused a contact region of ~0.07 cm and was made to slide slowly across the articulating surface. Microcinematography was facilitated via a Bausch and Lomb 250× stereo microscope. The flow patterns were visualized by 4-μm-DVB (divinylbenzene) microspheres of neutral bouyant density tracers, and they confirmed the self-generating, or circulatory flow, mechanism at the articular surface. These experiments provided direct visual evidence that articular cartilage, under slowly moving, moderate loading conditions (~2 MPa), can generate a fluid film estimated to be no more than 10 μm thick. However, it remains to be seen whether the quantity of exudate is sufficient to form a viable lubricant film under in vivo conditions.

In summary, basically there are only two distinct lubrication mechanisms: boundary and fluid film. The so-called lubricating glycoprotein fraction of synovial fluid, with its great affinity for the articulating surfaces, appears to be the boundary lubricant [26, 71]. This mechanism is important when the microsurface asperities of the surfaces come into contact, or when, under prolonged severe loading conditions, the fluid film is depleted. The HAP complex of synovial fluid is an effective boundary lubricant for the articular surface against the joint capsule [26]. Fluid-film lubrication could function in a number of different modes: hydrodynamic (light loads and high speeds), elastohydrodynamic (moderate loads and speeds), and squeeze film (impact loads with impinging surface conditions). The rheological properties of synovial fluid—its non-linear shear thinning effect and its viscoelastic response—tend to extend the duration and spread for fluid-film lubrication. The source of the fluid lubricant depends on the kinematics of the joint and magnitudes of joint loading; synovial fluid and cartilage exudate are both viable sources. The latter probably becomes most significant when elastohydrodynamic conditions prevail or when significant deformation of the cartilage exists. Normal articular cartilage yields its interstitial fluid by a circulatory flow pattern where both fluid exudation and imbibition occur simultaneously. This source also extends the range of fluid-film lubrication. Fluid-flow and deformation characteristics of synovial fluid and articular cartilage during diarthrodial joint function depend upon their respective material behaviors and the operating conditions of the joint—hip, knee, fingers, etc. Pathological conditions, such as osteoarthritis and rheumatoid arthritis, where the tissue properties of both synovial fluid and articular cartilage are profoundly affected, will affect the lubrication mechanisms within the joint.

5.6 EXPERIMENTAL EVIDENCE FROM MEASUREMENTS OF FRICTION

Friction is caused by energy-dissipating mechanisms that occur during relative sliding motions between contacting surfaces [81]. Frictional resistance can come from two sources:

1. *Surface:* adhesion of one surface to the other surface, topographical interference to motion by contacting asperities on the two surfaces, and viscosity of the sheared lubricant film separating the two surfaces.
2. *Bulk:* internal energy dissipation mechanisms within the bulk material. (The viscous dissipation within the synovial fluid, obviously, can also be considered within the latter category.)

For cartilage, the internal friction is principally due to the friction existing between the interstitial fluid and the porous organic solid matrix when the fluid moves through the matrix and the intrinsic viscoelastic properties of the solid proteoglycan-collagen matrix [41]. These energy-dissipating mechanisms occur as long as joint motion and load can cause shear deformation of the synovial fluid, contact of surface asperities of the cartilage surfaces, interstitial fluid flow through the matrix of cartilage, or shear deformation of the organic solid matrix [42]. By nature of diarthrodial joints, all these deformational effects exist in vivo. The dissipation relating to bulk internal energy dissipation when a load moves to and fro along the surface of a dissipative medium is known as *ploughing friction*. The ploughing friction of diarthrodial joints is difficult to assess and was first successfully analyzed by Linn in 1968 [119].

The conventional parameter used to quantify frictional characteristics of a material is the *coefficient of friction*, which is defined as the ratio of the tangential frictional resistance to the normal force pressing the two surfaces together. In general, there have been two experimental configurations used to measure the coefficients of friction between articular cartilage surfaces. First, whole intact joints were studied in specially designed "arthrotripsometers" [119, 120]. The coefficients of friction for intact synovial joints, obtained from these arthrotripsometers, which measure both ploughing friction and surface friction, ranged from 0.04 to as low as 0.003.

The most popular experimental configuration of such studies was to use the joint (hip or ankle) under investigation as the fulcrum of a simple pendulum where one of the joint surfaces rocked freely over the other (Fig. 5.12). Early on, many investigators felt that the time history of

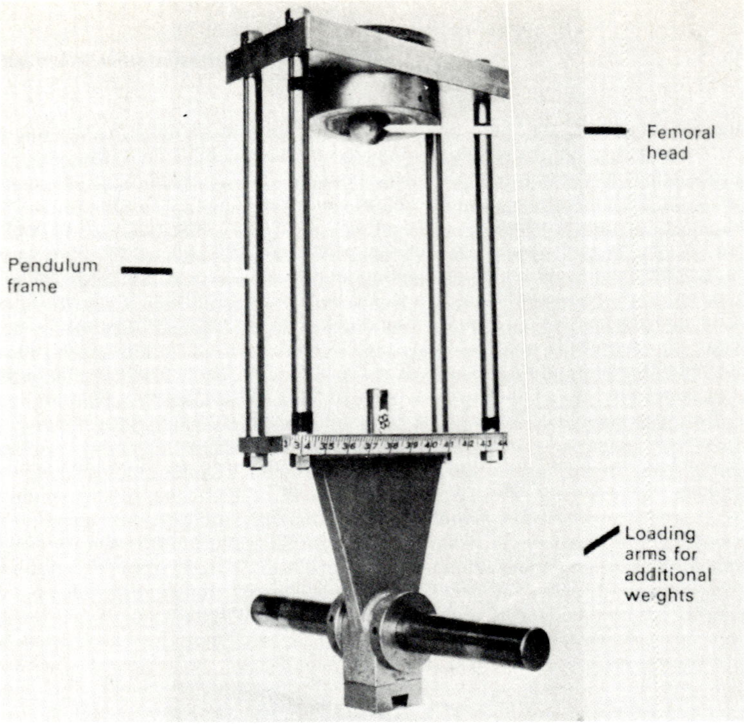

(a)

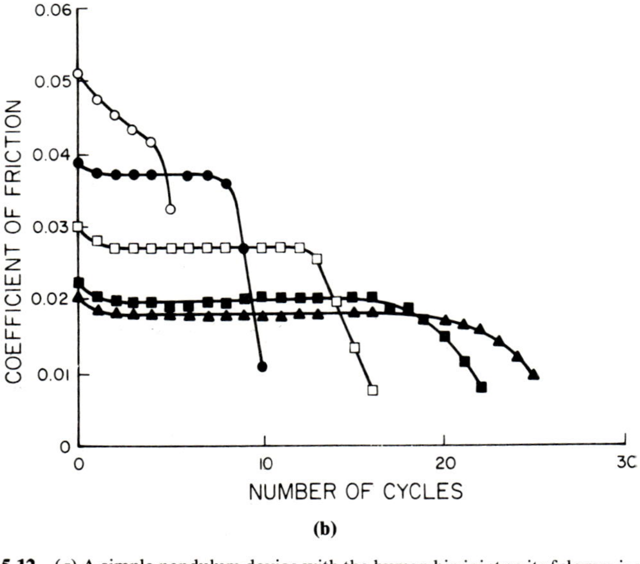

(b)

FIG. 5.12 (*a*) A simple pendulum device with the human hip joint as its fulcrum is used to measure the coefficient of friction between the femoral head (shown) and the acetabulum (not shown) by the decay of the amplitude of the pendulum motion [31]. (*b*) A typical set of curves for the coefficient of friction versus the number of cycles from one hip joint specimen under suddenly loaded condition—unlubricated with synovial fluid—showing a longer period of swing and a lower coefficient of friction with increasing load [31]; ○ = 133.5 N, ● = 213 N, □ = 375 N, ■ = 577 N, and ▲ = 1020 N. The greater efficiency with increasing load is believed to be due to the self-generating, or circulatory flow, mechanism of joint lubrication [55, 113].

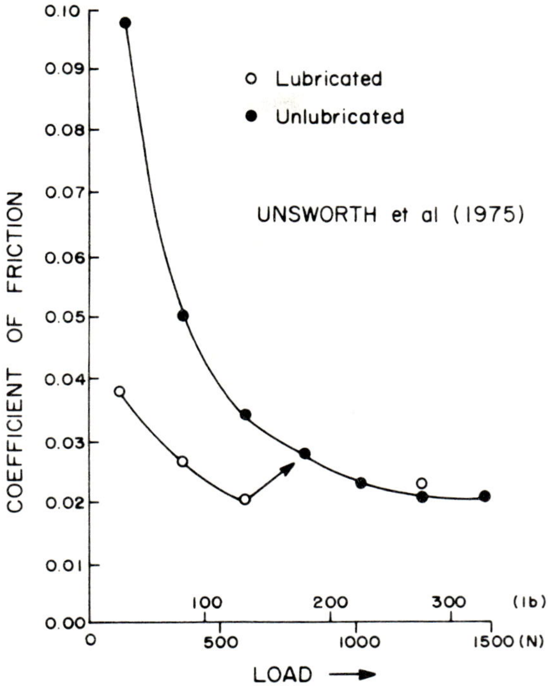

(c)

FIG. 5.12 (*c*) Differences of the coefficient of friction between unlubricated and synovial fluid–lubricated hip joints with varying applied loads. At a load > 90 N, no differences were observed—again indirectly demonstrating the importance of the self-generating mechanism of joint lubrication at physiological loads [31, 113].

the decay of the oscillation amplitude of the pendulum provided enough information to determine the mode of lubrication under the given loading conditions [6, 7, 10] (Fig. 5.12*b*). However, it was soon learned that such measurements alone are not sensitive enough to yield unambiguous distinctions to determine which mechanism of lubrication, fluid film or boundary, is operative [120]. By measuring the frictional torque directly, Unsworth and coworkers [120] obtained results that suggest the existence of a squeeze film when the load is applied suddenly. Frictional resistance was noted to increase over a few cycles of oscillation subsequent to loading. This effect was ascribed to the loss of squeeze films over time. When squeeze films were deliberately prevented by preloading the joint for a period of time before starting the oscillation, fluid-film actions seemed to persist at loads up to approximately body weight, indicating the possibility that some elastohydrodynamic and/or self-generating effects are in operation [120]. For tests where synovial fluids were removed, higher coefficients of friction were measured when the joints were lightly loaded. However, the difference due to the absence of synovial fluid became insignificant when a load greater than body weight was applied (Fig. 5.12*c*). Since normal loading across the hip joint is three to six times that of body weight, it seems that, under the experimental conditions, either the fluid lubricant film is being generated by articular cartilage (Fig. 5.11), or boundary lubrication is occurring at the articulating surfaces.

The second most frequently employed friction-testing configuration is to slide small pieces of cartilage over another surface, which can be either another piece of cartilage or an artificial material such as glass or rubber [19, 34, 53, 121]. Although under these conditions a certain amount of physiological reality is sacrificed, e.g., the elastohydrodynamic mechanism for lubrication would not be operative, this in vitro configuration does offer the means to separate purely surface frictional effects from ploughing effects. In these experiments, the entire surface of an excised "flat" specimen is loaded; therefore, no ploughing can occur. In one such experiment, the very flat conforming bovine shoulder joints were used [34] (Fig. 5.13*a*). Here, Malcom excised circular plugs from the humeral head and tested it against its mating glenoid surface in a continuously rotating articulation. Under these conditions, he found that the "interfacial" coefficient of friction (1) increased with time after application of load, (2) increased with magnitude of load, (3) was lower with synovial fluid as the lubricant than with buffer saline, and (4) was very sensitive to small vertical oscillations imposed on the rotating annular plug of cartilage (Fig. 5.13*b*). These "dynamic loads" dramatically decrease the coefficient of friction between the rotating surfaces [34]. Malcom ascribed such effects to the expression of interstitial fluid caused by the imposed oscillation. This frictional experiment is ingenious in that it permitted, as much as possible, a method to separate the effects of surface friction and bulk friction. It also demonstrated dramatically that when the interstitial fluid is caused to be expressed into the gap by the oscillation, a fresh fluid film is created, and thus frictional resistance is decreased; i.e., the lubrication mechanism changes from boundary to fluid film [34].

The role of synovial fluid in joint lubrication may vary under different conditions depending on the mode of lubrication operating at that moment. For example, in situations where fluid-film lubrication is operating, the shear-dependent viscosity of synovial fluid could be the determining factor in controlling lubricant efficiency. Under other conditions that favor boundary and boosted lubrication, presence of the lubricating glycoprotein (LGP) in the synovial fluid could be the determining factor regulating lubrication efficiency between the cartilaginous surfaces [37, 44].

By passing synovial fluid through filters with specific pore sizes, macromolecules of certain sizes can be separated from the fluid. The effects of these particular components on joint lubrication can then be investigated by using either the residue or the filtrate as the articular lubricant. For filters with 0.22-μm pores, McCutchen (1966) found that the residues lubricated the articular cartilage better than the filtrate [121]. On the other hand, if filters with 0.65-μm pores were used, the opposite occurred; i.e., the filtrate was the better lubricant. This means that a 0.22- to 0.65-μm macromolecule (hydrodynamic size) of the synovial fluid appears to enhance the lubricating property of the fluid. A reasonable guess as to the identity of this moiety is the hyaluronic acid protein complex. If this is true, the lubrication efficiency could be destroyed by treating synovial fluid with the enzyme hyaluronidase, which is known to disaggregate hyaluronic acid, which in turn reduces the fluid viscosity to almost that of saline. Thus, if hyaluronidase-digested fluid were used as the lubricant, one would expect a higher coefficient of friction. However, such effects were not observed experimentally [119, 122–124]. Instead, the hyaluronidase-treated synovial fluid was noted to lubricate as well as the normal synovial fluid; i.e., the coefficient of friction was not increased by hyaluronidase treatment of synovial fluid. On the other hand, when synovial fluid was treated with trypsin, thus degrading the glycoproteins, the corresponding coefficient of friction rose significantly, although the viscosity of such trypsin-treated synovial fluid was roughly the same as that of the untreated synovial fluid.

Results from these experiments imply that an important feature of synovial fluid is its biochemical composition. A protein moiety, LGP, which is adsorbed onto the articular surfaces, appears to be the boundary lubricant responsible for a low coefficient of friction between the contacting articulating surfaces [71]. This adsorbed layer greatly improves the frictional quality of the articular surfaces. Such results also seem to imply that the viscosity of synovial fluid is not significant in the frictional behavior of joints. This is not unreasonable when the dominant mode of lubrication is the boundary type or when the fluid film is too thin to be of any practical significance. These could be the existing in vivo conditions when joints are subjected to high loads for prolonged periods of time.

More recent results obtained by O'Kelly and coworkers (1978), using a static loading pendulum as well as a dynamic simulator which was designed to reproduce physiological loading patterns, have failed to confirm the previous assertion that hyaluronidase-treated synovial fluid was as good a lubricant as the normal synovial fluid [38]. In fact, O'Kelly and coworkers found

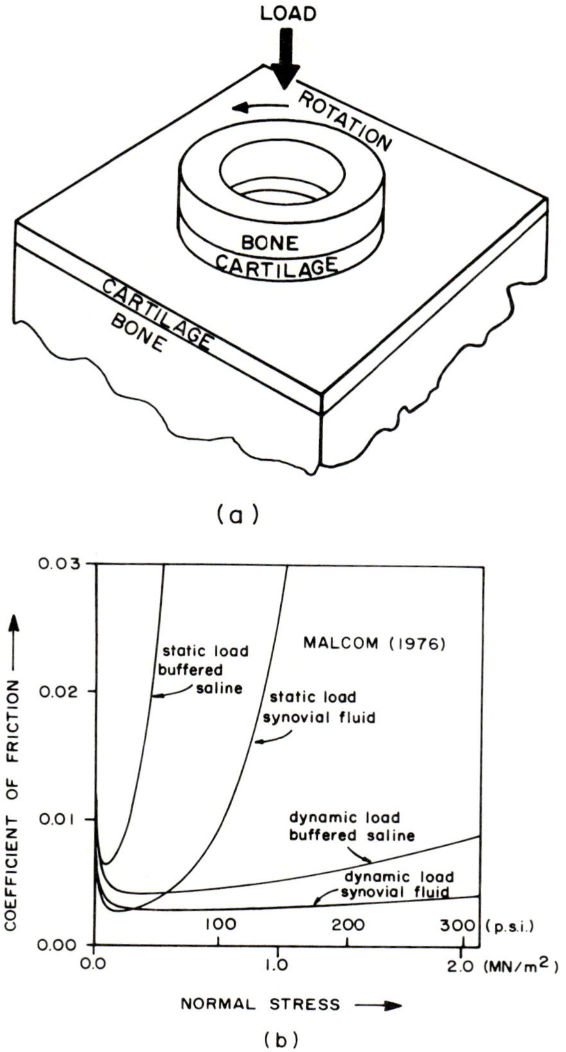

FIG. 5.13 (*a*) Specimens cut from the bovine humeral head and glenoid surface provide conforming surfaces for studies on interfacial friction. This configuration is believed to be effective in preventing ploughing friction from occurring [34, 37]. (*b*) Variation of the coefficient of friction corresponding to the test configuration depicted in (*a*), under various lubrication and loading conditions. Note the dramatic decrease of the coefficient of friction when the load is dynamically applied, demonstrating the effect of interstitial fluid flow in creating a lubricant at the surface.

TABLE 5.1 Coefficient of friction for articular cartilage and synovial joints

Investigator	Coefficient of friction μ	Joint tested
Charnley (1960) [8]	0.005–0.02	Human knee
Barnett and Cobbold (1962) [10]	0.015–0.03	Canine ankle
McCutchen (1962) [53]	0.02–0.35	Porcine shoulder
Linn (1967, 1968) [18, 119]	0.005–0.01	Canine ankle
Clarke et al. (1975) [32]	0.001–0.03	Human hip
Unsworth et al. (1975) [120]	0.01–0.04	Human hip
Malcom (1976) [34]	0.002–0.03	Bovine shoulder

that hyaluronidase digestion did have an effect in both the statically loaded and the dynamically loaded cases. In addition, quite contrary to the results of Swann and coworkers [124], the later investigators could find no indications from their results that suggested that friction rose in human joints when lubricated with synovial fluid that had been digested with trypsin to remove the proteins. These results suggested that the lubrication mechanism is of the mixed type even under static loading of up to 2.5 times body weight. Furthermore, when the loading was changed to the more physiological dynamic pattern, full fluid-film lubrication became effective in reducing joint friction. The discrepancies between the results of O'Kelly and coworkers [38] and those of Swann and coworkers [124] could very well be due to the differences in their experimental conditions, such as the magnitude of the applied loading and the prescribed kinematics, as well as the viscosity range of the lubricants used in their experiments.

Using an apparatus which was designed specially to discourage the hydrodynamic effects and to minimize the ploughing component of friction (thus the contribution from the interstitial fluid as well), Davis and coworkers (1979) studied the detailed mechanisms involved in boundary lubrication for excised disks of bovine nasal cartilage [36]. They hypothesized the existence of a thin, viscous, structured hydration shell at the surface, with the multiple layers of lubricating glycoprotein molecules held together by an alternating sequence of hydrophobic-hydrophobic and hydrophilic-hydrophobic bonds. These layers of lubricating, hydrated glycoproteins are believed to be adequate to separate the opposing surface asperities on the two articulating surfaces. These mutual electrostatic repulsive forces between the charged protein-polysaccharide components could enhance the boundary lubrication process [24].

In summary, we see that frictional measurements show that natural joints and excised pieces of articular cartilage have remarkably low coefficients of friction (see Table 5.1). These low frictional values persist with or without synovial fluid in the gap between the articulating surfaces and with or without the HAP complex. Static loading always results in an increase in the coefficient of friction. Dynamic oscillations or sliding articulations under physiological peak loads always result in a lowering of the coefficient of friction. At present, no comprehensive friction-lubrication theories exist. Frictional measurements and theories are at best empirical in nature.

5.7 WEAR MECHANISMS

Wear is a boundary degradation phenomenon involving loss of material from the surface of a solid medium as a result of mechanical interactions with its neighboring component [81]. There are basically two major categories of wear mechanisms. One, often referred to as *interfacial wear,* involves processes restricted mainly to the surface of the component parts *Fatigue wear,* on the other hand, is generally independent of the characteristics of the operating surfaces but involves the fatigue of the bulk material under cyclic stress and strain conditions. Formation of debris can be caused by coalescence of microscopic flaws within the material, which finally extend to the surface owing to deformational processes under the periodic loading conditions to which the material is subjected.

Interfacial wear can be further classified into several types on the basis of exact wear mechanisms. Adhesive wear occurs, even for very smooth surfaces, when a solid junction forms between the two contacting surfaces and when the junction is stronger then the cohesive strength of the bulk materials themselves. Surfaces will be torn from the weaker side as the bodies move relative to each other [125]. Abrasive wear may dominate if articulating surfaces of very dissimilar hardnesses are allowed to come into close contact. The hard asperities on the surface of the stiffer solid will cut into its softer counterpart and plough out debris as the asperities are being forced through the softer surface. Three-body wear can occur when a hard debris particle is wedged between two bearing surfaces, causing abrasive wear of the two softer bearing surfaces. Surfaces interacting with a moving fluid may also be significantly degraded by hydrodynamic erosion due to high pressures in the lubricant film. When chemical or electrochemical phenomena play important roles in enhancing the wear processes, corrosive wear is often used to describe those events.

Wear is usually measured as a rate in terms of the amount of material removed from the interacting surfaces per unit of time. The parameters which govern the wear rate depend very much on the mechanisms involved. For example, when fatigue processes are the predominant wear mode, factors such as the fracture toughness of the bulk material and the frequency and magnitude of the applied loads will determine the fatigue crack propagation velocity and hence the wear rate. In adhesive and abrasive interfacial wear, the volume of material lost from the surface generally increases with the applied pressures and with the distance over which sliding occurs, but decreases with increasing hardness of the softer component engaged in the articulation process [125].

Unlike ordinary engineering materials, biological tissues are quite capable of replacing the normally small amounts of material lost through an active biological repair process. Thus, under in vivo conditions, a dynamic equilibrium, or homeostasis, exists between the normal mechanical attrition process and the ability of the cartilage to repair itself within the joint. Thus, wear, as observed on the surface of biological tissue, should be regarded as a net result of such a balance. In spite of its relatively rough microtopology, articular cartilage in animal joints is well known for its extremely low wear rate under normal in vivo and in vitro conditions [126, 127] (Fig. 5.14). This indeed is reflective of the superb lubrication designs which safeguard the articulating surfaces from direct contact under a wide spectrum of physiological loading conditions. However, in pathological situations such as osteoarthritis, the surfaces of articular cartilage are often severely torn and fibrillated with splitting and fraying that might extend through the entire tissue thickness. This is frequently accompanied by a significant decrease of the proteoglycan content in the bulk cartilage, an increase in water content, and a disruption of the contiguous collagen fibrous network [48, 59]. Theories about what triggers such degradation events are still very speculative. The phenomena have motivated bioengineers to study the wear mechanisms for articular cartilage under various controlled in vitro conditions.

5.7.1 Measurement of Wear

Radin and coworkers (1971) studied bovine metacarpal-phalangeal joints lubricated by veronate saline under a static load of 1000 lb in an arthrotripsometer, and observed no significant wear even after 5000 h. However, when the static load was reduced to 500 lb, with an additional impact load of 500 lb applied periodically, the joint was noted to wear out in about 200 h [127]. This result agrees with the later work of Repo and Finlay (1977), who demonstrated that surface fractures could occur on the cartilage plugs when the applied impact exceeded a certain critical level [128]. Microscopically, such damage would correspond to ruptures in the collagen fibril network at the surface [59, 60]. These microflaws, once initiated, could propagate by fatigue to cause macroscopic fissures [84]. The possibility of tensile fatigue failure with age was confirmed by Weightman (1976) [61]. Significant tensile stresses can exist in the tissue even for compressively applied loads, particularly if the contact area is small [129] or when the region beneath the contact is softened [130]. The biphasic nature of the tissue tends to accentuate the effects of impact loading, because when a normal load is applied suddenly onto the tissue, a tremendous body drag force is induced upon the fibrillar network at the surface as the interstitial fluid tries to escape from the tissue [55].

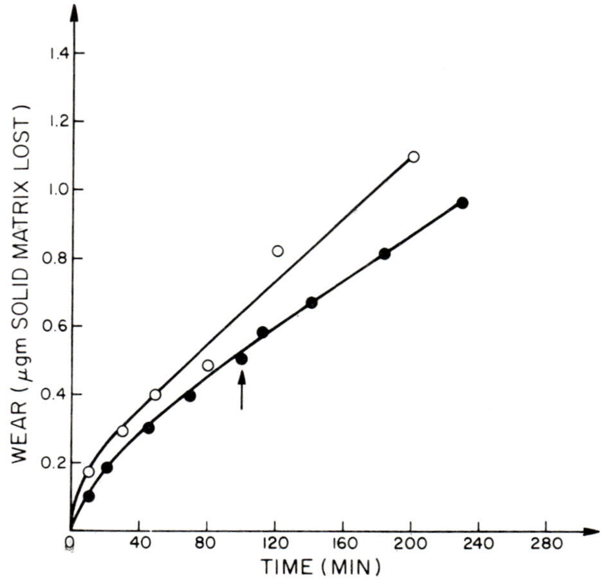

FIG. 5.14 Variation of wear rate as a function of time; each point is an average of between 12 and 15 runs (○ = 4.62 MPa and ● = 1.66 MPa). For the 1.66-MPa curve, the effect of imbibition is seen. At the point indicated by the arrow, the cartilage plugs were allowed to reimbibe fluid overnight. Note that the wear rate continues without the high wear rate of the initial period [126].

Simon (1971) reported the effects of various lubricants on the amount of wear when a steel rotary file was used to abrade cartilage [131]. Compared with saline, synovial fluid permits less wear in the tissue. Wear was significantly increased when the synovial fluid was digested with trypsin, but no significant change was observed if the synovial fluid was treated with hyaluronidase instead. This apparently agrees with the boundary lubrication concept proposed by Swann and coworkers [124]. It was also reported that trypsin-digested cartilage tended to wear out more than untreated tissues [131]. This did not come as a surprise either, since trypsin is known for its ability to denature, and thus rupture, the collagen fibril network and to digest proteoglycans. These simultaneous occurrences permit the otherwise trapped proteoglycan macromolecules to be lost from the tissue and thus increase wear.

Because of the extremely low wear rate under normal conditions, quantitative measurement of wear for articular cartilage has always been difficult. Lipshitz and Glimcher [126] used a standard colorimetric chemical method to carefully monitor the amount of wear in their experiments by quantifying (1) the changes in dimension and profile of the surface of the tissue after it had been worn, under various conditions, against a polished stainless steel plate and (2) the fraction of the hexosamine and hydroxyproline contents of the debris that dissolved during the wear experiment into the lubricant. Figure 5.14 shows some typical wear rate curves from these experiments. Further, they found that as much as 50 percent of the proteoglycans and 10 percent of the collagen worn from the surface existed in solution form in the lubricant. From these unique experiments, it was shown that (1) the wear rate apparently decreased with time until it attained a constant value; this steady wear rate depended on the characteristics of the opposing stainless steel surface and the applied pressure; (2) the wear rate increased with the applied normal pressure and the amplitude of the speed of relative motion; (3) the wear rate decreased by a factor of 5 if the tissue was fixed and stiffened by formaldehyde to retard the abrasive component of wear; and (4) synovial fluid could reduce wear by as much as 10 times

[126]. These investigators ascribed the initial decreasing wear rate over time to network breakdown and reorientation of fibrous macromolecules at the articular surface toward the wear direction and suggested that the subsequent constant wear rate was governed by the rate of crack initiation in the tissue.

Very little is known about the rate of wear in vivo. Even less is known about the mechanisms of wear for articular cartilage in vivo or in vitro. Much more research is needed in this area. However, it is known that collagen fibrils can fail in tension, proteoglycan aggregates can disaggregate and migrate from the tissue, and cartilage cells can respond, to some degree, to the natural attrition processes. How these changes in tissue composition and molecular architecture affect the material properties of the tissue and how these changes in material properties affect the function of the tissue remain unanswered. What has been demonstrated in vitro is that cartilage wear depends on the level of loading and the speed with which the articulating surfaces rub over each other. Theories on the wear of biological material, just as those on the wear of machine parts [81], remain in the realm of empirical science.

SUMMARY

Currently, there is no single, comprehensive theory that encompasses all the friction and lubrication phenomena existing within a joint in vivo. Probably many of the lubrication mechanisms discussed in this chapter can and do operate in vivo, depending on the particular nature of the motion, the load that the joint is subjected to, and the material properties constituting these biological bearings.

During high-speed, lightly loaded situations, such as the swing phase of walking or running, hydrodynamic and elastohydrodynamic lubrication modes appear to be reasonable. The resultant fluid film can be sustained for a brief period of time after the loading mode is suddenly changed and the squeeze-film mode is produced. The squeeze film in turn may be reinforced by boosting lubrication and the deformation of the cartilage surface. These latter conditions correspond to those that would occur during jumping and at heel strike or toe-off during walking. With increasing load and decreasing operating speed, the elastohydrodynamic effects may not be able to furnish a thick enough fluid film between the articular surfaces for efficient lubrication. At this point, the self-generating mechanism could be the dominant mechanism in developing the fluid-film lubricant. Under a fixed stationary load, as in standing for long periods of time, the fluid film could become so thin that the articulating surfaces would touch each other. Here, the adsorbed monolayer of LGP provides the boundary lubricant necessary to maintain a low coefficient of friction. This mechanism is enhanced by the ultrafiltrated synovial fluid gel remaining between the articular surfaces. Hence, the tribological efficiency of synovial joints is most probably the result of a series of complicated dynamic interactions occurring among the synovial fluid, its macromolecular components, the organic solid matrix of cartilage, and the interstitial water during joint function. Thus it is important to have accurate rheological models for both synovial fluid and articular cartilage.

Under normal conditions, such a comprehensive design of lubrication mechanisms can keep the amount of wear to a minimal, biologically acceptable level; any loss of material could be quickly replenished by the chondrocytes in the tissue. However, an imbalance could be produced under many pathological conditions. For example, absence of lubricating glycoproteins in the synovial fluid could greatly increase interfacial friction and accelerate the wear processes. Damage to the most highly hydrated superficial tangential zone of cartilage by the enhanced adhesive and abrasive mechanisms would render the tissue more vulnerable to excessive deformation, causing the fibrillar network to rupture and an accelerated loss of proteoglycans from the tissues to occur. Correspondingly, the aggregate modulus of cartilage would be significantly decreased, while its permeability would rise to an abnormal level. The increased fluid fluxes and the decreased modulus only encourage more surface and interior damage to the tissue and finally render the affected joint nonoperational as a bearing. Abnormal mechanical loading conditions could also initiate compositional changes and cause damage to the tissues. These events, if unchecked by the biological repair mechanisms of the tissue, could lead to a cyclic and progressive degradation process ultimately resulting in degenerate joint diseases and artificial joint replacements.

ACKNOWLEDGMENTS

This chapter is based on research sponsored by National Institute of Arthritis, Diabetes, Digestive and Kidney Diseases grants AM 26440 and AM 19094 and National Science Foundation grant MSM 82-11968. Any opinions, findings, and conclusions or recommendations expressed in this publication are those of the authors and do not necessarily reflect the views of the National Science Foundation. We thank Ms. C. M. Tanski for typing and Ms. R. A. Boshoff for editing this manuscript.

REFERENCES

1. O'Rahilly, R., and Gardner, E.: The embryology of movable joints, in L. Sokoloff (ed.), *The Joints and Synovial Fluid,* Academic Press, New York, 1978, vol. I, pp. 49–103.
2. Warwick, R., and Williams, P. L. (eds.): *Gray's Anatomy,* Saunders, Philadelphia, 1973.
3. Paul, J. P.: Joint kinetics, in L. Sokoloff (ed.), *The Joints and Synovial Fluid,* Academic Press, New York, 1980, vol. II, pp. 146–176.
4. Crowninshield, R. D., Johnston, R. C., Andrews, J. G., and Brand, R. A.: A biomechanical investigation of the human hip, *J. Biomech., 11,* 75–85, 1978.
5. MacConaill, M. A.: The function of intra-articular fibrocartilages, with special references to the knee and inferior radio-ulnar joints, *J. Anat., 66,* 210–227, 1932.
6. Jones, E. S.: Joint lubrication, *Lancet, 230,* 1043–1044, 1936.
7. Charnley, J.: The lubrication of animal joints, in *Symposium on Biomechanics,* Institution of Mechanical Engineers, London, 1959, pp. 12–22.
8. Charnley, J.: How our joints are lubricated, *Triangle, 4,* 175, 1960.
9. Lewis, P. R., and McCutchen, C. W.: Experimental evidence for weeping lubrication in mammalian joints, *Nature (London), 184,* 1284–1285, 1959.
10. Barnett, C. H., and Cobbold, A. F.: Lubrication within living joints, *J. Bone Jt. Surg., 44-B,* 662–674, 1962.
11. Dintenfass, L.: Lubrication in synovial joints, *Nature (London), 197,* 496–497, 1963.
12. Dintenfass, L.: Lubrication in synovial joints: A theoretical analysis, *J. Bone Jt. Surg., 45A,* 1241–1256, 1963.
13. Davies, D. V.: Synovial fluid as a lubricant, *Fed. Proc., 25,* 1069–1076, 1966.
14. Tanner, R. I.: An alternative mechanism for the lubrication of synovial joints, *Phys. Med. Biol., 11,* 119–127, 1966.
15. Fein, R. S.: Are synovial joints squeeze film lubricated?,'' *Proc. Inst. Mech. Eng., London, 181,* 125–128, 1967.
16. Maroudas, A.: Hyaluronic acid films, *Proc. Inst. Mech. Eng., London, 181,* 122–124, 1967.
17. Dowson, D.: Modes of lubrication in human joints, *Proc. Inst. Mech. Eng., London, 181,* 45–54, 1967.
18. Linn, F. C.: Lubrication of animal joints: I. The arthrotripsometer, *J. Bone Jt. Surg., 49-A,* 1079–1098, 1967.
19. Walker, P. S., Dowson, D., Longfield, M. D., and Wright, V.: Boosted lubrication in synovial joints by fluid entrapment and enrichment, *Ann. Rheum. Dis., 27,* 512–520, 1968.
20. Caygill, J. C., and West, G. H.: The rheological behavior of synovial fluid and its possible relation to joint lubrication, *Med. Biol. Eng., 7,* 507–516, 1969.
21. Little, T., Freeman, M. A. R., and Swanson, S. A. V.: Experiments on friction in the human hip joint, in V. Wright (ed.), *Lubrication and Wear in Joints,* Sector, London, 1969, pp. 110–114.
22. Mow, V. C.: The role of lubrication in biomechanical joints, *J. Lubr. Technol., 91,* 320–329, 1969.
23. Freeman, M. A. R., Little, T. D., and Swanson, S. A. V.: Lubrication of synovial joints: Possible significance of fat, *Proc. R. Soc. Med., 63,* 579–581, 1970.
24. Roberts, A. D.: The role of electrical repulsive forces in synovial fluid, *Nature (London), 231,* 434–436, 1971.

25. Radin, E. L., and Paul, I. L.: A consolidated concept of joint lubrication, *J. Bone Jt. Surg., 54-A*, 607–613, 1972.

26. Swann, D. A., and Radin, E. L.: The molecular basis of articular lubrication: I. Purification and properties of a lubricating fraction from bovine synovial fluid, *J. Biol. Chem., 247*, 8069–8073, 1972.

27. Ling, F. F.: A new model of articular cartilage in human joints, *J. Lubr. Technol., 96*, 449–454, 1974.

28. Higginson, G. R., and Norman, R.: The lubrication of porous elastic solids with reference to the functioning of human joints, *J. Mech. Eng. Sci., 16*, 250–257, 1974.

29. Cooke, A.: A study of the rheological characteristics of some fluids and tissues in synovial joint lubrication, Ph.D. thesis, Leeds University, 1974.

30. Gvozdanovic, D., Wright, V., and Dowson, D.: Formation of lubricating monolayers at the cartilage surface, *Ann. Rheum. Dis., 34*, suppl. 2, 100–101, 1975.

31. Unsworth, A., Dowson, D., and Wright, V.: The frictional behavior of human synovial joints: I. Natural joints, *J. Lubr. Technol., 97*, 369–376, 1975.

32. Clarke, I. C., Contini, R., and Kenedi, R. M.: Friction and wear studies of articular cartilage: A scanning electron microscope study, *J. Lubr. Technol., 97*, 358–368, 1975.

33. Lipshitz, H., Etheredge, R., and Glimcher, M. J.: *In vitro* wear of articular cartilage: I. Hydroxyproline, hexosamine, and amino acid composition of bovine articular cartilage as a function of depth from the surface; hydroxyproline content of the lubricant and the wear debris as a measure of wear, *J. Bone Jt. Surg., 57-A*, 527–537, 1975.

34. Malcom, L. L.: An experimental investigation of the frictional and deformational responses of articular cartilage interfaces to static and dynamic loading, Ph.D. thesis, University of California, San Diego, 1976.

35. Higginson, G. R.: Squeeze films between compliant solids, *Wear, 46*, 387–395, 1978.

36. Davis, W. H., Jr., Lee, S. L., and Sokoloff, L.: A proposed model of boundary lubrication by synovial fluid: Structuring of boundary water, *J. Biomech. Eng., 101*, 185–192, 1979.

37. Armstrong, C. G., and Mow, V. C.: Lubrication and wear of synovial joints, in R. Owen, J. Goodfellow, and P. Bullough (eds.), *Scientific Foundations: Orthopaedics and Traumatology*, William Heinemann Medical Books, London, 1980, pp. 223–232.

38. O'Kelly, J., Unsworth, A., Dowson, D., Hall, D. A., and Wright, V.: A study of the role of synovial fluid and its constituents in the friction and lubrication of human hip joints, *Eng. Med., 7*, 72–83, 1978.

39. Lai, W. M., Kuei, S. C., and Mow, V. C.: Rheological equations for synovial fluids, *J. Biomech., 100*, 169–186, 1978.

40. Mow, V. C., and Lai, W. M.: Mechanics of animal joints, *Ann. Rev. Fluid. Mech., 11*, 247–288, 1979.

41. Mow, V. C., Kuei, S. C., Lai, W. M., and Armstrong, C. G.: Biphasic creep and stress relaxation of articular cartilage in compression: Theory and experiments, *J. Biomech. Eng., 102*, 73–84, 1980.

42. Mow, V. C., Lai, W. M., and Holmes, M. H.: Advanced theoretical and experimental techniques in cartilage research, in R. Huiskes, D. Van Campen, and J. DeWijn (eds.), *Biomechanics: Principles and Applications*, Martinus Nijhoff, The Hague, 1982, pp. 147–174.

43. Higginson, G. R., and Unsworth, T.: The lubrication of natural joints, in J. H. Dumbleton (ed.), *Tribology Series*, vol. 3: *Tribology of Natural and Artificial Joints*, Elsevier, New York, 1981, pp. 47–73.

44. Dowson, D., Unsworth, A., Cooke, A. F., and Gvozdanovic, D.: Lubrication of joints, in D. Dowson and V. Wright (eds.), *An Introduction to the Biomechanics of Joints and Joint Replacement*, Mechanical Engineering, London, 1981, pp. 120–145.

45. Simon, W. H., Friedenberg, S., and Richardson, S.: Joint congruence: A correlation of joint congruence and thickness of articular cartilage in dogs, *J. Bone Jt. Surg., 55-A*, 1614–1620, 1973.

46. Muir, I. H. M.: The chemistry of the ground substance of joint cartilage, in L. Sokoloff (ed.), *The Joints and Synovial Fluid*, Academic Press, New York, 1980, vol. II, pp. 27–94.

47. Venn, M. F., and Maroudas, A.: Chemical composition and swelling of normal and osteoarthritic femoral head cartilage: I. Chemical composition, *Ann. Rheum. Dis., 36*, 121–129, 1977.

48. Armstrong, C. G., and Mow, V. C.: Variations in the intrinsic mechanical properties of human articular cartilage with age, degeneration and water content, *J. Bone Jt. Surg., 64-A*, 88–94, 1982.

49. Kempson, G. E.: Mechanical properties of articular cartilage, in M. A. R. Freeman (ed.), *Adult Articular Cartilage* (2d ed.), Pitman, London, 1979, pp. 333–414.

50. Roth, V., Mow, V. C., Lai, W. M., and Eyre, D. R.: Correlation of intrinsic compressive properties of bovine articular cartilage with its uronic acid and water content, *Trans. Orthop. Res. Soc.*, *6*, 49, 1981.

51. Obrink, B.: A study of the interactions between monomeric tropocollagen and glycosaminoglycans, *Eur. J. Biochem.*, *33*, 387–400, 1973.

52. Torzilli, P. A., Rose, D. E., and Dethmers, D. A.: Equilibrium water partition in articular cartilage, *Biorheology*, *19*, 519–537, 1982.

53. McCutchen, C. W.: The frictional properties of animal joints, *Wear*, *5*, 1–17, 1962.

54. Edwards, J.: Physical characteristics of articular cartilage, *Proc. Inst. Mech. Eng., London*, *181*, 16–24, 1967.

55. Mow, V. C., and Lai, W. M.: Recent developments in synovial joint biomechanics, *SIAM Rev.*, *22*, 275–317, 1980.

56. Kempson, G. E., Freeman, M. A. R., and Swanson, S. A. V.: Tensile properties of articular cartilage, *Nature (London)*, *220*, 1127–1128, 1968.

57. Kempson, G. E., Freeman, M. A. R., and Swanson, S. A. V.: The determination of a creep modulus for articular cartilage from indentation tests on the human femoral head, *J. Biomech.*, *4*, 239–250, 1971.

58. Wirth, C. R., Augello, F. A., Mow, V. C., and Roth, V.: Variation of tensile properties of human patella cartilage with age and histological indices, *Trans. Orthop. Res. Soc.*, *5*, 38, 1980.

59. Maroudas, A., and Venn, M. F.: Chemical composition and swelling of normal and osteoarthrotic femoral head cartilage: II. Swelling, *Ann. Rheum. Dis.*, *36*, 399–406, 1977.

60. Maroudas, A.: Swelling pressure versus collagen tension in normal and degenerate articular cartilage, *Nature (London)*, *260*, 808–809, 1976.

61. Weightman, B. O.: Tensile fatigue of human articular cartilage, *J. Biomech.*, *9*, 193–200, 1976.

62. Armstrong, C. G., Bahrani, A. S., and Gardner, D. L.: *In vitro* measurement of articular cartilage deformations in the intact human hip joint under load, *J. Bone Jt. Surg.*, *61-A*, 744–755, 1979.

63. Hoch, D. H., Grodzinsky, A. J., Koob, T. J., Eyre, D. R., and Albert, M. L.: Early changes in material properties in rabbit articular cartilage after meniscectomy, *Trans. Orthop. Res. Soc.*, *7*, 152, 1982.

64. McDevitt, C. A., and Muir, H.: Biochemical changes in the articular cartilage of the knee in experimental and natural osteoarthrosis in the dog, *J. Bone Jt. Surg.*, *58-B*, 94–101, 1976.

65. Armstrong, C. G., and Mow, V. C.: Biomechanics of normal and osteoarthrotic articular cartilage, in P. D. Wilson, Jr., and L. R. Straub (eds.), *Clinical Trends in Orthopaedics*, Thieme-Stratton, New York, 1982, pp. 189–197.

66. Ogston, A. G., and Stanier, J. E.: Further observations on the preparation and composition of hyaluronic acid complex of ox synovial fluid, *Biochem. J.*, *52*, 149–156, 1952.

67. Sundblad, L.: Studies of hyaluronic acid in synovial fluids, *Acta Soc. Med. Upsal.*, *58*, 113–238, 1953.

68. Sundblad, L.: Glycosaminoglycans and glycoproteins in synovial fluid, in E. A. Balazs and R. W. Jeanloz (eds.), *The Amino Sugars*, Academic Press, New York, 1965, vol. 2A, pp. 229–250.

69. Balazs, E. A., Watson, D., Duff, I. F., and Roseman, S.: Hyaluronic acid in synovial fluid: I. Molecular parameters of hyaluronic acid in normal and arthritic human fluids, *Arthritis Rheum.*, *10*, 357–376, 1967.

70. Sandson, J.: Human synovial fluid: Detection of a new component, *Science*, *155*, 839–841, 1967.

71. Swann, D. A.: Macromolecules of synovial fluid, in L. Sokoloff (ed.), *The Joints and Synovial Fluid*, Academic Press, New York, 1978, vol. I, pp. 407–435.

72. Laurent, T. C., and Ogston, A. G.: The interaction between polysaccharides and other macromolecules, *Biochem. J.*, *89*, 249–253, 1963.

73. Balazs, E. A., and Gibbs, D. A.: The rheological properties and biological function of hyaluronic acid, in E. A. Balazs (ed.), *Chemistry and Molecular Biology of the Intercellular Matrix*, Academic Press, New York, 1970, vol. 3, pp. 1241–1253.

74. Scott Blair, G. W., Williams, P. O., Fletcher, E. T. D., and Markham, R. L.: On the flow of certain pathological human synovial effusions through narrow tubes, *Biochem. J.*, *56*, 504–508, 1954.

75. Bloch, B., and Dintenfass, L.: Rheological study of human synovial fluid, *Aust. N. Z. J. Surg.*, *33*, 108–113, 1963.

76. Davies, D. V., and Palfrey, A. J.: Some of the physical properties of normal and pathological synovial fluids, *J. Biomech.*, *1*, 79–88, 1968.

77. Myers, R. R., Negami, S., and White, R. K.: Dynamic mechanical properties of synovial fluid, *Biorheology*, *3*, 197–209, 1966.

78. Gibbs, D. A., Merrill, E. W., and Smith, K. A.: Rheology of hyaluronic acid, *Biopolymers, 6,* 777–791, 1968.

79. Palfrey, A. J., and White, J. B.: The viscosity of synovial fluid during oscillatory movement, *Biorheology, 5,* 189–198, 1968.

80. Ferguson, J., Boyle, J. A., McSween, R. N. M., and Jasani, M. K.: Observations on the flow properties of the synovial fluid from patients with rheumatoid arthritis, *Biorheology, 5,* 119–131, 1968.

81. Bowden, F. P., and Tabor, D.: *Friction and Lubrication of Solids,* 2d ed., Oxford University Press, London, 1954.

82. Wismans, J., Veldpaus, F., Janssen, J., Huson, A., and Struben, P.: A three-dimensional mathematical model of the knee-joint, *J. Biomech., 13,* 677–685, 1980.

83. Longmore, R. B., and Gardner, M. J.: Development with age of human articular cartilage surface structure, *Ann. Rheum. Dis., 34,* 26–37, 1975.

84. Redler, I., and Mow, V. C.: Biomechanical theories of ultrastructural alterations of articular surfaces of the femoral head, in W. H. Harris (ed.), *The Hip,* Mosby, St. Louis, 1974, pp. 23–59.

85. Greenwald, A. S.: Joint congruence—a dynamic concept, in W. H. Harris (ed.), *The Hip,* Mosby, St. Louis, 1974, pp. 3–22.

86. Rushfeldt, P. D., Mann, R. W., and Harris, W. H.: Improved techniques for measuring in vitro the geometry and pressure distribution in the human acetabulum: I. Ultrasonic measurement of acetabular surfaces, sphericity and cartilage thickness, *J. Biomech., 14,* 253–260, 1981.

87. Gardner, D. L., and McGillivray, D. C.: Living articular cartilage is not smooth, *Ann. Rheum. Dis., 30,* 3–9, 1971.

88. Gardner, D. L.: The influence of microscopic technology on knowledge of cartilage surface structure, *Ann. Rheum. Dis., 31,* 235–258, 1972.

89. Ghadially, F. N., Moshurchak, E. M., and Thomas, I.: Humps on young human and rabbit articular cartilage, *J. Anat., 124,* 425–435, 1977.

90. Clarke, I. C.: The microevaluation of articular surface contours, *Ann. Biomed. Eng., 1,* 31–43, 1972.

91. Dowson, D., Longfield, M. D., Walker, P. S., and Wright, V.: An investigation of the friction and lubrication in human joints, *Proc. Inst. Mech. Eng., London, 182,* 68–76, 1968.

92. Eberhart, H. D., Inman, V. T., and Saunders, J. B. de C.: *Fundamental Studies of Human Locomotion and Other Information Relating to Design of Artificial Limbs,* University of California Press, Berkeley, 1947.

93. Frankel, V. H., and Burstein, A. H.: *Orthopaedic Biomechanics,* Lea & Febiger, Philadelphia, 1971.

94. Walker, P. S., and Hajek, J. V.: The load bearing areas in the knee joint, *J. Biomech., 5,* 581–589, 1972.

95. Seireg, A., and Arvikar, R.J.: The prediction of muscular load sharing and joint forces in the lower extremities during walking, *J. Biomech., 8,* 89–102, 1975.

96. Morrey, B. F., and Chao, E. Y.: Determination of three dimensional joint motion, *Trans. Orthop. Res. Soc., 1,* 74, 1976.

97. Walker, P. S.: *Human Joints and Their Artificial Replacements,* Thomas, Springfield, Ill., 1977.

98. Simon, S. R.: Biomechanics of joints, in W. N. Kelly, E. D. Harris, S. Ruddy, and C. B. Sledge (eds.), *Textbook of Rheumatology,* Saunders, Philadelphia, 1981, pp. 294–316.

99. Greenwald, A. S., and Haynes, D. W.: Weight bearing areas in human hip joint, *J. Bone Jt. Surg., 54-B,* 157–163, 1972.

100. Seedhom, B. B.: Transmission of the load in the knee joint with special reference to the role of the menisci: I. Anatomy, analysis and apparatus, *Eng. Med., 8,* 207–219, 1979.

101. Seedhom, B. B., and Hargreaves, D. J.: Transmission of the load in the knee joint with special reference to the role of the menisci: II. Experimental results, discussion and conclusions, *Eng. Med., 8,* 220–228, 1979.

102. Ling, F. F., Klaus, E. E., and Fein, R. S. (eds.): *Boundary Lubrication—An Appraisal of World Literature,* American Society of Mechanical Engineers, New York, 1969.

103. Martin, H. M.: Lubrication of gear teeth, *Engineering, 102,* 199, 1916.

104. Kapitza, P. L.: Hydrodynamic theory of lubrication during rolling, *Zh. Tekh. Fiz., 25,* 747–762, 1955.

105. Hamrock, B. J., and Dowson, D.: Elastohydrodynamic lubrication of elliptical contacts for materials of low elastic modulus: I. Fully flooded conjunction, *J. Lubr. Technol., 100,* 236–245, 1978.

106. Hooke, C. J., and O'Donoghue, J. P.: Elasticohydrodynamic lubrication of soft, highly deformed contacts, *J. Mech. Eng. Sci., 14,* 34–48, 1972.

107. Lewis, P. R., and McCutchen, C. W.: Lubrication of mammalian joints, *Nature (London), 185,* 920–921, 1960.

108. Maroudas, A.: Biophysical chemistry of cartilaginous tissues with special reference to solute and fluid transport, *Biorheology, 12,* 233–248, 1975.

109. Mansour, J. M., and Mow, V. C.: The permeability of articular cartilage under compressive strain and at high pressures, *J. Bone Jt. Surg., 58-A,* 509–516, 1976.

110. Grodzinsky, A. J., Roth, V., Myers, E. R., Grossman, W. D., and Mow, V. C.: The significance of electromechanical and osmotic forces in the nonequilibrium swelling behavior of articular cartilage in tension, *J. Biomech. Eng., 103,* 221–231, 1981.

111. Seller, P. C., Dowson, D., and Wright, V.: The rheology of synovial fluid, *Rheol. Acta, 10,* 2–7, 1971.

112. Lai, W. M., and Mow, V. C.: Ultrafiltration of synovial fluid by cartilage, *J. Eng. Mech. Div. Trans. ASCE, 104,* 79–96, 1978.

113. Mansour, J. M., and Mow, V. C.: On the natural lubrication of synovial joints: Normal and degenerate, *J. Lubr. Technol., 99,* 163–173, 1977.

114. Lai, W. M., and Mow, V. C.: Flow fields in a single layer model of articular cartilage created by a sliding load, in M. K. Wells (ed.), *1979 Advances in Bioengineering,* American Society of Mechanical Engineers, New York, 1979, pp. 101–104.

115. Maroudas, A., Bullough, P., Swanson, S. A. V., and Freeman, M. A. R.: The permeability of articular cartilage, *J. Bone Jt. Surg., 50-B,* 166–177, 1968.

116. Hori, R. Y., and Mockros, L. F.: Indentation tests for human articular cartilage, *J. Biomech., 9,* 259–268, 1976.

117. Hertz, H.: *Miscellaneous Papers,* D. E. Jones and G. A. Schott (trans.), Macmillan, London, 1896, pp. 146–183.

118. Mow, V. C., and Lai, W. M.: The optical sliding contact analytical rheometer (OSCAR) for flow visualization at the articular surface, in M. K. Wells (ed.), *1979 Advances in Bioengineering,* American Society of Mechanical Engineers, New York, 1979, pp. 97–99.

119. Linn, F. C.: Lubrication of animal joints: II. The mechanism, *J. Biomech., 1,* 193–205, 1968.

120. Unsworth, A., Dowson, D., and Wright, V.: Some new evidence on human joint lubrication, *Ann. Rheum. Dis., 34,* 277–285, 1975.

121. McCutchen, C. W.: Boundary lubrication by synovial fluid: Demonstration and possible osmotic explanation., *Fed. Proc., 25,* 1061–1068, 1966.

122. Wilkins, J. F.: Proteolytic destruction of synovial boundary lubrication, *Nature, (London), 219,* 1050–1051, 1968.

123. Radin, E. L., Swann, D. A., and Weisser, P. A.: Separation of a hyaluronate-free lubricating fraction from synovial fluid, *Nature, (London), 228,* 377–378, 1970.

124. Swann, D. A., Radin, E. L., Nazimiec, M., Weisser, P. A., Curran, N., and Lewinnek, G.: Role of hyaluronic acid in joint lubrication, *Ann. Rheum. Dis., 33,* 318–326, 1974.

125. Dowson, D.: Basic tribology, in D. Dowson and V. Wright (eds.), *Introduction to the Biomechanics of Joints and Joint Mechanics,* Mechanical Engineering, London, 1981, pp. 49–60.

126. Lipshitz, H., and Glimcher, M. J.: *In vitro* studies of the wear of articular cartilage: II. Characteristics of the wear of articular cartilage when worn against stainless steel plates having characterized surfaces, *Wear, 52,* 297–339, 1979.

127. Radin, E. L., and Paul, I. L.: Response of joints to impact loading: I. In vitro wear. *Arthritis Rheum., 14,* 356–362, 1971.

128. Repo, R. U., and Finlay, J. B.: Survival of articular cartilage after controlled impact. *J. Bone Jt. Surg., 59-A,* 1068–1076, 1977.

129. Askew, M. J., and Mow, V. C.: The biomechanical function of the collagen fibril ultrastructure of articular cartilage, *J. Biomech. Eng., 100,* 105–115, 1978.

130. Roth, V., and Mow, V. C.: Finite element analysis of contact problems for indentation of articular cartilage, in E. Grood and C. Smith, (eds.), *Advances in Bioengineering,* American Society of Mechanical Engineers, 1977, pp. 47–48.

131. Simon, W. H.: Wear properties of articular cartilage *in vitro, J. Biomech., 4,* 379–389, 1971.

132. Wright, V. (ed.): *Lubrication and Wear in Joints,* Lippincott, Philadelphia, 1969.

CHAPTER 6

Properties of Tendons and Ligaments

Andrus Viidik
Department of Connective Tissue Biology
Institute of Anatomy, University of Aarhus
DK-8000 Aarhus C, Denmark

6.1 FUNCTION OF TENDONS AND LIGAMENTS

The *physiological function* of muscle tendons is to transmit forces with precision from muscles to bones in order to execute joint motion, while that of joint ligaments is to provide stability to joints and limit their range of motion. Tendons and ligaments thus never function alone but are parts of functional units, i.e., bone-tendon-muscle-tendon-bone and bone-ligament-bone, respectively [1, 2], the functional requirement being to resist tensile forces. The structural component in them responsible for their tensile strength is a fibrous protein, collagen (see Sec. 6.6). This protein, which has an ultimate tensile strength in the range of 45 to 125 MPa (the corresponding strain value being 0.08 to 0.35) [3], also provides "skeletons" for most soft tissues and organs in the body, e.g. skin, organ capsules, and trabeculae, along with another fibrous protein, elastin, for some of these tissues, e.g., blood vessels, especially arteries, and the ligamentum nuchae. The study of tendon and ligament tissue is thus also of interest for the elucidation of the biomechanical properties of other soft tissues, since collagen is seldom found in such relatively pure and parallel-fibered form in these other tissues.

The *physiological working range* of tendons and ligaments is not firmly established. No information is available regarding ligaments, and that on tendons is based mostly on extrapolations. There is a strong correlation between the size of tendons and the force the corresponding muscles produce when stimulated to maximum tetanic contraction, with certain differences between fusiform and penniform muscles [4]. The ultimate tensile strength of a tendon is thus estimated to be about four times the maximum isometric tension its muscle can produce [5, 6]. The information on the mechanical behavior of tendons under strict in vivo conditions is scanty and possibly disturbed by the implantation of transducers. The load rate on a horse muscle tendon is reported to be up to 800 N/s during walking and may be doubled during trotting [7, 8]. Comparative in vivo and in vitro experiments [9] have recorded strain values for horse flexor tendons during walking in excess of 0.05, which suggests that the clearly viscoelastic part of the stress-strain behavior is reached under physiological conditions. The maximum strain in a sheep muscle tendon is, on the other hand, reported to be only 0.026 (strain rate 0.44 s^{-1}) [10]. Higher values should, however, be reached during more strenuous activities. Cats, when jumping vertically to a height of 120 cm, develop stresses 10 to 15 percent in excess of the maximum isometric force developed by the same muscles during tetanic contraction [11]. The assumption [4–6] that the physiological working range of a tendon is restricted to one-fourth its ultimate tensile strength might thus not hold true for every normal situation. It seems, however, reasonable to postulate that one-half this strength is not exceeded, which means that the safety factor is at least 2 (see below).

Considerably higher stress levels have been derived from indirect calculations based on anatomical data (weights, distances between joint axes, and centers of gravity for body parts) combined with kinesiological data (joint angles and angular accelerations) and at times force plate recordings. Alexander [12] calculated that the ankle extensor tendons of dogs in the act of jumping over various obstacles were subjected to stresses of the same magnitude as yield stress (84 MPa). He also reported similar data for hopping kangaroos (40 to 80 MPa) and galloping buffalos (14 to 48 MPa). These figures result in rather low safety factors, especially for the dog (1.0) and possibly the kangaroo (1.1 to 2.1). Such low figures suggest that the yield stress was underestimated or that some error was introduced into the assumptions for these calculations. Barfred [13], on the other hand, calculated the maximum stress on the human calcaneal tendon to be about 2000 N during ordinary running push-offs and 4340 N during a sudden alteration of running direction causing an accidental partial rupture. As the ultimate tensile strength of this tendon is, under laboratory in vitro "static" testing conditions, about 4000 N [14], this tendon appears to have a safety factor of about 2.0 under physiological conditions. This assumption agrees with the values recorded for the above-mentioned tendon rupture. Other investigators have, however, reached considerably higher figures in their calculations. Values up to 10,500 N on the two calcaneal tendons, i.e. more than 5,250 N on one of them in case of an unintentional asymmetrical loading, have been reported for the takeoff to the second half of a double somersault [15]. "Dynamic" loading locks to some extent the viscous elements in the material properties of the tendon and increases thereby the ultimate strength. Increase of the strain rate from 0.036 to 7.2 s^{-1} thus increases the ultimate tensile strength from 57 to 107 MPa and the energy absorption by a factor of about 3 [16]. The strain rate recorded at the above-mentioned tendon rupture was, on the other hand, only about 5 s^{-1}. It is doubtful whether the strain rates during strenuous sports activities are high enough to create increases sufficient to reestablish a reasonable safety margin (at least 1.5 to 2.0 for activities that do not habitually result in tendon injuries). Although these calculations have been limited to tensile forces as such, without taking into account energy absorption and storage, such indirectly derived data must be interpreted with caution until confirmed by actual force and strain measurements in a suitable animal model [cf. 17].

The *weakest point in the functional unit* is usually not the tendon or ligament, as shown by in vitro tensile testing along the main fiber axis until failure. The rupture of a functional unit seldom occurs in the pure tendon or ligament tissue but occurs most often as a tear-off fracture in one of the bone attachments or as a separation in the tendon-bone transitional zone. The calcaneal tendon tears off with one larger or several smaller bone fragments from the calcaneus in most species [1, 18, 19], except for possibly the rat, which experiences pure tendon ruptures [20]. In bone-ligament-bone specimens from most species, including the rat, separation occurs as avulsions of bone from the attachment site [2, 21, 22]. The exceptions for this type of specimen might be primates [23] and dogs [24], where ruptures have been reported to occur most frequently in the anterior cruciate ligament itself in the femur-ligament-tibia specimens used; here, however, the results might be influenced by the experimental design with the joints tested in 45 and 90° flexion, respectively (for the influence of the fiber arrangement geometry, see Sec. 6.2.1).

The type of injury in bone-ligament-bone or bone-tendon-muscle-tendon-bone units that results from supraphysiological forces acting on the human locomotive system in vivo is variable. The "pure" ligament ruptures seen clinically are probably due to uneven distribution of forces on them, as the violent external force in most cases is applied as a combination of bending and torque on the joint and its ligament. In some instances an avulsion of a bone attachment occurs, while in others combinations of ligament ruptures and bone avulsions are seen. The tensile strength of the anterior cruciate ligament decreases in *in vitro* experiments, when the tibia is rotated medially or laterally in relation to the femur [24]. Whether the "spontaneous" calcaneal tendon rupture, seen typically in middle-aged men during moderate sports activities, is a rupture in a healthy or previously diseased tendon is not yet firmly established.

6.2 STRESS-STRAIN CHARACTERISTICS

6.2.1 Ultimate Tensile Strength and Strain

Table 6.1 compares a number of soft tissues in the body; the muscle tendons are the strongest of these tissues.

TABLE 6.1 Ultimate tensile strength and elastic stiffness, and strain (dimensionless) at that point, for various human tissues [30] and inorganic materials [143]

	Strength, MPa	Strain	Elastic stiffness, N/mm^2
Ligamentum nuchae*	2.4	1.25	7.5
Skin (thoracical)†	15	1.1	35
Muscle fascia‡	15	1.17	340
Tendon	60	0.10	700
Compact bone	150	0.015	18,000
Aluminum alloy	210	0.1	7,000
Medium alloy steel	600	0.12	20,600

From cattle. Mean values for strongest (3.2) and weakest (1.6) parts (the corresponding strains are 1.6 and 0.9, and the elastic stiffnesses are 7.2 and 8.0, respectively).

† Tested in transversal direction.

‡ Tested along the main fiber direction.

The *geometrical arrangement of fibers* reasonably in parallel with each other in tendons is the basis for their conspicuous strength. Provided they are parallel-fibered, ligaments have comparable mechanical properties. Fasciae and skin also derive their tensile strength from collagen fibers; here the lower strength and higher strain values are primarily caused by geometrical differences in fiber arrangement [25, 26]. While most of the fibers in a tendon become, because of the relative parallelism in arrangement, stress bearing at about the same time, those in a fascia, with less pronounced predominant fiber direction, are subjected to a more uneven distribution of the stress bearing in uniaxial tensile testing. Therefore, the whole tissue is substantially weaker and more extensible, in spite of the fact that the collagen fibers in a fascia have basically the same mechanical properties as do tendons. This increase in extensibility is even more pronounced for skin, in which the amount of ground substance into which the fibers are embedded is larger. The role of elastic fibers is lesser in skin (only about 4 percent of dry weight is elastin) [27] than in ligamentum nuchae (about 71 percent) [28]. Elastic fibers add considerably more to the extensibility of the tissue than to its strength.

The *size of the cross section* of a tendon also has an influence on its strength. Harkness [29] calculated the tensile strength for whole tendons to be about 150 to 300 MPa for pure collagen (using the collagen content of wet tissue and assuming the density of collagen to be 1.4), while that of a collagen fascicle is about 500 MPa. The *elastic stiffness* (for a definition of this term see Sec. 6.2.2) of a whole limb muscle tendon is about 700 N/mm^2 [30], while that of a tendon fascicle is about 1000 N/mm^2 [31]. Brillouin light scattering measurements, which are confined to regions of dimension 100 μm (collagen fiber bundle level), have recorded values of about 5100 N/mm^2 for native tendon tissue [32]. These differences can be attributed to normal geometric "aberrations" in the macrostructure of the tendon [33]; i.e., the fibrils, fibers, fiber bundles, and fascicles on their respective levels of structural hierarchy are not perfectly aligned with each other with respect to parallelism and length between origin and insertion. They are therefore not loaded uniformly in tensile testing. This is seen as the gradual straightening out of the waviness of the fiber bundles in a relaxed tendon during the initial phase of tensile testing [34–36] and as the period length scattering during the lengthening of the 67- to 68-nm banding periodicity in the relaxed fibril [34, 37].

A number of other factors, varying with experimental procedures (for details see Sec. 6.5), have profound influence on the data recorded. Parameters based on stress values are especially affected by the method used to assess specimen cross-sectional area and by the design of the clamps used for securing the specimen in the testing machine. Parameters based on strain values are influenced by how the specimen's original length is defined, by the design of the clamps, by the methods to record specimen elongation, and by the strain history of the specimen. The water content of the specimen (overhydration and drying out during testing) influences the results. It should also be noted that the ultimate tensile strength and strain values are pertinent to the tendon or ligament tissue *only* if the failure site is within pure parallel-fibered collagenous tissue, or if an isolated tendon or ligament specimen is tested.

For these reasons the data in Table 6.1 were taken from studies performed in one laboratory and on isolated tissues, since the experimental protocol varies with respect to the above-mentioned factors between different laboratories, rendering direct comparisons difficult.

6.2.2 The Complete Stress-Strain Curve

The complete stress-strain curve for a tendon is much more important for the assessment of its mechanical properties than the isolated set of parameters for the point of ultimate tensile strength (or point of failure, which in most instances coincides with or occurs slightly after the first-mentioned point on this curve), as the physiological functioning range is limited to the first part of this curve.

The possibility of a *prestress* existing under in vivo conditions should also be considered. Such a tension in the tendon, however, probably varies somewhat with muscle tone changes and also with the degree of flexion or extension in the involved joint(s). It has been demonstrated by intravital microscopy techniques that the smooth surface of a "relaxed" tendon in situ becomes wavy after transection [38]. This observation suggests that the in vivo resting state is located somewhere within the "toe" part of the in vitro recorded stress-strain curve (see below). No information is available on the magnitude of this prestress. In ligaments, on the other hand, the possible presence of a prestress is dependent on whether the structure is relaxed in any position of its joint or not. Evidence suggests that there is no prestress in knee ligaments [39, 40], while it is pronounced for spinal ligaments, especially the ligamentum flavum [41]. There is at present no feasible method of measuring prestresses in tendons. The only method to establish the starting point of a stress-strain curve is to develop a reproducible empirical definition (see Sec. 6.5.1).

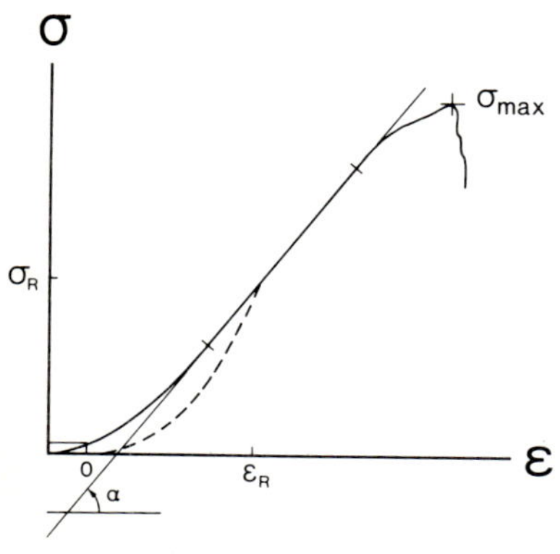

FIG. 6.1 Stress (σ)–strain (ε) curve for parallel-fibered collagenous tissue until failure (solid line); tan α is elastic stiffness. The beginning and end of the linear segment are marked with bars. The dashed curve is for unloading, if the cycle is reversed at the point (σ_R, ε_R). (*From [17].*)

The *stress-strain curve* for parallel-fibered collagenous tissue (Fig. 6.1) starts thus with an insidious "toe" part which gradually becomes steeper and merges into the linear part of the curve [42]. For tendons and ligaments, which consist of dense parallel-fibered collagenous tissue, the linear part is of considerable length [19], while for certain other soft tissues it is "compressed" into an inflection point on the curve [25]. The specimen can break directly at the end of this linear segment of the curve or, more commonly, after it has bent off somewhat toward the strain axis; the stress value can start to decline before failure occurs [19]. A number of parameters can be derived from this curve.

The *toe part* is most useful to characterize from a functional point of view, as most of the physiological functioning occurs within this part. Its increasing steepness can be expressed as the linear relationship between stress [43] or strain [44–46] and the corresponding tangent *modulus* (stress increment per unit strain increment, $d\sigma/d\varepsilon$). This approach was first used by Fung [43] for rabbit mesentery tissue in simple elongation, reducing the dynamic stress-strain curve to 1, which would have been obtained at an infinitely slow rate of loading (in my notation):

$$\frac{d\sigma_L}{d\lambda} = c_1\sigma_L(1 + c_2) \qquad (6.1)$$

where σ_L is lagrangian stress, λ is the extension ratio and c_1, c_2 are constants. For some other tissues it is necessary to represent the exponential stress response by two linear segments [47]. This linear relationship between stress and tangent modulus has also been found for the mouse tail tendon [48] and the canine anterior cruciate ligament [49] as well as for "tissue" reconstituted from highly purified type I collagen (the same type as in native tendons) [45, 46]. For the last-mentioned tissue, however, strain versus tangent modulus curves showed less dispersion between samples with different maturation times; the curves practically coincided when the tangent modulus was expressed in units of its maximum value for each curve. Whether this is also true for native tendon and ligament tissue is untested.

The beginning and end (L_B and L_E) of the *linear part* are easily determined, and from these coordinates the steepness of this curve segment (tan α in Fig. 6.1) can be determined. This parameter may be termed *elastic stiffness* [3, 25] to emphasize that it is not synonymous with *modulus of elasticity*, which would imply that the tissue in this region is perfectly elastic. All structures of soft, parallel-fibered collagenous tissue exhibit, even after meticulous preconditioning, a certain amount of viscosity throughout the stress-strain curve and, during initial tensile loading-unloading cycles, also some plasticity (see Sec. 6.2.3). It is thus futile to discuss phenomena such as yield stress and construct constitutive equations for these tissues on the assumption that they are perfectly elastic.

The *ultimate tensile strength and strain* (see Sec. 6.2.1) are the most commonly assessed parameters. Reporting them without reference to other parameters has little merit.

The *energy* required to break the specimen is measured as the area between the curve and the strain axis [19, 25]. Both the energy to reach the point of maximum stress and that to achieve failure should be calculated. This parameter has proved useful, especially in studies on the strength of healing wounds [50–53], but has not received the attention it merits from investigators of tendons and ligaments. It is important for the assessment of energy storage during physiological functioning [54] and energy absorption during violent stress impact. It must be realized that neither ultimate tensile strength nor strain can be used to estimate the capability to withstand impacts.

6.3 VISCOELASTICITY AND PLASTICITY

The stress-strain behavior described above (Secs. 6.2.1 and 6.2.2) is derived from single tensile loading experiments at rather low, finite strain rates. When the loading is reversed at a level of less than half the maximum stress, the unloading follows the dashed curve in Fig. 6.1. When such a loading-unloading cycle is repeated up to the same stress level, the curve is gradually shifted to the right, with that for the loading phase becoming steeper and that for the unloading phase remaining unchanged [2]. After a certain number of cycles the curve becomes identical with that from the previous cycle; i.e., the stationary phase of tissue behavior has been reached. The initial phase behavior is used for *preconditioning*, a procedure that might suggest that the stationary phase behavior is the "real" in vitro tissue behavior. At least the behavior in the stationary phase is easier to analyze [55, 56]. On the other hand, the fact that the changes occurring during the initial phase in vitro are irreversible does not imply that corresponding changes in vivo could not be reversed after an appropriate resting time, e.g., due to restoration of changes in fluid balance in the tissue by the blood circulation.

Similar in vitro initial phase behavior is seen in stress-relaxation and creep experiments. When the specimen is loaded until the stress value of σ_0 is reached, then at the constant strain level of ε_0 the stress relaxes with time (t) toward an asymptote (σ_A) (Fig. 6.2*a*). Similarly, the strain increases toward an asymptote (ε_A), when the stress is kept constant (σ_0) (Fig. 6.2*b*). The levels of these asymptotes increase and decrease, respectively, during initial phase behavior

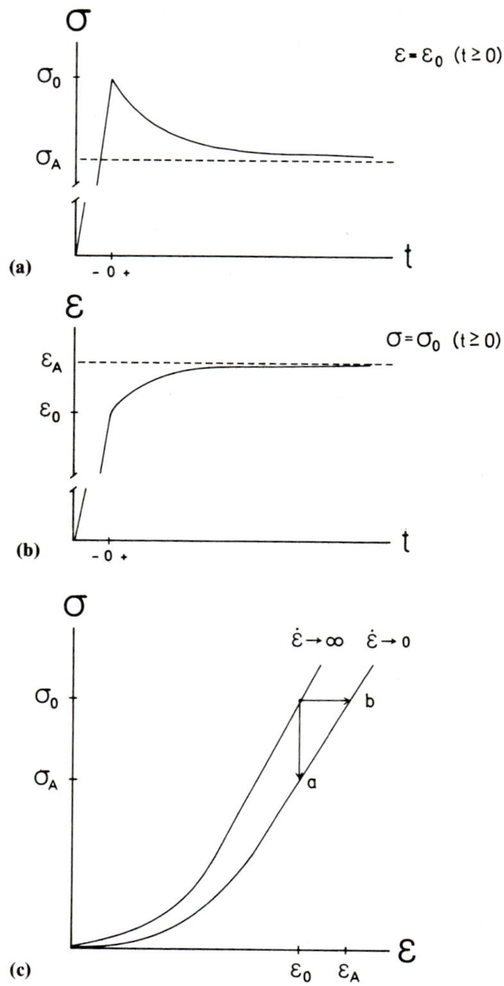

FIG. 6.2 Stress relaxation and creep. The loading is halted at time $t = 0$ at the point $(\sigma_0, \varepsilon_0)$. *a* shows stress relaxation toward an asymptote, σ_A; and *b* creep toward ε_A. *c* shows stress-strain curves for very high ($\dot{\varepsilon} \to \infty$) and very low ($\dot{\varepsilon} \to 0$) strain rates, with arrows *a* and *b* indicating stress relaxation and creep. (*From [17]*.)

until the stationary phase is reached [57]. The least time-consuming protocol (for preconditioning) to reach the in vitro stationary phase behavior is to use creep at a stress level somewhat above that used for analytical experiments.

The initial phase behavior in vitro thus includes strain hardening and an irreversible component of viscosity [55, 56]. The stationary phase behavior consists of nonlinear viscoelasticity, where the viscous components are less pronounced than during the initial phase. The movements of the locomotive system should thus be more precisely controlled during the stationary phase. It has been suggested that the initial phase corresponds to some extent to the limbering up of an athlete before a competition.

The nonlinear characteristics of the elastic components and the rate dependence of the viscous components can be studied in detail in the stationary phase only, and described by mathematical models (Sec. 6.4.1) or by constitutive equations based on discrete models (Sec. 6.4.2); the viscous component can further be studied by the T-jump technique (see Sec. 6.4.3). When the loading is performed at infinitely low speed ($\dot{\varepsilon} \rightarrow 0$), the viscous components are "exhausted" successively and neither stress-relaxation nor creep phenomena can occur, since the loading is halted at, for example, $(\sigma_0, \varepsilon_A)$ or $(\sigma_A, \varepsilon_0)$ in Fig. 6.2*c*. The stress-relaxation and creep phenomena increase in magnitude with increasing strain rate ($\dot{\varepsilon}$); i.e., the distance between the starting point (at $t = 0$) and the asymptote increases, until the strain rate (theoretically) goes toward infinity ($\dot{\varepsilon} \rightarrow \infty$). The stress-strain curve becomes steeper with increasing strain rate (Fig. 6.2*c*). A consequence of this is that at higher strain rates the specimen has absorbed more energy; when a certain strain level has been reached, the area under the curve ($\dot{\varepsilon} \rightarrow \infty$) up to the strain level of σ_0 is larger than the one under the curve ($\dot{\varepsilon} \rightarrow 0$) up to the same strain level. Higher ultimate tensile strengths and lower strains are thus recorded at increasing strain rates. The information available is not sufficient yet to determine whether this change with increasing strain rate has any practical influence on the tissue behavior in situations such as the above-mentioned double somersault.

6.4 MATHEMATICAL CHARACTERIZATION AND MODELING

6.4.1 Mathematical Models and Fung's Law

A number of more or less sophisticated, empirically found mathematical expressions have been suggested [see, e.g., 58, 59] since Wertheim in 1847 [60] suggested an exponential function for the description of the stress-strain curve for tendons:

$$\varepsilon^2 = c_3\sigma^2 + c_4\sigma \tag{6.2}$$

where ε is strain, σ is stress, and c_3 and c_4 are constants. Similarly, empirically found expressions have been used to describe stress-relaxation and creep phenomena [2]. These empirically found equations are useful only for the description of tissue properties under the limited experimental conditions (e.g., stress-strain *or* stress-relaxation conditions) from which they were derived. A broader approach to the mathematical characterization of tissue properties was introduced by Fung [43, 47, 61]. His studies of the response of different soft tissues to cyclic loading and unloading in uniaxial tensile experiments and to stress relaxation, and his creep-phenomenon testing, led to the formulation of a framework for quantitative research, a *quasi-linear viscoelasticity law* for soft biological tissues. It is valid for all the above-mentioned types of experiments and is written in the general form

$$\sigma(t) = f[\varepsilon(t)] + f'[\varepsilon(t - \tau); t, \tau] \tag{6.3}$$

where $\sigma(t)$ and $\varepsilon(t)$ are the stress and strain at time t, $f[\varepsilon(t)]$ is a function of ε at t, and $f'[\varepsilon(t - \tau); t, \tau]$ is a function of the entire strain history $\varepsilon(t - \tau)$. This emphasizes that the stress behavior of a specimen is dependent on the history of strain imposed on it. A *reduced relaxation function* $G(t)$ was incorporated so that the stress at time t is the sum of all past changes [47]:

$$\sigma(t) = \int_0^t G(t - t') \left\{ \frac{d\sigma_e[\lambda(t')]}{d\lambda} \right\} \left[\frac{d\lambda(t')}{dt'} \right] dt' \tag{6.4}$$

where $\sigma_e(t)$ has the same function as ε in the theory of linear viscoelasticity and σ_e is the elastic response. σ_e can be expressed in the equation for the nonlinear elasticity exhibited in simple elongation at an infinitely slow strain rate as

$$\sigma_L = \sigma_{L_\delta} e \exp\,[c_5(\lambda - \lambda_\delta)] \qquad (1 \leq \lambda \leq \lambda_1) \tag{6.5}$$

where σ_L is lagrangian stress, λ is the extension ratio $(\lambda l/l_0)$, and the subscript δ indicates a specific extension ratio and its corresponding stress value. As cyclic experiments show that biological tissues do not have the same rate dependence as linear viscoelasticity models, a continuous spectrum of relaxation time should be included. This broadens and flattens the internal damping peak characteristic of linear viscoelastic materials [47].

This mathematical representation was developed on the basis of experiments on other soft tissues not consisting of parallel-fibered connective tissue. It was later modified on the basis of the properties of rat tail tendons [49], approximating the tensile response to

$$\delta_c = c_6 \varepsilon^2 \tag{6.6}$$

and the relaxation function to

$$G(t) = A' \ln t + B' \tag{6.7}$$

and written as

$$\sigma(t) = c_7 \int_0^t [c_8 \ln (t - t') + 1]\varepsilon(t') \left[\frac{d\varepsilon(t')}{dt'}\right] dt' \tag{6.8}$$

where $c_8 = 2B'c_6$ and $c_7 = A'/B'$. Special forms of this equation can predict stress-strain curves for loading and unloading at constant strain rates as well as strain rate and hysteresis effects, while sinusoidal tests show poorer agreement.

Other mathematical representations have been developed for the description of the nonlinear viscoelastic behavior of parallel-fibered connective tissues, especially during simple elongation [62].

6.4.2 Discrete Models

Derivation of mathematical representation of tissue behavior on the basis of discrete models is an alternative to mathematical modeling based directly on phenomenological data (for review, see Ref. 44). This approach has inherent limitations since discrete models consist of "ideal" components of elasticity, viscosity, and plasticity, while biological tissues in general exhibit nonlinear behavior. A "true" discrete model would thus include a multitude of components. A model containing a limited number of components can, however, describe tissue behavior reasonably accurately and visualize tissue behavior in a way attractive for nonmathematically oriented biologists. This approach has been used for sea anemone body wall [63], bone [64, 65], chordae tendineae [66], and tendons and ligaments [55, 56, 67, 68]. Viidik [67] used a Kelvin element in series with a nonlinear spring array in which the springs come into function successively with increasing strain and a friction element to describe the stationary phase behavior of ligaments, and verified that its constants can be calculated from loading-unloading as well as stress-relaxation experiments [55, 56]. The constitutive equation is written

$$\frac{d\sigma}{dt} + \frac{\sigma(\phi + c_K)}{\eta_K} = \frac{d\varepsilon}{dt}\phi + \frac{\varepsilon c_K \phi}{\eta_K} - \frac{c_K \psi}{\eta_K} \tag{6.9}$$

where c_K is the spring constant, η_K is the constant of the dashpot in the Kelvin element, ψ is a step function, and ϕ is the step-function "constant" for the nonlinear spring array. ϕ may be written as

$$\phi(\varepsilon_\Sigma) = \sum_{\gamma=0}^{P} c_\gamma \mathcal{H}(\varepsilon_\Sigma - \Delta_\gamma) \tag{6.10}$$

where ε_Σ is the strain on the array, c is the spring constant, $\mathcal{H}$ is the Heaviside function, and Δ represents the strains before the springs come into action.

For the linear part of the curve the constants c_K, η_K, and linear ϕ can be evaluated from at least two special cases of this equation (i.e., infinitely slow loading, stress relaxation, and reversion from loading to unloading). Most of these expressions are easy to handle and therefore useful in the analysis of physiological as well as experimentally induced tissue changes [69]. For such purposes the tangent modulus constant c_1 [Eq. (6.1)] should be used for the characterization of the toe part rather than attempting to quantify the nonlinear spring array [Eq. (6.10)]. The more complex discrete model for the initial phase behavior [55, 56] is useful for the design of preconditioning protocols.

6.4.3 T-Jump Analysis

The activation energy for viscoelastic deformation can be used in the study of the temperature dependence of the viscoelasticity and determined by subjecting specimens to small and abrupt temperature changes (T jumps) during creep experiments [70]. This approach avoids the large scatter inherent in the superposition method [71], which requires repeated experiments; this scatter is large enough to mask the influence of temperature in the range 0 to 37°C [70]. The analysis is based on the theory of reaction rates [72]; when the temperature during a creep experiment is abruptly changed from T_1 to T_2 and the strain rates measured are $\dot{\varepsilon}_1$ and $\dot{\varepsilon}_2$ respectively, then

$$\frac{\dot{\varepsilon}_1}{\dot{\varepsilon}_2} = \exp\left[\Delta H\left(\frac{1}{T_1} - \frac{1}{T_2}\right)\right] \Delta H \tag{6.11}$$

where ΔH is the activation energy and R is the gas constant. With $\Delta T = T_1 - T_2$, the equation can be written as

$$\ln\frac{\dot{\varepsilon}_1}{\dot{\varepsilon}_2} = \frac{\Delta T}{RT_1T_2}\,\Delta H \tag{6.12}$$

which is a straight line, the slope being ΔH (kJ/mol) [70]. The activation energy is about 50 kJ/mol for the toe part [73] and about 155 kJ/mol for the linear part [74] of the stress-strain curve. This approach has been incorporated into a comprehensive model for the creep behavior of tendons [75] and is valuable for the determination of the roles of the collagen fibers and ground substance in the creep behavior of tendons (see Sec. 6.6).

6.5 TECHNICAL CONSIDERATIONS

6.5.1 Specimen-Related Sources of Error

The methods to measure specimen cross-sectional area vary considerably and thereby introduce difficulties in a comparison of the values for ultimate tensile strength from different reports. There are methodological problems with the measurement of the "true" cross-sectional area of a fresh tendon: (1) the size of the area is not uniform along its length from the muscle origin to the bone insertion, (2) fresh tendon tissue is somewhat compressible, and (3) the shape of the cross section (irregularly round to ovoid) and actual size limit the usefulness of some methods.

Some of these methods measure "physical" areas (for review, see Ref. 76). Planimetry of enlarged images of cross sections [77] is always accurate but requires the physical destruction of the tendon, since the accuracy of measuring only the areas at the two ends of the specimen to be tested is doubtful. A variety of calipers have been developed to measure the cross-sectional area at several sites along the specimen [78–80], but they lose their accuracy with decreasing specimen size.

Other methods measure "functional" cross-sectional areas. Commonly, these methods give a measure that is averaged for the whole length of the specimen. The use of Archimedes' principle [81] gives an average cross-sectional area for the whole length of the specimen. Wet weight per unit specimen length as an area measure [25, 82] is sensitive to loss or increase of water content caused by physiological processes in vivo (e.g., aging or change of hormonal balance) and to those caused by laboratory procedures. Dry weight per unit specimen length

[76] is a valid measure when all fractions of it contribute equally to the strength of the tissue, while amount of collagen per unit length [44] can be used if collagen is its only component with tensile strength.

The *methods to measure specimen original length* have inherent technical difficulties, and the differences in the various basic principles used introduce a source of variability between the strain values, especially at the point of ultimate tensile strength, in the various reports in the literature. The toe part of the stress-strain curve (Fig. 6.1) starts insidiously, and some of the waviness seen in incidental light microscopy (see Sec. 6.6) is straightened out before the recording equipment can measure any tension in the specimen.

The starting point for the stress-strain curve can be defined in several ways:

1. At the point of elongation reached from the relaxed state, when the minimum load the equipment can discriminate from zero is recorded [25]; care must be taken to define this load (or preferably stress) in absolute value or an error will be introduced by variation of specimen size or recorder amplification level.
2. The intercept between a tangent to the beginning of the toe part (the first few steps of the tangent modulus) and the elongation axis of the primary recorder curve (Viidik, to be published).
3. The point at which the first morphological changes occur in the specimen; this is difficult to determine in a reproducible way since the changes are insidious in character [42].
4. At the level of "prestress" present in vivo; this would be the logical choice for studies of physiological function ranges, provided this stress level could be determined unambiguously for a tendon (see Sec. 6.2.2).

The *previous strain history of the specimen*, e.g., that derived from specimen removal from its in situ location by dissection [25] or from a preconditioning procedure before the testing cycle to be analyzed, also has a significant influence on the results. For the standardization of experimental protocols in this respect it is useful to relate the procedures used to the discrete model for initial phase behavior (see Sec. 6.4.2) [55, 56, 67]. Special care should be taken when fragile specimens, e.g., wounds in early healing phases, are tested [50–53].

The *water content of the specimen* is important because its mechanical behavior—both overhydration (e.g., immersion in a fluid with low osmotic pressure before and/or during testing) and drying (e.g., prolonged testing in a too-dry atmosphere)—results in distortions of the results. It is noteworthy that specimens also swell when immersed in a solution of "physiological" saline buffered to pH 7.4; this results in enhanced initial phase phenomena (especially residual deformation) as well as more pronounced viscous components, and, in tensile testing, lesser "elastic stiffness" and larger maximum strain values [83, 84]. Also, storage in a 10% dextran solution or blood plasma increases the water content of the tissue [84].

Exposure to air, on the other hand, causes loss of water, most rapidly from specimens with very small cross-sectional areas (e.g., rat tail tendons). This stiffens the specimen significantly [14, 60, 81] and has the potential to introduce a systematic error in time-consuming experiments. The ideal experimental milieu would be air at 100% relative humidity at 37°C; for most experiments it suffices to maintain a humidity of more than 65% at 21°C [84] or to have the specimen wrapped in saline-moistened gauze [3]. There is no standard procedure in this respect; these factors must be evaluated with regard to the effects on the results of each experimental design [25].

The *storage of specimens prior to testing* up to 24 h does not require precautions other than prevention of changes in water content [19, 83–86]. For longer times storage at −20°C is advisable; lower storage temperatures do not seem to give additional advantages [86]. Available information shows that the main stress-strain parameters are not altered by storage in the frozen state, provided that both freezing and thawing are rapid to avoid protracted time periods with ice crystallization and high electrolyte concentration in the small amount of liquid water present [71, 83–85]. No information is available as to whether the viscoelastic properties are affected by freezing.

6.5.2 Equipment-Related Sources of Error

The clamping of specimens often poses technical problems, and numerous designs for clamps have been devised. The ideal clamp should hold the specimen firmly in its jaws without permitting slippage or producing abnormal stress concentrations which cause premature tissue failure at the edge of the jaw [25]. The clamp most efficient in preventing slippage has serrated jaws; it is useful for cyclic experiments but causes premature tissue failure when tensile strength testing is attempted [87]. The clamp with flat-surfaced jaws is the other extreme; it may be used for dried tendons [88] but is prone to slippage when fresh tissue is tested [78]. The intermediate way is to use rough surfaces and, if the specimens are large enough, a gradual decrease of pressure between the surfaces toward the jaw edges [82]. These surfaces may be of waterproof abrasive paper [82] or have a coating of rough metal [52, 89]. In order to avoid slippage as well as premature failure, the torque used for fastening the clamps must be determined empirically for each type and size of specimen; the practical test is that specimens failing with jaw breaks do so at the same stress level as the free-space breaks [25, 82]. Techniques of bedding specimen ends into hardened polymers or gluing them onto clamp surfaces, as well as "crushing" them between [90] or freezing them onto clamps [91], have also been used. These procedures create transitional zones with altered specimen properties and should therefore be avoided, except possibly in situations where specimen deformation is measured with a video technique (see below).

The clamping of whole functional units is easier, since the bones can be secured to clamps without undue problems. Pins through drilled holes can be used for this purpose [18, 92]; care must, however, be taken to avoid introducing fracture lines. This risk is eliminated when contour-shaped clamps especially made for each type and size of bone are used [1, 2, 21].

The methods to measure specimen deformation (elongation) most commonly used are to record machine crosshead movement (standard in most Instron and comparable machines) or changes in the distance between the clamps (typically with a coupled LVDT, or linear variable differential transformer). The crosshead movement recording is at times troubled by a mechanical play in the gear box which distorts the ends of the hysteresis loop in cyclic experiments. The LVDT measuring systems are more accurate in this respect but cannot eliminate that minor slippage of specimen tissue from the clamp jaws, and distorted deformation near the jaw edges are included into the recorded deformation. The best deformation measuring systems use video camera analysis of changes in the distance between two markers fixed onto the specimen at safe distances from the clamps [80, 93]. Such systems also allow the analysis of deformation in several segments of the same specimen.

The methods to measure tensile forces in specimens contain few problems. For specimens with short original lengths the elastic yield in the force transducer may distort deformation measurements with the crosshead movement and LVDT techniques. For high-speed testing, force transducers other than those based on piezoelectric elements may be inadequate.

The recording equipment poses no problems specific to the testing of tendons and ligaments. It should, however, be noted that for such experiments frequently used chart recorders with potentiometer-controlled chart feed for deformation recording are often not reliable enough for recording of cyclic loading-unloading tests [25].

6.6 STRUCTURE-FUNCTION INTERRELATION

Similar to the process of the vulcanization of rubber, which increases the density of cross-links and thereby the stiffness of the material, changes in the structural configuration of connective tissue components alter the mechanical properties of tendons and ligaments. The mechanical properties of rubber can be influenced by swelling in an organic liquid and the changes predicted by a general statistical-thermodynamic model. Such calculations are not possible for water-absorbing materials, e.g., wool and silk [94]. Connective tissues belong to this category; they are, further, composites in which the various components (collagen, elastin, proteoglycans, glycoproteins, and others) interact with each other. So far attempts to designate components in discrete models (see Sec. 6.4.2) to specific connective tissue components [68] have not produced conclusive results. The roles of collagen and elastin for the mechanical properties of the ligamentum nuchae have been analyzed by specific enzymatic removal of one tissue component between mechanical testing [95]. This approach has been used more extensively for aortic wall tissue [46, 96, 97]. Removal of collagen does not increase the extensibility of the tissue [46, 95], while removal of elastin does [46]; both removals decrease significantly the strength of the tissue. This agrees with the present concept of morphological interaction between collagen and elastin fibers; i.e., collagen fibers are wound around elastin cores [98]. The role of the proteoglycans, the main component of the ground substance, is less clear. Removal of only about half of them is possible by incubation in specific enzyme preparations but does not alter the mechanical properties of tendons [46]; the remaining half of them resist enzymatic attack, and their mechanical role is unknown. They could play a role for "gluing" the collagen fibrils into fibers. If the ground substance is removed by chelating agents, the fibrils are dispersed by dilute acid [99]; such agents, however, also alter the mechanical properties of "tissue" reconstituted from purified collagen [100].

Collagen is the main structural component of tendons and most ligaments (the main exception being ligamentum nuchae). The basic molecular unit in collagen is the left-handed helix, three of which are wound together into a right-handed superhelix, the collagen molecule (diameter 1.5 nm, length 300 nm) [101, 102]. Five rows of molecules are wound together into a microfibril, a left-handed "super-superhelix" (diameter 4 nm) [103, 104]; this model is in agreement with the distribution of hydrophobic residues in the helices [103]. A right-handed helical assembly of microfibrils into fibrils (diameter about 30 to 150 nm) is probably [105–107]. This alternate-handedness at consecutive hierarchial levels follows the principles for rope building; it converts axial stresses to lateral compression. In T-jump data this is seen as an orientation hardening mechanism; the dependence of isochronal stain rate on stress is of the form $\dot{\varepsilon} \propto \sigma^n$, $n < 1$ [70].

The collagen molecules are not coupled to each other by end-to-end links but are assembled into the microfibrils, guided by lateral intermolecular interactions and stabilized by cross-links to adjacent molecules in the same as well as neighboring microfibrils [108]. These cross-links are initially reducible, either aldimines or ketimines. With maturation they become nonreducible, possibly coupled to each other in the form of poly-chain links [108]. This results in an increased stability, seen mechanically as increased stiffness, and to a certain extent increased strength in the tissue [109]. With excessive cross-linking, especially when it is induced by artificial means, e.g., formaldehyde treatment, the tissue becomes even stiffer as well as brittle, and its tensile strength decreases [110]. The same effect is achieved by tanning [111] and embalming procedures [14, 112]; it follows from this that embalmed specimens are unsuitable for biomechanical testing. The opposite effect is achieved by in vivo treatment with cross-linking preventing and/or splitting agents such as β-aminopropionitrile or D-penicillamine [113–117], which results in a drastic decrease in stiffness and strength.

The fibrils seem to be packed in parallel into fibers and fiber bundles (diameter about 1 to 12 μm) and surrounded by sheaths of "woven fabric" which at least to some extent is collagenous [118, 119]. The mechanical role of these sheaths is not clear. The prominent feature of relaxed tendon fiber bundles, in light microscopy with crossed polaroids, is alternating dark and light banding (for review, see Refs. 42, 120); this banding is caused by a planar waveform in the bundle along its main longitudinal axis. There is no agreement whether this waveform consists of smooth quasi-sinusoidal undulations [25, 42] or is a crimp with sharp, kinking bends [121, 122]. In any case, this waveform is straightened out during the toe part of the stress-strain curve [34, 110, 123]. This waviness is probably created by collagen-interfibrillar interaction, similar to buckling of synthetic fibers, when the surrounding matrix is induced to shrink [124]. During the linear part of the curve the birefringence pattern seen in polarized light microscopy is not further altered. During this curve segment, however, the period length of the axial banding of 67 to 68 nm seen in electron microscopy [34], as well as the corresponding low-angle x-ray diffraction pattern [37, 68, 125], increases, indicating elongation of the microfibrillar structure. The changes in the diffraction patterns are larger at same strain levels for immature tissue with less dense cross-linking. Also, elongation of the wide-angle x-ray diffraction pattern, which reflects the winding of the triple helix, has been reported [125].

These observations suggest that the primary structural event during the toe part of the stress-strain curve is the straightening out of the light microscopic waviness and an improved paralleling of the fiber structure [36, 42, 113], while the fibrillar and molecular structures are stretched during the linear part of the curve. T-jump analyses (see Sec. 6.4.3) have produced similar conclusions. The activation energy during creep within the toe part (~50 kJ/mol) is in the same magnitude as that for hyaluronic acid gels (20 to 40), while that for the linear part of the curve (~135) is comparable to that of tape of pure steer collagen (~140). Creep from toe-part stress levels should thus be derived from interfibrillar sliding and shear of the interfibrillar ground substance, while that from linear-part levels should be attributed to fibrillar creep [75].

Some models for tendon behavior (not discussed above in Sec. 6.4) disregard known structural facts, e.g., that the fibrillar organization is planar and not helical [126] or that tendons such as rat tail tendons do not contain elastin [127]. Others emphasize the properties of sharp, crimping sites of collagen fibrils in their models or mathematical calculations [128, 129], when no macromolecular peculiarities have been demonstrated for these sites. Such studies seem to be mathematical exercises that at present provide no further information on the interdependence between structure and function of parallel-fibered collagenous tissues.

The obvious avenue for further research in this field is to base modeling on data collected from studies on in vivo structurally altered and in vitro enzymatically dissected tissues combined with studies on artificial tissues which have been reconstituted from purified and macromolecularly characterized connective tissue components [cf. 45, 46].

6.7 PHYSIOLOGICAL AND PATHOLOGICAL CHANGES

The *influence of physical exercise* on tendons and ligaments is not consistent, possibly because of age and species differences among various reports. Changes occur in the bone insertions of tendons and ligaments: increased strength is found in rabbits [1, 2] and male [130] but not female [131] rats. The strength of isolated tendons is increased in rabbits [132] but not in mice [89]. Also, morphological and biochemical changes in these tissues are induced by training, the general trend being a retardation of processes seen with maturation and aging.

Maturation and aging results in increased stability of connective tissues [133]. The stiffness of tendons increases with age; so does their tensile strength up to maturity, but it levels off [16] or starts to decrease somewhat later [109]. The amount of collagen also increases, while elastic fibers, especially in skin and arteries, tend to become brittle and split into fragments (for review, see Ref. 133).

The influence of hormonal disturbances. An increased level of plasma corticosteroids results in stiffness and tensile strength of some connective tissue structures, e.g., skin [134–137] and at times also of tendons [134, 137, 138]. Local treatment of tendons with moderate doses of cortisol increases their mechanical stability after short treatment periods [135] but not after longer periods [136], when at times a decrease can be seen [139]. Knee joint ligaments become weaker, though not less stiff, after corticosteroid injections into the joint cavity [135, 140]; since the failure always in these instances occurs as a tear-off fracture of the bone insertion, the possible explanation is corticosteroid-induced local osteoporosis. For review and discussion on the influence of corticosteroids connective tissues see Ref. 141.

Other changes of hormonal balance also influence the mechanical properties of tendons and ligaments, e.g., those during pregnancy [5, 21]. Disturbances in the hormonal balance, as by diabetes mellitus, may induce profound changes in connective tissues by increasing their stability both biochemically and mechanically [142].

Available information on the structure and function of connective tissues, especially the possibilities for a number of changes in them, thus emphasize that a tendon cannot be taken as "just a tendon." That is, information on the structural characteristics of tendon tissue from molecular to macroscopic levels should be sought and utilized when biomechanical data are analyzed and interpreted.

REFERENCES

1. Viidik, A.: Tensile strength properties of achilles tendon systems in trained and untrained rabbits, *Acta Orthop. Scand., 40,* 261–272, 1969.
2. Viidik, A.: Elasticity and tensile strength of the anterior cruciate ligament in rabbits as influenced by training, *Acta Physiol. Scand., 74,* 372–380, 1968.
3. Viidik, A.: Biomechanics and functional adaptation of tendons and joint ligaments, in F. G. Evans (ed), *Studies on the Anatomy and Function of Bone and Joints,* Springer-Verlag, Heidelberg, 1966, pp. 17–39.
4. Elliott, D. H., and Crawford, G. N. C.: The thickness and collagen content of tendon relative to the strength and cross-sectional area of muscle, *Proc. R. Soc., B162,* 137–146, 1965.
5. Harkness, R. D.: Mechanical properties of collagenous tissues, in B. S. Gould (ed.), *Treatise on Collagen,* vol. 2: *Biology of Collagen,* Academic Press, New York, 1968, pp. 247–310.
6. Hirsch, G.: Tensile properties during tendon healing, *Acta Orthop. Scand., Suppl., 153,* 1974.
7. Barnes, G. R. G., and Pinder, D. N.: In vivo tendon tension and bone strain measurement and correlation, *J. Biomech., 7,* 35–42, 1974.
8. Simon, P. J.: The mechanics of tendon function in health and disease, Department of Physiology and Anatomy, Massey University, Parlmerston North, New Zealand, Report, 1978.
9. Lochner, F. K., Milne, D. W., Mills, E. J., and Groom, J. G.: In vivo and in vitro measurement of tendon strain in the horse, *Am. J. Vet. Res., 41,* 1929–1937, 1980.
10. Kear, M., and Smith, R. N.: A method for recording tendon strain in sheep during locomotion, *Acta Orthop. Scand., 46,* 896–905, 1975.
11. Walmsley, B., Hodgson, J. A., and Burke, R. E.: Forces produced by medial gastrocnemius and soleus muscles during locomotion in freely moving cats, *J. Neurophysiol., 47,* 1203–1216, 1978.
12. Alexander, R. M.: Factors of safety in the structure of animals, *Sci. Prog. Oxf., 67,* 109–130, 1981.
13. Barfred, T.: Kinesiological comments on subcutaneous ruptures of the achilles tendon, *Acta Orthop. Scand., 42,* 397–405, 1971.
14. Stucke, K.: Über das elastische Verhalten der achillessehne im belastungsversuch, *Arch. Klin. Chir., 265,* 579–599, 1950.
15. Grafe, H.: Aspekte zur ätiologie der subkutanen achillessehnenruptur, *Zbl. Chir., 94,* 1073–1082, 1969.
16. Haut, R. C.: Age-dependent influence of strain rate on the tensile failure of rat-tail tendon, *J. Biomech. Eng., 105,* 296–299, 1983.
17. Viidik, A.: Elastomechanik biologischer gewebe, in H. Cotta et al. (eds.), *Die Belastungstoleranz des Bewegeapparates,* Georg Thieme, Stuttgart, New York, 1980, pp. 124–136.
18. Davidsson, L.: Über die subkutanen sehnen-rupturen und die regeneration der sehne, *Ann. Chir. Gynaecol. Fenn. Suppl., 6,* 1956.
19. Viidik, A., Sandqvist, L., and Mägi, M.: Influence of postmortal storage on tensile strength characteristics and histology of rabbit ligaments, *Acta Orthop. Scand. Suppl., 79,* 1965.
20. Barfred, T.: Experimental rupture of the achilles tendon, *Acta Orthop. Scand., 42,* 406–428, 1971.
21. Rundgren, Å.: Physical properties of connective tissue as influenced by single and repeated pregnancies in the rat, *Acta Physiol. Scand. Suppl., 417,* 1974.
22. Tipton, C. M., James, S. L., Mergner, W., and Tcheng, T-K.: Influence of exercise on the strength of medial collateral knee ligaments of dogs, *Am. J. Physiol., 218,* 894–902, 1970.
23. Noyes, F. R., Torvik, P. J., Hyde, W. B., and DeLucas, J. L.: Biomechanics of ligament failure, *J. Bone J. Surg., 56A,* 1406–1418, 1974.
24. Alm, A., Ekström, H., and Strömberg, B.: Tensile strength of the anterior cruciate ligament in the dog, *Acta Chir. Scand. Suppl., 445,* 15–23, 1974.
25. Viidik, A.: Biomechanical behavior of soft connective tissues, in N. Akkas (ed.), *Progress in Biomechanics,* Sijthoff & Noordhoff, Alphen aan den Rijn, 1979, pp. 75–113.
26. Nilsson, T.: Biomechanical studies of rabbit abdominal wall, *J. Biomech., 15,* 131–135, 1982.
27. Viidik, A.: Rheology of skin with special reference to age-related parameters and their possible correlation to structure, in L. Robert and B. Robert (eds.), *Frontiers of Matrix Biology,* Karger, Basel, 1973, vol. 1, pp. 157–189.

28. Jackson, D. S., Sandberg, L. B., and Cleary, E. G.: The swelling of bovine ligamentum nuchae as a function of pH, *Biochem. J., 96,* 813–817, 1965.

29. Harkness, R. D.: Biological functions of collagen, *Biol. Rev., 36,* 399–463, 1961.

30. Yamada, H.: *Strength of Biological Materials,* F. G. Evans (ed.), Williams & Wilkins, Baltimore, 1970.

31. Torp, S., Baer, E., and Friedman, B.: Effects of age and mechanical deformation on the ultrastructure of tendon, in E. D. T. Atkins and A. Keller (eds.), *Structure of Fibrous Biopolymers,* Colstron Papers no. 26, Butterworth, London, 1975, pp. 223–250.

32. Cusack, S.: Brillouin light scattering and the elastic constants of collagen, in A. Viidik and J. Vuust (eds.), *Biology of Collagen,* Academic Press, London, 1980, pp. 281–288.

33. Harley, R., James, D., Miller, A., and White, J. W.: Phonons and the elastic moduli of collagen and muscle, *Nature, 267,* 285–287, 1977.

34. Viidik, A., and Ekholm, R.: Light and electron microscopic studies of collagen fibers under strain, *Z. Anat. Entwicklungs-gesch., 127,* 154–164, 1968.

35. Viidik, A.: Simultaneous mechanical and light microscopic studies of collagen fibers, *Z. Anat. Entwicklungsgesch., 136,* 204–212, 1972.

36. Viidik, A.: Functional properties of collagenous tissues, in D. A. Hall and D. S. Jackson (eds.), *International Review of Connective Tissue Research,* Academic Press, London, 1973, vol. VI, pp. 127–215.

37. Nemetschek, Th., Jonak, R., Nemetschek-Gansler, H., and Riedl, H.: Über die bestimmung von langperioden-änderungen am kollagen, *Z. Naturforsch., 33,* 928–936, 1978.

38. Schatzker, J., and Brånemark, P.-I.: Intravital observation on the microvascular anatomy and microcirculation of the tendon, *Acta Orthop. Scand. Suppl., 126,* 1969.

39. Lewis, J. L.: A note on the application and evaluation of the buckle transducer for knee ligament force measurement, *J. Biomech. Eng., 104,* 125–128, 1982.

40. Kennedy, J. C., Hawkins, R. J., and Willis, R. B.: Strain gauge analysis of knee ligaments, *Clin. Orthop. Rel. Res., 129,* 225–229, 1977.

41. Åkerblom, B.: *Standing and Sitting Posture,* Nordiska Bokhandeln, Stockholm, 1948.

42. Viidik, A.: Interdependence between structure and function in collagenous tissues, in A. Viidik and J. Vuust (eds.), *Biology of Collagen,* Academic Press, London, 1980, pp. 257–280.

43. Fung, Y. C. B.: Elasticity of soft tissues in simple elongation, *Am. J. Physiol., 213,* 1532–1544, 1967.

44. Viidik, A.: Mechanical properties of parallel-fibred collagenous tissues, in A. Viidik and J. Vuust (eds.), *Biology of Collagen,* Academic Press, London, 1980, pp. 237–255.

45. Danielsen, C. C.: Mechanical properties of reconstituted collagen fibrils, *Connect. Tissue Res., 9,* 219–225, 1982.

46. Viidik, A., Danièlsen, C. C., and Oxlund, H.: On fundamental and phenomenological models, structure and mechanical properties of collagen, elastin and glycosaminoglycan complexes, *Biorheology, 19,* 437–451, 1982.

47. Fung, Y. C. B.: Stress-strain history relations of soft tissues in simple elongation, in Y. C. B. Fung et al. (eds.), *Biomechanics: Its Foundations and Objectives,* Prentice-Hall, Englewood Cliffs, N.J., 1972, pp. 181–208.

48. Stromberg, D. D., and Wiederhielm, C. A.: Viscoelastic description of a collagenous tissue in simple elongation, *J. Appl. Physiol., 26,* 857–862, 1969.

49. Haut, R. C., and Little, R. W.: Rheological properties of canine anterior cruciate ligaments, *J. Biomech., 2,* 289–298, 1969.

50. Holm-Pedersen, P., and Viidik, A.: Tensile properties and morphology of healing wounds in young and old rats, *Scand. J. Plast. Reconstr. Surg., 6,* 24–35, 1972.

51. Fogdestam, I., and Gottrup, F.: Biomechanical methods in wound-healing research with special reference to skin and gastrointestinal tract, in A. Viidik and J. Vuust (eds.), *Biology of Collagen,* Academic Press, London, 1980, pp. 363–371.

52. Fogdestam, I.: A biomechanical study of healing rat skin incisions after delayed primary closure, *Surg. Gynecol. Obstet., 153,* 191–199, 1981.

53. Gottrup, F.: Healing of incisional wounds in the stomach and duodenum, *Danish Med. Bull., 31,* 32–48, 1984.

54. Proske, U.: Energy conservation by elastic storage in kangaroos, *Endeavour, 4* (new series), 148–153, 1980.

55. Frisén, M., Mägi, M., Sonnerup, L., and Viidik, A.: Rheological analysis of soft collagenous tissue: I. Theoretical considerations, *J. Biomech., 2,* 13–20, 1969.

56. Frisén, M., Mägi, M., Sonnerup, L., and Viidik, A.: Rheological analysis of soft collagenous tissue: II. Experimental evaluations and verifications, *J. Biomech., 2,* 21–28, 1969.

57. Viidik, A.: On the correlation between structure and mechanical function of soft connective tissues, *Verh. Anat. Ges., 72,* 75–89, 1978.

58. Morgan, F. R.: The mechanical properties of collagen fibres: Stress-strain curves, *J. Soc. Leather Trades Chem., 44,* 170–182, 1960.

59. Elden, H. R.: Physical properties of collagen fibers, in D. A. Hall (ed.), *International Review of Connective Tissue Research,* Academic Press, London, 1968, vol. IV, pp. 283–348.

60. Wertheim, M. G.: Mémoire sur l'élasticité et la cohésion des principaux tissu du corps humain, *Chim. Phys., 21,* 385–414, 1847.

61. Fung, Y. C. B.: Biomechanics, its scope, history and some problems of continuum mechanics in physiology, *Appl. Mech. Rev., 21,* 1–20, 1968.

62. Ko, F. K., Pastore, A., and Cole, F.: Nonlinear viscoelastic behavior of human and canine flexor tendons in simple elongation, in Y. C. B. Fung and J. G. Pinto (eds.), *Proceedings of the 3d International Congress of Biorheology,* La Jolla, Calif., 1978, p. 193, and Philadelphia College of Textiles and Science, Philadelphia, Report, 1978.

63. Alexander, R. M.: Visco-elastic properties of the body-wall of sea anemones, *J. Exp. Biol., 39,* 373–386, 1962.

64. Sedlin, E. D.: A rheological model for cortical bone, *Acta Orthop. Scand. Suppl., 83,* 1965.

65. Frankel, V. H., and Burstein, A. H: *Orthopaedic Biomechanics,* Lea and Febiger, Philadelphia, 1970.

66. Anna, O., Arnold, G., Hartung, C., and Zech, M.: Biomechanik und simulation des zugverhaltens der chordae tendineae des menschlichen herzens, *Z. Werkstoffstech., 5,* 209–217, 1974.

67. Viidik, A.: A rheological model for uncalcified parallel-fibred collagenous tissue, *J. Biomech., 1,* 3–11, 1968.

68. Riedl, H., Nemetschek, Th., and Jonak, R.: A mathematical model for the changes of the long-period structure in collagen, in A. Viidik and J. Vuust (eds.), *Biology of Collagen,* Academic Press, London, 1980, pp. 289–296.

69. Viidik, A., Oxlund, H., and Andreassen, T.: Rheological analysis of collagen—With reference to morphology and experimental techniques, in Y. C. B. Fung and J. G. Pinto (eds.), *Proceedings of the Third International Congress of Biorheology,* La Jolla, Calif., 1978, pp. 187–188.

70. Cohen, R. E., Hooley, C. J., and McCrum, N. G.: Mechanics of the viscoelastic deformation of collagenous tissue, in E. D. T. Atkins and A. Keller (eds.), *Structure of Fibrous Biopolymers,* Colston Papers no. 26, Butterworth, London, 1975, pp. 251–254.

71. Rigby, B. J., Hirai, N., Spikes, J. D., and Eyring, H.: The mechanical properties of rat tail tendon, *J. Gen. Physiol., 43,* 265–283, 1959.

72. Glasstone, S., Laidler, K. J., and Eyring, H.: *The Theory of Rate Processes,* McGraw-Hill, New York, 1941.

73. Cohen, R. E., Hooley, C. J., and McCrum, N. G.: Viscoelastic creep of collagenous tissue, *J. Biomech., 9,* 175–184, 1976.

74. Hooley, C. J.: The viscoelastic behaviour of tendon, Ph.D. thesis, Oxford University, Oxford, 1977.

75. Hooley, C. J., and Cohen, R. E.: A model for the creep behaviour of tendon, *Int. J. Biol. Macromol., 1,* 123–132, 1979.

76. Ellis, D. G.: Cross-sectional area measurements for tendon specimens: A comparison of several methods, *J. Biomech., 2,* 175–186, 1969.

77. Cronkite, A. E.: The tensile strength of human tendons, *Anat. Rec., 64,* 173–186, 1936.

78. Walker, L. B., Jr., Harris, E. H., and Benedict, J. V.: Stress-strain relationship in human cadaveric plantaris tendon, *Med. Electron. Biol. Eng., 2,* 31–38, 1964.

79. Woo, S. L-Y., Ritter, M. A., Amiel, D., Sanders, T. M., Gomez, M. A., Kuel, S. C., Garfin, S. R., and Akeson, W. H.: The biomechanical and biochemical properties of swine tendons, *Connect. Tissue Res., 7,* 177–183, 1980.

80. Woo, S. L-Y.: Mechanical properties of tendons and ligaments: I. Quasi-static and nonlinear viscoelastic properties, *Biorheology*, *19*, 385–396, 1982.

81. Rollhäuser, H.: Die festigkeit menschlicher sehnen nach quellung und trockung in abhängigkeit von lebensalter, *Gegenbauers Morph. Jahrb.*, *90*, 180–191, 1950.

82. Viidik, A.: Experimental evaluation of the tensile strength of isolated rabbit tendons, *Biomed. Eng.*, *2*, 64–67, 1967.

83. Viidik, A., and Lewin, T.: Changes in tensile strength characteristics and histology of rabbit ligaments induced by different modes of postmortal storage, *Acta Orthop. Scand.*, *37*, 141–155, 1966.

84. Galante, J. O.: Tensile properties of the human lumbar annulus fibrosus, *Acta Orthop. Scand. Suppl.*, *100*, 1967.

85. Ridge, M. D., and Wright, V.: A rheology study of skin, in R. M. Kenedi (ed.), *Biomechanics and Related Bioengineering Topics*, Pergamon Press, Oxford, 1965, pp. 165–175.

86. Butler, D. L., Noyes, F. R., and Grood, E. S.: Measurement of the mechanical properties of ligaments, in B. N. Feinberg et al. (eds.), *CRC Handbook of Engineering in Medicine and Biology*, sec. B: *Instruments and Measurements*, CRC Press, Palm Beach, Fla., 1978, vol 1., pp. 279–314.

87. Wright, D. G., and Rennels, D. C.: A study of the elastic properties of plantar fascia, *J. Bone Jt. Surg.*, *46A*, 482–492, 1964.

88. Compton, E. D.: Effect of tannage on strength of collagen fibres, *J. Am. Leather Chem. Assoc.*, *44*, 140–151, 1949.

89. Kiiskinen, A.: Physical training and connective tissue in young mice—Physical properties of achilles tendons and long bones, *Growth*, *44*, 123–137, 1977.

90. Hornbech Svendsen, K., and Thomsen, G.: A new clamping and stretching procedure for determination of collagen fiber stiffness and strength relations upon maturation, *J. Biomech.*, *17*, 225–229, 1984.

91. Riemersa, D. J., and Schamhardt, H. C.: The cryo-jaw, a clamp designed for in vitro rheology studies of horse digital flexor tendons, *J. Biomech.*, *15*, 619–620, 1982.

92. Tipton, C. M., Schild, R. J., and Tomanek, R. J.: Influence of physical activity on the strength of knee ligaments in rats, *Am. J. Physiol.*, *212*, 783–787, 1967.

93. Lanir, Y., and Fung, Y. C. B.: Two-dimensional mechanical properties of rabbit skin: I. Experimental system, *J. Biomech.*, *7*, 29–34, 1974.

94. Treolar, L. R. G.: *The Physics of Rubber Elasticity*, Clarendon Press, Oxford, 1975.

95. Minns, R. J., Soden, P. D., and Jackson, D. S.: The role of fibrous components and ground substance in the mechanical properties of biological tissues, *J. Biomech.*, *6*, 153–165, 1973.

96. Hoffman, A. S., Grande, L. A., and Parks, J. B.: Sequential enzymolysis of human aorta and resultant stress-strain behavior, *Biomater. Med. Devices Artif. Organs*, *5*, 121–145, 1977.

97. Hoffman, A. S., and Daly, C. H.: A commentary on the use of enzyme probes to elucidate the contribution of individual components to soft-tissue biomechanics in vitro, in A. Viidik and J. Vuust (eds.), *Biology of Collagen*, Academic Press, London, 1980, pp. 297–312.

98. Serafini-Fracassini, A., Field, J. M., Smith, J. W., and Stephens, W. G. S.: The ultrastructure and mechanics of elastic ligaments, *J. Ultrastruct. Res.*, *58*, 244–251, 1977.

99. Steven, F. S.: The effect of chelating agents on collagen interfibrillar matrix interactions in connective tissue, *Biochim. Biophys. Acta*, *140*, 522–528, 1967.

100. Oxlund, H., and Andreassen, T. T.: The role of hyaluronic acid, collagen and elastin in the mechanical properties of connective tissues, *J. Anat.*, *131*, 611–620, 1980.

101. Kühn, K., and Glanvile, R. W.: Molecular structure and higher organization of different collagen types, in A. Viidik and J. Vuust (eds.), *Biology of Collagen*, Academic Press, London, 1980, pp. 1–14.

102. Miller, A.: Structural studies on connective tissue, in A. Viidik and J. Vuust (eds.), *Biology of Collagen*, Academic Press, London, 1980, pp. 39–52.

103. Trus, B. L., and Piez, K. A.: Molecular packing of collagen: Three-dimensional analysis of electrostatic interactions, *J. Mol. Biol.*, *108*, 705–732, 1976.

104. Miller, A., and Wray, J. S.: Molecular packing in collagen, *Nature*, *230*, 437–439, 1971.

105. Bouteille, M., and Pease, D. C.: The tridimensional structure of native collagenous fibrils, their proteinaceous filaments, *J. Ultrastruct. Res.*, *35*, 314–338, 1971.

106. Lillie, J. H., MacCallum, D. K., Scaletta, L. J., and Ouchino, J. C.: Collagen structure: Evidence for helical organization of the collagen fibril, *J. Ultrastruct. Res.*, *58*, 134–143, 1977.

107. Rayns, D. G.: Collagen from frozen fractured glycerinated beef heart, *J. Ultrastruct. Res., 48,* 59–66, 1974.

108. Light, N. D., and Bailey, A. J.: Molecular structure and stabilization of the collagen fibre, in A. Viidik and J. Vuust (eds.), *Biology of Collagen,* Academic Press, London, 1980, pp. 15–38.

109. Vogel, H. G.: Age dependence of mechanical properties of rat tail tendons, *Aktuel. Gerontol., 13,* 22–27, 1983.

110. Mohanaradhakrishnan, V., Muthiah, P. L., and Hadhanyi, A.: A few factors contributing to the mechanical strength of collagen fibres, *Arzneim. Forsch., 25,* 726–735, 1975.

111. Gustavson, K. H.: *The Chemistry and Reactivity of Collagen,* Academic Press, New York, 1956.

112. Blanton, P. L., and Biggs, N. L.: Ultimate tensile strength of fetal and adult human tendons, *J. Biomech., 3,* 181–189, 1970.

113. Nimni, M. E.: Collagen: Structure, function, and metabolism in normal and fibrotic tissues, *Semin. Arthritis Rheum., 13,* 1–86, 1983.

114. Viidik, A., and Busted, N.: Thermal shrinkage characteristics of collagen under isotonic conditions with no tension, in U. J. Schmidt et al. (eds.), *Proc. 5th Symp. Basic Res. Gerontol.,* Perimed, Erlangen, 1977, pp. 493–502.

115. Vogel, H. G.: Influence of desmotropic drugs on viscoelastic properties of rat tail tendons, *Arzneim. Forsch., 34,* 213–216, 1984.

116. Haut, R. C.: *Sensitivity of Lathyritic Tendon to Strain Rate,* Res. Pub. GMR-4662, General Motors Research Laboratories, Warren, Pa., 1984.

117. Haut, R. C.: *The Influence of Covalent Crosslinking in Collagen on the Sensitivity of Tendon to Strain Rate,* Res. Rep. GMG-4730, General Motors Research Laboratories, Warren, Pa., 1984.

118. Kwon, D. S., Mason, P., and Rigby, B. J.: Influence of the tendon membrane on swelling and thermal stability, *Nature, 201,* 159–160, 1964.

119. Balian, G. A., and Bowes, J. H.: Composition of membranous tissue obtained from tendon and skin, *Nature, 213,* 518, 1967.

120. Diamant, J., Keller, A., Baer, E., Litt, M., and Arridge, R. G. C.: Collagen: Ultrastructure and its relation to mechanical properties as a function of ageing, *Proc. R. Soc. London, B180,* 293–315, 1972.

121. Gathercole, L. J., and Keller, A.: Light microscopic waveforms in collagenous tissues and their structural implications, in E. D. T. Atkins and A. Keller (eds.), *Structure of Fibrous Biopolymers,* Colston Papers no. 26, Butterworth, London, 1975, pp. 153–187.

122. Betsch, D. F., and Baer, E.: Structure and mechanical properties of rat tail tendon, *Biorheology, 17,* 83–94, 1980.

123. Abrahams, M.: Mechanical behaviour of tendon in vitro, *Med. Biol. Eng., 5,* 433–443, 1967.

124. Dale, W. C., and Baer, E.: Fibre-buckling in composite systems: A model for the ultrastructure of uncalcified collagen tissue, *J. Mater. Sci., 9,* 369–382, 1974.

125. Cowan, P. M., North, A. C. T., and Randall, J. T.: X-ray diffraction studies of collagen fibres, *Symp. Soc. Exp. H. Biol., 9,* 115–126, 1955.

126. Beskos, D. E., and Jenkins, J. T.: A mechanical model for mammalian tendon, *J. Appl. Mech., 107,* 755–758, 1975.

127. Lanir, Y.: Structure-strength relations in mammalian tendon, *Biophys. J., 24,* 541–554, 1978.

128. Kastelic, J., Paley, I., and Baer, E.: A structural model for tendon crimping, *J. Biomech., 13,* 887–893, 1980.

129. Shah, J. S., Jayson, M. I. V., and Hampson, W. G.: Mechanical implications of crimping in collagen fibres of human spinal ligaments, *Eng. Med. (IME), 8,* 95–102, 1979.

130. Tipton, C. M., Matthes, R. D., and Sandage, D. S.: In situ measurement of junction strength and ligament elongation in rats, *J. Appl. Physiol., 37,* 758–761, 1974.

131. Booth, F. W., and Tipton, C. M.: Effects of training and 17-B estradiol upon heart rates, organ weights, and ligamentous strength of female rats, *Int. Z. Angew. Physiol., 27,* 187–197, 1969.

132. Viidik, A.: The effect of training on the tensile strength of isolated rabbit tendons, *Scand. J. Plast. Reconstr. Surg., 1,* 141–147, 1967.

133. Viidik, A.: Age-related changes in connective tissues, in A. Viidik (ed.), *Lectures on Gerontology,* Academic Press, London, 1982, vol. I, pt. A, pp. 173–211.

134. Oxlund, H.: Changes in the biomechanical properties of skin and aorta induced by corticotrophin treatment, *Acta Endocrinol., 94,* 132–137, 1980.

135. Oxlund, H.: The influence of a local injection of cortisol on the mechanical properties of tendons and ligaments and the indirect effect on skin, *Acta Orthop. Scand., 51,* 231–238, 1980.

136. Oxlund, H.: Long term local cortisol treatment of tendons and the indirect effect on skin, *Scand. J. Plast. Reconstr. Surg., 16,* 61–66, 1982.

137. Oxlund, H., Manthorpe, R., and Viidik, A.: The biomechanical properties of connective tissue in rabbits as influenced by short-term glucocorticoid treatment, *J. Biomech., 14,* 129–133, 1981.

138. Oxlund, H., and Manthorpe, R.: The biomechanical properties of tendon and skin as influenced by long term glucocorticoid treatment and food restriction, *Biorheology, 19,* 631–646, 1982.

139. Vogel, Beeinflussung der mechanischen eigenschaften der haut von ratten durch hormone, *Arzneim. Forsch., 20,* 1849–1857, 1970.

140. Noyes, F. R., Grood, E. S., Nussbaum, N. S., and Cooper, S. M.: Effect of intra-articular corticosteroids on ligament properties, *Clin. Orthop. Rel. Res., 123,* 197–209, 1977.

141. Oxlund, H.: Changes in connective tissues during corticotrophin and corticosteroid treatment, *Danish Med. Bull., 31,* 187–206, 1984.

142. Andreassen, T. T., Seyer-Hansen, K., and Bailey, A. J.: Thermal stability, mechanical properties and reducible cross-links of rat tail tendon in experimental diabetes, *Biochim. Biophys. Acta, 677,* 313–317, 1981.

143. Swanson, S. A. V.: Biomechanical characteristics of bone, in R. M. Kenedi (ed.), *Advances in Biomedical Engineering,* Academic Press, London, 1971, vol. 1, pp. 137–187.

CHAPTER 7

Muscle Mechanics

Thomas A. McMahon
Departments of Applied Mechanics and Biology
Harvard University
Cambridge, Massachusetts

INTRODUCTION

In many bioengineering problems, it is necessary to understand how muscle works. The heart is a muscular pump; its performance is determined, to an important extent, by muscle mechanics. Mathematical predictions for changes in the performance of the heart caused by disease or by a cardiac assist device depend on a quantitative description of the functioning of cardiac muscle. Similar statements may be made about the respiratory pump and the mechanics of walking and running.

This chapter consists of two major parts. The first section summarizes the fundamental observations underlying a physical model of whole muscle properties and is due mainly to the work of A. V. Hill and his collaborators (1938). This model has been found satisfactory for many types of engineering calculations, although it has important limitations. The second section takes up more detailed mathematical descriptions of crossbridge attachment, detachment, and filament sliding, and considers a model for a very important feature of muscle mechanics, the process by which the tensile force is developed at a crossbridge. All these discussions refer to mammalian and amphibian skeletal muscle, but the same principles have been shown to apply to cardiac muscle, and may be applied with some modifications to vertebrate smooth muscle and insect muscle. This chapter is a much condensed and abridged treatment; for a more detailed discussion, including the experimental evidence behind many of the assumptions, see McMahon (1984).

7.1 MECHANICS OF WHOLE MUSCLE

7.1.1 Structure of the Contractile Machinery

A diagram showing the levels of organization of a skeletal muscle is given in Fig. 7.1. Skeletal muscles generally work in pairs about a joint. In this illustration, the agonist is the biceps, which flexes the elbow, and the antagonist is the triceps, which extends the elbow. Within a whole

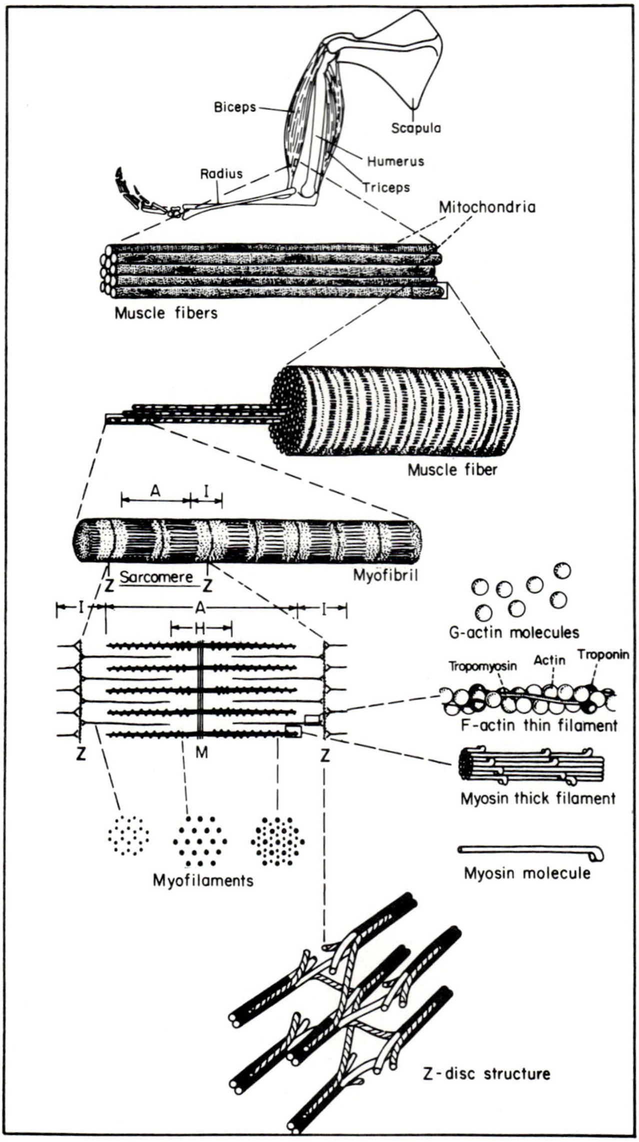

FIG. 7.1 Organization of striated muscle structure, showing the nomenclature applied to the bands and the proteins which make up the thick and thin filaments. (*Adapted in part from Bloom and Fawcett, 1968; Mackean, 1962; and Squire, 1981.*)

muscle, the individual muscle cells are the muscle fibers, surrounded by a membrane, the sarcolemma. Within the sarcolemma are the contractile proteins, bathed by an aqueous solution of organic and inorganic ions, the sarcoplasm. Specialized organelles called *mitochondria* have a role in oxidation of the foodstuffs and resynthesis of ATP, the fuel which the contractile proteins use directly. Myofibrils within a fiber are the smallest units which preserve the regular, repeating bands, or striations. The A band is anisotropic to polarized light; the I band is optically isotropic. In the center of each I band is the Z disk, a region of high refractive index. One sarcomere occupies the space between two Z disks.

The sarcomere is the basic contractile unit of muscle. Vertebrate striated muscle is made up of interdigitating thick and thin filaments. The thin filaments are connected at the Z disks; the thick filaments are connected somewhat more loosely at the M lines in the middle of the A bands. The thick filaments are composed mainly of myosin molecules arranged in such a manner that a bulbous region of each myosin molecule projects out from the thick filament. This projecting segment, which contains two heavy meromyosin S-1 subfragments joined to the main filament by a heavy meromyosin S-2 subfragment, is called a *crossbridge*. Present understanding is that crossbridges extending from the thick filaments combine momentarily with active sites on the thin filaments, creating a shearing force between the filaments which tends to draw the Z lines together. The nature of this shearing force is the main subject of Sec. 7.2.

7.1.2 Activation

The switch that turns the crossbridge activity on and off is located on the thin filaments in mammalian and amphibian striated muscle. The thin filament is made up of two helical strands of F-actin, each formed by polymerized globular G-actin molecules. Within the two grooves on either side of the thin filament are long, rodlike tropomyosin molecules. Troponin molecules located every seven G-actin monomers along the thin filament combine with calcium ions during the contractile event. When this happens, the troponin causes the tropomyosin rods to move slightly, removing a steric restraint at an active site on the thin filament and allowing the myosin crossbridges to attach to the actin.

The calcium ions which trigger the contractile event are normally bound to the sarcoplasmic reticulum, a membranous structure which runs among the myofibrils. Stimulation of the nerve supplying a muscle depolarizes the outer muscle membrane (sarcolemma). The wave of depolarization then runs deeply into the interior of the fiber, causing a release of calcium ions into the sarcoplasm. The calcium ions diffuse through the sarcoplasm until they reach the thin filaments, where they promote contraction by the troponin-tropomyosin mechanism discussed above. The calcium ions are reattached to the sarcoplasmic reticulum at the end of the contractile process, promoting inactivation. Figure 7.2 shows that the calcium "switch" is very sensitive; a change in calcium concentration from 10^{-7} to 10^{-6} M or vice versa is sufficient to turn the force-generating mechanism on and off completely.

7.1.3 Twitch and Tetanus

A whole muscle may be removed from an animal and mounted in an apparatus which allows its force to be measured at constant length (*isometric clamping*). The muscle is still capable of contracting many times, particularly if the blood supply is kept intact, or, if the muscle is small enough, it is maintained in an oxygenated bath. When an electrical stimulus is given by causing a brief current to flow through the bath, the muscle develops force according to a characteristic rising and falling curve called a *twitch* (Fig. 7.3). A series of stimuli produces a series of twitches; if the stimuli follow closely enough together, the force of adjacent twitches overlaps to give an *unfused tetanus*. The higher the frequency of stimulation, the greater the mean force of the unfused tetanus. At about 30 shocks per second in frog muscle at 0°C (50 or 60 per second in mammalian muscle at body temperature), the ripples fuse into a relatively steady force, the *tetanus*. Further increases in frequency above this level do not increase the mean force.

7.1.4 Tetanic Tension–Length Property

When a muscle is tetanized under isometric conditions, the tetanic tension changes as a function of length. In some whole muscles, the tetanic tension is a monotonically increasing function of length, while in others (Fig. 7.4*a*), the tetanic tension curve has a local maximum. The difference between the tetanized and passive (unstimulated) tension-length curves is called the *developed tension*. When a single fiber is removed from a muscle and mounted in a servo device

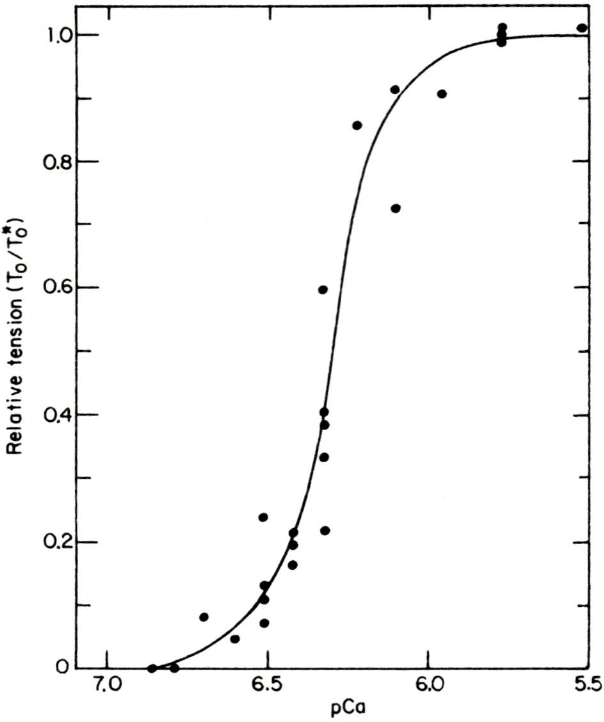

FIG. 7.2 Developed tension in a single fiber of a frog muscle as a function of pCa (negative log of free calcium concentration). This fiber was chemically skinned and held at a length which made the average sarcomere length during contraction between 2.10 and 2.22 μm. The developed tension T_0 is normalized by the tension T_0^* developed at pCa 5.49. (*From Julian and Moss, 1980.*)

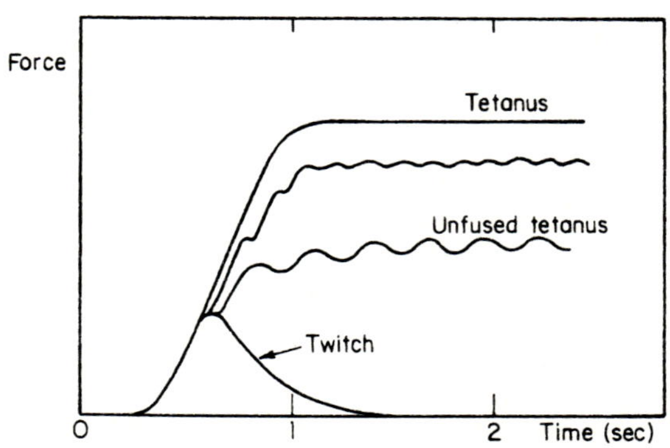

FIG. 7.3 Twitch and tetanus. When a series of stimuli is given, muscle force rises to an uneven plateau (unfused tetanus) which has a ripple at the frequency of stimulation. As the frequency is increased, the plateau rises and becomes smoother, reaching a limit as the tetanus becomes fused.

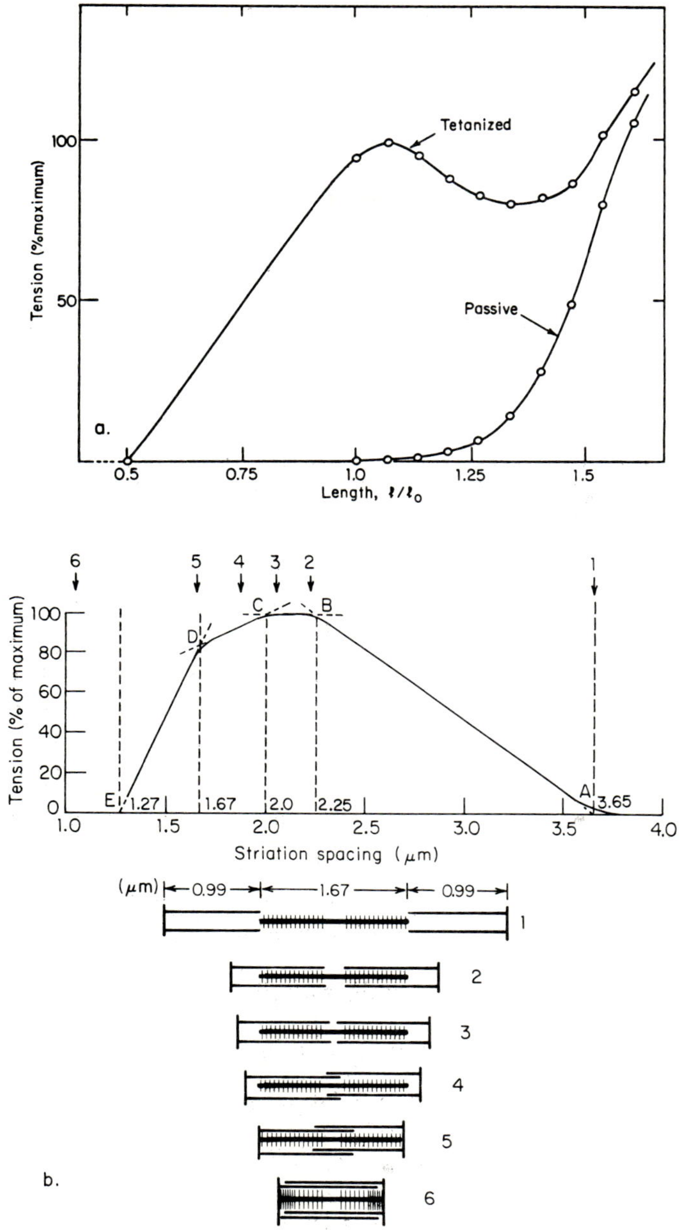

FIG. 7.4 (*a*) Tension-length curves for frog sartorius muscle at 0°C. The passive curve was measured on the resting muscle at a series of different lengths. The tetanized curve was measured at a series of constant lengths as the muscle was held in isometric contraction. At the rest length l_0 the muscle shows essentially no passive tension. (*From Aubert et al., 1951.*) (*b*) Developed tension versus length for a single fiber of frog semitendinosus muscle. The length of the segment was fixed for each measurement by a servo. The sliding-filament diagrams in the lower part of the figure show the appearance of the sarcomere striation pattern at the lengths corresponding to the numbers in the force-length diagram. (*Modified from Gordon et al., 1966.*)

which keeps a marked set of sarcomeres at constant length, the developed tension is found to have the sarcomere length dependence shown in Fig. 7.4*b*. Between points *A* and *B* in the figure, the tension is proportional to the width of the region of overlap between thick and thin filaments. This observation has been interpreted as evidence that the crossbridges are independent force generators working in parallel. Between points *B* and *C* in the figure, the force does not change with length because there exists a crossbridge-free region in the center of the thick filament. The force falls with decreasing length below the optimum length, at first because of interference between the thin filaments (*C–D*), and later because of buckling of the thick filaments (*D–E*) as they encounter the ends of the sarcomere (Z lines).

7.1.5 Quick-Release Experiments

Suppose a whole muscle is stimulated tetanically at constant length (Fig. 7.5). At a certain instant, a catch mechanism withdraws, so that after that instant the muscle is exposed to a constant force (isotonic condition). If the controlled isotonic force after the release is less than the isometric tension before the release, the muscle is observed to shorten in two stages.

The first stage (Δx_2) occurs very rapidly, and is attributed to a *series elastic element* functioning as an undamped spring. In muscles with long tendons, most of the series elasticity is present in the tendon, but there is evidence (see Fig. 7.13*a*) that the crossbridges also account for some of the series elasticity. A series elastic length-tension curve, obtained by many quick-release experiments, is shown in Fig. 7.5*d*.

After the rapid shortening due to series elasticity, a second stage of shortening occurs due to sliding of the thick with respect to the thin filaments. This slower length change (Δx_1) eventually brings the muscle to a new equilibrium point on the tetanic tension–length curve.

7.1.6 Force-Velocity Property

The rate of sarcomere shortening, dx_1/dt, immediately after Δx_2 has concluded, is found to be a function of the isotonic force (Fig. 7.6). By defining the muscle shortening velocity $v = -dx_1/dt$, the relation between force and steady shortening velocity may be given (empirically) by the *Hill equation* (Hill, 1938):

$$(T + a)(v + b) = (T_0 + a)b \tag{7.1}$$

or

$$\frac{v}{v_{max}} = \frac{1 - T/T_0}{1 + T/kT_0} \tag{7.2}$$

where $v_{max} = bT_0/a$ is the shortening velocity against no load, and $k = a/T_0 = b/v_{max}$. For most muscles, $0.15 < k < 0.25$. Here T_0 is the isometric tension against which the muscle neither shortens nor lengthens. The mechanical power produced by a shortening muscle is

$$\text{Power} = Tv = \frac{v(bT_0 - av)}{v + b} \tag{7.3}$$

When $k = 0.25$, the maximum power is delivered when $T = 0.309T_0$, $v = 0.309v_{max}$. Under those circumstances, the power is $0.096T_0v_{max}$.

It is important to note that Eqs. (7.1) and (7.2) apply only when the muscle is shortening, i.e., for positive v. For negative v, the force-velocity relation for small lengthening velocities is found to have a slope approximately six times steeper than for small shortening velocities.

When a muscle is forced to lengthen at moderate velocities, the force reaches a plateau near $1.8T_0$. This is called muscle *yielding*. At higher velocities of lengthening, the force may actually drop below the T_0 appropriate for the instantaneous muscle length.

The Hill equation [Eq. (7.1)] has been found to describe the relation between force and steady shortening velocity in nearly all muscles thus far examined, including cardiac and

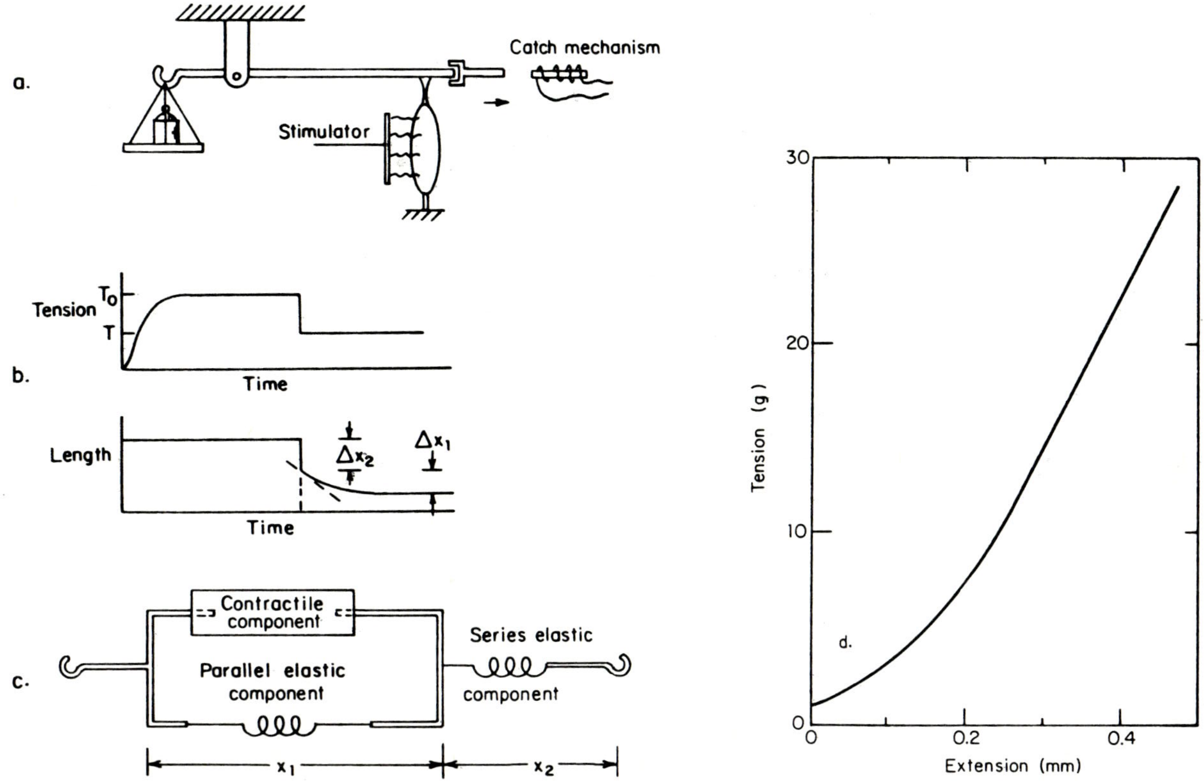

FIG. 7.5 (*a*) Quick-release apparatus. When the catch is withdrawn, the muscle is exposed to a constant force determined by the weight in the pan. (*b*) The muscle is stimulated tetanically. Upon release of the catch, the muscle shortens rapidly by an amount $\Delta\, x_2$ which depends on the difference in force before and after release. (*c*) A conceptual model of muscle. (*d*) Length-tension curve for the series elastic component (SEC) in a frog sartorius muscle. A change in force equal to the maximum isometric value produces a change in length of the SEC of about 2 percent of the muscle rest length. [*Part* (*d*) *modified from Jewell and Wilkie, 1958.*]

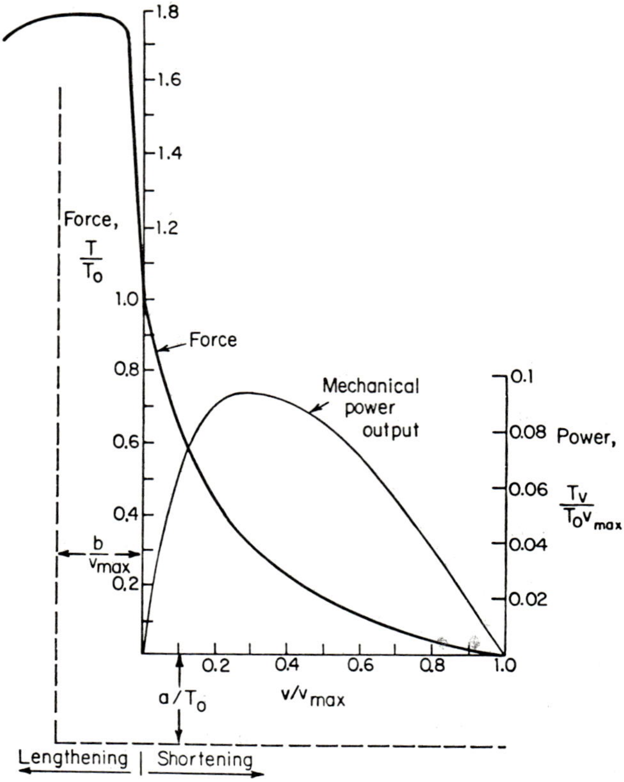

FIG. 7.6 Hill's force-velocity curve. The shortening part of the curve was calculated from Eq. (7.2) with $k = 0.25$. The asymptotes for Hill's hyperbola (broken lines) are parallel to the T/T_0 and v/v_{max} axes. Near zero shortening velocity, the lengthening part of the curve has a negative slope approximately six times steeper than the shortening part. The externally delivered power was calculated from the product of tension and shortening velocity.

smooth muscle as well as skeletal muscle. With Figs. 7.4*a* and 7.6, a three-dimensional surface may be specified for an individual muscle giving the instantaneous force at a particular length and velocity, assuming full (fused) tetanus.

7.1.7 Active-State Tension

Long before myosin crossbridges were identified as the force generators in muscle, Hill suggested that the contractile component in Fig. 7.5*c* might be represented by a force generator in parallel with a nonlinear damping element, as shown in Fig. 7.7. Modern evidence shows that the force generator, damping element, and series spring are all to a great extent properties of the crossbridges, and therefore may not be separated from one another as they are in Fig. 7.7. Nevertheless, many of the gross mechanical features of muscle may be modeled this way in good agreement with experimental observations.

Hill defined the *active-state tension* T_0 as the instantaneous force which the contractile component could bear while neither lengthening nor shortening. Suppose a muscle is held at a length near the optimum for developed tension, so that the spring K_{PE} bears little or no tension.

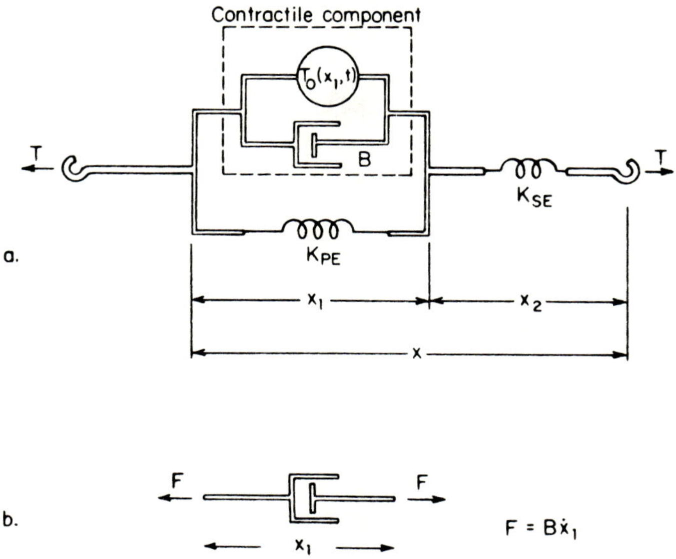

FIG. 7.7 (*a*) Active-state muscle model. The active state $T_0(x_1, t)$ is the tension developed by the force generator in the circle. (*b*) The dashpot element resists with a force proportional to the velocity.

In an isometric tetanus, active-state tension equals the developed isometric tension. For this reason, T_0 is a function of the length of the contractile component, x_1. In a twitch, T_0 rises to a peak and then falls. Because the damping element B subtracts some of the force whenever the contractile element has a nonzero shortening velocity, the measured tension T is equal to T_0 only when $\dot{x}_1 = 0$. Since $\dot{T} = K_{SE}\dot{x}_2$, and since $\dot{x}_1 = -\dot{x}_2$ under isometric conditions, $T = T_0$ whenever $\dot{T} = 0$ in such a muscle.

Using these arguments, Edman (1970) constructed a curve for $T_0(t)$ during an unfused tetanus (Fig. 7.8, broken line). A series of isometric stimulations produced a $T(t)$ curve with a set of maxima and minima. Subsequently the same muscle, receiving the same stimulation, began from a longer length and was released to shorten against a light load before becoming isometric again as it encountered a stop. The final length of the muscle was kept always the same. The broken line representing $T_0(t)$ was drawn through the maxima and minima of the several $T(t)$ curves.

Important conclusions are as follows: (1) The active-state tension $T_0(t)$ rises very rapidly. After a latency period lasting about 12 ms from the time of stimulation, the active-state tension rises sharply, requiring only 3 or 4 ms to increase from 25 to 65 percent of its maximum. (2) The active state T_0 reaches a much higher maximum force than the twitch tension $T(t)$ in a typical twitch. If the twitch tension is turned on and off as a square wave of duration C, if the damping element is a linear dashpot with damping constant B, if the muscle is held at its rest length so that K_{PE} is slack, and if K_{SE} is a linear spring, then the ratio of peak twitch tension to T_0 is

$$\frac{T_{\text{twitch}}}{T_0} = 1 - \exp \frac{-K_{SE}C}{B} \tag{7.4}$$

where K_{SE} and K_{PE} are spring constants of elements as shown in Fig. 7.7.

It is clear that this result depends on the ratio of C, the characteristic time the active state is "on," to B/K_{SE}, the muscle time constant. For frog muscle at room temperature, B/K_{SE} is roughly 30 ms; near 0°C, it is about 150 ms. Fast muscles, also called *twitch muscles,* have a T_{twitch}/T_0 ratio greater than 0.5.

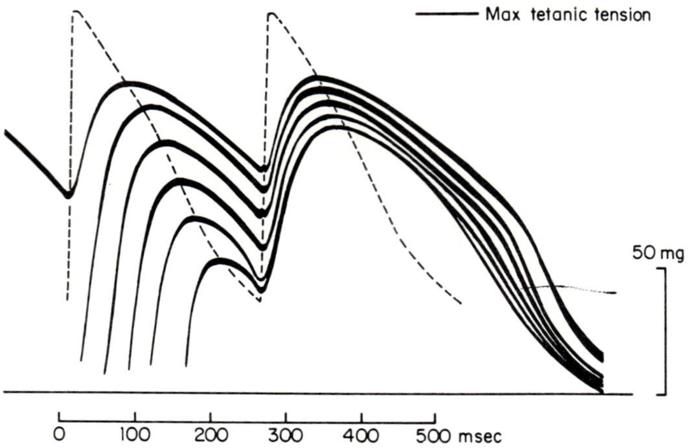

FIG. 7.8 Edman's experiment for determining both the rising and falling portions of the active-state curve. The top curve shows the tension in a series of twitches at constant length for a single fiber of frog semitendinosus muscle at 1°C. Lower curves show the tension when the muscle was released to the same length at various times during the next-to-last contraction cycle. The active-state curves (broken lines) are extrapolated from lines connecting the maxima and minima of the tension curves. These extrapolations intersect near the level of maximum tetanic tension for this fiber. (*From Edman, 1970.*)

7.2 MATHEMATICAL MODELS OF CROSSBRIDGE FUNCTION

7.2.1 Huxley's 1957 Model for Filament Sliding

Specifying the Model. In a classical paper published in 1957, Andrew F. Huxley proposed a mathematical model of crossbridge attachment, filament sliding, and crossbridge detachment designed to account for most of what was known about muscle then, including the energetic as well as the mechanical features. As shown in Fig. 7.9, the myosin crossbridge M was visualized as a socket element mounted on a spring suspension. Random thermal motions would cause the socket to vibrate. Occasionally when it came within range of an active site A on the thin filament, the crossbridge would attach itself to the filament. A shearing force between the thick and thin filaments would then act because of the distortion of the crossbridge spring. The displacement between the equilibrium position of an M site and the nearest A site on a thin filament was defined as x. When $x = 0$, no force would result if the crossbridge were attached. For $x > 0$, an attached crossbridge would generate a tensile force, i.e., one acting to shorten the sarcomere.

Let $n(x, t)$ be the probability that an arbitrarily chosen crossbridge, whose displacement is x, is attached. Assuming first-order rate kinetics,

$$\frac{Dn(x, t)}{Dt} = [1 - n(x, t)]f(x) - n(x, t)g(x) \tag{7.5}$$

where $Dn(x, t)/Dt$ is the convective derivative

$$\frac{Dn(x, t)}{Dt} = \frac{\partial n(x, t)}{\partial t} - v\frac{\partial n(x, t)}{\partial x} \tag{7.6}$$

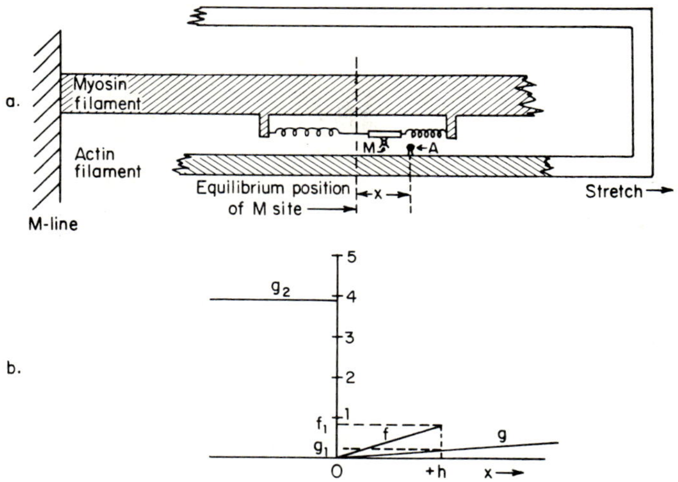

FIG. 7.9 A. F. Huxley's 1957 model for isotonic contraction of striated muscle. (*a*) The thick filament is fixed at the *M* line; the thin filaments move to the right if the muscle is stretched. (*b*) Rate constants for attachment and detachment are not symmetric about $x = 0$. The unit of the vertical scale is the value of $(f + g)$ at $x = h$. (*From Huxley, 1957.*)

with $v = -dx/dt$. Here $f(x)$ and $g(x)$ are rate constants

$$x < 0: \qquad f(x) = 0 \qquad g(x) = g_2 \tag{7.7}$$

$$0 \le x \le h: \qquad f(x) = \frac{f_1 x}{h} \qquad g(x) = \frac{g_1 x}{h} \tag{7.8}$$

$$x > h: \qquad f(x) = 0 \qquad g(x) = \frac{g_1 x}{h} \tag{7.9}$$

The assumption of this form for the rate constants ensures several necessary features: (1) There is a nonzero rate of attachment only for $x > 0$; hence the crossbridges will attach only in the positive force-producing range of x. (2) The attachment rate is zero for $x > h$. This reflects the assumption that there is a limited range of x for attachment. (3) The detachment rate constant is high in the region where $x < 0$ (see Fig. 7.9*b*), by comparison with the region where $x > 0$. Thus the rate of detachment, and consequently the rate of cycling of crossbridges and splitting of ATP, is slow unless the muscle is allowed to shorten.

As it stands, the model does not explicitly include the tension-length property. It must be modified slightly to be applied beyond the plateau region (C–B) in Fig. 7.4*b*.

Calculations of the force for a given time and shortening velocity are made in the following steps.

1. Eq. (7.5) is solved, using Eqs. (7.6) to (7.9), in each of the three regions $x < 0$, $0 \le x \le h$, and $x > h$, subject to the matching conditions

$$n(h^+, t) = n(h^-, t) \tag{7.10}$$

$$n(0^+, t) = n(0^-, t) \tag{7.11}$$

2. The tension per unit cross-sectional area T is obtained from

$$T = \frac{msk}{2l} \int_{-\infty}^{\infty} n(x, t)\, dx \tag{7.12}$$

where m is the number of crossbridges per unit volume, s is the length of one half sarcomere, k is the spring stiffness of the crossbridge, and l is the spacing distance of the crossbridges along a thick filament.

3. The rate of chemical energy utilization per unit volume may be calculated from

$$E = \frac{me}{l} \int_{-\infty}^{\infty} f(x)[1 - n(x, t)]\, dx \tag{7.13}$$

where e is the energy liberation per crossbridge in one cycle.

For a derivation of Eqs. (7.12) and (7.13), see McMahon (1984) or Huxley (1957).

Steady Shortening. When the muscle is shortening under isotonic (constant-force) conditions, neither the velocity nor the crossbridge distribution n is changing with time. Thus Eq. (7.6) becomes

$$\frac{Dn}{Dt} = -v\frac{dn}{dx} \tag{7.14}$$

and the solution for $n(x)$ in the three regions is

$x < 0$:
$$n = \frac{f_1}{f_1 + g_1}(1 - e^{-\phi/V})^{2xg_2/sV} \tag{7.15}$$

$0 \le x \le h$:
$$n = \frac{f_1}{f_1 + g_1}(1 - e^{(x^2/h^2 - 1)\phi/V}) \tag{7.16}$$

$x > h$:
$$n = 0 \tag{7.17}$$

where V is a normalized rate of shortening in muscle lengths per second defined by

$$v = \frac{sV}{2} \tag{7.18}$$

and

$$\phi = \frac{(f_1 + g_1)h}{s} \tag{7.19}$$

The crossbridge distributions for four values of the normalized shortening velocity, including the isometric condition $V = 0$, are plotted in Fig. 7.10. The tension per unit area T and rate of energy liberation per unit volume E are given by

$$T = \frac{msk}{2l}\left(\frac{f_1}{f_1 + g_1}\right)\frac{h^2}{2}\left\{1 - \frac{V}{\phi}(1 - e^{-\phi/V})\left[1 + \frac{1}{2}\left(\frac{f_1 + g_1}{g_2}\right)^2 \frac{V}{\phi}\right]\right\} \tag{7.20}$$

and

$$E = \frac{meh}{2l}\left(\frac{f_1}{f_1 + g_1}\right)\left[g_1 + f_1\frac{V}{\phi}(1 - e^{-\phi/V})\right] \tag{7.21}$$

These results, calculated from Eqs. (7.12) and (7.13) using Eqs. (7.15) to (7.17), are plotted in Fig. 7.11. The dimensionless ratios $g_1/(f_1 + g_1)$, $g_2/(f_1 + g_1)$, and $kh^2/2e$ were given the numerical values 3⁄16, 3.919, and 3⁄4, respectively, to make the result fit experimental observations. Here T_0 and E_0 are the isometric (active-state) tension per cross-sectional area and the isometric rate of energy utilization,

$$T_0 = \frac{mskh^2}{4l}\left(\frac{f_1}{f_1 + g_1}\right) \tag{7.22}$$

and

$$E_0 = \frac{mehg_1}{2l}\left(\frac{f_1}{f_1 + g_1}\right) \tag{7.23}$$

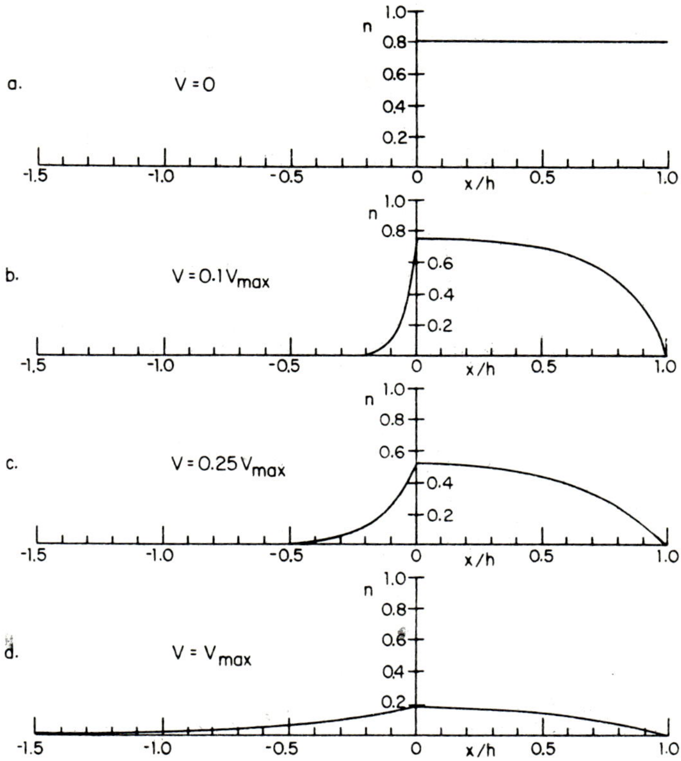

FIG. 7.10 Distributions of attached crossbridges over the dimensionless displacement x/h. (*a*) At zero shortening speed, an arbitrary crossbridge is equally likely to be attached in any positive tensile force–generating position up to the limit $x = h$. At successively higher speeds (*b*, *c*, and *d*) crossbridges which attach in the positive force region, $x/h > 0$, are swept into the negative force region, $x/h < 0$, before detaching. At the highest shortening speed, V_{max}, the force contributed by crossbridges attached in the region where $x/h > 0$ just balances the retarding influence of those still attached in the negative force-producing region, so that the net force is zero. (*From Huxley, 1957.*)

Steady Lengthening. When the muscle is forced to lengthen at constant velocity, dx/dt is positive and V is negative. The solution for $n(x)$ is as follows:

$x < 0$: $$n = 0 \tag{7.24}$$

$0 \le x \le h$: $$n(x) = \frac{f_1}{f_1 + g_1}\left(1 - e^{(\phi/V)(x^2/h^2)}\right) \tag{7.25}$$

$x > h$: $$n(x) = \frac{f_1}{f_1 + g_1}\left(1 - e^{\phi/V}\right)e^{(-g_1 h/sV)(1-x^2/h^2)} \tag{7.26}$$

The tension per cross-sectional area is given by evaluating Eq. (7.12), using Eqs. (7.24) to (7.26):

$$T = \frac{msk}{2l}\left(\frac{f_1}{f_1 + g_1}\right)\frac{h^2}{2}\left[1 - \frac{Vf_1}{\phi g_1}\left(1 - e^{\phi/V}\right)\right] \tag{7.27}$$

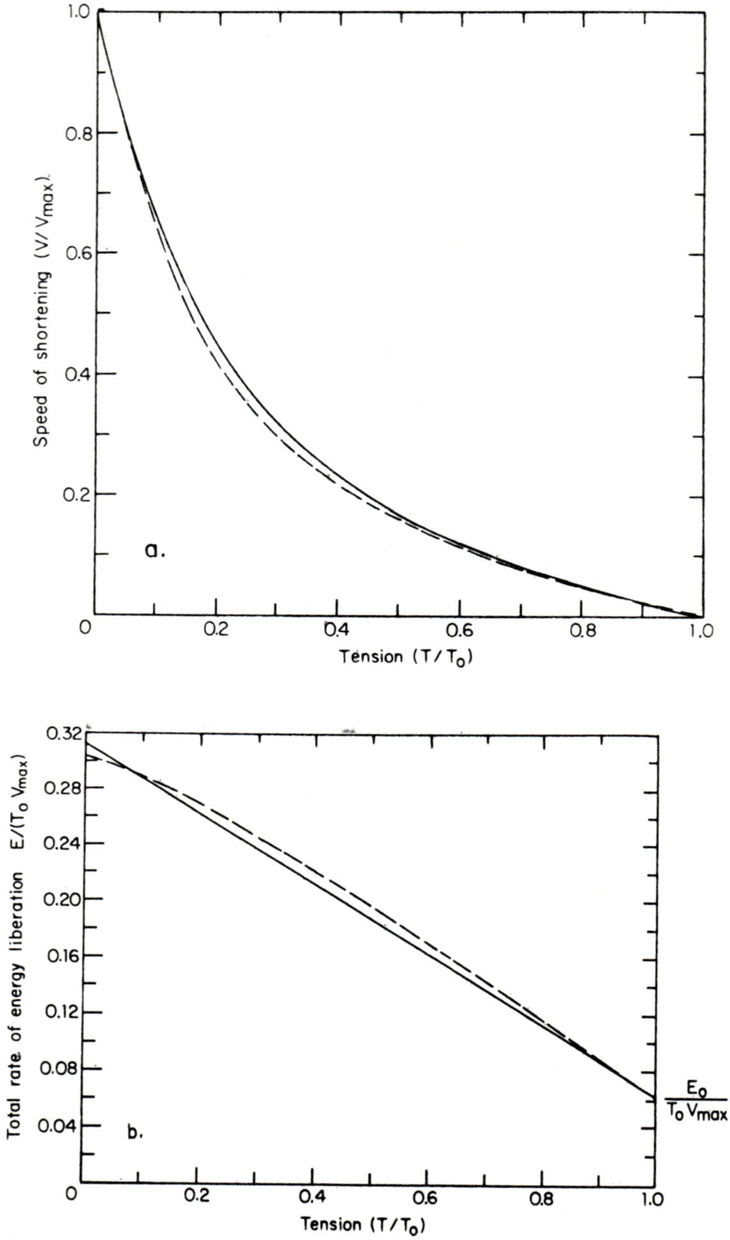

FIG. 7.11 (*a*) Speed of shortening versus tension. Solid line shows result of experiments [Hill equation, Eq. (7.1), with $a/T_0 = 1/4$]; broken line shows Huxley's 1957 theory with constants chosen to provide best fit. (*From Huxley, 1957.*) (*b*) Total rate of energy liberation versus tension. Solid straight line shows empirical results summarizing experiments. The maintenance heat rate E_0 is given by $T_0V_{max}/16$. The broken line shows the results calculated from Huxley's 1957 model. (*From Huxley, 1957.*)

The crossbridge distribution now includes no crossbridges attached in the negative force-producing region ($x < 0$), but as shown in Eq. (7.26), there is a "tail" of crossbridges extending into the region $x > h$. These bridges attached in the region $0 \leq x \leq h$ and were dragged into the region $x > h$ by the motion of the sliding filaments before detaching.

It is interesting to calculate the ratio of the slopes of the force-velocity curve appropriate for small lengthening and small shortening velocities. Evaluating dT/dV at $V = 0^-$ from Eq. (7.27), and dT/dV at $V = 0^+$ from Eq. (7.20), we obtain

$$\frac{dT/dV|_{V=0^-}}{dT/dV|_{V=0^+}} = \frac{f_1}{g_1} \tag{7.28}$$

For values of the parameters chosen by Huxley to plot Fig. 7.11, $f_1/g_1 = 4.33$. Since the experimental observation was that the discontinuity in slope of the force-velocity relation at zero velocity was about a factor of 6.0 (Fig. 7.6), the model makes a very successful prediction of this important effect.

When the velocity of stretching is large, Eq. (7.27) reaches a limit:

$$T|_{V \to -\infty} = \frac{T_0(f_1 + g_1)}{g_1} \tag{7.29}$$

For the parameters chosen, this asymptotic force is $5.33T_0$, which is larger than the $1.8T_0$ observed in the experiments used to draw Fig. 7.6, but still in good qualitative agreement with the yielding phenomenon. The yielding occurs in the model for the same reasons it is presumed to occur in actual muscle. As the stretching speed becomes larger, fewer crossbridges are attached because they are dragged quickly through the attachment region $0 \leq x \leq h$. Those bridges which do attach generate more force, however, because they are stretched to large x displacements before they can detach. Small modifications of the shape and magnitude of the rate constants (Fig. 7.9b) could result in a somewhat better prediction of the yielding phenomenon, at the expense of a poorer prediction of the other mechanical and energetic results.

7.2.2 The Crossbridge Force Generator

Tension Transient Experiments. Several years after the publication of Huxley's 1957 model, a new experimental result came along which required an extension of the theory (Huxley and Simmons, 1971). This extension was relevant only to the period when a crossbridge was attached; the attachment and detachment mechanics described by the 1957 model did not require changes.

A typical result of the experiment is shown in Fig. 7.12. The length of a small region of a single muscle fiber is controlled by a servo device of the same type used in the experiments depicted by Fig. 7.4b. A small negative step change in length (less than 0.5 percent of the initial length) is imposed upon the marked region of the muscle fiber in as little as 0.2 ms. A length change this small should not dislodge attached crossbridges. The tension drops synchronously with the length, as would be the case if the crossbridges were undamped springs. The minimum in the tension record is designated T_1. Just after the length change is complete, the tension rises again and remains close to a plateau level (called T_2) for some time before finally recovering the full value it had before the (extremely small) length change.

The observed relationship between T_1, T_2, and the amplitude of the length step is shown in Fig. 7.13a. When the length of the fiber is changed, the T_1 and T_2 curves have the same shape, as would be expected if fewer crossbridges were acting in parallel (broken curves). Figure 7.13b shows that the rate of tension rise from T_1 to T_2 is extremely length dependent, being much faster for releases than for stretches.

Two Attached States. In an important paper reporting the results of the tension transient experiments discussed above, Huxley and Simmons (1971) showed how their experimental data could be fitted by a theoretical model in which the crossbridge makes transitions between two attached states. Subsequently, biochemical evidence has appeared supporting the idea that there actually are two distinct attached states.

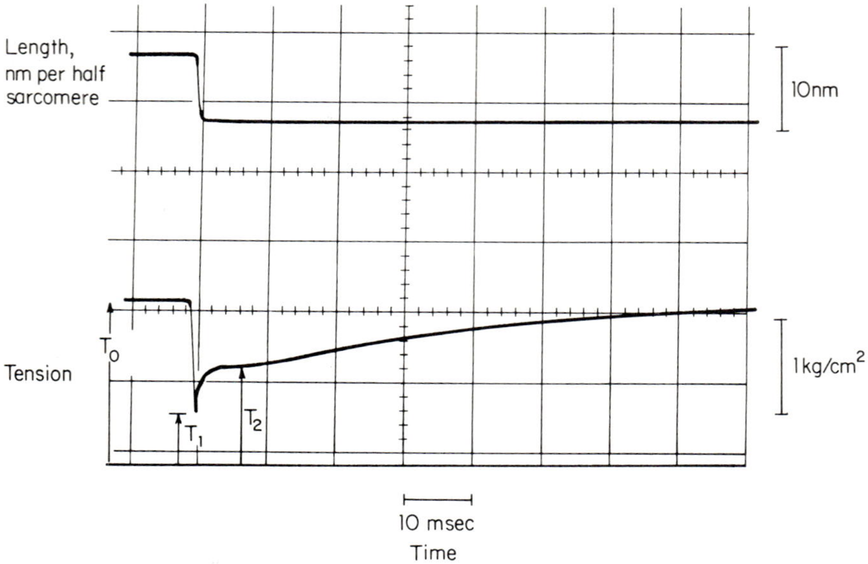

FIG. 7.12 Length (top) and tension (bottom) for a small step decrease in length of a tetanized single fiber from frog anterior tibialis at 0°C. Tension initially drops synchronously with the length change (T_1), then rapidly recovers, reaching an early plateau (T_2). Following the T_2 plateau, tension slowly returns to the original level. Only this final phase is thought to be associated with crossbridge attachment, filament sliding, and detachment. The length step in this experiment took approximately 1.0 ms. Later experiments using faster length steps (0.2 ms) produced the information shown in Fig. 7.13*a*. The T_1 for a given length step was higher using the slower step presumably because the early tension recovery process had begun before the length step was complete. (*From Huxley, 1974.*)

As shown in Fig. 7.14, the crossbridge head (subfragment S-1) can rock from state 1 to state 2, stretching the elastic link S-2, whenever the total energy of state 2 is lower than that of state 1. The displacement of the thin filament from its original position is y, so that positive y corresponds to a stretch, negative y to a release. The y displacement between the two states is h. The transitions between states are assumed to be controlled by Boltzmann statistics.

The potential energy stored in the (assumed linear) S-2 spring, U_s, is given by

$$U_s = \tfrac{1}{2}Kl^2 \tag{7.30}$$

where l is the length of the S-2 spring and K is the spring constant. Defining x as the S-1 link rotation parameter, where $x = -h/2$ in state 1 and $x = h/2$ in state 2,

$$l = y_0 + y + x \tag{7.31}$$

where y_0 is the length of the spring in the isometric condition before the quick change in y. If Eq. (7.31) is substituted into Eq. (7.30), the potential energy of the S-2 spring takes the form

$$U_s = \frac{K}{2}(y_0 + y + x)^2 \tag{7.32}$$

In Fig. 7.15*a*, U_s is shown for three values of y, assuming y_0 is fixed. In the initial isometric condition, $y = 0$, and the energy is a quadratic function of x with an intercept $U_s = Ky_0^2/2$ at $x = 0$. For positive y (a stretch), the curve is scaled upward.

A potential energy function is assigned arbitrarily to the rocking motion of the S-1 link (Fig. 7.15*b*). This includes two potential energy wells and three plateau segments.

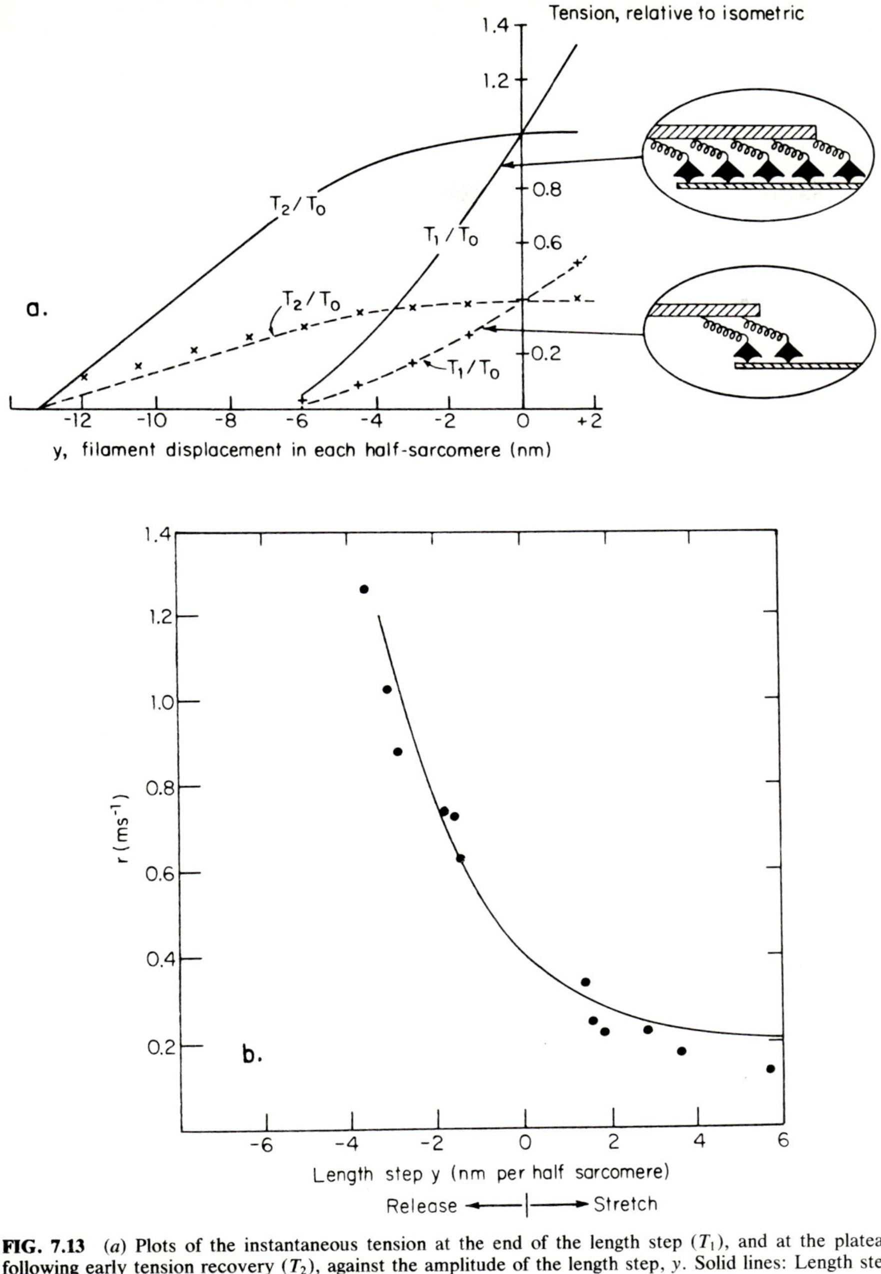

FIG. 7.13 (*a*) Plots of the instantaneous tension at the end of the length step (T_1), and at the plateau following early tension recovery (T_2), against the amplitude of the length step, y. Solid lines: Length step applied when the sarcomere length was 2.2 μm. Crosses: The same fiber has been stretched to sarcomere length 3.1 μm, reducing the overlap to only about 39 percent of what it was at 2.2 μm. The broken-line plots are scaled-down versions of the solid-line plots, multiplied by the factor 0.39. (*From Huxley, 1974.*) (*b*) Rate constant r (approximately the inverse of the time constant τ) for approach to the T_2 plateau. Points show results for frog muscle at 4°C; r has been estimated from $(\ln 3)/t_{1/3}$, where $t_{1/3}$ is the time required for the force to change from T_1 to $(2T_2 + T_1)/3$. Solid line shows a purely empirical fit to the data, $r = 0.2\,(1 + \exp - 0.5y)$. The rate of the tension recovery is substantially faster for releases than for stretches. (*From Huxley and Simmons, 1971.*)

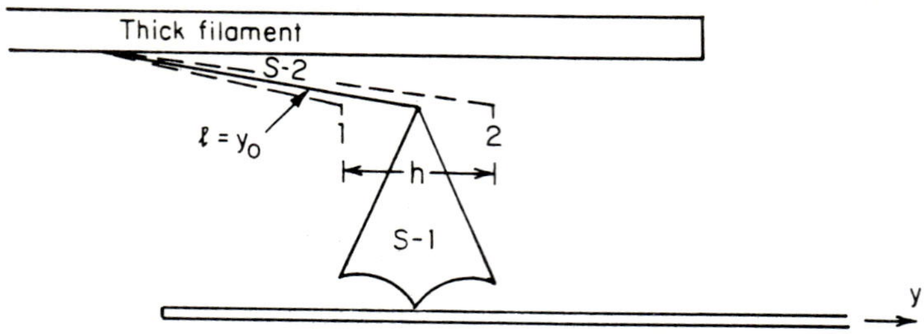

FIG. 7.14 Coordinate system for Huxley and Simmons model (1971). There are two stable attached states; here the S-1 element is shown midway between states 1 and 2. In this configuration, the length of the S-2 element is y_0; in general, its length is l. The rest (zero-force) length of the S-2 spring is zero; it does not buckle. The thick filament is assumed fixed. The displacement of the thin filament is measured by y, so that $y > 0$ corresponds to a stretch. The angle between the S-2 segment and the thick filament is taken to be nearly zero. A release of $-y = y_0$ brings the force to zero momentarily.

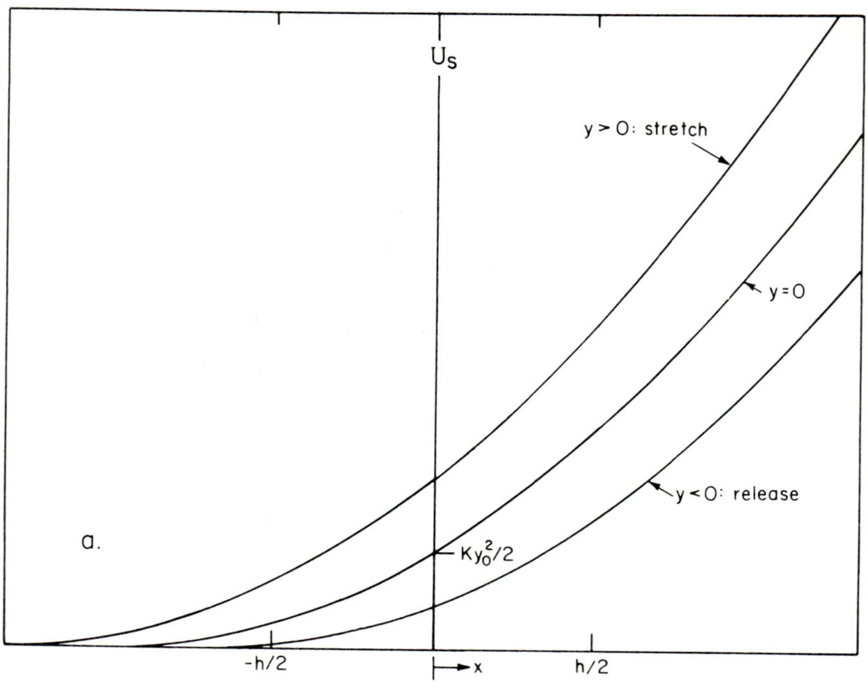

FIG. 7.15 Potential energy diagrams. (a) U_s, for the spring alone, is $U_s = \frac{1}{2}K(y_0 + y + x)^2$, shown for three values of y. The spring length, $l = y_0 + y + x$, has been written as the sum of a part due to filament sliding, y, and a part due to S-1 link rotation, x. (b) Potential energy of conformation of the S-1 link. (c) Total potential energy, sum of (a) plus (b), when $y = 0$. This corresponds to the isometric initial condition. Note that states 1 and 2 have the same total potential energy (at $-h/2$ and $h/2$).

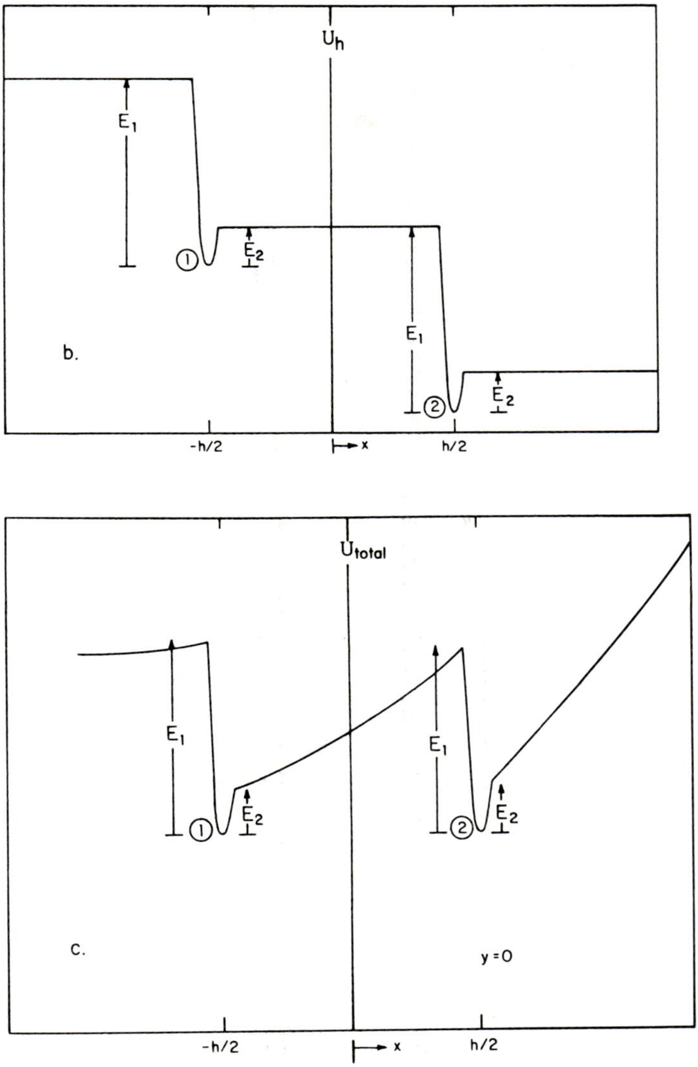

FIG. 7.15 ***(Continued)***

The total potential energy is the sum of U_s and U_h. It is assumed that in the initial condition of steady-state isometric stimulation, half the bridges are in either state. It follows that the total potential energy of a crossbridge at the bottom of the potential energy well is the same in state 1 and state 2. This is the condition shown in Fig. 7.15*c*.

After a quick stretch, the increase in potential energy contributed by stretching the S-2 spring is greater at $x = h/2$ than at $x = -h/2$. This means that the total energy well is lower for state 1 than for state 2 (Fig. 7.16*a*), and crossbridges tend to shift from state 2 to state 1, causing the tension to fall. The converse is true for a step decrease in length (Fig. 7.16*b*).

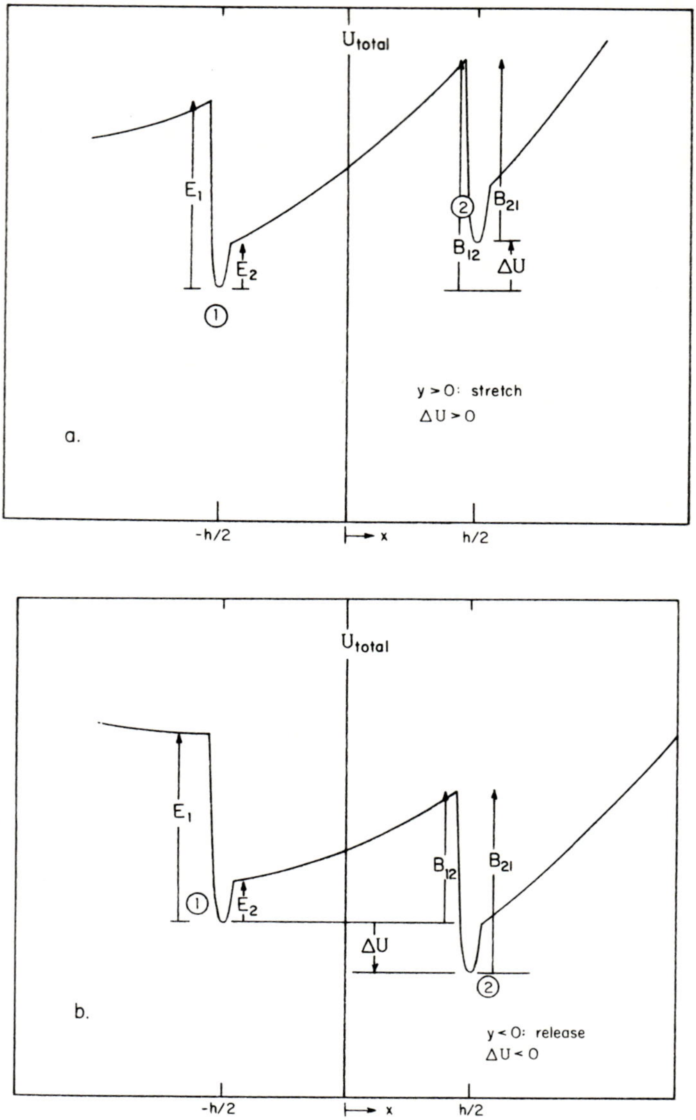

FIG. 7.16 Total potential energy versus S-1 link rotation parameter x. (*a*) For a stretch, $y > 0$, state 1 has a deeper potential energy well than state 2, and transitions tend to occur from 2 to 1. (*b*) After a release, $y < 0$, the transitions tend to occur from 1 to 2, stretching the S-2 spring and causing the tension to rise.

If k_+ and k_- are the rate constants for transitions from 1 to 2 and from 2 to 1,

$$\frac{k_+}{k_-} = Ce^{(B_{21}-B_{12})/\kappa\theta} \tag{7.33}$$

where C is a constant of proportionality to be determined and B_{21} and B_{12} are the activation energies

$$B_{21} = E_1 \tag{7.34}$$

and

$$B_{12} = \Delta U + E_1 \tag{7.35}$$

The difference in total potential energy between states 2 and 1 is ΔU. This ΔU is made up of two parts, a ΔU_h due to conformational energy,

$$\Delta U_h = E_2 - E_1 \tag{7.36}$$

plus a ΔU_s due to S-2 spring energy,

$$\Delta U_s = Kh(y_0 + y) \tag{7.37}$$

Summing Eqs. (7.36) and (7.37),

$$\Delta U = Kh(y_0 + y) + E_2 - E_1 \tag{7.38}$$

Using the assumption that during isometric tetanus each state is equally probable, $\Delta U = 0$ for $y = 0$. Applying this to Eq. (7.38),

$$E_2 - E_1 = -Khy_0 \tag{7.39}$$

Therefore, Eq. (7.38) becomes

$$\Delta U = Khy \tag{7.40}$$

Since $B_{21} - B_{12} = -\Delta U$, Eq. (7.33) becomes

$$\frac{k_+}{k_-} = Ce^{-\Delta U/\kappa\theta} = Ce^{-Khy/\kappa\theta} \tag{7.41}$$

From the isometric condition, when $\Delta U = 0$, k_+ and k_- are equal. Thus $C = 1$, and

$$\frac{k_+}{k_-} = e^{-Khy/\kappa\theta} \tag{7.42}$$

The Rate Equation. Let n_1 and $n_2 = 1 - n_1$ be the fractions of the total number of crossbridges attached in states 1 and 2. Assuming first-order kinetics,

$$\frac{dn_2}{dt} = k_+n_1 - k_-n_2 \tag{7.43}$$

Substituting $n_1 = 1 - n_2$,

$$\frac{dn_2}{dt} = -(k_+ + k_-)n_2 + k_+ \tag{7.44}$$

Subject to the initial condition for isometric tetanus,

$$n_2(0) = \tfrac{1}{2} \tag{7.45}$$

the solution is

$$n_2 = n_2^\infty + (\tfrac{1}{2} - n_2^\infty)e^{-t/\tau} \tag{7.46}$$

Both the time constant τ and the final value n_2^∞ depend on y, the magnitude of the quick length change.

$$\tau = \frac{1}{k_+ + k_-} = \left[k_-\left(1 + \exp\frac{-Khy}{\kappa\theta}\right)\right]^{-1} \tag{7.47}$$

$$n_2^\infty = \frac{k_+}{k_+ + k_-} = \frac{1}{2}\left(1 - \tanh\frac{Khy}{2\kappa\theta}\right) \tag{7.48}$$

Predicting the Tension Transients. Defining $F_1(y)$ and $F_2(y)$ as the force contributed by a single crossbridge in states 1 and 2, respectively, the total muscle force per crossbridge is

$$F(t) = n_1(t)F_1 + n_2(t)F_2 \tag{7.49}$$

Since

$$F_1 = K\left(y_0 + y - \frac{h}{2}\right) \tag{7.50}$$

and

$$F_2 = K\left(y_0 + y + \frac{h}{2}\right) \tag{7.51}$$

then

$$F(t) = K[y_0 + y + (n_2 - ½)h] \tag{7.52}$$

The force plateau T_2 at the end of the quick recovery period is found by evaluating Eq. (7.52) as $t \to \infty$:

$$T_2 = F(\infty) = K\left(y_0 + y - \frac{h}{2}\tanh\frac{Khy}{2\kappa\theta}\right) \tag{7.53}$$

The T_1 transient also follows from Eq. (7.49). Here the entire length step y is imposed on the S-2 spring while n_1 and n_2 remain at their original values (½):

$$T_1 = ½F_1(y) + ½F_2(y) = K(y_0 + y) \tag{7.54}$$

Notice that a release of amplitude y_0 brings the tension instantaneously to zero.

Fixing the Constants from Experiments. The experimental results for the rate constant for rapid tension recovery (Fig. 7.13*b*) may be fit by an equation of the form

$$r = \frac{r_0}{2}(1 + e^{-\alpha y}) \tag{7.55}$$

Since this is the same form as Eq. (7.47), with $\tau = 1/r$, Huxley and Simmons assigned

$$\alpha = \frac{Kh}{\kappa\theta} \tag{7.56}$$

and

$$\frac{r_0}{2} = k_- \tag{7.57}$$

In the experiments, α had a value close to 0.5 nm^{-1} and $r_0/2$ was approximately 0.4 ms^{-1}.

Since a release of magnitude y_0 drops the tension instantaneously to zero, the experimental T_1 curve should intersect the y axis at $-y_0$. In their original paper, which was based on early experiments with slower equipment, Huxley and Simmons (1971) estimated $y_0 = 8.0$ nm. This was the value used to calculate the data shown in Fig. 7.17. More recent experiments with improved equipment (Ford, Huxley, and Simmons, 1977) show $y_0 \approx 5.0$ nm.

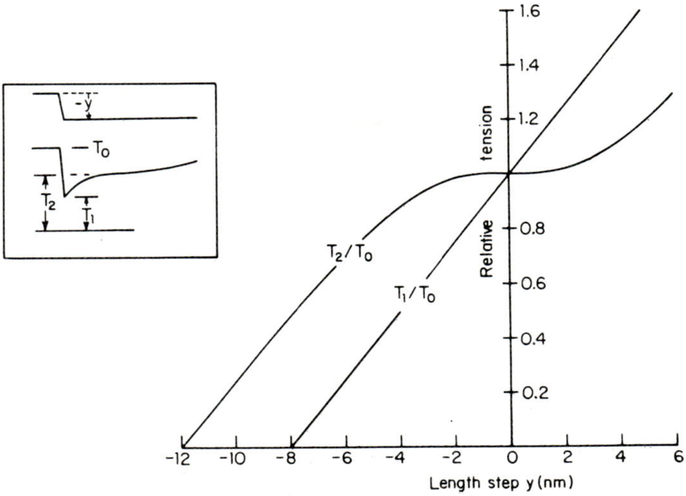

FIG. 7.17 Calculated curves of T_1 and T_2 versus length step y, showing predictions of the Huxley and Simmons model. (*From Huxley and Simmons, 1971.*)

The parameter h was determined by using Eq. (7.53) to match the shape of the experimental T_2 curve. The original paper gave $h = 8.0$ nm; later experiments required a revision to $h = 6.0$ nm.

7.2.3 Combined Model: Sliding and Flipping

The essential features of the Huxley 1957 model for filament sliding and the Huxley and Simmons 1971 model for crossbridge flipping may be combined into a single, self-consistent model, but not without loss of the simplicity which made possible the closed-form analytical solutions presented above. An example of such a combined model was presented by Eisenberg, Hill, and Chen (1980).

The model consists of four crossbridge states (Fig. 7.18), two attached (1 and 2) and two unattached (R and N). The unattached states are labeled refractory (R) and nonrefractory (N): the crossbridge must pass from the R to the N state before it can attach, but to do so it must overcome a free energy barrier. Because the rate constants for transitions between the N and 1 states and between the 1 and 2 states are higher than between the R and N states, the free energy barrier between the R and N states effectively establishes the rate of crossbridge attachment. Thus the rate constant α_{RN} is functionally equivalent to the f in Huxley's 1957 model.

The free energy diagram shown in Fig. 7.18 is in most respects analogous to the potential energy diagrams of Fig. 7.15. An important difference is that in the present model, crossbridge elasticity and change in conformational state are not independent. Attached states 1 and 2 are assumed to represent the crossbridge bound to the thin filament in a chemical state where the angle between the S-1 link and the thin filament under zero-force conditions is either 90° (state 1) or 45° (state 2). The x distance between these two states (same meaning as in Fig. 7.14) is taken as 80 Å (8 nm). Because the force-displacement properties of the crossbridge are assumed to be linear, the shape of the free energy curves corresponding to states 1 and 2 are parabolic. As shown by the vertical distance between the arrows, when the crossbridge executes one complete R-N-1-2-R cycle, the net free energy change is Δ, where Δ is the chemical potential of ATP hydrolysis.

The forward and reverse rate constants for each step around the cycle were estimated in such a way as to be consistent with known in vitro equilibrium constants and free energy values.

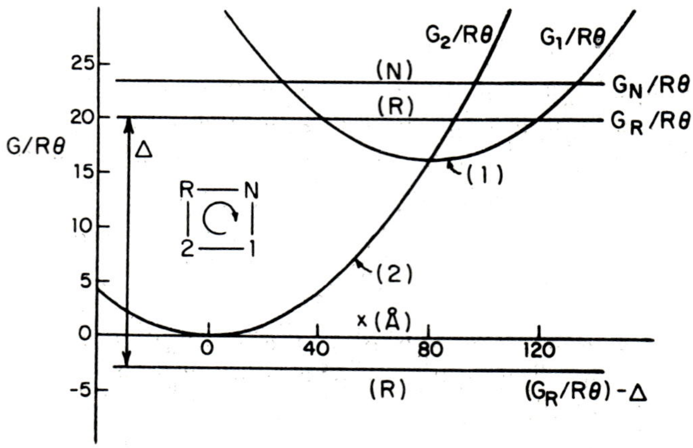

FIG. 7.18 Diagram showing the free energy versus the crossbridge rotation parameter x for the Eisenberg, Hill, and Chen model. All free energies are normalized by $R\theta$, where θ is absolute temperature. The energy of attached state 2 when $x = 0$ is arbitrarily set to zero, and all other energies and x displacements are given relative to this condition. (*Adapted from Eisenberg, Hill, and Chen, 1980.*)

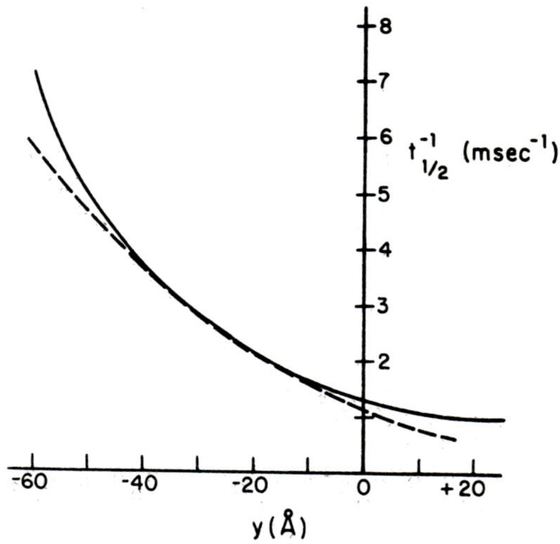

FIG. 7.19 Rate of force recovery in the isometric transient $T_1 - T_2$. The ordinate, $1/t_{1/2}$, is the inverse of time required for the force to rise one-half of the way from T_1 to T_2. Solid line shows computed values; broken line shows experimental measurements (Ford et al., 1977). (*Adapted from Eisenberg et al., 1980.*)

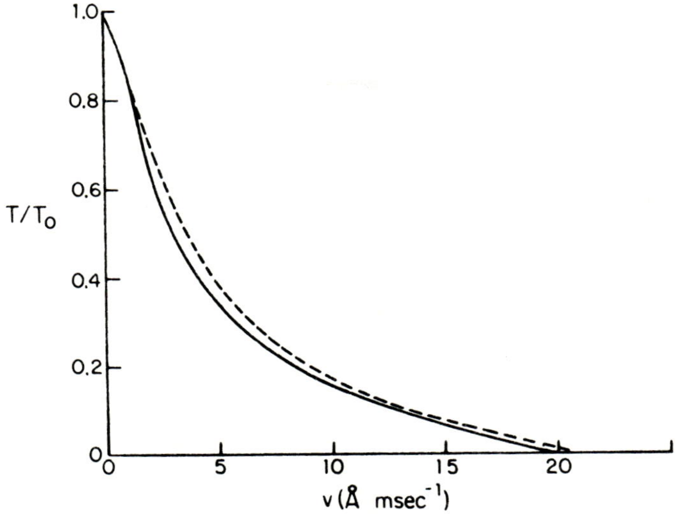

FIG. 7.20 Force-velocity curves for shortening. Solid curve shows computed results for the model described in Fig. 7.18. Broken curve shows experimental results for frog muscle at 1.2°C (Edman and Hwang, 1977). (*Adapted from Eisenberg et al., 1980.*)

Other rate constants were adjusted to make the predicted behavior of the model agree with one aspect or another of the experimental facts.

Probability distributions analogous to the $n(x, t)$ of Huxley's 1957 model were defined for each of the four states, and rate equations analogous to Eq. (7.5) for each of the states were solved simultaneously using numerical techniques. The tension per crossbridge and rate of energy flux were calculated from these probability distributions, using equations analogous to Eqs. (7.12) and (7.13).

Results calculated for quick length steps produced T_1 and T_2 curves which were in good agreement with recent experimental results (Ford et al., 1977). As shown in Fig. 7.19, the calculated rate constant $1/t_{1/2}$ [analogous to $1/\tau$ in Eq. (7.47)] was quite close to the experimental values for moderate shortening steps but somewhat different for large shortening steps and lengthening steps.

For steady shortening at constant force, the model predicted a force-velocity curve quite close to recently reported experimental results (Fig. 7.20). The agreement between predicted and observed energy flux values was not as good, however. The authors did not present any calculations testing their model against experimental results for the case of a muscle forced to lengthen while active.

CONCLUSION

At present, the experimental knowledge of muscle mechanics contains these salient points:

1. Muscle force appears to be due to crossbridge force generators working independently.
2. The fuel used by the crossbridges directly is ATP.
3. Steady shortening and lengthening are described by the force-velocity curve (Fig. 7.6), and there is a reasonably unique relation between force, length, and velocity at a given activation except when muscle yields.

4. For small length changes in an otherwise isometric muscle, the tension transients of Fig. 7.12 are observed.

Problems in biomechanics frequently involve several muscles operating together, often under circumstances where activation is changing and neither length nor velocity is held constant. Depending on the circumstances, a model similar to that shown in Fig. 7.7 may be sufficient, even though it is incapable of accounting for many features of real muscle behavior, including points 1, 2, and 4 above. The most comprehensive muscle models combine filament sliding and attached-state transitions to predict more of the major features seen experimentally, but even these models have some limitations. Their major disadvantage, i.e., their computational complexity, should become less of a drawback with each passing year as it continues to become faster and less expensive to perform numerical calculations.

REFERENCES

Aubert, X., Roquet, M. L., and Van der Elst, J. (1951). The tension-length diagram of the frog's sartorius muscle. *Arch. Intern. Physiol.*, *59*, 239–241.

Bloom, W., and Fawcett, D. W. (1968). *A Textbook of Histology* (9th ed). Saunders, Philadelphia.

Edman, K. A. P. (1970). The rising phase of the active state in single skeletal muscle fibres of the frog. *Acta Physiol. Scand.*, *79*, 167–173.

Edman, K. A. P., and Hwang, J. C. (1977). The force-velocity relationship in vertebrate muscle fibers at varied tonicity of the extracellular medium. *J. Physiol.*, *269*, 255.

Eisenberg, E., Hill, T. L., and Chen, Y. (1980). Crossbridge model of muscle contraction: Quantitative analysis. *Biophys. J.*, *29*, 195–227.

Ford, L. E., Huxley, A. F., and Simmons, R. M. (1977). Tension responses to sudden length change in stimulated frog muscle fibers near slack length. *J. Physiol.*, *269*, 441–515.

Gordon, A. M., Huxley, A. F., and Julian, F. J. (1966). The variation in isometric tension with sarcomere length in vertebrate muscle fibres. *J. Physiol.*, *184*, 170–192.

Hill, A. V. (1938). The heat of shortening and the dynamic constants of muscle. *Proc. R. Soc. London B*, *126*, 136–195.

Huxley, A. F. (1957). Muscle structure and theories of contraction. *Prog. Biophys. biophys. Chem.*, *7*, 255–318.

Huxley, A. F. (1974). Muscular contraction. *J. Physiol.*, *243*, 1–43.

Huxley, A. F., and Simmons, R. M. (1971). Mechanial properties of the cross-bridge of frog striated muscle. *J. Physiol.*, *218*, 59–60P.

Jewell, B. R., and Wilkie, D. R. (1958). An analysis of the mechanical components in frog's striated muscle. *J. Physiol.*, *143*, 515–540.

Julian, F. J., and Moss, R. L. (1980). Sarcomere length-tension relations of frog skinned muscle fibers at lengths above the optimum. *J. Physiol.*, *304*, 529–539.

Mackean, D. G. (1962). *Introduction to Biology*. Murray, London.

McMahon, T. A. (1984). *Muscles, Reflexes, and Locomotion*. Princeton University Press, Princeton, N.J.

Squire, J. (1981). *The Structural Basis of Muscular Contraction*. Plenum Press, New York.

CHAPTER 8

Noninvasive Blood Pressure Recording *and* the Genesis of Korotkoff Sound

Gary Drzewiecki
Julius Melbin
Abraham Noordergraaf
Cardiovascular Studies Unit
Departments of Bioengineering and Animal Biology
University of Pennsylvania
Philadelphia, Pennsylvania

INTRODUCTION

The Polish physician Józef Struś, in the Renaissance period, appears to have been the first to introduce the graphic display of the arterial pressure pulse to facilitate the discussion of its clinical significance (1555). Struś obtained the shape of the pulse by noninvasive means (see Bugiel, 1908). In recent years, invasive techniques have been emphasized (Li et al., 1976). In this chapter we discuss only noninvasive measurements of blood pressure.

First, methods are reviewed. These are classified in two categories: those that rely on an occlusive cuff and those that rely on transcutaneous pulse recording. Additionally we discuss, in detail, one method in each category: the auscultatory method of Korotkoff and the method known as arterial tonometry. Techniques associated with both categories have been described in detail by Smith and Bickley (1964) and by Geddes (1970). This chapter, however, deals primarily with analytical aspects and presents results from recent studies that have not been published previously.

The popularity of Korotkoff's auscultatory technique persists, despite a lack of accuracy, because of its simplicity, convenience, and repeatability, in addition to its being atraumatic and inexpensive. Invasive techniques however, sufficiently accurate to serve as standards, encourage development of improved noninvasive approaches (Noordergraaf, 1979).

Noninvasive measurements usually determine systemic arterial maximum (systolic) and minimum (diastolic) blood pressure. The temporal pattern of pressure, however, during a complete cardiac cycle has always been of interest. Utilization of the pulse shape includes that by Frank (1899), and subsequently by many others, to determine stroke volume; by Smith and Bickley (1964) in aerospace applications; by Lax et al. (1956), Tavel (1978), Busse et al. (1975), and Chlebus (1979) for clinical diagnosis; by Raines et al. (1973) to determine hemodynamic parameters; by Van der Hoeven et al. (1973) to measure systolic time intervals; and by Tursky (1974) in psychological and biofeedback studies.

8.1 REVIEW OF OCCLUSIVE CUFF METHODS

8.1.1 Background

Occlusive cuff methods all rely on the application of a known air pressure to a limb, most commonly the upper arm. The cuff is inflated to a pressure exceeding that of the arterial pressure and allowed to deflate slowly (2 to 3 mmHg/s). When cuff pressure exceeds arterial pressure, a segment of artery is forced to collapse, partially or completely. Ideally, if surrounding tissues completely transmit cuff pressure to the artery, and if this is all that is necessary for arterial collapse, then the arterial lumen begins to open when arterial pressure equals cuff pressure. This is usually assumed to occur.

Differences among cuff methods are primarily in the manner of detection of the arterial lumen opening. Detection schemes include a skin flush method (Gaertner, 1899), an oscillometric method (Marey, 1885; Erlanger, 1903), an auscultatory method (Korotkoff, 1905), a palpatory method (Norris, 1916), and an ultrasonic method (Ware, 1965). With each, the first signal detected during decreasing cuff pressure is taken to indicate systolic pressure. With continued cuff deflation, the signal alters in a manner characteristic to the particular method (Fig. 8.1). The significance of this alteration has not been fully analyzed.

Palpatory Method. The cuff is utilized in the standard manner and, simultaneously, the pulse is palpated at the brachial, the radial, or a digital artery. Detection of arterial opening is correlated with perception of the pulse distal to the cuff. The method is empirical and lacks a theoretical base, although the analysis can be accomplished (Noordergraaf, 1978; see also Sec. 8.4).

Since palpation is generally inadequate to detect the small alterations in the pulse, the method cannot detect diastolic pressure. It can, however, determine systolic pressure as long as a palpable pulse is present. Thus it finds application where other occlusive cuff methods fail, e.g., in children, with patients in shock, or with hypotensive patients.

Auscultatory Method. Arterial lumen opening is detected via a stethoscope. Sound generated in the (usually brachial) artery, and so detected, is termed *Korotkoff sound* after the Russian army physician Korotkoff (1905), who discovered it while studying the collateral circulation of the canine limb. Although he was concerned with the degree of occlusion imposed by an inflated cuff, he recognized the potential of the sound as an indicator of arterial pressure. An advantage of the auscultatory method is its ability to determine both systolic and diastolic pressures; this was appreciated by Korotkoff.

Figure 8.2 illustrates the Korotkoff sound superimposed upon the cuff pressure. Korotkoff sounds are divided in five phases, of which phases I, IV, and V have been of greatest interest. Phase I is clearly perceived and defines the onset of Korotkoff sound. Phases IV and V are defined as the onset of muffling of the sound and of its disappearance, respectively. In accordance with recommendations by the American Heart Association (Kirkendall et al., 1981), the phase I sound is interpreted as occurring when cuff pressure equals systolic blood pressure. The subsequent phases, IV and V, are associated with diastolic pressure; phase IV is the recommended gauge for children and phase V is that for adults.

For the auscultatory method, accuracy is reported to be -5 to -20 mmHg mean error for systolic pressure and $+12$ to $+20$ mmHg mean error for diastolic pressure (London and London, 1967). The errors appear to be more subject-related than observer-related (Weinman, 1972).

The width of the cuff is an important consideration in the auscultatory method (Karvonen et al., 1964; Simpson et al., 1965; Geddes and Whistler, 1978). Experimental results suggest a cuff-width–to–arm-circumference ratio of 0.4 (Kirkendall et al., 1981). The occlusive cuff was

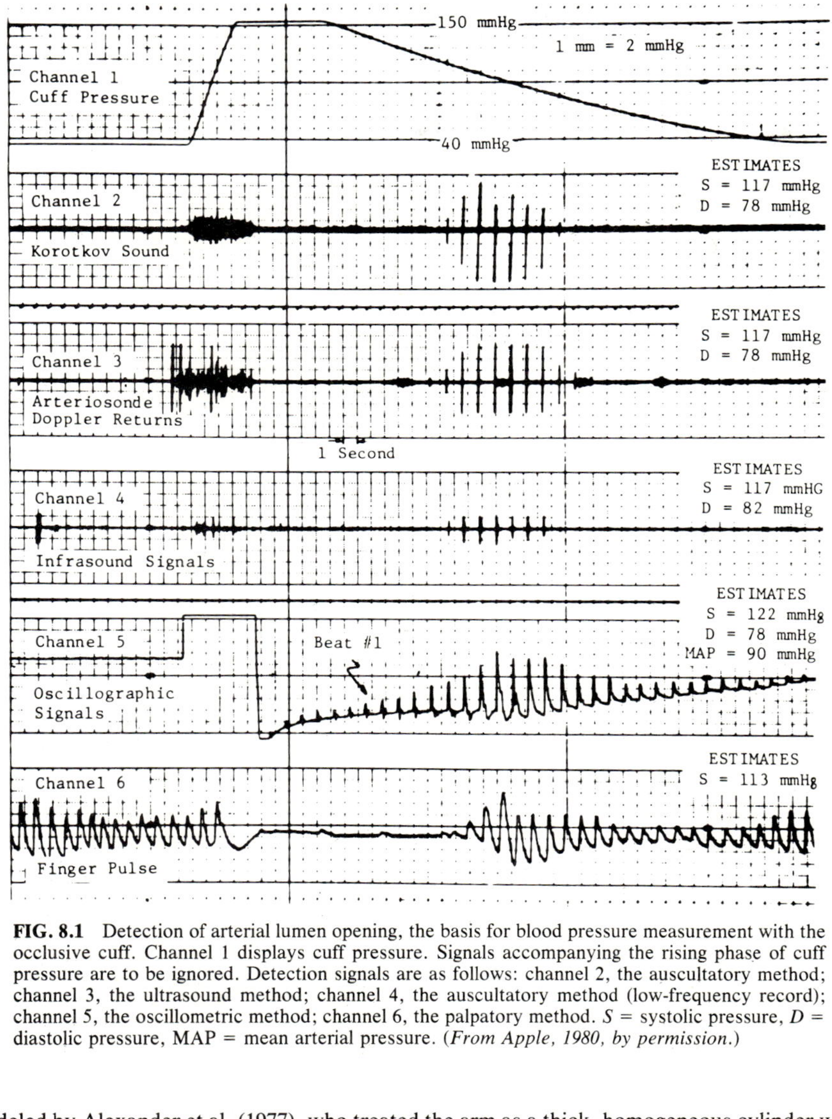

FIG. 8.1 Detection of arterial lumen opening, the basis for blood pressure measurement with the occlusive cuff. Channel 1 displays cuff pressure. Signals accompanying the rising phase of cuff pressure are to be ignored. Detection signals are as follows: channel 2, the auscultatory method; channel 3, the ultrasound method; channel 4, the auscultatory method (low-frequency record); channel 5, the oscillometric method; channel 6, the palpatory method. S = systolic pressure, D = diastolic pressure, MAP = mean arterial pressure. (*From Apple, 1980, by permission.*)

modeled by Alexander et al. (1977), who treated the arm as a thick, homogeneous cylinder with the artery at the axis of symmetry (Fig. 8.3). A wide cuff was found to be capable of accurate pressure transmission at the center of the cuff, with the pressure decreasing toward zero with distance from the center. Hence, a narrow cuff imposes an arterial stress below that of cuff pressure and results in overestimation of blood pressure. Although the theoretical model did not so indicate, experimental observations disclosed that there exists an optimal width for the cuff. Errors in estimation of blood pressure increase as cuff width exceeds the optimum. The reason for this remains unclear.

Other problems are peculiar to the auscultatory method. For example, hypotension and shock may be accompanied by complete loss of Korotkoff sound; the cause of this phenomenon is unknown, and measurements in small children may fail for this reason. Another problem

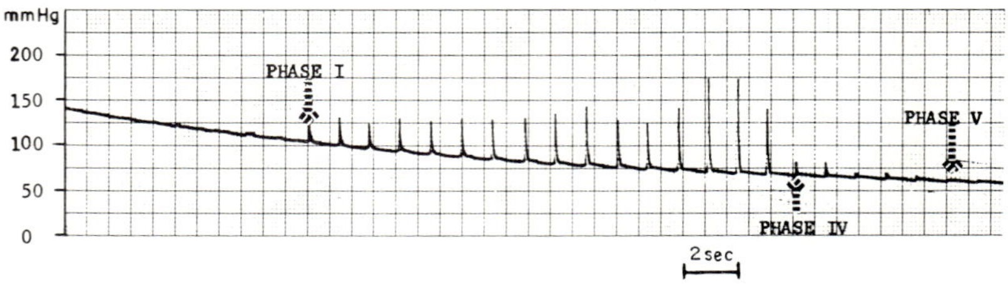

FIG. 8.2 Korotkoff sound intensity with decreasing cuff pressure. Phases I, IV, and V are marked. Phase I marks systolic blood pressure. Phases IV and V indicate diastolic pressure. (*From Maurer and Noordergraaf, 1976.*)

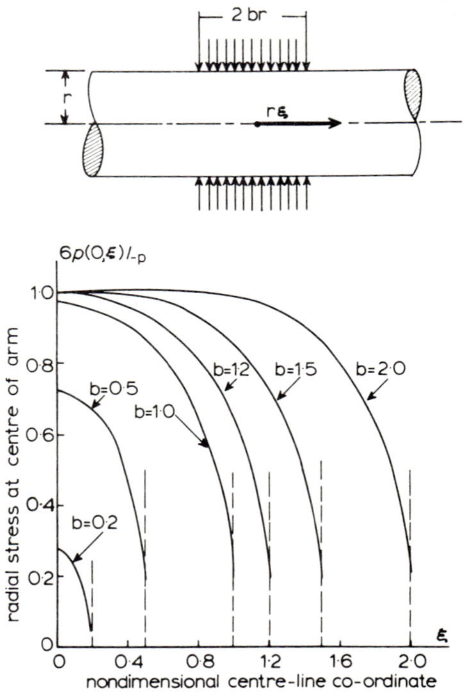

FIG. 8.3 Top: A uniform axisymmetric load, representing the occlusive cuff pressure, is applied to a solid cylindrical body model of the arm. b is the ratio of cuff width to arm diameter; ξ is the axial coordinate, nondimensionalized by the radius r, with its origin at the cuff center. Bottom: Computation of radial stress at the center axis of the arm, normalized by the applied load. Results indicate that the cuff pressure is transmitted exactly to the center of the arm when the cuff width is at least 1.2 times the diameter of the arm. (*From Alexander et al., 1977, by permission.*)

relates to whether, and under what conditions, phase IV or phase V is the best indicator of diastolic pressure (Levine, 1981).

There is evidence that the auscultatory method may measure more than just blood pressure. Burger, a pioneer of vectorcardiography, observed that Korotkoff sound was affected by the state of the arm's distal vasculature (1955). Cohn (1967) also observed large discrepancies in measurements when drugs altered the arm's peripheral resistance, and Whitcher (1969) observed sound alteration with the degree of venous filling. As perceived by Burger, such effects offer clues to the origin of the Korotkoff sound (see Secs. 8.4.1 and 8.4.2).

Oscillometric Method. Fundamental concepts of the oscillometric method remain unclear, since its development was purely empirical. Current interpretation is that volume expansion of the compressed arterial segment is transmitted to the occlusive cuff, causing a corresponding reduction in air volume and an increase in pressure.

Marey (1885) discovered that cuff pressure oscillated over a range of mean values of cuff pressure. Assuming that this range extended from systolic to diastolic pressure levels, Marey reasoned that these oscillations should be maximal when the mean transmural pressure is zero. He termed this *vascular unloading,* implying that the arterial wall is not stressed circumferentially. Animal experiments, accomplished by Posey et al. (1969), confirmed that the mean cuff pressure, at which oscillations are maximal, corresponds closely to mean arterial pressure. Similar observations were reported for human subjects (Ramsey, 1979). Oscillometry gained extensive application following modifications of the sphygmomanometer by von Recklinghausen (1940); the current instrument is very similar to his design.

Mauck et al. (1980) modeled the oscillometric system as an elastic vessel passing through a compression chamber. This was studied experimentally by substituting excised canine carotid arteries for the elastic vessel. They concluded that the magnitude of arterial pressure and of arterial wall elasticity tends not to affect the accuracy of the method. It was also found that as the volume of the compression chamber increased relative to the enclosed arterial volume, mean arterial pressure corresponded less to the level at which maximum oscillations occurred. The volume of deformable material surrounding the artery (muscle tissue, skin, air-filled cuff) may thus be an important feature. The size of the occlusive cuff was shown to effect accuracy in canine oscillometric experiments (Posey et al., 1969), but this has not been investigated in humans.

Renewed interest in oscillometry followed the excellent results reported by Ramsey (1979). In 17 human subjects, oscillometric measurements were compared with those from an indwelling catheter, for a variety of clinical states. For mean arterial pressure, Ramsey found an overall mean error of 0.23 mmHg. Although the oscillometric method yields accurate determination of systolic pressure (Van Bergen et al., 1954), it is not generally useful for diastolic pressure, a major disadvantage (Geddes et al., 1982). Concern with diastolic pressure relates to issues of cardiac afterload and hypertension (Noordergraaf and Melbin, 1978).

Ultrasound Method. Ultrasound was first used to detect arterial lumen opening by Ware (1965) and offers the most direct noninvasive detection of arterial motion. The conventional occlusive cuff is employed with an ultrasonic probe positioned over the artery just distal to the cuff. Local arterial wall and fluid motion is sensed by Doppler frequency shift in the reflected sound. Thus far, the method is empirical and lacks theoretical support.

The method's accuracy has been studied extensively in normal and in severely ill patients (Hochberg and Solomon, 1971). In comparison with intraarterial pressure, the ultrasonic method indicated higher values for mean systolic and diastolic pressures, the differences being 0.5 and 7.6 mmHg, respectively.

Circumstances in which blood pressure can be measured with ultrasound include hypotension and shock (Kirby et al., 1969; Buggs et al., 1973). The method has also measured pediatric blood pressure where the auscultatory method fails (Massie et al., 1973).

8.1.2 Automation

Occlusive cuff methods have been automated to execute the protocol normally accomplished manually after the occlusive cuff is placed (Ponomarenko et al., 1973). Criteria for blood pressure determination are most critical to such an instrument. Signal processing and detection techniques determine the cuff pressure at which selected criteria are satisfied.

For example, Korotkoff sound may be recorded via a microphone placed over the brachial artery. Typically, these signals are characterized by their relative frequency content (Golden et al., 1974; Wolthius et al., 1974). Detection can thus be accomplished using two bandpass filters, one detecting phase I sound in the 18- to 26-Hz range and the other detecting phase IV sound in the 40- to 60-Hz range. Wolthius et al. (1974) determined that the ratio of the amplitudes of filtered to unfiltered sound should be 0.45 for the decision threshold of the systolic channel. For the diastolic channel, the ratio between filtered phase IV sound to overall maximum filtered sound was optimal at a value of 0.17. Other techniques rely on a single bandpass filter (NASA, 1965; Allred and Johnson, 1966; Rogge et al., 1968; Maurer and Noordergraaf, 1976) or utilize more than two filters (Ware and Kahn, 1963).

Some automated systems are designed to measure blood pressure while the subject is active or in a noisy environment. An electronically timed "window" is set relative to the R wave of the ECG so that Korotkoff sound is recorded only when desired and interference by extraneous noise is minimized (Canzoneri et al., 1968; Wolthius et al., 1974; Weaver et al., 1978). Reducing the cuff deflation time also improves performance (Katona and Bolvary, 1975).

Automation was introduced mainly to permit continuous, unattended pressure monitoring and to improve accuracy. Portable instruments that monitor blood pressure throughout the day have been employed by Schneider and Costiloe (1975) and in studies of hypertensives by Kain et al. (1964) and Sokolow et al. (1973). The accuracy of such blood pressure measurements, acquired during exertion, is questionable (Wolthius et al., 1981). In general, automation has not increased the accuracy of the auscultatory method over that achieved by a trained technician (Weinman, 1972; Apple, 1980).

8.1.3 Summary

The development of occlusive cuff methods has been directed toward increased information, improved accuracy, and simplification of the technique. In the process, the auscultatory method of Korotkoff displaced earlier methods because it provides both systolic and diastolic pressures with a relatively simple apparatus. Recent technology (ultrasound, automation) does not truly challenge the auscultatory method except in special cases such as with neonatal and unattended monitoring. Nevertheless the desire for a more complete, noninvasive blood pressure measurement has prompted investigation of sophisticated methods capable of providing the entire pulse.

8.2 REVIEW OF NONINVASIVE ARTERIAL PULSE RECORDING METHODS

8.2.1 Transcutaneous Pulse Recording

Since the time of Struś, a variety of pulse-recording instruments have been developed. In an early approach, a tiny mirror was bonded to the skin slightly off the axis of a superficial artery. Light reflecting off the moving mirror revealed the pulse shape. Marey (1885), who contributed much to sphygmomanometry, also developed several instruments to record the pulse noninvasively. These instruments are continually being improved (Welkowitz and Deutsch, 1976).

It is convenient to distinguish three categories of transcutaneous pulse methods: the volume pulse method, the pressure pulse method, and other methods.

We consider first the volume pulse method. Volume pulsations are observed at the skin surface as a consequence of arterial wall distension. If the arterial wall is treated as an axially tethered, homogeneous, isotropic cylinder, a classic relation for internal pressure and radial displacement can be derived (Love, 1927). The relation for a thick wall cylinder of linear elastic material is (Noordergraaf, 1978)

$$\frac{dr}{dp} = \frac{3}{2}\frac{r(r+h)^2}{Eh(2r+h)} \tag{8.1}$$

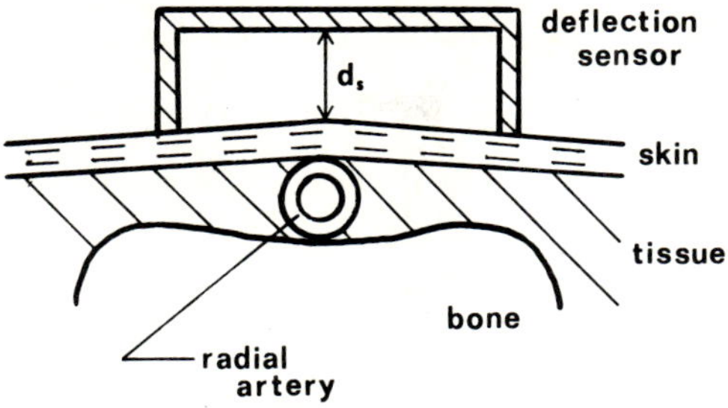

FIG. 8.4 Volume pulse recording. A rigid frame placed over the recording site allows the skin surface to displace freely. The surface deflection sensor mounted on the frame transduces the displacement u_r as a function of time. u_r is the change in stand-off distance, d_s. u_r results from local arterial wall, tissue, and skin deformations due to alteration of the vessel's internal pressure.

where p is arterial pressure, r is internal radius, h is wall thickness, and E is the modulus of elasticity of the wall material.

For small changes in arterial pressure, Δp, radial displacement u_r can be written as

$$u_r = \frac{dr}{dp}\,\Delta p \tag{8.2}$$

In the ideal case, the perceived skin surface displacements are directly proportional to changes in arterial pressure (Fig. 8.4).

Conversely, the pressure pulse method requires a surface device to restrain arterial distension below the sensor. The transducer is located over the vessel so that the artery is deformed while contact stress is measured (Fig. 8.5). This contact stress depends on the sensor contact area, the stresses related to initial deformation, and the arterial pressure. The arterial pressure is, therefore, a fraction of the total surface contact stress. With equilibrium conditions and no additional deformation stresses imposed after positioning the transducer, its signal will vary in proportion to the variation in arterial pressure.

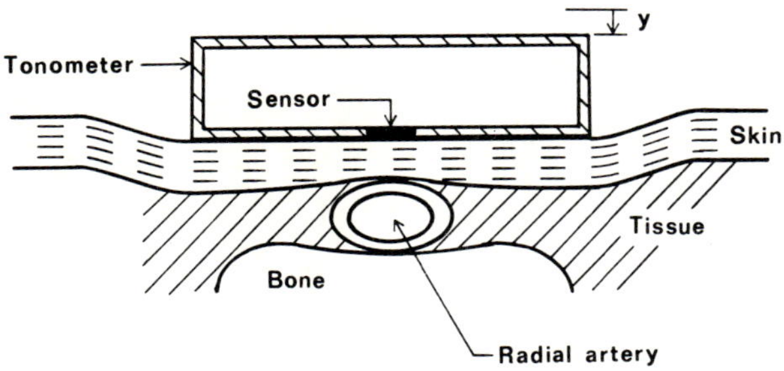

FIG. 8.5 Pressure pulse recording. A rigid frame placed on the skin surface indents the tissue a distance y. If the rigid surface maintains constant strain in the contact plane, the sensor records a stress containing both arterial pulse pressure and initial deformational stress. Arterial tonometry is a special case of this general method.

Pulse-recording instruments should not introduce significant nonlinearity or filtering. In a study that compared intraarterial pressure pulses versus transcutaneously recorded volume pulses, pulse fidelity was found to differ among subjects (Abbott, 1969), a finding for which no explanation was offered.

Proper placement of the pulse recorder is another problem that plagues its use. Minimal movement of the subject can cause the transducer to shift from its original site (Bahr and Clark, 1980). Consequently, pulse recorders are often unreliable and their signals poorly reproducible. These problems have been somewhat alleviated by enlarging the sensing area; this ensures that a portion of the transducer is always over the vessel of interest (Davis et al., 1963). Miniature transducer arrays enable arteries to be located electronically and have shown promising results (Weaver et al., 1978).

Attempts to calibrate pulse recorders have been only moderately successful (Correll, 1959; Stein and Blick, 1971, 1972). Initially, calibration was achieved using an alternate method to determine blood pressure. Once completed, however, it could be maintained for only short periods before recalibration was necessary. Local tissue "accommodation" and nonlinear and viscoelastic effects have been postulated as causes.

Since the auscultatory method provides calibrated blood pressure, investigators have utilized the method for transcutaneous pulse recording. It has been demonstrated that the pulse shape can be synthesized using the interval between the ECG R wave and Korotkoff sound data (Rodbard et al., 1957; Arzbaecher and Novotney, 1973; Johnston and Kakkar, 1974). The procedure requires several beats, so that beat-to-beat variations cannot be resolved. To alleviate this problem, a system was designed to monitor systolic pressure continuously, using the occlusive cuff. Here, a tracking system adjusts cuff pressure to that of systolic pressure by maintaining the phase I Korotkoff sound, this latter being detected with a microphone (Shapiro et al., 1981). This technique may be used for only a few minutes before the blood supply to the arm must be reestablished.

While fairly accurate recordings of arterial pulse shape can be obtained with transcutaneous methods, calibration problems remain formidable.

8.2.2 Arterial Distensibility Method

The arterial distensibility method was introduced by Correll (1959) as a calibration procedure for the transcutaneous volume pulse method. For calibration, Correll assumed that instantaneous arterial displacement is known and that the arterial wall is composed of linear elastic material. Assuming that Eq. (8.2) holds, it is formulated as the linear relation

$$u_r = k(p_1 - p_0) \tag{8.3}$$

where p_0 is reference pressure, p_1 is absolute arterial pressure, and k is dr/dp (a constant).

If k and p_0 are known and u_r can be measured, arterial pressure may be computed. To find k, Correll elevated the measuring device to two different heights, z_1 and z_2, relative to the heart. In terms of z_1 and z_2, k from Eq. (8.3) appears as

$$k = \frac{u_{r1} - u_{r2}}{\rho_b g(r_1 - r_2)} \tag{8.4a}$$

where ρ_b is blood density and g is the acceleration of gravity.

To determine p_0, the vessel is occluded with a cuff proximal to the measuring site. Then, from Eq. (8.3),

$$p_0 = p_a - \frac{u_{r,a}}{k} \tag{8.4b}$$

where p_a is the arterial pressure in the occluded vessel and $u_{r,a}$ is the measured displacement of the occluded vessel. p_a is assumed to equal atmospheric pressure. Experiments with human subjects indicate the method to be accurate only for variations in mean arterial pressure to 25 mmHg (Correll, 1959).

Sacks and Abbott (1962) analyzed the inability of Correll's method to monitor large pressure changes and found the problem to reside in the assumption of a linear elastic arterial wall. The arterial wall actually exhibits nonlinear elastic properties; i.e., k must be strain-dependent. Sacks and Abbott suggest piecewise linear approximations. To calibrate over a wide range of pressure, they propose that the subject exercise, avoiding perturbation with drugs or invasive interventions. Shifts in transducer placement, however, make the method cumbersome and impractical.

Since the pulse propagation velocity is a function of arterial pressure (King, 1947), it has often been proposed as a noninvasive means to calibrate pulse pressure (Gribbin et al., 1976; Sacks and Abbott, 1962). Inherent difficulties are similar to those of the arterial distensibility method. Ghista et al. (1978) and Ghista (1978) proposed that the above procedures together might provide sufficient information to eliminate the unknowns introduced by the nonlinear vessels during calibration. Ghista demonstrated, theoretically, that calibration can be achieved with measurement, at three pressure levels, of internal and external radii and of the pulse-wave velocity. These investigators indicate that a 99 percent accuracy in arterial dimensions is required to ensure a 90 percent accuracy in the determination of arterial pressure. For the radial artery, this is equivalent to a resolution of 0.0016 cm, a stringent requirement for the ultrasound equipment. This analysis used in vitro arterial distension data reported in the literature, but Ghista has not completed testing on human subjects.

8.2.3 Vascular Unloading Method

The vascular unloading method dates back to Marey (1885) (see "Oscillometric Method" in Sec. 8.1.1), who substituted a digit for the arm and discarded the cumbersome chamber that was used to enclose it. Similarly, Shirer (1962) attempted to measure digital arterial pressure by use of counterpressure in a digit chamber. These results were unsatisfactory; the measurements were apparently affected by viscoelastic properties of the digit and by the compliance of the chamber seal. To overcome these problems, Peñáz (1973) introduced a servo-controlled counterpressure instrument. The degree of vascular unloading was determined by a photoplethysmograph that electronically controlled the digit chamber's pressure to balance the arterial pressure. Feedback control is apparently effective in overcoming the early problems with the vascular unloading method (Wesseling et al., 1978; Yamakoshi et al., 1980; Smith and Beneken, 1980; Molhoek et al., 1981). A rigorous analytical treatment, however, has not yet been accomplished, such effort being hampered by only qualitative understanding of the photoplethysmograph itself.

The initial reference setting for the plethysmograph signal was found to be crucial for proper operation of the instrument (Yamakoshi et al., 1980). Yamakoshi et al. begin by opening the control loop (Fig. 8.6). Then cuff pressure is slowly decreased, under manual control, until the pulsations in the plethysmograph signal reach a maximum. The mean value of this plethysmograph signal is taken as the reference value. Yamakoshi et al. claim that this reference level is close to that recommended by Peñáz (1973) but simpler to obtain.

The vascular unloading method of Peñáz has been compared to direct intraarterial records in human studies (Fig. 8.7), but the method underestimates both systolic and diastolic pressures by 5 to 8 mmHg (Yamakoshi et al., 1980; Molhoek et al., 1981; Wesseling, 1981). Objections to the vascular unloading method arise as a result of its peripheral location. Digital arteries have a considerable amount of smooth muscle, and their state of contraction contributes to the circumferential stress that is to be nulled. If vascular tone is sufficiently large, measurement errors arise from violation of the assumption that the vessel is passive. Also, wave reflection is more likely to alter the pulse shape at the periphery in comparison to that of more central arteries (Westerhof et al., 1973).

8.2.4 Arterial Tonometry

Arterial tonometry, as a transcutaneous pulse-sensing method, requires an adequately sized, superficial artery that is supported by a bony structure. In human beings, several locations meet these criteria; the radial artery, just proximal to the wrist, is the popular choice.

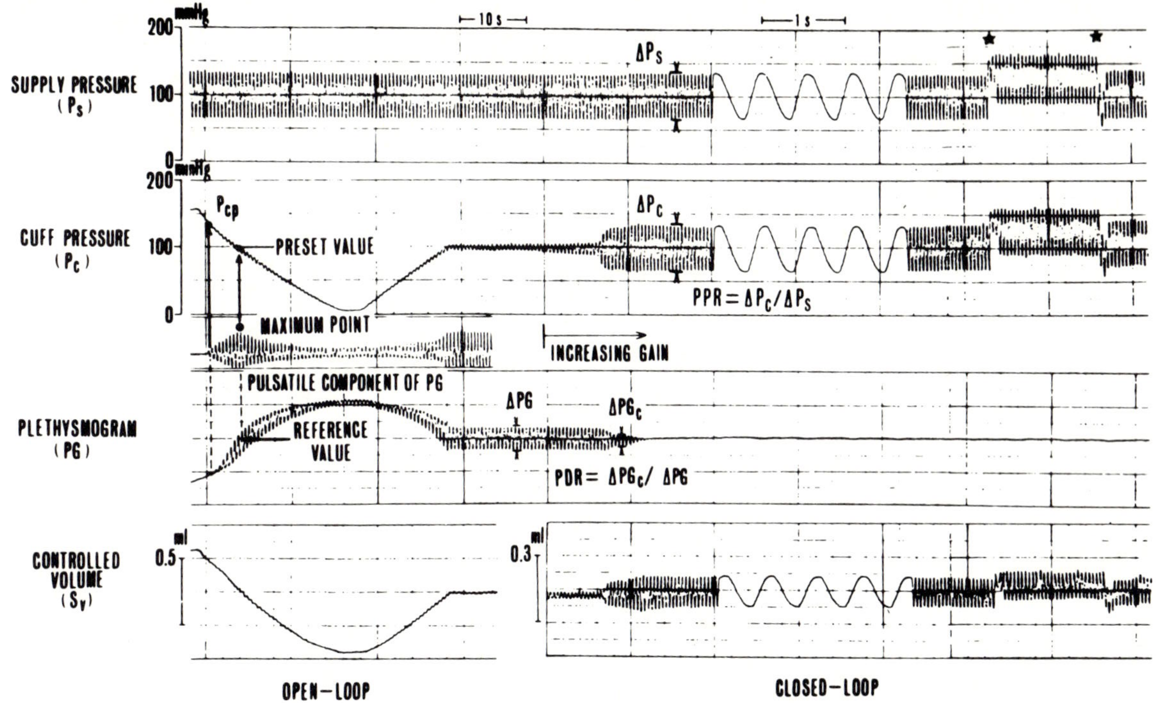

FIG. 8.6 Procedure by Yamakoshi et al. (1980) to set the reference level for the method of vascular unloading. Open-loop control is used to manually locate the maximum pulsatile component of the plethysmogram while the cuff pressure is adjusted. The mean value of the plethysmogram signal is taken as reference. When closed-loop conditions are restored, the system is assumed to maintain transmural pressure near zero. (*From Yamakoshi et al., 1980, by permission.*)

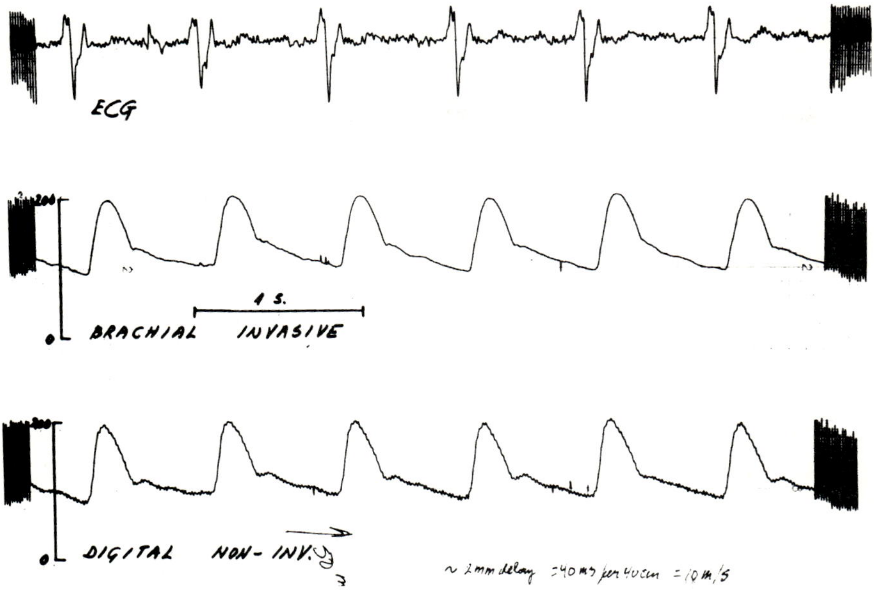

FIG. 8.7 Simultaneous records of brachial artery pressure, obtained invasively, and digital artery pressure, obtained noninvasively by vascular unloading. (*From Smith and Beneken, 1980, by permission.*)

A typical arrangement for a tonometric blood pressure determination is illustrated in Fig. 8.5. The tonometer is applied over the center of the artery, with sufficient force to distort the vessel. If this position is maintained, the normal contact stress between the skin and the tonometer surface, in a small area directly over the artery, should be close to the instantaneous intraluminal pressure. This contact stress is converted by a surface sensor into an electric signal providing, in principle, continuous pulse recording with direct calibration.

Arterial tonometry derives from the older technique of ocular tonometry. Pressman and Newgard (1963) found, in humans, good correlation of tonometric measurements of blood pressure with those by sphygmomanometer. Following this work Okino (1964) applied a variety of transducers to internally pressurized, elastic tubing. Despite the absence of intervening tissue, Okino's tonometer records indicated the presence of nonlinearity. Subsequent investigators confirmed the results of Pressman and Newgard (Stein and Blick, 1971, 1972). Stein and Blick also provided extensive data, in humans, comparing tonometry with invasive pressure measurements and verified the accuracy of the former. Later, other tonometers were developed to overcome some inherent problems. For example, the difficulty of precise positioning of the tonometer was mitigated by the introduction of automatic features such as a servo-controlled application force (Bahr and Petzke, 1973) and site positioning by means of a transducer array (Eckerle and Newgard, 1976; Weaver et al., 1978). Recently, the contradictory result of Okino (1964) was resolved by Drzewiecki (1979) and Drzewiecki and Noordergraaf (1979), who demonstrated analytically the need for appropriate arterial applanation.

Difficulties remain that deter tonometers from becoming viable clinical or research tools. First, there is the lack of an applied theory; results obtained by previous investigators vary widely probably because of design inconsistency and an incomplete analysis of the technique. Secondly, guidelines are unavailable for tonometer utilization or positioning, and thirdly, time-dependent tissue properties are thought to add untoward effects.

8.3 ANALYSIS OF ARTERIAL TONOMETRY

8.3.1 Transcutaneous Pulse Dynamics

We consider first the volume pulse recorder. The distension-pressure relation [Eq. (8.1)] contains, as one component, the modulus of elasticity of the arterial wall. In actuality, the arterial wall is nonlinear and viscoelastic. For the normal 40-mmHg pulse pressure, the modulus of elasticity may be approximated as constant, but viscoelasticity should be considered (Westerhof and Noordergraaf, 1970). This requires the modulus of elasticity to be represented in complex form $E(j\omega)$, where $j = (-1)^{1/2}$ and ω = angular frequency. The pressure-distension relation becomes

$$\Delta r(j\omega) = \frac{3}{2}\frac{r(r+h)^2}{|E|e^{j\phi(\omega)}h(2r+h)}|\Delta p(j\omega)|e^{j\omega t} \qquad (8.5)$$

so that Δr is directly proportional to Δp and is also frequency-dependent. With normal arterial pulses and wall material constants, a computed radial distension for the aorta is shown in Fig. 8.8. Frequency distortion is evident from the hysteresis loop of Δp versus Δr. Abbott (1969) observed such distortion in human subjects during a study of volume pulse recorders. Since tissue properties for a particular subject are not known a priori, this distortion cannot be canceled easily and remains a difficulty inherent in the volume pulse method.

By definition, the pressure pulse method requires that the displacement of the skin at the instrument's contact surface be held to zero. The arterial tonometer is a special case of a pressure pulse transducer; during measurement with this instrument the artery is partially flattened. Thus, if the pulse recorder compresses both tissue and artery, an advantage of tonometry will be realized. Drzewiecki et al. (1982a) have shown that, with arterial flattening, shear stress becomes negligible compared with the normal stress in the arterial wall and skin, and uniform contact stress is developed over the transducer-skin interface. At equilibrium, material properties contribute only negligibly so that nonlinearities and frequency distortion are minimized. This is the ideal state for dynamic pulse recording. Practically, it is difficult to achieve zero shear stress since it is introduced by tissue adjacent to the transmitting zone.

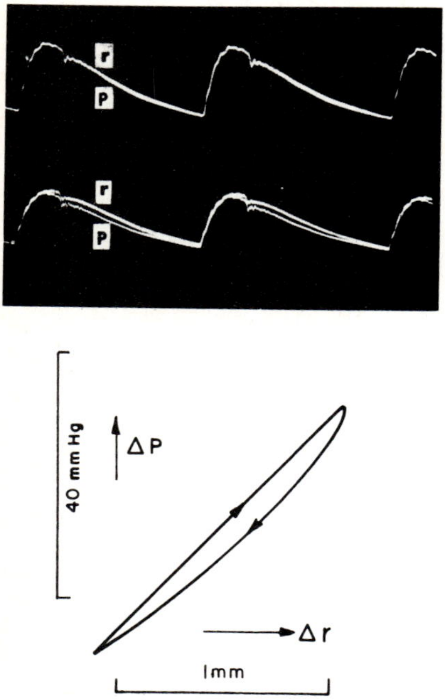

FIG. 8.8 Top curves: The computed radial distension and arterial pressure for a linear elastic wall, and, immediately below, for a viscoelastic wall. Bottom curve: The hysteresis loop obtained when plotting arterial distension and change in pressure for the viscoelastic wall. (*From Westerhof and Noordergraaf, 1970.*)

Drzewiecki et al. (1982a) used a finite element model to assess the feasibility of the pressure pulse method under different loading conditions. The inset in Fig. 8.9 illustrates a typical situation. The top surface of a volume of tissue is constrained to zero deflection. The sides located at a large distance from the loading site are also constrained to zero deflection. The bottom surface can displace freely, and, here, a spatially discontinuous load is applied, representing that imposed by the flattened artery. The stress in the y direction, σ_y, at the top surface is computed by finite element techniques. The contact stress is defined as $\sigma_c = \sigma_y$ for a pure slip boundary. Over the distance $x = \pm 0.02$ in (0.508 mm), σ_c approximately equals the magnitude of the applied unit

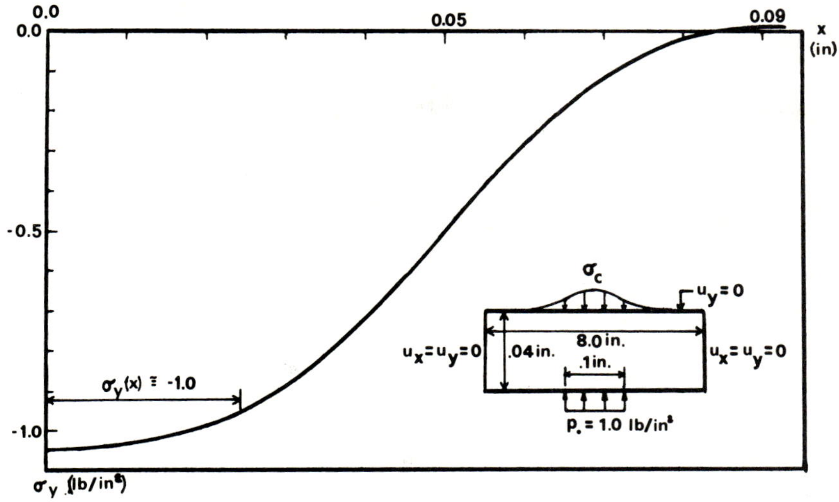

FIG. 8.9 The resulting contact stress, $\sigma_c(x) = -\sigma_y$, due to unit pressure applied over a restricted area of skin as computed by a finite element model. The dimensions and boundary conditions are shown in the inset. σ_y is plotted over half the surface since it is symmetric. Young's modulus was taken as 14.5 lb/in². Values are worst-case physiologic conditions.

load at the bottom surface. Hence, stress equilibrium can exist, suggesting that the pressure pulse method may be feasible under physiological conditions.

A number of researchers have obtained good results with the pressure pulse method (Van der Hoeven and Beneken, 1970; Stein and Blick, 1971). A study by Van der Hoeven and Beneken involving human subjects compares volume versus pressure pulse methods with direct arterial puncture (Fig. 8.10). Superior linearity with reduced hysteresis was consistently obtained with the pressure pulse method. These workers indicate that forcible compression, i.e., the tonomeric procedure, improved the fidelity of the instrument.

In a recent experiment performed by the authors, a pressure transducer was placed in contact with one side of a section of canine skin while a pressure source (capable of producing a step change) was in contact with the opposite side. The system response (Fig. 8.11) remained uniform in both amplitude and phase to 100 Hz, a range well above normal pulse frequency content.

Transcutaneous pulse models can be applied to a variety of pulse instruments. Mechanical considerations suggest a design based on minimal recorder sensor strain after positioning, with additional benefits accruing from arterial flattening. The models and experiments performed under these conditions indicate that excellent fidelity of the noninvasive arterial pressure record can be obtained.

8.3.2 Mechanical Modeling

In general, tonometric studies have made little use of modeling techniques. The first model for an arterial tonometer cast the entire artery-instrument complex as a series-parallel system of lumped linear elastic elements (Fig. 8.12) (Pressman and Newgard, 1963). This model served as a useful guide for the tonometer constructed by these workers, as well as for subsequent research.

To gain information additional to that provided by lumped elements, a model was recently developed that treats the artery-tonometer complex as a continuum. The components are

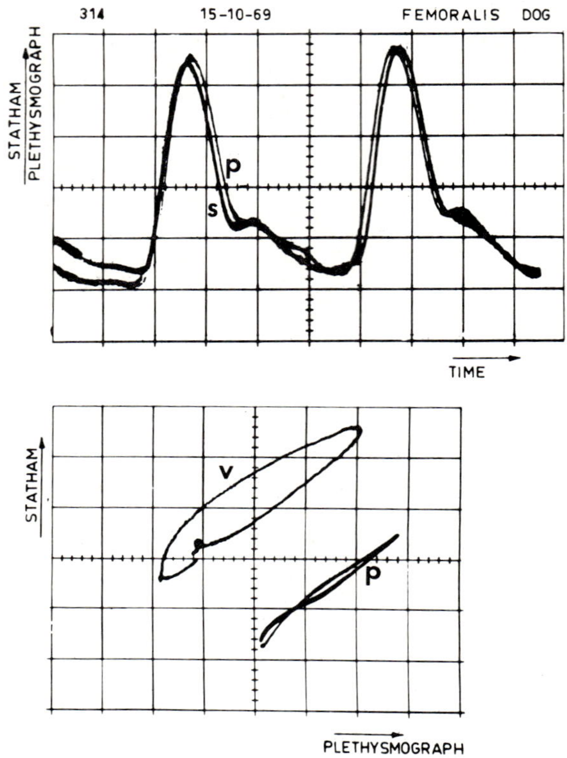

FIG. 8.10 Top: Invasive (curve *S*) and noninvasive volume pulse (curve *P*) recordings obtained simultaneously. Bottom: Direct comparison of both pressure records. Note the hysteresis loop for the volume pulse method (curve *V*) and that from the pressure pulse method (curve *P*). (*From Van der Hoeven and Beneken, 1970, by permission.*)

represented as solid mechanical structures which exhibit bending and shear deformation. These aspects were neglected in the past (Drzewiecki et al., 1983), as were others, including viscoelastic and nonlinear tissue properties, skin transmission phenomena, the determination of contact stress, and the characteristics of the instrumentation involved.

In order to derive arterial pressure from externally measured interfacial stresses, the contact stress problem is separated into two parts: that relating to point-contact deflection and that relating to line-contact deflection. As the initially circular vessel is deformed, point contact is maintained so long as the radius of curvature, ρ, at the contact point is finite. As $\rho \to \infty$, the arterial wall flattens and line contact occurs (Fig. 8.13). Hence, there exists a section of free vessel wall as well as a section whose curvature is zero. Solutions for such a model derive from curved beam mechanics.

The solution for the contact stress, $\sigma_c(x)$, was found to be (Drzewiecki et al., 1983)

$$\sigma_c(x) = V(b)\beta \frac{\cosh \beta x}{\sinh \beta t(b)} \tag{8.6}$$

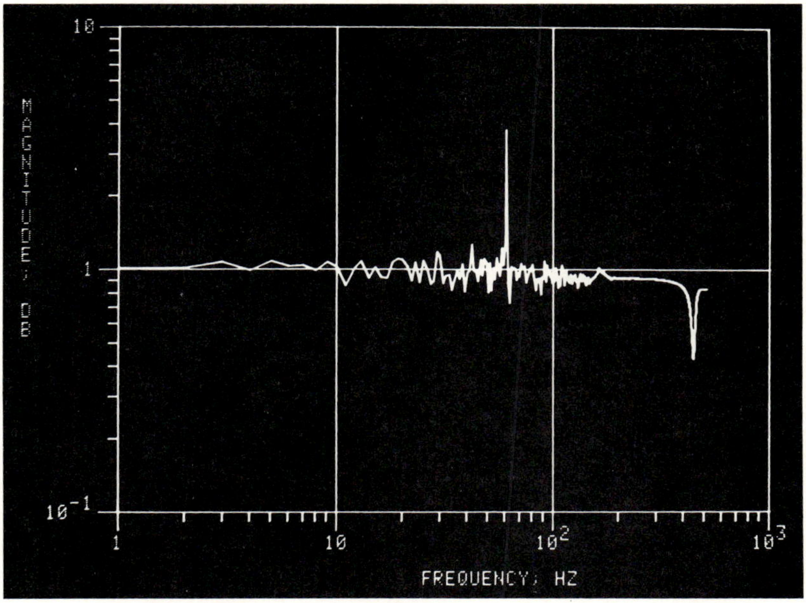

FIG. 8.11 A pressure pulse record of the response of canine dermal tissue to a pressure step function. A typical sample of the system response is shown to be uniform to more than 100 Hz (spikes indicate electrical noise).

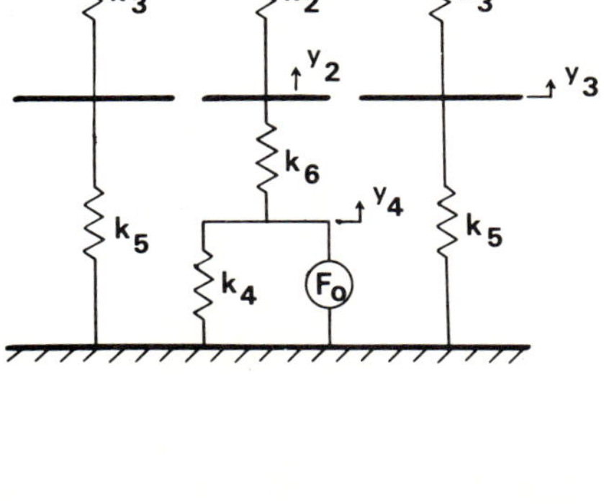

FIG. 8.12 The physical system model for an arterial tonometer proposed by Pressman and Newgard (1963). Only forces and deflections normal to the skin surface are considered. The linear elastic elements represent the following parameters: k_1 = mounting system; k_2, k_3 = the transducer and its frame, respectively; k_4 = the vessel wall; k_5 = adjacent tissue; and k_6 = tissue directly above the vessel. y_1, y_2, y_3, and y_4 are the displacements for these elements. F_0 is the force due to the vessel pressure, and F_1 is the total system positioning force.

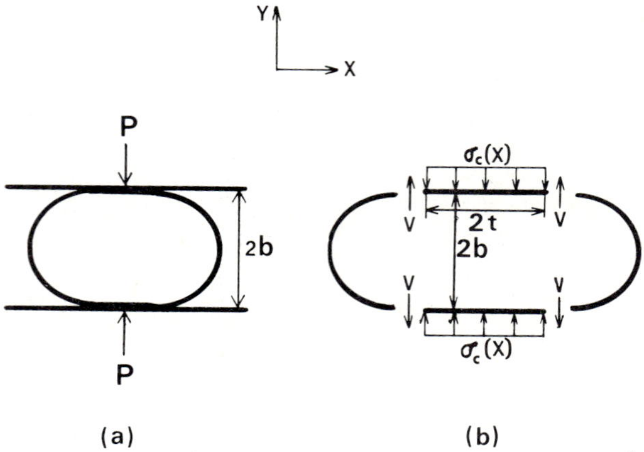

FIG. 8.13 (*a*) Vessel deflection shape such that the wall consists of free and surface constrained sections. (*b*) The free-body diagram for these sections. Contact stress $\sigma_c(x)$ is distributed along the flattened section length $2t$.

with
$$\beta = \frac{AG}{1.5EI} \tag{8.7}$$

and
$$V(b) = \frac{0.1757EI}{b^2} \tag{8.8a}$$

$$t(b) = \frac{\pi}{2}\left(r - \frac{b}{0.7185}\right) \tag{8.8b}$$

where $2b$ is plane separation distance, x is the position along wall contact, I is the moment area of arterial cross section, A is the cross-sectional area of arterial wall, E is Young's modulus, and G is the shear modulus.

The response of this model is illustrated for the radial artery in Fig. 8.14. The contact stress distribution along the flattened vessel wall was computed using Eqs. (8.6) to (8.8). The plots result from geometric and material descriptors for an isotropic approximation of the radial artery (Westerhof et al., 1969): $E = 8.0 \times 10^6$ dyn/cm^2, $G = 2.67 \times 10^6$ dyn/cm^2, $\nu = 0.5$ (Poisson's ratio), $r = 0.172$ cm, and $h = 0.043$ cm.

The computed contact stress, $\sigma_c(x)$, is plotted as a function of x along the contact length; $x = 0$ corresponds to a point directly over the center of the vessel. Each member of the family of curves represents a given deflection y of the vessel from its initially circular shape. For example, for a deflection of 50 percent of the initial diameter, σ_c is plotted over a distance $x = \pm 0.08$ cm, which is the flattened length $2t$ for that deflection (Fig. 8.13).

To interpret the graph, first consider p_i, the vessel's internal pressure, to be zero. The stress-strain relations imply that $\sigma_c(x)$ is related to bending and shear deformation. If $\sigma_c(x)$ approaches zero, bending and shear stress become negligible. If a fluid pressure Δp_i is added to the vessel lumen then, by superposition,

$$\sigma_c(x) = \sigma_c(x)|_{p_i=0} + \Delta p_i$$

for small Δp_i. Hence, when $\sigma_c(x)|_{p_i=0} \approx 0$ in Fig. 8.14, $\sigma_c(x) \approx \Delta p_i$, and the contact stress represents intraluminal pressure. If, however, $\sigma_c(x)|_{p_i=0}$ is appreciable (i.e., large deformational stress exists at the contact surface), such as with smaller deflections (e.g., 30 percent), one cannot expect to determine vessel pressure.

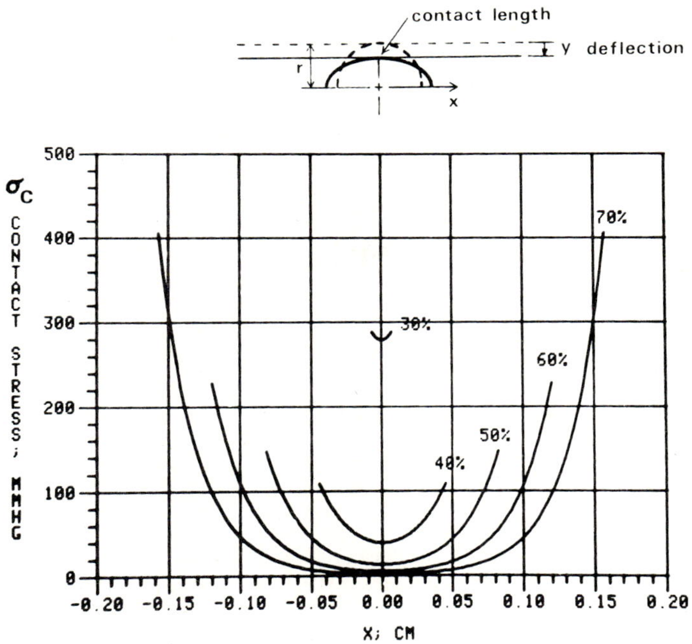

FIG. 8.14 Computed contact stress distribution, expressed in millimeters of mercury, from the model of the radial artery by Drzewiecki et al. (1983). Each curve represents the stress distribution for different deflections in the *y* direction, noted on the curves as fractions of the original vessel radius.

Criteria for tonometry follow from analysis of this model:

1. The artery should be deformed with a rigid, planar body until, by applanation, deformation contact stress becomes negligible (Fig. 8.14). An arbitrary degree of vessel flattening alone is insufficient.
2. The tonometer sensor should measure local contact stress rather than the contact force on the planar surface.
3. The sensor must record only over the flattened length of wall where deformation stresses are negligible, since there always exist regions of flattened wall where σ_c does not equal intraluminal pressure.

These design criteria impose severe requirements on available transducer technology.

The model also assists with positioning of the tonometer. Positioning can be associated with two components, radial (i.e., perpendicular to the surface) and tangential to the surface. For tangential positioning (note Fig. 8.14 and the 70 percent deflection curve), the contact stress will reflect arterial pressure most accurately over a range of $x = \pm 0.025$ cm. If the tonometer sensor dimension is exactly 0.05 cm, it must be centered on the artery. If smaller, it should lie within the range indicated.

The effect of radial positioning is illustrated in Fig. 8.15, where the vessel is shown deflected to three different degrees. The tonometer has a fixed sensor area and measures the total of the stresses bearing on its surface. Below each deflection is shown the contact stress distribution redrawn from Fig. 8.14.

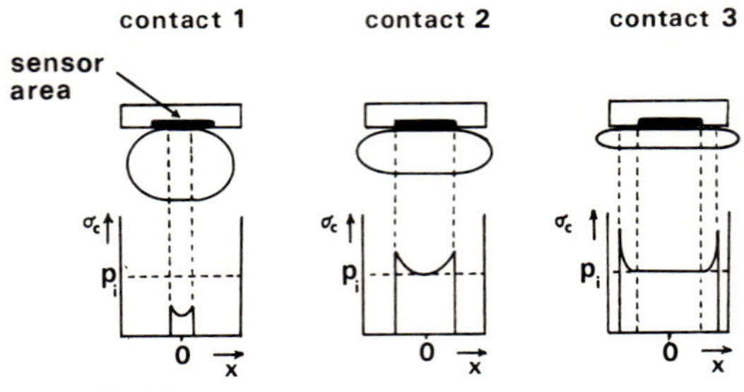

FIG. 8.15 Illustrative contact ranges and expected contact stresses for three radial positions; p_i = the vessel's internal pressure.

For contact 1, the tonometer signal will be a function of contact length, deformational stress, and p_i. The tonometer output may be less than predicted from the vessel's pressure, p_i, and depends on contact length. For contact 2, the contact length is equal to or greater than the sensor length, and the tonometer's output will be a function of deformation stress and p_i. Here the tonometer output exceeds the vessel's intraluminal pressure owing to the contribution of deformation stresses. Finally, for contact 3, the tonometer will measure p_i essentially since deformation stresses are negligible in the contact region. Further deformation requires an additional force so small that the tonometer output becomes independent of position. Hence, for the contact 3 range, a plateau will be observed in the tonometer output as it is displaced radially. This plateau denotes proper radial positioning and is a manifestation of the fact that the stress required to deform a vessel is inversely related to vessel eccentricity.

Since variation in material properties has been considered to be a problem for tonometers, the effect of a nonstatic modulus of elasticity was analyzed by using a range of values for E_s (the static modulus of elasticity) in Eqs. (8.6) to (8.8). Stress computed for a radial artery with a 60 percent deformation and three values of E_s is shown in Fig. 8.16. For some region over the

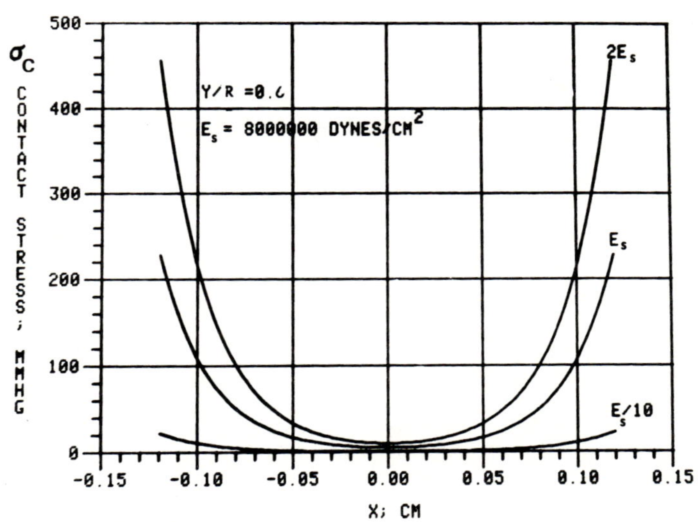

FIG. 8.16 Contact stress computed as a function of position x (the distance from the artery's center axis). Vessel deflection is 60 percent. Each curve represents one value of Young's modulus. Note that the range extends over an order of magnitude.

artery the measured stress $\sigma_c \approx 0$ for all values of E_s, the region narrowing as E_s increases. Where $d\sigma_c/dE_s \approx 0$, tonometer accuracy is not degraded by material property changes. Consequently, the tonometer should be designed to operate with the largest elastic modulus that may be encountered. These results agree with those from the more general stress transmission models (Sec. 8.3.1).

8.4 ORIGIN OF KOROTKOFF SOUND

The origin of Korotkoff sound is critical to the auscultatory method of blood pressure measurement, but existing theories that explain its generation are all inadequate. In this section a fluid-dynamic model is presented as the basis for a new theory that predicts many characteristics of the sound.

8.4.1 Existing Theories of Korotkoff Sound

The auscultatory method was first demonstrated to the Imperial Medical Academy in St. Petersburg by the physician Korotkoff (1905). An occlusive cuff was applied to a canine limb and the distal end of the femoral artery was exposed and severed to allow observation of blood flow. Korotkoff concluded that sound, which could be detected with the stethoscope, occurred when decreasing cuff pressure reached systolic pressure and ceased when cuff pressure reached diastolic pressure. Korotkoff disproved the postulate that the sound originated in the heart by perfusing an isolated canine limb from an independent source. Korotkoff believed that the sound originated from the arterial wall (knowing it was collapsed by cuff pressure) but had little supporting evidence.

Since Korotkoff's time, several theories have been proposed. These may be generally assigned to one of two categories (McQueen, 1976; Conrad et al., 1980):

1. *Pressure origin.* In this category the Korotkoff sound is attributed to a transient local disturbance of arterial pressure that is reflected in arterial wall motion. Examples include dispersion in finite-amplitude waves (Bramwell and Hickson, 1927; Beam, 1968; Tavel et al., 1969; Grigoryan et al., 1980) and arterial wall instability (Anliker and Raman, 1966; Raman, 1967; Cohen and Dinnar, 1975).
2. *Flow origin.* In this category a blood flow phenomenon is thought to be the source of Korotkoff sound. Examples include water hammer (Erlanger, 1916; Kositskii, 1958), disturbed flow (Chungcharoen, 1964; Gupta et al., 1975), jet noise (Erlanger, 1916), periodic wakes (Bruns, 1959), and self-excited oscillations of flow in a collapsible tube (Rodbard, 1953; Lighthill, 1972; Ur and Gordon, 1970; Conrad et al., 1978; Caro et al., 1979).

Considerable effort has been devoted to the determination of the origin of Korotkoff sound. A dissertation attempting to identify the best theory that explains the phenomena was inconclusive (McQueen, 1976). To quote von Recklinghausen (1940), the sound "had become even more obscure than before."

Nevertheless, McQueen weakened flow-related theories. He detected, noninvasively, motion of the arterial wall and of blood distal to the occlusive cuff with a Doppler ultrasonic probe. A similar experiment had been performed earlier by McCutcheon and Rushmer (1967), who used an ultrasonic flow probe and a microphone placed over the brachial artery just distal to the occlusive cuff. McCutcheon and Rushmer, as well as McQueen, found that blood flow had hardly commenced when sound was detected; the Doppler probe did detect transient arterial wall motion at this time (Fig. 8.17). They concluded that Korotkoff sound is generated when the arterial lumen opens, prior to any significant flow. Although they do not completely discount flow-based theories, these experiments favor pressure-origin theories. Flow-related phenomena are assumed to contribute when the sound is murmurlike, i.e., during phase II subsequent to the initial tapping sound when flow first develops (McCutcheon and Rushmer, 1967).

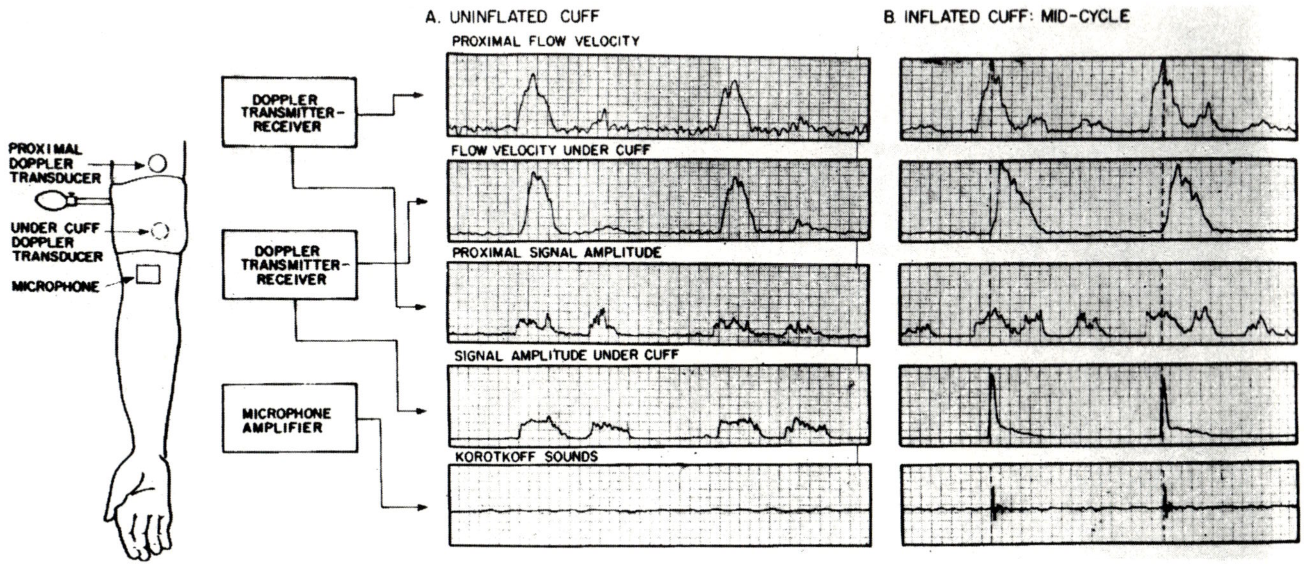

FIG. 8.17 Simultaneous Korotkoff sound microphone and Doppler ultrasound recordings. When the cuff is inflated (right tracings) and Korotkoff sounds occur, the sound tracing (bottom) coincides with the onset of the Doppler flow pulse (second from top). (*From McCutcheon and Rushmer, 1967, by permission.*)

Flow-related sources, such as disturbed flow, can no longer be viewed as responsible for Korotkoff sound, and theories that focus on arterial wall motion emerge as more appropriate candidates.

The first pressure-origin theory (see above) was based on pulse component dispersion. If the arterial pulse is to maintain its shape during transmission, all its frequency components must propagate at the same wave speed; i.e., there can be no dispersion. Proponents of the frequency dispersion theory believe that Korotkoff sound results from the arterial pulse altering to a transient wave, or "shock-wave," as it passes under the occlusive cuff. This pulse is thought to cause sudden arterial wall motion, audible as a tap.

The second theory attributes arterial wall instability to the pressurized cuff collapsing the underlying vessel. Proponents of this theory believe that noncircular cross sections amplify luminal flow or pressure disturbances. Oscillations in the acoustic frequency range may be generated by either induced resonance or buckling of the arterial wall. With buckling, rapid changes in the artery's bulk compliance are manifested, concomitant with various cross sections. Both theories require only small internal pressure changes for large wall deflections.

Evaluation of these theories depends on data on arterial wall motion that can be transduced to sound detectable by a stethoscope. Since this was not available, a quantitative cause-effect relationship was not developed to support the postulated mechanisms (McQueen, 1976).

8.4.2 Fluid-Dynamic Model

A recently developed pressure-related theory for the origin of Korotkoff sound provides a more quantitative base and predicts observed phenomena (Drzewiecki 1985). This theory relates both to fluid dynamics in a collapsible blood vessel and to sound transmission. For the first part, the artery and occlusive cuff system was modeled using information on arterial pulse dynamics (Noordergraaf, 1978). The system is illustrated in Fig. 8.18. Three segments of the artery are identified, i.e., proximal, distal, and collapsing; this last segment underlies the occlusive cuff. The proximal segment provides the normal pressure pulse, whereas the distal segment leads into the peripheral vasculature. The stethoscope is conventionally placed on the surface overlying the distal artery. The distal arterial pressure disturbance is first derived for a given pressure pulse, cuff pressure, and other local and distal vascular characteristics. This pressure disturbance may be analyzed for potential sound production.

The second part of the model expresses the transmission of sound to the ear. The arterial wall, skin, and stethoscope were analyzed and the computed sound compared for verification to actual recorded sounds.

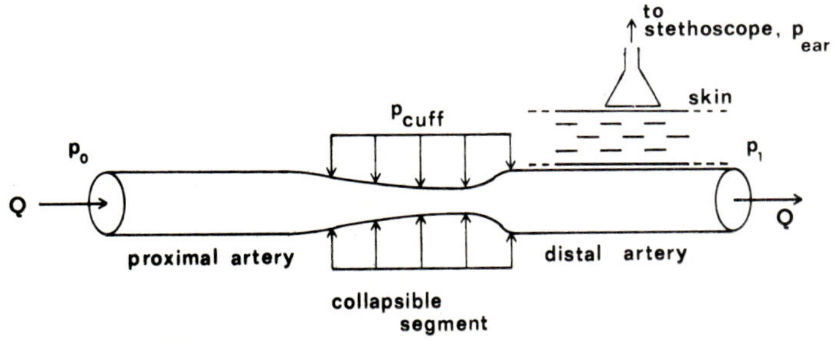

FIG. 8.18 Proposed Korotkoff sound–generating model representing the brachial artery during an auscultatory pressure measurement. The proximal artery receives pulsatile blood pressure and flow. The cuff collapsed segment is assumed to be under uniform radial load. Depending on the transmural pressure, the collapsible segment can collapse to various degrees. The distal artery connects the collapsible segment to the peripheral vasculature of the arm. Just distal to the cuff a stethoscope detects arterial motion through layers of tissue.

In order to solve the dynamic pressure and flow problem, the pulsatile characteristics of a collapsible vessel were formulated as the set of empirical relations derived by Brower and Noordergraaf (1973). These steady pressure-flow relations in a collapsed vessel can be represented by

$$p_0 - p_1 = \frac{Q_1}{Q_c} \frac{p_{\text{cuff}} - p_1}{[1 + (Q_1/Q_c)^n]^{1/n}} \quad \text{if} \quad p_0 - p_1 > R_r Q_1 \tag{8.9}$$

Otherwise

$$p_0 - p_1 = R_r Q_1 \tag{8.10}$$

where

$$Q_c = a + be^{-(p_{\text{cuff}} - p_1)/c} \tag{8.11}$$

R_r is the resistance of the collapsible segment with an open lumen, p_0 is proximal pressure, p_1 is distal pressure, p_{cuff} is external pressure, and Q_1 is flow. Q_c is an empirical quantity related to transmural pressure. The parameters a, b, c, and n were chosen to best represent the canine femoral artery as measured by Conrad et al. (1980). The parameters were found to be $a = -1.252 \times 10^{-3}$ cm³/s, $b = 8.819$ cm³/s, $c = 10.61$ mmHg, $n = 0.904$, and $R_r = 1.765 \times 10^3$ g/(cm⁴ · s).

The application to pulsatile flow of these steady-flow relations is justified by the fact that distributed inertial and compliant terms are small over the short collapsed length, as compared with the large distal mass of blood and total peripheral compliance. Furthermore, viscous resistance terms, which are included, dominate during most of the cardiac cycle and particularly during partial arterial collapse. Also, the empirical relation uses data from the canine femoral artery to approximate the human brachial artery. These assumptions become inapplicable only at the high-frequency end of the computed results.

The electrical analog of the fluid mechanical system is illustrated in Fig. 8.19. The distal arterial system is modeled as a modified *Windkessel* (Noordergraaf, 1978) whose element values are determined from the vessel's characteristic impedance R_0, the total peripheral resistance of the arm, R_s, and the total peripheral compliance of the arm, C_s. For the human they are approximated as $R_s = 1.2 \times 10^5$ g/(cm⁴ · s), $R_0 = 3.4 \times 10^3$ g/(cm⁴ · s), and $C_s = 10.1 \times 10^{-6}$ (cm⁴ · s²/g).

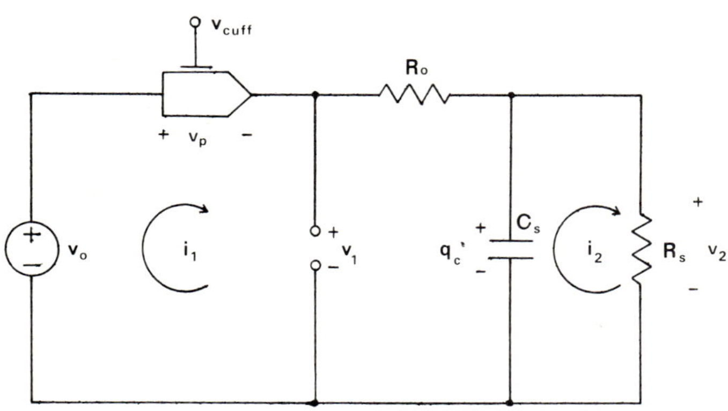

FIG. 8.19 Electrical analog of the fluid-dynamic part of the Korotkoff sound model. The voltage source V_0 represents proximal pressure, and V_1 represents the pressure distal to the cuff. The nonlinear voltage-controlled element represents the collapsible segment. R_s, C_s, and R_0 make up the modified *Windkessel* for the distal vasculature. q_c is the state variable of the system.

To solve for the electrical analog of distal arterial pressure, v_1, the state equation for the system is written as

$$\frac{q_c}{dt} = \frac{v_0}{R_0C_s} - \frac{v_p}{R_0C_s} - \frac{q_c}{C_s}\left(\frac{1}{R_s} + \frac{1}{R_0}\right) \tag{8.12}$$

where the state variable q_c is the voltage across the total peripheral capacitance and v_p is the voltage drop across the collapsible element. The loop voltage equations for both the open and collapsed vessels are

$$v_0 = R_r i_1 + R_0 i_1 + q_c \tag{8.13}$$

$$v_0 = R_0 i_1 + q_c + \frac{i_1}{Q_c}\,\frac{v_{\text{cuff}} - R_0 i_1 - q_c}{[1 + (i_1/Q_c)^n]^{1/n}} \tag{8.14}$$

The solution of these nonlinear, discontinuous equations was obtained by numerical methods. Finally, $v_1 = v_0 - v_p$.

The arterial pressure just distal to the occlusive cuff, computed from the model with cuff (external) pressure fixed at 100 mmHg, is illustrated in Fig. 8.20. The pulse appears distorted, and when this shape is filtered for frequencies in the acoustic range, these are found coincident with the foot of the ascending portion of the pulse. This is seen in Fig. 8.20 as an increased positive slope. Such distortions of the pulse have been recognized in direct human arterial recordings (Tavel et al., 1969) and in noninvasive recordings (Whitcher, 1969) (see Fig. 8.21*d*, no filter). Thus, the model indicates that frequencies in the acoustic range are introduced into the pulse during passage through the collapsible segment of artery.

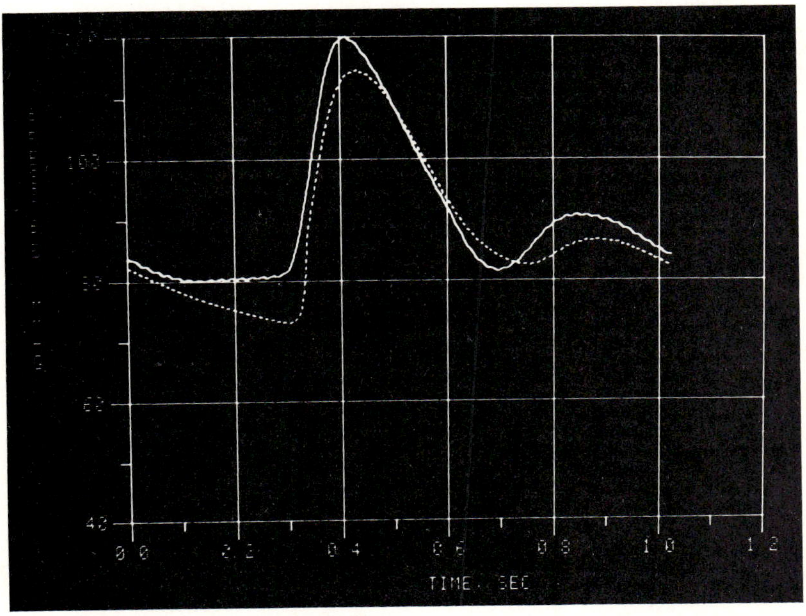

FIG. 8.20 Solid trace: proximal brachial pressure. Dotted trace: arterial pressure just distal to the collapsible artery segment computed from the model in Fig. 8.19. One cycle is shown with cuff pressure fixed at 100 mmHg and the system at steady state. The distal pulse is distorted from its normal shape, particularly in the early upstroke.

The literature concerning Korotkoff sound contains many observations from noninvasive procedures and a few observations from invasive procedures. These include microphone records, variation of sound intensity with cuff pressure, variation of sound spectra with cuff pressure, sound phase shift with cuff pressure, time of occurrence relative to distal pressure and flow pulse, and spatial variation of sound intensity with recording site. The fluid mechanical model is capable of predicting these phenomena, of which a few are presented here.

The dependence of the variation of intensity of Korotkoff sound on cuff pressure has been known since the method was first introduced. Distal pulses computed with the model were filtered for frequencies in the acoustic range and the resultant "sound" intensity plotted against cuff pressure (Fig. 8.21*a*). The expected behavior is evident, i.e., sound intensity increasing at systolic pressure and decreasing toward diastolic pressure. Similar curves have been observed by Whitcher (1969), Golden et al. (1974), and Maurer and Noordergraaf (1976) (Fig. 8.21*b* to *d*). It should be noted that the model overestimates blood pressure and that the accuracy for diastolic pressure determination is poor in comparison with that of systolic pressure. This has also been demonstrated when occlusive cuff measurements are compared with invasive measurements (London and London, 1967).

The model also provides the time course of arterial wall motion and hence Korotkoff sound, an accomplishment not achieved with previous theories. To illustrate, a phase II to III Korotkoff sound (region of loudest sound) was computed. Those who have investigated Korotkoff sound replace the stethoscope with a contact microphone (Maurer and Noordergraaf, 1976; Gupta et al., 1975). A dominant spike is evident in the recordings of these investigators and in the computed sound (Fig. 8.22). Discrepancies with the computed sound can be accounted for as arising from asymmetric microphone placement, the signal processing employed and subject-to-subject variation.

The flow pulse and Korotkoff sound are plotted simultaneously in Fig. 8.23. The model indicates that sound coincides with the onset of the flow pulse and conforms to that observed with Doppler ultrasound (McCutcheon and Rushmer, 1967).

For the purpose of automated instrument design (Sec. 8.1.2), the power spectra of the Korotkoff sound has been studied extensively (Ware and Anderson, 1966; McCutcheon and Rushmer, 1967; Gupta et al., 1975; Maurer and Noordergraaf, 1976). Power spectra computed

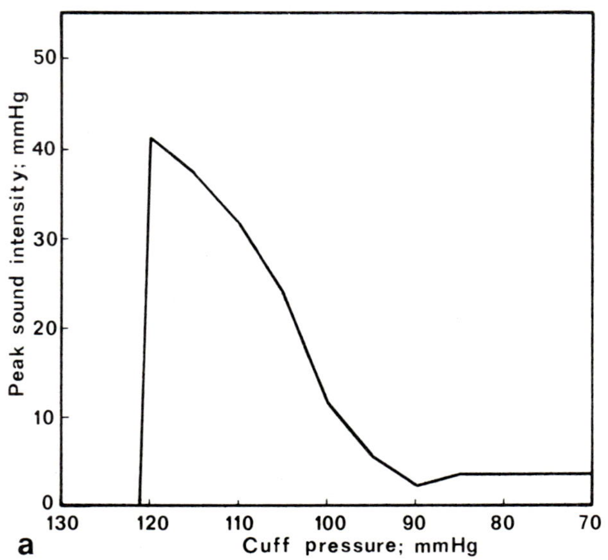

FIG. 8.21 Variation of Korotkoff peak sound intensity with cuff pressure. (*a*) Computed from model with blood pressure at 120/80 mmHg.

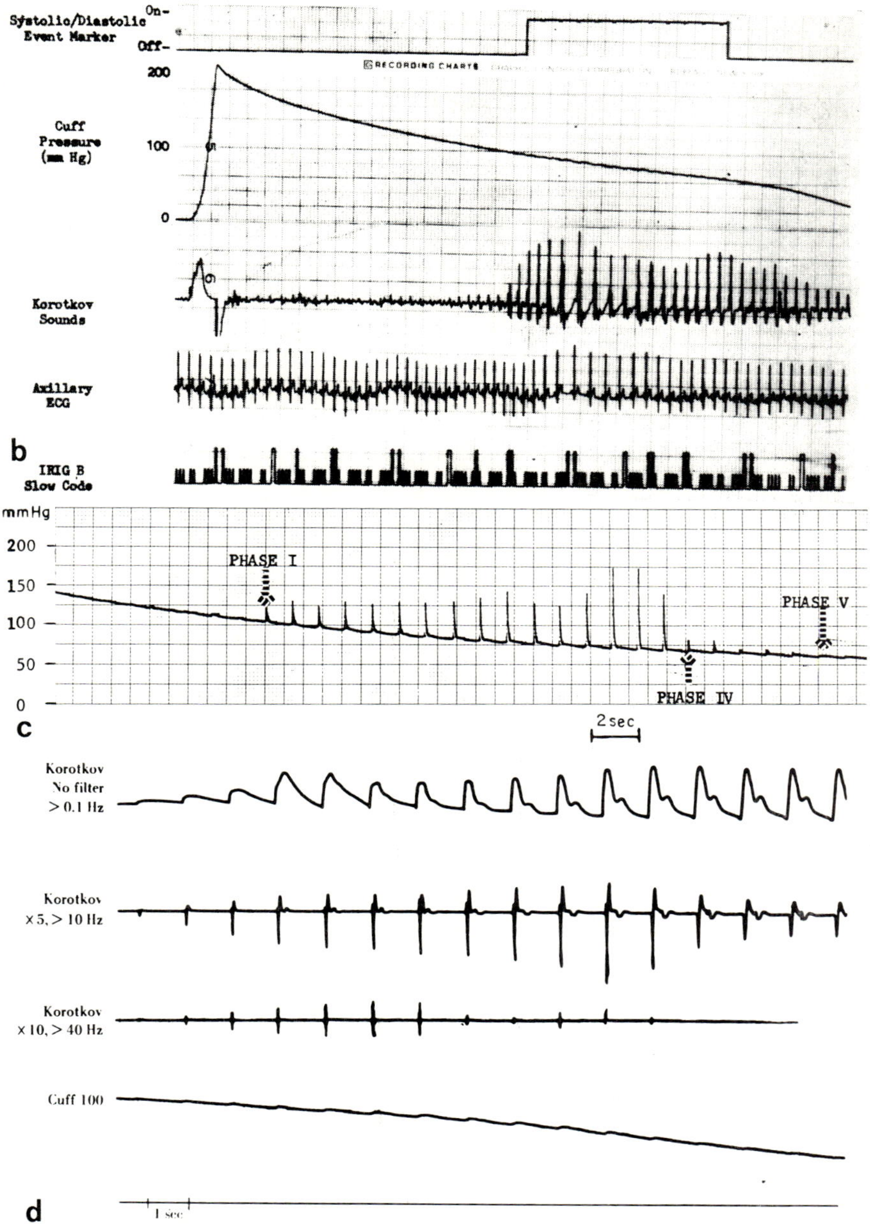

FIG. 8.21 (*b*) Microphone data. (*From Golden et al., 1974, by permission.*) (*c*) Microphone data. (*From Maurer and Noordergraaf, 1976.*) (*d*) Microphone data. (*From Whitcher, 1969, by permission.*) Observe the effect of different filters on the Korotkoff sound traces.

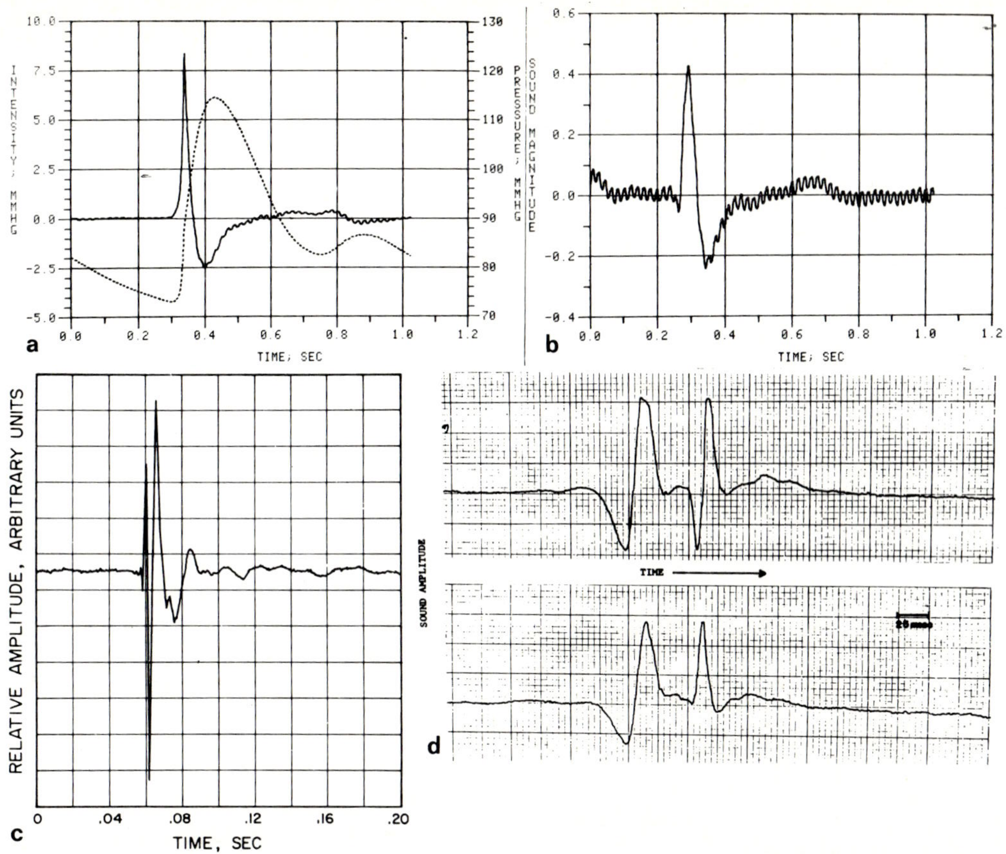

FIG. 8.22 (*a*) Korotkoff sound time course (solid) computed from the model for phase III. Cuff pressure is 100 mmHg. The computed brachial pulse just distal to the cuff (dotted) is shown in relation to the sound. (*b*) A Korotkoff sound recording at phase III (authors' experiment). (*c*) Korotkoff sound recording. (*From Gupta et al., 1975, by permission.*) (*d*) Korotkoff sound recordings. (*From Maurer and Noordergraaf, 1976.*)

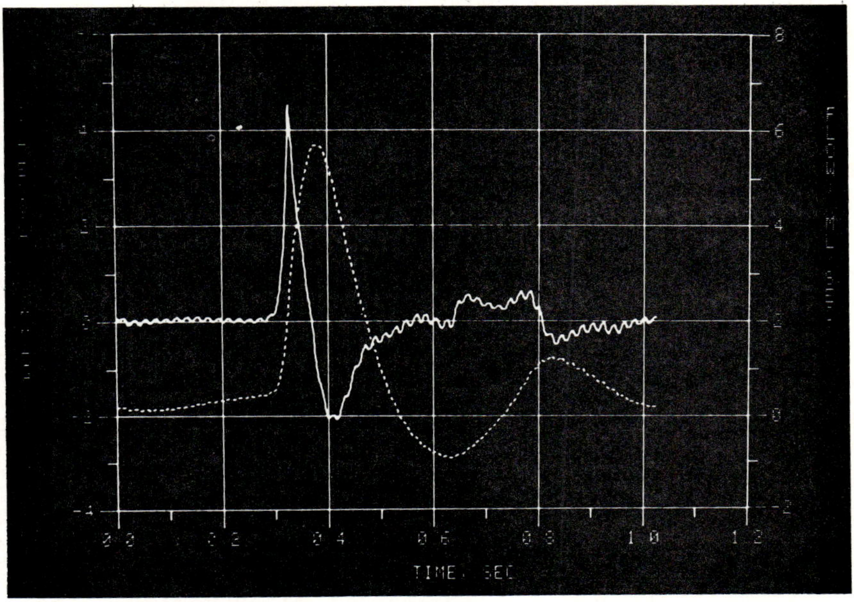

FIG. 8.23 Computed Korotkoff sound (solid trace) and the brachial flow pulse just distal to the cuff (dotted trace), with cuff pressure at 95 mmHg. The sound occurs at the onset of flow.

with the model at cuff pressures of 121, 100, and 95 mmHg (Fig. 8.24*a* to *c*) indicate loss of high-frequency content (“muffling”) as cuff pressure decreases. That sound energy decreases with increasing frequency has also been observed by others (Fig. 8.24*d* to *f*).

Unlike earlier models, the new model can synthesize sound similar to Korotkoff sound. To assess whether absolute magnitudes of the oscillations as presented to the ear are realistic, transmission through the artery, skin, and stethoscope has been analyzed by Drzewiecki et al. (1982b). The system is illustrated in Fig. 8.25 in terms of its electrical analog.

The electrical analog was arranged into two-port networks which are represented by the transmission matrices

$$\mathbf{T}_s = \begin{bmatrix} 1 & \dfrac{1}{j\omega C_s(\omega)} \\ 0 & 1 \end{bmatrix} \qquad \mathbf{T}_b = \begin{bmatrix} 1 + \dfrac{C_b}{C_d} & \dfrac{1}{j\omega C_d} \\ j\omega C_b & 1 \end{bmatrix}$$
$$\mathbf{T}_h = \begin{bmatrix} \cosh \gamma L & Z_0 \sinh \gamma L \\ \dfrac{\sin \gamma L}{Z_0} & \cosh \gamma L \end{bmatrix} \tag{8.15}$$

where $\mathbf{T}_s$ is the skin matrix; $\mathbf{T}_b$ is the bell and diaphragm matrix; $\mathbf{T}_h$ is the tubing transmission matrix for transfer functions h_b, h_e; Z_0 is the acoustic impedance of air; γ is the propagation constant of stethoscope tubing, experimentally determined by Latimer and van Vollenhoven (1981); L is the length of tubing; and $j\omega$ is the complex angular frequency. For a unique solution, the required input and output relations are

$$P_a = \frac{Q_1}{j\omega c_a(\omega)} + P_1 \tag{8.16}$$

$$Q_{e1} = -j\omega c_e P_{e1} \tag{8.17}$$

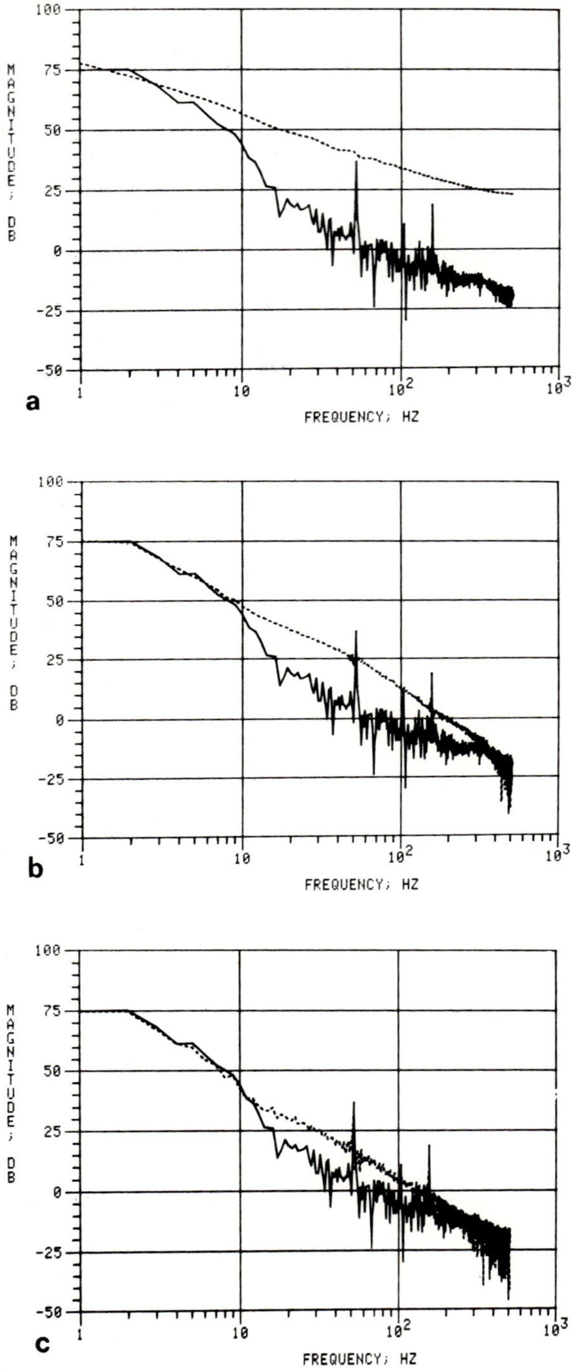

FIG. 8.24 Computed relative power frequency spectra of proximal (solid) brachial pulse and pulse just distal to the collapsed segment (dotted) for a cuff pressure of (*a*) 120, (*b*) 100, and (*c*) 95 mmHg.

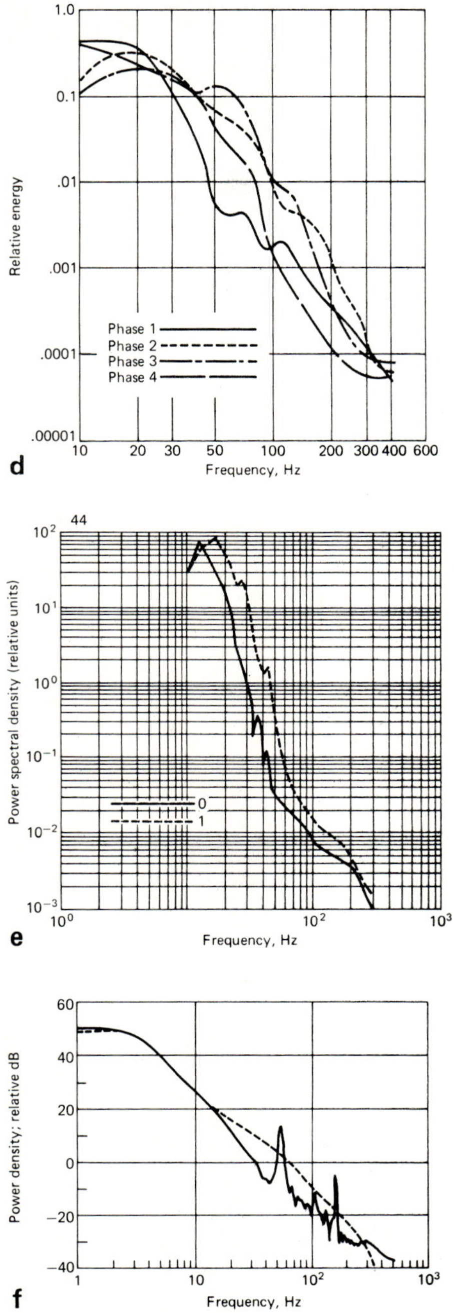

FIG. 8.24 (*d*) relative energy frequency spectra. (***From Ware and Anderson, 1966, by permission.***) (*e*) Measured power spectral density, phase IV. (*From Maurer and Noordergraaf, 1976.*) (*f*) **Computed power spectra, phase III (dotted curve) compared with recorded phase III (solid curve, authors' data). Both signals were normalized by (energy)$^{1/2}$ for relative power comparison.**

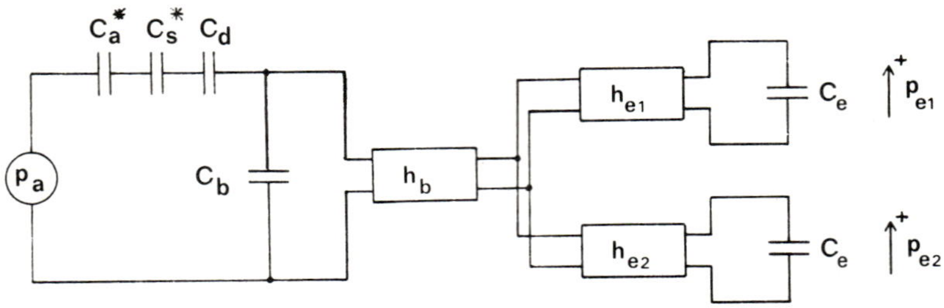

FIG. 8.25 Model of the transmission system from artery to ear. Input is computed arterial pressure P_a just distal to the cuff; output is computed Korotkoff sound pressures P_{e1} and P_{e2} at the ears. C_a^*, C_s^*, and C_d represent the arterial, skin, and stethoscope diaphragm compliances, respectively. The starred compliances have viscoelastic properties. C_b is the air compliance in the bell. h_b, h_{e1}, and h_{e2} are the acoustic transmission lines representing the stethoscope bell and ear tubing. C_e is the compliance of the air volume contained in the auditory canals.

where P_a is arterial pressure; P_{e1} is pressure in the auditory canal; and P_1, Q_1 are pressure and flow, respectively, at the artery-skin interface. Capital letters denote the variables to be complex phasors. For a given arterial pressure, Korotkoff sound at the ear was computed with a digital computer from Eqs. (8.16) to (8.18).

$$\begin{bmatrix} P_1 \\ Q_1 \end{bmatrix} = [\mathbf{T}] \begin{bmatrix} P_{e1} \\ -Q_{e1} \end{bmatrix} \tag{8.18}$$

where **T** is the complete transmission matrix

$$\mathbf{T} = [\mathbf{T}_s][\mathbf{T}_b][\mathbf{T}_{hb}] \begin{bmatrix} 1 & 0 \\ 0 & 2 \end{bmatrix} [\mathbf{T}_{he}] \tag{8.19}$$

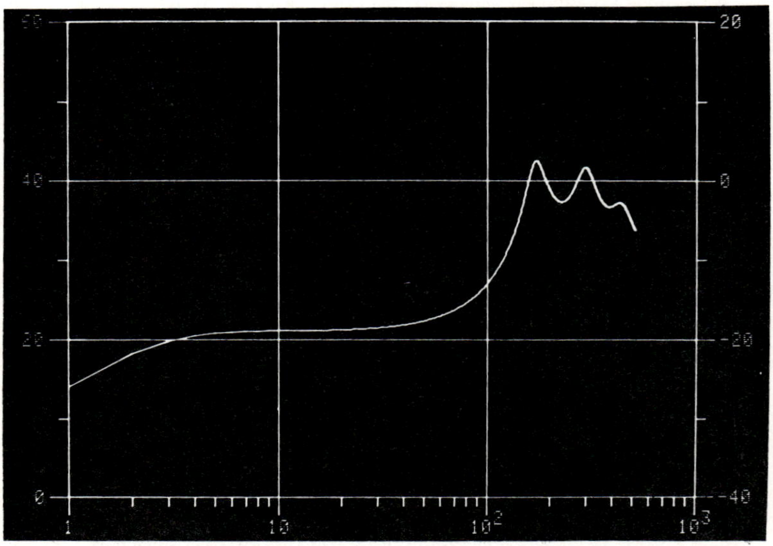

FIG. 8.26 The computed frequency response for the transmission system between the arterial lumen and the ear. Left ordinate: input pressure of 60 dB with a uniform frequency response. Right ordinate: system response magnitude in decibels. Abscissa: frequency in hertz.

The steady-state response of the overall linear transmission system is illustrated in Fig. 8.26. Amplitudes of the lower-frequency oscillations are reduced at the ear, and the characteristics of the stethoscope become evident at higher frequencies, manifesting a resonant frequency at 120 Hz. Losses greatly compromise energy transmitted to the ear at frequencies above 500 Hz. The overall system selectively attenuates frequency components of the incoming signal and exhibits a semblance of differentiation in the time domain. The stethoscope itself is capable of sound amplification only at its primary resonant frequency.

The transmission model indicates that at the lower end of the audible frequency range (< 20 Hz) the intensity of Korotkoff sound is 1.72×10^{-9} W/m^2. During generation, Korotkoff sound intensity averaged 2.73×10^{-8} W/m^2 consistent with auditory requirements; the audible threshold for the human ear is about 10^{-8} W/m^2 in this frequency range (Fletcher and Munson, 1933).

The fluid-dynamic model supports the postulate that Korotkoff sound derives from pressure pulse distortion. This distortion stems from nonlinear pressure-flow relations in the time-varying, collapsing segment of artery. Such distortions and losses overall shift pulse energy into the acoustic frequency range that reaches the ear at levels compatible with the threshold of hearing.

SUMMARY AND CONCLUSIONS

Since invasive methods measure pressure as a calibrated temporal function, a similar goal has been set for noninvasive methods. Achievement of this goal has been slow.

From the outset only one of two desirable options seemed feasible with any one method. For example, a tiny mirror, placed on the skin over a superficial artery, provided a sensitive record of the pressure oscillation but not of the absolute value of pressure. Alternatively, occlusive cuff methods obtain absolute pressure but only for the systolic and diastolic extremes.

This chapter summarizes methods that have received most of the attention with respect to these two options. In particular we focus on one of the oldest and most popular methods for the determination of systolic and diastolic pressures: the Korotkoff method. This empirical method relies on the production of audible sound, the genesis of which has long been a matter to be clarified. A fluid mechanical, acoustic model is proposed that provides a quantitative base for a new theory explaining the origin of Korotkoff sound. Results indicate that the Korotkoff sound arises from oscillations resulting from distortion of the arterial pulse under the cuff due to nonlinear pressure-flow relations in the collapsed segment of the artery. Although not yet categorically established, the theory appears to embody experimentally observed features and predicts the genesis of sound from the arterial pulse.

For the measurement of continuous pressure, we focus on arterial tonometry and present a continuum mechanical model of skin and artery. The analysis indicates that a tonometer, given suitable design characteristics, can provide a calibrated, temporal pressure signal. Design characteristics must consider positioning, calibration, and tissue variation; all are problems associated with tonometry.

General models are also presented that couple concepts of arterial tonometry with transcutaneous pulse dynamics and pulse recording. For minimum distortion in the recorded pulse, pressure pulse methods emerge as superior to volume pulse methods.

ACKNOWLEDGMENTS

This paper was supported, in part, by National Institutes of Health grants HL 24966, HL 22223, HL 10330, and GM 07627.

REFERENCES

Abbott, J. A. (1969). The fidelity of the externally recorded human pulse. *Am. J. Med. Sci., 258,* 40.

Alexander, H., Cohen, M., and Steinfeld, L. (1977). Criteria in the choice of an occluding cuff for the indirect measurement of blood pressure. *Med. Biol. Eng. Comput., 15,* 2.

Allred, J. E., and Johnson, J. B. (1966). Design and development of an automatic indirect blood pressure monitoring device. Tech. Rep. 66-80, U.S. Air Force School Aerospace Medicine, Brooks AFB, Texas.

Anliker, M., and Raman, K. R. (1966). Korotkoff sounds at diastole—A phenomena of dynamic instability of fluid-filled shells. *Int. J. Solids Struct., 2,* 467.

Apple, H. P. (1980). Automatic noninvasive blood pressure monitors: What is available. In J. S. Gravenstein, R. S. Newbower, A. K. Ream, N. T. Smith, and J. Barden (eds.), *Essential Noninvasive Monitoring in Anesthesia.* Grune & Stratton, New York.

Arzbaecher, R. C., and Novotney, R. L. (1973). Non-invasive measurement of the arterial pressure contour in man. *Bibl. Cardiol., 31,* 63.

Bahr, D. E., and Clark, K. R. (1980). Continuous arterial tonometry. In J. S. Gravenstein, R. S. Newbower, A. K. Ream, N. T. Smith, and J. Barden (eds.), *Essential Noninvasive Monitoring in Anesthesia.* Grune and Stratton, New York.

Bahr, D., and Patzke, J. (1973). The automatic arterial tonometer. *Proc. 26th ACEMB,* p. 259.

Beam, R. M. (1968). Finite amplitude waves in fluid-filled elastic tubes: Wave distortion, shock waves and Korotkoff sounds. NASA Tech. Note TND-4803, p. 52.

Bramwell, J. C., and Hickson, S. K. (1927). The relation of pulse form to sound production in arteries: 1. The form of the sphygmogram during decompression. *Heart, 13,* 109.

Brower, R. W., and Noordergraaf, A. (1973). Pressure-flow characteristics of collapsible tubes: A reconciliation of seemingly contradictory results. *Ann. Biomed. Eng., 1,* 333.

Bruns, D. L. (1959). A general theory of the causes of murmurs in the cardio-vascular system. *Am. J. Med., 27,* 360.

Buggs, H., Johnson, P. E., Gordon, L. S., Balguma, F. B., and Wettach, G. E. (1973). Comparison of systolic arterial blood pressure by transcutaneous Doppler probe and conventional methods in hypotensive patients. *Anesth. Analg., 52,* 776.

Bugiel, W. (1908). *Joseph Strus, Polish Physician of the Renaissance: His Life and Papers.* Nowiny Lekarskie, Poznan.

Burger, H. C. (1955). Unpublished research notes.

Busse, R., Wetterer, E., Bauer, R. D., Pasch, T., and Summa, Y. (1975). The genesis of the pulse contours of the distal leg arteries in man. *Pfluegers Arch., 360,* 63.

Canzoneri, J. III, Geddes, L. A., Vallbona, C., and Viscusi, R. L. (1968). Electronic system for detection of indirect systolic and diastolic blood pressure. *Biomed. Sci. Instr. Proc. 5th Nat. Biomed. Sci. Instr. Symp.* Plenum Press, New York, vol. 4, pp. 65–70.

Caro, C. G., Pedley, T. J., Schroter, R. C., and Seed, W. A. (1979). *The Mechanics of the Circulation.* Oxford University Press, Oxford.

Chlebus, H. (1979). Clinical application of carotid electrosphygmography: Noninvasive access to cardiovascular dynamics. *Exp. Appl. Bibl. Cardiol., 37,* 73.

Chungcharoen, D. (1964). Genesis of Korotkoff sounds. *Am. J. Physiol., 207,* 190.

Cohen, D., and Dinnar, U. (1975). Buckling of fluid-filled elastic cylindrical tubes under dynamic pressure. *Isr. J. Technol., 13,* 192, 1975.

Cohn, J. (1967). Blood pressure measurement in shock. *JAMA, 199,* 972.

Conrad, W. A., Cohen, M. L., and McQueen, D. M. (1978). Note on the oscillations of collapsible tubes. *Med. Biol. Eng. Comput., 16,* 211.

Conrad, W. A., McQueen, D. M., and Yellin, E. L. (1980). Steady pressure flow relations in compressed arteries: Possible origin of Korotkoff sounds. *Med. Biol. Eng. Comput., 18,* 419.

Correll, R. W. (1959). Theoretical analysis and preliminary development of an indirect blood pressure recording system. M.S. thesis, Massachusetts Institute of Technology, Cambridge, Mass.

Davis, M., Gilmore, B., and Freis, E. (1963). Improved transducer for external recording of arterial pulse wave. *IEEE Trans. Biomed. Elect., BME-10,* 173.

Drzewiecki, G. M. (1979). Arterial tonometry. M.S. thesis, University of Pennsylvania, Philadelphia.

Drzewiecki, G. M., and Noordergraaf, A. (1979). Arterial tonometry: Basic requirements. *Proc. 32d ACEMB.* Denver, p. 58.

Drzewiecki, G. M., Melbin, J., and Noordergraaf, A. (1982a). Analytical comparison of transcutaneous pulse recording methods. *Proc. Tenth Annu. NE Bioeng. Conf.* Dartmouth College, Hanover, N.H., p. 121.

Drzewiecki, G. M., Melbin, J., and Noordergraaf, A. (1982b). A model for auscultation. *Proc. 35th ACEMB.* Philadelphia, p. 76.

Drzewiecki, G. M. (1985). The origin of the Korotkoff sound and arterial tonometry. Ph.D. diss., University of Pennsylvania, Philadelphia.

Drzewiecki, G. M., Melbin, J., and Noordergraaf, A. (1983). Arterial tonometry: Review and analysis. *J. Biomech., 16,* 141.

Eckerle, J. S., and Newgard, P. M. (1976). A non-invasive transducer for the continuous measurement of arterial blood pressure. *Proc. 29th ACEMB*. Boston, p. 98.

Erlanger, J. (1903). A new instrument for determining the minimum and maximum blood pressures in man. *Johns Hopkins Hosp. Rep., 12,* 53.

Erlanger, J. (1916). Studies in blood pressure estimation by indirect method: II. The mechanism of the compression sounds of Korotkoff. *Am. J. Physiol., 40,* 82.

Fletcher, H., and Munson, W. A. (1933). Loudness, its definition, measurement and calculation. *J. Acoust. Soc. Am., 5,* 82.

Frank, O. (1899). Die Benutzung des Princips der Pitot'schen Röhrchen zur Bestimmung der Blutgeschwindigkeit. *Z. Biol., 19,* 1.

Gaertner, G. (1899). Ueber einen neuen Blutdruckmesser. *Wien. Med. Wochenschr., 49,* 1412.

Geddes, L. A. (1970). *The Direct and Indirect Measurement of Blood Pressure*. Year Book, Chicago.

Geddes, L. A., Voelz, M., Combs, C., Reiner, D., and Babbs, C. F. (1982). Characterization of the oscillometric method for measuring indirect blood pressure. *Ann. Biomed. Eng., 10,* 271.

Geddes, L. A., and Whistler, S. J. (1978). The error in indirect blood pressure measurement with incorrect size of cuff. *Am. Heart J., 96,* 4.

Ghista, D. N. (1978). Analysis for noninvasive monitoring of arterial blood pressure by (I) Phono cuff-sphygmomanometry, for discrete measurements and (II) arterial ultrasonic imaging, for continuous measurements. In *Noninvasive Cardiovascular Measurements. Soc. Photo-opt. Inst. Eng., 167,* 107.

Ghista, D. N., Jayaraman, G., and Sandler, H. (1978). Analysis for the noninvasive determination of arterial properties and for the transcutaneous continuous monitoring of arterial blood pressure. *Med. Biol. Eng. Comput., 16,* 715.

Golden, D. P., Jr., Wolthius, R. A., Hoffler, G. M., and Gowen, R. J. (1974). Development of a Korotkoff sound processor for automatic identification of auscultatory events: I. Specification of preprocessing bandpass filters. *IEEE Trans. Biomed. Eng., BME-21,* 114.

Gribbin, B., Steptoe, A., and Sleight, P. (1976). Pulse wave velocity as a measure of blood pressure change. *Psychophysiology, 13,* 86.

Grigoryan, S. S., Saakyan, Y. Z., and Tsaturyan, A. K. (1980). Mechanism of generation of Korotkoff sounds. *Sov. Phys. Dokl., 25,* 177.

Gupta, R., Miller, J. W., Yoganathan, A. P., Udwadia, F. E., Corcoran, W. H., and Kim, B. M. (1975). Spectral analysis of arterial sounds: A noninvasive method of studying arterial disease. *Med. Biol. Eng., 13,* 700.

Hochberg, H. M., and Solomon, H. (1971). Accuracy of an automated ultrasound blood pressure monitor. *Curr. Ther. Res., 13,* 129, 1971.

Johnston, K. W., and Kakkar, V. V. (1974). Noninvasive measurement of systolic pressure slope: A reliable index of the presence of peripheral arterial occlusive disease. *Arch. Surg., 108,* 52, 1974.

Kain, H. K., Hinman, A. T., and Sokolow, M. (1964). Arterial blood pressure measurements with a portable recorder in hypertensive patients: I. Variability and correlation with "casual" pressures. *Circulation, 30,* 882.

Karvonen, M. J., Telivuo, L. J., and Jarniven, E. J. K. (1964). Sphygmomanometer cuff sizes and the accuracy of indirect measurement of blood pressure. *Am. J. Cardiol., 13,* 688.

Katona, Z., and Bolvary, G. (1975). Automatic indirect sphygmomanometry. *Bio-med. Eng. (London), 10,* 405.

King, A. L. (1947). Waves in elastic tubes: Velocity of the pulse wave in large arteries. *J. Appl. Phys., 18,* 595.

Kirby, R. R., Kemmerer, W. T., and Morgan, J. L. (1969). Transcutaneous Doppler measurement of blood pressure. *Anesthesiology, 31,* 86.

Kirkendall, W. M., Feinleib, M., Freis, E. D., and Mark, A. (1981). Recommendations for human blood pressure determination by sphygmomanometers: Subcommittee of the AHA postgraduate education committee. *Hypertension, 3,* 510a.

Korotkoff, N. C. (1905). On the subject of methods of determining blood pressure. *Wien. Med. Wochenschr., 11,* 365.

Kositskii, G. T. (1958). Theoretical basis for the auditory method of blood pressure determination. *Sechenov Physiol. J. USSR, 44,* 1100.

Latimer, K. E., and van Vollenhoven, E. (1981). Function, physics, and calibration of stethoscopes. In D. N. Ghista (ed.), *Biomechanics of Medical Devices.* Dekker, New York, chap. 5.

Lax, H., Feinberg, A., and Cohen, B. M. (1956). The normal pulse wave and its modification in the presence of human atherosclerosis. *J. Chronic Dis., 3,* 618.

Levine, S. R., (1981). "True" diastolic blood pressure. *N. Engl. J. Med., 304,* 362.

Li, J. K-J., van Brummelen, A. G. W., and Noordergraaf, A. (1976). Fluid-filled blood pressure measurement systems. *J. Appl. Physiol., 40,* 839.

Lighthill, M. J. (1972). Physiological fluid dynamics: A survey. *J. Fluid Mech., 52,* 475.

London, S. B., and London, R. E. (1967). Comparison of indirect pressure measurements (Korotkoff) with simultaneous direct brachial artery pressure distal to the cuff. *Adv. Intern. Med., 13,* 127.

Love, A. E. H. (1927). *A Treatise on Mathematical Elasticity* (3d ed.). Cambridge University Press, Cambridge.

Marey, E. J. (1885). *La Méthode Graphique dans les Sciences Expérimentales et Principalement en Physiologie et en Médicine.* Masson, Paris.

Massie, H. L., Ziedonis, J. G., and Black, I. (1973). Ultrasonic measurement of infant blood pressure. *Med. Instr., 7,* 240.

Mauck, G. W., Smith, C. R., Geddes, L. A., and Bourland, J. D. (1980). The meaning of the point of maximum oscillations in cuff pressure in the indirect measurement of blood pressure: II. *J. Biomech. Eng., 102,* 28, 1980.

Maurer, A., and Noordergraaf, A. (1976). Korotkoff sound filtering for automated three-phase measurement of blood pressure. *Am. Heart J., 91,* 584.

McCutcheon, E. P., and Rushmer, R. F. (1967). Korotkoff sounds, an experimental critique. *Circ. Res., 20,* 149.

McQueen, D. M. (1976). The origins of Korotkoff sounds: Studies of arterial motion under an occlusive cuff. Ph.D. Dissertation, Stevens Institute of Technology, Castle Point Station, Hoboken, N.J..

Molhoek, P. G., Wesseling, K. H., Arntzenius, A. C., Settels, J. J. M., van Vollenhoven, E., Weeda, H. W. A., and de Wit, B. (1981). Evaluation of the Peñáz servo-plethysmo-manometer for continuous, noninvasive measurement of finger blood pressure. In A. C. Arntzenius, A. J. Dunning, and H. A. Snellen (eds.), *Blood Pressure Measurement and Systemic Hypertension.* Medical World Press, IMS Breda, Holland.

NASA. (1965). Blood pressure measuring system 54803, Project Gemini. Contract NAS9-2887, Air Res. Manu. Div., The Garrett Corp.

Noordergraaf, A. (1978). Circulatory System Dynamics. Academic Press, New York.

Noordergraaf, A. (1979). Invasive versus noninvasive measurements: Noninvasive access to cardiovascular dynamics, experimental and applied. *Bibl. Cardiol., 37,* 1.

Noordergraaf, A., and Melbin, J. (1978). Ventricular afterload: A succinct yet comprehensive definition (editorial). *Am. Heart J., 95,* 545.

Norris, G. W. (1916). *Blood Pressure and its Clinical Application.* Lea & Febiger, Philadelphia, p. 424.

Okino, H. (1964). Measurement of intraluminal pressure from external pressure with strain transducers. *J. Appl. Physiol., 19,* 546.

Peñáz, J. (1973). Photoelectric measurement of blood pressure, volume, and flow in the finger. *Dig. 10th Int. Conf. Med. Eng.* Dresden, German, p. 104.

Ponomarenko, V. M., Vorontsov, R. G., and Bobrov, B. S. (1973). Methods and instruments for automatic arterial pressure measurement. *Biomed. Eng. N.Y., 7,* 361.

Posey, J. A., Geddes, L. A., Williams, H., and Moore, A. G. (1969). The meaning of the point of maximum oscillations in cuff pressure in the indirect measurement of blood pressure: I. *Cardiovasc. Res. Cen. Bull., Houston, 8,* 15.

Pressman, G. L., and Newgard, P. M. (1963). A transducer for the continuous external measurement of arterial blood pressure. *IEEE Trans. Biomed. Eng., BME-10,* 73.

Raines, J. K., Jaffrin, M. Y., and Rao, S. (1973). Noninvasive pulse recorders: Development and rationale. *Med. Instr., 7,* 245.

Raman, K. R. (1967). Korotkoff sounds at systole: A phenomenon of dynamic instability. *J. Sci. Eng. Res. Indian Inst. Tech. (Kharagpur), 11,* 227.

Ramsey, M. III. (1979). Noninvasive automatic determination of mean arterial pressure. *Med. Biol. Eng. Comput., 17,* 11.

Rodbard, S. (1953). The significance of the intermediate Korotkoff sounds. *Circulation, 8,* 600.

Rodbard, S., Rubinstein, H. M., and Rosenblum, S. (1957). Arrival time and calibrated contour of the pulse wave determined indirectly from recordings of arterial compression sounds. *Am. Heart J., 53,* 205.

Rogge, J. D., Brown, W. K., and Brown, C. A. (1968). An automatic system for the indirect determination and recording of human arterial blood pressure. Tech. Rep. 68-25, U.S. Air Force School Aerospace Medicine, Brooks AFB, Texas.

Sacks, A. H., and Abbott, D. E. (1962). *Determination of the Physical Constants Relating Blood Pressure and Arterial-wall Displacements.* Vidya Rep. No. 65, Palo Alto, Calif.

Schneider, R. A., and Costiloe, J. P. (1975). Twenty-four hour automatic monitoring of blood pressure and heart rate at work and at home. *Am. Heart J., 90,* 695.

Shapiro, D., Greenstadt, L., Lane, J., and Rubinstein, E. (1981). Tracking-cuff system for beat-to-beat recording of blood pressure. *Psychophysiology, 18,* 129.

Shirer, H. W. (1962). Blood pressure measuring methods. *IRE Trans. Bio-med. Elect., BME-9,* 116.

Simpson, J. A., Jamieson, G., Dickhaus, D. W., and Grover, R. (1965). Effect of size of cuff bladder on accuracy of measurement of indirect blood pressure. *Am. Heart J., 70,* 208.

Smith, N. T., and Beneken, J. E. W. (1980). An overview of arterial pressure monitoring. In J. S. Gravenstein, R. S. Newbower, A. K. Ream, and N. T. Smith (eds.), *Essential Noninvasive Monitoring in Anesthesia.* Grune & Stratton, New York, p. 75.

Smith, C. R., and Bickley, W. H. (1964). *The Measurement of Blood Pressure in the Human Body.* NASA.

Sokolow, M., Perloff, D., and Cowan, R. (1973). The value of portably recorded blood pressure in the initiation of treatment of moderate hypertension. *Clin. Sci. Mol. Med., 45,* 195s.

Stein, P. D., and Blick, E. F. (1971). Arterial tonometry for the atraumatic measurement of arterial blood pressure. *J. Appl. Physiol., 30,* 593.

Stein, P. D., and Blick, E. F. (1972). The potential usefulness of arterial tonometry for the measurement of instantaneous changes of arterial blood pressure. *Med. Inst., 6,* 272.

Struś, J. (1555). Sphygmicae artis jam mille ducentos annos peritae et desideratae, Libri V a Josepho Struthio Posnanience, medico recens conscripti, Basel.

Tavel, M. E. (1978). *Clinical Phonocardiography and External Pulse Recording* (3d ed.). Year Book, Chicago.

Tavel, M. E., Faris, J., Nassar, W. K., Feigenbaum, H., and Fisch, C. (1969). Korotkoff sounds: Observation on pressure pulse changes underlying their formation. *Circulation, 39,* 465.

Tursky, B. (1974). The indirect recording of human blood pressure. In P. A. Obrist, A. H. Black, J. Brener, and L. V. DiCara (eds.), *Cardiovascular Psychophysiology: Current Issues in Response Mechanisms, Biofeedback, and Methodology.* Aldine, Chicago, p. 93.

Ur, A., and Gordon, M. (1970). Origin of Korotkoff sounds. *Am. J. Physiol., 218,* 524.

Van Bergen, F. H., Weatherhead, D. S., Treloar, A. E., Dobkin, A. B., and Buckley, J. J. (1954). Comparison of indirect and direct methods of measuring arterial blood pressure. *Circulation, 10,* 481.

Van der Hoeven, G. M. A., and Beneken, J. E. W. (1970). A reliable transducer for the recording of the arterial pulse wave. Prog. Rep. 2, Inst. Med. Phys. TNO, Utrecht.

Van der Hoeven, G. M. A., de Monchy, C., and Beneken, J. E. W. (1973). Studies on innocent praecordial vibratory murmurs in children: I. Systolic time intervals and pulse wave transmission times in normal children. *Br. Heart J., 35,* 669.

von Recklinghausen, H. (1940). Blutdruckmessung und Kreislauf in den Arterien des Menschen. Steinkopff, Dresden, Leipzig.

Ware, R. W. (1965). New approaches to the indirect measurement of human blood pressure. *Third Natl. Biomed. Sci. Inst. Symp. (ISA BM-65).* Dallas, Tex.

Ware, R. W., and Anderson, W. L. (1966). Spectral analysis of Korotkoff sounds. *IEEE Trans. Biomed. Eng., BME-13, 170.*

Ware, R. W., and Kahn, A. R. (1963). Automatic indirect blood pressure determinations in flight. *J. Appl. Physiol., 18,* 210.

Weaver, C. S., Eckerle, J. S., Newgard, P. M., Warnke, C. T., Angell, J. B., Terry, S. C., and Robinson, J. (1978). A study of noninvasive blood pressure measurement technique. In *Noninvasive Cardiovascular Measurements. Soc. Photo-opt. Inst. Eng., 167,* 89.

Weinman, J. (1972). An appraisal of some noninvasive cardiovascular techniques. *Med. Biol. Eng., 10,* 496, 1972.

Welkowitz, W., and Deutsch, S. (1976). *Biomedical Instruments: Theory and Design.* Academic Press, New York.

Wesseling, K. H. (1981). Noninvasive, continuous, calibrated blood pressure by the method of Peñáz. In A. C. Arntzenius, A. J. Dunning, and H. A. Snellen, (eds.) *Blood Pressure Measurement and Systemic Hypertension.* Medical World Press, IMS Breda, Holland.

Wesseling, K. H., de Wit, B., Snoeck, B., Weber, J. A. P., Hindman, B. W., Nijland, R., and Van der Hoeven, G. M. A. (1978). An implementation of the Peñáz method for measuring arterial blood pressure in the finger and first results of an evaluation. Prog. Rep. 6, Inst. Med. Phys. TNO, Utrecht.

Westerhof, N., Bosman, F., De Vries, C. F., and Noordergraaf, A. (1969). Analog studies of the human systemic arterial tree. *J. Biomech., 2,* 121.

Westerhof, N., Elzinga, G., and van den Bos, G. C. (1973). Influence of central and peripheral changes on the hydraulic input impedance of the systemic arterial tree. *Med. Biol. Eng., 11,* 710.

Westerhof, N., and Noordergraaf, A. (1970). Arterial viscoelasticity: A generalized model. Effect on input impedance and wave travel in the systemic tree. *J. Biomech., 3,* 357.

Whitcher, C. (1969). Blood pressure measurement. In W. J. Bellville and C. S. Weaver (eds.), *Techniques in Clinical Physiology.* Macmillan, New York.

Wolthius, R. A., Golden, D. P., Jr., and Hoffler, G. W. (1974). Development of a Korotkoff sound processor for automatic identification of auscultatory events: II. Decision logic specifications and operational verification. *IEEE Trans. Biomed. Eng., BME-21,* 119.

Wolthius, R. A., Hull, D., McAfoose, D., and Fischer, J. (1981). Portable blood pressure measurements: Performance of Korotkoff sound analysis techniques. *Hypertension, 3,* 596.

Yamakoshi, K., Shimazu, H., and Togawa, T. (1980). Indirect measurement of instantaneous arterial blood pressure in the human finger by the vascular unloading technique. *IEEE Trans. Biomed. Eng., BME-27,* 150.

CHAPTER 9

Thermal Properties of Tissues

John C. Chato
Departments of Mechanical Engineering and Bioengineering
University of Illinois at Urbana-Champaign
Urbana, Illinois

INTRODUCTION

The thermal behavior of biological materials, particularly living tissues, depends on passive or "intrinsic" properties as well as on active responses of the biological system to thermal stimuli. The intrinsic thermophysical properties influencing heat transfer in biological materials are basically the same as those in any other material—thermal conductivity and diffusivity, specific heat, and emissivity. However, the magnitudes of these properties tend to be quite variable not only from body to body but also within the same tissue. In many cases, these properties are directionally dependent, i.e., anisotropic. Thus, one must realize and accept the inherent inaccuracies in any thermal calculations made for living tissues.

9.1 ENERGY TRANSPORT

All life depends on the acquisition, storage, transformation, and transmission of energy. Two of the most important forms of energy transmission are physical work and heat. The latter is our chief concern, but we have to refer from time to time to all other aspects of energy in order to understand the reasons and needs for heat transfer in biological systems. The basic laws of thermodynamics state that energy is needed for any process to occur and that during the occurrence of any real process some heat will be liberated. Transfer of heat is accomplished by way of a temperature difference, or *gradient*. Thus, although biological systems are generally considered as isothermal, temperature differences, however small, must exist in them. Another reason that we cannot treat biological processes as entirely isothermal is that many of the most important processes are strongly temperature dependent. Thus, the normal temperature fluctuations occurring in life will affect the rate of these processes and, consequently, the amount of heat liberated.

Although slight temperature fluctuations and differences will always be found in them, biological systems in general must achieve remarkably close control of their temperature levels in order to function properly. This is achieved by the maintenance of a dynamic equilibrium in which the heat produced by the various internal processes of a living body is rejected to the environment.

We have said already that energy is needed for any process to occur. In addition, the energy must come from some available source, and in the ordinary course of life on our planet, it cannot appear or disappear spontaneously. That energy cannot be created or destroyed but only transformed is the essence of the first law of thermodynamics. Its quantification represents the major effort of most engineering studies of energy, including bioheat transfer. Since we are concerned primarily with heat transfer and not with the vast number of energy transforming processes occurring in life, we will limit our discussions to the magnitudes of heat liberated and their temperature dependence.

Before examining the various biological heat transfer processes, we must state another important aspect of heat transfer in general, which is perhaps instinctively obvious, namely, that a temperature difference is required to transfer heat, which will always flow from the higher to the lower temperature level. This is the essence of the second law of thermodynamics.

9.2 METABOLIC HEAT GENERATION

For our purposes, most internal energy-producing processes in living systems can be lumped into a single concept: *metabolism*. Although the word encompasses a great many complicated processes, we need be concerned only with the overall effect, that is, the location and the magnitude of the energy generated which must be removed as heat in order to maintain thermal equilibrium. Let us ascertain the orders of magnitude involved.

Since biological reactions are usually considered as isothermal, the energy quantities released or absorbed are equal in magnitude to the change of the Gibbs free energy. Its decrease means that energy is liberated. For the hydrolysis of typical phosphate compounds at pH = 7 and 25°C, the change of the standard free energy ranges from −9.6 to −53.6 kJ/(g · mol) [1, 2]. Perhaps the most notable of specialized molecules for the storage and transformation of biological energy is adenosine triphosphate (ATP), with a molecular weight of 459 and a standard free energy change of −29.3 kJ/(g · mol) or −63.8 J/g. Reference 3 gives somewhat higher values of −33.5 kJ/(g · mol) or −73 J/g under body conditions at 38°C. It also gives the free energy change for another important energy-producing molecule, creatine phosphate (CP), with a molecular weight of 211, as −39.8 kJ/(g · mol) or −188.5 J/g also under body conditions. The concentrations of these two primary energy-producing compounds in the cell are of the order of 5 mmol/L (or per kilogram) for ATP and about 10 times as much for CP [3]. Thus, the easily available energy content of the cells of muscles is about 2.16 J/mL (or per gram) of intracellular fluid. This amount allows a strong muscular contraction to last for perhaps 30 s. The cells of any biological system such as the human body are, of course, not isolated, but receive and reject materials. Thus, the maximum energy content is not as important as the rate at which the reactions occur.

Assuming that the metabolic heat generation is directly related to the oxygen consumption, Mitchell et al. [4] found the thermal energy generation for resting muscle and skin to be 0.17 W/kg. They arrived at this value by using the following physiological quantities:

Oxygen consumption of resting tissue = 4 mL 0_2 per 100 mL blood flow [5]
Representative blood flow = 0.21 mL/s per kilogram of tissue
Thermal energy generation = 20 J/mL 0_2 [3, 6]

Mitchell et al. [4] suggested the following relation describing temperature dependence for the metabolic heat generation in resting muscle:

$$q_m = 0.17(2)^{(T-37)/10} \text{ W/kg } (\approx \text{W/L}) \tag{9.1}$$

Students of biology will recognize the factor 2 as the Q_{10} factor, i.e., the magnitude of the change with 10°C. The temperature T in the exponent is, of course, in Celsius degrees.

Metabolic heat generation is not uniform in the body. The basal rate, at rest, is about 80 W. Most of this, about 70 W, is developed in the head and trunk and remains relatively constant. Most of the additional heat generated during exercise is released in the muscles. The total can reach 1 kW. Previous investigators [7–9] have used somewhat different sets of anatomical and physiological data, but the most complete set was reported by Stolwijk [7]; Table 9.1 is based primarily on this work. The table shows representative geometric parameters and some basal physiological data relevant to heat transfer. The so-called basal values indicate minimum magnitudes occurring when the body is at rest. During exercise these values increase. For example, let us assume that a person is performing work corresponding to a metabolic rate of 750 W (approximately 1 hp) and that the increase of heat generation beyond the basal rate, ~670 W, is evenly distributed among all limbs. Then the metabolic rate in the limbs becomes 680 W, and the volumetric heat generation rate rises to q_m = 23.4 kW/m³, an almost 60-fold increase. To estimate the corresponding increase in blood flow, we can use the proportionality factor suggested by Stolwijk [7] for muscles, i.e., 0.239-mL/s increase for each 1-W increase. For our example, the blood flow in the limbs becomes 164.8 mL/s and the volumetric rate becomes w_b = 5.656 kg/(s · m³), a 33-fold increase.

It should be mentioned here that the body surface area can be best estimated by the DuBois formula [12]:

$$A_D = 0.203\, W^{0.425} H^{0.725} \tag{9.2}$$

where the area A_D, weight W, and height H are in consistent SI units of square meters, kilograms, and meters, respectively.

TABLE 9.1 Typical anatomical and physiological parameters for the various parts of the human body

Body part	Weight or volume,* kg or L	Skin surface area, m²	Basal metabolism		Basal evaporative heat loss		Basal effective blood flow	
			W	q, kW/m³	W	q_e, W/m²	mL/s	w_b, kg/(s · m³)
Head	4.84	0.1222	13.56	2.802	0.71 Internal: 5.23	5.81	13.84	2.860
Trunk and central blood	40.41	0.6538	56.09	1.388	3.80 Internal: 5.23	5.81	67.44	1.669
Total head and trunk	45.25	0.7760	69.65	1.539¶	4.51 Internal: 10.46	5.81¶	81.28	1.796¶
Arms	6.93	0.2367	2.75	0.397	1.38	5.83	0.76	0.1097
Hands	0.67	0.0862	0.24	0.358	0.51	5.92	0.61	0.9104
Legs	20.56	0.5956	8.14	0.396	3.47	5.83	2.65	0.1289
Feet	0.98	0.1195	0.35	0.357	0.67	5.61	0.91	0.9286
Total limbs	29.14	1.0380	11.48	0.394¶	6.03	5.81¶	4.93	0.1692¶
Total body	74.39	1.8140	81.13	1.091	10.54† Internal: 10.56§	5.81‡	86.21	1.159

Source: Based on Refs. 7 to 11.
* Weight-volume conversions were made on the basis of a uniform density of 1 kg/L.
† = 4.38 × 10⁻⁶ kg/s or 15.8 g/h evaporation rate (latent heat = 2407 kJ/kg at 35°C).
‡ = 2.41 × 10⁻⁶ kg/(s · m²) or 8.68 g/(h · m²).
§ = 4.34 × 10⁻⁶ kg/s or 15.6 g/h.
¶ Overall average.
Note: The significant digits used do not imply corresponding accuracy.

9.3 THERMAL EFFECTS OF BLOOD FLOW

Another important physiological factor influencing the thermal behavior of biological systems is internal circulation of fluids. We will consider only blood circulation and perfusion in warmblooded animals, or homeotherms, where the blood supplied to various parts of the body is kept at a regulated high temperature level and, therefore, can be locally at a different level than the temperature of the surrounding tissue. Thus, blood flow has a dual influence on the energy balance: first, it can be a heat source or sink, depending on its relative local temperature, and second, it transports energy by flow. Undoubtedly, the thermal interactions between the complicated circulatory system and the various tissues represent the greatest difficulty in trying to develop quantifiable models for biological thermal systems. Different investigators [13–17] analyzed the problem in different ways and came up with very similar conclusions that blood vessels with diameters less than about 0.05 mm have such rapid heat transfer with the surrounding tissue that they can be assumed to be at the same temperature. Therefore, unless a direct shunt exists between a terminal artery and a terminal vein, all blood supplied by the arteries will achieve thermal equilibrium in the terminal branches before reaching the capillary bed and will enter the larger veins at the same temperature as the surrounding tissue. The larger arteries and veins with diameters over 0.05 mm, however, will carry blood which can be at a different temperature than the surrounding tissue. The complicated geometry of the vasculature poses another difficulty because in most cases it is impossible to define an overall, general diffusion pattern. Thus, heat transfer to a single blood vessel can be estimated with some confidence [13], but to model a tissue segment containing many different kinds of blood vessels requires careful analysis of the geometry of the system as well as determination of the flow and heat transfer parameters. The two most prevalent models will be described later.

In many applications a single parameter, the volumetric blood perfusion rate w_b, is used to characterize the effect of the blood circulation on the heat transfer. Typical values are given in Table 9.1, and a numerical example is given in Sec. 9.2.

9.4 INTRINSIC THERMAL PROPERTIES

Summaries of intrinsic thermal properties of tissues were published from time to time in the literature [18–22]. Additional data on a great variety of biological materials have been reported in agricultural literature. Qashou et al. [20] summarized thermophysical data on food products. Stoll [23] presented a summary on heat transfer in biotechnology which discussed many aspects of the thermal behavior of biological materials, particularly human tissues. Shizer [24] gave a thorough discussion of the mammalian thermoregulatory system, the thermal behavior of biological media, and some analytical studies of bioheat transfer.

9.4.1 Thermal Conductivity *k*

The magnitude of the thermal conductivity varies from the minimum values obtainable in vitro or for fat to the maximum value, that of water, which occurs in tissues with high water content. Spells [25] proposed the first general correlation for thermal conductivities of various biological materials. This correlation indicated a linear variation of the thermal conductivity with water content, with the maximum value corresponding to that of water. More recently, Poppendiek [26] and Cooper and Trezek [27] proposed more fundamental correlations based on the concept that tissue is composed primarily of three components: water, protein, and fat; consequently, its properties may be calculated as weighted averages of the properties of its three main components. For thermal conductivity, Cooper and Trezek [27] proposed the following correlation:

$$\bar{k} = \rho \sum \frac{k_n m_n}{\rho_n}$$

$$= \rho \times 10^{-3}(0.628 m_{\text{water}} + 0.117 m_{\text{protein}} + 0.231 m_{\text{fat}}) \text{ W/(m} \cdot {}^\circ\text{C)} \qquad (9.3)$$

where ρ is tissue density in kg/m^3 (g/L) and m is mass fraction. In the absence of reliable experimental data on a particular tissue, this correlation can certainly be used to obtain a reasonable estimate for the thermal conductivity.

Although a relatively large number of values have been reported for thermal conductivities of biological materials, a large majority of the data is not supported by analyses of the composition of the material. Also, unfortunately, some of the data have been obtained by faulty experimental techniques. In spite of these shortcomings, the magnitudes of the intrinsic thermal conductivities of many biological materials, such as mammalian and, in particular, human tissues, have been reasonably well established and can be used for estimates of thermal behavior.

9.4.2 Specific Heat c_p, Thermal Diffusivity α, and Thermal Inertia $k\rho c_p$

Since it is difficult to measure the specific heat of biological materials independently, and almost impossible to measure that of tissue in vivo, very little data are available on specific heat itself. Most of the measurements have been made on the volumetric specific heat, ρc_p, indirectly by determining either the thermal diffusivity, $\alpha = k/\rho c_p$, or the thermal inertia, $k\rho c_p$, and then dividing out the thermal conductivity. The diffusivity and inertia are related in that both are made up of the same three parameters. There are considerable data on these properties because the transient techniques used for measurement yield these combinations directly. In the conventional transient heat conduction problem, the diffusivity is the controlling parameter, and certain types of solutions to the transient heat conduction equation contain the thermal inertia.

Cooper and Trezek [27] suggest the following general correlations for the mean specific heat $\overline{c_p}$ and density $\bar{\rho}$:

$$\begin{aligned}\overline{c_p} &= \Sigma\, c_{pn} m_n \\ &= 4.2 m_{\text{water}} + 1.09 m_{\text{protein}} + 2.3 m_{\text{fat}} \text{ kJ/(kg} \cdot {}^\circ\text{C)}\end{aligned} \tag{9.4}$$

This expression can be approximated for human tissue by

$$\overline{c_p} \approx 4.2(0.4 + 0.6 m_{\text{water}}) \text{ kJ/(kg} \cdot {}^\circ\text{C)} \tag{9.5}$$

and

$$\bar{\rho} = \frac{1}{\Sigma\, m_n/\rho_n} = \frac{1}{m_{\text{water}}/1000 + m_{\text{protein}}/1540 + m_{\text{fat}}/815} \text{ kg/m}^3 \text{ (g/L)} \tag{9.6}$$

Table 9.2 shows selected experimental data for properties of tissues.

9.4.3 Emissivity ε

The emissivity of a surface determines its thermal interaction by radiation. Consequently, it is of importance almost exclusively for the skin. An exception is, for example, when a thermograph is used to determine the temperature distribution on the exposed tissues during surgery or during an experiment involving access to internal organs. In such cases the exposed tissue surfaces are almost always wet. Consequently, in these cases, the radiation properties are influenced by the liquid covering the tissue.

The emissivity for thermal radiation, i.e., in the infrared range, of dry skin is remarkably constant at $\varepsilon = 0.98$ to 0.99 for all kinds of human skins and virtually independent of the color or the presence of blemishes.

TABLE 9.2 Selected measured values of thermophysical properties of tissues

Description	Conductivity, W/(m · °C)	Diffusivity, $m^2/s \times 10^7$	Inertia, $W^2 \cdot s/(m^4 \cdot °C^2) \times 10^{-6}$
In vitro materials at room to body temperatures			
Most human internal, soft tissues	0.4–0.6	1.0–2.0	0.9–2.0
Dry bone and bone marrow	0.22		
Fat	0.1–0.4		0.3–0.5
Skin	0.2–0.4	0.8–1.2	1.2–2.2
Body fluids	0.5–0.6		
Water (0 to 100°C)	0.57–0.68	1.3–1.7	2.4–2.7
Materials below 0°C (property values generally increase with decreasing temperature)			
Frozen tissues	1.0–1.6		
Blood: −10, −60, −100°C*	1.6, 2.1, 2.7	8.7, 16.9, 23.7	
Plasma: −10, −40, −100°C*	2.0, 2.3, 3.2	9.7, 15.1, 26.9	
Packed cells: −10, −60, −100°C*	1.2, 1.7, 2.3	6.8, 14.1, 20.4	
In vivo materials (properties increase with increased blood perfusion and water content)			
Bone	0.3–3.1		
Cartilage	1.4–2.8		
Kidney and liver	0.6–1.2	2.0–4.3	
Muscle	0.7–1.0		
Skin	0.5–2.8	0.4–1.6	1.0–7.0

Source: Condensed from Refs. 18, 19, 21, 22.
* Values vary linearly between consecutive temperatures given.

9.5 MODELING TECHNIQUES

Conventional techniques [28] may be applied to tissues in vitro, but it has been shown that the thermal properties of such materials change rather rapidly from the instant of death. The change is not just a result of the cessation of blood perfusion but seems to be caused by chemical and, possibly, physical changes and by variations in water content. Consequently, data obtained with tissue in vitro cannot be used with any confidence to interpret the behavior of living tissues. Thus, measurements should be made in vivo. Such procedures, however, generally depend on interpretation of the experimental data in terms of a thermal model of the tissue. One reason is that steady-state conditions do not exist long enough in live tissues to enable the use of the standard steady-state methods. Consequently, transient techniques had to be developed which depend on the use of thermal models. Several models can be found in the literature which differ primarily in the handling of the thermal effects of blood perfusion and of the presence of relatively large blood vessels.

The most commonly used model is the so-called bioheat equation, first proposed by Pennes [29], which contains a separate blood perfusion term. In the rectangular, cartesian coordinate system (x_1, x_2, x_3) the equation is

$$\rho c_p \frac{\partial T}{\partial t} = \sum_{i=1}^{3} \frac{\partial}{\partial x_i}\left(k_i \frac{\partial T}{\partial x_1}\right) + w_b c_b (T_a - T_v) + q_m \qquad (9.7)$$

where T is the tissue temperature; t is time; T_a is the local blood supply, or arterial, temperature; c_b is the specific heat of blood; and q_m is the volumetric metabolic heat generation rate. The venous blood temperature T_v is usually assumed to be equal to the local tissue temperature, and the thermal conductivity is taken as isotropic, i.e., $k_1 = k_2 = k_3 = k$. This equation is currently probably the most widely used model for interpretation of thermal data in living tissues. The separate blood perfusion term potentially facilitates the quantitative evaluation of the volumetric blood perfusion rate w_b. However, practical methods to yield reasonably accurate values for w_b are still in the developmental stage. The bioheat equation is a relatively simple model. This fact represents both its greatest advantage, mathematical tractability, and its greatest shortcoming, oversimplification to a scalar form of an effect which has definite vectorial components.

The more complicated mathematical models include various degrees of vectorial representation of the blood flow. Some use averaging schemes, while others consider the detailed geometry of the blood vessels. Although such models are theoretically better than the bioheat equation, they generally require more detailed information than is available for most tissues. Thus they are useful only in limited cases (e.g., see Ref. 30). A more promising model was proposed by Chen and Holmes [14], who suggested two additional terms in the bioheat equation: one proportional to the local blood perfusion velocity and temperature gradient and the other representing an effective conductivity due to capillary perfusion.

The simplest model uses the standard, transient heat conduction equation and lumps all effects into a single, isotropic, "effective" thermal conductivity and corresponding thermal diffusivity. The values obtained depend on the blood perfusion rate and, to a lesser degree, on the local metabolic heat generation rate and can vary by a factor of about 5. The minimum values, corresponding to nearly zero blood perfusion and very low metabolic heat, can be taken as the intrinsic properties, whereas the higher values are measures primarily of the blood perfusion rate. Quantitative functional relationships between the increase in the "effective" properties and the blood perfusion rate have been established only recently [15–17].

Weinbaum, Jiji, and Lemons [15–17] developed a "new" bioheat equation on the basis of their observation that in the muscles near the skin surface the thermally significant arteries and veins run next to each other in a parallel, counterflow pattern. The local temperature differences between these paired vessels were found to be less than 0.2°C, indicating that most of the heat exchange occurred between the blood vessels with relatively little heat diffusion into the tissue. The average local temperature of the paired vessels was found to be the same as the local tissue temperature. They concluded that the principal mode of heat transfer between the blood and the tissue is an "incomplete countercurrent exchange in the thermally significant vessels," where most, but not all, of the heat conducted out of the artery is conducted locally into the adjacent vein. A detailed, three-layer model was developed first [16], followed by a simplified model which predicts a single, local, average blood-tissue temperature [17].

$$\rho c_p \frac{\partial T}{\partial t} = \sum_{i=1}^{3} \sum_{j=1}^{3} \left\{ \frac{\partial}{\partial x_i} \left[(k_{ij})_{\text{eff}} \frac{\partial T}{\partial x_j} \right] - \frac{\pi^2 n a^2 k_b^2}{4\sigma k} Pe^2 l_j \frac{\partial l_i}{\partial x_i} \frac{\partial T}{\partial x_j} \right\} + q_m \tag{9.8}$$

where the new symbols are as follows: n is the number density of the vessel pairs crossing the surface of the control volume per unit area; a is the vessel radius, assumed to be constant; k and k_b are the intrinsic thermal conductivities of the tissue and blood, respectively; $Pe = \text{Re}\, Pr$ is the mean Peclet number of the blood flow; l_i and l_j are direction cosines of the blood vessels;

$$(k_{ij})_{\text{eff}} = k \left(\delta_{ij} + \frac{\pi^2}{4\sigma k^2} n a^2 k_b^2 Pe^2 l_i l_j \right)$$

a tensor conductivity with the Kroneker delta δ_{ij}; and $\sigma = \pi/[\cos^{-1}(d/2a)]$, where d is the centerline distance between paired blood vessels.

For the simple, one-dimensional case where the blood vessels are in the same direction as the temperature gradient, Eq. (9.8) simplifies even further into a scalar equation:

$$\rho c_p \frac{\partial T}{\partial t} = \frac{\partial}{\partial x} \left(k_{\text{eff}} \frac{\partial T}{\partial x} \right) + q_m \tag{9.8a}$$

where
$$\frac{k_{\text{eff}}}{k} = 1 + \frac{\pi^2 n a^2 k_b^2 Pe^2}{4\sigma k^2} \tag{9.9}$$

Estimated magnitudes of this ratio are given in Table 9.3 for vessel pairs of varying size. The table indicates that for vessels with diameters greater than about 0.2 mm the countercurrent heat exchange dominates over conduction ($k_{\text{eff}}/k > 2$), whereas for vessels with diameters less than about 0.05 mm conduction dominates ($k_{\text{eff}}/k = 1$). Weinbaum and Jiji [17] estimated that for vessels with diameters between 0.07 and 0.15 mm the thermal effects of the directed perfusion through the thermally insignificant vessels and capillary beds are of the same magnitude as those of the countercurrent heat exchange, with a maximum enhancement in heat transfer of 20 to 60 percent due to each mechanism. Table 9.3 also indicates that the effective thermal conductivity is dominated by the largest thermally significant vessels in the region. Thus, the local average vascular properties may be approximated by the largest countercurrent vessels in the vicinity.

TABLE 9.3 Effective thermal conductivity

Diameter $2a$, mm	Velocity u, cm/s	Countercurrent pairs, n/cm^2	Peclet no., Pe = Re Pr	Eq. (9.9), k_{eff}/k
0.3	6	1	144	3.5
0.2	5	2	80	1.7
0.1	3	16	24	1.2
0.05	2	64	8	1.05

Source: Weinbaum and Jiji [17].

The application of Eq. (9.8) requires identification of a representative geometry for the countercurrent vessel system and a spatial distribution for the capillary bleed-off. Jiji et al. [16] describe some typical cases.

It is still to be determined by further studies to what extent this model should replace the "old" bioheat equation. The analysis of Chen and Holmes [14] suggested that both the increase of the effective thermal conductivity (the new model) and the blood perfusion term (the old model) may be needed, depending on the character of the vasculature. It seems that in those regions of the body where the thermally significant arteries and veins lie in the parallel, counterflow pattern described in Refs. 14 to 17, the new model is preferable. However, in other regions other models may be advantageous.

9.6 MEASUREMENT TECHNIQUES

Detailed discussion of various measurement techniques was given by the author [22, 31] in other publications dealing exclusively with thermal effects in biological systems. The reader is referred there for more details.

Measurements made on the skin surface, because of its accessibility, can be easier to carry out than those made inside the tissue. Thus, more methods are available for measurements of properties on the skin than of those in deep tissues. Steady-state temperature data on the skin can provide useful information on the pathological conditions near the surface. For example, excess blood perfusion generates a local high temperature zone, whereas reduced blood perfusion creates low temperatures. Thus, asymmetrical high temperature regions on a thermograph of the breast can indicate the presence of tumors.

Transient temperature data on the skin can provide further information on the conditions within the tissues underlying the surface. Several early investigators [18, 19, 21] used the temperature change of the skin surface due to an applied heat source to estimate the thermal inertia $k\rho c_p$ of the tissue. Chen, Pedersen, and Chato [32] suggested that the variation of skin temperatures following a carefully controlled change in the heat transfer rate at the surface can provide three-dimensional information on the underlying tissue, such as the location and extent of a region with very high local blood perfusion rate.

Conventional methods of measuring temperatures on the surface or inside tissues, such as by thermocouples and thermistors, require careful applications of the technique with special precautions exercised to eliminate the conduction errors along the leads and to minimize the thermal disturbance created by the bead and the leads. The following rules should be observed:

1. The bead and leads should be as small as possible within required strength limitations. Cain and Welch [33] developed a microthermocouple consisting of thin films deposited on a quartz filament, and recently Chato [32] suggested a glass- or quartz-encased thermocouple junction that can be made in the 10- to 50-μm-diameter range with reasonable strength.
2. The wire leads should be placed along the isotherm going through the measuring junction for as long a distance as practicable.
3. For surface temperature measurements, the surrounding air should be kept at or near the same temperature as the surface, and the surface should be dry.

Infrared radiation (IR) techniques such as thermography can provide distributed as well as time-varying temperature data which can be utilized in various ways. Near normal body temperature, the bulk of the radiation from the dry human skin will occur in the infrared region, where the emissivity is $\varepsilon = 0.98$ to 0.99 independent of color or pigmentation. Even though this is a very narrow range, the measurement protocol for thermography must be carefully developed to take into account both convective and radiative interactions between the skin and its environment, or significant errors could develop. Modern thermographic equipment can provide built-in capability for showing isothermal contours, for showing temperature profiles along a line across the field of vision, and even for color coding the temperature levels. For surfaces other than dry skin, the radiative properties, specifically the emissivity, will determine the efficacy of the thermograph as an accurate temperature-measuring device. For example, the grooves between adjacent teeth tend to show on the thermographic image as being at a higher temperature than the rest of the teeth due to the "blackbody" effect, i.e., the increased effective emissivity of an enclosure. Wet surfaces will also make the evaluation of thermographic data difficult because of the varying emissivity of the liquid cover as well as the temperature disturbances created by evaporative cooling.

Another, less expensive, although less accurate, method of obtaining distributed temperature patterns on the skin is the application of liquid crystals which change color from red to violet as their temperature rises within a prescribed range. Liquid crystal solutions can be applied directly to the skin over a black sublayer, a somewhat messy procedure, or they can be incorporated into flexible sheets [35] or semirigid plates [36] pressed on the skin. Because these applications tend to reduce heat flow, evaporation, and, with the applied pressure, even the local blood perfusion rate, they cannot be used for accurate temperature measurements. They can be useful, however, for showing relative temperature patterns which can disclose the presence of vascular disorders.

Other methods of temperature measurements are under development, and some of them, such as some fiber-optic devices, are beginning to be used in commercially available units. A detailed discussion of the currently considered methods is given in the book cited in Ref. 22.

Reviews of current measuring techniques for thermal properties of in vivo tissues [21, 31] show that all useful methods employ transient techniques which require very short measurement times lasting from a few seconds to a few minutes. Such methods have the additional advantage of providing not only the thermal conductivity k (which is the only parameter obtainable from a steady-state situation), but also the volumetric specific heat ρc_p in terms of the diffusivity or inertia, as discussed earlier. The devices can be classified according to geometry or according to the method of operation. The most important geometric configurations are as follows:

1. One-dimensional, rectangular heat flow perpendicular to a surface, used only at the skin
2. Spherical measuring device with a spherically symmetric, one-dimensional heat-flow pattern

3. Long, cylindrical measuring device with essentially one-dimensional, cylindrical symmetry of the heat flow around it

All useful operating methods assume an initial, uniform, quasi-steady-state condition followed by:

1. A step change in the temperature of the measuring device. The decay of the power required to maintain the changed temperature level yields the thermal properties in conjunction with an appropriate thermal model of the tissue.
2. A constant power input to the measuring device. The increase of the temperature in the device and, possibly, also in the tissue at known locations provides the data used for interpretation.
3. A temperature pulse at the measuring device generated by a known heat input. The subsequent decay of temperature yields the desired data.

Chato [37] suggested first the single spherical bead configuration with a step change in temperature. Balasubramaniam and Bowman [38] improved this method by accounting for the nonuniform temperature distribution inside the measuring bead. The analysis of the data involves essentially the evaluation of the power input to the bead as a function of $t^{-1/2}$. The intercept is proportional to the thermal conductivity of the surrounding tissue, and the slope of the function yields $(k\rho c_p)^{1/2}$. The method requires careful calibration of the bead diameter by performing measurements on materials with accurately known properties, e.g., agar solution.

Chen et al. [39] proposed the use of the pulse technique with a small spherical bead. The known, initial heat input, together with the temperature measured at a given time, yields the value for $k(\rho c_p)^{-1/3}$. Since the size of the bead does not enter into this analysis, the bead can be very small and both thermal disturbance and trauma are minimal.

Current efforts with both these methods include the quantitative determination of the blood perfusion rate or the presence of large blood vessels by interpreting the deviation of the measured data from the analytically predicted behavior. For example, in the constant-temperature method the presence of a large blood vessel or a highly perfused region at some distance from the bead will cause leveling off of the power curve. The time at which this effect takes place is an indication of the distance from the bead. In the pulse method, the variation of the measured thermal conductivity with time is a function of the blood perfusion rate in the tissue.

The cylindrical method has been used by Trezek and Cooper with constant heat input [40] and as a simple temperature decay method [41]. The cylindrical, constant-power technique has been used extensively in food products [42–44]. This geometry requires a larger sampling volume than the spherical beads. This method also requires a relatively long measuring time because of an initial, nonlinear transient which is generally not used in the analysis of data.

9.7 THERMOREGULATION

It must always be kept in mind that biological systems respond actively to various stimuli such as changes in temperature. These responses must be known and taken into account in any attempt to describe thermal behavior. Our interest is the human thermoregulatory system, but we must remember that different biological systems achieve thermal balance in different ways.

Homeotherms, including humans, strive to maintain a fixed internal thermal state, i.e., a constant temperature in the "core." This temperature ranges rather narrowly, between 35 and 40°C, for all mammals. Although it is kept remarkably constant under a given set of steady conditions, it undergoes a controlled change with time of day (circadian rhythm), with changes in activity (increases with exercise), with disease (fever), and, as extreme cases, with seasonal changes in hibernators or estivators (winter or summer sleepers). It is a currently accepted idea that this core temperature is controlled centrally by the hypothalamus [45, 46]. Two distinct areas of this organ are involved. The anterior portion generates responses to increased temperatures in the body such as peripheral vasodilatation, which allows more warm blood to diffuse toward the skin, sweating, and panting. The posterior portion generates responses to decreased temperatures such as peripheral vasoconstriction, shivering, and increased metabolism. These two areas operate generally in a mutually inhibitory fashion.

Just what constitutes the feedback signals and how these are utilized by the central "controller" are under constant study. The various signals represent thermal inputs from the core of the body as well as from its periphery, but there are other influences apparent such as exercise level, the time of day, sleep, and state of well-being. The general, overall purpose of the system seems to be the maintenance of some constant internal thermal environment. For details, the reader is referred to the extensive literature, the most recent summaries of which are in the book cited in Ref. 22.

SUMMARY

The thermal behavior of biological tissues depends on the passive, intrinsic thermal properties, in particular, the conductivity, diffusivity, and inertia, as well as on the active physiological responses such as blood perfusion and metabolic heat generation. In this chapter we discuss the magnitudes of these properties, the various thermal models used in analysis, and the currently used methods of measurement.

The available data exhibit considerable uncertainty because of the variability of the properties even for the same tissue owing to differences in composition (e.g., protein, fat, and water) and microstructure. Various degrees of anisotropy can also occur. The values of the properties reported represent almost exclusively locally averaged measurements. In spite of these difficulties, the ranges of the magnitudes have been well enough established, as shown in Tables 9.1 and 9.2, to allow reasonable estimates of the temperature patterns and heat transfer rates within biological tissues. It must be kept in mind, however, that in most biological applications the temperature variations are very small, generally less than 4°C. Consequently, errors which may be considered acceptable in ordinary heat transfer (say, 0.5°C) can be significant in biological problems. At present there seems to be no way to circumvent this dilemma. Therefore, critical work involving thermal effects must be performed with carefully designed instrumentation which measures temperatures at as many points as needed without significantly disturbing the thermal field.

Although measurements of the intrinsic properties are still being conducted for various tissues and for other biological materials in order to obtain new or more accurate data, the emphasis has been shifting toward the development of methods to measure the physiological responses, particularly the blood perfusion rates. Thermal methods for such purposes are in various stages of development at present and need further work to become successful.

For further applications and refinements of the mathematical models, a careful "mapping"

TABLE 9.4 Conversion factors

Parameter	To convert from	To	Multiply by
k	W/(m · °C)	W/(cm · °C)	10^{-2}
		cal/(s · cm · °C)	0.239×10^{-2}
		Btu/(h · ft · °F)	0.5778
c_p	J/(kg · °C)	cal/(g · °C)	2.388×10^{-4}
		Btu/(lb · °F)	2.388×10^{-4}
α	m^2/s	cm^2/s	10^4
		ft^2/h	3.875×10^4
$k\rho c_p$	$W^2 \cdot s/(m^4 \cdot °C^2)$	$W^2 \cdot s/(cm^4 \cdot °C^2)$	10^{-8}
		$cal^2/(s \cdot cm^4 \cdot °C^2)$	5.712×10^{-10}
		$Btu^2/(h \cdot ft^4 \cdot °F^2)$	8.616×10^{-6}
Energy	J	cal	0.2388
		Btu	0.9478×10^{-3}

of the various parts of the body will be required which would identify the local microvasculature and, thus, allow the determination of the appropriate model for each region.

For the reader's convenience, unit conversion factors are provided in Table 9.4.

REFERENCES

1. Peusner, L.: *Concepts in Bioenergetics,* Prentice-Hall, Englewood Cliffs, N.J., 1974.
2. Lehninger, A.: *Bioenergetics,* W. A. Benjamin, New York, 1965.
3. Guyton, A. C.: *Basic Human Physiology,* Saunders, Philadelphia, 1971.
4. Mitchell, J. W., Galvez, T. L., Hangle, J., Myers, G. E., and Siebecker, K. L.: Thermal response of human legs during cooling, *J. Appl. Physiol., 29,* 859–865, 1970.
5. Selkurt, E. E. (ed.): *Physiology,* Little, Brown, Boston, 1966.
6. *NASA Life Sciences Data Book,* Webb, Yellow Springs, Ohio, 1962.
7. Stolwijk, J. A. J.: Mathematical model of thermoregulation, in J. D. Hardy, A. P. Gagge, and J. A. J. Stolwijk (eds.), *Physiological and Behavioral Temperature Regulation,* Charles C Thomas, Springfield, Ill., 1970, pp. 703–721.
8. Huckaba, C. E., Hansen, L. W., Downey, J. A., and Darling, R. C.: Calculation of temperature distribution in the human body, *AIChE J., 19,* 527–532, 1973.
9. Wissler, E. H.: A review of thermal control in the human, *Chem. Eng. Med., Adv. Chem. Ser., 118,* 236–255, 1973.
10. Aschoff, J., and Wever, R.: Kern and Schale im Wärmehaushalt des Menschen, *Naturwissenschaften, 45,* 477–485, 1958.
11. Stolwijk, J. A. J., and Hardy, J. D.: Partitional calorimetric studies of responses of man to thermal transients, *J. Appl. Physiol., 21,* 967–977, 1966.
12. DuBois, D., and DuBois, E. F.: A formula to estimate the approximate surface area if height and weight are known, *Arch. Int. Med., 17,* 863–871, 1961.
13. Chato, J. C.: Heat transfer to blood vessels, *Trans. ASME, J. Biomech. Eng., 102,* 110–118, 1980.
14. Chen, M. M., and Holmes, K. R.: Microvascular contributions in tissue heat transfer, in *Thermal Characteristics of Tumors, Ann. N.Y. Acad. Sci., 355,* 137–150, 1980.
15. Weinbaum, S., Jiji, L. M., and Lemons, D. E.: Theory and experiment for the effect of vascular microstructure on surface tissue heat transfer: I. Anatomical foundation and model conceptualization, *Trans. ASME, J. Biomech. Eng., 106,* 321–330, 1984.
16. Jiji, L. M., Weinbaum, S., and Lemons, D. E.: Theory and experiment for the effect of vascular microstructure on surface tissue heat transfer: II. Model formulation and solution, *Trans. ASME, J. Biomech. Eng., 106,* 331–341, 1984.
17. Weinbaum, S., and Jiji, L. M.: A new simplified bioheat equation for the effect of blood flow on local average tissue temperature, *Trans. ASME, J. Biomech. Eng., 107,* 131–139, 1985.
18. Chato, J. C.: A survey of thermal conductivity and diffusivity data on biological materials, ASME Pap. No. 66-WA/HT-37, 1966.
19. Chato, J. C.: Heat transfer in bioengineering, in B. T. Chao (ed.), *Advanced Heat Transfer,* University of Illinois Press, Urbana, 1969, pp. 404–412.
20. Qashou, S., Vachon, R. I., and Touloukian, Y. S.: Thermal conductivity of foods, *ASHRAE Trans., 78-1,* 165–183, 1972.
21. Bowman, H. F., Cravalho, E. G., and Woods, M.: Theory, measurement, and application of thermal properties of biomaterials, *Annu. Rev. Biophys. Bioeng., 4,* 43–80, 1975.
22. Chato, J. C.: Measurement of properties related to thermal behavior of biological systems, in A. Shitzer and R. C. Eberhart (eds.), *Heat Transfer in Medicine and Biology: Analysis and Applications,* Plenum Press, New York, 1985, vol. 1, chap. 8, pp. 167–192 and vol. 2, App. 2, pp. 413–418.
23. Stoll, A. M.: Heat transfer in biotechnology, in J. P. Hartnett and T. F. Irvine, Jr. (eds.), *Advances in Heat Transfer,* Academic Press, New York, 1967, vol. 4, pp. 65–141.
24. Shitzer, A.: Studies of bioheat transfer in mammals, in C. Gutfinger (ed.), *Topics in Transport Phenomena,* Halsted Press, New York, 1975, chap. 3.
25. Spells, K. E.: The thermal conductivities of some biological fluids, *Phys. Med. Biol., 5,* 139–153, 1960.

26. Poppendiek, H. F., et al.: Thermal and electrical conductivities of biological fluids and tissues, in *Reports,* under Contract No. ONR4095(00), Bioscience, Solana Beach, Calif., specifically DDC No. Ad. 630712, 1966; also *Cryobiology,* 1966, vol. 3, pp. 318–217.

27. Cooper, T. E., and Trezek, G. J.: Correlation of thermal properties of some human tissues with water content, *Aerosp. Med., 42,* 24–27, 1971.

28. Eckert, E. R. G., and Goldstein, R. J.: *Measurements in Heat Transfer,* Hemisphere, Washington, D.C., 1976.

29. Pennes, H. H.: Analysis of tissue and arterial blood temperatures in the resting human forearm, *J. Appl. Physiol., 1,* 93–122, 1948.

30. Keller, K. H., and Seiler, L., Jr.: An analysis of peripheral heat transfer in man, *J. Appl. Physiol., 30,* 779–786, 1971.

31. Chato, J. C.: Measurement of thermal properties of growing tumors, in *Thermal Characteristics of Tumors, Ann. N.Y. Acad. Sci., 335,* 67–85, 1980.

32. Chen, M. M., Pedersen, C. O., and Chato, J. C.: On the feasibility of obtaining three-dimensional information from thermographic measurements, *Trans. ASME, J. Biomech. Eng., 99,* 58–64, 1977.

33. Cain, C. P., and Welch, A. J.: Thin film temperature sensors for biological measurements, *IEEE Trans. Biomed. Eng., BME-21,* 421–423, 1974.

34. Chato, J. C.: A miniature thermocouple for in vivo and in vitro applications, *Proceedings of the 3d International Conference on Mechanics in Medicine and Biology,* Compiegne, France, 1982, pp. 141–142.

35. Katz, S., and Masterfano, T.: A preliminary report on the clinical testing of a new elastomeric liquid crystal film, in *Thermal Characteristics of Tumors, Ann. N.Y. Acad. Sci., 335,* 484–488, 1980.

36. Widder, B., Auskultation und Plattenthermographie: Sich ergänzende Methoden in der Früherkennung von drohenden Schlaganfällen, *Nervenarzt, 49,* 189–191, 1978.

37. Chato, J. C.: A method for the measurement of thermal properties of biological materials, in *Thermal Problems in Biotechnology, ASME Symp. Ser.,* 1968, pp. 16–26.

38. Balasubramaniam, T. A., and Bowman, H. F.: Thermal conductivity and thermal diffusivity of biomaterials: A simultaneous measurement technique, *Trans. ASME, J. Biomech. Eng., 99,* 148–154, 1977.

39. Chen, M. M., Holmes, K. R., and Rupinskas, V.: Pulse-decay method of measuring the thermal conductivity of living tissues, *Trans. ASME, J. Biomech. Eng., 103,* 253–260, 1981.

40. Trezek, G. J., and Cooper, T. E.: Analytical determination of cylindrical source temperature fields and their relation to thermal diffusivity of brain tissue, in *Thermal Problems in Biotechnology, ASME Symp. Ser.,* 1968, pp. 1–15.

41. Cooper, T. E., and Trezek, G. J.: A probe technique for determining the thermal conductivity of tissue, *Trans. ASME, J. Heat Transfer, 94,* 133–140, 1972.

42. Van Der Held, E. F. M., and Van Drunen, F. G.: A method of measuring the thermal conductivity of liquids, *Physica, 15,* 865–881, 1949.

43. Nix, G. H., Lowery, G. W., Vachon, R. I., and Tanger, G. E.: Direct determination of thermal diffusivity and conductivity with a refined line source technique, in *Progress in Aeronautics and Astronautics: Thermophysics of Spacecraft and Planetary Bodies,* Academic Press, New York, 1967, vol. 20, pp. 865–878.

44. Sweat, V. E.: Experimental values of thermal conductivity of selected fruits and vegetables, *J. Food Sci., 38,* 1080–1083, 1975.

45. Guyton, A. C.: *Organ Physiology, Structure and Function of the Nervous System,* Saunders, Philadelphia, 1972.

46. Bazett, H. C.: The regulation of body temperatures, in L. H. Newburgh (ed.), *Physiology of Temperature Regulation and the Science of Clothing,* Hafner, New York, 1968, pp. 109–192.

CHAPTER 10

Mechanics of the Uterus in Pregnancy and Labor

Joseph Mizrahi
Zvi Karni*
Department of Biomedical Engineering
Technion-Israel Institute of Technology
Haifa, Israel

INTRODUCTION

The mechanics of the uterus during labor and delivery is of great interest in obstetrics. Basic subjects of a mechanical nature include questions on the manner of production and propagation of uterine deformation and on the mechanism of normal uterine activity. An understanding of these, as well as other related processes, is of fundamental importance in defining and evaluating uterine dysfunction, such as incoordinate uterine contractions, reversed uterine polarity, or asymmetrical uterine action at an early stage of labor. Such an understanding may play an important role in clinical obstetrics in preventing unnecessary fetal distress and prolonged labors.

In this chapter the mechanical parameters taking part in uterine activity during labor and delivery are discussed, with emphasis on tying together theory and experiment.

10.1 MECHANICAL PARAMETERS OF THE UTERUS IN PREGNANCY AND LABOR

10.1.1 Structural Aspects

The main functions of the uterus are to provide a safe and adequate environment for the developing ovum and to deliver its contents at term by the process of parturition. During pregnancy considerable changes take place in the shape and dimensions of the uterus, as well as in the structure of its muscle [1]. The uterine muscle hypertrophies despite its relative inactivity. This is unusual for a muscular organ, and it is made possible by the pregnancy-maintaining

* Deceased.

hormonal influence of progesterone [2]. At the end of pregnancy the uterus consists of 45 percent muscle tissue and 55 percent connective tissue. The uterine cervix consists, on the other hand, of only 15 percent muscle and 85 percent connective tissue. The muscle fibers of the uterus are integrated into bundles of connective tissue, the compliance of which enables the uterus to adapt to any required volume.

It was claimed [3] that distension produced by the growing fetus is a major factor in stimulating growth of connective tissue in the rat. Moreover, following parturition, 85 percent of the collagen was found to be resorbed within 4 days in the rat [4]. Mechanical factors associated with the presence of the material contained within the uterine cavity have been reported to play a major role in maintaining the structure and composition of the uterine wall. This can be demonstrated by artificially inhibiting involution, i.e., the rapid decrease of the size of the uterus following parturition, as was done by several investigators. The various methods to demonstrate this included distension of the rabbit uterus by a balloon after parturition [5], surgical retention of the fetus in the uterine rat [6], and substitution of silicone rubber for the fetus [7].

10.1.2 Kinematic and Dynamic Aspects

Mosler [2] lists two different kinds of tonus in the uterus during pregnancy. One is the plastic tonus, for gradual adaptation of the uterus during its growth process, and the other is the contractile tonus, to enable short-time adaptation of the contents during the birth process. As opposed to the irregular, uncoordinated, and low-amplitude elementary contractions existing in the nonpregnant uterus, local contractions of considerable magnitude appear during the early stage of pregnancy. These contractions are usually not perceivable despite their high frequency of approximately one contraction per minute [8]. The amniotic pressure rise during these contractions is 2 to 4 mmHg.

From the thirtieth week of pregnancy, the contractions begin to have a certain coordination of sequence and are called the *Braxton Hicks contractions*. These contractions spread to a larger area but are painless and incapable of performing external work [2]. They were monitored by several investigators [9–15], and their frequency was reported to be one contraction per hour approximately with an intraamniotic pressure increase of 10 to 15 mmHg and tonus of 3 to 8 mmHg. Before onset of labor, the intensity and frequency of the contractions increase as they become well-coordinated, phasic, and more regular. Some values of amniotic pressures during contractions at various stages of pregnancy are given in Table 10.1.

The significance of existence of frequent Braxton Hicks contractions to the fetal environment was pointed out by Heyns [16]. He explained that contractions of the uterus cause lowering of the pressure in the choriodecidual space, resulting in an increase in volume of the space over and above that occupied by its blood. These pressure changes might accelerate blood flow to the fetus during a contraction, thus improving its otherwise poor oxygenation. It

TABLE 10.1 Typical values of amniotic pressure intensity during contractions, frequency of contractions, activity of the uterus, tonus pressure, and cervical dilatation in various stages of pregnancy and during labor

Stage	Intensity, mmHg	Frequency, contractions per 10 min	Activity, Montevideo units	Tonus, mmHg	Cervical dilatation, cm
Throughout pregnancy	2	10	20	3	0
> 30th week (Braxton Hicks)	20	0.16	3.2	5	0
Prelabor	20	1	20	7	0
Onset of labor	28	3	84	8	2
1st stage	41	4.2	172	10	10
2d stage	47	5	235	12	Fully dilated

Source: Mosler [2] and Caldeyro-Barcia and Poseiro [8].

seems, therefore, important to develop methods of stimulating reasonably frequent Braxton Hicks contractions [16]. A mechanical method to stimulate uterine contractions at normal midpregnancy by applying stretching to the uterus was recently reported [17].

10.2 UTERINE ACTIVITY MEASUREMENTS

The experimental methods used to investigate the mechanical aspects of the gravid human uterus—and here we do not refer to reports concerned with the changes in uterine electrical activity which result from the muscle contraction [e.g., 18–20]—fall into three groups, as discussed in Secs. 10.2.1 to 10.2.3.

10.2.1 Intramyometrial Measurements

The intramyometrial method of detecting uterine activity was suggested and applied by Caldeyro-Barcia and Poseiro [8]. Amniotic pressure is measured by means of a catheter introduced into the amniotic cavity through the muscle, while intramyometrial pressure is measured by means of microballoons penetrating the mass of the myometrium, each of which is connected by a catheter to an electromanometer, as schematically shown in Fig. 10.1. This rather bold method includes many inconveniences; however, its use precipitated many important investigations which yielded a better understanding of uterine contractions, particularly with respect to the simultaneous recording of the intramyometrial pressure in different areas of the uterus.

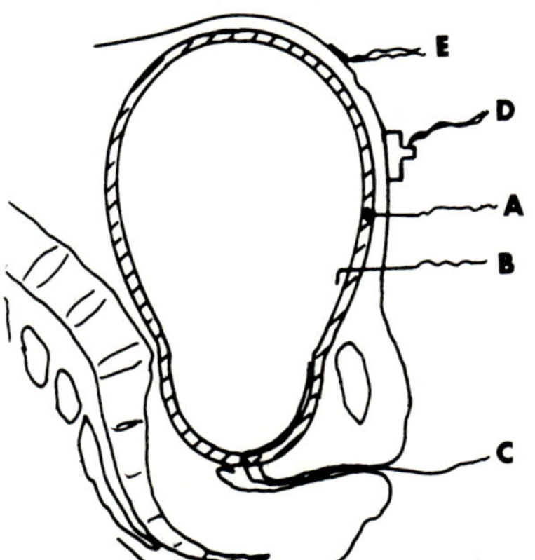

FIG. 10.1 Schematic illustration of techniques of uterine activity measurements. (*a*) Intramyometrial microballoons penetrated into the myometrium and connected by a catheter to an electromanometer, for measuring intramyometrial pressure [8]. (*b*) Catheter introduced to amniotic cavity through the muscle to measure amniotic pressure [8]. (*c*) Open-ended needle or catheter placed between muscle and amniotic membrane through the cervical canal to measure pressure in the amniotic sac [24–27]. (*d*) Tocodynamometer, in which part of the abdominal wall area and underlying uterine wall is turned into a flat diaphragm by applying force to it [9, 10, 32–34]. (*e*) Strain gauges and strain rosettes attached to the abdominal wall [40, 43, 54, 70].

The Uterine Activity Measure. In order to make possible a quantitative analysis of the tracings of the uterine contractions, Caldeyro-Barcia defined a *uterine activity unit* as the product of the intensity of the amniotic pressure of a given contraction multiplied by the frequency of the contractions. When the intensity is measured in millimeters of mercury and the frequency is expressed as the number of contractions per 10 min, the uterine activity is then expressed in *Montevideo units*. Typical values of uterine activity are illustrated in Table 10.1.

In terms of the uterine activity just defined, a classification of the different stages of labor could be made. However, this quantitative index by itself could be applied only to the activity of the organ as a whole; it did not supply any information on the separate action of the different parts: fundus, medium part, lower segment, and so forth. For this action, the intramuscular pressure was recorded simultaneously in several parts of the uterus by microballoons inserted into the wall of the myometrium, as described above. The contraction of the muscle which surrounded each microballoon caused a local rise in the intramyometrial pressure which was recorded by a pressure transducer connected to each of the microballoons. With this method, study of the propagation of the contractile wave throughout the uterus became possible, and the intensity and duration of the contraction in different areas of the organ could be compared. These data therefore permitted evaluation of the coordination of the contractions.

Pacemaker and the Propagation of a Contraction. Of the findings that emerge from Caldeyro-Barcia's studies, two are of major importance to the course of normal labor. The first refers to the existence of a pacemaker. This point has been a subject of dispute among investigators. Some researchers supported its existence [18, 19], whereas others rejected it [21]. Caldeyro-Barcia and Poseiro [8] support the existence of one dominant pacemaker area, situated near the uterine end of the uterine tubes. From this pacemaker nearly all the contractile waves originate and spread throughout the uterus in a descending propagation. The point emphasized by Caldeyro-Barcia in this connection was that under normal conditions only one pacemaker is concerned with the initiation of each contractile wave. This is in contradiction to earlier conclusions [22], which supported the assumption that two synchronized contractile waves originate simultaneously from areas near each end of both tubes.

The second significant conclusion from Caldeyro-Barcia's work was that, despite the fact that the wave reaches different parts of the uterus at different times, the peak of the contractions is attained almost at the same time in all these parts. Consequently, the farther the area is from the pacemaker, the shorter the duration of the systolic phase of the contraction.

Caldeyro-Barcia's experiments were never repeated later on. In very exceptional cases, only the general method was adopted [23], and uterine contractions were recorded through a catheter inserted transparietally into the amniotic cavity, to monitor the effect of a progesterone-type of hormone infusion on uterine activity.

While the theory which resulted from Caldeyro-Barcia's investigations could neither be confirmed nor definitely disproved by other investigators, several questions about the method arise:

1. The signals detected by the intramyometrial microballoons, while undoubtedly correlated with the uterine contractions, are difficult to interpret in terms of a definite mechanical identity. It is not clear what they actually represent: pressure, stress, or deformation.
2. The catheters are made of thin polyethylene tubes that are constantly exposed to blocking and kinking; this could distort the recordings.
3. Another source of distortion is the possible presence of a physiological local reaction around the catheter due to the insertion of the catheter itself.
4. The patient is subjected to a considerable degree of inconvenience and possibly a slight risk when this method is applied.

10.2.2 Semi-Internal Measurements

The method of semi-internal measurement was developed in the early thirties, when the measurements of intraamniotic pressure became of value in obstetrics for the diagnosis and controlled treatment of dystocia [24]. The hydrostatic pressure in the amniotic sac is measured by means of an open-ended needle or a catheter placed between muscle and amniotic membrane through the cervical canal, with pressure gauges attached to them [24–27], as shown schematically in Fig. 10.1. The following parameters were measured by this method: (1) intraamniotic pressure and frequency of contractions in the uterus [11–13, 15], (2) pressure between fetal head and the surrounding parts of the lower uterine segment [28], and (3) pressure between fetal head and cervix [26, 29, 30]. More recent studies suggest a modified version of the above technique in recording the intraamniotic pressure by duplication of catheters to overcome difficulties of blocking and kinking [31]. Despite its scientific perfection, this method causes some degree of inconvenience to the patient and demands special attendance by the obstetrician. Moreover, it can measure only intraamniotic pressure.

10.2.3 External Measurements

Tocodynamometer. To overcome the inconvenience of the intrauterine apparatus for measuring intraamniotic pressure, abdominal tocodynamometers have been used extensively [9, 10, 32–34]. In these instruments, part of the abdominal wall area and the underlying uterine wall is turned into a flat diaphragm by applying external forces to it (Fig. 10.1). Three main versions of this measurement technique have been developed. In the first, a flat plate is used to apply a given force and the resulting contact area with the abdominal wall is measured. Another version includes a plunger which indents the surface of the abdominal wall, and the third consists of a plunger which gently rests on the surface and follows the changes in curvature of the abdominal wall.

For various reasons inherent in their design, these instruments could not always record true intraamniotic pressure. Indeed, it has been questioned [35] which of the following three is the real meaning of what the tocodynamometer reads: (1) uterine curvature change, (2) uterine wall rigidity change, or (3) some measure of the inner fluid pressure. Typical errors in these instruments' readings can originate from skin and uterine wall tension, thickness of the subcutaneous layers, posture of the patient, and position of the fetus in the uterus. Nevertheless, these instruments are in routine use to time contractions and note some changes in uterine tonus during the treatment of a patient.

A new tocodynamometer including a strain-gauge circuit was later developed and reported upon [36]. This new type is reported to overcome the defects of earlier instruments and to make possible absolute measurement of intraamniotic pressure through the abdominal wall. It consists of a guard ring which supports a flat, pressure-sensitive plate by which the external surface of the abdominal wall is pressed. Measurements of tonus made by this tocodynamometer revealed pressures of 15 g/cm^2 at the umbilical level with the patient in the supine position. At the cervical level, the tonus was higher by 12 g/cm^2 in the supine position and by 30 g/cm^2 in the standing position. In a later study, this apparatus was combined with an ultrasonic Doppler recorder to measure continuously the fetal heart rate and uterine contractions for monitoring the fetal condition during contraction [31].

Smyth's claim [29] of measuring absolute intraamniotic pressure was reexamined by Duggan et al. [37], who found that the pressure readings obtained by this tocodynamometer were dependent upon the force of application on the abdominal wall. No plateau was found where the transducer output was constant and thus representative of the intrauterine pressure. This evidence suggested, therefore, that pressure results obtained using Smyth's instrument are questionable and that the interpretation of uterine function based on such experimental findings should be considered with caution. Consequently, Jilek et al. [38] rejected the tocodynamometer as a quantitative method for measuring uterine activity, since it was able to give only relative information on the intensity, duration, and frequency of contractions. Instead of this method, they used an intrauterine catheter and measured the actual intrauterine pressure. To quantitate uterine activity, they measured the area under the intrauterine pressure curve. A similar approach, which also correlated the area under the intrauterine pressure waveform to amplitude, was recently reported [39].

Multistrain-Gauge and Strain-Rosette Measurements. In order to investigate the nature of the normal uterine contractions and the patterns of deformation the uterus undergoes (both questions of basic importance), multichannel recording techniques were applied to monitor uterine activity simultaneously at different sites of the uterus. The strain-gauge technique applied in multistrain measurements, which was first reported by Karni et al. [40], presents definite advantages compared with the earlier work done with tocodynamometers [41] or microballoons [8]. This method is based on multistrain measurements of uterine activity with the use of high-extension electrical resistance rubber strain gauges (Fig. 10.1). Electrical strain gauges had not been used before in obstetrical monitoring except for secondary purposes. They were, however, introduced for the purpose of recording (1) intrauterine pressures in different parts of the uterus [28], (2) dilatation of the cervix with the use of a tweezerlike instrument which contained a strain gauge [29], (3) spring deflection in the guard-ring tocodynamometer [36], and (4) bending of clamps which were attached to the internal margin of the external cervical os, to indicate the magnitude of the forces which dilate the cervix during labor. In this latter use cervical tocometry was combined with external tocography, so that the cervical forces were synchronized with the uterine contractions. Dilating forces of up to 9 N for the fully dilated os were obtained [42].

The original multistrain measurements of uterine activity were later supplemented by strain-rosette measurements (Fig. 10.2) and incorporated into a mechanical shell model, for the uterus, set up by the present authors [43], to be discussed in Secs. 10.3 and 10.4. The strain gauges used

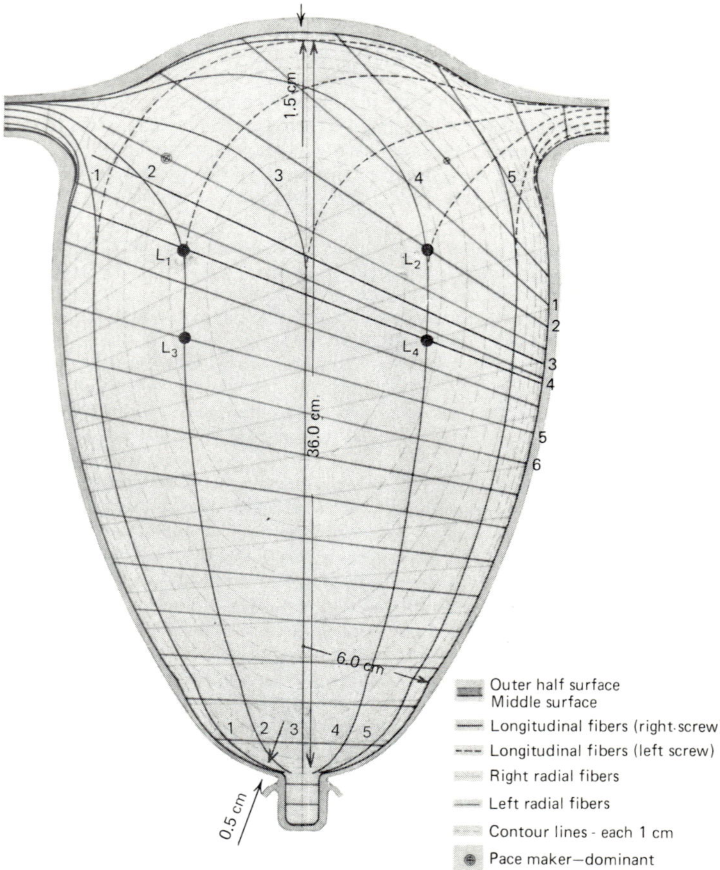

FIG. 10.2 Prototype of the average uterus, before labor, with families of longitudinal and circular fibers, used in the thin shell model [43]. To measure isotropy, strain rosettes were attached at locations L1 to L4. The rosette is an arrangement of four strain gauges in four different directions, 45° apart, about a given centroid. The angles between the axes being known, the readings of any three gauges are necessary and sufficient, in the case of perfect isotropy, to permit the prediction of the fourth strain, but the latter is also independently measured so that the computed strain can be checked against that which is measured. The entire rosette is then rotated about its centroid and the whole process repeated. If the computed values from three arbitrary gauges coincide with the measured values of the fourth, for all orientations of the rosette, complete isotropy is established.

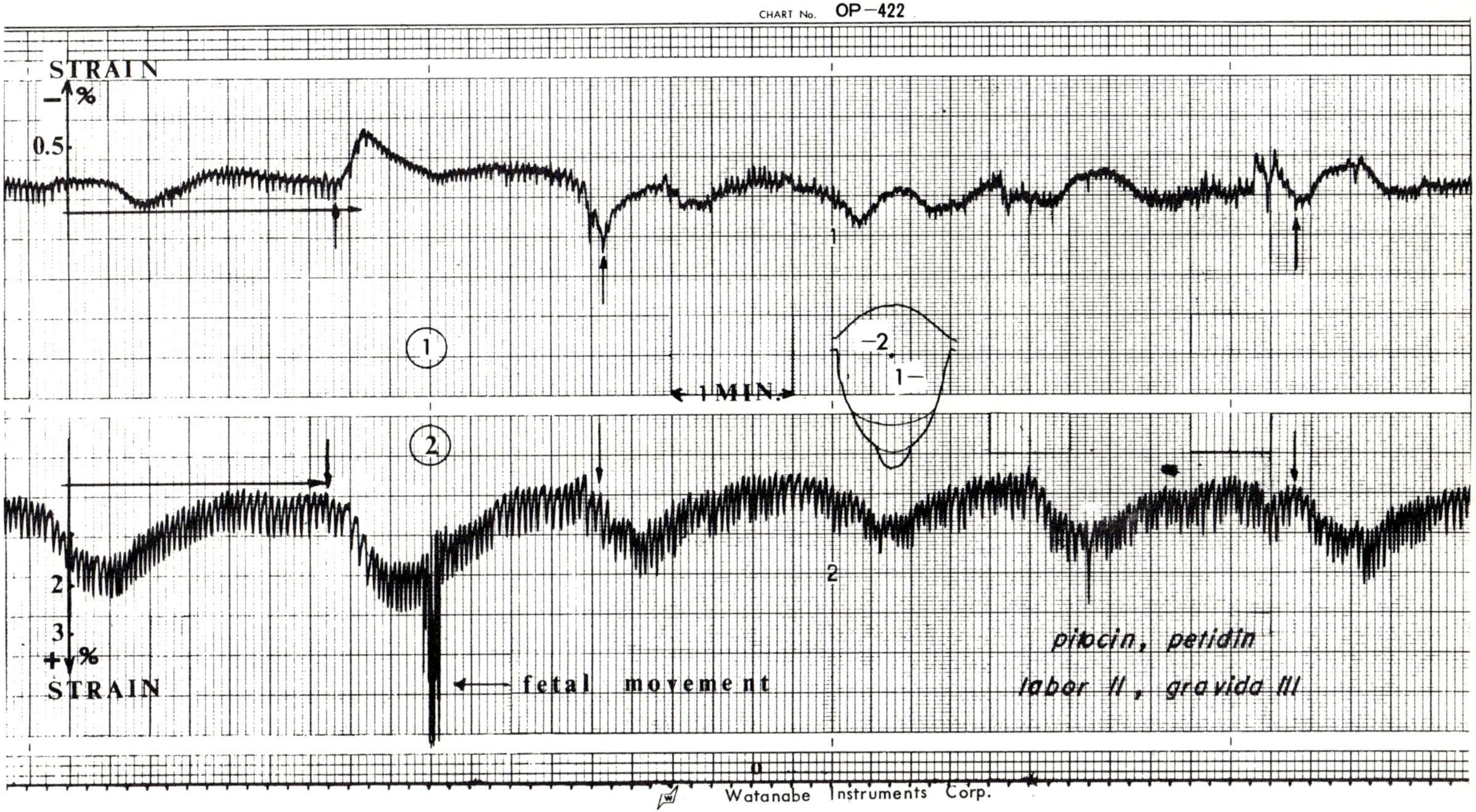

FIG. 10.3 A two-gauge strain curve. The upper gauge (no. 2) underwent extension, whereas the lower gauge (no. 1) underwent compression. The rapid oscillations correspond to the breathing of the patient, the transient accelerations or waveforms reflect the contraction of the uterine muscle, and the rapid accelerations or spikes represent sudden movement of the patient or fetal movements.

had a maximum extensibility of 20 percent and compressibility of 15 percent, an active length of 13 mm, and a pickup sensitivity of 2 μm (0.01 percent strain).

The advantages of the strain-gauge technique can be listed as follows. First, it fits in with the precise nature of the recorded quantity, strain, being a differential rather than an integral entity. Second, because of its high sensitivity and reliability, it offers a deeper probe into the more refined behavior of the uterus in labor. Third, no inconvenience is caused to the patient, nor would local traumas follow which might disturb the genuine physiology of that organ. Finally, the interpretation of the results covers a broad spectrum and may answer queries which relate to the anatomy and physiology, normal and pathological, of the uterus. A typical strain record is shown in Fig. 10.3. Both multistrain-gauge and strain-rosette measurements were introduced as boundary conditions into the thin shell model for the uterus, as discussed later in Secs. 10.3 and 10.4.

Fetal Movements. The strain measurements also yielded information about fetal movements, breathing changes, and the effect of analgesics on contractions throughout the various stages of labor [43]. (See Fig. 10.3.) Monitoring of fetal movements objectively is of considerable importance since these movements have been reported to reflect the well-being of the fetus in utero [44–46]. The reduction of these movements compared with the normal could thus be used as an alarm signal, indicating impending intrauterine death of the fetus and requiring immediate delivery.

Introduction of the strain-gauge method as an external and harmless method for monitoring fetal movements was described by Mizrahi et al. [47]. A differential recording technique was described which served to uncouple the signals due to fetal movements from those of the respiratory movements of the patient. This method has clear advantages of being inexpensive and safe, as compared with the existing ultrasonic [48–50] and electromagnetic [44] methods.

10.3 MECHANICAL MODELING

10.3.1 The Pressure Vessel Model

A number of workers investigated the role of mechanical factors in uterine activity and regulation by setting up mathematical formulations based on physical principles. Sellheim [51] suggested that physical function of the uterus could be best understood in hydrodynamic terms. Reynolds [41], following Samuel Haughton's "On the Muscular Forces Employed in Parturition," tried to estimate the quantitative characteristics of the gravid uterus, such as volume, surface, tensile stress, and greatest inner pressure. The relationships among the same parameters were also investigated by Mosler [2]. Special importance was attributed by him to the role of wall tension of the uterus in regulating the contractions. Wall tension was related to the amniotic pressure and the radius of curvature of the uterus, as in the case of a pressure vessel. Along the same lines, Reynolds [41] explained that the process of expulsion is irreversible since the radius of curvature of the fundus exceeds that of the cervix, thus producing a higher tension in the fundal region.

Role of Wall Tension in Pregnancy and During Labor. According to Mosler [2], two forces oppose each other during pregnancy: the uterine power (due to wall tension) and the resistance of the closed cervix. In early and midpregnancy, the volume is small; therefore the wall tension is insignificant. The uterine power is low, and the resistance of the cervix relatively high. Uterine tonus in this period is kept low by adaptation of the uterus to its growing volume. In fact, the low-magnitude resting intrauterine pressure during pregnancy may be attributed to the intraabdominal pressure itself. As pregnancy advances, there is a progressive increase in volume and a gradual rise in the wall tension, resulting in greater uterine power and a decrease in the cervical resistance. When the wall tension reaches a critical value, the two forces will be unbalanced and labor will begin [2, 52]. At this stage every local contraction normally triggers off a generalized uterine contraction. An increased tension improves labor performance.

Mechanism of Expulsion of the Fetus. The first stage of labor is isometric, until the cervix opens. The power of the myometrium is just sufficient to dilate the cervix and rupture the membranes which envelop the fetus [41]. In normal labor the minimum active pressure amplitude to rupture the membranes is 40 mmHg [2]. During the expulsion period, contractions normally consist of two parts. One is isometric, with an almost constant volume, during which the energy is converted into pressure and deformation of the uterus. The other part is characterized by a volume decrease and can be considered as isotonic, often with considerably high internal pressure of up to 80 mmHg. Calculations have shown [41] that the force required to expel the fetus is higher than that which the myometrium can develop. In most cases expulsion of the fetus in the second stage of labor is aided by the reflex action of the voluntary abdominal muscles, called the *bearing-down effort*. The combined pressure increase during a contraction may amount to 110 to 130 mmHg [8]. Alternatively, external intervention such as use of forceps is required for providing the extra force necessary for fetal expulsion.

10.3.2 The Thin Shell Model

An extensive investigation of the deformation of the uterus was carried out by the present authors [43, 53]. Questions related to strain of the muscle fibers, and to variations in curvatures and in the fiber angles of the deforming uterus, were treated within the framework of a comprehensive kinematic model for the uterus based on in vivo strain measurements [53, 54], described earlier in Sec. 10.2. On the basis of the anatomical and physiological data available [1, 55–57], the myometrium of the pregnant uterus at term was regarded as a closed, axisymmetrical thin shell structure, the thickness of which is small in comparison with the other dimensions. Strictly speaking, the pearlike shape of the uterus is nonsymmetrical [58], but its closeness to a surface of revolution makes the assumption of axial symmetry a good approximation which has been extensively used in the literature on obstetrics. It is almost an isolated muscular system, free to deform in the abdominal cavity, and is subjected to the constraints of its ligament attachments and the cervix. In the static state, the shell is subjected to internal pressures only. Later, upon the release of the trigger mechanism of labor, the muscle fibers undergo an induced strain along their trajectories in the surface of the shell, acting against the inner pressure, and produce a deformational pattern of the uterus which varies with time. A more complete description of the thin shell model and its solution are given in Sec. 10.4. Summary of the main mechanical models used for describing uterine contractions is given in Table 10.2.

TABLE 10.2 Summary of principal models used in mechanics of uterine contractions and in cervical dilatation

Model and investigators	Parameters correlated
Mechanics of uterine contractions	
Pressure vessel model (Sellheim [51], Reynolds [41], Mosler [2])	Pressure, radius of sphericity, tensile stress; and volume, work, pressure
Thin shell model (Mizrahi, Karni, et al. [43, 53, 54, 57])	Strain, curvature, twist, displacement, volume
Extended Hill model for constitutive equation (Mizrahi, Karni, et al. [87, 89])	Stress, strain rate, strain rate invariants
Cervical dilatation	
Extrusion model (Lindgren and Holmlund [113, 115])	Pressure, friction, radius of head, radius of cervix
Hydraulic axisymmetric model (Rice et al. [116])	Pressure, radius of head, radius of cervix, tensile stress, hoop force, elastic stretch
Two-stage model (Egan and Richardson [119])	Pressure, radius of head, radius of cervix, tensile stress, hoop force, time after effacement of cervix

10.4 KINEMATICS OF THE UTERUS DURING LABOR CONTRACTIONS

10.4.1 Basic Assumptions of the Thin Shell Model

In setting up the thin shell model [43, 53] the following assumptions were made:

1. Axial symmetry of a pearlike shape.
2. Uniformity in the thickness variation of the uterine wall, from the thinnest part in the inferior part of the uterus to the thickest part in the fundal region.
3. In keeping with the symmetry assumption, the longitudinal systems of muscle fibers which thread out from the tubes downward are regarded as both right-handed and left-handed at each tube when viewed from the inside.
4. Planar systems of circular muscle fibers, seen as families of straight lines in Fig. 10.2. The trajectories of the longitudinal and of the circular fibers are not lines of curvature, and shearing of surface elements in the direction of these fibers may take place.

The above-listed assumptions enable one to formulate the kinematic equations of the deforming thin shell, and these equations can be solved by means of measured boundary strains. In order to study the dynamic aspects of the deforming shell, material properties of the uterine muscle have to be incorporated into the model. Although some work has been reported on this subject (see Chap. 6), much more work remains to be done.

Following the anatomy of the myometrium, the system of reference was set along the two primary families of nonorthogonal muscle fibers [43, 53, 54]. As a result, oblique coordinates were used throughout the analysis, for which the notation of physical components of tensors was adopted [59–63]. The time-dependent solution of the deforming shell was sought as a sequence of incremental states of deformation [64]. Each is sufficiently small by itself so that the infinitesimal relations for the small deformation were employed. Two separate approaches were considered: the continuum approach and the network approach.

10.4.2 Continuum Approach

According to the continuum approach the uterine muscle was assumed to behave as a continuum, for which the Codazzi-Gauss equation [65], or the vanishing of the Riemann-Christoffel tensor [66] for the deformed middle surface, hold. These equations are the kinematic compatibility conditions, which express the deformational continuity of the middle surface of the shell. By introducing the measured strain components and their derivatives as boundary conditions into these equations, the complete pattern of strain, curvature, twist, and displacement were obtained [43].

10.4.3 Network Approach

By the network approach, the uterine muscle was assumed to be formed of noninteracting families of muscle fibers. The uterus was thus pictured as a wicker basket made of soft fibers, the geometrical changes of which reflected the deformation of the basket. In this case a set of equations, expressing the noninteraction of the families of muscle fibers, replaced the compatibility equations of the continuum model [43].

10.4.4 Boundary Conditions

The earlier-described measurements of uterine strains were introduced as boundary conditions into the shell equations. The reduction from outer surface to midsurface readings was performed [54] by means of the Kirchhoff-Love assumption [65, 67–69], which underlies the thin shell theory.

10.4.5 Features of Uterine Contractions

The following parameters were inferred directly from the strain records [70] (see Fig. 10.3): (1) polarity of uterine contractions, (2) dominance of the pacemaker, (3) intensity of contractions, (4) frequency of contractions, and (5) tonus of the myometrium during relaxation. The findings suggested that there was one dominant pacemaker at the uterine fundus from which nearly all uterine contractions originated and spread in a descending fashion. The calculated speed of propagation was 5.7 ± 0.234 cm/s.

The existence of more than one peak during uterine contractions was a common feature, and it indicated either multiple firing of a single pacemaker or the combined effect of several pacemakers, one of which is dominant. The common finding of two peaks, with the first larger than the second, lent support to the theory of Caldeyro-Barcia and Poseiro [8] that there are two pacemakers, one dominant and one subordinate, on the anterior wall of the uterine fundus about halfway between the umbilicus and each of the fallopian tubes. Electrical potential recordings, made simultaneously with the strain measurements, would shed more light on the subject of pacemaker activity.

Another finding reported by Mizrahi et al. [70] was that there exists a considerable time lag between the maximal points of the contraction at different locations on the uterus, in contradiction to previous reports [8]. The strains measured in the long fibers were of positive sign (extension) in the fundal region and of negative sign (contraction) in the lower part of the uterus, with a transition zone of zero strain in a horizontal strip below the umbilicus. This transition was a feature seen in all patients, although it differed in location.

10.4.6 Deformational Patterns of the Uterus

Time variations of the radii of curvature along the longitudinal and circular fibers, as well as the relative torsion during a uterine contraction, were obtained by solving the thin shell continuum model and are reported by Mizrahi et al. [54]. The average increase in longitudinal curvature from the onset of contraction to its peaks was 200 percent. The curvature along the circular fibers, on the other hand, increased on the average by only 20 percent. It thus followed that the combined effect of curvature changes implies that the initial pearlike shape of the uterus becomes more spherical during the contractions, with the radii of curvatures attaining almost similar values.

The deformational changes occurring in the uterus during labor contractions were discussed by Heyns [16, 71] in relation to abdominal decompression manipulation, obtained by placing the patient in appropriate vacuum chambers. It was argued that the rounding of the gravid uterus, by which the pearlike shape becomes more spherical, involves a shortening of the axial diameter and an inevitable lengthening of the transverse diameter, since the inner content of the uterus stays incompressible, at least to a first approximation. The transverse push of the abdominal wall is one cause of labor pains. By increasing the abdominal volume and overcoming the resistance of its walls through the decompression technique, the labor pains should be reduced. Moreover, the myometrium is also free to contract longitudinally, making cervix dilatation and expulsion of the fetus possible. It should, however, be noted that application of the decompression technique has been discontinued, after questions related to fetal safety remained unanswered.

10.5 MECHANICAL BEHAVIOR OF THE UTERINE MUSCLE

The mechanical behavior of the uterine muscle can be studied in various ways: in vitro or in vivo, by using specimens or on the whole organ, and in the active or the passive state.

10.5.1 In Vitro Methods

A technique for the examination of uterine contractions in vitro was proposed by Magnus [72] and developed by Kehrer [73]. By this technique, it has been shown that the isolated uterus or strips from a human pregnant uterus can develop laborlike contraction, in spite of their being detached from any neural influence. The strips of muscle had to be kept in an oxygen-saturated solution at body temperature, and its mechanical activity was recorded by means of a lever mechanism attached to the specimen. Continuous supply of fresh nutritive solution was later found to prolong the survival time of the isolated organs being studied. Improvement of the Magnus-Kehrer apparatus was reported by Mosler [2], by which both electrical and mechanical activities of longitudinal and annular muscle families could be made. This investigator, however, favored whole organ to muscular strip investigation. Two main reasons were given for this: (1) the difficulty of reproducing identical strips, in which the fibers run in a predetermined direction, and (2) the fact that nourishment of the strips is not natural and homogeneous. The characteristics thus determined did not provide reliable information on the actual behavior of the segment in the intact organ.

10.5.2 Applicability of Hill's Equation in the Active State

Significant similarity between the behavior of the uterine muscle in the active state and that of smooth muscle was reported [74–76]. This was based on observations that (1) the uterine muscle contracts relatively slowly and relaxes even more slowly, (2) contraction is tetanic, and (3) there exists a direct correlation between the frequency of the action potentials and the tension developed in the muscle. Other experiments have shown a similarity of the myometrium to skeletal muscle, for strips of rabbit uteri [77, 78] and for human uteri [79, 80].

One-dimensional investigations on smooth muscles, uterine as well as nonuterine [77, 81–84], indicated that smooth muscles displayed the same type of hyperbolic relationship between the force and velocity of shortening as do striated muscles [85]. The Hill model of muscle dynamics was also the basis for the description of uterine contractility and uterine function in human parturition [86].

10.5.3 In Vivo Methods and the Question of Isotropy and Anisotropy

In relation to isotropy and anisotropy of the uterine muscle, the anatomical sandwich-like separation into definite layers would suggest an anisotropic behavior of this muscle. On the other hand, because of the intricate interlacing of the fiber bundles through the various layers of the myometrium, one would expect the muscle to show isotropy rather than anisotropy. To resolve this problem the present authors made strain-rosette measurements of the uterus during deformation [43, 87], as shown schematically in Fig. 10.2.

The above-described rosette analysis showed that the functional behavior of the uterine muscle is basically anisotropic in character [87]. However, during the early stages of labor, when the deformations are still small and the different groups of muscle fibers, owing to their interlacing, still act as an ensemble, to a close approximation, isotropy prevails. With advancement of labor, deviations from isotropy build up, indicating that each group of fibers becomes more and more functionally independent of the others. They even exhibit different material properties, the longitudinal fibers showing a higher resilience than the circular ones [88]. A pattern of a delayed recovery from deformation during the contractional relaxations is a common feature of the various fiber groups, which is even more pronounced at the final stage of labor.

10.5.4 Constitutive Equation for Uterine Muscle

A constitutive equation for smooth muscles, with special reference to the myometrium, was proposed by the authors [89]. It is based on Hill's equation [85], found to be valid for smooth muscles in one-dimensional experiments, and assumes a mathematically isotropic relationship* between stress and strain rate, two interrelated physical quantities in Hill's equation [90].

The basic relation used in deriving the constitutive equation was the generalized principle of dimensions [93, 94], according to which an isotropic relation can be expressed as a sum of symmetric tensors of second rank, multiplied by scalars. Next, by use of the Cayley-Hamilton equation [92], and by a comparison with the series expansion of Hill's equation, a constitutive equation relating the stress and strain rate tensors, through the principal invariants of the strain rate, was obtained.

10.5.5 Passive Properties of Uterine Specimens

Passive mechanical properties of uterine muscle in vitro were reported by several investigators. Reynolds [41], after comparing the forces necessary to rupture the uterus with those actually developed by the organ, stated that, during its activity, the uterus cannot destroy itself.

Data on passive stress-strain relations in tension were obtained by loading samples in 20-g increments each half minute until the tissue ruptured [95]. From these data the true stress at rupture was calculated and found to be approximately 0.5 MPa [88]. The elastic modulus at stress of 0.07 MPa was found to be 1.2 MPa. Dead-weight loading of 100 g for 1 min was also used to isolate the elastic and plastic component of deformation, but no analysis of the time-dependent behavior of deformation was made [95]. From step-loading tensile measurements [80], the following elastic moduli were calculated [88] at 0.07-MPa stress and found to be 0.6 and 1 MPa for pregnant and nonpregnant uterus, respectively. These values should, however, be treated with caution since some stress relaxation was permitted between loading steps.

Pearsall and Roberts [88] characterized the range of passive mechanical behavior of the human myometrium in terms of true stresses versus true strain, in both tension and compression at the low strain rate of 0.5 min^{-1}. The specimens were taken from the fundal region and tested in air soon after excision. The small statistical scatter of results as compared to variability between specimens tended to support that testing of myometrium in air was satisfactory. The data were found to obey a logarithmic relationship between true strain and true stress, for stress values above 7 kPa. Earlier, Fung [96, 97] suggested that this relationship was typical of many biological materials which can be regarded as springs that stiffen under load. The mechanical values found by Pearsall and Roberts [88] were 0.5 to 1.4 MPa for elastic modulus (corresponding to 0.07 MPa stress) and tensile strength of 0.56 to 2.1 MPa. Summary of tensile properties of uterine specimens is given in Table 10.3. The modulus of tension was the same as that in

* It should be noted that an *isotropic relationship* refers to invariance requirements [91, 92], one of which is that constitutive functions must be absolute tensor point functions of their variables. Such functions are invariant with respect to coordinate transformations. Otherwise, a change of description would produce a different response in the same material subjected to the same disturbances. The coordinate invariance is achieved by writing the equations in tensorial forms. Another requirement is the material invariance, by which a constitutive equation must be invariant with respect to the transformation of material coordinates. If there exists no preferred direction with respect to a certain constitutive property of the material, we say that the material is *isotropic* with respect to that property. Otherwise, the material is said to be *anisotropic*.

TABLE 10.3 Summary of passive tensile properties tested in vitro

Investigators	Maximal stress, MPa	Modulus, MPa (at 0.07 MPa stress)	Mode of loading
Wood [95]	0.5	. . . .	Step loading
Conrad and Kuhn [80]	. . . .	0.6 (pregnant) 1.0 (nonpregnant)	Stress relaxation and step loading
Pearsall and Roberts [88]	0.5–2.1	0.5–1.4	Dynamic loading at 0.5 min^{-1}

compression for the same stress values but higher for the same strain values. Graphically, this was expressed by the fact that the compression curve was shifted to the right of the tension curve. This difference was attributed to the fact that in tension the collagen network can be stretched taut; this was not so in compression. Differences in stiffness were found between the inner and outer layers of the myometrium. The outer tissue was less stiff in compression but more stiff in tension. No attempt was made to correlate these findings with the differences in fiber orientations of the layers of the myometrium.

10.5.6 Response to Indentation In Vivo

A finite element approach was used to predict the stress and deformation of uterine muscle tissue, caused by a flexible shield indentation intrauterine contraceptive device [98]. The analysis undertaken was that of plane strain, and tissue behavior was approximated as piecewise linear; the analysis was aimed at detecting possible damages caused by the shield to the uterus.

An instrument was designed and constructed for the measurement of the force-displacement relation of the uterine cervix in vivo [99]. The penetrated cervix was indented on its walls by a measured force, and the angle formed between the walls of the impressed area was monitored. The significance of the results obtained in this study cannot be established because of misuse of mechanical concepts by the author.

10.6 CERVICAL DYNAMICS

10.6.1 Mechanical Characteristics and Geometry of the Cervix during Labor

The quantitative relation between intrauterine pressure and cervical dilatation is of major significance in describing the mechanics of labor and the process of delivery. Important factors in describing this relation are the geometry of the birth system and the mechanical characteristics of the cervical tissue.

During pregnancy, the cervix must resist any tendency to dilate, and the circular musculature of the lower part of the uterus assists to preserve pregnancy, acting as a sphincter. The cervical tissue consists mainly of closely packed collagenous fibers, arranged circumferentially, with only a small amount of muscle (10 to 15 percent). Toward the end of pregnancy, significant transformation in the structure of the cervical tissue occurs to prepare it for its dilatation during labor. The structure of the collagen fibers is destroyed, resulting in small collagen fibrils randomly dispersed in a glycosaminoglycan matrix, rendering the cervical tissue more extensible [100, 101].

At the onset of labor, yielding of the cervical opening is achieved in each contraction by thickening of the upper part of the uterus and thinning of its lower part. Distension of the genital tract was reported to be the main cause of pain in each uterine contraction [8]. Although each contraction can dilate the cervix by up to 30 percent, most of this stretching is reversed, and the gradual opening of the cervix from 2 cm at the onset of labor to 10 cm during delivery is a result of a ratchet mechanism, through which the net gain in each contraction is approximately 2 percent. The progressive dilatation of the cervix is assisted by the viscoelastic behavior of cervical tissue during parturition [102]. The mean duration of the dilatation process of the cervix was reported to be 6 h for multiparas and 9 h for primiparas [103]. It was found [104] that about 2 contractions per 10 min are necessary to maintain a given dilatation and that 2½ to 3 contractions per 10 min is the optimum frequency for rapid cervical dilatation.

Before expulsion of the fetus the intrauterine pressure is homogeneous [104]. After engagement, however, the pressure between the presenting part of the fetus and uterine wall in the girdle of contact becomes three to four times higher than the intraamniotic pressure [104, 105]. Descent of the fetal head is oscillating and not uniform. It progresses during each contraction and represses partly when the uterus relaxes due to the elastin retraction of the vagina and perineum [8].

10.6.2 Theories of Cervical Dilatation

Early Models. It has been proposed that radial forces from the lower ligaments were responsible for a major part of the cervical dilatation [106, 107]. This, however, contradicted earlier observations [108, 109], indicating that the relative movements of the parts of the system during labor were too small to allow the ligaments a major role in cervical dilatation. Another theory suggested that dilatation is due to forces developed from the wedging action of the fetal head [30, 110–112].

Frictional Forces. Fetal motion in the cervix is resisted by frictional forces. Several factors contribute to the reduction of these frictional forces: the waxy substance covering the fetal skin, amniotic fluid, blood, and other secretions, and the reciprocating motion of the fetus in the cervix which keeps these lubricants. It was found, however, that the normal forces between head and cervix were considerably higher than the driving force due to the intrauterine pressure, increasing the relative importance of friction in the dynamics of the cervix [113, 114]. An attempt to calculate the coefficient of friction between the head and cervix was made [115], assuming that the frictional force is that force which is required to balance the axial force on the head resulting from intrauterine pressure; and a coefficient of friction of 0.20 was found for five normal labors. Rice et al. [116], however, pointed out that these calculations neglected the axial component of the head-cervix pressure, which opposes the axial force from intrauterine pressure.

The Hydraulic Model. Rice et al. [116] proposed a model for cervical dilatation, based on the assumption that the hydraulic force on the fetus produced by the contraction is the primary mechanism for dilatation, the frictional and inertial forces being negligible (Table 10.2). Geometrically, it was assumed that the uterus, cervix, and fetal head are axisymmetrical and coaxial. The cervix was assumed to be passive. Force-length characteristics for the cervix were taken from existing reports [117, 118].

The Two-Stage Model. A two-stage dilatation process was proposed by Egan and Richardson [119] (Table 10.2). The first stage is the effacement process of the cervix, which occurs shortly before the onset of labor during which the tissue fibers reorient themselves, resulting in an increase in stiffness in the direction of deformation. The second stage of the dilatation process occurs as the fetal head is pushed into the cervix during contraction. This model made use of the equilibrium equations for axisymmetric membranes, neglecting frictional and dynamic effects during dilatation. A constitutive equation assuming isotropy, homogeneity, and incompressibility in the uneffaced state was thereafter introduced, by uncoupling the relaxation time–dependent behavior of the tissue following effacement from the instantaneous response occurring during effacement. The solution of this model included intrauterine pressure and dilatation data, earlier reported by Richardson et al. [120].

10.7 CONCLUSIONS AND SUGGESTIONS FOR FUTURE RESEARCH

The large amount of data on the mechanical parameters of uterine activity collected through numerous years of investigations need to be introduced into a consistent model if the mechanics of the uterus, normal or pathological, are to be properly understood. To the knowledge of the present authors, the only reported comprehensive mechanical model of the uterus as an organ is the thin shell model (see Secs. 10.3 and 10.4). The kinematic part of this model has been already solved. The dynamic part, however, requires investigation.

The following points are some suggestions for future research.

1. Simultaneous measurement of electric potentials of the uterine muscle with the strain measurements in order to shed more light on the subject of pacemaker activity.

2. Mechanical properties of the uterine muscle, both in the active and passive states, as function of location and direction of the specimens, as well as of time during pregnancy and labor. The following are points of special interest: (1) experimental validation of the three-dimensional constitutive equation, based on Hill's model, (2) correlation of mechanical properties with behavioral differences between the upper tensile and the lower compressive zones of the uterus, and (3) establishment of material isotropy of the uterus during labor.
3. Formulation of the complete equations of the thin shell model and their solution by means of (1) the constitutive equation for uterine muscle and (2) in vivo measured boundary conditions of strains and pressures, to yield the displacement and stress patterns of the uterus during labor contractions.

REFERENCES

1. Greenhill, J. P. (ed.): *Obstetrics,* 13th ed., Saunders, Philadelphia, 1969.
2. Mosler, K. H.: The dynamics of uterine muscle, *Adv. Obstet. Gynaecol., 35,* 1–88, 1968.
3. Harkness, M. L. R., and Harkness, R. D.: The distribution of the growth of collagen in the uterus of the pregnant rat, *J. Physiol., 132,* 492–501, 1956.
4. Harkness, R. D., and Moralee, B. E.: The time-course and route of loss of collagen from the rat's uterus of the pregnant rat, *J. Physiol., 132,* 502–508, 1956.
5. Csapo, A., Erdos, T., De Mattos, C. R., Cromms, E., and Moscowitz, C.: Stretch-induced uterine growth, protein synthesis and function, *Nature, 207,* 1378–1379, 1965.
6. Yamey, A. R.: Localized inhibition of collagen resorption in the post-partum rat uterus, *J. Physiol., 267,* 29, 1977.
7. Wray, S.: The role of mechanical and hormonal stimuli on uterine involution in the rat, *J. Physiol., 328,* 1–9, 1982.
8. Caldeyro-Barcia, R., and Poseiro, J. J.: The powers and the mechanism of labor, in J. P. Greenhill, (ed.), *Obstetrics,* 13th ed., Saunders, Philadelphia, 1969, chap. 14.
9. Schwalm, H.: Das Pastogramm: Hilfsmittel der Praktischen und wissenschaftlichen Geburtsluffe, *Geburts. Frauenheilk., 17,* 105–121, 1957.
10. Stock, W.: Tokographische untersuchungen zum problem des Geburtsbeginnes, inaugural dissertation, Med. Fak. Univ., Würzburg, 1965.
11. Bengtsson, I. Ph.: Oxytocin response, withdrawal and reinforcement of defence mechanism of the human uterus at midpregnancy, *Am. J. Obstet. Gynecol., 83,* 1083–1093, 1962.
12. Csapo, A.: Defense mechanism of pregnancy, *Ciba Found. Coll. Endocrinol., 9,* 3–26, 1961.
13. Csapo, A.: The effects of oxytocic substances on the excitability of the uterus, in (R. Caldeyro-Barcia and H. Heller) (eds.), *Oxytocin,* Pergamon Press, London, 1961.
14. Caldeyro-Barcia, R.: A quantitative study of the action of synthetic oxytocin on the pregnant human uterus, *J. Pharmacol. Exp. Ther., 121,* 18–31, 1958.
15. Lindgren, L.: The influence of anaesthetics and analgesics on different types of labour, *Acta Anaesth. Scand., 2,* 49–56, 1959.
16. Heyns, O. S.: The effect of the human placenta of change of uterine shape during contraction, *S. Afr. J. Med. Sci., 23,* 157–166, 1958.
17. Manabe, Y., Sakaguchi, M., and Nakajima, A.: Stretch-induced uterine activity at mid-trimester, *Int. J. Biol. Res. in Pregnancy 2*(2), 63–69, 1981.
18. Steer, C. H., and Hertsch, G. J.: Electrical activity of the human uterus in labor: The electrohysterograph, *Am. J. Obstet. Gynecol., 59,* 25, 1950.
19. Larks, S. D., Dasgupta, K., Assali, N. S., Morton, D. G., and Bellarmy, A. W.: The human electrohysterogram: Electrical evidence for the existence of pacemaker function in the parturient uterus, *J. Obstet. Gynaecol. Br. Emp., 66,* 229–238, 1959.
20. Larks, S. D.: *Electrohysterography,* Charles C Thomas, Springfield, Ill., 1960.
21. Schulman, H. H., and Romney, S. L.: Variability of uterine contractions in normal human parturition, *Obstet. Gynecol., 36,* 215, 1970.

22. Ivy, A. C.: The functional anatomy of labor, with special reference to the human being, *Am. J. Obstet. Gynecol., 44,* 952, 1942.

23. Scommegna, A., Burd, L., Goodman, C., and Bierniarz, J.: The effect of pregnenolone sulfate on uterine contractility, *Am. J. Obstet. Gynecol., 108*(7), 1023–1029, 1970.

24. Moir, J. C.: Recording the contractions of the human pregnant and non-pregnant uterus, *Trans. Edinb. Obstet. Soc., 92,* 93, 1934.

25. Karlson, S.: A contribution to the methods of recording of the mobility of the human uterus, *Acta Obstet. Gynecol. Scand., 24,* suppl. 4, 1944.

26. Carey, H. M.: An improved technique for the induction of labour, *J. Obstet. Gynaecol. Br. Emp., 61,* 59, 1954.

27. Williams, E. A., and Stallworthy, J. A.: A simple method of internal tocography, *Lancet, 1,* 330, 1952.

28. Ingelman-Sundberg, A., Lindgren, L., and Ljungstrom, T.: An electronic method for intra-uterine measurements of pressure during labour, *J. Obstet. Gynaecol. Br. Emp., 60,* 321, 1953.

29. Smyth, C. N.: *Exposition Scientifique,* Sandoz, Basel, 1954.

30. Lindgren, L.: The lower parts of the uterus during the first stage of labour in occipito-anterior vertex presentation, *Acta Obstet. Gynecol. Scand., 34,* suppl. 2, 1–79, 1955.

31. Thomas, D. L., Hansen, S., Torbet, T. E., and Hay, D. M.: A comprehensive system for monitoring the foetal heartrate and uterine contraction, *Med. Biol. Eng., 11,* 703, 1973.

32. Frey, E.: Der Hysterotonograph: Eine neue Aparatur mit automatischer Schreibvorrichtung fur die klinische Wehenkontrolle, *Zentralbl. Gynaekol., 57,* 545, 1933.

33. Lorand, S.: Tokometer, Wehenmessinstrument geburtshilflichen Praxis, *Gynaecologia, 124,* 98, 1947.

34. Reynolds, S. R. M., Harris, J. S., and Kaiser, I. H.: Clinical measurements of uterine forces in pregnancy and labor, Charles C Thomas, Springfield, Ill., 1954.

35. Murphy, D. P.: *Uterine Contractility,* Lippincott, Philadelphia, 1946.

36. Smyth, C. N.: The guard-ring tocodynamometer, *J. Obstet. Gynaecol. Br. Emp., 64,* 59, 1957.

37. Duggan, T. C., Laughland, A. W., and MacGregor, J.: Measurement of intrauterine pressure, *Lancet, February 1966,* 488–489, 1966.

38. Jilek, J., Hon, E. H., and Yeh, S. H.: A technique for the measurement of uterine activity, *Med. Res. Eng., 11*(5), 4–5, 1972.

39. Seitchik, J.: Measure of contraction strength in labor: Area and amplitude, *Am. J. Obstet. Gynecol., 138,* 727–730, 1980.

40. Karni, Z., Polishuk, W. Z., and Segal, S.: Multi-strain measurements of uterine activity, *Proc. 24th Annu. Conf. Eng. Med. Biol.,* Las Vegas, Nev., Oct. 31, Nov. 4, 1971, p. 101.

41. Reynolds, S. R. M.: *Physiology of the Uterus,* 2d ed., Hafner, New York, 1965.

42. Siener, H. H.: Cervical dynamometry: A new method in obstetrical research, *Am. J. Obstet. Gynecol., 89,* 579, 1964.

43. Mizrahi, J.: Deformation analysis of the uterine muscle during labor and delivery, D.Sc. thesis, Technion-Israel Institute of Technology, Haifa, 1975.

44. Sadovsky, E., Polishuk, W. Z., Mahler, Y., and Malkin, A.: Correlation between electromagnetic recording and material assessment of fetal movement, *Lancet,* 1973, 1141–1143, 1973.

45. Sadovsky, E., and Yaffe, H.: Daily fetal movement recording and fetal prognosis, *Obstet. Gynecol., 41,* 845–850, 1973.

46. Pearn, J. H.: Fetal movements and Werdnig Hoffman disease, *J. Neurol. Sci. (Amst.), 18,* 373–379, 1973.

47. Mizrahi, J., Karni, Z., and Polishuk, W. Z.: Fetal movement recording in late pregnancy, *S. Afr. J. Sci., 73,* 385–386, 1977.

48. Reinold, E.: Fetal movements in early pregnancy, *Z. Geburtsh. Gynaekol., 174,* 220–225, 1971.

49. Reinold, E.: Observation of active fetal movement by ultrasonic process in the first months of pregnancy, *Pediatr. Paedol., 6,* 274–279, 1971.

50. Reinold, E.: A technique for the observation of fetal movements, *Wien. Klin. Wochenschr., 84,* 747–749, 1972.

51. Sellheim, H.: Die Geburt des Menschen, in *Deutsche Frauenheilkunde,* Bergmann, Wiesbaden, 1913.

52. Kamina, P., de Tourris, H., and Rideau, Y.: Anatomie fonctionnelle du Perinée obstetrical: Le bassin mou, *Rev. Fr. Gynecol. Obstet.*, *67*(1), 17–23, 1972.

53. Mizrahi, J., and Karni, Z.: A mechanical model for uterine muscle activity during labor and delivery, *Isr. J. Technol.*, *13*, 185–191, 1975.

54. Mizrahi, J., Karni, Z., and Polishuk, W. Z.: A kinematic analysis of uterine deformation during labor, *J. Franklin Inst.*, *306*(2), 119–132, 1978.

55. Bumm, E.: *Grundiss zum Studium der Geburtshilfe,* 13th ed., Bergmann, Wiesbaden, 1921.

56. Gillespie, E. C.: Principles of uterine growth in pregnancy, *Am. J. Obstet. Gynecol.*, *59*, 949–956, 1950.

57. Karni, Z., and Polishuk, W. Z.: The uterus as a thin shell structure, *Research Report of the Julius Silver Institute of Biomedical Engineering*, No. TSI 70-05, Technion-Israel Institute of Technology, Haifa, 1970.

58. Netter, F.: *Reproductive System,* Ciba Collection of Medical Illustrations, New York, 1965, vol. 11.

59. McConnell, A. J.: *Applications of the Absolute Differential Calculus,* Blackie, London, 1931.

60. Ericksen, J. L.: Tensor fields, in S. Flugge (ed.), *Encyclopedia of Physics,* Springer-Verlag, New York, 1960, vol. III, pt. 1, pp. 794–858.

61. Novozhilov, V. V.: *Theory of Elasticity* (in Russian), National Shipbuilding Industry, Leningrad, 1958.

62. Karni, Z.: The physical tensor and application, *Final Scientific Report,* USAF AFOSR AF-EOOAR 60-024, 1969.

63. Ollendorf, F.: *Die Welt der Vektoren,* Springer-Verlag, Vienna, 1950.

64. Biot, M. A.: *Mechanics of Incremental Deformations,* Wiley, New York, 1965.

65. Novozhilov, V. V.: *The Theory of Thin Shells,* Nordhoff, Groningen, 1959.

66. Murnaghan, F. D.: *Finite Deformation of an Elastic Solid,* Wiley, New York, 1951.

67. Flügge, W.: *Statik und Dynamik der Schalen,* Springer-Verlag, Berlin, 1934.

68. Goldenveizer, A. L.: *Theory of Elastic Thin Shells,* Gostekhizdat, Moscow, 1953.

69. Green, A. E., and Zerna, W.: *Theoretical Elasticity,* Oxford University Press, London, 1954.

70. Mizrahi, J., Karni, Z., and Polishuk, W. Z.: Strain uterography in labour, *Br. J. Obstet. Gynaecol.*, *84*(12), 930–936, 1977.

71. Heyns, O. S.: Abdominal decompression in the first stage of labor, *J. Obstet. Gynaecol. Br. Emp.*, *66*, 220–228, 1959.

72. Magnus, R.: Die Beziehung des Darmnervensystemes zur automatischen Darmbewegung, *Pfluegers Arch. Ges. Physiol.*, *102*, 349–467, 1904.

73. Kehrer, E.: Physiologische und Pharmakologische Untersuchungen an den überlebenden und lebenden inneren Genitialien, *Arch. Gynaekol.*, *81*, 161–210, 1907.

74. Jung, H.: Uber die Aktionspontentiale am schwageren und nichtswangeren Uterus, *Pfluegers Arch. Ges. Physiol.*, *262*, 13–24, 1955.

75. Jung, H.: Erregungsleitung und Erregungsbildung am Uterus, *Z. Geburtsh. Gynaekol.*, *147*, 51–71, 1956.

76. Jung, H.: Uber die Beziehung der Aktionspotentiale des Uterus zur mechanischen Leistung, *Pfluegers Arch. Ges. Physiol.*, *263*, 419–426, 1956.

77. Csapo, A., and Goodall, M.: Excitability, length-tension relation and kinetics of uterine muscle contraction in relation to hormonal status, *J. Physiol.*, *126*, 384–395, 1954.

78. Schofield, B. M., and Wood, C.: Length-tension relation in rabbit and human myometrium, *J. Physiol.*, *175*, 125–133, 1964.

79. Wood, C.: Physiology of uterine contractions, *J. Obstet. Gynaecol. Br. Commonw.*, *71*, 360–373, 1964.

80. Conrad, J. T., and Kuhn, W. K.: The active length-tension relationship in human uterine muscle, *Am. J. Obstet. Gynecol.*, *97*, 154–160, 1967.

81. Katz, B.: The relation between force and speed in muscular contraction, *J. Physiol.*, *96*, 45–64, 1939.

82. Aberg, A. K., and Axelsson, J.: Some mechanical aspects of an intestinal smooth muscle, *Acta Physiol. Scand.*, *64*, 15–27, 1965.

83. Stephans, N. L., Kroeger, E., and Mehta, J. A.: Force-velocity characteristics of respiratory airway smooth muscle, *J. Appl. Physiol.*, *26*, 685–692, 1969.

84. Gordon, A. R., and Siegman, M. J.: Mechanical properties of smooth muscle, *J. Physiol., 221,* 1243–1254, 1971.

85. Hill, A. V.: Heat of shortening and dynamic constants of muscle, *Proc. R. Soc. London Ser. B., 126,* 136, 1938.

86. Barclay, M. L., Dettart, W., and Mercer, J. E.: Wave-form analysis of intrauterine pressure curves with methods and models developed in cardiac research, *Am. J. Obstet. Gynecol., 128*(3), 242–250, 1977.

87. Mizrahi, J., Karni, Z., and Polishuk, W. Z.: Isotropy and anisotropy of uterine muscle during labor contractions, *J. Biomech., 13,* 211–218, 1980.

88. Pearsall, G. W., and Roberts, V. L.: Passive mechanical properties of uterine muscle (myometrium) tested in vitro, *J. Biomech., 11,* 167–176, 1978.

89. Mizrahi, J., and Karni, Z.: A constitutive equation for isotropic smooth muscle, *Isr. J. Technol., 19,* 143–146, 1981.

90. Buchtal, F., and Kaiser, E.: The rheology of the cross striated fibre, *Dan. Biol. Med., 21*(7), 1–318, 1951.

91. Eringen, A. C.: *Nonlinear Theory of Continuous Media,* McGraw-Hill, New York, 1962, p. 135.

92. Eringen, A. C.: *Mechanics of Continua,* Wiley, New York, 1967, p. 232.

93. Reiner, M.: A mathematical theory of dilatancy, *Am. J. Math., 67,* 350–362, 1945.

94. Reiner, M.: Elasticity beyond the elastic limit, *Am. J. Math., 70,* 433–466, 1948.

95. Wood, C.: The expansile behavior of the human uterus, *J. Obstet. Gynaecol. Br. Commonw., 71,* 615–620, 1964.

96. Fung, Y. C.: Elasticity of soft tissues in simple elongation, *Am. J. Physiol., 213,* 1532–1544, 1967.

97. Fung, Y. C.: Mathematical representation of the mechanical properties of the heart muscle, *J. Biomech., 3,* 381–404, 1970.

98. Wilson, J. F., Tsui, Y., and Roberts, V. L.: Plane penetration of uterine muscle by intrauterine shields, *J. Bioeng., 2,* 139–157, 1978.

99. Bakke, T.: A new mechanical instrument for the measurement of fibro-elasticity, with special reference to its use in the assessment of the consistency of the uterine cervix, *Acta Obstet. Gynacol. Scand., 52,* 277–287, 1973.

100. Bryant, W. M., Greenwell, J. E., and Weeks, P. M.: Alterations in the collagen organisation during dilatation of the cervix uteri, *Surg. Gynecol. Obstet., 126,* 27–39, 1968.

101. Danforth, D. N., Buckingham, J. C., and Roddick, J. W.: Connective tissue changes incident to cervical effacement, *Am. J. Obstet. Gynecol., 80,* 939–945, 1960.

102. Harkness, M. L. R., and Harkness, R. D.: Changes in the physical properties of the uterine cervix of a rat during pregnancy, *J. Physiol., 148,* 524–547, 1959.

103. Woraschk, H. J., Ehinger, B., and Ropke, F.: Dynamics of the dilatation of os uteri during birth, *Zentralbl. Gynaekol., 100*(18), 1173–1177, 1978.

104. Lindgren, L.: The physiology and pathology of the cervix during labor, *Acta Obstet. Gynecol. Scand., 47,* suppl. 1, 37–56, 1968.

105. Strada-Sáenz, G., Cobo, E., Gómez Rogers, C., and Sala, N.: Trabajo uterino en parto espontáneo, *Third Uruguayan Cong. Obstet. Gynecol., 3,* 392, 1960.

106. Sciapades, E.: Die normale Beckenbodenstrukhur und ihre Wiederherstellung nach Schädigungen, *Arch. Gynaekol., 175,* 54, 1944, cited by L. Lindgren, The lower parts of the uterus during the first stage of labor in occipito-anterior vertex presentation, *Acta Obstet. Gynecol. Scand., 34,* suppl. 2, 1–79, 1955.

107. Wolf, W. (1946): Cited by L. Lindgren, the lower parts of the uterus during the first stage of labour in occipito-anterior vertex presentation, *Acta Obstet. Gynecol. Scand., 34,* suppl. 2, 1–79, 1955.

108. Wahl, F. A.: Kritik an der derzeitigen Auffassung über die Wehenarbeit und ihre Messung, *Arch. Gynaekol., 157,* 1, 1934.

109. Moloy, H. C.: In discussion of L. A. Calkins, Second stage of labor—The descent phase, *Am. J. Obstet. Gynecol., 48,* 798, 1944.

110. Rydberg, E.: *The Mechanism of Labor,* Charles C Thomas, Springfield, Ill., 1954.

111. Csapo, A.: The theoretic, diagnostic and prognostic value of intrauterine pressure, *Bibl. Gynecol., 42,* 93–124, 1966.

112. Lindgren, L.: Der Biomechanismmus der Zervixdilation wahrend der Geburt, *Gynaekol. Runds. (Basel), 11,* 237–276, 1971.

113. Lindgren, L.: The causes of fetal head moulding in labor, *Acta Obstet. Gynecol. Scand., 39,* 46–62, 1960.

114. Schwarcz, R. L., Strada-Saenz, G., Althabe, O., Fernandez-Funes, J., and Caldeyro-Barcia, R.: Pressure exerted by uterine contractions on the head of the human foetus during labor, in *Perinatal Factors Affecting Human Development,* Scientific Pub. No. 185, Pan American Sanitary Bureau, Washington, 1969, p. 115.

115. Holmlund, D., and Lindgren, L.: Friction between the fetal head and uterine wall during normal labor and lower uterine spasm, *Am. J. Obstet. Gynecol., 103,* 939–941,1969.

116. Rice, D. A., Yang, T. Y., and Stanley, P. E.: A simple model of the human cervix during the first stage of labor, *J. Biomech., 9,* 153–163, 1976.

117. Rice, D. A.: Mechanism and measurement of cervical dilatation, Ph.D. thesis, Purdue University, W. Lafayette, Ind., 1974.

118. Romero-Salinas, G., Pantle, J., Aramburú, G., Figueroa, J. G., Bieniarz, J., and Caldeyro-Barcia, R.: Registro continuo de dilatacion cervical en el parto humano con método electronico directo, *Fourth Uruguayan Cong. Obstet. Gynecol., 2,* 718, 1964.

119. Egan, J. M., and Richardson, J. D.: Biomechanical analysis of the deformation of the lower uterus during labour, in R. Huiskes, D. van Campen, and J. DeWijn (eds.), *Biomechanics: Principles and Applications,* Martinus Nijhoff, The Hague, 1982, pp. 403–409.

120. Richardson, J. D., Sutherland, I. A., Allen, D. W., and Dore, F.: The development of an instrument for monitoring dilatation of the cervix during labour, *Biomed. Eng., 2,* 311–313, 1976.

CHAPTER 11

Skin Mechanics

Y. Lanir
Department of Biomedical Engineering
Technion-Israel Institute of Technology
Haifa, Israel

INTRODUCTION

The skin forms a barrier between the body and the external environment. Its functions [1, 2] are to contain and support internal organs, to protect them from unfavorable external stimuli (chemical, thermal, mechanical, biological, and radiative), to participate in maintaining a proper internal milieu, and to serve as a sensory interface with the external world.

Mechanically, the most important functions of the skin are to support internal organs and protect them (from abrasions, blunt impact, cutting, and penetration), while at the same time allowing considerable mobility. The skin's ability to perform this variety of functions results from a unique combination of mechanical properties which are closely related to its microstructure. Both properties and structure change with body site, age, sex, weight, exposure, pregnancy, and disease, and under the influence of drugs and chemicals.

Skin mechanics, apart from being an interesting subject in itself, has considerable clinical impact, especially in plastic surgery. The present chapter will be devoted to the mechanical and structural characteristics of the skin and their significance in plastic surgery.

11.1 CONSISTENCY AND STRUCTURE OF THE SKIN

The skin accounts for about 16 percent of the body weight in human adults. Its surface area is 1.5 to 2.0 m^2 in adults, and it varies in thickness from 0.2 mm (eyelid) to 6.0 mm (sole of foot) [3]. The skin consists of an outer layer, the epidermis, and the dermis. The latter contains collagen, elastin, reticulin, fibrocytes, blood and lymph vessels, nerve endings, hair and hair follicles, and glands (sebaceous, sweat, and apocrine) together with their associated ducts. All are embedded in what is called the *ground substance*—a gelatinous matrix consisting of water, mucopolysaccharides (notably hyaluronic acid and chondroitin sulfate), proteins (mostly soluble collagen), enzymes, and electrolytes.

The mechanical and barrier functions of the epidermis stem primarily from the stratum corneum—an outer layer of keratinized cells. It is from ~10 to 200 μm thick (sole of foot) and is organized as a multicellular membrane in which the cells (~0.8 μm thick) are stacked in columns and interdigitate with cells of adjacent columns [4].

The predominant components in the dermis are the ground substance and collagen fibers. The collagen accounts for 60 to 80 percent of the dry weight of the skin depending on age, sex, and site [5, 12]. The collagen (primarily of type I, with lesser amount of type III) appears as a three-dimensional, apparently disordered, network of wavy or coiled fibers. They are structured in several layers of primarily planar networks [6], with some fibers running between them [7]. Scanning electron microscopic observation [8] suggests that the collagen fibers near the epidermis (papillary layer) are finer and randomly oriented, while at the midzone they are coarser and densely packed, and may show preferred orientation. In the deep zone they are coarse again but are loosely packed.

The ground substance matrix accounts for 70 to 90 percent of the skin's volume [9]. Although it was traditionally considered to be amorphous, there is evidence [10, 11] which suggests that some components of the ground substance are incorporated in the structure of the collagen fiber.

Elastin accounts for about 4 percent of the dry weight [12, 13]. It appears as fine fibers which intertwine around the thicker collagen fibers in the deep layers of the dermis, but are rather straight close to the epidermis [6, 14–16]. Dick [16] observed that the elastin fibers in the main dermis layer are coarser than those close to the epidermis.

Reticulin fibers account for 0.4 percent of the dry weight of the skin [17]. They are located around blood vessels and hair, close to the epidermis [9]. Their chemical composition is similar to that of collagen [18]. Because of their minute quantity, they have little effect on the skin's overall response.

In response to uniaxial stretch the skin contracts in the lateral direction and the collagen fibers gradually align themselves in the direction of stretch. Their packing becomes denser and they subsequently stretch [6, 19, 20]. The fibers' alignment increases the strength and stiffness in the direction of stretch—an important merit in a tissue which may be stretched in various directions.

Stretch also affects the surface pattern of the skin. When skin is uniaxially stretched, the surface grooves align in the direction of stretch while the surface roughness in this direction decreases [21].

11.2 THE MECHANICAL ROLE OF THE SKIN COMPONENTS

Collagen fibers are the major mechanical elements in the skin. When these fibers are stretched, their effects predominate over those of all the other components. They are strong (tensile strength of 1.5 to 3.5 × 10^2 MPa) and stiff (Young's modulus in the linear region is approximately 1 GPa), and they stretch reversibly up to 2 to 4 percent. In the native state they are viscoelastic [22, 71, 113]. Thermodynamic studies have revealed that collagen behaves like a crystalline rather than a rubberlike material [23].

Elastin is considerably less stiff than collagen but can be reversibly stretched to more than 100 percent [24]. Its elastic properties are related to configurational entropic changes as in rubber and rubberlike materials [25]. On the basis of mechanoenzymatic studies [26], mechanohistological observations [16, 27, 28], and comparative mechanical data [15], it is apparent that the elastin fibers are the first to be stretched when the tissue is strained. Their effect on the response of the whole skin is thus significant at low levels of strain when the collagen fibers are still crimped. This notion is further strengthened by thermomechanical data [29] which show that at low strain levels skin has a negative thermal expansion coefficient (as in elastin) while at higher strain levels it is positive (as in collagen).

The role of the ground substance matrix in the response of the skin to tension is not yet clear. Studies [155] on the effects of temperature and humidity on the spectrum of relaxation times in very low strain levels (0.6 percent) show similar responses of skin to that of gels consisting of hyaluronic acid and water. This suggests that the ground substance is responsible for the viscoelastic behavior in this low strain range. It was also suggested that the fluidlike ground substance exudates from the interfiber space while the fibers become reoriented and densely packed upon stretch. This "flow" of ground substance may be associated with the viscous part of the skin's behavior, primarily in the "toe" region (the nonlinear region of the stress-strain curve), in which the collagen fibers are not yet fully aligned and stretched.

There exists, however, other evidence which suggests a lesser significance of the matrix. Study of the in vitro response of rat skin in which the hyaluronic acid (regarded as the predominant mechanical component in the ground substance) was differentially digested [30] revealed little effect on the rate of creep as compared with untreated controls. Similarly, Vlasblom [31] found no effect of the same treatment (hyaluronidase) on the rate of creep in torsional tests on human skin in vivo. Daly [32] found that in the initial reorientation phase (in which the migration of collagen fibers is thought to squeeze the fluid away), the response of the skin is approximately elastic. The viscous effect is highest at the final linear region when the collagen fibers are closely packed and straight.

Compressive tests of the skin (indentometry and skin-fold compressibility) show significant effects of the ground substance. The speed of compression [33–35], the amount of expressible fluid [34], and the skin's permeability [36] all increase upon application of hyaluronidase. These effects are thought to be related to the water affinity of the matrix mucopolysaccharides.

The epidermis contributes little to the skin's resistance to stretch [37, 38], but its contribution to the frictional resistance of the skin is predominant [39–41]. Its effect on the compressive indentation response of the skin is not clear [35, 42]. More detailed review of the role of the epidermis is given in Sec. 11.5.8.

11.3 IN VIVO STRESSES AND STRAINS

One of the most important features of the skin is its state of tension. Skin, in vivo, is normally under tension. This tension (expressed as force per unit length) varies in the range of 0 to 20 N/m [1, 27], depending on site, direction, and body posture. Upon excision, the skin will retract by 5 to 30 percent, again depending on site, direction, and posture.

Directional effects in the skin tension have since long been reported [43] and investigated in detail [20, 44]. It was found that circular punctures produce elliptical holes (splits) in the skin. The trajectories of the splits' major axes (cleavage lines) are called *Langer's lines* (Fig. 11.1). They are of considerable importance in plastic surgery.

Langer's cleavage lines are in fact intimately related to the visible crease and wrinkle lines of the skin, except on the palms, soles, and some areas on the limbs [44]. Flint [45] concluded that some earlier contradictory claims (stating that Langer's lines do not follow the same pattern as the wrinkles) were based on grossly inaccurate reproduction of Langer's original drawings and their misinterpretation.

Data on directional variability of in vivo tension and deformation [46–48] show that the highest stresses and strains are in the direction of Langer's lines. In addition, retraction of linear wounds is higher when the wounds are cut across Langer's lines rather than along them [44, 48, 49].

The skin's mechanical anisotropy follows the same directional pattern as Langer's lines: Its extensibility is lowest in the direction of the lines, both in vivo [50] and in vitro [20, 49, 51]. The skin's stiffness was shown to be highest in the direction of Langer's lines [51], but detailed observations show this to be true only at the low strain levels. At higher strains the slope in the linear region of the stress-strain curve is similar in all directions [20, 47–49].

11.3.1 Structural Aspects of Langer's Lines

Langer noticed that the cleavage lines correspond to the direction of preferred disposition of collagen fibers. Later, from histological observation [44, 47], it was concluded that both the collagen and elastin fibers are preferentially oriented in the lines' direction. Similar conclusions were drawn from scanning electron microscopic studies [7, 52].

Apart from the preferential disposition along Langer's lines, it was also found that the elastin [45] and the collagen [47] fibers along these lines are more stretched than those across the lines.

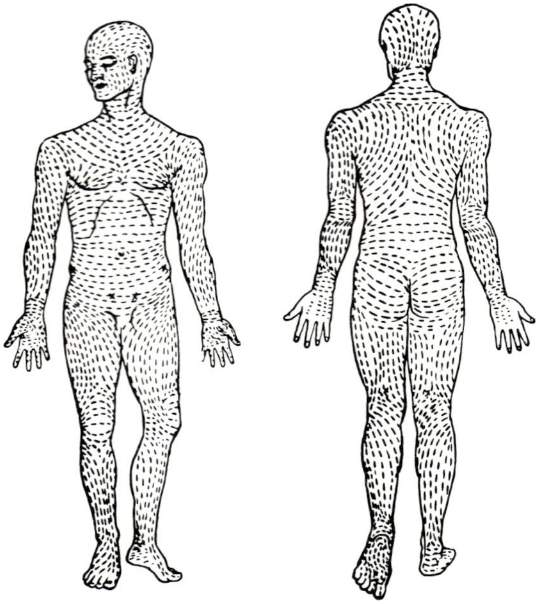

FIG. 11.1 Langer's cleavage lines. (*From Cox [44], with permission.*)

Consequently, the former are under higher tension and higher stretch, thus giving rise to a lower extensibility and to a higher initial stiffness in this direction—features which indeed characterize the skin in vivo.

It was suggested that the elastic fibers are the ones mainly responsible for the in vivo tension field. This hypothesis is supported by histological observation [16, 27, 45].

A recent study of the directional variability in the epidermis surface pattern [21] indicates that the major grooves (highest roughness) are along Langer's lines.

11.4 METHODS IN SKIN MECHANICS RESEARCH

Several engineering methods have been applied in the investigation of the mechanical properties of the skin, both in vivo and in vitro. An ideal procedure should consist of in vivo multiaxial tests with uniform strain field, such that the skin's comprehensive properties (the constitutive equations) can be readily associated with the results. This ideal has not yet been achieved. Under in vivo tests it is exceedingly difficult to obtain a uniform strain field throughout the specimen or to control the conditions along the boundaries. In addition, the resting tension and deformation [46, 53, 54], as well as the skin's thickness [55, 56], must be measured. Under in vitro conditions these experimental problems are easier to deal with. The disadvantage of in vitro tests is that the effects of blood and blood pressure, lymphatic drainage, in vivo metabolism, and nervous and hormonal controls are not present, and their significance cannot be evaluated. The most visible result of the altered environment in vitro is that the skin specimen swells [29].

Another problem common to many in vivo and in vitro tests is the attachment via grips. They may introduce local stress concentration or induce nonuniformity in the strain field. If slippage occurs, then the strain measurement based on the grips' displacement is inaccurate. The effect of the grips is particularly significant under in vivo tests in which stretching is performed via tabs which are glued to the epidermis [57]. Some of the preceding problems can be overcome by direct measurement of the tissue strain at sites sufficiently remote from the grips [29].

Although no overall superior experimental method is available, some techniques have distinguishing features which make them better suited for specific studies. Uniaxial in vivo tests [14, 57, 109, 114] are well suited for studies of directional effects, but the tissue's strain is not uniform. Improved strain uniformity can be achieved under either in vitro uniaxial tests [47, 51, 75, 98] or strip biaxial tests both in vivo [54] and in vitro [123]. In vivo suction tests with racetrack-shaped cups [53] and in vitro biaxial tests [29] are also associated with uniform strain fields and are well suited for studies of the anisotropic constitutive behavior of the skin. In vitro torsional tests [31, 126, 129, 130] and suction tests with circular cups, both in vivo [31] and in vitro [27], are easy to perform, but their interpretation in terms of the skin's properties is very difficult due to the highly nonuniform strain field. Skin-fold compressibility tests [153] and indentation tests [33, 39, 76, 100–102] are probably the easiest to perform in vivo. Although they cannot be used for investigation of the skin's constitutive behavior (owing to a highly nonuniform strain field), they have found extensive use in parametric studies of the compressive response of the skin, especially with regard to the effect of diseases and medicocosmetic agents.

Other details on experimental procedures can be found in previous reviews which have concentrated on in vivo techniques [58], on pharmacological aspects [59], and on hardware considerations [60].

Investigations of the epidermis and its stratum corneum were carried out with similar techniques to those used for the whole skin. Under in vivo conditions it is difficult to isolate the effects of the epidermis from those of the dermis. The stratum corneum can, however, be separated from the dermis and tested by itself. The separation can be done by a variety of methods which include heating to 60°C [61], suction blistering [62], chemical blistering with ammonia [63] or with cantharidin [64], and by enzymatic digestion [65]. It has been suggested [66], however, that some epidermal damage may be caused by these treatments. This topic requires further clarification.

11.5 GENERAL CHARACTERISTICS OF THE SKIN'S MECHANICAL RESPONSE

The functionally important mechanical properties of the skin are extensibility, resistance to friction, and response to lateral compressive loading. The first is associated mainly with the dermis and the second mainly with the epidermis, and the last is due to the combined effects of both.

Skin properties are highly variable, depending on species, age, exposure, hydration, obesity, disease, and biological difference between individuals. In the same individual, the skin's properties vary with site and orientation and may be altered by irradiation, drugs, and chemicals. Space limitations prevent the inclusion of an appropriate account of all the above effects in this chapter. Attention will be concentrated primarily on the general mechanical characteristics of the skin.

In the following, the important mechanical features of the dermis and epidermis will be outlined. A more quantitative outlook will be provided in the review of skin models.

11.5.1 In Vivo versus in Vitro Response

It is difficult to compare in vivo and in vitro responses of the same specimen both because of the poor control of boundary conditions in vivo and because of the physiological differences between in vivo and in vitro conditions. Reported results of such comparisons do not always agree: Cook et al. [53] found that in vivo tension is between approximately 10 percent (at low strain levels) and 20 percent (at high strain levels) higher than in vitro tension for the same strain in rat skin. Kenedi et al. [67] observed slightly higher loads in detached human skin specimens than in in situ specimens, but the differences were within their experimental error.

Vogel and Denkel [57] and Vogel [156] show that if the same gripping method is used both in vivo and in vitro, then the stress-strain relations are similar, with a slight difference in the strain-dependent anisotropic response.

11.5.2 Equilibrium Configuration and Preconditioning

Because of the skin's considerable slackness at low strain levels, it is difficult to establish a stress-free configuration for skin. If the specimen floats on a physiological solution, then the stress-free configuration can be established [29]. It depends, however, on the specimen's strain history. It has been recognized for some time that skin behaves in an approximately repetitive mode only after several test cycles. The stress in the second cycle is lower than in the first one, then it drops somewhat less in the third cycle, and so on until eventually a stable response is obtained. This is called *preconditioning*. It is accompanied by a parallel increase in the gage length in uniaxial tests, and by configurational changes in biaxial tests. In the case of biaxial stretching it was found [29] that each biaxial test requires its own preconditioning and is associated with its own equilibrium configuration. The effects of preconditioning are reversible if no damage has been caused. Recovery of both initial configuration and initial response will occur provided the specimen has not contracted in any planar direction [29].

The time constant for recovery from preconditioning is measured in hours. This is orders of magnitude higher than the characteristic viscoelastic relaxation time. It is thus possible to separate the two effects. This should, however, be done with caution, since skin may exhibit a wide spectrum of relaxation times.

Preconditioning affects the in vivo response as well. This was observed in uniaxial [68] and torsional tests [31, 69]. Preconditioning exists in other tissues [70], as well as in isolated collagen fibers [71].

From the modeling point of view, the effect of preconditioning must be included in any general constitutive formulation. This has not been done so far. Moreover, there seems to be no relevant data on this topic.

11.5.3 Homogeneity

Skin is obviously not a homogeneous material because of the fibrous, cellular, vascular, granular, and amorphous components of which it consists.

In practical problems we are usually interested in dimensions of specimens which are orders of magnitude larger than any of the skin's components. In many clinical problems, the response of the entire skin thickness is of importance. In these cases we may consider the skin as statistically homogeneous.

11.5.4 Compressibility

The question of whether skin can be considered incompressible is important since the constitutive formulation of incompressible materials is significantly simpler than in the general case. The evidence is, however, conflicting. Upon direct compressive loading [34], the skin will lose more than half its water at ~0.3 MPa. If however, hydrostatic pressure is used [72], then the compressibility (expressed by the bulk modulus) is 0.33 GPa, close to that of water (~0.25 GPa). Vossoughi and Vaishnav [73] point out that it is the relative compressibility (ratio of bulk to shear modulus) which determines whether a material can be considered incompressible. They showed this ratio in the skin to be of order 10^3, thus confirming its incompressibility.

Stretching tests in vitro show different results. Several investigators observed fluid secretion from specimens stretched in air [14, 74, 75].

The relevance of these in vitro observations to the in vivo state is not clear. Direct evidence of in vivo compressibility can be found in impedance tests [76]. The ratio of the bulk to the shear modulus was found to be of order 10^6. Although this points to in vivo incompressibility, more evidence is needed before conclusions can be drawn.

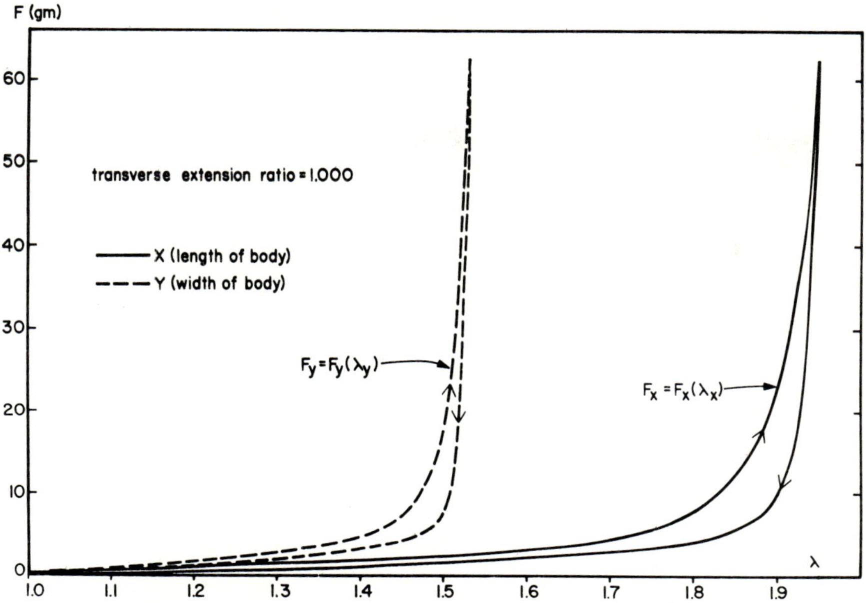

FIG. 11.2 Load-strain relations of a square (3.5 × 3.5 cm) specimen of rabbit abdominal skin. The tissue is stretched in one direction while the transverse dimension is kept constant. Solid line: stretch along the body length. Dashed line: stretch along the body width. (*From Lanir and Fung, [29], with permission.*)

11.5.5 Stress-Strain Relations

The stress-strain relations of the skin are both nonlinear and anisotropic (Fig. 11.2). Initially the skin is slack, and small stresses produce large strains. With further stretch the stiffness increases up to a maximum constant level in the linear region. Further stress will result in failure. The skin's anisotropy is evident from the different responses in different directions, both in vitro and in vivo, and under uniaxial as well as biaxial stretching.

Under biaxial homogeneous stretch along and across Langer's lines, both in vitro [29] and in vivo [46], no shear strains were observed. This points to some symmetry in the skin's response, namely, orthotropy or transverse isotropy.

As mentioned earlier, the nonlinearity of the skin's response can be readily accounted for by the rotation and gradual straightening of the collagen fibers. In the high-strain region they are all straight, and the stress-strain curve becomes linear. Under uniaxial stretch it is often observed [20] that the linear slope is equal in all directions. This is probably associated with a complete alignment of the fibers in the direction of stretch.

11.5.6 Viscoelasticity

Skin is a viscoelastic tissue. Its stress-strain relations are rate-dependent and exhibit considerable hysteresis. Skin tissue shows stress relaxation under constant strain and creep under constant stress. The skin's viscoelastic nature has been demonstrated in uniaxial, biaxial, torsional, and various dynamic compressive tests (penetration, elevation, skin-fold compression, ballistometry, and acoustic impedance).

The viscoelastic response of the skin is not linear: the modes of relaxation and creep depend on the corresponding levels of strain or stress in both uniaxial [15, 77, 78] and biaxial testing

(Fig. 11.3*a*). At low strain levels the skin may behave in an approximately elastic manner both in vitro [26] and in vivo [54]. Under in vitro creep testing, it was shown that following a transient time of several minutes the extension rate becomes constant and is proportional to the stress [79].

In vitro biaxial stress-relaxation tests reveal [29, 80] that two principal stress components relax in different modes (Fig. 11.3*b*). The skin is thus anisotropically viscoelastic. The above tests also reveal that under uniaxial stress-relaxation tests, the lateral dimension continuously contracts following the application of stress.

11.5.7 Damage and Strength

Although total rupture of healthy skin is not common, increased tension may cause various degrees of damage. Ruptures due to penetration of sharp objects occur in animals [2] and in humans during accidents. The skin's penetration failure under moderately sharp edges was found to be of a tensile (rather than shear) character and independent of edge shape and rate of loading [81].

Partial damage to the skin is more common, as demonstrated by the striae. *Striae* are stretch lines that occur under moderate tensions (not high enough to cause arrest of blood supply—blanching) which persist for a long time. They are common in pregnancy and may also occur over the muscles in excessive body-building exercises and in rapidly enlarging breasts [82]. At higher tension levels, blanching of the skin and subsequent necrosis will occur. Some skin damage in the form of permanent stretch may occur under tensions lower than those which cause striae.

The tensile strength of skin has typical values in the range of 2.5 to 16 MPa [83]. It varies with site [84], is higher in the main fiber direction [85–87], is higher in males than in females [85], and is believed to increase with age [83, 85, 88, 89], but contrary trends have also been reported [90]. Skin may also break under considerably lower stresses (as low as one-fifth of the tensile strength) under long creep tests [79, 87, 89].

11.5.8 Mechanical Properties of the Epidermis

The epidermis is exposed to a wide range of temperature and humidity, whose effects on its mechanical response are of prime interest.

Epidermal Stress-Strain Relations. The stress-strain relations are in fact significantly influenced by both humidity and temperature [91]. At low humidities (25 to 30% relative humidity), the curve is linear up to rupture, with a slope (Young's modulus) of order 1 to 10 GPa. But at 100% relative humidity the curve has between two [92] and three [93] linear regions with an initial slope which is 1000-fold smaller than in the dry state [92, 94]. A similar drop of the slope is caused by higher temperatures [92, 95]. The stress-strain data have a considerable scatter which can be reduced by a suitable moistening regime [94]. Under in vivo conditions the epidermis is hydrated only on its dermal side. Its properties are then different than in the fully hydrated state [96].

Epidermal Strength. The tensile strength was found to decrease, and the breaking strain to increase, with increasing humidity and temperature [91–93]. The strength is of order 10^2 MPa at low (26%) humidities and 10 MPa in the wet state. There is a similar drop with temperature in the 25 to 60°C range [92]. The breaking strain increases with humidity by close to two orders of magnitude, from approximately 2 percent at 26% relative humidity to 150 to 200 percent if wet [92].

The overall contribution of the thin stratum corneum is small compared with that of the thick dermis [37, 38]. In rupture tests the stratum corneum was observed to tear before the dermis [83].

Epidermal Viscoelasticity. The stratum corneum is viscoelastic. Its stress-strain relations are rate-dependent, becoming stiffer with increasing strain rate [4]. In dynamic in vitro tests the

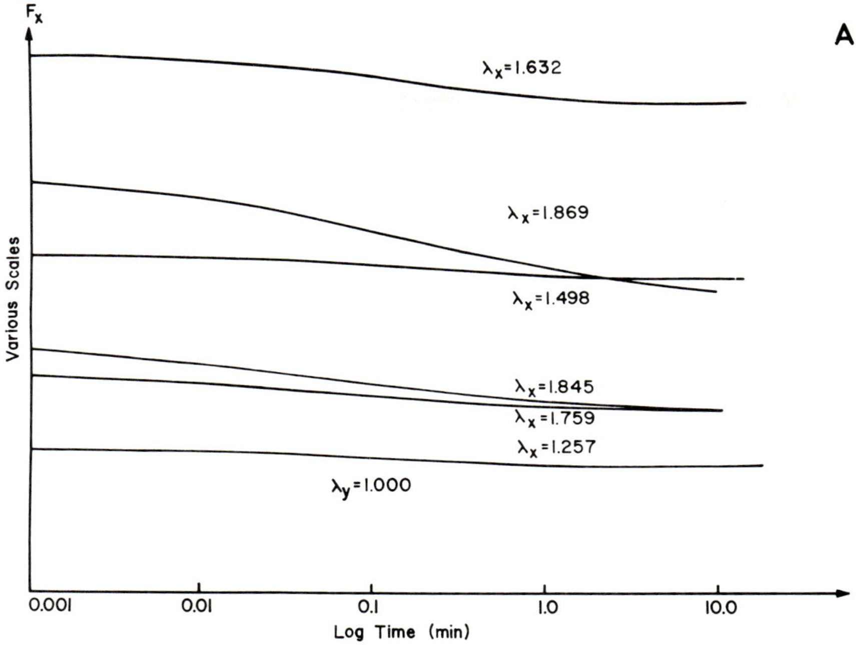

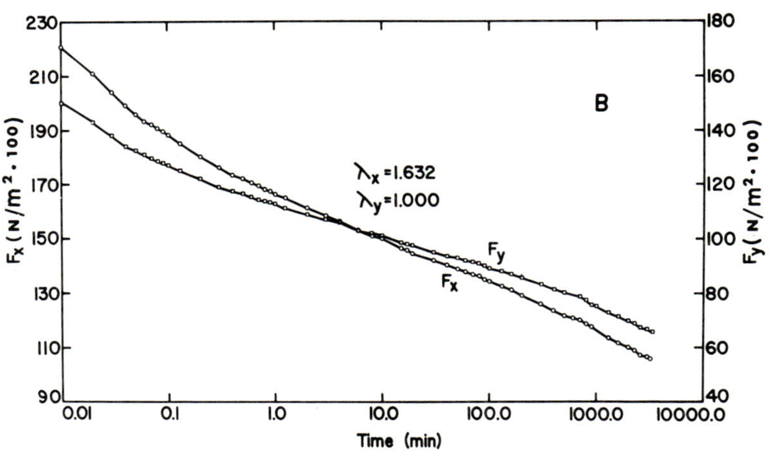

FIG. 11.3 (*A*) Mode of stress-relaxation in a square (3.5 × 3.5 cm) specimen of rabbit abdominal skin following a step stretch in the *x* (body length) direction, at various stretch ratios (λ_x). The transverse dimension is kept constant (λ_y = 1.000). (*B*) A comparison between the simultaneous relaxations of two principal stress components following a step stretch. The stretch ratios are indicated. (*From Lanir and Fung [29], with permission.*)

isolated stratum corneum shows storage and loss properties. Both depend on strain, frequency, humidity, and temperature. The loss (E'') and damping (tan δ) moduli were found to be highest in the 50 to 80% relative humidity range in human specimens [97], while the storage modulus (E') drops continuously with increasing humidity. Temperature has a dramatic effect on the storage and damping moduli in the physiological range but much less effect on the loss modulus. There are damping peaks at about 10 and 40°C. Frequency has only a small effect. Both E' and E'' increase slightly with frequency, while tan δ is relatively unaffected in the 3.5 to 110-Hz range [98]. The dynamic modulus E' of human stratum corneum under 72% relative humidity, at 30°C, and at 200 Hz was found [157] to increase with strain initially, peak at a strain level of ~0.20, and drop gradually thereafter. E' is unaffected by strain rate and by preconditioning cycles and remains constant throughout the stress-relaxation process following a step change in the static strain.

Frictional Properties. The frictional properties of the epidermis are of considerable interest in many fields (e.g., the textile and shaving industries). Unfortunately, the available results are few and at times conflicting. Well-controlled in vivo tests [41] revealed linear relations between frictional force and lateral load (Amonton's law) and no difference between dynamic and static coefficients of friction (0.5 with polyethylene). In vitro tests, however, showed a lower dynamic coefficient and a reduction of the coefficient of friction with load for several textile materials [99].

The frictional properties of the epidermis depend on several environmental factors. The coefficient was shown to increase by sweat or moderate moistening and by the removal of grease, but was reduced by wetting, by dry talc powder, and by addition of oil [41]. The effect of wetting in vivo varies with time [39].

11.5.9 The Skin's Response to Lateral Loading

The response of the skin to lateral compressive loading is of considerable functional and clinical importance. Studies of this subject are based primarily on indentation and skin-fold compressibility tests. Elevation tests have been used as well [121], but they are associated primarily with stretching deformation and consequently yield different results [100]. Most of the compressive studies on the skin were done in vivo. The main results are discussed in the following subsections.

Effects of Dermis and Epidermis on Compressive Response. The skin's response to compressive loading involves both the dermis and the epidermis. In some previous reports it was observed that certain cosmetic agents (including moisturizers) modified the tissue's response under specific procedures. It was thus claimed that these tests are associated primarily with the epidermis. Tests which involve the removal of the stratum corneum yielded conflicting results. In some [42] the denuded skin had a different response from the intact specimen, while in a recent study [35] no difference was observed. It can be seen that the roles of the dermis and epidermis under compressive loadings are not yet clear and this area of study requires further investigation.

Compressive Viscoelastic Effects. The compressive response of the skin is highly viscoelastic [33, 101, 102]. It is also nonlinear: the stiffness was found to increase with indentation [100]. Both the viscous and the nonlinear aspects of the skin response were found to diminish considerably by preload [103]. The immediate (elastic) and delayed (viscous) responses of the skin depend on age, sex, site, hydration, and obesity. A treatment with hyaluronidase was found to increase both the magnitude [35] and the speed [33] of indentation. The magnitude also increased after treatment with elastase but decreased after injection of saline. It did not change following treatment with collagenase [35]. These results indicate that the tighter elastin fibers and the ground substance may be the predominant components in the response to compressive loading and that collagen has little or no effect.

Compressive Dynamic Response. The dynamic response of the skin is related to its capacity to absorb impact energy in accidents. Most of the previous studies used the normal vibrational impedance characteristics of the skin as the key to its dynamic behavior. Theory and experiments show [76, 104] that shear waves predominate at frequencies up to 5 kHz while compressive waves are more important thereafter. Surface waves which can be observed on the skin are a combination of both.

The impedance characteristics were found to depend on frequency, humidity, probe area, and site on the body, as well as on the static pressure [104]. If the shear impedance is expressed as a complex number ($\mu = \mu_1 + i\mu_2 w$, where w is frequency and $i = \sqrt{-1}$), then typical values were $\mu_1 = 2.5 \times 10^3$ Pa, $\mu_2 = 15$ Pa · s [76]. In another study [158] the corresponding values at 600 to 1000 Hz were 1.5×10^3 Pa and 5.0 Pa · s. The impedance and phase angle were found [105] to increase linearly with the logarithm of frequency, but the precise relations depend on the preload. Increased humidity was found [159] to increase significantly the absorbed energy of human skin in vivo in the frequency range of 700 to 800 Hz.

In Vitro Compressive Response. A compressive study at physiological pressure ranges [106] revealed linear stress-strain relations up to a pressure of 7.8×10^3 Pa and an exponential relation (measured up to a pressure of 1.75×10^5 Pa). The behavior under in vitro cyclic compressive loading was found in the same study to be linearly viscoelastic at low pressures. At higher pressures the viscoelastic parameters decayed continuously and linearly with the logarithm of time (or cycle number).

11.5.10 Binding Forces in the Skin

The binding forces between different layers of the skin are of significance in its in vivo response, but little is known about this topic. The binding of dermis to subcutaneous tissues varies in different sites on the body (e.g., auditory canal versus penis). Cook et al. [53] found that the binding of dermis to subcutaneous tissues in the torso of the rat affects the results of suction tests by 10 to 20 percent. This was concluded from a comparison of in vivo and in vitro tests of the same specimen. Vlasblom [31] observed that a sliding load of 8×10^{-3} N was required to displace 1.0 cm^2 of human forearm skin by 1.0 mm. This introduced only a small (< 10 percent) error in his in vivo torsion tests.

The dermal-epidermal adhesion was found to be of a predominantly viscous nature [107]. The product of suction pressure and time to blister was found in this work to be approximately constant and equal to 90 ± 30% MPa · s in the lower abdominal human skin.

The binding forces between layers of the stratum corneum vary with site and sex. They are of order 2×10^4 Pa [108].

11.6 MODELS OF SKIN MECHANICS

Three major classes of models have been developed for the skin. In the continuum models, general material theories were specialized for the skin and used to simulate its multiaxial response. The phenomenological models consist of mathematical formulas describing the skin's behavior under specific deformational schemes. Some of them consist of mechanical elements such as springs and dashpots. In the structural models the skin's behavior is analyzed in terms of the combined effects of its components.

General multiaxial theories for viscoelastic behavior are available. They are, however, much too complex for practical use. In some cases, simpler approaches are adequate. One simplification widely used in skin research is based on the fact that under constant rate of stretch the stress-strain relations, in both the extension (ascending) and contraction (descending) phases, are unique for a preconditioned specimen. The skin can thus be regarded as pseudoelastic in these widely used tests.

The pseudoelastic stress-strain relations can be expressed as follows:

$$S^{ij} = \frac{\partial W}{\partial \varepsilon_{ij}} \qquad i, j = 1, 2, 3 \tag{11.1}$$

where $W(\varepsilon_{ij})$ is a function of the strain components and is known as the *strain energy function* (per unit original volume). S^{ij} and ε_{ij} are Kirchhoff's stress and Green's strain components, respectively. They are referred to the original dimensions of the specimen. W can also be expressed in terms of the three principal stretch ratios $-\lambda_i$.

TABLE 11.1 Pseudoelastic models for the skin under stretch

Source	Model*	Tests, results, and remarks
Alexander and Cook [46]	$N = \frac{C}{\lambda}\left[\lambda^2 - \frac{1}{(\lambda\lambda p)^2}\right] \exp K(I_1 - 3)^2$ where λp = transverse λ	Isotropic model; rectangular suction cup; in vivo human skin; C, K depend primarily on sun exposure Typical values: $C = 10^4$ Pa, $K = 100$
Allaire et al. [114]	$W = C_1(I_1 - 3) + C_2(I_2 - 3) + C_3(I_3 - 1)^2 - (C_1 + 2C_2)(I_3 - 1)$	Isotropic model; uniaxial, in vivo, human forearm skin $C_1 = 2.16\text{–}5.90 \times 10^4$ Pa $C_2 = (-2.46)\text{–}(-6.27) \times 10^4$ Pa $C_3 = (-0.88)\text{–}(1.65) \times 10^4$ Pa
Danielson [115]	$N^{\alpha\beta} = A^{\sigma\beta\lambda\mu}\gamma_{\lambda\mu} \exp (B^{\kappa\delta}\gamma_{\kappa\delta} + C\gamma^{\rho\phi}\gamma_{\rho\phi})$	Skin considered as membrane; $A^{\alpha\beta\lambda\mu}$, $B^{\kappa\delta}$ = tensors of elastic constants
Danielson and Natarajan [116]	Tension field theory for finite deformation of the skin	Consider skin as isotropic membrane where one principal stress is large and positive, and the others are negligible. Theory requires knowledge of uniaxial stress-strain relations; shows agreement with data of Z-plasty transformation [141]
Demiray [117]	$W = \frac{B}{2(I_1 - 3)}$	Assumes incompressibility and isotropy
Elden [118]	$\sigma = K\left(\frac{L}{L_0} - 1\right)^2$	Ring uniaxial, rat tail skin, in vitro; K depends on body weight and cross-sectional area Typical value of $K = 1.5$ MPa
Glaser et al. [119]	$0 \le \varepsilon \le 0.3$: $\sigma = A\varepsilon^n$ $0.3 \le \varepsilon$: Linear $\sigma(\varepsilon)$ relations	Uniaxial, in vitro, guinea pig $A = 138$ MPa; $n = 2.47$ Linear slope = 46.2 ± 10.3 MPa
Gou [120]	$W = C\{\exp [KI_1(I_1^2 - 3I_2)] - 1\} + W_1(I_3)$	Assumes isotropy
Kenedi et al. [67]	$\sigma = A\varepsilon^n$	Uniaxial, in vitro and in vivo human skin, stepwise stretch. A, n = function of age and direction; A scattered in a wide range while n = 6–8
Lanir [110, 111]	$N^{\alpha\beta} = \int_0^{\pi} U(\theta)\left(\frac{\partial\varepsilon}{\partial\varepsilon_{\alpha\beta}}\right) d\theta + N_p A^{\alpha\beta}$ $U(\theta) = \sum_{i=1}^{2} S_i \cdot F_i(\varepsilon) \cdot R_i(\theta)$ $\varepsilon = \varepsilon_{11} \cos^2\theta + 2\varepsilon_{12} \sin\theta \cos\theta + \varepsilon_{22} \sin^2\theta$	A structural biaxial model; considers the skin's response as due to elastin, collagen, and ground substance h_0 = initial thickness J = area ratio ε = fiber strain S = fiber density F_i = fiber stress-strain relations

		R_i = fibers' orientational distribution $N_p A^{\alpha\beta}$ = hydrostatic pressure term 1 = elastin; 2 = collagen
Peng et al. [54]	$W = E \sum_{i=1}^{3} \frac{\lambda_i^n - 1}{n}$	Strip biaxial, in vivo human skin, stepwise stretch $E = 3.9$ Pa; $n = 18.99$
Ridge and Wright [122]	$0 < \sigma < 14.7$ Pa: $\varepsilon = a + b \log \sigma$ $14.7 < \sigma < 98.1$ Pa: $\sigma = C + A\varepsilon^n$	Uniaxial, in vitro human skin; n, A = function of age, sex, and site; A depends on direction and cross-sectional area as well Typical values: $A = 2 \times 10^4$ Pa; $n = 3.3$
Snyder and Lee [123]	$W = \frac{k}{a^2} \exp[a(\beta - 1)] - \left(\frac{k}{a}\right)\beta$ $\beta = \frac{I_1 + (I_1^2 - 3I_2)^{1/2}}{I_2}$	Assumes incompressibility and isotropy; pure shear (strip biaxial) and uniaxial, in vitro frog skin; considerable variation of a, k among animals in uniaxial tests
Tong and Fung [124]	$W = \frac{1}{2}(\alpha_1\varepsilon_1^2 + \alpha_2\varepsilon_2^2 + \alpha_3\varepsilon_{12}^2 + 2\alpha_4\varepsilon_1\varepsilon_2) + \frac{1}{2}C \exp(a_1\varepsilon_1^2 + a_2\varepsilon_2^2 + a_3\varepsilon_{12}^2 + 2a_4\varepsilon_1\varepsilon_2 + \gamma_4\varepsilon_1^2\varepsilon_2 + \gamma_5\varepsilon_1\varepsilon_2^2)$	Assumes plane stress, orthotropy, and coincidence of X_1, X_2 with axes of material symmetry. In vitro biaxial stretching ($\varepsilon_{12} = 0$); rabbit abdominal skin [29]. Assumes $\alpha_1 = \alpha_2$. Found parameters to change considerably in a single specimen from one test to another
Veronda and Westmann [75]	$W = C_1\{\exp[\beta(I_1 - 3)] - 1\} + C_2(I_2 - 3) + g(I_3)$	Assumes isotropy; uniaxial, in vitro cat skin, stepwise loading; measured λ_1, λ_2, λ_3 $C_1 = 27$ Pa; $C_2 = 137$ Pa; $\beta = 5.03$
Vlasblom [31]	For $\varepsilon \to 0$: Hooke's law	Assumes linear elastic response under small deformations. Torsion, stepwise, and intermittent creep test in vivo, human arm skin, twist angle < 0.03 rad. Linear behavior with shear modulus $G = 1.9 \pm 0.2 \times 10^4$ Pa
Wijn [109, 125]	$\frac{d\sigma}{d\varepsilon} = E_0 + K\sigma$	A structural model for small deformations; considers the effect of elastin alone; torsional and uniaxial, in vivo human skin; small deformations. E_0 depends on age and direction of test Typical values: $E_0 = 2$ MPa; $K = 20$

* Parameters are defined as follows:

ε, λ = uniaxial strain and stretch ratio, respectively
σ, N = uniaxial stress and stress resultant, respectively
$N^{\alpha\beta}$, $\gamma_{\alpha\beta}$ ($\gamma^{\alpha\beta}$) = biaxial stress resultant and strain components, respectively
$\varepsilon_{\alpha\beta}$, ε_a ($\equiv \varepsilon_{\alpha\alpha}$) = biaxial strain components in cartesian coordinates
λ_1, λ_2, λ_3 = principal stretch ratios
I_1, I_2, I_3 = strain invariants
W = strain energy function

TABLE 11.2 Viscoelastic models for the skin under stretch

Source	Model*	Tests, results, and remarks
Barbenel et al. [126]	$\sigma(t) = \sigma(\infty) + \sigma(0) \int_{-\infty}^{\infty} H(\tau) \exp \frac{-t}{\tau} \, d \ln \tau$	Linear viscoelastic model; torsion creep, uniaxial relaxation and cyclic test; human skin in vivo. For torsion test under constant torque: $H(\tau) = 8 \times 10^{-2}$ mN · m $0 < \tau < 35$ s 0 $\tau > 35$ s
Galford and McElhaney [77]	$\frac{\varepsilon(t)}{\sigma_o} = C_1 + C_2 \ln t$	Uniaxial creep tests, monkey scalp, in vitro. Stress in range of 690 to 4140 Pa $C_1 = 6.0 \times 10^{-5}$ Pa^{-1}; $C_2 = 9.4 \times 10^{-7}$ Pa^{-1}
Jamison et al. [78]	See Fig. 11.4*a*	Uniaxial creep and constant rate tests, guinea pig skin, in vitro. E_2, η_2, η_2 depend on stress level and direction of stress
Kenedi et al. [127]	$\sigma(t) = \sigma(0) \left(a_0 + \sum_{i=1}^{n} a_i \exp \frac{-t}{\tau_i} \right)$	Linear viscoelastic model. Uniaxial relaxation, human skin, in vitro; n, a_i, and τ_i vary with age, sex, and site on body
Lanir [80]	$\sigma(t) = B + \frac{A}{(1 + t/\tau)^N}$	Biaxial relaxation tests, rabbit skin, in vitro [29]. B, A, N depend on stress and are different for different stress components
Lanir [111, 128]	$N^{\alpha\beta}(t) = \int_0^{\pi} U(\theta, t)(\partial\varepsilon/\partial\varepsilon_{\alpha\beta}) \, d\theta + N_p A^{\alpha\beta}$ $U(\theta, t) = \sum_{i=1}^{2} S_i \mathcal{F}_i[\varepsilon(\overset{t}{\underset{0}{\tau}}), t] R_i(\theta)$ $\varepsilon(t) = \varepsilon_{11}(t) \cos^2 \theta + 2\varepsilon_{12}(t) \cdot \sin \theta \cos \theta + \varepsilon_{22}(t) \sin^2 \theta$	A structural model which considers the contribution of elastin (elastic behavior), collagen, and ground substance. Symbols as in Table 11.1 $\mathcal{F}_i$ = rheological law for elastin ($i = 1$) and collagen ($i = 2$) fibers $\mathcal{F}_p A^{\alpha\beta}$ = time-dependent pressure term
Sanders [129]	$V(t) = V_e + V_v \left(1 - \exp \frac{-t}{\tau}\right) + At$	Plastoviscoelastic model. Torsion creep test, human skin in vivo, small twist angles V_e, V_v, V = elastic, viscous, and total deformation At = plastic (permanent) deformation
Vlasblom [31]	$J(t) = J(\infty) - [J(\infty) - 1] \exp \frac{-t}{\tau}$ $\tau \approx 30$ s	Linear viscoelastic model. Torsion creep tests, human skin in-vivo
Wijn et al. [130]	See Fig. 11.4*b* I.† $E_s = 1.5 \times 10^6$; $E_p = 4.0 \times 10^6$; $\eta_p = 2.0 \times 10^7$; $\eta_s = 3.0 \times 10^8$ II. $E_s = 3.8 \times 10^5$; $E_p = 4.0 \times 10^5$; $\eta_p = 2.0 \times 10^7$; $\eta_s = 3.0 \times 10^8$ III. $G_s = 2.8 \times 10^4$; $G_p = 1.0 \times 10^5$; $\gamma_p = 5.0 \times 10^5$; $\gamma_s = 9.0 \times 10^6$	Linear viscoelastic model. Uniaxial and torsion creep tests, human skin, in vivo I = uniaxial along Langer's lines II = uniaxial across Langer's lines III = torsion

In the case of isotropic materials W is solely a function of the three strain invariants I_1, I_2, and I_3. They are related to the strain and stretch ratio components as follows:

$$I_1 = 3 + 2\varepsilon_{rr} = \lambda_1^2 + \lambda_2^2 + \lambda_3^2 \tag{11.2}$$

$$I_2 = 3 + 4\varepsilon_{rr} + 2(\varepsilon_{rr}\varepsilon_{ss} - \varepsilon_{rs}\varepsilon_{sr})$$

$$= \lambda_1^2\lambda_2^2 + \lambda_1^2\lambda_3^2 + \lambda_2^2\lambda_3^2 \tag{11.3}$$

$$I_3 = \det(\delta_{rs} + 2\varepsilon_{rs}) = \lambda_1^2\lambda_2^2\lambda_3^2 \tag{11.4}$$

where Einstein's summation rule over repeated indices applies. Equation (11.1) for isotropic materials is transformed to:

$$S^{ij} = \frac{\partial W}{\partial I_n}\frac{\partial I_n}{\partial \varepsilon_{ij}} \qquad W = W(I_1, I_2, I_3) \tag{11.5}$$

In several studies, the skin is considered to be a membrane. The membrane tension $N^{\alpha\beta}\alpha$, $\beta = 1, 2$ (expressed as force per unit original length) is a function of the planar strain components $\varepsilon_{\alpha\beta}$.

The pseudoelastic models for the skin are listed in Table 11.1. Some are expressed in terms of direct stress-strain relations. Others apply the strain energy function. They differ from each other primarily in their functional form and underlying assumptions (isotropy versus orthotropy, incompressibility versus compressibility, uniaxial versus multiaxial stretching). Two functional forms are common: the power and the exponential forms.

The time-dependent models for the skin are listed in Table 11.2 and illustrated in Fig. 11.4. Here again the differences are primarily in the functional form (e.g., logarithmic or exponential time effect) and underlying assumptions (uniaxial, multiaxial, linear, or nonlinear viscoelasticity).

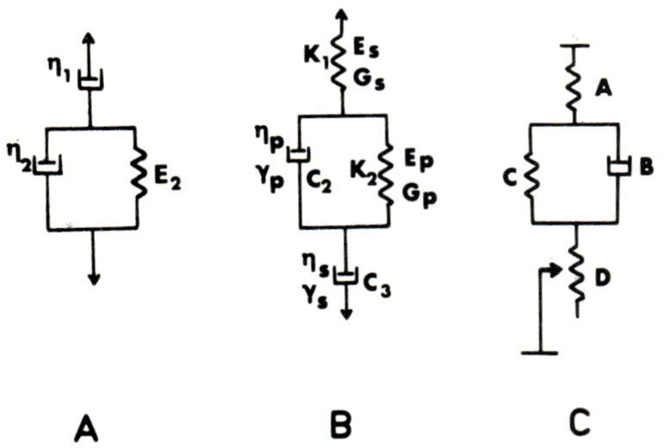

FIG. 11.4 Mechanical models for the skin as proposed by (*A*) (Jamison et al. [78], (*B*) Dinnar [112] and Wijn et al. [130], and (*C*) Sakata et al. [106].

The compressive response of the skin has been modeled as well. The models range from purely elastic [100, 106], through purely viscous [33] to linear viscoelastic [112] and plastoviscoelastic [106]. They are listed in Table 11.3.

11.7 SKIN MECHANICS AND PLASTIC SURGERY

Plastic surgeons have known for some time that the tension existing in sutured wounds affects the resulting scar [131]. The directional aspect of the skin's tension and the skin's extensibility have also been known. It is thus natural to expect that by application of standard methods from

TABLE 11.3 Models for compressive response of the skin

Source	Model	Tests, results, and remarks
Dikstein and Hartzshtark [100]	$p = kI^2$	In vivo indentation under constant loads; human forehead skin; 0.2-cm² flat cylindrical indenter $k = 60$ Pa p = pressure, Pa I = indentation, m
Dinnar [112]	See Fig. 11.4*b*	Assumes linear viscoelasticity. From fit to data of Hickman et al. [101] found: $k_1 = 2.5$; $k_2 = 10$; $C_2 = 1.8 \times 10^4$; $C_3 = 10^4$ Parameters change in edema
Sakata et al. [106]	$0 < F < 0.1$ N: linear behavior $0.1\ \text{N} < F < 2.25\ \text{N}$: $\frac{\Delta h}{ho} = a + b \ln F$	Pseudoelastic model. In vitro, constant rate of indentation loading; sheep abdominal skin; indenter is flat cylinder of diameter 0.397 cm F = load, N Typical values: $a = -0.49$; $b = -0.16$
Sakata et al. [106]	See Fig.11.4*c*	Plastoviscoelastic model. In vitro cyclic test superimposed on constant load $A = a_1 + b_1 \ln t$; $B = a_2 + b_2 \ln t$; $C = a_3 + b_3 \ln t$; $D = b_4 \ln t$ Observed logarithmic decay of viscoelastic response
Tregear and Dirnhuber [33]	$h^2(t) = \frac{h^2(0)}{1 + kFt}$	In vivo indentation under constant load, human skin over tibia and dorsal arm skin $h(t)$ = thickness at time t F = the constant load k = constant Assumes compression involves only viscous effects

engineering mechanics, it would be possible to improve the incisions so that the resulting scars will be less conspicuous and of minimal functional restraint.

It is fair to state that this goal has only partially been achieved. In the following sections, past efforts will be reviewed and topics which require further study will be outlined.

11.7.1 The Significance of Langer's Lines

Various guidelines have been used for the choice of the optimal incisional direction. Probably the most widely used and yet controversial is the one associated with Langer's lines [20]. Other methods rely alternatively on the crease or wrinkle lines, on the direction of hair or hair follicles, and on skin tension [132–135]. The approach using Langer's lines stipulates that the incision should be made along these lines in order to minimize the tension across the suture, thus ensuring the least conspicuous scar.

Past criticism of the use of Langer's lines as incisional guides [136, 137] was based primarily on the mismatch between their pattern and that of the normal crease or wrinkle lines in several sites, particularly some areas of the face. It has also been stated that Langer's lines represent a static state of the skin while in practical application the kinetic effects associated with the whole range of body motion should be considered [138]. Kraissl [132] suggested that the wrinkle lines are the proper guides for incisions since they run uniformly perpendicular to, and are indeed formed by, the underlying muscles. Wrinkle lines thus represent dynamic forces acting on the skin.

Flint [45] has reviewed these and other past criticisms of Langer's patterns as incisional guides and observed that much of the criticism was based on grossly inaccurate and misinterpreted reproduction of Langer's original patterns. If the correct data of Langer and the later figures of Cox [44] are used, it becomes apparent, that, with few exceptions (palms, soles, and skin near knee, elbow, and ankle), the wrinkle line patterns coincide with Langer's lines [44].

The validity of Langer's lines as incisional guides is further strengthened by later studies which showed them to coincide with directions of maximum tension and deformation. It is not surprising then that wounds retract more if cut across rather than along Langer's lines [44, 47–49] and that the tension needed to close a wound (the suture tension) is considerably higher in a perpendicular wound than in a parallel one [48].

11.7.2 Stresses in Sutures

The use of Langer's lines in skin surgery is the most successful application of mechanics and mechanical concepts for clinical purposes. With a sense of humbleness we realize that this application is based on the work of Langer reported in 1861 [20].

In the last three decades much effort has been made to establish a quantitative guide for incisions so that the suture stresses can be predicted. This ongoing endeavor faces several difficulties, not the least of which is the absence of sufficient quantitative data on the preexisting tension field in the skin. Another major obstacle is that the existing models for skin mechanics need further development because either they cannot simulate the behavior of skin under all deformational schemes or their validity in the in vivo state has not been demonstrated.

Different methods have been utilized in the design and mechanical analysis of skin incisions. A comprehensive work on several patterns of incision was carried out by Limberg [139] on the basis of inextensible paper models. A similar study was done by McGregor [140] on the Z-plasty procedure. This approach, although of some value as a preliminary tool, is not realistic since it ignores the skin's extensibility and thus provides incorrect results.

Furnas and Fischer [141] performed an in vivo experimental investigation of the biomechanical aspects in the transformation of regular and compound Z-plasty. Z-plasty [142] is a surgical technique in which the incision is shaped like a Z. This prevents excessive scar formation in cases of long incisions positioned across Langer's lines, along joints, or perpendicular to free edges (e.g., lips, eyelids). Furnas and Fischer measured the closure force, the distortion of surrounding strain field, and the changes in the central length of the Z-plasty as a function of its parameters (overall length, number of flaps, tip angle, and asymmetry in angles). Their results provide a useful guide for the design of Z-plasty, but the specific state of pretension which existed in their study (on dogs) was not specified.

Theoretical analyses of tension patterns in sutures were carried out for a few cases. Danielson and Natarajan [116] used the tension field theory to solve the tension pattern in a suture resulting from the closure of an elliptical hole along its major axis. They also analyzed the forces needed to transform a Z-plasty of similar characteristics to those used by Furnas and Fischer [141]. By adjusting parameters they obtained good agreement with the experimental data. Danielson [115] used the same theory to determine the tractions required to close a rectangular hole. Bradsley [160] and DeHoff and Key [143] used the finite element method with the skin mechanical model of Veronda and Westmann [75] to determine both the tension which develops in the closure of an elliptical hole as a function of its axis ratio, and the total force needed to transform an 8-cm Z-plasty of various tip angles. Their results show the same trend as the experimental data of Furnas and Fischer [141].

11.7.3 Mechanical Aspects of Wound Healing

The healing of skin wounds involves a variety of complex biological mechanisms. Yet the functionally most important feature of the healing process is the development of tensile strength in the scar. Other significant characteristics are the rate of contraction of open wounds and the rate of epithelization.

The healing of a wound has, however, some other important mechanical aspects. For example, the effect of various mechanical parameters on the healing process is of considerable clinical as well as scientific significance.

The healing process has three stages [144]. In the latent period (up to approximately 5 days) there is no gain in strength. It is followed by a period of fibroplasia (5 days or more) in which there is a rapid increase in strength. This is related to an increase in the collagen content [145]. In the maturation period, which can last for many weeks or even months [146], the collagen content reaches a plateau, and the further increase in strength is associated with changes within the collagen structure. This may be related to the increase in fibril size, which is in turn controlled by the ratio of collagen to mucopolysaccharide contents [147, 148].

The development of the scar's ultimate mechanical properties with time has been studied by several investigators using experimental wounds in animals. Tensile strength, strain at rupture, maximum stiffness, and energy absorption at rupture are all increasing with time after experimental wounding [90, 119, 149], and this process continues even after 1 year [146]. The tensile stress and stiffness increase from approximately zero at 6 days up to levels close or equal to the unwounded control after 1 year. The energy absorption after 1 year is about half its normal value. The strain at rupture increases from about 25 percent of its normal value to about 90 percent in the same period [146]. These results indicate that skin may eventually regain its initial strength, though earlier reports claim that this is never achieved [144].

The stress-strain relations of the scar tissue change significantly with time after wounding. The curve for a scar was found to be below the curve of unwounded control at 10 days, but above it after 20 days [149]. Brody et al. [68] found that the scar on human forearm skin exhibits nonlinear stress-strain relations which are similar in shape but of higher stiffness than those of normal skin.

Delayed primary closure of skin wounds (a common procedure in military surgery by which, for reasons of minimizing infections, the wound is left open for a few days and then sutured) was found [161] to result in wounds which, 10 days after wounding, are as strong as those subjected to primary (immediate) closure. Later on, however, the former becomes twice as strong as the wounds receiving primary closure.

The Effect of Stress on Healing. The effect of stress on wound healing is a topic of considerable clinical significance. Unfortunately, little study has been reported thus far. It has been long recognized that stress has an important effect on the healing of bone (Wolff's law), but its role in the healing of soft tissues has not been studied systematically. Accumulating clinical experience has guided surgeons to perform incisions along, rather than across, Langer's lines. At the same time Berard et al. [150] found parallel scars to be thicker and weaker than those cut across the wrinkle line at scar ages of 14 and 21 days.

There is some evidence to suggest that stress may have a beneficial effect on wound healing [162], even if applied intermittently [151, 152], but the available data are not sufficient and this topic requires further study.

Stress was found to help prevent contracture (shrinking) of skin grafts [154]. The extent of contracture of human split skin grafts is lower at higher tension sites on the body. At a specific site it is least along Langer's lines as compared with other directions.

11.8 SUGGESTIONS FOR FUTURE RESEARCH

Skin mechanics is a topic of considerable interest in several fields of science and medicine. Significant progress has already been made, especially in the domain of scientific research. Much can be done in the future to increase the impact of our knowledge in the clinical world.

For engineers, many of the clinical problems in plastic surgery can be classified as boundary value problems. Their solution requires the knowledge both of the boundary conditions and of the constitutive behavior of the skin. In terms of the boundaries, it will be useful to develop a fast, simple, and convenient method with which one could measure the skin's in vivo tension under clinical conditions. Another subject which needs further clarification is the binding forces to the subcutaneous tissues. They have not been adequately studied in terms of their effect on the skin's extensibility.

Much progress had been made in the constitutive formulation of the skin's response, but two important subjects have hardly been dealt with: (1) the effects of preconditioning and (2) the relation between in vitro and in vivo behavior, particularly under biaxial stretching. The structurally based constitutive formulations seem to have considerable potential, but they must be tested and, if necessary, improved on the basis of well-controlled biaxial tests.

In the field of clinical research, an important and yet little understood subject is the effect of stress on wound healing. Systematic quantitative information on the time course and range of stresses which facilitate optimal healing would be useful.

In epidermal research it is necessary to develop an experimental method under which the effects of dermis and epidermis can be distinguished. The preliminary work on the relations between Langer's lines on the one hand and the epidermal groove pattern on the other is promising. Further study and validation of these relations are important in terms of clinical potential.

REFERENCES

1. Harkness, R. D.: Functional aspects of the connective tissues of skin, in E. A. Balazs, (ed.), *Chemistry and Molecular Biology of the Intercellular Matrix,* Academic Press, New York, 1970, vol. 3, pp. 1309–1340.
2. Harkness, R. D.: Mechanical properties of skin in relation to its biological function and its chemical components, in H. R. Elden (ed.), *Biophysical Properties of the Skin,* Wiley, New York, 1971, pp. 393–436.
3. Whitaker, L. A.: Characteristics of skin of importance to the surgeon, *Int. Surg., 57,* 877–879, 1972.
4. Wilkes, G. L., Nguyen, A., and Wildnauer, R.: Structure-property relations of human and neonatal rat stratum corneum: I. Thermal stability of the crystalline lipid structure as studied by x-ray diffraction and differential thermal analysis, *Biophys. Acta, 304,* 267–275, 1973.
5. Bottoms, E., and Shuster, S.: Radiology: Effect of ultraviolet light on skin collagen, *Nature (London), 199,* 192–193, 1963.
6. Finlay, B.: Scanning electron microscopy of the human dermis under uniaxial strain, *Bio-Med. Eng., 4,* 322–327, 1969.
7. Millington, P. F., Gibson, T., Evans, J. H., and Barbenel, J. C.: Structural and mechanical aspects of connective tissue, in R. M. Kenedi (ed.), *Advances in Biomedical Engineering,* Academic Press, New York, 1971, vol. 1, pp. 189–248.

8. Brown, I. A.: Scanning electron microscopy of human dermal fibrous tissue, *J. Anat., 113,* 159–168, 1972.

9. Tregear, R. T.: *Physical Functions of Skin,* Academic Press, London, 1966.

10. Reed, R.: Freeze-etched connective tissue, *Int. Rev. Connect. Tiss. Res., 6,* 257–305, 1973.

11. Myers, D. B., Highton, T. C., and Rayns, D. G.: Ruthenium red–positive filaments interconnecting collagen fibrils, *J. Ultrastruct. Res., 42,* 87–92, 1973.

12. Hult, A. M., and Goltz, R. W.: The measurement of elastin in human skin and its quantity in relation to age, *J. Invest. Dermatol., 44,* 408–412, 1965.

13. Weinstein, G. D., and Boucek, R. J.: Collagen and elastin of human dermis, *J. Invest. Dermatol., 35,* 227–230, 1960.

14. Gibson, T., Kenedi, R. M., and Craik, J. E.: The mobile microarchitecture of dermal collagen: A bioengineering study, *Br. J. Surg., 52,* 764–770, 1965.

15. Gibson, T., and Kenedi, R. M.: The structural components of the dermis and their mechanical characteristics, in W. Montagna, J. P. Bentley, and R. L. Dobson (eds.), *The Dermis,* Appleton-Century-Crofts, Educational Division Corporation, New York, 1968, pp. 19–38.

16. Dick, J. C.: Observation on the elastic tissue of the skin with a note on the reticular layer at the junction of the dermis and epidermis, *J. Anat., 81,* 201–211, 1947.

17. Montagna, W.: *The Structure and Function of Skin,* Academic Press, London, 1962.

18. Bowes, J. H., and Kenten, R. H.: The amino-acid distribution of collagen, elastin and reticular tissue, *Biochem. J., 43,* 358–365, 1949.

19. Brown, I. A.: A scanning electron microscope study of the effects of uniaxial tension on human skin, *Br. J. Dermatol., 89,* 383–393, 1973.

20. Langer, K.: Zur Anatomie und Physiologie der Haut: I. Uber die Spaltbarkeit der Cutis, *Sitzungber. Akad. Wiss. Wien., 44,* 19–46, 1861.

21. Ferguson, J., and Barbenel, J. C.: Skin surface pattern and the directional mechanical properties of the dermis, in R. Marks and P. A. Payne (eds.), *Bioengineering and the Skin,* MTP Press, Lancaster, 1981, pp. 83–92.

22. Haut, R. G., and Little, R. W.: A constitutive equation for collagen fibers, *J. Biomech., 5,* 423–430, 1972.

23. Hall, R. H.: Energy and entropy effects on the elasticity of collagen fibers, *J. Soc. Leather Trades Chem., 36,* 137–148, 1952.

24. Carton, R. W., Dainauskas, J., and Clark, J. W.: Elastic properties of single elastic fibers, *J. Appl. Physiol., 17,* 547–551, 1962.

25. Meyer, K. H., and Ferri, C.: Die Elastischen Eigenschaften der Elastischen und der Kollagenen Fasern und ihre molekulare Deutung, *Arch. Physiol., 238,* 78–90, 1936.

26. Daly, C. H.: The role of elastin in the mechanical behaviour of human skin, *Proc. 8th Int. Conf. Med. Biol. Eng.,* Chicago, 1969, p. 18/7.

27. Dick, J. C.: Tension and resistance to stretching of human skin and other membranes, *J. Physiol., 112,* 102–113, 1951.

28. Lee, M. M. C.: Physical and structural age changes in human skin, *Anat. Rec., 129,* 473–493, 1957.

29. Lanir, Y., and Fung, Y. C.: Two-dimensional mechanical properties of rabbit skin: II. Experimental results, *Biomechanics, 7,* 171–182, 1974.

30. Harkness, M. L. R., and Harkness, R. D.: Effect of enzymes on mechanical properties of tissues, *Nature, 183,* 1821–1822, 1959.

31. Vlasblom, D. C.: Skin elasticity, Ph.D. thesis, University of Utrecht, The Netherlands, 1967.

32. Daly, C. H.: Viscoelastic properties of human skin, *Proc. 21st Annu. Conf. Eng. Med. Biol. (ACEMB),* Houston, Tex., 1968, pp. 18–21.

33. Tregear, R. T., and Dirnhuber, P.: Viscous flow in compressed human and rat skin, *J. Invest. Dermatol., 45,* 119–125, 1965.

34. Hvidberg, E.: Investigations into the effect of mechanical pressure on the water content of isolated skin, *Acta Pharmacol. (Kobenhavn), 16,* 245–249, 1960.

35. Dikstein, S., and Hartzshtark, A.: What does low-pressure indentometry measure, *Arzt. Kosmetol., 16,* 327–328, 1983.

36. Day, T. D.: The effect of hyaluronidase on the flow of water in connective tissue, in A. Frey-Wyssling (ed.), *Deformation and Flow in Biological Systems,* North-Holland, Amsterdam, 1952, pp. 503–505.

37. Daly, C.: Mechanical properties of skin, Ph.D. thesis, University of Strathclyde, Glasgow, 1966.

38. Harkness, M. L. R., and Harkness, R. D.: Functional alterations in the mechanical properties of collageneous frameworks, in S. F. Jackson, R. D. Harkness, S. M. Partridge, and G. R. Trisran (eds.), *Structure and Function of Connective and Skeletal Tissue,* Butterworth, London, 1965, pp. 376–381.

39. Highley, D. R., Coomey, M., Den Beste, M., and Wolfram, L. J.: Frictional properties of skin, *J. Invest. Dermatol., 69,* 303–305, 1977.

40. Naylor, P. F. D.: The measurement of epidermal strength, *St. John's Hosp. Trans. Dermatol. Scc., 31,* 29–33, 1952.

41. Naylor, P. F. D.: The skin surface and friction, *Br. J. Dermatol., 67,* 239–248, 1955.

42. Guibara, E., Nicholls, S., and Marks, R.: Measurement of force required for point indentation of stratum corneum: The effect of hydration, *Bioeng. Skin., 2,* 29–30, 1979.

43. Dupuytren, G.: *Traité Théorique et Pratique des Blessures par Armes de Guerre,* Maison, Paris, 1834, vol. 1, p. 66.

44. Cox, H. T.: The cleavage lines of the skin, *Br. J. Surg., 29,* 234–240, 1942.

45. Flint, M. A.: Langer's lines: A commentary, *Surg. Annu., 8,* 25–46, 1976.

46. Alexander, H., and Cook, T. H.: Accounting for natural tension in the mechanical testing of human skin, *J. Invest. Dermatol., 69,* 310–314, 1977.

47. Ridge, M. D., and Wright, V.: The directional effect of skin—A bioengineering study of skin with particular reference to Langer's lines, *J. Invest. Dermatol., 46,* 341–346, 1966.

48. Ksander, G. A., Vistnes, L. M., and Rose, E. H.: Excisional wound biomechanics, skin tension lines, and elastic contraction, *Plast. Reconstr. Surg., 59,* 398–406, 1977.

49. Flint, M. H.: The biological basis of Langer's lines, in J. J. Longacre (ed.), *The Ultrastructure of Collagen,* Charles C Thomas, Springfield, Ill., 1973.

50. Gibson, T., Stark, H., and Evans, J. H.: Directional variation in extensibility of human skin in-vivo, *J. Biomech., 2,* 201–204, 1969.

51. Jansen, L. H., and Rottier, P. B.: Elasticity of human skin related to age, *Dermatologica, 115,* 106–111, 1957.

52. Brown, I. A.: Structural and mechanical studies on human skin, Ph.D. thesis, University of Strathclyde, Glasgow, 1971.

53. Cook, T., Alexander, H., and Cohen, M.: Experimental method for determining the 2-dimensional mechanical properties of living human skin, *Med. Biol. Eng. Comput., 15,* 381–390, 1977.

54. Peng, S. T. J., Landel, R. F., and Brody, G. S.: In-vitro study of human skin rheology, *Proc. 6th N.E. Bioeng. Conf.,* Pergamon Press, New York, 1978, pp. 350–354.

55. Black, M. M.: A modified radiographic method for measuring skin thickness, *Br. J. Dermatol., 81,* 661–666, 1969.

56. Payne, P. A., Grove, G. L., Alexander, H., Quilliam, R. M., and Miller, D. L.: Cross-sectional ultrasonics scanning of skin using plastic film transducers, *Bioeng. Skin, 3,* 241–246, 1982.

57. Vogel, H. G., and Denkel, K.: Methodological studies on biomechanics of rat skin comparing in-vivo and in-vitro results, *Bioeng. Skin, 4,* 71–79, 1982.

58. Barbenel, J. C., and Payne, P. A.: In-vivo mechanical testing of dermal properties, *Bioeng. Skin, 3,* 8–38, 1981.

59. Hartzshtark, A., and Dikstein, S.: Mechanical testing of human skin in-vivo, *Rev. Pure Appl. Pharmacol. Sci., 3,* 83–122, 1982.

60. Stevens, J. C., and Jones, N. B.: A review of apparatus for investigating the mechanical properties of soft animal tissue in-vitro, *Eng. Med., 6,* 112–119, 1972.

61. Middleton, J. D.: The mechanism of water binding in stratum corneum, *Br. J. Dermatol., 80,* 437–450, 1968.

62. Kiistala, U.: A suction blister device for separation of viable epidermis and dermis, *J. Invest. Dermatol., 50,* 129–137, 1968.

63. Baumberger, J. P., Suntzeff, V., and Cowdry, E. G.: Methods for the separation of epidermis from dermis and some physiological and chemical properties of isolated epidermis, *J. Nat. Cancer Inst., 11,* 413–423, 1941.

64. Einbinder, J. M., and Welzer, R. A.: Separation of epidermis from dermis by use of disodium cantharides, *J. Invest. Dermatol., 41,* 109–122, 1963.

65. Ogura, R., Knox, J. M., and Griffin, A. C.: Separation of epidermis for the study of epidermal sulphydryl, *J. Invest. Dermatol., 35,* 239–243, 1960.

66. Rietschel, R. L., and Abers, W. A.: Effects of harvesting techniques on hydration dynamics: Gravimetric studies of stratum corneum, *J. Soc. Cosmet. Chem., 2,* 777–782, 1978.

67. Kenedi, R. M., Gibson, T., and Daly, C. H.: Bio-engineering studies of the human skin: II, in R. M. Kenedi, (ed.), *Biomechanics and Related Bio-Engineering Topics,* Pergamon Press, Oxford, 1965, pp. 147–158.

68. Brody, G. S., Peng, T. J., and Landel, R. F.: The rheological properties of human skin and scar tissue, in R. Marks and P. A. Payne (eds.), *Bioengineering and the Skin,* MTP Press, Lancaster, 1981, pp. 147–158.

69. Finlay, B.: Dynamic mechanical testing of human skin in vivo, *J. Biomech., 3,* 557–568, 1970.

70. Chu, B. M., Frasher, W. G., and Wayland, H.: Hysteretic behavior of soft living animal tissue, *Ann. Biomed. Eng., 1,* 182–203, 1972.

71. Morgan, F. R.: The mechanical properties of collagen fibers: Stress-strain curves, *J. Soc. Leather Trades Chem., 44,* 170–182, 1960.

72. North, J. F., and Gibson, F.: Volume compressibility of human abdominal skin, *J. Biomech., 11,* 203–207, 1978.

73. Vossoughi, J., and Vaishnav, R. N.: A comment, *J. Biomech., 12,* 481, 1979.

74. Beckwith, T. G., Brody, G. S., Glaser, A. A., Prevenstik, T., and White, W. L.: Standardization of methods of measuring the mechanical properties of wounds, *ASME Human Factors,* Div. Pap. No. 63 Wa 276, American Society of Mechanical Engineers, New York, 1963, p. 1.

75. Veronda, D. R., and Westmann, R. A.: Mechanical characterization of skin—Finite deformations, *J. Biomech., 3,* 111–124, 1970.

76. von Gierke, H. E., Oestreicher, H. L., Franke, E. K., Parrack, H. O., and von Wittern, W. W.: Physics of vibrations in living tissues, *J. Appl. Physiol., 4,* 886–900, 1952.

77. Galford, J. E., and McElhaney, J. H.: A viscoelastic study of scalp, brain and dura, *J. Biomech., 3,* 211–221, 1970.

78. Jamison, C. E., Marangoni, R. D., and Glaser, A. A.: Viscoelastic properties of soft tissues by discrete model characterization, *J. Biomech., 1,* 33–46, 1968.

79. Fry, P., Harkness, M. L. R., Harkness, R. D., and Nightingale, M.: Mechanical properties of tissues of lathyritic animals, *J. Physiol. London, 164,* 77–89, 1962.

80. Lanir, Y.: Biaxial stress relaxation in skin, *Ann. Biomed. Eng., 4,* 250–270, 1976.

81. Gadd, C. W., Lange, W. A., and Peterson, F. J.: *Strength of Skin and its Measurement with Special Applications to Trauma,* Biomechanics Monograph ASME, American Society of Mechanical Engineers, New York, 1967, pp. 184–191.

82. Gibson, T., and Kenedi, R. M.: Biomedical properties of skin, *Surg. Clin. North Am., 47,* 279–941, 1967.

83. Rollhäuser, H.: Die Zugfestigkeit der Menschlichen haut; *Gegenbaurs Morphol. Jahrb, 90,* 249–61, 1950.

84. Fazekas, I. G., Kosa, F., and Basch, A.: On the tensile strength of skin in various body areas, *Dtsch. Z. Gesamte Gerichtl. Med., 64,* 62–68, 1968.

85. Mendoza, S. A., and Milch, R. A.: Age variations of nominal tensile strength of wistar rat skins, *Gerontologia, 10,* 42–46, 1964.

86. Jansen, L. H., and Rottier, P. B.: Some mechanical properties of human abdominal skin measured on excised strips, *Dermatologica, 117,* 65–83, 1958.

87. Wenzel, H. G.: Untersuchungen einiger Mechanischer Eigenschaften der Hant, insbesondere der Striae Cutis Distensae, *Virchows Arch. Pathol. Anat., 317,* 654–706, 1950.

88. Vogel, H. G., Kobelt, D., Korting, G. W., and Holzmann, W.: Prüfung der Festigkeitseigenschaften von Rattenhaut in Abhangigkeit von Lebensalter und Geschlecht, *Arch. Klin. Exp. Dermatol., 239,* 296–306, 1970.

89. Fry, P., Harkness, M. L. R., and Harkness, R. D.: Mechanical properties of the collagenous framework of skin in rats of different ages, *Am. J. Physiol., 206,* 1425–1429, 1964.

90. Sussman, M. D.: Aging of connective tissue: Physical properties of healing wounds in young and old rats, *Am. J. Physiol., 224,* 1167–1171, 1973.

91. Middleton, J. D.: The influence of temperature and humidity of stratum corneum and its relation to skin chapping, *J. Soc. Cosmet. Chem., 24,* 239–243, 1973.

92. Papir, Y. S., Hsu, K. H., and Wildnauer, R. H.: The mechanical properties of stratum corneum: I. The effect of water and ambient temperature on the tensile properties of newborn at stratum corneum, *Biochim. Biophys. Acta, 399,* 170–180, 1975.

93. Wildnauer, R. H., Bothwell, J. W., and Douglass, A. B.: Stratum corneum biochemical properties: I. Influence of relative humidity on normal and extracted human stratum corneum, *J. Invest. Dermatol., 56,* 72–78, 1971.

94. Park, A. C., and Baddiel, C. B.: Rheology of stratum corneum: II. A Physico-chemical investigation of factors influencing the water content of the corneum, *J. Soc. Cosmet. Chem., 23,* 13–21, 1972.

95. Wilkes, G. L., Wildnauer, R. H., and Brown, I. A.: Structure-property relationship of skin—A Polymer Composite, *CRC Crit. Rev. Bioeng., 1,* 453–495, 1973.

96. Agache, P., Boyer, J. P., and Laurent, R.: Biomechanical properties and microscopic morphology of human stratum corneum incubated on a wet pad in-vitro, *Arch. Dermatol. Forsch., 246,* 271–283, 1973.

97. Takahashi, M., Kawasaki, K., Tanaka, M., Ohta, S., and Tsuda, Y.: The mechanism of stratum corneum plasticization with water, in R. Marks and P. A. Payne (eds.), *Bioengineering and the Skin,* MTP Press, Lancaster, 1981, pp. 67–73.

98. Wilkes, G. L., and Wildnauer, R. H.: Structure-property relationships of the stratum corneum of human and neonated rat: II. Dynamic mechanical studies, *Biochem. Biophys. Acta, 304,* 276–289, 1973.

99. Comaish, S., and Bottoms, E.: The skin and friction: Deviations from Amonton's laws, and the effects of hydration and lubrication, *Br. J. Dermatol., 84,* 37–43, 1971.

100. Dikstein, S., and Hartzshtark, A.: In-vivo measurement of some elastic properties of human skin, in R. Marks and P. A. Payne (eds.), *Bioengineering and the Skin,* MTP Press, Lancaster, 1981, pp. 45–53.

101. Hickman, K. E., Lindan, O., Reswick, J. B., and Scanlan, R. H.: Deformation and flow in compressed skin tissue, *Biomedical Fluid Mechanics Symposium,* American Society of Mechanical Engineers, New York, 1966, pp. 127–147.

102. Schade, H.: Untersuchungen zur Organfunktion des Bindegewebes: I. Die Elasticitats function des Bindesgewebes und die initiate Messung, *Z. Exp. Pathol. Ther., 11,* 369–399, 1912.

103. Ziegert, J. C., and Lewis, J. L.: In-vivo mechanical properties of soft tissue covering bony prominences, *J. Biomech. Eng., 100,* 194–201, 1978.

104. Moore, T. J.: A survey of the mechanical characteristics of skin and tissue in response to vibratory stimulation, *IEEE Trans. Man-Mach. Sys., MMS-11*(1), 79–84, 1970.

105. Thompson, D. E., Hussein, H. M., and Perritt, R. Q.: Point impedance characterization of soft tissues in-vivo, in R. Marks and P. A. Payne (eds.), *Bioengineering and the Skin,* MTP Press, Lancaster, 1981, pp. 103–111.

106. Sakata, K., Parfitt, G., and Pinder, K. L.: Compressive behavior of physiological tissue, *Biorheology, 9,* 173–184, 1972.

107. Lowe, L. B., and Van der Leun, J. C.: Suction blisters and dermal-epidermal adherence, *J. Invest. Dermatol., 50,* 308–314, 1968.

108. Marks, R., Nicholls, S., and Fitzgeorge, D.: Measurement of intracorneal cohesion in man using in-vivo techniques, *J. Invest. Dermatol., 69,* 299–302, 1977.

109. Wijn, P. F. F.: A fibre model description of the alinear viscoelastic properties of the human skin in-vivo for small deformation, *1st Int. Conf. Mech. Med. Biol.,* Aachen, Germany, 1978.

110. Lanir, Y.: A structural theory for the homogeneous biaxial stress-strain relationship in flat collageneous tissues, *J. Biomechan., 12,* 423–436, 1979.

111. Lanir, Y.: Constitutive equation for fibrous connective tissues, *J. Biomech., 16,* 1–12, 1983.

112. Dinnar, U.: A note on theory of deformation in compressed skin tissues, *Math. Biosci., 8,* 71–82, 1970.

113. Viidik, A.: A rheological model for uncalcified parallel fibered collagenous tissue, *J. Biomech., 1,* 3–11, 1968.

114. Allaire, P. E., Thacker, J. G., Edlich, R. F., Rodenheaver, G. J., and Edgerton, M. T.: Finite deformation theory for in-vivo human skin, *J. Bioeng., 1,* 239–249, 1977.

115. Danielson, D. A.: Human skin as an elastic membrane, *J. Biomech., 6,* 539–546, 1973.

116. Danielson, D. A., and Natarajan, S.: Tension field theory and the stress in stretched skin, *J. Biomech., 8,* 135–142, 1975.

117. Demiray, H.: A note on the elasticity of soft biological tissues, *J. Biomech., 5,* 309–311, 1972.

118. Elden, H. R.: Biophysical properties of aging skin, in W. Montagna, J. P. Bentley, and R. L. Dobson (eds.), *Advances in Biology of Skin,* vol. X: *The Dermis,* Meredith, New York, 1970, pp. 231–252.

119. Glaser, A. A., Marangoni, R. D., Must, J. S., Beckwith, T. G., Brody, G. S., Walker, G. R., and White, W. L.: Refinements of the methods for the measurement of the mechanical properties of unwounded and wounded skin, *Med. Electron. Biol. Eng., 3,* 411–419, 1965.

120. Gou, P. F.: Strain-energy functions for biological tissues, *J. Biomech., 3,* 547–550, 1970.

121. Pierard, G. E., and Lapiere, C. M.: Physiopathological variations in the mechanical properties of skin, *Arch. Dermatol. Res., 260,* 231–239, 1977.

122. Ridge, M. D., and Wright, V.: A rheological study of skin, in R. M. Kenedi (ed.), *Biomechanics and Related Bioengineering Topics,* Pergamon Press, Oxford, 1965, pp. 165–175.

123. Snyder, R. W., and Lee, L. H. N.: Experimental study of biological tissue subjected to pure shear, *J. Biomech., 8,* 415–419, 1975.

124. Tong, P., and Y. C. Fung: The stress-strain relationship for the skin, *J. Biomech., 9,* 649–657, 1976.

125. Wijn, P. F. F.: The alinear viscoelastic properties of human skin in vivo for small deformations, Ph.D. thesis, Katholieke Universteit Nijmegen, The Netherlands, 1980.

126. Barbenel, J. C., Evans, J. H., and Finlay, J. B.: Stress-strain-time relations for soft connective tissue, in R. M. Kenedi (ed.), *Perspectives in Biomedical Engineering,* Macmillan, London, 1973, pp. 165–172.

127. Kenedi, R. M., Gibson, T., Daly, C. H., and Abrahams, M.: Biomedical characteristics of human skin and costal cartilage, *Fed. Proc., 25,* 1084–1087, 1960.

128. Lanir, Y.: The rheological behavior of the skin—experimental results and a structural model, *Biorheology, 16,* 191–202, 1979.

129. Sanders, R.: Torsional elasticity of human skin in vivo, *Pfluegers Arch., 342,* 255–260, 1973.

130. Wijn, P. F., Brakkee, A. J. M., Stienen, G. J. M., and Vendrick, A. J. H.: Mechanical properties of human skin in-vivo for small deformations: A comparison of uniaxial strain and torsion measurements, in R. M. Kenedi, J. M. Cowden, and J. T. Scales (eds.), *Bedsore Biomechanics,* Macmillan, London, 1976, pp. 103–108.

131. Kocher, T. H.: The direction of skin incision, in T. H. Kocher (ed.), *Textbook of Operative Surgery,* 5th ed., Black, London, 1911, p. 30.

132. Kraissl, C. J.: The selection of appropriate lines for elective surgical incisions, *Plast. Reconstr. Surg., 8,* 1–28, 1951.

133. Webster, J. P.: Deforming scars: Their causes, prevention and treatment, *Pa. Med. J., 38,* 929–937, 1935.

134. Borges, A. F., and Alexander, J. E.: Relaxed and skin tension lines: Z-plasties on scars, and fusiform excision of lesions, *Br. J. Plast. Surg., 15,* 242–247, 1962.

135. Gibson, T.: Biomechanics in plastic surgery, in R. M. Kenedi (ed.), *Biomechanics and Related Bioengineering Topics,* Pergamon Press, New York, 1965, pp. 129–134.

136. Kraissl, C. J., and Conway, H.: Excision of small tumors of skin and face with special reference to wrinkle lines, *Surgery, 25,* 592–598, 1949.

137. Rubin, L. R.: Langer's lines and facial scars, *Plast. Reconstr. Surg., 3,* 147–151, 1948.

138. Thacker, J. G., Iachetta, F. A., Alliare, P. E., Edgerton, M. T., Rodeheaver, G. T., and Edlich, R. R.: Biomedical properties of skin: Their influence on planning surgical excisions, *Basic Science in Plastic Surgery,* Mosby, 1976, pp. 72–79.

139. Limberg, A. A.: *Planning Local Plastic Operations on the Surface of the Body: Theory and Practice Manual for Surgeons,* NTIS Trans. AD-621-014, Leningrad Technical Institute, 1929.

140. McGregor, I. A.: The theoretical basis of Z-plasty, *Br. J. Plast. Surg., 9,* 256–259, 1957.

141. Furnas, D. W., and Fischer, G. W.: The Z-plasty: Biomechanics and mathematics, *Br. J. Plast. Surg., 24,* 144–160, 1971.

142. Hazrati, E.: Compound right angle Z-plasty, *Plast. Reconstr. Surg., 10,* 133–139, 1952.

143. DeHoff, P. H., and Key, J. E.: Application of the finite element analysis to determine forces and stresses in wound closing, *J. Biomech., 14,* 549–554, 1981.

144. Viljanto, J.: Tensile strength of healing wounds, in H. R. Elden (ed.), *Biophysical Properties of the Skin,* Wiley, New York, 1971, pp. 451–483.

145. Dunphy, J. E., and Udupa, K. N.: Chemical and histochemical sequences in the normal healing of wounds, *N. Engl. J. Med., 253,* 847–51, 1955.

146. White, W. L., Brody, G. S., Glaser, A. A., Maragnoni, R. D., Beckwith, T. C., Must, J. S., and Lehman, J. A.: Tensiometric studies of unwounded and wounded skin: Results using a standardized testing method, *Ann. Surg., 173,* 19–25, 1971.

147. Fessler, J. H.: A structural function of mucopolysaccharide in connective tissue, *Biochem. J., 76,* 124–132, 1960.

148. Bryant, W. M., and Weeks, P. M.: Secondary wound tensile strength gain: A function of collagen and mucopolysaccharide interaction, *Plast. Reconstr. Surg., 39,* 84–91, 1967.

149. Forrester, J. C., Zederfeldt, H. B., Hayes, T. L., and Hunt, T. K.: Mechanical, biochemical and architectural features of repair, in J. E. Dunphy and W. Van Winkle, Jr. (eds.), *Repairs and Regeneration,* McGraw-Hill, New York, 1968, pp. 71–85.

150. Berard, C. W., Woodward, S. C., Herrmann, J. B., and Pulaski, E. J.: Healing of incisional wounds in rats: The relationship of tensile strength and morphology of the normal skin wrinkle lines, *Ann. Surg., 159,* 260–270, 1964.

151. Arem, A. J., and Madden, J. W.: Effects of stress on healing wounds: I. Intermittent noncyclical tension, *J. Surg. Res., 20,* 93–102, 1976.

152. Langrana, N. A., Alexander, H., Straucher, I., Mehta, H., and Ricci, J.: Effect of mechanical load in wound healing, *Bioeng. Skin, 3,* 210–224, 1982.

153. Lewis, H. E., Meyer, J., and Paudiscio, A. A.: Recording skinfold calipers for determination of subcutaneous edema, *J. Lab. Clin. Med., 66,* 154–160, 1965.

154. Sawhney, C. P.: Contracture of skin grafts and its relation to cleavage line of skin, *Br. J. Plast. Surg., 24,* 233–237, 1971.

155. Potts, R. O., and Breuer, M. M.: The low-strain viscoelastic properties of skin, *Bioeng. Skin, 4,* 105–114, 1983.

156. Vogel, H. G.: Mechanical properties of rat skin as compared by in-vivo and in-vitro measurement, *Bioeng. Skin 3,* 198–209, 1982.

157. Rasseneur, L., de Rigal, J., and Leveque, J. L.: Simultaneous determination of the static and dynamic modulus of elasticity of stratum corneum, *Bioeng. Skin, 4,* 52–64, 1982.

158. Potts, R. O., Chrisman, D. A., Jr., and Buras, E. M., Jr.: The dynamic mechanical properties of human skin in-vivo, *J. Biomech., 16,* 365–372, 1983.

159. Torgalkar, A. M., A resonance frequency technique to determine energy absorbed in stratum corneum in-vivo, in R. Marks, and P. A. Payne (eds.), *Bioengineering and the Skin,* MTP Press, Lancaster, 1981, pp. 55–65.

160. Bradsley, T. J.: A finite element stress analysis of biological tissue, M.Sc. thesis, Bucknell University, Lewisburg, Pa., 1978.

161. Fogdestam, I.: Delayed primary closure: An experimental study of the healing of skin incisions, doctoral thesis, University of Aarhus, Denmark, 1980.

162. Brunius, U., and Ahren, C.: Healing of skin incisions during reduced tension of the wound area, *Acta Chir. Scand., 135,* 383–390, 1969.

CHAPTER 12

Properties of Red Blood Cells

Robert M. Hochmuth
Professor of Biomedical Engineering
Department of Biomedical Engineering
Duke University
Durham, North Carolina

The disk-shaped mammalian red blood cell has a very thin membrane (~10 nm) that encapsulates a viscous, liquid hemoglobin solution (cytoplasm). Since blood is a concentrated suspension of red cells, the ability of the cells to deform largely determines the ease with which blood flows through the vessels of the body. In general, cell deformation is a complex phenomenon caused by external forces acting over the surface of the membrane. For a given distribution of forces, the rate and extent of deformation depend on the *intrinsic* and *extrinsic* properties of the cell. The intrinsic properties are inherent to the membrane and cytoplasm and are mechanical in nature. They include various membrane elasticities and viscosities and a cytoplasm viscosity. The extrinsic properties pertain to the surface area and volume of the cell. The role the properties play in governing cellular deformation depends on their relative magnitudes and the distribution of forces acting over the membrane surface.

The extrinsic properties of the red cell are measured directly with the light microscope, using either conventional transmitted light or interference holography. With conventional microscopy the accuracy of measurement of surface area and volume is improved by aspirating the disk-shaped cell into a micropipet and in this way converting the disk shape of the cell into portions of spheres and cylinders. The intrinsic viscosity of the cytoplasmic hemoglobin solution is measured simply by removing and isolating the cytoplasm from the membrane and then measuring its viscosity with conventional techniques. However, although the size and shape of the cell and its cytoplasmic viscosity are measured by obvious methods, the methods for the measurement of the *intrinsic properties of the membrane* are not so obvious.

In general, the intrinsic material properties of a membrane are determined by applying a known force and then measuring the resulting deformation. Either a precise suction force applied with a micropipet or a stretching force created by a fluid shear can be used to deform the membrane. Aspiration with a micropipet has been shown to be an extremely versatile and accurate technique. *All* known elastic and viscous membrane properties have been measured with it.

12.1 HISTORICAL BACKGROUND

Rand and Burton [1] were the first to use a micropipet to apply a suction pressure to the (human) red cell membrane. They found that the aspiration pressure required to deform a portion of the membrane did not vary over the surface, and therefore the membrane surface is isotropic. In a closely related study, Rand [2] showed that a swollen red cell will lyse when the membrane tension is about 20 mN/m.

The analysis of membrane surface stresses (stress resultants) produced by micropipet aspiration, osmotic swelling, and fluid shear requires the rational application of the equations of mechanical equilibrium. This was first done by Fung [3]. Using simple, basic principles of membrane mechanics, Fung showed clearly that no pressure gradient could be maintained across the membrane if the bending resistance is small, and his estimate of the bending rigidity showed that it was negligible.* Shortly after the analysis of Fung [3], Fung and Tong [4] systematically analyzed the osmotic sphering of the red cell with the equations of membrane equilibrium and a general, nonlinear stress-strain law for the membrane surface. They showed in general that if the membrane behaves as a three-dimensional, homogeneous material, then either the elasticity or the thickness of the membrane must vary in a systematic way over the surface in order to permit the cell to swell without the membrane buckling (wrinkling). However, following the general analysis of Fung and Tong, experimental evidence began to accumulate that the membrane was not a three-dimensional material but a special kind of two-dimensional material—one in which deformation occurs only in the plane of the membrane, and in which the local surface area remains constant during deformation and swelling. Canham and Parkinson [5] showed in a series of careful measurements with the light microscope that little, if any, area change occurs during the cell swelling process. With dual-beam interference holography, Evans and Fung [6] improved the resolution of the light microscope by an order of magnitude and accurately measured the shape, surface area, and volume of approximately 50 red cells at three osmolarities. The surface area of the cells at the three osmolarities was essentially the same.† Image holography of single cells in isotonic and hypotonic solutions proved conclusively that there was no area change in the swelling process [7]. Finally, extensional deformation of red cells attached to a surface in a fluid shear flow indicated that cells stretch at constant surface area [8].

With the principle of constant-surface-area deformation firmly in mind, Evans [9] proposed that the red cell membrane be modeled as a two-dimensional, incompressible material. That is, the membrane is a material that deforms at constant membrane surface area and constant volume. Thus, since the area and volume remain constant during deformation, the thickness does also. Evans developed a simple but general stress-strain law that describes the finite (large) deformation of such a material. With this model and the equations of membrane equilibrium, Evans analyzed cell swelling [9] as well as the fluid shear deformation and micropipet aspiration of red cell membrane [10]. This latter analysis showed clearly that the two-dimensional stretching of red cell membrane, such as that which occurs in micropipet aspiration or fluid shear deformation of flaccid red cells, is characterized by a single elastic constant or shear modulus [10].

At about the same time as Evans's work, Skalak et al. [11] independently proposed a similar model for membrane deformation. However, their model had one very important generalization: the membrane was assumed to be slightly compressible. Thus, two elastic constants are required to characterize the plane, elastic deformation of the membrane—a relatively small shear modulus that characterizes the membrane's resistance to extension or shear, and a relatively large two-dimensional bulk or area modulus that characterizes the membrane's strong resistance to an increase in area.

---

* Rand and Burton measured a small but significant pressure difference across the membrane of 2.3 mmH_2O. Recent measurements indicate that this value for the pressure difference is at least an order of magnitude too large.

† The surface area at the lowest osmolarity, 131 mosmol, was 7.5 percent larger. Either the cells with smaller surface areas had hemolyzed at this osmolarity, or some stretching of the membrane surface area had taken place.

In summary, the original experiments and analysis indicate that the red cell can be mechanically modeled as a thin elastic membrane* surrounding a viscous liquid [3, 4, 9–11]. The membrane has little resistance to bending [3] and a very large resistance to an increase in area [11]. It readily elongates in shear at constant surface area [8, 10, 11]. It is "totally anisotropic: a continuum in two dimensions, molecular character in the third dimension" [9]. It "behaves like a two-dimensional version of a soft, but nearly incompressible material" [11].

12.2 SIZE AND SHAPE

The human red cell in particular and mammalian red cells in general are axisymmetric, biconcave, disk-shaped bodies.† The human red cell in cross section and its approximate dimensions are shown in Fig. 12.1. Traditionally, the diameter of the red cell is measured with the light microscope, with the cell lying flat on a glass slide, and the size and shape of the cell are measured with the cell hanging on edge [12]. Measurements by different observers of the mean diameter of human and other mammalian red cells are summarized by Ponder [12].

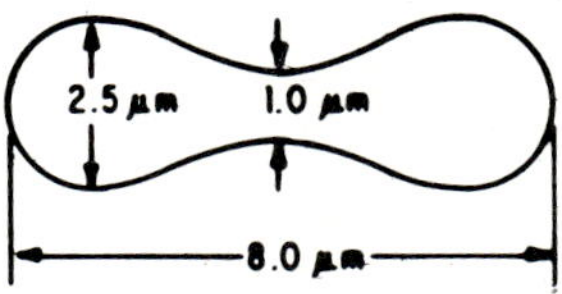

FIG. 12.1 The red cell in cross section and its nominal dimensions.

Since measurements with the light microscope have an uncertainty of approximately 0.5 μm, Evans and Fung [6] used a dual-beam interference microscope to create image holograms of red cells, thereby improving the resolution of the light microscope by an order of magnitude. In their studies the following equation was used to describe the shape of the red cell:

$$D(r) = \left[1 - \left(\frac{r}{R_0}\right)^2\right]^{1/2}\left[C_0 + C_2\left(\frac{r}{R_0}\right)^2 + C_4\left(\frac{r}{R_0}\right)^4\right] \tag{12.1}$$

where $D(r)$ is the thickness of the cell, r is the radius from the axis of symmetry, R_0 is the maximum radius, C_0 is the minimum thickness, and C_2 and C_4 are additional constants describing the shape of the cell. Values for these constants, as well as values for the surface area, volume, and maximum thickness, are given in Table 12.1 for 50 red cells at three different

TABLE 12.1 Values for the parameters R_0, C_0, C_2, C_4 in Eq. (12.1), as well as surface areas and volumes, at three osmolarities for approximately 50 red cells from one donor

Parameter	Osmolarity, mosm 300	217	131
Diameter $2R_0$, μm	7.82	7.59	6.78
Minimum thickness C_0, μm	0.81	2.10	
Maximum thickness, μm	2.58	3.30	
C_2, μm	7.83	7.58	0
C_4, μm	−4.39	−5.59	0
Surface area, μm²	135	135	145
Volume, μm³	94	116	164

Source: Data compiled from Evans and Fung [6].

* Later work showed that the membrane can behave as a viscoelastic or a viscoplastic material.

† There is a great deal of speculation about why the red cell has its biconcave shape. Fung simply says that "the cell takes up the biconcave shape whenever it is completely relaxed, for much the same reason [that] a rubber glove has a natural shape" [3].

osmolarities. All cells are from a single donor and are suspended in an Eagle-albumin solution. Measurements under isotonic conditions of various dimensions, surface areas, and volumes by Evans and Fung [6], Fung and Tsang (as cited by Fung in Ref. 13), Linderkamp and Meiselman [14], and Ponder [12] are given in Table 12.2. The results by Fung and Tsang as cited by Fung

TABLE 12.2 Dimensions, area, and volume of human red cells in isotonic solutions

	Diameter, μm	Minimum thickness, μm	Maximum thickness, μm	Surface area, μm^2	Volume, μm^3	Reference*
	7.82	0.81	2.58	135	94	6
	7.65	1.44	2.84	130	98	13
	8.5	1.0	2.4	163	87	12
	7.88	. . . .	. . . .	134	90	14
Average	8.0	1.1	2.6	140	92	

* References 6 and 13 detail the use of image holography for 50 cells from one donor [6] or 1581 cells from 14 donors [13]. Reference 12 gives the classical measurements by Ponder with the light microscope. Reference 14 gives the method for using the light microscope and micropipet aspiration to measure accurately the surface area and volume of 200 cells from 5 donors.

[13] represent pooled data from 14 individuals and 1581 cells. The results by Linderkamp and Meiselman represent data from 200 cells (40 from each of five donors). Note that Linderkamp and Meiselman used the light microscope to measure accurately the surface area and volume of cells by aspirating individual cells into a pipet until the portion of the cell outside the pipet forms a sphere, and the portion of the cell inside the pipet forms a cylinder and a spherical cap.

The size of the red cell decreases as the cell becomes more dense with age because of an increase in hemoglobin concentration [14]. The diameter, surface area, and volume of the least dense 5 percent and most dense 5 percent of 200 cells (40 from each of five donors) are shown in Table 12.3.

TABLE 12.3 Variation of cell size with density

Density fraction	Diameter, μm	Surface area, μm^2	Volume, μm^3
Top 5 percent	8.5	148	98.1
Unfractionated	7.9	134	89.8
Bottom 5 percent	7.4	118	78.0

Source: Linderkamp and Meiselman [14].

It is interesting to speculate about how large a human red cell can be. Chen and Fung [15] used image holography to measure the largest cell in samples of 100 cells. They studied 35 to 55 samples per donor from five different donors and found that the largest 1 percent of the red cell population has an average diameter, area, and volume of about 9.25 μm, 180 μm^2, and 128 μm^3, respectively. They predict that the largest cell in a population of 10^8 red cells will have a diameter between 15.7 and 17.1 μm.

The data shown in Table 12.1 illustrate that the red cell swells as the osmolarity decreases. If it is assumed that the number of osmotically active particles within a red cell remains constant as the cell swells, and that only a fraction R of the isotonic volume is osmotically active, then the following relation between V, the volume of the cell relative to the isotonic volume, and T, the osmolarity of the suspending medium relative to the isotonic osmolarity (~300 mosmol), is easily derived [12]:

$$V = \frac{R}{T} + 1 - R \tag{12.2}$$

The value for R is about 0.57 [6, 14]. The value for R decreases with an increase in cell density; i.e., the cell becomes less osmotically active. $R = 0.52$ for the most dense 5 percent of a sample of cells, while $R = 0.60$ for the least dense 5 percent [14].

12.3 MEMBRANE ELASTICITY

Since the cytoplasm of the red cell is a newtonian liquid, the unique biconcave shape of the cell (Fig. 12.1) comes from the natural shape of its membrane. The deformed cell remembers its undeformed shape and recovers it within a fraction of a second when the forces of deformation are removed.* Thus, the cell membrane behaves as an elastic solid. Although in general cell and membrane deformation can be quite complex, for a two-dimensional, incompressible, elastic material such as the red cell membrane, a deformation will always consist of one or more of three fundamental, independent deformations [16], as illustrated in Fig. 12.2. These deforma-

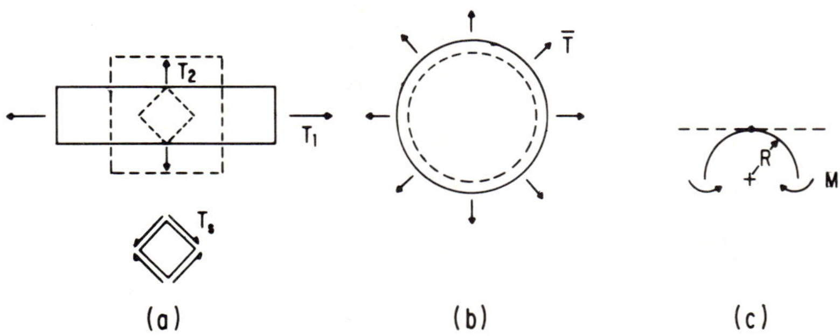

FIG. 12.2 Independent deformations of the red cell membrane. (*a*) Elongation and shear (at 45° to the principal axis of elongation). T_1 and T_2 are surface stress resultants, and T_s is the shear resultant $[T_s = (T_1 - T_2)/2]$. (*b*) Isotropic dilatation of the membrane. $\bar{T}$ is the isotropic tension. (*c*) Bending of the membrane. M is the bending moment (per unit length), and R is the radius of curvature.

tions are (1) an elongation or "shear" of the membrane without either increasing the surface area or bending it, (2) a uniform (isotropic) dilation (expansion) of the membrane surface without either shearing or bending it, and (3) a bending of the membrane without either shearing or expanding it. Each of these fundamental elastic deformations is characterized by an elastic modulus: (1) a shear modulus μ with units of N/m, (2) an area expansion modulus K with units of N/m, and (3) a bending modulus B with units of N · m. Figure 12.3 illustrates three pipet aspiration experiments for measuring the three elastic moduli. In general, a larger modulus indicates a greater resistance to that particular form of deformation.

---

* It will be shown in Sec. 12.4 that relatively large forces applied for long time periods can result in a permanent deformation of the membrane and a new stress-free state.

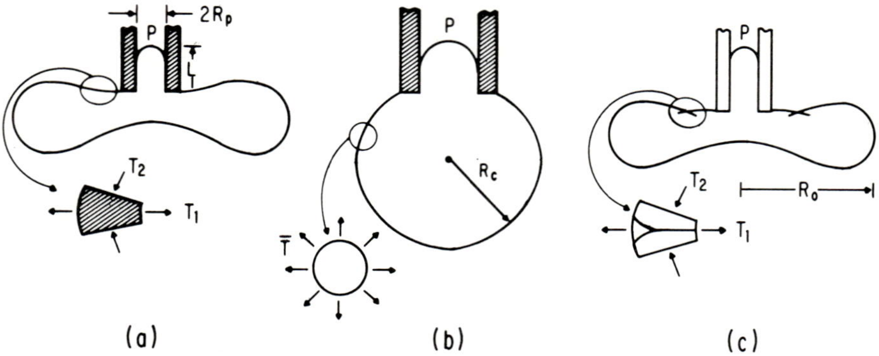

FIG. 12.3 Three micropipet aspiration experiments. (*a*) Micropipet aspiration of a flaccid cell for the measurement of the shear modulus μ. P is the aspiration pressure, R_p is the pipet radius, L is the aspiration length, and T_1 and T_2 are the stress resultants acting on the membrane surface. (*b*) Micropipet aspiration of a spherical cell for the measurement of the area modulus K. R_c is the cell radius, and $\bar{T}$ is the isotropic tension. (*c*) Buckling of a flaccid cell for the measurement of the bending modulus B. R_0 is the maximum cell radius.

12.3.1 Shear Elasticity

For the constant-area extension of a two-dimensional material, the maximum shear resultant T_s occurs along a line at 45° to the principal axis system (Fig. 12.2*a*). Shear and extension represent the same phenomenon and are characterized by a single elastic constant—the shear modulus μ. The relation between the shear resultant T_s and the extension ratio λ (extended length/original length) is [10, 16]

$$T_s \equiv \frac{T_1 - T_2}{2} = \frac{\mu}{2}(\lambda_1^2 - \lambda_1^{-2}) \tag{12.3}$$

For the pipet aspiration experiment, the relation between the aspiration pressure P (relative to the pressure outside the cell) and the extension up the pipet, L, as shown in Fig. 12.3*a*, is [17, 18]

$$P\frac{R_p}{\mu} = \frac{L}{R_p} - 1 + \ln\frac{2L}{R_p}$$

$$\approx 2.45\left(\frac{L}{R_p}\right) - 0.63 \qquad 1 \leq \frac{L}{R_p} \leq 4 \tag{12.4}$$

From Eq. (12.4) it is readily seen that μ is proportional to the slope of the P versus L line:

$$\mu = \frac{R_p^2}{2.45}\frac{\Delta P}{\Delta L} \tag{12.5}$$

The slope $\Delta P/\Delta L$ can be measured accurately. The major uncertainty in the determination of a value for μ is the uncertainty in the measurement of a value for the pipet radius R_p. A 20 percent error in the measurement of R_p (say, 0.2 out of 1.0 μm) will give about a 40 percent error in the value for μ.

On the basis of a single datum point for two different cells, Evans [10] first obtained a value for μ of 5×10^{-3} mN/m. [Note that the value for μ as defined by Eq. (12.3) and given in this paper is exactly one-half of the value given in Evans's original work.] This value agrees closely with the value for μ obtained from a study of the shear deformation of point-attached red cells

[8, 10]. Since the original measurement [10] was made, many investigators have measured values for μ [14, 17–24]. Values range from about 4×10^{-3} mN/m [17, 19] to 8×10^{-3} mN/m [20]. The typical value is $\mu = 6$ to 7×10^{-3} mN/m [14, 18, 21–24]. Recently, however, Evans et al. [25] accurately measured the inner diameter of their pipets from the insertion depth of a tapered microneedle that had been calibrated with the scanning electron microscope. Their value is $\mu = 9 \pm 1.7 \times 10^{-3}$ mN/m. Since this value is based on an accurate measurement of R_p, it must be given greater weight. Thus, at this time,

$$\mu \approx 6\text{–}9 \times 10^{-3} \text{ mN/m} \tag{12.6}$$

For a given pipet and a given sample of cells from the same donor, the standard deviation in the measurement of μ is about ± 20 percent. This is caused either by cell-to-cell variability or by errors in the measurement of the slope of the P versus L line [Eq. (12.5)]. All experiments to date indicate that μ, as defined by Eq. (12.3), is a constant. (Measurements are made at room temperature.)

12.3.2 Area Elasticity

An isotropic dilatation of the red cell membrane is produced by aspirating a preswollen cell into a micropipet until the portion of the cell outside the pipet forms a sphere (Figs. 12.2*b* and 12.3*b*). The isotropic tension $\bar{T}$ is given by [2, 26]

$$\bar{T} = \frac{PR_p}{2(1 - R_p/R_c)} \tag{12.7}$$

where P is the aspiration pressure, R_p is the pipet radius, and R_c is the cell radius. As the aspiration pressure is increased, the area of the cell expands because of an increase in the isotropic tension in the membrane. The expansion of the surface area is measured accurately by measuring the movement up the pipet of the membrane "tongue" within the pipet [26, 27]. Extremely small area dilatations can be measured in this way.

Evans et al. [26] propose a simple linear relation between the isotropic tension in the membrane and the relative area expansion

$$\bar{T} = K \frac{\Delta A}{A_0} \tag{12.8}$$

where ΔA is the increase in surface area, A_0 is the original area, and K is the area expansion modulus. The value for K at room temperature is [18, 27]

$$K = 450 \text{ mN/m} \tag{12.9}$$

with a standard deviation of about ± 15 to 20 percent [18]. (Measurements are made at room temperature.)

12.3.3 Bending Elasticity

Although small, the membrane's resistance to bending stabilizes the bioconcave shape of the cell during the early phases of the swelling process [28, 29] and during pipet aspiration [30]. For a shear elastic modulus of 6×10^{-3} mN/m, the bending modulus has to be about 10^{-19} N · m (10^{-12} erg) for the cell to swell smoothly (without popping) from a biconcave shape into a sphere [28, 29]. Also, this value for the bending modulus implies that bending does not contribute significantly to the pressure required to aspirate the membrane into a pipet [30].

The bending modulus B and shear modulus μ can be combined to form a dimensionless group [28]: $B/\mu R^2$, where R is a characteristic radius of curvature of the bend. Thus, bending is important if the radius of curvature is small.

The bending energy per unit area, F, relative to the stress-free state, is given by [16, 31–33]

$$F = \left(\frac{B}{2}\right)(C_1 + C_2 - C_0)^2 \tag{12.10}$$

where C_1, C_2 are the principal curvatures (the reciprocals of the principal radii of curvature), and C_0 is the chemically induced, or "spontaneous," curvature. The change in the free energy with respect to curvature gives the bending moment per unit length M:

$$M = B(C_1 + C_2 - C_0) \tag{12.11}$$

Evans [34] has designed a micropipet aspiration experiment (Fig. 12.3*c*) to measure the bending modulus B. In this experiment the cell is aspirated into a pipet until the circumferential compressive load on the membrane (Fig. 12.3*c*) causes it to buckle. A perturbation analysis of the onset of buckling shows that

$$\frac{B}{PR_p^3} = f\left(\frac{R_p}{R_0}\right) \tag{12.12}$$

where f is a given function of the ratio of the pipet radius R_p to the radius of the rim of the cell R_0. As can be seen in Eq. (12.12), the aspiration pressure P at which buckling occurs varies strongly and inversely with the pipet radius. The value for the bending elasticity is the specific value for B that gives the best fit of Eq. (12.12) to the experimental (P versus R_p) data. It is [25, 34]

$$B = 1.8 \pm 0.2 \times 10^{-19}\ \text{N} \cdot \text{m} \tag{12.13}$$

This measured value for B is two to three times larger than previous estimates [28–30]. (Measurements are made at room temperature.)

12.4 MEMBRANE VISCOSITY

The elasticity of the membrane characterizes its resistance to deformation. On the other hand, the viscosity of the membrane characterizes its resistance to a *rate* of deformation. For all experiments discussed in this section, the *rate* of deformation is one of shear (extension), as shown in Fig. 12.2*a*, and, thus, a shear viscosity is measured. Three types of viscous behavior will be discussed: (1) viscoelastic solid behavior, (2) membrane yield and viscoplastic flow, and (3) force relaxation and creep.

12.4.1 Viscoelastic Solid Behavior

When the membrane of a flaccid red cell is elongated (stretched) by micropipet aspiration (Fig. 12.3*a*) or fluid shear deformation, and then the force of deformation is suddenly removed, the cell rapidly recovers its undeformed shape. For this recovery process, Evans and Hochmuth [35] calculated that the viscous dissipation in the membrane is about two orders of magnitude greater than that in the internal hemoglobin solution. Thus, the membrane is the dominant source of viscous dissipation during the recovery process.

Evans and Hochmuth [35] accounted for the dissipation in a "viscoelastic" solid membrane by the addition to Eq. (12.3) of a type of relationship governed by Newton's law of viscosity:

$$T_s = \left(\frac{\mu}{2}\right)(\lambda_1^2 - \lambda_1^{-2}) + 2\eta\,\frac{\partial \ln \lambda_1}{\partial t} \tag{12.14}$$

where t is time, and η is the coefficient of surface viscosity. A characteristic time constant t_c is obtained in Eq. (12.14) simply by dividing η by μ:

$$t_c \equiv \frac{\eta}{\mu} \tag{12.15}$$

Hochmuth et al. [36] devised a simple and accurate way to measure t_c. In their experiment illustrated in Fig. 12.4, a red cell is elongated by pulling at diametrically opposite points on the

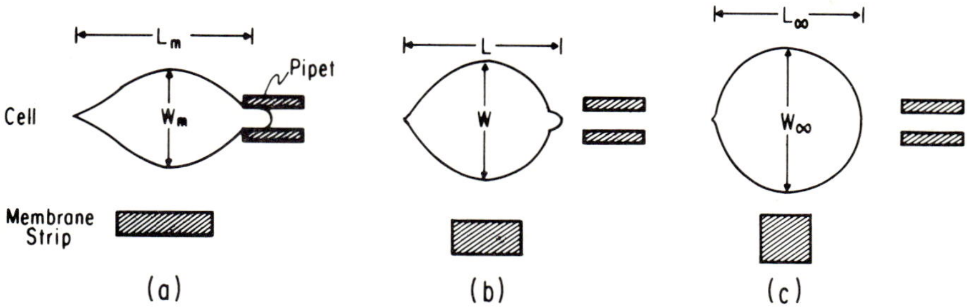

FIG. 12.4 Red cell recovery experiment showing the cell and a "membrane strip" of constant area cut from the central portion of the cell. (*a*) Extension of the cell prior to release. L_m is the maximum length of the cell, and W_m is the maximum width. (*b*) Cell at some time after its release. (*c*) Fully recovered cell.

red cell rim. Upon release, the red cell recovers its biconcave shape in about 0.35 s. This recovery process is described by the integration of Eq. (12.14) with $T_s = 0$ [35, 36]:

$$\frac{L/W - (L/W)_\infty}{L/W + (L/W)_\infty} \cdot \frac{(L/W)_m + (L/W)_\infty}{(L/W)_m - (L/W)_\infty} = \exp \frac{-t}{t_c} \tag{12.16}$$

where L/W, the length-width ratio, is the extension ratio squared (λ_1^2). The subscript m stands for the initial (maximum) value at $t = 0$, and the subscript ∞ stands for the final equilibrium value. The best fit of the experimental data to Eq. (12.16) gives a value for t_c. Since the original recovery experiments of Hochmuth et al. [36] were made, several investigators [14, 23–25] have used this technique to measure a value for t_c. The range of values for t_c measured in the recovery experiments is [14, 23–25, 36]

$$t_c = 0.1 - 0.13 \text{ s} \tag{12.17}$$

with a standard deviation of about ±20 percent. (Measurements are made at room temperature.) With the range of values for μ [Eq. (12.6)] and the range of values for t_c [Eq. (12.17)], the range of values for η is calculated with Eq. (12.15):

$$\eta = 0.6\text{–}1.2 \times 10^{-3} \text{ mN} \cdot \text{s/m} \tag{12.18}$$

Chien et al. [17] aspirated a portion of the red cell membrane into a pipet (Fig. 12.3*a*) and then studied the recovery of the membrane within the pipet upon removal of the aspiration pressure. Their analysis of this experiment, using the viscoelastic solid model of Evans and Hochmuth [Eq. (12.14)], gave a recovery time constant of 0.13 ± 0.05 s for 100 cells. Additional experiments by Chien et al. [17], in which cells were rapidly aspirated into a pipet, indicate that some "shear thinning" of the membrane may occur at high rates of deformation, since the time constant decreased by about a factor of 5. However, in these experiments it is possible that an unstable "buckling" of the membrane occurred at higher aspiration pressures [34]. Such a buckling of the membrane would allow the cell to move up the pipet at a faster rate than expected, and this would lead to the calculation of smaller values for t_c.

12.4.2 Yield and Continuous Flow

Macroscopic flow of red cell membrane has been observed by Hochmuth et al. [37]. In these experiments, cells are allowed to adhere to a glass surface and then are subjected to a fluid shear force. When the fluid shear force exceeds a certain critical value ($\sim 10^{-8}$ mN), the cells begin to move slowly downstream as though attached to the surface at a single point by a "tether." Under the action of the constant fluid shear force, this long, thin, apparently hollow membrane filament (tether) steadily increases in length as membrane material flows from the flaccid cell body to the tether. An increase in the fluid shear force results in an increase in the rate of growth of the tether and, therefore, an increase in the flow of membrane material from cell body to the tether.

Evans and Hochmuth [38] modeled this membrane-flow process by assuming that the membrane behaves as a two-dimensional version of a "Bingham plastic":

$$\frac{\partial \ln \lambda_1}{\partial t} = 0 \qquad T_s < T_y$$
$$T_s = T_y + 2\eta_p \frac{\partial \ln \lambda_1}{\partial t} \qquad T_s > T_y \tag{12.19}$$

where T_y is the value for the shear resultant at the yield point where the membrane begins to flow, and η_p is the viscosity for plastic flow. The relation between T_y and the force on the cell at the point where the tether is formed is [38]

$$T_y = \frac{f}{4\pi R_t} \tag{12.20}$$

where R_t is the tether radius, and f is the force (calculated as $f = \tau_s A_c$, where τ_s is the fluid shear stress at the wall of the channel and A_c is the projected area of the cell). From observations with the light microscope and the scanning electron microscope, R_t is estimated to be between 20 and 50 nm for a flaccid cell. Thus, when $f = 10^{-8}$ mN,

$$T_y = 1.6\text{–}4.0 \times 10^{-2} \text{ mN/m} \tag{12.21}$$

The value for η_p [Eq. (12.19)] is inversely proportional to the tether growth rate (i.e., the cell velocity) and directly proportional to the force on the cell (and tether) at yield [38]. However, it is independent of the tether radius. Its value is [21, 38–40]

$$\eta_p \approx 10^{-2} \text{ mN} \cdot \text{s/m} \tag{12.22}$$

However, this value for η_p is only a good estimate since the analysis on which the calculation for η_p is based [38] assumes an axisymmetric configuration as an approximation of the complicated disklike geometry.

Hochmuth and Evans [41] and Hochmuth et al. [42] have extracted tethers from axisymmetric spherical cells held in a pipet (Fig. 12.3*b*). A small bead adhering to the membrane at a point opposite to the point of aspiration is translated at a constant rate (Fig. 12.5). A reservoir of membrane material is created within the pipet, and this material is depleted as the tether is formed at constant membrane surface area and volume. Thus, since the cell is essentially spherical, the value for the tether radius is readily calculated from conservation principles [41]:

$$R_t = \left(1 - \frac{R_p}{R_c}\right)(R_p)\left(\frac{-dL_p}{dL_t}\right) \tag{12.23}$$

where R_p is the radius of the pipet, R_c is the maximum radius of the cell, $-dL_p$ is the decrease in the aspirated portion of the membrane within the pipet, and dL_t is the increase in tether length. Since all quantities on the right-hand side of Eq. (12.23) can be measured, the tether radius R_t can be calculated as the tether is formed. Typically, the value for R_t is about 15 nm [42]. Its minimum value is about 6 nm [43].

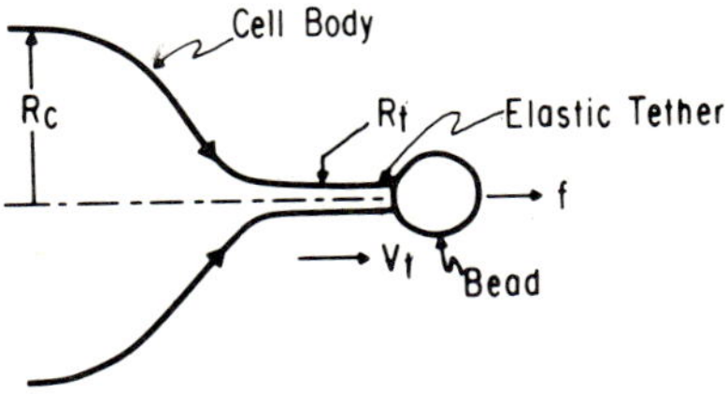

FIG. 12.5 Flow of cell membrane from a nearly spherical cell body to a cylindrical, elastic tether. R_c is the maximum radius of the cell body, R_t is the tether radius, and V_t is the tether (or cell) velocity.

The membrane-flow, tether-formation process shown in Fig. 12.5 is analyzed with the equations of membrane mechanics and Newton's law of viscosity [i.e., T_y in Eq. (12.19) is neglected]. This analysis [41] shows that the flow of membrane from cell body to tether causes a tension difference between the axial tension in the tether, $f/2\pi R_t$, and the isotropic tension $\bar{T}$ in the cell membrane. (The quantity f is the axial force that pulls on the tether.) The result is

$$R_t = \frac{f}{2\pi\bar{T}} - \frac{2\eta_p V_t}{c\bar{T}} \tag{12.24}$$

where η_p is the coefficient of surface viscosity, V_t is the tether velocity, and $c = 1.6$. The value for η_p in Eq. (12.24) is obtained by measuring the tether radius as a function of the isotropic tension $\bar{T}$ and the velocity of the tether V_t:

$$\eta_p \approx 3 \times 10^{-3}\ \text{mN} \cdot \text{s/m} \tag{12.25}$$

Note that the recovery experiments in which the membrane behaves as a viscoelastic solid give a smaller value for the viscosity of 0.6 to 1.2×10^{-3} mN · s/m [Eq. (12.18)], while the studies of membrane flow from a flaccid cell give a larger value for the viscosity of about 10^{-2} mN · s/m [Eq. (12.22)].

12.4.3 Creep and Force Relaxation

Small forces applied for periods of time greater than a few minutes result in permanent deformation of the membrane [19]. This process of creep and force relaxation in red cell membrane has been modeled by placing an elastic membrane component [Eq. (12.3)] in series with a linear (newtonian) viscous component [44]. With this model in mind, Markle et al. [45] studied the permanent deformation processes in red cell membrane with two techniques: micropipet aspiration of a portion of a flaccid cell (Fig. 12.3*a*) and extension of a whole cell between two micropipets. In both experiments the extent of residual membrane deformation is negligible when the time of extension is less than several minutes. However, extensions maintained for longer time periods result in significant permanent deformation. The magnitude of the permanent deformation is proportional to the total time period of extension and the level of the applied force. When these results are analyzed according to the analytical model of this long-term deformation process [44], a value for a "creep viscosity" can be calculated. However, the results are strongly influenced by the concentration of albumin in the suspending fluid. For albumin concentrations of 0.0, 0.01, 0.1, and 1 g %, the creep viscosity is equal to 3.6, 14.8, 25.6, and 51.9 mN · s/m, respectively. These values for the creep viscosity are approximately four orders of magnitude greater than those values given in Eqs. (12.18), (12.22), and (12.25). These extremely large values for the membrane creep viscosity probably represent a slow and permanent molecular reorganization of the structural membrane components (e.g., "spectrin").

12.5 HEMOGLOBIN VISCOSITY

The cytoplasm within the human red cell is a concentrated solution of hemoglobin. In normal cells, hemoglobin concentrations range from 0.31 to 0.38 kg/L [46]. The viscosity of hemoglobin at 37°C and 0.31 kg/L is about 6 mN · s/m^2 [47].

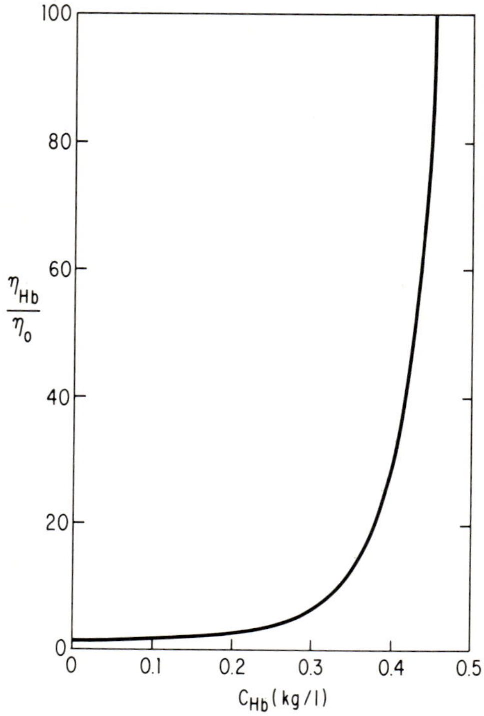

FIG. 12.6 Graph of Eq. (12.26) [48]. Variation of hemoglobin viscosity η_{Hb}. η_0 is the viscosity of the suspending phase (e.g., saline).

Ross and Minton [48] use the following equation to fit the hemoglobin viscosity data of Chien et al. [49] and Charache et al. [50]:

$$\eta_{Hb} = \eta_0 \exp \frac{A_1 C_{Hb}}{1 - A_2 C_{Hb}} \tag{12.26}$$

where η_{Hb} is the hemoglobin viscosity, η_0 is the viscosity of the suspending medium (i.e., saline), C_{Hb} is the hemoglobin concentration, and A_1 and A_2 are constants: $A_1 = 3.6$ L/kg, and $A_2 = 0.40 \times 3.6 = 1.4$ L/kg. A plot of Eq. (12.26) is shown in Fig. 12.6. Equation (12.26) is valid for hemoglobin solutions at 37°C, although it is probably accurate at other temperatures as long as η_0 is evaluated at the appropriate temperature. It is also accurate for abnormal hemoglobin such as Hb S (sickle-cell hemoglobin) or Hb C as long as the hemoglobin molecules remain in solution [49, 50].

12.6 CHANGES OF MATERIAL PROPERTIES

12.6.1 Variation with Species

Waugh and Evans [20] have measured the shear elasticity of several mammals, birds, amphibians, and reptiles. In addition, they estimated the viscosity for several species from a study of

TABLE 12.4 Values for the shear elasticity μ and viscosity η for several species

Species	$\mu \times 10^3$ mN/m	$\eta \times 10^4$ mN · s/m	Reference
Human	8.0	. . . .	20
Human	5.4	. . . .	22
Opossum	7.7	3.4	20
Rabbit	3.6	. . . .	22
Turkey	40	58	20
Bullfrog	37	67	20
Conga snake	68	130	20
Toadfish	34	87	20
Pond turtle	120	370	20
Painted turtle	97	. . . .	20
King snake	104	120	20

the recovery time of small "bumps" produced by micropipet aspiration. Also, Nash and Wyard [22] measured the elasticity of red cells from rabbits. The results of these studies are given in Table 12.4. Note that the values for μ given by Waugh and Evans [20] are divided by 2 before being listed in Table 12.4 in order to make them consistent with the current definition [Eq. (12.3)]. Note also that nucleated cells have extremely large values for the shear elasticity and viscosity.

12.6.2 Effect of Temperature

Table 12.5 shows the variation with temperature of the shear elasticity, area elasticity, viscosity for viscoelastic recovery η, and the viscosity for viscoplastic flow η_p. Note that the shear modulus changes only slightly with temperature while the values for the viscosities depend strongly on the temperature. The change in μ with temperature is -6×10^{-5} mN/(m · °C), and the change in K with temperature is -6 mN/(m · °C) [18]. The thermal area expansivity for the red cell membrane is 1.2×10^{-3}/°C, and the heat of extension for unit extension is 2×10^{-2} mN · m/m² [18]. The viscosity data in Table 12.5 can be expressed as $\eta = \exp(-E/RT)$, where E is the "activation energy" and RT is the gas constant times the absolute temperature. For η and η_p in Table 12.5, the values for E are 10.0 and 18.6 kcal/(g · mol) [39]. Unlike the elastic and viscous properties, the yield shear resultant T_y, surprisingly, does not appear to change with temperature [39].

TABLE 12.5 Variation of membrane properties with temperature*

Temperature, °C	$\mu \times 10^3$ mN/m [18]	K, mN/m [18]	t_c, s [51]	$\eta \times 10^4$ mN · s/m [51]	$\eta_p \times 10^3$ mN · s/m [39]
5–6	8.0	~560	0.27	21	
12	. . . .	530	. . . .	. . . .	29
15–16	6.8	~500	0.20	14	10
25	6.6	440	0.11	7.4	7.2
35–37	6.1	375	0.06	3.6	2.7
40	. . . .	360	. . . .	. . . .	1.3
46	5.2	~320	. . . .	~0	

* μ = shear elasticity, K = area modulus, t_c = recovery time constant, η = surface viscosity during recovery, η_p = plastic viscosity. Data compiled from references given in parentheses.

Heat treatment of cells at high temperatures results in an irreversible increase in the elastic shear modulus μ [52]. Cells heated at 47° for 6 min and then rapidly cooled (quenched) to room temperature experience about a threefold increase in μ [52]. A similar increase occurs within 3 to 4 min when cells are heated at 48.8 instead of 47°C [52].

12.6.3 Biochemical Treatments

Small concentrations of membrane-permeable, bifunctional, SH-reagents (e.g., diamide, tetrathionate) decrease the elongation of human red cells freely suspended in a shear field [53] and, therefore, increase the shear modulus. Small concentrations of the lectin *wheat germ agglutinin* (WGA) in the presence of albumin can cause a threefold increase in the value for η without changing the value for μ [54]. ATP depletion via incubation at 37°C for 24 h has *no* effect on μ, K, or η_p [55]. Finally, cholesterol enrichment and depletion of human red cell membrane have *no* effect on μ or η [56].

12.6.4 Changes in Hemoglobin Concentration

Since hemoglobin binds to the membrane of intact red blood cells [57], it is possible that the elastic and viscous properties of the membrane will depend on cellular hemoglobin concentration. Cells with higher hemoglobin concentrations are denser and older. Cells can be separated into different density fractions either by simple centrifugation [14, 22, 24] or by suspending them in solutions with discontinuous density gradients, i.e., Stractan layers [25]. The latter is more precise if sufficiently small density increments are used.

For normal cells, the value for the shear elastic modulus μ is independent of hemoglobin concentration [14, 22, 24, 25]. Also, the bending elasticity B is independent of hemoglobin concentration except for concentrations greater than 0.46 kg/L [25]. However, the recovery time constant t_c [Eqs. (12.15) and (12.16)] increases with density [14, 24] and hemoglobin concentration [25]. Thus, η increases in direct proportion with t_c, since μ is a constant. Red cells from the bottom 5 percent of a centrifuge tube have a value for t_c that is about 50 percent larger than those from the top 5 percent [14]. Studies in which cells were precisely separated into density fractions with Stractan solutions reveal a dramatic increase in t_c with an increase in hemoglobin concentration [25]. For example, at 0.42 kg/L (an abnormally large value), the value for t_c is about 10 times larger than its value at a normal hemoglobin concentration of 0.32 kg/L [25]. The increase in the bulk hemoglobin viscosity (Fig. 12.6) caused by this increase in hemoglobin concentration should increase t_c by only approximately 10 percent, on the basis of estimates of the dissipation in bulk hemoglobin relative to that in the membrane [14, 23, 35, 36, 51]. Thus, one may conclude that the in-plane shear viscosity of the membrane is increased because of an increase in the amount of membrane-bound hemoglobin [14, 23–25]. However, the elasticity of the membrane remains unaffected by this postulated increase in membrane-bound hemoglobin.

12.6.5 Oxygenated Sickle Cells

The densest of the sickle cells, the so-called irreversibly sickled cell (ISC), has a value for μ that is about twice that of normal cells [24]. When separated by Stractan gradients, the shear elasticity of sickle cells begins to increase when the hemoglobin concentration exceeds about 0.38 kg/L [25]. At 0.45 kg/L, μ is about 2½ times that of normal cells [25]. The bending elasticity does not change [25]. The value for t_c increases with hemoglobin concentration just as it does for normal cells. Thus, the viscosity of the densest sickle cells is very large because of an increase in both μ and t_c [24]. Swelling these cells does not alter μ but dramatically decreases t_c so that their values for η approach normal values [24].

12.6.6 Hereditary Spherocytes

In hereditary spherocytosis, the value for the yield shear resultant T_y is about one-third that of normal [21]. The value for μ is about two-thirds that of normal [21]. The value for K is also reduced by as much as a factor of 2 (unpublished observations). The plastic-flow viscosity η_p appears to be normal [21].

12.6.7 Other Cells

Ghost cells have a greater value for μ than intact cells, but their value for η is normal [23]. The elasticity of red cells from males with Duchenne's muscular dystrophy and female carriers is only slightly elevated [58].

ACKNOWLEDGMENTS

Almost half of the references (27 out of 58) appeared in the *Biophysical Journal*. The author gratefully acknowledges the support of this journal. This work is supported by grant number HL 23728 from the National Heart, Lung, and Blood Institute.

REFERENCES

1. Rand, R. P., and Burton, A. C.: Mechanical properties of the red cell membrane: I. Membrane stiffness and intracellular pressure, *Biophys. J., 4,* 115–135, 1964.
2. Rand, R. P.: Mechanical properties of the red cell membrane: II. Viscoelastic breakdown of the membrane, *Biophys. J., 4,* 303–316, 1964.
3. Fung, Y. C.: Theoretical considerations of the elasticity of red cells and small blood vessels, *Fed. Proc., 25,* 1761–1772, 1966.
4. Fung, Y. C. B., and Tong, P.: Theory of the sphering of red blood cells, *Biophys. J., 8,* 175–198, 1968.
5. Canham, P. B., and Parkinson, D. R.: The area and volume of single human erythrocytes during gradual osmotic swelling to hemolysis, *Can. J. Physiol. Pharmacol., 48,* 369–376, 1970.
6. Evans, E., and Fung, Y-C.: Improved measurements of the erythrocyte geometry, *Microvasc. Res., 4,* 335–347, 1972.
7. Evans, E. A., and Leblond, P. F.: Geometric properties of individual red blood cell discocyte-spherocyte transformations, *Biorheology, 10,* 393–405, 1973.
8. Hochmuth, R. M., and Mohandas, N.: Uniaxial loading of the red-cell membrane, *J. Biomech., 5,* 501–509, 1972.
9. Evans, E. A.: A new material concept for the red cell membrane, *Biophys. J., 13,* 926–940, 1973.
10. Evans, E. A.: New membrane concept applied to the analysis of fluid shear– and micropipette-deformed red blood cells, *Biophys. J., 13,* 941–954, 1973.
11. Skalak, R., Tözeren, A., Zarda, R. P., and Chien, S.: Strain energy function of red blood cell membranes, *Biophys. J., 13,* 245–264, 1973.
12. Ponder, E.: *Hemolysis and Related Phenomena,* Grune & Stratton, New York, 1948.
13. Fung, Y. C.: *Biomechanics: Mechanical Properties of Living Tissue,* Springer-Verlag, New York, 1981.
14. Linderkamp, O., and Meiselman, H. J.: Geometric, osmotic, and membrane mechanical properties of density-separated human red cells, *Blood, 59,* 1121–1127, 1982.
15. Chen, P. C. Y., and Fung, Y. C.: Extreme-value statistics of human red blood cells, *Microvasc. Res., 6,* 32–43, 1973.
16. Evans, E. A., and Hochmuth, R. M.: Mechanochemical properties of membranes, in F. Bronner and A. Kleinzeller (eds.), *Current Topics in Membranes and Transport,* Academic Press, New York, 1978.

17. Chien, S., Sung, K-L. P., Skalak, R., Usami, S., and Tözeren, A.: Theoretical and experimental studies on viscoelastic properties of erythrocyte membrane, *Biophys. J., 24,* 463–487, 1978.

18. Waugh, R., and Evans, E. A.: Thermoelasticity of red blood cell membrane, *Biophys. J., 26,* 115–132, 1979.

19. Evans, E. A., and LaCelle, P. L.: Intrinsic material properties of the erythrocyte membrane indicated by mechanical analysis of deformation, *Blood, 45,* 29–43, 1975.

20. Waugh, R., and Evans, E. A.: Viscoelastic properties of erythrocyte membranes of different vertebrate animals, *Microvasc. Res., 12,* 291–304, 1976.

21. Waugh, R. E., and LaCelle, P. L.: Abnormalities in the membrane material properties of hereditary spherocytes, *J. Biomech. Eng., 102,* 240–246, 1980.

22. Nash, G. B., and Wyard, S. J.: Erythrocyte membrane elasticity during in vivo ageing, *Biochim. Biophys. Acta, 643,* 269–275, 1981.

23. Nash, G. B., and Meiselman, H. J.: Red cell and ghost viscoelasticity: Effects of hemoglobin concentration and in vivo ageing, *Biophys. J., 43,* 63–73, 1983.

24. Nash, G. B., Johnson, C. S., and Meiselman, H. J.: Mechanical properties of oxygenated red blood cells in sickle cell (HbSS) disease, *Blood, 63,* 73–82, 1984.

25. Evans, E., Mohandas, N., and Leung, A.: Static and dynamic rigidities of normal and sickle erythrocytes: Major influence of cell hemoglobin concentration, *J. Clin. Invest., 73,* 477–488, 1984.

26. Evans, E. A., Waugh, R., and Melnik, L.: Elastic area compressibility modulus of red cell membrane, *Biophys. J., 16,* 585–595, 1976.

27. Evans, E. A., and Waugh, R.: Osmotic correction to elastic area compressibility measurements on red cell membrane, *Biophys. J., 20,* 307–313, 1977.

28. Zarda, P. R., Chien, S., and Skalak, R.: Elastic deformations of red blood cells, *J. Biomech., 10,* 211–221, 1977.

29. Fischer, T. M., Haest, C. W. M., Stohr-Liesen, M., Schmid-Schönbein, H., and Skalak, R.: The stress-free shape of the red blood cell membrane, *Biophys. J., 34,* 409–422, 1981.

30. Evans, E. A.: Minimum energy analysis of membrane deformation applied to pipet aspiration and surface adhesion of red blood cells, *Biophys. J., 30,* 265–284, 1980.

31. Helfrich, W.: Elastic properties of lipid bilayers: Theory and possible experiments, *Z. Naturforsch., 28c,* 693–703, 1973.

32. Evans, E. A.: Bending resistance and chemically induced moments in membrane bilayers, *Biophys. J., 14,* 923–931, 1974.

33. Deuling, H. J., and Helfrich, W.: Red blood cell shapes as explained on the basis of curvature elasticity, *Biophys. J., 16,* 861–868, 1976.

34. Evans, E. A.: Bending elastic modulus of red blood cell membrane derived from buckling instability in micropipet aspiration tests, *Biophys. J., 43,* 27–30, 1983.

35. Evans, E. A., and Hochmuth, R. M.: Membrane viscoelasticity, *Biophys. J., 16,* 1–11, 1976.

36. Hochmuth, R. M., Worthy, P. R., and Evans, E. A.: Red cell extensional recovery and the determination of membrane viscosity, *Biophys. J., 26,* 101–114, 1979.

37. Hochmuth, R. M., Mohandas, N., and Blackshear, P. L., Jr.: Measurement of the elastic modulus for red cell membrane using a fluid mechanical technique, *Biophys. J., 13,* 747–762, 1973.

38. Evans, E. A., and Hochmuth, R. M.: Membrane viscoplastic flow, *Biophys. J., 16,* 13–26, 1976.

39. Waugh, R. E.: Temperature dependence of the yield shear resultant and the plastic viscosity coefficient of erythrocyte membrane: Implications about molecular events during membrane failure, *Biophys. J., 39,* 273–278, 1982.

40. Hochmuth, R. M., Evans, E. A., and Colvard, D. F.: Viscosity of human red cell membrane in plastic flow, *Microvasc. Res., 11,* 155–159, 1976.

41. Hochmuth, R. M., and Evans, E. A.: Extensional flow of erythrocyte membrane from cell body to elastic tether: I. Analysis, *Biophys. J., 39,* 71–81, 1982.

42. Hochmuth, R. M., Wiles, H. C., Evans, E. A., and McCown, J. T.: Extensional flow of erythrocyte membrane from cell body to elastic tether: II. Experiment, *Biophys. J., 39,* 83–89, 1982.

43. Hochmuth, R. M., Evans, E. A., Wiles, H. C., and McCown, J. T.: Mechanical measurement of red cell membrane thickness, *Science, 220,* 101–102, 1983.

44. Evans, E. A., and Skalak, R.: *Mechanics and Thermodynamics of Biomembranes,* CRC Press, Boca Raton, Fla., 1980.

45. Markle, D. R., Evans, E. A., and Hochmuth, R. M.: Force relaxation and permanent deformation of erythrocyte membrane, *Biophys. J., 42,* 91–98, 1983.

46. Williams, A. R., and Morris, D. R.: The internal viscosity of the human erythrocyte may determine its lifespan in vivo, *Scand. J. Haematol., 24,* 57–62, 1980.

47. Cokelet, G. R., and Meiselman, H. J.: Rheological comparison of hemoglobin solutions and erythrocyte suspensions, *Science, 162,* 275–277, 1968.

48. Ross, P. D., and Minton, A. P.: Hard quasispherical model for the viscosity of hemoglobin solutions, *Biochem. Biophys. Res. Commun., 76,* 971–976, 1977.

49. Chien, S., Usami, S., and Bertles, J. F.: Abnormal rheology of oxygenated blood in sickle cell anemia, *J. Clin. Invest., 49,* 623–634, 1970.

50. Charache, S., Conley, C. L., Waugh, D. F., Ugoretz, R. J., Spurell, J. R., and Gayle, E.: Pathogenesis of hemolytic anemia in homozygous hemoglobin C disease, *J. Clin. Invest., 46,* 1795–1811, 1967.

51. Hochmuth, R. M., Buxbaum, K. L., and Evans, E. A.: Temperature dependence of the viscoelastic recovery of red cell membrane, *Biophys. J., 29,* 177–182, 1980.

52. Rakow, A. L., and Hochmuth, R. M.: Effect of heat treatment on the elasticity of human erythrocyte membrane, *Biophys. J., 15,* 1095–1100, 1975.

53. Fisher, T. M., Haest, C. W. M., Stohr, M., Kamp, D., and Deuticke, B.: Selective alteration of erythrocyte deformability by SH-reagents: Evidence for an involvement of spectrin in membrane shear elasticity, *Biochim. Biophys. Acta, 510,* 270–282, 1978.

54. Smith, L., and Hochmuth, R. M.: Effect of wheat germ agglutinin on the viscoelastic properties of erythrocyte membrane, *J. Cell Biol., 94,* 7–11, 1982.

55. Meiselman, H. J., Evans, E. A., and Hochmuth, R. M.: Membrane mechanical properties of ATP-depleted human erythrocytes, *Blood, 52,* 499–504, 1978.

56. Chabane, A., Flamm, M., Sung, K. L. P., Lee, M. M., Schachter, D., and Chien, S.: Influence of cholesterol content on red cell membrane viscoelasticity and fluidity, *Biophys. J., 44,* 171–176, 1983.

57. Eisinger, J., Flores, J., and Salhany, J. M.: Association of cytosol hemoglobin with the membrane in intact erythrocytes, *Proc. Nat. Acad. Sci. U.S.A., 79,* 408–412, 1982.

58. Nash, G. B., and Wyard, S. J.: Measurement of erythrocyte membrane elasticity as a diagnostic aid in Duchenne muscular dystrophy, *J. Med. Genet., 19,* 262–265, 1982.

CHAPTER 13

Rheology of Leukocytes

Geert W. Schmid-Schönbein
Department of Applied Mechanics and Engineering Sciences
University of California, San Diego
La Jolla, California

INTRODUCTION

The term *leukocyte,* or *white blood cell,* stands for a class of five morphologically distinct cells which are circulating in the blood and have many functions in normal and diseased tissue. They play a key role in the protective mechanisms of the body against diseases, in tissue inflammation, in wound healing, and in other physiological and pathological processes.

Among the formed elements of the blood, the leukocytes constitute the smallest population, with a ratio of one leukocyte to about 1000 erythrocytes. Therefore, in many blood rheological considerations leukocytes have traditionally been neglected. This assumption of their insignificance may be satisfactory for blood flow in large vessels, but in the microcirculation, where the diameters of the blood vessels and cells are comparable, leukocytes can exert a strong hemodynamic influence. Whereas leukocytes in the bloodstream are carried passively by the plasma, leukocytes migrating across the endothelium into the tissue become active and require chemical energy for active deformation.

In this short review we discuss some morphological and rheological properties of leukocytes and explore implications with respect to their motion in the microcirculation. The remarks will be limited to the circulating leukocytes. For classical subjects of leukocyte physiology, the reader is referred to the following books and papers: for the development in the bone marrow, thymus, and lymph, [1–4]; for the biochemistry and cellular physiology, [5–9]; for immunology, [10–12]; for chemotaxis, [13, 14]; and for phagocytosis, [15]. No reference will be made to diseased cells (see, e.g., the book by Rapaport [16], or a textbook in hematology). Several aspects of leukocyte rheology have recently been reviewed [17–19].

13.1 MORPHOLOGY

Cell volume is needed to analyze cell motion in the circulation. The several different methods existing today for visualizing leukocytes have led to disagreement about the cell dimensions, and have also yielded different morphological pictures. Thus, there exists a need to evaluate each method and to seek a unifying picture of leukocyte morphology.

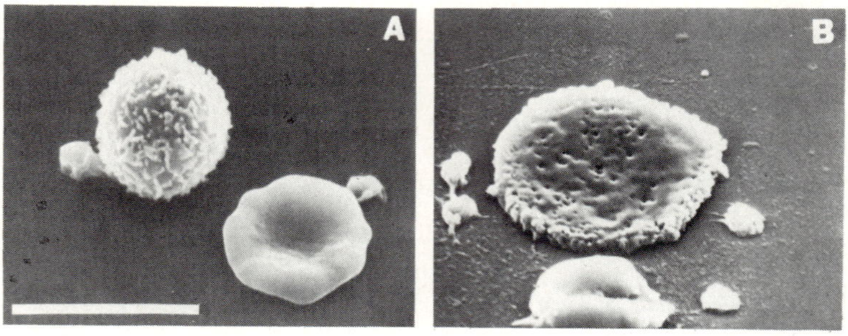

FIG. 13.1 Scanning electron microscopic image of (*a*) human leukocyte in its undeformed spherical state (crossbar represents a length of 10 μm) and (*b*) leukocyte on a hematological blood film after deformation into a pancake shape. The thickness is about 1 μm and its diameter is greater than 12 μm. One erythrocyte and several platelets are shown for comparison.

13.1.1 The Blood Film

Since the introduction of histological dyes about 100 years ago [20, 21], leukocytes have been investigated on blood films (*Ausstrich* in German). Blood-film investigation is the most common technique in hematology today, and many modifications exist. In a typical experiment a small drop of blood is placed on an objective glass slide, and with a second slide the blood is gently spread out to a thin film with a monolayer of cells. However, in the process of spreading, the leukocytes are compressed from their undeformed spherical shape to a flattened pancake shape (Fig. 13.1). The cells remain on the blood film in this shape because the time required for recovery of the deformed cell to the spherical shape is longer than the time which most blood films need in order to dry. The diameters of these spread-out cells are determined not only by the cell volume but also by the cell membrane area [22]. The values of the diameters of the flattened cells (10 to 20 μm) are often cited in hematology textbooks, but they exceed those values for the diameters of the undeformed spherical cells and are incompatible with normal flow through capillaries.

On a stained blood film five classes of circulating leukocytes can be recognized: neutrophils, eosinophils, and basophils—sometimes collectively referred to as *granulocytes*—and lymphocytes and monocytes. Each granulocyte is recognized by the color and size of its granules. Lymphocytes and monocytes have small granules, few of which can be resolved with light microscopy; thus they are usually distinguished by their coloration, size, and nuclear shape. Many detailed color photographs of leukocytes on a blood film have been published, and the reader is referred to Refs. 23 and 24 or to other hematology textbooks.

13.1.2 Light Macroscopic Investigation

In the undeformed state leukocytes in aqueous suspension are spherical and have many fine membrane folds on their surfaces (Fig. 13.2*b*). In a hypertonic medium the cells shrink (Fig. 13.2*a*), showing larger membrane folds, and in a hypotonic medium they swell and the membrane unfolds (Fig. 13.2*c*), maintaining constant area [25]. After the membrane is unfolded to a smooth sphere, further swelling leads to membrane rupture and hemolysis. In an isotonic medium the leukocyte diameter distribution of a normal human being (without hematological disease symptoms) shows two separate peaks (Fig. 13.3), one peak at about 6 μm due to the lymphocyte population and a second peak at about 7.5 μm. This distribution represents a superposition of the neutrophil, basophil, eosinophil, and monocyte populations, each of which has a relatively narrow gaussian diameter distribution [25]. These distributions are altered under

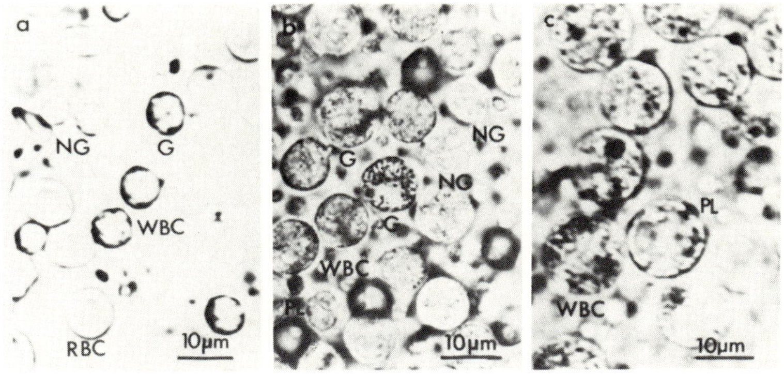

FIG. 13.2 Human leukocyte in the passive state in aqueous solution at (*a*) 525, (*b*) 305, and (*c*) 100 mosmol. Granular leukocytes (G) and nongranular leukocytes (NG; mostly lymphocytes) may be distinguished. (*From Schmid-Schönbein et al.* [25], *with permission.*)

conditions of disease [26]. The average population diameter is shown in Fig. 13.4 as a function of the solution osmolality. These values agree with in vivo measurements in humans [27].

13.1.3 Scanning Electron Microscopic Investigation

The use of the scanning electron microscope (SEM) allows observation of cell surface features in three dimensions. However, SEM images are not always readily interpreted without consideration of the cell dehydration. When leukocytes are prepared for the SEM, they are chemically fixed and then dehydrated. In many cases the dehydration is achieved with the critical-point drying method [28]. During the dehydration process the cells shrink but preserve the surface folds which were present after chemical fixation. Swollen leukocytes (in 100 mosmol) have no membrane fold, and appear in the SEM as smooth spheres (Fig. 13.5). Leukocytes in an isotonic (300 mosmol) or hypertonic (500 mosmol) medium exhibit the usual membrane folds (Fig. 13.5), but their population diameter is about the same as that of the swollen cells when it is measured from SEM images (Fig. 13.4). Thus, the state of cell hydration before the preparation for microscopy is not reflected on SEM images. Nevertheless, the shape of the cell surface folds in SEM imaging is strongly dependent on the degree of water uptake prior to fixation.

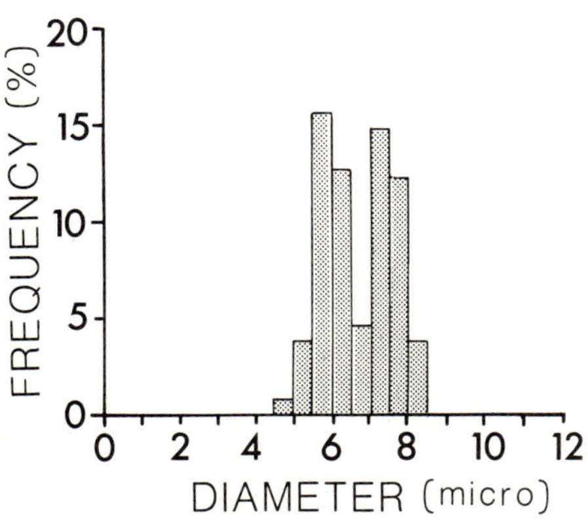

FIG. 13.3 Diameter histogram of human leukocyte population in 300-mosmol suspension. The average diameter is 6.7 μm. Diameters were measured across the inner ring of the light interference ring demonstrating the edge of the leukocyte. If the diameter is measured across the outer edge of the interference band, about 0.45 m has to be added to the diameter values. (*From Schmid-Schönbein et al.* [25], *with permission.*)

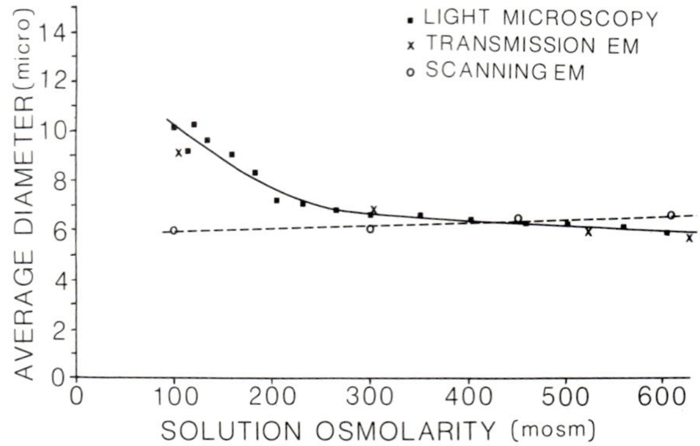

FIG. 13.4 Average cell diameter of human leukocyte population as a function of the osmolality of the suspending medium. The light microscopic values (■) represent measurement on the inner edge of the light interference ring. The transmission electron microscopic measurements (*x*) represent values obtained from stereological analysis of random thin sections. The scanning electron microscopic measurements (○) represent diameters across the outer edge of the surface folds. (*From Schmid-Schönbein et al.* [25], *with permission.*)

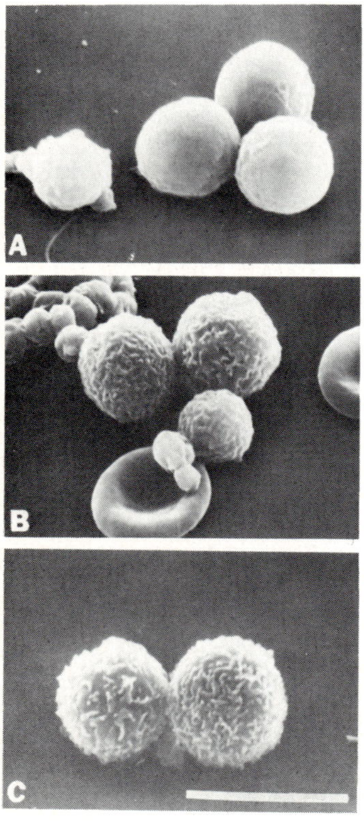

FIG. 13.5 Scanning electron microscopic image of human leukocytes in suspending medium of (*A*) 100, (*B*) 300, and (*C*) 500 mosmol. The length of the crossbar represents 10 μm. The leukocytes at 100 mosmol have no surface folds, and the cell diameters are about equal in all three cases.

The above investigative results explain the claim that, with the SEM, human T, or thymus-derived, lymphocytes may be distinguished from the B lymphocytes by the absence of the surface folds [29]. While some investigators support this finding [30, 31], others see no difference [32, 33] or report conversely that B lymphocytes have a smooth membrane and T lymphocytes retain their folds [34, 35]. The explanation for these contradictory findings probably lies in the differing preparation of the cells for the SEM. For example, the investigators who observe T lymphocytes with a smooth surface have obtained them by filtering the cells through a nylon filter. The T lymphocytes remain attached to the nylon filters and must be washed off at high shear stress. This washing process damages the cells, causing them to undergo partial or complete swelling and a smoothing of membrane folds. Following the dehydration procedure in SEM preparation, these swollen cells appear on SEM images with few or no surface folds, but their diameters no longer differ from those of the normal unswollen cells in the SEM (Fig. 13.4). The surface folds of freely suspended neutrophils consist of crestlike elevations, whereas those of lymphocytes often appear to be spikelike extensions. Temperature also affects the shape and size of the folds [30].

13.1.4 Transmission Electron Microscopic Investigation

When leukocytes are prepared for the transmission electron microscope (TEM), the water in the cell is replaced by a polymer resin. This replacement process can be performed with such care that cell shrinking or swelling remains below the resolution of light microscopy [25]. However, with the conventional TEM one observes the image of only an extremely thin sur-

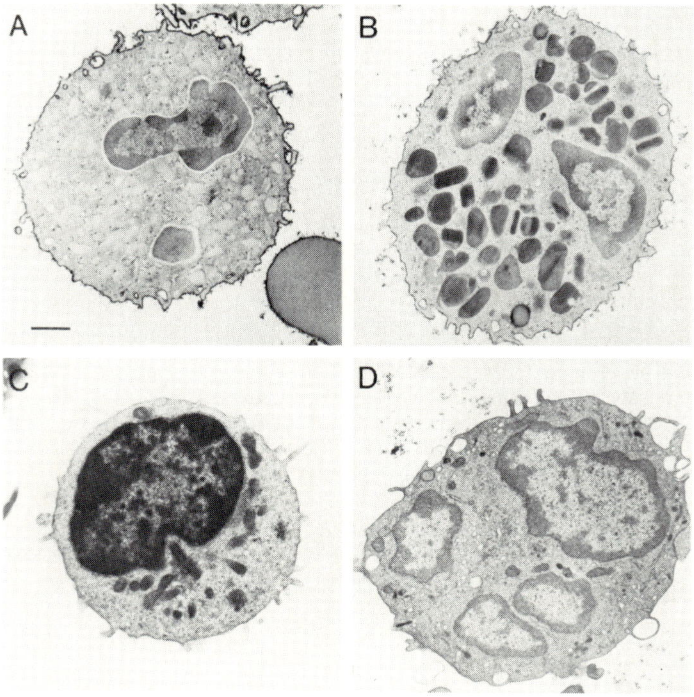

FIG. 13.6 Transmission electron microscopic images of human leukocytes: (*A*) neutrophils, (*B*) eosinophils, (*C*) lymphocytes, (*D*) monocytes. The length of the crossbar represents 1 μm.

faced electron-opaque stain on the section. In order to derive from these plane sections three-dimensional cell features such as volume, membrane area, and others, a stereological reconstruction is necessary [25, 36–38].

Figure 13.6 shows a TEM image of a neutrophil, eosinophil, lymphocyte, and monocyte, each at 300 mosmol. All cells have in common a folded cytoplasmic membrane, and a double membrane surrounding the nucleus. There are, in addition, numerous membrane-bound vacuoles in the Golgi region of the cells. The endoplasmic reticulum is also largely composed of membranes; each cell granule is surrounded by a bilipid membrane. The volume ratios of cell nucleus, granules, and mitochondria are distinctly different among the leukocyte types (Table 13.1). The granules of the eosinophils and basophils show a fine, crystal-like structure [39, 40]. Other features of the fine structure of cell organelles have also been reviewed [1, 41–43].

TABLE 13.1 Leukocyte volume distribution*

	Nucleus, %	Granules, %	Mitochondria, %	Cytoplasm, %
Lymphocyte	44.4	0.2	3.8	51.6
Neutrophil	21.3	15.4	0.6	62.7
Monocyte	25.9	2.6	3.5	68.1
Eosinophil	17.9	23.0	0.7	58.4

* At 310 mosmol.

The volume and membrane surface areas of the cell and nucleus are listed in Table 13.2. All leukocytes in isotonic medium have considerable excess membrane above that needed to cover their volume into a smooth sphere. During osmotic swelling and dehydration of the cells, the membrane areas remain constant within error of measurement [25]. Thus the available membrane area limits the degree of swelling of the cell or any of its organelles. It also limits the extent of cell deformation during migration on a substrate.

13.1.5 Leukocyte Volume Distribution Measured with Coulter-Counter Techniques

As a method of rapidly determining the volume distribution of a leukocyte population, the Coulter-counter sizing technique has been used frequently [26, 44–48]. The method is based on an electrical impedance measurement as the cells pass through a narrow orifice. Although it yields the typical dual-peak volume distribution of human leukocytes, the absolute volume estimates are often higher than direct light microscopic or TEM-stereological estimates. The reasons for the discrepancy are not resolved presently; it may be that (1) the TEM stereological values indicate the actual cell mass, whereas the Coulter-counter values may include some volume of plasma entrapped between the leukocyte surface folds; (2) the plastic or glass spheres used to calibrate the Coulter-counter device may have different electrical properties than leukocytes. Further investigation is necessary to resolve this discrepancy.

TABLE 13.2 Leukocyte morphometry*

	Neutrophil		Eosinophil		Lymphocyte		Monocyte	
Average cell volume, μm^3	182	(2.02)†	206	(2.29)	116	(1.29)	234	(2.6)
Average cell membrane area, μm^2	295	(2.10)	324	(2.31)	264	(1.88)	436	(3.11)
Average nuclear volume, μm^3	40		37		59		60	
Average nuclear membrane area, μm^2	112		99		97		129	

* At 310 mosmol.

† The numbers in parentheses indicate cell volume and membrane area, respectively, relative to that of average human erythrocytes (volume = 90 μm^3, area = 140 μm^2).

13.2 LEUKOCYTE DENSITY

Leukocytes have different mass densities. This difference forms the basis for separation of the cell types in a density gradient produced with high-molecular-weight substances such as dextran, albumin, or others [49–53]. Direct estimates of the cell densities shows that lymphocytes and monocytes have an overlapping density distribution with a mean value of about 1.063 g/cm^3 (range 1.055 to 1.070), whereas neutrophils and eosinophils are heavier with an average density of about 1.080 g/cm^3 (range 1.075 to 1.085) [46, 53, 54]. Both populations are lighter than erythrocytes, which have an average density of about 1.100 g/cm^3, and heavier than plasma, which has a density of about 1.035 g/cm^3 [55], and platelets [56].

13.3 LEUKOCYTE CYTOPLASMIC BIOCHEMISTRY

As background, a short summary of the cell chemical constituents is provided. The cell matrix (protoplasm), i.e., the material in which all cell organelles are embedded, comprises in the case of the neutrophil about 60 percent of the cell volume (Table 13.1). Therefore, it is the most important component in determining the rheological properties of the cell. Among the molecules that have been identified in the cell matrix are actin, both in monomeric (polymerized) and polymeric (depolymerized) form, myosin, and actin-binding protein [57–62]. All proteins are involved in the polymerization of the cytoplasm. Actin extracts in the presence of actin-binding protein can be induced under isothermal conditions to polymerize from a viscous solution to a viscoelastic gel with a strong elastic component [63]. The gelation process is Ca^{2+}-dependent. If the gel is mechanically sheared, it may assume viscous properties but resets spontaneously to a gel. There are also several molecules such as gelsolin and acumentin which may influence the actin filament length [64]. The ratio between actin and myosin is higher in leukocytes than in muscle tissue. In the unpolymerized state there is no specific fiber orientation of the molecules that can be visualized with the conventional transmission electron microscope. In the polymerized state, a fine fiber orientation (microfilaments) can be seen in the pseudopodium of the cell, suggesting molecular rearrangement of actin molecules. Electron microscopic observation of the cytoplasm after removal of the plasma membrane, critical-point drying, and rotary shadowing show a filament network with relatively small volume ratio [64].

Neutrophils play an important role in *phagocytosis,* a process of engulfment and digestion of foreign microorganisms. It requires release of enzymes from the granules [7]. Since the first neutrophil granules were isolated by differential centrifugation [65], and their enzymatic activity was analyzed, more than 30 different enzymes have been isolated [7, 66]. These enzymes serve a digestive function [65, 67]. The basophil granules appear to be carriers of heparin and histamine [8], substances which are associated with anticoagulation, smooth muscle stimulation, vascular permeability, and other physiological functions.

13.4 RHEOLOGICAL PROPERTIES OF SINGLE LEUKOCYTES

The rheological properties of the leukocyte cytoplasm are markedly different from those of the red blood cells, although the mechanical properties of the cytoplasmic membrane have similar features. A leukocyte in free suspension may be observed in the passive state, where it is spherical (Figs. 13.2, 13.5, and 13.6), or in an active state, where it forms spontaneous pseudopodia, i.e., regions of polymerized protoplasm (Fig. 13.8). In both cases the leukocyte may undergo large deformation.

13.4.1 The Passive State

The cell may be kept in the passive state by the removal of bivalent ions, e.g., with ethylenediaminetetraacetic acid (EDTA), by inhibitors of actin polymerization such as cytochalasin B, or by other agents [68]. A direct method to test the mechanical properties of the leukocytes is aspiration into a hollow glass micropipet. This method can be implemented in different ways. Some investigators [68–72] aspirate the entire cell into the pipet and monitor the required pressure for full entry of the cell. The aspiration pressure in these experiments is not only determined by the mechanical properties of the cell but also by its size and by the degree of adhesion to the inner micropipet wall.

A somewhat different test to investigate the cytoplasmic viscoelastic properties can be performed by applying a step aspiration pressure at the cell surface and observing the time course of a small surface deformation into the pipet [73–75]. Under these conditions adhesion to glass is negligible. All body forces on the cell are small so that in equilibrium

$$\frac{\partial \sigma_{ij}}{\partial x_j} = 0 \qquad (13.1)$$

where σ_{ij} is the Cauchy stress tensor, and x_i are spacial coordinates ($i = 1, 2, 3$) in index notation [76]. If the cell is assumed to be a homogeneous, incompressible isotropic sphere, its material properties have been tested for small strains e_{ij} such that

$$e_{ij} = \frac{1}{2}\left(\frac{\partial u_i}{\partial x_j} + \frac{\partial u_j}{\partial x_i}\right) \qquad (13.2)$$

where μ_i is the displacement. The observed deformation histories can be well approximated by a standard solid viscoelastic model (Fig. 13.7):

$$\sigma'_{ij} + \frac{\mu}{k_2}\frac{\partial \sigma'_{ij}}{\partial t} = k_1 e'_{ij} + \mu\left(1 + \frac{k_1}{k_2}\right)\frac{\partial e'_{ij}}{\partial t} \qquad (13.3)$$

where σ'_{ij} is the stress deviation, e'_{ij} is the strain deviator, and $\partial/\partial t$ is the partial derivative with respect to time. μ, k_1, and k_2 are three material constants to be determined by the experiment [73]. Typical values for neutrophils in an isotonic saline-albumin solution at room temperature are

$$\mu = 130 \pm 54 \text{ dyn} \cdot \text{s/cm}^2$$

$$k_1 = 275 \pm 119 \text{ dyn/cm}^2$$

$$k_2 = 737 \pm 346 \text{ dyn/cm}^2$$

The values for monocytes, lymphocytes, and eosinophils are of the same order of magnitude. The material coefficients are a function of several chemicophysical factors. They are tempera-

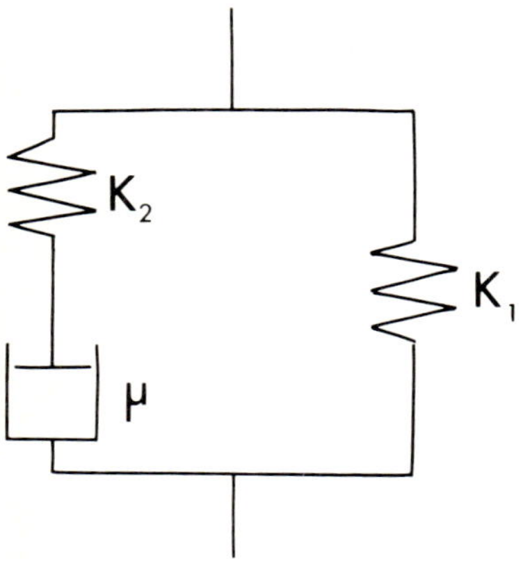

FIG. 13.7 Standard solid model. (*From Schmid-Schönbein et al. [73], with permission.*)

ture- and pH-dependent, and all three coefficients are exponential functions of the cell-suspending osmolarity in the range between 200 and 650 mosmol. In the range between 100 and 200 mosmol the coefficients are relatively constant for small strains [77].

For a stress of short duration the leukocytes show a stiff elastic response with an elastic shear modulus $G = (k_1 + k_2)/2$ of about 506 dyn/cm^2. This is several orders of magnitude smaller in the case of the erythrocytes. The viscosity coefficient μ is about 2000 times larger than the viscosity of the hemoglobin solution in erythrocytes. The model and the viscoelastic coefficients in the case of large strains are currently undetermined.

If the leukocyte is swollen in a hypotonic medium, the cell cytoplasm becomes diluted and the membrane unfolds. If a fully swollen cell (e.g., at 180 mosmol) is subjected to a pipet aspiration test, only a small deformation is observed and no viscoelastic creep follows [77]. In this case the applied stress is balanced by an isotropic tension in the membrane. For aspiration pressures between 4000 and 20,000 dyn/cm^2, which are associated with an increase of membrane tension by 0.127 dyn/cm, the change of membrane area is measured to be about 0.02 percent. The resulting membrane-area compressibility modulus is 636 dyn/cm. This value is similar to the one reported for red cell membranes [78]. The viscous and elastic properties during in-plane shearing of leukocyte membranes or during bending are currently unexplored.

The assumption of linearity [Eq. (13.2)] restricts the generality of the standard solid model [Eq. (13.3)] to small strains. Experiments on leukocytes entering into a stenosis of a glass capillary, where the cell is subjected to a large deformation, indicate that the three-element standard solid is still the simplest model to explain the deformation history and recovery of the cells [79].

The physical interpretation of the standard solid model may be stated tentatively as follows: The Maxwell fluid element (μ, k_2) can be ascribed mostly to the cell interior, particularly the ungelled actin suspension of the cytoplasm. The stress contribution of the cytoplasmic membrane in its folded shape is probably small in light of the high flexibility to bending of such biological membranes. The elastic modulus k_1 may be associated with the cell membrane or the cortical cytoplasm. A small membrane stress acting over a long period of time could restore the undeformed spherical shape of leukocytes. Also, random brownian motion of the granules may act to restore the undeformed shape. Since the granules undergo continuous brownian motion in the cortical region of the cytoplasm, no permanent cell-wide structure, such as a larger network of microtubules or the endoplasmic reticulum, seems to be there to restore the original spherical shape.

13.4.2 The Active State

When a leukocyte is freely suspended in autologous plasma or a Ringer's solution at body or room temperature, it will deform spontaneously. This deformation is asymmetric, and it occurs when the cell adheres to a substrate or is in free suspension because only a uniform hydrostatic pressure is acting on its membrane [68, 75]. Therefore, in the latter case it is clear that all mechanical energy needed for this deformation is provided by the cell itself. During this deformation the cell projects part of its cytoplasm beyond the perimeter of the undeformed cell, forming a structure sometimes called a *pseudopodium* or a "protopod." The pseudopodium is membrane-covered and varies in shape and size (Fig. 13.8*a*). It forms and retracts in a cyclic motion with typical growth velocity of about 5 μm/min. During formation the region of the pseudopodium is devoid of granules and other cell organelles (Fig. 13.8*b*), and a fine fibrillar structure can be observed in the protoplasm [75]. This suggests that pseudopodium formation is due to polymerization of the cell actin network associated with a shape change of the cytoplasm [80]. Tests with the micropipet show that the polymerized material in the pseudopodium shows relatively stiff elastic properties whereas the remainder of the cell shows the typical viscoelastic response as in the passive state [75]. This behavior is consistant with the gelation of actin monomers to actin polymers, which has been observed in vitro with leukocyte extracts or the cross-linking of actin-polymer fragment [63]. The formation of a pseudopodium is the mechanism by which the cell projects its membrane and cytoplasm when adhering to a substrate and thereby displacing itself in an amoeboid motion.

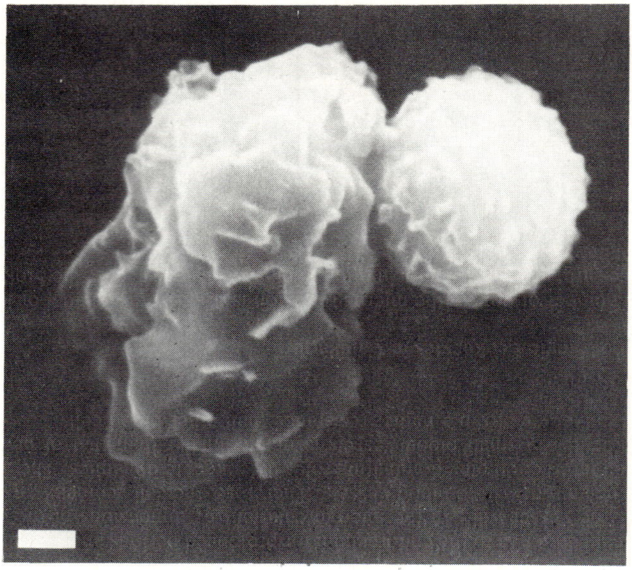

(a)

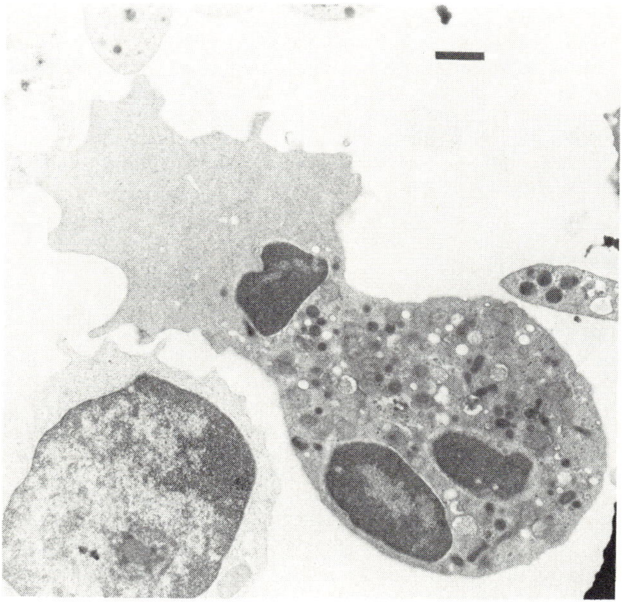

(b)

FIG. 13.8 Human leukocyte in the active state during pseudopodium formation. The cells were freely suspended before and during chemical fixation without adhesion to a substrate. (*a*) Scanning electron microscopic image. (*b*) Transmission electron microscopic image. The length of the crossbar represents 1 μm.

13.5 RHEOLOGY OF LEUKOCYTE SUSPENSIONS

At the usual low concentration (about 5000 cells per cubic millimeter at an equivalent leukocrit of about 0.1 percent) of leukocytes in whole blood, the influence on the blood viscosity in large blood vessels is negligible. However, if this concentration is increased, as in leukemic patients, the leukocytes cause a larger increase in apparent blood viscosity than would an equivalent volume of red cells [70, 81–83]. A blood suspension with high fraction of leukocytes shows less dependency on shear rate than a normal blood suspension with the usual low leukocrit (Fig. 13.9*a*). The apparent viscosity at high shear rates increases exponentially with the leukocrit (Fig. 13.9*b*) [83]. This behavior approaches that of suspensions of rigid spheres, which exhibit a shear rate-independent apparent viscosity that increases exponentially with volume fraction. The relatively stiff leukocytes appear to be deformed only a small degree by the plasma shear stresses in such a flow. However, leukocyte suspensions at high volume fractions (greater than 45 percent) exhibit a strong non-Newtonian viscosity.

13.6 BLOOD FILTRATION EXPERIMENTS AND LEUKOCYTE PORE PLUGGING

Whereas the role of leukocytes at the normal low leukocrits is negligible in large blood vessels, they have a significant effect on the flow in narrow channels such as the pores of a polycarbonate filter. When a whole-blood suspension is filtered through a filter with 5-μm-diameter pores at a constant flow rate, a monotonically increasing filtration pressure and resistance across the filter are observed. The increase in resistance is due to transient or permanent plugging of the pores by leukocytes [84–86]. It is not observed in the absence of leukocytes, and pore plugging is eliminated when the pore diameter exceeds the leukocyte diameters. In filtration experiments the flow at the level of the pores is unsteady [87]. Among different types of leukocytes the neutrophils seem to participate most often in plugging. The results of these filtration experiments underscore the importance of blood leukocytes in obstruction of single-file capillaries.

13.7 LEUKOCYTES IN THE MICROCIRCULATION

Circulating leukocytes are usually observed in the passive state. The cells may be deformed by plasma fluid stresses or an adhesive stress to the endothelium, but they return to the spherical shape once these stresses are removed. Pseudopodium formation is initiated, however, when leukocytes remain attached to the endothelium for prolonged periods and when they eventually migrate across the vessel wall. In the microcirculation the dimensions of the cells are of the same order of magnitude as the blood vessels. Therefore, the cell geometry, its deformability, and the cell wall interaction play dominating roles in leukocyte dynamics. The degree of deformation of leukocytes in small blood vessels is much smaller than that of erythrocytes [88], indicating that leukocytes form a higher resistance than other blood cells.

13.7.1 Motion of Single Leukocytes in Capillaries

The Reynolds number of flow in the microcirculation is of the order of 10^{-1} or less. Therefore, all inertial forces are negligible and the motion of plasma is governed by the Stokes' approximation of the equations of motion [76]:

$$0 = -\nabla p + \mu_p \nabla^2 \mathbf{u} \tag{13.4}$$

The plasma is assumed to be incompressible, so

$$\nabla \mathbf{u} = 0 \tag{13.5}$$

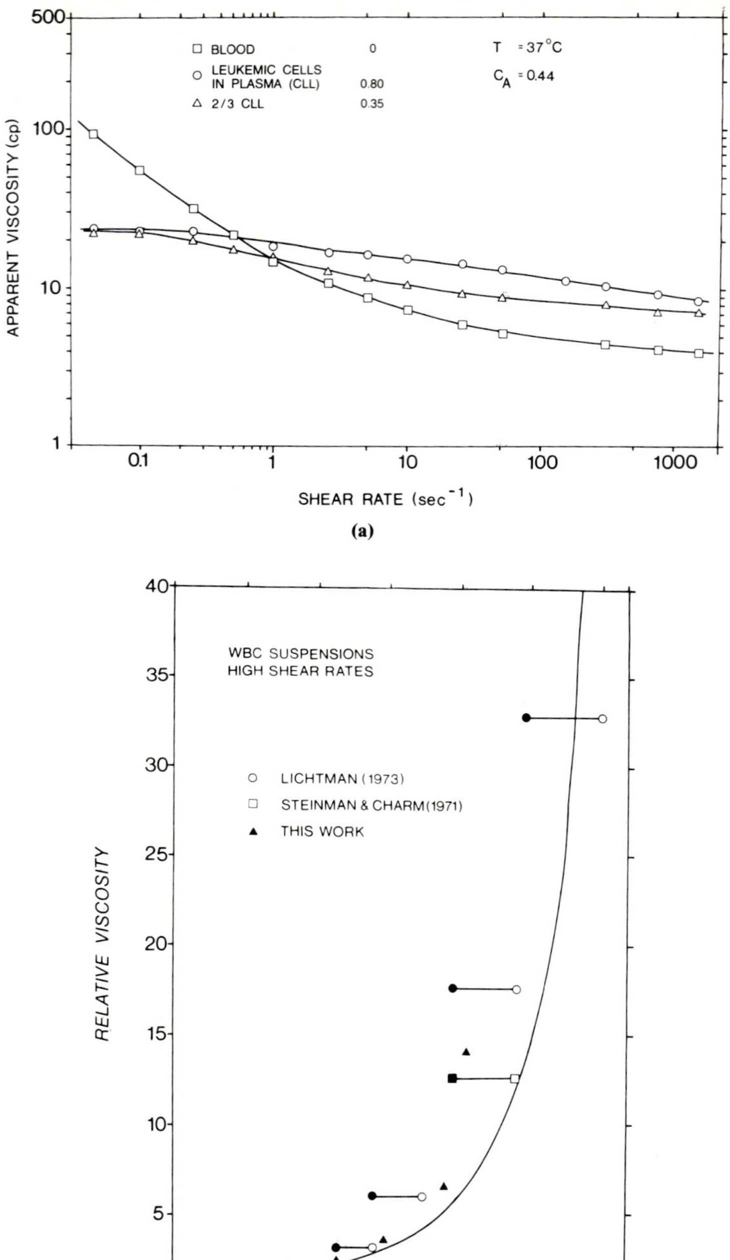

FIG. 13.9 (*a*) Apparent viscosity of human blood as a function of shear rate for normal blood and elevated leukocyte volume from a patient with lymphocytic leukemia (CLL). (*b*) Relative apparent viscosity of blood as a function of leukocyte concentration at high shear rate of the order of 1000 s^{-1}. C_A is the total cell (lymphocyte plus erythrocyte) volume fraction. The curve in (*b*) represents the apparent viscosity for a suspension of solid spheres. (*From Cohelet and Lichtman* [83], *with permission.*)

where **u** is the velocity vector, p is the pressure, ∇ is the gradient operator, and μ_p is the coefficient of plasma viscosity. Neglecting the effect of gravity and assuming that the cell is freely suspended in plasma, then the equation of motion for the leukocyte requires that the net force and net moment on each cell vanish. In addition, the deformation of each cell must be compatible with the stresses or velocities of the plasma surrounding it.

For elastic particles in a narrow capillary, solutions to this problem have been obtained by lubrication theory, using a series solution for the stress and displacement distribution inside individual cells [89]. These solutions show that cells located on the axis of a cylindrical tube move faster than the mean fluid velocity. The velocity is strongly dependent on the diameter ratio $\lambda = a/r$, where a is the cell radius in the deformed state and r is the tube radius. The relative apparent viscosity is given by

$$\eta = \frac{\Delta p}{16\mu_p \overline{V} a/r^2} \tag{13.6}$$

where Δp is the pressure drop across the cell, $16\mu_p\overline{V}a/r^2$ is the Poiseuille pressure drop over the same length of tube, and $\overline{V}$ is the mean velocity. For elastic particles the apparent viscosity is a function of the elastic shear modulus of the cell and the pressure drop across the cell. If the initial diameter ratio of the cell (i.e., before being deformed in the tube) is greater than 1, then the apparent viscosity tends to infinity as the pressure drop and the velocity tend to zero [89].

13.7.2 Hydrodynamic Interaction between Leukocytes and Erythrocytes in Capillaries

In capillaries with a single file of cells the higher stiffness and larger volume of the leukocytes in comparison to erythrocytes lead to a hydrodynamic interaction which results in the regular attachment of leukocytes to the endothelium of postcapillary venules.

The axial velocity of the leukocytes in single-file capillaries is lower than that of erythrocytes because of a smaller deformation during flow. For example, in a 6.8-μm-diameter capillary the ratio between leukocyte and erythrocyte velocity is 0.88 ± 0.06 [90]. The ratio is somewhat higher in larger capillaries, but is still different from 1, and it differs among leukocyte types. This difference causes an axial redistribution of a train of erythrocytes in the presence of a leukocyte. Downstream from the leukocyte, a plasma gap develops which increases linearly with time and with the velocity difference between leukocyte and the next erythrocyte downstream. Upstream from the leukocyte, red cells will accumulate and form a region of higher hematocrit [27, 91, 92]. This cell-train configuration is broken up on entering a postcapillary venule. The erythrocyte immediately upstream from the leukocyte will pass it and, in the process, push the leukocyte laterally against the endothelium. At that instant of time, two cells are side by side in a capillary vessel, and the leukocyte attaches to the endothelium. The ensuing adhesive bond between the leukocyte and endothelium membrane will retard the motion of the leukocyte, which may then form a contact area with the endothelium and roll on its surface. The erythrocytes which follow will pass, and hence will close the plasma gap downstream [92]. The magnitude of the adhesive force and the net fluid shear force and moment will determine whether the leukocyte will remain attached or will roll slowly downstream on the endothelium [93] (see below). The leukocytes can also be observed to crawl slowly upstream on the endothelium against the bloodstream. This process requires pseudopodium formation.

The hydrodynamic interaction with the red cells leads to regular endothelial attachment of all leukocytes passing through capillaries. It occurs in postcapillary venules with diameters about twice that of the leukocytes. This interaction explains the rapid shift of circulating cells in capillaries to marginating cells in venules [94]. The shifting process is dependent upon the presence of the erythrocytes. Without erythrocytes leukocytes make no regular attachment to the venous endothelium [92, 95].

13.7.3 Leukocyte Motion in Arterioles and Venules

In larger blood vessels with multiple layers of cells, the leukocytes, like other blood cells, are subject to longitudinal and radial random displacements about a mean position. This is due to random collisions with red cells in a shear field. The amplitude of the lateral dispersions increases toward the wall, where a higher velocity gradient exists and more frequent collisions with the red cells occur. The lateral dispersions cause random impacts of the leukocytes against the vessel wall [96]. The average radial position is also determined by erythrocyte rouleaux in the stream. In the presence of rouleaux the leukocytes are displaced toward the tube wall, and in the absence of rouleaux they tend toward the tube center [97–99]. Therefore, if the flow rate and shear rate in a blood vessel increase and rouleaux break up into single dispersed cells [97], the leukocytes tend to migrate toward the tube center [100, 101].

At diverging vascular bifurcations, the distribution of leukocytes into the daughter branches is greatly influenced by the flow rates into the downstream channels and by the radial position of the cells upstream from the branch [102, 103]. At a venous converging bifurcation where two single-file capillaries join together, the leukocyte may be displaced to a position close to the wall by the neighboring streamlines. Leukocytes that roll on the vascular wall across larger venous junctions usually remain attached to the endothelium [104].

13.7.4 Leukocyte-Endothelium Interaction

Following attachment of the leukocyte to the endothelium, a common contact area is formed with a typical intramembrane gap thickness of about 60 nm [105]. Whereas the endothelial cells deform little, the cell surface of the leukocyte undergoes a large deformation (Fig. 13.10). In the initial period of the attachment the leukocyte protrudes far into the bloodstream, but if the cell remains attached, it progressively flattens under the action of the adhesive stress [93, 105]. Among the leukocytes, the neutrophils and monocytes regularly adhere to the venous endothelium, under conditions of tissue inflammation. Lymphocytes are usually not involved, but they adhere to the endothelium of lymph nodes [107, 108].

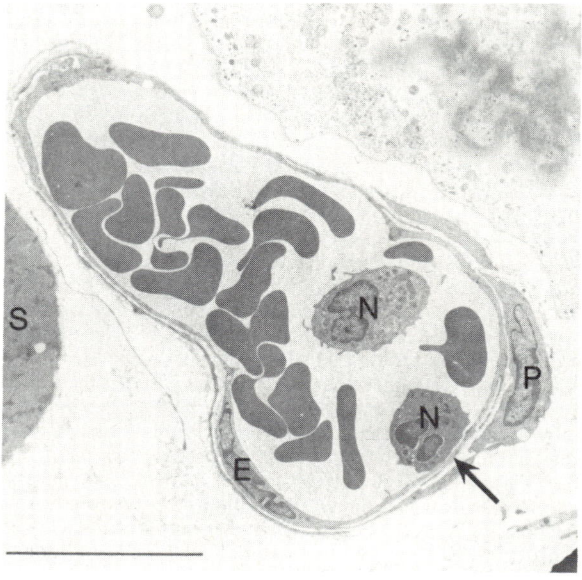

FIG. 13.10 Transmission electron microscopic view of a neutrophil (N) adhering to the endothelium (E) of a postcapillary venule in skeletal muscle of the rat. The pericyte cell layer (P) and skeletal muscle (S) with collagen fibers are shown. The contact area between neutrophil and endothelium can be seen (arrow).

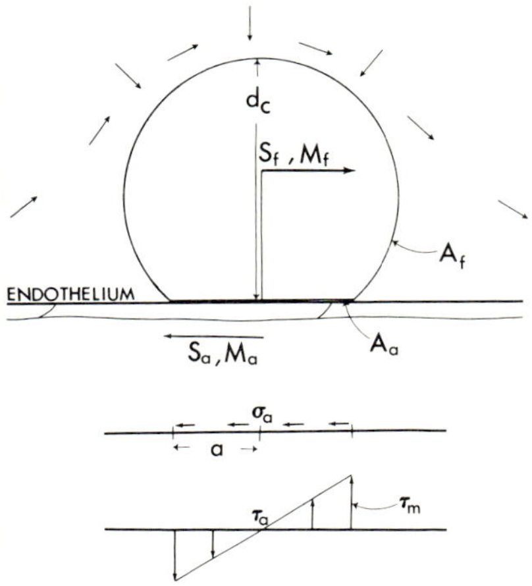

FIG. 13.11 Schematic of leukocyte adhering to vascular endothelium. For definition of forces and moments see text.

Due to the low Reynolds number of the flow field around an attached leukocyte, viscous shear forces dominate [93, 109]. Let S_f be the resultant fluid shear force and M_f the resultant moment over the surface A_f of the cell exposed to the bloodstream (Fig. 13.11). When the cell is stationary or rolling with constant velocity, S_f and M_f are in equilibrium with the resultant shear stress S_a and resultant moment M_a, due to an adhesive shear σ_a and normal τ_a distribution at the contact area A_a with the endothelium. The motion of the cell is determined by the magnitude of these net forces and moments and by the maximum values of S_a and M_a when the cell breaks away, that is, $S_{a,\max}$ and $M_{a,\max}$, respectively. If $M_f > M_{a,\max}$, the cell will start rolling about the axis of vorticity; if $M_f \leq (M_a)_{\max}$, no rolling will occur. If $S_f > (S_a)_{\max}$, the cell will be sheared away, and if $S_f \leq S_{a,\max}$, it will remain attached. S_f is a function of the mean blood velocity in the blood vessel, cell diameter d_c, plasma viscosity μ_p; so

$$S_f = \mu_p \overline{V} d_c C_s \tag{13.7}$$

where C_s is a dimensionless shear coefficient. For a stationary cell of spherical shape, C_s is a function of the ratio between cell and tube diameter, d_c/d_t, and the tube Reynolds number Re = $\overline{V} \cdot d_t \cdot \rho_p/\mu_p$ [93]. C_s is also dependent upon hematocrit, which renders the shear force a stochastic variable, owing to random hydrodynamic interactions with passing red cells. In the case of a large blood vessel, so that $d_c/d_t \approx 0$, the shear force and moment can be expressed in terms of the local wall strain rate $\dot{\gamma}$ as [110]

$$S_f = 1.7005(6\pi\mu_p)\left(\frac{d_c}{2}\right)^2 \dot{\gamma} \tag{13.8}$$

and

$$M_f = 0.944(4\pi\mu_p)\left(\frac{d_c}{2}\right)^3 \dot{\gamma} \tag{13.9}$$

A case of special interest is the maximum shear and normal stress between the white cell and the endothelium at which they break apart, i.e., at the net shear force $S_{a,\max}$ and net moment $M_{a,\max}$. It is a measure of the strength of the membrane-membrane bond, and it can be measured by increasing S_f and M_f until the leukocyte is sheared away. The maximum shear stress is of the same order of magnitude as the maximum normal stress, and in the rabbit omentum has been found to be between 265 and 1060 dyn/cm^2 [93]. The range of values may reflect differences among leukocyte types, heterogeneities on the venular wall, and dependence on concentration of the biochemical agents modulating the adhesive stress (see Sec. 13.8 below).

In arterioles the fluid shear force usually dominates the adhesive forces so that no leukocyte adhesion is observed under normal physiological conditions. However, in the presence of an inflammation the adhesive stress becomes strong enough to retard the leukocytes [111], although under conditions of low flow the fluid forces are not sufficient to wash away the leukocytes [112]. On the other hand, in venules the adhesive forces usually dominate, and so leukocyte margination is a common phenomenon. During an inflammatory reaction the number of marginating neutrophils is enhanced, which largely reflects an increase of the adhesive force.

Leukocytes can roll for long distances on the venous endothelium. During this motion the cells preserve the contact area with the endothelium, thus making new bonds downstream and breaking them upstream. The axial leukocyte velocity during rolling is a linear function of the erythrocyte velocities in the adjacent bloodstream [113], but its magnitude is much smaller

[114]. The rolling velocity is rapid in noninflamed tissues and is lowered in areas of inflammation.

13.7.5 Margination and Vascular Resistance

When leukocytes are adhering to the vascular endothelium, they usually protrude into the bloodstream. This protrusion causes an elevation of the hemodynamic resistance in the blood vessel [115, 116]. The effect is pronounced in narrow vessels and negligible in large ones. In the presence of a small number (n) of equal-sized, spherically shaped leukocytes, the ratio η between resistances with and without leukocytes is a linear function of n [116]:

$$\eta = 1 + n\left(\frac{C_s d_c}{4\pi l} - \frac{A_a}{\pi d_t l}\right) \tag{13.10}$$

where l is the vessel length, d_t is its diameter, and A_a is the leukocyte contact area. The increase of resistance in vessels of the microcirculation, even in the presence of only a few adhering leukocytes, is large in comparison with a corresponding effect produced by erythrocytes.

13.7.6 Margination and Endothelial Cell Damage

When larger numbers of leukocytes adhere to the vascular wall, damage may occur to the endothelium and subendothelial tissue [117, 118] and may lead to edema formation. In the lung the accumulation of leukocytes is accompanied by an increase in microvascular permeability and increase in protein-rich lymph flow [119]. The cell injury is probably mediated by release of oxygen-free radicals and lysosomal enzymes from the leukocytes [120, 121].

13.7.7 Leukocyte Capillary Plugging

When leukocytes enter into capillaries, they have to be deformed. The deformation takes place in a tapered portion of the vessel, usually at the entrance or at a protruding endothelial cell nucleus. The leukocyte temporarily blocks the blood flow in the capillary [122, 123]. The time necessary for the leukocytes to enter the capillary is typically 1000 to 2000 times longer than that required for the red cells to enter [124]. The resultant squeezing force P_3A_3 (Fig. 13.12, top), which compresses the leukocyte into an elongated shape, is linearly related to the driving pressure [79] as

$$P_3 A_3 = \frac{1}{\sin \theta}(P_1A_1 - P_2A_2) \tag{13.11}$$

For a typical taper angle $\theta = 6°$, the squeezing force P_3A_3 is 9.56 times larger than the driving force $P_1A_1 - P_2A_2$. Under normal physiological driving pressure $P_1 - P_2$, this squeezing force is sufficient to preserve a thin fluid film between leukocyte and endothelium and prevent adhesion. However, if the driving pressure gradient $P_1 - P_2$ is reduced, the cell will not be deformed sufficiently. Its membrane comes into close proximity with the endothelial membrane so that molecular adhesion forces take over. The leukocyte forms a large contact area with the endothelium and obstructs the capillary lumen (Fig. 13.13). Permanent trapping of the leukocyte in a capillary requires the action of an adhesive force. Without endothelial adhesion even a reduced pressure will eventually deform the viscoelastic cell until it can pass through the capillary.

Leukocyte capillary plugging has been first observed in human skin chambers and in mesentery [27, 124, 125] and during hemorrhagic shock in skeletal muscle capillaries [126]. In the ischemic skeletal muscle each capillary without flow contains at least one leukocyte, usually located at a protruding endothelial nucleus. When leukocytes are occasionally dislodged, circulation in that capillary starts again immediately. Obstruction of flow by leukocytes occurs only in capillaries, but not all capillaries are affected and no arterioles or venules are occluded.

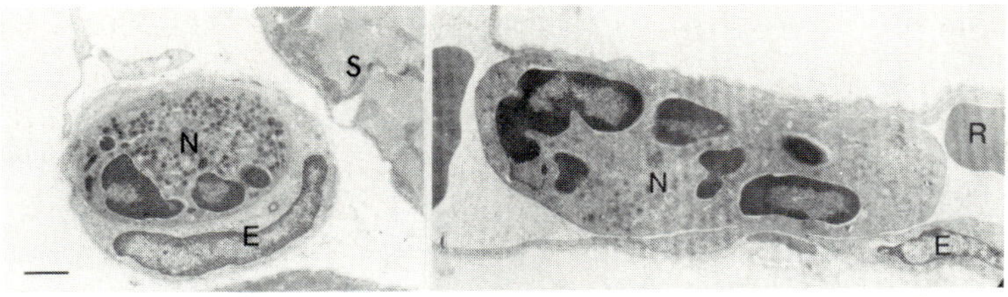

FIG. 13.12 Transmission electron microscopic image of neutrophils (N) occluding a capillary in skeletal muscle of the rat. Left side shows a cross section, right side a longitudinal section. There is a common contact area with the endothelium (E) with a gap thickness of about 600 Å (R = red cell, S = skeletal muscle cell.)

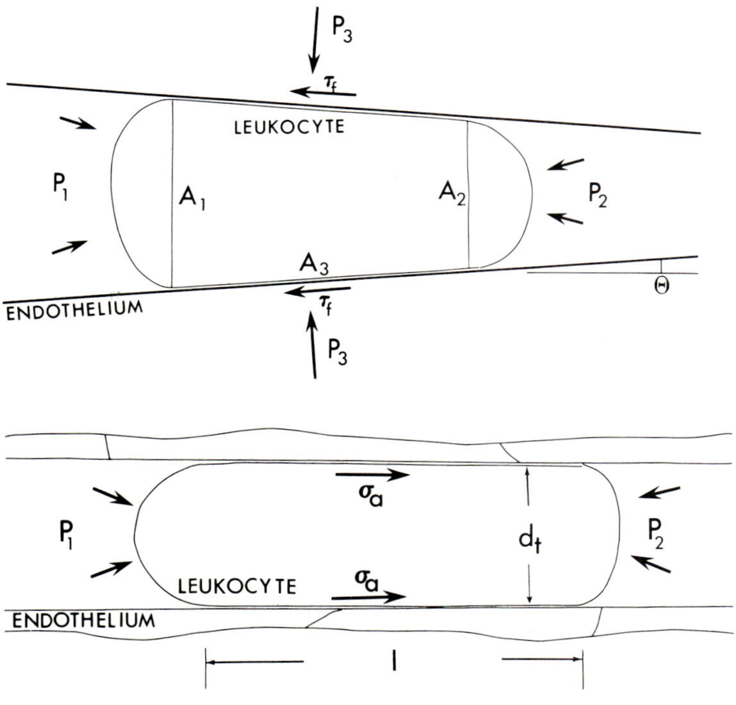

FIG. 13.13 Schematic of a leukocyte being compressed in a locally tapering vessel (top) and during entrapment in a capillary (bottom).

The pressure drop ($P_1 - P_2$) across a trapped leukocyte in a straight capillary is balanced by the adhesive shear stress (Fig. 13.13, bottom) via the equilibrium condition

$$(P_1 - P_2)\pi \left(\frac{d_c}{2}\right)^2 = \pi d_t l \sigma_a \tag{13.12}$$

where l is the length of contact area and d_t is the tube diameter. As long as the adhesive force [right-hand side of Eq. (13.12)] exceeds the pressure force (left-hand side), the leukocyte cannot be pushed out of the capillary. On the other hand, if σ_a is elevated because of the action of inflammatory chemical products, even a normal or elevated physiological capillary pressure drop may not be sufficient to remove the leukocyte.

Leukocyte capillary plugging is associated with an increase of vascular resistance [127]. This plugging has been observed in ischemic skeletal muscle and in myocardial ischemia after 1 to 5 h occlusion of the descending coronary artery in the dog heart. On the average, one leukocyte obstructs one unbranched capillary. Many capillaries with leukocytes (mostly neutrophils) in the ischemic tissue cannot be reperfused after restoration of normal arterial pressure [128]. Under these ischemic conditions leukocytes also adhere to the endothelium in arterioles and venules, but without occluded flow, and this adherence may be the cause of the "no-flow" and the "no-reflow" phenomena that have been observed in several organs [127, 128]. Similar events of leukocyte capillary plugging have been reported in the cat brain cortex [129]. Leukocyte accumulation during ischemia has also been observed in the lung and liver [130, 131].

Under conditions of a reduced perfusion pressure or during elevation of the leukocyte adhesive stress, an organ or part of an organ may become a filter for certain types of leukocytes. This filtering leads to hypoxia, followed by a sequence of biochemical reactions, fluid shifts across the vascular wall, and eventual damage of the endothelium and tissue parenchyma. Thus leukocyte capillary plugging may be a fundamental pathological mechanism in many diseases.

13.8 SOME ASPECTS OF THE BIOCHEMISTRY OF ADHESION

The adhesive stress necessary to separate a leukocyte from its substrate appears to depend upon a number of biochemical agents and their concentration. Several aspects of this problem have been reviewed [132, 133].

Various tests have been designed to obtain a measure for the adhesive bond, including in vivo and in vitro conditions. In vivo tests usually consist of counting the number of adhering white cells per length of vessel segments [115]. In vitro tests consist of counting the number of adhering leukocytes after filtration through a glass bead column [134] or through nylon fibers [135, 136]. Other tests consist of sedimenting leukocytes on a surface, washing them in a "standard procedure," and then counting the adhering cells. This has been performed with several artificial surfaces [136–138] and endothelial culture cell surfaces [139]. In many of these tests the fluid forces acting on the cells are not well defined. A quantitative test of leukocyte interaction in a two-body collision in a free fluid shear field has been suggested [140].

The biochemical conditions controlling adhesion appear to be type-specific to leukocytes. Most investigations have been performed on neutrophils, and remarks will be limited to these cells. In a normal, noninflamed tissue, such as the rabbit ear chamber or bat wing, the number of neutrophils adhering to the vascular wall reaches a minimum, but a few cells are usually observed to marginate on the venous endothelium. Upon perfusion with saline or other ionic buffers, a transient increase of adhering leukocytes is observed [141]. This indicates that among several chemicals a soluble substance on the endothelium appears to modulate adhesion. If an injury is inflicted on the tissue, a significant increase of adhering cells is observed. This affects mostly granulocytes, whereas lymphocytes pass readily and monocytes are observed to adhere to the injured tissue only if the inflammation progresses. Basophils appear to be attracted to sites of injury due to cold [8]. Endothelial adhesion of neutrophils in inflammation is dependent upon the presence of Ca^{2+} or other ions [114, 142, 143], but independent of the anticoagulant heparin [144, 145]. Treatment of the endothelium with neuraminidase to cleave sialic acid residues on the cell membranes leads to a significant increase of adhesion [141]. The elevated adherence in inflammation is mediated through a soluble material in plasma which can be suppressed by anti-inflammatory substances such as corticosteroids, aspirin, or others [146–149]. An elevation of the adhesion can also be induced in several other ways, for example, by infusion of an animal with bacterial endotoxins [114], by infusion with complement-activated plasma [143, 150, 151], and with many chemotactic agents, such as *N*-formylmethionyl peptide (FMLP) [143, 152, 153]. FMLP has a dose-dependent effect on granulocyte adhesion with a monotonically increasing function [153]. Both complement C5a and FMLP not only affect the strength of the adhesion but also convert the cell into the active state with pseudopodium formation [154, 155].

Although a large number of empirical results relating to leukocyte adhesion are available today, physicochemical analysis of the bonds between leukocytes and endothelial cells is still incomplete. The molecular basis of leukocyte adhesion appears to be an important subject for further research in light of the strong influence of adhesion on motion of leukocytes in the circulation and in the tissues.

ACKNOWLEDGMENTS

This work was supported by USPHS grant HL-10881 and by a grant from the Office of Graduate Studies, University of California, San Diego.

REFERENCES

1. Bessis, M.: *Living Blood Cells and Their Ultrastructure,* R. E. Weed (trans.), Springer-Verlag, New York, 1973.
2. Cline, M. J.: *The White Cell,* Harvard University Press, Cambridge, Mass., 1975.
3. Bainton, D. F.: Differentiation of human neutrophilic granulocytes: Normal and abnormal, in T. J. Greenwalt and G. A. Jamieson (eds.), *The Granulocyte: Function and Clinical Utilization,* Alan R. Liss, New York, 1977, pp. 1–27.
4. Keleman, E., Calvo, W., and Fliedner, T. M.: *Atlas of Human Hemopoietic Development,* Springer-Verlag, Berlin, 1979.
5. Braunsteiner, H.: *The Physiology and Pathology of Leukocytes,* rev. ed., D. Zucker-Franklin (ed.), Grune & Stratton, New York, 1962.
6. Murphy, P.: *The Neutrophil,* Plenum Press, New York, 1976.
7. Klebanoff, S. J., and Clark, R. A.: *The neutrophil: Function and Clinical Disorder,* North Holland, Amsterdam, 1978.
8. Parwaresch, M. R.: *The Human Blood Basophil,* Springer-Verlag, Berlin, 1976.
9. Beeson, P. B., and Bass, D. A.: The eosinophil, in L. H. Smith, Jr. (ed.), *Major Problems in Internal Medicine,* Saunders, Philadelphia, 1977, vol. XIV.
10. Elves, M. W.: *The Lymphocytes,* 2d ed., Year Book, Chicago, 1972.
11. Loor, F., and Roelants, G. E. (eds.): *Band T Cells in Immune Recognition,* Wiley, London, 1977.
12. Marchalonis, J. J. (ed.): *The Lymphocyte Structure and Function,* Dekker, New York, 1977, pts. I, II.
13. Wilkinson, P. C.: *Chemotaxis and Inflammation,* Churchill Livingstone, Edinburgh, 1974.
14. Gallin, J. I., and Quie, P. C. (eds.): *Leukocyte Chemotaxis: Methods, Physiology and Clinical Implications,* Raven Press, New York, 1978.
15. Elsbach, P.: Phagocytosis in the inflammatory process, in B. W. Zweifach, L. Grant, and R. T. McCluskey (eds.), *The Inflammatory Process,* 2d ed., Academic Press, New York, 1974, vol. 1, pp. 363–410.
16. Rapaport, S. I.: *Introduction to Hematology,* Harper & Row, New York, 1971.
17. Bagge, U., Brånemark, P-I., and Skalak, R.: Measurement and influence of white cell deformability, in G. D. O. Lowe, J. C. Barbenel, and C. D. Forbes (eds.), *Clinical Aspects of Blood Viscosity and Cell Deformability,* Springer-Verlag, New York, 1981.
18. Bagge, U., Born, G. V. R., and Gaehtgens, P. (eds.): *White Blood Cells Morphology and Rheology as Related to Function,* Martinus Nijhoff, The Hague, 1982.
19. Meiselman, H. J., Lichtman, M. A., La Celle, P. L.: *White Cell Mechanics: Basic Science and Clinical Aspects,* Alan R. Liss, New York, 1984.
20. Ehrlich, P.: Beiträge zue Kenntnis der Anilinfärbung und ihrer Verwendung in der mikroskopischen Technik, *Arch. Mikr. Anat., 13,* 267, 1877.

21. Ehrlich, P.: Methodologische Beiträge zur Physiologie und Pathologie der verschiedenen Formen der Leukozyten, *Z. Klin. Med., 1,* 553–560, 1880.

22. Schmid-Schönbein, G. W., Jan, K-M., Skalak, R., and Chien, S.: Deformation of leukocytes on a hematological blood film, *Biorheology, 21,* 767–781, 1984.

23. Bessis, M.: *Blood Smears Reinterpreted,* Springer International, New York, 1977.

24. Kapff, C. T., and Zande, J. H.: *Blood: Atlas and Source Book of Hematology,* Little, Brown, Boston, 1981.

25. Schmid-Schönbein, G. W., Shih, Y. Y., and Chien, S.: Morphometry of human leukocytes, *Blood, 56,* 866–875, 1980.

26. Humphries, R. K., and Miller, R. G.: Volume analysis of human peripheral blood leukocytes, *Ser. Haematol., 5,* 142–162, 1972.

27. Bagge, U., and Brånemark, P-I.: White blood cell rheology: An intravital study in man, *Adv. Microcirc., 7,* 1–17, 1977.

28. Anderson, T. F.: Technique for the preservation of three-dimensional structure in preparing specimens for the electron microscope, *Trans. N.Y. Acad. Sci., 13,* 130–134, 1951.

29. Polliack, A., Lampen, N., Clarkson, B. D., deHarven, E., Bentwich, Z., Siegal, F. P., and Kunkel, H. G.: Identification of human B and T lymphocytes by scanning electron microscopy, *J. Exp. Med., 138,* 607–624, 1973.

30. Lin, P. S., Cooper, A. G., and Wortis, H. H.: Scanning electron microscopy of human T-cell and B-cell rosettes, *N. Engl. J. Med., 289,* 548–551, 1973.

31. Renau-Piqueras, J.: Morphological differences in human peripheral blood lymphocytes: A freeze-etching and scanning electron microscopy study, *Acta Haematol. (Basel), 59,* 19–24, 1978.

32. Alexander, E. L., and Wetzel, B.: Human Lymphocytes: Similarity of B and T cell surface morphology, *Science,* 188, 732–734, 1975.

33. Newell, D. G., Roath, S., and Smith, J. L.: The scanning electron microscopy of normal human peripheral blood lymphocytes, *Br. J. Haematol., 32,* 209–316, 1976.

34. Thontwaite, J. T., and Leif, R. C.: The utilization of the plaque cytogram assay for the study of spontaneous rosette forming and complement independent plaque forming cells, *J. Reticulo. Endothel. Soc., 16* (suppl.), 2a, 1974.

35. Kay, M. M., Belohradsky, B., Yee, K., Vogel, J., Butcher, D., Wybran, J., and Fudenberg, H. H.: Cellular interactions: Scanning electron microscopy of human thymus-derived rosette-forming lymphocytes, *Clin. Immunol. Immunopathol., 2,* 301–309, 1974.

36. Underwood, E.: *Quantitative Sterology,* Addison-Wesley, Reading, Mass., 1970.

37. Ito, T., and Abe, K.: Cytometric analysis of thymic small lymphocytes: Studies by a stereological method in electron microscopy, in E. Yamada, V. Misuhira, K. Kurosumi, and T. Nagano (eds.), *Recent Progress in Electron Microscopy of Cells and Tissue,* University Park Press, Baltimore, 1976, pp. 253–265.

38. James, V.: Stereological analysis of leukaemic cells, *Br. J. Haematol., 39,* 17–24, 1978.

39. Miller, F., DeHarven, E., and Palade, G. E.: The structure of eosinophil leukocyte granules in rodents and in man, *J. Cell Biol., 31,* 349–362, 1966.

40. Terry, R. W., Bainton, D. F., and Farquhar, M. G.: Formation and structure of specific granules in basophilic leukocytes of the guinea pig, *Lab. Invest., 21,* 65–76, 1969.

41. Watanabe, I., Donahue, S., and Hoggatt, N.: Method for electron microscopic studies of circulating human leukocytes and observations on their fine structure, *J. Ultrastruct. Res., 20,* 366–382, 1967.

42. Porter, K. R., and Bonneville, M. A.: *Fine Structure of Cells and Tissue,* Lea & Febiger, Philadelphia, 1968.

43. Tanaka, Y., and Goodman, J. R.: *Fine Structure of WBCs: Electron Microscopy of Human Blood Cells,* Harper & Row, New York, 1972.

44. Gauthier, J., and Harel, P.: Human leukocytes: Their size distribution and mean corpuscular volume, *Can. Med. Assoc. J., 97,* 793–796, 1967.

45. Westring, D. W., Ladinsky, J. L., and Feick, P.: The volume distribution of human lymphocytes, *Proc. Soc. Exp. Biol. Med., 131,* 1077–1083, 1969.

46. Zucker, R. M., and Cassen, B.: The separation of normal human leukocytes by density and classification by size, *Blood, 34,* 591–600, 1969.

47. Ben-Sasson, S., Patinkin, D., Grover, N. B., and Daljanski, F.: Electrical sizing of particles in suspension: IV. Lymphocytes, *J. Cell Physiol., 84,* 205–214, 1974.

48. Segel, G. B., Cokelet, G. R., and Lichtman, M. A.: Measurement of lymphocyte volume: Importance of reference particle deformability and counting solution tonicity, *Blood, 57,* 894–899, 1981.

49. Böyum, A.: Isolation of leukocytes from human blood, *Scand. J. Clin. Lab. Invest., 21* (suppl. 94), 9–29, 1968.

50. Böyum, A.: Isolation and leukocytes from human blood: Further observations, *Scand. J. Clin. Lab. Invest., 21* (suppl. 97), 31–49, 1968.

51. Böyum, A.: A one-stage procedure for isolation of granulocytes and lymphocytes from human blood, *Scand. J. Clin. Lab. Invest., 21* (suppl. 97), 51–75, 1968.

52. Böyum, A.: Isolation of mononuclear cells and granulocytes from human blood, *Scand. J. Clin. Lab. Invest., 21* (suppl. 97), 77–89, 1968.

53. Shortman, K., Haskill, J. S., Szenberg, A., and Legge, D. G.: Density distribution analysis of lymphocyte populations, *Nature, 216,* 1227–1228, 1968.

54. Williams, N., Moore, M. A. S., Shortman, K., Condon, L., Pike, B., and Nossal, G. J. V.: Density separation and characterization of human blood leukocytes, including B and T lymphocytes, *Aust. J. Exp. Biol. Med. Sci., 52,* 491–503, 1974.

55. Merrill, E. W., and Wells, R. E.: Flow properties of biological fluids, *Appl. Mech. Rev., 14,* 663–673, 1961.

56. Bessis, M.: Une méthode permettant l'isolement des differents éléments figurés du sang, *Sang, 14,* 262–264, 1940.

57. Hartwig, J. H., and Stossel, T. P.: Isolation and properties of actin, myosin, and a new actin-binding protein in rabbit alveolar macrophages, *J. Biol. Chem., 250,* 5696–5705, 1975.

58. Stossel, T. P., and Hartwig, J. H.: Interactions between actin, myosin, and actin-binding protein from rabbit alveolar macrophages, *J. Biol. Chem., 250,* 5706–5712, 1975.

59. Shibata, N., Tatsumi, N., Tanaka, K., Okamura, Y., and Senda, N.: Leukocyte myosin and its location in the cell, *Biochim. Biophys. Acta, 400,* 222–243, 1975.

60. Senda, N., Shibata, N., Tatsumi, N., Kondo, K., and Hamada, K.: A contractile protein from leukocytes: Its extraction and some of its properties, *Biochim. Biophys. Acta, 181,* 191–200, 1969.

61. Boxer, L. A., Richardson, S., and Floyd, A.: Identification of actin-binding protein in membrane of polymorphonuclear leukocytes, *Nature, 263,* 259–261, 1976.

62. Rikihisa, Y., and Mizuno, D.: Demonstration of myosin on the cytoplasmic side of plasma membranes of guinea pig polymorphonuclear leukocytes with immunoferritin, *Exp. Cell Res., 110,* 87–92, 1977.

63. Brotschi, E. A., Hartwig, J. H., and Stossel, T. P.: The gelation of actin by actin-binding protein, *J. Biol. Chem., 253,* 8988–8993, 1978.

64. Stossel, T. P.: The structure of cortical cytoplasm, *Phil. Trans. R. Soc. London (Ser. B), 299,* 275–289, 1982.

65. Cohn, Z. A., and Hirsch, J. G.: The isolation and properties of the specific cytoplasmic granules of rabbit polymorphonuclear leukocytes, *J. Exp. Med., 112,* 983–1004, 1960.

66. Baggiolini, M.: The enzymes of the granules of polymorphonuclear leukocytes and their functions, *Enzyme, 13,* 132–160, 1972.

67. Hirsch, J. G., and Cohn, Z. A.: Degranulation of polymorphonuclear leukocytes following phagocytosis of microorganisms, *J. Exp. Med., 112,* 1005–1014, 1960.

68. Lichtman, M. A., Santillo, P. A., Kearney, E. A., Roberts, G. W., and Weed, R. I.: The shape and surface morphology of human leukocytes in vitro: Effect of temperature, metabolic inhibitors and agents that influence membrane structure, *Blood Cells, 2,* 507–531, 1976.

69. Lichtman, M. A., and Weed, P. I.: Alteration of the cell periphery during granulocyte maturation: Relationship to cell function, *Blood, 39,* 301–316, 1972.

70. Lichtman, M. A.: Rheology of leukocytes, leukocyte suspensions, and blood in leukemia, *J. Clin. Invest., 52,* 350–358, 1973.

71. Miller, E. M., and Myers, K. A.: Cellular deformability of the human peripheral blood polymorphonuclear leukocyte: Methods, normal variation, effect of physical and chemical alteration, *Res. J. Reticuloendothel. Soc., 18,* 337–345, 1975.

72. LaCelle, P. L., Bush, R. W., and Smith, B. D.: Viscoelastic properties of normal and pathologic human granulocytes and lymphocytes, in U. Bagge, G. V. R. Born, and P. Gaehtgens (eds.), *White Blood Cells: Morphology and Rheology as Related to Function,* Martinus Nijhoff, The Hague, 1982.

73. Schmid-Schönbein, G. W., Sung, K. L. P., Tözeren, H., Skalak, R., and Chien, S.: Passive mechanical properties of human leukocytes, *Biophys. J., 36,* 243–256, 1981.

74. Chien, S., Sung, K. L. P., Skalak, R., and Schmid-Schönbein, G. W.: Viscoelastic properties of leukocytes in passive deformation, in U. Bagge, G. V. R. Born, and P. Gaehtgens (eds.), *White Blood Cells: Morphology and Rheology as Related to Function,* Martinus Nijhoff, The Hague, 1982.

75. Schmid-Schönbein, G. W., Skalak, R., Sung, K. L. P., and Chien, S.: Human leukocytes in the active state, in U. Bagge, G. V. R. Born, and P. Gaehtgens (eds.), *White Blood Cells: Morphology and Rheology as Related to Function,* Martinus Nijhoff, The Hague, 1982.

76. Fung, Y. C.: *A First Course in Continuum Mechanics,* 2d ed., Prentice-Hall, Englewood Cliffs, N.J., 1977.

77. Sung, K. L. P., Schmid-Schönbein, G. W., Skalak, R., Schuessler, G. B., Usami, S., and Chien, S.: Influence of physicochemical factors on rheology of human neutrophils, *Biophys. J., 39,* 101–106, 1982.

78. Evans, E. A., and Wangh, R.: Osmotic correction to elastic area compressibility measurements on red cell membrane, *Biophys. J., 20,* 307–313, 1977.

79. Bagge, U., Skalak, R., and Attefors, R.: Granulocyte rheology: Experimental studies in an in-vivo microflow system, *Adv. Microcirc., 7,* 29–48, 1977.

80. Schmid-Schönbein, G. W., and Skalak, R.: Continuum mechanical model of leukocytes during protopod formation, *J. Biomech. Eng., 106,* 10–18, 1984.

81. Dintenfass, L.: *Blood Microrheology—Viscosity Factor in Blood Flow, Ischemia and Thrombosis,* Butterworth, London, 1971.

82. Steinman, M. H., and Charm, S. E.: Effects of high concentration of leukocytes on whole blood viscosity, *Blood, 38,* 299–301, 1971.

83. Cokelet, G. R., and Lichtman, M. A.: Rheology of leukocyte suspensions, in U. Bagge, G. V. R. Born, and P. Gaehtgens (eds.), *White Blood Cells: Morphology and Rheology as Related to Function,* Martinus Nijhoff, The Hague, 1982.

84. Lingard, P.: Capillary pore rheology of erythrocytes: I. Hydroelastic rheology of human erythrocytes in microscopic pores, *Microvasc. Res., 8,* 53–63, 1974.

85. Lichtman, M. A., and Kearney, E. A.: The filterability of normal and leukemic human leukocytes, *Blood Cells, 2,* 491–506, 1976.

86. Chien, S., Schmalzer, E. A., Lee, M. M. L., Impelluso, T., and Skalak, R.: Role of white blood cells in filtration of blood cell suspensions, *Biorheology, 20,* 11–22, 1983.

87. Skalak, R., Impelluso, T., Schmalzer, E. A., and Chien, S.: Theoretical modeling of filtration of blood cell suspension, *Biorheology, 20,* 41–56, 1983.

88. Brånemark, P-I., and Lindstrom, J.: Shape of circulating blood corpuscles, *Biorheology, 1,* 139–142, 1963.

89. Tözeren, H., and Skalak, R.: The steady flow of closely fitting incompressible elastic spheres in a tube, *J. Fluid Mech., 87,* 1–16, 1978.

90. Mayrovitz, H. N.: The relationship between leukocytes and erythrocytes velocity in arterioles, in U. Bagge, G. V. R. Born, and P. Gaehtgens (eds.), *White Blood Cells: Morphology and Rheology as Related to Function,* Martinus Nijhoff, The Hague, 1982.

91. Brånemark, P-I.: *Intravascular Anatomy of Blood Cells in Man,* Karger, Basel, 1971.

92. Schmid-Schönbein, G. W., Usami, S., Skalak, R., and Chien, S.: The interaction of leukocytes and erythrocytes in capillary and postcapillary vessels, *Microvasc. Res., 19,* 45–70, 1980.

93. Schmid-Schönbein, G. W., Fung, Y. C., and Zweifach, B. W.: Vascular endothelium-leukocyte interaction: Sticking shear force in venules, *Circ. Res., 36,* 173–184, 1975.

94. Athens, J. W., Haab, O. P., Raab, S. O., Mauer, A. M., Ashenbrucker, H., Cartwright, G. E., and Wintrobe, M. M.: Leukokinetic studies: IV. The total blood circulating and marginal granulocyte pools and the granulocyte turnover rate in normal subjects, *J. Clin. Invest., 40,* 989–995, 1961.

95. Bagge, U., Blixt, A., and Strid, K-G.: The initiation of post-capillary margination of leukocytes: Studies in vitro on the influence of erythrocyte concentration and flow velocity, *Int. J. Microcirc. Clin. Exp., 2,* 215–222, 1983.

96. Goldsmith, H. L., and Karino, T.: Microscopic considerations: The motion of individual particles, *Ann. N.Y. Acad. Sci., 293,* 241–255, 1977.

97. Vejlens, G.: The distribution of leukocytes in the vascular system, *Acta Pathol. Microbiol. Scand. Suppl., 33,* 11–239, 1938.

98. Palmer, A. A.: Platelet and leukocyte skimming, *Bibl. Anat., 9,* 300–393, 1967.

99. Chien, S.: Biophysical behaviour of red cells in suspension, in *The Red Blood Cell,* 2d ed., Academic Press, New York, 1975, vol. II, chap. 26.

100. Goldsmith, H. L., and Spain, S.: Margination of leukocytes in blood flow through small tubes, *Microvasc. Res., 27,* 204–222, 1984.

101. Nobis, U., Pries, A. R., Cokelet, G. R., and Gaehtgens, P.: Radial distribution of white cells during blood flow in small tubes, *Microvasc. Res., 29,* 295–304, 1985.

102. Schmid-Schönbein, G. W., Skalak, R., Usami, S., and Chien, S.: Cell distribution in capillary networks, *Microvasc. Res., 19,* 18–44, 1980.

103. Mayrovitz, H. N., and Rubin, R.: Leukocyte distribution to arteriolar branches: Dependence to microvascular blood flow, *Microvasc. Res., 29,* 282–294, 1985.

104. Bagge, U., and Karlsson, R.: Maintenance of white cell margination at the passage through small venular junctions, *Microvasc. Res., 20,* 92–95, 1980.

105. Marchesi, V. T., and Florey, H. W.: Electron micrographic observations on the emigration of leukocytes, *Q. J. Exp. Physiol., 45,* 343–347, 1960.

106. Mayrovitz, H. N., Wiedeman, M. P., and Tuma, R. F.: Factors in influencing leukocyte adherence in microvessels, *Thromb. Haemostasis, 4,* 823–830, 1977.

107. Marchesi, V. T., and Gowans, J. L.: The migration of lymphocytes through the endothelium of venules in lymph nodes: An electron microscopic study, *Proc. R. Soc. London (Biol. Sci.), 159,* 283–290, 1964.

108. Gowans, J. L., and Knight, E. J.: The route of recirculation of lymphocytes in the rat, *Proc. R. Soc. London (Biol. Sci.), 159,* 257–282, 1964.

109. Doroszewski, I., Golab-Meyer, Z., and Guryn, W.: Adhesion of cells in flowing suspensions: Effects of shearing force and cell kinetic energy, *Microvasc. Res., 18,* 421–433, 1979.

110. Tözeren, A., and Skalak, R.: Stress in a suspension near rigid boundaries, *J. Fluid Mech., 82,* 289–307, 1977.

111. Mayrovitz, H. N., Tuma, R. F., and Wiedeman, M. P.: Leukocyte adherences in arterioles following extravascular tissue trauma, *Microvasc. Res., 20,* 264–274, 1980.

112. Allison, F., Jr., Smith, M. R., and Wood, W. B., Jr.: Studies on the pathogenesis of acute inflammation, *J. Exp. Med., 102,* 655–676, 1955.

113. Mayrovitz, H. N., and Wiedeman, M. P.: Leukocyte adhesiveness as influenced by blood velocity, in J. Grayson and W. Zingg (eds.), *Microcirculation,* Plenum Press, New York, 1976, pp. 128–130.

114. Atherton, A., and Born, G. V. R.: Quantitative investigations of the adhesiveness of circulating polymorphonuclear leukocytes to blood vessel walls, *J. Physiol., 222,* 447–474, 1972.

115. Eriksson, E., and Robson, M. C.: New pathophysiological mechanism explaining post-burn oedema, *Burns, 4,* 153–156, 1978.

116. Lipowsky, H. H., Usami, S., and Chien, S.: In-vivo measurements of "apparent viscosity" and microvessel hematocrit in the mesentery of the cat, *Microvasc. Res., 19,* 297–319, 1980.

117. Stewart, G. F., Ritchie, W. G., and Lynch, P. R.: Venous endothelial damage produced by massive sticking and emigration of leukocytes, *Am. J. Pathol., 74,* 507–521, 1974.

118. Wedmore, C. V., and Williams, T. J.: Control of vascular permeability by polymorphonuclear leukocytes in inflammation, *Nature, 289,* 646–650, 1981.

119. Flick, M. R., Perel, A., and Staub, N. C.: Leukocytes are required for increased lung microvascular permeability after microembolization in sheep, *Circ. Res., 48,* 344–351, 1981.

120. Weening, R. S., Wever, R., and Roos, D.: Quantitative aspect of the production of superoxide radicals by phagocytizing human granulocytes, *J. Lab. Clin. Med., 85,* 245–252, 1975.

121. Sacks, T., Moldow, F. M., Craddock, P. R., Bowers, T. K., and Jacob, H. S.: Oxygen radicals mediate endothelial cell damage by complement stimulated granulocytes, *J. Clin. Invest., 61,* 1161–1167, 1978.

122. Sandison, J. C.: Contraction of blood vessels and observations in the transparent chamber in the rabbit's ear, *Anat. Rec., 54,* 105–127, 1932.

123. Clark, E. R., Clark, E. L., and Rex, R. O.: Observation on polymorphonuclear leukocytes in the living animal, *Am. J. Anat., 57,* 123–173, 1936.

124. Bagge, U., Johansson, B. R., and Olofson, J.: Deformation of white blood cells in capillaries: A combined intravital and electron microscopic study in the mesentery of rabbits, *Adv. Microcirc., 7,* 18–28, 1977.

125. Adell, R., Skalak, R., and Brånemark, P-I.: A preliminary study of rheology of granulocytes, *Blut, 21,* 91–105, 1970.

126. Bagge, U., Amundson, B., and Lauritzen, C.: White blood cell deformability and plugging of skeletal muscle capillaries in hemorrhagic shock, *Acta Physiol. Scand., 108,* 159–163, 1980.

127. Braide, M., Amundson, B., Chien, S., and Bagge, U.: Quantitative studies on the influence of leukocytes on the vascular resistance in a skeletal muscle preparation, *Microvasc. Res., 27,* 331–352, 1984.

128. Engler, R., Schmid-Schönbein, G. W., and Pavelec, R.: Leukocyte capillary plugging in myocardial ischemia and reperfusion in the dog, *Am. J. Pathol., 111,* 98–111, 1983.

129. Yamakawa, T.: Dynamics of white blood cells and red blood cells in microcirculatory networks of the cat brain cortex during hemorrhagic shock: Intravital microscopic study, *Microvasc. Res., 24,* 218A, 1982.

130. Wilson, J. W.: Leukocyte sequestration and morphologic augmentation in the pulmonary network following hemorrhagic shock and related forms of stress, *Adv. Microcirc., 4,* 197–232, 1972.

131. Schlag, G., and Redl, H.: Leukostasis in the lung during hypovolemic traumatic shock, *Anesthetist, 29,* 606–612, 1980.

132. MacGregor, R. R.: Granulocyte adherence, in L. E. Glynn, J. C. Houck, and G. Weissmann (eds.), *Handbook of Inflammation,* Elsevier–North Holland, New York, 1980, chap. 9.

133. Harlan, J. M.: Leukocyte-endothelial interactions, *Blood, 65,* 513–525, 1985.

134. Spitler, L. E., Petz, L., Spath, P., and Fundenberg, H. H.: Acquired lazy leukocyte syndrome, *Clin. Res., 19,* 568A, 1971.

135. MacGregor, R. R., Spagnuolo, P. J., and Lentrek, A. L.: Inhibition of granulocyte adherence by ethanol, prednisone, and aspirin, measured with a new assay system, *N. Engl. J. Med., 291,* 642–646, 1974.

136. Bryant, R. E., and Sutcliffe, M. C.: A method for quantitation of human leukocyte adhesion to glass, *Proc. Soc. Exp. Biol. Med., 141,* 196–202, 1972.

137. Banks, D. C., and Mitchell, J. R. A.: Leukocytes and thrombosis: I. A simple test of leukocyte behavior, *Thromb. Diath. Haemorrh., 30,* 36–46, 1973.

138. Neumann, A. W., Absolom, D. R., Van Oss, C. J., and Zingg, W.: Surface thermodynamics of leukocyte and platelet adhesion to polymer surfaces, *Cell Biophys., 1,* 79–92, 1979.

139. Lackie, J. M., and deBono, D.: Interactions of neutrophil granulocytes (PMNs) and endothelium in vitro, *Microvasc. Res., 13,* 107–112, 1977.

140. Goldsmith, H. L., Lichtarge, O., Tessier-Lavigne, M., and Spain, S.: Some model experiments in hemodynamics: VI. Two-body collision between blood cells, *Biorheology, 18,* 531–555, 1981.

141. Görog, P., and Born, G. V. R.: Increased adhesiveness of granulocytes in rabbit ear chamber blood vessels perfused with neuraminidase, *Microvasc. Res., 23,* 380–384, 1982.

142. Thompson, P. L., Papadimitrion, J., and Walkers, M.: Suppression of leukocyte sticking and emigration by chelation of calcium, *J. Pathol. Bacteriol., 94,* 389–396, 1967.

143. Hoover, R. L., Folger, R., Haering, W. A., Ware, B. R., and Karnovsky, M. J.: Adhesion of leukocytes to endothelium: Roles of divalent cations surface charge, chemotactic agents and substrate, *J. Cell Sci., 45,* 73–86, 1980

144. Garvin, J. E.: Factors affecting the adhesiveness of human leukocytes and platelets in vitro, *J. Exp. Med., 114,* 51–73, 1961.

145. Grant, L., Palmer, P., and Sanders, A. G.: The effect of heparin on the sticking of white cells to endothelium in inflammation, *J. Pathol. Bacteriol., 83,* 127–133, 1962.

146. Lentrek, A. L., Schrieber, A. D., and MacGregor, R. R.: The induction of augmented granulocyte adherence by inflammation: Mediation by a plasma factor, *J. Clin. Invest., 57,* 1098–1103, 1976.

147. MacGregor, R. R.: The effect of anti-inflammatory agents and inflammation of granulocyte adherence: Evidence for regulation by plasma factor, *Am. J. Med., 61,* 597–607, 1976.

148. Wiedeman, M. P.: Movement of leukocytes in flowing blood, in U. Bagge, G. V. R. Born, and P. Gaehtgens (eds.), *White Blood Cells: Morphology and Rheology as Related to Function,* Martinus Nijoff, The Hague, 1982.

149. Hammerschmidt, D. E., White, J. G., Craddock, P. R., and Jacob, H. S.: Corticosteroids inhibit complement induced granulocyte aggregation, *J. Clin. Invest., 63,* 798–803, 1979.

150. Fehr, J., and Jacob, H. A.: In-vitro granulocyte adherence and in-vivo margination: Two associated complement-dependent functions, *J. Exp. Med., 146,* 641–652, 1977.

151. Hammerschmidt, D. E., Harris, P. D., Wayland, J. H., Craddock, P. R., and Jacob, H. S.: Complement-induced granulocyte aggregation in vivo, *Am. J. Pathol., 102,* 146–150, 1981.

152. Schiffman, E., Corcoran, B. A., and Wahl, S. M.: *N*-formyl-methionyl peptides as chemoattractants for leukocytes, *Proc. Nat. Acad. Sci., 72,* 1059–1062, 1975.

153. Fehr, J., and Dahinden, C.: Modulating influences of chemotactic factor induced cell adhesiveness on granulocyte function, *J. Clin. Invest., 64,* 8–16, 1979.

154. Craddock, P. R., Hammerschmidt, D. E., White, J. G., Dalmasso, A. P., and Jacob, H. S.: Complement (C5a)-induced granulocyte aggregation in vitro, *J. Clin. Invest., 60,* 260–264, 1977.

155. Cheung, A. T. W., Miller, M. E., and Keller, S. R.: Movement of human polymorphonuclear leukocytes: A video tape analysis, *J. Reticul. Soc., 31,* 193–205, 1982.

CHAPTER 14

The Rheology and Tube Flow of Blood

Giles R. Cokelet
Department of Radiation Biology and Biophysics
University of Rochester
School of Medicine and Dentistry
Rochester, New York

14.1 RHEOLOGY OF BLOOD

14.1.1 General Characteristics

Human blood is a concentrated suspension of various types of near-colloidal-size cells in a complex aqueous solution. Since the cells are of various shapes, deformabilities, and electric charge densities, and since the suspending medium (plasma) contains molecular species ranging from small ionic and nonionic species to large proteins, one anticipates that blood will exhibit the typical complex rheological properties of a suspension of deformable, aggregable particles.

Typical characteristics for the blood cells of a healthy adult are shown in Table 14.1 The red cells (erythrocytes) form quite a uniform class of cells, in spite of the fact that they have a normal life span of 121 days, during which their characteristics change; they dominate the cell population on the basis of number and volume concentration. As a crude analogue, each of these cells can be considered to be an incompletely filled envelope (membrane) of a solution which contains all the blood's oxygen-carrying protein, hemoglobin (about 33 g/100 mL). The physical characteristics of these easily deformable cells are described in Chap. 12.

The white cells (leukocytes) are a diverse collection of nucleated, complex cells. They are less easily deformed than erythrocytes (see Chap. 13 for their characteristics). The platelets are very small and show Brownian motion; their mechanical properties have not been quantitated. Because the leukocytes and platelets normally occupy such a small fraction of the blood volume, their effect on the macroscopic rheology of blood is generally negligible.

While the cell populations shown in Table 14.1 are considered normal, they can vary over a wide range under normal physiological or pathological conditions (as well as in human-made suspensions of some engineering importance). The erythrocyte volume fraction of the blood (hematocrit) may be very high (50 to 60 percent in neonates, 65 to 75 percent in premature

TABLE 14.1 Representative characteristics of healthy adult human blood cells

Cell	Density, g/mL	Blood count, no./μL	Volume, fL	Dimensions, μm	Unstressed shape
Erythrocytes:					
Mature	1.09	5×10^6	87	8.5×2	Biconcave disk, unnucleated
Reticulocyte		30,000	300	8.5×5	Irregular, nucleated
Leukocytes*:	1.07				Wrinkle-surfaced sphere, nucleated
Neutrophil		4,200	440	9.4	
Eosinophil		170	440	9.4	
Basophil		50	440	9.4	
Lymphocyte		2,200	210	7.4	
Monocyte		460	460	9.5	
Platelets	1.03	300,000	15	3	Irregular disk

* See Chap. 13 for other values of leukocyte dimensions.

infants, 60 percent in adults living at very high altitudes, 75 percent in those with polycythemia) or very low (20 to 25 percent in those with sickle-cell anemia or thalassemia, or even lower in cases of high blood loss or during hypothermic surgery). White cell volume fraction (leukocrit) may become over 30 percent in hyperleukocytic leukemia patients (although in these patients the sum of the hematocrit and leukocrit remains roughly constant). The rheological properties of blood over a wide range of hematocrit and leukocrit values are therefore of interest.

The composition of the suspending medium for these blood cells can vary extensively, both naturally in the human body and in human-made suspensions. This fluid (plasma) is usually an aqueous solution, although it may contain small noncellular particles (e.g., emulsified fat particles). A list of the major components and typical concentrations in plasma from healthy individuals is given in Table 14.2. Two influences of these constituents are of direct concern to the blood's flow properties: (1) The membranes of the blood cells contain both simple and complex

TABLE 14.2 Major constituents of plasma from healthy adult humans

Solute	Typical concentration, g/L
Proteins	73
Albumin	37
Globulins	
α_1	4
α_2	7
β	9
γ	13
Fibrinogen	3
Lipids	5
Glucose	0.8
Amino acids	0.5
Urea	0.3
Lactate	0.1
Inorganic ions	
Na^+	3.2
K^+	0.2
Ca^{++}	0.1
Mg^{++}	0.01
Cl^-	3.6
$HCO_{3_=}^-$	1.7
$SO_{4_=}$	0.2
PO_4	0.1

molecular transport mechanisms; in particular, water passes rapidly across the red cell membrane, and therefore the volume and shape of these cells is rapidly influenced by changes in the chemical potential of water in the plasma (osmotic pressure). (2) The larger molecules in plasma (most importantly, fibrinogen) can cause erythrocyte aggregation, which may be broken up by sufficiently strong shear stresses; since the cell surfaces as well as the aggregation-causing macromolecules have net negative charges, the ionic environment in the suspending medium influences erythrocyte aggregation.

The volume and resting shape of a human erythrocyte as a function of the osmotic pressure of the suspending medium are quite well known. For example, Evans and Fung [1] report that the average volume is 94 f L at 300 mosmol (isotonic medium), where the cell is a biconcave discoid shape; 164 f L at 131 mosmol (hypotonic medium), where the cell is a sphere and further lowering of the osmotic pressure causes cell membrane rupture (hemolysis); and 116 f L at 217 mosmol, where the cell is still a biconcave discoid shape but where its major diameter is less and its minimum thickness is considerably increased over its isotonic value. Increasing tonicity over isotonic conditions results in decreased cell volume and transformation into a crenated shape (echinocytes). Also, erythrocyte density increases during the 121-day life span of red cells in the body; this density behavior forms the basis for the methods of fractionation of red cells according to their age.

Erythrocyte volume and/or shape may be abnormal under a variety of conditions. In diseases, cell shape may be altered; the example of sickle-cell disease is well known. Some sickled erythrocytes undergo repeated cyclic transformation from discoid (deformable) to sickle (rigid) to discoid shape as the cell travels cyclically through high-oxygen and low-oxygen regions, while other cells in the same blood are permanently rigid and sickle-shaped (irreversibly sickled cells); yet other cells always remain discoid. In microcythemia, the erythrocytes are abnormally small. Alteration of shape may occur even in healthy people: it has been reported [2] that during decompression from simulated deep-sea dives, appreciable fractions (6 to 12 percent) of the red cells in humans are converted to the echinocytic shapes. A large number of agents are known to cause crenation (e.g., 2,4-dinitrophenol) and spheration (e.g., chlorpromazine HCl) at low concentrations. (Reference 3 contains useful information on these agents.) It is important to note that metabolically depleted erythrocytes transform into echinocytes and to consider this fact in an interpretation of rheological data obtained with blood containing such cells (e.g., stored blood, "outdated" blood).

Many macromolecules can cause red cell aggregation: fibrinogen, globulins, and plasma expanders such as dextran (a neutral macromolecule) and polyvinylpyrrolidone. The arrangement of erythrocytes within the primary aggregate (rouleau) is regular: the cells aggregate discoid face to discoid face rather than by random clumping (as occurs in agglutination). The cells deform when they aggregate so that the adjoining faces are locally parallel and have maximum contact; the separation distance between adjacent cell faces increases with increase in the length of the aggregating macromolecule [4]. Since the cells have a net negative surface charge, they are surrounded by a diffuse layer of counterions; this gives rise to an electrostatic repulsion between cells which is not overbalanced by van der Waals forces. This repulsion prevents red cells from aggregating when in a simple aqueous medium such as isotonic saline. However, a macromolecule which can simultaneously adsorb on the surfaces of two neighboring cells could cause aggregation if it is long enough to bridge the cells while they are far enough apart so that the electrostatic repulsive forces are small. This mechanism is consistent with the observation that dextran molecules with average molecular weights below about 45,000 cannot cause red cell aggregation at any concentration [5]. However, larger dextran molecules do not cause aggregation at very low or very high concentrations. The behavior at lower concentrations reflects too few cross-links between cells to overcome electrostatic repulsion. The behavior at higher concentrations is explained as follows: As the concentration of dextran is increased in the suspending medium, more dextran molecules are adsorbed on the red cell surface, resulting in an increase in the dextran volume fraction in the fluid layer adjacent to the cell surface. This causes the diffuse counterion layer around the cell to expand until the counterion layer thickness is so great that the dextran molecules can no longer bridge the gap between adjacent erythrocyte surfaces. While this model explains the effects of dextran on red cell aggregation, it is important to note that for fibrinogen, the most important red cell aggregating agent in human plasma, no high concentration limit for red cell aggregation has been found [6].

On the basis of this mechanism of red cell aggregation, one would expect characteristics of erythrocyte aggregation to be influenced by the characteristics of the bridging macromolecules (and how they adsorb on the cell surface), the ionic composition of the suspending medium, the charge density of the cell surface, and the shape and deformability of the erythrocytes [7, 8].

14.1.2 Characteristic Times for Cell Deformation and Aggregation

Since blood contains deformable and aggregable cells, blood should be a non-Newtonian fluid, at least under those flow conditions where the degrees of cell deformability and aggregation are functions of shear rate, time, etc. Since these two cell properties provide a means of storing energy during flow (e.g., by deformation of the cell membrane and by dispersion of the aggregable particles), blood, in principle, should be viscoelastic.

Cell deformation and aggregation do not occur instantaneously. It is therefore helpful, in interpreting and utilizing blood rheological data, to consider the relationships between the characteristic times for these processes and for the stages through which blood passes as it flows in various situations. For example, in an assessment of unsteady flow the validity of the quasi-steady-state approximation can be considered (at each instant of time, the steady-state stress-strain rate relationship can be used).

The characteristic times for red cell deformation, aggregation, and disaggregation have been previously reviewed [9]. The characteristic time for red cell deformation is very short (on the order of 0.06 s or less), while that for white cells is very long (order of tens of seconds or longer). The characteristic time for red cell aggregation is a function of many parameters (hematocrit, shear rate, plasma composition, and, to a lesser extent, red cell shape and deformability) since it depends on the forces of attraction between cells and the collision frequency of cells. The characteristic time is on the order of 10 s in going from a completely dispersed cell state to an equilibrium aggregation state at a shear rate of about 10 s^{-1}, while the comparable time in the absence of superimposed shear is on the order of 1 min. The mechanism for breakup of aggregates, by an increase in shear rate, does not depend on cell collision but rather on how fast the disruptive stress is generated and how fast cells move apart; this process is almost instantaneous. Blood under different pathological conditions may have increased characteristic times for both aggregation and disaggregation.

When red cell aggregation occurs, two additional effects are encountered: (1) increased erythrocyte sedimentation and (2) red cell syneresis. Both these processes are slow, not only because the aggregation process is slow but also because the density difference between red cells and plasma is small (about 4 percent), in the case of sedimentation, and the net attractive forces between cells are small, in the case of syneresis. Consequently, the full development of the effects of these processes, even in small flow channels, may require tens of seconds to minutes to develop.

It therefore would appear that for flows which are of a predominantly high shear rate, so that cell aggregation does not occur, and which occur in flow channels with a characteristic dimension that is large compared with the cell dimensions (say, about 400 μm or larger), the steady-state approximation probably will be acceptable; in smaller channels, an effect due to nonuniform distribution of red cells across the flow-channel lumen and migration of red cells transverse to the flow may be important. In unsteady flows, where appreciable aggregation and disaggregation occurs, the secondary processes of sedimentation and syneresis may be the "rate-limiting" mechanisms, while all other flows can be satisfactorily represented through use of the quasi-steady-state approximation. This area has not been extensively explored.

14.1.3 Plasma Rheology

Until the early 1960s, there was considerable debate over whether plasma from healthy individuals exhibited Newtonian or non-Newtonian behavior. Since that time, it has become generally accepted that such plasma is Newtonian and that the earlier data showing apparently non-Newtonian behavior contain an experimental artifact. Some constituents of plasma are surfactants and form an air-plasma interfacial film which has sufficient mechanical cohesion to transmit a significant torque, measured by a rotational viscometer, between those surfaces which move at different angular velocities, and to support a significant pressure drop, measured by a capillary viscometer. When this torque or pressure drop artifact is removed, plasma from healthy humans is found to be Newtonian [20].

Harkness [10] reports that most plasmas show a 2 to 3 percent decrease in viscosity with each 1°C increase in temperature over the temperature range of 15 to 40°C. He reports for normal plasmas viscosity ranges of 1.70 to 1.915 mPa · s at 20°C, 1.50 to 1.72 mPa · s at 25°C, and 1.16 to 1.35 mPa · s at 37°C; these values show a temperature-insensitive ratio with water viscosity at the same temperature (1.68 to 1.92 times water viscosity). At temperatures above about 42°C, irreversible changes occur to some plasma constituents, resulting in increased viscosities upon return to lower temperatures.

The contributions of the plasma constituents to plasma viscosity have been estimated. The small molecular components of plasma (e.g., ionic salts, urea, glucose) contribute only 1 to 2 percent to the increase of plasma viscosity over water's viscosity, while the remaining part of the increase is due to the proteins. Each individual type of protein has a nonlinear viscosity-concentration curve, with the curve's slope increasing with concentration. A protein's contribution to plasma viscosity depends not only on its concentration but also on its molecular weight and nonsphericity. Thus, as has been shown [10], while the normal protein concentrations are in the ratio of albumin to globulin to fibrinogen as 4.47 : 3.03 : 0.3 g/100 mL, the contribution to plasma viscosity (over that of water) is about 36 : 41 : 22. Therefore, any prediction of the effect of variations in plasma protein composition on plasma viscosity must recognize the different nature of the contributions of the various proteins. In some diseases (e.g., Raynaud's syndrome [11]), other macromolecules can reach concentrations at which they affect plasma viscosity. Also, in pathological cases, some of the gamma globulins can cause abnormal temperature dependency for plasma viscosity at lower temperatures. High levels of lipoproteins can also cause increased plasma viscosity, especially if they are present in the form of chylomicrons (emulsified fat particles with diameters as large as 1 μm). A linear relationship has been reported [12] between plasma viscosity and chylomicron concentration, with a slope of 1.1×10^{-4} mPa · s per 1 mg/dL triglyceride.

Harkness [10, 13] has thoroughly reviewed the subject of human blood plasma viscosity, and his studies should be consulted for more extensive information on this subject.

14.1.4 Blood and Red Blood Cell Suspensions

General Features. The general rheological behavior of red cell suspensions is a strong function of the rate of deformation, since for a given suspension this dictates the degree of cell aggregation and deformation. Figure 14.1 is a typical diagram used to show the importance of red cell deformation and aggregation. The apparent viscosities of three different red cell suspensions are shown as a function of shear rate at 37°C. The three suspensions are of normal red cells in plasma ("blood"), normal red cells in isotonic saline containing 11% albumin, and hardened, discoid red cells suspended in the same saline solution. All hematocrits are 45 percent, and the suspending media all have the same viscosity (1.2 mPa · s). The blood shows both cell deformation and aggregation, the normal red cells in the saline medium show deformation but no aggregation, and the hardened cells in saline show neither deformation nor aggregation.

The red blood cell–plasma suspension (blood) shows the great range in apparent viscosity which this suspension has over a wide shear rate range; at high shear rates (above about 1000 s^{-1}) this fluid can be considered to be Newtonian (apparent viscosity independent of shear rate), although the critical shear rate at which blood is Newtonian depends mainly on such variables as hematocrit. The blood and red blood cell–saline suspension have identical rheological behavior at shear rates above about 4 s^{-1}, indicating that the non-Newtonian behavior shown by blood at shear rates above 4 s^{-1} is due to red cell deformational changes with shear rates. The constant apparent viscosity behavior shown at the highest shear rate is taken to mean that cell shape changes with shear rate changes are minimal or uninfluential. The difference in apparent viscosity between these two suspensions, at a given shear rate below 4 s^{-1}, is attributed to red cell aggregation in the blood.

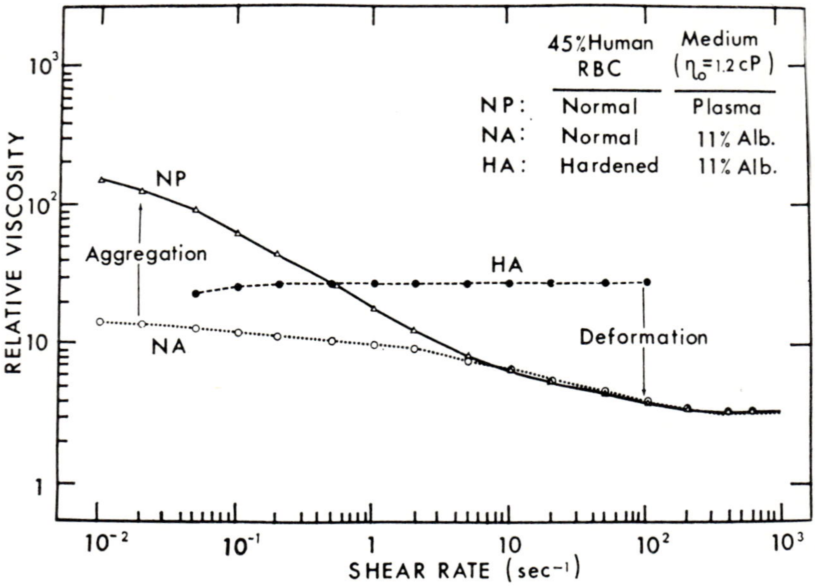

FIG. 14.1 The apparent viscosities of suspensions of human red cells, divided by the suspending medium viscosity (1.2 mPa · s), as a function of shear rate. NP = normal red cells in plasma, NA = normal cells in isotonic saline containing 11% albumin (to make suspending medium viscosity the same as plasma viscosity), and HA = hardened discoid red cells in the same saline. (*Reproduced with permission from S. Chien, Shear dependence of effective cell volume as a determinant of blood viscosity, Science, 168, 977, 1970.*)

The red cell–saline suspension appears to be approaching Newtonian behavior at lower shear rates, suggesting that the cells in this suspension are tending to act as rigid particles. Goldsmith and Marlow [14] have shown that below a shear stress of about 10^{-1} Pa, individual red cells will undergo a flipping motion with no evidence of cell deformation. (See Keller and Skalak [15] for a detailed mathematical analysis of this case, including prediction of as yet experimentally unconfirmed circulation of cell membrane and contents, called "tank treading," even when the cell undergoes flipping motion.) The shear stresses "seen" by a cell at shear rates below 1 s^{-1} are below this critical level.

Thus, it is interesting to compare in Fig. 14.1 the behavior of the red cell–saline suspension to the behavior of rigid, discoid red cells suspended in the saline medium (where the cells do not aggregate). While these two suspensions approach identical behavior as the shear rate is decreased, the normal red cell–saline suspension is slightly non-Newtonian at the lowest shear rates shown and has a slightly lower apparent viscosity at a given shear rate, even when the shear stress is well below 10^{-1} Pa. Perhaps the small difference at the lowest shear rates reflects tank treading even at these low shear stresses or intercellular interactions which cause small cellular deformations or orientational effects. Suspensions of hardened discoid erythrocytes have generally been found to be Newtonian, at least for shear rates below several hundred s^{-1}; at higher shear rates and hematocrits above about 30 percent, these suspensions become dilatant (apparent viscosity increases with increasing shear rate), perhaps because the characteristic time for cell rotation becomes longer than the characteristic time between collisions. In any event, the difference between rigid cell suspension behavior and normal cell suspension behavior at shear rates above about 4 s^{-1} is attributed to the effect of the differences in deformability of normal and hardened erythrocytes.

Descriptive Equations. It is convenient to have analytical expressions to represent rheological data. Because of their importance, hematocrit and shear rate are the primary independent variables for describing the apparent viscosity of blood and other red cell suspensions. Other independent variables, such as temperature and concentrations of macromolecules which can cause red cell aggregation at low shear rates, are usually of less practical importance in a descriptive equation because they are constant in a given problem, but the effect of their variation from case to case must be considered.

Easthope and Brooks [16] made a comparison of 11 equations which have been used in the literature to represent blood's rheological behavior; the equation which represented their data with the least error was the four-constant equation proposed by Walburn and Schneck [17]:

$$\tau = a_1 \exp\left(a_2 H + \frac{a_3}{H^2}\right) \gamma^{-1-a_4 H} \tag{14.1}$$

where τ = the shear stress, γ = the shear rate, H = hematocrit, and a_1, a_2, a_3, a_4 are constants. This equation was developed by starting with the power law relationship between shear stress and shear rate, and using regression curve fitting of trial equations built by successive inclusion of variables of decreasing influence. While this equation contains the two most influential independent variables, the constants are functions of temperature, macromolecular composition of the continuous phase, etc. It does not allow for a yield stress.

There are advantages of interpretation to be gained if the descriptive rheological equations are derived from a mechanistic model of blood flow. Because of the complexity of blood, especially due to its particle high volume fraction, it has not been possible to develop such equations without introduction of some empiricism or the use of approximations. An equation of this type, which has been extensively used to represent blood rheological data (first used for this purpose by Scott-Blair [18]), is that derived by Casson [19]:

$$\tau^{1/2} = a_0^{1/2} + a_1 \gamma^{1/2} \tag{14.2}$$

The derivation of this equation models the suspension as containing particles which can aggregate at low shear rates to form rodlike aggregates whose length increases as shear rate decreases; the similarity between these aggregates and rouleaux makes the model attractive for use with blood. This model has some inherent defects (e.g., the yield stress a_0 does not become zero except in the case where particle concentration is zero). However, the equation has been extensively used because it can represent blood rheological data over a limited shear rate range with one set of constant values. But extrapolation of this equation beyond its demonstrated range of suitability is not safe. The experimental determination of the yield stress a_0 is complicated by the effect of erythrocyte syneresis at low shear rates [20, 21].

Another equation which fits into this category of being partially based on a mechanistic model is that derived by Quemada [22–24]:

$$\eta_r = (1 - 0.5kH)^{-2} \tag{14.3}$$

where $$k \equiv \frac{k_0 + k_\infty \gamma_r^{1/2}}{1 + \gamma_r^{1/2}} \quad \text{and} \quad \gamma_r \equiv \frac{\gamma}{\gamma_c}$$

In this equation, the three coefficients which reflect the suspension properties are k_0 and k_∞, which are the intrinsic viscosities at zero and infinite shear rates, respectively, of the flow particles which predominate at those shear rates, and γ_c, the critical shear rate, which can be considered to be the inverse of the relaxation time for the dominant structural unit causing the suspension to be non-Newtonian. These coefficients are functions of hematocrit, suspending medium composition, etc.

The properties of this equation which are pertinent to blood rheology are the following:

1. k_0 and k_∞ should be positive, with k_0 larger than k_∞ for shear-thinning suspensions. If the suspension is Newtonian, the equation reduces to

$$\eta_r = (1 - 0.5Hk)^{-2} \tag{14.4}$$

where $k = k_0$ if $\gamma \ll \gamma_c$, $k = k_\infty$ if $\gamma \gg \gamma_c$, or $k = k_0 = k_\infty$ if γ_c is within the range of γ for which the suspension is Newtonian. For red cell suspensions which are Newtonian, the latter form of the equation is valid.

2. The equation reduces to Casson's equation if $\gamma \gg \gamma_c$:

$$\tau^{1/2} = (\eta_0 \gamma_c)^{1/2} \frac{H}{2} \frac{k_0 - k_\infty}{(1 - 0.5k_\infty H)^2} + \frac{\eta_0^{1/2}}{1 - 0.5k_\infty H} \gamma^{1/2} \tag{14.5}$$

This is fortuitous since at such high shear rates red cell aggregation would not be expected and Casson's equation is built on a model of particle aggregation. It is therefore not realistic to interpret the constant term in this special version of Quemada's equation as the square root of a yield stress.

3. When the relative viscosity reaches infinity as the hematocrit is increased, the equation reduces to

$$k_0 = \frac{2}{H_m} \tag{14.6}$$

where H_m is the maximum volume fraction physically permissible in the suspension if the particles are in the association form prevalent when γ is very much smaller than γ_c.

Because Quemada's equation represents the rheological data for blood and related red cell suspensions over a wide range of conditions (Easthope and Brooks [16] found it to be almost as good as the Walburn-Schneck equation) and because its coefficients have some physical interpretation, it is used here.

Suspensions of Normal Erythrocytes. The general features of the rheological properties of blood and other red blood cell suspensions have been depicted in Fig. 14.1. Besides the shear rate influence shown there, hematocrit is the most influential independent variable, and generally the apparent viscosity at a given shear rate increases rapidly as hematocrit increases. This is illustrated in Fig. 14.2 for blood at 37°C, where the logarithm of apparent viscosity is plotted as a function of hematocrit with curves for constant shear rate.

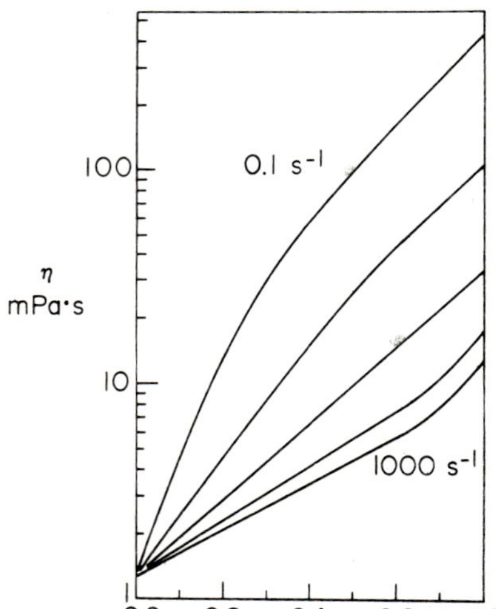

FIG. 14.2 The apparent viscosity of blood as a function of hematocrit, with each curve for a constant shear rate of 1000 (lowest curve), 100, 10, 1, or 0.1 s^{-1} (uppermost curve). Temperature is 37°C.

Data for blood at various hematocrits and two temperatures, and for normal erythrocytes suspended in saline containing 0.5% albumin, have been fitted to Quemada's equation. The coefficients resulting from this data fitting are shown in Figs. 14.3 to 14.5.

The coefficient k_0, the intrinsic viscosity at near-zero shear rate, is shown in Fig. 14.3. Data for blood at both 23 and 37°C fall on one curve. The best-fit polynomial equation is close to the equation $k_0 = 2/H$ (shown by dashed curve), which indicates that H_m is close to H. The k_0 curve for erythrocytes suspended in saline approaches the blood curve for H greater than about 0.65, indicating that above that H value, red cell crowding is so great that flow structures built by cell aggregation cannot fit within the suspension. In this respect, Quemada [24] has defined an *aggregation index* as $A = k_0/k_0^0$, where k_0^0 is for red cells suspended in saline. This index

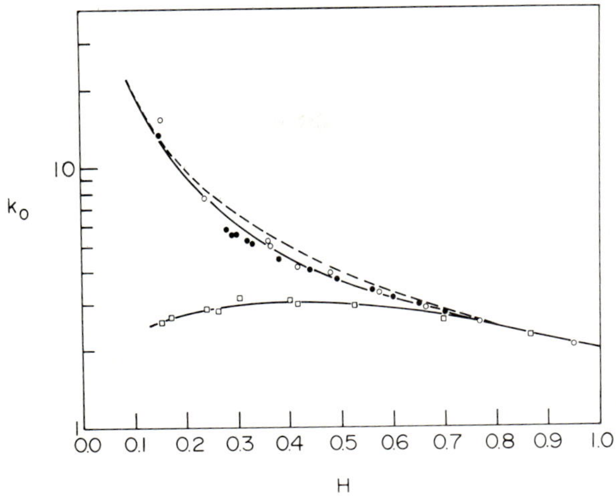

FIG. 14.3 The low shear rate intrinsic viscosity coefficient, k_0, for Quemada's equation as a function of hematocrit. Dashed curve is the equation $k_0 = 2/H$. Open circles are blood at 37°C (author's data); filled circles are blood at 23°C. (*From J. Dufax, D. Quemada, and P. Mills, Determination of rheological properties of red blood cells by Couvette viscometry, Rev. Phys. Appl. 15, 1367, 1980.*) Squares are red cells in saline with 0.5% albumin at 37°C (author's data). The curve for blood can be represented by $\ln k_0 = 3.874 - 10.41H + 13.80H^2 - 6.738H^3$, and the curve for red cells in saline $\ln k_0 = 0.6534 + 2.350H - 3.451H^2 + 1.1200H^3$.

compares qualitatively as a measure of aggregability with the index developed by Brooks et al. [25], $R = \eta\,(\text{saline})/\eta$, where η (saline) is the apparent viscosity for red cells in saline at a given shear rate and hematocrit, and η is for red cells in another medium at the same shear rate and hematocrit. It should be noted that defibrinated plasma (serum) still contains macromolecules which can cause very weak red cell aggregation, so that k_0 (serum) > k_0 (saline) except at high hematocrits (Quemada [26] gives a k_0 (serum) value of about 5.3 at $H = 0.10$).

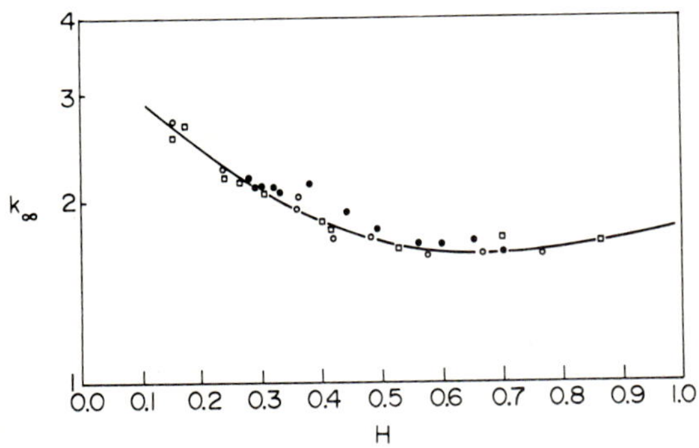

FIG. 14.4 The high shear rate coefficient k_∞ for Quemada's equation. Symbols same as for Fig. 14.3. The curve can be represented by the expression $\ln k_\infty = 1.3435 - 2.803H + 2.711H^2 - 0.6479H^3$.

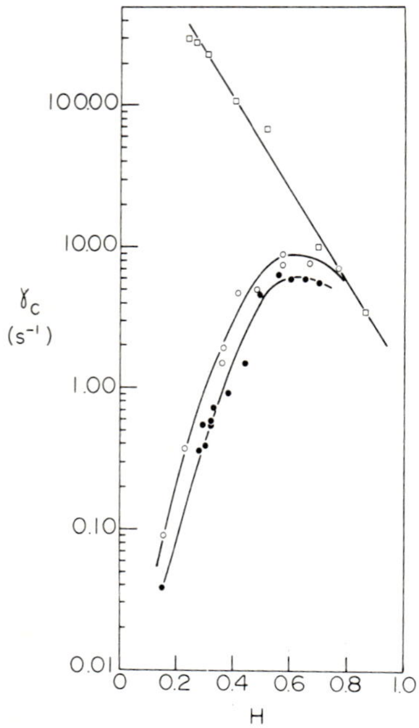

FIG. 14.5 The critical shear rate γ_c for Quemada's equation; symbols are as in Fig. 14.3. The curves can be represented by $\ln \gamma_c = -7.5411H + 7.7381$ (red cells in saline); $\ln \gamma_c = -6.1508 + 27.923H - 25.60H^2 + 3.697H^3$ (blood, 37°C); $\ln \gamma_c = -6.7597 + 25.626H - 18.988H^2$ (blood, 23°C).

The intrinsic viscosity at very high shear rates, k_∞, is shown in Fig. 14.4. Because at very high shear rates normal erythrocytes in all suspensions (continuous phase as plasma, serum, or saline) are dispersed, deformed, and circulating, k is expected to be the same for all suspensions of the same hematocrit. The slope of this curve changes its sign at a hematocrit close to that shown in Fig. 14.3, where the nature of the suspending medium stops influencing the value of k_0, presumably because of cell crowding. This coefficient is temperature-independent.

The third coefficient, γ_c, shows a more complex dependence on hematocrit, suspending phase composition, and temperature. This is illustrated in Fig. 14.5. It is reasonable to expect γ_c for blood and for normal erythrocytes in saline (at the same temperature) to approach each other at high hematocrits, because of cell crowding.

Miscellaneous Variables. The intent here is to provide information about the effects of some variables which have not been discussed, so that the interested reader will have an entry point to the pertinent literature.

Erythrocyte shape: Through the use of shape-changing drugs, the effect of red cell shape on the rheology of red cell–saline suspension has been studied. Transformation into more spheric shapes (stomatocytes) is reported to reduce suspension apparent viscosity at a constant hematocrit, while transformation into spiculated forms (echinocytes) increases low shear rate viscosity without affecting high shear rate viscosity [27, 28].

Erythrocyte deformability: A decrease in red cell deformability is expected to decrease low shear rate apparent viscosity of blood (because of decreased contact area between cells) and increase high shear rate apparent viscosity.

This expectation is supported by the study of Chien et al. [29], in which the time courses of the rheological properties of saline suspensions of human red cells were studied during cell hardening with acetaldehyde. At a constant true hematocrit, loss of cell deformability is first seen at the lowest shear rates, but as hardening progresses, the effect of rigidification is seen at progressively higher shear rates.

Data for suspensions of completely hardened discoid red cells [30], as well as for mixtures of normal and hardened red cells [31], have been reported. Sickle-cell blood is an interesting example of such a mixture of cells. Suspensions of erythrocyte ghosts have been studied [32].

Plasma proteins: Of the proteins normally found in plasma, fibrinogen has the greatest influence on low shear viscosity of blood, followed by the globulins and albumin [21]. Abnormal protein composition can lead to increased blood viscosity (e.g., IgM or cryoglobulins [33]).

White blood cells: White blood cells, being less deformable than erythrocytes, show higher suspension viscosities than red cell suspensions of comparable cellular volume fraction. In leukemia, the sum of red cell and white cell volume fractions in blood tends to remain constant, so that the blood viscosity effect of increased white cell count is reduced. The rheology of red cell–white cell plasma mixtures has been studied [34].

14.1.5 Conclusion

For an excellent statement about the correct understanding of blood rheology, the reader is referred to a chapter by S. Chien [35].

14.2 BLOOD FLOW IN TUBES

The use of macroscopic rheological data and the continuum model to predict blood flow behavior in large vessels seems satisfactory as long as the shear stresses are large enough to prevent appreciable red cell aggregation. If aggregation occurs, sedimentation and/or syneresis effects may become important factors. For example, for normal blood flow in horizontal tubes, sedimentation effects are seen if the bulk average velocity divided by vessel diameter ($\overline{U}$) is below about 1 s^{-1} (as shown, for example, in Figs. 7 and 8 in Ref. 36). The effect is expected to occur at higher flow rates if red cell aggregation is enhanced. In blood flow through vertical vessels, red cell aggregation leads to syneresis so that a wall layer of plasma forms, reducing the pressure gradient needed for a given flow rate (see Ref. 37 for an exaggerated example of this effect). With these cautions, simple continuum mechanics is satisfactory for vessels with diameters above about 500 μm.

14.2.1 Fahraeus Effect

Because of its particulate nature, blood fails to act as a continuum as the characteristic dimension of the vessel containing it approaches cellular dimensions. This discrepancy can be noted in several different ways.

One evidence of the deviation arises from a difference in the abilities of cells and the suspending medium to enter a vessel. For example, when blood flows from a stirred reservoir into a small tube, the cells may be retarded during entry into the tube because of collision with the entrance rim, collision between cells, etc., while plasma flows easily into the tube. This "screening effect" in steady-flow experiments of this type is indicated by the hematocrit of the blood discharging from the tube (H_D) being less than the hematocrit in the feed reservoir ($\underline{H_F}$). This effect is seen only for tubes with diameters below about 20 μm at high flow rates ($\overline{U} > 100\ s^{-1}$) [38] and at lower velocities [39]. Of course, if the blood "presented" to the vessel entrance is not representative of the average blood in the feed container, then H_D and H_F can differ for reasons other than screening at the vessel entrance.

Under steady-state conditions, H_D represents the composition of the blood actually entering the vessel. But in vessels with diameters smaller than about 500 μm, the average hematocrit in the tube, H_T, is less than H_D. Gaehtgens [40] prepared the graphical summary in Fig. 14.6 of data showing how the ratio H_T/H_D varies with tube diameter when H_F is constant at about 40 percent. This phenomenon of H_T/H_D being less than 1 is generally referred to as the *Fahraeus effect*. The explanation of this effect lies in two factors: (1) the red cells are not uniformly distributed across the vessel lumen, but, instead, the local hematocrit is low near the vessel wall and high in the central region of the lumen, and (2) the velocity is highest at the center of the vessel and decreases toward zero at the wall. Thus, locally in the tube high-hematocrit blood flows rapidly through the tube while low-hematocrit blood flows slowly, resulting in an effluent which when mixed has H_D greater than H_T.

The reason for the dispersion of data in Fig. 14.6 is not known. In many cases, H_F, H_T, and H_D were not all measured simultaneously, and, in particular, H_T and H_D were not determined in the same experiment. It is possible that H_D was not equal to H_F in some experiments, as assumed. The two curves for tubes with diameters between 8 and 30 μm are claimed to be limits of data; in this region, H_T/H_D is said to depend on flow velocity, with the upper curve representing low velocities, where cell deformation is small, and the lower curve representing high velocities, where cell deformation is large. The limit at 2.7 μm arises from the observation that the red cell surface area is relatively fixed and that a cylindrical vessel of this size is the smallest tube which can accommodate a human red cell.

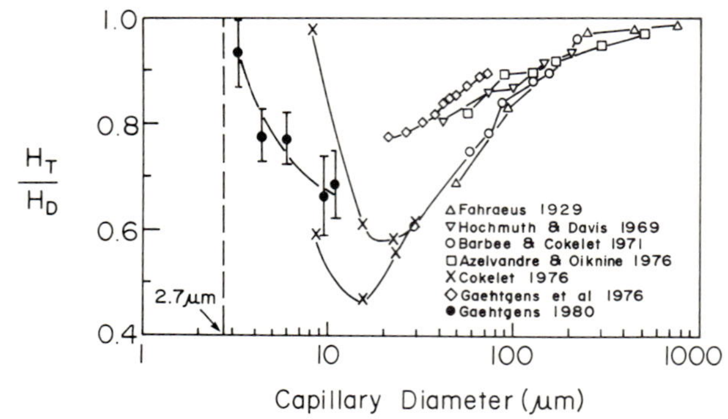

FIG. 14.6 The Fahraeus effect. (See Ref. 40 for data source identity.) *(Data from various authors as collected by Gaehtgens [40]. Reproduced with permission.)*

The ratio H_T/H_F is a linear function of H_F for a fixed value of D [38]. The slope of this function is zero for D greater than about 110 μm and increases as D decreases below that value. At a fixed H_F value, H_T decreases as D decreases. Thus, the minimum in Fig. 14.6 at 15 to 20 μm occurs because, below that diameter, H_D begins to decrease as D decreases.

14.2.2 Fahraeus-Lindqvist Effect

When the experimental relationship between pressure drop per unit length ($\Delta P/L$) and flow rate (Q) is determined for blood flow through a tube, a region of high Q is found where $\Delta P/L$ is proportional to Q; i.e., blood acts as a Newtonian fluid. Within this Newtonian flow region, if one decreases tube diameter D while keeping H_F constant, one discovers that when D is less than about 500 μm the flow does not obey Poiseuille's equation for flow of Newtonian fluids through tubes. Solved for viscosity η, this equation is

$$\eta = \frac{(\pi D^4)(\Delta P/L)}{128Q} \tag{14.7}$$

It has become traditional to use this equation to calculate an "apparent" blood viscosity from experimental tube-flow data, even when D is so small that the continuum model of blood is obviously inappropriate. In very large tubes ($D > 500$ μm), this equation is obeyed (i.e., η is constant for a given blood with a fixed H_F), but in smaller tubes, η is less than the large-tube value. This is illustrated in the summary diagram of Fig. 14.7, where the ratio of η to the value of η in a very large tube (all at same H_F) is plotted as a function of D. While the shape of this curve is similar to that of Fig. 14.6, the minimum in this curve occurs at a lower D value: 4 to 6 μm. While H_T decreases continuously as D decreases, and so η might be expected to decrease also, a point is reached in very small tubes where the cells barely squeeze through the tube, with very little clearance between cell and vessel wall. This situation results in large pressure losses per cell, which increase rapidly as D is decreased (the viscosity ratio in Fig. 14.7 could exceed unity as D approaches 2.7 μm).

The dispersion of the data in Fig. 14.7 has not been explained, although some of the cause may be the same as that causing data dispersion in Fig. 14.6. Additionally, the viscosity ratio, calculated with data collected in the non-Newtonian flow regime, is expected to be a function of flow rate. It has been proposed as an empirical observation that one can predict the flow behavior of blood in small tubes by using macroscopic (large-tube) rheological data and considering the hematocrit in the tube to be uniform at a value of H_T. This was proposed for tubes with D greater than 29 μm [41] and for D as small as 8 μm [38]; in vivo observations [42] are consistent with this simplistic pragmatic method. However, some researchers [43] have not found this technique to be acceptable.

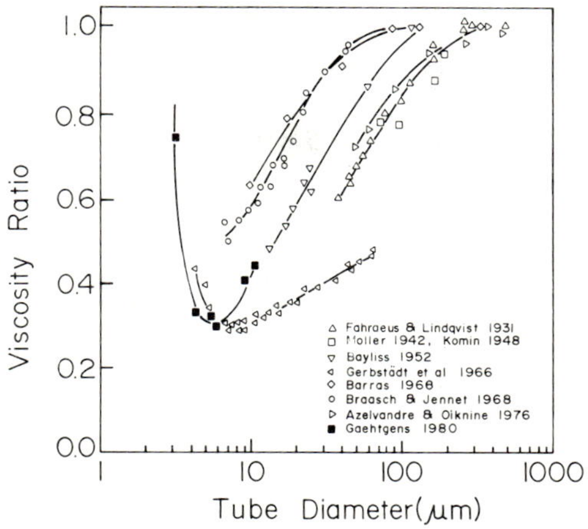

FIG. 14.7 The Fahraeus-Lindqvist effect. *(Data gathered by Gaehtgens [40]. Reproduced with permission.)*

14.2.3 Radial Distribution of White Cells and Platelets

Besides the nonuniform radial distribution of red cells in blood flow in tubes, it has been found that white cells and platelets also have nonuniform radial distributions.

The radial distribution of white cell concentration is a function of flow conditions [44, 45]. At high flow rates, white cell concentration in blood is highest at the tube center, but as the flow rate is decreased, the white cells shift out to regions next to the vessel wall. Under flow conditions allowing red cell aggregation, the white cells are displaced to the periphery of the flow.

The radial distribution of the local platelet concentration during blood flow in tubes has also been investigated [46]. In blood, the platelet concentration is highest near the vessel wall, while when only platelet-rich plasma flows in a tube, the platelet concentration is uniform.

All these observations are consistent with the hypothesis that during blood flow in a vessel, the largest particles, or "flow units," move toward the central region of the vessel and the smallest particles are concentrated nearer the wall.

14.2.4 Flow in Very Small Capillaries

In capillaries with diameters below 15 μm, red cells may travel either in single- or multifile manners. Figure 14.8 summarizes the nature of such flows in terms of two important parameters: vessel diameter and tube hematocrit. The details of how cells arrange themselves in capillary flow are also strongly influenced by cell deformability: a relatively poorly deformable cell (a white cell or less deformable erythrocyte) will tend to travel more slowly and other cells will accumulate behind it.

The shape of red cells traveling down capillaries is complex [47, 48]. In larger capillaries (about 8 to 12 μm diameter), the nonaxisymmetric shape results from a shift of internal hemoglobin solution into the leading-edge part of the cell from the trailing-edge region, so that the forward region of the cell bulges and the trailing edge is thinned. In smaller capillaries, the cell seems folded along a major axis which is parallel to the vessel axis, followed by a shift forward of internal hemoglobin solution. This complexity of shape makes mathematical modeling, even of single-file flow, difficult.

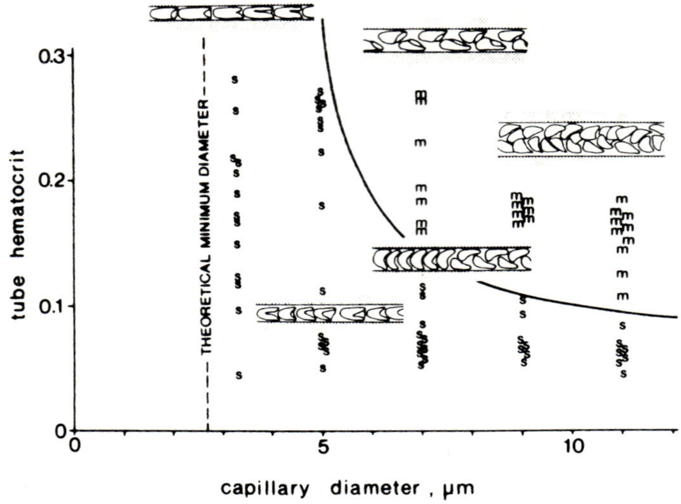

FIG. 14.8 The arrangement of human red cells in flow through glass tubes. The symbol *s* refers to single-file flow, *m* to multifile flow. (*Reproduced with permission from Gaehtgens, Duehrssen, and Albrecht [47].*)

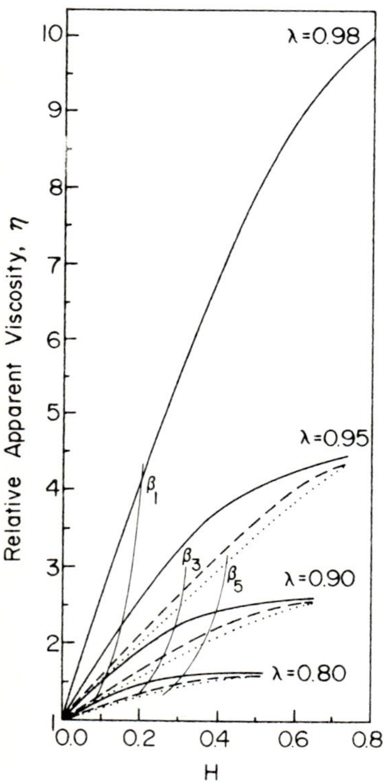

FIG. 14.9 The relative apparent viscosity of suspensions of model red cells for flow in small tubes; γ is the ratio of cell to tube diameters, β refers to points where the gap between cells or rouleaux is one tube diameter (subscript refers to number of cells in a rouleau). Solid curve is for single cells, dashed curve is for three-cell rouleaux, and dotted curve is for five-cell rouleaux. (*Modified from Skalak, Chen, and Chien [40], with permission.*)

Nevertheless, research with simple physical and mathematical models of axisymmetric particle flow in cylindrical tubes has provided some very useful generalizations. A good example is provided by a mathematical model of a suspension of rigid discoids flowing down a tube with their major planes perpendicular to the vessel axes [49]. In Fig. 14.9, the ratio of the suspension viscosity, as calculated from Poiseuille's equation, to the viscosity of the suspending medium (relative viscosity η) is plotted against H_T as a function of two variables: (1) λ, the ratio of cell diameter to tube diameter, and (2) how the cells are spaced along the tube axis—individual cells evenly spaced; evenly spaced rouleaux, each containing three cells; or evenly spaced five-cell rouleaux. A number of generalizations can be drawn from such a plot: (1) the value of λ must approach 0.80 before increased flow resistance due to the presence of cells is appreciable, but when λ becomes greater than 0.90, the resistance increases rapidly as λ approaches unity; (2) when the spacing between individual cells or rouleaux in the suspending medium is equal to or greater than the tube diameter, the additional pressure loss due to the cells is independent of the presence of the other cells (η is proportional to H); (3) at closer cell spacing, η tends to become less sensitive to H as H increases, because the suspending medium between cells tends to become immobilized and moves as a body with the cells. The ratio of particle velocity to average velocity of the suspension bulk has also been experimentally and mathematically studied; this ratio varies from 2 for λ of zero to unity when λ is unity, and is insensitive to particle shape. Using empirical λ versus cell velocity data [38, 50], one can use these model predictions to get estimates of values of parameters for capillary blood flow.

Modeling has been extended to replacing the rigid particles with deformable red cell models which consist of elastic membranes enclosing a viscous fluid. See Skalak and Tözeren [51] for a review of this more realistic modeling work. The modeling of nonaxisymmetric cells is beginning [52], and the analytical study of the flow of tightly fitting cells in capillaries continues [53].

REFERENCES

1. Evans, E., and Fung, Y-C.: Improved measurements of the erythrocyte geometry, *Microvasc. Res., 4,* 335–347, 1972.
2. Carlyle, R. F., Nichols, G., and Rowles, P. M.: Abnormal red cells in blood of men subjected to simulated dives, *Lancet,* 114–116, May 26, 1979.
3. Deuticke, B.: Transformation and restoration of biconcave shape of human erythrocytes induced by amphiphilic agents and changes in ionic environment, *Biochim. Biophys. Acta, 163,* 494–500, 1968.
4. Chien, S., and Jan, K-M.: Ultrastructural basis of the mechanism of rouleaux formation, *Microvasc. Res., 5,* 155–166, 1973.
5. Brooks, D. E.: The effects of neutral polymers on the electrokinetic potential of cells and other charged particles: IV. Electrostatic effects in dextran-mediated cellular interactions, *J. Colloid Interface Sci., 43,* 714, 1973.
6. Brooks, D. E.: Red cell interactions in low flow states, in, J. Grayson and W. Zingg (eds.), *Microcirculation,* Plenum Press, New York, 1976, vol. I, pp. 33–52.
7. Jan, K-M.: Role of hydrogen bonding in red cell aggregation, *J. Cell Physiol., 101*(1), 49–55, 1979.
8. Brooks, D. E., Greig, R. G., and Janzen, J.: Mechanisms of erythrocyte aggregation, in G. R. Cokelet, H. J. Meiselman, and D. E. Brooks (eds.), *Erythrocyte Mechanics and Blood Flow,* Alan R. Liss, New York, 1980, pp. 119–140.
9. Cokelet, G. R.: Rheology and hemodynamics, *Annu. Rev. Physiol., 42,* 311–324, 1980.
10. Harkness, J.: Measurement of plasma viscosity, G. D. O. Lowe, J. C. Barbenel, and C. D. Forbes (eds.), in *Clinical Aspects of Blood Viscosity and Cell Deformability,* Springer-Verlag, New York, 1981, pp. 79–87.
11. Teitjen, G. W., Chien, S., Leroy, E. C., Gavras, I., Gavras, H., and Gump, F. E.: Blood viscosity, plasma proteins and Raynaud syndrome, *Arch. Surg., 110,* 1343–1346, 1975.
12. Seplowitz, A. H., Chien, S., and Smith, F. R.: Effects of lipoproteins on plasma viscosity, *Atherosclerosis, 38,* 89–95, 1981.
13. Harkness, J.: The viscosity of human blood plasma: Its measurement in health and disease, *Biorheology, 8,* 171–193, 1971.

14. Goldsmith, H. L., and Marlow, J.: Flow behavior of erythrocytes: I. Rotation and deformation in dilute suspensions, *Proc. R. Soc. London Ser. B., 182,* 351–384, 1972.

15. Keller, S. R., and Skalak, R.: Motion of a tank-treading ellipsoidal particle in a shear flow, *J. Fluid Mech., 120,* 27–47, 1982.

16. Easthope, P. L., and Brooks, D. E.: A comparison of rheological constitutive functions for whole human blood, *Biorheology, 17,* 235–247, 1980.

17. Walburn, F. J., and Schneck, D. J.: A constitutive equation for whole human blood, *Biorheology, 13,* 201–210, 1976.

18. Scott-Blair, G. W.: An equation for the flow of blood, plasma and serum through glass capillaries, *Nature, 183,* 613–614, 1959.

19. Casson, N.: A flow equation for pigment-oil suspension of the printing ink type, in C. C. Mills (ed.), *Rheology of Disperse Systems,* Pergamon Press, New York, 1959, pp. 84–104.

20. Cokelet, G. R.: The rheology of human blood, in Y. C. Fung, N. Perrone, and M. Anliker (eds.), *Biomechanics, Its Foundations and Objectives,* Prentice-Hall, Englewood Cliffs, N.J., 1972, pp. 63–103.

21. Chien, S., Usami, S., Dellenback, R. J., and Gregersen, M. I.: Shear-dependent interactions of plasma proteins with erythrocytes in blood rheology, *Am. J. Physiol., 219*(1), 143–153, 1970.

22. Quemada, D.: Rheology of concentrated disperse systems and minimum energy dissipation principle: I. Viscosity-concentration relationship, *Rheol. Acta, 16,* 82–94, 1977.

23. Quemada, D.: Rheology of concentrated dispersed systems: II. A model for non-Newtonian shear viscosity in steady flows, *Rheol. Acta, 17,* 632–642, 1978.

24. Quemada, D.: Rheology of concentrated dispersed systems: III. General features of the proposed non-Newtonian model: Comparison with experimental data, *Rheol. Acta, 17,* 643–653, 1978.

25. Brooks, D. E., Goodwin, J. W., and Seaman, G. V. F.: Rheology of erythrocyte suspensions: Electrostatic factors in the dextran-mediated aggregation of erythrocytes, *Biorheology, 11,* 69–77, 1977.

26. Quemada, D.: Red cell aggregation and thrombus formation: A rheological approach, *Topics in Hematology, Excerpta Med. Int. Cong.,* ser. no. 415, 733–736, 1976.

27. Suda, T., Maeda, N., Shimizu, D., Kamitsubo, E., and Shiga, T.: Decreased viscosity of human erythrocyte suspension due to drug-induced spherostomatocytosis, *Biorheology, 19,* 555–565, 1977.

28. Meiselman, H. J., and Baker, R. F.: Flow behavior of ATP-depleted human erythrocytes, *Biorheology, 14,* 111–126, 1977.

29. Chien, S., Usami, S., Dellenback, R. J., Bryant, C. A., and Gregersen, M. I.: Change of erythrocyte deformability during fixation in acetaldehyde, in H. H. Hartert and A. L. Copley (eds.), *Theoretical and Clinical Hemorheology,* Springer-Verlag, New York, 1971, pp. 136–143.

30. Chien, S., Usami, S., Dellenback, R. J., and Gregersen, M. I.: Shear-dependent deformation of erythrocytes in rheology of human blood, *Am. J. Physiol., 219*(1), 136–142, 1970.

31. Carr, R., and Cokelet, G. R.: Rheology of suspensions of normal and hardened erythrocytes and their mixtures, *J. Rheol., 25*(1) 67–82, 1981.

32. Usami, S., and Chien, S.: Shear deformation of red cell ghosts, *Biorheology, 10,* 425–430, 1973.

33. Mackenzie, M. R., and Babcock, J.: Studies of the hyperviscosity syndrome: II. Macroglobulinemia, *J. Lab. Clin. Med., 85,* 227–234, 1975.

34. Cokelet, G. R., and Lichtman, M. A.: Rheology of leukocyte suspensions, in U. Bagge, G. V. R. Born, and P. Gaehtgens (eds.), *White Blood Cells: Morphology and Rheology as Related to Function,* Martinus Nijhoff, The Hague, 1982, pp. 46–56.

35. Chien, S.: Biophysical behavior of red cells in suspensions, in *The Red Blood Cell,* 2d ed., Academic Press, New York, 1975, vol. II, pp. 1031–1133.

36. Merrill, E. W., Benis, A. M., Gilliland, E. R., Sherwood, T. K., and Salzman, E. W.: Pressure-flow relations of human blood in hollow fibers at low flow rates, *J. Appl. Physiol, 20*(5), 954–967, 1965.

37. Palmer, A. A., and Jedrzejczyk, H. J.: The influence of rouleaux on the resistance to flow through capillary channels at various shear rates, *Biorheology, 12,* 265–270, 1975.

38. Cokelet, G. R.: Macroscopic rheology and tube flow of human blood, in J. Grayson and W. Zingg (eds.), *Microcirculation,* Plenum Press, New York, 1976, vol. I, pp. 9–31.

39. Pries, A. R., Albrecht, K. H., and Gaehtgens, P.: Model studies of phase separation at a capillary orifice, *Biorheology, 18,* 355–367, 1981.

40. Gaehtgens, P.: Flow of blood through narrow capillaries: Rheological mechanisms determining capillary hematocrit and apparent viscosity, *Biorheology, 17,* 183–189, 1980.

41. Barbee, J. H., and Cokelet, G. R.: Prediction of blood flow in tubes with diameters as small as 29 μ, *Microvasc. Res., 3,* 17–21, 1971.

42. Lipowsky, H. H., Usami, S., and Chien, S.: *In vivo* measurement of "apparent viscosity" and microvessel hematocrit in the mesentery of the cat, *Microvasc. Res., 19*(3), 297–319, 1980.

43. Azelvandre, F., and Oiknine, C.: Effet Fahraeus et effet Fahraeus-Lindqvist: Résultats Expérimentaux et Modèles Théoriques, *Biorheology, 13,* 325–335, 1976.

44. Nobis, U., Pries, A. R., and Gaehtgens, P.: Rheological mechanisms contributing to WBC-margination, in U. Bagge, G. V. R. Born, and P. Gaehtgens (eds.), *White Blood Cells: Morphology and Rheology as Related to Function,* Martinus Nijhoff, The Hague, 1982, pp. 57–65.

45. Goldsmith, H. L., and Spain, S.: Margination of leukocytes in blood flow through small tubes, *Microvasc. Res., 27*(2), 204–222, 1984.

46. Beck, M. R., and Eckstein, E. C.: Preliminary reports on platelet concentration in capillary tube flows of whole blood, *Biorheology, 17,* 455–464, 1980.

47. Gaehtgens, P., Duehrssen, C., and Albrecht, K. H.: Motion, deformation and interaction of blood cells and plasma during flow through narrow tubes, *Blood Cells, 6,* 799–812, 1980.

48. Bagge, U., Branemark, P-I., Karlsson, R., and Skalak, R.: Three-dimensional observations of red blood cell deformation in capillaries, *Blood Cells, 6,* 231–237, 1980.

49. Skalak, R., Chen, P. H., and Chien, S.: Effect of hematocrit and rouleaux on apparent viscosity in capillaries, *Biorheology, 9,* 67–82, 1972.

50. Hochmuth, R. M., Marple, R. N., and Sutera, S. P.: Capillary blood flow: I. Erythrocyte deformation in glass capillaries, *Microvasc. Res., 2,* 409–419, 1970.

51. Skalak, R., and Tözeren, H.: Flow mechanics in the microcirculation, in J. F. Gross and A. Popel (eds.), *Mathematics of Microcirculation Phenomena,* Raven Press, New York, 1979, pp. 17–40.

52. Secomb, T. W., and Skalak, R.: A two-dimensional model for capillary flow of an asymmetric cell, *Microvasc. Res., 24,* 194–203, 1982.

53. Secomb, T. W., and Gross, J. F.: Flow of red blood cells in narrow capillaries: Role of membrane tension, *Int. J. Microcirc. Clin. Exp., 2,* 229–240, 1983.

CHAPTER 15

Mechanical Hemolysis

Perry L. Blackshear
Gertrude L. Blackshear
Department of Mechanical Engineering
University of Minnesota
Minneapolis, Minnesota

INTRODUCTION

A number of reviews which include material covering mechanical hemolysis have appeared in 1972 [1, 2] and in 1977 [3]. Most studies of hemolysis continue to support a hypothesis first enunciated by Rand [4] that hemolysis occurs when biaxial membrane *strain* exceeds a threshold value where pores open or the membranes tear. Meanwhile, great strides have been made in understanding the relationship between imposed stress and resulting strain in the red blood cell (RBC) membrane [5–15] and in the sieving process that occurs across membrane when it is strained slowly [16–20] or rapidly [21, 22]. Hemolysis which occurs intravascularly in vivo and in various flow systems in vitro is thought to be due at least in part to flow and physical factors which produce the hemolysis threshold strain in the RBC membrane. When the membrane strain is not enough to permit hemoglobin escape, smaller solute exchange can lead to cell changes which affect the process of lysis and can lead to colloid lysis.

In preparing this chapter we had the biographical retrieval services of Corporation Park, Scotia, New York, do a search under mechanical hemolysis back to 1965. Of the 1016 citations, we selected approximately 100 on which to base this handbook data. The first part of this chapter provides physiological background material and a summary of observed hemolysis in vivo; this material is assembled from four sources, themselves compendia [23–26]; in the second part hemolysis measured in relatively simple in vitro experiments is modeled in an effort to develop an understanding of the mechanical forces that lead to hemolysis. Here the effort will be to identify flow phenomena (that may be present in blood handling equipment) which may cause lysis, and examine how the RBCs interact with these flow phenomena to produce the lytic condition. Because many of the conclusions are tentative, we have included references in this second part.

15.1 HEMOLYTIC MECHANISMS

15.1.1 In Vivo Hemolysis

Background. Hemolysis can be equated with a shortened red cell lifetime in the circulation. A background summary of pertinent red cell physiology follows.

Red cells are formed in the bone marrow from pluripotential precursors. In the process Fe^{3+} is required for the manufacture of hemoglobin, which is made up of heme (ferroprotoporphyrin 9) and a protein, globin. There is a daily rate of red cell release to balance the normal daily destruction of 2×10^{11} cells. Normal red cell life span in the circulation is about 120 days.

The spleen has been modeled as a testing site for aging cells; all cells unable to deform in their passage through 3-μm openings in the splenic circulation are delayed, their environment becomes increasingly poor in nutrients, they become more spherical, and they are eventually phagocytized. Fragments of cells or whole cells are also taken up from the circulation by cells of other organs of the reticuloendothelial system, liver, bone marrow, and lymph nodes. In the cells, hemoglobin is broken down, giving off CO in the process, its iron is reused or stored, and the residue of the heme is converted to unconjugated bilirubin. This then reenters the circulation, is picked up by liver cells and conjugated with glucuronide, and discharged in the bile into the small intestine. In the small intestine, the conjugated bilirubin is converted by bacteria into fecal urobilinogen, which is found chiefly in the feces, although small amounts are absorbed and excreted in the urine.

With a shortened red cell lifetime (increased rate of destruction) the rate of red cell development (*erythropoiesis*) is increased in the bone marrow because of stimulation by plasma erythropoietin, produced in the kidney. If red cell lifetime has fallen to less than 15 to 20 days, the bone marrow cannot compensate adequately, and anemia, usually normocytic, develops.

Erythropoiesis has been classified as *effective,* which refers to formation of cells with normal life spans, and *ineffective,* which refers to cells that are destroyed in the marrow or soon after release.

Hemolysis can be due to *intracorpuscular* defects (defective cells) or *extracorpuscular* defects (other factors). If the cells are defective, they are destroyed when injected into a normal circulation or into the patient's own circulation. If the pathological condition is extracorpuscular, the patient's cells will not be destroyed in a normal circulation but normal cells will be destroyed in the patient's circulation.

Hemolysis (red cell destruction) can be either *extravascular,* which is the normal pathway (see above) and involves cells with mild or moderate defects and uptake and breakdown in reticuloendothelial cells, or *intravascular,* which occurs with more severe defects and in which breakdown of cells takes place in the bloodstream with release of hemoglobin into the plasma.

With increased (above-normal) extravascular hemolysis, as well as in intravascular hemolysis, there is shortened red cell lifetime and an increase in plasma-free bilirubin. This is strong evidence of increased red cell destruction. Plasma conjugated bilirubin may also increase, as does fecal, plasma, and urinary urobilinogen. Other general markers of increased red cell destruction are increased carbon monoxide production, formed in the catabolism of hemoglobin, and increased iron content of feces.

In intravascular hemolysis, the liberated serum hemoglobin is bound to haptoglobin, a plasma protein, to the amount of 100 to 140 mg hemoglobin per 100 mL plasma. The hemoglobin-haptoglobin complex is cleared by the reticuloendothelial system, and hemoglobin is broken down in the cells. The fraction broken down in the liver cells yields hematin, which is released into the plasma and binds to albumin to form methemalbumin. Its presence indicates recent intravascular hemolysis. Another fraction is filtered by the kidney; some of this is reabsorbed. Hemoglobinuria occurs after the resorptive capacity of the kidney is exceeded. Hemoglobin absorbed by renal tubular cells is broken down, ferritin and hemosiderin are formed, and the latter is found in the urine within a few days after onset of hemolysis.

Mechanisms of Hemolysis. The mechanisms of hemolysis are usually listed as follows; in many diseases there is more than one mechanism operating.

1. Colloid osmotic lysis. There is increased permeability of the red cell membrane with disturbance of the cation gradient. As a result, water enters the cell, which becomes more spherocytic. With marked distension, the membrane pores are enlarged and there is hemoglobin loss. Congenitally spherocytic cells are more easily lysed in this manner (increased osmotic fragility), as are red cells exposed to high temperatures and certain antibodies.

2. Perforation of a cell. This leads to hemoglobin loss without previous cell swelling. The defect must have a radius greater than 3.25 nm. This type of hemolysis occurs with chemical lysins and certain antigen-antibody reactions.

3. Fragmentation of red cells, with the fragments free of or containing associated hemoglobin. The donor cells become spheroidal and smaller. Mechanical trauma such as artificial heart valves or fibrin strands traversing blood vessels can cause this fragmentation. Abnormal hemoglobins can make the cells more rigid and thus more liable to fragmentation.

4. Erythrophagocytosis: red cell (or fragment) uptake by reticuloendothelial cells. Certain antibodies coat red cells and predispose them to phagocytosis.

Methods of Investigation

1. Analysis of the blood.

a. Morphology of red cells. Many unusual forms can be found in hemolytic disease depending on the etiology: spherocytes, cells with Heinz bodies, sickled cells, fragmented cells, as well as burr, spur, contracted, stippled, triangular, and budded cells, pyknocytes, and target forms.

b. Reticulocyte count. The reticulocyte count gives an approximation of red cell production. It is increased in hemolytic disease.

c. Serum chemistry. With increased hemolysis, the free hemoglobin is above normal, there is methemalbumin present, the unconjugated and conjugated bilirubin are above normal (as measured by indirect and direct van den Bergh's tests), and there is increased urobilinogen and lactic dehydrogenase, and decreased haptoglobin.

2. Osmotic and mechanical fragility tests detect intracorpuscular abnormalities.

3. Analysis of hemoglobin breakdown products.

a. Bilirubin. Bilirubin is believed to be formed solely from heme catabolism. Isotopically labeled bilirubin clears from the circulation as the sum of three exponentials; these plus the unconjugated serum bilirubin concentration allow calculation of the bilirubin production rate, which is elevated with hemolysis. The calculation requires samples for 24 to 48 hr.

b. Carbon monoxide excretion. The sole endogenous source of CO is the catabolism of heme. The rate of production is determined from blood carboxyhemoglobin or by using a gas-tight chamber over the head and measuring CO concentration. The method is cumbersome, but the lifetime of the red cell can be calculated from the data.

4. Red cell lifetime determination.

a. Cohort method. The cohort method utilizes a label incorporated in the developing red cells, usually ^{15}N, ^{14}C, or ^{59}Fe. These labeled red cells are destroyed simultaneously at about 120 days. The method requires samples for 150 to 180 days.

b. Random label methods. These methods utilize ^{50}Cr or ^{51}Cr and ^{32}P, ^{3}H, or ^{14}C DFP, (diisopropyl fluorophosphate). Problems include the need for a constant total red cell volume and, in the Cr method, elution of label. Normally, red cell survival shows a steady linear decline from 100 percent to 0 at about 120 days; a normal $t_{1/2}$ for ^{51}Cr is 27 to 35 days. Increased destruction shows an exponential decline with a shortened halftime. This method requires samples for 30 to 60 days. ^{51}Cr, a gamma emitter, can give additional information with organ scanning on sites of red cell destruction.

5. Analysis of urine. Hemoglobinuria or hemosiderinuria can be present.

6. Analysis of stool. Increased urobilinogen or increased iron can be present.

Hemolytic Diseases. Hemolytic anemia is due to many different causative agents. The red cells, themselves, may be abnormal due either to a hereditary defect, e.g., sickle-cell anemia, or to an acquired condition, e.g., nutritional deficiency. Or the fault may lie outside the red cell per se, as in immune disease, e.g., erythroblastosis fetalis, or as a result of exposure to toxic agents (e.g., heavy metals), physical agents (e.g., mechanical trauma), or infectious agents (e.g., as in malaria). At times there is both an intracorpuscular and an extracorpuscular abnormality (e.g., lead poisoning).

Finally, hemolytic anemia is associated with a wide variety of chronic diseases such as tumors, chronic inflammation, and uremia.

Pathological Side Effects. Hemolysis can lead to other disturbances of the organism. Above a certain level it will cause a secondary anemia. Release of red cell contents can stimulate clotting factors and has been suggested as a possible cause of disseminated intravascular coagulation. Increased red cell phagocytosis can lead to splenomegaly of such magnitude that surgery is indicated. Increased iron loss overloads the body's capacity to handle iron and can cause generalized siderosis, reticuloendothelial system depression, gallstones, and kidney failure.

Compensatory mechanisms include increased erythropoiesis and increased formation of haptoglobin.

Laboratory Values of Interest in the Study of Hemolysis

1. *Blood*

Average volume: 5000 mL.

Average red cell mass: 2000 mL.

Hematocrit: 46% (male), 41% (female).

Total red cell volume (mg/kg): 30 (male), 25 (female).

Red cell count ($\times 10^6/\mu$L blood): 5.1 (male), 4.5 (female).

Reticulocyte count (0.5 to 1.5 per 100 RBCs): $>$ 4% is abnormal (should be corrected for hematocrit).

Hemoglobin: 15 g/100 mL: 750 g in circulating red cells, 2.5 mg free hemoglobin in 100 mL normal plasma. Serum methemalbumin is found with large amounts of free hemoglobin in blood. Normal daily destruction of 0.83% releases 5.6 to 6.7 g hemoglobin; 100 g hemoglobin yields 3.6 g bilirubin; 1 g hemoglobin yields 35 mg urobilinogen.

Haptoglobin: binds 100 to 140 mg hemoglobin per 100 mL plasma. The complex is cleared 15 mg/100 mL per hour by the reticuloendothelial system.

Serum bilirubin: alcohol-soluble, indirect van den Bergh's test, normal $<$ 0.2 mg/100 mL; and conjugated to glucuronide, direct van den Bergh's test, normal $<$ 0.11 mg/100 mL.

2. *Red cell*

Density: 1.10

Diameter (dry film): 7.7 μm

Thickness: 1.95 μm

Volume (μm^3 per red cell): 90.1 (male), 90.4 (female)

Mean corpuscular hemoglobin concentration (g/100 mL RBCs): 30.2 (male), 30.2 (female)

Lifetime: 100 to 130 days

Normal daily destruction: $2 \times 10^{11} \approx 0.83\%$

Reticulocyte lifetime: $t_{1/2}$ = 1 to 2 days

3. *Fecal excretion*

Urobilinogen: normal, 100 to 200 mg/day

Iron: 1 mg/day

4. *Urine excretion*

Hemoglobinuria, when plasma value $>$ 135 mg/100 mL

Urobilinogen: normal, 0.5 to 1.5 mg

15.1.2 Mechanical Hemolysis In Vitro

Complex Blood Handling Systems. A number of studies have been published on pumps [27–29], heart-lung machines [28, 30], and prosthetic valves [31], in which rates of hemolysis were correlated with rates of flow and temperature [28]. The differences noted between prosthetic valves [31] and pumps [27] in vitro reflect the in vivo performance of the valves in some instances [31]. These relatively complicated systems involve an assortment of potentially traumatic stresses. The simultaneous presence of foreign materials, injurious wall contact, flow through apertures, high shear stress, turbulent flow, accelerating flow, pressure gradients, and, in some instances, blood-gas interfaces, frequently makes it difficult to place the blame on one particular lytic agent, or to draw inferences that would be of general utility.

In the following sections relatively simple flow-related mechanisms which lead to lysis are modeled to find a relationship between membrane properties, imposed forces, and lysis.

Membrane Viscoelastic Properties. Elsewhere in this handbook the membrane structure and its relationship to the mechanical properties exhibited by a red blood cell under stress have been summarized. If we relate the membrane tensions T_{11}, T_{22} to the principal stretch ratios λ_1 and λ_2, we may write

$$T_{11} = \mu \frac{\lambda_1^2 - \lambda_2^2}{2} + \eta_m \frac{\dot{\lambda}_1}{\lambda_1} + \sigma \tag{15.1}$$

$$T_{22} = -\mu \frac{\lambda_1^2 - \lambda_2^2}{2} - \eta_m \frac{\dot{\lambda}_2}{\lambda_2} + \sigma \tag{15.2}$$

where μ is the shear modulus, η_m is the coefficient of viscosity for the membrane surface in shear flow, and σ is the biaxial tension causing areal dilation of a cell surface under stress. For $\lambda_1 < 3$, μ, the shear modulus, and η_m, the membrane viscosity, are dependent upon the cytoskeleton and do not appear to be influenced by membrane lipid properties. The values for normal RBCs are $\mu = 0.0061 \pm 0.0012$ dyn/cm and $\eta_m = (6 \pm 2) \times 10^{-4}$ P. cm [11]. On the other hand, the response of the membrane area to changes in σ is thought to depend upon the lipid bilayer properties. Lysis is expected to occur when an increase in area is greater than some threshold value [4]. Rand [4] shows that when a biaxial tension σ is suddenly imposed on an RBC in a micropipet, the value of σ at lysis depends upon the duration of the imposed stress. These results are shown in Fig. 15.1. From these results Rand constructed a model of dilational properties of the RBC membrane, on the assumption that hemolysis occurs at a critical areal strain. Seeman [22] followed the spherically shaped cell volumes of RBCs osmotically lysing at differing rates and temperatures and confirmed Rand's hypothesis: lysis occurred at a fixed areal strain in osmotic lysis as well. Rand's model will be used below in interpreting hemolysis for other, more complex systems. The equation obtained by Rand [4] relating the biaxial tension σ_c that produces the critical strain S_c in time t, based on a viscoelastic model consisting of a Voight body in series with a spring and dashpot, is:

$$\frac{1}{\sigma_c} = \frac{1}{S_c h}\left[\frac{1}{Y_2} + \frac{1}{Y_1}\left(1 - \exp\frac{-Y_1}{\eta_1}t\right) + \frac{t}{\eta_2}\right] \tag{15.3}$$

where S_c is linear strain $(\lambda - 1)$, h is membrane thickness, and Y_2 is Young's modulus.

The values of the constants from Eq. (15.4) can be derived from the following values:

$$S_c h Y_2 = 28.6 \text{ dyn/cm}$$

$$S_c h Y_1 = 34.5 \text{ dyn/cm}$$

$$\frac{Y_1}{\eta_1} = 8^{-1}\ \text{s}^{-1}$$

$$\eta_2 S_c h = 666.7 \text{ dyn}\cdot\text{s/cm}$$

Recently, Evans and Waugh (for a full account of these properties, see Chap. 12 in this handbook) have measured the areal modulus from which hY_2 can be inferred to be 900 dyn/cm^2.

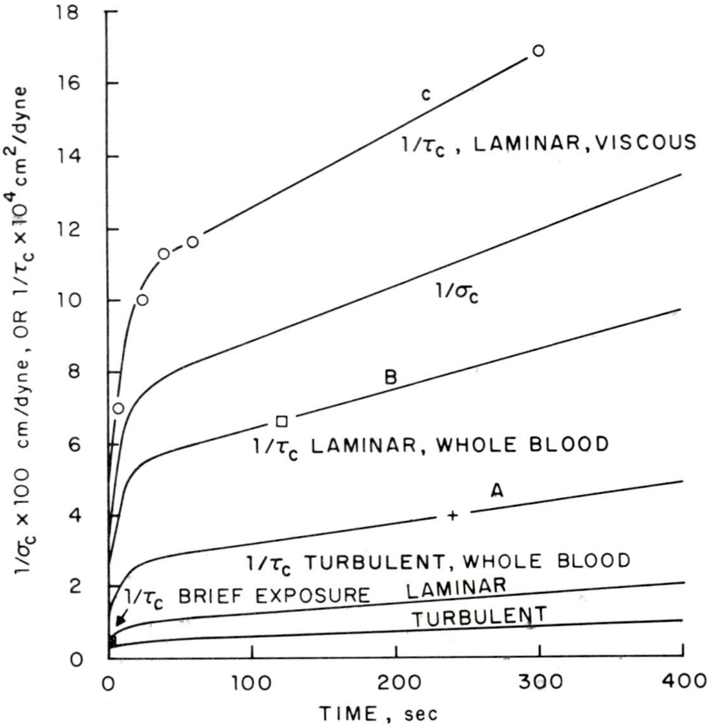

FIG. 15.1 The reciprocal of the critical membrane tension and the reciprocal of the critical shear stress versus time. The plot of $1/\sigma_c$ versus t is from Eq. (15.3). Rand's measurements [4] (not shown) fall directly on this curve. Curve C is a plot of $1/\tau_c = 1.42 \times 10^{-2}/\sigma_c$. The experimental values of $1/\tau_c(t)$ of Williams [37] fit this curve well, as shown. The shortest time of exposure examined is 5 s. Curve C is thought to represent the $1/\tau_c$ that characterizes RBCs sheared in a highly viscous (20-cP) medium for 5 s or longer. Curve B is $1/\tau_c = 7.25 \times 10^{-3}/\sigma_c$. A single datum from Everett et al. [2] located on this curve is thought to represent the behavior of whole blood sheared between confining surfaces for prolonged (2-min) periods. Curve A, $1/\tau_c = 3.6 \times 10^{-3}/\sigma_c$, with its single datum from Sutera and Mehrjardi [53], is thought to represent response to prolonged (4-min) *turbulent* shear of whole blood. The band embracing $7.25 \times 10^{-4}/\sigma_c \leq 1/\tau_c \leq 1.45 \times 10^{-3}/\sigma_c$ from Forstrom [51], Coakley et al. [60], and Keshaviah [69] represents the lytic region for short exposures (<1 s) to laminar and turbulent high fluid shear stress.

Thus, the S_c value is smaller than the one employed by Rand. The corresponding values of the other constants are (for $h = 10^{-6}$ cm) $Y_1 = 8 \times 10^8$ dyn/cm^2, $\eta_1 = 10^8$ dyn · s/cm^2, and $\eta_2 = 2.08 \times 10^{11}$ dyn · s/cm^2. Rand calculated different values of these moduli, on the basis of the then assumed critical areal strain of 17 percent. The moduli tabulated above are based on more recent direct measurements of biaxial stress versus strain. The predictions of Rand for the dynamics of lysis resulting from imposed *stress* remain unaltered. For example, after the sudden imposition of a biaxial tension of σ_c, lysis will occur after a time t, as shown in Fig. 15.1. The basic premise of Rand now reads as follows: Areal change occurs when the cell membrane is subjected to stress; hemolysis occurs when the area increases by approximately 6.4 percent. More important in a practical sense, Rand's model gives a convenient σ_c versus t behavior for predicting lysis.

Two modes of hemolysis are recognized in which the ghosts reseal: one in which an array of pores open and reseal rapidly with time constants of 10 ms [32], and a second in which large rents appear and reseal with a time constant of 20 [22].

When *biaxial* stress is clearly applied, as in the case of osmotic lysis, lysis in micropipet aspiration, and in microporous media filtration, then [from Eqs. (15.1) and (15.2)] $T_{11} = T_{22}$ and $\lambda_1\lambda_2 - 1$ must reach approximately 0.064. For shorter duration of membrane tension (less than 1 s), $T_{11} = T_{22} = 28.6$ dyn/cm at lysis. For a greater duration of greater tension (more than 10 s), $T_{11} = T_{22}$, less than 15 dyn/cm can cause lysis. We may consider the more complicated cases in the light of this model.

When the lysis results from what appears to be an imposed *uniaxial* stress, as in the case of shear flow or flow into an abrupt reduction in cross section, lysis can be again related to the condition that the critical $\sigma_{c,t\to 0}$ of Fig. 15.1 has been reached by application of Eqs. (15.1) and (15.2) in the following way. If T_{11} is the uniaxial tension and $T_{22} = 0$, we see that $\frac{1}{2}(T_{11} + T_{22}) = \sigma$ [33, 34]. Thus a critical uniaxial tension $T_{11,c} = 2\sigma_c$. For example, when $T_{11,c} = 2\sigma_c$, the criterion for lysis by briefly imposed uniaxial tension [4] is $T_{11} = 2\sigma_{c,t\to 0} = 58$ dyn/cm.

In the following sections the results of relatively well defined tests will be reviewed and, where possible, compared on the basis of Eqs. (15.1) to (15.3) and the prediction based on the Rand hypothesis, Fig. 15.1 [4].

Fluid Shear Stress. Frequent use is made of the cone-plate or the Couette flow viscometer to put a suspension of cells under stress for a prescribed period of time after which (or during which) the increase in suspending phase hemoglobin is determined [22, 33, 34, 36–44]. In his 1972 review, Leverette et al. [2] called attention to the time dependence of hemolysis in the shear apparatus and extended the comparison to other devices for stressing the red cell: as imposed stress increased, time to lysis decreased. Richardson [34] points out the inconsistencies that are found in investigations of simple shear flow hemolysis tests: although time for hemolysis to a given extent is reduced with increasing shear stress, the absolute levels of this time varied tremendously from one experimental apparatus to another. A most important issue is raised by Bessis and Mohandas [45, 46] and independently by Morris and Williams [49]. In transparent-walled viscometers, these investigators measured cell defraction patterns that show cell shape as shear rate is altered. They found that at a particular shear stress, the cell becomes elongated as the suspending phase viscosity increases. Hemolysis, over a prescribed time for the same shear stress, shows a similar increase with increasing suspending phase viscosity; Fig. 15.2 shows how these quantities increase as suspending phase viscosity goes from 3 to 20 cP. Williams [37] shows that the *shear stress alone* is not sufficient to predict hemolysis rate or thresholds, but cell shape and cell tumbling (or lack of it) play a role as well.

Bessis and Mohandas [45, 46] show that tank treading in a Couette flow viscometer cannot be made to occur at any shear stress when the suspending phase viscosity is less than the cytoplasm viscosity (~8 cP). Goldsmith [47] has studied the orientation of RBCs in a ghost suspension designed to behave like normal blood. He found that RBC tumbling is impeded. Some tank tread–like behavior may thus be expected at the conditions of normal circulation owing to the interaction of the cells with each other. It seems clear that the membrane shear is far less than when the RBCs are suspended in higher-viscosity media.

Fischer [45], in his counterrotating, transparent-walled, cone-plate viscometer, has shown that when the spacing between the two surfaces approaches cell dimensions, cells suspended in isotonic saline (where viscosity is about 1 cP) can be forced into the elongated tank-treading mode, whereas cells in the wider part of the cone-plate viscometer merely tumbled. The fact that Leverett [2] (who used saline or plasma as the suspending phase) finds such a weak influence of suspending phase viscosity on hemolysis threshold (and no influence as hematocrit is varied from 0.3 to 60 percent), and whose threshold corresponds to that which Morris and Williams [49] (who used a viscous suspending phase) give for partially elongated cells, suggests that the hemolysis may have been taking place in the narrow part of the cone-plate portion of the viscometer; i.e., tank treading and some cell elongation was forced by containing walls rather than by suspending phase viscosity. The turbulent shear stress data reported by Sutera and Mehrjardi [53] are accompanied by photographs of elongated cells obtained in turbulent flow, for a medium, however, in which the viscosity is that of isotonic saline. We conclude then, that the interaction between the local turbulence and the cell produces an effect that causes elongation similar to that experienced by cells in a viscous suspending phase. However, the cell does not appear as vulnerable to an imposed turbulent shear stress as it does in a viscous suspending phase at the same stress [33].

Fischer [48] suggests that in tank treading, viscous energy dissipation in an elongated ellip-

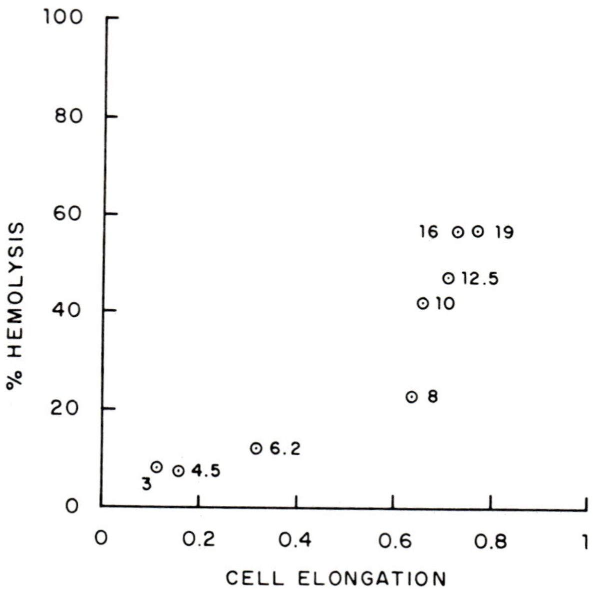

FIG. 15.2 Hemolysis and cell elongation in laminar shear stress of 1500 dyn/cm² as viscosity is increased from 3 to 20 cP [37]. Here the index of hemolysis is hemoglobin released during 5 min at 25°C divided by total hemoglobin. The elongation is (cell length − cell width) ÷ (cell length + cell width). Viscosity is shown with each point. Hemolysis and elongation are seen to increase with increasing suspending phase viscosity until a viscosity of 12 cP is reached. Thereafter, very little increase in either index is noted. An elongation of 0.75 for a human RBC gives an increase in length to a value approximately 1.6 times the initial cell diameter.

soidal cell should depend strongly on the membrane viscosity. Neither Richardson [33] nor Keller and Skalak [50] consider the influence of membrane viscosity in shear flow on the stress distribution in the ellipsoidal tank-treading cell, although the latter paper predicts the *shape* of the tank-treading cell [50]. At present, it is not possible to predict the membrane tension distribution of a tank-treading cell as a function of suspending phase viscosity, membrane viscosity shear modulus, fluid shear stress, and cytoplasm viscosity.

Earlier models of deforming droplets relate components of membrane tension to the shear stress in the suspending phase [33, 49]. Judging from the shape of the tank-treading cell, it appears that the force acting on the surface is essentially uniaxial. If this force $T_{11,c}$ is set proportional to the fluid stress $T_{11,c} = a\tau_c l$ in the suspending phase and to a cell length (which in turn depends upon shear stress and suspending phase viscosity [49]), we can contrast this T_{11} with the prediction of Eq. (15.3). We would expect for this case that $T_{22} = 0$, and that $\sigma_c = \frac{1}{2}T_{11,c}$.

We see from Fig. 15.2 that when $\eta = 20$ cP, the RBC elongation is virtually complete (at $\tau = 1500$ dyn/cm²) so that length $l = 1.6d_0$, where d_0 is initial RBC diameter. Then, if we set

$$\sigma_c = \tfrac{1}{2}T_{11,c} = a\tau_c\ (\text{dyn/cm}^2) \times 1.6 \times 8 \times 10^{-4}\ (\text{cm})$$

where σ_c is from Eq. (15.3) and τ_c from the experiments of Williams [49], we obtain $a = 11.1$.

Thus, in order to find the threshold for lysis in laminar shear flow where the cells are fully extended, we simply set

$$\tau_c = \frac{\sigma_c}{1.42 \times 10^{-2}\ \text{cm}}$$

If we use the Leverett data at 2-min exposure times to evaluate a comparable value for blood, we find

$$\tau_c = \frac{\sigma_c}{7.25 \times 10^{-3}\ \text{cm}}$$

These two limiting curves are shown in Fig. 15.1 For brief exposures to laminar flow in capillary tubes, Keshaviah [69] and Coakley et al. [60] found $\tau_c = 20{,}000$ dyn/cm^2, yielding $\tau_c = \sigma_c/(1.45 \times 10^{-3})$, as shown in Fig. 15.1.

The case for turbulent flow is more demanding. The data of Forstrom [51] and Bernstein et al. [52] show that for very brief exposures ($\sim 10^{-5}$ s) in a turbulent jet, lysis occurs at $\tau_c = 40{,}000$ dyn/cm^2. This gives for the limit $t \to 0$,

$$\tau_c = \frac{\sigma_c}{7.25 \times 10^{-4}\ \text{cm}}$$

This curve is also shown in Fig. 15.1. Sutera and Mehrjardi [53] found that in turbulent shear stress sustained for 4 min, significant hemolysis and membrane alteration occurred above $\tau =$ 2500 dyn/cm^2. From these data we estimate, for prolonged turbulent exposure,

$$\tau_c = \frac{\sigma_c}{3.6 \times 10^{-3}\ \text{cm}}$$

This curve is shown in Fig. 15.1.

Richardson [33] and Williams [43] both estimate $\tau_c = \sigma_c/10^{-3}$ to $\sigma_c/10^{-4}$. The data for all sustained shear flows (i.e., 10 s or longer) produce lysis at values of τ_c *smaller* than would be predicted on the basis of a $\sigma_c(t)$ obtained from nonsheared RBCs.

All the tests in whole blood in tubes and jets in which the shear stress duration is less than 1 s show thresholds of lytic $\tau_c > 10{,}000$ dyn/cm^2. These results are more nearly in accord with a reasonable multiplier of Rand's σ_c, say,

$$\tau_c = \frac{\sigma_c}{1.4 \times 10^{-3}}$$

These data suggest that the effect of shear in reducing τ_c probably takes more than 1 s. Most high–shear stress episodes experienced by an RBC last a few milliseconds so that from a practical standpoint the bar in the lower left-hand part of Fig. 15.1 depicts the safe operating zone. For some reason, when prolonged exposure to shear stress is imposed, the lytic region appears to jump for $t > 0.1$ s from the shaded region in the lower left-hand corner of Fig. 15.1 to curve *A* for turbulent flow, curve *B* for laminar flow at normal hematocrits, and curve *C* for laminar flow and hematocrit ≈ 70.

Osmotic Lysis. Osmotic lysis occurs when the membrane area increases by approximately 6.4 percent because of volume swelling [4, 22]. The mechanics of lysis depend on the rate of swelling. There are three general forms of membrane behavior reported, depending upon the rapidity with which the membrane is osmotically stressed.

Fast hemolysis: Here osmolarity of suspending media is reduced faster than water can enter the cell (in about 0.05 s) [32]. Cell swelling occurs in about 0.6 s [32]. A rapid hemoglobin release results, which is then halted within 50 ms by restoring isotonicity [32].

These experiments suggest that small pores open and close rapidly, allowing portions of the cell contents to escape; fast hemolysis is not an all-or-nothing phenomenon [32]; cells reseal with varying amounts of hemoglobin retained. Anderson and Lovrien [32] did not measure the accompanying K^+ flux. Because the sudden straining and relaxation of the membrane (in fast hemolysis [32]) should be similar to the membrane stress history in cells forced through polycarbonate sieves, we may expect there to be no time for prelytic K^+ release, but approximately twice the K^+ as Hb release during hemolysis [21]. The data from Refs. 21 and 22 are summarized in Fig. 15.3. The sequence of events is depicted in Fig. 15.4.

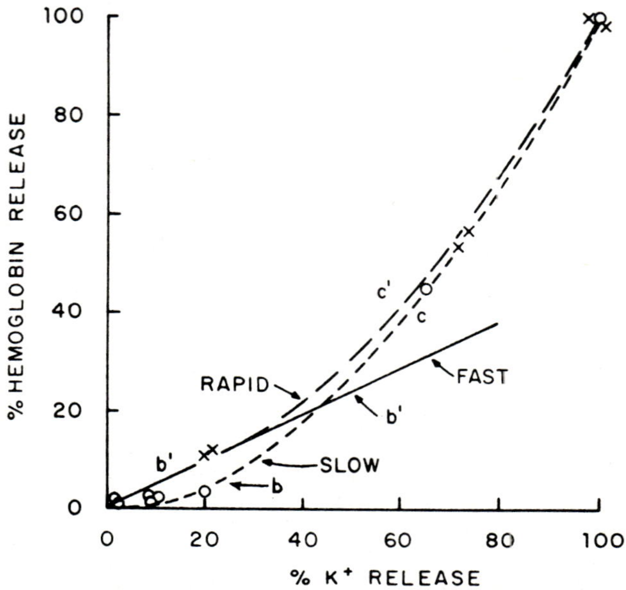

FIG. 15.3 Percent hemoglobin release versus percent K^+ release for fast [21, 32], rapid [22], and slow [22] hemolysis. The letters associated with the curves correspond to the membrane behavior sketched in Fig. 15.4. The slope of the Hb–K^+ line for fast hemolysis [21] corresponds to the earliest stages of rapid hemolysis and is thought to be associated with multiple ~6-nm pores depicted by *b* in Fig. 15.4. The nearly horizontal part of the slow hemolysis curve represents state *b* in Fig. 15.4, where multiple K^+ leaks occur. When the single large pores open, the cell contents flow out as depicted by the steep slopes of the slow and rapid hemolysis curves. In fast hemolysis, the small multiple pores close rapidly (0.04 s) when stress is removed [32]. Presumably sudden stress relief accompanies the formation of the single large pore and permits the closing of the multiple small pores as well.

"Rapid" hemolysis: Here the schedule of rate of dilution of suspending phase is sufficiently fast to reach the hemolytic salt concentration in about 5 s. A single large pore (~1 μm) opens for approximately 10 s and reseals. Hemolysis is an all-or-nothing phenomenon. The initial percent rate of loss for K^+ is roughly twice the initial percent rate of loss for hemoglobin, as seen in Fig. 15.3.

Slow hemolysis: Here the schedule of dilution is such that hemolysis occurs in approximately 10 min. A prelytic K^+ loss of 20 percent precedes the release of Hb. As a result the lytic salt concentration is lower than in "rapid" hemolysis. Again, a single large pore opens (~1 μm) and can reseal grossly in 10 to 20 s, with further closure (sufficient to make the ghosts behave as perfect osmometers) in 1 h (at 37°C). Recent studies suggest that resealing can be manipulated. This behavior is also shown in Figs. 15.3 and 15.4.

Figure 15.4 shows the sequence of events in rapid and slow lysis. In fast lysis, the pores are thought to be smaller and to reseal more rapidly.

Seeman et al. [22] show that reduced temperature also increases the osmotic fragility of the RBC, shifting the lytic threshold to higher salt concentrations. Again the shift is seen to be due to prelytic K^+ release at the high temperatures and to its reduced release at low temperatures.

The fact that in slow hemolysis K^+ release preceded Hb release suggests the formation of porelike structures in the membrane, such that sieving would be expected. Other evidence of sieving in osmotic hemolysis has been reported: MacGregor and Tobias [18], in what seems to

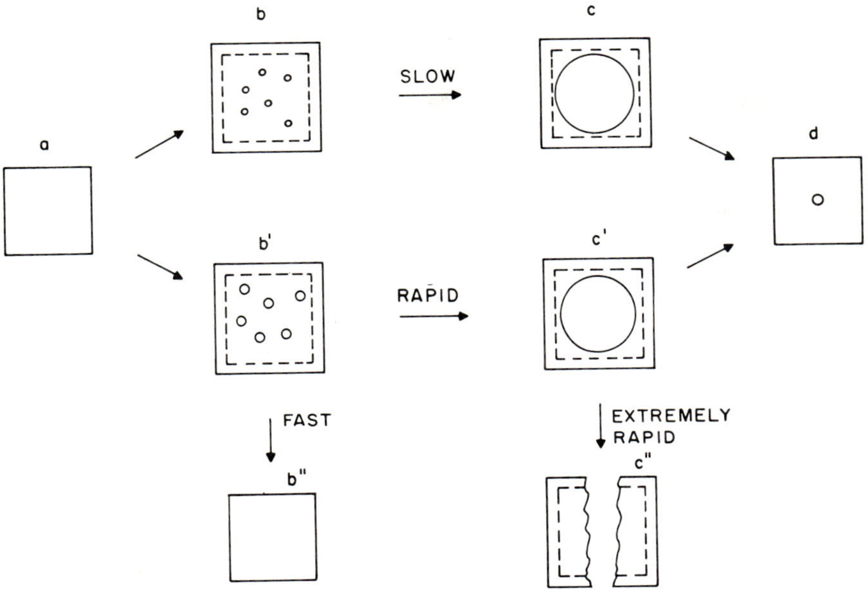

FIG. 15.4 Membrane behavior during lysis. These sequences are inferred from Refs. 11, 21, 51, 71, 73, and 74. The element of membrane is shown strained uniformly in the two principal directions. This kind of straining is thought to occur when a curved membrane supports a pressure difference. A second, similar sequence may be pictured resulting from a rapid shear in which the area again is made to increase [39, 49]. (*a*) An element of membrane area A is initially relatively impermeable to K^+. (*b*) As the membrane is strained biaxially, the area increases an amount ΔA. In slow hemolysis there is a prelytic period when $0 < \Delta A/A < 0.064$ such that K^+ leaks rapidly but Hb does not. (*c*) Multiple pores increase in size but collapse when one large pore per RBC (diameter, 1 μm) opens and hemolysis occurs. (*d*) The pore can reseal grossly in 10 to 20 s [22]. This resealing process can be halted and manipulated, leaving a single pore per ghost [73, 74]. A sudden straining of the membrane to $\Delta A/A = 0.064$ opens multiple pores of apparently the same size (~6 nm) such that hemoglobin leaks at half the rate of K^+ over a range of leak rates [21] (*b*′). If time (<1 to 10 s) permits, a single large pore, one per cell, opens, and hemolysis ensues (*c*′). *c*′ differs from *c* primarily in the K^+ and Hb content when the large pore opens. *c*′ goes to *d* in the same manner as *c* to *d*. If stress is relaxed on *b*′ before the large pore opens, K^+ and Hb leaks cease within approximately 40 ms [32] and the ghost reseals, *b*″. If the overwhelming stress is combined with fluid shear, the cell shatters, as indicated in *c*″, releasing cell contents and membrane fragments.

be a very important paper, demonstrated molecular sieving by measuring release of potassium, adenylate kinase, and the hemoglobin monomer, dimer, and tetramer. They inferred that as membrane strain progressively increased, the potassium loss commenced first, then the hemoglobin monomer, followed by adenylate kinase, hemoglobin dimer, and finally the hemoglobin tetramer. When the external fluid contained sodium chloride or lithium chloride as the primary solute, ingress of the cation, at the time when potassium leaks were first noted, led to colloid osmotic swelling and the loss of hemoglobin. On the other hand, when the larger dextrose molecule was used as the primary external solute, the delay after the first appearance of potassium loss was extended, and only when the dextrose was able to equilibrate across the membrane did the swelling occur and lysis ensue. It should be mentioned here that colloid osmotic pressure was also cited as the primary cause for lysis when an electrostatic potential of 1 V over the membrane was introduced, causing cation permeability increases [17]. These results support the observations of Seeman [22] and Tatsumi and Kimura [54], who, while using an ion-flux detector of cell volume, noted a short-circuiting effect when the cell pores opened up, suggesting that the cell volume rapidly shrank. Seeman [22] calls attention to the possibility that this is more easily understood if we suppose that the ion flux could occur through the pores. Tatsumi and Kimura [54] show that critical volume at the onset of this drop in apparent volume occurred at 1.9 ± 0.11 times the initial cell volume and was weakly dependent on the mean cell hemoglobin concentration. It should be added that the shift to lower salt concentration in slow hemolysis [22] is modeled more successfully by assuming that K^+ escapes with a negligible ingress of Na^+.

Hemoglobin Loss and Pore Formation in Red Cell Filtration. Hemolysis in the filters used in transfusion and other perfusion processes has been discussed by Linco [55], who showed that an elevated temperature reduced hemolysis in the filtration process. The use of filters intentionally to cause hemolysis and at the same time to separate cytoplasm from membrane components is discussed by Hultquist et al. [56]; here attention is called to the fact that the cells held up by the filter and exposed to the pressure differential are gently lysed with their membranes resealing. A similar result was obtained by Chien and coworkers [21], using larger filters through which the cells could pass, having lost some of their hemoglobin, and then could reseal. Evidence of pore formation in the membrane was presented by Chien et al. [21]. This evidence is given in Fig. 15.3. It shows the index of potassium loss as the pressure differential across the filters is increased and as the rate of hemoglobin loss is increased. Chien et al. [21] call attention to the fact that the index of potassium loss is roughly twice that of hemoglobin, suggesting that the pores that form must be approximately the size of the hemoglobin molecule or a little larger. The consistency of this slope for a wide range of loss rates led Chien et al. [21] to suggest that only pores of one size open and that increased stress merely causes an increased number of pores of the same size to open (see Figs. 15.3 and 15.4). Blackshear and Anderson [57] extended this study to cover other filter sizes and produced the hemolysis map shown in Fig. 15.5. In modeling lysis in filter pores, they assumed that the lubricating layer of fluid between cell membrane and pore produced a cylinder of cell membrane material of a diameter initially smaller than the filter pore. Then in order for the cell to be able to enter the pore, the cell must either lose volume or fill the pore. At the driving pressures where hemolysis occurred, they inferred that volume loss occurred first.

This effect has still not been adequately elucidated. Chien et al. [21] have reported apparent membrane viscosities, derived from RBCs entering a micropipet, that are significantly smaller than those obtained from cells expelled from the micropipet following a greater-than-20-s dwell time. If we assume the membrane viscosity is independent of rate of strain, these data are consistent with the existence of a lubricating layer between the membrane and pipet wall.

Blackshear et al. [58] and, more recently, Skalak et al. [59] have modeled flow through a filter as consisting of parallel pores through which either suspending fluid only, or cells only, pass. From these considerations, the velocity of the cell and the dwell time of the cell in its stressed state can be inferred. At the condition for lysis, the stress duration is about 10^{-2} s.

It seems reasonable, then, to relate the membrane pore-forming mechanisms in flow through microporous media to fast hemolysis and the early stages of rapid hemolysis. In both cases, the model proposed by Chien et al. [21] fits the data presently available: a rapid imposition of a lytic membrane stress causes pores of roughly hemoglobin diameter to appear in the cell membrane. An elevation of the stress over the threshold value merely causes more such pores to open. In time (order, 1 s), these pores would grow. But if stress is released before about 1 s, the pore reseals rapidly [32].

Hemolysis in Tube Flow. A number of papers have been published recently concerning hemolysis in tube flow [35, 51, 60–65]. Some concern themselves with lysis which occurs when a cell touches a wall [30, 35, 40, 41, 65–67].

Wall roughness on a scale of the RBC correlates with hemolysis [35]. Surface material may chemically interact [67], and bubbles trapped in surface imperfections can produce lysis. Even when these potentially trauma-producing surfaces are present, their effect is mediated by flow processes [68]. A number of papers discuss the relationship between the containing surfaces and superimposed shear stress on hemolysis [30, 35, 40, 41, 65–67]. Solin et al. [40, 41] show that over a wide range of shear rates, the particulate matter formed in a conical, cylindrical test system was independent of shear rate, but strongly a function of surface-to-volume ratio. Hemolysis, on the other hand, was linearly dependent upon shear rate as well as upon surface-to-volume ratio. This type of hemolysis and the accompanying particle generation occur at fluid stress thresholds below the stress required for lysis by fluid shear. The specific mechanism of

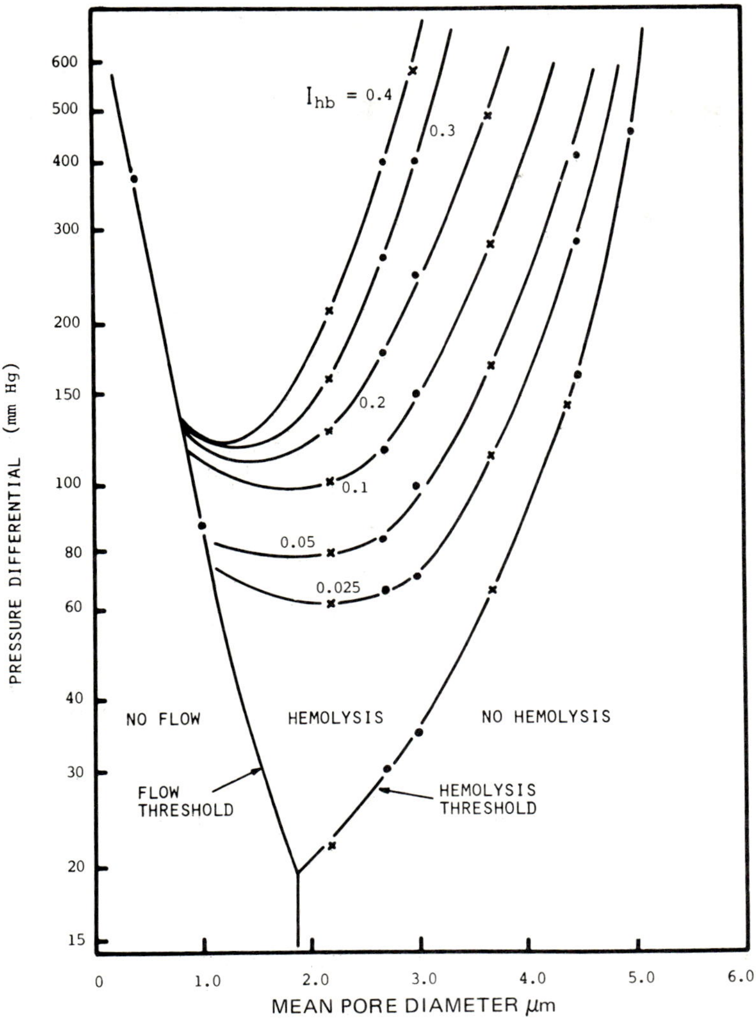

FIG. 15.5 Hemolysis and hemolysis thresholds in microporous structures. Circles are taken from Ref. 57; *X*'s are from Ref. 21. The hemolysis threshold agrees well with the value predicted by method of Ref. 57 and the Rand value of $\sigma_{c,t\to 0}$.

hemolysis depends on the surface. Morgan et al. [67] show that the hemolytic activity of some fibrous amphiboles depends on the strong adherence of some of the surface-related proteins on the red cell to the mineral, followed by osmotic lysis. The hemolytic effect of silica, on the other hand, is attributed to hydrogen bonding of silicic acid groups with constituents of the erythrocyte membrane. Glass contact hemolysis has been recently studied by Barreto [66]; Bacher and Williams [35] and Wielogorski et al. [65] show that surface roughness plays a role in hemolysis. In the latter case, it was found that only during the first hour of blood handling in tubes did the roughness make a difference. When pumping of blood was continued for longer periods, the rate of further hemolysis was independent of the tube rugosity. It seems clear that fragmentation and surface-induced osmotic hemolysis make it possible to hemolyze red cells by wall contact at flow conditions much more benign than those required to cause hemolysis in accelerating or shear flows through flow-induced membrane strain. Even benign surfaces become lytic with the presence of lipid particles, the chylomicra that are transiently elevated following a fat-rich meal [52]. Fortunately, surfaces appear to become passivated by wall coatings of plasma proteins with time. Although under some circumstances wall interaction may play a role, the most important practical limit on the hemolysis threshold in tube flow seems to be governed by the fluid mechanics of entry and exit effects in the tube.

Hemolysis in sharp-edged entrances of tubes on the order of 1 mm in diameter or less occurs at an entrance velocity of 600 cm/s [69]. Because hemolysis at an abrupt increase in cross section commences at a tube velocity of 700 cm/s [69], the entrance to, say, a conventional hypodermic needle represents the locus of hemolysis [70].

In determining the effects of suspending phase viscosity, tube size, and tube entry configuration on the hemolysis threshold, Keshaviah [69] found that hemolysis occurred when the maximum "elongational" stress, $\eta\, du/dz$, exceeded 10^4 dyn/cm^2, where η is suspending phase viscosity and du/dz is the velocity gradient in the flow direction. Rand's criterion [4] that membrane strain exceeds a critical value can, in this case where exposure to stress is brief, be equated to the requirement that the membrane biaxial tension exceed about 29 dyn/cm.

If we again postulate that a uniaxial membrane tension T_{11} generates a biaxial tension of $\sigma_c = \frac{1}{2}T_{11}$, then Rand's criterion for rapid lysis to occur is again $T_{11} \geq 58$ dyn/cm. Keshaviah shows that (for a range of assumed elongated shapes) $T_{11} \propto \eta\, du/dz$ and is of order 10 to 100 dyn/cm when $\eta\, du/dz = 10^4$ dyn/cm^2 [69].

In practical terms, for capillaries of diameter ~1 mm and blood with $\eta = 4$ cP, this means an upper limit on tube mean velocity of 600 cm/s with a sharp entry, and 1700 cm/s with a carefully fared entry [69].

If the flow in a tube is laminar, hemolysis caused by fluid shear within the tube is most likely to occur to the cell nearest the wall. Here both the stress and the dwell time will be greatest. If we use Bugliarello's dimensions [70] of the cell-free zone near the wall, the velocity will be $\dot{\gamma} \times 10^{-4}/H$ cm/s, where H is the cell fraction of blood and η is wall shear rate.

In fully developed laminar flow of a newtonian fluid in a circular tube, $\dot{\gamma} = 8\, U/d$, where U is mean velocity and d is tube diameter. The dwell time for the slowest cell is, thus, tube length L divided by slow cell velocity; that is, $t = HdL/10^{-4}\, U$ (in seconds).

For example, a tube with shear rate $\dot{\gamma} = 104\ \text{s}^{-1}$ and length L cm, with $H = 0.4$, will have a dwell time for the slowest cell of $0.4L$ s.

If we once again employ Rand's criterion, we may map the lytic threshold for whole blood in tube flow for the values of σ_c/τ_c derived above. We did this and found that a tube length less than 1 m produced very short dwell times such that $\sigma_c \approx \sigma_{c,t\to 0}$. Note that to compare velocities, $\dot{\gamma} = 8\, U/d$.

$U = \tau_c d/8\eta$; thus, for a 3-mm tube and whole blood ($\eta = 0.04$), *if* $\tau_c = 40{,}000$, hemolysis will occur in the entrance before it will occur in the tube. Data from Refs. 60 and 69 suggest that when $t \approx 1$ s, $\sigma_c \geq 20{,}000$ dyn/cm^2, as shown in Fig. 15.1.

Finally, at the sudden expansion from the tube flow into a stationary reservoir, the turbulent jet studies of Forstrom [51] should apply. To a good approximation, the limiting stress is independent of viscosity and hematocrit and occurs at a jet velocity of 1700 cm/s.

In comparing these thresholds, it is seen that the sharp entrance offers the greatest trauma, the exit the next greatest trauma, and only for very long, very small tubes would hemolysis occur in the tub itself due to shear stress alone.

Extremely Rapid Hemolysis. The shear stress required to cause lysis in very short times is found to be on the order of 10^4 to 10^5 dyn/cm^2 [51]. In a recent editorial following an entire issue devoted to lysis by ultrasound [71], the author points out that ultrasonic hemolysis is an all-or-nothing phenomenon proceeding at a logarithmic rate in which the membrane bursts into fragments spontaneously and the erythrocyte contents are lost. This is quite similar to the results shown in the Forstrom jet tests [51], but with a difference discussed below. The gradual stretching and opening of pores in the membrane is thought not to be a factor here. What seems to be a factor is the existence of sufficient time during the brief exposure to the very high shear stress for the cell to distort into a shape which would tear the membrane. Smaller cells or more viscous cells appear to be protected [51]. The influence of size is given in Table 15.1. Studies

TABLE 15.1 Jet fragility test results*

Mammal (no.)	Volume, μm^3	U_c, cm/s
Human cord (12)	103	1930
Human female (14)	84	2120
Human male (26)	84	2240
Hamster (10)	62	2260
Dog (19)	72	2310
Raccoon (1)	44	2310
Rabbit (3)	70	2330
Squirrel monkey (2)	63	2380
Brown bear (1)	52	2390
Guinea pig (1)	76	2460
Rat (10)	64	2470
Pig (1)	58	2700
Mouse (10)	48	2800
Siberian tiger (1)	53	3070
Domestic cat (1)	39	3150
Horse (1)	37	3510
Cow (2)	32	3850
Sheep (4)	31	3880
Goat (1)	24	5800

Source: Forstrom [51].

* These results show that the critical shear stress ($\tau_c = 0.008U^2$ dyn/cm^2) in the jet test increases dramatically with a reduction in cell volume.

involving the distortion of fluid drops combined with membrane-tearing theories may be more useful for the very high shear rate and very fast hemolysis; see Ultman [72] and Makowski [19].

There are a number of unresolved issues in extremely fast hemolysis. Forstrom [51] showed that, at the threshold velocity for lysis, cells appeared either healthy or as ghosts. At slightly higher than threshold velocities fragments appear. At velocities twice threshold velocity where stress is four times the threshold, only small fragments (or healthy cells) *but no ghosts* were observed. These considerations have contributed to the fragmentation sequence shown in Fig. 15.4.

When we attempt to contrast these given results with the characteristics of fast lysis in flow through microporous media and in fast osmotic lysis, we are again struck with the role membrane shear must play. Membrane biaxial tension four times the threshold value was applied (briefly) in the case of the filter lysis experiments [57], and could well have been applied in the fast osmotic lysis [32] experiments, but the pronounced fragmentation did not occur; the evidence suggests instead that more ~6-nm pores opened as stress increased [21]. At this point, we can only infer that when shear is present the pores do form at the threshold strain, but in a strong shear field they may be torn.

CONCLUSION

The recent research on the properties of the pores in ghosts resealing from osmotic lysis strongly suggests that these are ordered structures in the lipid bilayer. They can be manipulated and are remarkably stable [73, 74]. Lieber and Steck [74] point out that when the pore was fixed at 1 μm, where it could be visualized in the light microscope, it was circular and always appeared at the rim of the ghost. This and the sensitivity of the pore dynamics to a variety of lipid-intercalated molecules (e.g., cholesterol) suggest that the pore may be a lipid structure (and not a membrane failure) throughout the lysis process. In the material we have presented here, we feel the data support the Rand hypothesis [4] that lysis occurs when the membrane area increases by 6.4 percent, no matter whether the source of stress is osmotic, fluid shear, or flow through apertures.

The data also support the existence of different types of pores in the membrane: many small pores of uniform size when stress exposure is brief (less than 1 s) and one single, large pore in each cell when time to lysis is 1 to 10 s^{-1} or greater. If lipid structures are involved in both pore types, a rearrangement of the lipids about the pore could explain both the following phenomena: the σ_c versus t for biaxial tension lysis (osmotic, micropipet, filtration, etc.) and the apparent acceleration of the pore-forming process in the presence of membrane shear. Such a perspective suggests potentially fruitful research.

A reexamination of the stress histories required to cause mechanical lysis and the stresses encountered by RBCs in the vasculature reaffirm the conclusion [57] that the mechanisms that govern lysis in microporous media could also be responsible for lysis in vivo. In brief, an RBC forced into a pore that restrains it (but just) can show membrane tensions in excess of 29 dyn/cm for pores up to 4.5 μm in diameter at pressure drops that may be expected in the microvasculature and across heart valves and patches. Thus, lysis of spherocized cells may occur in normal vasculature and normal cells lysed in abnormally small flow channels, as suggested by Fig. 15.5. Prosthetic valves which have such a pressure differential when closure is at cell dimensions should also cause lysis. It seems unlikely to us that shear stress of normal RBCs in the absence of wall effects would be the source of lysis in vivo. Postalimentary chylomicron population or other plasma factors that predispose the cell to lysis may make the shear stress threshold lower [51].

Finally, the engineering applications of mechanical lysis seem very exciting indeed. The prospects of controlling content and pore size in the resealed ghost suggest applications in health-related drug encapsulation and delivery [74, 75], as well as in the broader chemical engineering application of encaged reagents or catalysts with controlled access.

REFERENCES

1. Blackshear, P. L. Jr.: Mechanical hemolysis in flowing blood, in Y. C. Fung, N. Perrone, and M. Anliker (eds.), *Biomechanics: Its Foundations and Objectives,* Prentice-Hall, Englewood Cliffs, N.J., 1972, pp. 501–528.
2. Everett, L. B., Hellums, J. D., Alfrey, C. P., and Lynch, E. C.: Red blood cell damage by shear stress, *Biophys. J., 12,* 257–273, 1972.
3. Sutera, S. P.: Brief reviews: Flow-induced trauma to blood cells, *Circ. Res., 41*(1), 2–8, 1977.
4. Rand, R. P.: Mechanical properties of the red cell membrane: II. Viscoelastic breakdown of the membrane, *Biophys. J., 4,* 303–316, 1964.
5. Brailsford, J. D., Korpman, R. A., and Bull, B. S.: The aspiration of red cell membrane into small holes: New data, *Blood Cells, 3,* 25–38, 1977.
6. Bull, B. S., Braillsford, J. D., and Korpman, R. A.: Red cell membrane deformability: An examination of two apparently disparate methods of measurement, *Blood Cells, 3,* 39–54, 1977.
7. Chien, S.: Blood rheology, in N. H. C. Hwang, D. R. Gross, and D. J. Patel (eds.), *Quantitative Cardiovascular Studies: Clinical and Research Applications of Engineering Principles,* University Park Press, Baltimore, 1979, chap. 6, pp. 241–287.
8. Chien, S., Sung, K-L. P., Skalak, R., Usami, S., and Tözeren, A.: Theoretical and experimental studies on viscoelastic properties of erythrocyte membrane, *Biophys. J., 14,* 463–487, 1978.
9. Hochmuth, R. M., Buxbaum, K. L., and Evans, E. A.: Brief communication: Temperature dependence of the viscoelastic recovery of red cell membrane, *Biophys. J., 29,* 177–182, 1980.
10. Hochmuth, R. M., Mohandas, N., and Blackshear, P. L., Jr.: Measurement of the elastic modulus for a red cell membrane using a fluid mechanical technique, *Biophys. J., 13,* 747–762, 1973.
11. Hochmuth, R. M., Worthy, P. R., and Evans, E. A.: Red cell relaxation and the determination of membrane viscosity, *Biophys. J., 26,* 111–114, 1979.
12. Hochmuth, R. M., Wiles, H. C., Evans, E. A., and McCown, J. T.: Extensional flow of erythrocyte membrane from cell body to elastic tether, *Biophys. J., 39,* 83–89, 1982.

13. Makowski, L.: A physical basis for red cell membrane elasticity, *J. Theor. Biol., 61,* 27–45, 1976.
14. Shiga, T., Maeda, N., Suda, T., Kon, K., Sekiya, M., and Oka, S.: Rheological and kinetic dysfunctions of the cholesterol-loaded, human erythrocytes, *Biorheology, 16,* 363–369, 1979.
15. Weed, R. I.: Membrane structure and its relation to haemolysis, *Clin. Haematol., 4*(1), 3–28, 1975.
16. Jay, A. W. L., and Rowlands, S.: The stages of osmotic haemolysis, *J. Physiol., 252,* 817–832, 1975.
17. Kinosita, K., Jr., and Tsong, T. Y.: Formation and resealing of pores of controlled sizes in human erythrocyte membrane, *Nature, 268,* 438–441, 1977.
18. MacGregor, R. D. II, and Tobias, C. A.: Molecular sieving of red cell membranes during gradual osmotic hemolysis, *J. Membrane Biol., 10,* 345–356, 1972.
19. Makowski, L.: A phase transition model for hemolysis, *J. Theor. Biol., 61,* 47–53, 1976.
20. Saari, J. T., and Beck, J. S.: Hypotonic hemolysis of human red blood cells: A two-phase process, *J. Membrane Biol., 23,* 213–226, 1975.
21. Chien, S., Luse, S. A., and Bryant, C. A.: Hemolysis during filtration through micropores: A scanning electron microscopic and hemorheologic correlation, *Microvasc. Res., 3,* 1183–1203, 1971.
22. Seeman, P.: Ultrastructure of membrane lesions in immune lysis, osmotic lysis and drug-induced lysis, *Fed. Proc., 33*(10), 2116–2124, 1974.
23. Harris, J. W., and Kellermeyer, R. W.: *The Red Cell,* rev. ed., Harvard University Press, Cambridge, Mass., 1970.
24. Leavell, B. S., and Thorup, O. A.: *Fundamentals of Clinical Hematology,* 4th ed., Saunders, Philadelphia, 1976.
25. Ponder, E.: *Hemolysis and Related Phenomena,* Grune & Stratton, New York, 1948.
26. Williams, W. J., Beutler, E., Erslev, A. J., and Rundles, R. W.: *Hematology,* 2d ed., McGraw-Hill, New York, 1977.
27. Gyde, O. H. B., and Harris, B. R.: A laboratory study of the haemolysis caused by different blood pumps, *J. Clin. Pathol., 29,* 87–88, 1976.
28. Han, N., Lehle, H. L., and Brieden, H.: Hämolysewirksame Faktoren im extrakorporalien Kreislauf: In vitro Untersuchungen, *Thoraxchirurgie, 21,* 123–130, 1973.
29. Kletschka, H. D., Rafferty, E. H., Olsen, D. A., Clausen, E. W., Hubbard, L. C., Robinson, A. R., and Larson, W. H.: Artificial heart, *Minn. Med.,* October 1975, pp. 757–781.
30. Mortensen, J. D.: Evaluation of tests for blood damage produced by oxygenators, *Trans. Am. Soc. Artif. Intern. Organs, 23,* 747–755, 1977.
31. Bleifeld, W., Kramer, C., and Pelzer, H.: Strömungsdynamische Einflussgrössen der mechanischen Hämolyse bei Herzklappenfehlern, *Thoraxchirurgie, 21,* 130–135, 1973.
32. Anderson, P. C., and Lovrien, R. E.: Human red cell hemolysis rates in the subsecond to seconds range, *Biophys. J., 20,* 181–191, 1977.
33. Richardson, E.: Deformation and haemolysis of red cells in shear flow, *Proc. R. Soc. London Ser. A, 338,* 129–153, 1974.
34. Richardson, E.: Applications of a theoretical model for haemolysis in shear flow, *Biorheology, 12,* 27–37, 1975.
35. Bacher, R. P., and Williams, M. C.: Hemolysis in capillary flow, *J. Lab. Clin. Med., 76,* 485–496, 1970.
36. MacCallum, R. N., Lynch, E. C., Hellums, J. D., and Alfrey, C. P., Jr.: Fragility of abnormal erythrocytes evaluated by response to shear stress, *J. Lab. Clin. Med., 85*(1), 67–74, 1975.
37. Williams, A. R.: Viscoelasticity of the human erythrocyte membrane, *Biorheology, 10,* 313–319, 1973.
38. Rooney: J. A.: Hemolysis near an ultrasonically pulsating gas bubble, *Science, 169,* 869–871, 1970.
39. Sanders, M. F., and Coakley, W. T.: Factors affecting disruption of biological material by an ultrasonically vibrating wire, *Exp. Cell Res., 73,* 410–444, 1972.
40. Solen, K. A., Whiffen, J. D., and Lightfoot, E. N.: The effect of shear; specific surfaces, and air interface on the development of blood emboli and hemolysis, *J. Biomed. Mat. Res., 12,* 381–399, 1978.
41. Solen, K. A., Whiffen, J. D., and Lightfoot, E. N.: Plasma glucose concentrations and the development of blood emboli and hemolysis," *J. Lab. Clin. Med., 98*(2), 206–216, 1981.
42. Sutera, S., Mehrjardi, M., and Mohandas, N.: Deformation of erythrocytes under shear, *Blood Cells, 1,* 369–374, 1975.

43. Williams, A. R.: The effect of cell volume and sugars on haemolysis by hydrodynamic shear stress, *Biorheology, 11,* 331–336, 1974.

44. Williams, A. R.: The effect of bovine and human serum albumins on the mechanical properties of human erythrocyte membranes, *Biochim. Biophys. Acta, 307,* 53–54, 1973.

45. Bessis, M., and Mohandas, N.: A diffractometric method for the measurement of cellular deformability, *Blood Cells, 1,* 307–313, 1975.

46. Bessis, M., and Mohandas, N.: Laser diffraction patterns of sickle cells in fluid shear fields, *Blood Cells, 3,* 229–239, 1977.

47. Goldsmith, H. L.: Red cell motions and wall interactions in tube flow, *Fed. Proc., 30*(5), 1578–1588, 1971.

48. Fischer, T. M.: Brief communication: On the energy dissipation in a tank-treading human red blood cell, *Biophys. J., 32,* 863–868, 1980.

49. Morris, D. R., and Williams, A. R.: The effects of suspending medium viscosity on erythrocyte deformation and haemolysis in vitro, *Biochim. Biophys. Acta, 550,* 288–296, 1979.

50. Keller, S. R., and Skalak, R.: Motion of a tank-treading ellipsoidal particle in a shear flow, *J. Fluid Mech., 120,* 27–47, 1982.

51. Forstrom, R. J.: A new measure of erythrocyte membrane strength: The jet fragility test, Ph.D. thesis, University of Minnesota, Minneapolis, St. Paul, 1969.

52. Bernstein, E. F., Blackshear, P. L., Jr., and Keller, K. H.: Factors influencing erythrocyte destruction in artificial organs, *Am. J. Surg., 114*(1), 126–138, 1967.

53. Sutera, S. P., and Mehrjardi, M. H.: Deformation and fragmentation of human red blood cells in turbulent shear flow, *Biophys. J., 15,* 1–10, 1975.

54. Tatsumi, N., and Kimura, E.: Expansibility of erythrocytes during the course of hypotonic hemolysis, *Experientia, 37,* 210–212, 1981.

55. Linco, K.: In-line blood warming and microfiltration devices: II. Influence of blood temperature on flow rate and hemolysis during pressure transfusion through microfilters and transfusion sets, *Acta Anaesth. Scand., 23,* 46–50, 1979.

56. Hultquist, D. E., Vasko, M. R., and Gray, R. H.: The use of filtration techniques for the lysis and study of red blood cells, *Am. J. Hematol., 3,* 231–234, 1977.

57. Blackshear, P. L., Jr., and Anderson, R. J.: Hemolysis thresholds in microporous structures, *Blood Cells, 3,* 377–395, 1977.

58. Blackshear, P. L., Jr., Christianson, T. J., Majerle, R. J., and Vargas, F. F.: Resistance of erythrocyte flow into pores, *J. Rheol., 23*(6), 681–702, 1979.

59. Skalak, R., Impelluso, T., Schmalzer, E. A., and Chien, S.: Theoretical modeling of filtration of blood cell suspensions, *J. Rheol., 20,* 41–56, 1983.

60. Coakley, W. T., James, C. J., and Macintosh, I. J. C.: Haemolysis of human erythrocytes in dextran solutions during rapid flow in capillaries, *Biorheology, 14,* 91–97, 1977.

61. Eurenius, S., and Smith, R. M.: Hemolysis in blood infused under pressure, *Anesthesiology, 39*(6), 650–651, 1973.

62. Herrera, A. J.: Blood transfusions: Effect of speed of infusion and of needle gauge on hemolysis, *J. Pediatr., 99*(5), 757–758, 1981.

63. Sherwood, G. K., and Schmidt, P. J.: Needles and hemolysis, *N. Engl. J. Med., 10,* 283, *11,* 598, 1970.

64. Moss, G., and Staunton, C.: Needles and hemolysis, *N. Engl. J. Med., 10,* 283, *11,* 598, 1970.

65. Wielogorski, J. W., Davy, T. J., and Regan, R. J.: The influence of surface rugosity on haemolysis occurring in tubing, *Biomed. Eng., 10,* 91–94, March 1976.

66. Barreto, D. M.: The early stages of glass-contact haemolysis, *Acta Haematol., 45,* 55–64, 1971.

67. Morgan, A., Holmes, A., and Talbot, R. J.: The haemolytic activity of some fibrous amphiboles and its relation to their specific surface areas, *Ann. Occup. Hygiene, 20,* 39–48, 1977.

68. Steinbach, J.: Hemolysis at tube walls, M.S. thesis, University of Minnesota, Minneapolis, St. Paul, 1970.

69. Keshaviah, P. R.: Hemolysis in the accelerated flow region of an abrupt contraction, Ph.D. thesis, University of Minnesota, Minneapolis, St. Paul, 1974.

70. Bugliarello, G., Kapur, C., and Hgiao, G.: The profile viscosity and other characteristics of blood flow in

a nonuniform shear field, in E. H. Lee and A. L. Copley (eds.), *Proc. 4th Int. Cong. Rheol.,* Interscience, New York, 1964, vol. 4, p. 351.

71. Summary, *Scand. J. Hematol., 29,* 52–53, 1976.

72. Ultman, J. S.: The recoverable membrane energy as the criterion for hemolysis of the red blood cell, in H. L. Gabelnick and M. Litt (eds.), *Rheology of Biological Systems,* Charles C Thomas, Springfield, Ill., 1973, pp. 85–100.

73. Lieber, M. R., and Steck, T. L.: A description of the holes in human erythrocyte membrane ghosts, *J. Biol. Chem., 257*(19), 11,651–11,659, 1982.

74. Lieber, M. R., and Steck, T. L.: Dynamics of the holes in human erythrocyte membrane ghosts, *J. Biol. Chem., 257*(19), 11,660–11,666, 1982.

CHAPTER 16

Static Elastic Properties of Blood Vessels

Thomas R. Canfield
Philip B. Dobrin
Departments of Radiotherapy and Surgery
Hines Veterans Administration Hospital and
Loyola University Medical Center
Hines, Illinois

INTRODUCTION

The present chapter discusses the mechanical behavior of arteries and the mathematical method required for quantification of such data. The following discussion presents static stress-strain data based on static equilibria making no assumptions regarding elastic properties of the tissue, except that such equilibria can be maintained. Computation of elastic properties of tissues is much more complex and is developed in detail below. Although this chapter is entirely concerned with the *elastic* (Bergel, 1961; Patel and Vaishnav, 1980) or *pseudoelastic* (Fung et al., 1979) behavior of blood vessels, it should be emphasized that the arterial wall also exhibits distinctive *inelastic* behavior. This includes viscosity, creep, stress relaxation, and length-force or pressure-diameter hysteresis. This deviation from elastic behavior is markedly increased by activation of vascular smooth muscle. Such behavior has been variously called viscoelastic (Remington, 1955) or plastoelastic (Alexander, 1971).

16.1 LONGITUDINAL ELASTIC BEHAVIOR

In vivo pressurized arteries are deformed simultaneously in all directions. However, experimental studies (Patel and Fry, 1969) have demonstrated that arteries deform orthotropically. Therefore, arterial deformations may be examined in three orthogonal directions, i.e., the longitudinal, circumferential, and radial directions. These are illustrated in Fig. 16.1. In situ, vessels are held under longitudinal tension. This develops in utero and increases gradually during early postnatal life (Dobrin et al., 1975). Longitudinal extension reaches the maximum

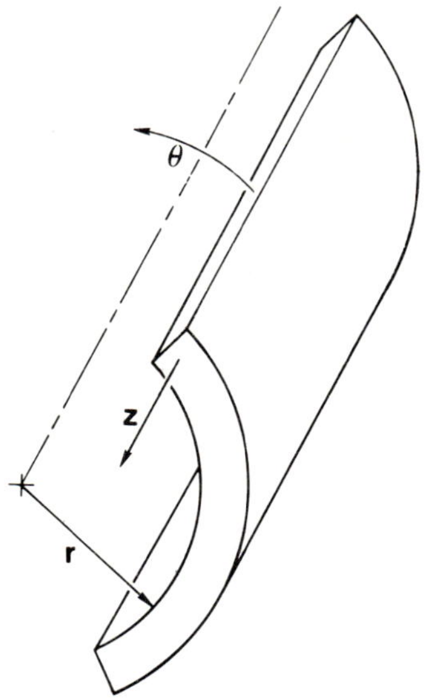

FIG. 16.1 Illustration of three orthogonal directions in an artery: the circumferential (θ), the longitudinal (z), and the radial (r). (*From Dobrin, 1978. By permission of the American Physiological Society.*)

during adulthood, then declines with age (Band et al., 1972; Learoyd and Taylor, 1966). The degree of longitudinal extension depends upon body position. For example, in dogs, extension of the neck is associated with changes in the length of the common carotid artery, ranging from 20 percent with maximum cervical flexion to 60 percent with extreme cervical extension. These percentage changes are given with respect to the unloaded length of the vessel (Dobrin, 1983). Similarly, with flexion of the knee joint the length of the popliteal artery is increased 20 percent in human beings and 9.5% in dogs (Browse et al., 1979). However, in spite of these changes in length with body position, arteries change negligibly in vivo with each pressure pulse. For example, the descending thoracic aorta lengthens about 1 percent during each cardiac cycle, while the abdominal aorta simultaneously shortens by the same amount (Lawton, 1957; Patel and Fry, 1964; Patel et al., 1964). An exception to this rule is found in the longitudinal motion of the ascending aorta and pulmonary arteries, which exhibit 5 to 11 percent changes in length with each cardiac cycle (Patel et al., 1964), but these changes in length result from overall motion of the heart during rapid ventricular ejection. The negligible changes in length of most vessels result from the longitudinal extension imposed upon them, the presence of perivascular connective tissue, and the arterial side branches which tend to fix the position of vessels. The tethering effect of these factors increases with increasing distance from the heart (Patel and Fry, 1966). In addition, the longitudinal extension (traction) imposed on vessels acts to buffer changes in length that might result

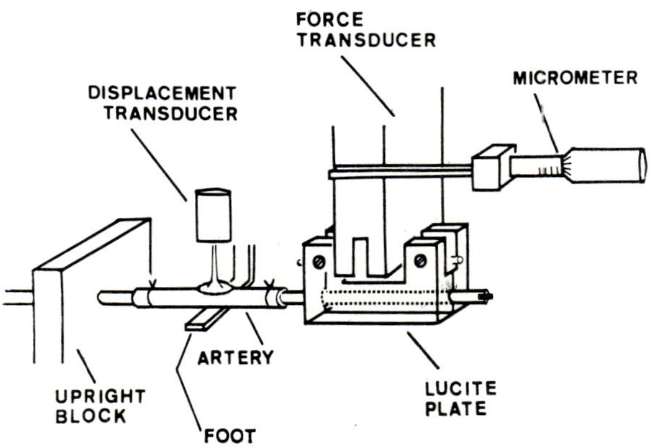

FIG. 16.2 Technique for studying vessels in vitro. Vessels were pressurized through the catheter at the left. Diameter was measured with displacement transducer. Nonrotating blade micrometer permitted vessel elongation without twisting. (*By permission of the American Heart Association, Inc.*)

from alterations in pressure. In order to study these interactions, experiments were performed on cylindrical segments of dog common carotid arteries in vitro. These vessels were mounted in a tissue bath filled with physiological salt solution where they were extended to in situ length (Fig. 16.2). The arterial segments were pressurized while longitudinal force was measured with a force transducer, and diameter was measured with a linear displacement transducer. Experimental results are shown in Fig. 16.3. The net longitudinal force is the sum of the forces

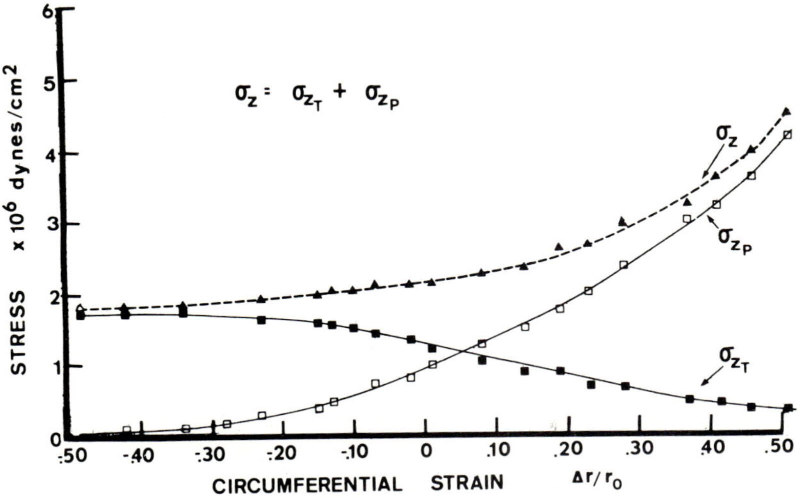

FIG. 16.3 Longitudinal stresses for a dog carotid artery plotted as a function of circumferential strain, where strain is computed with reference to the dimensions of the fully unloaded vessel. Data illustrate the relative constancy of net longitudinal stress (σ_z) because of the interaction of the longitudinal stress due to traction (σ_{zT}) and the longitudinal stress due to pressure (σ_{zP}). *(From Dobrin and Doyle, 1970. By permission of the American Heart Association, Inc.)*

resulting from traction and pressurization. As pressure increases, the component due to pressure increases, but this tends to be counterbalanced by a comparable fall in tractional force. Therefore net longitudinal force tends to remain constant over a wide range of pressures, thereby stabilizing vessel length.

16.2 CIRCUMFERENTIAL ELASTIC BEHAVIOR

Early descriptions of vascular mechanics (Roy, 1880) revealed a markedly biphasic length-force relationship with tissue extension. This property has been confirmed by numerous studies on strips, rings, and cylindrical segments of blood vessels (Bergel, 1961). Figure 16.4 shows pressure-radius curves obtained from experiments performed on cylindrical segments of arteries in vitro. The experimental arrangement was similar to that shown in Fig. 16.2. Vessels were pressurized in small steps over the full pressure range in order to relax them. The labeled curves show data obtained after relaxation, but before treatment (PRE), after excitation of the vascular muscle with norepinephrine (NEpi), and after poisoning of the muscle with potassium cyanide (KCN). Both the PRE and the KCN curves display biphasic pressure-radius curves, with little hysteresis. The curves obtained following excitation of the muscle exhibits marked distension-

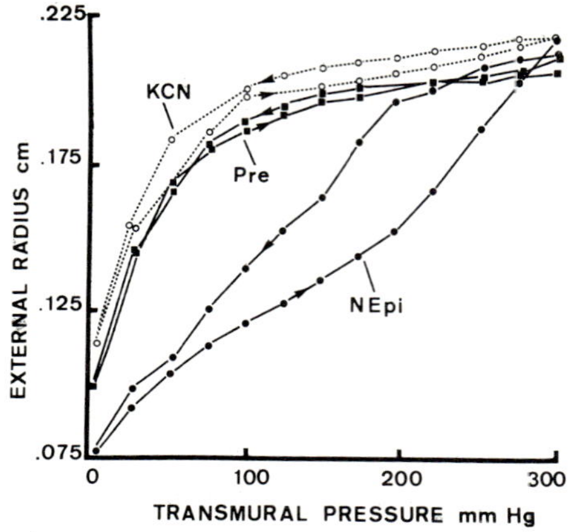

FIG. 16.4 Pressure-diameter curve for a dog carotid artery in relaxed state (Pre), after excitation of the muscle (NEpi), and after metabolic poisoning of the muscle (KCN). See text for discussion. (*From Dobrin and Rovick, 1969. By permission of the American Physiological Society.*)

retraction hysteresis. Chemical and enzymatic degradation studies (Reuterwall, 1921; Krafka, 1939; Remington et al., 1945; Roach and Burton, 1957) demonstrated that the extensibility exhibited by vascular tissue at small extensions is associated with unwrinkling, orientation, and gradual extension of elastin, whereas the stiffness observed at large extensions is associated with the progressive loading of elastin and of collagen fibers. Elastin is an elastomeric protein capable of undergoing extensions of up to 130 percent (Carton et al., 1962), whereas collagen is a protein consisting of a triple helix and is capable of undergoing extensions of only 2 to 4 percent (Stromberg and Wiederhielm, 1969). The elastic modulus of elastin is about 4.0×10^5 Pa, whereas the elastic modulus of collagen is about 1.0×10^9 Pa. The elastic modulus for the intact vessel wall computed by various means indicates that the stiffness of the intact structure is negligible at minimum extension, but at maximum extension is approximately 1.0×10^6 Pa. Therefore it is likely that only a small portion of wall collagen is actually load-bearing. Biochemical studies by Fischer and Llaurado (1966) showed that elastin represents between 15 and 41 percent of the dry defatted tissue of arteries, the lowest being found in the coronary arteries and the greatest being found in the ascending aorta, while collagen represents between 19 and 47 percent of the weight of the dry defatted tissue, the lowest being found in the aorta and the greatest being found in the coronary artery. Wolinsky and Glagov (1967) demonstrated a relationship between the number of lamella and circumferential tension in a large number of arteries. However, the content or ratio of collagen to elastin correlated poorly with vessel stiffness (Cox, 1978), implying an important role of architecture. Histological studies by Ho and coworkers (1972) showed a correlation between the amount of elastin in vessel walls and the mechanical stiffness. Cox (1978) estimated that in highly distended vessels, as little as 8 percent of dog carotid artery stiffness, and 20 to 25 percent of mesenteric, renal, iliac, and coronary artery stiffness, is attributable to collagen. It should be emphasized that a significant portion of the collagen in vessel walls is located in the adventitia and is not easy to remove completely. Several histological studies have presented evidence that adventitial collagen is not normally load-bearing except in vessels distended to an extreme degree. The studies of Wolinsky and Glagov (1964) demonstrated that the elastic lamella of the media are corrugated in appearance at low pressure until approximately 80 mmHg; they gradually become more straightened with further distension. On the other hand, medial collagen fibers do not become oriented circumferentially until pressures higher than physiological values are achieved. Adventitial collagen fibers are not fully stretched or fully oriented even at very high pressures in normal vessels, but may be stretched and loaded in aneurysmal vessels (Sumner et al., 1970; Dobrin et al., 1984).

Vascular smooth muscle also contributes to elastic properties in the circumferential direction. Some of these muscle cells form rings, being joined end to end, and are located between concentric elastic lamellae; other muscle cells appear to bridge segments of elastin. The latter organization is found in the aorta; however, very complex architectural arrangements are found when the aorta is examined in detail with electron microscopy (Clark and Glagov, 1979). Activation of vascular smooth muscle may be manifested mechanically in two ways: it may cause a decrease in vessel diameter, i.e., vasoconstriction, and it may cause an increase in stiffness so that the vessel can withstand increased transmural pressure and distending force at the same diameter. The latter mode of contraction is comparable to isometric contraction in striated muscle. Clearly, then, activation of the muscle increases the mechanical stiffness of the tissue. Figure 16.5 illustrates a sequence of isometric and isobaric contractions in a cylindrical segment of dog carotid artery studied in a tissue bath in vitro. Contraction was elicited by stimulation with a supramaximal dose of norepinephrine (NEpi).

Isometric contraction proceeded gradually, requiring elevation in the transmural pressure by the investigator. After completion of isometric contraction, the pressure was returned to its original level to permit "isobaric" contraction. This permitted the vessel to decrease in diameter. Because the distending force is the product of pressure, internal diameter, and vessel length

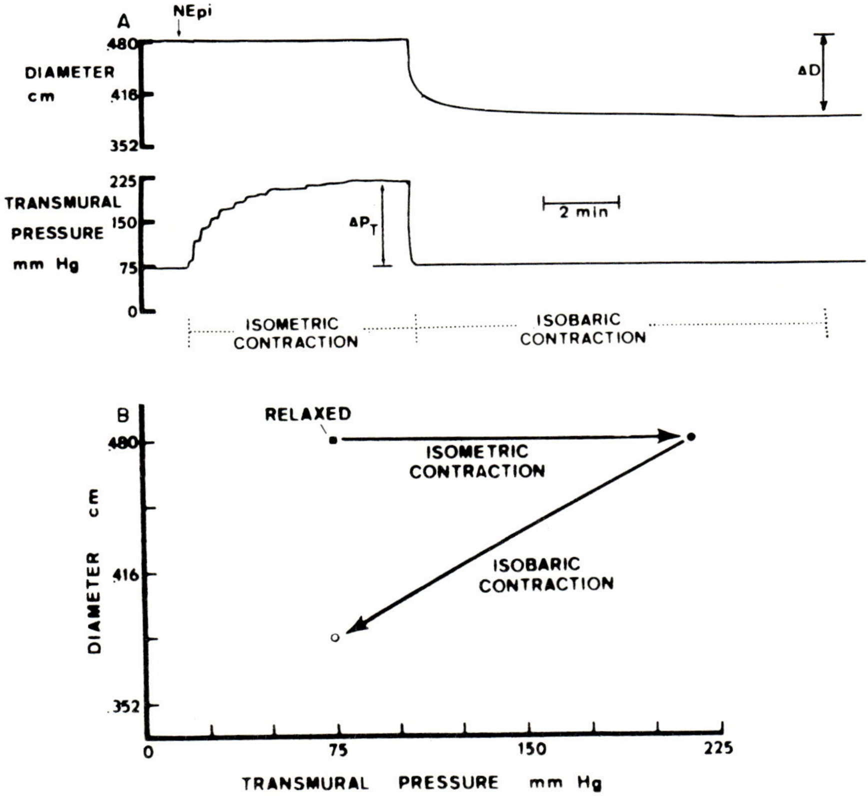

FIG. 16.5 Top: Polygraph illustrating isometric and isobaric contraction of vascular muscle in a dog carotid artery. Bottom: Same data depicted in pressure diameter coordinates. (*From Dobrin, 1973. By permission of the American Physiological Society.*)

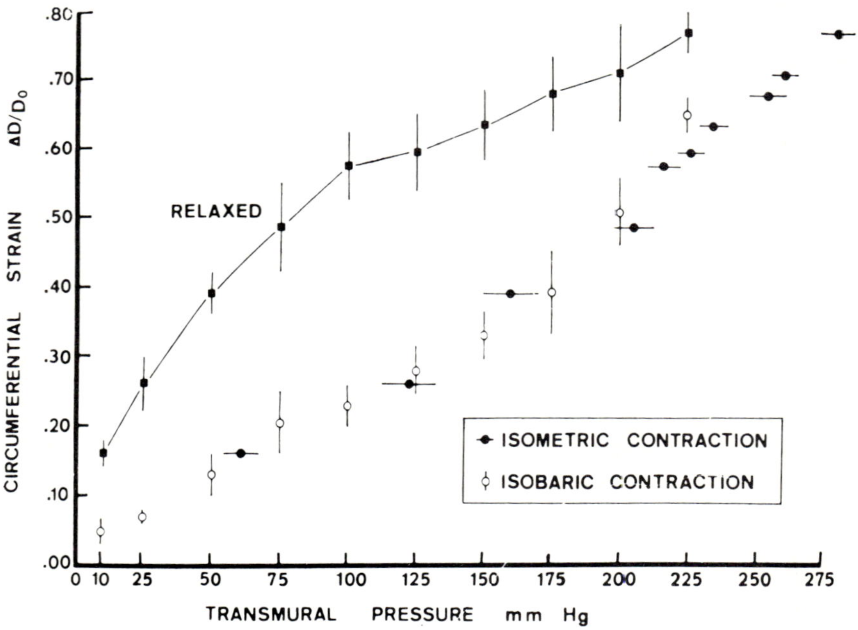

FIG. 16.6 Summary of isometric and isobaric contractions for 160 dog carotid arteries; see text for details. (*From Dobrin, 1973. By permission of the American Physiological Society.*)

(see Fig. 16.7), isobaric contractions are neither isometric nor isotonic. Figure 16.6 summarizes isometric and isobaric contraction data for 160 segments of dog carotid artery subjected to isometric contraction (constant diameter) associated with an increase in transmural pressure and then permitted to constrict at the original pretreatment pressure (isobaric contraction). The data show that when vessels are excited in this sequential fashion, the isometrically and isobarically contracted vessels tend to fall along a single pressure-diameter curve, thus indicating comparability of the two modes of contraction. Studies of a large number of different vessels indicate that isobaric contraction can decrease the diameter of blood vessels up to 50 percent, although in most cases it may be no more than half this amount. Moreover, in most vessels the pressure at which maximum contraction occurs appears at the usual in vivo transmural pressure to which the vessels are subjected (Dobrin, 1978). The mechanics of vasoconstriction are quite complex because with decreased diameter there is decreased distending force, even at constant pressure. At the same time, vasoconstriction is associated with shortening of the vascular muscle cells, and this is associated with decreased active contractile stress.

During isometric contraction the vascular muscle exerts contractile force, but not enough to overcome the applied load. Therefore it cannot cause vascular constriction (Fig. 16.5). However, because the muscle generates active force, all the elements in series with the force-generating apparatus are extended covertly. The degree of covert extension depends upon the elastic properties of those elements. These elements are collectively termed the *series elastic component,* or SEC. By subjecting the SEC to small, rapid changes in load during (Dobrin, 1974; Cox, 1977) and after (Dobrin and Canfield, 1973; Herlihy and Murphy, 1974; Cox, 1977; Peiper et al., 1978) isometric and isobaric contraction, it is possible to estimate the compliance of the SEC. Such studies have shown that the SEC is extended 7 to 11 percent during maximum isometric contraction. The precise identity of the SEC is unknown, although temperature variation studies and enzymatic degradation experiments (Dobrin and Canfield, 1977) suggest that much of the SEC is located in elastin. Studies using large variations in temperature

(Mulvany, 1979) suggest that there may be both damped and undamped components to the SEC.

Activation of vascular smooth muscle also alters the resistance of the wall to deformation. Although the early literature reported conflicting effects of vascular muscle on vessel stiffness, much of this conflict appears to have resulted from the manner by which the data were presented. Those authors who presented stiffness data as a function of transmural pressure or force generally found that activation of the muscle caused decreased vessel stiffness, whereas those who presented stiffness as a function of volume or diameter generally reported increased stiffness (see Dobrin and Rovick, 1969). Insight may be gained into the effects of the muscle by examining the circumferential stress-strain curves presented in Fig. 16.4. The slopes of these curves are directly related to vessel stiffness. Examining these slopes with respect to strain shows that activation of the muscle increased vessel stiffness at small diameters. When the curves are examined as a function of constant pressure, activation of the muscle caused constriction to smaller diameters, unloading the connective tissue and generally decreasing vessel stiffness at most pressures. In vivo, small vessels constrict to smaller dimensions during sympathetic nerve activation, but large vessels constrict slightly or not at all. Indeed, in the intact circulation, activation of vascular muscle by means of a generalized pressor response may cause an *increase* in the diameter of the conduit arteries due to the increase in pressure (Gerová and Gero, 1967). Under these circumstances one may expect activation of the muscle to increase vessel stiffness, and, of course, distension will increase the stiffness of the connective tissue as well. Examination of Fig. 16.4 also indicates that a vessel in which the muscle has been activated exhibits marked pressure-diameter hysteresis in that the activated muscle can bear much greater loads than it can generate. For many years the mechanism of this behavior was unexplained. Recently, Seigman et al. (1976) showed that tone in smooth muscle reflects calcium-dependent persistence of contractile crossbridges, while Dillon et al. (1981) demonstrated that in vascular muscle, dephosphorylation of the myosin light chain stops crossbridge cycling, but maintains persistent crossbridges.

16.3 RADIAL ELASTIC BEHAVIOR

The third direction in the arterial wall is radial. Unlike the tensile deformations and stresses in the longitudinal and circumferential directions, deformations and stresses in the radial direction are compressive and are only about 5 percent of that in the circumferential direction and 10 percent of that in the longitudinal direction. Very little is known regarding the behavior of arteries in the radial direction per se because such behavior is inevitably coupled with deformations and much larger stresses in the remaining two directions. What is known regarding mechanical properties in the radial direction has been obtained by Patel et al. (1970). Their work shows that the stiffness in the radial direction is generally comparable to that observed longitudinally and circumferentially. Moreover, the stiffness in the radial direction may not be very important in determining the overall behavior of intact arteries because of the relatively small stress in this direction, as compared with those in the remaining directions. An important feature of the radial behavior of vessels concerns the compression to which it is subjected by pressurization. It has been suggested that compressive stress or associated tissue distortion may compress the vasa vasorum which feed the vessel wall, possibly predisposing to abnormal metabolism of lipids. The pathophysiological implications of this phenomenon relative to atherosclerosis are discussed in Chap. 22 of this handbook (Goldstick and Dobrin, 1987) and elsewhere (Heistad and Marcus, 1979; Dobrin, 1983).

16.4 ELASTICITY OF ARTERIES

The remainder of this chapter concerns itself with analytic methods for determining the elastic properties of blood vessels. These techniques may be used on arteries, veins, and other tubular structures encountered in biomechanics. However, attention will be given only to the elastic properties of arteries. The elastic properties of arteries have been computed on the basis of the following assumptions:

1. Arteries are cylindrical.
2. The arterial wall is incompressible.
3. Deformations are orthotropic.
4. The wall is homogeneous.
5. The wall is mechanically isotropic.
6. Usual analytic methods may be employed.

There is good evidence that arteries are cylindrical (Sobin and Tremer, 1980), that the arterial wall is incompressible (Carew et al., 1968; Dobrin and Rovick, 1969), and that wall deformations are orthotropic (Patel and Fry, 1969). However, the last three assumptions are much less secure. There is much evidence to indicate that arteries are not isotropic, their departure from isotropy depending upon the vessel considered as well as the degree of deformation imposed on it (Patel et al., 1969; Apter et al., 1970; Dobrin and Doyle, 1970; Cox, 1985). The arterial wall is clearly *not* homogeneous, being composed of elastin, collagen, ground substance, and smooth muscle cells arranged in a highly complex architecture (Clark and Glagov, 1979). However, because of its incompressibility at any given level of deformation, the overall mechanical properties of the wall may be assumed to reflect the appropriately weighted average of the properties of its load-bearing components at their respective levels of deformation. There is evidence that arteries exhibit uniform resistance to compression in the radial direction at several zones across the wall thickness (Dobrin, 1983). Uniformity of properties in the longitudinal and circumferential directions has not been determined. The assumption that arterial mechanics may be described by conventional analytic methods requires further consideration. Classical infinitesimal strain theory assumes very small changes in dimensions with deformation and linearity of properties over these small dimensional changes. Blood vessels undergo very large changes in dimension—as much as 70 percent in the longitudinal and circumferential directions relative to their respective unloaded states. Two analytic methods have been employed to circumvent these violations of classical elasticity theory. The first is the use of incremental methods (Bergel, 1961), i.e., the use of linear elasticity over small increments of deformation; the second is the use of finite deformation theory, a method originally developed to study the behavior of elastomeric materials such as rubber (Green and Adkins, 1970). The development of such analyses is given below. The results of analytic solutions then are evaluated using real values obtained from in vitro experimental studies. The experimental studies were carried out as shown in Fig. 16.2, but the vessels were pressurized at four lengths, i.e., 20, 40, 60, and 80 percent longitudinal extensions.

16.4.1 Strain Measures

Consider a thin tissue specimen undergoing a uniaxial extension. Its original length is given by l_0. When a load is applied, it will elongate to a new length, l. A principal extension ratio λ may be computed to be the ratio of this new deformed length l to the reference, or undeformed, length l_0:

$$\lambda = \frac{l}{l_0} \tag{16.1}$$

Principal extension ratios are convenient measures for large deformations, but they are not universally employed. In particular, small deformations are not suited for this type of representation when the changes in length are orders of magnitude less than the dimensions of the body being deformed. For these small deformations one usually employs engineering strains, ε, which are derived by direct comparison of changes in dimension with the overall dimensions of a body prior to deformation. For uniaxial extension, the engineering strain is given by

$$\varepsilon = \frac{\Delta l}{l_0} = \lambda - 1 \tag{16.2}$$

In continuum mechanics two additional types of strain are frequently used to describe large deformations of materials. They are quadratic and natural strain measures. Both reduce to engineering strains when the deformations are small and are expressible in terms of the principal extension ratios. For the uniaxial extension test the quadratic strain is given by

$$e = \tfrac{1}{2}(\lambda^2 - 1) \tag{16.3}$$

whereas a natural strain has the form

$$\eta = \log \lambda \tag{16.4}$$

The reduction to engineering strain can be appreciated when one expresses these strains in terms of the engineering strain ε. The quadratic strain is then

$$e = \varepsilon + \tfrac{1}{2}\varepsilon^2 \tag{16.5}$$

When $\varepsilon \ll 1$, the second term on the right-hand side may be neglected and the two strains are nearly equal. The natural strain may be written as

$$\eta = \log (1 + \varepsilon) = \varepsilon - \tfrac{1}{2}\varepsilon^2 + \cdots + (-1)^{n-1}\frac{\varepsilon^n}{n} + \cdots \tag{16.6}$$

where the series expansion is valid when $-1 < \varepsilon < 1$. Using similar arguments, the natural strain approaches the engineering strain when $\varepsilon \ll 1$, and terms of order ε^2 and higher may be ignored. Either strain is a valid measure of deformation. Quadratic strains are more easily incorporated into a strain energy density function, a potential for description of an elastic material. The quadratic form may be a desirable property because it conforms to measures of length in the euclidean space of our experience (Truesdell and Toupin, 1960). Natural strains may be interpreted in a different way. They are a measure of relative deformation which can be appreciated by considering the expression equivalent to Eq. (16.4),

$$\eta = \int_{l_0}^{l} \frac{dl}{l} \tag{16.7}$$

The natural strain is the integral sum of the relative changes in the dimensions of a body.

16.4.2 Static Deformation of Cylindrical Vessels

In the analysis of deformation of cylindrical blood vessels, all the material points may be located and identified by their initial positions in a cylindrical coordinate system whose components are given by (R, Θ, Z). This set of coordinates may be defined as the reference configuration, and all other configurations may be described in terms of them (Fig. 16.1). For example, a particle originally located at a particular position (R, Θ, Z) may be deformed to a new position, say (r, θ, z), at some later time t. These deformed coordinates (r, θ, z) would, in general, be functions of the original coordinate positions (R, Θ, Z) and t. For static deformation of elastic materials the explicit dependence upon t may be dropped. And in the restricted case of static internal pressurization and uniform longitudinal extension of an orthotropic blood vessel, the deformation may be described by

$$\begin{aligned} r &= r(R) \\ \theta &= \Theta \\ z &= \mu Z + c_1 \end{aligned} \tag{16.8}$$

where (r, θ, z) and (R, Θ, Z) are given above, c_1 is an arbitrary constant, and μ is a constant-longitudinal-extension ratio. In this particular deformation cylindrical geometry is maintained. The actual forces that produce longitudinal extension of blood vessels in situ result from connections to surrounding tissue, internal distending pressure, and viscous drag due to blood flow. These forces would not be expected to produce constant longitudinal extension, there would be shearing strains, and μ would have functional dependence on R and perhaps Θ. If these effects were large enough, they could invalidate assumptions of pure orthotropy in vessel mechanics. However, their effect appears to be small relative to the 20 to 60 percent overall longitudinal extensions to which blood vessels may normally be subjected. When this shearing can be ignored, the principal axes of strain lie along the directions of the (r, θ, z) coordinate axes, and the principal distension ratios are given by

$$\lambda_r = \frac{dr}{dR} \qquad \lambda_\theta = \frac{r}{R} \qquad \lambda_z = \frac{dz}{dZ} \tag{16.9}$$

These numbers express the relative dimensions of the sides of a block of material that was originally a unit cube in the reference configuration. The volume J of this cube is the product of the three principal extension ratios:

$$J = \lambda_r \lambda_\theta \lambda_z \tag{16.10}$$

In the case of an incompressible vessel, $J \equiv 1$, indicating that the volume is constant. Employing Eqs. (16.8) and (16.9) in (16.10) allows one to write

$$\mu \frac{r}{R}\frac{dr}{dR} = 1 \tag{16.11}$$

and since R is always greater than zero in the cylindrical blood vessel,

$$r\,dr = \frac{1}{\mu} R\,dR \tag{16.12}$$

Integration of this expression results in

$$r^2 = \frac{1}{\mu} R^2 + c_2 \tag{16.13}$$

where c_2 is an integration constant; this is different for each deformed configuration. For a particular deformed configuration, the integration constant c_2 may be evaluated as follows:

$$c_2 = r_e^2 - \frac{1}{\mu} R_e^2 = r_i^2 - \frac{1}{\mu} R_i^2 \tag{16.14}$$

where the subscript e is used to denote the external radii and the subscript i is used to denote the internal radii. As a consequence, the principal extension ratios may be expressed in terms of either the deformed radial coordinate r or the reference radial coordinate R. Therefore,

$$\lambda_r = \sqrt{\frac{1}{\mu} - \frac{c_2}{\mu r^2}} = \frac{R}{\sqrt{\mu(R^2 + \mu c_2)}} \qquad \lambda_\theta = \frac{r}{\sqrt{\mu(r^2 - c_2)}} = \sqrt{\frac{1}{\mu} + \frac{c_2}{R^2}} \qquad \lambda_z = \mu \tag{16.15}$$

For this orthotropic deformation the quadratic strains defined in the previous section are given by

$$e_r = \tfrac{1}{2}(\lambda_r^2 - 1)$$
$$e_\theta = \tfrac{1}{2}(\lambda_\theta^2 - 1) \qquad (16.16)$$
$$e_z = \tfrac{1}{2}(\lambda_z^2 - 1)$$

while the corresponding natural strains are

$$\eta_r = \log \lambda_r$$
$$\eta_\theta = \log \lambda_\theta \qquad (16.17)$$
$$\eta_z = \log \lambda_z$$

16.4.3 Static Equilibrium of Cylindrical Vessels

Blood vessels retract both longitudinally and circumferentially when excised. Restoration to natural dimensions requires the application of internal pressure p_i and a longitudinal tether force F_T. The net effect of these combined tractions can be expressed in the following two equations:

$$T = p_i r_i \qquad (16.18)$$

and

$$F_R = F_T + p_i \pi r_i^2 \qquad (16.19)$$

The first equation is the familiar law of Laplace for a right circular cylinder with internal radius of r_i. It indicates that the force due to internal pressure p_i must give rise to a tensile force (per unit length) within the wall, T. This tensile force is the integral of the force intensity, or stress σ_θ, across the wall (see Fig. 16.7):

$$T = \int_{ri}^{re} \sigma_\theta \, dr = \bar{\sigma}_\theta h \qquad (16.20)$$

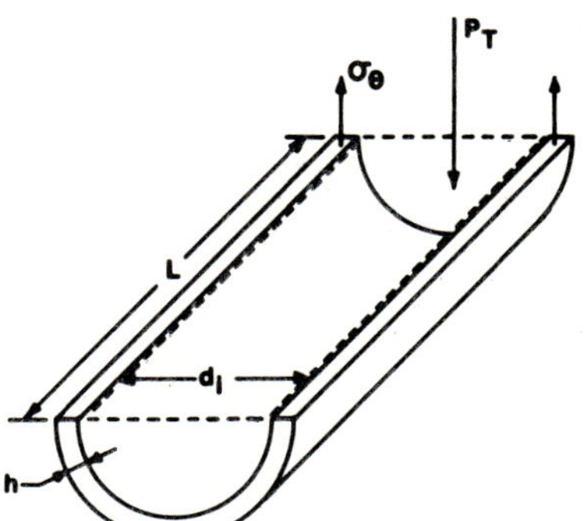

FIG. 16.7 Schematic free-body diagram of a blood vessel sliced longitudinally. The internal pressure P_T acts over the area Ld_i, which is exactly balanced by the integral of the circumferential stress σ_θ over the wall area $2Lh$. *(From Dobrin, 1983. By permission of the American Physiological Society.)*

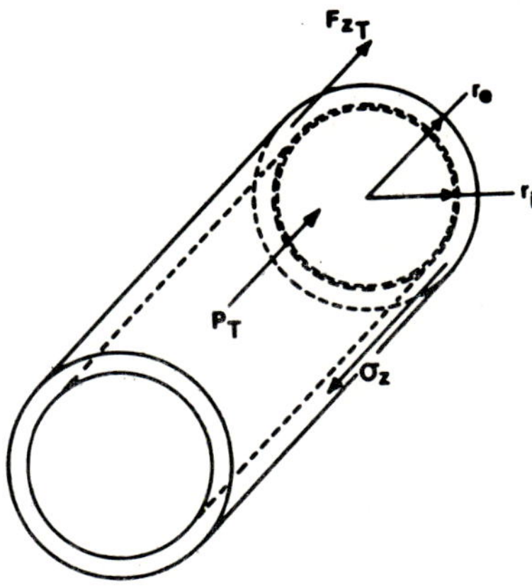

FIG. 16.8 Schematic free-body diagram of a blood vessel loaded longitudinally. The internal pressure P_T acting over the area πr_i^2, plus an external traction force F_T, must be exactly balanced by the integral of the longitudinal stress σ_z over the wall area $\pi(r_e^2 - r_i^2)$. *(From Dobrin, 1983. By permission of the American Physiological Society.)*

where $\bar{\sigma}_\theta$ is the mean value of stress and h is the wall thickness. Equilibrium is satisfied in Eq. (16.19) with the development of a longitudinal retractive force F_R, which is equal to the combination of longitudinal distending force due to internal pressurization acting on the enclosed circular area, $p_i \times \pi r_i^2$, plus the longitudinal tether force F_T, applied externally to maintain extension (see Fig. 16.8). This retractive force is the integral sum of σ_z over the annular cross section of the blood vessel,

$$F_R = 2\pi \int_{ri}^{re} \sigma_z r\, dr = \bar{\sigma}_z \pi h(r_e + r_i) \tag{16.21}$$

where $\bar{\sigma}_z$ is the mean stress, h is the wall thickness, r_e is the external radius, and r_i is the internal radius.

The mean stresses are good approximations for thin vessels where the variation through the thickness is small. In general, the stress distributions can be determined only by solution of a set of partial differential equations known as the *equilibrium equations,* with the imposition of boundary conditions, compatibility conditions, and known relationships between stresses and strains. However, for sufficiently thin walled vessels, $r \approx r_e \approx r_i$, and the stresses through the thickness may be approximated by

$$\begin{aligned} \bar{\sigma}_\theta &= p_i \frac{r}{h} \\ \bar{\sigma}_z &= \frac{F_T}{2\pi hr} + \frac{p_i}{2}\frac{r}{h} \end{aligned} \tag{16.22}$$

The validity of this so-called membrane assumption is dependent upon material properties and geometry. For example, when a tube of linear elastic material is pressurized internally and the strains are restricted to small deformations, variations in circumferential stress σ_θ across the wall are less than 5 percent for $r/h > 20$. When the material is nonlinear or the deformation is large, the variation in stress across the wall can be more severe (Mirsky, 1973; Doyle and Dobrin, 1973). For arteries it may be that the membrane assumption is somewhat limited. However, other factors must be considered. Arteries are not homogeneous, and the effect of this inhomogeneity is difficult to assess. The material properties may vary significantly as one traverses the wall for imbedded structures such as elastic lamellae and may result in localized stress concentrations, if the initial prestress is unknown. Though the true stress distribution may never be determined, the mean values are known and can be determined by using Eqs. (16.19) to (16.21).

When a straight cylindrical vessel is subjected to constant internal pressurization p_i and uniform longitudinal extension e, it is expected that the shearing stresses would be zero and that there would be no longitudinal or circumferential variation in the remaining three principal stresses, σ_r, σ_θ, and σ_z. These stresses would be functions of r alone. The internal stress field may then be determined, when the differential equation

$$\frac{1}{r}\frac{d}{dr}(r\sigma_r) - \frac{\sigma_\theta}{r} = 0 \tag{16.23}$$

is satisfied. This equilibrium equation relates the two stresses σ_θ and σ_r, and provides a means of assessing the unknown hydrostatic pressure p. Complete determination of the internal stress field also requires known stress-strain relationships for the material comprising the vessel wall and the boundary conditions. The condition of uniform longitudinal extension prescribes the longitudinal strain, and the internal pressurization results in the two boundary conditions for the radial stress:

$$\begin{aligned} \sigma_r(r_i, \mu) &= -p_i \\ \sigma_r(r_e, \mu) &= 0 \end{aligned} \tag{16.24}$$

16.4.4 Small-Strain Analysis

As an introduction, consider the deformation of a thick-walled, homogeneous, incompressible, isotropic, elastic tube subjected to internal pressurization p_i and constant longitudinal extension $1 + e$. When the deformations are small, the displacement $u = r - R$ is also small, and higher-order terms in the displacement gradients may be ignored. The infinitesimal strains associated with the axisymmetric deformation are approximated by

$$\varepsilon_r = \frac{du}{dr} \qquad \varepsilon_\theta = \frac{u}{r} \qquad \varepsilon_z = e \tag{16.25}$$

The stresses for the vessel are obtained from the following expressions:

$$\sigma_r = 2G\varepsilon_r - p \qquad \sigma_\theta = 2G\varepsilon_\theta - p \qquad \sigma_z = 2G\varepsilon_z - p \tag{16.26}$$

where G is the shear modulus and p is an additional hydrostatic pressure. The infinitesimal incompressibility condition is given by

$$\varepsilon_r + \varepsilon_\theta + \varepsilon_z = 0 \tag{16.27}$$

In general, p is a function of position, and that can only be determined when the stresses given in Eq. (16.26) and the boundary conditions given in Eq. (16.24) are used to satisfy the equilibrium equation, Eq. (16.23). This combination results in the inhomogeneous ordinary differential equation

$$\frac{d^2u}{dr^2} + \frac{1}{r}\frac{du}{dr} - \frac{u}{r^2} = + \frac{1}{2G}\frac{dp}{dr} \tag{16.28}$$

When the incompressibility constraint is satisfied, it is found that p is constant through the thickness. The resulting solution is

$$p = -Ge - \frac{p_i r_i^2}{r_e^2 - r_i^2} \qquad u = -\frac{e}{2}r + \frac{p_i}{2G}\frac{r_e^2 r_i^2}{r_e^2 - r_i^2}\frac{1}{r} \tag{16.29}$$

The strains are computed to be

$$\varepsilon_r = -\frac{e}{2} - \frac{p_i}{2G}\frac{r_e^2 r_i^2}{r_e^2 - r_i^2}\frac{1}{r^2} \qquad \varepsilon_\theta = -\frac{e}{2} + \frac{p_i}{2G}\frac{r_e^2 r_i^2}{r_e^2 - r_i^2}\frac{1}{r^2} \qquad \varepsilon_r = e \tag{16.30}$$

and the following stresses result:

$$\sigma_r = p_i \frac{r_i^2}{r_e^2 - r_i^2}\left(1 - \frac{r_e^2}{r^2}\right)$$

$$\sigma_\theta = p_i \frac{r_i^2}{r_e^2 - r_i^2}\left(1 + \frac{r_e^2}{r^2}\right) \tag{16.31}$$

$$\sigma_z = 3Ge + p_i \frac{r_i^2}{r_e^2 - r_i^2}$$

The first two stresses, σ_r and σ_θ, are identical to those determined for a compressible tube (Love, 1927), and the third is found as limiting case for a material approaching incompressibility. The limited applicability of these expressions for blood vessels is appreciated when one observes that the displacement u_r is a *linear* function of the internal pressure p_i and longitudinal extension e. This is clearly inconsistent with experimental data, as shown in Fig. 16.3, indicating that linear infinitesimal elasticity theory is not applicable.

16.4.5 Large-Strain Analysis

Large-strain solutions for the homogeneous, isotropic, elastic vessels can be obtained by following the approach of Chu and Oka (1973). In an isotropic elastic material there are three invariants of strain which are given by

$$I_1 = \lambda_r^2 + \lambda_\theta^2 + \lambda_z^2$$

$$I_2 = \lambda_r^2\lambda_\theta^2 + \lambda_\theta^2\lambda_z^2 + \lambda_z^2\lambda_\theta^2 \tag{16.32}$$

$$I_3 = \lambda_r^2\lambda_\theta^2\lambda_z^2$$

where λ_r, λ_θ, and λ_z are the principal extension ratios described previously. These strain invariants are measures of deformation which are completely independent of the orientation of the material. The vessel material is assumed to possess a strain energy density function W, which is a function of these invariants and is a potential for determining the stresses. For an incompressible, isotropic material, it has a general form

$$W = W(I_1, I_2) \tag{16.33}$$

since I_3 is constant. A material that is *linear* with regard to the two strain invariants was first proposed by Mooney (1940) to approximate the properties of rubberlike elastomers. It has the form

$$W = \frac{G}{2}[(I_1 - 3) + k(I_2 - 3)] \tag{16.34}$$

where G can be interpreted to be the shear modulus and k is a second material parameter. The stresses are determined by

$$\sigma_r = \lambda_r \frac{\partial W}{\partial \lambda_r} + p = G[1 + k(\lambda_\theta^2 + \lambda_z^2)]\lambda_r^2 + p$$

$$\sigma_\theta = \lambda_\theta \frac{\partial w}{\partial \lambda_\theta} + p = G[1 + k(\lambda_r^2 + \lambda_z^2)]\lambda_\theta^2 + p \tag{16.35}$$

$$\sigma_z = \lambda_z \frac{\partial w}{\partial \lambda_z} + p = G[1 + k(\lambda_r^2 + \lambda_\theta^2)]\lambda_z^2 + p$$

Again, an unknown hydrostatic pressure p has been introduced because of material incompressibility. When the parameter k is zero, the material is referred to as *neohookean,* and Eqs. (16.35) resemble Eqs. (16.26).

If we let $\lambda = \lambda_\theta$ and $\mu = \lambda_z$, the second expression in Eq. (16.15),

$$\lambda = \frac{r}{\sqrt{\mu(r^2 - c_2)}} \tag{16.36}$$

may be differentiated to allow transformation of the independent variable from r to λ by

$$r\frac{d\lambda}{dr} = \lambda - \mu\lambda^3 \tag{16.37}$$

The transformed equilibrium equation is then

$$(\lambda - \mu\lambda^3)\frac{d\sigma_r}{d\lambda} + \sigma_r - \sigma_\theta = 0 \tag{16.38}$$

Substitution of the expressions for σ_r and σ_θ into Eq. (16.38) results in the first-order ordinary differential equation

$$\frac{dp}{d\lambda} = -G(1 + k\mu^2)\left(\frac{1}{\mu\lambda} - \frac{1}{\mu^2\lambda^3}\right) \tag{16.39}$$

for p in terms of λ. In this case the hydrostatic pressure term can be integrated exactly to give

$$p = -G(1 + k\mu^2)\left(\frac{\log\lambda}{\mu} + \frac{1}{2\mu^2\lambda^2}\right) + c_0 \tag{16.40}$$

where c_0 is a constant that may be determined from the second boundary condition on σ_r. The final results can be written as

$$\begin{aligned}
\sigma_r &= G\left[\frac{1}{\lambda^2\mu^2} + k\left(\frac{1}{\mu^2} + \frac{1}{\lambda^2}\right) - (1 + k\mu^2)\left(\frac{\log\lambda}{\mu} + \frac{1}{2\mu^2\lambda^2}\right)\right] + c_0 \\
\sigma_\theta &= G\left[\lambda^2 + k\left(\frac{1}{\mu^2} + \lambda^2\mu^2\right) - (1 + k\mu^2)\left(\frac{\log\lambda}{\mu} + \frac{1}{2\mu^2\lambda^2}\right)\right] + c_0 \\
\sigma_z &= G\left[\mu^2 + k\left(\lambda^2\mu^2 + \frac{1}{\lambda^2}\right) - (1 + k\mu^2)\left(\frac{\log\lambda}{\mu} + \frac{1}{2\mu^2\lambda^2}\right)\right] + c_0
\end{aligned} \tag{16.41}$$

The constant c_0 may be eliminated by subtraction of the first from the second boundary condition on σ_r. This yields the relationship

$$p_i = \sigma_r(r_e, \mu) - \sigma_r(r_i, \mu) = p_i(r_e, \mu) \tag{16.42}$$

because r_i can be expressed in terms of r_e, R_e, R_i, and μ. Since p_i, r_e, and μ are quantities which can be measured directly, this equation is useful in the determination of material constants and allows the prediction of expected pressure-radius behavior. Since the radial stress has a low relative value when compared with the other two stresses, this argues against the exclusive use of an expression involving only that term in the determination of material constants. An alternate expression involving the longitudinal tether force F_T may be deduced from Eqs. (16.19) and (16.21),

$$F_T = 2\pi\int_{ri}^{re} \sigma_z r\, dr - p_i\pi r_i^2 \tag{16.43}$$

This is useful when measured values of longitudinal retractive force are available.

These mathematical models then were tested and compared using experimental data. Experiments were performed on cylindrical segments of arteries in vitro (Fig. 16.2) in order to obtain vessel dimensions and longitudinal retractive force values for a physiological range of pressures and vessel lengths. Results are shown in Fig. 16.9*a* and *b* for 24 relaxed dog carotid arteries at four lengths, $\mu = 1.2$, 1.4, 1.6, and 1.8. In each figure the lowest curve corresponds to $\mu = 1.2$, and the next curve up represents $\mu = 1.4$, up to $\mu = 1.8$. The data shown in these figures are empirical values. By contrast, Fig. 16.10*a* and *b* presents the predicted values for longitudinal retractive forces and vessel dimensions at each pressure based on a Mooney model. The disagreement between the observed values shown in Fig. 16.9*a* and *b* and the predicted values shown in Fig. 16.10*a* and *b* shows that the Mooney model is not very satisfactory. For example,

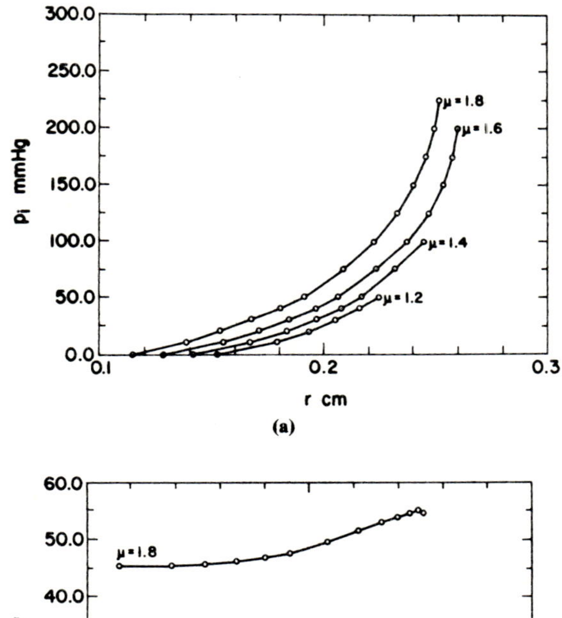

(a)

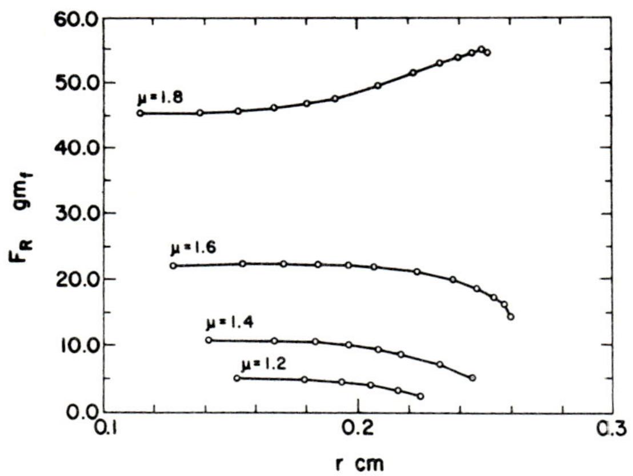

(b)

FIG. 16.9 (*a*) Pressure-radius data for 24 carotid arteries at various degrees of longitudinal extension μ. The initial dimensions in the unloaded fully retracted state were given by $R_e = 0.166$ cm and $R_e - R_i = 0.084$ cm. (*b*) Longitudinal distending force as a function of radius for 24 carotid arteries corresponding to (*a*).

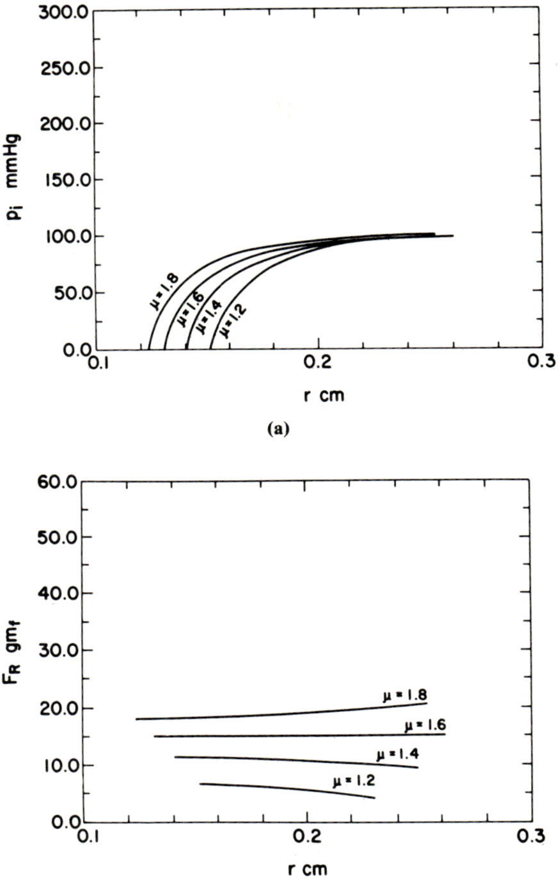

FIG. 16.10 (*a*) Pressure-radius curves predicted for a tube of Mooney material, Eq. (16.34), with $G = 30.0$ kPa and $k = 0.1$ at various degrees of longitudinal extension μ. (*b*) Longitudinal distending force as a function of radius corresponding to (*a*).

the carotid artery becomes less compliant when held at fixed length and subjected to increasing levels of internal pressure. In contrast, the Mooney tube becomes more compliant at a fixed length and subject to increasing levels of internal pressure. This is not a result of poor selection of material constants. The Mooney material exhibits strain softening when one principal extension ratio is held fixed.

More reasonable agreement can be achieved with an alternate form of an isotropic model. A strain energy density function of the exponential form

$$W = \frac{G_0}{2k_1} \exp\left[k_1(I_1 - 3) + k_2(I_2 - 3)\right] \tag{16.44}$$

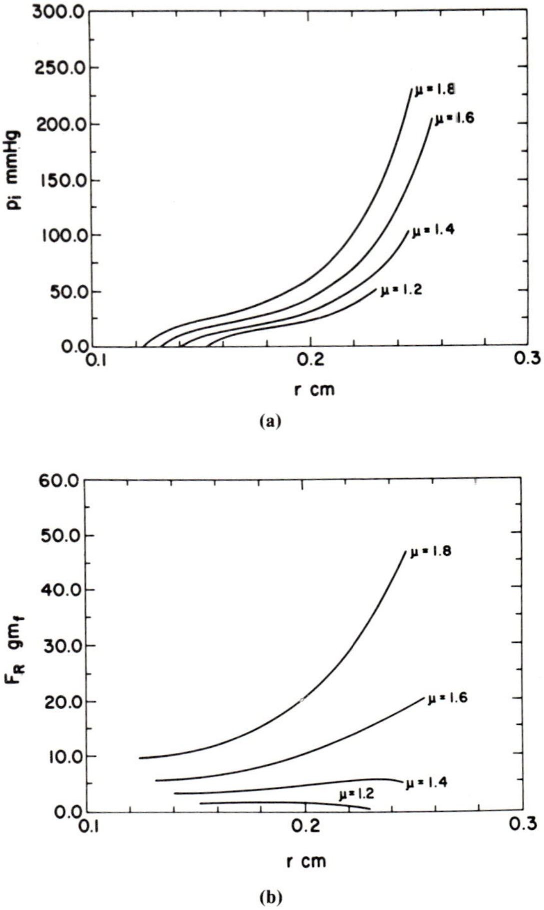

FIG. 16.11 (*a*) Pressure-radius curves predicted by using the isotropic exponential material, Eq. (16.44), with $G_0 = 0.42$ kPa, $k_1 = 0.6$, and $k_2 = 0.01$ at various degrees of longitudinal extension μ. (*b*) Longitudinal distending force as a function of radius corresponding to (*a*).

also may be employed. Figure 16.11*a* and *b* presents results obtained by using the experimental data given in Fig. 16.9*a* and *b* to obtain a strain energy function of an exponential form. Although the fit obtained does not involve a sophisticated minimization of the squared error terms, the general features of the experimental data are reproduced quite well. With this model, expressions similar to Eq. (16.41) may be used to predict stress distributions across the thickness of the wall for a homogeneous vessel. These are shown in Fig. 16.12*a* to *c* for pressure p_i = 20, 40, 60, 80, 100, and 120 mmHg at a longitudinal extension $\mu = 1.5$, i.e., approximately in vivo length. Finally, Figure 16.13 shows a comparison of the distributions in radial stress at 80 mmHg, i.e., at a typical diastolic pressure. The lowest curve is predicted by linear elasticity, the one immediately above it is the Mooney material, and the curved one above that is the exponential model.

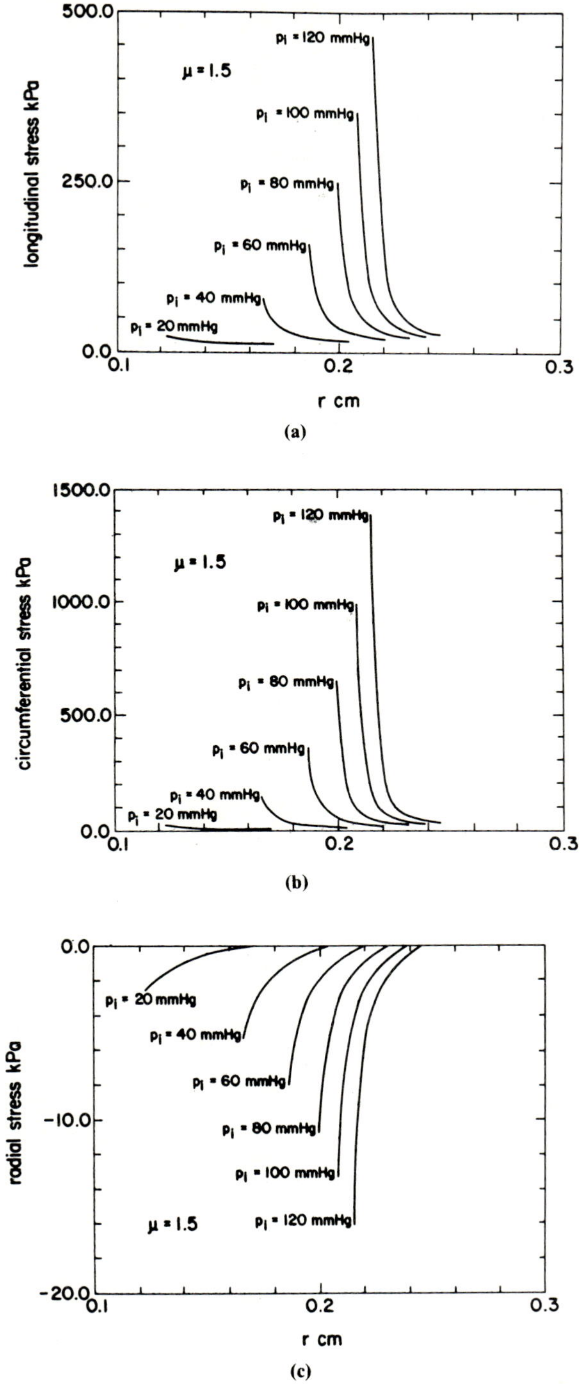

FIG. 16.12 (*a*) Longitudinal stress distributions across the thickness of the vessel wall predicted with the isotropic exponential material, Eq. (16.44), at in vivo length $\mu = 1.5$ and various internal pressures. (*b*) Circumferential stress distributions. (*c*) Radial stress distributions.

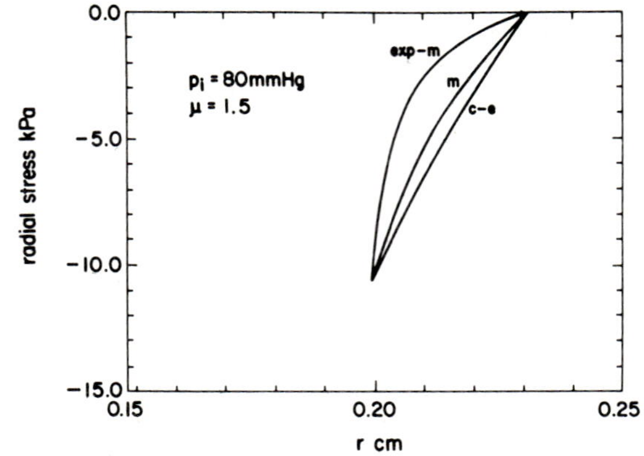

FIG. 16.13 A comparison of radial stress distributions across the thickness of the vessel wall predicted with classical elasticity ($c - e$), for the Mooney material (m) and the exponential material ($\exp - m$).

16.4.6 Incremental Elasticity of Blood Vessels

An alternate approach to vascular mechanics has been to apply infinitesimal elasticity theory to local or small incremental deformations. Because large arteries undergo only small changes in dimension with each cardiac cycle, the degree of local nonlinear behavior should be small and could be approximated with locally linear elastic theory. Small incremental changes in vessel strain would result in proportional incremental changes in stress. The proportionality factors or elastic parameters would not be constants but would, in fact, be functions of the current state of stress and strain. With this incremental approach one would retain some of the familiar nomenclature of infinitesimal elasticity theory. The elastic parameters could be expressed as *incremental elastic moduli* and *incremental Poisson's ratios*. Discussion of changes in the measured values of these parameters for blood vessels in various degrees of deformation and vascular muscle excitation have been the subject of numerous articles in the literature on physiology (Bergel, 1961; Dobrin and Doyle, 1970; Cox, 1975).

The following presents some of the analytic aspects of incremental elasticity relevant to the study of blood vessel mechanics. In it an increment in the quadratic strain is introduced to facilitate presentation of the method. This choice is not unique. One might just as well use any other rigorously defined measure of strain and obtain similar results. However, practical comparisons of derived elastic moduli and Poisson's ratios should be based on the single method. There are relationships between the various methods, but they will not be discussed here. A detailed account of the relationship between finite strains, incremental strain formulations, and various moduli may be found in the recent work of Vaishnav and Vassoughi (1984).

Let the initial and final states of an incremental deformation of blood vessel be denoted by 1 and 2, respectively. The final stress and strain states are determined by the incremental equations

$$
\begin{aligned}
\sigma_r^{(2)} &= \sigma_r^{(1)} + \Delta\sigma_r \\
\sigma_\theta^{(2)} &= \sigma_\theta^{(1)} + \Delta\sigma_\theta \\
\sigma_z^{(2)} &= \sigma_z^{(1)} + \Delta\sigma_z \\
e_r^{(2)} &= e_r^{(1)} + \Delta e_r \\
e_\theta^{(2)} &= e_\theta^{(1)} + \Delta e_\theta \\
e_z^{(2)} &= e_z^{(1)} + \Delta e_z
\end{aligned}
\qquad (16.45)
$$

in the absence of rotations. For a material that exhibits isotropy, the incremental changes in strain are related to the incremental changes in stress by the set of equations

$$\Delta e_r = \frac{1}{E}(\Delta\sigma_r - \nu\,\Delta\sigma_\theta - \nu\,\Delta\sigma_z)$$
$$\Delta e_\theta = \frac{1}{E}(-\nu\,\Delta\sigma_r + \Delta\sigma_\theta - \nu\,\Delta\sigma_z) \tag{16.46}$$
$$\Delta e_z = \frac{1}{E}(-\nu\,\Delta\sigma_r - \nu\,\Delta\sigma_\theta + \Delta\sigma_z)$$

where E is the elastic modulus and ν is Poisson's ratio. These have limited usefulness for blood vessels because of their isotropy. Even materials that are initially isotropic undergo strain-induced anisotropy that is not addressed by these equations. A more general set of equations having a similar form which allows cylindrical orthotropy is

$$\Delta e_r = \frac{1}{E_r}(\Delta\sigma_r - \nu_{r\theta}\,\Delta\sigma_\theta - \nu_{rz}\,\Delta\sigma_z)$$
$$\Delta e_\theta = \frac{1}{E_\theta}(-\nu_{\theta r}\,\Delta\sigma_r + \Delta\sigma_\theta - \nu_{\theta z}\,\Delta\sigma_z) \tag{16.47}$$
$$\Delta e_z = \frac{1}{E_z}(-\nu_{zr}\,\Delta\sigma_r - \nu_{z\theta}\,\Delta\sigma_\theta + \Delta\sigma_z)$$

In the above expressions there are nine elastic parameters: the three elastic moduli, E_r, E_θ, and E_z, and six Poisson's ratios, $\nu_{r\theta}$, ν_{rz}, $\nu_{\theta r}$, $\nu_{\theta z}$, ν_{zr}, and $\nu_{z\theta}$. Only six of the parameters are independent because of the symmetry relationships

$$\frac{\nu_{r\theta}}{E_r} = \frac{\nu_{\theta r}}{E_\theta}$$
$$\frac{\nu_{rz}}{E_r} = \frac{\nu_{zr}}{E_z} \tag{16.48}$$
$$\frac{\nu_{\theta z}}{E_\theta} = \frac{\nu_{z\theta}}{E_z}$$

The additional condition of incompressibility,

$$\Delta e_r + \Delta e_\theta + \Delta e_z = 0 \tag{16.49}$$

may also be imposed to further reduce the number of parameters. This equation is always true when increments in natural strains are employed. Since the stress increments are assumed to be independent, incompressibility may be invoked in three orthogonal cases to yield

$$1 - \nu_{r\theta} - \nu_{rz} = 0$$
$$1 - \nu_{\theta r} - \nu_{\theta z} = 0 \tag{16.50}$$
$$1 - \nu_{zr} - \nu_{z\theta} = 0$$

When these are coupled with the symmetry relations, it is concluded that there are only three independent parameters characterizing vessel orthotropy. It is convenient to choose the three elastic moduli, E_r, E_θ, and E_z. All the Poisson's ratios may then be expressed in terms of these parameters, that is,

$$
\begin{aligned}
\nu_{r\theta} &= \frac{1}{2}\left(1 - \frac{E_r}{E_z} + \frac{E_r}{E_\theta}\right) \\
\nu_{\theta r} &= \frac{1}{2}\left(1 - \frac{E_\theta}{E_z} + \frac{E_\theta}{E_r}\right) \\
\nu_{rz} &= \frac{1}{2}\left(1 - \frac{E_r}{E_\theta} + \frac{E_r}{E_z}\right) \\
\nu_{zr} &= \frac{1}{2}\left(1 - \frac{E_z}{E_\theta} + \frac{E_z}{E_r}\right) \\
\nu_{\theta z} &= \frac{1}{2}\left(1 - \frac{E_\theta}{E_r} + \frac{E_\theta}{E_z}\right) \\
\nu_{r\theta} &= \frac{1}{2}\left(1 - \frac{E_z}{E_r} + \frac{E_z}{E_\theta}\right)
\end{aligned} \tag{16.51}
$$

Studies have been performed in which the radial stress component has been ignored because of its small relative magnitude. In this case one need only consider a reduced set of equations (Dobrin and Doyle, 1970),

$$
\begin{aligned}
\Delta e_\theta &\approx \frac{1}{E_\theta}\left(\Delta\sigma_\theta - \nu_{\theta z}\,\Delta\sigma_z\right) \\
\Delta e_z &\approx \frac{1}{E_z}\left(-\nu_{z\theta}\,\Delta\sigma_\theta + \Delta\sigma_z\right)
\end{aligned} \tag{16.52}
$$

The condition of symmetry still holds; hence

$$\frac{\nu_{\theta z}}{E_\theta} = \frac{\nu_{z\theta}}{E_z} \tag{16.53}$$

The two elastic moduli can be measured independently, and it follows from Eq. (16.52) that

$$\nu_{z\theta} = \frac{\Delta\sigma_z}{\Delta\sigma_\theta} \tag{16.54}$$

The three elastic material parameters determined by this technique were sufficient to characterize these orthotropic tubes. Dog carotid arteries were studied as cylinders in vitro (Fig. 16.2). They were subjected to distending pressure while held at a single in situ length. Values computed for Poisson's ratios [Eq. (16.54)] using these data are shown in Fig. 16.14. These data show that the Poisson's ratio is about 0.3, not 0.5 as would be predicted for an isotropic material. These data also show that, in most cases, excitation of the vascular muscle did not greatly alter the value for the Poisson's ratio.

16.4.7 Finite Strain Orthotropy of Blood Vessels

It is well established that blood vessels exhibit anisotropic mechanical behavior (Apter, 1970; Dobrin and Doyle, 1970; Patel et al., 1969) and that this anisotropy is due in part to the vascular structure (observe Fig. 16.15). The alternate layering of concentric cylinders of elastic lamellae and smooth muscle cells within the media suggests one degree of material anisotropy. In addition, orientation of smooth muscle and the means of interconnection between fibers of elastin and collagen may also introduce anisotropy. Incremental methods are able to treat this type of anisotropy, but the equations must still be integrated. This can result in difficulty in the determination of a total energy integral that is both single-valued and path-independent. For a

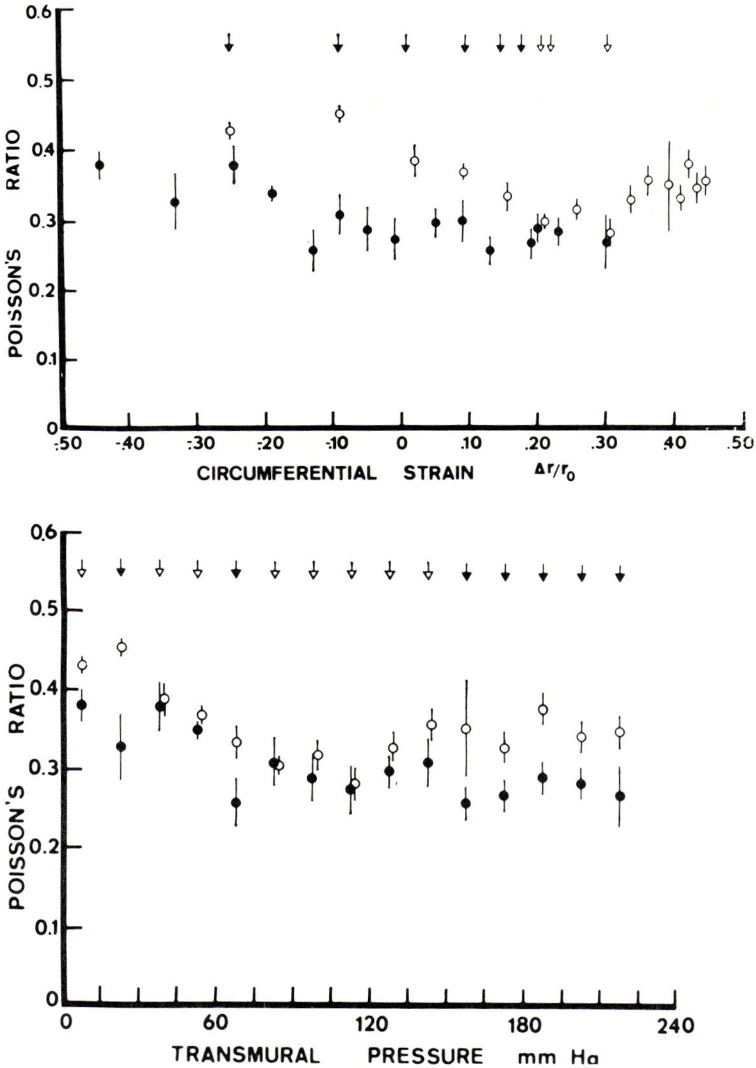

FIG. 16.14 Poisson's ratio computed with respect to circumferential strain and pressure in both relaxed (open symbols) and contracted (closed symbols) dog carotid arteries. Closed arrows denote statistical difference $P < 0.05$ between relaxed and contracted vessels; open arrows denote no statistical difference. (*From Dobrin and Doyle, 1970. By permission of the American Heart Association.*)

vessel to exhibit true elasticity, it must possess a strain energy density function under either isothermal or adiabatic conditions. If this potential exists, the stress-strain relationships are a form of its total derivative. Uniqueness may be satisfied when stresses are expressed as functions of strains, and these are the types of relationships usually obtained from experimental data. These relationships should be integrable, and there are general restrictions that should be imposed which would affect their form and that of the elastic potential. This potential should be

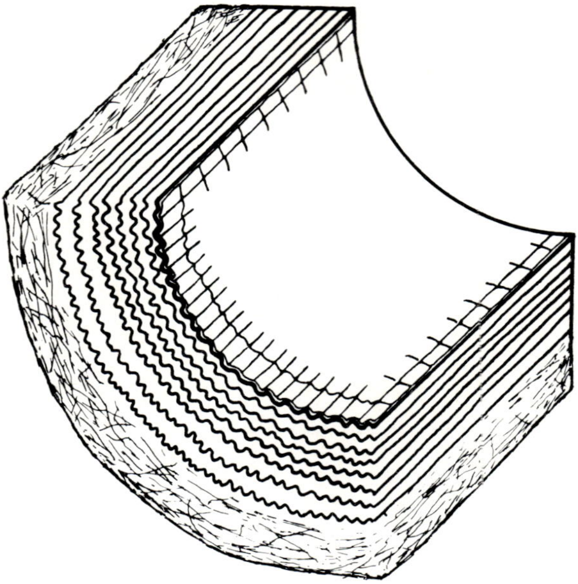

FIG. 16.15 Cross section of a blood vessel showing the layered structure of elastic lamellae with intervening smooth muscle cells in the media. The outer third of the vessel, the adventitia, is composed largely of disorganized collagen fibers.

positive definite and should allow a stress-free state. Also deformation from this stress-free state should not result in the release of energy (Truesdell and Toupin, 1960). The latter conditions are difficult to justify since the stress-free state may never exist in blood vessels. However, a vessel free of all external tractions is in a state that minimizes its internal energy.

Patel and Fry (1969) demonstrated that shear strains were small and could be ignored when cylindrical segments of arteries were subjected to internal pressurization and longitudinal extension. Thus, vessels could be treated as orthotropic. As a result, a partial strain energy density function could be constructed in the form

$$W = W(\lambda_r, \lambda_\theta, \lambda_z) \tag{16.55}$$

Of course, incompressibility could be added as an additional constraint and the stresses determined by

$$\begin{aligned} \sigma_r &= \lambda_r \frac{\partial W}{\partial \lambda_r} + p \\ \sigma_\theta &= \lambda_\theta \frac{\partial w}{\partial \lambda_\theta} + p \\ \sigma_z &= \lambda_z \frac{\partial w}{\partial \lambda_z} + p \end{aligned} \tag{16.56}$$

where p is the additional hydrostatic pressure term. This partial strain energy density function may also be expressed in the alternate form

$$W = W(e_r, e_\theta, e_z) \tag{16.57}$$

Doyle and Dobrin (1971) obtained exponential relationships to describe the membrane stresses for the dog carotid artery at in vivo lengths,

$$\bar{\sigma}_\theta = A\left(1 - \frac{\lambda_0}{\bar{\lambda}_\theta}\right) e^{K\bar{\lambda}_\theta}$$
$$\bar{\sigma}_z = Be^{C\bar{\lambda}_\theta^{-3}} \tag{16.58}$$

where A, B, and K are constants for the particular extension. Using these expressions, they were able to deduce that the carotid exhibits true material anisotropy. A convincing piece of evidence is that the strain invariants, I_1 and I_2, may assume the same values at two different radii, but the stored energy increases monotonically with increased mean circumferential extension ratio $\bar{\lambda}_\theta$. Doyle and Dobrin (1973) were able to use their expression to compute stress distributions by assuming that the vessel was homogeneous and that the average material properties held throughout the wall. Hayashi et al. (1974) and Sato et al. (1979) also suggested exponential forms for both the circumferential and longitudinal membrane stresses in blood vessels. Hayashi et al. (1975) proposed a symmetric form which is equivalent to

$$\bar{\sigma}_\theta - p = (Ae_z + 1)e^{Be_\theta} - 1$$
$$\bar{\sigma}_z - p = (Ce_\theta + 1)e^{De_z} - 1 \tag{16.59}$$

where e_θ and e_z are quadratic measures of strain and A, B, C, and D are material constants. From these expressions and those of Doyle and Dobrin (1973), one cannot infer a total differential since additional terms are required. A possible form of a strain energy density function that has no explicit dependence upon e_r and that would lead to stress-strain relations similar in form to Eqs. (16.59) is

$$W = (Ae_z + 1)\int_0^{e_\theta} \frac{e^{B\psi}}{2\psi + 1}\,d\psi + (Ce_\theta + 1)\int_0^{e_z} \frac{e^{D\eta}}{2\eta + 1}\,d\eta$$
$$- \tfrac{1}{2}\log(2e_z + 1) - \tfrac{1}{2}\log(2e_\theta + 1) \tag{16.60}$$

The straightforward, but less intuitive, approach is to assume a form of the strain energy density function and derive expressions for the stress-strain relationships for comparison with experimental data. Successful attempts have been made to model the elastic properties of blood vessels with this method. In these studies, only a membrane approximation was sought, and it was assumed that

$$W = W(e_\theta, e_z) \tag{16.61}$$

No explicit dependence of W on e_r was assumed because of incompressibility. In the earlier study the strain energy density function was assumed to be in the form of a truncated Taylor series in the strains (Vaishnav et al., 1972)

$$W = Ae_\theta^2 + Be_\theta e_z + Ce_z^2 + De_\theta^3 + Ee_\theta^2 e_z + Fe_\theta e_z^2$$
$$+ Ge_z^3 + He_\theta^4 + Ie_\theta^3 e_z + Je_\theta^2 e_z^2 + Ke_\theta e_z^3 + Le_z^4 + \cdots \tag{16.62}$$

where A, B, C, etc., are all material constants. They compared results obtained when the series was truncated to the second, third, and fourth orders. Higher-order polynomial approximations gave more accurate agreement with the experimental data; however, it was felt that the improvement in accuracy offered by increasing the degree of the polynomial from third to fourth order was only slight and did not merit the additional computational investment. More recently, Fung and colleagues (1979) chose an exponential representation with a quadratic polynomial in strain as its argument,

$$W = \frac{C}{2}\exp\left[a_1(e_\theta^2 - e_\theta^{*2}) + a_2(e_z^2 - e_z^{*2}) + 2a_4(e_\theta e_z - e_\theta^* e_z^*)\right] \tag{16.63}$$

where C, a_1, a_2, and a_4 are material constants. In addition, they introduced what they called *standard strains* [indicated by the asterisk (*)] which correspond to standard reference stresses. These stresses can be chosen arbitrarily to facilitate determination of material constants. They chose values equivalent to those applicable to in vivo conditions, i.e., in situ length and diameters observed since these are easily reproduced in the laboratory. Fung (1972) has long advocated exponential stress-strain relationships for biological tissue because of the approximately linear relationships usually found between tissue stiffness and stress. Large blood vessels, such as the carotid and iliac arteries, exhibit this strong stiffening response to increased levels of stress. For this reason exponential forms have often been used successfully to model their behavior.

REFERENCES

Alexander, R. S. (1971). Contribution of plastoelasticity to the tone of the cat portal vein. *Circ. Res., 28,* 461–469.

Apter, J. T., Marquez, E., and Janas, M. (1970). Dynamic viscoelastic anisotropy of canine aorta correlated with aortic wall composition. *J. Assoc. Adv. Med. Inst., 4,* 15–21.

Band, W., Goedhard, W. J. A., and Knoop, A. A. (1972). Effect of aging on dynamic viscoelastic properties of the rat's thoracic aorta. *Pfluegers Arch., 331,* 357–364.

Bergel, D. H. (1961). The static elastic properties of the arterial wall. *J. Physiol. London, 156,* 445–457.

Browse, N. L., Young, A. W., and Thomas, M. L. (1979). The effect of bending on canine and human arterial walls and on blood flow. *Circ. Res., 45,* 41–48.

Carew, T. E., Vaishnav, R. N., and Patel, D. J. (1968). Compressibility of the arterial wall. *Circ. Res., 23,* 61–68.

Carton, R. W., Dainauskas, J., and Clark, J. W. (1962). Elastic properties of single elastic fibers. *J. Appl. Physiol., 17,* 547–551.

Chu, B. M., and Oka, S. (1973). Influence of longitudinal tethering on the tension in thick-walled blood vessels in equilibrium. *Biorheology, 10,* 517–525.

Clark, J. M., and Glagov, S. (1979). Structural integration of the arterial wall: I. Relationships and attachments of medial smooth muscle cells in normally distended and hyperdistended aortas. *Lab. Invest., 40,* 587–602.

Cox, R. H. (1975). Anisotropic properties of the canine carotid artery *in vitro*. *J. Biomech., 8,* 293–300.

Cox, R. H. (1977). Determination of series elasticity in arterial smooth muscle. *Am. J. Physiol.,* 233, H248–H255.

Cox, R. H. (1978). Passive mechanics and connective tissue composition of canine arteries. *Am. J. Physiol., 234,* H533–H541.

Dillon, P. F., Aksoy, M. O., Driska, S. P., and Murphy, R. A. (1981). Myosin phosphorylation and the cross-bridge cycle in arterial smooth muscle. *Science, 211,* 495–497.

Dobrin, P. B. (1973). Isometric and isobaric contraction of carotid arterial smooth muscle. *Am. J. Physiol., 225,* 659–663.

Dobrin, P. B. (1974). Vascular muscle series elastic element stiffness during isometric contraction. *Circ. Res., 34,* 242–250.

Dobrin, P. B. (1978). Mechanical properties of arteries. *Physiol. Rev., 58,* 397–460.

Dobrin, P. B. (1983). Vascular mechanics. In J. T. Shepherd and F. Abboud (eds.), *Handbook of Physiology,* vol. 3: *The Cardiovascular System.* American Physiological Society, Washington, D.C., sec. 2, pp. 65–104.

Dobrin, P. B., Baker, W. H., and Gley, W. C. (1984). Elastin, collagen and the mechanical behavior of arterial aneurysms. *Arch. Surg., 119,* 405–409.

Dobrin, P. B., and Canfield, T. R. (1973). Series elastic and contractile elements in vascular smooth muscle. *Circ. Res., 33,* 454–464.

Dobrin, P. B., and Canfield, T. R. (1977). Identification of smooth muscle series elastic component in intact carotid artery. *Am. J. Physiol., 232,* H122–H130.

Dobrin, P. B., Canfield, T., and Sinha, S. (1975). Development of longitudinal retraction of carotid arteries in neonatal dogs. *Experientia, 31,* 1295–1296.

Dobrin, P. B., and Doyle, J. M. (1970). Vascular smooth muscle and the anisotropy of dog carotid artery. *Circ. Res., 27,* 105–119.

Dobrin, P. B., and Rovick, A. A. (1969). Influence of vascular smooth muscle on contractile mechanics and elasticity of arteries. *Am. J. Physiol., 217,* 1644–1652.

Doyle, J. M., and Dobrin, P. B. (1971). Finite deformation of the relaxed and contracted dog carotid artery. *Microvasc. Res., 3,* 400–415.

Doyle, J. M., and Dobrin, P. B. (1973). Stress gradients in the walls of large arteries. *J. Biomech., 16,* 631–639.

Fischer, G. M., and Llaurado, J. G. (1966). Collagen and elastin content in canine arteries selected from functionally different vascular beds. *Circ. Res., 19,* 394–399.

Fung, Y. C., Fronek, K., and Patitucci, P. (1979). Pseudoelasticity of arteries and the choice of its mathematical expression. *Am. J. Physiol., 237,* H620–631.

Gerová, M., and Gero, J. (1967). Reflex regulation of smooth muscle tone of conduit vessel. *Angiologica, 4,* 348–358.

Goldstick, T. K., and Dobrin, P. B. (1987). Arterial wall transport and its relationship to atherogenesis. In R. Skalak and S. Chien (eds.), *Handbook of Bioengineering.* McGraw-Hill, New York.

Green, A. E., and Adkins, J. E. (1970). *Large Elastic Deformations* (2d ed.). Clarendon Press, Oxford.

Hayashi, K., Sato, M., Handa, H., and Moritake, K. (1974). Biomechanical study of the constitutive laws of vascular walls. *Exp. Mech., 14,* 400–404.

Hayashi, K., Sato, M., Niimi, H., Handa, H., Moritake, K., and Okumura, A. (1975). Analysis of constitutive laws of vascular walls by finite deformation theory. *Jpn. J. Med. Elec. Biol. Eng., 13,* 293–298.

Heistad, D. D., and Marcus, M. L. (1979). Role of vasa vasorum in nourishment of the aorta blood vessels. *16,* 225–238.

Herlihy, J. T., and Murphy, R. A. (1974). Force-velocity and series elastic characteristics of smooth muscle from the hog carotid artery. *Circ. Res., 34,* 461–466.

Ho, K-J., Lin, C-Y., Gaylash, F. T., Patel, A. S., Ling, L. B., and Taylor, C. B., (1972). Aortic compliance: Studies on its relationship to aortic constituents. *Arch. Pathol., 94,* 537–546.

Krafka, J., (1939). Comparative study of the histophysics of the aorta, *Am J. Physiol., 125,* 1–14.

Lawton, R. W. (1957). Some aspects of research in biological elasticity: Introductory remarks. In R. W. Remington (ed.), *Tissue Elasticity.* American Physiological Society, Washington, D.C., pp. 1–11.

Learoyd, B. M., and Taylor, M. G. (1966). Alterations with age in the viscoelastic properties of human arterial walls. *Circ. Res., 18,* 278–292.

Love, A. E. H. (1927). *A Treatise on the Mathematical Theory of Elasticity* (4th ed.). Cambridge University Press, Cambridge.

Mirsky, I. (1973). Ventricular and arterial wall stresses, based on large deformation analysis. *Biophys. J., 13,* 1141–1159.

Mooney, M. (1940). A theory of large elastic deformation. *J. Appl. Physiol., 11,* 582–592.

Patel, D. J., and Fry, D. L. (1964). In situ pressure-radius-length measurements in ascending aorta of anesthetized dogs. *J. Appl. Physiol. 19,* 413–416.

Patel, D. J., and Fry, D. L. (1966). Longitudinal tethering of arteries in dogs. *Circ. Res., 19,* 1011–1021.

Patel, D. J., and Fry, D. L. (1969). The elastic symmetry of arterial segments in dogs. *Circ. Res., 24,* 1–8.

Patel, D. J., Greenfield, J. C., and Fry, D. L. (1964). In vivo pressure-length-radius relationship of certain blood vessels in man and dog. In E. O. Attinger (ed.), *Pulsatile Blood Flow.* McGraw-Hill, New York, pp. 293–306.

Patel, D. J., Janicki, J. S., and Carew, T. E. (1969). Static anisotropic elastic properties of the aorta in living dogs. *Circ. Res., 25,* 765–779.

Patel, D. J., Tucker, W. K., and Janicki, J. S. (1970). Dynamic elastic properties of the aorta in the radial direction. *J. Appl. Physiol., 28,* 578–582.

Patel, D. J., and Vaishnav, R. N. (1980). *Hemodynamics and Its Role in Disease Processes.* University Park Press, Baltimore.

Peiper, U., Klemt, P., and Schleupner, R. (1978). The temperature dependence of parallel and series elastic elements in the vascular smooth muscle of the rat portal vein. *Pfluegers Arch., 378,* 25–30.

Remington, J. W. (1955). Hysteresis loop behavior of the aorta and other extensible tissues. *Am. J. Physiol., 180,* 83–95.

Remington, J. W., Hamilton, W. F., and Dow, P. (1945). Some difficulties involved in the prediction of the stroke volume from the pulse wave velocity. *Am. J. Physiol., 144,* 536–545.

Reuterwall, O. P. (1921). Über die Elastizitat der Gefaswond und die Methode ihrer naheren Prufung. *Acta Med. Scand. Suppl. 2,* 1–175.

Roach, M. R., and Burton, A. C. (1957). The reason for the shape of the distensibility curves of arteries. *Can. J. Biochem. Physiol., 35,* 681–690.

Roy, C. S. (1880). The elastic properties of the arterial wall. *J. Physiol. London, 3,* 125–129.

Sato, M., Hayashi, K., Niimi, H., Moritake, K., Okumura, A., and Handa, H. (1979). Axial mechanical properties of arterial walls and their anisotropy. *Med. Biol. Eng. Comput., 17,* 170–176.

Seigman, M. J., Batter, T. M., Mooers, S. U., and Davies, R. E. (1976). Calcium-dependent resistance to stretch and stress relaxation in resting smooth muscles. *Am. J. Physiol., 231,* 1051–1058.

Sobin, S., and Tremer, H. M. (1980). Cylindricity of the arterial tree in the dog and cat. *Fed. Proc., 39,* 269.

Stromberg, D. D., and Wiederhielm, C. A. (1969). Viscoelastic description of a collagenous tissue in simple elongation. *J. Appl. Physiol., 26,* 857–862.

Sumner, D. S., Hokanson, D. E., and Strandness, D. E., Jr. (1970). Stress-strain characteristics and collagen-elastin content of abdominal aortic aneurysms. *Surg. Gynecol. Obstet., 130,* 459–466.

Truesdell, C., and Toupin, R. (1960). The classical field theories. In S. Flugge (ed.), *Handbuch der Physik.* Springer-Verlag, Berlin, vol. 3, pp. 226–858.

Vaishnav, R. N., and Vassoughi, J. (1984). Incremental formulations in vascular mechanics. *J. Biomech. Eng., 106,* 105–111.

Vaishnav, R. N., Young, J. T., Janicki, J. S., and Patel, D. J. (1972). Nonlinear anisotropic elastic properties of the canine aorta. *Biophys. J., 12,* 1008–1027.

Wolinsky, H., and Glagov, S. (1964). Structural basis for the static mechanical properties of the aortic media. *Circ. Res., 14,* 400–413.

Wolinsky, H., and Glagov, S. (1967). A lamellar unit of aortic medial structure and function in mammals. *Circ. Res., 20,* 99–111.

CHAPTER 17

Nonlinear Mathematical Models of the Arterial System: Effects of Bifurcations, Wall Viscoelasticity, Stenoses, and Counterpulsation on Pressure and Flow Pulses

J. C. Stettler
P. Niederer
M. Anliker
Institute of Biomedical Engineering of the University of Zurich and the Swiss Federal Institute of Technology
Zurich, Switzerland

INTRODUCTION

The pressure and flow pulses generated by the intermittent ejection of blood from the left ventricle of the heart undergo characteristic nonlinear shape changes as they propagate along the different branches of the arterial tree [1, 2]. These changes are primarily due to the geometrical arrangement and the taper of the arteries, as well as the mechanical properties of the vessel wall and the blood [3]. Accordingly, disease or aging of the arterial conduits may noticeably influence the propagation characteristics of the pulses.

Modern ultrasound technology [4, 5] allows one to quantify noninvasively the flow patterns in larger blood vessels within a range of about 10 cm from the skin. With the aid of appropriately designed ultrasonic instruments, it is therefore possible to detect vascular diseases in various parts of the circulatory system [6–8, 68]. While the diagnostic significance of these methods is widely recognized today, they lead to a rather empirical understanding of basic hemodynamics. It is the task of a mathematical analysis of arterial blood flow to provide a physically sound basis to interpret the observed phenomena. Accordingly, a number of theoretical studies of pulse propagation in arteries have been carried out in order to gain some insight into the degree to which the different properties of the arterial system are responsible for the characteristic features of the pressure and flow pulses. [3, 9–17].

A lack of anatomical knowledge as well as of computational power prevents the formulation of a comprehensive mathematical model of the entire arterial tree, including three-dimensional pulsatile flow. However, the unidirectional nature of blood flow justifies the attempt to apply a one-dimensional approximation to long arterial conduits. In some of the early models [3, 11] even discrete side branches at various locations in an arterial conduit were simulated within the framework of a one-dimensional model by a suitably chosen leakage of the arterial wall which varied appropriately with pressure, flow, and distance from the heart. In part, satisfactory results were obtained with these models. However, from an anatomical and fluid-dynamic point of view, they were not sufficiently good approximations of reality.

In order to better approximate the pertinent features of the arterial tree in normal as well as in pathological situations, it was found to be advantageous to divide the arterial system into segments which are individually modeled in terms of standardized elements representing tapered tubes, bifurcation regions, and pathological sections, respectively. The feasibility of such an approach is assessed by a comparison of the predicted flow patterns in healthy and pathological cases with ultrasound Doppler measurements.

The requirement is that the blood be treated as an incompressible Newtonian fluid and the flow as one-dimensional. This implies that only spatial averages of the pressure and velocity at a given cross section are calculated. Small curvatures of the blood vessels can be neglected. The numerical solution utilized for the long segments is based on the method of characteristics using specified time intervals [12, 19, 20]. Numerical solutions utilizing the finite differences method have been discussed by Gerrard and Taylor [21].

One-dimensional models with discrete branches have been analyzed by various authors. For example, Clark and Kufahl [14] considered the canine brain perfusion, Hillen et al. [17] developed a model for the flow in the posterior communicating arteries, while Raines et al. [22] investigated the pulse propagation in the human leg. A model for the systemic arterial pulse wave transmission using simplified idealized branch conditions was presented by Schaaf and Abbrecht [9]. Finally, two-dimensional models of the flow through tapered arterial segments were presented by Ling and Atabek [23], Kimmel and Dinnar [24], and Imaeda and Goodman [25].

A recent comparison of predicted and measured flow patterns [13, 26] has indicated that the viscoelastic damping properties of arteries [27–31] should be realistically included in the mathematical model if accurate theoretical results are to be obtained for greater distances from the heart.

Viscoelastic damping mechanisms should also be included if a comprehensive analysis of shock formation is undertaken [32, 43]. Such shock transitions can occur in the mammalian circulation in cases of severe aortic valve insufficiency and have been investigated theoretically by Anliker et al. [3].

Whenever the Navier-Stokes equations are specialized to one-dimensional tube flow, an assumption has to be made regarding the velocity profile. The resulting expression for the viscous drag essentially depends on this assumption. In arterial pulse propagation models, a Poiseuille approximation is usually made which, under certain conditions, may lead to a severe underestimation of viscous damping [38, 40].

17.1 MATHEMATICAL MODEL OF THE ARTERIAL SYSTEM

17.1.1 Geometrical Approximations

The anatomical features of arterial conduits suggest that a mathematical model be built upon the basis of three different types of elements:

1. Segments with no or only insignificant side branches
2. Bifurcation regions
3. Segments with pathological features

Each of these elements requires different fluid-dynamic approximations which will be described in the following sections.

Segments with No or Only Insignificant Side Branches. In 1775, Euler [33] developed the inviscid hydrodynamic equations for one-dimensional flow in elastic tubes. Later on, these equations were expanded to include frictional effects as well as seepage through the wall [34].

$$\frac{\partial S}{\partial p}\frac{\partial p}{\partial z}v + \frac{\partial S}{\partial z}v + S\frac{\partial v}{\partial z} + \psi + \frac{\partial S}{\partial p}\frac{\partial p}{\partial t} = 0 \tag{17.1}$$

$$\frac{\partial v}{\partial t} + v\frac{\partial v}{\partial z} + \frac{1}{\rho}\frac{\partial p}{\partial z} - \frac{f}{S\rho} = 0 \tag{17.2}$$

In these equations, $v(z, t)$ and $p\,(z, t)$ are the spatially averaged velocity and pressure, z is the axial coordinate (usually identical with the distance from the heart), and t denotes the time (Fig. 17.1*a*). ρ denotes the density of the blood, f is an approximate expression for the frictional force due to blood viscosity, and $\psi(p, v, z)$ simulates a continuous pressure and flow-dependent seepage.

The skin friction f is specified in a simplified manner [3, 12]:

$$f = -8\pi\nu v \tag{17.3}$$

Equation (17.3) assumes a parabolic velocity profile at every instant. This is a crude approximation of the real flow conditions, as has been shown by Witzig [35], Paulsen and Hasenkam [36], Kimmel and Dinnar [24], and, for pulsatile flow in curved tubes, Chandran et al. [37].

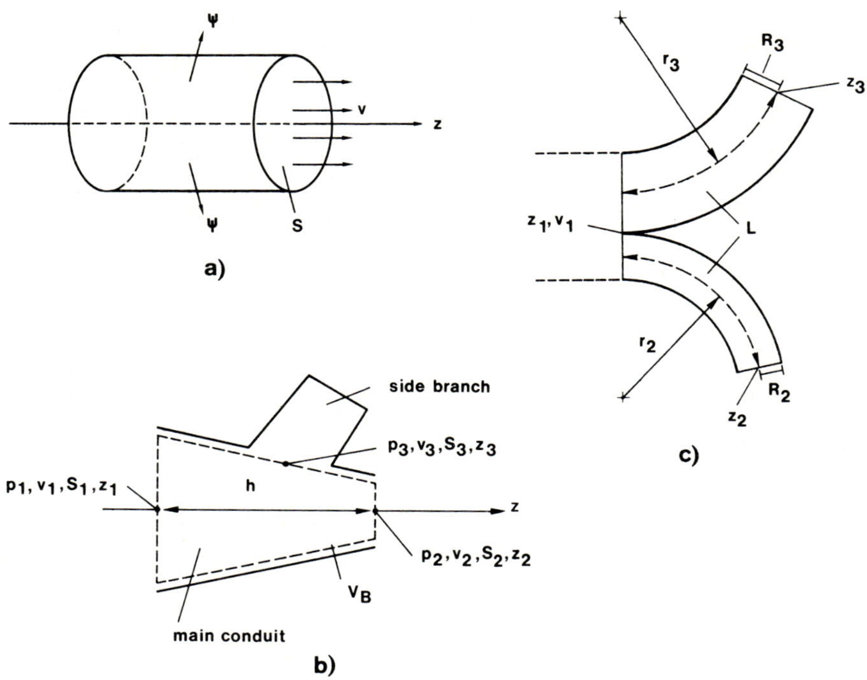

FIG. 17.1 Models of arterial elements, from which a conduit is formed: (*a*) Segment with continuous outflow (only small side branches); (*b*) bifurcation segment; (*c*) geometrical approximations of the branching of the flow in a bifurcation.

The effect of the skin friction on the solution of the one-dimensional equations of pulsatile flow in distensible tubes has been investigated by Gerrard [38] for a straight tube without bifurcations. He suggests replacing Eq. (17.3) by a more realistic approximation proposed by Trikha [39]. Zielke [40] and Brown [41] have proposed a method to calculate the skin friction f more accurately by utilizing a convolution method. A study concerning the effect of various skin-friction expressions on the pulse-wave propagation has been performed by Niederer [32].

Either pressure-dependent [3, 11, 20] or flow-dependent [18, 42] expressions for ψ have been considered. On the basis of detailed modeling of branching, a generalized term was used [12, 26]:

$$\psi(p, z) = \phi_p(z)(p - p_c)S^2 + \phi_q(z)Sv \tag{17.4}$$

In this relation $\phi_p(z)$ and $\phi_q(z)$ serve as outflow parameters, which are specified in Ref. 12.

Bifurcation Regions. Whenever a large vessel branches off from the conduit of interest, we define a *bifurcation region* as shown in Fig. 17.1*b*. Its form is approximated by a truncated cone with height h and surfaces S_1 and S_2. For the volume V_B of the bifurcation region the mass balance requires:

$$v_1S_1 = v_2S_2 + v_3S_3 + \frac{d}{dt}\left[\frac{h}{3}(S_1 + S_2 + \sqrt{S_1S_2})\right] \tag{17.5}$$

In formulating the momentum equation between the points z_1 and z_j ($j = 2, 3$) we make use of a modified Bernoulli relation:

$$p_1 - p_j = \frac{\rho}{2}(v_j^2 - v_1^2) + \rho\int_{z1}^{zj}\frac{\partial v}{\partial t}\,ds + \Delta p_j \tag{17.6}$$

The term Δp_j incorporates the pressure loss due to viscous friction of the blood, as well as the influence of curvature of the vessels in an approximate fashion. An empirical expression for Δp_j has been given by Prandtl [44]:

$$\Delta p_j = \lambda_j \frac{L}{4R_j} v_1^2 \tag{17.7}$$

where

$$\lambda_j = 0.37\left(\frac{1}{2}Re_j\sqrt{\frac{R_j}{r_j}}\right)\left(0.36\cdot\frac{64}{Re_j}\right) \quad \text{for} \quad Re_j\sqrt{\frac{R_j}{r_j}} > 10^{1.6}$$

$$= \frac{64}{Re_j} \quad \text{for} \quad Re_j\sqrt{\frac{R_j}{r_j}} < 10^{1.6}$$

and

$$Re_j = \frac{2R_j\rho v_1}{\mu}$$

This approximation of a bifurcation (Fig. 17.1*c*) and the resulting equation [Eq. (17.7)] for Δp_j describe the real flow situation through a bifurcation only in an approximate way [45–47]. However, in view of the low viscosity of blood and the relatively small area exhibiting the complicated spatial flow patterns prevailing in a bifurcation area, the chosen approximations appear to be justified.

Segments with Pathological Features. Among the pathological alterations of an arterial segment, short stenosis represents the most frequent and therefore one of the most important. Various features of flow through stenosed arteries have been investigated experimentally by Young and Tsai [48] and by Younchareon and Young [49], while Daly [50] uses a finite difference method to model the pulsatile flow through an axisymmetric contraction. Sabbah and Stein

[51] conducted an experiment regarding the dynamics of retrograde coronary flow in aortic valvular stenosis in an in vitro pulse-duplicating system.

Often, a stenosis is short and has the form of an abrupt and short narrowing of the blood vessel (Fig. 17.2). The length L of the stenosis is less than the computational axial increment Δz defined in Sec. 17.2.

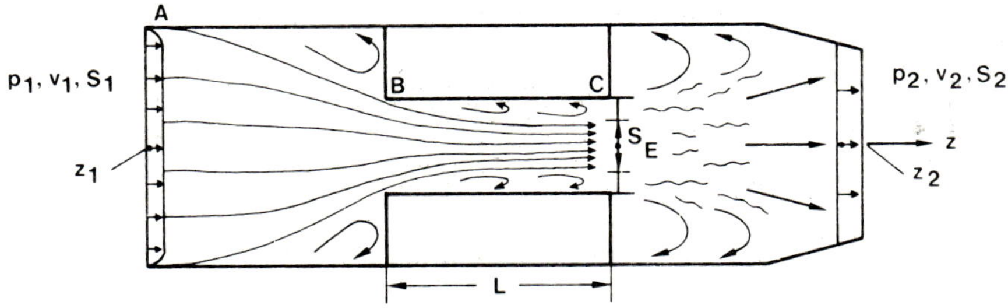

FIG. 17.2 Model of a short stenosis.

Frequently, these stenoses are accompanied by fat and calcium depositions in the arterial wall (see, e.g., Bollinger [52]) which result in a local stiffening of the wall material. The cross-sectional change with pressure of the entire segment can thus be neglected in formulating the mass balance equation between z_1 and z_2 (Fig. 17.2):

$$v_1 S_1 = v_2 S_2 \tag{17.8}$$

For the pressure drop $p_1 - p_2$ across the stenosis, an equation similar to Eq. (17.6) can be derived:

$$p_1 - p_2 = \frac{\rho}{2}(v_2^2 - v_1^2) + \rho \int_{z1}^{z2} \frac{\partial v}{\partial t} \Delta z + \Delta p \tag{17.9}$$

Flow separation in points A and B (Fig. 17.2) is equivalent to an additional narrowing of the cross section. Hence, the stenosis has an effective area S_E. Furthermore, flow separation in C causes an increased pressure drop which is taken into account by using the relation for the so-called Borda-Carnot effect (e.g., see Bennett and Myres [53]).

Accordingly we have

$$\Delta p = \frac{\rho}{2} v_1^2 \left(\frac{S_1}{S_E} - 1\right)^2 + \frac{8\pi\mu L v_1 S_1}{S_E^2} \tag{17.10}$$

where the first term on the right-hand side of the equation represents the Borda-Carnot effect and the second term represents the friction between B and C (Fig. 17.2).

Fukushima et al. [54] have calculated numerically the velocity profiles in an arterial model with one or two constrictions based on the assumption of quasi-steady flow [which is also assumed in Eq. (17.10)]. Young and Tsai [48] have proposed an expression for calculating the pressure drop over a stenosis that includes the effects of the unsteadiness of the heart pulse, while Rooz et al. [55] have developed a finite element model for pulsatile flow in flexible tubes. The applicability of their model is documented by comparing predicted flow and pressure waveforms with corresponding in vitro measurements for pulsatile flow in a straight, flexible, partially obstructed tube. However, the synthesis of such refined models of stenoses with models of the arterial tree remains to be done.

17.1.2 Mechanical Aspects

Distensibility of Blood Vessels. For elastic arterial walls the cross section S at a given location z_0 is a function of the pressure $p(z_0, t)$ alone. The explicit form of $S(p, z_0)$ can best be determined from the dependence of the phase velocity of small-amplitude pressure waves, as this velocity is amenable to measurements under physiological conditions [1, 56]:

$$c^2(p, z) = \frac{S(p, z)}{\rho \dfrac{\partial S}{\partial p}(p, z)} \tag{17.11}$$

The available results of wave-speed measurements indicate that a relation of the form

$$c(p, z) = (c_0 + Bp)g(z) \tag{17.12}$$

can be used [3]. The two constants c_0 and B, as well as the function $g(z)$, are chosen according to the measurements. At present, our knowledge of vessel elasticity under physiological conditions could not support a more refined choice of $c(p, z)$. Upon integrating Eq. (17.11), one obtains, for the pressure dependence of $S(p, z)$,

$$S(p, z) = S_0(z) \exp\left[\frac{p - p_0}{c(p_0, z)\rho c(p, z)}\right] \tag{17.13}$$

The application of the outlined mathematical model with elastic vessel walls to normal and certain pathological situations has yielded realistic results [12]. A comparison between the computed pressure and flow curves with in vivo measurements, however, revealed that in the distal parts of the arterial tree the calculated wavefronts are steeper and the peaks sharper than observed in the measurements. This discrepancy may be attributed to the fact that the viscoelastic properties of the vessel wall have been neglected and the influence of blood viscosity has not been properly taken into account.

Holenstein et al. [13] developed a new method to simulate the nonlinear viscoelastic properties of the wall material with the aid of a convolution integral of the creep function and the pressure history. Following their analysis, one obtains, for the cross-sectional area S,

$$S(t, z) = S^e(p(t), z) + \int_0^\infty J(\tau)S^e(p(t - \tau), z)\, d\tau = S^e + S^i \tag{17.14}$$

where the integral represents S^i, $S^e(p, z)$ denotes the instantaneous elastic response, and $J(\tau)$ denotes the creep function with the normalization $J(0) = 1$. As a first approximation, the creep function can be assumed to be independent of z, because available experimental results to date would not allow such a dependency to be determined. Using Eq. (17.14), the phase velocity (Eq. 17.11)] can be expressed as

$$c^2 = \frac{S^e + S^i}{\rho\,(\partial S^e/\partial p)} \tag{17.11a}$$

and is therefore a function of the pressure history.

Outflow Conditions. In the case of a parent artery branching into two or more smaller vessels, the distribution of blood flow in the individual branches depends on the geometry of the bifurcation as well as on the geometric structure of the branching arteries and their mechanical wall properties. Detailed information on these parameters is reported by Schönfeld et al. [57], who investigated the elastic response as well as the geometry of the aortic-iliac junction of 10 dogs.

In the long segments, described in the preceding section, emanating from the heart or a bifurcation, the method of characteristics is utilized to integrate the pulse-propagation equations. Consequent application of this procedure for all first- and higher-order side branches would, however, result in a computer program which could no longer be applied efficiently. Therefore, it is desirable to replace side branches of minor interest by suitably chosen, lumped outflow resistance elements, which relate the pressure p_3 and flow Q_3 (Fig. 17.1b) at the entrance of a branch to the pressure p_1 and flow Q_1 in the main conduit.

On the basis of heuristic considerations, Stettler et al. [12] derived the following outflow condition:

$$Q_3 = R_p(p_1 - p_c)S_3^2 + R_q Q_1 \tag{17.15}$$

Its applicability has been shown by direct comparison of results obtained with the method of characteristics and with Eq. (17.15), respectively.

17.2 NUMERICAL SOLUTION

17.2.1 Application of the Method

Equations (17.1) and (17.2), which govern the pulse propagation along segments with no or only insignificant side branches, form a system of quasi-linear hyperbolic partial differential equations. The numerical solution is found by utilizing the method of characteristics and specified time intervals [19]. Since in our model the bifurcations are treated separately, the functions $S_0(z)$ and ψ exhibit only a relatively weak dependence of z. Therefore, a first-order integration scheme is applicable [18].

Upon transforming the system of Eqs. (17.1) and (17.2) and by using Eq. (17.3), the two characteristic directions

$$\frac{dz}{dt} = v \pm c \tag{17.16}$$

are obtained, whereby c is given by Eq. (17.12). In the viscoelastic case, Eq. (17.12) cannot be applied directly, since the wave velocity depends on the pressure history. Since Eq. (17.12) describes the propagation velocity of small-amplitude waves, however, a linearized form of the viscoelastic equations can be derived whereby the phase velocity is represented as a function of pressure and frequency.

The characteristic equations are

$$dv \pm \frac{dp}{\rho c} = \left(-\frac{8\pi\mu v}{\rho S} \mp \frac{vc}{S}\frac{\partial S}{\partial z} \mp \frac{c}{S}\psi \mp \frac{c}{S}\frac{\partial S^i}{\partial t} \right) dt \tag{17.17}$$

with $S^i = 0$ for elastic walls

and $S^i = \int_0^\infty J(\tau) S^e \, d\tau$ for viscoelastic walls [67]

The equation one gets using the upper (lower) sign in Eq. (17.17) is valid along the characteristic direction using the upper (lower) sign in Eq. (17.16).

For the solution to be numerically stable, the Courant stability criterion [19] for the step sizes Δz and Δt,

$$\frac{\Delta z}{\Delta t} > |v| + c \tag{17.18}$$

has to be fulfilled for all integration points z_j and time steps t_j.

17.2.2 Boundary Conditions

Consider an arbitrary arterial system which is built up from two of the segment types mentioned in Sec. 17.1.1 (see Fig. 17.3). While the coordinate z applies to the main conduit, at each bifurcation a new axial coordinate $z_{B,n}$ for the nth side branch originating at that location has to be defined. Each branch and each segment may have its own spatial and temporal step sizes $\Delta z_{B,n}$ and $\Delta t_{B,n}$.

In order to have a unique solution for the characteristic Eq. (17.17), a complete set of boundary conditions has to be specified. At the root of the aorta ($z = 0$ in the main conduit), we prescribe the ejection pattern of the left ventricle in the form of the volume flow rate $Q_0(t)$ as measured by Spencer and Denison [58] and Wetterer and Kenner [59]. At the distal end of the main conduit and of each side branch which is not replaced by the lumped outflow model, either

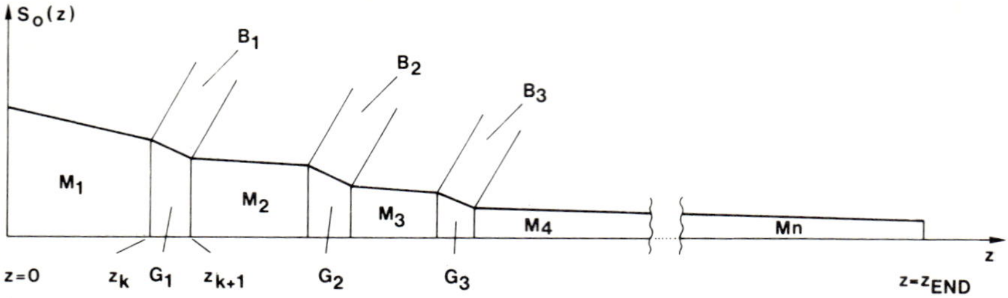

FIG. 17.3 Principle of subdividing an arterial conduit into long segments M_i, bifurcation G_i, and side branches B_i.

a constant capillary pressure p_c, a constant venous pressure, or a peripheral resistance is given. Models from the type presented here are intended to simulate arterial pulse propagation and are as such relatively insensitive to the details of the flow and pressure distributions in the microvasculature. Therefore, no attempt is made here to incorporate the results from the extensive studies of blood flow in capillaries and veins which have been carried out. For references the reader is referred to Skalak et al. [60] and Shapiro [61]. The boundary conditions used for all segments ending or originating in a bifurcation are described for the bifurcation shown in Fig. 17.4. We assume the velocities and pressures at the boundary of the bifurcation volume at time

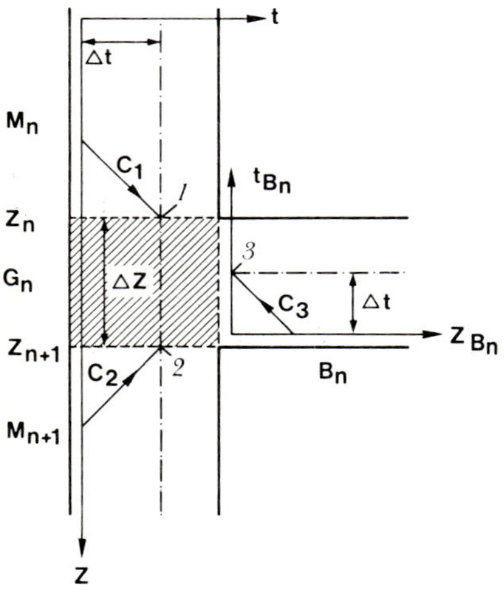

FIG. 17.4 Boundary conditions for bifurcation $G_n : C_1$, C_2, and C_3 denote characteristic directions in the coordinate system z, t (points 1 and 2), and z_{B_n}, t_{B_n} (point 3), respectively.

t_0 to be known, namely, $v(z_n)$ and $p(z_n)$, $v(z_n + \Delta z)$, and $p(z_n + \Delta z)$, as well as $v(z_{B,n} = 0)$. If we restrict ourselves to the simple case $\Delta t_{B,n} = \Delta t$, the six unknowns—namely, velocities and pressures at the boundaries of the bifurcation volume at time $t_0 + \Delta t$—are calculated with the aid of the following relations:

- The characteristic Eq. (17.17) along the characteristic directions C_1 and C_2 in the main conduit
- Either the characteristic equation along the characteristic direction C_3 in the $z_{B,n}$, $t_{B,n}$ plane of the side branch, or the outflow condition [Eq. (17.15)] in case of replacement of the side branch by outflow parameters
- Mass balance Eq. (17.5) for the bifurcation volume G_n
- Integrated momentum Eq. (17.6) with Eq. (17.7) between points z_n and $z_n + \Delta z$; z_n and $z_{B,n} = 0$, respectively

In order to make Eqs. (17.6) and (17.7) amenable to numerical solution, the term

$$\rho' \int_{z1}^{z2} \frac{\partial v}{\partial t}\, dz$$

is approximated by

$$\rho \frac{\Delta z}{2\,\Delta t} \{v(z_n, t_0 + \Delta t) + v(z_n + \Delta z, t_0 + \Delta t) - [v(z_n, t_0) + v(z_n + \Delta z, t_0)]\} \tag{17.19}$$

and the quadratic terms in v are linearized with respect to the increments within the time interval Δt.

In case of unequal integration step size for the adjoining branches, interpolation procedures are applied, as described in detail by Stettler [62]. The boundary conditions for pathological segments are treated in the same manner as shown for the bifurcations.

The computational accuracy of the computer method was checked by verifying the mass and energy balance equations for a control volume surrounding the entire arterial conduit. This method is of particular help during program development, and it also offers additional insight into energy dissipation phenomena [12].

17.3 STANDARD CASE OF HUMAN ARTERIAL TREE

17.3.1 General Layout

In the following discussion, the pulse propagation in a human arterial conduit which extends from the root of the aorta to the foot (Fig. 17.5) is examined. Only a few authors have performed calculations of the pulse propagation along the entire length of such a relatively long and variable conduit [9, 12, 13]. Models based on the linearized fluid-dynamic equations [1, 2] are more suited for analyzing the pulse propagation along a shorter segment.

A standard case of the arterial tree is defined by choosing the physiological and geometric quantities such that they correspond to an average healthy human adult. Thereby, extensive use was made of the work by Westerhof et al. [63], McDonald [1], and Snyder et al. [64]. The model consists of a main conduit and eight side branches. The side branches 1, 2, 3, and 6, corresponding to coronary arteries and truncus brachiocephalicus, left carotid and subclavian arteries, truncus coeliacus, and internal iliac artery, respectively, are replaced by outflow parameters.

All parameters which determine the cross section, the outflow conditions, the boundary conditions, and the phase velocity, as well as the parameters of the wall viscoelasticity model, are given in Refs. 12 and 13. Except as noted in Fig. 17.6, all the calculated pressure and flow pulses in this chapter are for the mathematical model with elastic vessel walls.

The computed pressure and flow pulses shown in Figs. 17.5 and 17.7 to 17.9 exhibit a good overall agreement with published measurements [1, 6, 52, 59]. The characteristic features of the arterial pressure and flow pulses as they are observed in humans are clearly reproduced. In particular, the theoretical results document the presence of the incisura up to the bifurcation at $z = 42.5$ cm. Furthermore, they predict a gradual decrease of the diastolic pressure and increase of the systolic pressure with increasing distance from the heart. Besides this, they show a steepening of the wavefront up to a distance of about 140 cm. Beyond the bifurcation the development of a dicrotic pressure wave is observed. In addition, the flow pulse shows a progressively pronounced backflow during diastole with increasing distance from the heart until the diameter is reduced to approximately 0.3 cm.

The difference between the curves in Fig. 17.5 and the one obtained with the viscoelastic model becomes apparent in the distal region, as shown in Fig. 17.6. With the viscoelastic model, the wavefronts remain smooth even in the distal region. It appears therefore that the tendency to generate shock waves [65] is diminished in the viscoelastic wall model [43]. It is known that shock waves do not occur in humans under normal physiological conditions. Accordingly, the viscoelastic model seems to describe the pulse propagation over a long distance in a more realistic way than the elastic model. It is, moreover, known that a realistic representation of the

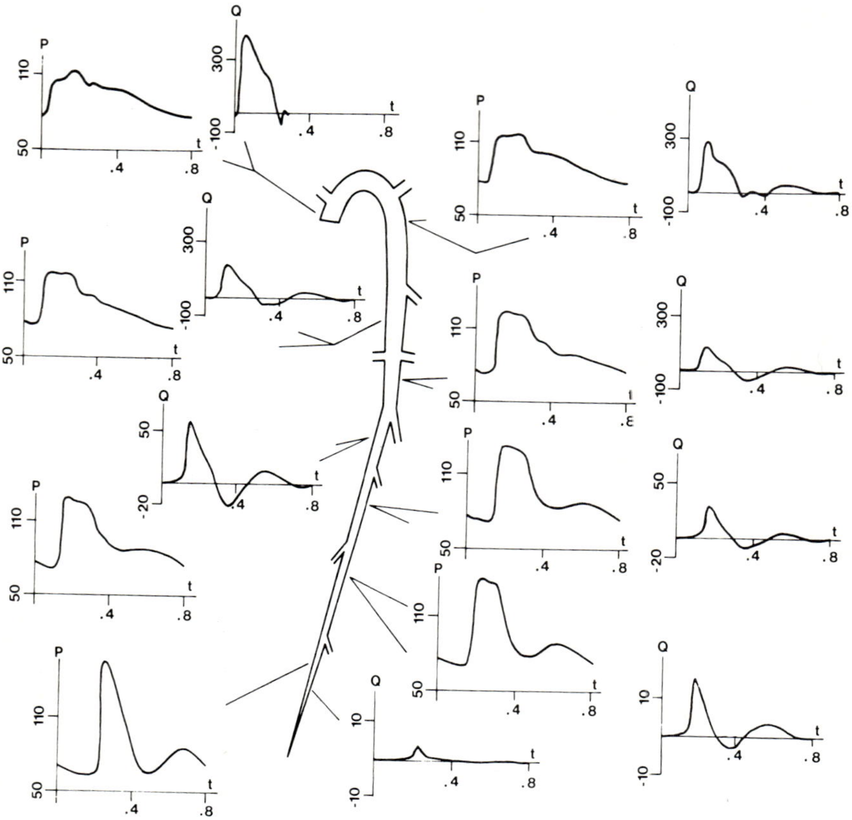

FIG. 17.5 Calculated pressures and flow patterns at different locations for standard case. Pressures are given in mmHg, and flows in mL/s.

effects of blood viscosity leads to damping characteristics which differ in their frequency dependence from those induced by viscoelasticity. It is, however, not known at present to what extent the peripheral pulse shapes are influenced by this additional damping.

17.4 COMPARISON OF SIMULATION RESULTS WITH PRESSURE AND FLOW MEASUREMENTS IN NORMAL ADULTS

In the following discussion, the computed flow and pressure pulses are compared with measured data for various locations along the arterial conduit of the standard case. First a comparison of the pressure pulses with those adopted from Wetterer [66] (Fig. 17.7) shows that the model correctly reproduces the characteristic changes of the pulse shapes with increasing distance from the heart. The calculated systolic and diastolic pressures fall well within the range of physiological values as measured in healthy human adults [72].

Also, the theoretical flow curves agree, within the limits of biological variability among normal adults, with those measured by pulsed Doppler ultrasound. This is documented in Figs. 17.8 to 17.10 for the femoral, abdominal, popliteal, and carotid arteries.

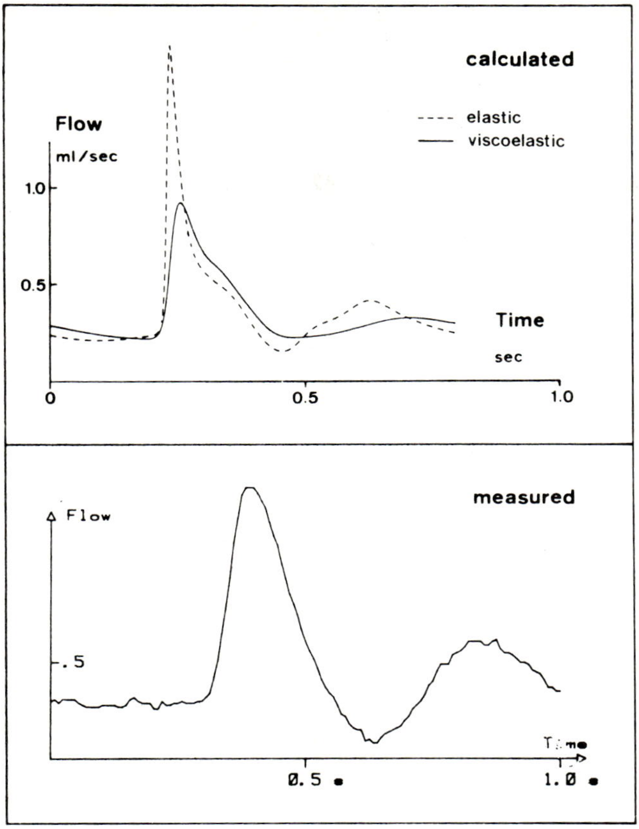

FIG. 17.6 Predicted and measured flow pulses in the arteria dorsalis pedis (z = 130 cm) of a young healthy adult. Note that the heart rate of the measured subject is lower than is assumed in the standard case.

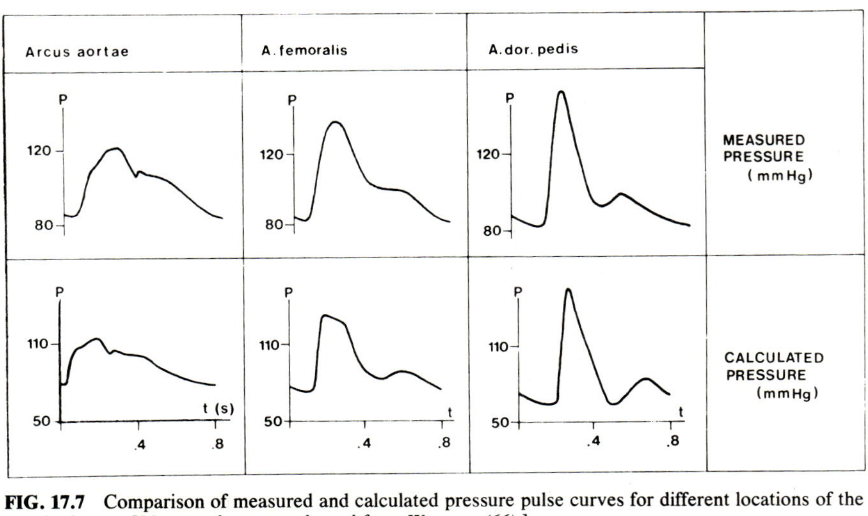

FIG. 17.7 Comparison of measured and calculated pressure pulse curves for different locations of the standard case. [*Measured curves adopted from Wetterer (66).*]

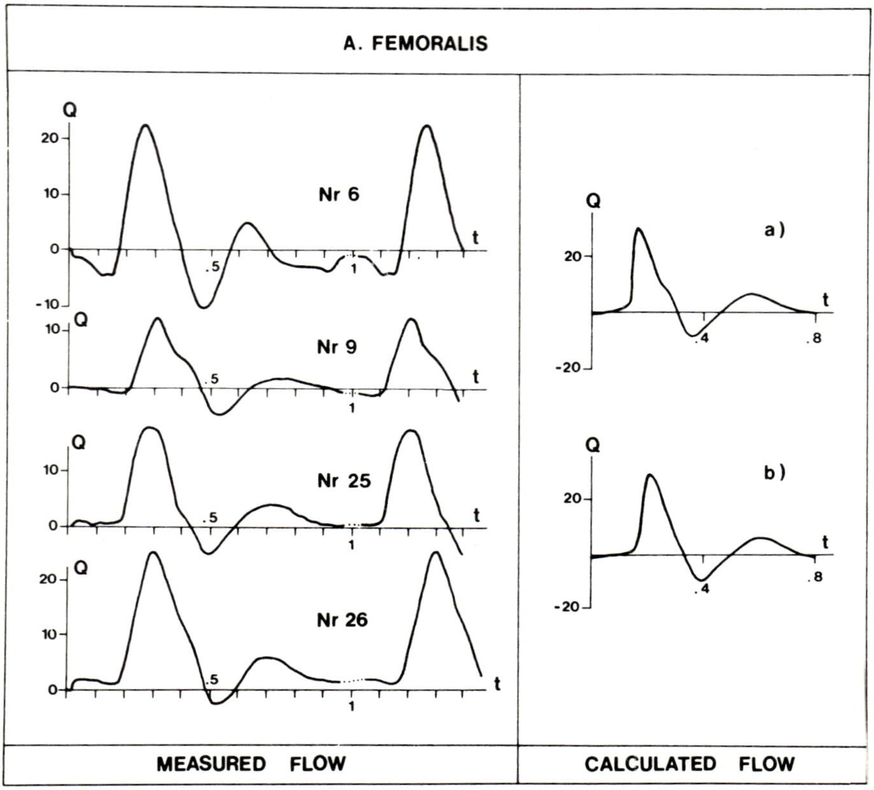

FIG. 17.8 Comparison of measured and calculated flow pulse in A. femoralis (Q: mL/s; t: sec). Left: Flow curves in four healthy volunteers, obtained with pulsed US-Doppler technique and averaged over eight heart cycles. Right: Calculated curves: (*a*) standard case; (*b*) increased systolic rise time of proximal boundary condition.

In Fig. 17.8, the curves measured in four healthy volunteers indicate the range of normal biological variability in the case of the femoral artery. This variability is even more pronounced in the carotid artery (Fig. 17.10). Note that the pulse calculated in branch 2 exhibits a pulse amplitude which is substantially higher than that recorded in the common carotid artery (Fig. 17.10). This is attributed to the fact that in the model branch 2 includes the left subclavian artery. The variability in the pulse shapes observed in the noninvasive flow measurements and documented in part in Figs. 17.8 and 17.10 are manifestations of typical individual features of the arterial system of the subjects examined and are not caused by temporary aberrations of cardiovascular functions. The variability of the theoretical model due to parameter variations can be found in Refs. 12 and 62.

17.5 FLOW AND PRESSURE PULSES IN THE PRESENCE OF A STENOSIS

17.5.1 Stenosis with Collateral Vessels in the Femoral Artery

The gradual growth of an arterial stenosis represents a pathological process which may develop over several years. It is often accompanied by the formation of small collateral vessels which

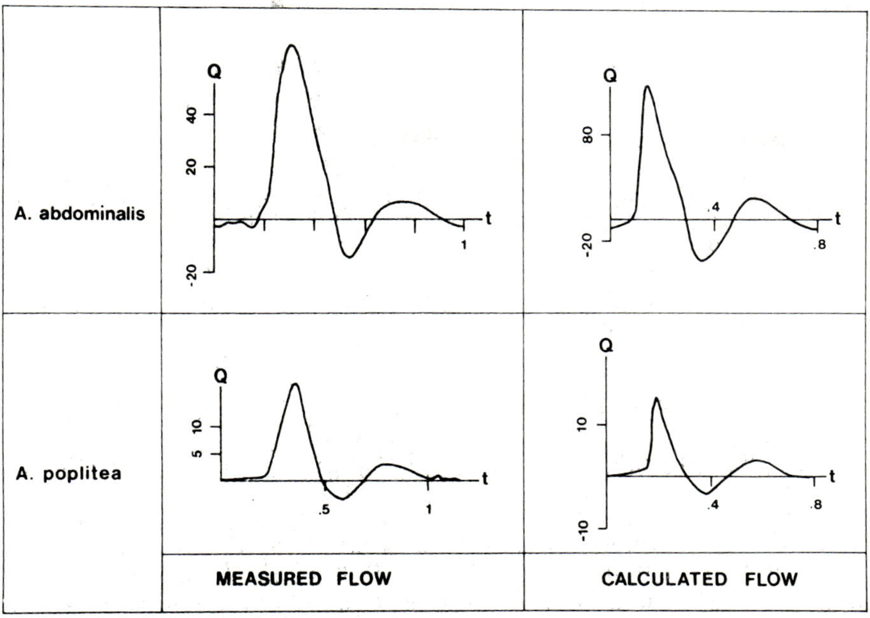

FIG. 17.9 Comparison of measured and calculated flow pulse in A. abdominalis and A. poplitea (Q: mL/s; t: sec). Measured curves are obtained with pulsed US-Doppler technique and averaged over eight heart cycles.

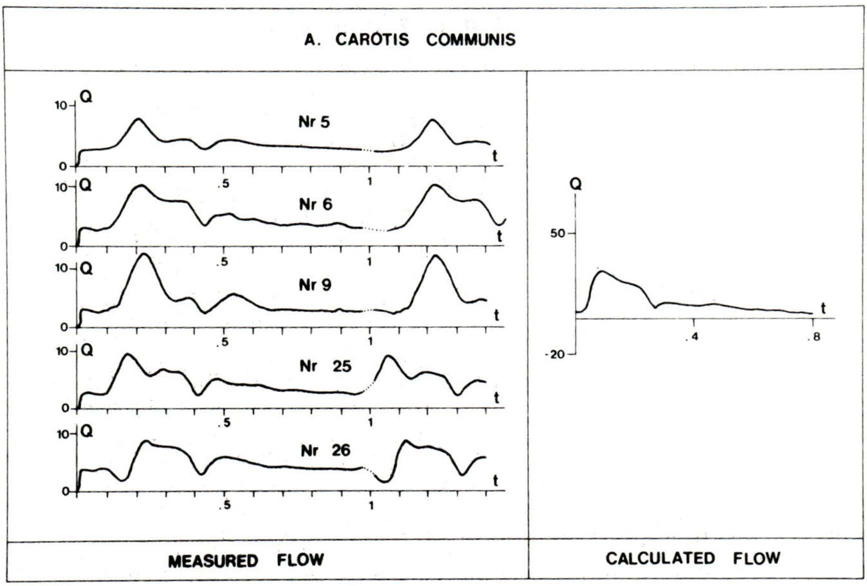

FIG. 17.10 Comparison of measured and computed flow in A. carotis communis. Left: Averaged (eight heart cycles) flow curves in five healthy volunteers, US-Doppler measurements (Q: mL/s; t: sec).

compensate to some degree the loss in peripheral perfusion of the affected organs. Stenoses are predominantly seen in older patients. This latter fact is taken into account in the model by a general increase in the wave speed of 20 percent.

A specific case is examined here in which a total occlusion with collateral vessels occurred in the femoral artery, as shown in Fig. 17.11. On the basis of radiographs the pathological

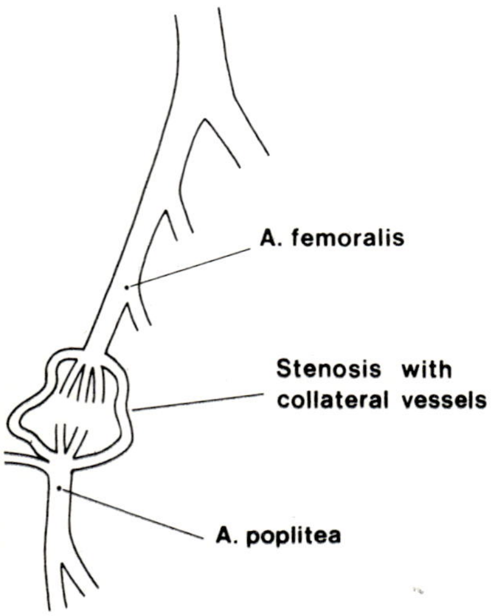

FIG. 7.11 Model of a total stenosis in the A. femoralis with collateral vessels.

changes between the femoral and popliteal arteries could be measured and approximated in our model. At $z = 70$ cm the artery subdivides into five identical small vessels which recombine at $z = 90$ cm. The total cross section of the five collaterals is 20 percent smaller than the corresponding cross section of the original vessel at this location in the standard case. The results of the simulation and the measurements are seen in Fig. 17.12.

The main feature of the computed and measured flow pulses (see Fig. 17.12) proximal to the stenosis is the lack of backflow during early diastole. Both curves are in excellent agreement. At

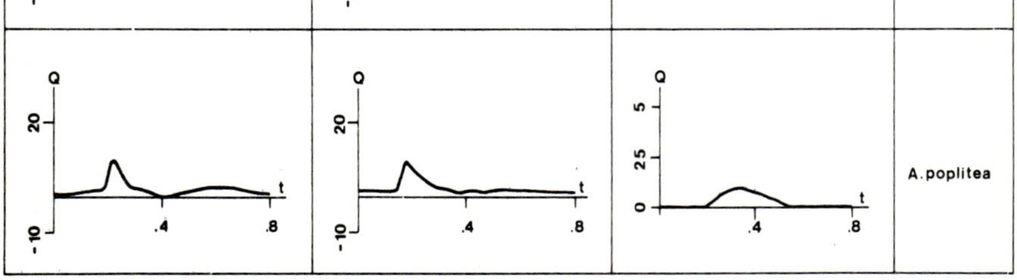

FIG. 17.12 Computed and measured flow curves for stenosis with collateral flow as shown in Fig. 17.11 (Q: mL/s; t: sec).

the distal site of the stenosis, where the pulse is markedly damped, the computed and measured curves differ insofar as the measured pulse is smoother and its amplitude smaller. This is due to our assumption that all collaterals have recombined at $z = 90$ cm, whereas in reality the anatomical details are more complicated. The lack of smoothing in the model can additionally be explained by the neglect of the damping due to vessel wall viscoelasticity as well as by the insufficient damping provided by the Poiseuille viscosity model.

17.5.2 The Effect of Short Stenoses on the Flow and Pressure Pulses

Whenever a stenosis grows rapidly in time, we expect the collateral flow to be insignificant and the resulting flow conditions to be described by the mathematical model (Eq. (17.9)] (see Fig. 17.2). Two different cases of such a short stenosis are documented in Figs. 17.13 and 17.14. Note that the mathematical model presented makes it possible to determine the effect of such a pathological vessel constriction not only on the flow conditions inside the main conduit, but also

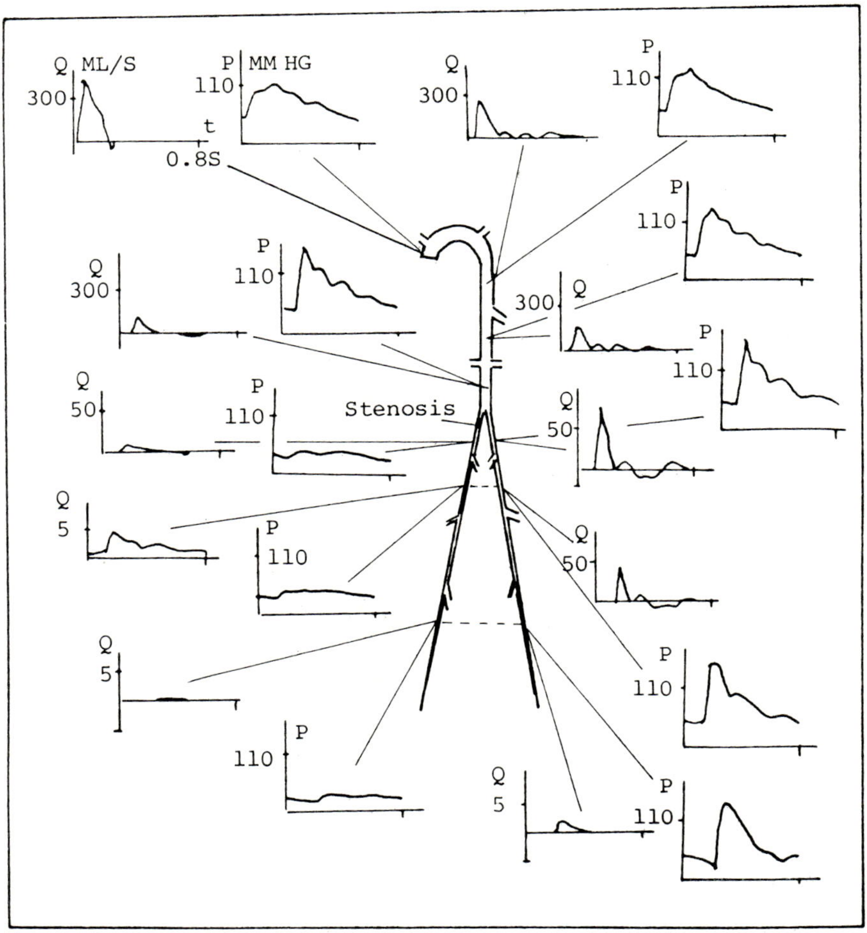

FIG. 17.13 Computed pressure and flow pulse curves for a short 95% stenosis in the iliac artery of the left leg. The pressure drop induced by the stenosis is approximated by expression (17.10).

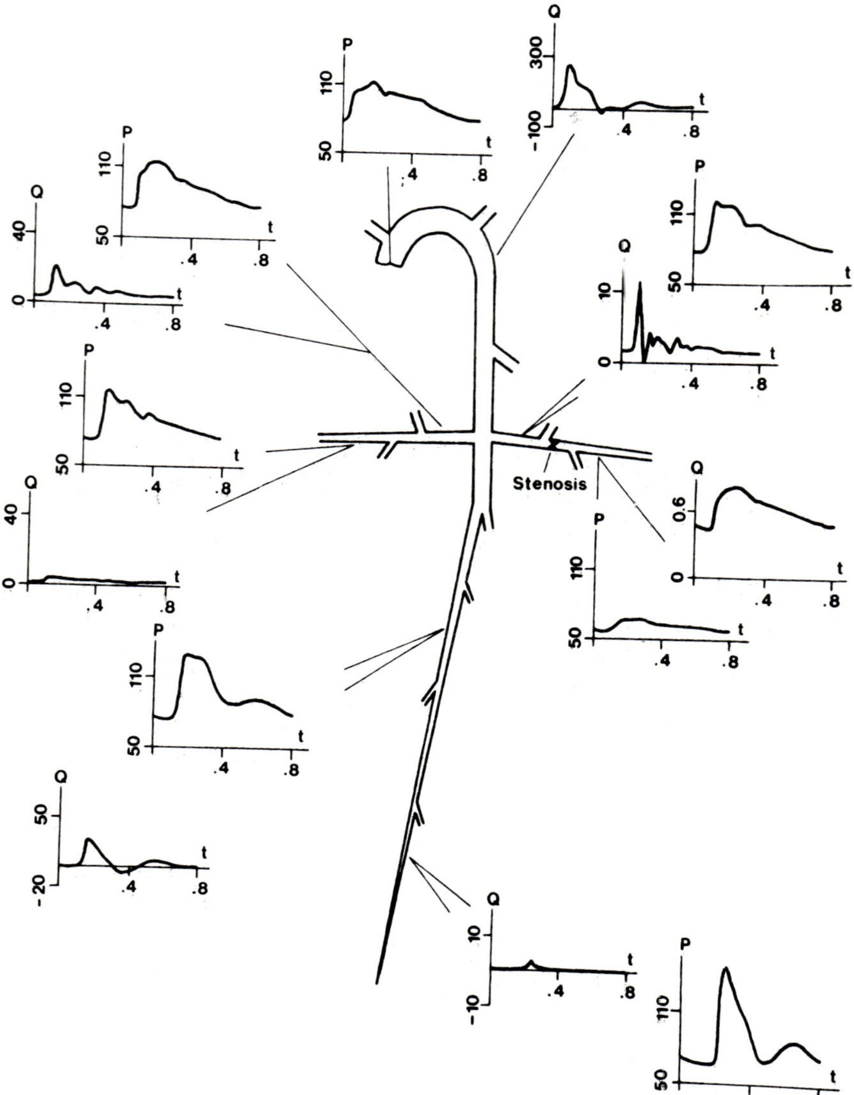

FIG. 17.14 Pulse curves obtained for a stenosis in the right renal artery. A 90% stenosis of length L = 0.45 cm is assumed to be in the right renal artery, 3.0 cm away from its bifurcation in the main conduit. Note that the lengths of the two renal arteries and the length of the main conduit are not to scale (P: mmHg; Q: mL/s).

simultaneously on the pulse propagation inside the various side branches of interest. For a one-sided renal or iliac stenosis, the flow and pressure pulses in the healthy renal or iliac artery will then be of special interest (see Figs. 17.13 and 17.14). Figure 17.13 shows a 95 percent stenosis of short length, $L = 2$ cm, located in the iliac artery of the left leg. Because of the nonsymmetry of the two arterial conduits distal to the iliac bifurcation, the pulse propagation in both legs has to be calculated with the method of characteristics to evaluate realistically the effect of this stenosis on the flow and pressure pulses.

Strong reflections of the heart pulse at the stenosis cause several peaks to occur in the pressure and flow pulses during one cycle. This phenomenon of several peaks is not observed for the standard case. Because of the stenosis, the pulse curves at various locations in the left leg are very much damped, whereas those in the right leg show very high peaks during the systolic phase. The predicted strong damping of the flow pulses behind the stenosis and the disappearance of the backflow have been confirmed by ultrasonic measurements of the velocity in the femoral arteries of patients with aortoiliac diseases [7, 68, 69]. Contrary to the standard case, for which the flow into both legs (integrated over one period) is equal (approximately 6.3 mL), the stenosis causes the total flow into the left leg to be reduced to only 3.8 mL and the one into the right leg to 5 mL. This is due to an increase in the mean pressure of approximately 2.5 mmHg in the main conduit for $z \leq 40$ cm. Consequently, the total flow into the side branches 1 to 4 is increased.

A realistic simulation of a one-sided renal stenosis makes it also necessary to calculate the pulse-wave propagation in both renal arteries with the help of the MOC. In contrast to the iliac stenosis discussed above, a one-sided renal stenosis does not significantly alter the characteristic *shape* of the pressure and flow pulses in the main conduit. On the other hand, the pressure in the main conduit *averaged* over one period, as well as the outflow conditions into the various side branches and the external heart work, is altered in the following way: An increase of the mean pressure of approximately 3 mmHg is observed for the entire main conduit. As a consequence, the flow into the first two side branches, as well as the flow into the left renal artery, is increased compared with that of the standard case. During one period only 2.4 mL of blood is flowing into the right stenosed renal artery, whereas 7.6 mL is flowing into the renal artery without the stenosis (standard case: 7 mL in each renal artery). The external heart work is increased due to the stenosis by approximately 5 percent, which is considerably more than for the above-discussed iliac stenosis. Clinical data show that a one-sided renal stenosis can indirectly produce a high pressure (hypertension) in the arterial system, with diastolic pressures above 120 mmHg [70]. Such a strong increase of the pressure cannot occur in the present simulation, because under physiological conditions hypertension in these cases is caused by hormone (renin-angiotensin) action, which, of course, is not part of a mechanical model.

17.6 OPTIMAL OPERATION OF INTRAAORTIC BALLOON PUMPS

Optimal timing of inflation and deflation of intraaortic balloon pumps (IABPs) is essential for their effectiveness in assisting failing left ventricles [71, 73]. To date, no generally accepted criterion for optimal pump performance has been established. One reason for this may be that not all the various performance indices to be considered, such as coronary artery flow, left ventricular stroke work, end-diastolic pressure, cardiac output, and tension-time index, can be optimized concurrently [83]. Also, in a relative weighting of these indices, one would presumably have to take the status of the individual patient into account. Comprehensive solutions to this problem are difficult to obtain with the aid of animal experiments. The mathematical model of the human arterial system and its hemodynamics, discussed in Sections 17.1 to 17.3, appear to offer an alternate approach and thus additional help in establishing criteria for optimal timing of IABPs. Accordingly, an extension of the model outlined so far was developed to include the desired features [76].

17.6.1 Governing Equations

The IABP will occupy part of the blood volume inside the aorta. The arterial cross section will then be reduced by the IABP cross section over a certain length in the descending aorta (see Figs. 17.15 and 17.16).

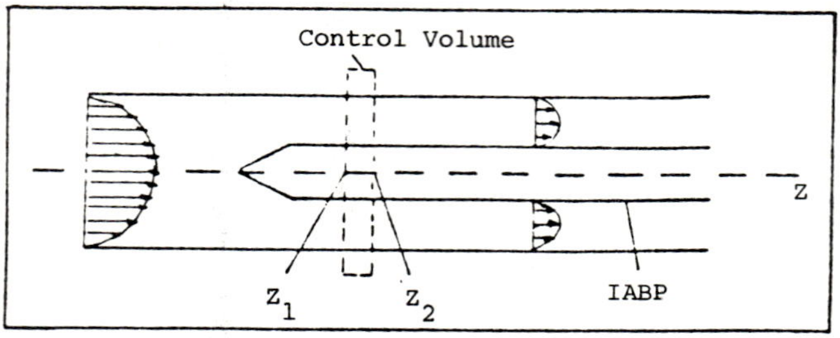

FIG. 17.15 Schematic drawing of the blood flow in the aorta with inserted IABP.

To calculate the pulse-wave propagation alongside the IABP, we have to modify the mass and momentum equations [Eqs. (17.1) and (17.2)] accordingly. This can be achieved by replacing the arterial cross section $S(p, z)$ in Eqs. (17.1) and (17.2) with the reduced cross section $S(p, z) - S_B(p_B, z)$. $S_B(p_B, z)$ is the IABP cross section, and p_B denotes the pressure difference between the pressure inside the IABP, P_I, and the outside pressure, which is the arterial pressure p. In addition, the frictional force f alongside the IABP has to be calculated for a new velocity profile (see Fig. 17.15). In a first approximation, we again assume quasi-steady laminar flow:

$$f = \frac{-12\pi\mu v(S - S_B)}{(\sqrt{S} - \sqrt{S_B})^2} \tag{17.20}$$

The modified system of quasi-linear differential equations is solved with the method of characteristics. Equation (17.11) then becomes

$$c_B = \sqrt{\frac{S - S_B}{(\partial S/\partial p - \partial S_B/\partial p)\rho}} \tag{17.21}$$

and for the characteristic equations [Eq. (17.17)] results

$$dv \pm \frac{dp}{\rho c_B} = \left[-\frac{12\pi\mu v}{(\sqrt{S} - \sqrt{S_B})^2} \mp \frac{v c_B}{S - S_B}\frac{\partial(S - S_B)}{\partial z} \mp \frac{C_B}{S - S_B}\left(\psi \pm \frac{\partial S_B}{\partial t}\right)\right] dt \tag{17.22}$$

along the two characteristic directions

$$\frac{dz}{dt} = v \pm c_B \tag{17.23}$$

c_B denotes the wave speed alongside the IABP inside the descending aorta (see Fig. 17.17). For this purpose the pressure dependency of the cross section of the IABP had to be determined experimentally. As in Anliker et al. [56, 74] the wave speed c_B of small sinusoidal pressure variations inside the IABP was measured, which for this purpose was filled with water at a given pressure p_B. In this model the IABP behaves like a fluid-filled elastic tube. Consequently, Eq. (17.11), relating wave speed, cross-sectional area, and pressure dependency, respectively, is valid:

$$c(p_B) = \sqrt{\frac{S_B(p_B)}{\rho_w(\partial S_B/\partial p_B)}} \tag{17.24}$$

where ρ_w is the density of water; and $p_B = p_I - p_A$, where p_I is the pressure inside the IABP and p_A is the pressure outside the IABP. Integration of Eq. (17.24) yields, according to Eq. (17.13),

$$S_B(p_B) = S_{B,0} \exp\left[\frac{p_B - p_0}{c(p_0)c(p_B)\rho_w}\right] \tag{17.25}$$

where $S_{B,0}$ denotes the cross-sectional area at a reference pressure p_0. For a clinically used IABP [75] the wave speed turned out to be pressure-independent:

$$c(p_B) = c(p_0) = c_{\text{IABP}} = 1200 \text{ cm/s}$$

and therefore

$$S_B(p_B) = S_{B,0} \exp \frac{P_B - P_0}{\rho_w c^2_{\text{IABP}}}$$

Finally, for the IABP being inserted into the artery, we get, for the pressure difference p_B,

$$p_B = p_I - p$$

with p being the arterial pressure, as introduced in Sec. 17.1. Therefore

$$S_B(p) = S_{B,0} \exp \frac{P_I - P_0 - P}{\rho_w c^2_{\text{IABP}}} \tag{17.26}$$

and

$$\frac{\partial S_B}{\partial p} = \frac{-S_B}{\rho_w c^2_{\text{IABP}}} \tag{17.27}$$

Equations (17.26) and (17.27) can then be used to evaluate f and c_B in Eqs. (17.20) and (17.21), respectively.

17.6.2 Geometric Properties of the IABP

Figure 17.16 shows the geometry of the IABP for a given z-independent transmural pressure. As can be seen from Fig. 17.16, large variations in $S_B(z)$ are present between z_B and z_1 as well as

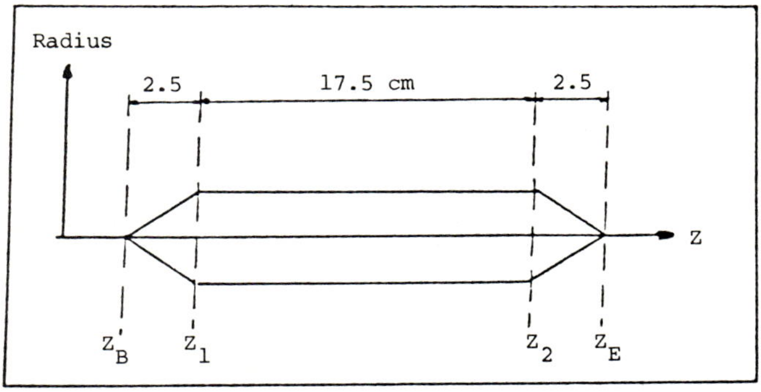

FIG. 17.16 Simplified geometry of the IABP at p_B equal to 100 mmHg. As for arteries, axisymmetry of the balloon is assumed.

between z_2 and z_E. This can lead to numerical instabilities if the pulse-wave propagation is calculated with the help of the MOC along these domains. Special segments are therefore required to model the flow situation between z_B and z_1 and between z_2 and z_E. For this purpose, the same principle is applied as in describing a bifurcation segment or a short stenosis. The governing equations for calculating pressure and velocity values at the new segment boundaries will therefore be mass and momentum equations, as well as two characteristic equations along the two characteristic directions with their endpoints on the segment boundaries. Pressure losses due to flow separation between z_2 and z_E can, as with a short stenosis, be included in the momentum equation. It finally follows (see Fig. 17.17) that the position of the IABP in the

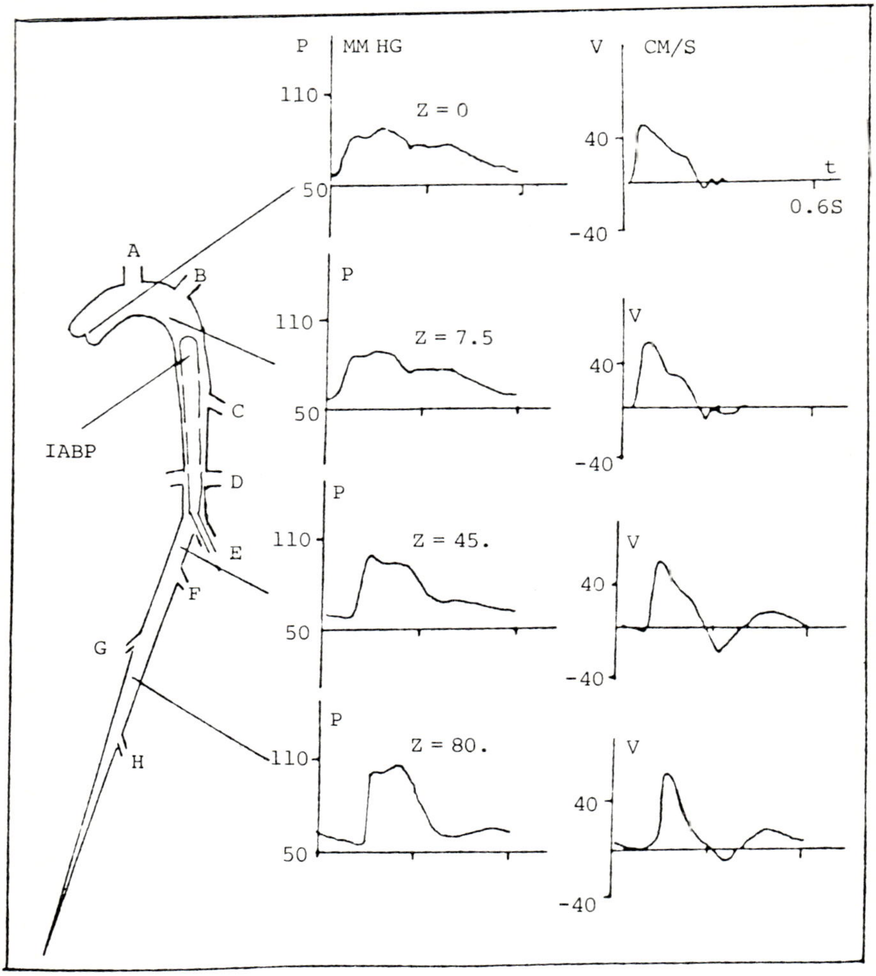

FIG. 17.17 Calculated pressure and velocity patterns for modified standard case. Also shown is the location of the IABP in the descending aorta.

arterial tree overlaps over the two bifurcations of the truncus coeliacus and renal arteries. Accordingly, mass and momentum equations [Eqs. (17.5) and (17.6)], valid for a bifurcation segment, are adapted to this simulation by replacing the cross-sectional area of the descending aorta, S, by the reduced area $S - S_B$ [62].

Inflation and Deflation of the IABP. Inflation and deflation of the IABP are crucial parameters in determining the effectiveness of the IABP as a heart-assisting device. Of special importance are, therefore, the length of the inflation and deflation periods, as well as their timing with respect to the heart cycle [76]. Additional but not less important parameters are the end volume and the geometric shape of the IABP (see, e.g., the "bubble-blowing phenomenon" as described by Shapiro et al. [77] and Weikel et al. [78]).

In the following discussion, the effectiveness of IABPs is investigated with regard to the timing sequence. To do so, an ideal balloon is assumed for which a linear change of volume during inflation and deflation can be prescribed. At the end of, and also during, the inflation period, the geometric shape of the balloon is assumed to be the one shown in Fig. 17.16, with a total end volume of 25 mL. MaMahon et al. [79] have filmed the motions of a balloon in a hydraulic model by a high-speed motion camera. They found the balloon inflation and deflation to be much more complicated than assumed in our model. This was also reported by Schilt et al. [75], who observed clinically a possible drastic reduction of the effectiveness of an IABP due to spatially uneven inflation of the balloon. In a future study, the assumed idealized inflation-deflation patterns of the balloon should be modeled more realistically. Note that the pressure inside the IABP, p_I, required to achieve the prescribed end volume can be calculated by

$$P_I = P_0 + \bar{p} + \rho_w c_B^2 \ln \frac{S_{B,\max}}{S_{B,0}} \tag{17.28}$$

where $\bar{p}$ is the mean arterial pressure inside the artery between z_B and z_E, at the end of the balloon inflation:

$$\bar{p} = \frac{1}{z_2 - z_1} \int_{z1}^{z2} p(z)\, dz \tag{17.29}$$

Note that, according to Eq. (17.27), the cross-sectional area S_B of the balloon depends on the pressure difference $p_I - p$. Keeping p_I constant between inflation and deflation periods will then result in an increase of the balloon cross-sectional area if the arterial pressure p is decreased [minus sign in Eq. (17.27)]. The cross-sectional area of the artery, however, will decrease with decreasing p [see Eq. (17.13)]. Therefore, we have to check during the calculation whether the physical condition $S_B \leq S$ is fulfilled all the time.

17.6.3 Results of the IABP Model

The model of the standard case of the human arterial tree, defined in Sec. 17.3, was adapted to analyze the hemodynamics of the system when an IABP with a length of 22.5 cm is placed at a distance of 12.5 cm from the aortic root ($z = 0$). For this purpose an elderly patient was considered with moderately stiff arteries [compared with the standard case, the wave speed $c(p_0, z)$ was increased by 60 percent], a heart rate of 100, and a stroke volume of 40 mL. For this modified standard case, the calculated pressure and velocity patterns along the arterial tree are shown in Fig. 17.17 for the case where the pump is not in operation (even so, its location is indicated in Fig. 17.17). Note that for all the various pump actions investigated (Fig. 17.18), the same ejection pattern of the left ventricle is prescribed as a proximal boundary condition. Thus, the flow rate at $z = 0$ is independent of the output pressure, which, for the failing heart, is not a true approximation of the physiological conditions [80].

Figure 17.18 and Table 17.1 summarize the results obtained for various timings of the IABP actions. T_1 and T_2 denote the beginnings of inflation and deflation respectively, in reference to the beginning of ventricular ejection. ΔT_F and ΔT_E refer to time periods required for filling and emptying the balloon. In the parametric study described here, it was assumed that $\Delta T_F = \Delta T_E$.

For four different pump timings, the pressure data obtained at $z = 0$, the aortic root, are shown in Fig. 17.18. For comparison, the pressure curve for no pump action is also shown in this figure. Initiation of pump inflation and deflation in Fig. 17.18 are clearly marked by an abrupt pressure increase or decrease. Such sudden pressure variations associated with the pump action are typical and have also been measured in animal as well as laboratory experiments [81, 82].

In cases 2, 3, and 4 the beginnings of inflation (T_1) and deflation (T_2) are assumed to be triggered on the basis of the R wave of the ECG. In case 5, however, they are based on the initial pressure rise caused by cardiac ejection and detected by a hypothetical cathether tip manometer at the leading edge of the IABP. For none of the cases evaluated did branch A exhibit a backflow. The results for the reduction of the external heart work, and the peak as well mean values of the aortic root pressure (bifurcation of the coronary arteries) during systolic and diastolic phases, are summarized in Table 17.1.

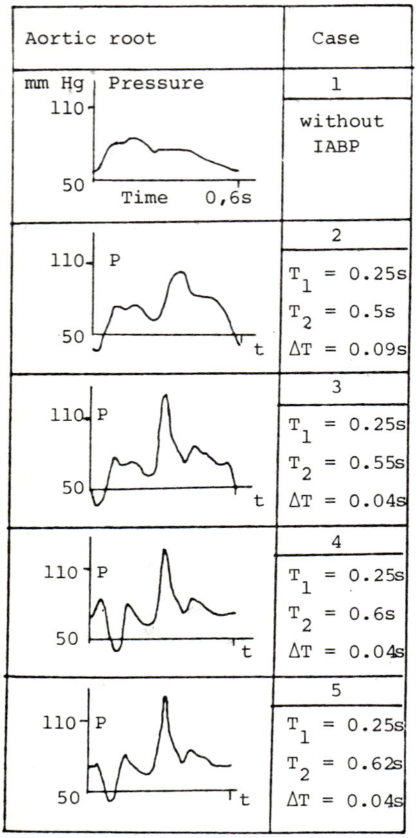

FIG. 17.18 Pressure patterns at the aortic root for different timings of balloon inflation and deflation.

TABLE 17.1 Theoretical prediction of IABP timing effects on external heart work and pressure at aortic root ($z = 0$)

		Systolic phase pressure,* mmHg			Diastolic phase pressure,* mmHg		
No.	$-\Delta E$, %	P_{min}	P_{max}	P	P_{ed}	P_{max}	P
1	. . . .	58	88	79	59	76	69
2	16	36	76	65	41	105	80
3	21	36	80	62	52	132	82
4	21	41	80	66	70	132	83
5	19	38	85	67	70	130	82

*P_{min} = minimal pressure; P_{max} = maximal pressure; P_{ed} = end-diastolic pressure; P = mean pressure.

The results show that the chosen pump actions produce reductions in external heart work ΔE from 16 to 21 percent. The mean diastolic pressure at the aortic root is increased by 11 to 14 mmHg, which results in an increase of the flow into the coronary arteries. Furthermore, a reduction in the mean systolic pressure seen by the ventricle by 12 to 17 mmHg is also achieved through the pump action. Comparison of cases 2 and 3 show that it is advantageous to keep the inflation and deflation periods T short. This is in agreement with the results of Jaron et al. [84], who developed a lumped-parameter model to represent the effects of cardiac assistance by intraaortic balloon pumping. Furthermore, the data given in Table 17.1 indicate that case 4, for which the balloon deflation can be triggered by the R wave of the ECG, displays the best performance.

REFERENCES

1. McDonald, D. A.: *Blood Flow in Arteries,* 2d ed., Williams & Wilkins, Baltimore, 1974.
2. Caro, C. G., Pedley, T. J., Schroter, R. C., and Seed, W. A.: *The Mechanics of the Circulation,* Oxford University Press, Oxford, 1978.
3. Anliker, M., Rockwell, R. L., Ogden, E.: Nonlinear analysis of flow pulses and shock waves in arteries: I. Derivation and properties of mathematical model; II. Parametric study related to clinical problems, *J. Appl. Math. Phys. (ZAMP), 22,* 217 (Part I), 563 (Part 5), 1971.
4. Renemann, R. S.: *Cardiovascular Application of Ultrasound,* North-Holland, Amsterdam, 1974.
5. Anliker, M.: Current and future aspects of biomedical engineering, *Triangle, 16,* 3, 1977.
6. Anliker, M., Casty, M., Friedli, P., Kubli, R., and Keller, H.: Noninvasive measurement of blood flow, in N. H. C. Hwang and N. A. Normann (eds.), *Cardiovascular Flow Dynamics and Measurements,* University Park Press, Baltimore, 1977.
7. Fulton, T. J., Hamilton, W. A. P., Graham, J. C., Roberts, C. V.: On-line analysis of the femoral artery flow velocity wave form and its application in the diagnosis of arterial disease, *J. Biomed. Eng., 5,* 51, 1983.
8. Jenni, R., Hübscher, W., Casty, M., Anliker, M., and Krayenbühl, H. P.: Quantitation of aortic regurgitation by a percutaneous 128-channel, digital ultrasound Doppler instrument, 3d Symposium on Echocardiology, Erasmus University, Rotterdam, June 1979.
9. Schaaf, B. W., and Abbrecht, P. H.: Digital computer simulation of human systemic arterial pulse wave transmission: A nonlinear model, *J. Biomech., 5,* 345, 1972.
10. Jones, E.: A mathematical model for nonlinear analysis of flowpulses utilizing an integral technique, *J. Appl. Math. Phys. (ZAMP), 24,* 565, 1973.
11. Rumberger, J. A., and Nerem, R. M.: A method-of-characteristics calculation of coronary blood flow, *J. Fluid Mech., 182,* 425, 1977.
12. Stettler, J. C., Niederer, P., and Anliker, M.: Theoretical analysis of arterial hemodynamics including the influence of bifurcations, pts. I, II, *Ann. Biomed. Eng., 9,* 145 (Part I), 165 (Part II), 1981.
13. Holenstein, R., Niederer, P., and Anliker, M.: A viscoelastic model for use in predicting arterial pulse waves, *ASME J. Biomech. Eng., 102,* 318, 1980.
14. Clark, M. E., and Kufahl, R. H.: Simulation of the cerebral macrocirculation, in J. Baan and A. Noordergraaf (eds.), *Cardiovascular System Dynamics,* MIT Press, Boston, 1978.
15. Van der Werff, T. J.: Significant parameters in arterial pressure and velocity development, *J. Biomech., 7,* 437, 1974.
16. Gerrard, J. H.: Numerical analysis and linear theory of pulsatile flow in cylindrical deformable tubes: The testing of a numerical model for blood flow, *Med. Biol. Eng. Comput., 20,* 49, 1982.
17. Hillen, B., Tae Gaasbeek, and Hoogstraten, H. W.: A mathematical model of the flow in the posterior communicating arteries, *J. Biomech., 15,* 441, 1982.
18. Elsner, J.: Mathematische Modellstudien der Druck—und Flusspulse in menschlichen Arterien, Diss. No. 5582, Swiss Fed. Inst. of Tech., Zürich, 1975.
19. Lister, M.: The numerical solution of hyperbolic differential equations by the method of characteristics, in *Mathematics for Digital Computers,* Wiley, New York, 1960.
20. Streeter, V. L., Keitzer, W. R., and Bohr, O. F.: Pulsatile pressure and flow through distensible vessels, *Circ. Res., 13,* 3, 1963.

21. Gerrard, J. H., and Taylor, L. A.: Mathematical model representing blood flow in arteries, *Med. Biol. Eng. Comput.*, *15*, 611, 1977.

22. Raines, J., Jaffrin, M., and Shapiro, A. H.: A computer simulation of arterial dynamics in the human leg, *J. Biomech.*, *7*, 77, 1974.

23. Ling, S. C., and Atabek, H. B.: A nonlinear analysis of pulsatile flow in arteries, *J. Fluid Mech.*, *55*, 493, 1972.

24. Kimmel, E., and Dinnar, U.: Pulsatile flow in tapered tubes: A Model of blood flow with large disturbances, *Trans. ASME, J. Biomech. Eng.*, *105*, 112, 1983.

25. Imaeda, K., and Goodman, F. O.: Analysis of non linear pulsatile blood flow in arteries, *J. Biomech.*, *13*, 1007, 1980.

26. Anliker, M., Stettler, J. C., Niederer, P., and Holenstein, R.: Prediction of shape changes of propagating flow and pressure pulses in human arteries, in R. D. Bauer and R. Busse (eds.), *The Arterial System*, Springer-Verlag, New York, 1978.

27. Remington, J. W.: Hysteresis loop behavior of the aorta and other extensible tissues, *Am. J. Physiol.*, *180*, 83, 1955.

28. Bergel, D. H.: The dynamic elastic properties of the arterial wall, *J. Physiol.*, *156*, 458, 1961.

29. Hardung, V.: Propagation of pulse waves in viscoelastic tubing, in W. F. Hamilton (ed.), *Handbook of Physiology*, sec. 2: *Circulation*, American Physiological Society, Washington, D.C., 1963.

30. Patel, D. J., Janicko, J. S., Vaishnav, R. N., and Young, J. T.: Dynamic anisotropic viscoelastic properties of the aorta in living dogs, *Circ. Res.*, *32*, 93, 1973.

31. Gow, B. S., and Taylor, M. G.: Measurement of viscoelastic properties of arteries in the living dog, *Circ. Res.*, *23*, 111, 1968.

32. Niederer, P.: Propagation and damping of shock-like transitions in the human arterial tree, *J. Appl. Math. Phys. (ZAMP)*, *36*, 204, 1985.

33. Euler, F.: Principia pro motu sanguinis per arterias determinado: Opera posthuma mathematica et physica anno 1844 detecta, in P. H. Fuss and N. Fuss (eds.), *Petropoli Apud Eggers et Socios*, 1862, vol. 2.

34. Rudinger, G.: Review of current mathematical methods for the analysis of blood flow, *Biomedical Fluid Mechanics Symposium*, American Society of Mechanical Engineers, New York, April, 1966.

35. Witzig, K.: Ueber die Wellenbewegung zäher Flüssigkeiten in elastischen Röhren, Ph.D. dissertation, University of Bern, 1914.

36. Paulsen, P. K., and Hasenkam, J. M.: Three-dimensional visualization of velocity profiles in the ascending aorta in dogs, measured with a hot-film anemometer, *J. Biomech.*, *16*, 3, 1983.

37. Chandran, K. B., Yearwood, T. L., and Wieting, D. W.: An experimental study of pulsatile flow in a curved tube, *J. Biomech.*, *12*, 793, 1979.

38. Gerrard, J. H.: The effect of the skin friction on the solution of the one dimensional equations of pulsatile flow in distensible tubes, *Med. Biol. Eng. Comput.*, *19*, 79, 1981.

39. Trikha, A. K.: An efficient method for simulating frequency dependent friction in transient liquid flow, *Trans. ASME Ser. I., J. Fluid Eng.*, *97*, 97, 1975.

40. Zielke, W.: Frequency-dependent friction in transient pipe flow, *ASME J. Bas. Eng.*, *90*, 109, 1968.

41. Brown, F. T.: A quasi-method of characteristics with application to fluid lines with frequency dependent wall shear and heat transfer, *ASME J. Bas. Eng.*, *91*, 217, 1969.

42. Skalak, R., and Stathis, T.: A porous tapered elastic tube model of a vascular bed, *Symposium on Biomechanics*, American Society of Mechanical Engineers, New York, 1966.

43. Holenstein, R., Nerem, R. M., and Niederer, P.: On the propagation of a wave front in viscoelastic arteries, *ASME J. Biomech. Eng.*, *106*, 115, 1984.

44. Prandtl, L.: *Führer durch die Strömungslehre*, 3d ed., Braunschweig, 1948.

45. Walburn, F. J., and Stein, P. D.: Flow in a symmetrically branched tube simulating the aortic bifurcation: The effects of unevenly distributed flow, *Ann. Biomed. Eng.*, *8*, 159, 1980.

46. Cheng, L. C., Clark, M. E., Robertson, J. M., and Chao, N. H.: Effects of flow division on bifurcation flow characteristics, in D. J. Schneck (ed.), *Proceedings First Mid-Atlantic Conference on Bio-Fluid Mechanics*, Virginia Polytechnic Institute and State University, Blackburg, 1978.

47. Karino, T., Kwong, N. H. M., and Goldsmith, H. L.: Particle flow behavior in models of branching vessels: I. Vortices in 90° T-junctions, *Biorheology*, *16*, 231, 1979.

48. Young, D. F., and Tsai, F. Y.: Flow characteristics in models of arterial stenosis: I. Steady Flow; II. Unsteady flow, *J. Biomech., 6,* 395 (I), 547 (II), 1973.

49. Youngchareon, W., and Young, D. F.: Initiation of turbulence in models of arterial stenosis, *J. Biomech., 12,* 185, 1979.

50. Daly, B. J.: A numerical study of pulsatile flow through stenosed canine femoral arteries, *J. Biomech., 9,* 465, 1976.

51. Sabbah, H. N., and Stein, P. D.: Effect of aortic stenosis on coronary flow dynamics: Studies in an in-vitro pulse duplicating system, *ASME J. Biomech. Eng., 104,* 221, 1982.

52. Bollinger, A.: *Funktionelle Angiologie,* Thieme, Stuttgart, 1979.

53. Benett, C. O., and Myres, J. E.: *Momentum, Heat, and Mass Transfer,* 2d ed., McGraw-Hill, New York, 1974.

54. Fukushima, T., Azuma, T., and Matsuzawa, T.: Numerical analysis of blood flow in the vertebral artery, *ASME J. Biomech. Eng., 104,* 143, 1982.

55. Rooz, E., Young, D. F., and Rogge, T. R.: A finite element simulation of pulsatile flow in flexible obstructed tubes, *ASME J. Biomech. Eng., 104,* 119, 1982.

56. Anliker, M., Histand, M. B., and Ogden, E.: Dispersion and attenuation of small artificial pressure waves in the canine aorta, *Circ. Res., 23,* 539, 1968.

57. Schönfeld, D., Atabek, H. B., and Patel, D. J.: Geometry and elastic response of the aortic-iliac junction, *J. Biomech., 12,* 483, 1979.

58. Spencer, M. P., and Denison, A. B.: The squarewave electro-magnetic flowmeter: Theory of operation and design of magnetic probes for clinical and experimental applications, *IRE Trans. Med. Elect., 6,* 220, 1959.

59. Wetterer, E., and Kenner, T.: *Dynamik des Arterienpulses,* Springer-Verlag, Berlin, 1968.

60. Skalak, R., Keller, S. R., and Secomb, T. W.: Mechanics of blood flow, *ASME J. Biomech. Eng., 103,* 102, 1981.

61. Shapiro, A. H.: Steady flow in collapsible tubes, *ASME J. of Biomech. Eng., 99,* 126, 1977.

62. Stettler, J. C.: Theoretische Untersuchungen der arteriellen Hämodynamik unter Berücksichtigung von Verzweigungen, Diss. Swiss Fed. Inst. of Tech. 6376, 1979.

63. Westerhof, N., Bosmann, F., DeVries, C. J., and Noordergraaf, A.: Analog studies of the human systemic arterial tree, *J. Biomech., 2,* 121, 1969.

64. Snyder, M. F., Rideout, V. C., and Hillestad, R. J.: Computer modeling of the human systemic arterial tree, *J. Biomech., 1,* 341, 1968.

65. Forbes, L. K.: A note on the solution of the one-dimensional unsteady equations of arterial blood flow by the method of characteristics, *J. Aust. Math. Soc. Ser. B, 21,* 45, 1979.

66. Wetterer, E.: Bau und Funktion des Gefässsystems, in W. D. Kreidel (ed.), *Physiologie,* Thieme, Stuttgart, 1975.

67. Holenstein, R.: Viskoelastizität von Arterien: mathematische Beschreibung und Einfluss auf die Druck—und Flusspulse, Diss. Swiss Fed. Inst. of Tech. 6801, 1981.

68. Casty, M.: Differenzierung zwischen normalen und pathologischen Strömungsverhältnissen im Arteriellen Kreislauf mittels gepulstem Ultraschall, Diss. Swiss Fed. Inst. of Tech. 6626, 1980.

69. Anliker, M.: The instantaneous velocity profile in large arteries as a diagnostic indicator of abnormalities of the vessel wall surface, in S. Effert and J. D. Meyer-Erkelenz (eds.), *Blood Vessels,* Springer-Verlag, Berlin, 1976.

70. Ganong, X.: *Lehrbuch der medizinischen Physiologie,* 3d ed., Springer-Verlag, Berlin, 1900.

71. Nerz, A. R., Myerowitz, P. D., and Blackshear, P. L., Jr.: A simulation of the dynamics of counterpulsation, *ASME J. Biomech. Eng., 101,* 105, 1979.

72. Brubakk, A. O., and Aaslid, R.: Use of a model for simulating individual aortic dynamics in man, *Med. Biol. Eng. Comput., 16,* 231, 1978.

73. Philippe, E., Clark, J. W., Lande, A., and Ellis, J. R.: Microprocessor control of intra-aortic balloon pumping, *Ann. Biomed. Eng., 8,* 209, 1980.

74. Anliker, M., Moritz, W. E., and Ogden, E.: Transmission characteristics of axial waves in blood vessels, *J. Biomech., 1,* 235, 1968.

75. Schilt, W., Freed, P. S., Ghassa, K., and Kantrowitz, A.: Temporary non-surgical intra-arterial cardiac assistance, *Trans. Am. Soc. Artif. Intern. Organs, 13,* 322, 1967.

76. Anliker, M., Stettler, J. C., Niederer, P., and Schilt, W.: Optimal operation of intra-aortic balloon pumps, in *Conference Digest, European Conference of Artificial Organs*, Zürich, 1978.

77. Shapiro, A. H., Jaffrin, M., Clark, C., Murthy, V. S., and McMahon, T. A.: Fluid mechanics of intra-aortic balloon counterpulsation, in *Proc. Artif. Heart Program Conf.*, Washington D.C., 1969, U.S. Department of Health and Human Services, 1970.

78. Weikel, A. M., Jones, R. T., Dinsmore, R., and Petschek, H. E.: Size limits and pumping effectiveness of intra-aortic balloons, *Ann. Thorac. Surg.*, *12,* 45, 1971.

79. McMahon, T. A., Clark, C., Murthy, V. S., and Shapiro, A. H.: Intra-aortic balloon experiments in a lumped-element hydraulic model of the circulation, *J. Biomech.*, *4,* 335, 1971.

80. Jones, R. T., and Laird, J. D.: Simulation of the human aorta for the purpose of studying artificial blood pumps, presented at the *Symposium on Prospects for Simulation and Simulators of Dynamic Systems*, Baltimore, 1966.

81. Clark, J. W., Philippe, E., Peterson, R., Lande, A., and Ellis, J. R.: Automatic control of a series-parallel mechanical circulatory assist system in severe uni- or biventricular failure, *Ann. Biomed. Eng.*, *8,* 57, 1980.

82. Martin, P. J., and Jaron, D.: New controller for in-series cardiac-assist devices, *Med. Biol. Eng. Comput.*, *16,* 243, 1978.

83. Williams, M. J., Jr., Rubin, J. W., and Ellison, R. G.: Experimental determination of optimum performance of counterpulsation assist pumping under computer control, *Comput. Biomed. Res.*, *10,* 545, 1977.

84. Jaron, D., Moore, T. W., and Ping, H.: Theoretical considerations regarding the optimization of cardiac assistance by intraaortic balloon pumping, *IEEE Trans. Biomed. Eng.*, 30, 177, 1983.

CHAPTER 18

Mechanics of Blood Flow in the Microcirculation

Herbert H. Lipowsky
Department of Physiology
College of Physicians and Surgeons
Columbia University
New York, New York

INTRODUCTION

> *Facts are necessary, of course, but unless fertilized by ideas, correlated with other facts [and] illuminated by thought, I consider them as material only for science. . . . Too many experiments and observations are being made and published and too little thought bestowed upon them.*
>
> AUGUST KROGH, 1929

In the five decades following this caveat by the eminent microvascular physiologist August Krogh, over 500 publications have presented both qualitative and quantitative observations on intravascular pressures and flows in the microcirculation. Furthermore, with the advent of new techniques and sophisticated instrumentation for direct intravital microscopy, about 90 percent of these studies have appeared within the last 15 years. Attendant to this exponential growth of physiological studies have been many contributions from the physical sciences which have provided many tools for acquiring new data, understanding old and new concepts, and correlating numerous observations. This amalgam of physiology and engineering, embodied in the concept of bioengineering, has brought together two partners with seemingly disparate and sometimes discordant methodologies, as evidenced by the almost uncontrolled avalanche of independent physiological variables and the requisite rigor and orderliness of engineering. Nonetheless, the direct application of the analytical tools of fluid mechanics, applied mechanics, and electrical engineering, as well as of many other disciplines, has resulted in placing the mechanics of microvascular blood flow on a firm foundation.

As a first step in unraveling the mysteries of microvascular blood flow, numerous in vitro studies have concentrated on evaluating the intrinsic properties of blood as a viscous fluid, with the aim of providing a frame of reference for analysis of in vivo hemorheological behavior at both the macro-and microcirculatory levels [1–4]. On the basis of in vitro experiments in tube, cone-plate, and Couette viscometers, the non-Newtonian properties of blood have been characterized in terms of increases in apparent viscosity with increasing hematocrit, red cell aggregation attendant to a decrease in shear rates, and decreased red cell deformability. In vitro investigations into the rheological behavior of blood at the microcirculatory level by simulations in small-bore tubes, and both direct and indirect in vivo methods, have served to highlight the departure of blood from a Newtonian fluid and have generally precluded the acceptance of equations, such as the well-known Poiseuille-Hagen relationship [5, 6], describing Newtonian fluid behavior in small-bore tubes. For example, it was reported by Fahraeus [7] that the hematocrit of blood flowing in small-bore tubes diminished in comparison to its value in the feeding reservoir as tube diameter was decreased to about 40 μm. This effect has been attributed to changes in the ratio of average red cell to plasma velocity, such that the dynamic tube hematocrit can be substantially less than the feed value. As tube diameter is decreased, red cell velocity increases with respect to the mean velocity of plasma, while the conservation of red cell flux is maintained, as represented by the equality of the discharge hematocrit at the tube exit and the feed reservoir value. This finding was followed by the discovery of Fahraeus and Lindqvist [8] that the apparent viscosity of blood was reduced as well in small-bore tubes in comparison with values obtained in relatively large scale viscometric instruments.

In vivo observations on blood rheological behavior by Whittaker and Winton [9] provided the first indirect evidence on how such properties of suspensions might affect microvascular resistance. By perfusing isolated hind limbs of dogs with suspensions of erythrocytes, they found that the in vivo apparent viscosity of blood was almost half that measured in macroscale viscometers (large-bore tubes). Similar results have been obtained within the last 20 years for the isolated rabbit ear [10, 11], for the tail artery of the rat [12], and for the isolated calf muscle of the cat [13, 14], to cite a few. In an attempt to resolve these differences between in vitro and in vivo determinations of apparent viscosity, studies of isolated organs have been conducted which attribute this disparity to the presence of inertial losses outside the microcirculation proper [15, 16]. In these experiments, a measure of inertial effects was obtained by analysis of the nonlinearity in pressure drop (ΔP) versus flow (Q) curves of an isolated vascular segment, by expressing their relationship as $\Delta P = aQ + bQ^2$, where the latter term reflects inertial losses. The results suggested the presence of significant inertial losses (as much as 40 percent of ΔP, for perfusion with cell-free media), which diminished (to 5 percent of ΔP) as the hematocrit was increased to 0.5 [16]. Other studies provided evidence that the nonlinearity of the pressure-flow curves could be eradicated when maximum vasodilatation was pharmacologically induced [17], thus ascribing this nonlinear relationship to vascular dimensional changes with pressure.

Based upon the pioneering work of Fahraeus [7], Fahraeus and Lindqvist [8], and Whittaker and Winton [9], an understanding of the basic factors governing the in vivo rheological behavior of blood in the microcirculation has evolved over the years. It is now clearly recognized that blood is a complex suspension of discrete particles which are free to interact kinematically and chemically with and within the confines of the vascular compartment. The resultant resistance to blood flow stems from both vascular (topographical and geometric) as well as intravascular (rheological) factors. Also, these two sets of factors are free to interact with one another to yield a state of rheological equilibrium throughout the terminal vascular network, under specified arteriovenous (AV) boundary conditions in pressure and/or flow.

Within this framework, intravascular factors encompass the relationship between blood velocity (shear rates), pressure gradients, hematocrit, and white cell count, as they apply to single unbranched microvessels and at bifurcation points. The determinants of this relationship have been shown to be red cell deformability and aggregation, leukocyte deformability, and adhesion of leukocyte to endothelial cell, as well as the very nature of the blood-endothelium contact surface itself.

Vascular or topographical features of the microvascular network can influence this relationship both directly and indirectly at any level of the hierarchy of microvessels from large arterioles to large venules. Direct effects on these intravascular factors may be derived principally through alterations in microvessel diameter by regulatory behavior evoked by a variety of physiological mechanisms. Furthermore, the branching pattern of the network itself and the deployment of microvessels with specific lengths and diameters play the major roles in determining the local portion of the overall AV pressure drop which appears along the length of any individual microvessel. This local pressure gradient will also be affected by the combination of vascular and intravascular factors within all other parts of the network. Hence, with all factors considered, the task of evaluating the in vivo rheological behavior of blood within the microvascular network is indeed quite complex.

The first approach to this problem was presented in the pioneering investigations of Landis [18]. To separate the interaction between network topography and hemorheology, Landis used the studies of Poiseuille to compute pressure gradients in individual microvessels of the frog mesentery, using values of apparent viscosity from in vitro measurements and in situ measurements of red cell velocity. On comparison of these computed pressure gradients with those measured in his earlier studies [19], he concluded that the Poiseuille-Hagen law was not directly applicable to the microcirculation. While the results of these early studies have been superseded by studies employing newly developed instrumentation for in vivo experiments, his methodology remains today a prime example of the means by which one may examine the rheological basis of microvascular blood flow.

In the following sections, some of the advances in intravital microscopy will be described briefly, with particular reference to their application to the description of topographical effects and the rheological behavior of blood throughout the microvascular network and within individual microvessels.

18.1 INSTRUMENTATION AND METHODS

The ability to perform direct in situ measurements of pressure and flow relationships in the microcirculation has grown steadily over the past 10 to 20 years with the development of electromechanical and optical techniques. With these new tools, a wealth of data on intravascular pressure and red cell velocity has been obtained for microvessels ranging in size from about 100 μm for arterioles and venules down to 5 to 7 μm for true capillaries. The development and application of these techniques is briefly summarized as follows.

Techniques of intravital microscopy have been applied to a wide variety of tissues in many species of animals. The basic approach has been to expose a thin region of tissue with a minimum of dissection and surgical trauma, and, with suitable mechanical support, to view the tissue through the microscope under either transmitted or incident illumination. To ensure the viability of the exposed tissue, the preparation is usually irrigated with a physiological salt solution which has been heated to body temperature (37°C). An excellent survey of these methods may be found in Ref. 20.

The measurement of intravascular pressure in vessels comparable in diameter to the size of an individual red blood cell has become an almost routine procedure with the development of the servo-null technique. Today's readily available electromechanical techniques were developed from the early techniques of Landis [19], who employed finely drawn micropipets with external tip diameters on the order of 5 to 10 μm which were filled with a dye solution. Following intubation of a microvessel, static equilibrium of the dye-plasma interface was used to indicate the equilibration of intravascular pressure and that of a known pressure applied to the back end of the pipet. This technique was mechanized some 30 years later by Wiederhielm et al. [21], who used pipets filled with electrically conducting 1-*M* NaCl. Upon initially aspirating a small amount of the lesser conducting 0.155-*M* plasma into the pipet tip, the electrical conductivity could be monitored as a function of the pressure changes external to the tip. The resultant alterations in conductivity were fed to a bridge circuit, and deviations from the initial balance at atmospheric pressure were then amplified to drive a pump connected to the back end of the micropipet. The resulting servomechanism generated a pressure equal and opposite to that external to the tip of the pipet by maintaining a constant pipet conductivity. Through this servo-null approach, intravascular pressures could be monitored with a fidelity which satisfactorily provided both steady and pulsatile pressure measurements within an accuracy of a few percent.

Further development of the servo-null system by Intaglietta et al. [22] facilitated the routine use of this method in many laboratories and its subsequent application to the measurement of pressure drops along single unbranched microvessels for rheological studies [23]. In this latter approach, two servo-null systems were coupled by a highly sensitive differential pressure transducer in order to permit the measurement of upstream-to-downstream pressure drops to within an accuracy of 0.05 cmH_2O.

Methods for the measurement of red cell velocity have also undergone substantial developments and refinements over the years. Utilizing the techniques of cross-correlation, Wayland and Johnson [24] demonstrated that red cell velocity could be easily obtained under the microscope by monitoring the fluctuations in light intensity caused by the passage of red cells at two axial stations along the length of a given microvessel. This procedure was accomplished by projecting the image of the vessel on the surface of a screen or mirror which had two slits cut out at an effective spacing of about 7 μm. The temporal fluctuations in light intensity transmitted through the slits were then converted to electric signals with the use of two photodetectors. The resultant photometric signatures were then subjected to on-line cross-correlation to determine the transit time of the red cells from one slit to the other. With the transit time and distance known, the velocity could then be computed with appropriate analog circuitry. Additional refinements in the design of the electronic system for performing the cross-correlation [25] have made this tool readily available for the measurement of steady and pulsatile velocities to within an accuracy of 5 percent, and with a frequency response flat to about 5 Hz. Further developments have demonstrated that the measured red cell velocity can be empirically related to the mean velocity of red cells plus plasma (V_m) [26–30]. In these studies, it was shown that the mean velocity of the suspension (V_m) was related to the velocity of red cells measured along the vessel centerline (V_{rbc}) by the expression $V_m = V_{rbc}/1.6$.

To complete the instrumentation needed for analysis of blood rheological behavior, optical techniques have been developed which permit the in situ measurement of microvessel hematocrit and vessel luminal diameter. It has been demonstrated [31] that the optical density of blood flowing in small-bore tubes may be uniquely related to the intraluminal hematocrit. Optical density, defined as the logarithm of the ratio of incident to transmitted light intensity, i.e., log (I_0/I), can be measured with the same photodetectors used for the velocity measurements [32–34], or by video densitometry [35, 36]. Microvessel hematocrits have also been measured by direct cell counting of the number of red blood cells (RBCs) per vessel volume [37, 38] or by sampling the cell count with fluorescently tagged tracer RBCs [39].

Measurements of microvessel diameter have also become routine to within the limits of resolution offered by the light microscope (about 0.5 μm). By use of the concept of image shearing devised by Baez [40], techniques using closed-circuit video systems have been developed [41]. This approach relies upon electronically displacing, or "shearing," a portion of the video image to produce an analog signal proportional to the displacement of two landmarks within the microscopic field of view. By use of the inner walls of a microvessel as the targets for the shearing, luminal diameter measurements can be accurately and quickly made simultaneously with measurements of the hemodynamic parameters.

18.2 NETWORK DISTRIBUTIONS OF HEMODYNAMIC PARAMETERS

18.2.1 Intravascular Pressures

Direct in situ measurements of intravascular pressure and red cell velocity have served to characterize their distribution throughout successive microvascular divisions in many tissues. Evaluation and interpretation of the results of these studies have usually pursued the relationship between network structure (branching patterns) and functional behavior, in light of the needs of the parenchymal tissue (i.e., convective transport and transcapillary exchange of water and solutes). To achieve the basic goal of relating network structure to function, the principal theme has been to seek a quantitative evaluation of network hemodynamics in light of commonalities between networks of various configurations and their departures from the norm under pathophysiological conditions.

To demonstrate the general features of network hemodynamics, Fig. 18.1 presents the AV distributions of intravascular pressures and red cell velocities versus luminal diameter for two splanchnic and two skeletal muscle beds. Luminal diameter may be interpreted as an index of

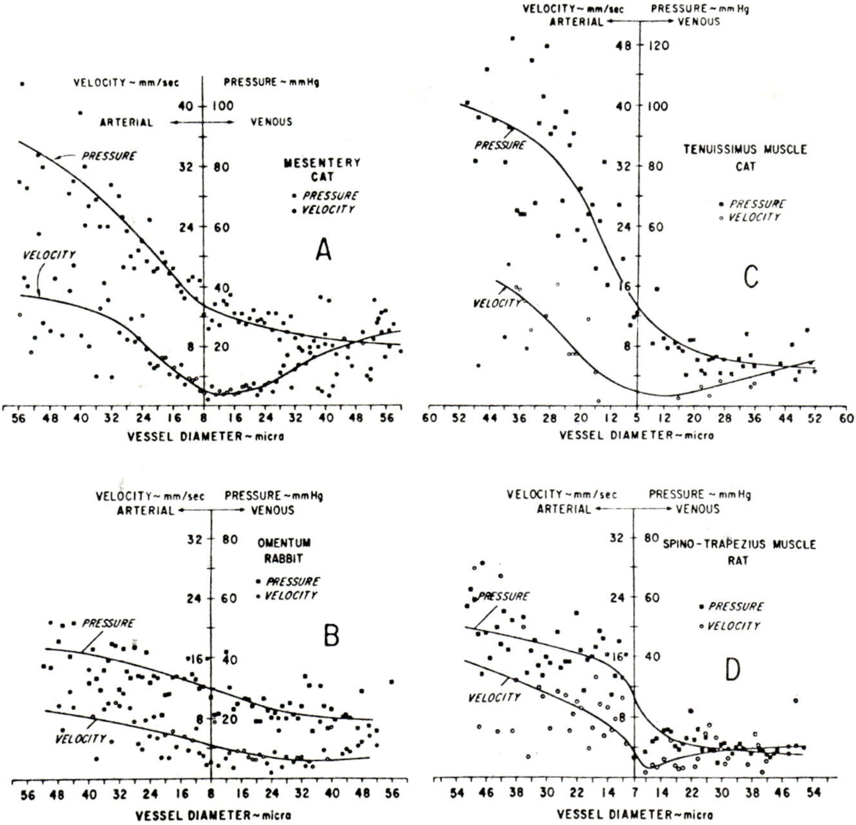

FIG. 18.1 Arteriovenous distributions of intravascular pressure and red cell velocity in four representative networks. The solid curves are piecewise cubic spline fits. *(From Zweifach and Lipowsky [42], with permission.)*

position within the overall hierarchy of microvessels, for the mesentery (cat), omentum (rabbit), spinotrapezius muscle (rat), and tenuissimus muscle (cat) [42]. All measurements have been obtained for the normal resting state of these tissues, and individual pressures and velocities have been averaged over three to five cardiac cycles. From these distributions, it is evident that each network has a hemodynamic pattern characteristic of its topography, overall AV pressure drops, flows, and blood rheological properties. Similarities in the function of each network may be gleaned from these distributions by a suitable normalization of the data. Specifically, under the assumption that the distribution of intravascular pressures has been sampled in sufficiently large numbers (which appears satisfied by the 300 to 500 measurements obtained for each tissue) one may address the attendant variation of intravascular pressure with diameter as being indicative of how the total throughput of the network is apportioned by a succession of hypothetical serial resistances between each functional division (arterioles, precapillaries, capillaries, etc.). Normalization of the intravascular pressure (P) with respect to the arteriolar and venular pressures (P_a and P_v, respectively) has been proposed to represent the fraction of the total overall network resistance distal to a microvessel of a specific size (position in the network), namely, $\overline{P} = (P - P_v)/(P_a - P_v)$ [43].

The resultant AV distributions of $\overline{P}$ computed for P_a and P_v corresponding to pressures in

50-μm arterioles and venules, respectively, shown in Fig. 18.2, demonstrate the similar manner in which pressure decreases throughout each network. The fraction of total network resistance required to attenuate arteriolar pressure to its true-capillary values ranges from about 0.2 for mesentery and tenuissimus muscle to about 0.45 for the omentum and spinotrapezius preparations. The magnitude of pressure P at the true-capillary level is maintained, presumably to assure that capillary hydrostatic pressure will be in equilibrium with tissue hydrostatic pressure and oncotic forces to preserve blood-to-tissue fluid balance. Hence, capillary pressure will not vary widely from one tissue to another, although the means by which these reductions are incurred is affected by the network topography.

The arteriovenous distribution of the resistance to the total throughput of each network can be visualized more clearly by examining the slopes of P and the normalized pressure $\bar{P}$ in Fig. 18.2. Specifically, if one treats the abscissas of Fig. 18.2 as rightward-running coordinates by taking the diameters of the arterioles as negative and those of the venules as positive (i.e., transforming to a coordinate system $x = D_{\text{cap}} - D$ for the arterioles, and $x = D - D_{\text{cap}}$ for the venules), then the slope of pressure or flow may be computed from $d\bar{P}/dx$ or $d\bar{Q}/dx$, respectively, with $d/dx = d/dD$. Presented in Fig. 18.3 are the AV distributions of $d\bar{P}/dD$ as obtained for these four networks. (The negative of $d\bar{P}/dD$ is plotted since the monotonically decreasing pressure distribution yields a negative slope.) This representation of the network gradients in pressure delineates the specific location of the major sites of resistance to flow throughout the ensemble of microvessels and agrees fairly well with topographical features. For example, the mesentery shows its maximum network resistance occurring in the precapillary microvessels ranging in diameter from 15 to 35 μm, which is consistent with observations on the arcading modular structure of this network [44]. The omentum, however, shows a maximum pressure gradient at the true-capillary level, consistent with its nearly dichotomous branching pattern [45]. The two skeletal muscle beds indicate a maximum falloff in pressure in the immediate vicinity of the precapillary vessels, which may be identified with small arterioles which run transversely to the muscle fibers prior to the final ramification at the true capillaries [46, 47].

18.2.2 Volumetric Flow Rates

While the AV distributions of P and $\bar{P}$ are indicative of the relative deployment of network resistances, in terms of serial and/or parallel elements, the AV distribution of flow can be used

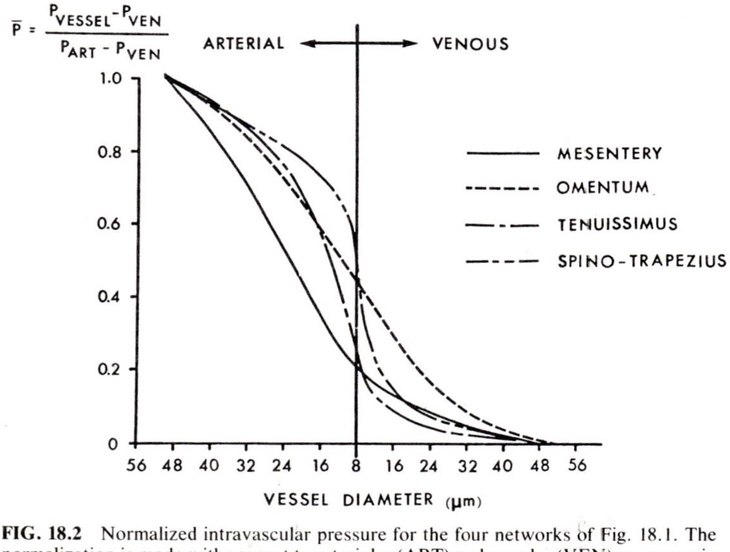

FIG. 18.2 Normalized intravascular pressure for the four networks of Fig. 18.1. The normalization is made with respect to arteriolar (ART) and venular (VEN) pressures in 50-μm vessels. *(From Zweifach and Lipowsky [42], with permission.)*

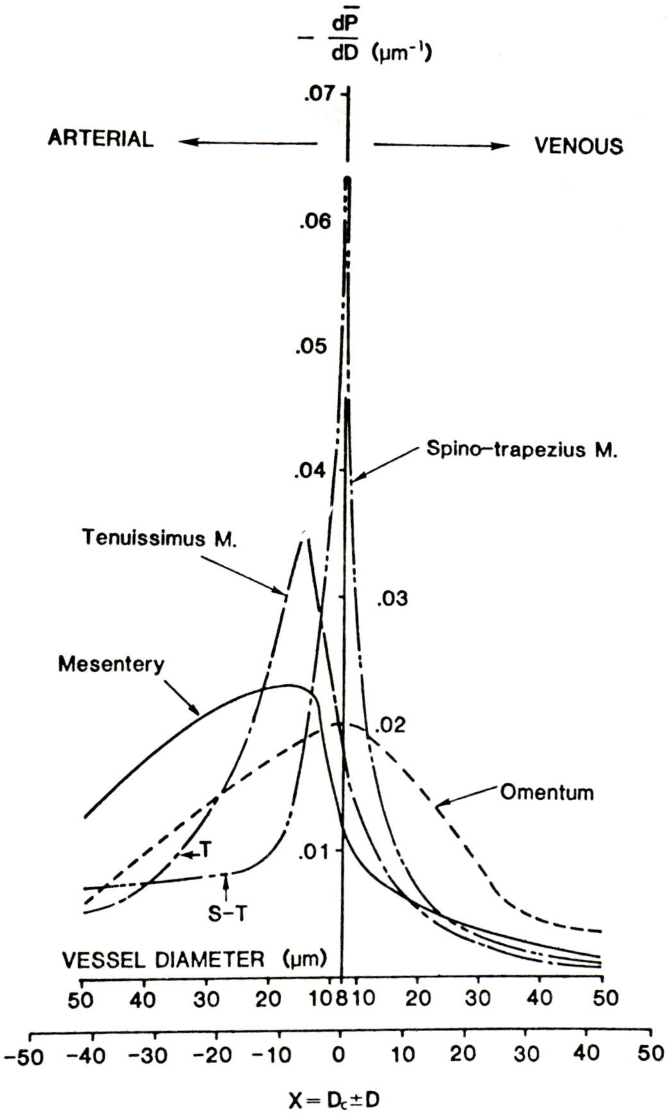

FIG. 18.3 The slopes of the normalized pressure distributions of Fig. 18.2 serve to demonstrate the rate of pressure attenuation throughout successive network divisions. The slopes were obtained by arbitrarily transforming vessel diameter into a rightward-running abscissa with $x = D_{cap} - D$ for the arterioles and $x = D - D_{cap}$ for the venules.

to gain insight into the relative number of microvessels which are effectively in parallel at each division to carry off the total throughput of the network. Using empirical relationships between red cell velocity (V_{rbc}) and mean velocity of cells plus plasma (V_m) [26, 27, 48], the bulk volumetric flow can be computed as $Q = V_m \pi D^2/4$, and its AV distribution plotted as in Fig. 18.4 for the mesenteric circulation [49]. In light of the limited variation of V_{rbc} (Fig. 18.1*a*) relative to D^2, it appears that microvessel cross-sectional area dominates the AV distribution of

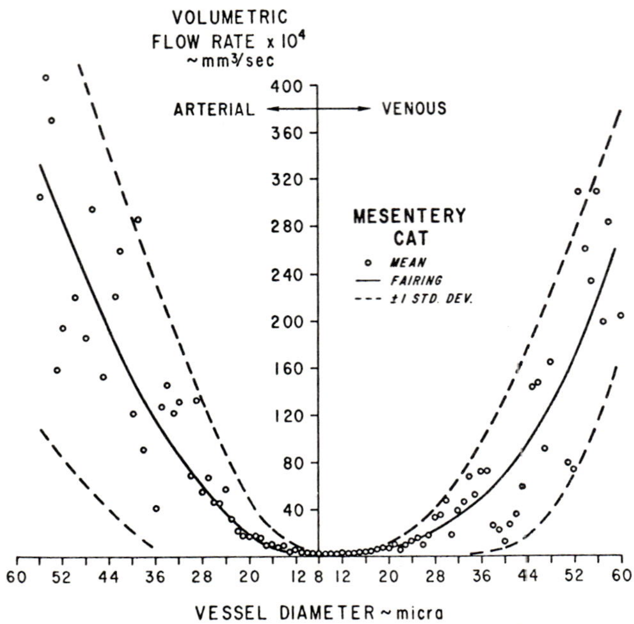

FIG. 18.4 Arteriovenous distribution of volumetric flow rate for the mesenteric microvasculature. The piecewise spline fit of the data (solid curve) is bounded by the standard deviations of the data (dashed curves). *(From Zweifach and Lipowsky [49], with permission.)*

flow, thus yielding a parabolic distribution. In most preparations, the minimum flow appears to fall in the immediate postcapillary microvessels following the true capillaries.

Analyses of the relationship between network topography and the AV distribution of volumetric flow have been made to explore the possibility of an optimization in network design [50, 51]. On the basis of the hypothesis that the energy expended in perfusing successive divisions of the vasculature is proportional to blood volume within a segment [52], the principle of minimum work was applied to examine the relationship between flow (Q) and luminal diameter (D). By use of the Poiseuille-Hagen equation to calculate the viscous dissipation within specific microvessels, and by minimization of the total work performed as blood traverses successive divisions (if vascular volume is taken as proportional to D^2), it was computed that Q is inversely proportional to D^3. Power law regressions of in vivo data ($Q = kD^m$) revealed that m varied from 2.88 to 2.99 for the cremaster muscle, omentum, and tenuissimus muscle [51]. Also, paired measurements of Q and D in the cremaster muscle of both normal and hypertensive rats, with presumably subtle differences in vascular topography attendant to this disease process, demonstrated an exponent of $m = 3.01$ for arteriole diameters ranging from 6 to 108 μm [52]. These findings have led to the conclusion that $Q = kD^3$ represents a common functional theme of the microvascular network.

As noted therein, such a relationship may also arise from a state of uniform viscous dissipation throughout successive segments. For example, for a Newtonian fluid flowing in a tube, the energy due to viscous dissipation (E) is given as $\eta\dot{\gamma}^2$, where the Newtonian wall shear rate ($\dot{\gamma} = 8V_m/D$) may be related to flow as $\dot{\gamma} = 32Q/\pi D^3$. From in situ measurement of $\dot{\gamma}_w$ throughout the mesenteric network (see Fig. 18.11), it appears that $\dot{\gamma}_w$ varies by only a factor of 3 whereas D^3 may vary over three decades. This behavior suggests that $\dot{\gamma}_w$ and hence E (for $\eta \approx$ constant) are relatively invariant throughout the network.

18.2.3 Transit Times

Further attempts to relate microhemodynamics to network topography have employed tracer dilution techniques based upon the Stewart-Hamilton principle for measuring regional blood flow [53]. The basic approach entails the computation of mean transit time ($\bar{t}$) of blood from an upstream site, e.g., a small artery where an indicator is injected, to specific locations within the microvascular network. The value of $\bar{t}$ is usually determined from in situ measurements of indicator concentration within individual microvessels from the relationship

$$\bar{t} = \frac{\int_0^\infty c(t)t\,dt}{\int_0^\infty c(t)\,dt}$$

A variety of indicators have been used such as saline [54], fluorescently labeled plasma [55], plasma labeled with Evans blue dye [56–58], fluorescently tagged red cells [59], and sulfuronated red cells which are significantly darker than normal red cells [60]. The results of these studies have addressed the relative velocities of red cells and the suspending medium within individual microvessels, as well as the topographical distribution of $\bar{t}$ throughout successive microvascular segments.

In the context of describing network hemodynamics, these studies have clearly demonstrated the departure of the network pattern from an orderly succession of serial segments, each comprised of numerous parallel elements. To illustrate this point, Fig. 18.5 presents the frequency distribution of $\bar{t}$ measured from the superior mesenteric artery to 41 arterioles and 49 venules with diameters from 10 to 70 μm [54]. Although the average value of $\bar{t}$ for the venule (5.6 s) is longer than that for the arteriole (3.8 s), the substantial overlap of the two frequency distributions is indicative of the intertwining of the constituent vessels. Anatomical views of the mesenteric network topography support this finding, as evidenced by the arteriole-to-arteriole and venule-to-venule anastomoses typical of its arcading pattern of vessels [44].

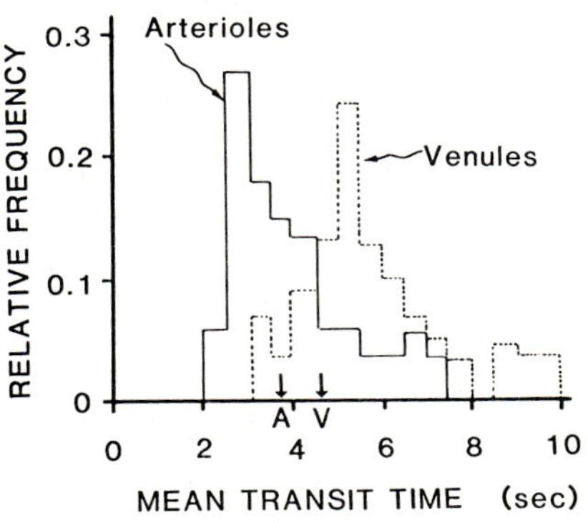

FIG. 18.5 Frequency distribution of mean transit time of red cells from the superior mesenteric artery to arterioles and venules within the mesenteric network. The substantial overlap of the two classes of microvessels vividly demonstrates the great extent of vascular intertwining and departure from a serial deployment of microvessels. *(Redrawn from Nellis and Lee [54].)*

Application of transit-time techniques to individual microvessels have served to quantitate the differences between mean red cell velocity (V_{rbc}) and mean plasma velocity (V_{pl}) [56–58]. Upon comparison of red cell $\bar{t}$ with that of a bolus of dye solution, in vivo measurements of V_{rbc}/V_{pl} were found to average 1.38 ± 0.18 SD (standard deviation) for capillaries ranging in diameter from 6.2 to 11.6 μm [56]. Under comparatively higher flow rates, in vitro studies in 11-μm glass tubes resulted in a ratio of 1.84, suggesting a possible shear rate dependency arising from red cell deformation [58]. Studies aimed at addressing the disparity of red cell and plasma transit times throughout successive microvascular segments, employing labeled plasma and red cells, have shown consistently shorter transit times of red cells compared with those of plasma at all divisions of the cremasteric network [59].

18.3 HEMODYNAMICS IN INDIVIDUAL MICROVESSELS

18.3.1 Pressure Gradients

While it is clearly recognized that the pressure gradient along the length of an individual microvessel ($\Delta P/l$) serves to determine the flow for a given luminal diameter and magnitude of apparent viscosity of blood, with the possibility of numerous interconnections throughout the network, the relationship between single-vessel pressure gradients and those throughout the network remains obscure. The network gradient of pressure presented in the preceding section is a functional parameter which expresses the variability of an intravascular pressure with respect to specific anatomical landmarks (arterioles, precapillaries, etc.). Measurements of pressure gradients within individual microvessels vividly demonstrate these differences, as illustrated in Fig. 18.6, where the upstream-to-downstream pressure drop, divided by vessel length, is presented for the mesenteric network [61]. Also shown is the corresponding AV distribution of network pressure gradient from Fig. 18.3. Figure 18.6*b* reveals that for the single vessels, $\Delta P/l$ increases steadily to attain a maximum value at the true-capillary level, as indicated by the piecewise spline fit of these in vivo data. In contrast, the network gradient achieves its maximum in the precapillary to small arteriolar vessels ranging in diameter from 15 to 35 μm. While this disparity can be attributed, in part, to differences in length scale $l(D)$ in formulating the network gradient, doing so yields unreasonable estimates. For example, the value of $dP/dD = dP/dx$ can be corrected for length scale by taking advantage of the fact that $dP/dD = (dP/dl)(dl/dD)$. In order to achieve agreement between dP/dD and $\Delta P/l$, one must demonstrate a decrease in vessel length by a factor of 4 in going from 30-μm arterioles to true capillaries ($D = 7$ μm), to facilitate a rise in network pressure gradient to attain a maximum at the true-capillary level.

Analyses of unbranched microvessel lengths as a function of luminal diameter do not reveal trends of sufficient magnitude to support this hypothesis [45, 62], although some evidence to the contrary may be found for the conjunctiva network [63]. Vessel length between branch points of the mesenteric network has been found to vary from about 200 to 500 μm, without any systematic trend as a function of diameter [62]. The differences between network and individual vessel gradients appear to lie in the inability to characterize the network topography in terms of either purely serial and/or parallel elements.

18.3.2 Resistance to Blood Flow

From simultaneous measurements of the upstream-to-downstream pressure drop (ΔP) and red cell velocity, the resistance to flow in individual microvessels of the mesenteric network has been obtained from the analogy to Ohm's law, $R = \Delta P/Q$ [23, 42]. Presented in Fig. 18.7*a* is the AV distribution of R obtained for the mesenteric network, where R is plotted as a function of luminal diameter [61]. It is evident that single-vessel resistance rises about five decades as blood traverses the arteriolar network, to achieve a maximum at the true-capillary level. This distribution in resistance represents the variation of intravascular resistance commensurate with the variation in topographical lengths and attendant blood rheological behavior. Power law regressions of the form $R = aD^{-m}$ (solid curves) reveal exponents of 3.65 and 3.94 for arteriolar and venular vessels, respectively.

To further elucidate the dependency of resistance on luminal diameter, the resistance per unit of vessel length (R/l) is presented in Fig. 18.7*b* [61]. The power law regressions of these data yield exponents of 4.04 and 3.94 for arteriolar and venular vessels, respectively. These results are quite remarkable in that if blood behaved as a Newtonian fluid, with a fully developed flow (parabolic velocity profile) in a smooth-walled tube of circular cross section, then these regression exponents would be 4.00 as predicted by the Poiseuille-Hagen relationship, where $R/l = 128\eta/\pi D^4$. Evidently, by virtue of the large range in vessel diameters encountered as blood traverses the network, the fourth-power relationship appears as the dominant factor in specifying intravascular resistance, in the normal flow state. On a global or network scale, deviations from this behavior appear to be comparatively minor, and may be attributed to local variations in apparent viscosity, which arise from its dependency on hematocrit, red cell aggregation, and obstruction of the lumen by leukocytes adhering to the endothelium.

18.3.3 Microvessel Hematocrit

The importance of hematocrit as a determinant of the apparent viscosity of blood and in the transport of oxygen to tissues has prompted numerous macro- and microcirculatory studies

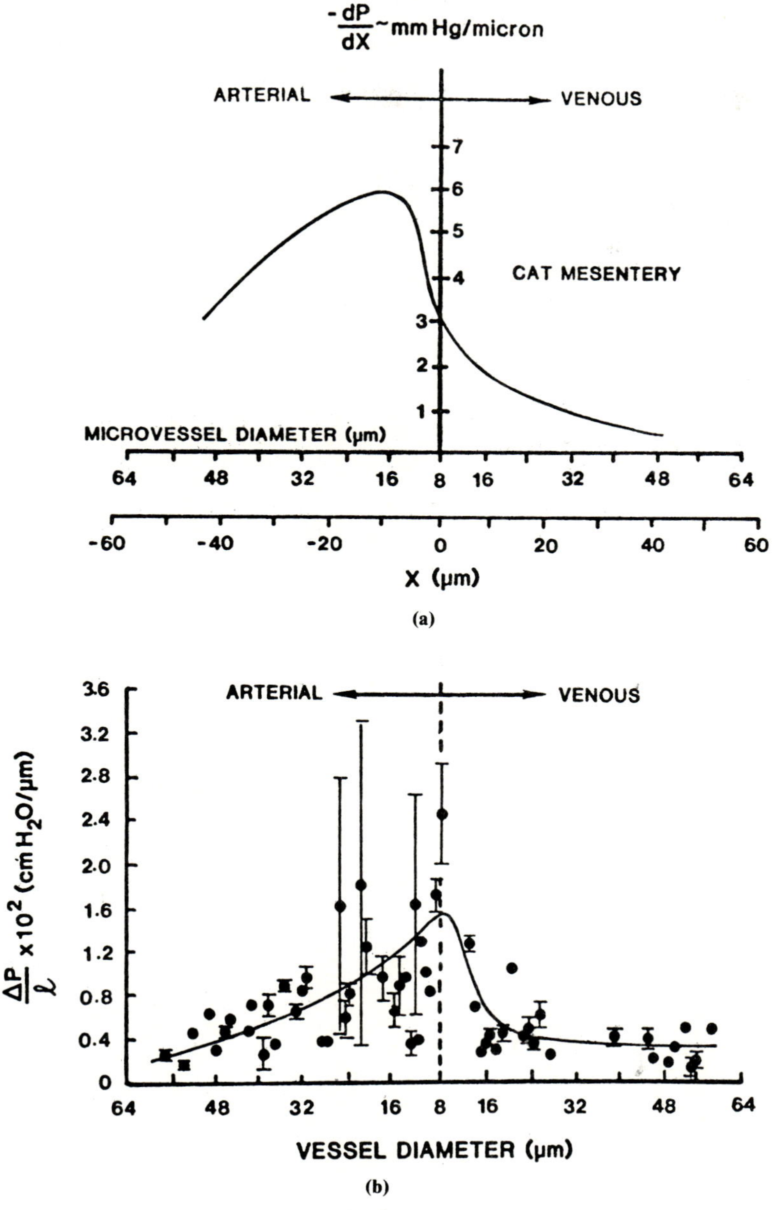

FIG. 18.6 Gradients of intravascular pressure throughout the mesenteric network: (*a*) Network gradient, computed from the slope of the AV pressure distribution. (*b*) Direct measurements of upstream-to-downstream pressure drop divided by vessel length for individual microvessels. *(From Zweifach and Lipowsky [42], with permission.)*

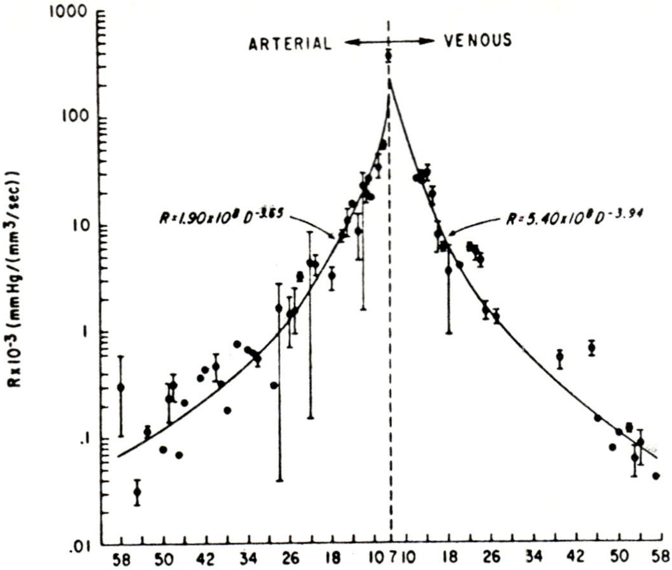

(a)

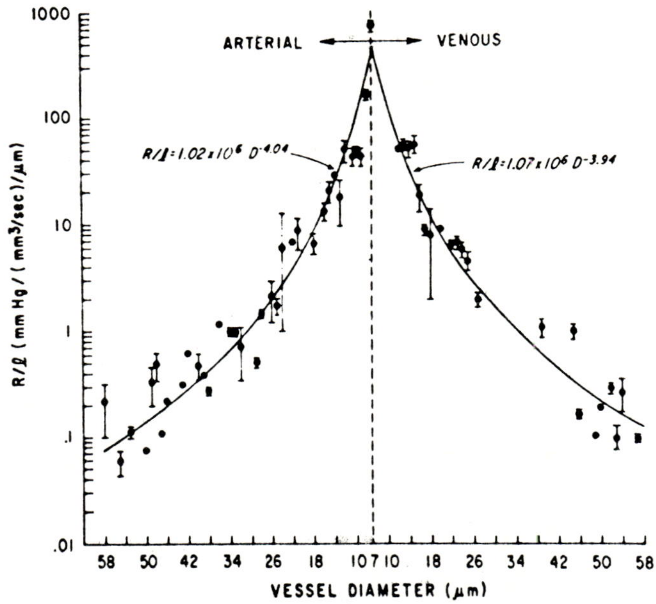

(b)

FIG. 18.7 (*a*) Resistance to blood flow (R) computed from the simultaneous measurements of flow (Q) and pressure drop (ΔP), in single unbranched vessels of mesentery, where $R = \Delta P/Q$. (*b*) Resistance per unit vessel length (R/l) plotted against luminal diameter (D). The solid curves are power law regressions of the form $R = aD^m$. *(From Lipowsky et al. [61], with permission.)*

aimed to evaluate its systemic and microvascular distributions. Essential to an understanding of these distributions is the recognition that two specific hematocrits are associated with the flow of blood: First is the actual microvessel or dynamic hematocrit [64] (H_{micro}), which corresponds to that value determined by instantaneously stopping the flow in a vessel and measuring the packed cell fraction, either directly by centrifugation or indirectly from erythrocyte and plasma labeling by radioisotopic methods. The second hematocrit of importance is the discharge (H_{disch}), or outflow, hematocrit, which is that value obtained by collecting the effluent of a vessel in a hypothetical "mixing cup" and determining its packed cell fraction [65]. Differences between H_{micro} and H_{disch} arise principally because of the relative increase in the mean velocity of erythrocytes above that of the mean velocity of plasma as suggested by the early investigations of Fahraeus [7] and subsequent in vitro and in vivo studies [56–58].

Application of these techniques to the delineation of the distribution of H_{micro} throughout successive microvascular divisions has demonstrated a steady fall in H_{micro} throughout the network, to the point where average capillary values on the order of 20 percent of H_{sys} may be found in some tissues. For example, cinephotometric measurements of H_{micro} in omentum capillaries have revealed average values of 17.2 percent in the 8-μm capillaries, which correspond to H_{micro}/H_{sys} equal to 0.45 [66]. On the basis of the microocclusion technique, mesenteric capillaries smaller than 10 μm in diameter were found to average 8.2 percent (H_{micro}/H_{sys} = 0.23) [32], and comparable values of 10.4 percent have been shown in the cremaster muscle with H_{micro}/H_{sys} = 0.2 [67]. In larger microvessels, the optical density technique has been applied to determine average mesenteric arteriolar H_{micro} for $35 < D < 110$ μm equal to 30.5 percent (H_{micro}/H_{sys} = 0.87) [31]. Fairly complete AV distributions of H_{micro} have been obtained for the mesentery of the cat [68], mesentery of the rat [34], and the hamster cheek pouch [39], as illustrated in Fig. 18.8.

As shown therein for the cat mesentery, microhematocrits normalized with respect to H_{sys} (average systemic hematocrit equals 35 percent for cat, 42 percent for rat, and 50 percent for hamster) fall from about 0.8 in the 70-μm arterioles to 0.2 at the true-capillary level (D = 7 μm), with a subsequent rise in the venous circuitry to reach a value of 0.9 in the 70-μm venules. From these trends it appears that the minimum hematocrit occurs in the immediate postcapillary vessels of 12 μm diameter. For microvessels ranging in diameter from 10 to 60 μm, values of H_{micro}/H_{sys} averaged 0.46 ± 0.24 SD in arterioles and 0.53 ± 0.25 SD in venules.

Similar trends have been observed in the rat mesentery and hamster cheek pouch networks (Fig. 18.8), with differences attributed to variations in network topography. Average capillary values of H_{micro}/H_{sys} were for rat mesentery 0.36 ± 0.23 SD and for the cheek pouch 0.31 ± 0.18 SD. It should be emphasized that although H_{micro} falls dramatically in the capillary network, H_{micro} is not a parameter which is indicative of conservation of red cell flux. Instead, it is the product of flow and H_{disch} which must be conserved, and is determined by the Fahraeus effect and plasma skimming.

Measurements of microvessel hematocrit in the nonresting state have also received attention in light of the delivery of oxygen in the face of increased metabolic demand. Estimates of capillary hematocrit by red cell counting in 5-μm capillaries of cremaster muscle [67] yielded values of H_{micro} equal to 10.4 ± 2.0 (standard error, or SE) percent in the resting state in contrast to values of H_{sys} on the order of 50 percent (hamster). When the muscle was exercised with 1-Hz contractions, capillary hematocrits were found to rise to 18.5 ± 2.4 (SE) percent. Also when the tissue was suffused with Ringer's solutions of high Po_2 (40 mmHg) and low Po_2 (8 mmHg), modulations in H_{micro} could be achieved between 6.3 ± 1.8 (SE) percent and 10.4 ± 2.0 (SE) percent, respectively, principally due to arteriolar vasodilatation [67]. Such variations in H_{micro} have been explored in part by macrocirculatory studies aimed to examine the interplay between hematocrit, blood viscosity factors, and resistance.

Indirect macrocirculatory experiments have investigated whole-organ perfusion in light of the relationship of increased blood viscosity with elevated hematocrit and the flux of red cells to the capillary bed. Theoretical considerations of the resistance-viscosity relationship make it conceivable that increases in viscosity attendant to elevations in perfusion hematocrit might negate the advantage of higher red cell concentrations (oxygen-carrying capacity) by reducing perfusion. This concept has been explored in the framework of an optimal hematocrit for maximum red cell flux and hence oxygen delivery to tissue [69–73]. For values of H_{sys} (i.e., H_{disch}) ranging from 0.3 to 0.4, the volumetric flux of red cells through the entire network may be maximized by virtue of the nonlinear relationship between blood viscosity and hematocrit [71].

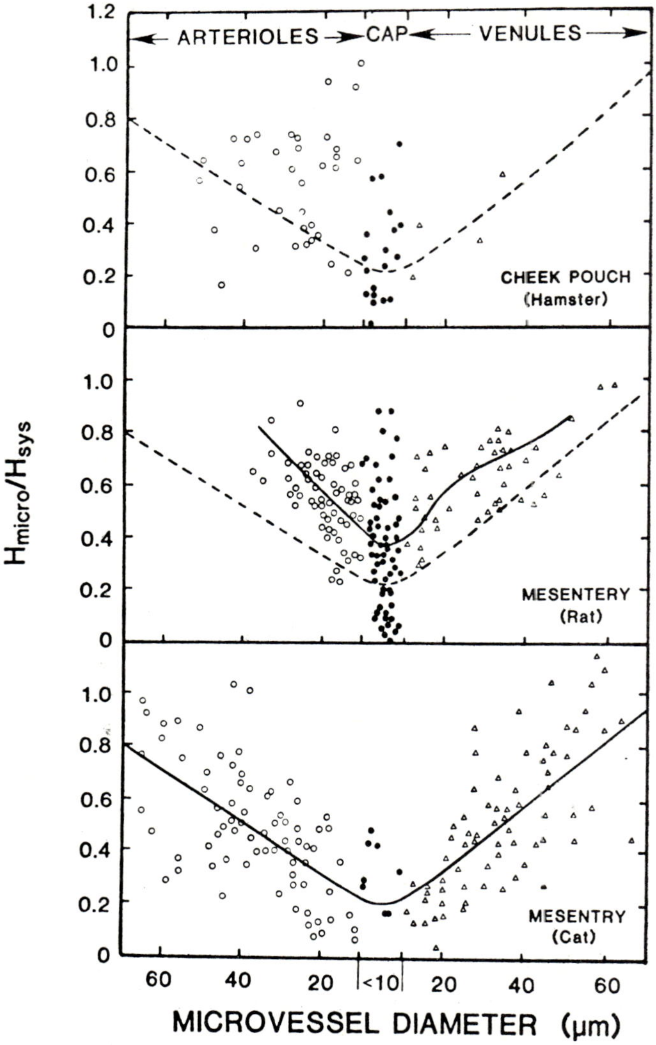

FIG. 18.8 AV distributions of microvessel hematocrit normalized with respect to systemic values. The cat mesentery data [68] are fitted by a cubic spline which also appears as the dashed curves above, for comparison of data from the rat mesentery [34] and hamster cheek pouch [39].

Microcirculatory studies aimed to delineate the flux of red cells during systemic hemodilution and/or concentration have revealed a general invariance of red cell velocity over a wide range of H_{sys} from 0.3 to 0.6 in the mesentery of the rat [74]. The precise relationships among microvascular hematocrit, red cell flux, and systemic hematocrit remain to be determined.

18.3.4 Apparent Viscosity

The apparent viscosity of blood (η) flowing in individual mesenteric vessels has also been evaluated by computing η as that value which satisfies the Poiseuille-Hagen relationship, $\eta =$

$(\Delta P\, D^2)/kV_m l$, where V_m is the mean (bulk) velocity of cells plus plasma and k is a constant dependent on the units employed [23, 60]. Examination of the in situ variability of apparent viscosity η with microvessel hematocrit H_{micro} under high shear rate conditions (2000 s^{-1}, associated with systemic hemodilution) has shown that within the physiological range of mesenteric hematocrits η varies linearly with H_{micro} [66], as shown in Fig. 18.9. Comparison of a linear regression of the in vivo measurements of η versus H_{micro} (solid line bounded by dashed ±95 percent confidence limits) with those measurements obtained in vitro in a cone-plate viscometer at comparable shear rates shows no significant difference between the two. This agreement tends to substantiate the hypothesis by Barbee and Cokelet [75], established in vitro for small-bore tubes ranging in lumen from 29 to over 800 μm, that, given comparable hematocrits and shear rates, the apparent viscosity of blood will be the same regardless of the size of the tube or microvessel.

Measurements of the AV distribution of η throughout the mesenteric network in the resting flow state are summarized in Fig. 18.10 (upper curve), where η is plotted as a function of luminal diameter (network position) [61]. Values of η averaged 3.6 ± 1.9 (SD) cP for arterioles and 5.2 ± 2.7 (SD) cP for venules. The relatively greater viscosities in the venules have been attributed to both higher hematocrits in venules of comparable arteriolar diameter, as well as shear rate–related phenomena (e.g., red cell aggregation and/or leukocyte-endothelium adhesion). These values of η are considerably higher than those obtained for blood by the macrocirculatory whole-organ approach (~2 cP in the dog hind limb [9]; 2.35 cP for cat calf muscle [14]), and correspond to values obtained in vitro for hematocrits of 40 to 50 percent [67]. The apparent viscosity deduced by macrocirculatory studies may be considerably lower owing to the inability of such studies to evaluate the effective hematocrit at the microcirculatory level, as suggested by Whittaker and Winton [9]. The relatively high values of η obtained in situ may also be biased somewhat since vessels with zero values of H_{micro} (plasma only) were not amenable to velocity and, hence, viscosity measurements, or since vessels of low hematocrit (and, hence, viscosity) may have been inaccessible.

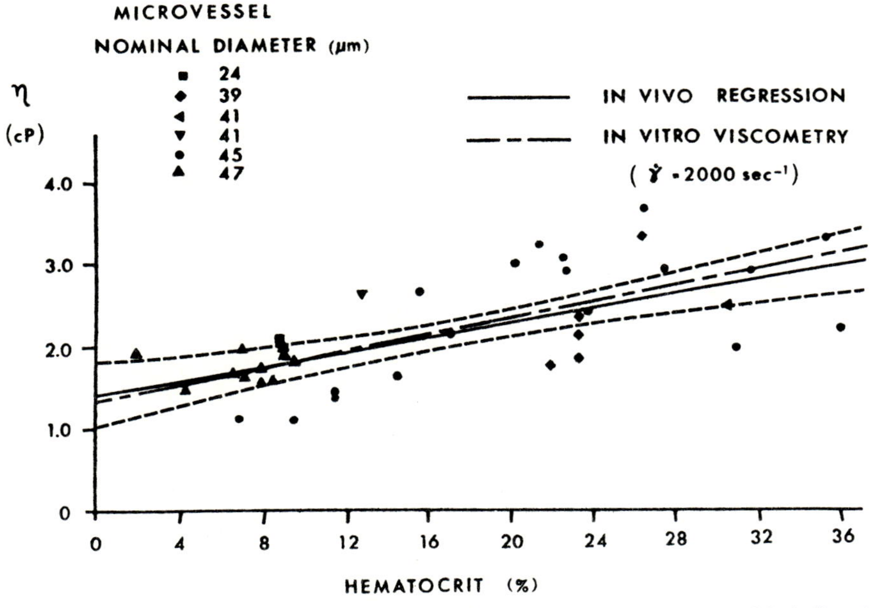

FIG. 18.9 Apparent viscosity (η) of blood versus microvessel hematocrit for arterioles of the indicated diameter. Values of η were computed as those values which satisfy the Poiseuille-Hagen equation for the measured ΔP and Q. A linear regression of the data (solid line) is bounded by ±95 percent confidence limits (dashed curves). Shown for comparison are measurements from a cone-plate viscometer at comparable shear rates. *(From Lipowsky et al. [68], with permission.)*

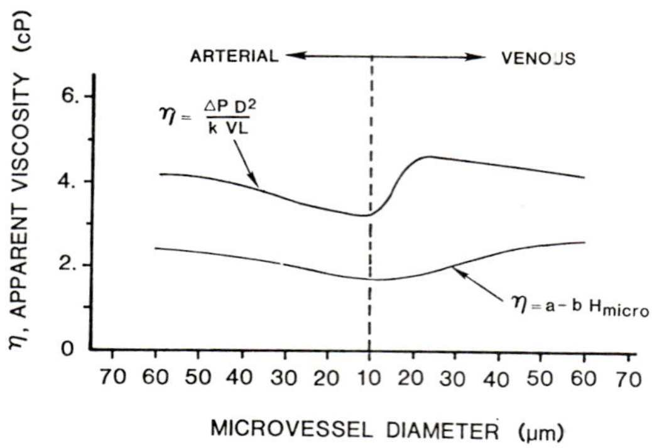

FIG. 18.10 AV distribution of blood apparent viscosity. The upper curve represents values computed from the Poiseuille-Hagen equation for the measured pressure drop and flow. The lower curve represents η computed from the AV distribution of H_{micro} and the linear relationship of Fig. 18.9, that is, $\eta = a + bH_{\text{micro}}$.

Comparison of the network distribution of η with those values calculated from the measured AV distribution of H_{micro} (by use of the linear relationship of η versus H_{micro} depicted in Fig. 18.9) reveals a substantial disparity, as shown in Fig. 18.10 (lower curve). This difference may arise from the fact that the measurements of η versus H_{micro} were made under high wall–shear rate conditions with an absence of leukocyte-to-endothelium adhesion. It has been shown that adhesion of leukocytes to the endothelium is strongly shear rate–dependent [37, 76–79]. In cases where shear rates result in leukocyte-to-endothelium adhesion, the subsequent obstruction of the microvessel lumen may raise intravascular resistance to blood flow [37]. If this increase in resistance is attributed entirely to an elevation in apparent viscosity, increases on the order of 100 percent or more may occur, depending on vessel diameter [67].

18.3.5 Shear Rate–Related Phenomena

Characterization of the rheological behavior of blood, both in vitro and in vivo, requires the specification of shear rates imposed on the suspension. To satisfy this requirement, it has become common practice to characterize microvessel shear rates in terms of the wall shear rate of a Newtonian fluid flowing with an equivalent mean velocity for a given tube diameter. Under these conditions, the wall shear rate $\dot{\gamma}$ may be derived from the Navier-Stokes equations as $8V_m/D$. From a rheological standpoint, the representation of effective shear rates imposed on a fluid can be improved somewhat for a homogenous, non-Newtonian fluid [2]; however, in view of the particulate nature of blood, such refinements have given way to the empiricism of an equivalent Newtonian wall shear rate. Although some blood rheologists have preferred the characterization of blood shear rates for tube flow in terms of the "reduced velocity" V_m/D [75, 80], the utility of using $8V_m/D$ appears equally justified by the conceptual advantages of an effective Newtonian value.

Wall shear rates throughout the microvasculature can be easily calculated from the AV distribution of red cell velocity with diameter by use of the empirical relationship between V_{rbc} and V_m (i.e., $V_{\text{rbc}}/V_m = 1.6$ [26, 27]), as given in Fig. 18.11. Upon doing so for the mesenteric network, one finds that arteriolar values of $\dot{\gamma}$ range from about 1400 s^{-1} in 50-μm arterioles, reaching a maximum value of about 1800 s^{-1} in 25-μm precapillary vessels and then steadily declining to a minimum of 500 s^{-1} in the 20-μm postcapillary venules, followed by a rise to 800 s^{-1} in 50-μm venules [61]. Average values of $\dot{\gamma}$ for microvessels ranging in size between 10 and 60 μm were found to be 1200 s^{-1} in the arterioles and 850 s^{-1} in the venules, with an overall

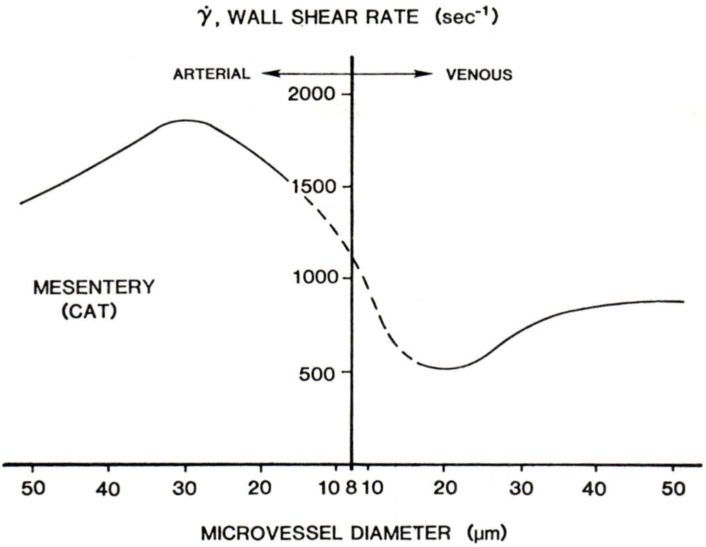

FIG. 18.11 AV distribution of Newtonian wall shear rate for the mesenteric network computed from the data of Fig. 18.4 on the basis of $\gamma = 8V_{\text{mean}}/D$.

average of 1330 s^{-1}. These wall shear rates are considerably higher than those usually studied in vitro [2], mainly because the in vivo relationship between apparent viscosity and $\dot{\gamma}$ becomes invariant (asymptotically) for $\dot{\gamma} > 1000\ s^{-1}$ at physiological hematocrits.

It should be emphasized that these shear rates are derived from continuum concepts as the slope of the velocity profile at the tube wall. In situ observations on the velocity profile of red cells flowing in 16- to 43-μm arterioles and venules (omentum) [66] have revealed considerable blunting of the radial velocity profile. These studies have also demonstrated that for red cells flowing in close proximity to the endothelium (less than 1 μm away), with velocities on the order of 0.5 mm/s, shear rates as high as 16,400 s^{-1} can be calculated on the assumption of a linear shear field between the cell and vessel wall.

On the basis of the in vitro behavior of η versus $\dot{\gamma}$, the high shear rates in the mesenteric circulation suggest an invariance of in vivo apparent viscosity. In contrast, there appears to be a 40 percent rise in η corresponding to the 30 percent decrease in in vivo $\dot{\gamma}$. To evaluate the potential for velocity (shear rate) effects in individual microvessels, curves of flow (Q) versus upstream-to-downstream pressure drop (ΔP) have been obtained in vivo [61], as shown in Fig. 18.12. A nonlinear variation of Q with ΔP (obtained by occlusion of the superior mesenteric artery) is suggestive of dramatic elevations of η as the mean velocity of the flow is reduced. The shape of these curves, as well as the zero flow intercept on the ΔP axis, is similar in nature to whole-organ measurements of Q versus ΔP obtained for the isolated intestinal loop [81]. For example, for the 39-μm arteriole shown here, a rise in η from about 3 cP in the normal flow state to 20 cP is obtained as flow velocity is reduced to 0.3 mm/s.

These elevations in apparent viscosity of blood during reductions in shear rates may arise either from red cell aggregation, as suggested by in vitro studies [82, 83] or by leukocyte-endothelium adhesion, as evidenced by in vivo observations [37]. The study of the effects of red cell aggregation has, for the most part, relied upon the whole-organ approach because of a lack of suitable instrumentation for its in situ quantitation. However, conflicting assessments of the effect of red cell aggregation on in vivo η have not yielded definitive results. Studies using high-molecular-weight dextrans have shown both no significant rise in η [10] as well as a substantial increase [13, 84, 85] which may be due to an attendant rise in plasma viscosity with increased dextran concentration.

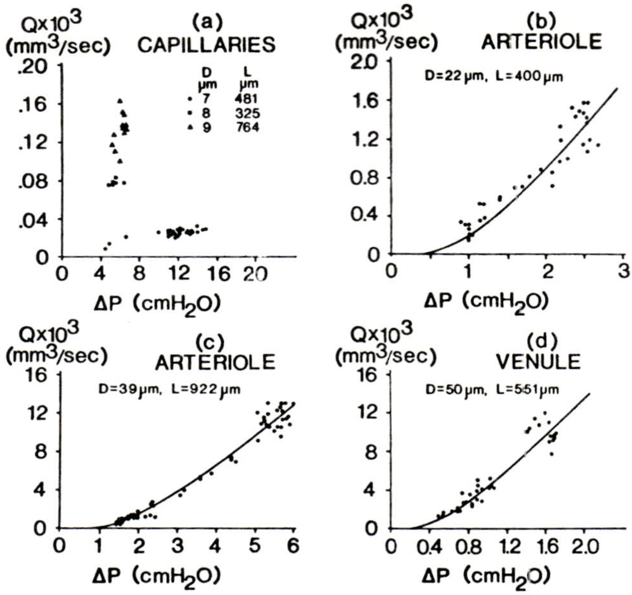

FIG. 18.12 Microvessel volumetric flow rate (Q) versus upstream-to-downstream pressure drop (ΔP) for representative individual microvessels. *(From Lipowsky and Zweifach [23], with permission.)*

18.3.6 Branch Points and Bifurcations

The variability of hemodynamic parameters throughout the microvascular network, examined in the preceding sections on the basis of increasing vascularity (i.e., numbers of microvessels), has also been examined in light of the flow behavior of blood at a branching site. In most networks configured as a succession of branching tubes (e.g., muscle and splanchnic beds), the sequential division of arterioles into smaller-diameter branches may be characterized as a bifurcation involving a parent and two daughter branches. Although substantial departures from this schema are evident in many tissues, most analyses of branch-flow processes have centered around bifurcations with the goal of exploring (1) attenuation of intravascular pressure, (2) intermittency of flow, and (3) hematocrit variations.

The reduction of intravascular pressure from parent to daughter branches in the arteriolar network has been measured for parent arterioles ranging in diameter from 50 to 20 μm [86], as shown in Fig. 18.13. Parent-to-daughter pressure drop was found to increase substantially with diminishing ratio of their respective diameters (D_d/D_p), and rose from 5 to 30 percent as D_d/D_p decreased from 0.9 to 0.1. Such pressure losses may arise from irregularities at the daughter branch orifice, conceptualized as a precapillary sphincter [87] which may obstruct the inlet by as much as 80 percent. Theoretical analysis of the flow field at a branch point by the two-dimensional finite element approach has predicted that, even in the low Reynolds number regime of the microcirculation ($\sim 10^{-2}$), substantial pressure drops may be incurred at a constricted orifice [88]. Because of difficulties in evaluating these geometric factors throughout the network by intravital microscopy, the relative contribution of this process to the overall AV distribution of intravascular pressure remains obscure.

The intermittency of flow at a branch has also been studied, although mostly on a qualitative basis, and has been noted as a result of the transient plugging of capillaries by the less deformable leukocytes [89, 90] and transients in plasma skimming arising from randomly placed plasma gaps in the parent vessel [90]. The degree of intermittency has been shown to increase dramati-

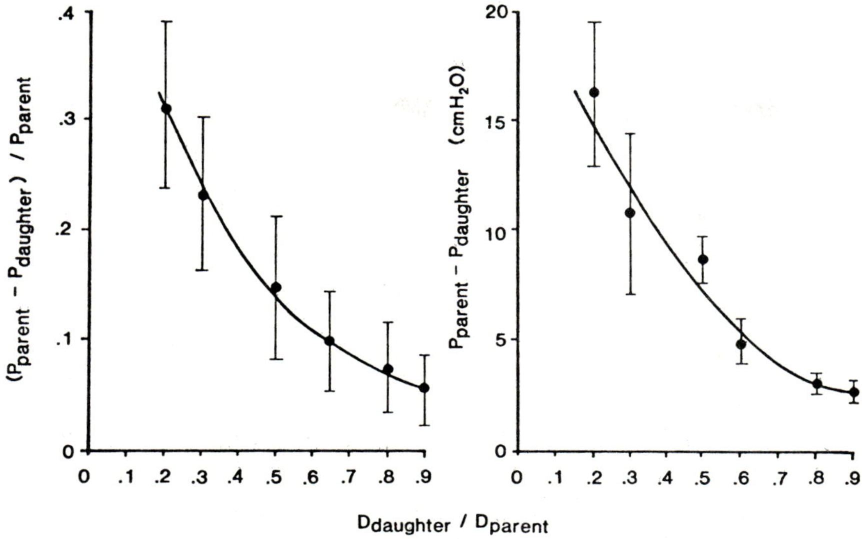

FIG. 18.13 Parent-to-daughter branch pressure drop at bifurcations in the mesenteric microvasculature. Left: fractional difference in pressure drop; right: pressure drop in cmH_2O. *(From Zweifach [86], with permission.)*

cally in the region of the true capillaries [91, 92], where stochastic effects may influence the apportionment of red cells from parent to daughter microvessels [93, 94]. On theoretical grounds, it has been illustrated that the distribution of cells to either of two daughter branches at a bifurcation is a function of the balance of hydrodynamic forces on the particles, and that red cells will be drawn into the branch with the greatest flow [93]. This situation can result in an intermittency of flow by virtue of increases in η within a branch which has received additional red cells. The resultant increase in branch resistance may serve to lower the pressure gradient and flow at the orifice of the branch, and thus swing the balance of forces to favor admission of cells to the contralateral branch [94]. Analysis of stochastic events at the capillary level, in terms of their frequencies of occurrence, has shown that they are greatly overshadowed by the pulsatility, which is synchronous with the heart rate [95].

The apportionment of red cells at a bifurcation has received considerably more attention than either pressure drop or transient flow variations. The concept of plasma skimming introduced by Krogh [96], and the resultant low hematocrits for capillaries, have prompted an examination of the laws of conservation governing red cell flux at a bifurcation. Specifically, conservation of volumetric flow at a bifurcation must satisfy the following relationships for bulk flow of cells plus plasma in the daughter and parent vessels, Q_p and Q_d, respectively, as well as for the volumetric flux of red cells, Q_{rbc}:

$$Q_{d1} + Q_{d2} = Q_p$$

and

$$Q_{\text{rbc},d1} + Q_{\text{rbc},d2} = Q_{\text{rbc},p}$$

where the volumetric flux of red cells is given by the product of bulk flow and the discharge hematocrit (H_{disch}) for each branch, that is, QH_{disch}. These relationships must be satisfied regardless of whether or not plasma skimming or other factors favor a preferential distribution of red cells into a branch. However, comparison of the ratio of QH_{disch} between daughter and parent branches and the flow ratio Q_d/Q_p should reveal a linear relationship only if the red cells are uniformly distributed without preferential shunting into a branch.

Within this framework, two regimes of the ratio of daughter to parent vessel diameter have been considered: first, that of a sizable arteriole feeding a capillary side branch ($D_d/D_p \ll 1$), and second, that of the bifurcation of vessels of near capillary diameters ($D_d/D_p \approx 1$). In the arteriole-capillary branch, nonuniformities in the radial distribution of hematocrit and the presence of a marginal zone of plasma along the wall of the arteriole have been ascribed as the principal factors to be considered in the dilution of the side-branch hematocrit [97]. Theoretical [98] and in vitro studies [99] of blood flow in small-bore glass tubes have stressed the importance of daughter branch diameter and flow velocity as determinants of branch hematocrit. The larger the branch diameter and the greater the branch flow, the farther into the parent stream the dividing streamline will protrude relative to the parent vessel plasma layer. Such behavior has been examined in the arteriolar network of the mesentery, where measurements of preferentially higher hematocrits have been correlated with higher velocities in a daughter branch [100].

Detailed studies of blood cell distributions at a bifurcation, with all branches comparable to capillary diameters, have emphasized the roles of both the balance of hydrodynamic forces and blood cell deformability (both erythrocytes and leukocytes) as a determinant of the cell concentrations in the daughter branch [94, 101]. The net hydrodynamic forces acting to steer a cell into a particular side branch have been shown to be a direct function of the relative ratio of the branch flow to the parent flow, i.e., the sum of the two daughter branch flows [102]. In situ observations utilizing the rabbit ear chamber [94] have shown that, in the case of parent vessels with sparse cell concentrations, the position of the blood cell (either erythrocytes or leukocytes) with respect to the dividing streamline, i.e., radial eccentricity of the cell position, is the principal determinant of which branch the cell may enter. The position of the dividing streamline is in turn affected by the ratio of daughter to parent vessel bulk flow. This situation may lead to a strikingly nonuniform distribution of red cells as depicted in Fig. 18.14*a*, where the ratio of daughter to parent red cell flux is plotted against the corresponding bulk-flow ratio [94].

Although such nonuniform distributions can occur at an individual bifurcation, it appears that when one considers the averaging of many branch points the effect is less striking. To illustrate, Fig. 18.14*b* presents red cell flux ratios for over 200 branch points in the cremaster muscle [104]. The deviation of the capillary distributions from a linear relationship is small but statistically significant. In contrast, the deviation from a linear relationship in comparatively larger arterioles and venules with parent and daughter vessels of comparable size, ranging from 20 to 70 μm (Fig. 18.14*c*), was not statistically significant [103]. Hence, it appears that the effects of plasma skimming do not contribute greatly to the distribution of blood throughout successive microvascular divisions when parent vessels feed daughter vessels of comparable size.

In addition to velocity factors, the relative distribution of cells at a bifurcation is also dictated by the forces required to deform a blood cell in order for it to enter a capillary.

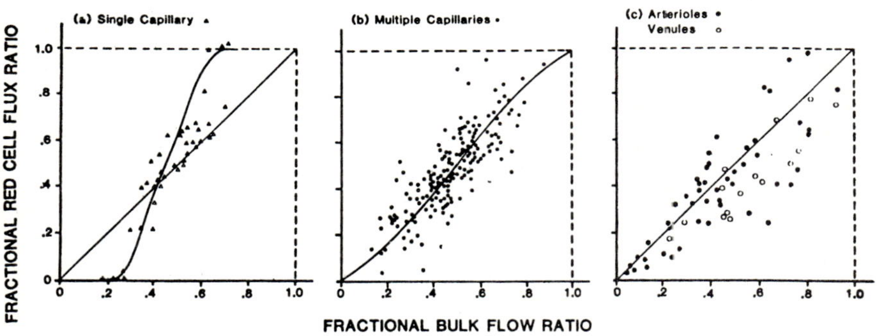

FIG. 18.14 Fractional red cell volumetric flux ratio from daughter to parent vessels versus corresponding bulk flow ratio: (*a*) A single capillary bifurcation, where the preferential distribution of red cells is clearly evident [94]. (*b*) Multiple measurements for over 200 capillary bifurcations, which on the average show a small but significant departure from a linear relationship. (*c*) Arterioles and venules ($20 < D < 70$ μm), which do not significantly depart from a linear cell distribution [103].

Modeling experiments have demonstrated that if the size of an undeformed blood cell is larger than the branch diameter, then it is possible for seizure of cells to occur in the capillary network [101]. The seizure of cells in the capillary branches may thus lead to an increase in capillary hematocrit above that of the feeding arteriole, and hence an inversion of the Fahraeus effect.

CONCLUSIONS

The task of describing the mechanics of microvascular blood flow revolves around the intimate relationship between topographical and blood rheological factors. This recurrent theme remains in the forefront of efforts to delineate the mechanisms which contribute to progression of the disease process. Numerous pathophysiological states, such as hypertension, diabetes, and hematological disorders, present subtle departures from the norm in the chain of physical processes which link network structure and function. In addition, many of the experimental tools and analytical techniques being used to explore the microcirculation have developed from the physical sciences, which have often been unified with physiology under the label of *bioengineering*. One of the most promising areas of future collaboration between physiology and engineering may be that of mathematical modeling and computer simulations. Given the constitutive equations which describe the rheological behavior of blood and a description of the network topography and its determinants, network hemodynamics may be modeled by traditional techniques of network analysis. By coupling such studies with simulations of the mechanisms of microvascular control and regulation, e.g., convective and diffusive transport of nutrients and metabolites, as well as the mechanics of vascular smooth muscle, it will be possible to establish an integrated approach to gain a thorough understanding of the functional behavior of the microcirculation in health and disease.

ACKNOWLEDGMENTS

This work was funded in part by NIH research grants HL-16851 and HL-28381 and research career development award HL-00594.

REFERENCES

1. Bayliss, L. E.: Rheology of blood and lymph, in A. Frey-Wyssling (ed.), *Deformation and Flow in Biological Systems*, Interscience, New York, 1952, pp. 355–418.
2. Cokelet, G. R.: The rheology of human blood, in Y. C. Fung, N. Perrone, and M. Anliker (eds.), *Biomechanics: Its Foundations and Objectives*, Prentice-Hall, Englewood Cliffs, N.J., 1972.
3. Meiselman, H. J., and Cokelet, G. R.: Blood rheology, *Adv. Microcir.*, *5*, 32–61, 1973.
4. Cokelet, G. R.: Rheology and hemodynamics, *Annu. Rev. Physiol.*, *42*, 311–324, 1980.
5. Poiseuille, J. L. M.: Récherches expérimentales sur le mouvement des liquides dans les tubes de tres petits diamètres, *C. R. Acad. Sci.*, *11*, 961–967, 1840.
6. Hagen, G.: Uber die Bewegung das Wassers in Engen Zylindrischen Rohren, *Pogg. Ann.*, *46*, 423, 1839.
7. Fahraeus, R.: The suspension stability of blood, *Physiol. Rev.*, *9*(2), 241–274, 1929.
8. Fahraeus, R., and Lindqvist, T.: The viscosity of the blood in narrow capillary tubes, *Am. J. Physiol.*, *96*, 562–568, 1931.
9. Whittaker, S. R. F., and Winton, F. R.: The apparent viscosity of blood flowing in the isolated hindlimb of the dog and its variation with corpuscular concentration, *J. Physiol.*, *78*, 339–369, 1933.
10. Hint, H. C.: The flow properties of erythrocyte suspensions in isolated rabbit's ear; the effects of erythrocyte aggregation, hematocrit and perfusion pressure, in *2d European Conference on Microcirculation*, Pavia, 1962, Bibl. Anat., 4, 112–118, 1964.

11. Skovborg, F., Nielsenand, A. V., and Schlichtkrull, J.: Blood viscosity and vascular flow rate: Blood viscosity measured in a cone-plate viscometer and flow rate in an isolated vascular bed, *Scand. J. Clin. Lab. Invest., 21,* 83–88, 1968.

12. Kurland, G. S., Charm, S. E., Brown, S. L., and Tousignant, P.: A comparison of blood flow in a living vessel and in glass tubes, in A. L. Copley (ed.), *Hemorheology, Proceedings of the First International Conference,* Pergamon Press, New York, 1968, pp. 607–615.

13. Djojosugito, A. M., Folkow, B., Oberg, B., and White, S.: A comparison of blood viscosity *in vitro* and in a vascular bed, *Acta Physiol. Scand., 78,* 70–84. 1970.

14. Baeckstrom, P., Folkow, B., Kendrick, E., Lofving, B., and Oberg, B.: Effects of vasoconstriction on blood viscosity *in vivo, Acta Physiol. Scand., 81,* 376–384, 1971.

15. Benis, A. M., Usami, S., and Chien, S.: Effect of hematocrit and inertial losses on pressure-flow relations in the isolated hindpaw of the dog, *Circ. Res., 27,* 1047–1068, 1970.

16. Benis, A. M., Usami, S., and Chien, S.: A reappraisal of Whittaker and Winton's results on the basis of inertial losses, *Biorheology, 11,* 153–161, 1974.

17. Eliassen, E., Folkow, B., and Oberg, B.: Are there any significant inertial losses in the vascular bed, *Acta Physiol. Scand., 87,* 567–569, 1973.

18. Landis, E. M.: Poiseuille's law and the capillary circulation, *Am. J. Physiol., 103,* 432–443, 1933.

19. Landis, E. M.: The capillary pressure in frog mesentery as determined by micro-injection, *Am. J. Physiol., 75,* 548–570, 1926.

20. Shepro, D., Fulton, G. P., and Zweifach, B. W. (eds.): *Proceedings of The Microcirculatory Approach to Peripheral Vascular Function,* Tucson, Ariz., 1971, *Microvasc. Res., 5*(3), 229–437, 1973.

21. Wiederhielm, C. A., Woodbury, J. W., Kirk, W., and Rushmer, R. F.: Pulsatile pressures in the microcirculation of the frog's mesentery, *Am. J. Physiol., 23,* 684–696, 1957.

22. Intaglietta, M., Pawula, R. F., and Tompkins, W. R.: Pressure measurements in the mammalian microvasculature, *Microvasc. Res., 2,* 212–220, 1970.

23. Lipowsky, H. H., and Zweifach, B. W.: Methods for the simultaneous measurement of pressure differentials and flow in single unbranched vessels of the microcirculation for rheological studies, *Microvasc. Res., 14,* 345–361, 1977.

24. Wayland, H., and Johnson, P. C.: Erythrocyte velocity measurement in microvessels by a two-slit photometric method, *J. Appl. Physiol., 22,* 333–337, 1967.

25. Tompkins, W. R., Monti, R., and Intaglietta, M.: Velocity measurements by self-tracking correlator, *Rev. Sci. Instr., 45,* 647–649, 1974.

26. Baker, M., and Wayland, H.: On-line volumetric flow rate and velocity profile measurement for blood in microvessels, *Microvasc. Res., 7,* 131–143, 1974.

27. Lipowsky, H. H., and Zweifach, B. W.: Application of the two-slit photometric technique to the measurement of microvascular volumetric flow rates, *Microvasc. Res., 15,* 93–101, 1978.

28. Harper, S. L., and Bohlen, H. G.: In vitro and in vivo measurements of red cell velocity with epi- and transillumination, *Microvasc. Res., 25,* 186–193, 1983.

29. Rosenblum, W. I., and El-Sabban, F.: Measurement of red cell velocity with a two-slit technique and cross-correlation: Use of reflected light and either regulated or unregulated power supplies, *Microvasc. Res., 22,* 225–227, 1981.

30. Lee, T. Q., Schmid-Schönbein, G. W., and Zweifach, B. W.: The application of an improved dual slit photometric analyzer for volumetric flow measurements in microvessels, *Microvasc. Res., 26,* 351–361, 1983.

31. Jendrucko, R. J., and Lee, J. S.: Hematocrit measurement in cat arterioles by microphotometry, *Fed. Proc., 33,* 313, 1974.

32. Lipowsky, H. H., Usami, S., Chien, S., and Pittman, R. N.: Hematocrit determination in small bore tubes from optical density measurements under white light illumination, *Microvasc. Res., 20,* 51–70, 1980.

33. Lipowsky, H. H., Usami, S., Chien, S., and Pittman, R. N.: Hematocrit determination in small bore tubes by differential spectrophotometry, *Microvasc. Res., 24,* 42–55, 1982.

34. Kanzow, G., Pries, A. R., and Gaehtgens, P.: Analysis of the hematocrit distribution in the mesenteric microcirculation, *Int. J. Microcirc. Clin. Exp., 1,* 67–80, 1982.

35. Pittman, R. N., and Duling, B. R.: A new method for the measurement of percent oxyhemoglobin, *J. Appl. Physiol., 38*(2), 315–320, 1975.

36. Pittman, R. N., and Duling, B. R.: Measurement of percent oxyhemoglobin in the microvasculature, *J. Appl. Physiol.*, *38*(2), 321–327, 1975.

37. Schmid-Schönbein, G. W., Fung, Y. C., and Zweifach, B. W.: Vascular endothelium-leukocyte interaction, sticking shear force in venules, *Circ. Res.*, *36*, 173–184, 1975.

38. Klitzman, B., and Johnson, P. C.: Capillary network geometry and red cell distribution in hamster cremaster muscle, *Am. J. Physiol.*, *242*, H211–H219, 1982.

39. Sarelius, I. H., and Duling, B. R.: Direct measurement of microvessel hematocrit, red cell flux, velocity and transit time, *Am. J. Physiol.*, *243*, H1018–H1026, 1982.

40. Baez, S.: A method for in-line measurement of lumen and wall of microscopic vessels *in vivo*, *Microvasc. Res.*, *5*, 299–308, 1973.

41. Intaglietta, M., and Tompkins, W. R.: Microvascular measurements by video shearing and splitting, *Microvasc. Res.*, *5*, 309–312, 1973.

42. Zweifach, B. W., and Lipowsky, H. H.: Pressure-flow relations in blood and lymph microcirculation, in E. M. Renkin and C. C. Michel (eds.), *Handbook of Physiology*, vol. 4: *The Cardiovascular System*, American Physiological Society, Bethesda, Md., 1984, pt. 1, pp. 251–307.

43. Richardson, D. R., and Zweifach, B. W.: Pressure relationships in the macro and microcirculation of the mesentery, *Microvasc. Res.*, *2*(4), 478–488, 1970.

44. Frasher, W. G., and Wayland, H.: A repeating modular organization of the microcirculation of cat mesentery, *Microvasc. Res.*, *4*, 62–76, 1972.

45. Intaglietta, M., and Zweifach, B. W.: Geometrical model of the microvasculature of rabbit omentum from in vivo measurements, *Circ. Res.*, *28*, 593–600, 1971.

46. Erikson, E., and Myrhage, R.: Microvascular dimensions, and blood flow in skeletal muscle, *Acta Physiol. Scand.*, *86*, 211–222, 1972.

47. Fronek, K., and Zweifach, B. W.: Microvascular pressure distribution in skeletal muscle and the effect of vasodilation, *Am. J. Physiol.*, *228*(3), 791–796, 1975.

48. Lalone, B. J., and Johnson, P. C.: Estimation of arteriolar volume flux from dual-slit red cell velocity: An in vivo study, *Microvasc. Res.*, *16*(3), 327–339, 1978.

49. Zweifach, B. W., and Lipowsky, H. H.: Quantitative studies of microvascular structure and function: III. Microvascular hemodynamics of at mesentery and rabbit omentum, *Circ. Res.*, *41*, 380–390, 1977.

50. Mayrovitz, H. N.: Microvascular blood flow: Functional relation to vessel size, in *Proceedings of 34th Annual Conference on Engineering in Medicine and Biology*, *23*, 9, 1981.

51. Mayrovitz, H. N., and Roy, J.: Microvascular blood flow: Evidence indicating a cubic dependence on arteriolar diameter, *Am. J. Physiol.*, *245*, H1031–H1038, 1983.

52. Murray, C. D.: The physical principle of minimum work: I. The vascular system and the cost of blood volume, *Proc. Nat. Acad. Sci.*, *12*, 207–213, 1926.

53. Meir, P., and Zierler, K. L.: On the theory of the indicator-dilution method for measurement of blood flow and volume, *J. Appl. Physiol.*, *6*, 731–744, 1954.

54. Nellis, S. H., and Lee, J. S.: Dispersion of indicator measured from microvessels of cat mesentery, *Circ. Res.*, *35*, 580–591, 1974.

55. Baker, C. H., Davis, D. L., and Sutton, E. T.: Arteriolar capillary and venular FITC-dextran time concentration curves and plasma flow velocities, *Proc. Soc. Exp. Biol. Med.*, *161*, 370–377, 1979.

56. Starr, M. C., and Frasher, W. G.: A method for the simultaneous determination of plasma and cellular velocities in the microvasculature, *Microvasc. Res.*, *10*, 95–101, 1975.

57. Starr, M. C., and Frasher, W. G.: In vivo cellular and plasma velocities in microvessels of the cat mesentery, *Microvasc. Res.*, *10*, 102–106, 1975.

58. Gaehtgens, P., Benner, K. U., Schickendantz, S., and Albrecht, K. H.: Method for simultaneous determination of red cell and plasma flow velocity *in vitro* and *in vitro*, *Pfluegers Arch.*, *361*, 191–195, 1976.

59. Baker, C. H., Wilmoth, F. R., Sutton, E. T., and Takach, K.: Red blood cell and plasma distribution in SHR cremaster muscle microvessels, *Am. J. Physiol.*, *242*, H381–H391, 1982.

60. Baker, C. H., Sutton, E. T., and Davis, D. I.: Microvessel mean transit time and blood flow velocity of sulfhemoglobin RBC, *Am. J. Physiol.*, *238*, H745–H749, 1980.

61. Lipowsky, H. H., Kovalcheck, S., and Zweifach, B. W.: The distribution of blood rheological parameters in the microcirculation of cat mesentery, *Circ. Res.*, *43*, 738–749, 1978.

62. Lipowsky, H. H., and Zweifach, B. W.: Network analysis of microcirculation of cat mesentery, *Microvasc. Res., 7,* 73–83, 1974.
63. Fenton, B. M., and Zweifach, B. W.: Microcirculatory model relating geometric variations to changes in pressure and flow rate, *Ann. Biomed. Eng., 9,* 303–321, 1981.
64. Wayland, H.: Rheology and the microcirculation, *Gastroenterology, 52,* 343–355, 1967.
65. Cokelet, G. R.: Blood rheology interpreted through the flow properties of the red cell, in J. Grayson and W. Zingg (eds.), *Microcirculation,* Plenum Press, New York, 1976, vol. 1, pp. 9–32.
66. Schmid-Schönbein, G. W., and Zweifach, B. W.: RBC velocity profiles in arterioles and venules of the rabbit, *Microvasc. Res., 10*(2), 165–179, 1975.
67. Klitzman, B., and Duling, B. R.: Microvascular hematocrit and red cell flow in resting and contracting striated muscle, *Am. J. Physiol., 237*(4), H481–H490, 1979.
68. Lipowsky, H. H., Usami, S., and Chien, S.: *In vivo* measurements of hematocrit and apparent viscosity in the microvasculature of cat mesentery, *Microvasc. Res., 19,* 297–319, 1980.
69. Crowell, J. W., and Smith, E. E.: Determinant of the optimal hematocrit, *J. Appl. Physiol., 22,* 501–504, 1967.
70. Gaehtgens, P., Kreutz, F., and Albrecht, K. H.: Optimal hematocrit for canine skeletal muscle during rhythmic isotonic exercise, *Eur. J. Physiol., 41,* 27–39, 1979.
71. Jan, K. M., and Chien, S.: Effect of hematocrit variations on coronary hemodynamics and oxygen utilization, *Am. J. Physiol., 233,* H106–H113, 1977.
72. Messmer, K., Sunder-Plassman, L., Klovekorn, W. P., and Holper, K.: Circulatory significance of hemodilution: Rheological changes and limitations, *Adv. Microcirc., 4,* 1–77, 1972.
73. Meiselman, H. J.: In vivo viscometry: Effect of hemodilution, in K. Messmer and H. Schmid-Schönbein (eds.), *Hemodilution,* Karger, Basel, 1972, pp. 143–159.
74. Driessen, G. K., Heidtmann, H., and Schmid-Schönbein, H.: Effect of hemodilution and hemoconcentration on red cell flow velocity in the capillaries of the rat mesentery, *Pfluegers Arch., 380*(1), 1–6, 1979.
75. Barbee, J. H., and Cokelet, G. R.: Prediction of blood flow in tubes with diameters as small as 29 μm, *Microvasc. Res., 3,* 17–21, 1971.
76. Atherton, A., and Born, G. V.: Effect of blood flow velocity on the rolling of granulocytes in venules, *J. Physiol. (London), 231*(1), 35P–36P, 1973.
77. Atherton, A., and Born, G. V.: Relationship between the velocity of rolling granulocytes and that of the blood flow in venules, *J. Physiol. (London), 233,* 157–165, 1973.
78. Grant, L.: The sticking and emigration of white blood cells in inflammation, in B. W. Zweifach, L. Grant, and R. T. McClusky (eds.), *The Inflammatory Process,* 2d ed., Academic Press, New York, 1973, vol. II, pp. 205–249.
79. Mayrovitz, H. N., and Wiedeman, M. P.: Leukocyte adhesiveness as influenced by blood velocity, in J. Grayson and W. Zingg (eds.), *Microcirculation,* Plenum Press, New York, 1976, vol. I, pp. 128–129.
80. Barbee, J. H., and Cokelet, G. R.: The Fahraeus effect, *Microvasc. Res., 3,* 6–16, 1971.
81. Selkurt, E. W., Scibetta, M. P., and Cull, T. E.: Hemodynamics of intestinal circulation, *Circ. Res., 6,* 92–99, 1958.
82. Schmid-Schönbein, H.: Blood rheology in the microcirculation, *Pfluegers Arch., 336,* 84–87, 1972.
83. Schmid-Schönbein, H.: Microrheology of erythrocytes, blood viscosity and the distribution of blood flow in the microcirculation, *Int. Rev. Physiol., 9,* 1–62, 1976.
84. Gustafsson, L., Appelgren, L., and Myrvold, H. E.: The effect of polycythemia on blood flow in working and non-working skeletal muscle, *Acta Physiol. Scand., 109,* 143–148, 1980.
85. Gustafsson, L., Appelgren, L., and Myrvold, H. E.: Blood flow and apparent viscosity in working and non-working skeletal muscle of the dog after high and low molecular weight dextran, *Circ. Res., 48,* 465–469, 1981.
86. Zweifach, B. W.: Quantitative studies of microcirculatory structure and function: I. Analysis of pressure distribution in the terminal vascular bed in cat mesentery, *Circ. Res., 34*(6), 843–857, 1974.
87. Wiedeman, M. P., Tuma, R. F., and Mayrovitz, H. N.: Defining the precapillary sphincter, *Microvasc. Res., 12,* 71–75, 1976.
88. Vawter, D. L., Fung, Y. C., and Zweifach, B. W.: Distribution of blood flow and pressure from a microvessel into a branch, *Microvasc. Res., 8,* 44–52, 1974.

89. Olofsson, J., Bagge, U., and Branemark, P. I.: Influence of white blood cells on the distribution of blood in microvascular compartments, *Bibl. Anat., 11,* 405–410, 1973.

90. Palmer, A. A.: A study of blood flow in minute vessels of the pancreatic region of the rat with reference to intermittent corpuscular flow in individual capillaries, *Q. J. Exp. Physiol., 44,* 149–159, 1959.

91. Asano, M.: On plasma skimming observed in cutaneous microcirculation in man and rabbits, *J. Physiol. Soc. Jpn., 35*(8), 424–425, 1973.

92. Hauck, G., and Schroer, H.: Stability of the postcapillary bloodstream, *Pfluegers Arch., 312*(1), R38, 1969.

93. Fung, Y. C.: Stochastic flow in capillary blood vessels, *Microvasc. Res., 4,* 34–48, 1973.

94. Schmid-Schönbein, G. W., Skalak, R., Usami, S., and Chien, S.: Cell distribution in capillary networks, *Microvasc. Res., 19,* 18–44, 1980.

95. Tyml, L. K., and Groom, A. C.: Regulation of blood flow in individual capillaries of resting skeletal muscle in frogs, *Microvasc. Res., 20,* 346–357, 1980.

96. Krogh, A.: *The Anatomy and Physiology of Capillaries,* Yale University Press, New Haven, Conn., 1922.

97. Palmer, A. A.: Axial drift of cells and partial plasma skimming in blood flowing through glass slits, *Am. J. Physiol., 209,* 1115, 1965.

98. Papenfuss, H. D., and Gaehtgens, P.: Effect of bifurcations on hematocrit reduction in the microcirculation: I. Fluid dynamic concepts of phase separation, *Bibl. Anat., 18,* 50–52, 1979.

99. Gaehtgens, P., and Papenfuss, H. D.: Effect of bifurcations on hematocrit reduction in the microcirculation: II. Experimental studies in narrow capillaries, *Bibl. Anat., 18,* 53–55, 1979.

100. Johnson, P. C., Blaschke, J., Burton, K. S., and Dial, J. H.: Influence of flow variation on capillary hematocrit in mesentery, *Am. J. Physiol., 221,* 105–112, 1971.

101. Yen, R. T., and Fung, Y. C.: Inversion of Fahraeus effect and effect of mainstream flow on capillary hematocrit, *J. Appl. Physiol., 42*(4), 578–586, 1977.

102. Yen, R. T., and Fung, Y. C.: Effect of velocity distribution on red cell distribution on capillary blood vessels, *Am. J. Physiol., 235*(2), H251–H257, 1978.

103. Lipowsky, H. H., Rofe, S., Tannenbaum, L., Firrell, J. C., Usami, S., and Chien, S.: Microvessel hematocrit distributions at bifurcations, *Microvasc. Res., 21,* 249–250, 1981.

104. Klitzman, B., and Johnson, P. C.: Capillary network geometry and red cell distribution in the hamster cremaster muscle, *Am. J. Physiol., 242,* H211–H219, 1982.

CHAPTER 19

Interstitial Fluid Pressure and Lymph Flow

Alan R. Hargens
Departments of Surgery and Bioengineering
University of California, San Diego
La Jolla, California

INTRODUCTION

In terms of colorful controversies and important issues, the topics of interstitial fluid pressure and lymph formation have gained much attention among physiologists and bioengineers during the past 20 years. Prior to this period, information regarding interstitial pressures and lymphatic drainage was accumulated slowly because of limited availability of quantitative techniques. However, foundations for our present understanding of fluid exchange between the blood and tissue spaces were firmly laid by two pioneering physiologists, Carl Ludwig and Ernest Starling. Ludwig hypothesized that interstitial fluid is formed by filtration of blood through the capillary wall during periods of elevated capillary blood pressure [1]. Starling proposed that blood volume is maintained and edema is prevented by the colloidal osmotic pressure of blood proteins [2]. These basic theories provide the framework for our present knowledge of fluid exchange between blood and interstitial fluids.

Various ideas regarding the magnitude of interstitial fluid pressure and the mechanism of lymph formation have been argued forcefully by many investigators during the past two decades. Frequently, these controversies dominated many studies of transport dynamics between blood, interstitial fluid, and peripheral lymph. Sometimes arguments concerning whether interstitial fluid pressure was positive or negative (with reference to atmospheric or ambient pressure) led to lively debates regarding the validity and utility of one pressure-measuring technique over another. Most workers in this field today, however, agree that some tissues (subcutaneous, peritoneal, synovial, and lung tissues) normally have negative interstitial fluid pressures, whereas other tissues (muscle tissue and tissue of filtering organs such as the liver and kidney) normally have positive interstitial fluid pressures. During the past 20 years, our understanding of interstitial fluid pressure and lymph flow has advanced markedly, partly because of the above-mentioned controversies but also as a result of the availability of new technology (macro- and microtechniques) for measuring pressures in blood, tissue, and peripheral lymph.

The importance of the topics of interstitial fluid pressure and lymph flow may be inferred from the number of reviews related to this subject [3–17], some of which are very recent. For details concerning past controversies regarding mechanisms and techniques for measuring interstitial fluid pressure, readers are referred to these reviews. In the present chapter, a summary of interstitial fluid pressure and lymph flow will encompass the following subjects: interstitial and lymphatic anatomy (dimensional relationships and functional properties), transcapillary exchange (parameters of the Starling equilibrium, techniques for measuring transcapillary pressures with special emphasis on interstitial pressures), regulation of interstitial fluid volume, and formation of lymph.

19.1 ANATOMY AND FUNCTION OF INTERSTITIUM AND LYMPHATIC SYSTEM

The interstitium consists of structural components [collagen fibrils and glycosaminoglycans (GAGs)] and fluid components (water, proteins, ions, and other metabolic substrates and products) which surround single cells or clusters of cells. These components comprise the ground substance of the so-called connective tissue that forms the milieu around which all cells and microcirculatory elements function (Fig. 19.1). Collagen fibrils and GAGs interpenetrate each other to provide elasticity, to oppose interstitial deformation, and to maintain interstitial volume [18, 19]. Fibroblasts synthesize collagen precursors (procollagen), which move into the interstitium to form tropocollagen polymers, which in turn, polymerize to collagen fibrils (Fig. 19.2). Similarly, hyaluronic acid and proteoglycans are transported extracellularly and contained by collagen fibrils, thus forming an interstitial matrix.

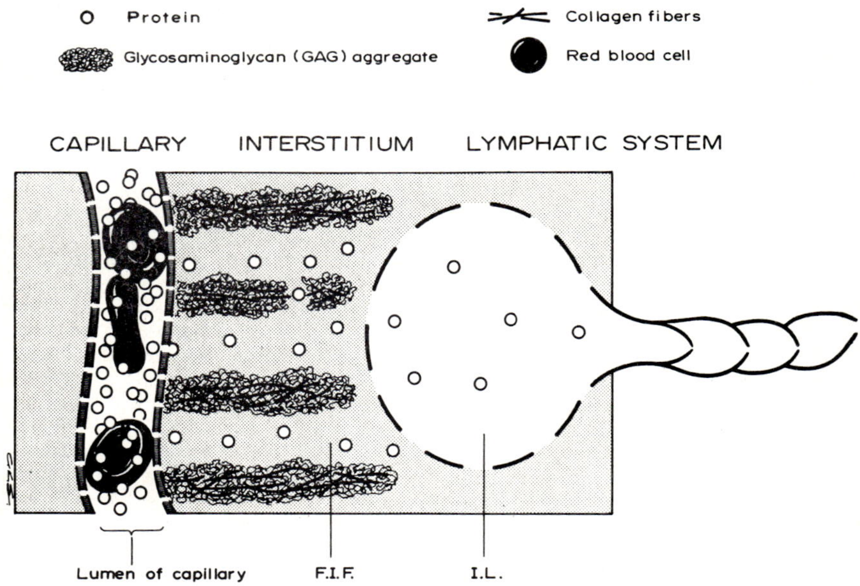

FIG. 19.1 Structure and flow dynamics of the interstitium. Ultrafiltration of plasma across the capillary wall yields a gradient of plasma protein and hydrostatic pressure from the blood to interstitial fluid. Capillary walls with a high protein reflection coefficient σ_p retain most protein in the blood. Tissue can be considered as a three-phase system with cells (not shown) surrounded by collagen-GAG aggregates and microscopic channels of free interstitial fluid (F.I.F.). As fluid moves from blood through the interstitium and into the initial lymphatic (I.L.), its composition is altered by exclusion phenomena and chromatography effects. Structural elements are not drawn to scale (refer to text for dimensions).

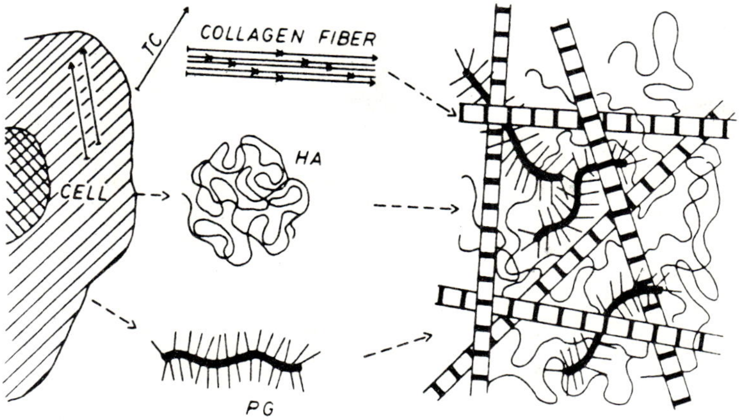

FIG. 19.2 Interstitial matrix components are formed intracellularly, transported to interstitial spaces, and polymerized to form a framework of interpenetrating macromolecules. Collagen fibers are formed from tropocollagen (TC), whereas glycosaminoglycans form from hyaluronic acid (HA) and proteoglycan (PG). (*From Laurent [19].*)

19.1.1 Dimensions, Composition, and Properties of the Interstitium

The volume of interstitial fluid, based upon equilibration of various extracellular and intravascular isotopes, ranges from 12 to 20 percent of body weight [12]. Certain tissues contain larger reservoirs of interstitial fluid, and thus these organs are relatively more important in overall fluid balance within the body (Table 19.1). Interstitial fluid volume is about four times plasma volume

TABLE 19.1 Interstitial fluid volume (IFV) in various organs

Tissue	Tissue weight, % of body weight	IFV, (mL/100 g wet tissue)	Total tissue IFV, % of total body IFV
Skeletal muscle	~40	8–20	25–40
Skin	10–18	40–60	30–40
Viscera	12–15	20–30	20–30
Other organs	....	....	< 10
Whole body	100	12–20	100

Source: Aukland and Nicolaysen [12].

(two to three times blood volume with hematocrit of 40). Although interstitial structures are well defined and relatively uniform biochemically, interstitial anatomy varies greatly from one tissue to another and even within a given tissue. For example, some regions are devoid of collagen fibrils, GAGs, or both [20]. As depicted in Fig. 19.1, large macromolecules such as globulins are preferentially and rapidly transported within channels of free fluid so that large protein molecules experience a "chromatography effect" as they are transported from the blood to peripheral lymph [19, 21–23]. Free fluid channels are estimated to range between 2 and 10 μm in diameter, on the basis of brownian motion studies in subcutaneous tissue of the bat wing [24]. More recently, however, Casley-Smith and Vincent determined that these tissue fluid channels are 100 to 200 nm in diameter for various tissues [24a].

As indicated earlier, the interstitium consists of a gel-like matrix of collagen fibrils and GAGs which interpenetrate and attract one another primarily by hydrogen and electrostatic bonds. Whereas collagen fibrils function as structural support for the matrix, negatively charged GAGs (hyaluronate and proteoglycan) function importantly to imbibe interstitial fluid and to exclude proteins and other large molecules. Thus, the transport of fluid and solutes from blood to lymph is primarily confined within the interstitial channels of free fluid and represents a complex biorheological phenomenon [25–27].

Hyaluronate is an unbranched macromolecule with a diameter, length, and molecular weight of 1 nm, 3 μm, and 10^6 to 10^7 daltons, respectively. When placed in solution, it assumes a random coil configuration with a 500-nm diameter and imbibes a great deal of saline by virtue of its high density of negative charge. For example, when human umbilical connective tissue (a structure rich in hyaluronate) is placed in Tyrode solution in absence of external compression, it swells to twice its usual volume [28]. At a normal water–dry weight ratio of 10, the umbilical cord has an imbibition pressure of −12 mmHg. During dehydration, this imbibition pressure increases as a power-series expansion so that at water–dry weight ratio of 5, the imbibition pressure reaches −80 mmHg. Repulsive forces within a matrix of high anionic charge density cause the tissue to exert large negative fluid pressures [29]. When water or proteinaceous fluid is available for entry into the tissue, fluid is imbibed and the tissue swells. Granger and coworkers [28] conclude that the swelling behavior of the interstitial matrix is characterized by (1) a low compliance (dV/dP) state at normal hydration and (2) a nonlinear compliance falloff to very low levels with dehydration.

According to studies of imbibition pressure in isolated samples of umbilical tissue, elastic recoil forces primarily cause the interstitial matrix to swell when excess fluid is available. With elimination of electrostatic and associated Donnan effects between anionic charges in the matrix by placing the tissue in a solution with high salt concentration, only a small shift of the volume-pressure relationship occurs [28]. This suggests that elastic recoil of *GAG macromolecules* within the interstitial matrix provides the primary mechanism for imbibition (Fig. 19.3). In

ELASTIC RECOIL AND FLUID PRESSURE OF THE EXTRACELLULAR MATRIX

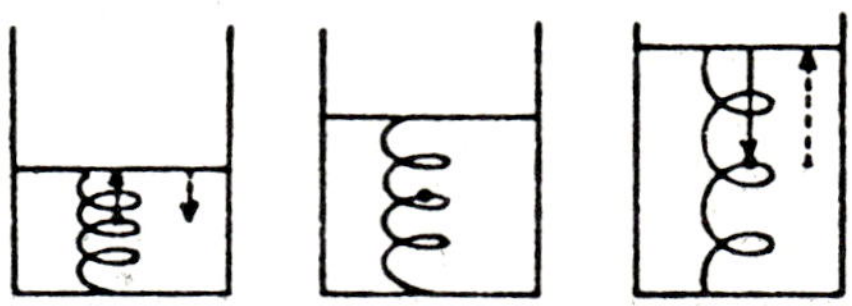

	COMPACTED MATRIX	UNSTRESSED MATRIX	EXPANDED MATRIX
Elastic Recoil Force (——▸)	Positive	Zero	Negative
Fluid Pressure (---▸)	Negative	Zero	Positive

FIG. 19.3 Relationship between elastic recoil and interstitial fluid pressures in tissue matrix during dehydration or compaction (left diagram), unstressed state (central diagram), and edematous or excess fluid conditions (right diagram). (*From Granger et al. [28].*)

an unstressed matrix (central diagram), these macromolecules are not compressed or stretched, so that recoil pressure and hydrostatic pressure are zero. If fluid is lost by dehydration (left-hand diagram of Fig. 19.3), the matrix is compressed, elastic recoil force becomes positive, and interstitial fluid pressure becomes negative to balance the recoil force. On the other hand, if fluid is injected into the tissue (e.g., in a filtering organ or under conditions of elevated capillary filtration), the positive interstitial fluid pressure induces a negative elastic recoil pressure (right-hand diagram). Edema is a condition which exemplifies the third situation. It is important to note that certain tissues (subcutaneous, lung) normally have negative fluid pressure in their interstitium and represent a compressed matrix (left-hand diagram), while other tissues (filtering organs such as the kidney) normally have positive interstitial fluid pressure in their expanded matrix. During abnormal conditions of severe dehydration or edema, any tissue can assume the compacted or expanded state, respectively. Not all investigators, however, agree with this description of the mechanism for tissue swelling. For example, Meyer and Silberberg [30, 31] suggest that, under normal in vivo conditions, the *collagen framework* assumes an unstressed state and that dehydration or swelling is opposed by the network of collagen fibrils, which tends to restore the tissue to its steady state.

GAG and hyaluronate macromolecules, which interpenetrate the framework of collagen fibrils in the interstitium, will diffuse out of the tissue (e.g., articular cartilage, umbilical tissue) over relatively long time periods. This process is considerably more rapid if the collagen fibrils are digested away by collagenase [30]. Recently, lymph was found to contain detectable hyaluronate concentrations [32], which suggests that this large mucopolysaccharide passes across the interstitial-lymph barrier. Apparently, however, this is a very slow process since hyaluronate and other GAG macromolecules are much more highly concentrated in tissue spaces than in lymph [33–35]. Hyaluronate is cleared from plasma primarily by the liver at a rate of 10 to 30 mg/day per kilogram of body weight [36].

19.1.2 Interstitial Flow Phenomena

Early studies by Ogston and associates [35, 37–39] demonstrated that GAG and related macromolecules profoundly affect the transport and distribution of proteins and other macromolecules within a tissue matrix. Thus, interstitial matrix excludes any solute whose diameter is greater than the pore diameter of the entangled matrix of collagen fibrils and GAG. Recently, Granger summarized these relationships for the interstitium in quantitative terms [21]. If a tissue matrix of equivalent fibrils is assumed, the fraction F_E of fluid space excluded by the matrix is

$$F_E = 1 - \exp\,[-(r_p + r_f)^2\overline{V}_f C_f] \tag{19.1}$$

where r_p is protein radius, r_f is fibril radius, $\overline{V}_f$ is partial specific volume of fibril polymers, and C_f is fibril concentration. This relationship indicates that exclusion increases with protein size and fibrillar density such that dehydration raises F_E. In addition to these steric considerations, electrostatic repulsion between the negatively charged proteins and negatively charged GAG matrix at physiological pH values offers further exclusion to protein penetration. High charge densities within articular cartilage, intervertebral disks, meniscus, and corneal stroma are particularly important for exclusion phenomena in these tissues.

Confinement of protein molecules to larger channels within the interstitium increases their colloidal osmotic pressure Π_t [21, 38, 40] to a level higher than in a pure saline solution:

$$\Pi_t = RTn_p[(1 - F_E)V_t]^{-1} \tag{19.2}$$

where Π_t is interstitial fluid colloidal osmotic pressure (sometimes referred to as *interstitial fluid oncotic pressure*), R is the universal gas constant, T is absolute temperature, n_p is number of moles of protein for a number-averaged molecular weight, and V_t is tissue fluid volume. As will be discussed in Sec. 19.2, which is concerned with transcapillary exchange, any condition which adds protein-poor fluid to the interstitium (e.g., elevation of capillary blood pressure or reduction of plasma protein concentration) will tend to reduce Π_t and wash out interstitial proteins so that edema is prevented [12]. Moreover, since the exclusion phenomenon amplifies changes in Π_t with interstitial dilution, Granger suggests that this "exclusion amplification extends the margin of safety against edema formation beyond the limit provided by simple dilution of the extravascular protein pool" [21]. Granger summarizes albumin exclusion fractions for various tissues (Table 19.2).

After passing from the capillary wall into interstitial fluid, solutes are transported by diffusion, by convection, or by both. Molecular diffusion through the interstitium is a random process which depends on the gradient of solute concentration, solute size, and other parameters specific to the tissue matrix. Ogston and coworkers [39] suggest the ratio of diffusion in a

TABLE 19.2 Albumin exclusion from various tissues

Tissue	F_E for albumin	Reference(s)
Subcutis and skeletal muscle	0.2–0.4	Bell et al. [41]; Rutili [42]
Liver	0.3	Goresky [43]
Lung	0.3–0.5	Selinger et al. [44]
		Parker et al. [45]
Intestine	0.4	Granger et al. [46]
Umbilical cord	0.5	Granger and Shepherd [47]
Smooth muscle	0.6	Barr and Malvin [48]
'Corneal stroma	0.8	Maurice [49]
Articular cartilage	0.9	Maroudas [50]

Source: Granger [21].

matrix, D_m, to free diffusion in water, D_w, is an exponential function:

$$\ln \frac{D_m}{D_w} = -\left(\frac{r_s + r_f}{r_f}\right)(\bar{V}_f C_f)^{1/2} \tag{19.3}$$

where symbols are as defined in Eq. (19.1). Carman [51] provides a relationship for convective (volume) flow J_v through porous media:

$$J_v = \frac{Kr_f^2}{4\eta}\left[\frac{(1 - C_f)^3}{C_f^2}\right]\frac{dP}{dX} \tag{19.4}$$

where K is a constant, η is a fluid viscosity, and dP/dX is the gradient of hydrostatic pressure across the matrix. With reference to transport across the blood-joint barrier (J. R. Levick, personal communication), synovial interstitial conductivity (K_s) is probably more important than capillary wall conductivity [K_f in Eq. (19.9)]. Granger [21] indicates that if the umbilical cord is hydrated to 50 percent of normal, hydraulic conductivity decreases to 7 percent of its normal value [10^{-11} cm^4/(dyn · s)]. Conversely, conductivity increases 20 times when the tissue swells by only a factor of 3. In terms of irreversible thermodynamics [52], interstitial convective flow J_v is a function of hydrostatic ΔP and colloidal osmotic $\Delta\Pi_p$ gradients:

$$J_v = L_p(\Delta P - \sigma_p \, \Delta\Pi_p) \tag{19.5}$$

where L_p is hydraulic permeability and σ_p is the protein reflection coefficient. If $\sigma_p = 1$, all proteins are excluded from the interstitial matrix. If $\sigma_p = 0$, all proteins move through the interstitial matrix without restriction. On the basis of a formulation by Anderson and Malone [53],

$$\sigma_p = \left[1 - \left(1 - \frac{r_p}{r_m}\right)^2\right]^2 \tag{19.6}$$

Granger [21] calculated albumin reflection coefficients σ_A and equivalent pore radii for various tissues (Table 19.3).

Because of retardation of large proteins by the interstitial matrix, these macromolecules build up in interstitial fluid by a sieving effect [12, 18, 55]. Experimental evidence for interstitial sieving is available from the relatively high concentrations of globulin obtained by the implanted wick technique [56] and from protein transport in rabbit lung tissue [57]. However, chromatography effects between the matrix phase and fluid-channel phase would produce more rapid transport of large proteins compared with that of small proteins [21]. A chromatography effect is supported by the finding that fibrinogen equilibrates more rapidly in hindpaw lymph than albumin [55, 58]. However, direct evidence for protein gradients within the interstitium is not available.

TABLE 19.3 Albumin reflection coefficient σ_A and equivalent pore radii for various tissues

Tissue	σ_A	Pore radius, Å	Reference
Subcutis	0.12	193	Granger, unpublished data
Capsule lining:			Granger and Taylor [54]
Outer	0.14	177	
Inner	0.23	133	
Umbilical matrix	0.28	118	Granger and Shepherd [47]
Corneal stroma	0.93	46	Granger, unpublished data

Source: Granger [21].

19.1.3 Lymphatic Transport

This section will deal solely with mechanisms of lymph transport after interstitial fluid has entered the initial lymphatics (also referred to as *terminal lymphatics*). Later discussions will evaluate possible mechanisms of lymph formation. Previously, initial lymphatics have been referred to as *lymphatic capillaries*, but the latter term is a misnomer since the smallest lymph vessels have diameters 5 to 10 times larger than blood capillaries and since functionally these vessels are very different. The entry sites of interstitial fluid into the lymphatic system are usually difficult to visualize even in thin tissues such as mesentery, but dye can be injected to delineate the boundaries of these vessels [59]. As depicted in Fig. 19.1, the primary function of the lymphatic system is to drain off excess fluid and metabolic end products from the interstitium. Another important function is to return extravasated proteins to blood, thus preventing a buildup of interstitial proteins and maintaining a relatively low Π_t. After entry into the initial lymphatics, lymph is mainly propelled by spontaneous, rhythmic contractions of smooth muscle cells which envelop the vessel [59–65], although in some cases lymph flow is facilitated by exercise or voluntary contraction of limb muscles [65–67].

Intralymphatic pressure rises from slightly negative to ambient values in initial lymphatics up to 30 mmHg or higher in large (200- to 300-μm) collecting vessels. Depending on the contractile cycle of an initial lymphatic, intraluminal pressures are negative (about −1 mmHg) or positive (about +1 mmHg) as determined by direct micropuncture [68, 69]. Earlier studies of micropressure in initial lymphatics detected only ambient or positive pressures [59], but it is thought that this was due to suffusion of the mesentery with saline and subsequent edema formation within the tissue [68]. After entry into the 25- to 50-μm initial lymphatics, intraluminal pressure increases downstream in each successive lymphatic segment (Fig. 19.4). Pressure increases in the downstream direction of lymph flow and is maintained by one-way valves which are prominent features of all mammalian lymphatics [70].

In a normally functioning lymphatic segment, lymph flow is generated by an upstream contraction. This flow is associated with increased hydrostatic pressure in the segment and a progressive distension of the collecting lymphatic channel (Fig. 19.5). After attaining a particular threshold pressure and corresponding diameter specific to that particular segment, the vessel contracts. During contraction, a further transient rise in pressure occurs, closing the upstream valve by a momentary retrograde or bidirectional flow and opening the downstream valve. As soon as the downstream valve opens, intraluminal pressure falls even though the contractile phase continues. As the contiguous, downstream segment contracts, pressure in the feeding segment falls to its minimum baseline value, and the valve separating the two segments closes. The whole cycle then repeats itself. As a result of these pressure-volume dynamics, a contractile wave propagates downstream from the initial lymphatic to the collecting lymphatic channels.

When intralymphatic pressure is reduced artificially by upstream occlusion or lymph withdrawal, spontaneous contractions are significantly decelerated to the point of quiescence [64]. Since increased intraluminal pressure stretches the lymphatic wall, these results support the concept that contractions are induced by development of increased tension in smooth muscle

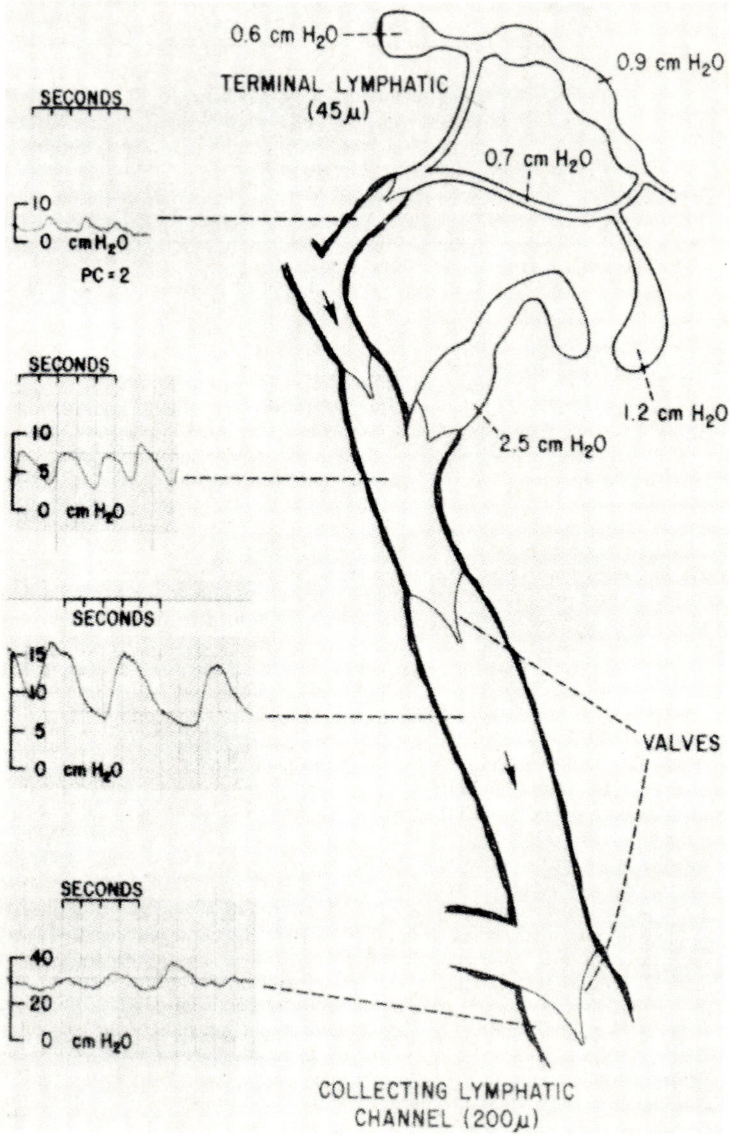

FIG. 19.4 Intralymphatic pressures increase from initial ("terminal") lymphatics, where lymph is formed to collecting lymphatic channels. Within a given intervalvular segment, pressures are equal. (*From Zweifach and Prather [59].*)

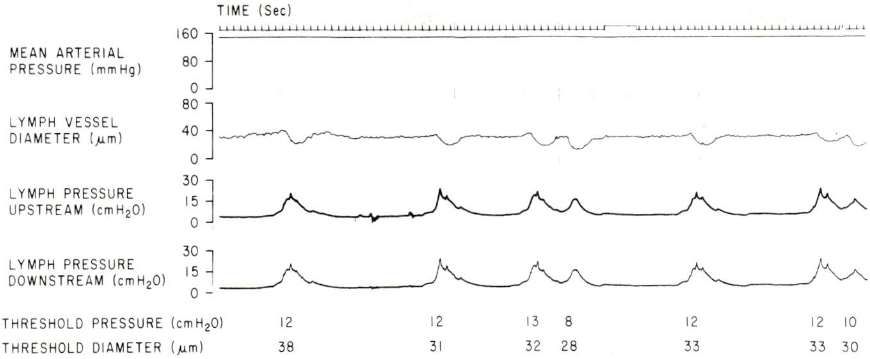

FIG. 19.5 Normal sequence of contraction in collecting lymph vessel of rat mesentery. Two micropressure recordings (one upstream and one downstream) are presented for the same intervalvular segment. Vessel diameter is measured near the downstream micropipet. Mean contraction velocity is 4 Hz. Contraction occurs at a threshold pressure and vessel diameter of 12 cmH_2O and 32 μm, respectively. However, some contractions occur at 8 and 10 cmH_2O. (*From Hargens and Zweifach [64].*)

which envelops the vessel. Lymphatic wall tension T_l can be calculated from pressure-diameter results obtained during normal, artificially increased, and artificially decreased transmural pressures P_l by use of the Laplace relationship:

$$T_l = P_l \cdot r_l \qquad (19.7)$$

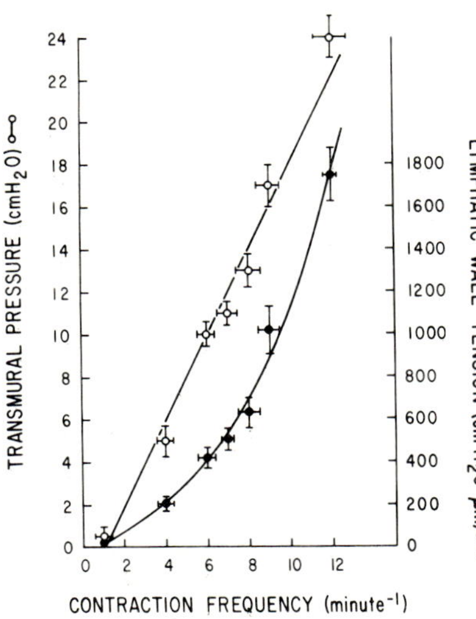

FIG. 19.6 Myogenic contraction of collecting lymph vessels. Contraction frequency is a linear function of transmural pressure. Wall tension, calculated from the Laplace equation, correlates well with its corresponding contraction frequency. Values represent means ±SE (standard error) for at least 28 pressure-radius correlations at each frequency. (*From Hargens and Zweifach [64].*)

where r_l is internal radius of the lymphatic segment (Fig. 19.6). Although the Laplace equation assumes cylindrical vessel geometry, the lymphatic segment is probably cylindrical only during full distension by high P_l. Contraction frequency is a linear function of P_l through a frequency range of 1 to 12 Hz. Thus, a stretch or myogenic mechanism appears to be a more plausible cause of lymphatic contraction than a neurogenic mechanism.

In terms of lymphatic permeability, some evidence exists that initial lymphatics are more permeable to proteins than larger collecting lymphatic channels. When dye-tagged protein is injected into an initial lymphatic, it leaks out through discrete sites along the thin endothelial cell membrane [59, 71]. Similar injections within collecting vessels show no such leakage. In addition, direct protein analyses in succeeding lymphatic segments suggest that lymph is concentrated during its passage from the periphery to the pre- and postnodal collecting lymphatics [71]. Thus, the lymphatic system itself may function in a Starling-type framework, with lymph being concentrated by ultrafiltration as it courses through the extensive network of collecting vessels. A possible driving force for such a mechanism is provided by the fact that lymphatic pressure P_l is substantially greater than interstitial fluid pressure P_t. However, evidence of lymph concentration from prenodal to postnodal vessels is suspect in light of recent studies which demonstrate significant changes in lymph composition related to microcirculatory and metabolic processes occurring within the lymph node itself [72–76].

19.2 TRANSCAPILLARY EXCHANGE

19.2.1 Early Foundations

Based on prior work by Ludwig [1], Starling's experiments [2] led to our present understanding of pressure parameters which govern fluid balance between blood and interstitial fluid. Two recent reviews [16, 77], clearly indicate that Starling never presented a mathematical formulation of the so-called Starling equilibrium. It wasn't until the late 1920s that the first formal mathematical relationships of transcapillary exchange were developed [78–80]. However, Starling is rightly credited with providing the basic concepts and mechanisms for fluid balance between the microcirculation and interstitium. Basically, Starling [2] indicated that the level of capillary blood pressure P_c determined the amount of ultrafiltration into tissue, but this had been reported previously by Ludwig [1]. Importantly, Starling suggested for the first time that blood colloidal osmotic pressure Π_c was the primary factor in resorption of interstitial fluid, and that P_c was nearly balanced by Π_c so that fluid balance was maintained across the capillary wall. Starling suggested that elevated interstitial fluid pressure P_t did not aid fluid resorption. Also, he indicated that interstitial fluid colloidal osmotic pressure Π_t falls at high P_c (high filtration levels) and that this response prevents edema. Starling gave lymphatic return a very minor role in his model because of his belief that P_c was closely balanced by Π_c. Today, however, it is apparent that lymph flow is an important adjunct to the maintenance of transcapillary fluid balance and prevention of edema.

19.2.2 Starling Equilibrium

Recent development of techniques which measure exchange across single capillaries [81–83], interstitial fluid pressure [84–89], blood colloidal osmotic pressure [90, 91], and interstitial fluid colloidal osmotic pressure [92–97] offers the opportunity to investigate directly the forces which maintain fluid balance across the capillary wall. Today these techniques are sufficiently developed in experimental animals and clinical diagnoses in humans that extensive data are available on normal and pathological states of tissue [12, 14, 98–107].

Direct measurement of capillary blood pressure in nail folds of human toes demonstrates that approximately two-thirds of the hydrostatic pressure gradient form the heart to toe is effectively added to capillary pressure [108]. In an erect-standing subject, therefore, capillary blood pressure in the toe is 80 to 90 mmHg, compared with only 30 mmHg when the toe is at heart level. Humans, as well as other terrestrial animals, probably avoid edema development in their feet by myogenic vasoconstriction of their microcirculation, which reduces capillary blood flow and transcapillary fluid filtration in dependent tissues [109]. For example, when precapillary smooth muscles in the feet are exposed to high arterial blood pressure during standing, these sphincter muscles contract to raise arteriolar resistance and to lower capillary pressure P_c and capillary filtration coefficient K_f, according to the Starling equation:

$$J_c = K_f[(P_c - P_t) - \sigma_p(\Pi_c - \Pi_t)] \\ = J_l \tag{19.8}$$

where J_c = net transcapillary fluid transport, with dimensions of volume per time per capillary surface area [e.g., mL/(min · cm²)]
K_f = capillary filtration coefficient, with dimensions of volume per time per capillary surface area per pressure [e.g., mL/(min · cm² · mmHg)
P_c = capillary blood hydrostatic pressure, mmHg

P_t = interstitial fluid pressure, mmHg
σ_p = capillary membrane reflection coefficient (dimensionless; ranges between 0 and 1)
Π_c = capillary blood colloid osmotic pressure, mmHg
Π_t = interstitial fluid colloidal osmotic pressure, mmHg
J_l = lymph flow

Shifts of fluid between blood and tissue are primarily determined by the pressure parameters described in this equation. In filtering organs, lymphatic drainage provides the primary path for the return of interstitial fluid to the blood. Equation (19.8) requires that all Starling pressures are in the steady state and that K_f and σ_p are constants for this steady-state condition. These pressures and associated transcapillary fluid flows regulate tissue fluid balance and prevent edema formation (Fig. 19.7).

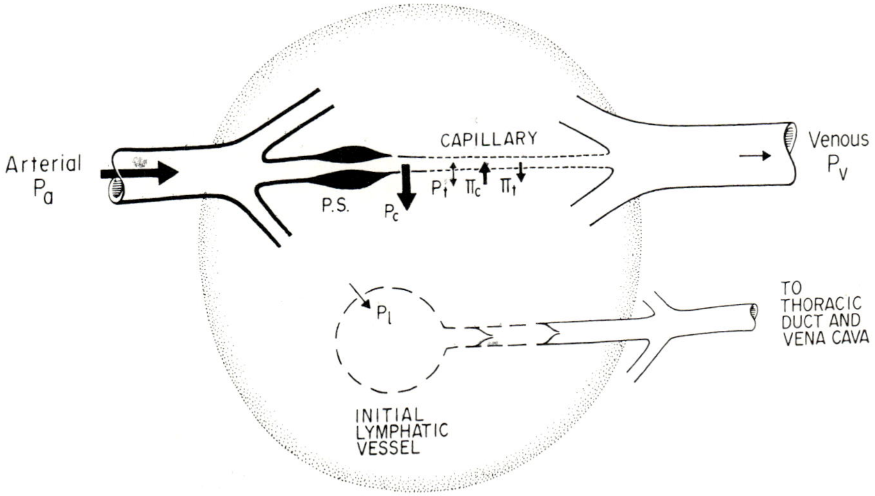

FIG. 19.7 Starling pressures which regulate transcapillary fluid balance. Pressure parameters which determine direction and magnitude of transcapillary exchange include capillary blood pressure P_c, interstitial fluid pressure P_t (directed into capillary when positive or directed into tissue when negative), plasma colloidal osmotic pressure Π_c, and interstitial fluid colloidal osmotic pressure Π_t. Precapillary sphincters (P.S.) regulate P_c, capillary flow, and capillary filtration coefficient K_f. It is generally agreed that a hydrostatic pressure gradient (P_t > lymph pressure P_l) drains off excess interstitial fluid under conditions of net filtration. Relative magnitudes of pressures are depicted by the size of arrows.

Capillary Filtration Coefficient K_f. K_f is the product of the hydraulic conductivity for the capillary basement membrane L_p and capillary surface area A:

$$K_f = L_p A \tag{19.9}$$

where L_p depends primarily on the number and size of capillary pores as given by

$$L_p = \frac{\psi N r^4}{\eta t} \tag{19.10}$$

where ψ is a constant, N is pore number per unit area, r is pore radius, η is viscosity of the ultrafiltrate, and t is capillary membrane thickness [16]. As previously indicated, K_f is primarily regulated by precapillary sphincter muscles which (1) constrict to close off capillaries during some conditions (high P_c or high Po_2) or (2) dilate to "recruit" a greater capillary surface area A. Moreover, some tissues have higher capillary densities than others (thus higher A and higher K_f) even though L_p may be the same for the different tissues.

Values for K_f are obtained by measuring changes in weight of an isolated organ following perturbation of P_c. Measured by this technique, K_f is expressed in units of mL/(min · mmHg) per 100 g tissue, and typical values are 0.010 to 0.015 for skeletal muscle, 0.08 to 0.24 for the intestine, 0.2 to 0.3 for the colon, and 0.07 to 0.25 for lung capillaries [16]. Hydraulic conductivity of the blood-joint barrier L_p increases with elevated intraarticular pressure [110], but acute saline expansion does not affect capillary membrane properties [111] even though proteins are shifted from plasma to tissue spaces [112]. Studies of heart capillaries, however, indicate that plasma protein and blood cells reduce macromolecular permeability [113]. Recent theoretical work suggests that convective and diffusive transports across heteroporous membranes such as the capillary wall can be described by equations for two sets of pores with greatly different radii [114]. Finally, an increased capillary filtration coefficient may be an important cause of edema following arterial reconstructive surgery [115].

Capillary Blood Pressure P_c. Capillary blood pressure P_c is measured indirectly by the isogravimetric technique [116] or directly by dye microinjection [117] or servo-null micropressure techniques [85, 118]. Although the indirect, isogravimetric procedure provides an average value of P_c for the entire tissue or organ, this value may be functionally more relevant than direct, local measures of P_c. On the other hand, isogravimetric preparations are unphysiological in that the tissue is completely immobile, lymph flow is slower than normal, and thus P_c is probably underestimated. Representative values for isogravimetric P_c range between 9 and 20 mmHg in isolated limbs (normal hydration) of dogs and cats [119–121]. Directly determined P_c values are commonly higher than those determined by isogravimetry. For example, P_c in skeletal muscle of cats and dogs ranges between 20 and 40 mmHg [122, 123]. However, these values may overestimate true P_c since micropuncture procedures may select capillaries with highest flow and P_c properties. Furthermore, vasomotion produces intermittent flow through capillaries that may reduce P_c to venular pressure (10 to 20 mmHg), providing a functionally lower P_c [124]. Such a phenomenon would increase capillary resorption of interstitial fluid during low- or no-flow conditions. In summary, true P_c is inherently variable, and its magnitude depends on the organ under study (microcirculatory anatomy) and on the technique of tissue preparation.

Capillary Blood Colloidal Osmotic Pressure Π_c. It is reasonable to assume that Π_c is identical to Π_{systemic} in most tissues. Therefore, a blood sample obtained from a superficial vein is probably representative of capillary blood in terms of plasma protein concentration, the factor which determines colloidal osmotic (oncotic) pressure. The recent development of membrane osmometers for macro- and microsamples [90, 91] makes the measurement of Π_c fairly routine. Human Π_c normally ranges between 20 to 35 mmHg [94, 96, 97]. Interestingly, Π_c in ocean fish ranges from 8 mmHg in temperate species [93] to over 200 mmHg in polar species living near sea ice [125].

It should be emphasized that plasma proteins move through the capillary wall at different rates in different tissues. Some organs have high rates of protein flux from plasma to interstitial fluid (e.g., liver, lung), while other organs have relatively low rates (e.g., muscle, kidney); therefore, when considering Π_c for a given tissue, these differences should be kept in mind. In this regard, the protein reflection properties of the osmometer membrane should be matched with those present in the tissue under study.

Capillary Reflection Coefficient σ_p. Since capillary membranes do not usually operate as perfect osmometers where $\sigma_p = 1$, calculated or measured colloidal osmotic pressures in blood and interstitial fluid (Π_c and Π_t, respectively) must be multiplied by a factor $\sigma_p < 1$ which characterizes a given tissue [see Eq. (19.8)]. Estimates for several organs are now available, based on lymph-to-plasma ratios of protein concentrations (C_l/C_p) at various lymph flows [16]:

$$\frac{C_l}{C_p} = 1 - \frac{\sigma_p}{1 - \sigma_p e^{-x}} \tag{19.11}$$

where

$$x = (1 - \sigma_p)\frac{J_l}{PS} \tag{19.12}$$

in which J_l is lymph flow and PS is the product of the permeability coefficient and surface area [126]. Calculated σ_p values for various tissues range between 0.70 and 0.95 for hindpaw [127], intestine [126], colon [128], lung [129], and synovium [130]. σ_p in liver sinusoids approaches zero [131].

Interstitial Fluid Pressure P_t. Our knowledge of osmosis and fluid resorption by gels is important in understanding mechanisms of pressure generation in interstitial fluids and blood-interstitium-lymph transport. Although classical thermodynamics offers a rigorous mathematical approach to osmotic pressures within solutions, gels, and tissue matrix [34, 38, 132, 133], this approach offers no mechanism for understanding the origins and mechanisms of osmosis, gel imbibition, and tissue swelling [134–137]. In this regard, the viewpoint of a negative solvent (fluid) pressure mechanism has proved valuable for understanding and predicting osmotic flows, imbibition phenomena, and other colligative properties (Fig. 19.8). The unifying concept is that water is drawn into a solution or matrix down a pressure gradient from water in the dish (ambient pressure) to the solution or matrix (negative water pressure). The "springlike" mechanism which induces water tension in the solution or gel resides in the kinetic dispersal pressure of solute molecules or mechanical elasticity of the gel matrix, respectively [134].

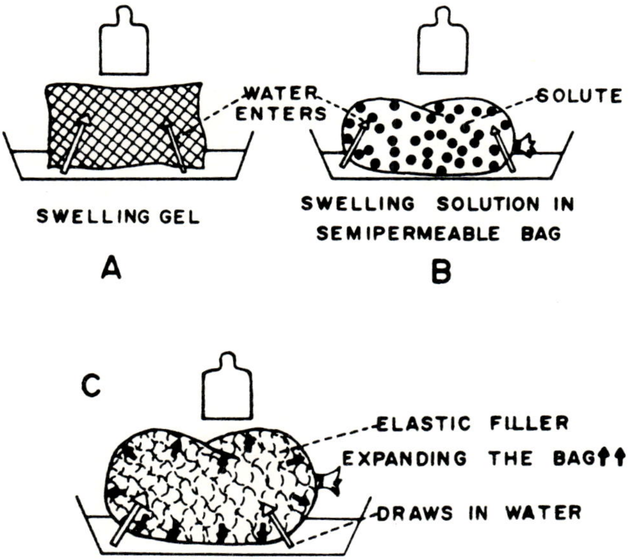

FIG. 19.8 Negative fluid pressure mechanism for imbibition and osmosis. Block of moist gelatin placed in dish of water (A), and slack dialyzing bag filled with protein solutes also placed in water dish (B). Both display five identical responses to water: (1) initial lower vapor pressure; (2) swelling when placed in water; (3) ability to lift weight; (4) when swelling force is balanced by weight, vapor pressure returns to that of ambient water; and (5) if water is substituted by a permeant solution, the solution is drawn into gel (A) or bag (B) irrespective of water potential. A simple physical model (C) explains both mechanisms: If a dialyzing bag is stuffed with a springlike substance (e.g., shredded rubber sponge), placed in a water dish, some water is squeezed out, and the bag is closed; the compressed filler puts the water under tension and lowers vapor pressure. Water is drawn in from the dish and the bag swells. The bag with compressed sponge will lift a weight. When the spring can lift no more, water pressure becomes ambient and vapor pressure returns to normal. If this bag is placed in a permeant solution of any strength, the negative pressure in the bag draws in the bulk solution. (*From Hammel and Scholander [134].*)

Guyton and coworkers [4] dichotomize total tissue pressure into two separate and additive components: interstitial fluid pressure and solid tissue pressure. According to their viewpoint, solid pressure adds to fluid pressure to give total pressure (Fig. 19.9). In tissues, fluid pressure determines capillary exchange as defined by the Starling formulation [see Eq. (19.8)]. Solid pressure is caused by cells, collagen fibers, and gel elastic forces that balance fluid pressure and osmotic pressure; and total pressure combines the fluid and solid components and is expressed as surface pressure in skin [138]. Total tissue pressure can be measured by flaccid balloons or cannulated veins since both solid and fluid pressure combine to cause vessel collapse.

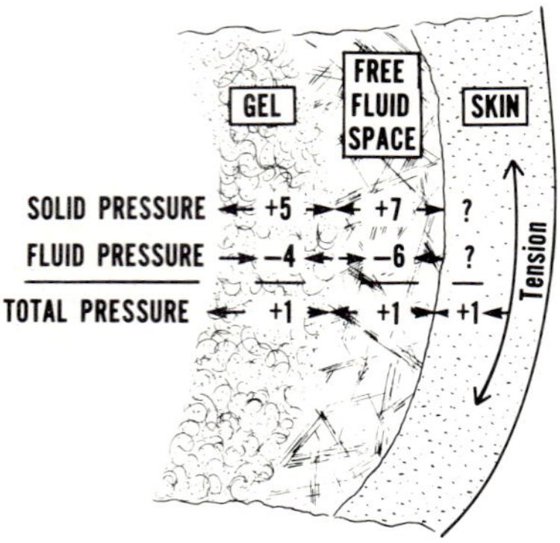

FIG. 19.9 Summary of Guyton's balance of pressures at interfaces between tissue gel, free fluid space, and inner skin surface (*From Guyton et al. [138].*)

Negative fluid pressures are measurable by a push technique or a pull technique (Fig. 19.10). To create a system with obvious negative pressure, a compressed spring is inserted into a syringe which applies a constant pressure P against the syringe plunger (Fig. 19.10*a*). The syringe is filled with water, and a cotton plug is inserted into the nozzle for capillary action to prevent air from entering the syringe. Under these conditions the water has a permanent negative pressure $-P$. This negative pressure is measured either by pushing the plunger with pressure $+P$ until the water is freed above the cotton plug (Fig. 19.10*b*) or by pulling (via suction pressure $-P$ applied to the nozzle) the water until it is free (Fig. 19.10*c*). Both techniques measure the same pressure, namely, the matrix elastic pressure or spring pressure which induces tension in the water (Fig. 19.10*a*). In practice, the push technique is applicable to a negative fluid pressure of any magnitude, whereas the pull technique (e.g., wick technique at bottom of Fig. 19.10) is usable only for weak tensions (i.e., negative fluid pressures < 2 mH_2O). In the wick technique, a simple way to measure negative fluid pressure is to determine the distance H of the glass capillary tube below the wick when the capillary meniscus is stationary while adding the tube's capillarity C [134]. At the bottom of Fig. 19.10, the negative fluid pressure equals $H + C$ cmH_2O.

Since Guyton pioneered the concept of a negative interstitial fluid pressure P_t [84], controversy has developed over the validity of various techniques for measuring P_t. Guyton found that P_t in chronically implanted capsules (those implanted for several weeks) was typically -6 to -7 mmHg in subcutis where inflammation was minimized [138]. Prior to 1960, P_t was usually measured by needles which routinely found $P_t > 0$ (for review, see Ref. 3). Later studies of P_t in subcutaneous, peritoneal, and muscle tissue of several animals by the wick catheter [86] confirmed Guyton's finding that P is often negative. However, intramuscular fluid pressure may reach as high as several hundred millimeters of mercury during maximum isometric contractions in humans. Moreover, this pressure increases with muscular depth in accordance with the Laplace law [see Eq. (19.7)]. Subsequent to Guyton's pioneering paper [84], P_t was also determined by the servo-null micropressure system [69, 85, 89], the wick-in-needle technique [87], and the slit-catheter technique [88]. Recently, the capsule, wick, and micropipet techniques for measuring P_t were described, compared, and evaluated in detail [139]. While chronic animal studies are ideal for the capsule technique [139–141], acute and human studies require a method such as that offered by the wick catheter [94, 142]. Micropipet procedures are also limited to acute studies and to surface (exposed) tissues. However, smaller pieces of tissue and more localized pressure gradients can be investigated by the micropipet [69, 89, 143–147].

Interstitial Fluid Colloidal Osmotic Pressure Π_t. The final component of the Starling formulation includes colloidal osmotic (oncotic) pressure of interstitial fluid [see Eq. (19.8)]. In nonedematous tissue it is difficult to obtain more than a few microliters of interstitial fluid for studies of

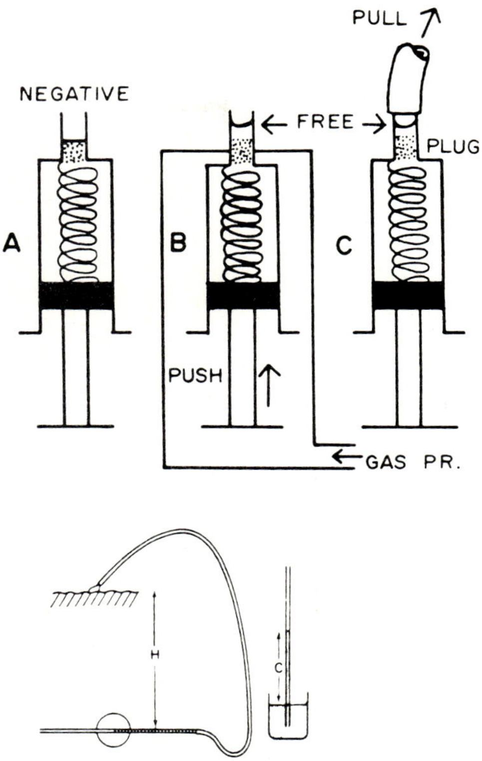

FIG. 19.10 To create negative fluid pressure, an expansion spring is inserted into syringe barrel with porous cotton plug in nozzle (A). Fluid pressure is measured by push (B) or pull (C) technique, which frees the fluid meniscus in the nozzle. Either technique measures the spring compression, which creates the negative fluid pressure. Usually the push technique (B) is performed in a pressure chamber by introducing high gas pressures. One example of the pull method is the wick catheter technique at bottom, where equilibrium, negative fluid pressure equals H (height of capillary tube with stationary meniscus) + C (capillarity of tube). (*From Hammel and Scholander [134].*)

Π_t. Several procedures have been used to collect interstitial fluid samples or to measure $P_t + \Pi_t$ directly. These techniques include implanted wick fibers [92, 96], empty wick catheters [93, 94], an implantable colloidal osmometer to measure $P_t + \Pi_t$ [95], chronically implanted capsules [148, 149], liquid paraffin cavities [150], and micropipets [151–153]. These techniques are considered in detail elsewhere [12, 14]. Also, clinical techniques for measuring P_t and Π_t have been recently compared and evaluated [97] (see Fig. 19.11).

In general, these techniques yield interstitial protein concentrations which are 40 to 60 percent of plasma protein concentration. Because of the nonlinear relationship between Π_t and protein concentration, Π_t/Π_c is usually greater than 0.5 in subcutaneous tissue of the human thorax but less than 0.5 in subcutaneous tissue of the human ankle. Noddeland recently found a Π_t gradient in subcutaneous tissue of normal humans having a mean $\Pi_c \pm$ SD of 29 ± 4 mmHg:

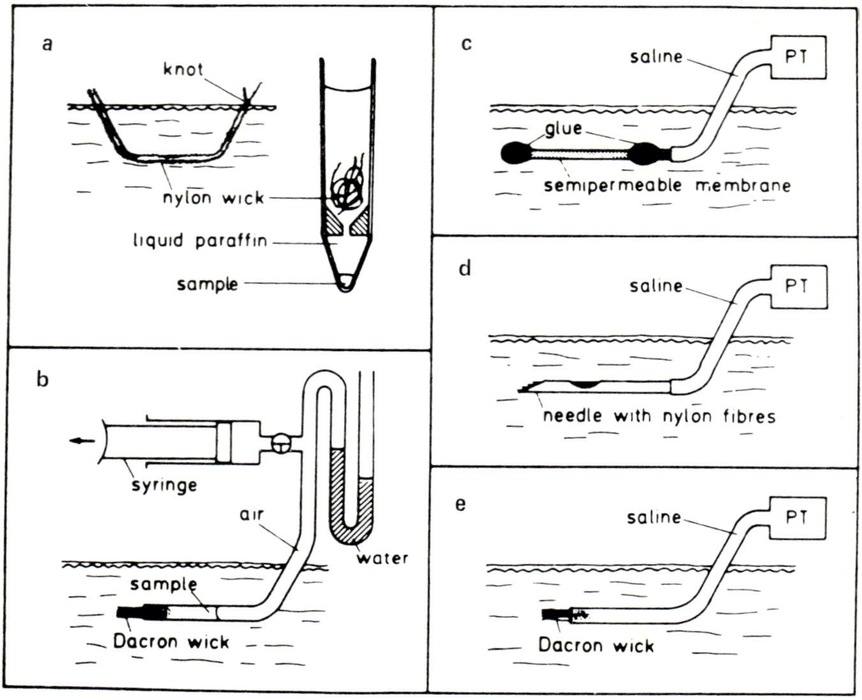

FIG. 19.11 Clinical methods for collection of interstitial fluid (*a, b*), measurement of $P_t + \Pi_t$ (*c*), and measurement of P_t (*d, e*) in human subcutaneous tissue: (*a*) Implanted nylon wick, (*b*) empty wick catheter, (*c*) implantable colloidal osmometer, (*d*) wick-in-needle, and (*e*) wick catheter. PT represents pressure transducer. Suction in empty wick catheters should not be over 3 cmH_2O. (*From Noddeland et al. [97].*)

Π_t in the throax was 15 ± 2 mmHg, while Π_t in the ankle was 10 ± 2 mmHg [154]. Although data remain contradictory, most investigators believe that peripheral lymph is an excellent representation of interstitial fluid [12, 14].

19.3 REGULATION OF INTERSTITIAL FLUID VOLUME

Interstitial compliance (dV_t/dP_t) is a very important parameter which characterizes a given tissue's ability to counteract edema formation or to prevent dehydration. High compliance indicates that large changes in interstitial fluid volume V_t occur with only small changes in P_t. Conversely, low compliance indicates that small changes in V_t occur with large changes in P_t. The basic shape of the compliance curve was determined by Guyton in 1965 [155]. In a review that examines local regulatory mechanisms of interstitial fluid volume, Aukland and Nicolaysen [12] compile inverse dP_t/dV_t curves for several tissues by various techniques (Fig. 19.12). These studies indicate that normal limb tissues have a relatively low compliance region at low P_t and V_t due to compaction of cells, collagen, GAG, and other structural components; an intermediate zone of somewhat higher compliance; and, at high V_t, another region of low compliance where fascia or skin prevent further interstitial expansion [12, 86, 123, 156–158]. Similarly, compliance in a joint space obeys this sigmoidal relationship [159, 160]. Synovial fluid is considered interstitial fluid that normally exists in a subatmospheric pressure state [161].

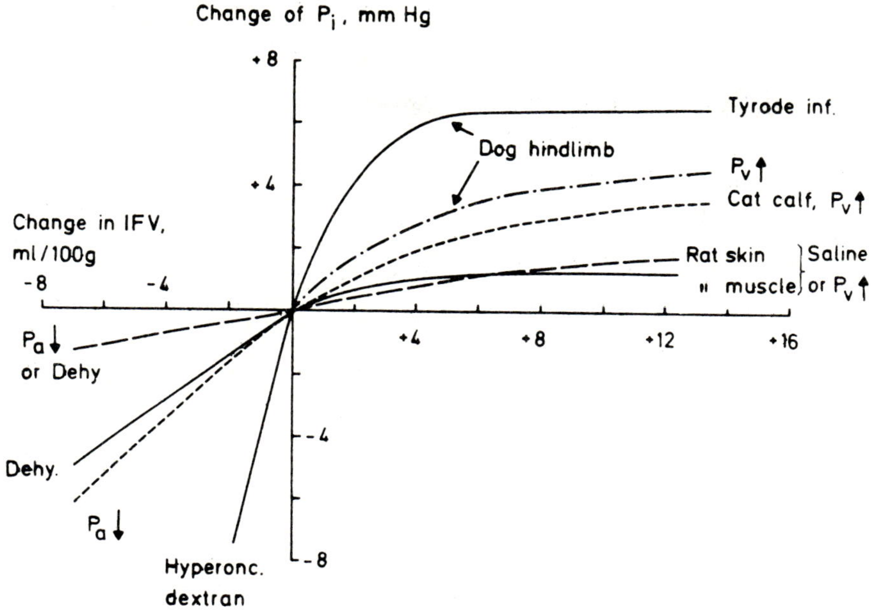

FIG. 19.12 Change in interstitial fluid pressure dP_i with change in interstitial fluid volume *d*IFV in dog hind limb (total limb volume versus subcutaneous capsule P_i), isolated calf muscle (total volume versus P_i calculated by isovolumetry), intact rat muscle and skin (local ^{51}Cr-EDTA space minus plasma volume versus micropipet P_i). IFV was varied by reduced arterial blood pressure ($Pa\downarrow$); dehydration (Dehy.); increased venous blood pressure ($P_v\uparrow$); or intravenous infusion of hyperoncotic dextran, saline, or Tyrode solution. (*From Aukland and Nicolaysen [12].*)

Taylor and collaborators [162] suggest that interstitial resistance plays a role in opposing edema formation. Although lymph flow increases in all organs with elevated P_c, it reaches a plateau (*maximum lymphatic carrying capacity*) which is characteristic for a given tissue. Since lymph flow increases more with lowered Π_c than with elevated venous pressure or P_c in intestine [163] and lung [164], Taylor and coworkers postulate that lymph flow is a function of fluid conductance through the interstitium [162]. In this respect, lymph formation could be enhanced and may provide an extra margin of safety against severe edema during the early stages of tissue swelling when interstitial fluid channels are enlarged. Recent evidence in support of their hypothesis is provided by rapid clearance of tissue edema and P_t reduction following hyaluronidase treatment of compartment syndromes [165]. Of course, the interstitial resistance safety factor would complement other acknowledged mechanisms of edema prevention (increased P_t, elevated Π_c, and decreased Π_t).

19.4 FORMATION OF LYMPH

The mechanism of lymph formation remains controversial and unknown. Specific mechanisms may vary for different tissues, but all probably involve hydrostatic and/or osmotic pressure gradients across the wall of the initial lymphatic vessel. Few direct measurements of such gradients are available. Using the servo-null micropressure technique, Hogan found that a pressure gradient ($P_t > P_l$) for lymphatic filling exists during part of the contractile cycle of the initial lymphatic, despite the fact that average P_l exceeds adjacent P_t [69]. In the bat wing, P_t is slightly higher than P_l in adjacent lymphatics [24, 166]. In order for fluid to move from an interstitium with $P_t < 0$ into peripheral lymph, the initial lymphatic must create a P_l more negative than P_t. Guyton and colleagues [7, 138, 139], Reddy and associates [167], and Hogan [69] propose that initial lymphatics pump fluid from the interstitium. In effect, this pumping mechanism would also create the negative P_t which is evident in some tissues. Presumably smooth muscle contraction of the initial lymphatic propels lymph beyond the first one-way valve of the collecting lymphatic channel, while interstitial anchoring filaments and flaplike valves along the lymphatic endothelium [168] allow the suction mechanism.

Support for an osmotic mechanism of lymph formation is primarily based upon theoretical concerns and electron microscopic data of Casley-Smith [17]. Briefly, various tracer and microdensitometric techniques indicate that average protein concentration in the initial lymphatic is about three times that in interstitial (free-fluid) channels [17, 169, 170]. An oncotic gradient ($\Pi_l > \Pi_t$) is produced by ultrafiltration of protein-free lymph from the initial lymphatic during changes in total tissue pressure. Presently, however, the complicated process of lymph formation proposed by Casley-Smith lacks general support by workers in this field.

It must also be emphasized that tissue motion is essential for lymph formation processes. In a resting limb of an anesthetized animal, lymph flow rapidly falls to zero [171]. Intramuscular P_t is a linear function of contractile force and increases to values as high as 600 mmHg during maximal voluntary contractions in human subjects [172, 173]. Pressure gradients in muscles with curved fibers conform to the Laplace law [174] and are sufficient to generate lymph formation and flow. Finally, recent studies indicate that initial lymphatics of fascia [175] and muscle [176] are filled and purged by arteriolar vasomotion.

SUMMARY

Mechanisms for the generation of interstitial fluid pressure and lymph flow have undergone intense scrutiny during the past 20 years, sparking considerable controversy with reference to techniques employed and conclusions drawn. Interstitial pressure and transport phenomena are intimately related to the matrix of collagen fibrils and glycosaminoglycans which constitute interstitial ground substance. Negative fluid pressure is caused by elastic recoil and electrostatic repulsion of GAG and can be related to osmotic mechanisms in solutions. Exclusion and possible chromatography effects determine the movement of plasma proteins from blood to peripheral lymph. Exact details of lymph formation are still unknown, but probably involve a hydrostatic pressure gradient induced by pumping of initial lymphatic vessels, tissue motion, and vasomotion. Flow through larger lymphatic channels is generated by myogenic contraction of smooth muscle in the lymphatic wall coupled with a system of one-way valves.

Transcapillary fluid exchange is regulated primarily by the Starling pressures: capillary hydrostatic and oncotic pressures P_c and Π_c, and interstitial fluid hydrostatic and oncotic pressures P_t and Π_t. Interstitial fluid pressure is positive in filtering organs such as the kidney and negative in other tissues (subcutis, lung, and sometimes muscle). The choice of a technique for measuring P_t is determined by the preparation or experimental condition under study. Finally, edema or dehydration of tissues is prevented by adjustments in Starling pressures and lymph flow, constraints of interstitial compliance, and interstitial resistance.

ACKNOWLEDGMENTS

I thank Drs. Knut Aukland and J. Rodney Levick for their critiques and helpful suggestions.

This work was supported in part by the Veterans Administration and by NIH grants AM-25501 and AM-26344 and a research career development award AM-00602.

REFERENCES

1. Ludwig, C. F.: *Lehrbuch der Physiologie des Menschen,* C. F. Winter, Leipzig, 1861, vols. I, II.
2. Starling, E. H.: On the absorption of fluids from the connective tissue spaces, *J. Physiol. (London), 19,* 312–326, 1896.
3. Landis, E. M., and Pappenheimer, J. R.: Exchange of substances through the capillary walls, in W. F. Hamilton (ed.), *Handbook of Physiology,* sec. 2: *Circulation,* American Physiological Society, Washington, D.C., 1963, vol. II, chap. 29, pp. 961–1034.
4. Guyton, A. C., Granger, H. J., and Taylor, A. E.: Interstitial fluid pressure, *Physiol. Rev., 51,* 527–563, 1971.
5. Wiederhielm, C. A.: The interstitial space, in Y. C. Fung et al. (eds.), *Biomechanics: Its Foundation and Objectives,* Prentice-Hall, Englewood Cliffs, N.J., 1972, pp. 273–286.
6. Staub, N. C.: Pulmonary edema, *Physiol. Rev., 54,* 678–811, 1974.
7. Guyton, A. C., Taylor, A. E., and Granger, H. J.: *Circulatory Physiology,* vol. II: *Dynamics and Control of the Body Fluids,* Saunders, Philadelphia, 1975.
8. Guyton, A. C.: Interstitial fluid pressure and dynamics of lymph formation, *Fed. Proc., 35,* 1861–1862, 1976.
9. Gauer, O. H., and Kirsch, K.: The interstitial tissue pressure and microcirculation, in H. Meessen (ed.), *Handbuch der allgemeinen Pathologie,* vol. III: *Mikrozirkulation,* Springer-Verlag, Berlin, 1977, chap. 7, pp. 385–409.
10. Nicoll, P. A., and Taylor, A. E.: Lymph formation and flow, *Annu. Rev. Physiol, 39,* 73–95, 1977.
11. Zweifach, B. W., and Silberberg, A.: The interstitial-lymphatic flow system, in A. C. Guyton and D. B. Young (eds.), *International Review of Physiology III,* University Park Press, Baltimore, 1979, vol. 18, pp. 215–260.
12. Aukland, K., and Nicolaysen, G.: Interstitial fluid volume: Local regulatory mechanisms, *Physiol. Rev., 61,* 556–643, 1981.
13. Brace, R. A.: Progress toward resolving the controversy of positive vs. negative interstitial fluid pressure, *Circ. Res., 49,* 281–297, 1981.
14. Hargens, A. R.: *Tissue Fluid Pressure and Composition,* Williams & Wilkins, Baltimore, 1981.
15. Rodbard, S.: The capsular barrier between the interstitial fluid and the source of lymph, *Curr. Mod. Biol., 3,* 27–34, 1969.
16. Taylor, A. E.: Capillary fluid filtration: Starling forces and lymph flow, *Circ. Res., 49,* 557–575, 1981.
17. Casley-Smith, J. R.: Mechanisms in the formation of lymph, in A. C. Guyton and J. E. Hall (eds.), *International Review of Physiology IV,* University Park Press, Baltimore, 1982, vol. 26, pp. 147–187.
18. Laurent, T. C.: The structure and function of the intercellular polysaccharides in connective tissues, in C. Crone and N. A. Lassen (eds.), *Capillary Permeability,* Munksgaard, Copenhagen, 1970, pp. 261–277.
19. Laurent, T. C.: The ultrastructure and physical-chemical properties of interstitial connective tissue, in O. H. Gauer et al. (eds.), *Capillary Exchange and the Interstitial Space, Pfluegers Arch. Suppl., 336,* 21–42, 1972.
20. Merker, H. J., and Günther, T.: Morphologic observations, in O. H. Gauer et al. (eds.), *Capillary Exchange and the Interstitial Space, Pfluegers Arch. Suppl., 336,* 33–34, 1972.
21. Granger, H. J.: Physicochemical properties of the extracellular matrix, in A. R. Hargens (ed.), *Tissue Fluid Pressure and Composition,* Williams & Wilkins, Baltimore, 1981, chap. 5, pp. 43–61.
22. Preston, B. N., Laurent, T. C., Comper, W. D., and Checkley, G. J.: Rapid polymer transport in concentrated solutions through the formation of ordered structures, *Nature, 287,* 499–503, 1980.
23. Hauck, G.: The connective tissue space in view of the lymphology, *Experientia, 38,* 1121–1122, 1982.
24. Wiederhielm, C. A.: The interstitial space and lymphatic pressures in the bat wing, in A. P. Fishman and H. H. Hecht (eds.), *The Pulmonary Circulation and the Interstitial Space,* University of Chicago Press, Chicago, 1969, pp. 29–41.

24a. Casley-Smith, J. R., and Vincent, A. H.: The quantitative morphology of interstitial channels in some tissues of the rat and rabbit, *Tissue Cell, 10,* 571–584, 1978.

25. Katz, M. A.: Interstitial space—The forgotten organ, *Med. Hypoth., 6,* 885–898, 1980.

26. Johnson, J. A., and Bloom, G.: Permeability and reflection coefficients from osmotic transients—Extravascular factors, *Microvasc. Res., 22,* 80–92, 1981.

27. Silberberg, A.: The mechanics and thermodynamics of separation flow through porous molecularly disperse, solid media—The Poiseuille Lecture 1981, *Biorheology, 19,* 111–127, 1982.

28. Granger, H. J., Dhar, J., and Chen, H. I.: Structure and function of the interstitium, in J. T. Sgouris and A. Rene (eds.), *Proceedings of the NIH Symposium on Albumin,* National Institutes of Health, Bethesda, Md., 1975, pp. 114–124.

29. Hargens, A. R.: Interstitial osmotic pressures associated with Donnan equilibria, in A. R. Hargens (ed.), *Tissue Fluid Pressure and Composition,* Williams & Wilkins, Baltimore, 1981, chap. 8, pp. 77–85.

30. Meyer, F. A., and Silberberg, A.: *In vitro* study of the influence of some factors important for any physicochemical characterization of loose connective tissue in the microcirculation, *Microvasc. Res., 8,* 263–273, 1974.

31. Meyer, F. A.: Distribution and transport of fluid as related to tissue structure, in A. R. Hargens (ed.), *Tissue Nutrition and Viability,* Springer-Verlag, New York, 1986, chap. 2, pp. 25–46.

32. Laurent, U. B. G., and Laurent, T. C.: On the origin of hyaluronate in blood, *Biochem. Int., 2,* 195–199, 1981.

33. Wiederhielm, C. A.: Dynamics of capillary fluid exchange: A nonlinear computer simulation, *Microvasc. Res., 18,* 48–82, 1979.

34. Comper, W. D., and Laurent, T. C.: Physiological Function of connective tissue polysaccharides, *Physiol. Rev., 58,* 255–315, 1978.

35. Ogston, A. G., and Phelps, C. F.: The partition of solutes between buffer solutions containing hyaluronic acid, *Biochem. J., 78,* 827–833, 1961.

36. Fraser, J. R. E., Laurent, T. C., Pertoft, H., and Baxter, E.: Plasma clearance, tissue distribution and metabolism of hyaluronic acid injected intravenously in the rabbit, *Biochem. J., 200,* 415–424, 1981.

37. Ogston, A. G., and Sherman, T. F.: Effects of hyaluronic acid upon diffusion of solutes and flow of solvent, *J. Physiol., 156,* 67–74, 1961.

38. Laurent, T. C., and Ogston, A. G.: The interaction between polysaccharides and other macromolecules: 4. The osmotic pressure of mixtures of serum albumin and hyaluronic acid, *Biochem. J., 89,* 249–253, 1963.

39. Ogston, A. G., Preston, B. N., and Wells, J. D.: On the transport of compact particles through solutions of chain polymers, *Proc. R. Soc. London, 333,* 297–316, 1973.

40. Wiederhielm, C. A., Fox, J. R., and Lee, D. R.: Ground substance mucopolysaccharide and plasma proteins: Their role in capillary water balance, *Am. J. Physiol., 230,* 1121–1125, 1976.

41. Bell, D. R., Watson, P. D., and Renkin, E. M.: Exclusion of plasma proteins in the interstitium of tissues from the dog hindpaw, *Am. J. Physiol., 239,* H532–H538, 1980.

42. Rutili, G.: Transport of macromolecules in subcutaneous tissue studied by FITC-dextran, *Acta Univ. Upsaliensis Suppl., 306,* 1978.

43. Goresky, C. A.: The nature of transcapillary exchange in the liver, *Can. Med. Assoc. J., 92,* 517–522, 1965.

44. Selinger, S. L., Bland, R. D., Demling, R. H., and Staub, N. C.: Distribution volumes of ^{131}I-albumin, ^{14}C-sucrose, and ^{36}C1 in sheep lung, *J. Appl. Physiol., 39,* 773–779, 1975.

45. Parker, J. C., Falgout, H. J., Parker, R. E., Granger, D. N., and Taylor, A. E.: The effect of fluid volume loading on exclusion of interstitial albumin and lymph flow in the dog lung, *Circ. Res., 45,* 440–450, 1979.

46. Granger, D. N., Mortillaro, N. A., Kvietys, P. R., Rutili, G., Parker, J. C., and Taylor, A. E.: Role of the interstitial matrix during intestinal volume absorption, *Am. J. Physiol., 238,* G183–G189, 1980.

47. Granger, H. J., and Shepherd, A. P.: Dynamics and control of the microcirculation, in J. H. U. Brown (ed.), *Advances in Biomedical Engineering,* Academic Press, New York, 1979, vol. 7, pp. 1–63.

48. Barr, L., and Malvin, R. L.: Estimation of extracellular spaces of smooth muscle using different sized molecules, *Am. J. Physiol., 208,* 1042–1045, 1965.

49. Maurice, D. M.: The physical state of water in the corneal stroma, in M. Langham (ed.), *The Cornea: Macromolecular Organization of a Connective Tissue,* Johns Hopkins Press, Baltimore, 1969, pp. 193–204.

50. Maroudas, A.: Distribution and diffusion of solutes in articular cartilage, *Biophys. J., 10,* 365–379, 1970.

51. Carman, P. C.: Fluid flow through granular beds, *Trans. Inst. Chem. Eng., 15,* 150–166, 1937.

52. Katchalsky, A., and Curran, P. F.: *Nonequilibrium Thermodynamics in Biophysics,* Harvard University Press, Cambridge, Mass., 1965.

53. Anderson, J. L., and Malone, D. M.: Mechanism of osmotic flow in porous membranes, *Biophys. J., 14,* 957–982, 1974.

54. Granger, H. J., and Taylor, A. E.: Permeability of connective tissue linings isolated from implanted capsules: Implications for interstitial pressure measurements, *Circ. Res., 36,* 222–228, 1975.

55. Watson, P. D., and Grodins, F. S.: An analysis of the effects of the interstitial matrix on plasma-lymph transport, *Microvasc. Res., 16,* 19–41, 1978.

56. Aukland, K., and Johnsen, H. M.: Protein concentration and colloid osmotic pressure of rat skeletal muscle interstitial fluid, *Acta Physiol. Scand., 91,* 354–364, 1974.

57. Bill, A.: Plasma protein dynamics: Albumin and IgG capillary permeability, extravascular movement, and regional blood flow in unanesthetized rabbits, *Acta Physiol. Scand., 101,* 28–42, 1977.

58. Watson, P. D., Bell, D. R., and Renkin, E. M.: Early kinetics of large molecule transport between plasma and lymph in dogs, *Am. J. Physiol., 239,* H525–H531, 1980.

59. Zweifach, B. W., and Prather, J. W.: Micromanipulation of pressure in terminal lymphatics of the mesentery, *Am. J. Physiol., 228,* 1326–1335, 1975.

60. Smith, R. O.: Lymphatic contractility, a possible mechanism of lymphatic vessels for the transport of lymph, *J. Exp. Med., 90,* 497–509, 1949.

61. Lee, J. S., Motility, lymphatic contractility and distention pressure in intestinal absorption, *Am. J. Physiol., 208,* 621–627, 1965.

62. Yoffey, J. M., and Courtice, F. C.: *Lymphatics, Lymph, and the Lymphomyeloid Complex,* Academic Press, New York, 1970.

63. Nicoll, P. A.: Excitation-contraction of single vascular smooth muscle cells and lymphatics *in vivo, Immunochemistry, 12,* 511–515, 1975.

64. Hargens, A. R., and Zweifach, B. W.: Contractile stimuli in collecting lymph vessels, *Am. J. Physiol., 233,* H57–H65, 1977.

65. Olszewski, W. L., and Engeset, A.: Intrinsic contractility of prenodal lymph vessels and lymph flow in human leg, *Am. J. Physiol., 239,* H775–H783, 1980.

66. Olszewski, W., Engeset, A., Jaeger, P. M., and Sokolowski, J., and Theodorsen, T.: Flow and composition of leg lymph in normal men during venous stasis, muscular activity and local hyperthermia, *Acta Physiol. Scand, 99,* 149–155, 1977.

67. Olszewski, W. L., Engeset, A., and Sokolowski, J.: Lymph flow and protein in the normal male leg during lying, getting up and walking, *Lymphology, 10,* 178–183, 1977.

68. Clough, G., and Smaje, L. H.: Simultaneous measurement of pressure in the interstitium and the terminal lymphatics of the cat mesentery, *J. Physiol., 283,* 457–468, 1978.

69. Hogan, R. D.: The initial lymphatics and interstitial fluid pressure, in A. R. Hargens (ed.), *Tissue Fluid Pressure and Composition,* Williams & Wilkins, Baltimore, 1981, chap. 16, pp. 155–163.

70. Rusznyak, I. M., Földi, M., and Szabo, G.: *Lymphatics and Lymph Circulation,* Pergamon Press, London, 1967.

71. Hargens, A. R., and Zweifach, B. W.: Transport between blood and peripheral lymph in intestine, *Microvasc. Res., 11,* 89–101, 1976.

72. Quin, J. W., and Shannon, A. D.: The influence of the lymph node on the protein concentration of efferent lymph leaving the node, *J. Physiol., 264,* 307–321, 1977.

73. Adair, T. H., Moffatt, D. S., Paulsen, A. W., and Guyton, A. C.: Quantitation of changes in lymph protein concentration during lymph node transit, *Am. J. Physiol., 243,* H351–H359, 1982.

74. Zweifach, S. S., Kramer, G. C., Gunther, R. A., and Renkin, E. M.: Direct comparison of pre- and postnodal lymph in unanesthetized sheep, *Microvasc. Res., 25,* 263, 1983.

75. Maron, M. B.: Modification of lymph during passage through the lymph node: Effect of histamine, *Am. J. Physiol., 245,* H553–H559, 1983.

76. Adair, T. H., and Guyton, A. C.: Modification of lymph by lymph nodes: II. Effect of increased lymph node venous blood pressure, *Am. J. Physiol., 245,* H616–H622, 1983.

77. Staub, N. C.: Lung fluid and solute exchange, in N. C. Staub (ed.), *Lung Biology in Health and Disease,* Marcel Dekker, New York, 1978, vol. 9, pp. 3–16.

78. Landis, E. M.: The relationship between capillary pressure and the rate at which fluid passes through the walls of single capillaries, *Am. J. Physiol., 82,* 217–238, 1927.

79. Landis, E. M.: Micro-injection studies of capillary permeability: III. The effect of lack of oxygen on the permeability of the capillary wall to fluid and the plasma proteins, *Am. J. Physiol., 83,* 523–542, 1928.

80. Iverson, P., and Johansen, E. H.: Pathogenese und Resorption von Trans- und Exudation in der Pleura, *Klin. Wochenschr., 8,* 1311–1319, 1929.

81. Curry, F. E., Mason, J. C., and Michel, C. C.: Osmotic reflexion coefficients of capillary walls to low molecular weight hydrophilic solutes measured in single perfused capillaries of the frog mesentery, *J. Physiol. (London), 261,* 319–336, 1976.

82. Michel, C. C.: The measurement of permeability in single capillaries, *Arch. Int. Physiol. Biochem., 86,* 657–667, 1978.

83. Gore, R. W., and McDonagh, P. F.: Fluid exchange across single capillaries, *Annu. Rev. Physiol., 42,* 337–357, 1980.

84. Guyton, A. C.: A concept of negative interstitial pressure based on pressures in implanted perforated capsules, *Circ. Res., 12,* 399–414, 1963.

85. Wiederhielm, C. A., Woodbury, J. W., Kirk, S., and Rushmer, R. F.: Pulsatile pressures in the microcirculation of frog's mesentery, *Am. J. Physiol., 207,* 173–176, 1964.

86. Scholander, P. F., Hargens, A. R., and Miller, S. L.: Negative pressure in the interstitial fluid of animals, *Science, 161,* 321–328, 1968.

87. Fadnes, H. O., Reed, R. K., and Aukland, K.: Interstitial fluid pressure in rats measured with a modified wick technique, *Microvasc. Res., 14,* 27–36, 1977.

88. Rorabeck, C. H., Castle, G. S. P., Hardie, R., and Logan, J.: Compartmental pressure measurements: An experimental investigation using the slit catheter, *J. Trauma, 21,* 446–449, 1981.

89. Wiig, H., Reed, R. K., and Aukland, K.: Micropuncture measurement of interstitial fluid pressure in rat subcutis and skeletal muscle: Comparison to wick-in-needle technique, *Microvasc. Res., 21,* 308–319, 1981.

90. Prather, J. W., Gaar, K. A., Jr., and Guyton, A. C.: Direct continuous recording of plasma osmotic pressure of whole blood, *J. Appl. Physiol., 24,* 602–605, 1968.

91. Aukland, K., and Johnsen, H. M.: A colloid osmometer for small fluid samples, *Acta Physiol. Scand., 90,* 485–490, 1974.

92. Fadnes, H. O., and Aukland, K.: Protein concentration and colloid osmotic pressure of interstitial fluid collected by the wick technique: Analysis and evaluation of the method, *Microvasc. Res., 14,* 11–25, 1977.

93. Hargens, A. R., Millard, R. W., and Johansen, K.: High capillary permeability in fishes, *Comp. Biochem. Physiol., 48-A,* 675–680, 1974.

94. Hargens, A. R., Cologne, J. B., Menninger, F. J., Hogan, J. S., Tucker, B. J., and Peters, R. M.: Normal transcapillary pressures in human skeletal muscle and subcutaneous tissues, *Microvasc. Res., 22,* 177–189, 1981.

95. Reed, R. K.: An implantable colloid osmometer: Measurements in subcutis and skeletal muscle of rats, *Microvasc. Res., 18,* 83–94, 1979.

96. Noddeland, H.: Colloid osmotic pressure of human subcutaneous interstitial fluid sampled by nylon wicks: Evaluation of the method, *Scand. J. Clin. Lab. Invest., 42,* 123–130, 1982.

97. Noddeland, H., Hargens, A. R., Reed, R. K., and Aukland, K.: Interstitial colloid osmotic and hydrostatic pressures in subcutaneous tissue of human thorax, *Microvasc. Res., 24,* 104–113, 1982.

98. Mubarak, S. J., and Hargens, A. R.: *Compartment Syndromes and Volkmann's Contracture,* Saunders, Philadelphia, 1981.

99. Reddy, N. P., Palmieri, V., and Cochran, G. V. B.: Subcutaneous interstitial fluid pressure during external loading, *Am. J. Physiol., 240,* R327–R329, 1981.

100. Stranden, E., and Myhre, H. O.: Pressure-volume recordings of human subcutaneous tissue: A study in patients with edema following arterial reconstruction for lower limb atherosclerosis, *Microvasc. Res., 24,* 241–248, 1982.

101. Wiig, H., Tveit, E., Hultborn, R., Reed, R. K., and Weiss, L.: Interstitial fluid pressure in DMBA-induced rat mammary tumours, *Scand. J. Clin. Lab. Invest., 42,* 159–164, 1982.

102. Noddeland, H., Riisnes, S. M., and Fadnes, H. O.: Interstitial fluid colloid osmotic and hydrostatic pressures in subcutaneous tissue of patients with nephrotic syndrome, *Scand. J. Clin. Lab. Invest., 42,* 139–146, 1982.

103. Qvarfordt, P., Christenson, J. T., Eklöf, B., Ohlin, P., and Saltin, B.: Intramuscular pressure, muscle blood flow and skeletal muscle metabolism in chronic anterior tibial compartment syndrome, *Clin. Orthop. Rel. Res., 179,* 284–290, 1983.

104. Stranden, E., and Myhre, H. O.: Transcapillary forces in patients with lower limb ischemia, *Scand. J. Clin. Lab. Invest., 43,* 233–239, 1983.

105. Stranden, E.: Transcapillary forces in subcutaneous tissue of patients following operation for lower limb atherosclerosis, *Scand. J. Clin. Lab. Invest., 43,* 381–388, 1983.

106. Qvarfordt, P., Eklöf, B., and Ohlin, P.: Intramuscular pressure in the lower leg in deep vein thrombosis and phlegmasia cerulea dolens, *Ann. Surg., 197,* 450–453, 1983.

107. Tønder, K. J. H., and Kvinnsland, I.: Micropuncture measurements of interstitial fluid pressure in normal and inflamed dental pulp in cats, *J. Endodontics, 9,* 105–109, 1983.

108. Levick, J. R., and Michel, C. C.: The effects of position and skin temperature on the capillary pressures in the fingers and toes, *J. Physiol. (London), 274,* 97–109, 1978.

109. Mellander, S., Öberg, B., and Odelram, H.: Vascular adjustments to increased transmural pressure in cat and man with special reference to shifts in capillary fluid transfer, *Acta Physiol. Scand., 61,* 34–48, 1964.

110. Knight, A. D., and Levick, J. R.: Changes in hydraulic conductivity of the blood-joint barrier with pressure, *J. Physiol. (London), 324,* 47P, 1981.

111. Mullins, R. J., and Bell, D. R.: Permeability of rabbit skin and muscle microvasculature after saline infusion, *J. Surg. Res., 32,* 390–400, 1982.

112. Mullins, R. J., and Bell, D. R.: Changes in interstitial volume and masses of albumin and IgG in rabbit skin and skeletal muscle after saline volume loading, *Circ. Res., 51,* 303–313, 1982.

113. McDonagh, P. F.: Both protein and blood cells reduce coronary microvascular permeability to macromolecules, *Am. J. Physiol., 245,* H698–H706, 1983.

114. Groome, L. J., Kinasewitz, G. T., and Diana, J. N.: Diffusion and convection across heteroporous membranes: A simple macroscopic equation, *Microvasc. Res., 26,* 307–322, 1983.

115. Stranden, E.: Transcapillary fluid filtration in patients with leg edema following arterial reconstruction for lower limb atherosclerosis, *VASA, 12,* 219–224, 1983.

116. Pappenheimer, J. R., and Soto-Rivera, A.: Effective osmotic pressure of the plasma proteins and other quantities associated with the capillary circulation in the hindlimbs of cats and dogs, *Am. J. Physiol., 152,* 471–491, 1948.

117. Landis, E. M.: The capillary pressure in frog mesentery as determined by micro-injection methods, *Am. J. Physiol., 75,* 548–570, 1926.

118. Intaglietta, M., Pawula, R. F., and Tompkins, W. R.: Pressure measurement in the mammalian microvasculature, *Microvasc. Res., 2,* 212–220, 1970.

119. Eliassen, E., Folkow, B., Hilton, S. M., Öberg, B., and Rippe, B.: Pressure-volume characteristics of the interstitial fluid space in the skeletal muscle of the cat, *Acta Physiol. Scand., 90,* 583–593, 1974.

120. Chen, H. I., Granger, H. J., and Taylor, A. E.: Interaction of capillary, interstitial and lymphatic forces in the canine hindpaw, *Circ. Res., 39,* 245–254, 1976.

121. Brace, R. A., Guyton, A. C., and Taylor, A. E.: Determinants of isogravimetric capillary pressure in the isolated dog hindlimb, *Am. J. Physiol., 233,* H130–H135, 1977.

122. Fronek, K., and Zweifach, B. W.: Microvascular pressure distribution in skeletal muscle and the effect of vasodilation, *Am. J. Physiol., 228,* 791–796, 1975.

123. Hargens, A. R., Akeson, W. H., Mubarak, S. J., Owen, C. A., Evans, K. L., Garetto, L. P., Gonsalves, M. R., and Schmidt, D. A.: Fluid balance within the canine anterolateral compartment and its relationship to compartment syndromes, *J. Bone Joint Surg., 60A,* 499–505, 1978.

124. Intaglietta, M.: Vasomotor activity, time-dependent fluid exchange and tissue pressure, *Microvasc. Res., 21,* 153–164, 1981.

125. Hargens, A. R.: Freezing resistance in polar fishes, *Science, 176,* 184–186, 1972.

126. Granger, D. N., and Taylor, A. E.: Permeability of intestinal capillaries to endogenous molecules, *Am. J. Physiol., 238,* H457–H464, 1980.

127. Rutili, G., Granger, D. N., Mortillaro, N. A., Parker, J. C., and Taylor, A. E.: Analysis of blood-lymph transport in resting and passively flexed hind-leg, *Physiologist, 22,* 110, 1979.

128. Richardson, P. D. I., Granger, D. N., Kvietys, P. R., and Mailman, D.: Permeability characteristics of colonic capillaries, *Am. J. Physiol., 239,* G300–G305, 1980.

129. Parker, J. C., Roselli, R. J., and Brigham, K. L.: Effects of prolonged left atrial pressure elevation on lung microvascular protein sieving in unanesthetized sheep, *Fed. Proc., 39,* 279, 1980.

130. Levick, J. R.: Blood flow and mass transport in synovial joints, in E. M. Renkin and C. C. Michel (eds.), *Handbook of Physiology,* sec. 2: *The Cardiovascular System,* vol. IV: *Microcirculation,* American Physiological Society, Bethesda, Md., 1984, chap. 19, pp. 917–947.

131. Granger, D. N., Miller, T., Allen, R., Parker, R. E., Parker, J. C., and Taylor, A. E.: Permselectivity of cat liver blood-lymph barrier to endogenous macromolecules, *Gastroenterology, 77,* 103–109, 1979.

132. Lewis, G. N., and Randall, M.: *Thermodynamics,* rev. by K. S. Pitzer and L. Brewer, McGraw-Hill, New York, 1961.

133. Silberberg, A.: The role of matrix mechanical stress in swelling equilibrium and transport through networks, *Macromolecules, 13,* 742–748, 1980.

134. Hammel, H. T., and Scholander, P. F.: *Osmosis and Tensile Solvent,* Springer-Verlag, Berlin, 1976.

135. Hammel, H. T.: Hulett's enhanced solvent tension: An alternative to Lewis' diminished solvent activity, *Phys. Chem. Liq. 13,* 25–36, 1983.

136. Hammel, H. T.: Solution solvent couples solute molecules and enhances their internal tension, *Phy. Chem. Liq., 13,* 255–268, 1983.

137. Katz, M. A., and Bresler, E. H.: Osmosis, in N. C. Staub and A. E. Taylor (eds.), *Edema,* Raven Press, New York, 1984, chap. 2, pp. 39–60.

138. Guyton, A. C., Barber, B. J., and Moffatt, D. S.: Theory of interstitial pressures, in A. R. Hargens (ed.), *Tissue Fluid Pressure and Composition,* Williams & Wilkins, Baltimore, 1981, chap. 2, pp. 11–19.

139. Adair, T. H., Hogan, R. D., Hargens, A. R., and Guyton, A. C.: Techniques in the measurement of tissue fluid pressures and lymph flow, in R. J. Linden (ed.), *Techniques in the Life Sciences,* vol. P3: *Cardiovascular Physiology,* Elsevier, County Clare, Ireland, 1983, chap. 1, pp. 1–27.

140. Katz, M. A.: Validity of interstitial fluid hydrostatic pressure measurement in hollow porous polyethylene capsules, *Microvasc. Res., 16,* 316–326, 1978.

141. Parker, J. C., and Taylor, A. E.: Comparison of capsular and intra-alveolar fluid pressures in the lung, *J. Appl. Physiol.: Respir. Environ. Exercise Physiol., 52,* 1444–1452, 1982.

142. Hoffman, E. A., Lai-Fook, S. J., Wei, J., and Wood, E. H.: Regional pleural surface expansile forces in intact dogs by wick catheters, *J. Appl. Physiol.: Respir. Environ. Exercise Physiol., 55,* 1523–1529, 1983.

143. Lai-Fook, S. J.: Mechanical factors determining pulmonary interstitial fluid pressure, *Lung, 160,* 175–186, 1982.

144. Lai-Fook, S. J.: Perivascular interstitial fluid pressure measured by micropipettes in isolated dog lung, *J. Appl. Physiol.: Respir. Environ. Exercise Physiol., 52,* 9–15, 1982.

145. Wiig, H., and Reed, R. K.: Rat brain interstitial fluid pressure measured with micropipettes, *Am. J. Physiol., 244,* H239–H246, 1983.

146. Wiig, H., and Noddeland, H.: Interstitial fluid pressure in human skin measured by micropuncture and wick-in-needle, *Scand. J. Clin. Lab. Invest., 43,* 255–260, 1983.

147. Bhattacharya, J., Gropper, M. A., and Staub, N. C.: Interstitial fluid pressure gradient measured by micropuncture in excised dog lung, *J. Appl. Physiol.: Respir. Environ. Exercise Physiol., 56,* 271–277, 1984.

148. Taylor, A. E., and Gibson, H.: Concentrating ability of lymphatic vessels, *Lymphology, 8,* 43–49, 1975.

149. Senay, L. C., Jr., Mitchell, G., McClain, E., Mitchell, D., and Connors, N.: A subcutaneous capsule for long-term studies, *Am. J. Physiol., 241,* F687–F691, 1981.

150. Haljamäe, H., and Freden, H.: Comparative analysis of the protein content of local subcutaneous tissue fluid and plasma, *Microvasc. Res., 2,* 163–171, 1970.

151. Haljamäe, H., Linde, A., and Amundson, B.: Comparative analyses of capsular fluid and interstitial fluid, *Am. J. Physiol., 227,* 1199–1205, 1974.

152. Rutili, G., and Arfors, K-E.: Protein concentration in interstitial and lymphatic fluids from the subcutaneous tissue, *Acta Physiol. Scand., 99,* 1–8, 1977.

153. Rutili, G., and Arfors, K-E.: Collections of interstitial fluid: A comparison between wick and micropuncture techniques, *Microvasc. Res., 15,* 107–109, 1978.

154. Noddeland, H.: Influence of body posture on transcapillary pressures in human subcutaneous tissue, *Scand. J. Clin. Lab. Invest., 42,* 131–138, 1982.

155. Guyton, A. C.: Interstitial fluid pressure: II. Pressure-volume curves of interstitial space, *Circ. Res.*, *16,* 452–460, 1965.

156. Hargens, A. R.: Pressure-volume relationships in dehydrating animal tissues, *Comp. Biochem. Physiol.*, *56A,* 363–367, 1977.

157. Reed, R. K., and Wiig, H.: Compliance of the interstitial space in rats: I. Studies on hindlimb skeletal muscle, *Acta Physiol. Scand.*, *113,* 297–305, 1981.

158. Wiig, H., and Reed, R. K.: Compliance of the interstitial space in rats: II. Studies on skin, *Acta Physiol. Scand.*, *113,* 305–315, 1981.

159. Knight, A. D., and Levick, J. R.: Pressure-volume relationships above and below atmospheric pressure in the synovial cavity of the rabbit knee, *J. Physiol. London*, *328,* 403–420, 1982.

160. Levick, J. R.: Joint pressure-volume studies: Their importance, design and interpretation, *J. Rheumatol.*, *10,* 353–357, 1983.

161. Levick, J. R.: An investigation into the validity of subatmospheric synovial-fluid pressure recordings and their dependence on joint angle, *J. Physiol. (London)*, *289,* 55–68, 1979.

162. Taylor, A. E., Parker, J. C., and Rippe, B.: Edema and the tissue resistance safety factor, in A. R. Hargens (ed.), *Tissue Nutrition and Viability*, Springer-Verlag, New York, 1986, chap. 9, pp. 185–195.

163. Granger, D. N., Parker, R. E., Quillan, E. W., Brace, R. A., and Taylor, A. E.: Lymph transients, in P. Malick et al. (eds.), *Lymphology*, Georg Thieme, Stuttgart, 1979, pp. 61–63.

164. Kramer, G. C., Harms, B. A., Gunther, R. A., Renkin, E. M., and Demling, R. H.: The effects of hypoproteinemia on blood-to-lymph fluid transport in sheep lung, *Circ. Res.*, *49,* 1173–1180, 1981.

165. Gershuni, D. H., Hargens, A. R., Lieber, R. L., O'Hara, R. C., Johansson, C. B., and Akeson, W. H.: Decompression of an experimental compartment syndrome in dogs with hyaluronidase, *Clin. Orthop. Rel. Res.*, *197,* 295–300, 1985.

166. Wiederhielm, C. A., and Weston, B. V.: Microvascular, lymphatic, and tissue pressures in the unanesthetized mammal, *Am. J. Physiol.*, *225,* 992–996, 1973.

167. Reddy, N. P., Krouskop, T. A., and Nevell, P. H.: A note on the mechanisms of lymph flow through the terminal lymphatics, *Microvasc. Res.*, *10,* 214–216, 1975.

168. Leak, L. V.: The structure of the lymphatic capillaries in lymph formation, *Fed. Proc.*, *35,* 1863–1871, 1976.

169. Casley-Smith, J. R.: Mechanisms in the formation of lymph, in A. C. Guyton and J. E. Hall (eds.), *Cardiovascular Physiology IV,* vol. 26 of *International Review of Physiology*, University Park Press, Baltimore, 1982, pp. 147–200.

170. Casley-Smith, J. R.: Varying total tissue pressures and the concentration of initial lymphatic lymph, *Microvasc. Res.*, *25,* 369–379, 1983.

171. Brace, R. A., Guyton, A. C., and Taylor, A. E.: Time course of lymphatic protein concentration in the dog, *Microvasc. Res.*, *14,* 243–249, 1977.

172. Hargens, A. R., Mubarak, S. J., Owen, C. A., Garetto, L. P., and Akeson, W. H.: Interstitial fluid pressures on muscle and compartment syndromes in man, *Microvasc. Res.*, *14,* 1–10, 1977.

173. Sejersted, O. M., Hargens, A. R., Kardel, K. R., Blom, P., Jensen, Ø., and Hermansen, L.: Intramuscular fluid pressure during isometric contraction of human skeletal muscle, *J. Appl. Physiol.: Respir. Environ. Exercise Physiol.*, *56,* 287–295, 1984.

174. Skalak, R.: Approximate formulas for myocardial fiber stresses, *J. Biomech. Eng.*, *104,* 162–163, 1982.

175. Intaglietta, M., and Gross, J. F.: Vasomotion, tissue fluid flow and the formation of lymph, *Int. J. Microcirc. Clin. Exp.*, *1,* 55–65, 1982.

176. Skalak, T. C., Schmid-Schönbein, G. W., and Zweifach, B. W.: Lymph transport in skeletal muscle, in A. R. Hargens (ed.), *Tissue Nutrition and Viability*, Springer-Verlag, New York, 1986, chap. 12, pp. 243–261.

CHAPTER 20

Network Models of Peripheral Circulation

Aleksander S. Popel
Department of Biomedical Engineering
Johns Hopkins University
Baltimore, Maryland

INTRODUCTION

Blood flow in peripheral or microcirculatory networks is a complex phenomenon affected by a great variety of factors including network topography, blood rheology, and blood cell–vessel wall interactions. Much new quantitative information pertaining to the peripheral circulation has been accumulated in the last two decades. The key factors in this progress were the development of new experimental techniques which made it possible to measure velocity, pressure, and blood cell concentration in the small vessels, and the development of imaging techniques permitting better vessel visualization for topographical measurements. Many recent quantitative studies of microcirculatory networks are summarized in Refs. 1, 2, and 96.

The need to intepret the complex interrelations between experimentally measured hemodynamic variables, as well as to understand the mechanisms of local blood flow regulation and to describe quantitatively the transport of oxygen and nutrients in the system, is behind the efforts of mathematical modeling of the microcirculation [3]. A number of mathematical models have been developed that take into account factors characteristic of the microcirculatory networks. Among these factors are the particulate structure of blood that is manifested in the Fahraeus effect (diminution of hematocrit in small blood vessels), the Fahraeus-Lindqvist effect (diminution of the apparent viscosity in small vessels), and the preferential flow of red blood cells (RBCs) at vascular bifurcations. Another important factor is the heterogeneity of the topographical and hemodynamic variables in the microcirculatory networks. Mathematical models of the microcirculation which take into account these factors significantly improve our understanding of the complex blood flow phenomena. These mathematical models are discussed in this chapter, together with related experimental data on topographical and hemodynamic characteristics of the microvascular beds.

20.1 TOPOGRAPHICAL DESCRIPTION OF CIRCULATORY NETWORKS

20.1.1 Topological Characteristics of Vascular Networks

The first step in quantifying the topography of a circulatory network is describing the interconnections between different vessels and segments, i.e., defining the network topology. It is clear from simply looking at examples of microvascular networks (e.g., [2]) that the topological structure differs from one network to another. For example, some networks contain vascular arcades, or closed loops within the arterial or venous networks, whereas other do not. Graph theory is a very suitable mathematical tool for the description of network topology [4, 5]. Several definitions and concepts from graph theory are introduced here, and then applied to the vascular networks.

A *graph* is a set of points (nodes) and edges (segments) that connect some of the nodes. A connected graph is one in which any two nodes are connected by a path of segments. A *circuit* is a closed path in the graph, and a *tree* is a connected graph that contains no circuits. Graph theory is applicable to vascular networks if the junctions of vascular segments are identified as nodes of the graph, and the vascular segments as edges. Closed paths have been quantitatively described in capillary networks [6, 7, 99], and in arteriolar and venular networks [97–99]; the methods of graph theory were used in Refs. 97 to 99 to analyze vascular arcades.

In studies of the circulatory networks it is common to divide vessels into groups according to their structural and functional characteristics, thus introducing branching orders. Traditionally, the orders are assigned according to the morphological characteristics of the vessel and its location in the network, e.g., capillaries, precapillary arterioles, postcapillary venules, terminal arterioles, distributing arterioles. An early quantitative study of the vascular topography that used this classification, including the measurements of diameters and the number of vessels in each branching order, was published by Mall in 1888 [8]. This classic work remains a major source of data on vascular topography quoted in many textbooks on cardiovascular physiology. A very detailed topographical study of the bat wing microvasculature was done by Wiedeman [9]. In most investigations, a numerical order is assigned to vessels according to either centrifugal or centripetal schemes, as illustrated in Fig. 20.1. In the centrifugal scheme, which has been

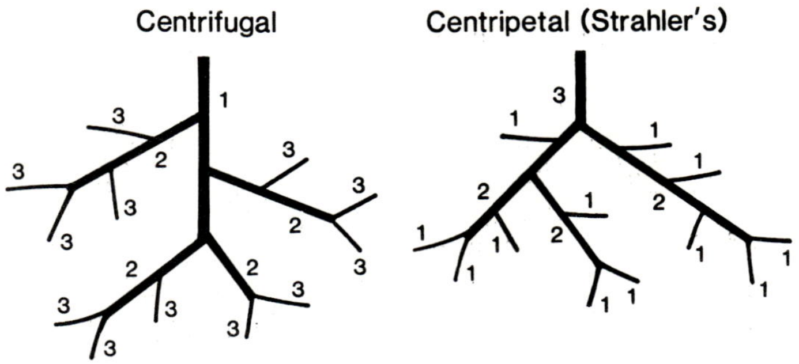

FIG. 20.1 Comparison of centrifugal and centripetal methods of ordering.

used in most studies, the feeding arterial vessel (the trunk) is assigned order 1, while the daughter branches are assigned order 2 or higher depending on their size, morphological structure, and other factors. The ordering proceeds to the capillaries in the arterial tree until all arterioles have been assigned branching orders. In the venous tree, the ordering starts with the largest venous vessel, which is assigned order 1, and proceeds in the same fashion toward the capillaries. In a centripetal scheme, the ordering starts at the level of capillaries, which are assigned order 1, and proceeds toward larger vessels. The centripetal scheme was initially introduced in a geomorphological study of rivers by Horton [10], and has been modified by [illegible] Topographical and hemodynamic measurements in the peripheral circulatory networks, keyed to references

Tissue	Topographical measurements: Topology	Topographical measurements: Diameter	Topographical measurements: Length	Hemodynamic measurements: Velocity	Hemodynamic measurements: Pressure	Hemodynamic measurements: Hematocrit or RBC flux	Mathematical model
Cat mesentery	22,* 23*	22–25	22	23, 25	22, 24	25	22, 105
Rat mesentery	26, 101*	26, 27	101	....	27	26	
Cat omentum	....	28	....	28	28		
Rabbit omentum	21,* 23,* 29*	21, 23, 29	29, 21	23, 30	23, 30	....	21, 30
Rat intestinal muscle and muscosal villi	31	31	....	....	31		
Rabbit mucosal villi	32*	32	32	....	....	....	32
Dog small intestine	8*	8					
Cat tenuissimus muscle	33, 34	33, 34	....	33, 34	34		
Cat sartorius muscle	35, 104*, 106	35, 104	104	35	106	....	107, 108
Rat gracilis muscle	36	36	....	....	36		
Rat cremaster muscle	37–39	37–39	....	....	37–39		
Rat spinotrapezius muscle	97*–99*	97–99	97–99	110	110	....	52
Human bulbar conjunctiva	21*	21	21	....	....	....	21
Rat pia mater	40	40	....	40			
Hamster cheek pouch	41, 111	41, 111	....	42	41, 111	42	
Bat wing	9*	9	9	43	44	....	43, 45
Frog sartorius muscle capillary network	7,* 46*	....	7, 46	46			
Rat glomerular capillary network	47,* 48*	47, 48	47, 48	....	....	....	47–49
Pulmonary circulation	102*, 103*	102, 103	102, 103	....	....	....	112

* These references contain quantitative topological data; unmarked references contain primarily qualitative topological data.

whereas the data on lengths are scarce. One of the reasons for this deficiency is that the geometrical data, in most cases, have been collected not with subsequent modeling in mind but rather with the intent to characterize the physiological state of the tissue. Since the vessel diameters can be altered by various experimental manipulations, such as the change of systemic pressure, or can be altered chronically, as in the case of hypertension, the vessel diameters carry important physiological information about the state of the vasculature. The vessel lengths, on the other hand, have not been reported to vary in these experiments, and thus their values are of considerably less interest to the physiologist. Additionally, vessel length measurements are also more complex to make, for they require traversing the entire vessel; this is difficult in part because of the difficulty of determining the beginning and the end of the vessel.

It should be noted that vessel length is a necessary variable in calculating the vessel resistance and, therefore, is required for network modeling. All mathematical models referred to in the last column of Table 20.2 are developed for those tissues in which the data on lengths are available. In recent years it has become more common to collect the network topographical data specifically for subsequent incorporation into a theoretical model [21, 22, 32, 43, 47, 48, 50–52].

An important question related to the problems discussed in the preceding paragraph is how to group vessels together so that the mean diameter and length for a given group of vessels are meaningful parameters. An analysis of statistical dispersion of the geometrical (and hemodynamic) variables for vessels of the same order under different ordering schemes for a given tissue would serve as an objective criterion in evaluating the utility of various ordering schemes; such an analysis has yet to be performed.

If vessels are grouped together with the help of a certain ordering scheme and the sample size is sufficiently large, then the geometrical parameters can be presented in the form of frequency distributions, or histograms. Such distributions are routinely obtained in experiments, but only the mean values and standard deviations of diameters and lengths are commonly reported. In a few cases, the histograms have been published [6, 7, 53]. These distributions carry valuable information necessary for more detailed mathematical modeling.

An important task in a quantitative topographical description of vascular networks is correlating the mean vessel diameters and lengths with the branching orders. Table 20.2 contains a number of data sets which represent the mean lengths and, in more cases, diameters as functions of the vessel order. For conventional centrifugal ordering, no universal relationships between these quantities have been reported.

Using a centripetal ordering scheme, Horton [10] established a law of stream lengths for rivers, which states essentially that the mean lengths of streams of different orders form a geometric sequence. Thus, a length ratio RL can be defined in the same manner as the bifurcation ratio, so that, on the average,

$$L_\alpha = L_1(RL)^{\alpha-1} \tag{20.2}$$

Here RL is independent of order α. In a similar way, it could be conjectured that vessel diameters in a network can be expressed in terms of a diameter ratio RD,

$$D_\alpha = D_1(RD)^{\alpha-1} \tag{20.3}$$

Relationships (20.2) and (20.3) imply that the logarithms of L_α or of D_α plotted versus α should fall on a straight line. The validity of Eqs. (20.2) and (20.3) have been tested for such biological systems as botanical trees [14] and bronchial trees (Table 20.1). Fenton and Zweifach [21] tested the data on human conjunctiva in normal and diabetic subjects, and the data on rabbit omentum for compliance with the geometric sequences (20.2) and (20.3) and found that there was, indeed, a definite trend for the mean diameters and lengths to follow the sequences. The ratios RL and RD, calculated in Ref. 21, are shown in Table 20.1 together with data on other circulatory networks.

The Horton type of relationships should be looked at with caution, however. Vessel diameters will vary with a change in vascular tone (owing to changes in arterial pressure or oxygen environment). The number of vessels containing flowing blood may similarly vary [35]. It is known that the responses of arterioles in response to arterial pressure changes are uneven, with small arterioles experiencing larger relative changes [54]. Whether or not the vessel diameters and the number of flowing vessels would satisfy Horton-type relationships under the condition of altered vascular tone—even if for a control state these relationships are satisfied—is an unanswered question. An experimental study devoted to testing the validity of Horton-type equations for different states of the vasculature would be an important contribution to the present knowledge of the vascular network topography.

Up to this point, only such metric parameters as diameters and lengths of vessels have been discussed; the data on other variables are very limited. There are no systematic data on distribution throughout a microvascular network of such variables as branching angles and shape of the vascular cross section, although in recent studies [55, 56] the parameters to characterize branching angles have been introduced.

20.2 EXPERIMENTAL DATA ON HEMODYNAMICS OF CIRCULATORY NETWORKS

20.2.1 Distribution of Flow and Pressure

The experimental data on flow and pressure distributions in microvascular networks are compiled in Table 20.2. In most cases, only mean values of flow, velocity, and pressure in groups of vessels (grouped either by branching order or by diameter range) have been reported. Several such distributions of pressure and velocity in the arteriolar networks are shown in Fig. 20.3; no universal behavior is evident from these graphs. When some of these data are plotted versus branching order by a centrifugal ordering scheme, the mean pressures show a linear dependence on the order, with different slopes in pre- and postcapillary networks (see Fig. 8.6 in Ref. 2). Histograms of velocity distribution in the capillaries are reported in Refs. 46 and 53. The values of standard deviations do not give a complete description of the corresponding frequency distributions if these distributions are not gaussian. Thus, despite a large number of studies on velocity and pressure distributions in networks, a systematic analysis of histograms of these variables in different branching orders has not been performed.

The distribution of the shear rate and the apparent viscosity of blood was obtained for the cat mesentery by Lipowsky et al. [57]. Recent results on the rheology of blood, particularly on

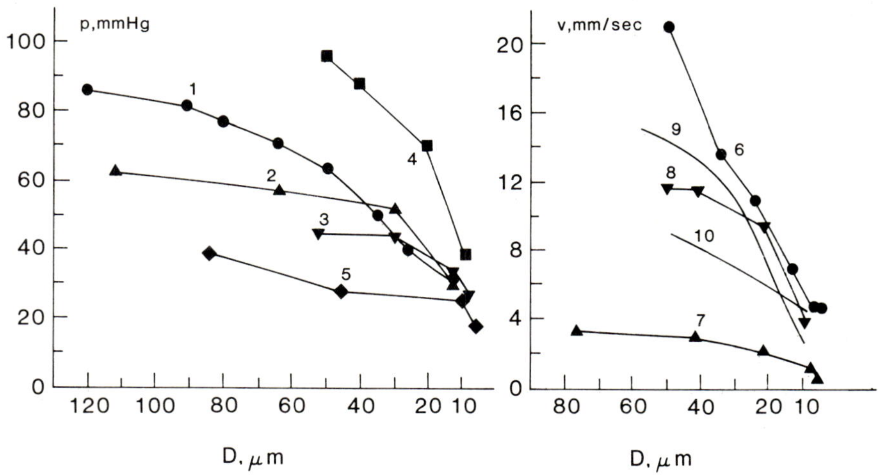

FIG. 20.3 Distribution of the mean pressure and velocity throughout the arterial network in different tissues. Pressure: 1—cat mesentery [24]; 2—hamster cheek pouch [41]; 3—rat intestinal muscle [31]; 4—cat tenuissimus muscle [34]; 5—rat cremaster muscle [38]. Velocity: 6—rat pia mater [40]; 7—bat wing [43]; 8—cat tenuissimus muscle [34]; 9—cat mesentery [23]; 10—rabbit omentum [23].

distribution of rheological parameters in the microvascular networks, are summarized in Ref. 58.

20.2.2 Red Blood Cell Distribution

The distribution of red blood cells in the microcirculation has an important effect on the oxygen transport to tissue and the rheology of blood in vivo [25]. Because of the technical difficulty of making the hematocrit measurements, the hematocrit distributions through the microcirculation (including pre- and postcapillary vessels) have been studied only in three tissues: cat mesentery [25], rat mesentery [26], and hamster cheek pouch [42]. Figure 20.4 shows the ratio of the vessel

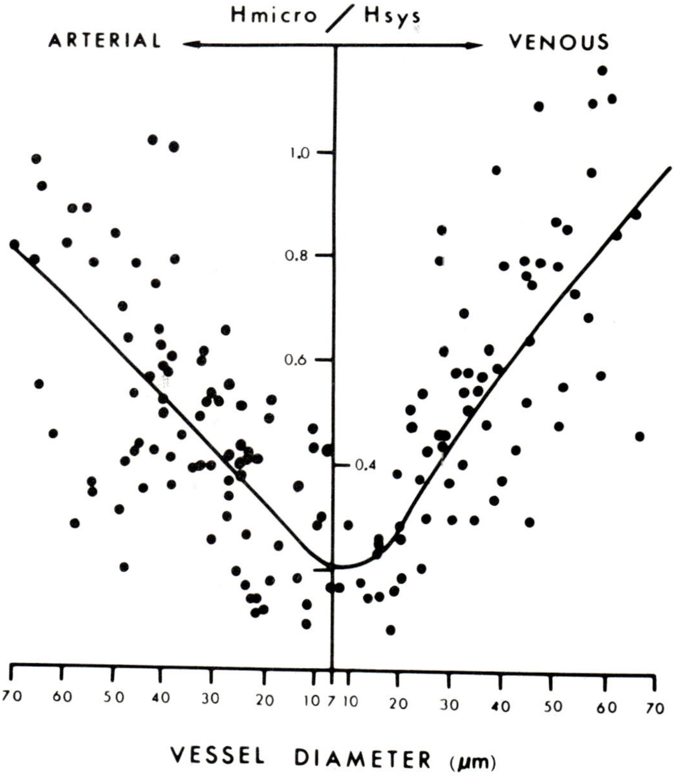

FIG. 20.4 Arteriovenous distributions of microvessel hematocrit normalized with respect to its corresponding systemic value (H_{micro}/H_{sys}) in the cat mesentery. The solid curve represents a piecewise cubic spline fit of the data. The minimum microvessel hematocrit appears to lie within the immediate postcapillary venules, approximately 10 μm in diameter. *(From Lipowsky et al. [25], by permission.)*

hematocrit H_{micro} to the systemic hematocrit H_{sys} [25]. In the rat mesentery the distribution of discharge hematocrit obtained from the vessel hematocrit distribution measured in vivo and from a correlation between vessel and discharge hematocrits established in vitro was reported in Ref. 26. Despite the significant scatter of experimental data, it is evident that the average tube and discharge hematocrits gradually decrease through the arterial network and increase through the venous network. The reported values for the ratio of the average minimum hematocrit, which is found in or near the capillaries, to the systemic hematocrit H_c/H_{sys} are 0.45 in the rabbit omentum [59]; 0.20 [60], 0.24 [53], and 0.26 [61] in the hamster cremaster muscle; 0.26 in the cat mesentery [25]; and 0.36 in the rat mesentery [26].

The mechanisms governing the hematocrit distribution are not clearly understood at the present time. In the following section, some hypothetical mechanisms will be discussed together with additional experimental results on RBC distribution at the vascular bifurcations.

20.3 MATHEMATICAL MODELING OF BLOOD FLOW IN PERIPHERAL CIRCULATORY NETWORKS

20.3.1 Models of Steady Flow in Arteriovenous Networks

One of the first analyses of pressure and flow distribution in a vascular bed was inspired by the necessity to assess the effect of rheological factors in different parts of the circulation. Schmid-Schönbein and Devendran [62] combined the data of Mall [8] for dog small intestine and of Wiedeman [9] for the bat wing, which resulted in a full "network" for which numbers of vessels, diameters, and lengths in different functional orders were specified. A hydraulic analysis of the network was done under the assumption of parallel connection of vessels of like orders and series connection of different orders. A constant viscosity of fluid was also assumed. The calculations predicted the distribution of resistance, intravascular pressure, and wall shear stress throughout the network. These calculations indicated that about 70 percent of the total pressure drop will occur between the small arteries with an average diameter of 0.2 mm and the postcapillary venules.

Lipowsky and Zweifach [22] obtained very detailed experimental data on modules of the cat mesentery including the vessel arrangement, lengths, and diameters. Then a mathematical model was constructed with the network topology and geometry closely following the experimentally observed configuration. In this study (and in studies discussed below, unless specifically noted) a constant apparent viscosity of blood was assumed. A linear set of equations was solved, yielding the pressure at each node of the network and flow in each segment. The calculated pressures were then grouped according to diameter ranges, and the results compared with the experimental data. The model has recently been extended to include shear rate dependency of blood viscosity [105].

Levitt et al. [32] studied the network architecture of rabbit small intestine mucosa by use of vascular casts. A simplified arcading lattice-shaped network was used as a geometrical model for the mucosal circulation, with a more complex geometry for the premucosal circulation. The predicted level of the mucosal capillary pressure agreed with known experimental values.

Mayrovitz et al. [43, 45] introduced a different approach to modeling of large networks. The network topological arrangement was a regular branching, diverging arterial tree connected through the capillaries to a branching, converging venous tree (Fig. 20.5). The model was applied to simulate the flow distribution in the bat wing microvasculature; the input parameters for the model were taken from Ref. 9. The distribution of impedances in the network was computed, and, subsequently, the pressures and velocities were determined. The results were presented in the form of average pressure and velocity for vessels of different branching orders, and the pressure distribution predicted by the model compared favorably with the experimental data [44]. The predicted average pressure was shown to vary linearly with the vascular order (but not with vessel diameter). The model was also employed to study the effect of diameter variation for different vascular orders on the network hemodynamics [63]. Blood flow in branching and anastomosing networks was further analyzed in Refs. 2 and 109.

Intrinsic in the asymmetric branching topography of a network is the heterogeneity of flow and pressure distribution. The analysis of heterogeneities in the microcirculatory system is attracting increased attention from researchers because it appears that many important effects could be explained through a coherent description of system heterogeneity. Popel [65] followed the approach of Mayrovitz et al. [45] but specifically addressed the question of distribution of pressure and flow within each branching order. Vessels of the same order were assumed

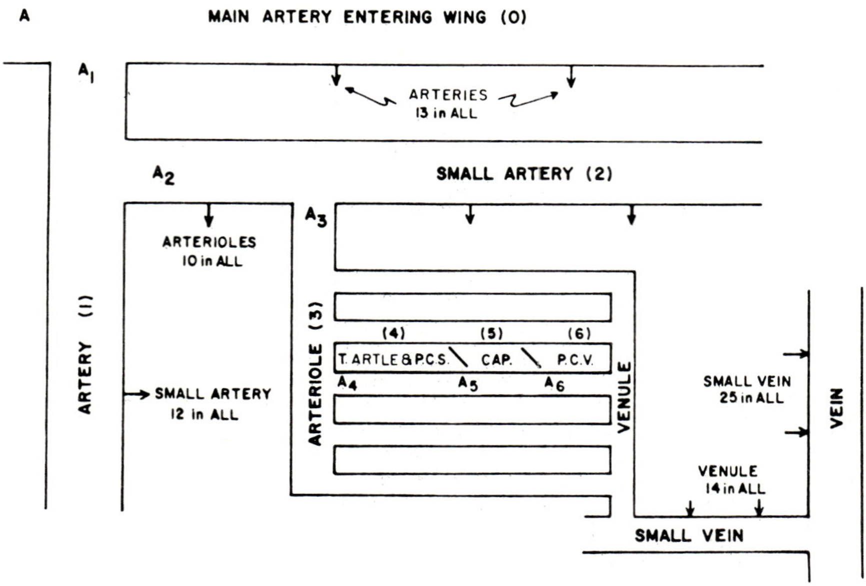

FIG. 20.5 Idealized topographical model based on the microvasculature of the bat wing. (*From Mayrovitz et al. [63]; reproduced by permission.*)

identical; thus, the source of heterogeneity was in the branching structure of the network. The results of calculations were presented in the form of pressure and flow histograms for different arterial branching orders.

Fenton and Zweifach [21] made use of the extensive literature on geological objects, botanical trees, neural networks, airways, and macrovascular beds, and applied the methods developed in these fields to a microvascular study of the human bulbar conjunctiva and rabbit omentum. Using Strahler's scheme of vessel ordering, and Horton's relationships, the authors obtained bifurcation ratios and length and diameter ratios, and employed these results in the mathematical modeling. The computer generation of a network started with a fixed number of first-order vessels—the capillaries. First, two capillaries were combined to form a segment of a second-order vessel; then a choice of whether or not to add another capillary was based on the following procedure. If the bifurcation ratio was, for example, 2.7, by use of random number generator the number 1 was drawn randomly with probability of 0.7 and the number 0 with probability of 0.3; in the former case a third capillary was joined with the newly formed second-order vessel, whereas in the latter case the formation of the second-order vessel was complete. This procedure was repeated until all the capillaries were used, and then a similar procedure was repeated with second-order vessels, and so forth, until only a single vessel of a higher order (the trunk) remained. A similar generation was done for the venous tree, again starting with capillaries, each of which was divided in half, with one half attributed to the arterial tree and the other half to the venous tree. After the arterial and venous trees were generated separately, they were combined by randomly matching arterial and venous capillaries. This completed the network generation. The next step was to assign a diameter and a length to each segment; it was accomplished, basically using Horton's laws, by assigning the same values to all vessels within a given branching order. Thus, the geometrical parameters of the network were deterministic, whereas the topology was stochastic. The introduction of a stochastic element into the microcirculatory model was a principal novelty compared with previous modeling schemes. The flow and pressure distribution was then computed in all nodes and segments of the network by solving a set of simultaneous linear algebraic equations for pressures in the nodes. Since the networks were topologically random, the results were averaged over an ensemble of networks. In Fig. 20.6, adapted from Ref. 21, the calculated pressure and flow in different branching orders are compared with experimental data for rabbit omentum. An extended analysis of the hemodynamics of the computer-generated networks can be found in Ref. 18. In this study the number, lengths, and diameters of vessels were altered, and the effect of these variations on the network hemodynamics was investigated. Also, extensive computations were made for the human conjunctival microvasculature on the basis of topographical data in normal and diabetic subjects.

Popel and Levin [66], Dawant et al. [67], and Levin et al. [68] extended the above approach by permitting heterogeneity of geometrical variables within each vascular order because there was ample experimental evidence of significant dispersion of lengths and diameters for vessels of a given branching order. Of course, the quantitative character of this dispersion depends on the ordering scheme used (in other words, on the grouping of vessels). The analysis of some published results [21, 53] indicates that the coefficient of variation (i.e., the standard deviation divided by the mean) is around 20 percent for the diameters and 80 percent for lengths. These characteristics can certainly change with changes of the vascular tone, and they probably depend on the type of tissue. The goal in Refs. 67 and 68 was to investigate, in principle, the effect of dispersion of lengths and diameters on the hemodynamic variables. The input parameters of the model were specified in the form of probability density distributions of geometrical variables, and the calculated hemodynamic variables were also presented in the form of probability density distributions. The computer network generation was similar to the scheme proposed in Ref. 21 with the exception that the number of daughter vessels was drawn from a Poisson distribution; as a result, the networks possess a more pronounced topological variability. After the topological structure of the network was defined, the diameters and lengths of vessels were assigned randomly according to certain probability distributions; a normal and log-normal distribution for diameters and lengths, respectively, were chosen in this work. A general formulation was also given which allowed the parameters of different vessels to be statistically correlated. A sensitivity analysis was then performed to investigate the number of different geometrical and topological samples necessary for a statistical convergence. Figure 20.7 shows normalized histograms of dimensionless capillary flow and midcapillary pressure calculated by ensemble-averaging the distributions over 2500 random networks each containing 20 capillaries and, typically, three orders (Strahler's) of arterioles and of venules. The top-panel histograms correspond to the case of zero dispersions of diameters and lengths (i.e., the model similar to that of Ref. 21). The lower-panel histograms were obtained for the same average diameters and lengths as in the case of zero dispersion, but in this case the coefficients of variation of diameters and lengths for all branching orders were 0.2 and 0.75, respectively. The numbers next to the histograms show the mean, minimum, maximum, and coefficient of variation of the corresponding dimensionless variable. The presence of reverse capillary flow (in about 3 percent of capillaries) is noted. The distribution of flow and pressure is, obviously, strongly affected by the geometrical dispersion. On the other hand, the distribution of the pressure averaged over vessels of a given branching order is almost insensitive to the geometrical dispersion. Thus, pressure distribution could be predicted fairly well without taking the dispersion into account.

An important aim in the mathematical modeling of peripheral circulatory networks is the quantitative prediction of RBC distribution, and a proper interpretation of the presently available experimental data on hematocrit distribution in the microvasculature (Fig. 20.4). Several hypothetical mechanisms have been proposed that may govern the intravascular RBC distribution. Klitzman and Duling [60] suggest that the observed low values of the capillary hematocrit may be attributed to a thin, completely or partially immobile plasma layer adjacent to the endothelium. The physical mechanisms responsible for the formation of such a layer have yet to be experimentally investigated [69]. Shunting of red blood cells through arteriovenous shunts was discussed as a factor in the hematocrit reduction in cat [25] and rat [26] mesentery.

Preferential flow of red blood cells at vessel bifurcations has been observed in vitro [70, 71, 113, 114] and in vivo [53, 72]; the results were recently summarized by Cokelet [73, 115]. Figure 20.8 shows the ratio of daughter to parent vessel RBC fluxes, F_i/F_o, versus the ratio of the corresponding flow rates, Q_i/Q_o. The data in Ref. 53 were fitted using sigmoid curves

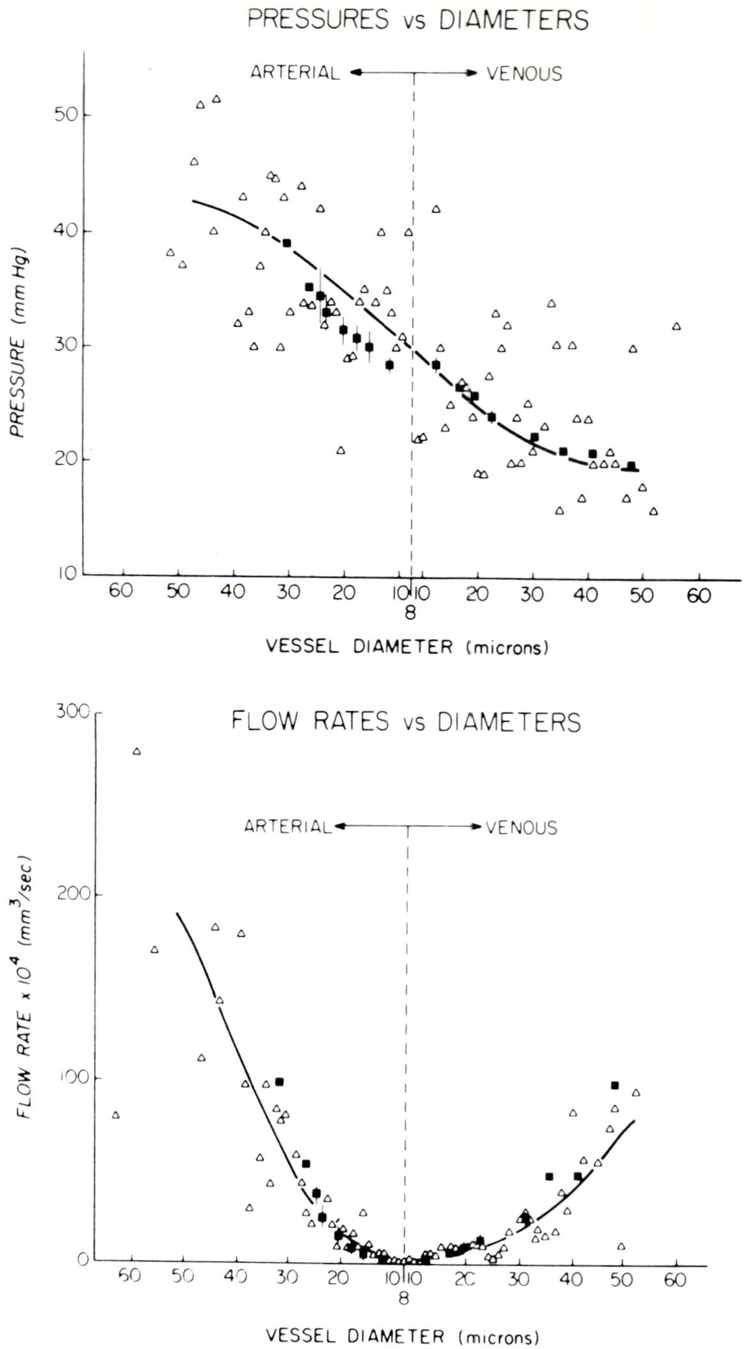

FIG. 20.6 Experimental and calculated values for intravascular pressure and volumetric flow rate versus diameter in the rabbit omentum. Δ represents in vivo values [23]; ■ represents calculated values (±SD) [21]; solid line represents the cubic spline [23]. (*From Fenton and Zweifach [21], by permission.*)

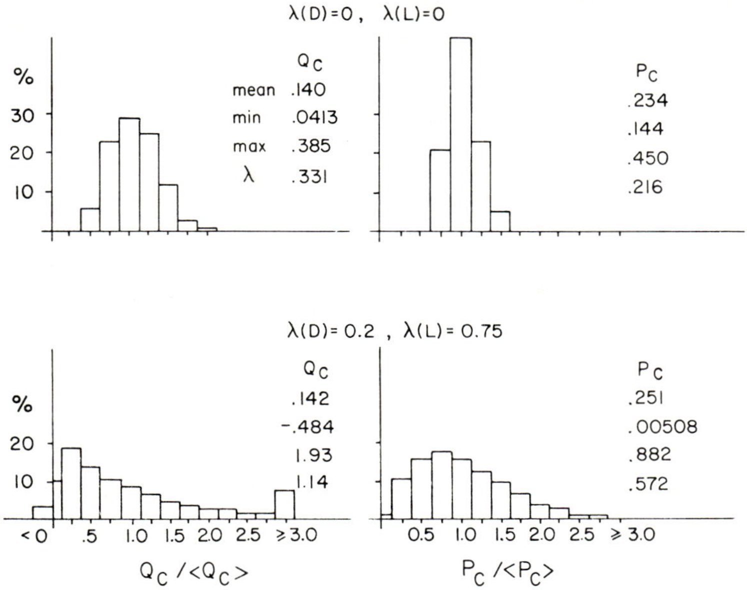

FIG. 20.7 The effect of dispersion of vessel diameters D and lengths L on histograms of capillary flow Q_C and midcapillary pressure P_C. The histograms are normalized by ensemble averaged values denoted by brackets, < >. Reverse flow in a group of capillaries is present in the case when the geometric variables are dispersed.

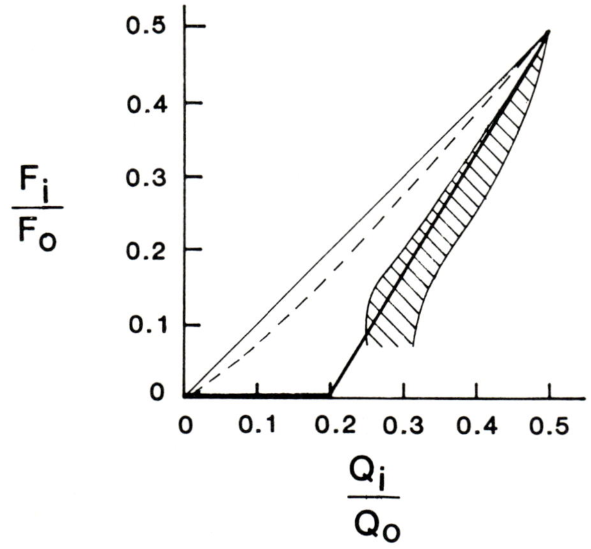

FIG. 20.8 The ratio of RBC fluxes versus the ratio of blood flow rates found by Klitzman and Johnson [53] (short dashes), Schmid-Schönbein et al. [72] (thick solid line), and Yen and Fung [70] (shaded area). This plot is half the entire plot; the upper part is generated by reflecting each point on this plot through the point (0.5, 0.5) to get F_i/F_o values for Q_i/Q_o between 0.5 and 1.0. (*From Cokelet [73], p. 8; by courtesy of Marcel Dekker, Inc.*)

$$\frac{F_o - F_i}{F_i} = \left(\frac{Q_o - Q_i}{Q_i}\right)^b \tag{20.4}$$

with the mean value $b = 1.15$ obtained for all capillary bifurcations, and $b = 1.35$ for capillary bifurcations with the flow ratio $Q_i/Q_o < 0.25$ or > 0.75. In vitro experiments give $b = 1.10$ to 1.58 depending on the feed hematocrit value [71]. The data from Refs. 70 and 72 shown in Fig. 20.8 correspond to yet higher values of the parameter b. Lipowsky et al. [74], however, found no systematic preferential distribution of red cells at arteriolar bifurcations in vessels of 20 to 99 μm diameter. It was also discovered [72] that the eccentricity of the cell pathway in the parent vessel is an important determinant of the cell behavior at the bifurcations. Figure 20.8 makes it clear that an uncertainty exists in *FQ* correlations at vessel bifurcations; additional research in this area is required. Several authors have suggested that the preferential flow of RBCs at bifurcations is a possible mechanism of the hematocrit reduction beyond the level attributed to the Fahraeus effect [26, 53, 73, 75]. To test this hypothesis, Cokelet [73] considered a diverging series-parallel network with six bifurcations along each pathway. At each bifurcation the ratio of flows in daughter segments was chosen randomly from five specified numbers, and the discharge hematocrit was calculated in the terminal branches by use of Eq. (20.4). It was shown that, under reasonable assumptions, the predicted mean vessel (tube) hematocrit can be as small as 23 percent of the inlet hematocrit. Levin et al. [68] also used Eq. (20.4) at each bifurcation, but in their model the topological and geometrical parameters were based on experimental data [21], and the flow in each vascular segment was a calculated variable.

An example of the predicted discharge hematocrit distribution based on this model is shown in Fig. 20.9, upper panel. The graph depicts the discharge hematocrit distribution obtained by averaging over an ensemble of networks each containing 20 capillaries for the case of zero dispersions of diameters and lengths, and for the case of coefficients of variation 0.2 and 0.75, respectively. The preferential flux index in Eq. (20.4), $b = 1.35$, was chosen for all bifurcations, and the inlet hematocrit $H_{\text{in}} = 0.5$ was assumed. Clearly, the geometrical dispersion in the network is responsible for the hematocrit reduction, whereas a topological stochasticity present in the case represented by the upper curve does not lead to any appreciable deviation of the discharge hematocrit from its inlet value. In fact, in this case the predicted average capillary discharge hematocrit exceeds the inlet hematocrit. Thus it appears that the heterogeneity of vessel diameters and lengths contributes to the observed hematocrit variation in the network; however, it is also clear that for the adopted value of b the predicted discharge hematocrit reduction due to the preferential RBC flow at bifurcations is only a fraction of the experimentally observed hematocrit reduction, and thus additional mechanisms must be considered. For example, for $b = 1.15$ found by Klitzman and Johnson [53], the predicted value of the ratio of the ensemble-averaged capillary hematocrit, $\langle H_c \rangle$, to the inlet hematocrit, H_{in}, is $\langle H_c \rangle / H_{\text{in}} = 0.9$, that is, only 10 percent reduction. Figure 20.9, lower panel, depicts $\langle H_c \rangle / H_{\text{in}}$ as a function of the parameter b. The steep drop in hematocrit occurs for the values of b between 1 and 2; the maximum reduction of $\langle H_c \rangle$ is about 40 percent. Subsequent increase of hematocrit is due to gradual filling of some capillaries to the maximum hematocrit of 1. It can be concluded that both the mechanisms and the quantitative description of the phenomenon require further careful analysis.

20.3.2 Models of Blood Flow in Capillary Networks

Shea [47] presented a detailed analysis of a rat glomerular capillary network. The vascular topography was reconstructed from ultrathin sections; a graph showing capillary interconnections in a glomerulus is shown in Fig. 20.10. The hemodynamic calculations were done in a way similar to those of Lipowsky and Zwiefach [22]. The nodal pressures corresponding to two experimentally measured values of the total flow were compared with available micropressure measurements.

Papenfuss and Gross [76, 108] formulated a general model of flow in a capillary network not based on the assumption of constant viscosity. The known theoretical and experimental results on single-file particle motion in small tubes were utilized to express the apparent viscosity and vessel hematocrit in terms of the tube diameter, feed hematocrit, and velocity of blood. At the

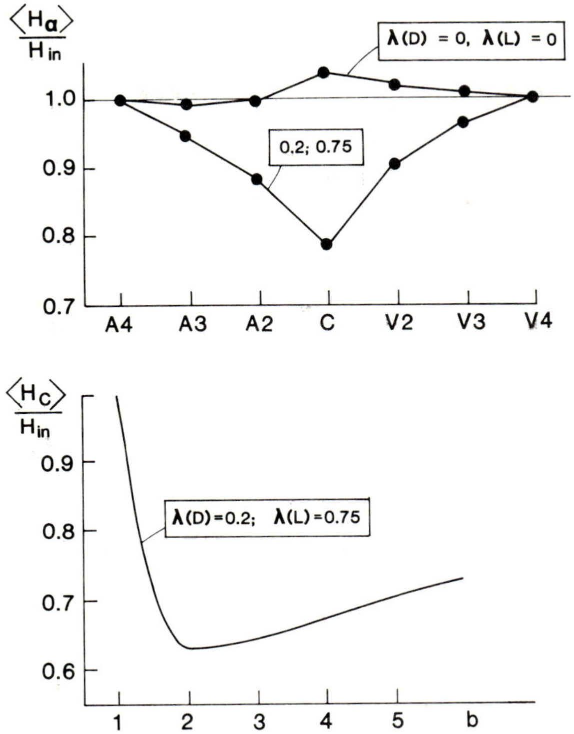

FIG. 20.9 Upper panel: Distribution of the calculated average discharge hematocrit (solid circles) throughout the network for an ensemble of 2500 networks for $b = 1.35$. The upper curve is for the case of zero disperson of diameters and lengths; the lower curve is for the case where the disperson is present. Lower panel: The normalized average capillary discharge hematocrit as a function of the preferential flux index b, for the coefficients of variation $\lambda(D) = 0.2$, $\lambda(L) = 0.75$.

vascular bifurcations a relationship similar to Eq. (20.4) was used. The resulting nonlinear set of equations was solved and the effect of nonlinearity was demonstrated. An elaborate model of the rat glomerular capillary network was developed by Lambert et al. [48]. The topological and geometrical parameters of the network were obtained by reconstructing from serial sections of a rat glomerulus. In addition to the features described in Ref. 76, the transcapillary filtration was taken into account. A detailed account of the computer program and the numerical techniques used in the solution of a nonlinear set of equations is given by Van Damme et al. [49]. Schmid-Schönbein et al. [72] developed a network model which described the time-dependent motion of individual red and white blood cells. A computer program was written for a general network configuration and applied to the case of flow through a capillary bifurcation and through a small capillary network. The flow in the network instantly responded to such events as the appearance of a white blood cell in the network, or the passage of a white blood cell through a capillary bifurcation. A stochastic element was introduced into the flux-flow relationship at bifurcations. A strong localized effect of white blood cells was noted. The importance of nonaxisymmetric position of the cells was established. This model was extended by Fenton et al. [116].

In a recent study Wieringa et al. [77] constructed a three-dimensional capillary network model of the myocardial microcirculation. The authors considered a set of hexagonally stacked,

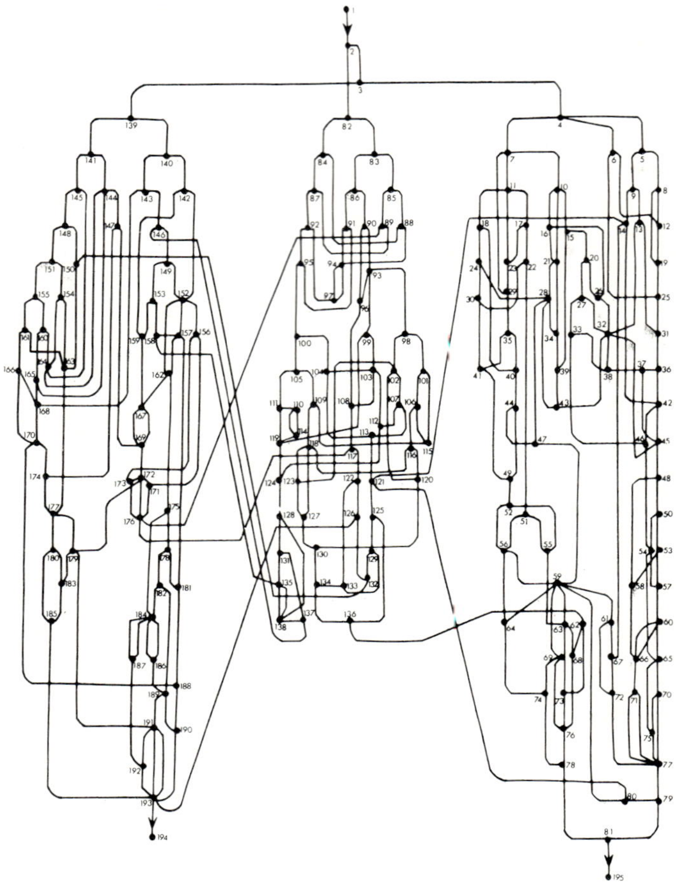

FIG. 20.10 Graph showing the connections of the capillaries of a rat glomerulus. The afferent and efferent arterioles are identified by arrows and black circles numbered 1 (afferent) and 194, 195 (efferent). Other numbered black circles represent points of branching or confluence of capillaries. The apparent lengths of the arcs connecting nodes are not realistically portrayed, but are diagrammatically adjusted to reveal the branching pattern. (*From Shea [47]; reproduced by permission.*)

parallel capillaries and randomly distributed a specified number of interconnections between the capillaries. In addition, the venous ends of capillaries were connected sequentially to the arterial ends of other capillaries. A number of arteriolar sources and venular sinks were randomly distributed in the network. The flow through the network was simulated under the assumption of constant viscosity, and statistical characteristics of flow were obtained on an ensemble of eight networks. An example of the flow pattern is shown in Fig. 20.11, where capillaries lying in one plane are depicted and flow directions and magnitudes are identified. The results of simulation were discussed in relation to experimental information available on the myocardial blood flow.

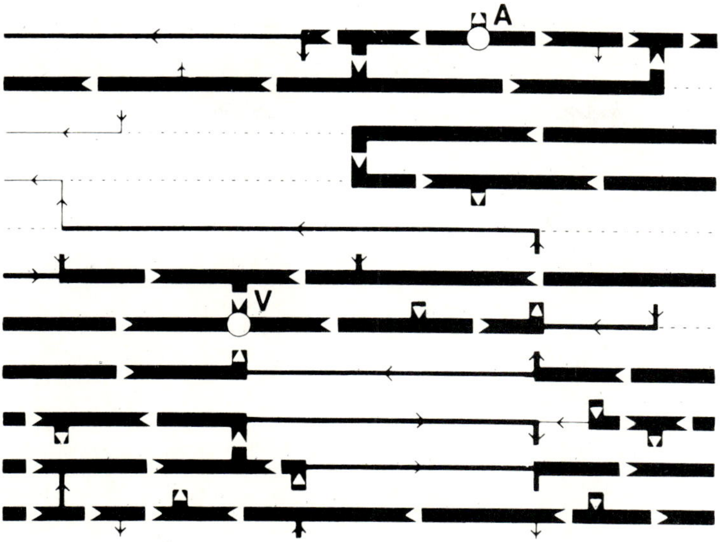

FIG. 20.11 The flow pattern in a typical longitudinal section through the capillary network in a model of the myocardial microcirculation. The flow direction is marked by arrowheads. The flow magnitude is divided into four classes shown by dark lines (largest flow), heavy lines, thin lines, and broken lines (smallest flow). Interconnections perpendicular to the capillaries can be seen as bars. A short bar indicates an interconnection to a capillary in another longitudinal plane above or below the plane shown. A-arteriolar source, V-venular sink. (*From Wieringa et al. [77], p. 203; by courtesy of Marcel Dekker, Inc.*)

20.3.3 Models of Time-Dependent Flow in Peripheral Circulation

A mathematical model of pulsatile flow in a microvascular network was formulated by Gross et al. [30]. The vessel compliance was taken into account, as was the transcapillary filtration of fluid, and a linear diffusion type of equation was obtained for the pulsatile pressure. Inertia was neglected, and the diffusion coefficient for each vessel was expressed in terms of the vessel resistance and compliance. The solution of the equation was found in the form of an exponentially attenuating (with distance) propagating wave. The model was applied to study pulsatile flow in the rabbit omentum microvasculature; nine vessel generations were identified and a series-parallel arrangement of the network was assumed. The input parameters for the model were taken from the experiment, and the distribution of pulsatile pressure, flow amplitudes, and phase angles were calculated for all vascular generations. An extensive discussion of different aspects of pulsatile flow in the microcirculation can be found in a review paper by Gross [78].

Mayrovitz et al. [79] addressed the question of time-dependent blood flow in a microvascular network resulting from the arterial vasomotion, periodic activity of precapillary sphincters, and venomotion. The network was represented by "equivalent" time-dependent resistances and compliances. A system of ordinary differential equations was derived, and its solution was obtained with the help of a hybrid computer. Some examples of time-dependent flow caused by vessel periodic activity are presented in Ref. 79 and additional results are given in Ref. 64. A time-dependent capillary flow model [72, 116] has been already discussed in the preceding section.

20.3.4 Optimal Models of Vascular Networks

In the preceding sections we reviewed methods of quantitative description of the peripheral circulatory networks and hemodynamic simulations. These studies do not reveal the principles that may govern the networks' structure and behavior. There have been attempts to formulate general optimality principles applicable either to the entire network, or to a single vessel or bifurcation. An optimality principle for network structure was mentioned by Thompson [80]; however, a rigorous formulation of the principle of minimum power was first given by Murray [81]. He suggested that the sum of the power required to move the fluid and the power required to maintain a given volume of blood is minimized in the circulation. Using Poiseuille's law to calculate the first term and assuming that the second term is proportional to the blood volume, Murray concluded that vessel diameters at a bifurcation must satisfy a relationship $D_0^3 = D_1^3 + D_2^3$, where D_0 is the parent vessel diameter and D_1, D_2 are the diameters of daughter vessels. Murray [82] applied this relationship to derive optimal branching angles of vessels of different size. If the relationship is applied to a series of branching vessels such that the flow through a given branching order of daughter vessels is equal to the total flow, then the quantity ΣD_i^3, in which the sum is taken over all vessels in a given order, is independent of order. If vessels in a given branching order are identical, then the ratio of vessel diameters for consecutive orders equals $2^{1/3}$, the result first given by Thompson [80]. Another important consequence of the optimality principle is the proportionality between the flow Q and D^3. This relationship is often referred to as Murray's law. Recently, Sherman [83] discussed in detail the derivation, implications, and limitations of Murray's law. In particular, he has shown that in an optimal network the velocity in a full-flow branching order should be proportional to D, the pressure gradient to D^{-1}, the Reynolds number to D^2, and the resistance to D^{-1}. The topographical relationships were validated by using Mall's data [8]. Mayrovitz and Roy [84] correlated values of flow and diameter for individual vessels of the rat cremaster muscle in the form $Q = kD^m$, and obtained a value of m close to 3. In addition, the data were divided into five groups corresponding to five arterial branching orders, and the values of Q and D averaged within each order were correlated. This calculation was performed for the rat cremaster muscle by use of the original data, and also for cat mesentery, cat tenuissimus muscle, and rabbit omentum by use of data from the literature; in all cases the values of the exponent m were close to 3. These results appear to support the applicability of Murray's law to the microcirculation.

Other optimality principles have been proposed, such as minimum drag force [85] and minimum lumen volume [86]. Zamir [87], Zamir et al. [88], Zamir and Medeiros [89], Uylings [90], and Roy and Woldenberg [91] applied optimality principles to the analysis of branching angles in the arterial system. Additional information on this subject can be found in Refs. 19, 92, and 93, where the principles are discussed and applied to the bronchial trees.

CONCLUSION

In conclusion, some additional comments on the existing theoretical and experimental results pertaining to the peripheral circulatory network will be given and the areas where an adequate model or an important set of experimental data are lacking will be pointed out.

With regard to network topography, considerable work has to be done on developing methods of topographical description, particularly for the arcading networks. The problem of the quantitative description of arcading structures, common to the microvasculature, has not been adequately addressed in the literature. Although quantitative work on different methods of ordering in the vascular networks has begun in recent years [21] a systematic study applying different schemes to the same tissue and analyzing the obtained characteristics is lacking.

There are only several peripheral networks for which the geometrical parameters necessary for mathematical modeling are known (see Table 20.2). Information about vessel lengths, taper, and branching angles is especially scarce. The available data are, in most cases, in the form of mean values for different branching orders, with ordering schemes that are sometimes ambiguous. For more detailed network models, the input information should be obtained in the form of probability distributions (histograms) of variables. The important problem of correlations between the geometrical characteristics of neighboring vessels, or between the length and diameter of the same vessel, has not been addressed either theoretically or experimentally. It is clear, however, that if such correlations exist, they probably have an impact on the network hemodynamics.

When the network topography is specified, simulation of the network hemodynamics is hampered by several factors. First, for a given vessel segment the relationship between the apparent viscosity of blood and the vessel diameter, hematocrit, and shear rate has to be known. Significant amounts of in vitro data are presently available, but in vivo data are very limited [58] and more research in this area is necessary. There is evidence that microvessels are not generally circular in cross section; thus caution should be exercised in applying glass-tube data to in vivo situations. There are also arguments which question the applicability of in vitro rheological data to in vivo rheology because of the possible existence of a plasma layer adjacent to the endothelium where the flow is restricted [60, 69]. Second, for the mathematical modeling of network hemodynamics, the mechanics of blood cells at bifurcations needs to be understood. It has been shown that the distribution of red blood cells at bifurcations is affected by such factors as the position of cells in the parent vessels [72], and the ratio of flow, shear rate, or velocity in daughter vessels [53, 70–72]. An analysis of the problem suggests that there are several additional characteristic parameters, among which are the bifurcation angles and the ratio of daughter to parent vessel diameters; hence a more complete knowledge of the relationship between the flux of red blood cells at a bifurcation and the geometrical and dynamic parameters characterizing the bifurcation is necessary for inclusion into the mathematical models [115].

To simulate the network response to variations of arterial or venous pressure, it is necessary to know the dependence of vessel diameters on the intravascular pressure. Some data for arterioles are available [24], but additional research has to be done to establish reliable constitutive equations for arterioles and venules.

Modeling of time-dependent flow presents considerable technical and principal problems. There are several causes of time dependence in the microvasculature: variation of the systemic arterial and venous pressures or the tissue pressure, spontaneous or induced variations of vessel diameters, and rheological factors due to interactions among red and white blood cells and the vascular wall. It is known that the vascular wall possesses viscoelastic properties; however, an adequate constitutive relationship between the circumferential stress and strain and their derivatives with respect to time (or alternatively, between the pressure and diameter and their time derivatives) has not been formulated. Next, very limited qualitative and quantitative information has been accumulated regarding spontaneous vessel activity (vasomotion); the mechanism of the phenomenon is poorly understood, and parameters characterizing the oscillations have not been properly identified. Apparently, the conventional Fourier analysis of vasomotive patterns is insufficient for identification of the key variables which could be used to describe the phenomenon [94]. A recent study [95] demonstrates an alternative approach to the problem. Some of the rheological factors in capillary networks have been dealt with in Ref. 72; however, the effects in arterioles and venules remain to be considered. Future developments in time-dependent modeling will pose a special requirement to presenting the experimental data, in that characterization of the temporal behavior will become necessary, whereas presently the data are commonly represented in a time-averaged form; in particular, time correlation of different parameters, e.g., diameters of neighboring vessels, might be important.

ACKNOWLEDGMENTS

The author takes pleasure in thanking Drs. Bruce Klitzman and Dean C. Winter for reading the manuscript and making critical comments, Ms. Leigh Hollyer for her help in preparation of the manuscript, and Ms. Mary Lamparty for typing it. The author acknowledges the support from the National Institutes of Health through grants HL-17421, HL-26901, and HL-18292.

REFERENCES

1. Johnson, P. C. (ed.): *Peripheral Circulation,* Wiley, New York, 1981.
2. Wiedeman, M. P., Tuma, R. F., and Mayrovitz, H. N.: *An Introduction to Microcirculation,* Academic Press, New York, 1981.
3. Popel, A. S.: Mathematical modeling of convective and diffusive transport in the microcirculation, in J. F. Gross and A. Popel (eds.), *Mathematics of Microcirculation Phenomena,* Raven Press, New York, 1980, pp. 63–88.
4. Carre, B.: *Graphs and Networks,* Clarendon Press, Oxford, 1979.
5. Wilson, R. J., and Beineke, L. W. (eds.): *Applications of Graph Theory,* Academic Press, New York, 1979.
6. Honig, C. R., Feldstein, M. L., and Frierson, J. L.: Capillary lengths, anastomoses, and estimated capillary transit times in skeletal muscle, *Am. J. Physiol., 233,* H122–H129, 1977.
7. Plyley, M. J., Sutherland, G. J., and Groom, A. C.: Geometry of the capillary network in skeletal muscle, *Microvasc. Res., 11,* 161–173, 1976.
8. Mall, F. P.: Die Blut und Lymphwege im Dunndarm des Hundes, *Ber. Sachs. Ges. Akad. Wiss., 14,* 151–200, 1888.
9. Wiedeman, M. P.: Dimensions of blood vessels from distributing artery to collecting vein, *Circ. Res., 12,* 375–378, 1963.
10. Horton, R. E.: Erosional development of streams and their drainage basins; hydrophysical approach to quantitative morphology, *Bull. Geol. Soc. Am., 56,* 275–370, 1945.
11. Strahler, A. N.: Quantitative analysis of watershed geomorphology, *Trans. Am. Geophys. Union, 38,* 913–920, 1957.
12. Woldenberg, M. J.: Hierarchical systems: Cities, rivers, alpine glaciers, bovine livers and trees, *Harvard Papers in Theoretical Geography, Geography and the Properties of Surfaces,* pap. 19, Harvard University Press, Cambridge, Mass., 1968.
13. Haggett, P., and Chorley, R. J.: *Network Analysis in Geography,* Arnold, London, 1969.
14. Barker, S. B., Cummings, G., and Horsfield, K.: Quantitative morphology of the branching structure of trees, *J. Theor. Biol.,* 40, 33–43, 1973.
15. Berry, M., Hollingworth, T., Anderson, E. M., and Flinn, R. M.: Application of network analysis to the study of the branching patterns in dendritic fields, in G. W. Kreutzberg (ed.), *Advances in Neurology,* Raven Press, New York, 1975, vol. 12, pp. 217–245.
16. Uylings, H. B. M., Smit, G. J., and Veltman, W. A. M.: Ordering methods in quantitative analysis of branching structure of dendritic trees, in G. W. Kreutzberg (ed.), *Advances in Neurology,* Raven Press, New York, 1975, vol. 12, pp. 247–254.
17. Horsfield, K.: Some mathematical properties of branching trees with application to the respiratory system, *Bull. Math. Biol., 38,* 305–315, 1976.
18. Fenton, B. M.: Topographical simulation of the blood vessels of the human bulbar conjunctiva and application to pressure-flow relationships, Ph.D. dissertation, University of California, San Diego, La Jolla, Calif., 1980.
19. Horsfield, K., and Thurlbeck, A.: Volume of the conducting airways calculated from morphometric parameters, *Bull. Math. Biol., 43,* 101–109, 1981.
20. Singhal, S., Henderson, R., Horsfield, K., Harding, K., and Cumming, G.: Morphometry of human pulmonary arterial tree, *Circ. Res., 33,* 190–197, 1973.
21. Fenton, B., and Zweifach, B. W.: Microcirculatory model relating geometrical variation to changes in pressure and flow rate, *Ann. Biomed. Eng., 9,* 303–321, 1981.
22. Lipowsky, H. H., and Zweifach, B. W.: Network analysis of microcirculation of cat mesentery, *Microvasc. Res., 7,* 73–83, 1974.
23. Zweifach, B. W., and Lipowsky, H. H.: Quantitative studies of microcirculatory structure and function: III. Microvascular hemodynamics of cat mesentery and rabbit omentum, *Circ. Res., 41,* 380–390, 1977.
24. Gore, R. W.: Pressures in cat mesenteric arterioles and capillaries during changes in systemic arterial blood pressure, *Circ. Res., 34,* 581–591, 1974.
25. Lipowsky, H. H., Usami, S., and Chien, S.: *In vivo* measurements of "apparent viscosity" and microvessel hematocrit in the mesentery of the cat, *Microvasc. Res., 19,* 297–319, 1980.

26. Kanzow, G., Pries, A. R., and Gaehtgens, P.: Analysis of the hematocrit distribution in the mesenteric microcirculation, *Int. J. Microcirc. Clin. Exp., 1,* 67–79, 1982.

27. Richardson, D. R., and Coates, F.: Effects of norepinephrine infusion (IV) on microvasculature pressures and capillary blood flow in the mesentery, *Microvasc. Res., 9,* 166–181, 1975.

28. Intaglietta, M., Richardson, D. R., and Tompkins, W. R.: Blood pressure, flow, and elastic properties in microvessels of cat omentum, *Am. J. Physiol., 221,* 922–928, 1971.

29. Intaglietta, M., and Zweifach, B. W.: Geometrical model of the microvasculature of rabbit omentum from in vivo measurements, *Circ. Res., 28,* 593–600, 1971.

30. Gross, J. F., Intaglietta, M., and Zweifach, B. W.: Network model of pulsatile hemodynamics in the microcirculation of the rabbit omentum, *Am. J. Physiol., 226,* 1117–1123, 1974.

31. Gore, R. W., and Bohlen, H. G.: Microvascular pressures in rat intestinal muscle and mucosal villi, *Am. J. Physiol., 233,* H685–H693, 1977.

32. Levitt, D. G., Sircar, B., Lifson, N., and Lender, E. J.: Model for mucosal circulation of rabbit small intestine, *Am. J. Physiol., 237,* E373–E382, 1979.

33. Eriksson, E., and Myrhage, R.: Microvascular dimensions and blood flow in skeletal muscle, *Acta Physiol. Scand., 86,* 211–222, 1972.

34. Fronek, K., and Zweifach, B. W.: Microvascular blood flow in cat tenuissimus muscle, *Microvasc. Res., 14,* 181–189, 1977.

35. Sullivan, S. M., and Johnson, P. C.: Effect of oxygen on arteriolar dimensions and blood flow in cat sartorius muscle, *Am. J. Physiol., 241,* H547–H556, 1981.

36. Henrich, H. N., and Hecke, A.: A gracilis muscle preparation for quantitative microcirculatory studies in the rat, *Microvasc. Res., 15,* 349–356, 1978.

37. Smaje, L., Zweifach, B. W., and Intaglietta, M.: Micropressure and capillary filtration coefficients in single vessels of the cremaster muscle of the rat, *Microvasc. Res., 2,* 96–110, 1970.

38. Bohlen, H. G., Gore, R. W., and Hutchins, P. M.: Comparison of microvascular pressures in normal and spontaneously hypertensive rats, *Microvasc. Res., 13,* 125–130, 1977.

39. Meininger, G. A., Harris, P. D., Joshua, I. G., Miller, F. N., and Wiegman, D. L.: Microvascular pressures in skeletal muscle of one-kidney one-clip renovascular hypertensive rats, *Microcirculation, 1,* 237–254, 1981.

40. Ma, Y. P., Koo, A., Kwan, H. C., and Chen, K. K.: On-line measurement of the dynamic velocity of erythrocytes in the cerebral microvessels in the rat, *Microvasc. Res., 8,* 1–13, 1974.

41. Joyner, W. L., Davis, M. J., and Gilmore, J. P.: Intravascular pressure distribution and dimensional analysis of microvessels in hamsters with renovascular hypertension, *Microvasc. Res., 22,* 190–198, 1981.

42. Sarelius, I. H., and Duling, B. R.: Direct measurement of microvessel hematocrit, red cell flux, velocity, and transit time, *Am. J. Physiol., 243,* H1018–H1026, 1982.

43. Mayrovitz, H. N., Tuma, R. F., and Wiedeman, M. P.: Relationship between microvascular blood velocity and pressure distribution, *Am. J. Physiol., 232,* H400–H405, 1977.

44. Wiederhielm, C. A., and Weston, B. V.: Microvascular, lymphatic, and tissue pressures in the unanesthetized mammal, *Am. J. Physiol., 225,* 992–996, 1973.

45. Mayrovitz, H. N., Wiedeman, M. P., and Noordergraaf, A.: Analytical characterization of microvascular resistance distribution, *Bull. Math. Biol., 38,* 71–82, 1976.

46. Tyml, K., Ellis, C. G., Safranyos, R. G., Fraser, S., and Groom, A. C.: Temporal and spatial distributions of red cell velocity in capillaries of resting skeletal muscle, including estimates of red cell transit time, *Microvasc. Res., 22,* 14–31, 1981.

47. Shea, S. M.: Glomerular hemodynamics and vascular structure, *Microvasc. Res., 18,* 129–143, 1979.

48. Lambert, P. P., Aeikens, B., Bohle, A., Hanus, F., Pegoff, S., and Van Damme, M.: A network model of glomerular function, *Microvasc. Res., 23,* 99–128, 1982.

49. Van Damme, M., Pegoff, S., and Lambert, P. P.: The pressure and flow distribution within a filtering capillary network, *Comput. Programs Biomed., 13,* 239–250, 1981.

50. Schmid-Schönbein, G. W., Zweifach, B. W., and Kovalcheck, S.: The application of stereological principles to morphometry of the microcirculation in different tissues, *Microvasc. Res., 14,* 303–317, 1977.

51. Sullivan, S. M., and Pittman, R. N.: Hamster retractor muscle: A new preparation for intravital microscopy, *Microvasc. Res., 23,* 329–335, 1982.

52. Skalak, T. C.: A mathematical hemodynamic network model of the microcirculation in skeletal muscle, using measured blood vessel distensibility and topology, Ph.D. dissertation, University of California, San Diego, La Jolla, Calif., 1984.

53. Klitzman, B., and Johnson, P. C.: Capillary network geometry and red cell distribution in hamster cremaster muscle, *Am. J. Physiol., 242,* H211–H219, 1982.

54. Sullivan, S. M., and Johnson, P. C.: Effect of oxygen on blood flow autoregulation in cat sartorius muscle, *Am. J. Physiol., 241,* H807–H815, 1981.

55. Chen, I. I., and Prewitt, R. L.: A mathematical representation for vessel network, *J. Theor. Biol., 97,* 211–219, 1982.

56. Zamir, M.: Three-dimensional aspects of arterial branching, *J. Theor. Biol., 90,* 457–476, 1981.

57. Lipowsky, H. H., Kovalcheck, S., and Zweifach, B. W.: The distribution of blood rheological parameters in the microvasculature of cat mesentery, *Circ. Res., 43,* 738–749, 1978.

58. Usami, S.: Physiological significance of blood rheology, *Biorheology, 19,* 29–46, 1982.

59. Schmid-Schönbein, G. W., and Zweifach, B. W.: RBC velocity profiles in arterioles and venules of the rabbit omentum, *Microvasc. Res., 10,* 153–164, 1975.

60. Klitzman, B., and Duling, B. R.: Microvascular hematocrit and red cell flow in resting and contracting striated muscle, *Am. J. Physiol., 237,* H481–H490, 1979.

61. Sarelius, I. H., Damon, D. N., and Duling, B. R.: Microvascular adaptations during maturation of striated muscle, *Am. J. Physiol, 241,* H317–H324, 1981.

62. Schmid-Schönbein, H., and Devendran, T.: Blood rheology in the microcirculation, *Pflügers Arch. (Suppl), 336,* 84–87, 1972.

63. Mayrovitz, H. N., Wiedeman, M. P., and Noordergraaf, A.: Microvascular hemodynamic variations accompanying microvessel dimensional changes, *Microvasc. Res., 10,* 322–339, 1975.

64. Mayrovitz, H. N.: The microcirculation: Theory and experiment, Ph.D. dissertation, University of Pennsylvania, Philadelphia, 1974.

65. Popel, A. S.: A model of pressure and flow distribution in branching networks, *J. Appl. Mech., 102,* 247–253, 1980.

66. Popel, A. S., and Levin, M.: A model of blood flow in microvascular networks with stochastic geometry, *Int. J. Microcirc. Clin. Exp., 1,* 310–311, 1982.

67. Dawant, B., Levin, M., and Popel, A. S.: Effect of dispersion of vessel diameters and lengths in stochastic networks: I. Modeling of microcirculatory flow, *Microvasc. Res., 31,* 203–222, 1986.

68. Levin, M., Dawant, B., and Popel, A. S.: Effect of dispersion of vessel diameters and lengths in stochastic networks: II. Modeling of microvascular hematocrit distribution, *Microvasc. Res., 31,* 223–234, 1986.

69. Duling, B. R., Sarelius, I. H., and Jackson, W. F.: A comparison of microvascular estimates of capillary flow with direct measurements of total striated muscle flow, *Int. J. Microcirc. Clin. Exp., 1,* 409–424, 1982.

70. Yen, R. T., and Fung, Y. C.: Effect of velocity distribution on red cell distribution in capillary blood vessels, *Am. J. Physiol., 235,* H251–H257, 1978.

71. Dellimore, J. W., Dunlop, M. J., and Canham, P. B.: Ratio of cells and plasma in blood flowing past branches in small plastic channels, *Am. J. Physiol., 244,* H635–H643, 1983.

72. Schmid-Schönbein, G. W., Skalak, R., Usami, S., and Chien, S.: Cell distribution in capillary networks, *Microvasc. Res., 19,* 18–44, 1980.

73. Cokelet, G. R.: Speculation on a cause of low vessel hematocrits in the microcirculation, *Microcirculation, 2,* 1–18, 1982.

74. Lipowsky, H. H., Rofe, S., Tannenbaum, L., Firrell, J. C., Usami, S., and Chien, S.: Microvessel hematocrit distributions at bifurcations, *Microvasc. Res., 21,* 249–250, 1981.

75. Popel, A. S.: Effect of heterogeneity of capillary flow on the capillary hematocrit, *1979 Biomechan. Symp., Proc. Am. Soc. Mech. Eng., Appl. Mech. Div.,* American Society of Mechanical Engineers, New York, 1979, vol. 32, pp. 83–84.

76. Papenfuss, H. D., and Gross, J. F.: Microhemodynamics of capillary networks, *Biorheology, 18,* 673–692, 1981.

77. Wieringa, P. A., Spaan, J. A. E., Stassen, H. G., and Laird, J. D.: Heterogeneous flow distribution in a three dimensional network simulation of the myocardial microcirculation—A hypothesis, *Microcirculation, 2,* 195–216, 1982.

78. Gross, J. F.: The significance of pulsatile microhemodynamics, in G. Kaley and B. M. Altura (eds.), *Microcirculation,* University Park Press, Baltimore, 1977, vol. 1, pp. 365–390.

79. Mayrovitz, H. N., Wiedeman, M. P., and Noordergraaf, A.: Interaction in the microcirculation, in J. Baan, A. Noordergraaf, and J. Raines (eds.), *Cardiovascular System Dynamics,* MIT Press, Cambridge, Mass., 1978, pp. 194–204.

80. Thompson, D. W.: *On Growth and Form,* Cambridge University Press, Cambridge, 1981 (first published in 1917), pp. 125–131.

81. Murray, C. D.: The physiological principle of minimum work: I. The vascular system and the cost of blood volume, *Proc. Nat. Acad. Sci. U.S.A., 12,* 207–214, 1926.

82. Murray, C. D.: The physiological principle of minimum work applied to the angle of branching of arteries, *J. Gen. Physiol., 9,* 835–841, 1926.

83. Sherman, T. F.: On connecting large vessels to small, *J. Gen. Physiol., 78,* 431–453, 1981.

84. Mayrovitz, H. N., and Roy, J.: Microvascular blood flow: Evidence indicating a cubic dependence on arteriolar diameter, *Am. J. Physiol., 242,* H1031–H1038, 1983.

85. Zamir, M.: The role of shear forces in arterial branching, *J. Gen. Physiol., 67,* 213–222, 1976.

86. Kamiya, A., and Togawa, T.: Optimal branching structure of the vascular tree, *Bull. Math. Biophys., 34,* 431–438, 1972.

87. Zamir, M.: Nonsymmetrical bifurcations in arterial branching, *J. Gen. Physiol., 72,* 837–845, 1978.

88. Zamir, M., Medeiros, J. A., and Cunningham, T. K.: Arterial bifurcations in the human retina, *J. Gen. Physiol., 74,* 537–548, 1979.

89. Zamir, M., and Medeiros, J. A.: Arterial branching in man and monkey, *J. Gen. Physiol., 79,* 353–360, 1982.

90. Uylings, H. B. M.: Optimization of diameters and bifurcation angles in lung and vascular tree structures, *Bull. Math. Biol., 39,* 509–520, 1977.

91. Roy, A. G., and Woldenberg, M. J.: A generalization of the optimal models of arterial branching, *Bull. Math. Biol., 44,* 149–360, 1983.

92. Horsfield, K., and Cumming, G.: Angles of branching and diameters of branches in the human bronchial tree, *Bull. Math. Biophys., 29,* 245–259, 1967.

93. Horsfield, K., and Thurlbeck, A.: Relation between diameter and flow in branches of the bronchial tree, *Bull. Math. Biol., 43,* 681–691, 1981.

94. Borders, J. L.: Vasomotion patterns in skeletal muscle in normal and hypertensive rats, Ph.D. dissertation, University of California, San Diego, La Jolla, Calif., 1980.

95. Colantuoni, A., Bertuglia, S., and Intaglietta, M.: Variation of rhythmic diameter changes at the arterial microvascular bifurcations, *Pflügers Arch., 403,* 289–295, 1985.

96. Popel, A. S., and Johnson, P. C. (eds.): *Microvascular Networks: Experimental and Theoretical Studies,* Karger, Basel, in press.

97. Engelson, E. T., Skalak, T. C., and Schmid-Schönbein, G. W.: The microvasculature in skeletal muscle: I. Arteriolar network in rat spinotrapezius muscle, *Microvasc. Res., 30,* 29–44, 1985.

98. Engelson, E. T., Schmid-Schönbein, G. W., and Zweifach, B. W.: The microvasculature in skeletal muscle: III. Venous network anatomy in normotensive and spontaneously hypertensive rats, *Int. J. Microcirc. Clin. Exp., 4,* 229–248, 1985.

99. Schmid-Schönbein, G. W., Skalak, T. C., Engelson, E. T., and Zweifach, B. W.: Microvascular network anatomy in rat skeletal muscle, in A. S. Popel and P. C. Johnson (eds.), *Microvascular Networks: Experimental and Theoretical Studies,* Krager, Basel, in press.

100. Woldenberg, M. J.: Quantitative analysis of biological and fluvial networks, in A. S. Popel and P. C. Johnson (eds.), *Microvascular Networks: Experimental and Theoretical Studies,* Karger, Basel, in press.

101. Ley, K., Pries, A. R., and Gaehtgens, P.: Topological structure of rat mesenteric microvessel networks, *Microvasc. Res.,* in press.

102. Yen, R. T., Zhuang, F. Y., Fung, Y. C., Ho, H. H., Tremer, H., and Sobin, S. S.: Morphometry of cat's pulmonary arterial tree, *J. Biomech. Eng., 106,* 131–136, 1984.

103. Yen, R. T., Zhuang, F. Y., Fung, Y. C., Ho, H. H., Tremer, H., and Sobin, S. S.: Morphometry of cat pulmonary venous tree, *J. Appl. Physiol., 55,* 236–242, 1983.

104. Koller, A., and Johnson, P. C.: Methods for in vivo mapping and classifying microvascular networks in skeletal muscle, in A. S. Popel and P. C. Johnson (eds.), *Microvascular Networks: Experimental and Theoretical Studies*, Karger, Basel, in press.

105. Lipowsky, H. H.: Network hemodynamics and the shear rate dependency of blood viscosity, in A. S. Popel and P. C. Johnson (eds.), *Microvascular Networks: Experimental and Theoretical Studies*, Karger, Basel, in press.

106. House, S. D., and Johnson, P. C.: Microvascular pressures in venules of skeletal muscle during arterial pressure reduction, *Am. J. Physiol.*, in press.

107. Popel, A. S., and Dawant, B.: Flow heterogeneity in microvascular networks: Comparison of theoretical predictions with experimental data, in A. S. Popel and P. C. Johnson (eds.), *Microvascular Networks: Experimental and Theoretical Studies*, Karger, Basel, in press.

108. Papenfuss, H. D., and Gross, J. F.: Mathematical simulation of blood flow in microcirculatory networks, in A. S. Popel and P. C. Johnson (eds.), *Microvascular Networks: Experimental and Theoretical Studies*, Karger, Basel, in press.

109. Mayrovitz, H. N.: Hemodynamic significance of microvascular arteriolar anastomosing, in A. S. Popel and P. C. Johnson (eds.), *Microvascular Networks: Experimental and Theoretical Studies*, Karger, Basel, in press.

110. Zweifach, B. W., Kovalcheck, S., Delano, F., and Chen, P. Micropressure-flow relationships in a skeletal muscle of spontaneously hypertensive rats, *Hypertension*, *3*, 601–614, 1981.

111. Davis, M. J., and Gore, R. W.: Pressure distribution in the microvascular network of the hamster cheek pouch, in A. S. Popel and P. C. Johnson (eds.), *Microvascular Networks: Experimental and Theoretical Studies*, Karger, Basel, in press.

112. Zhuang, F. Y., Fung, Y. C., and Yen, R. T.: Analysis of blood flow in cat's lung with detailed anatomical and elasticity data, *J. Appl. Physiol.*, *55*, 1341–1348, 1983.

113. Fenton, B. M., Carr, R. T., and Cokelet, G. R.: Nonuniform red cell distribution in 20 to 100 micron bifurcations, *Microvasc. Res.*, *29*, 103–126, 1985.

114. Chien, S., Tventestrand, C. D., Farrell Epstein, M. A., and Schmid-Schönbein, G. W.: Model studies on distributions of white and red blood cells at microvascular bifurcations, *Am. J. Physiol.*, *248*, H568–H576, 1985.

115. Cokelet, G. R.: Blood flow through arterial microvascular bifurcations, in A. S. Popel and P. C. Johnson (eds.), *Microvascular Networks: Experimental and Theoretical Studies*, Karger, Basel, in press.

116. Fenton, B. M., Wilson, D. W., and Cokelet, G. R.: Analysis of the effects of measured white blood cell entrance times on hemodynamics in a computer model of a microvascular bed, *Pflügers Arch.*, *403*, 396–401, 1985.

CHAPTER 21

Fluid Mechanics in Atherosclerosis

Robert M. Nerem
Murina J. Levesque
Physiological Fluid Mechanics Laboratory
Department of Mechanical Engineering
University of Houston
Houston, Texas

INTRODUCTION

In recent years there has been a resurgence of interest in the study of blood flow in the major mammalian arteries. This interest and the associated activity are a direct result of the possible role of hemodynamics and, in a broader sense, fluid mechanics in atherosclerosis, a disease which is the chief cause of death in the United States. This disease occurs in the large and medium-sized arteries. It is primarily concentrated in the intima, i.e., the inner part of the arterial wall, and it involves the accumulation of lipids, smooth muscle proliferation, and the synthesis of connective tissue components. Over a period of time lesions develop which typically have been classified as one of three types—the fatty streak, the fibrous plaque, and the so-called complicated lesion.

Severe atherosclerosis interferes with blood flow. This is particularly important for the heart and the brain, with the result being myocardial or cerebral ischemia or even a myocardial infarction or stroke. However, even in the early stages of the disease process, i.e., during atherogenesis, there may be an important relationship between the disease and the characteristics of the blood flowing through the arteries. This is not because the disease has altered the blood flow properties and thus influenced the health of the organs being perfused. Rather, it is because of the possibility that there may be arterial blood flow or hemodynamically related factors involved in the initiation and early progression of the disease.

In this context, there are two fundamental questions concerning atherosclerosis and its genesis, to which fluid mechanics may hold at least part of the answers. First, why does the disease occur in the large and central arteries and not on the venous side or even in the pulmonary arteries? Second, why is the development of lesions localized so that there appears to be a very definite, focal pattern to the disease? To address these two questions, we will first review the state of our knowledge with regard to the evidence for such a hemodynamic or fluid mechanical involvement. This evidence includes both the pattern of the disease and the role of hypertension as a risk factor.

As indirect as this evidence is, it has provided the primary motivation for the resurgence of interest over the last 15 years in the study of blood flow through large arteries. The detailed characteristics of arterial flows thus will be considered next. Blood flow in the major mammalian arteries presents a number of interesting and challenging problems. Such flows are pulsatile and pass through vessels which are complex both in geometry and in their elastic nature. The former is of particular interest with regard to atherosclerosis. It now appears that this geometric complexity, as represented by vessel branching and the general tortuosity of the vasculature, may be all important in determining not only the detailed characteristics of arterial flows but also the spatial nature of any hemodynamic involvement in the disease process.

The possible mechanisms through which any such hemodynamic involvement might manifest itself have also been the subject of extensive study. The vascular endothelium, as the inner lining of a blood vessel, is the interface between the arterial wall and the flowing blood; this monolayer of cells has been the focus of investigations of hemodynamics, and, in a broader sense, fluid mechanics, as a factor in atherogenesis. There has been a primary interest in the influence of mechanical forces, as imposed by the flow of blood, on endothelial morphology and function; these factors will be reviewed here.

We will then discuss the role of hemodynamics in the disease process and, in particular, during atherogenesis. In the initiation and early development of the disease, the endothelium is believed to be intact and the vascular geometry unaltered from its normal state. Although this normal geometry may vary considerably from individual to individual, if there is a role of hemodynamics or arterial fluid mechanics in the early development of atherosclerosis, it is the fluid mechanics of the normal cardiovascular system and its role in the biology of the arterial wall which are of interest.

This is not meant in any way to deny the possibility of a role of hemodynamics in later disease stages, nor the importance of thrombosis in atherosclerosis. The latter is certainly a factor in the more advanced stages of the disease, and hemodynamic, or fluid mechanical, factors are believed to be important in platelet adherence and related processes. However, whatever this hemodynamic role is in the later stages of disease, it more than likely is completely different from that played by hemodynamics in the genesis of the disease. Furthermore, in the later stages of atherosclerosis, the geometry of the now-diseased arteries are so changed that it is the fluid mechanics of an altered cardiovascular system which must be considered. At this point in the development of the disease there is a coupling between the hemodynamics and the disease process; as the disease continues to progress so as to further alter the local arterial geometry, the local fluid mechanics will undergo change, possibly influencing its role in and the further course of the disease.

Finally, it should be emphasized that, given the characteristic time of this disease, the role of hemodynamics is quite possibly an extremely subtle one. This suggests that the rather detailed hemodynamic characteristics may be important, and the evidence for such a subtle involvement, the detailed flow characteristics themselves, and mechanisms at a cellular level which may be participatory are discussed here and are used as a means to speculate on the role of fluid mechanics in atherosclerosis and specifically in lesion localization.

21.1 EVIDENCE FOR A FLUID MECHANICAL INVOLVEMENT

Although the possible importance of hemodynamics in atherogenesis has provided a requirement for a more detailed knowledge of the properties of arterial blood flow, there is some question as to what is the evidence for the involvement of fluid mechanics in the atherosclerotic process. The answer to this question centers on the pattern of the disease [1–4]. Not only is it large, thick-walled arteries such as the aorta and the iliac, femoral, coronary, and cerebral arteries which are most commonly affected (see Fig. 21.1), but, more specifically, it appears that it is often regions of arterial branching and sharp curvature which have the greatest predilection for the development of atherosclerosis. These are also regions where the flow will assume unusual characteristics or at least deviate from what otherwise might be considered a well-behaved arterial flow. The indictment provided by this indirect evidence, particularly as it

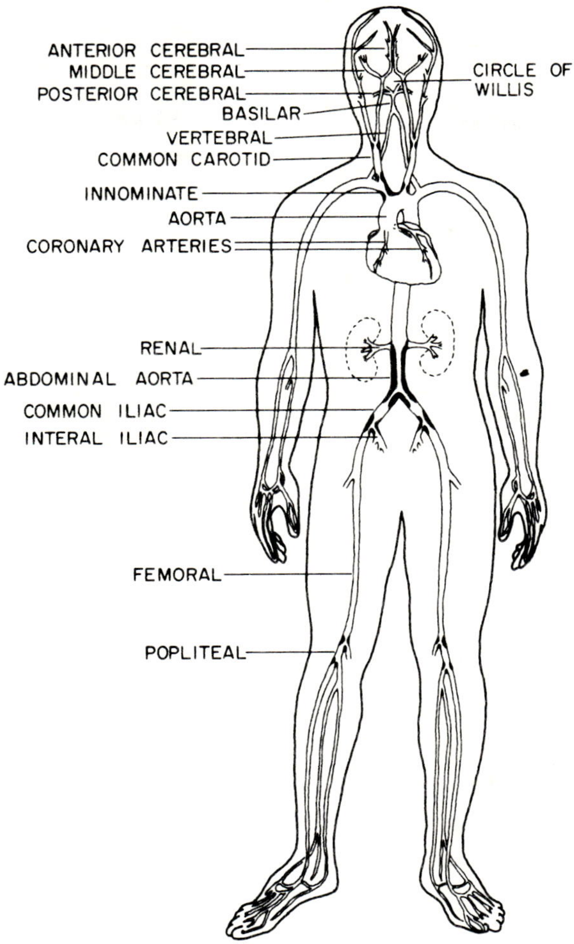

FIG. 21.1 Pattern of atherosclerosis in humans, showing a concentration of disease in the large and medium-sized arteries. (*From Spain [4].*)

relates to the bifurcations and geometrical contortions of the arterial vasculature, motivates much of the present interest in arterial fluid dynamics.

Unfortunately, in terms of understanding fluid mechanics as a factor in the atherosclerotic process, it is not enough to identify the general regions of branching and sharp curvature. This is because such general regions may include locations of flow separation, localized positions of accentuated secondary motion, areas where the wall shear stress will be high and others where it will be low, and portions of the wall itself where the stresses within will be intensified and others where stress levels may be lessened. The net result is that any indictment of fluid mechanics as a disease factor which is based on the pattern of the disease only raises questions relative to the details.

There is, of course, some information available on the specifics of the pattern of the disease in humans. Caro et al. [3] provided an early review of such observations and indicate that, with regard to the effect of branching, in the visceral branches of the abdominal aorta the two principal features are a relative sparing of the flow dividers and relatively heavy involvement of the "hips," or outer walls of the junctions. This type of distribution is also reported for Y junctions by Texon [5] and by Mustard and Packham [6]. Similar findings are reported by Mitchell and Schwartz [1] for the intercostals, where a relative sparing around and distal to intercostal ostia was shown within the general aortic posterior fatty streaking. Kjaernes et al. [7] and Svindland and Walløe [8] have investigated the localization of early lesions, including both carotid and coronary bifurcations. For the bifurcation of the brachiocephalic artery into the subclavian and common carotid arteries and for the bifurcation of the right carotid artery, early lesions form initially on the outer walls of the bifurcation, while the inner walls downstream of the flow divider and the lateral walls are spared. For the left main coronary bifurcation, the lesions are again found preferentially on the outer walls, with the inner walls downstream of the bifurcation being left free. The lesions also were predominantly located at the myocardial side of the vessels, i.e., along the inner curvature of the curved vessels. This is in general agreement with the coronary data of Fox and Seed [9], although they suggested the possibility of a spiral pattern in the left anterior descending (LAD) coronary artery. Fox and Seed also looked at the right coronary artery and concluded that lesions were concentrated on the inner walls of major curvature.

This general picture of lesion distribution in the vascular system is somewhat "muddied" by the data of Sinzinger et al. [10], obtained from fetuses, neonates, and infants, and by the mixed observations in animals. With regard to the former, fatty streaking was observed to occur preferentially on the distal side of aortic intercostal ostia. In the studies of Cornhill and Roach [11] of the hypercholesterolemic rabbit, a similar pattern was observed. However, in the white Carneau pigeon, where atherosclerosis occurs spontaneously, it has been found that even more advanced lesions have almost no distal components and are of a relatively pure proximal nature [12].

The results thus seem to divide into two groups. In human fetuses, neonates, and infants [10], and in experimental studies of diet-induced disease in animals [11], the disease appears to favor the occurrence of lesions in regions which, as we will see in the next section, are believed to be high-shear regions. However, in adult human disease [3, 7–9, 13], as well as in the white Carneau pigeon [12], where atherosclerosis occurs spontaneously, the disease appears to favor low-shear regions. This pattern of disease, reflected by differences in spatial location among lesions, raises important questions, both with regard to the possible importance of fluid mechanics in atherogenesis and in terms of differences in the development of experimental as opposed to spontaneous disease.

In addition to the pattern of the disease, another piece of evidence may have potential hemodynamic or fluid mechanical implications: hypertension is a risk factor in coronary artery disease. A basic question, the answer to which has considerable implication for hemodynamics, is what is the role of blood pressure? At one time it was thought that blood–arterial wall lipid transport was due to a pressure-driven, bulk-flow process. Although data now available tend to refute any major importance of such a pressure effect for intact endothelium [14], there still could be a major effect of pressure on water movement through the endothelium, on endothelial cell function, and/or on cell turnover. Furthermore, if mean blood pressure is important in terms of a hypertension effect, then it may be due to its effect on arterial geometry or on wall metabolism. The former would affect local flow details, while the latter would represent a very indirect type of hemodynamic effect. On the other hand, if an elevated pulse pressure is of importance in hypertension being a risk factor, then another possibility presents itself. Associated with an increased pulse pressure, there would be an altered velocity waveform, an increased peak velocity, and an increased level of arterial wall shear stress [15].

With regard to a pressure effect on arterial geometry, it has been reported that gross elevation of aortic pressure distorts the geometry of the aortic tree to the point where angles of branching are markedly changed [16] (see Fig. 21.2). This could have the effect of changing a relatively streamlined flow system into one considerably more tortuous, which then would tend to accentuate the type of complicated fluid-dynamic phenomena already noted. There are also data showing an accelerated occurrence of atherosclerosis in subjects with a human coronary geometry deviating from that normally found [17]. Furthermore, aortic coarctation, which represents a major change in vessel geometry, results in a significant alteration in the uptake of

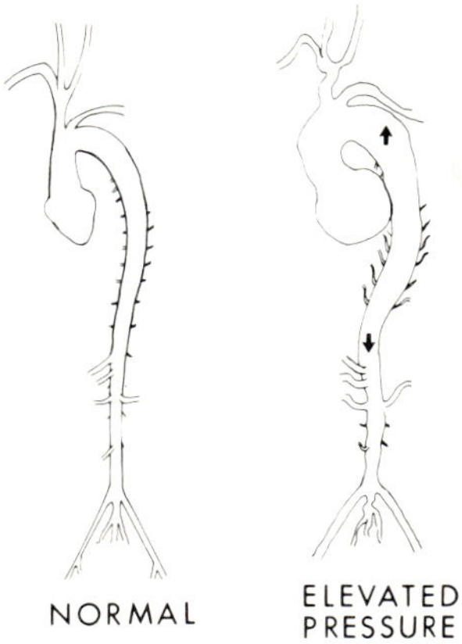

FIG. 21.2 Illustration of the effect of an elevation in blood pressure on the geometry of the aorta, showing the distortion which results. (*From Fry [16].*)

Evans blue dye [18], and our own recent studies have demonstrated a similar effect on endothelial geometry and orientation, as will be seen in a later section.

In fact, in considering the various factors involved in atherogenesis, a possible common thread is vascular geometry, i.e., the geometry of the arterial system. The physiological phenomena that would appear to be most directly influenced by arterial geometry are the detailed hemodynamic or fluid mechanical characteristics. This thus suggests that vascular geometry and the way it influences the local detailed flow properties may be a primary determinant of a fluid mechanical role in atherosclerosis, and this will be explored further in the next few sections.

21.2 DETAILED CHARACTERISTICS OF ARTERIAL FLOWS

Until recently, little information existed about the detailed nature of viscous flow phenomena in the arterial system. This is at first sight surprising since the subject of viscous flow has been of continuing interest and importance in a wide range of engineering contexts. However, the problem in the arterial system in general has been uniquely difficult, not only because of the unsteady nature of the basic flow but also because of limitations on instrumentation and access.

Partly as a result of the motivation provided by a possible role of hemodynamics in atherogenesis, much has been learned about arterial flows in the last 15 years. A major achievement was the development of the ability to make point-velocity measurements within blood vessels. Such measurements have become possible with the hot-film constant-temperature anemometer and the pulsed ultrasonic Doppler and the laser Doppler anemometer systems. These techniques have provided, for the first time, insight into the temporal and spatial velocity variations present locally in the vascular system. In vivo studies using such point-velocity systems of measurement have been carried out in a variety of animals and in humans [19–26]. Furthermore, these approaches also have been used in physical models [13]. From this combination of approaches a much clearer picture of arterial fluid mechanics has emerged.

The classical picture used to explain the detailed characteristics of flowing blood is that of laminar Poiseuille flow. In this the flow is likened to that in a long, rigid, circular pipe with a constant flow rate. The velocity pattern, or velocity profile, i.e., the variation in velocity as one moves radially across the lumen of the vessel from one wall to the other, is characterized as being parabolic. This is illustrated in Fig. 21.3. However, we now know that flow in the larger arteries is, in general, not Poiseuille flow; i.e., it is not a fully developed viscous flow characterized by a parabolic profile. There are, of course, many reasons for this. Primary among these is the unsteady nature of the flow in large arteries. In addition, there is the possibility of transition to or at least transitory bursts of turbulence [23], at least in the aorta. The concept of Poiseuille flow, strictly speaking, applies only under laminar, steady-state flow conditions.

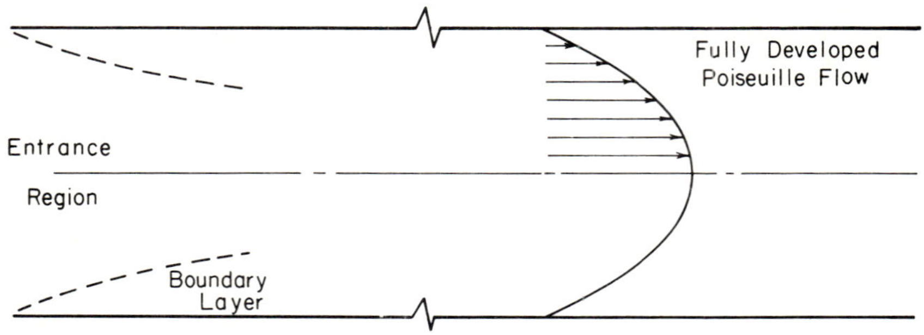

FIG. 21.3 Laminar pipe flow illustrating both the entrance region and the downstream region of fully developed, Poiseuille flow.

In the context of our interests here, however, one of the very major complexities of blood flow in the arterial system is that the vessels involved are often curved, and they branch repeatedly. This is illustrated in Fig. 21.4 for the coronary system. As can be seen, the geometry is extremely tortuous. However, it is also highly individualistic. For example, in a study of human coronary casts [27], it was found that the average angle of bifurcation of the left main coronary artery into the left circumflex and the LAD coronary arteries was 72°; however, individual values ranged from 50 to 95°. Similar individual differences were found in comparing

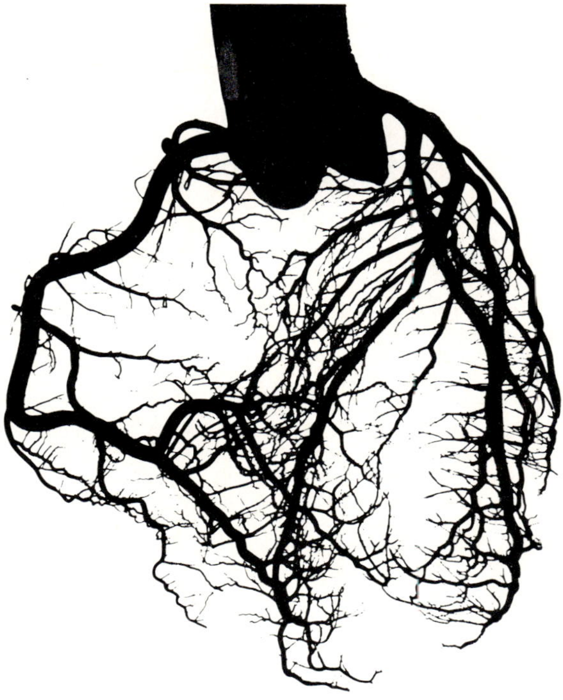

FIG. 21.4 A cast of human coronary vessels illustrating the complexity of its geometry. (*From Nerem and Seed [27].*)

such geometric characteristics as vessel length, diameter, and percent of taper. The branching, asymmetric nature of the vascular geometry means that the flow will be characterized fluid mechanically by entry phenomena, asymmetries in the velocity patterns, complicated secondary motions, and even flow separation, all of which are far more difficult to analyze than simple steady-state, fully developed Poiseuille flow [15]. The individualistic nature of this geometry means that the specific detailed characteristics of the flow will differ from one individual to the next. Even so, it is possible to describe in a general way the nature of blood flow in the arterial system.

The entrance effect, or the existence of an entry length, is a well-known problem in fluid dynamics [28]. In general, when a flow passes from a large-diameter vessel into a smaller one, the velocity distribution at the entrance to the smaller vessel is found to be rather blunt. A wall boundary layer, initially very thin at the entrance of the vessel, increases in thickness as the flow proceeds distally (see Fig. 21.3). At a distance from the entrance called the *entry length*, the boundary layer closes in on itself, and at this point the frictional effect of a viscous fluid passing through the vessel has spread across the entire lumen of the vessel. If one calculates the entry length for flow in the human aorta [15], the result is a length of at least 150 cm, which is far greater than the length of the aorta. The flow in the aorta thus cannot be characterized as fully developed, and this has been shown experimentally [20, 22–23]. In fact, all the larger arteries of the circulatory system, including the epicardial coronary vessels [27], are subject to entrance effects.

An additional complexity of the arterial system is that its asymmetric geometry, including vessel branching and curvature, quite naturally produces asymmetries in the velocity patterns. For example, at a bifurcation the flow in the upstream parent vessel divides into the two daughter vessels so as to bring relatively high velocity blood at the center of the parent vessel in close proximity to the wall of the flow divider, and this has been demonstrated both for aortic flows [20, 22] and for coronary flows [24]. Skewing of the velocity profile also occurs in a curved vessel. The nature of this skewing depends on whether the flow is largely inviscid or fully viscous. By *largely inviscid* is meant a flow such as is present in the ascending aorta of large animals, in which there is an entrance region where viscous effects are confined to a thin-walled boundary layer and thus the core flow behaves in an inviscid manner [22]. In the case of fully viscous flow, on the other hand, the flow is skewed toward the outer wall. An example of this is the flow in the left common coronary artery [24, 29]. Furthermore, the bifurcation of the left common into the LAD and the left circumflex coronary arteries represents a case where there is skewing due to both branching and curvature. Our own studies demonstrate that the skewing produced by the curving of these epicardial vessels over the heart in many cases dominates that associated with the bifurcation process, depending, of course, on the exact geometry [29].

Associated with the flow in a bifurcation or a curved vessel is the secondary motion induced by the curved path the fluid must follow. As is illustrated in Fig. 21.5, secondary flows are characterized by a swirling, helical component superimposed on the main streamwise velocity along the tube axis. The most familiar example is that of steady flow in a pipe [28], a relevant example, since the arch of the aorta is in effect a pipe bend. Secondary motions are also produced by branchings, where, just as in a curved pipe, the curvature associated with the change in flow direction is accompanied by a centrifugal pressure gradient. Although there have been numerous fluid-dynamic studies of secondary flow phenomena [15], instrumentation limitations have prevented any in vivo observations.

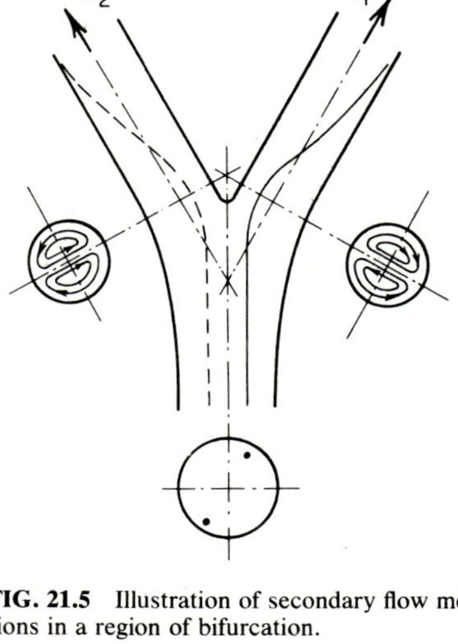

FIG. 21.5 Illustration of secondary flow motions in a region of bifurcation.

An additional complication introduced by the geometry of the arterial system is that the flow may actually separate from the wall, giving rise to a recirculating, dead-water region containing slow-moving, almost stagnant fluid. For example, for steady flow in a bifurcation the flow may separate because of the adverse pressure gradient associated with the deceleration of the fluid as it passes from the parent vessel into daughter vessels whose combined cross-sectional area exceeds that of the parent vessel. Downstream of the bifurcation the flow will reattach to the wall. The bubble, or recirculation region, located between the points of separation and reattachment, contains recirculating fluid. Needless to say, the occurrence of flow separation in pulsatile flows is an extremely complex phenomenon, for there the recirculation region is itself unsteady in nature, and flow separation and reattachment points, if present, can change location or even disappear and then reappear as the flow pulses. Flow patterns for oscillatory flows obviously are far more complicated than for steady flows, and very little is known yet about the quantitative details of many of the phenomena discussed here for arterial blood flow conditions.

The picture is further complicated by the fact that, although it is the detailed local flow properties which are believed important to the atherogenic process, these are themselves determined by the flow properties of the system. Thus the division of flow at the bifurcation of a vessel is primarily determined by the peripheral resistance downstream of each of the two daughter vessels. As an example, Rabinovitz et al. [30] have shown that the flow division at the LAD–left circumflex coronary bifurcation is markedly different in the presence of hypertension as compared to control conditions, and that it in fact depends on the mechanism by which hypertension has been produced.

Now so far we have concerned ourselves with the complexities of the local flow details; however, associated with the local velocity pattern, there will be a mechanical, hemodynamic force exerted by the flowing blood on the arterial wall. The effect of this force has been of particular interest in terms of understanding the fluid mechanics of an interaction with the endothelium. The total force per unit wall area, i.e. the total stress, imposed on the arterial wall by the flow of blood is manifested through both a normal and a tangential component. The normal component of the stress is essentially equal to the blood pressure, while the tangential component is a frictional force per unit area and is termed the *wall shear stress*. For a newtonian fluid, the wall shear stress is equal to the product of the fluid viscosity and the shear rate S evaluated at the wall. The latter is by definition equal to the rate of change in the streamwise velocity component as one moves away from the wall, i.e., across the lumen of the vessel, and this in and of itself suggests an importance of the local flow pattern and detailed characteristics.

Whereas a blood pressure of 100 mmHg represents a normal force slightly in excess of 10^5 dyn/cm^2, wall shear stresses are believed to be several order of magnitudes lower. Limited experimental data (some of which are suspect), as well as theoretical determinations, indicate peak shear stresses at the level of 100 to 200 dyn/cm^2 [15]. Although the shear stress levels are obviously much smaller than the stress due to normal pressure, the latter is borne by the entire arterial wall, while the shear stress is largely taken up by the endothelium.

Finally, it is through our understanding of arterial blood flow, as summarized here, that we believe we can characterize vascular regions as high-shear or low-shear regions. For example, in a region of branching, in general, we would expect the flow divider to experience higher shear, while immediately proximal to the branch, the shear should be comparatively lower. However, one must use caution in doing this, for in complicated flow regions the shear stress can vary considerably over a very short distance. For example, in the region of the iliac bifurcation one might speculate that the shear would be low on the outer walls. However, there are now data suggesting that as the flow passes into the daughter vessels, there is a significant increase in the shear stress level [31]. Thus, depending exactly on where you are located on the outer wall, it may be a region of low shear stress or one of high shear stress, and any attempt to characterize the pattern of the disease as being associated with low shear or high shear as done in the preceding section involves a certain element of risk.

21.3 MECHANICAL INFLUENCES ON THE ENDOTHELIUM

Early studies of a possible influence of hemodynamically related events on the endothelium focused on transendothelial transport. Fry's results [32] suggested that the wall shear stress influences the rate of transport of macromolecules between blood and the arterial wall, and the results of Caro and Nerem [33] using ^{14}C-4-cholesterol have shown that it is not an effect of wall shear on diffusion boundary layer–controlled transport, but one considerably more subtle in terms of an effect of shear stress on the properties of the arterial wall. Similar steady-flow results have been obtained using ^{131}I-albumin, and data from experiments using both radioactively labeled albumin and cholesterol indicate that it is not only the mean component of arterial pressure and flow but also the oscillatory nature of the phenomena which has an important influence on the blood–arterial wall transport of macromolecules [34, 45]. These results suggest that at low levels of wall shear stress there is at best a very weak shear dependence; however, at shear stress levels greater than 50 dyn/cm^2, there is a much stronger dependence of the transport rate on shear stress [15]. The more recent results of Thibault and Fry [36] confirm this observation, but raise a question as to whether excised vessels, particularly at high shear stress levels, have an intact endothelium.

Vaishnav et al. [37] have attempted to quantify the strength of the endothelium in the canine aorta. In these experiments jets of physiological saline were impinged normally on the endothelium of freshly excised segments of canine middle descending thoracic aortas. A characteristic type of annular lesion was found which showed that the endothelium could withstand large normal stresses where the jet impinged, but was easily eroded by the shear stress resulting from the jet efflux. In general, the size of the lesion increased with an increase in the jet flow rate, an increase in the duration of jet impingement, and an increase in the temperature of the bath. These results thus demonstrate that the erosion stress τ_e, that is, the arterial shear stress required to cause endothelial denudation, is strongly dependent on the duration of exposure and on environmental factors such as temperature. For short time durations, τ_e increases to the order of several thousand dynes per square centimeter, and this also has been borne out by the experiments of Languille [38]. It thus appears that peak physiological wall shear stress levels are far below those required to cause denudation, and the influence of fluid mechanics in the atherogenic process must occur in the presence of an intact endothelium.

What then is the mechanism by which any fluid mechanical effect on the endothelium becomes manifest? Is it a hemodynamically related effect on transendothelial transport? Is it an effect on endothelial cell function? Or are endothelial cell function and transendothelial permeability in fact related to one another? If it is a transport problem, what are the transport mechanisms? Assuming that the intercellular junctions are too tight to allow for the passage of large-molecular-weight macromolecules, how do lipids and other such molecules move across the endothelium from blood into the arterial wall? One possibility is by some vesicular process in which material at the luminal side of the endothelial cell is taken into the cell by an invagination of the cell membrane. The resulting vesicle then detaches and through brownian motion moves across the cell, and at the other side the vesicle attaches and the particle is released. Vesicles in endothelium have been observed [39]; however, Weinbaum's calculations [40] suggest that vesicular transport may not be the answer.

Weinbaum [40], in fact, has investigated a number of possibilities for the movement of molecules across the endothelial interface. Of particular interest has been his theoretical calculations of the effect of a damaged cell and the influence of the concomitant "leaky" junctions on the enhancement of macromolecule transport. These results, summarized in Fig. 21.6, indicate that as the number of damaged cells increases from 1 in a 1000 to 5 in a 1000, the flux or uptake ratio increases by 67 percent. This suggests that the endothelium indeed can have a rate-controlling effect on the transport of macromolecules, as had been indicated earlier by the experiments of Caro and Nerem [33], but one which is related to cell damage or death, i.e., the life expectancy of a cell, and, as such, possibly to cell function. Weinbaum likens this to the process of cell turnover, and on the basis of his calculations, an enhanced lipid influx would occur in regions of high cell turnover. In vivo data [41] indicate that endothelial cell turnover rates will be higher in regions of more rounded endothelial cells as opposed to those of more elongated ones. As we will see shortly, this suggests that low-shear regions have higher cell turnover rates than high-shear regions.

The movement of water across the arterial wall is another area which has been under investigation. The consolidation theory of Kenyon [42] and the confirming experiments of Wijesinghe [43] predict that arteries in vivo will experience substantial oscillatory fluid transport between the lumen and the subluminal tissue if resistance to water flow at the endothelium is not excessive. Blackshear et al. [44] have estimated the endothelial contribution to the total

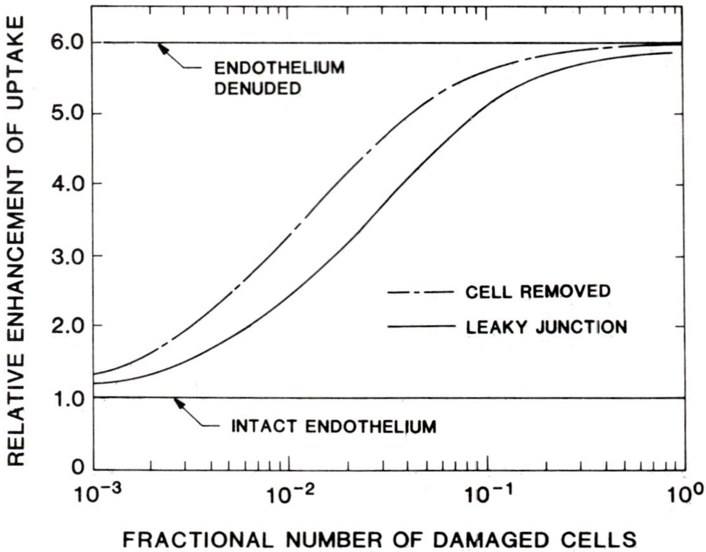

FIG. 21.6 Relative enhancement of flux or uptake as a function of cell turnover, i.e., fractional number of cells damaged, for the leaky junction model (width of intercellular junction, 200 Å; wall thickness, 1 mm). Dashed curve equivalent would result if entire endothelial cell were removed. (*From Weinbaum [40].*)

resistance to water flow to be 10 percent in the human aorta, and in this case the movement of water across the arterial wall may be important to an understanding of macromolecule transport within and the distribution of such molecules across the arterial wall. The experiments of Caro et al. [45] confirm this, and furthermore suggest that, even if the endothelium is controlling the rate of blood–arterial wall macromolecule transport, the movement of molecules within the wall is also important.

One of the conclusions which comes out of the theoretical work of Weinbaum and his coworkers is that the resistance of the arterial wall to the passage of macromolecules is important. This is not in any way to deny that the endothelium is a major barrier to the uptake of molecules such as lipids. However, once such large molecules have entered the intima, the transport properties of the intima and the media will determine the distribution of such molecules across the arterial wall. To be more specific, once a molecule has penetrated the endothelium, in whatever manner, there are then two important pathways within the arterial wall to be considered. One is the radial pathway, which may be characterized by a resistance R_r, that is, the resistance of the wall to the transport of molecules radially or directly across it. The other is the longitudinal pathway, which may be characterized by a resistance R_z, that is, the resistance of the wall to the transport of molecules along the wall or in a direction parallel to the axis of the artery.

Caro et al. [45] were the first to show strong exclusion by the media and thus strong medial resistance. Studies by Smith and Staples [46] on the composition of plasma and on interstitial fluid from the intima and media of the arterial wall support the view of a high transport resistance of the underlying wall. Furthermore, wall transport appears to depend on the flow of water across the wall influencing not only the transported species [47] but also the transport properties of the wall itself. Wall transport appears to depend, moreover, on the state of contraction of the smooth muscle of the media [48].

The problem of the transport of liquids in soft tissue also has been investigated in the context of articular cartilage [49]. It was found that the strain field induced by the drag of the permeating liquid, i.e., *drag-induced compression,* is nonuniform with respect to depth, causing a nonuniform distribution of permeability throughout the tissue. Thus, for a uniform sample without flow, there is uniform compression; but the same sample with flow will have nonuniform compression and thus nonuniform transport properties. Although these experiments were done for cartilage, the same phenomenon should be exhibited in the arterial wall and thus could contribute to an increased medial resistance. It thus has been shown that the resistance of the media is certainly not negligible and appears to make a major contribution to R_r, and thus to the transport within the wall itself. If R_r is in fact locally much greater than R_z, then the movement of molecules will be preferentially in a longitudinal direction, i.e., along the wall. For example, for molecules entering in a region of high cell turnover, not only will they accumulate below the location of endothelial cell turnover, but they will also move longitudinally so as to occupy that general region of the wall.

The interest in arterial wall transport stems from its being the link between luminal hemodynamic events and biochemical processes within the arterial wall. However, there are at least two other mechanisms which lend themselves to a shear-dependent, transport-related hypothesis. The first of these is that hemodynamics is a controlling factor in determining the availability of oxygen to the intima [15, 50]. There have been a number of observations correlating hypoxic conditions and the incidence of atherosclerosis, and assuming that the diffusion boundary layer is limiting the rate of oxygen transport, then it is low-shear regions of the wall which would have the greater predilection for tissue hypoxia.

The other way in which hemodynamic considerations would favor a low-shear hypothesis is if shear in fact determines platelet deposition, i.e., the transport of platelets to the wall. As inferred earlier, there may be many ways in which endothelial injury can occur, possibly on a continuing and random basis. Such injury might then be followed by platelet deposition, release reactions, and smooth muscle cell proliferation—in some cases leading to repair of the wall and in some cases leading to atherogenesis. If platelet deposition is enhanced in low-shear regions [51], then a disease pattern could emerge which would be consistent with a low-shear hypothesis. However, if the genesis of atherosclerosis occurs in the presence of an intact endothelium, then this would speak against a platelet adherence–aggregation hypothesis since it is believed that platelets do not adhere to intact endothelium. Platelets still could experience transient endothelial encounters, possibly releasing platelet factors and stimulating smooth muscle cell proliferation.

As noted earlier, thrombosis is believed to be, and more than likely is, an important element in the later disease stages, and if the influence of fluid mechanics on platelet deposition is such as to favor a pattern similar to that of early lesion localization, then this might serve to compound the hemodynamic effect and possibly further augment the pattern earlier established during atherogenesis. Unfortunately, as the disease progresses, there are changes which occur in arterial geometry, and thus there will also be changes in the detailed characteristics of blood flow, e.g., in the shear stress levels and in the distribution of shear stress within the arterial system.

There is also a body of accumulating data which indicates that endothelial cell geometry and orientation are determined by hemodynamically related events. If the functioning of a cell is influenced by its shape, then this could be a means whereby a hemodynamic effect becomes manifest. Of particular interest in our own studies of arterial endothelium has been the influence of hemodynamic forces, i.e., the effect of pressure and of wall shear stress. With the vascular casting technique [52], endothelial cell geometry and orientation have been studied under conditions corresponding to acutely elevated pressure and in regions of hemodynamic interest. Such cell properties as length, width, area, perimeter, angle of orientation, and shape index have been determined on a cell-by-cell basis and statistically representative values calculated. Of these, the shape index, which is a measure of the elongation of a cell and defined as $4\pi A/P^2$ (where A is cell area and P is cell perimeter), is the most interesting. Note that a decrease in shape index corresponds to cell elongation. These studies, although largely conducted with the rabbit as an animal model, have looked at various animals ranging from the rat to the dog. The data obtained provide striking evidence that the in vivo endothelium is intact, i.e., no regions of denudation were found, and that in general the shape index has a range from 0.2 to 0.6 for physiological conditions. Furthermore, studies of rabbit abdominal aortic endothelium indicate that, over a pressure range from 65 to 150 mmHg, there is a 42 percent increase in shape index; i.e., with increased pressure there is a significant rounding of the endothelial cells [53]. Whether this effect would also be shown for chronic hypertensive conditions and whether it is due to a change in cell shape or to an adjustment in cell junctional overlap are not known.

We have also examined endothelial cell patterns in the region of intercostal ostia, as determined from casts of rabbit aortas; these have been shown to be very suggestive of the complicated flow patterns that one might expect to be present, as will be discussed in the next section [54]. In a more recent experiment, aortic stenosis has been produced in the dog by banding the abdominal aorta. By use of the vascular casting technique, endothelial cell geometry and orientation have been determined as a function of the position relative to the stenosis (see Fig. 21.7). As may be seen, far proximal to the stenosis the shape index is normal and equal to that for control conditions. However, as one enters into the stenosis, where the cross-sectional area decreases, the cells become more elongated and achieve their greatest elongation, i.e., smallest shape-index value, right at the throat of the stenosis, where the shear stress would be expected to be the highest. Immediately downstream of the throat, one finds very round endothelial cells. This is presumably a region where there is a jet and the flow is separated from the walls. However, as the jet grows and ultimately reattaches to the wall, there is a gradual return of the shape index to control conditions. This then represents an experiment whereby changing the geometry of the aorta to produce a specific hemodynamic situation results in an endothelial cell pattern that is consistent with what one would expect from the fluid dynamics of the situation.

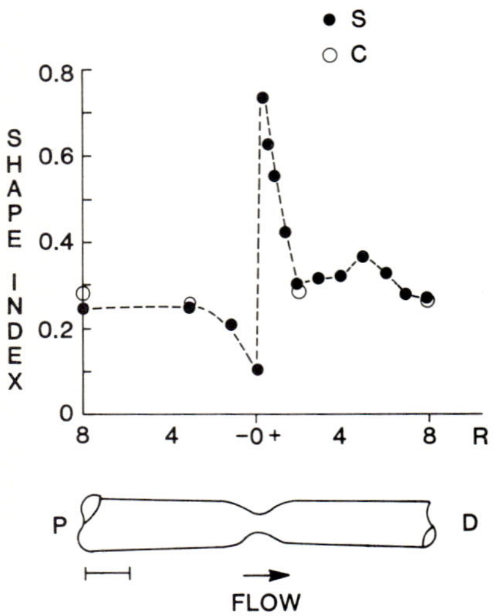

FIG. 21.7 Endothelial shape index as a function of position through a stenosed abdominal aorta of a dog (S); open circles correspond to unstenosed aortas of controls (C).

Recently, in vitro studies of the endothelial cellular dynamics have been initiated using cultured populations of vascular cells. Dewey et al. [55] have reported the use of a rotating cone-plate viscometer to study cultured bovine aortic endothelial cells under conditions of a uniform fluid shear stress. These studies suggest not only a shear stress effect on cell orientation but also an effect on endothelial cytoskeletal organization [56]. Stress fibers, i.e., contractile proteins such as myosin and actin, were more prominent and showed orientation with the direction of flow. The extracellular matrix beneath the cell monolayer also appeared to be influenced, with collagen on the coverslips having an oriented fibrillar pattern with flow as opposed to a basket-weave pattern for unstressed, static controls. It should be noted that in vivo the occurrence and distribution of stress fibers have been examined in mouse and rat endothelial cells in perfusion-fixed large arteries and veins [57]. For example, in the lower thoracic aorta the fibers were strongly aligned with the flow, while in the inferior vena cava there were very few stress fibers with no apparent preferred orientation. Since the latter might be expected to have lower shear stresses than the former, these results suggest that hemodynamic factors influence both the abundance and orientation of such elements.

Our own laboratory efforts also have been directed mostly at the culture of bovine endothelial cells. Various projects are now in progress. In one of these, confluent cultured endothelial cells are being exposed to known levels of shear stress, using a parallel-plate, channel-flow device. From the results obtained, there is a clear indication of the presence of a shear stress effect on orientation, but one which is substrate-dependent. For example, using a glass substrate with only a 1-h exposure to a shear stress of 8 dyn/cm^2, there is almost a total orientation of the cells with the direction of flow. Furthermore, the more closely a cell is oriented with the flow, the more elongated it is, as evidenced by a lower shape index. This is contrasted with endothelial cells on a plastic substrate (Thermanox coverslips) exposed to a shear stress of 85 dyn/cm^2, where there was virtually no tendency to orient the cells in the direction of the flow after 1 h, even though the shear was an order of magnitude higher. However, cells grown on the same plastic substrate also exposed to a shear stress of 85 dyn/cm^2, but for an extended period of 24 h, again showed a strong orientation. Here again there was a correlation such that the more oriented a cell was with the flow, the more elongated it was.

Others have also documented the effect of flow on the orientation of cultured endothelial cells. In addition, data are beginning to be accumulated on the influence of shear stress on cell function. For example, Davies et al. [58] have reported data on endocytosis gathered by using the soluble enzyme tracer molecule horseradish peroxidase (HRP). In a series of experiments [58, 59] rates of endocytosis have been measured for endothelial cells, with and without shear, and with changing shear. Where endothelial cells were exposed to two different levels of steady stress, 0 and 8 dyn/cm^2, for 48 h, the rates of endocytosis were equal, i.e., independent of the level of shear stress. However, in a companion experiment where the shear stress was alternated between 1 and 8 dyn/cm^2 every 15 min over a 2-h period, the rate of endocytosis was twice that for static, no-flow controls (1 dyn/cm^2 is not sufficient to produce endothelial orientation). To investigate the response of endothelial cells to an oscillating shear stress as occurs in vivo, the cone axis was tilted relative to the axis of rotation, thus producing a "wobble" in the plate. The shear oscillated at 1 Hz, with a peak-to-peak amplitude of 10 dyn/cm^2, and with a mean of 8 dyn/cm^2. Over a wide range of combinations of different shear stress levels and times of application, there was no observable increase of uptake from the level of static controls. This perhaps suggests that, at least for HRP, temporal effects are important, but not on a time scale characteristic of the normal heartbeat, i.e., the time associated with the oscillating shear stress in a large artery.

There also have been studies of prostacyclin (PGI_2) production by cultured endothelial cells, but with somewhat different results. In vitro PGI_2 production was measured under no-flow conditions, steady shear stress, and pulsatile shear stress [60]. Cultured human umbilical vein endothelial cells were placed in a parallel-plate flow chamber where pulsatility was introduced by periodically varying the circuit's resistance. At the onset of a steady shear stress of 16 dyn/cm^2, PGI_2 production peaked, but then leveled off after 1.5 h. However, upon the superimposition of a pulsatile shear stress at a frequency of 1 Hz and an amplitude of 10 percent of the mean value, PGI_2 production rose quickly to a level a factor of 3 higher than the previous peak and continued to increase over the next few hours. Similar results were obtained by Grabowski et al. [61], and these, taken together with the earlier noted results, suggest that, depending on the molecule and the particular cell process of interest, there may be a wide variety of time scales associated with the influence of a flow-imposed shear stress on cell function.

A different technique for studying the response of cultured endothelial cells to transient mechanical phenomena is now under way in our own laboratory. This involves exposing cells to an oscillatory pressure field. In a pilot study, cells grown on polystyrene were exposed to different levels of static or oscillatory pressures at frequencies of 1 and 2 Hz. In general, the endothelial cells responded to an increase in pressure by changing their geometry. As an example, endothelial cells exposed to a static pressure of 100 mmHg for 5 h showed significant change in surface area, perimeter, length, and shape index (i.e., 50, 42, 54, and −23 percent, respectively). This response is dependent on the pressure level, duration of exposure, and degree of confluence. The results showed no significant difference between cells exposed to an oscillatory pressure versus those exposed to a constant pressure. However, with an oscillatory pressure, the endothelial cells tended to lose their degree of confluence and were characterized by enhanced migration and increased mitotic activity. It is of interest that, in general, observations of cultured endothelial cells undergoing a change in mechanical environment, either through flow or pressure, demonstrate enhanced activity as measured by such variables as migration, mitosis, HRP endocytosis, and PGI_2 production.

These data suggest that fluid mechanical forces play an important role in determining cellular structure and function, both of which are important elements in maintaining vascular integrity. Furthermore, the morphology of an endothelial cell and its response to its mechanical environment may depend on the cell membrane's mechanical properties. Thus, experiments recently have been initiated in our laboratory to study endothelial cell membrane deformation.

21.4 HEMODYNAMICS, THE ENDOTHELIUM, AND LESION LOCALIZATION

It is appropriate at this point to examine the relationship of endothelial changes to the detailed local flow pattern as well as to the pattern of early lesions. Our knowledge is too limited—perhaps the interactions are too complicated—to conduct such an examination in a general way; however, it is possible to do so for the aorta and some of its major branch vessels.

With regard to the local hemodynamics, Fig. 21.8 illustrates what the flow pattern should be like in the vicinity of an aortic branch vessel, e.g., an intercostal or the point where the celiac

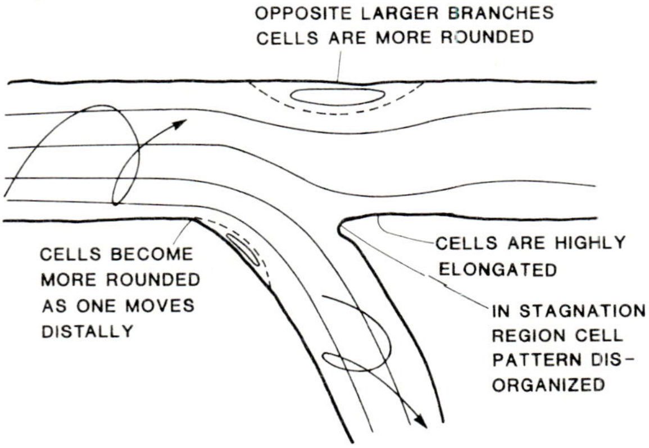

FIG. 21.8 Illustration of aortic flow in the vicinity of a branch vessel, with selected notations on endothelial cell characteristics.

artery branches off the aorta. The flow is unsteady and laminar, although possibly with turbulent bursts at peak systole in larger animals. The velocity profile proximal to the branch point may be approximately symmetric, although it is not a parabolic Poiseuille flow because of entrance effects. However, immediately distal to the branch point the velocity profile will be asymmetric as streamlines move toward the wall from which the branch vessel emanates (see Fig. 21.8). Farther downstream the flow will adjust and tend to return to an approximately symmetric velocity profile. There will be secondary motions induced by the streamline curvature associated with the branching process and in the thoracic aorta, due to the curvature of the aortic arch.

There is also the possibility of flow separation occurring. Although there are no in vivo observations to confirm this, the two most likely locations are illustrated in Fig. 21.8. One location is on the proximal side of a branching vessel as the flow turns to enter the branch, and the other location is on the wall opposite the flow divider. The likelihood of separation occurring depends on how sharp a turn the flow must negotiate for the first location and the amount of flow drawn away through the branch for the second location. Even if flow separation does not occur, both these regions would be characterized as generally low in shear, followed by a rapid increase in shear to produce a locally elevated wall shear stress.

If flow separation does occur, this local increase in shear is due to the reattachment of the flow on the wall. At the flow divider, there is a low-shear stagnation region on the leading edge. However, once one is away from the leading-edge region, there should be a rather high shear both on the aortic surface and on the flow-divider surface within the branch. This elevation in shear stress should persist for a distance on the order of a vessel diameter.

In the region of the iliac bifurcation, qualitatively the flow field would be similar to that just described. The leading edge of the flow divider would be a stagnation-point region, and immediately distal on the flow-divider surface a high shear stress would be expected. On the lateral walls, the shear will be low. However, there is a rapid increase in shear near the entrance to the daughter vessels which is not necessarily associated with the reattachment of a separated flow.

In the coronary arteries, the Reynolds numbers are much lower, the viscous effects are more dominant, and the flow is laminar. The velocity profile in many regions will be more like a parabolic Poiseuille flow, except that there will be skewing of this profile due to vessel curvature and branching. There also still may be significant entrance effects, resulting in a blunting of the velocity profile. With regard to the level of shear stress, the velocity profile in general would suggest a range of 10 to 50 dyn/cm^2 for mean stress levels [27] and a factor of 3 higher for peak stresses. In the left main coronary artery there will be a higher shear stress on the outer wall as it curves over the myocardium. At the LAD-LCFX (left circumflex coronary artery) bifurcation the flow divider will have a higher shear stress, and on the outer walls of this bifurcation, the shear stress may be lower.

A picture of the endothelium is also beginning to emerge. First, as already noted, it appears to be intact, with desquamation [62] occurring without denudation. The endothelial cells are of an elongated character and aligned with the direction of the blood flow. In regions of branching where the flow pattern may be quite complicated, as just discussed and as illustrated in Fig. 21.9, the cells will not necessarily be aligned with the axis of the blood vessel. Furthermore, cells will be more rounded in some regions than others. This appears to be related to the local flow conditions and may be of some importance, since the recent paper by Repin et al. [63] suggests that atherosclerosis in humans is more likely to occur in regions having more rounded endothelial cells.

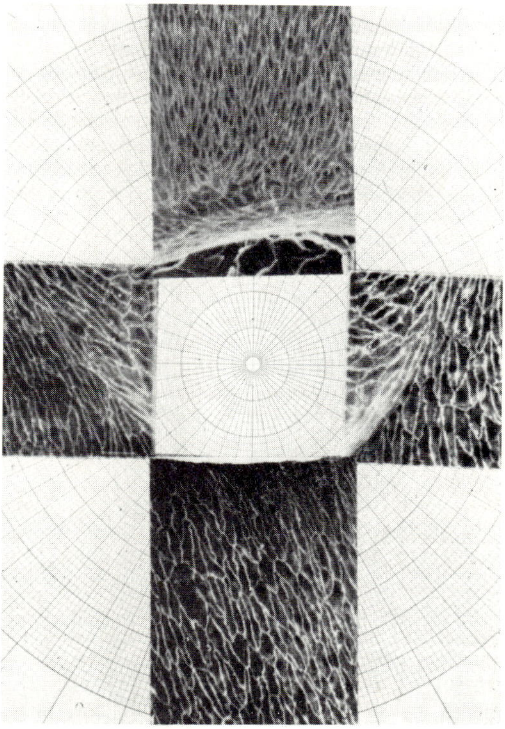

FIG. 21.9 En face montage of endothelial cell photographs in the vicinity of the intercostal branch of a rabbit; the endothelial cell pattern reflects the local streamline pattern of the flow in the vicinity of the wall.

To be more specific about the shape of aortic endothelial cells in a region of branching (such as illustrated in Fig. 21.9), comments on the endothelial cell pattern are overlaid on Fig. 21.8. As is noted, as one approaches the branch point from the proximal side, the cells on the surface from which the branch emanates will become increasingly more round, indicating the possibility of a decreasing wall shear stress. On the leading edge of the flow divider, the cells will also be of a generally round shape; however, here, in contrast to other locations, the pattern will be highly disorganized, demonstrating no particular preferred orientation, and presumably indicative of a stagnation region. On the aortic flow-divider surface immediately distal to the leading-edge region, there is a region a few cell lengths long in which the cells are highly elongated, particularly when compared with the shorter, more rounded cells proximal to the branch point. This suggests a comparatively higher shear stress on the distal side of the branch than on the proximal side. For the celiac artery, which is a large branch vessel, there also is a region of more rounded cells on the aortic surface opposite the flow divider; and in the region of the iliac bifurcation, endothelial cells appear to be more rounded on the outer, lateral walls and more elongated on the flow divider.

Collectively, these facts suggest that endothelial cells are more elongated in regions of high shear and more rounded in regions of low shear. Although this must be considered only a preliminary assessment because our knowledge of some of the details of the flow field is still incomplete, the initial results from in vitro studies of cultured endothelial cells exposed to a known flow condition support the existence of such a relationship between the shape of an endothelial cell and the level of wall shear stress to which it is exposed.

Although there is limited information on the pattern of early lesions for spontaneous disease, the existence of a general pattern indicating low-shear regions to be of higher predilection for atherogenesis has been confirmed in general in studies of the adult human as well as in studies using the white Carneau pigeon. It is also consistent with the suggestion of Repin et al. [63] that regions of more rounded cells have a much higher predilection for atherogenesis. More rounded cells do appear to be associated with regions of low shear, and thus one can hypothesize the following sequence of events as being intrinsic to the process of atherogenesis. In regions of low shear, which appear to have a higher predilection for the disease, the endothelium may be characterized as having comparatively rounder cells. These cells appear to have a shorter lifetime, i.e., a higher cell turnover rate, which may be associated with the influence of hemodynamic forces or other related events on cell function or dysfunction. During cell turnover, intercellular junctions become leaky, allowing for an enhanced influx of lipids and other macromolecules. Thus, regions of low shear and the concomitant rounder endothelial cells have a higher cell turnover rate, and as a result demonstrate a greater permeability. It is this enhanced permeability and the associated increased influx of lipid macromolecules and other plasma constituents which we believe to be an essential ingredient in the initiation of atherosclerosis.

However, in addition to the multitude of questions relating to any hemodynamic influence on endothelial cells, there are two other cautionary notes. First, as noted earlier, the pattern of lesions in diet-induced atherosclerosis in animals is, in general, not in agreement with the concept of low-shear regions having an enhanced predilection for atherogenesis. The same may be said for the pattern of early lesions in fetuses, neonates, and infants. Although this observation can possibly be explained on the basis that the lipid environment and thus the disease is not the same as in the adult human, further investigation is necessary. Second, although the evidence surveyed here provides an indictment for low-shear regions as having the highest predilection for early lesions, it must be reemphasized that our knowledge is still too incomplete to be certain. In fact, whether it is shear, rapid changes in shear as reflected in the shear pulse amplitude, or some other related hemodynamic event remains to be demonstrated.

Finally, hypertension is a risk factor in heart disease. Furthermore, it has been demonstrated, as noted earlier, that with increased pressure, endothelial cells become more round. This, taken together with the suggestion of Repin et al. [63], provides a possible explanation for why hypertension is a risk factor. The influence of hypertension may be multifactorial in that vascular geometry, the velocity field, and endothelial cell shape are all affected. In addition, it may be precisely because of this multifactorial influence that hypertension is so significant a risk factor.

CONCLUSION

Let us now return to the two questions posed in the introduction: (1) Why is atherosclerosis an arterial disease which is not seen either on the venous side of the system or in the pulmonary arteries, and (2) why is atherosclerosis a focal disease with the striking feature of lesions being localized in regions of sharp curvature and branching?

Before attempting to answer these questions, let us briefly summarize what we believe is known. First, the disease occurs in large arteries with at least moderately thick walls and not in thin-walled vessels such as veins or the pulmonary arteries. Second, an important event in the development of a lesion is the accumulation of lipids and/or other plasma constituents within the intima. Third, on the basis of our present knowledge of arterial fluid dynamics, lesions occur in regions which we would categorize as being of low shear. Fourth, the disease is initiated in the presence of an intact endothelium. Fifth, endothelial cells are more rounded in regions of low shear as well as apparently in regions more prone to the development of lesions. Sixth, although the endothelium offers major resistance to the passage of macromolecules from blood into the arterial wall, once these macromolecules are inside, their transport within is influenced by the wall properties, particularly the relative resistance of the media versus the intima.

With these facts in mind, let us return to the first question. Why is atherosclerosis an arterial disease which occurs in relatively thick walled arteries and not in veins or on the pulmonary side (except during pulmonary hypertension, when the wall of the pulmonary artery does thicken and atherosclerotic lesions have been observed)? All these vessels—large arteries and veins in both the systemic and pulmonary circulations—are exposed to blood with a comparable lipid composition and to a comparable fluid mechanical environment. The Reynolds number, the Womersley parameter, the vessel geometry, and the wall shear stresses are all approximately the same. Only the pressure, and the thickened wall with the properties and the effects on flow which go with it, are different. It thus appears to be a disease of thick-walled blood vessels, and, at least from the viewpoint of fluid mechanics, this suggests the possible importance of transport phenomena within the wall. It is hypothesized that intimal lipid accumulation is an important event and that this occurs due to an imbalance between influx and efflux. If the former is primarily controlled by transport across the endothelium and the latter by transport across the wall, with removal either through the vasa vasorum or the lymphatics, then it can easily be seen that an imbalance and subsequent accumulation could occur. In this sense, a necessary, although not sufficient, condition for accumulation is an enhanced permeability of the endothelium; i.e., the higher the permeability—whether it be by increased cell turnover, enlarged intercellular gaps, or some other mechanism—the greater the passage of macromolecules such as lipids into the wall. However, whether or not accumulation actually occurs depends on the resistance of the underlying arterial wall, particularly the media, to transport, i.e., on the movement of such molecules across to the advential side where they will be removed. If it is the relatively thick walled vessels—the large and medium-sized arteries or the pulmonary artery in the case of chronic pulmonary hypertension—which have an increased wall resistance R_r, then this is where atherosclerotic lesions should be found, explaining why atherosclerosis is an arterial disease. Of course, this hypothesis ignores the details of lipid biochemistry and the number of voids in our knowledge of wall transport; however, it does represent a possible answer to the first question and points the way to additional studies which should be carried out.

The second question posed is why there is a focal pattern to the disease, with lesions being localized in regions of sharp curvature and branching. We believe that it is not just such general regions, but more specifically those parts of the wall exposed to low shear, which are in jeopardy. The answer to this question, as suggested in the previous paragraph, seems to reside with the endothelium and its resistance to the passage across it of macromolecules such as lipids, i.e., transendothelial transport.

Not only does a region characterized as being of low shear have a higher predilection for the disease, but also its endothelium has comparatively rounder cells. These cells appear to have larger gaps around their endothelial cell borders which may make possible and even enhance lipid transport. These cells also appear to have a shorter lifetime, i.e., a higher cell turnover rate, which also may be associated with the influence of hemodynamic forces or other related events on cell function or dysfunction. Thus, regions of low shear and the concomitant rounder endothelial cells, because of either a larger intercellular gap or a higher cell turnover rate, demonstrate a greater permeability. It is this enhanced permeability and the associated increased influx of macromolecules such as lipids which we believe to be an essential ingredient in the initiation of atherosclerosis. However, as noted earlier, accumulation occurs only when the underlying arterial wall resistance is comparatively high, as in thick-walled arteries.

It should be noted that, though these regions are characterized as being of low shear, we do not know whether this is shear stress, changes in shear possibly reflected in the shear pulse amplitude, or some other related hemodynamic event. Studies of cultured endothelial cells, as noted earlier, suggest that cells undergoing a change in mechanical environment demonstrate enhanced activity, as measured by such variables as migration, mitosis, HRP endocytosis, and PGI_2 production. If this is the case, then perhaps regions characterized as low shear should be looked at as regions of greatest alternating shear stress. This would be true because, for a given flow pulsatility, there would be a greater possibility of flow reversal—and thus the presence of an oscillating or alternating shear stress—in a region of low mean shear stress than in a region of a high mean shear stress. However, we do not know either the mechanisms by which cell function is influenced by such a mechanical event or the time constants for the various cellular processes which may be of interest. We also do not know the answer to such a basic question as why cell turnover is different in regions of low shear versus those of high shear, i.e., what determines the lifetime of a cell.

Finally, much of the evidence for a hemodynamic involvement presented here can be construed to be in fact related to geometry. As illustrated in Fig. 21.10, this thus suggests that

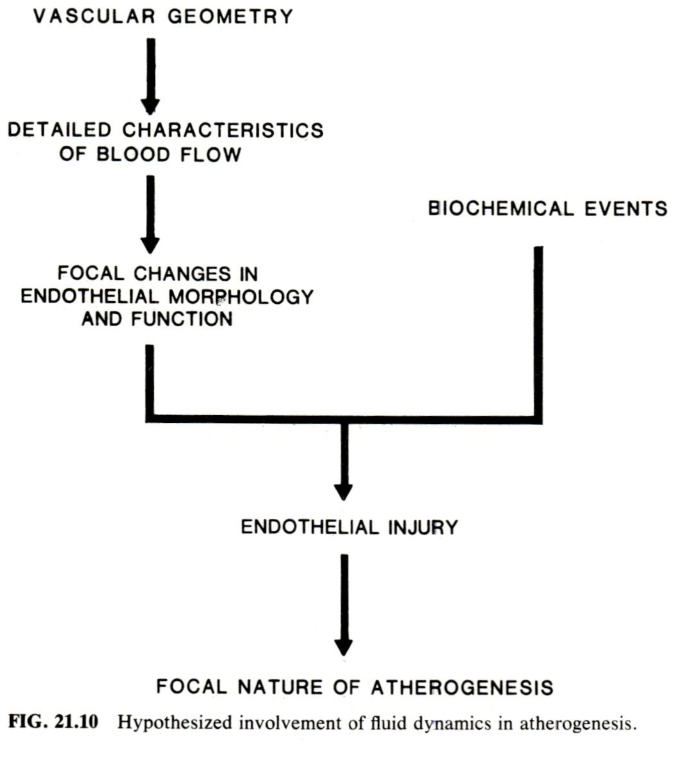

FIG. 21.10 Hypothesized involvement of fluid dynamics in atherogenesis.

vascular geometry, and the way it influences the local detailed flow properties, may be the primary determinant of a hemodynamic effect on the initiation of atherosclerosis. Under such a hypothesis, whatever affects vascular geometry would alter the local, detailed flow characteristics and correspondingly influence endothelial morphology and function. For an individual subject, that geometry might be inherited, i.e., a family history or genetic effect, and as such one's geometry may in itself be a risk factor. Alternatively, a subject's vascular geometry may have been significantly altered by a risk factor such as hypertension. However, whatever the effect, the primary determinant of the pattern of atherogenesis would be vascular geometry, and whether or not an individual has an abnormal predilection for the disease would depend not only on biochemically related factors but also on the exact nature of the geometry, e.g., branching pattern and degree of curvature and/or tortuosity, and the resulting detailed hemodynamic characteristics.

Within this view of the disease process it is the influence of hemodynamically-related events on the endothelium which is believed to be a factor in atherogenesis and thus in lesion localization. The endothelium represents a selective barrier to movement of materials between blood and the arterial wall. It is selective in the sense that it impedes the movement of certain substances more than others, and in controlling the movement of materials, the endothelium at least in part is influencing the biochemical processes deeper within the wall. The properties of this barrier may be altered by both mechanical and chemical influences, i.e., hemodynamics and blood chemical constituents. Of these it is the hemodynamic or mechanical events that have a primary focal or localizing effect. The detailed flow properties and thus the exact nature of a hemodynamic effect are intimately related to local vascular geometry; i.e., an individual's vascular geometry is the primary determinant of any hemodynamic effect, with the endothelium being the mediator of that effect. In this context what is of interest is the influence of hemodynamically-related events on endothelial morphology, endothelial cell function, cell turnover rate, and transendothelial transport. Furthermore, any such effect is quite possibly of an extremely subtle nature; i.e., a relatively small imbalance over a prolonged time period could produce a major change. Such an imbalance might initially occur within the blood, e.g., its lipid profile or various hormone concentrations. In the transfer of this imbalance to the arterial wall, hemodynamics serves only as a possible high-gain filter, and the effect will be amplified more in some regions than in others because of differences in local hemodynamics. In this the exact nature of any hemodynamic effect would appear to be highly dependent on blood chemistry.

This hypothesized subtle influence of hemodynamics on the endothelium is not meant to totally rule out a more traumatic effect such as endothelial cell denudation; however, such gross injury would appear to be nonphysiological and is not likely to occur under normal flow conditions. It is also not meant to rule out a role for thrombosis in atherosclerosis. In fact, it is believed that thrombosis is a very important factor in the later stages of the disease.

A few years ago the prevailing view was that if one could produce a statistically significant picture of the pattern of the disease and a correspondingly statistically significant picture of the flow pattern, then one could draw conclusions about the relationship of hemodynamics to atherogenesis. However, it now appears that neither the normal pattern of disease (in the sense of mean data representing the statistically normal situation) nor the normal flow pattern is of primary interest. Rather, the deviations from the normal are the important interests, and it is in this context that vascular geometry may be a risk factor. From this point of view, vascular geometry, whether inherited or as altered by the influence of a factor such as elevated blood pressure, would determine the local, detailed hemodynamics and the resulting influence on endothelial morphology and function. This, together with other, biochemically related factors, would determine the focal nature of the disease.

It is thus geometry—and through geometry, hemodynamics—which we believe to be an important factor in lesion localization. If we can understand the exact influence of vascular geometry on the local, detailed characteristics of arterial flows, and the specific role of hemodynamics in the biologic processes which lead to the genesis of atherosclerosis, then as vascular imaging techniques become more advanced, we may be able to tell individuals at an early age whether or not their vascular geometry puts them at an above-normal risk for the development of atherosclerosis.

ACKNOWLEDGMENTS

The authors thank S. A. Altobelli, C. G. Caro, C. J. Schwartz, and W. A. Seed for their contributions to the work and ideas reflected in this paper. The authors also are appreciative of the support provided by the National Science Foundation through grants CME-80-01701 and MEA-8200344 and the National Institutes of Health through grant HL-26890.

REFERENCES

1. Mitchell, J. R. A., and Schwartz, C. J.: *Arterial Disease,* Blackwell, Oxford, 1965.
2. Montenegro, M. R., and Eggen, D. A.: Topography of atherosclerosis in the coronary arteries, *Lab. Invest., 18,* 126, 1968.
3. Caro, C. G., Fitz-Gerald, J. M., and Schroter, R. C.: Atheroma and arterial wall shear: Observation, correlation and proposal of a shear dependent mass transfer mechanism for atherogenesis, *Proc. R. Soc. London (Biol.), 177,* 109, 1971.
4. Spain, D. M.: Atherosclerosis, *Sci. Am., 215*(2), 48–56, 1966.
5. Texon, M.: The role of vascular dynamics in the development of atherosclerosis, in M. Sandler and G. H. Bourne (eds.), *Atherosclerosis and Its Origin,* Academic Press, New York, 1963.
6. Mustard, J. F., and Packham, M. A.: Thrombosis and the development of atherosclerosis, in R. W. Wissler and J. C. Geer (eds.), *The Pathogenesis of Atherosclerosis,* Williams & Wilkins, Baltimore, 1972.
7. Kjaernes, M., Svindland, A., Walløe, L., and Wille, S. Ø.: Localization of early atherosclerotic lesions in an arterial bifurcation in humans, *Acta Pathol. Microbiol. Scand. Sect. A, 89,* 35–40, 1981.
8. Svindland, A. D., and Walløe, L.: Localization of early atherosclerotic lesions in carotid and coronary bifurcations: A refutation of the high shear stress hypothesis, in G. Schettler, et al. (eds.), *Proceedings of a Symposium on Fluid Dynamics as a Localizing Factor in Atherosclerosis,* Springer-Verlag, Heidelberg, in press.
9. Fox, B., and Seed, W. A.: Localization of early atheroma in human coronary arteries, *ASME J. Biomech. Eng., 103*(3), 208–212, 1981.
10. Sinzinger, H., Silberbauer, K., and Auerswald, W.: Quantitative investigation of sudanophilic lesions around the ostia of human fetuses, newborn, and children, *Blood Vessels, 17,* 44, 1980.
11. Cornhill, J. F., and Roach, M. R.: A quantitative study of the localization of atherosclerotic lesions in the rabbit aorta, *Atherosclerosis, 23,* 489, 1976.
12. Cornhill, J. F., Levesque, M. J., and Nerem, R. M.: Quantitative study of the localization of sudanophilic and coeliac lesions in the white Carneau pigeon, *Atherosclerosis, 35,* 103, 1980.
13. Friedmann, M. H., Hutchins, G. M., Bargeron, C. B., Deters, O. J., and Mark, F. F.: Correlation of human arterial morphology with hemodynamic measurements in arterial casts, *ASME J. Biomech. Eng., 103,* 204, 1981.
14. Caro, C. G.: Mechanical factors in atherogenesis, in N. H. C. Hwang and N. A. Norman (eds.), *Cardiovascular Flow Dynamics and Measurements,* University Park Press, Baltimore, 1975, p. 473.
15. Nerem, R. M.: Arterial fluid dynamics and interactions with vessel walls, in C. J. Schwartz and N. Werthesson (eds.), *Structure and Function of the Circulation,* Plenum Press, New York, 1981, vol. 2, p. 719.
16. Fry, D. L.: Hemodynamic forces in atherogenesis, in P. Scheinberg (ed.), *Cerebrovascular Diseases,* Raven Press, New York, 1976, p. 77.
17. Velican, D., and Velican, C.: Accelerated atherosclerosis in subjects with some minor deviations from the common type of distribution of human coronary arteries, *Atherosclerosis, 40,* 309, 1981.
18. Somer, J. B., Evans, G., and Schwartz, C. J.: Influence of experimental aortic coarctation on the pattern of aortic Evans blue uptake in vivo, *Atherosclerosis, 16,* 127, 1972.
19. Baker, D. W.: Pulsed ultrasonic Doppler blood-flow sensing, *IEEE Trans. Sonics Ultrasonics, Su-17,* 170, 1970.
20. Schultz, D. L., Tunstall-Pedoe, D. S., Lee, G. J., Gunning, A. J., and Bellhouse, B. J.: Velocity distribution and transition in the arterial system, in G. E. W. Wolstenholme and J. Knight (eds.), *Circulatory and Respiratory Mass Transport,* Churchill, London, 1969, p. 172.

21. Peronneau, P., Hinglais, J., Pellet, M., and Leger, F.: Velocimètre sanguin par éffet Doppler à émission ultrasonore pulse, *L'Onde Electrique, 50,* 369, 1970.

22. Seed, W. A., and Wood, N. B.: Velocity patterns in the aorta, *Cardiovasc. Res., 5,* 137, 1971.

23. Nerem, R. M., Seed, W. A., and Wood, N. B.: An experimental study of the velocity distribution and transition to turbulence in the aorta, *J. Fluid Mech., 52,* 137, 1972.

24. Nerem, R. M., Rumberger, J. A., Gross, D. R., Muir, W. W., and Geiger, G. L.: Hot-film coronary artery velocity measurements in horses, *Cardiovasc. Res., 10,* 301, 1976.

25. Wells, M. K., Winter, D. C., Nelson, A. W., and McCarthy, T. C.: Blood velocity patterns in coronary arteries, *ASME J. Biomech. Eng., 99,* 26, 1977.

26. Kajiya, F., Hoki, N., Tomonago, G., and Nishihara, H.: A laser Doppler velocimeter using optical fiber and its application to local velocity measurements in the coronary artery, *Experentia, 37,* 1171–1173, 1981.

27. Nerem, R. M., and Seed, W. A.: Coronary artery geometry and its fluid mechanical implications, in G. Schettler et al. (eds.), *Proceedings of a Symposium on Fluid Dynamics as a Localizing Factor in Atherosclerosis,* Springer-Verlag, Berlin, 1983, pp. 51–59.

28. Pedley, T. J.: *The Fluid Mechanics of Large Blood Vessels,* Cambridge University Press, Cambridge, 1980.

29. Altobelli, S. A., and Nerem, R. M.: Experimental studies of coronary artery fluid mechanics, *Symposium on the Mechanics of the Coronary Circulation,* ASME AMD/BED/FED Conference, Houston, Tex., June 20–22, 1983.

30. Rabinovitz, R., Levesque, M. J., and Nerem, R. M.: Coronary flow redistribution due to aortic pressure changes, *Proceedings of the 36th Annual Conference on Engineering in Medicine and Biology,* Columbus, Ohio, Sept. 12–14, 1983.

31. Walburn, F. J., and Stein, P. D.: Shear rate at the wall in a symmetrically branched tube simulating the aortic bifurcation, *Biorheology, 19,* 307, 1982.

32. Fry, D. L.: Certain histological and chemical responses of the vascular interface to acutely induced mechanical stress in the aorta of the dog, *Circ. Res., 24,* 83, 1969.

33. Caro, C. G., and Nerem, R. M.: Transport of ^{14}C-4-cholesterol between serum and wall in perfused dog common carotid artery, *Circ. Res., 32,* 197, 1973.

34. Nerem, R. M., Mosberg, A. T., and Shwerin, W. D.: Transendothelial transport of ^{131}I-albumin, *Biorheology, 12,* 31, 1975.

35. Chien, S., Lee, M. M. L., Laufer, L. S., Handley, D. A., Weinbaum, S., Caro, C. G., and Usami, S.: Effects of oscillatory mechanical disturbance on macromolecular uptake by arterial wall, *Arteriosclerosis, 1,* 326, 1981.

36. Thibault, L. E., and Fry, D. L.: Hydrodynamically induced wall shear stress effects of Evans blue dye uptake, in R. M. Nerem and J. R. Guyton (eds.), *Hemodynamics and the Arterial Wall,* University of Houston Press, Houston, Tex., 1980, p. 34.

37. Vaishnav, R. N., Patel, D. J., Atabek, H. B., Deshpande, M. D., Plowman, F., and Vossoughi, J.: Determination of the local erosion stress of the canine endothelium using a jet impingement method, *ASME J. Biomech. Eng., 105*(1), 77–83, 1983.

38. Languille, B. L.: Integrity of endothelium following exposure to high shear stress, in G. Schettler et al. (eds.), *6th International Symposium on Atherosclerosis Abstracts,* FRG Press, Berlin, 1982, p. 616.

39. Simionescu, M., Simionescu, N., and Palade, G. E.: Morphometric data on the endothelium of blood capillaries, *J. Cell Biol., 60,* 128–152, 1974.

40. Weinbaum, S.: The endothelial interface between tissue and blood, *Proceedings of the New York Academy of Sciences International Conference on Surface Phenomena in Hemorheology: Their Theoretical, Experimental and Clinical Aspects,* held Oct. 26–29, 1982, New York Academy of Sciences, New York, 1983, pp. 92–114.

41. Caplan, B. A., and Schwartz, C. J.: Increased endothelial cell turnover in areas of in vivo Evans blue uptake in the pig aorta, *Atherosclerosis, 17,* 401–417, 1973.

42. Kenyon, D. E.: The theory of an incompressible solid-flux mixture, *Arch. Rat. Mech. Anal., 62,* 131, 1976.

43. Wijesinghe, A. M.: *Deformation and Fluid Flow in Porous Viscoelastic Solids,* Ph.D. thesis, Massachusetts Institute of Technology, Cambridge, Mass., 1978.

44. Blackshear, P. L., Jr., Vargas, F. F., Vargas, C. B., Pribyl, J. G., and Blackshear, G. L.: Hydraulic

conductivity of intact endothelium, permeability of media, and consolidation properties in the rabbit aorta, in R. M. Nerem and J. F. Cornhill (eds.), *The Role of Fluid Mechanics in Atherogenesis, Proceedings from a Specialists Meeting,* Ohio State University, Columbus, Ohio, *17,* 1–17, 6, 1978.

45. Caro, C. G., Lever, M. J., Laver-Rudick, Z., Meyer, F., Kiron, N., Ebel, W., Parker, K. H., and Winlove, C. P.: Net albumin transport across the wall of the rabbit common carotid artery perfused in situ, *Atherosclerosis, 37,* 497, 1980.

46. Smith, E. B., and Staples, E. M.: Intimal and medial plasma protein concentrations and endothelial function, *Atherosclerosis, 37,* 497, 1980.

47. Bratzler, R. L., Chisholm, G. M., Colton, C. K., Smith, K. A., Zilversmit, D. B., and Lees, R. S.: The distribution of labelled albumin across the rabbit thoracic aorta in vivo, *Circ. Res., 40,* 182, 1977.

48. Caro, C. G., and Lever, M. J.: Effect of vasoactive agents and applied stress on the albumin space of excised rabbit carotid arteries, *Atherosclerosis, 46,* 137, 1983.

49. Lai, W. M., and Mow, V. C.: Drag-induced compression of articular cartilage during a permeation experiment, *Biorheology, 17,* 111, 1980.

50. Crawford, D. W., Back, L. H., and Cole, M. A.: In vivo measurements of the boundary layer oxygen gradient in the normal dog femoral artery, in R. M. Nerem and J. R. Guyton (eds.), *Hemodynamics and the Arterial Wall,* University of Houston Press, Houston, Tex., 1980, p. 67.

51. Forstrom, R. J., Voss, G. O., and Blackshear, P. L., Jr.: Fluid dynamics of particle (platelet) deposition for filtering walls: Relationship to atherosclerosis, *ASME Fluids Eng., 96,* 168, 1974.

52. Levesque, M. J., Cornhill, J. F., and Nerem, R. M.: Vascular casting: A new method for the study of arterial endothelium, *Atherosclerosis, 34,* 457, 1979.

53. Cornhill, J. F., Levesque, M. J., and Nerem, R. M.: The effect of arterial pressure on endothelial cell morphology, *2d Austrian Atherosclerosis Conference,* Vienna, Apr. 7–12, 1980.

54. Nerem, R. M., Levesque, J. M., and Cornhill, J. F.: Vascular endothelial morphology as an indicator of blood flow, *ASME J. Biomech. Eng. 103,* 172, 1981.

55. Dewey, C. F., Bussolari, S. R., Gimbrone, M. A., and Davies, P. F.: The dynamic response of vascular endothelial cells to fluid shear stress, *ASME J. Biomech. Eng., 103,* 177, 1981.

56. White, C. E., Fujiwara, K., Shefton, E. J., Dewey, C. F., Jr., and Gimbrone, M. A., Jr.: Fluid shear influences on cell shape and cytoskeletal organization in cultured vascular endothelium, *Fed. Proc., 41,* 321, 1982.

57. White, C. E., Gimbrone, M. A., Jr., and Fujiwara, K.: Factors influencing the expression of stress fibers in vascular endothelial cells in situ, *J. Cell Biol., 97,* 416–424, 1982.

58. Davies, P. F., Dewey, C. F., Jr., Bussolari, S. R., Gordon, E. J., and Gimbrone, M. A., Jr.: Influence of hemodynamic forces on vascular endothelial function: In vitro studies of shear stress and pinocytosis in bovine aortic endothelial cells, *J. Clin. Invest., 73,* 1121–1129, 1984.

59. Dewey, C. F., Jr.: Effects of fluid flow on living vascular cells, *ASME J. Biomech. Eng., 106,* 31, 1984.

60. Frangos, J. A., Ives, C. L., Eskin, S. G. and McIntire, L. V.: Effect of pulsatile flow on PGI_2 production by endothelial cells, *Abstracts of the 56th Scientific Sessions,* American Heart Association, III-73, 1983.

61. Grabowski, E. F., Jaffe, E. A., and Weksler, B. B.: Prostacyclin production by cultured endothelial cell monolayers exposed to step increases in shear stress, submitted for publication.

62. Schwartz, S. M., Gajdusek, C. M., and Selden, S. C.: Vascular wall growth control: The role of the endothelium, *Arteriosclerosis, 1,* 107, 1981.

63. Repin, V. S., Dolgov, V. V., Zaikina, O. E., Novikov, J. D., Antonov, A. S., Nikolaeva, M. A., and Smirnov, V. N.: Heterogeneity of endothelium in human aorta, *Atherosclerosis, 50,* 35, 1984.

CHAPTER 22

Arterial Wall Oxygen Transport and Its Relationship to Atherogenesis

Thomas K. Goldstick
Departments of Chemical Engineering,
Biomedical Engineering, and Neurobiology/Physiology
Northwestern University
Evanston, Illinois

Philip B. Dobrin
Department of Surgery
Hines Veterans Administration Hospital and Loyola University Medical Center
Hines, Illinois

22.1 BLOOD FLOW TO THE ARTERIAL WALL

The arterial wall is nourished by diffusion from the lumen and from the vasa vasorum. Most vessels greater than 0.5 mm in wall thickness possess vasa vasorum, but these are located only in the adventitia and outer media (Geiringer, 1951; Wolinsky and Glagov, 1967). India-ink studies of the pig aorta (Jaeger, 1964) show that some vasa may derive directly from the lumen of the vessel, but that most originate from other vessels, such as intercostal arteries, to enter from the adventitia. Schoenenberger and Muller (1960) termed those arising from the lumen *vasa interna,* and those arising from an external vessel *vasa externa.* According to Jaeger, in the pig aorta, about one-third of wall blood flow is carried by the vasa interna and about two-thirds is carried by the vasa externa. Other investigators have not described the vasa interna, and raise the possibility that these elements may be artifacts, perhaps representing focal leaks in excised vessels.

Most of the blood supply to the aortic wall derives from intercostal arteries. Heistad et al. (1979) found that, in dogs, sympathetic nerve stimulation decreased flow through the vasa. Wilens et al. (1965) and Heistad et al. (1981) found that ligation of the intercostal arteries caused a significant decrease in aortic wall flow with the development of medial necrosis. Thus, the wall nutrition provided by the vasa vasorum appears to be critical to wall viability.

Using microsphere methods to quantitate flow, Heistad et al. (1978) reported that, under control conditions, flow to the outer two-thirds of the dog thoracic aorta in vivo was about 10 mL/100 g · min, while flow to the inner third of the same vessel was only about 2 mL/100 g · min.

Total flow to the abdominal aorta was markedly less. Flows to the inner and outer regions of the media of the abdominal aorta were both about 1 mL/100 g · min. Infusion of adenosine increased flow through the outer media of the thoracic aorta to 18 mL/100 g · min, while hemorrhagic hypotension decreased it to 2 mL/100 g · min. Acute hypertension did not increase flow because there was an associated increase in hydraulic resistance.

Detailed discussions of the role of hemodynamic and mechanical factors that may limit wall nutrition are reviewed elsewhere (Heistad and Marcus, 1979; Dobrin, 1983). Briefly, Schoenenberger and Muller (1960) suggested that wall deformation and wall compressive stress may limit flow through the vasa vasorum. Simon et al. (1973) investigated this possibility by simulating wall behavior with a finite deformation model. Their analysis indicated that elevated transmural pressure should increase the pressure perfusing the vasa vasorum, but also should exert increased compressive stress against these intramural vessels. This study predicted increased flow through the wall arterioles but decreased flow through the series-coupled venules because of wall compression, thereby limiting flow. Simon et al. (1973), Vaishnav et al. (1973), and

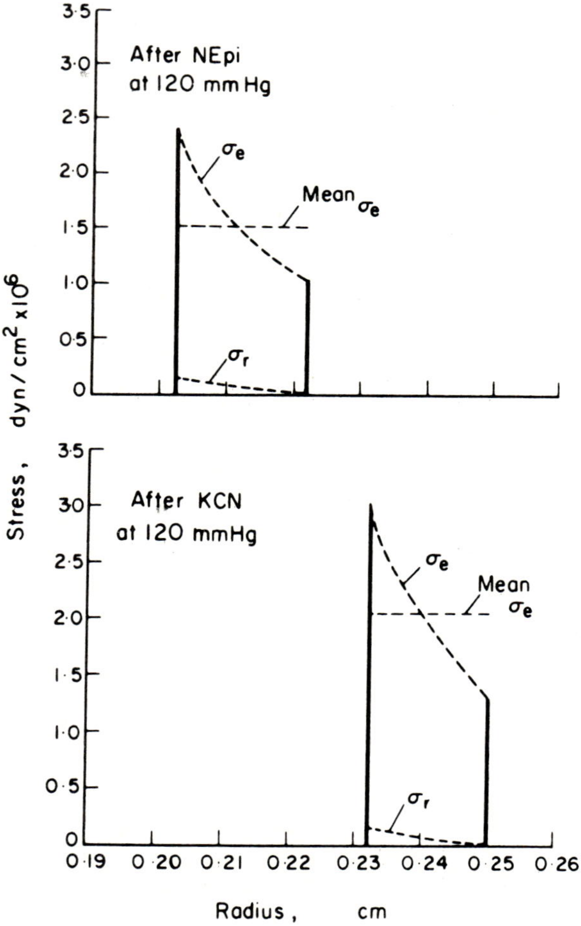

FIG. 22.1 Stress distribution in the walls of arteries. σ_e is circumferential tensile stress; σ_r is radial compressive stress. Both stresses are high at the intima and decline toward the adventitia. (*From Doyle and Dobrin, 1973.*)

Doyle and Dobrin (1973) all found that both circumferential tensile stress and radial compressive stresses are distributed curvilinearly across vessel walls. For both stresses, the largest values are found at the lumen. The tensile circumferential stress declines across the wall to give somewhat lower values at the adventitia, while the compressive stress declines to zero at the adventitia (Fig. 22.1). Strains also are distributed, with greatest values found at the lumen and lesser values found at the adventitia (Doyle and Dobrin, 1973). Therefore, tissue distortion and compression should be most pronounced at the intima and inner regions of the media. However, the vasa usually are located only in the middle and outer regions of the media, so that vasa-dependent wall nutrition should be only moderately impeded. Finally, the vasa vasorum participate in generalized vasoconstriction brought about by increased sympathetic nerve activity (Heistad et al., 1979). This also may limit flow though the vasa. Some forms of chronic hypertension are mediated by sympathetic nerve activity (Buñag and Takeda, 1979), suggesting that these forms of hypertension may be especially associated with decreased wall nutrition.

22.2 THEORY OF ARTERIAL WALL OXYGEN TRANSPORT

The delivery of oxygen to the arterial wall depends on a coupled system. On the luminal side, oxygen comes from the luminal blood; on the adventitial side, it comes from the vasa vasorum. But the vasa vasorum extend only to the outer regions of the media, as illustrated schematically in Fig. 22.2. The most significant feature of this nutrient transport system is the relatively thick avascular annulus adjacent to the lumen. This region is about 500 μm in the thoracic aorta of a human and is supplied with oxygen only by radial diffusion. Standard transport equations may be applied to yield a computer simulation of oxygen transport to, and through, the avascular annulus (Schneiderman and Goldstick, 1978a, 1978c; Schneiderman et al., 1979, 1982). Although this simulation was for one geometrically simple vessel, the human thoracic aorta, it included virtually all the important physiological features, as well as several factors thought to be important in the genesis of arterial disease. Factors examined were blood flow pulsatility, blood shear rate, luminal narrowing (stenosis), and carboxyhemoglobin levels.

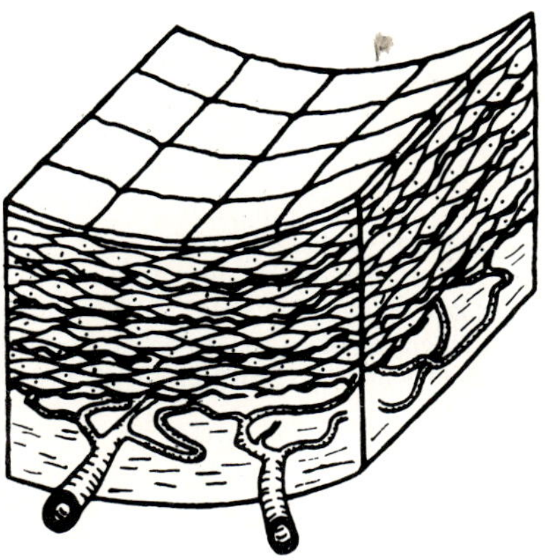

FIG. 22.2 Diagram of vessel wall showing intima, media, and adventitia. Vasa vasorum penetrate only to outermost region of media.

Oxygen transport through the thoracic aorta was modeled as five axially symmetric regions (Fig. 22.3). From the center of the vessel outward these are:

1. Blood. This fluid is almost universally modeled as a homogeneous Newtonian fluid having a uniform hemoglobin concentration in local equilibrium with oxygen.
2. Cell-free plasma. The very thin layer (10 μm) next to the intima where oxygen is in simple solution.
3. Intima. The avascular tissue region next to the lumen (approximately 100 μm thick).
4. Avascular media. The next outermost avascular tissue region (approximately 500 μm thick).
5. Vascular media and adventitia. The outermost vascular portion where the tissue oxygen tension is assumed to be maintained constant.

Although the effects of physiological pulsatile flow on wall nutrition have not been investigated, extensive studies of a simplified sinusoidal model similar to that described by Womersley (Schneiderman et al., 1982) suggest that pulsatility has a minimum effect on oxygen transport in

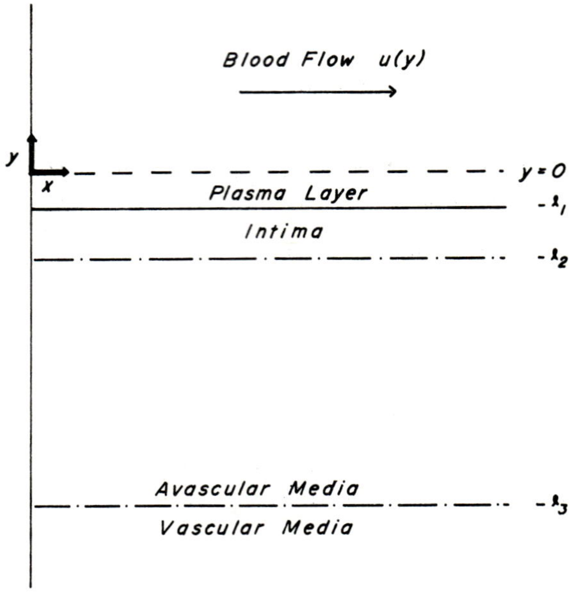

FIG. 22.3 Mathematical model of human thoracic aorta in steady flow. All layers are to scale except the plasma layer. The blood is considered to be completely mixed at $x = 0$.

the *near-wall* region. Accordingly, only steady flow was considered. For straight vessels of constant cross section, the model employed (Schneiderman and Goldstick, 1978a, 1978c) assumes mass transport to be restricted to a boundary layer, and the radius of curvature to be so large compared to the thickness that the system resembles flow over a plate (Fig. 22.2). If blood is Newtonian and flow occurs with a steady Poiseuille velocity profile, the governing partial differential equation in the oxygen boundary layer (Schneiderman and Goldstick, 1978c) is

$$u(y)\frac{\partial P}{\partial x} = \frac{D_B}{\Phi(P)}\frac{\partial^2 P}{\partial y^2} \tag{22.1}$$

where $\Phi(P) = 1 + (dC/dP)/k_B$ (dimensionless), P is the oxygen partial pressure or oxygen fugacity (torr), and the remainder of the symbols are defined in the list of symbols at the end of the chapter. In the stagnant plasma layer the governing equation is

$$\frac{d^2P}{dy^2} = 0 \tag{22.2}$$

And, finally, in both the tissue layers the equation (Goldstick, 1973) is

$$\frac{d^2P}{dy^2} = \frac{Q_T}{D_T k_T} \tag{22.3}$$

At the beginning of the thoracic aorta it was assumed that the blood is fully mixed, i.e., that

$$P(x, y) = P_a \qquad x \le 0,\ y \ge 0 \tag{22.4}$$

It also was assumed that, over the length of the thoracic aorta, the oxygen transport boundary layer is relatively thin compared with the vessel radius, i.e., that

$$P(x, y) = P_a \qquad x \geq 0,\ y \to \infty \tag{22.5}$$

At the interface between any two layers, the Po_2 and the flux must be continuous, giving the following boundary conditions:

$$P(x, l_j^+) = P(x, l_j^-) \qquad x \geq 0 \tag{22.6}$$

$$\left(Dk \frac{\partial P}{\partial y}\right)_{l_j^+} = \left(Dk \frac{\partial P}{\partial y}\right)_{l_j^-} \qquad x \geq 0 \tag{22.7}$$

Finally, with virtually no experimental information available regarding the vasa vasorum except that they are absent from the lumen to about 500 μm from the lumen, it was assumed that they maintain a constant boundary condition at the avascular-vascular tissue junction such that

$$P(x, -l_3) = P_o \qquad x \geq 0 \tag{22.8}$$

Simultaneous numerical solutions to Eqs. (22.1) to (22.3) subject to the boundary conditions of Eqs. (22.4) to (22.8) were obtained for a number of different cases. Equation (22.1) was solved numerically by using an analogue to the Crank-Nicholson method while coupling it to analytical solutions to Eqs. (22.2) and (22.3). For example, Fig. 22.4 shows the effect on the arterial wall O_2 profile of increasing levels of carboxyhemoglobin (COHb) (Schneiderman and Goldstick, 1978a). Levels as high as 20% COHb are associated with smoking of three to four packages of cigarettes per day (40 to 60 cigarettes per day), and also have been found in nonsmoking parking garage attendants, traffic officers, and fire fighters. These data show a marked and systematic depression of wall Po_2 with increasing levels of COHb.

Another example of the results from this kind of analysis are shown in Figs. 22.5 and 22.6. These diagrams illustrate the simulated effect of a stenosis on oxygen transport to the wall.

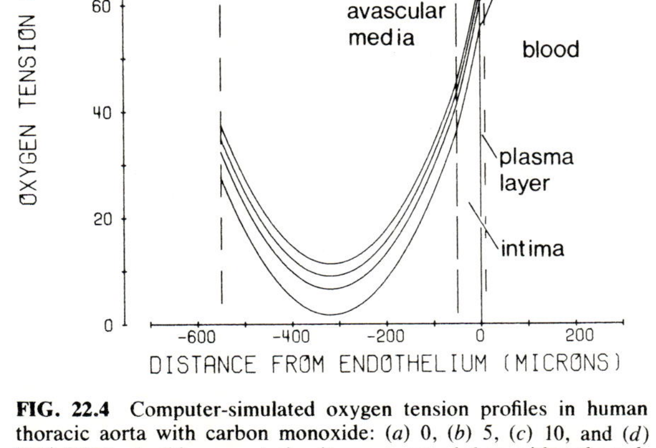

FIG. 22.4 Computer-simulated oxygen tension profiles in human thoracic aorta with carbon monoxide: (*a*) 0, (*b*) 5, (*c*) 10, and (*d*) 20% COHb. In this computation it was assumed that neither the pulmonary system nor the vasa vasorum autoregulated in response to the altered COHb. (*From Schneiderman and Goldstick, 1978a.*)

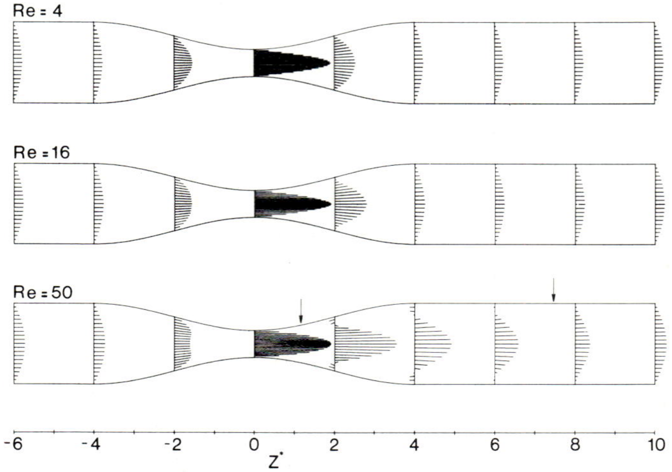

FIG. 22.5 Velocity vectors in the region of an 89 percent constriction. These vectors were drawn by a Calcomp attached to a CDC 6600 computer, using the Fourier series previously fitted to the results of Morgan and Young (1974). The arrows represent the points of flow separation and reattachment. The position variable z^* is the ratio of the axial location z to the unconstricted radius. (*From Schneiderman et al., 1979.*)

Figure 22.5 shows the velocity vectors near an 89 percent sinusoidal constriction. These results were originally obtained by Morgan and Young (1974) and were subsequently fitted by Schneiderman et al. (1979) to a Fourier series having 40 harmonics. The three Reynolds numbers shown were below, at, and above that required for flow separation. At Re = 50 the arrows indicate the location of flow separation and flow reattachment. Within the experimental possibilities of testing this result it has been found to be accurate and to represent adequately a stenosis in vivo. In analyzing this particular system, Eq. (22.1) was modified to include "radial" as well as axial flows. In this case, upstream differencing was used on the modified form of Eq. (22.1) and was solved for the appropriate boundary conditions. The axial diffusion term was included because of the possibility that axial diffusion would dominate convection in regions of low flow. For the finite difference scheme, the Gauss-Seidel method was employed. The wall flux obtained under conditions of stenotic constriction is shown in Figure 22.6 (peaked curves), and may be compared with the wall flux for a tube with constant luminal area and the same Reynolds number (smooth curve). Although it is difficult to know precisely how to interpret the effect of the region of reduced flux, it is clear that disturbed flow leads to alterations in flux.

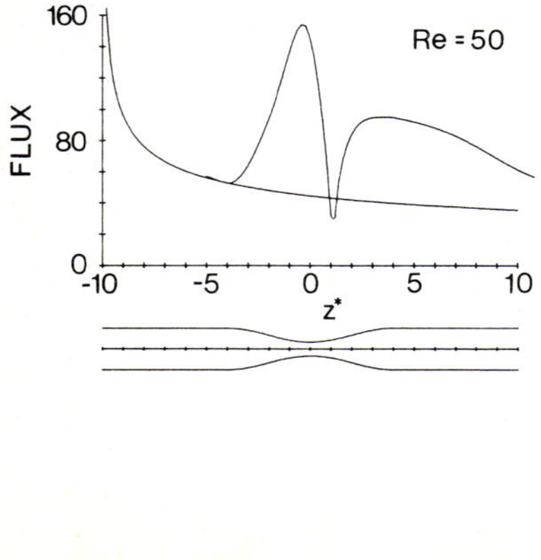

FIG. 22.6 Oxygen flux to the wall. The flux is compared with that to the wall of an unconstricted vessel (smooth curve), both with Re = 50. The position variable z^* is the ratio of z to the unconstricted radius. Flux is in arbitrary units. This figure was generated by a Calcomp attached to a CDC 6600 computer. (*Modified from Schneiderman et al., 1979.*)

22.3 EXPERIMENTAL MEASUREMENTS

Moss et al. (1968) used 125-μm-diameter platinum wire electrodes to measure the oxygen levels in the dog femoral arteries in vivo. Niinikoski et al. (1973) studied the rabbit aorta using needle-type microcathodes. Both these investigators found a gradual decline in Po_2 as the electrodes were advanced inward from the adventitia, with Po_2 values increasing abruptly as the electrodes entered the blood. Crawford et al. (1980) found similar results in the rabbit femoral artery and modeled the wall as a thick-walled structure with uniform oxygen consumption. This model predicted a mean oxygen consumption by the media of 99×10^{-6} mL/(mL · s), varying linearly with wall stress. Jurrus and Weiss (1977) measured the Po_2 gradients in in vitro sections of rabbit aortic arch. They found that, in normal vessels, the Po_2 fell from ambient levels on the two surfaces to a minimum of 51 torr at the midmedia. The minimum was reduced to 27 torr in vessels with early atherosclerotic disease, and to 0 torr with advanced atherosclerotic lesions. Buerk and Goldstick (1982) used microelectrodes to measure Po_2 levels in the intact dog aorta in vivo. This vessel more nearly resembles large human arteries than dog and rabbit femoral arteries or rabbit aorta, because it possesses distinct vasa vasorum. These investigators found that wall Po_2 was high at the adventitia, declining to 30 to 35 torr at the inner regions of the media. As the electrodes were advanced toward the lumen, the Po_2 rose steeply to reach a steady value in the blood. The Po_2 of 30 torr recorded in the media of the dog aorta is considerably greater than the experimentally determined critical Po_2 required to sustain tissue viability, that is, 5 to 12 torr (Gore and Whalen, 1968). However, as shown by Jurrus and Weiss (1977), minimum Po_2 levels in the arterial wall decline in the presence of atherosclerotic disease. Figure 22.7 presents an example of an in vivo wall Po_2 recording.

All direct measurements of wall Po_2 are technically very difficult, and the use of electrodes in this setting introduces serious problems. Large electrodes are electrically stable, but may consume a significant portion of tissue oxygen. In this way the electrodes themselves may lower measured Po_2 values. Large electrodes also do not permit precise localization of tissue oxygen,

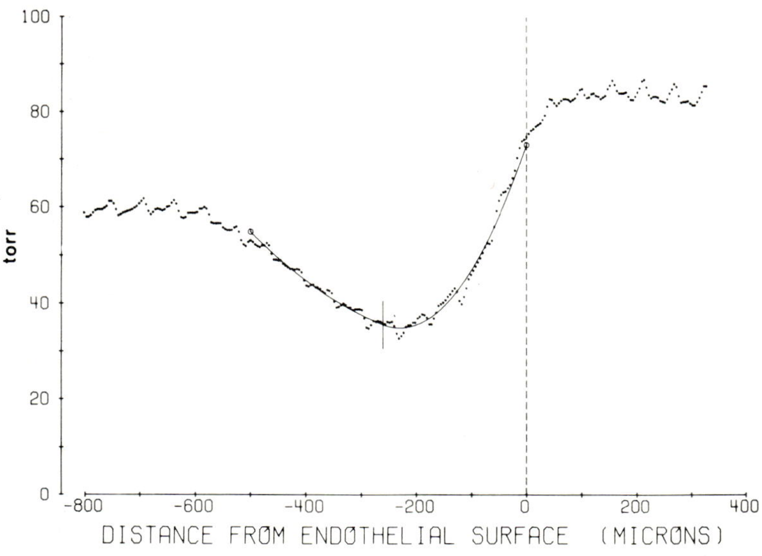

FIG. 22.7 Typical oxygen tension profile in the arterial wall. An oxygen electrode was slowly passed from the adventitial side through the wall of the thoracic aorta of an anesthetized dog. Dots are experimental data; line is two-layer fit predicted by theory, i.e., Eqs. (22.3), (22.6), and (22.7). (*Unpublished data of Buerk and Goldstick.*)

as, for example, when the electrode used occupies a significant fraction of the wall thickness. Use of a microelectrode avoids some of these problems but introduces a new set of difficulties. It may be difficult to penetrate the stiff adventitia with small microelectrodes, and bending may occur. Similarly, the wall may be dimpled or compressed at the site of contact with incomplete electrode insertion. Vibrating devices have been used to overcome this problem (Kanabus et al., 1977). Microelectrodes are difficult to calibrate (Schneiderman and Goldstick, 1978b) and "poisoning" of the electrode surface may occur. Finally, use of all electrodes suffers from problems in identifying the precise location of the measurement within the tissue, and particularly the blood-tissue interface.

22.4 PATHOPHYSIOLOGICAL IMPLICATIONS OF IMPAIRED WALL NUTRITION

It has long been postulated that tissue hypoxia is an intervening variable linking diverse epidemiological factors and arterial wall disease. Hueper (1944) suggested that inhibition of oxidative metabolism of the vascular wall could lead to degenerative changes. A current theory of atherogenesis (Ditzel, 1976) proposes that intermittent hypoxia in the media of the wall of arteries triggers the cellular proliferation observed in this disease. Moreover, recent evidence emphasizes the exacerbation of hypoxia by atherosclerotic lesions (Jurrus and Weiss, 1977). Perhaps the reason that the arterial wall is susceptible to atherosclerotic degeneration is its relatively avascular structure. Indeed, the avascular inner zone in the human aorta may be at the limit for maintenance of cell viability by diffusion, for only a moderate increase in thickness may lead to tissue hypoxia (Kirk and Laursen, 1955; Schneiderman and Goldstick, 1978a). The avascular zone of the abdominal aorta is significantly thicker than that in the thoracic aorta, perhaps explaining why atherosclerosis and the development of aneurysms are much more common in the abdominal aorta. Nakata (1967) and Nakata and Kamiya (1970) studied the effect of selective obstruction of the vasa vasorum in the presence of a cholesterol-rich diet. These investigators found that obstruction produced vascular disease in the dog aorta, a tissue normally resistant to atherosclerosis. These studies agree with those of Kjeldsen et al. (1968), Garbarsch et al. (1969), and Vesselinovitch et al. (1974), all of whom showed that respiratory hypoxia in rabbits accelerates atherogenesis while hyperoxia inhibits the process.

Studies of the metabolism of the arterial wall in vitro lend support to the hypothesis that hypoxia interferes with lipid transport and degradation. Low-density lipoproteins (LDLs), the cholesterol-rich lipoproteins, are relatively small particles which normally pass through the intima to penetrate the arterial wall. Several studies suggest that the permeability of LDLs may be increased by hypoxia (Getz et al., 1969; Kjeldsen et al., 1968; Garbarsch et al., 1969; Vesselinovitch et al., 1974). LDLs accumulate if the rate of LDLs entering the wall exceeds the rate of elimination (Adams and Bayliss, 1969), a situation enhanced by high plasma LDL concentrations, by hypertension, and by any factor which impedes metabolic clearance or transport of lipids from the wall. It is believed that hypoxia may impair LDL metabolism and transport. In excised human aorta, both atherosclerosis and hypoxia inhibit phospholipid synthesis, causing micellar lipids to be transformed into globular fats (Filipovic and Rutemoller, 1976). In turn, the accumulation of lipid-laden cells in the wall tends to further increase local hypoxia, thereby further inhibiting local synthesis of phospholipids. Other investigators (Robertson, 1963; Whereat, 1964; Robertson, 1968; Scott et al., 1969; Morrison et al., 1972) found that the early fatty streaks of atherosclerosis showed increased oxygen uptake. Although the mechanism is poorly understood at present, it seems likely that altered wall metabolism observed in atherosclerosis may reflect changes in the mural enzymes and biochemical pathways. For example, different metabolic pathways have been observed in atherosclerosis-susceptible and atherogenesis-resistant pigeons (Kalra and Brodie, 1974). Biochemical differences also have been observed in atherosclerotic human aortas (Filipovic and Rutemoller, 1976), revealing that atherosclerotic lesions are much more susceptible to hypoxia. Similarly, when cultures of arterial smooth muscle cells taken from the human aorta are exposed to hypoxia, these cells proliferate and increase their uptake of lipids while decreasing their degradation of lipids (Robertson, 1968; Albers and Bierman, 1976). These cholesterol-choked cells take on the appearance of "foam cells," a classic precursor to frank atherosclerotic lesions. Thus, inhibition of wall metabolism by atherosclerosis and by hypoxia may perpetuate, if not accelerate, atherogenesis. Further discussions of this subject may be found in the extensive reviews by Wissler et al. (1976) and St. Clair (1976).

ACKNOWLEDGMENTS

This chapter was based on the Ph.D. thesis research of Gary Schneiderman and Donald G. Buerk at Northwestern University. Preparation of the manuscript was supported by a Veterans Administration clinical investigatorship and NIH grants HL-26056 and EY-04085. This manuscript was typed by Ms. Virginia Thiel.

LIST OF SYMBOLS

$C(P)$ = concentration of oxygen combined with hemoglobin as defined by the oxygen-hemoglobin equilibrium curve, mL O_2 [STP]/mL blood
D = oxygen diffusivity, cm^2/s
k = oxygen solubility, mL O_2 [STP]/(mL · torr)
l_j = y coordinate of a tissue interface, cm
P = oxygen tension, Po_2, torr
Q = oxygen consumption rate of tissue, mL O_2 [STP]/(mL · s)
$u(y)$ = axial blood velocity profile, cm/s
y = boundary layer coordinate, distance from plasma layer (positive in blood direction), cm
x = axial coordinate, distance from entry into arterial segment, cm
$\Phi(P)$ = $1 + (dC/dP)/k_B$, dimensionless

Subscripts

a = mixed arterial
B = whole blood
o = outer vascularized tissue
T = tissue, either intimal or medial

REFERENCES

Adams, C. W. M., and Bayliss, O. B. (1969). The relationship between diffuse intimal thickening, medial enzyme failure and intimal lipid deposition in various human arteries. *J. Atherosclerosis Res., 10,* 327–339.

Albers, J. J., and Bierman, E. L. (1976). The effect of hypoxia on uptake and degradation of low density lipoproteins by cultured human arterial smooth muscle cells. *Biochim. Biophys. Acta, 424,* 422–429.

Buerk, D. G., and Goldstick, T. K. (1982). Arterial wall oxygen consumption rate varies spatially. *Am. J. Physiol., 243,* H948–H958.

Buñag, R. D., and Takeda, K. (1979). Sympathetic hyper-responsiveness to hypothalamic stimulation in young hypertensive rats. *Am. J. Physiol., 237,* R39–R44.

Crawford, D. W., Back, L. H., and Cole, M. A. (1980). In vivo oxygen transport in the normal rabbit femoral arterial wall. *J. Clin. Invest., 65,* 1498–1508.

Ditzel, J. (1976). The problem of tissue oxygenation in diabetes mellitus as related to the development of diabetic angiopathy. In J. Grayson and W. Zingg (eds.), *Microcirculation.* Plenum Press, New York, pp. 263–274.

Dobrin, P. B. (1983). Vascular mechanics. In J. T. Shepherd and F. M. Abboud (eds.), *Handbook of Physiology*, sec. 3: *The Cardiovascular System*. American Physiological Society, Bethesda, Md, pp. 65–102.

Doyle, J. M., and Dobrin, P. B. (1973). Stress gradients in the walls of large arteries. *J. Biomech.*, *6*, 631–639.

Filipovic, I., and Rutemoller, M. (1976). Comparative studies of fatty acid synthesis in atherosclerotic and hypoxic human aorta. *Atherosclerosis*, *24*, 457–469.

Garbarsch, G., Matthiessen, M. E., Helin, P., and Lorenzen, I. (1969). Arteriosclerosis and hypoxia: I. Gross and microscopic changes in rabbit aorta induced by systemic hypoxia: Histochemical studies. *J. Atherosclerosis Res.*, *9*, 283–294.

Geiringer, E. (1951). Intimal vascularization and atheromatosis. *J. Pathol. Bacteriol.*, *63*, 201–211.

Getz, G. S., Vesselinovitch, D., and Wissler, R. W. (1969). A dynamic pathology of atherosclerosis. *Am. J. Med.*, *46*, 657–673.

Goldstick, T. K. (1973). Oxygen transport. In J. H. U. Brown and D. S. Gann (eds.), *Engineering Principles in Physiology*. Academic Press, New York, vol. 2, pp. 257–282.

Gore, R. W., and Whalen, W. J. (1968). Relations among tissue Po_2, Qo_2, and resting heat production of frog sartorius muscle. *Am. J. Physiol.*, *214*, 277–286.

Heistad, D. D., and Marcus, M. L. (1979). Role of vasa vasorum in nourishment of the aorta. *Blood Vessels*, *16*, 225–238.

Heistad, D. D., Marcus, M. L., Larsen, G. E., and Armstrong, M. L. (1981). Role of vasa vasorum in nourishment of the aortic wall. *Am. J. Physiol.*, *240*, H781–H787.

Heistad, D. D., Marcus, M. L., Law, E. G., Armstrong, M. L., Ehrhardt, J. C., and Abboud, F. M. (1978). Regulation of blood flow to the aortic media in dogs. *J. Clin. Invest.*, *62*, 133–140.

Heistad, D. D., Marcus, M. L., and Martins, J. B. (1979). Effects of neural stimuli on blood flow through vasa vasorum in dogs. *Circ. Res.*, *45*, 615–620.

Hueper, W. C. (1944). Arteriosclerosis: The anoxemia theory. *Arch. Pathol.*, *38*, 162–181, 245–285, 350–364.

Jaeger, M. (1964). The flow through the arterial wall. In E. O. Attinger (ed.), *Pulsatile Blood Flow*. McGraw-Hill, New York, pp. 307–322.

Jurrus, E. R., and Weiss, H. S. (1977). In vitro tissue oxygen tensions in the rabbit aortic arch. *Atherosclerosis*, *28*, 223–232.

Kalra, V. K., and Brodie, A. F. (1974). Metabolic differences between the arteries of atherosclerotic susceptible and resistant pigeons. *Biochem. Biophys. Res. Comm.*, *61*, 1372–1378.

Kanabus, E. W., Feldstein, C., and Crawford, D. W. (1980). Excursion of vibrating microelectrodes in tissue. *J. Appl. Physiol.*, *48*, 737–741.

Kirk, J. E., and Laursen, T. J. S. (1955). Diffusion coefficients of various solutes for human aortic tissue with special reference to variation in tissue permeability with age. *J. Gerontol.*, *10*, 288–302.

Kjeldsen, K., Wanstrup, J., and Astrup, P. (1968). Enhancing influence of arterial hypoxia on the development of atheromatosis in cholesterol-fed rabbits. *J. Atherosclerosis Res.*, *8*, 835–845.

Morgan, B. E., and Young, D. P. (1974). An integral method for the analysis of flow in arterial stenoses. *Bull. Math. Biol.*, *36*, 39–53.

Morrison, A. D., Clements, R. S., Jr., and Weingrad, A. T. (1972). Effects of elevated glucose concentrations on the metabolism of the aortic wall. *J. Clin. Invest.*, *51*, 3114–3123.

Moss, A. J., Samuelson, P., Augell, C., and Minken, S. L. (1968). Polarographic evaluation of transmural oxygen availability in intact muscular arteries. *J. Atherosclerosis Res.*, *8*, 803–810.

Nakata, Y. (1967). An experimental study on the vascular lesions caused by obstruction of vasa vasorum. *Jpn. Circ. J.*, *31*, 275–287.

Nakata, Y., and Kamiya, K. (1970). An experimental study of vascular lesions caused by obstruction of the vasa vasorum, II. *Jpn. Circ. J.*, *34*, 1029–1034.

Niinikoski, J., Heughan, C., and Hunt, T. K. (1973). Oxygen tensions in the aortic wall of normal rabbits. *Atherosclerosis*, *17*, 353–359.

Robertson, A. L., Jr. (1963). Hypoxia and sterol synthesis in human atheromatous plaques. *Fed. Proc.*, *22*, 161.

Robertson, A. L., Jr. (1968). Oxygen requirements of the human arterial intima in atherogenesis. *Prog. Biochem. Pharmacol.*, *4*, 305–316.

St. Clair, R. W. (1976). Metabolism of the arterial wall and atherosclerosis. In R. Paoletti and A. M. Gotto, Jr. (eds.), *Atherosclerosis Reviews*. Raven Press, New York, vol. I, pp. 61–117.

Schneiderman, G., Ellis, C. G., and Goldstick, T. K. (1979). Mass transport to walls of stenosed arteries: Variation with Reynolds number and blood flow separation. *J. Biomech., 12,* 869–877.

Schneiderman, G., and Goldstick, T. K. (1978a). Carbon monoxide-induced arterial wall hypoxia and atherosclerosis. *Atherosclerosis, 30,* 1–15.

Schneiderman, G., and Goldstick, T. K. (1978b). Oxygen electrode design criteria and performance characteristics: Recessed cathode. *J. Appl. Physiol., 45,* 145–154.

Schneiderman, G., and Goldstick, T. K. (1978c). Significance of luminal plasma layer resistance in arterial wall oxygen supply. *Atherosclerosis, 31,* 11–20.

Schneiderman, G., Mockros, L. F., and Goldstick, T. K. (1982). Effect of pulsatility on oxygen transport to the human arterial wall. *J. Biomech., 15,* 849–858.

Schoenenberger, F., and Muller, A. (1960). Über die Vaskularisation der Rinderaortenwand. *Helv. Physiol. Acta, 18,* 136–150.

Scott, R. F., Morrison, E. S., and Kroms, M. (1969). Aortic respiration and glycolysis in the pre-proliferative phase of diet-induced atherosclerosis in swine. *J. Atherosclerosis Res., 9,* 5–16.

Simon, B. R., Kobayashi, A. S., Wiederhielm, C. A., and Strandness, D. E. (1973). Deformation of the arterial vasa vasorum at normal and hypertensive arterial pressure. *J. Biomech., 6,* 349–359.

Vaishnav, R. N., Young, J. T., and Patel, D. J. (1973). Distribution of stresses and of strain-energy density through the wall thickness in a canine aortic segment. *Circ. Res., 32,* 577–583.

Vesselinovitch, D., Wissler, R. W., Fischer-Dzoga, K., et al. (1974). Regression of atherosclerosis in rabbits: I. Treatment with low-fat diet, hypoxia, and hypolipidemic agents. *Atherosclerosis, 19,* 259–275.

Whereat, A. F. (1964). Lipid biosynthesis in aortic intima from normal and cholesterol fed rabbits. *J. Atherosclerosis Res., 4,* 272–282.

Wilens, S. L., Malcom, J. A., and Vazquez, J. M. (1965). Experimental infarction (medical necrosis) of the dog's aorta. *Am. J. Pathol., 47,* 695–711.

Wissler, R. W., Vesselinovitch, D., and Getz, G. (1976). Abnormalities of the arterial wall and its metabolism in atherogenesis. *Prog. Cardiovasc. Dis., 18,* 341–369.

Wolinsky, H., and Glagov, S. (1967). Nature of species differences in the medial distribution of aortic vasa vasorum in mammals. *Circ. Res., 20,* 409–421.

CHAPTER 23

Flow through Collapsible Tubes

Roger D. Kamm
Department of Mechanical Engineering
Massachusetts Institute of Technology
Cambridge, Massachusetts

23.1 THE PHYSIOLOGICAL IMPORTANCE OF FLOW THROUGH COLLAPSIBLE TUBES

23.1.1 Normal Physiological Occurrences of Vessel Collapse

The fluid-carrying vessels of the body are prone to collapse for a variety of reasons. Often, collapse is a normal and even useful physiological function which occurs as a result of normal variations in pressure acting either internal or external to the vessel. Perhaps the best-known case of vessel collapse as a normal course of events is found in the return flow of blood to the right heart. As Guyton [1] demonstrated, the rate of venous return to the heart increases with decreasing right atrial pressure, but only up to the point at which the pressure inside the inferior and superior venae cavae is roughly equal to that acting external to these vessels. When right atrial pressure is further reduced, the vessels partially collapse, thereby limiting further increases in the rate of venous return. This illustrates one of the more fascinating characteristics of collapsible tube flow, the ability of the tube to limit the rate of flow passing through it despite changes in downstream pressure. This phenomenon, termed *flow limitation*, will be discussed more fully later in this chapter.

Even the fluctuations in intrathoracic pressure associated with normal breathing can cause veins to partially collapse at the point where they enter the thoracic cage, producing a flow situation of considerable complexity owing to the superposition of respiratory and cardiac pulsations. The veins of the upper extremities and neck are also subject to frequent collapse. This can be readily observed in the superficial veins when, for example, the hand is raised above the level of the heart.

The deep veins of the lower extremity also collapse during walking or any activity causing contraction of the leg muscles. Arnoldi [2] has demonstrated that intramuscular pressures in excess of 200 mmHg are common and that these high pressures, when compressing a muscular vein, cause it to collapse. With collapse, the contained blood is expelled in the direction of the heart, thereby aiding venous return from the muscular sinuses of the lower leg.

The pulmonary vasculature also exhibits phenomena characteristic of flows in collapsible tubes. The collapse of veins in the lung leads to a nonuniform distribution of blood flow to different regions of the lung parenchyma which depends on the local hydrostatic head with respect to the left side of the heart [3, 4].

The respiratory airways exhibit flow limitation and collapse during a maneuver of forced expiration due to the external pressures acting on the airways resulting from intense muscular effort [5–8]. Flow limitation in the lung has been likened to that which occurs in gas-dynamic duct flows and is apparently due to the acceleration of the gas flow to the local speed of wave propagation [8, 9] (see Sec. 23.4.1).

Other situations involving vessel collapse are thought to occur in the ureter [10–12], where there exists the possibility of flow-induced oscillations [13], in the flow of aqueous humor [14] or venous blood [15] from the eye, in the cerebral circulation [16] and the circulation of the heart [17], and in the renal tubules [18].

23.1.2 Externally Induced Collapse

Collapse is sometimes induced for the purpose of medical diagnosis, prophylaxis, or treatment. One of the most familiar examples is the measurement of blood pressure by the auscultatory method by which the physician determines systolic and diastolic arterial pressures through subtle changes in the character of the Korotkoff sounds which are produced at the point of vessel collapse (see, e.g., Ref. 19). Other uses include the application of external pressure to the lower extremities for the prevention of deep vein thrombosis [20], or as a means of external cardiac assist [21, 22].

These collapsible tube flows exhibit all the phenomena associated with flows in distended tubes, but, in addition, they offer a wealth of other interesting characteristics. These include autoregulation, flow limitation, and flow-induced flutter, to name but a few. Because of this variety of interesting features, collapsible tube flows have been widely studied by numerous investigators, and the results have consequently been presented in varied forms. In this chapter, the literature will be presented in a format which attempts to coalesce the experiences of many investigators.

23.2 THE STARLING RESISTOR

No discussion of flow in collapsible tubes is complete without a thorough analysis of the Starling resistor. This device, used for decades to simulate various physiological flow phenomena, consists of a compliant tube segment mounted between two rigid tubes and surrounded by a rigid enclosure that can be pressurized (see Refs. 23 and 24 for comprehensive reviews). It exhibits many of the same characteristics as physiological collapsible tube flows and is quite useful for the purpose of illustrating such phenomena as flow limitation and flow-induced oscillations.

Flows in a Starling resistor (Fig. 23.1) are driven by the pressure difference $p_1 - p_2$ given in

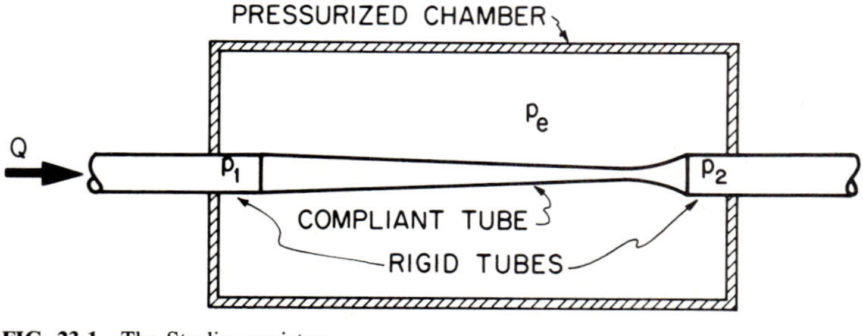

FIG. 23.1 The Starling resistor.

the figure, while some pressure p_e acts external to the tube. The behavior of the Starling resistor can be characterized by two independent variables and one dependent variable, as was first pointed out by Brower and Noordergreaf [24]. The dependent variable that is typically selected is the volumetric rate of flow passing through the tube, Q. The two independent variables are most conveniently expressed as pressure differences, selected from $p_1 - p_2$, $p_e - p_2$, and $p_1 - p_e$, depending on the phenomenon one wishes to demonstrate.

One of the most interesting and useful characteristics of a Starling resistor is its ability to exhibit flow limitation. This can best be demonstrated in the Starling resistor by the following experiment. Let p_e and p_1 be fixed (with $p_1 > p_e$) while p_2 is slowly lowered. If p_2 and p_1 are equal, obviously no flow passes through the device. As p_2 is decreased, Q begins to rise, at first in a linear fashion (Fig. 23.2) [25]. When p_2 approaches and falls below p_e, the flow rate reaches

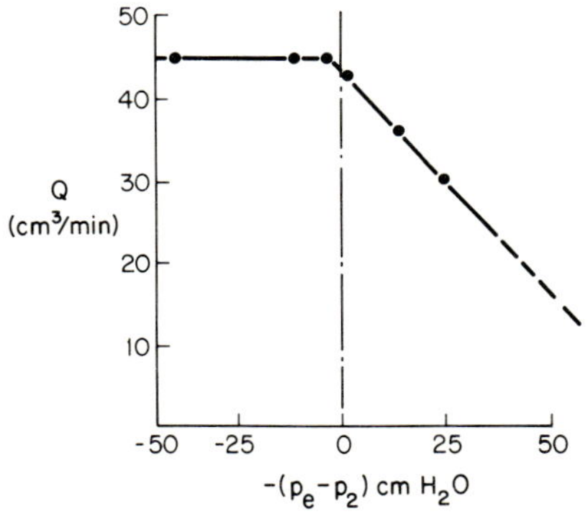

FIG. 23.2 Flow through a Starling resistor as a function of $p_e - p_2$ with $p_1 - p_e$ held constant. *(From the data of Holt [25].)*

a maximum value such that further reductions in p_2 have no perceptible effect on Q. Depending on the particular conditions of the experiment, this flow limitation can be a result of either of two mechanisms—wave speed limitation or viscous flow limitation. In the former, flow speed inside the tube at the downstream end comes to equal the speed at which area waves propagate through the tube. Hence, when p_2 is lowered, the reduction in pressure affects the flow so that it is not able to propagate upstream beyond the point where the two speeds are equal. We can say that, in this situation, the flow is choked in the same sense that a gas-dynamic flow is choked in a De Laval nozzle (see also Sec. 23.4.1).

The second mechanism of flow limitation relies exclusively upon the viscous properties of the fluid [9, 26–28]. When the downstream segment of the tube is collapsed, any further reduction in p_2 has two effects: to increase the pressure driving the flow, and to increase the resistance to flow owing to a reduction in the cross-sectional area of the tube. These two effects drive the flow rate in opposite directions and tend to balance one another, producing the observed flow limitation. The mechanism which regulates flow in any particular situation is that which limits flow to the lowest value.

This property of flow limitation has been demonstrated in two different types of experiments which, as can be inferred from the form of Fig. 23.2, are equivalent. In one, p_2 is lowered while the other pressures are held constant; in the other, p_1 and p_e are raised simultaneously (holding $p_1 - p_e$ constant) by connecting the external chamber to the inlet tube. The latter device was used in experiments conducted by Rodbard [29, 30] and was termed a *capillaron*.

The results of Fig. 23.2 suggest that, over a certain range of parameters, the rate of flow through a Starling resistor is independent of the pressure difference $p_e - p_2$. This can be made even more evident if we plot Q versus $p_1 - p_e$ as in Fig. 23.3. Here, even though $p_e - p_2$ varies over a considerable range, the flow rate depends only upon $p_1 - p_e$ and, in fact, varies roughly linearly with $p_1 - p_e$. This is in stark contrast to all rigid flow devices in which the flow depends only upon $p_1 - p_2$.

Oscillations of a Starling resistor have been studied by various investigators [23, 31, 32]. It

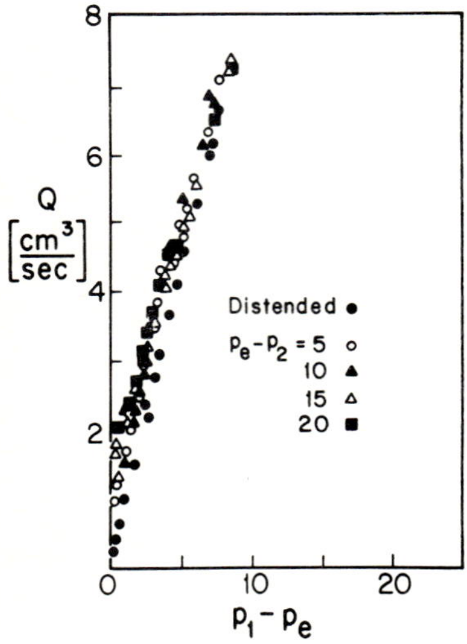

FIG. 23.3 Flow through a Starling resistor as a function of $p_1 - p_e$ for several different values of $p_e - p_2$. Pressures in cm H_2O. (*Modified from Fry et al. [66].*)

has been shown, however, that the character of these oscillations is governed largely by the properties of the system downstream of the device [33, 34] and is also likely to be influenced by the nonphysiological attachment between the compliant and rigid tubes at the downstream end. Since oscillations of similar character have not yet been demonstrated by hardware experiments in which this problem of a nonphysiological boundary condition has been eliminated, direct application of such results to physiological vessels remains in question.

23.3 THE MECHANICS OF THE TUBE WALL

An essential element of all theoretical analyses of flow in collapsible tubes is the constitutive relation which defines the response of the tube to changes in transmural pressure ($p_e - p$, where p is the internal vessel pressure), often referred to as the *tube law*. Although all vessels found in physiological systems are imbedded in other tissues, we will at first neglect the influence of external tethering. Furthermore, we will assume, for now, that the vessel is being subjected to a uniform transmural pressure which does not vary in either the circumferential or axial direction, and that the tube circumference remains of constant length despite changes in cross-sectional area. The tube wall will be treated as a homogeneous, isotropic structure, the walls of which are subject to both tensile and bending stresses.

The response of the tube to changes in transmural pressure has been treated by several investigators [35–37], each using somewhat different analytical and numerical techniques. Flaherty and co-workers [37] wrote the structural equations for a small element of the tube wall at a circumferential position s (Fig. 23.4) as follows:

$$\sum F_n = \frac{dN}{ds} - TK + P = 0$$

$$\sum F_T = \frac{dT}{ds} + NK = 0 \tag{23.1}$$

$$\sum M = \frac{dM}{ds} + N = 0$$

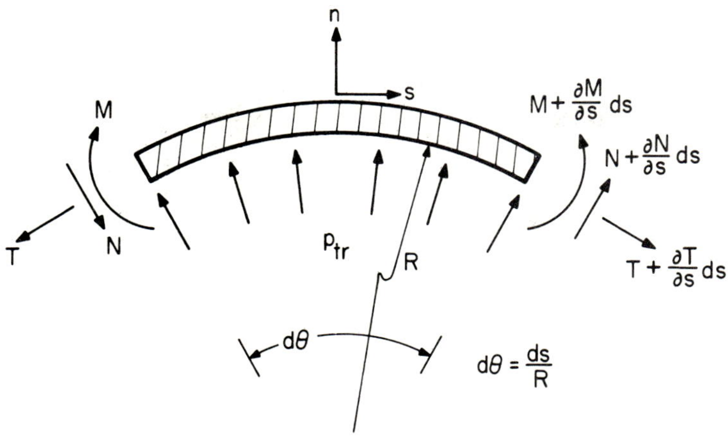

FIG. 23.4 Free-body diagram for a small segment of tube wall.

where P is a dimensionless transmural pressure [$P = (p - p_e)/K_p$, with $K_p = E(h/R)^3/12(1 - \sigma^2)$, where h is wall thickness, E is Young's modulus, and σ is Poisson's ratio], $N(s)$ and $T(s)$ are dimensionless components of force in the outward normal and tangential directions, respectively, and $M(s)$ is the local dimensionless bending moment. Equation (23.1) could be reduced to the following differential equation in terms of the local curvature k when the bending moment is assumed to be proportional to the difference between the curvature of a circular tube and the actual curvature.

$$k'' + \tfrac{1}{2}k^3 - \bar{c}k^3 + P = 0 \tag{23.2}$$

Here $\bar{c}$ is an integration constant given by

$$\bar{c} = T + \tfrac{1}{2}k^2 \tag{23.3}$$

evaluated at $s = 0$. The numerical solution of this equation was obtained in three different regimes: (1) before the two opposing walls come into contact, (2) when contact occurs at a single point (or line in a uniformly pressurized tube), and (3) when a line of contact exists between the two opposing surfaces. The numerical solution is carried out for a single quadrant of the tube using boundary conditions that are appropriate for each of the three different regimes just defined. The resulting relationship between transmural pressure and cross-sectional area is given in Fig. 23.5 with the two transition points identified: no contact to point contact, $\alpha = A/A_0 = 0.27$, where A_0 is the cross-sectional area when the transmural pressure is zero; and point contact to line contact, $\alpha = 0.21$. Thus in the case of uniform compression of a long tube, the tube law takes on the simple form

$$\alpha = \frac{A}{A_0} = f(P) \qquad \text{or} \qquad P = P(\alpha) \tag{23.4}$$

The essential features of vessel collapse are contained in these results. Additional effects which can be important in certain conditions have also been considered. Kresch [36] determined the collapse behavior for tubes of elliptical cross section at zero transmural pressure when the vessel walls were allowed to be distensible. Swidler [38] used an analysis similar to that of Flaherty and colleagues [37] and included the influence of longitudinal curvature and tension. Her predictions were used to help in the development of a simple analytical model for axial tension [39] which could be easily incorporated into the steady-flow analysis.

The predictions of numerical analyses have been compared with experimental measurements of the tube law using long segments of latex rubber tubing, suspended in liquid to negate the effects of hydrostatic pressure. The results of one set of experiments [40] are shown in Fig. 23.5 and can be seen to be in good agreement with the theoretical predictions. Because of the difficulty in obtaining similar measurements on physiological vessels, only qualitative or rough quantitative comparisons can be made (see, e.g., Ref. 41).

Another important property of the tube, the velocity of long area waves, can be directly inferred from the tube law. These are related through the expression (see Sec. 23.5.1)

$$c = \left(\frac{\alpha}{\rho}\frac{dp}{d\alpha}\right)^{1/2} \tag{23.5}$$

Because of the nonlinear behavior of the tube law, the wave speed, as derived from experimental data using Eq. (23.5), can be seen to vary considerably as the tube collapses from a circular state (Fig. 23.5). The abrupt jump in c at a value of α of about 0.27 corresponds to the two walls of the tube coming into contact, with a consequent increase in tube stiffness.

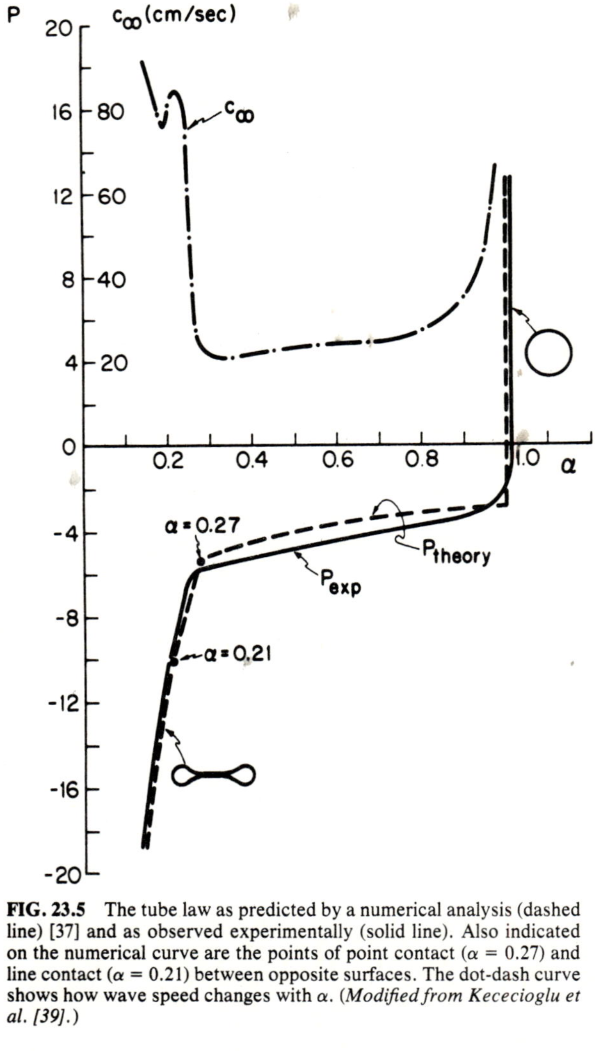

FIG. 23.5 The tube law as predicted by a numerical analysis (dashed line) [37] and as observed experimentally (solid line). Also indicated on the numerical curve are the points of point contact ($\alpha = 0.27$) and line contact ($\alpha = 0.21$) between opposite surfaces. The dot-dash curve shows how wave speed changes with α. (*Modified from Kececioglu et al. [39].*)

All the foregoing assumes that the structural characteristics of physiological collapsible tubes are determined exclusively by the properties of the vessel wall. Although this is often justifiable for vessels of circular cross section, in which case the vessel wall is relatively "stiff," the partially collapsed vessel is much more compliant, and its resistance to collapse might therefore be more subject to the influence of surrounding tissues. Furthermore, the change in vessel shape during collapse causes considerable distortion of the surrounding tissue and will be met with a greater resistance than simple changes in vessel diameter for a tube of circular cross section. These influences are rarely taken into account but are of obvious importance in measurements of vessel compliance made in vivo [42]. As has been suggested in a study of venous collapse induced by external compression [43], this influence can be accounted for by a change in slope of the rapid-collapse portion of the tube law to reflect the increased stiffness due to external tethering to surrounding tissues.

23.4 STEADY FLOW

23.4.1 The Governing Equations

Several theories have been developed for flow through a collapsible tube, with different objectives in mind [8, 9, 44, 45]. Here we will follow the derivation of Shapiro [9].

The differential equations governing the steady, one-dimensional flow of an incompressible fluid are as follows:

Conservation of Mass

$$\frac{dA}{A} + \frac{du}{u} = 0 \tag{23.6a}$$

Conservation of Momentum

$$-A\,dp - \tau_w s\,dx - \rho g A\,dz = \rho A u\,du \tag{23.6b}$$

where τ_w is the wall shear stress, s is the tube perimeter, z is the elevation in a gravity field, ρ is the fluid density, and g is the acceleration due to gravity. With the introduction of dimensionless variables, and allowing for axial variations in the resting tube area A_0, external pressure, tube properties (as reflected by changes in K_p), and the dimensionless form of the tube law, each of the dependent variables can be expressed in differential form as a sum of terms representing the independent variables. Two useful dependent variables are the speed index (S) and the dimensionless cross-sectional area (α). For the speed index, one obtains

$$\frac{dS^2}{S^2} = \frac{1}{1 - S^2}\left\{\frac{-2 + (2 - M)S^2}{A_0}\frac{dA_0}{dx} + \frac{M}{\rho c^2}\frac{d(p_e + \rho g z)}{dx} + \frac{4M\tau_w}{\rho c^2 D_e} + \left[\frac{MP}{\rho c^2} - \frac{(1 - S^2)}{K_p}\right]\frac{dK_p}{dx} + \frac{MK_p}{\rho c^2}\frac{\partial P}{\partial x}\right\}dx \tag{23.7}$$

where D_e is the hydraulic diameter, and the important tube law property M is defined by

$$M = 3 + \frac{\alpha P''}{P'} \tag{23.8}$$

Equation (23.7) includes all the independent variables mentioned above and allows for axial variations in A_0, K_p and in the form of the tube law. An equation in terms of the area ratio (including only the effect of wall friction for illustration) has the form

$$\frac{d\alpha}{\alpha} = \frac{S^2}{S^2 - 1} \frac{4\tau_w}{\rho u^2 D_e} dx$$

Equation (23.7) is particularly useful for the purpose of exploring the different types of flows that are possible in collapsible tubes. The appearance of $1 - S^2$ on the right in Eq. (23.7) suggests that subcritical flows ($S < 1$) and supercritical flows ($S > 1$) are intrinsically different. Consider, for example, the effects of friction alone such that the right-hand side of Eq. (23.7) is comprised only of the third term with all others equal to zero. For $S < 1$, friction can be seen to cause the speed index to increase in the axial direction, driving a subcritical flow toward $S = 1$. For $S > 1$, the term $1 - S^2$ changes sign, thereby causing the speed index in supercritical flows to decrease, again driving the flow toward $S = 1$. By the same reasoning, it can be shown that friction causes the area ratio to decrease in the axial direction for subcritical flows and to increase if the flow is supercritical.

This example leads us to two special types of behavior that occur in collapsible tube flows: (1) choking and (2) continuous passage through $S = 1$. Continuing with the example above, in which only friction acts to alter the flow, consider the outcome of allowing either a sub- or supercritical flow to come to the condition $S = 1$. Since the sub- and supercritical flow behaviors are exactly opposite in their influence on S, it is impossible, in the presence of friction alone, to pass through $S = 1$. Since all flows with only friction acting are driven toward $S = 1$, and since no solution exists after the critical point has been reached, then a critical flow is possible only if the tube terminates at the point where the critical state is achieved. In practice, if a flow is slowly accelerated by a gradual reduction in downstream pressure, once the critical state is reached at the downstream end of the tube, the flow can be said to be choked and can be accelerated no farther. This choking behavior is analogous to choked flows in gas dynamics and is the same as the wave speed flow limitation phenomenon described earlier.

By examination of Eq. (23.7) it can be seen that there is only one situation in which a continuous passage through $S = 1$ might be possible. That is, only if the collection of terms on the right-hand side goes to zero precisely at the point where S passes through unity could a smooth transition occur. This might be produced, for example, by applying a nonuniform external pressure to the tube such that the point of maximum pressure coincides with $S = 1$. Although it might seem difficult to obtain such a unique coincidence, in practice it is relatively easy to produce a sub- to supercritical transition simply by pinching the tube with your fingers or by using a mechanical sphincter. Figure 23.6 shows a region of supercritical flow preceded by a mechanical sphincter and followed by a shocklike transition to subcritical flow. The *shock,* or *elastic jump,* shown in this figure is described next.

23.4.2 The Elastic Jump

If a sub- to supercritical flow transition is possible, is it also possible to decelerate the flow from a super- to a subcritical state? The theory developed above draws no distinction between sub- to

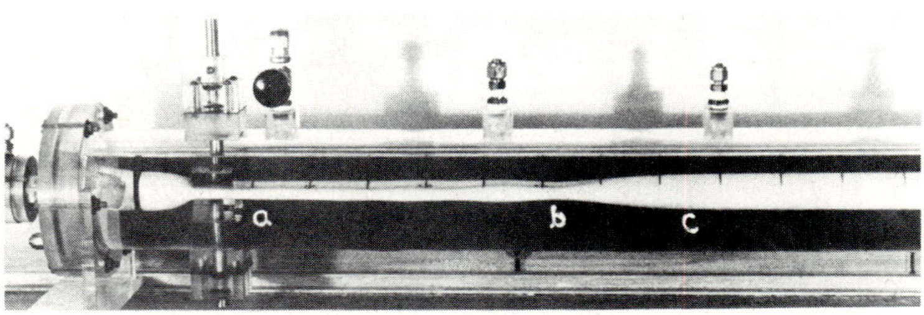

FIG. 23.6 Photograph of an elastic jump in a compliant tube suspended in water. Flow from left to right. (*a–b*) Supercritical region with precursor waves emanating from jump. (*b–c*) Elastic jump. (*c*–end) Subcritical region. (*From Kececioglu et al. [39], with permission.*)

supercritical transitions and super- to subcritical transitions. Therefore we have no reason to believe that either is impossible. Unlike transitions of the former type, however, experiments have not yet been able to demonstrate the existence of a smooth super- to subcritical transition. Instead, the flow tends to decelerate to subcritical speed by means of an elastic jump, across which the cross-sectional area, flow velocity, and pressure each undergo an abrupt change (over a distance of several tube diameters) while the speed index goes from greater than 1 to less than 1. The close analogy between an elastic jump and either a shock wave in gas-dynamic duct flow or an hydraulic jump in free surface flow can readily be seen.

Figure 23.6 depicts an elastic jump as it occurs in a collapsible latex tube suspended in water. We generally refer to the region of rapid area expansion (b to c) as the jump, although the train of waves seen standing upstream (a to b) originate at the jump and are closely associated with it.

The jump (b to c) has a length of from one to five tube diameters depending on its strength and the imposed longitudinal tension. As in the gas-dynamic shock, energy is dissipated by the elastic jump. This occurs by several mechanisms, including viscous dissipation and the release of energy in the form of a stationary wave train propagating upstream from the jump. These "precursor waves" appear to be stationary because their phase speed equals the local fluid velocity. They are able to propagate upstream because their group velocity exceeds the local flow speed (see, e.g., Sec. 23.5.1).

Several analyses have appeared in the literature to predict the nature of the transition across the elastic jump [9, 44, 46, 47]. To model the jump realistically, however, proves to be an extremely difficult task owing to several complicating factors, including: (1) the essential coupling between the pressure distribution created by the flow and the geometry of the tube wall, (2) the importance of the three-dimensional aspects of the jump, and (3) the flow separation that is likely to occur in the region of rapid area expansion. For all these reasons, present analyses must rely on various simplifying assumptions and can therefore be thought of as only approximate models of the true elastic jump. Perhaps the most rigorous analysis is that of Cowley [47], who has examined two cases: (1) a jump with internal dissipation (associated, for example, with turbulence) but negligible wall shear stress, and (2) a jump with no internal dissipation but including the effects of wall shear stress. Only the former will be considered here since that analysis seems most applicable to the current experimental evidence.

Considering case (1) above and denoting the area, velocity, and internal pressure by A, u, and p, respectively, at the upstream (b) and downstream (c) locations, Cowley obtains, from conservation of mass and momentum,

$$A_b u_b = A_c u_c = Q \tag{23.9}$$

$$A_b u_b^2 - \rho A_c u_c^2 = p_c A_c - p_b A_b + \int_{S_{b-c}} p\mathbf{n} \cdot \mathbf{z}\, dS \tag{23.10}$$

where S_{b-c} is the internal surface area between points b and c, and $\mathbf{n}$ and $\mathbf{z}$ are unit vectors normal to the tube wall and pointing in the axial direction, respectively.

If one assumes that there exist points upstream and downstream of the shock at which the tube area is uniform, then the area and transmural pressure are uniquely related there by the tube law. Cowley goes on to show that if the external pressure is uniform and the shock is quasi-steady, then

$$\int_{S_{b-c}} p\mathbf{n} \cdot \mathbf{z}\, dS = -\int_{A_b}^{A_c} P(A)\, dA \tag{23.11}$$

This expression is valid whether or not the tube law is strictly satisfied at all points along the tube. Combining Eqs. (23.9) to (23.11), one obtains the following relationship between conditions upstream and downstream of the shock:

$$\Phi(A_b) = \Phi(A_c) \tag{23.12}$$

where $$\Phi(A) \equiv \frac{Q^2}{A} + \int_A \frac{A}{\rho} \frac{d(p - p_e)}{dA}\, dA$$

This is the same result obtained by Beam [46] and Oates [44], but Cowley's analysis is less restrictive in that he does not need to assume that the tube law is valid throughout the shock. Cowley does assume, however, that the tube walls have no curvature at points b and c (upstream and downstream of the jump), and that wall shear stress is negligible within the control volume. These assumptions prove to be difficult to satisfy under typical experimental conditions and probably account for some differences between the theoretical predictions and the experimental results reported next.

Griffiths [48, 49] was probably the first to demonstrate experimentally the existence of a stationary elastic jump, although the lack of precise measurement capabilities prevented him from proving that the transition was indeed from a super- to subcritical flow state. Later experiments [39, 40, 50] were able to demonstrate clearly the existence of the jump and to characterize it in terms of upstream and downstream conditions. McClurken and colleagues [40] were able to determine the flow velocity, cross-sectional area, and pressure upstream and

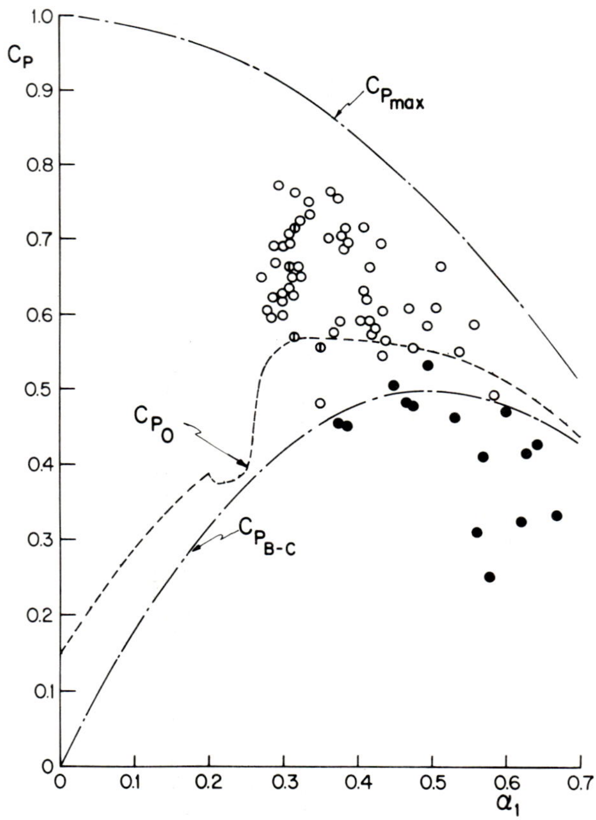

FIG. 23.7 Experimental values for the jump pressure recovery coefficient [$C_p = 2(p_c - p_b)/\rho u_b^2$] compared to various calculated values: $C_{p,\max}$ based on the assumption of no loss of kinetic head; $C_{p,B\text{–}C}$ based on the Borda-Carnot solution; and C_{pO} based on the predictions of Eq. (23.12). All calculated values were determined by setting $\alpha_c = 1$, and the curve for C_{pO} is for $S_b = 3$. ● = long shocks, ○ = intermediate shocks, ⦶ = short shocks. (*From Kececioglu et al. [39], with permission.*)

downstream of the shock and compared their measurements with several theoretical predictions. They found that the dimensionless pressure recovery coefficient generally fell between the two limiting curves of a jump free of viscous dissipation and of one with total loss of dynamic head in the jump (Fig. 23.7). In those jumps for which the pressure recovery fell below the lower curve (generally for very long jumps), it was thought that wall shear stress played a dominant role. The analysis of Oates (and also of Cowley) represented an improvement over either of these extreme cases, but still failed to yield a good prediction.

The lack of close agreement between theory and experiment can probably be attributed to the two assumptions identified above. For very long shocks, wall shear stress clearly plays an important role, causing the pressure recovery to fall below the prediction based on a total loss of kinetic energy. For shorter shocks, it is not possible to identify a point just upstream of the shock where the wall curvature is zero or insignificant as required by Cowley. Therefore the neglect of the effects of tension at points b and c is often not justified.

McClurken and co-workers [40] also predicted the length of the jump by using the simple notion that the jump was merely the first wave in the train of waves propagating upstream. A prediction of wavelength including the effects of longitudinal tension (see Sec. 23.5.1) is in reasonably good agreement with the measured lengths of the jump.

23.5 UNSTEADY-FLOW PHENOMENA

23.5.1 Wave Propagation

Although studies of steady flow provide useful insights into collapsible tube flow behavior, most physiological flows involve some degree of unsteadiness. Two important features of unsteady flow will be discussed here: (1) wave propagation and (2) flows induced by the application of nonuniform pressures applied external to the tube. Both features can be analyzed by use of the appropriate unsteady, one-dimensional equations of continuity for an incompressible fluid,

$$\frac{\partial}{\partial x}(uA) + \frac{\partial A}{\partial t} = 0 \tag{23.13}$$

and for conservation of momentum,

$$\frac{\partial u}{\partial t} + u\frac{\partial u}{\partial x} + \frac{1}{\rho}\frac{\partial}{\partial x}(p - p_e) + \frac{1}{\rho}\frac{\partial p_e}{\partial x} - G_x + \frac{4}{D_e}\frac{\tau_w}{\rho} = 0 \tag{23.14}$$

where G_x is the x component of all body forces per unit mass.

If the fourth, fifth, and sixth terms in Eq. (23.14) are set equal to zero, the equations are hyperbolic. Consequently, changes in state are transmitted along the tube by means of propagating waves which travel at a speed $u + c$ or $u - c$ depending on whether the waves are propagating with or against the flow, respectively, with the wave speed c given by Eq. (23.5).

Brower and Scholten [31] have measured wave speed in long tubes resting on a flat surface over a range of frequency and transmural pressure, and have found that predictions based on Eq. (23.5) are in good agreement with the low-frequency limit from the direct measurement. There was, however, a noticeable change in the observed wave speed as frequency increased which cannot be described by the simple hyperbolic equations above. This frequency dependence of wave speed, or wave speed dispersion, can be due to a variety of factors, one of which is the stiffness associated with the combination of longitudinal tension and curvature. As the wavelength decreases (with increasing frequency), the axial wall curvature increases, tending to make the tube seem less compliant, thereby increasing the wave speed. Wall mass, or the mass of surrounding fluid or tissue, can have a similar effect.

The effects of longitudinal tension have been studied by McClurken and colleagues [40] with a simple geometrical representation of the tube cross section in the partially collapsed state. Assuming that the effects of circumferential bending and longitudinal tension could be superimposed for waves of small amplitude, they obtained the following dispersion relationship:

$$c^2 = c_\infty^2 + \frac{\pi^2 T D_0 \bar{\alpha}}{\rho\lambda^2(1 - \bar{\alpha})^{1/2}} + \frac{\pi^4 K_p D_0^4 \bar{\alpha}}{2\rho\lambda^2(1 - \bar{\alpha})^{1/2}} \tag{23.15}$$

where c_∞ is given by Eq. (23.5), T is the axial tension, $D_0 = (A_0/\pi)^{1/2}$, λ is the wavelength, and $\bar{\alpha}$ is the mean area ratio. This prediction was compared with experimental measurements obtained from the precursor waves preceding stationary elastic jumps under conditions in which the third term on the right in Eq. (23.15) was negligible. Since, for the waves to be stationary in the laboratory reference frame, the phase speed of the wave must equal the flow velocity, the phase speed could be inferred from measurements of both the steady rate of flow passing through the tube and the local cross-sectional area. Good agreement was obtained between the theory and the experimental measurement, as indicated in Fig. 23.8.

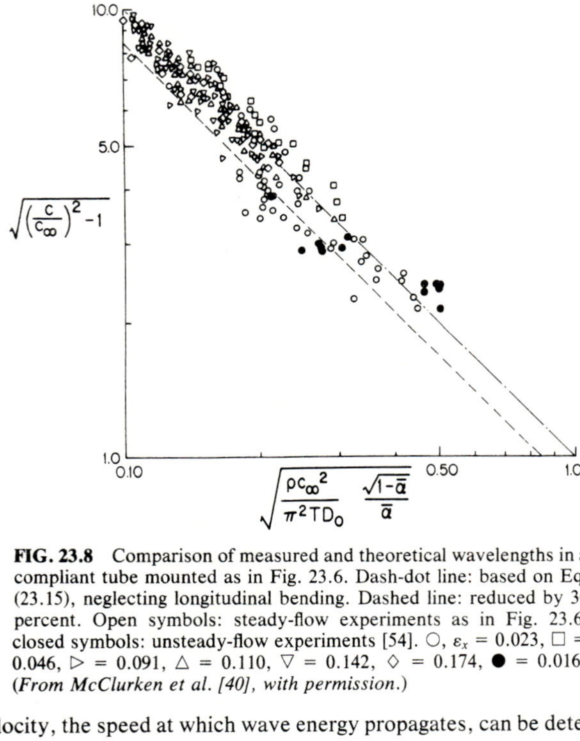

FIG. 23.8 Comparison of measured and theoretical wavelengths in a compliant tube mounted as in Fig. 23.6. Dash-dot line: based on Eq. (23.15), neglecting longitudinal bending. Dashed line: reduced by 30 percent. Open symbols: steady-flow experiments as in Fig. 23.6; closed symbols: unsteady-flow experiments [54]. ○, ε_x = 0.023, □ = 0.046, ▷ = 0.091, △ = 0.110, ▽ = 0.142, ◇ = 0.174, ● = 0.016. *(From McClurken et al. [40], with permission.)*

The group velocity, the speed at which wave energy propagates, can be determined from the relationship

$$c_g = c - \lambda \frac{dc}{d\lambda} \tag{23.16}$$

For a dispersion relationship of the type given in Eq. (23.15) for which the wave speed is monotonically increasing for decreasing wavelength, the group velocity always exceeds the phase velocity. This explains the appearance of standing waves upstream of the jump as seen in Fig. 23.6. Energy emanates from the jump and is transmitted at the group velocity. For waves above a certain frequency, this group velocity is greater than the supercritical flow speed and the waves can therefore propagate upstream against the flow. Those waves for which the phase velocity equals the flow velocity are therefore able to send energy upstream, but appear stationary to an observer. Such waves are analogous to capillary waves in free surface flows.

23.5.2 Changes in Wave Shape

For the purpose of studying the effects of tube law nonlinearity on the propagation of large-amplitude waves, Eqs. (23.13) and (23.14) can be transformed by using standard techniques [51] to the characteristic form given in the equations

$$\frac{dx}{dt} = u \pm c \tag{23.17}$$

$$du \pm c\frac{dA}{A} + H\,dt = 0 \tag{23.18}$$

$$H \equiv \frac{\partial}{\partial x}\left(\frac{p_e}{\rho} - \int G_x\,dx\right) + \frac{4\tau_w}{\rho D_e} \tag{23.19}$$

for a tube with uniform resting area and stiffness (constant A_0 and K_p). The curves in x–t space defined by Eq. (23.17) map the wave trajectories and identify the speed of wave propagation in the rightward ($+c$) and leftward ($-c$) directions with respect to the fluid motion. Along these physical characteristics, changes in state occur as determined by Eq. (23.18).

Waves propagating in a compliant tube can be shown either to steepen or broaden, depending upon whether the wave is one of compression or rarefaction and upon the characteristics of the tube as expressed by the tube law. To illustrate, consider the propagation of a rightward-running wave into an otherwise quiescent fluid. If we assume that the tube is uniform, that frictional effects are insignificant, and that there are no gradients in external pressure, then H is zero and the state characteristics [Eq. (23.18)] can be integrated along the physical characteristics [Eq. (23.17)] to yield [52]

$$u_R = -\int \frac{c}{A}\,dA + \text{constant} \tag{23.20}$$

$$u_L = \int \frac{c}{A}\,dA + \text{constant} \tag{23.21}$$

where the subscripts R and L denote rightward-running and leftward-running waves, respectively.

Since only rightward-running waves (R waves) are present, all leftward-running characteristics have the same integration constant in Eq. (23.21). Therefore, all points along a single R wave satisfy the same two equations and have the same solution. If u and A have constant values, then $u + c$ must also have constant values, and each R-wave trajectory is a straight line in x–t space. These are called *simple waves* [53]. It follows then that changes in state associated with a simple wave of a particular family are mapped onto a state characteristic of opposite family. Therefore, in a simple R wave, all properties are constant on the straight lines $dx/dt = u + c$, and changes in state from one such line to the next are related through

$$du = \frac{c}{\alpha}\,d\alpha = \frac{dp}{\rho c} \tag{23.22}$$

To determine whether a simple wave will either steepen or broaden, it is sufficient to consider the sign of

$$\frac{D}{Dt}\frac{\partial u}{\partial x} = \frac{\partial}{\partial t}\frac{\partial u}{\partial x} + (u + c)\frac{\partial}{\partial x}\frac{\partial u}{\partial x} \tag{23.23}$$

representing the change in velocity gradient in a lagrangian frame of reference. By incorporating previous equations, this expression can be rewritten as [52]

$$\frac{D}{Dt}\frac{\partial u}{\partial x} = -\frac{M}{2}\left(\frac{\partial u}{\partial x}\right)^2 \tag{23.24}$$

Expressions similar to Eq. (23.24) can be derived for A and p as well. It can be seen therefore, that compression waves ($du/dx > 0$) will either steepen [$(D/Dt)(du/dx) > 0$] or broaden [$(D/Dt)(du/dx) < 0$] depending on the sign of M. If $M > 0$, compressive waves will steepen; if $M < 0$, compressive waves will broaden. Rarefaction waves steepen for $M < 0$ and broaden for $M > 0$.

Direct calculations from an experimentally determined tube law suggest that generally $M > 0$ [52]. However, there is some evidence to suggest that M falls below zero when the two opposing walls of the tube come into point contact, in which case the steepening and broadening behavior reverses.

Experiments in which a wave is generated at one end of a very long tube [54] show that, if the wave is of sufficient amplitude, it will steepen into a traveling elastic jump or shock wave, the general character of which is in agreement with the predictions of the characteristic equations derived above. The detailed structure of the wave, however—in particular, the precursor waves that travel out in front of the shock—depends upon the effects of longitudinal tension and other factors producing wave speed dispersion. A theoretical analysis has also demonstrated the creation of an elastic jump by the steepening of a large-amplitude pressure wave introduced at one end of a long vessel, although the detailed structure, in these calculations, was dependent on the viscoelastic properties of the tube wall [55] rather than on longitudinal tension.

23.5.3 The Collapse of a Tube by External Compression

The problem of vessel collapse by external compression is important in a variety of situations ranging from the normal operation of the calf muscle pump to the methods of periodic external compression for the prevention of deep vein thrombosis [43, 56] or for external cardiac assist by lower limb compression [57]. Largely because of the strong nonlinearities in the tube law for a partially collapsed tube, the process of dynamic collapse exhibits several interesting and unique characteristics.

Some of the concepts developed in connection with the analysis of simple waves can also be applied to the problem of tube collapse by a spatial step in external pressure. If the tube is pressurized as shown in Fig. 23.9*a*, the regions both to the left and right of the step change in external pressure are regions of simple waves and can be treated as above. The step in pressure acts as the source of pressure disturbances which propagate to both the left and right away from the pressure discontinuity in external pressure. It remains, then, only to identify the matching conditions at the step in order to obtain a complete solution. These conditions are that the flow rate be continuous across the step, and that the stagnation pressure ($p + \rho u^2/2$) be constant. This, of course, neglects any loss in head that might occur across the discontinuity owing, for example, to flow separation or turbulence. It is convenient now to take the unique relationship that exists between u and A for simple waves, and express it instead in terms of $u\alpha$ versus α (Fig. 23.9*b*) so that the two states—to the right and to the left of the pressure discontinuity—fall on the intersection points between the state characteristics and a horizontal line. The "elevation" of the horizontal line is determined by the constraint on stagnation pressure. This results in a unique relationship between the applied pressure (the magnitude of the pressure step) and rate of flow passing the discontinuity at $x = 0$, as illustrated in Fig. 23.9*c*.

With the application of a time-varying pressure step, the tube area just upstream of the step diminishes, displacing fluid into the downstream segment and thereby inflating the tube to the right of the step. These waves of collapse and distension propagating to the left and right, respectively, cause the tube to collapse over an increasing length as time progresses.

Of particular note is the fact that the curve of $u\alpha$ in Fig. 23.9*c* has a maximum value beyond which the flow cannot be accelerated. This occurs at precisely the point at which the flow is critical ($S = 1$) and is therefore choked. As in the earlier description of choked flow, the fluid to the left of the pressure discontinuity can no longer be accelerated by waves propagating upstream that originate, in this case, at the pressure step.

In real flow situations, if pressures greater than the minimum required to produce choking are applied to the tube, a shock develops either within the region of varying external pressure or downstream of it, thus violating the previously stated matching condition of constant stagnation pressure.

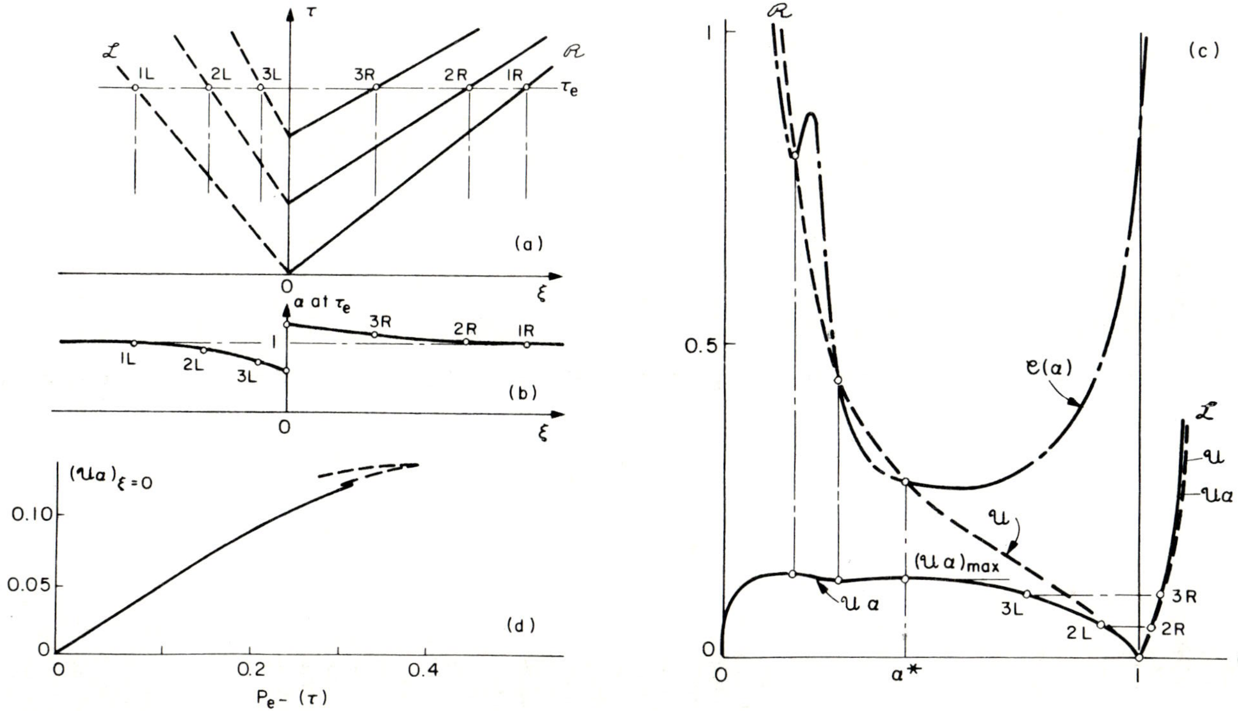

FIG. 23.9 Simple wave solution for time-varying step change in external pressure over the range $\xi = x/L < 0$, where L is a characteristic length of the tube. (*a*) ξ–τ wave diagram, showing R compression waves (solid lines) and L rarefaction waves (dashed lines) ($\tau = tc_0/L$). (*b*) Area distribution at the time τ_e. (*c*) Characteristic curves $u/c_0 \equiv U(\alpha)$ defined in Eqs. (23.20) and (23.21) and corresponding curves of $U\alpha$ versus α. Also curve of $C = c/c_0$ derived from experimental data. (*d*) Dimensionless flow at $\xi = 0$ as a function of applied external pressure, $P_e \equiv p_e/\rho c_0^2$.

23.6 OSCILLATIONS IN A COMPLIANT TUBE

Experimentally, oscillations have most frequently been observed at the downstream end of the tube, whether the tube is rigidly constrained at its ends (the Starling resistor) or is untethered. The latter case can be demonstrated by allowing the air to escape freely from a balloon, producing what is often referred to as the "Bronx cheer" type of oscillation. In only one instance of in vitro experiments have oscillations been reported that originate at a region far from the end of a tube and that are virtually free from any influence of the external rigid system [58]. This situation is, however, more representative of physiological examples, where rigid attachments are rare. These oscillations were observed to originate in the expansion region of an elastic jump and can be either random or more purely sinusoidal in character, depending on the local flow conditions [58]. It has been postulated [58, 59] that this type of tube flutter may be due to periodic vortex shedding similar to that observed in a two-dimensional diffusor with a moderate angle of divergence [60].

As suggested by the experiments described above, the flow-induced oscillations observed in a compliant tube are probably of multiple origin. The oscillations of a Starling resistor, as mentioned previously, are likely to depend not only on the properties of the flow and the compliant tube, but also on the capacitance and inertia of the several flow elements farther downstream.

Other oscillations are thought to be due to fluid-structure interactions, similar in character and origin to those that arise in flows past a compliant panel. The analysis of panel "flutter" is generally based on the linearized, inviscid flow equations and the structural equation for the flexible wall or vessel [61]. The latter variably includes the effects of wall mass, either axial bending stiffness or a "local" stiffness (as due, for example, to a series of springs attached to the outer surface of the wall), and structural damping.

In recent studies [62–64] the potential for instabilities of this type in a collapsible tube or channel conveying an incompressible fluid have been examined. Contrary to experimental observation, these analyses have predicted that the initial instability is one of static divergence if only a small amount of wall damping is included in the analysis. That is, the tube will either increase or decrease in area monotonically rather than oscillate about some mean value, at least while the constraints of the linearized model are satisfied. More recently, however, Grotberg [66] has shown that if fluid viscous effects, as well as wall damping, are included in the analysis, the initial instability is one of flutter.

In spite of the considerable advances made in the analysis of flutter, it is not yet possible to predict its onset and character in a typical experimental situation. This is no doubt partly due to the many assumptions that are necessary to make the analysis tractable. The most recent study [65] still deals with a two-dimensional channel and therefore neglects the effects of the tube sidewalls and the constraint they apply to the flow. Furthermore, the analysis assumes that the state of the tube prior to the imposition of infinitesimal deformations is perfectly uniform in the axial direction. Consequently, all the effects associated with axial variations in cross-sectional area—the possibility of flow separation or vortex growth, for example—are absent from the analysis. Further studies will clearly be necessary before this extremely complex problem is totally resolved.

REFERENCES

1. Guyton, A. C.: Venous return, in W. F. Hamilton and P. Dow (eds.), *Handbook of Physiology,* sec. 2: *Circulation,* American Physiological Society, Washington, D.C., 1963, vol. II.
2. Arnoldi, C. C.: Venous pressures in the leg of healthy human subjects at rest and during muscular exercise in the nearly erect position, *Acta Chir. Scand., 130,* 570, 1965.
3. Banister. J., and Torrance, R. W.: The effects of the tracheal pressure upon flow: Pressure relations in the vascular bed of isolated lungs, *Q. J. Exp. Physiol., 45,* 352–367, 1960.
4. Permutt, S., Bromberger-Barnea, B., and Bane, H. N.: Hemodynamics of collapsible vessels with tone: The vascular waterfall, *J. Appl. Physiol., 18,* 924–932, 1963.

5. Hyatt, R. E., Schilder, D. P., and Fry, D. L.: Relationship between maximum expiratory flow and degree of lung inflation, *J. Appl. Physiol., 13,* 331–336, 1958.

6. Pride, N. B., Permutt, B., Riley, R. L., and Bromberger-Barnea, B.: Determinants of maximal expiratory flow from the lungs, *J. Appl. Physiol., 16,* 77–84, 1967.

7. Lambert, R. K., and Wilson, T. A.: Flow limitation in a collapsible tube, *J. Appl. Physiol., 33*(1), 150–153, 1972.

8. Dawson, S. V., and Elliott, E. A.: Wave-speed limitation on expiratory flow—A unifying concept, *J. Appl. Physiol., 43,* 498–515, 1977.

9. Shapiro, A. H.: Steady flow in collapsible tubes, *J. Biomech. Eng., 99,* 126–147, 1977.

10. Griffiths, D. J.: Urethral elasticity and micturition hydrodynamics in females, *Med. Biol. Eng., 7,* 201–215, 1969.

11. Griffiths, D. J.: Hydrodynamics of male micturition: I. Theory of steady flow through elastic-walled tubes, *Med. Biol. Eng., 9,* 581–588, 1971.

12. Griffiths, D. J.: The mechanics of the urethra and of micturition, *Br. J. Urol., 45,* 497–507, 1973.

13. Holmberg, J. L., and Wilson, T. A.: Mechanics of the flow in an elastic tube extending from an orifice into a pressure vessel, Pap. No. 70-APM-ZZ, *J. Appl. Mech.,* 1970, pp. 1–5.

14. Johnson, M. C., and Kamm, R. D.: The role of Schlemm's canal in aqueous outflow from the eye, *Invest. Ophthalmol., 24,* 320–325, 1983.

15. Moses, R. A.: Hydrodynamic model eye, *Ophthamologica, 146,* 137–142, 1963.

16. Kety, S. S., Shenkin, H. A., and Schmidt, C. F.: The effects of increased intracranial pressure on cerebral circulatory functions in man, *J. Clin. Invest., 27,* 493–499, 1948.

17. Downey, J. M., and Kirk, E. S.: Inhibition of coronary blood flow by vascular waterfall mechanism, *Circ. Res., 36,* 753–760, 1975.

18. Winton, F. R.: Hydrostatic pressures affecting the flow of urine and blood in the kidney, *Harvey Lect., 47,* 21–52, 1951.

19. McCutcheon, E. P., and Rushmer, R. F.: Korotkoff sounds: An experimental critique, *Circ. Res., 20*(2), 149–161, 1967.

20. Cotton, L. T., and Roberts, V. C.: The prevention of deep vein thrombosis, with particular reference to mechanical methods of prevention, *Surgery, 81,* 228–235, 1977.

21. Birtwell, W., Giron, F., Soroff, H., Ruiz, U., Collins, J., and Deterling, R.: Support of the systemic circulation and left ventricular assist by synchronous pulsation of extramural pressure, *Trans. Am. Soc. Artif. Intern. Organs, 11,* 43–48, 1965.

22. Sanders, C. A., Buckley, M. J., Leinbach, R. C., Mundth, E. D., and Austen, W. G.: Mechanical circulatory assistance, *Circulation, 45,* 1292–1313, 1972.

23. Conrad, W. A.: Pressure-flow relationships in collapsible tubes, *IEEE Trans. Bio-Med. Eng., BME-16*(4), 284–295, 1969.

24. Brower, R. W., and Noordergraaf, A.: Pressure flow characteristics of collapsible tubes: A reconciliation of seemingly contradictory results, *Ann. Biomed. Eng., 1,* 333–335, 1973.

25. Holt, J. P.: The collapse factor in the measurement of venous pressure: The flow of fluid through collapsible tubes, *Am. J. Physiol., 134,* 292–299, 1941.

26. Rubinow, R. I., and Keller, J. B.: Flow of a viscous fluid through an elastic tube with applications to blood flow, *J. Theor. Biol., 35,* 299–313, 1972.

27. Shapiro, A. H.: Physiologic and medical aspects of flow in collapsible tubes, *Proc. 6th Can. Cong. Appl. Mech.,* Vancouver, B.C., 1977, pp. 883–906.

28. Lyon, C. K., Scott, J. B., and Wang, C. Y.: Flow through collapsible tubes at low Reynolds numbers, *Circ. Res., 45*(1), 68–73, 1980.

29. Rodbard, S.: A hydrodynamic mechanism for auto-regulation of flow, *Cardiologia, 48,* 532–535, 1966.

30. Rodbard, S., and Takacs, L.: Hydrodynamics of auto-regulation, *Cardiologia, 48,* 433–440, 1966.

31. Brower, R. W., and Scholten, C.: Experimental evidence on the mechanism for the instability of flow in collapsible vessels, *Med. Biol. Eng., 13,* 839–845, 1975.

32. Griffiths, D. J.: Oscillations in the outflow from a collapsible tube, *Med. Biol. Eng. Comput., 15,* 357–362, 1977.

33. Conrad, W. A., Cohen, M. L., and McQueen, D. M.: Note on the oscillations of collapsible tubes, *Med. Biol. Eng. Comput., 16,* 211–214, 1978.

34. Bertram, C. D., and Pedley, T. J.: A mathematical model of unsteady collapsible tube behavior, *J. Biomech.*, *15*(1), 39–50, 1982.

35. Kresch, E., and Noordergraaf, A.: Cross-sectional shape of collapsible tubes, *Biophys. J.*, *12,* 274–294, 1972.

36. Kresch, E.: Compliance of flexible tubes, *J. Biomech.*, *12,* 825–839, 1979.

37. Flaherty, J. E., Keller, J. B., and Rubinow, S. I.: Post buckling behavior of elastic tubes and rings with opposite sides in contact, *SIAM J. Appl. Math.*, *23*(4), 446–455, 1972.

38. Swidler, S. C.: Effect of longitudinal tension and curvature on pressure-area relationships of collapsible tubes, M.S. thesis, Department of Mechanical Engineering, Massachusetts Institute of Technology, Cambridge, Mass., 1980.

39. Kececioglu, I., McClurken, M. E., Kamm, R. D., and Shapiro, A. H.: Steady, supercritical flow in collapsible tubes: 1. Experimental observations, *J. Fluid Mech.*, *109,* 367–389, 1981.

40. McClurken, M. E., Kececioglu, I., Kamm, R. D., and Shapiro, A. H.: Steady, supercritical flow in collapsible tubes: 2. Theoretical studies, *J. Fluid Mech.*, *109,* 391–415, 1981.

41. Moreno, A. H., Katz, A., Gold, L. D., and Reddy, R. V.: Mechanics of distension of dog veins and other very thin-walled tubular structures, *Circ. Res.*, *27,* 1069–1080, 1970.

42. Morris, T. W., Abbrecht, P. H., and Leverett, S. D., Jr.: Diameter-pressure relationships in the unexposed femoral vein, *Am. J. Physiol.*, *227*(4), 782–788, 1974.

43. Kamm, R. D.: Bioengineering studies of periodic external compression as prophylaxis against deep vein thrombosis: I. Numerical studies, *J. Biomech. Eng.*, *104,* 87–95, 1982.

44. Oates, G. C.: Fluid flow in soft walled tubes: I. Steady flow, *Med. Biol. Eng.*, *13,* 773–778, 1975.

45. Bonis, M., and Ribreau, C.: Pressure-flow relationships in collapsible tubes, *Cardiovasc. Pulmon. Dyn.*, *92,* 459–466, 1977.

46. Beam, R. M.: Finite amplitude waves in fluid-filled elastic tubes: Wave distortion, shock waves, and Korotkoff sounds, NASA. Tech. Note D-480, Washington, D.C., 1968.

47. Cowley, S. J.: Elastic jumps on fluid-filled elastic tubes, *J. Fluid Mech.*, *116,* 459–473, 1982.

48. Griffiths, D. J.: Hydrodynamics of male micturition: I. Theory of steady flow through elastic walled tubes, *Med. Biol. Eng. Comput.*, *13,* 785–790, 1971.

49. Griffiths, D. J.: Negative-resistance effects in flow through collapsible tubes: I. Relaxation oscillations, *Med. Biol. Eng. Comput.*, *13,* 785–790, 1975.

50. Elliott, E. A., and Dawson, S. V.: Fluid velocity greater than wavespeed and transition from supercritical to subcritical flow in elastic tubes, *Med. Biol. Eng. Comput.*, *17,* 192–198, 1979.

51. Abbott, M. B.: *An Introduction to the Method of Characteristics,* Elsevier, New York, 1966.

52. Kamm, R. D., and Shapiro, A. H.: Unsteady flow in a collapsible tube subjected to external pressure or body forces, *J. Fluid Mech.*, *95*(1), 1–78, 1979.

53. Whitham, G. B.: *Linear and Nonlinear Waves,* Wiley, New York, 1974, pp. 164–167.

54. Jan, D. L., Kamm, R. D., and Shapiro, A. H.: Filling of partially-collapsed compliant tubes, *J. Biomech. Eng.*, *105,* 12–19, 1982.

55. Kivity, Y., and Collins, R.: Nonlinear wave propagation in visco-elastic tubes: Application to aortic rupture, *J. Biomech.*, *7,* 67–76, 1974.

56. Olson, D. A., Kamm, R. D., and Shapiro, A. H.: Bioengineering studies of periodic external compression as prophylaxis against deep vein thrombosis: II. Experimental studies on a simulated leg, *J. Biomech. Eng.*, *104*(2), 96–104, 1982.

57. Lueptow, R. M., Karlen, J. M., Kamm, R. D., and Shapiro, A. H.: Circulatory model studies of external cardiac assist by counterpulsation, *Cardiovasc. Res.*, *15*(8), 443–455, 1981.

58. Lichter, S.: Oscillation of shocks on compliant tubes, Ph.D. thesis, Massachusetts Institute of Technology, Cambridge, Mass., 1982.

59. Matsuzaki, Y., and Fung, Y. C.: On separation of a divergent flow at moderate Reynolds numbers, *J. Appl. Mech.*, *98*(2), 227–231, 1976.

60. Reneau, L. R., Johnston, J. P., and Kline, S. J.: Performance and design of straight, two-dimensional diffusers, *Trans. ASME D, J. Bas. Eng.*, *89,* 141–148, 1967.

61. Dugundji, J., Dowell, E., and Perkin, B.: Subsonic flutter of panels on continuous elastic foundations, *AIAA J.*, *1,* 1146–1154, 1963.

62. Weaver, D. S., and Paidoussis, M. P.: On collapse and flutter phenomena in thin tubes conveying fluid, *J. Sound Vibration, 50,* 117–132, 1977.

63. Matsuzaki, Y., and Fung, Y. C.: Stability analysis of straight and buckled two-dimensional channels conveying an incompressible flow, *J. Appl. Mech., 44,* 548–552, 1977.

64. Grotberg, J. B., and Davis, S. H.: Fluid-dynamic flapping of a collapsible channel: Sound generation and flow limitation, *J. Biomech., 13*(3), 219–230, 1980.

65. Grotberg, J. B.: A subsonic flutter anomaly, *J. Sound Vibration, 80*(3), 444–446, 1982.

66. Fry, D. L., Thomas, L. J., and Greenfield, J. C.: Flow in collapsible tubes, in D. J. Patel and R. N. Vaishnav (eds.), *Basic Hemodynamics and Its Role in Disease Processes,* University Park Press, Baltimore, 1980.

CHAPTER 24

Elasticity of the Lung

Donald L. Vawter
Jay D. Humphrey
School of Engineering
Georgia Institute of Technology
Atlanta, Georgia

INTRODUCTION

In this chapter we will look at our current knowledge of the mechanical behavior of the lung. We will be primarily interested in the elasticity of the lung but will also discuss the importance of surface tension. The presentation will be divided into three sections:

1. Experimental observations of the mechanical behavior of *lung tissue*.
2. Determination of the constitutive behavior of the lung tissue.
3. Numerical studies of the behavior of the *whole lung*.

In general, the approach of bioengineers working on lung mechanics has been to conduct experiments at the tissue level and then postulate a constitutive law which adequately describes the observations. These constitutive laws can then be used in numerical studies to predict overall organ response. The numerical predictions can then be compared with experimental observations to assess the adequacy of the constitutive law. In using this approach, however, it is implicitly assumed that one has not significantly altered the material behavior of the lung by obtaining specimens which can be studied at the tissue level. In reality, the behavior probably will be altered. The ramifications of this will be discussed subsequently.

24.1 EXPERIMENTAL OBSERVATIONS

Prior to 1968 most observations of the mechanical behavior of the lung were based on the behavior of the whole organ (e.g., pressure-volume relationships). It has been only recently that engineers and physiologists have attempted to describe the behavior of lung tissue as a material.

Four methods have been used in an attempt to better understand the mechanical behavior of the lung tissue:

1. Uniaxial tests on strips of alveolar wall
2. Biaxial tests on slabs of lung parenchyma
3. Triaxial tests on cubes of lung tissue
4. Indentation tests performed at the lobar level

24.1.1 Uniaxial Tests

In 1968 Fukaya et al. performed mechanical tests on individual alveolar walls. Typical specimen size was 30 × 30 × 200 μm. A sketch of the experimental approach is seen in Fig. 24.1. The

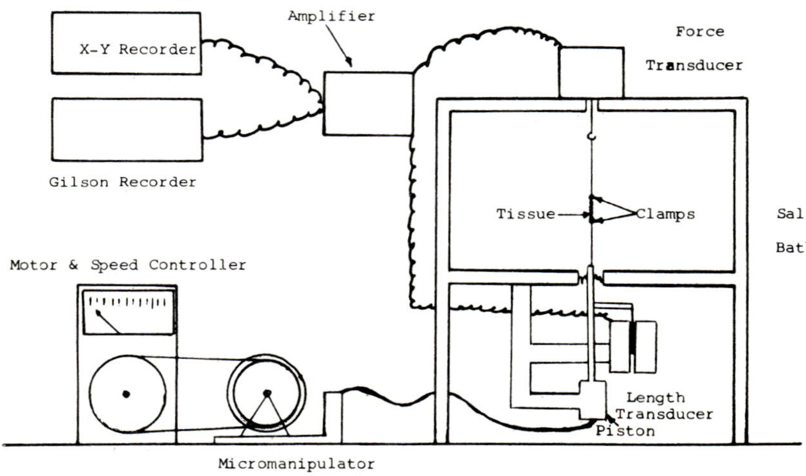

FIG. 24.1 Experimental apparatus for uniaxial microscopic tests on individual alveolar walls. Length is controlled by a micromanipulator. (*From Fukaya et al., 1968, with permission.*)

tissue is clamped at each end and immersed in a saline bath. Force-deformation relationships were measured using a Sanborn FTA1-1 force transducer and a Sanborn 7 dc DTO-50 displacement transducer. Peak force readings were under 10 mg. Measurements were quasi-static with recorder hysteresis apparent when the loading-unloading cycle was less than 3 s duration.

A typical response curve is shown in Fig. 24.2. Notice that the tissue exhibits highly nonlinear behavior, being very flexible at low loads and quite stiff when stretched. The primary goal of Fukaya's study was not to determine the constitutive behavior of the tissue but rather to have a method for comparing the response of many different specimens having diverse histories. He was, for example, concerned with the effect of aging on the response of the alveolar walls.

In using the results of the one-dimensional experiments in postulating a constitutive law, one must be aware of the experimental limitations of the uniaxial test. As seen in Fig. 24.1, it is not the specimen deformation that is measured but rather the spacing between the end clamps. A technical difficulty of nearly all tests on excised specimens is reliable gripping of the tissue. This is especially important when grip spacing is used to infer elongation. There can, of course, be large differences between the spacing measurements and the specimen deformation if there is any slippage or tear-out near the clamps. The importance of these edge conditions on the interpretation of uniaxial tests has been discussed in detail by Krueger and Pollack (1975) and by

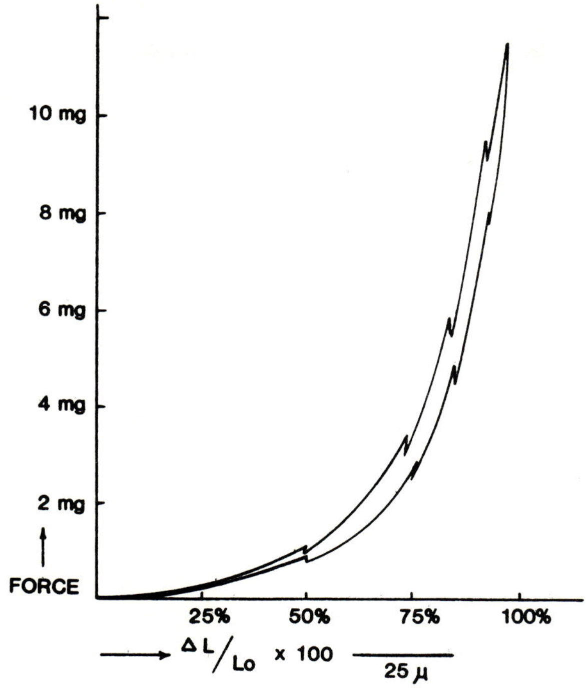

FIG. 24.2 Typical response curve for an alveolar wall stretched at a rate of 300 μm/min with pauses at five places on loading and four on unloading. The time-dependent hysteresis was almost eliminated by allowing stress relaxation. (*From Fukaya et al., 1968, with permission.*)

Pinto (1973). In the references cited the discussions concern cardiac muscle, but the principle remains the same.

Another difficulty connected with the uniaxial tests on microscopic specimens is in determining the stress and strain, given the force and deformation. In order to do this one needs, of course, a reliable measurement of both the length in the unstressed state and the cross-sectional area of the specimen. The cross-sectional area varies as the tissue is stretched. Because the alveolar walls are generally considered incompressible, the area of the specimen can be expressed as a function of length, given the value of the area in the unloaded state. Because of the extreme flexibility of lung tissue, it is difficult to measure this initial length and area reliably. Fukaya (1968) determined the length of the specimen and area not in the unstressed state but when a small preload had been applied. This eliminated curling of the specimen.

24.1.2 Bixial Tests

It has been shown (see Sec. 24.3) that measuring triaxial deformations of a slab of tissue loaded biaxially (in plane) provides sufficient information for determining a macroscopic constitutive relationship.

Vawter et al. (1978) have developed an experimental system for biaxial tests patterned after that of Lanir and Fung (1974) and have used the system for the study of lung parenchyma. The following is a brief description of the system with illustrative results. A more complete description of the system may be found in Vawter et al. (1978) and an analysis of the results in Vawter et al. (1979).

An optical-electromechanical system was developed for measuring the force-deformation behavior of biaxially loaded slabs of lung tissue whose size was approximately 5.0 × 5.0 × 0.5 cm. The slabs were stretched in plane by two orthogonal mechanical drive systems, one driven at a constant velocity and the other controlled by a servo which could be operated to keep either lateral displacement or load constant. A pictorial representation of the system is given in Figs. 24.3 and 24.4.

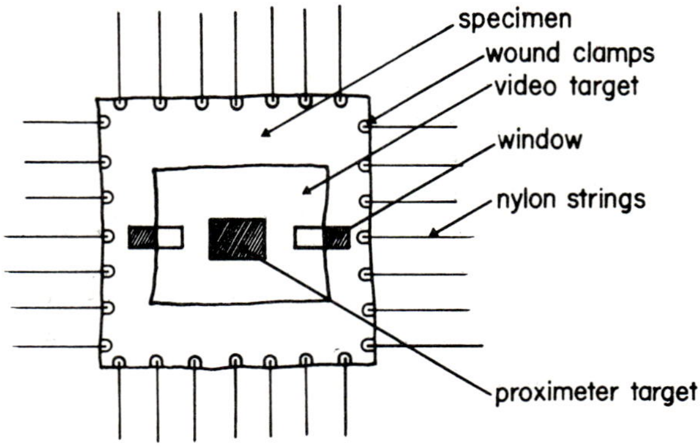

FIG. 24.3 Schematic of the video scene showing video targets, electronic windows, and specimen fixtures.

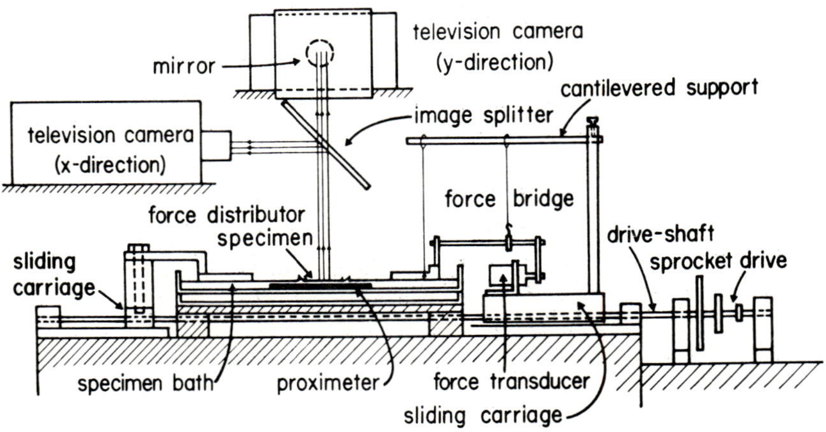

FIG. 24.4 Schematic of the overall testing device. For clarity, the upper proximeter is not shown.

Both in-plane forces and deformations as well as tissue thickness were measured during stress-strain tests and sampled on-line by a PDP-8E computer. The data acquired on digital magnetic tape were processed on a Burroughs 6700 computer. The system was designed to measure deformation in the center of the specimen where end conditions were shown to be unimportant.

In-Plane Dimensional Measurement. The in-plane dimensions were measured by a video technique previously used in the testing of skin (Lanir and Fung, 1974). The technique can be summarized as follows. Two sensing elements, called *windows*, which are displayed on the monitor, are placed at a given vertical position in the television screen. The height, length, and spacing of the windows may be independently chosen (Fig. 24.3). If, anywhere within the left-hand window, a certain predetermined gray level is reached, an electronic counting circuit is triggered. The counter is stopped when a second predetermined gray level is attained in the right-hand window. The two gray levels need not be the same. Knowing the scan rate and the time between starting and stopping of the counting circuit, one can compute the distance. The video analyzer does this electronically and gives an output voltage proportional to distance. The frequency response of the system is sufficiently high (20 Hz) for use in stress-strain tests (0.05 Hz).

In order to measure the in-plane dimensions of the lung specimens, a white rectangular target was painted in the center of the tissue. The target consisted of a suspension of titanium dioxide in saline, and covered about 40 percent of the length of the specimen and the central 15 percent of the area of the specimen.

The interface between the lung and the white target served as a contrast variation which triggered the counting circuit of the video analyzer. As the tissue stretched, the length of the target was measured continuously. Measuring the midportion of the specimen minimized the boundary effects.

Force Measurement and Mechanical Drive System. On each of the four sides of the specimen, seven strings were attached using 9-mm Clay-Adams surgical wound clips. The strings were fastened to four lucite force distributors (one to a side). On two perpendicular sides, the distributors were mounted to sliding carriages, while on the opposing sides, the distributors were connected rigidly by a force bridge to force transducers which were mounted to movable motor-controlled carriages. Each string was individually adjusted so that the video marker remained rectangular under tension as well as at rest.

Measurement of Specimen Thickness. The specimen cannot be assumed to be incompressible, and it was necessary to include specimen thickness as a measured parameter. The distensibility of the tissue precluded the use of conventional contacting measurement devices.

In order to measure thickness, a proximeter system was designed which consisted of two Bently proximeters and suitable targets. Proximeters are gap-to-voltage transducers which function by radiating radio-frequency (RF) signals from the tip of the probe to the surrounding area as a magnetic field. The presence of a conductor in the magnetic field sets up an eddy current and causes a loss of RF power. It was necessary to have proximeters mounted above and below the specimen to cancel out rigid-body motion of the specimen.

Cutting the Specimen. The most successful procedure found for cutting a smooth specimen was to defrost the lobe for 20 min at room temperature and then slice it on an electric rotary meat slicer. (The lobe was originally degassed, filled with saline, and frozen.) A thin slice was first taken to remove the pleura, and then a thick slab (4 to 5 mm) was cut and trimmed to 5 cm^2 per side.

Illustrative Results of the System. In order to illustrate the information which can be obtained from the system, we will show typical force-deformation curves for slabs of lung parenchyma, of which a typical specimen is shown in Fig. 24.5. The data presented are not filtered or smoothed in any way, and only every fourth data point is shown. In the following figures x refers to the direction which is driven at constant velocity, y is the direction of servo-controlled dimension, and z is the direction of specimen thickness. The results shown are from operation in the force mode, i.e., the lateral force (y direction) is held constant; λ_x, λ_y, and λ_z are the stretch ratios in the x, y, and z directions respectively; and T_x, T_y are the lagrangian stresses. T_z is, of course, zero throughout.

Figure 24.6 plots T_x versus λ_x for four different lateral loads. For clarity, only the extension portion of the cycles are shown. Notice that the curves are highly nonlinear, with the tissue being quite flexible at low stresses and quite stiff at higher stresses. Figure 24.7 shows the lateral

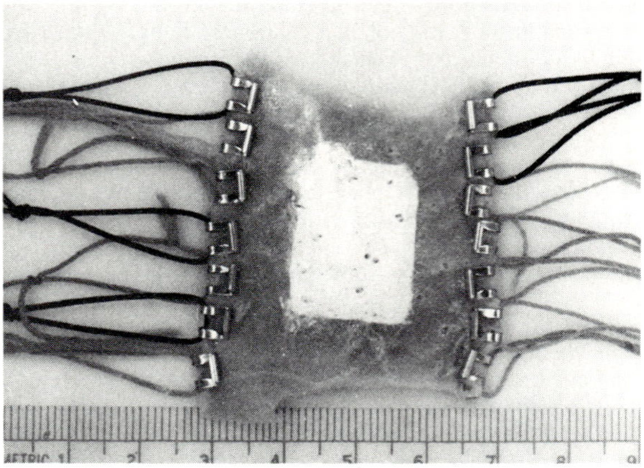

FIG. 24.5 Parenchymal specimen showing clamps and video target. Lateral clamps are not shown. Notice that the specimen is relatively free of large airways.

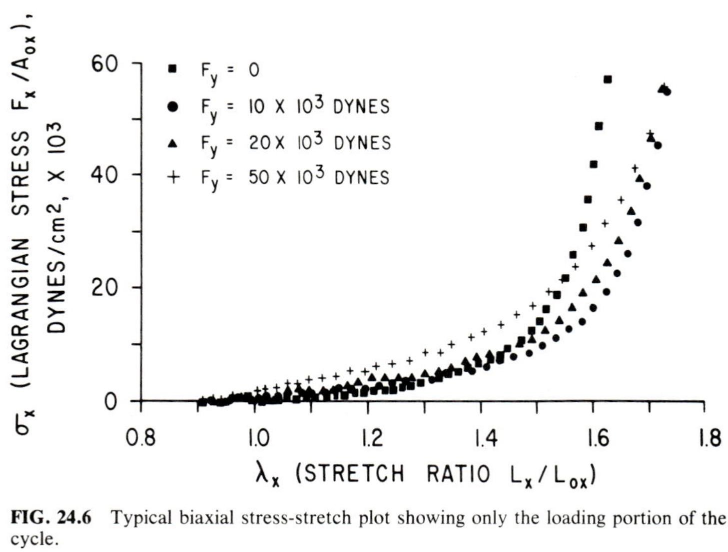

FIG. 24.6 Typical biaxial stress-stretch plot showing only the loading portion of the cycle.

dimension, λ_y, versus λ_x for the above lateral loads. As the tissue is stretched farther and, hence, is stiffer, the lateral extension required to keep constant decreases. In Fig. 24.8, we see thickness λ_z versus λ_x. Notice the curve is nonmonotonic, showing the competition between increasing λ_x and decreasing λ_y. The former is dominant for small λ_x, while λ_y dominates for large λ_x.

24.1.3 Triaxial Tests

On the surface, triaxial testing of parenchyma appears to offer advantages over biaxial testing in that the deformations induced are more representative of the in vivo situation than are those

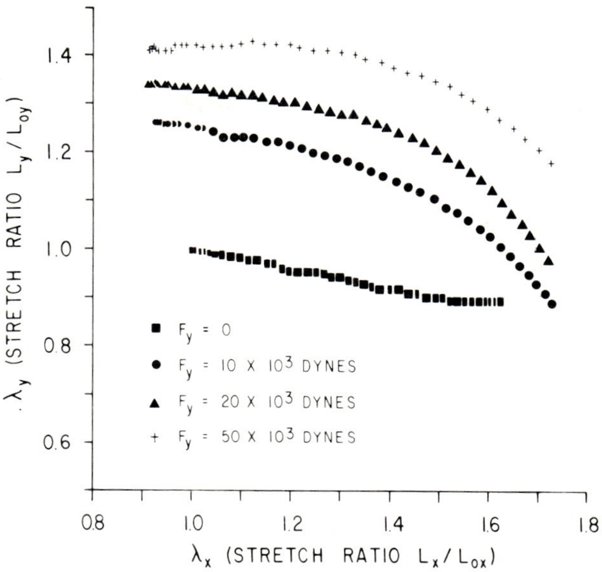

FIG. 24.7 Variation of the laterial extension λ_y for the loading test shown in Fig. 24.6.

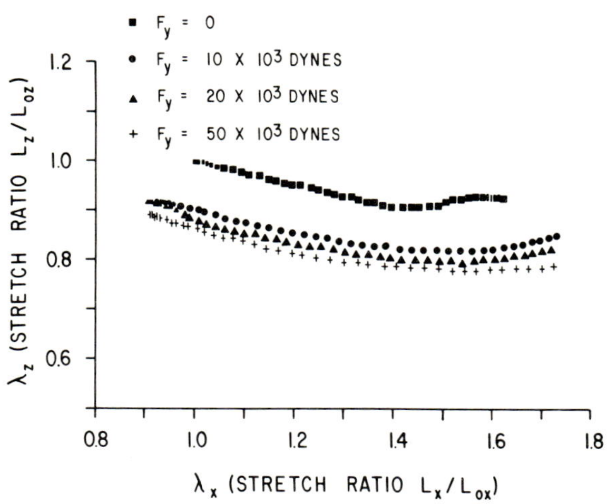

FIG. 24.8 Variation of the thickness λ_z for the loading test shown in Fig. 24.6.

induced in the biaxial tests. In the biaxial tests the out-of-plane dimension is always in compression, a situation occurring infrequently in vivo (Vawter et al., 1978). As we will discuss subsequently, this potential advantage is at least partially offset by technical difficulties encountered in triaxial testing.

Hoppin et al. (1975) have performed triaxial tests on both air-filled and saline-filled cubes of lung parenchyma with edge length about 1 cm.

Application of Force. Sixteen tiny fishhooks (no. 24) were hooked into each face of the cube. The fishhooks were connected to a lucite plate via 14-cm-long 4-0 silk threads. Figure 24.9 shows a schematic sketch of the system. Forces were applied by hanging dead weights.

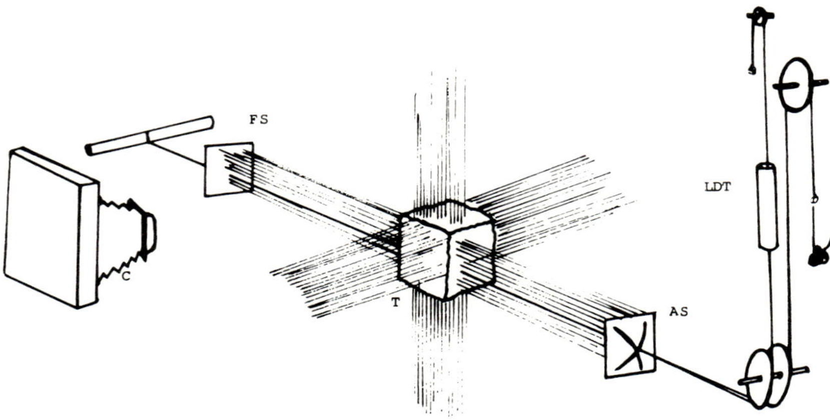

FIG. 24.9 Schematic drawing of the triaxial testing apparatus showing the tissue specimen (T) suspended by fixed supports (FS) and active supports (AS). Active supports lead to calibrated weights (W) and to a linear differential transducer (LDT) readout of extension. Camera (C) records extension for purposes of calibration and baseline length measurement. Connections of only one axis are shown. (*From Hoppin et al., 1975, with permission.*)

Displacement Measurements. One of each pair of lucite plates was fixed, with the other movable. The displacement of the movable plate was measured with a linear differential transducer (LDT) and by a camera for verification.

Results. The authors have presented results on only two saline-filled cubes and three air-filled cubes. Figure 24.10 shows typical force-extension ratio curves for a saline-filled specimen. Note that this is *not* a stress-free extension ratio curve. Also, the extension ratio definition is unconventional, since the reference is not the edge length in the stress-free state. Rather, Hoppin et al. (1975) have defined the reference length by

$$l_{x0} = l_{x,\max}\left(\frac{m}{\rho_0}\right)^{1/3}(l_{x,\max}l_{y,\max}l_{z,\max})^{-1/3}$$

where $l_{x,\max}$, $l_{y,\max}$, $l_{z,\max}$ are the maximum edge lengths, m is the mass of the cube, and ρ_0 is the "assumed density in the collapsed state." The present authors were unable to find any record for the value of this assumed density.

Limitations of Triaxial Tests. The triaxial test has a problem in common with the biaxial test. In both tests the lung is frozen and cut before testing. Although there is little evidence that freezing and thawing the lung alters the mechanical behavior significantly, it clearly is not a desirable procedure. One cannot, of course, compare the triaxial or biaxial results of a specimen that has never been frozen with one that has been frozen. To the authors' knowledge a comparison of specimens frozen once and those frozen and thawed many times has not been made.

Another difficulty with the triaxial test is strain measurement. Although the authors attempted to distribute the loads evenly, there must be significant distortion near the point of attachment. If one measures tissue deformation by measuring the distance between testing

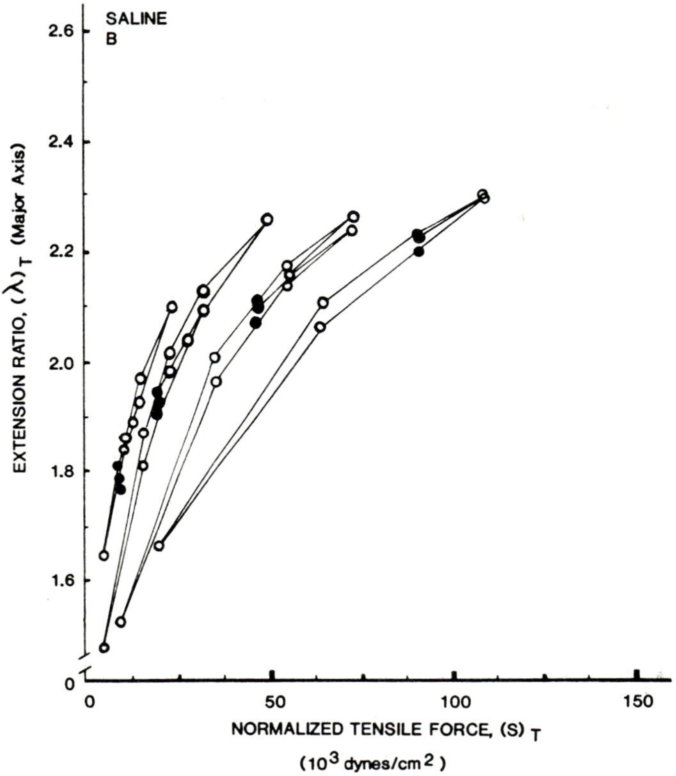

FIG. 24.10 Typical force-extension ratio curves for a saline-filled cube under points of asymmetrical loading (open circles) and symmetrical loading (closed circles). (*From Hoppin et al., 1975, with permission.*)

grips, any "pullout" near the contacts is necessarily included. Hoppin et al. (1975) have estimated that 80 percent of the alveoli in their preparation were open.

Another problem common to all tests on excised tissue is that the tissue is cut into pieces prior to testing. The magnitude of the importance of this fact, although intuitively it would appear severe, is unknown.

24.1.4 Other Macroscopic Tests

Because of the potential change in material properties which may occur when the lung tissue is cut into slabs or cubes, some investigators have attempted to determine mechanical behavior from tests on either the whole lung or an individual lobe.

The major difficulty in intact experiments is interpretation of the results. This is the classic experimental stress analysis dilemma: "If you can do the experiment, you can't solve the boundary value problem, and if you can solve the boundary value problem, you can't perform the experiment." The truth of this statement will be apparent to anyone who attempts to obtain a homogeneous specimen of lung tissue of regular geometric shape.

Lai-Fook and colleagues (1976, 1980) have performed experiments which attempt to simulate tractable boundary value problems. Two problems have been studied: (1) indentation of an elastic half space and (2) the behavior of a flexible disk bonded to an elastic half space.

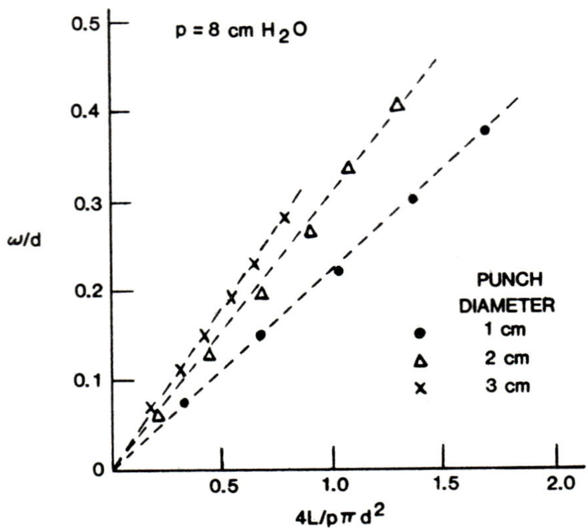

FIG. 24.11 Dimensionless displacements versus dimensionless applied pressure of 1-, 2-, and 3-cm-diameter punches at inflation pressure of 8 cmH_2O. The punch displacement is normalized with respect to punch diameter, whereas the applied load is made dimensionless by dividing by the inflation pressure multiplied by the contact area. (*From Lai-Fook et al., 1976, with permission.*)

In 1976 Lai-Fook, Wilson, Hyatt, and Rodarte performed cylindrical punch indentation tests on lobes of dog lungs at a fixed indentation pressure. The punch displacement versus applied load was measured using punch diameters of 1, 2, and 3 cm. The results shown in Fig. 24.11 are for an inflation pressure of 8 cmH_2O. The authors calculated the material properties using the results of classical linear elasticity for the deformation of a semi-infinite elastic solid (Timoshenko and Goodier, 1970). It should be noted that the solution used is based on the assumptions of small strain and a hookean stress-strain relationship, neither of which is likely to be valid for the problem studied. Shields (1979) has analyzed the experiments by use of the finite element method and the constitutive law proposed by Fung (1974), which will be discussed later. The effect of the pleural membrane on the test was studied subsequently by Hajji (1978). In 1979 Hajji, Wilson, and Lai-Fook extended these results and proposed a method for measuring not only the material properties but also the tension in the pleural membrane. These later results showed that it is absolutely essential to take account of the pleural membrane when analyzing indentation tests.

Lai-Fook et al. (1980) have also bonded small flexible disks to the lung's surface and then inflated or deflated the lung while measuring the deformation of the disk. They again have used a linear analysis in an attempt to determine the material properties, in particular Poisson's ratio. It is shown elsewhere (Vawter, 1983) that Poisson's ratio loses its physical meaning in finite elasticity.

24.2 MEASUREMENT OF THE "REFERENCE STATE" ON THE LUNG

As we will soon see, a description of the mechanical behavior of the lung must be based on a known reference state. In classical elasticity the logical reference state is the stress-free state. Because of the rigidity of engineering materials, measurement of this state is relatively simple.

For soft tissues such as the lung parenchyma, a stress-free state is difficult to attain and even more difficult to measure. The reason for this is apparent from Figs. 24.2, 24.6, and 24.10. At low stress levels the lung is quite deformable and a measurement of the length of zero load is difficult. Given the deformability at low loads, it seems that any measurement should necessarily be a noncontacting measurement.

There have been three approaches used to overcome the above difficulties:

1. For one-dimensional tests a small preload was applied and the resultant length was used as the reference length. Ideally one would like to apply a constant "prestress" rather than a preload. Since those testing one-dimensional specimens generally used specimens of almost the same size, the use of a constant preload probably approaches applying a constant prestress. However, difficulties arise when comparing the results of tests performed on specimens of varying size.

2. The stress-free state of the tissue can be estimated if the volume of the whole lung can be measured in a stress-free environment. One way to approximate the stress-free state is to degas the lung and then refill it with saline, nearly eliminating surface tension effects. In order to minimize gravitational effects the lung should be immersed in saline (whose density is near that of the lung tissue). If one then allows the saline within the lung to communicate with the bath, the lung volume should equilibrate to the stress-free volume (if such a volume exists). Studies by Vawter (1977b) of five lungs revealed that a unique stress-free state exists and that the volume of the lung divided by the volume of the lung tissue has an average value of 2.74 in the stress-free state. If one assumes the lung is made up of alveoli, which are regular polyhedra, then the above value implies a ratio of diameter (of an inscribed sphere) to thickness (of the alveolar wall) of 7.1.

3. Using another reference state may also be helpful. If one bases the reference state at total lung capacity (TLC), then one is using a reference state that lies on the steep portion of the stress-strain curve. In this region large changes in stress cause relatively small changes in strain. One practical difficulty is that the stress-strain curve is never truly vertical and hence there really is no such thing as TLC. This is usually circumvented by defining TLC as the volume attained when the inflating pressure is 25 cmH_2O.

One may conclude that no one measurement of a reference state is entirely satisfactory. Given the importance of comparing experimental results from different laboratories, it would seem prudent to quantify the volume (or lengths) at as many of the commonly used reference states as possible. This would eliminate the necessity of inferring calculations from figures in the literature (which, given size restrictions, often do not portray the accuracy of the underlying measurements).

24.3 DETERMINATION OF THE CONSTITUTIVE BEHAVIOR OF THE LUNG

In the preceding section we discussed a variety of experiments which measured the deformation of the lung during prescribed loading. In this section we discuss attempts to model the mechanical behavior of the lung. If successful, a model will accurately predict the mechanical response of the lung which has been experimentally observed.

The pioneering work on modeling the mechanical behavior of the lung is the paper by Mead, Takishima, and Leith in 1970. It was in this paper that the basic idea of modeling the lung parenchyma as a mesh of membranes was presented. In order to describe qualitatively the interaction among alveoli, they modeled the lung as a network of springs connected in a hexagonal array. Under a variety of loads the enclosed hexagons vary in size throughout the structure. Mead et al. described this nonuniformity as *interdependence,* a term widely used in the recent literature. They stopped short of a mathematical formulation; however, this was carried out shortly afterward by Wilson (1972). Wilson analyzed the behavior of the two-dimensional spring model, using a linearized approach valid only for small perturbations from the equilibrium configuration.

A further extension was made by Lambert and Wilson (1973), in which the lung tissue is pictured as number of interconnected, randomly oriented, plane, elastic membranes. Again, this model was linearized for small incremental variations of stresses and strains about a state of uniform inflation. The incremental stresses and strains obey Hooke's law, for which the concepts of Young's modulus and Poisson's ratio apply. However, these constants vary with the state of equilibrium on which the perturbations are imposed. In particular, unless the state of equilibrium is isotropic (as in a uniform inflation of a lung with material isotropy), the incremental law is anisotropic. Only the isotropic case is analyzed by Lambert and Wilson.

The journey into a description of the three-dimensional structure of the lung parenchyma began in 1974 with the work of Frankus and Lee and of Fung. Frankus and Lee used a dodecahedron as the geometric form for a typical alveolus. One can fill a three-dimensional space with dodecahedrons without creating voids; however, only for the simplest geometries will regular dodecahedrons suffice.

Frankus and Lee assumed that the orientation of elastic fibers within the alveolar wall was random and hence that the behavior of the alveolar wall was isotropic. They used a constitutive equation for the strain energy density (W) in terms of the strain invariants (I_1, I_2, I_3) of the form

$$W = a_1 I_1 + a_2(I_1^2 - 2I_2) + a_3(I_1^3 - 3I_1 I_2 + 3I_3) + a_4(I_1^4 - 4I_1^2 I_2 + 2I_2^2 + 4I_1 I_3)$$
$$+ b_1 I_2 + b_2(I_2^2 - 2I_1 I_3) + c_1 I_3 + c_2(I_1 I_2 - 3I_3) + c_3(I_1^2 I_2 - 2I_2^2 - I_1 I_3)$$

where typical values of the constants are (as determined using the pressure-volume curve):

$a_1 = -21.06$	$b_1 = 2.673$	$C_1 = 1.324$
$a_2 = 19.76$	$b_2 = -0.350$	$C_2 = -1.94$
$a_3 = -7.88$		$C_3 = 0.943$
$a_4 = 1.062$		

Fung used the concept of the "average alveolus" and determined the macroscopic stress state by first considering the histological geometry of the alveoli and then deriving certain integrals which represent the macroscopic stress. He then derived a statistical relationship between the macroscopic strains and the strains in the individual interalveolar septa by means of a mathematical concept of a statistical "mean alveolus." Linearization is unnecessary in Fung's approach. Fung gives analytical results for cases when the mean alveolus is spherical (and deforms to an ellipsoid) and cubic (deforms to a rectangular parallelepiped). By far the simplest model for the mean alveolus is the cube. Fung found that the final constitutive equation does not change very much with the shape of the mean alveolus. For this reason the cubic alveolus was used in a later analysis of the atelectasis of the lung (Fung, 1975) and by Vawter et al. (1979), who used the model in combination with the observations on biaxial slabs of lung discussed above.

Vawter et al. (1979) used the cubic model of the alveolus discussed above in order to formulate a relationship between macroscopic stress in the lung and macroscopic strain. A macroscopic relationship is formed by first describing the behavior of an individual alveolar membrane and then summing the contributions of all membranes. For an individual membrane it was hypothesized that the strain energy of the membrane can be described by

$$M_0 W^{(1)} = \frac{C'}{2} \exp\,(a_1 E_x^2 + a_2 E_y^2 + 2a_4 E_x E_y)$$

where M_0 is the mass of the interalveolar septa per unit area of the membrane in the resting (unstressed) state; $W^{(1)}$ is the strain energy per unit mass of the alveolar septa; E_x, E_y are the strains in the x and y directions, respectively, as defined by Green; and C, a_1, a_2, and a_4 are material constants.

Combining the contributions of all membranes leads to the following equation for the langrangian stress (T_x) in the x direction:

$$T_x = C\lambda_x(a_1E_x + a_4E_y)\exp(a_1E_x^2 + a_2E_y^2 + 2a_4E_xE_y)$$
$$+ C\lambda_x(a_1E_x + a_4E_z)\exp(a_1E_x^2 - a_2E_z^2 + 2a_4E_xE_z)$$

and to a macroscopic strain energy function given by

$$\rho_0W = \tfrac{1}{2}C\exp(a_1E_x^2 + a_2E_y^2 + 2a_4E_xE_y) + \text{symmetrical terms by permutation}$$

If the model is restricted to the isotropic case (i.e., $a_1 = a_2$), then the description of the strain energy is much simpler and is given by

$$\rho_0W = \tfrac{1}{2}C\exp(\alpha I_1^2 + \beta I_2)$$

where I_1 and I_2 are the strain invariants

$$I_1 = E_{xx} + E_{yy} + E_{zz}$$
$$I_2 = E_{xx}E_{yy} + E_{yy}E_{zz} + E_{zz}E_{xx} - E_{xy}^2 - E_{yz}^2 - E_{zx}^2$$

and C, α, β are material constants.

Values for the material constants are given in Table 24.1 with the data from the biaxial experiments of Vawter et al. (1978). Figure 24.12 shows a typical comparison between the experimental data and the model predictions. Good correlation was seen as long as uniaxial tests were not involved. Biaxial and uniaxial tests appear to have some fundamental differences which the above approach does not correctly describe.

Once one attempts to model the parenchyma structure explicitly rather than statistically, the generation of constitutive equations based on experimental observations becomes much more difficult. In general, some numerical technique, such as the finite element method, is required to determine the material constants. Primarily this is necessary because, with a complicated alveolar geometry, it is not possible to predict analytically the response of the structure to prescribed loading even when the material behavior is known.

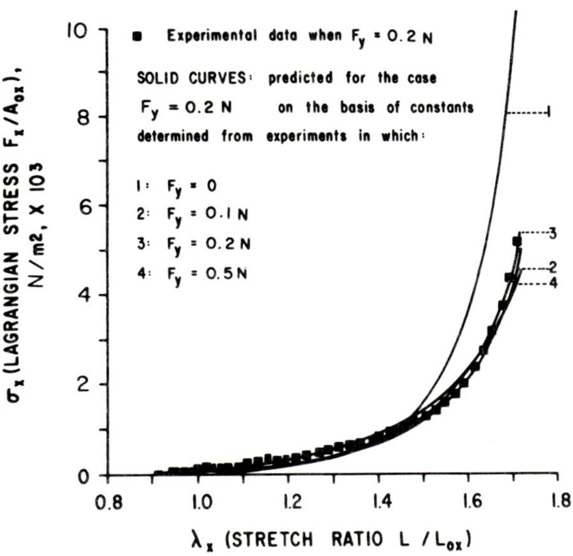

FIG. 24.12 Prediction based on the biaxial loading response with a laterial load of 0.1 N. Note that curves 2, 3, and 4 fit well. Curve 1, however, which is calculated with constants derived from the uniaxial case, does not fit well.

TABLE 24.1 Best-fit material constants of dog's lung tissue elasticity in loading process (stretching), without surface tension, in normal saline bath at 20°C, pH 6.7

		Tissue considered anisotropic					Tissue considered isotropic				
						Correlation based on				Correlation based on	
Species no.	F_y, N	C, Pa	a_1	a_2	a_4	T_x, T_y	C, Pa	$a_1 = a_2$	a_4	T_x	T_x, T_y
4081	0	74	2.84	*	*	0.995	74.0	2.840	*	0.995	*
	0.1	147	1.44	4.67	0.522	0.996	88.0	1.705	5.049	0.997	0.932
	0.2	225	1.27	3.71	0.569	0.995	141.3	1.512	1.688	0.996	0.934
	0.5	1210	0.481	0.860	0.177	0.972	571.3	0.732	0.427	0.992	0.766
4181	0	58.4	3.51	*	*	0.998	58.4	3.514	*	0.998	*
	0.1	86.7	2.18	4.71	0.920	0.997	50.2	2.587	4.063	0.999	0.973
	0.2	100	2.12	3.89	0.976	0.996	122.5	1.962	0.879	0.997	0.976
	0.5	289	1.33	2.28	0.427	0.966	313.2	1.259	0.613	0.998	0.863
5191	0	56.4	2.954	*	*	0.997	56.4	2.954	*	0.997	*
	0.1	420	0.807	1.95	0.641	0.995	120.5	1.397	4.084	0.999	0.935
	0.2	802	0.524	1.37	0.477	0.999	199.9	1.093	1.978	0.998	0.898
	0.5	741	0.762	2.60	1.02	0.989	909.4	0.688	0.785	0.999	0.769
6021	0	160.7	3.75	*	*	0.999	160.7	3.749	*	0.999	*
	0.1	274	1.46	4.42	0.613	0.994	233.8	1.560	4.486	0.998	0.982
	0.2	565	0.945	2.43	0.788	0.995	582.9	0.917	0.985	0.999	0.974
	0.5	938	0.697	1.94	0.673	0.990	611.6	0.941	0.769	0.996	0.782

* a_2 and a_4 cannot be determined from uniaxial loading experiments.

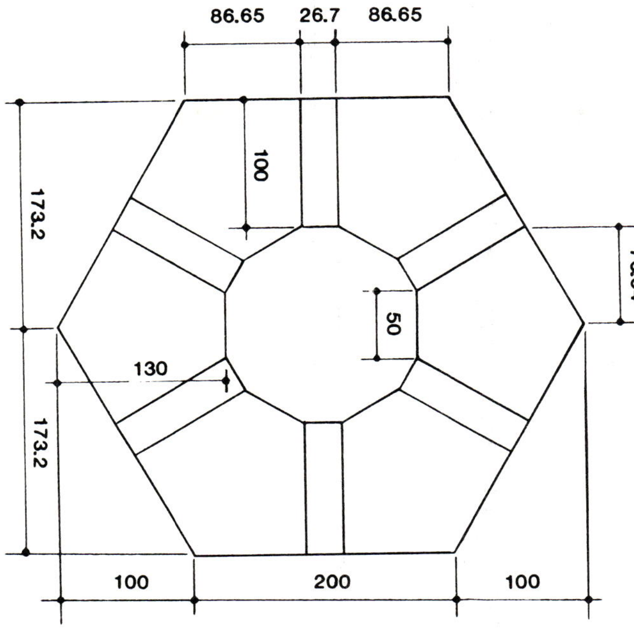

FIG. 24.13 The proposed geometric model of a cluster of alveoli around an air duct: cross section normal to the axis of the duct. All dimensions are for the state of zero pressure and are in micrometers. (*From Karakaplan et al., 1980, with permission.*)

Karakaplan et al. (1980, 1982) have modeled the lung parenchyma and attempted to take into account the morphology of the parenchyma structure. They have used a honeycomb type of structure to model the geometry, after the description of Weibel (1963), who describes alveoli as small lateral outpouchings of alveolar ducts and who finds a honeycomb bent around the cylindrical surface of the alveolar duct an acceptable model of a cluster of alveoli. Figures 24.13 and 24.14 illustrate their model. The alveolar duct is terminated at one end by an alveolus and connected to a larger airway at the other end. Larger expanses of parenchyma are formed by joining these alveolar duct clusters.

Their model for the constitutive behavior of the alveolar wall sheets is similar to that of Fung and is given by

$$W = \tfrac{1}{2}(\lambda\delta_{ij}\delta_{kl} + \mu\delta_{ki}\delta_{lj})e_{ij}e_{kl}\exp(aI_1^2 + bI_2)$$

where λ, μ, a, b are material constants; δ_{ij} is Kronecker's delta; and the indices assume values of 1 and 2.

In order to describe the behavior of the parenchyma, or at least the duct model, Karakaplan et al. (1982) make use of the finite element method. Basically they model the structure by the finite element method and then determine a set of material constants (λ, μ, a, b) which gives the best agreement with experimental results either on strips of parenchyma or with pressure-volume measurements. Typical values of the constants, obtained by using pressure-volume curves, are

$$\lambda = 7355 \text{ dyn/cm}^2$$

$$\mu = 1.2258 \text{ dyn/cm}^2$$

$$a = 0.50$$

$$b = 0.50$$

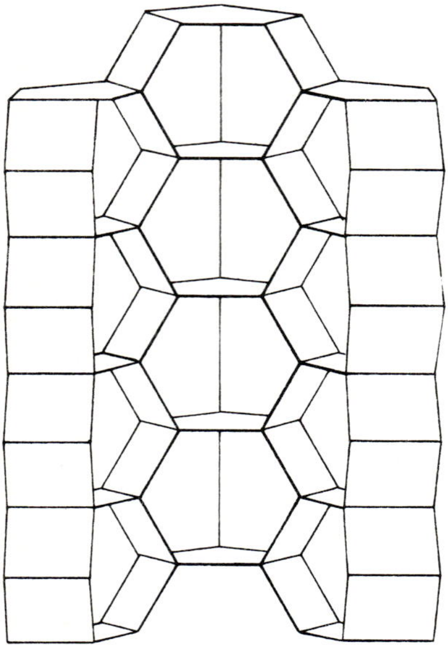

FIG. 24.14 Interior of the alveolar duct shown in Fig. 24.13. (*From Karakaplan et al., 1980, with permission.*)

Comparison of their model with a variety of experimental data is included in the cited references.

Most models of the lung parenchyma assume that the alveolar sheets are the source of mechanical strength. Recently this view has been challenged by Wilson and Bachofen (1982), who have proposed a model in which the alveolar rims of the alveolar openings serve as a source of strength. Their model can be described succinctly from the abstract of the cited reference:

> There are two networks of force-bearing elements. The first is an interdependent part of the peripheral connective tissue system that starts from the pleura and extends into the interlobar and interlobular fissures. At the sublobular level, its geometry is not yet fully clear. This network is extended by changes in lung volume and is insensitive to surface tension. The second network is composed of the line elements that form the rims of the alveolar openings. This network is the terminal part of the axial fiber system that surrounds bronchi, bronchioli, and arteries. The line elements of this network are extended by the outward force of surface tension. The two-dimensional alveolar walls that form the alveoli are negligible mechanical components except as platforms for surface tension at the air-liquid interface.

24.4 SURFACE TENSION IN THE LUNG

An important, perhaps dominant, force in the constitutive behavior of the lung is that of surface tension. A discussion of the surface-active behavior is beyond the scope of this article. Suffice it to say here that there is general agreement that the value of the surface tension varies with alveolar surface area but that there is no general agreement as to the quantitative description of the variation. The reader is referred to the literature for a discussion (Brown et al., 1959; Clements, 1962; Flicker and Lee, 1974; Scarpelli, 1965; Scarpelli and Gabbay, 1965; Shields, 1979; Shields and Vawter, 1980; Slama et al., 1973; Vawter and Shields, 1982; Wilson, 1980).

24.5 MATHEMATICAL MODELS OF WHOLE-LUNG BEHAVIOR

In the earlier sections we have found that the constitutive behavior of the lung is quite complex and that any discussion of the behavior of the whole lung is likely to involve numerical modeling rather than a purely analytical approach. Because the lung is so flexible at low levels of stress, it is dramatically affected by gravitational and other forces. This flexibility causes a nonuniform distribution of air within the lung and hence affects gas exchange. There have been a number of numerical studies on the response of the whole lung to gravitational and other forces. Most of these studies make use of the finite element method. Rather than describe in detail the finite element models of each researcher, we will discuss only the assumptions used in the analysis. Each cited reference contains a more complete description of the model used.

The pioneering finite element analysis of the lung was done by West and Matthews in 1972. This important paper was instrumental in showing the importance of gravity in the deformation of the lung and was probably the catalyst for more refined analyses. Because of limited knowledge of the material behavior of the lung, an elastic model based on an extension of uniaxial stress-straining behavior was assumed. From the viewpoint of continuum mechanics such an extension to finite elasticity is inconsistent, and hence the quantitative results are subject to question. In particular, the analysis implies that tissue behavior is isotropic even when tissue is distorted, which is not consistent with experimental observations.

The grid used was rather coarse, and a demonstration of convergence of the solution was not presented. Only elastic behavior was modeled, although it should be noted that the material properties were determined by using air-filled pressure-volume curves. The effects of surface tension, then, are implicit in the analysis. The necessity of modeling surface tension behavior explicitly has been shown in papers to be discussed shortly.

The mathematical basis for the West and Matthews analysis (1972) is found in the paper by Mathews and West (1972). Extensions of the method to assess the effect of chest size and shape are presented by Vawter, Matthews, and West (1975). This later work allows for distortion of the lung due to the chest cavity but makes no improvement on the modeling of the tissue behavior.

Recently Liu and Lee (1978) have presented a three-dimensional analysis of the finite deformation of the lung under gravity. The analysis deals with nonlinear elasticity and finite deformation consistently. The major drawback is that the effects of surface tension are not considered explicitly. The authors used tetrahedral elements and a total of 40 nodes. No mention is made of a study of the effect of grid size on convergence.

Pao et al. (1978) have presented a description of the variation of strain for an excised lobe of canine lung. Although termed a finite element analysis by the authors, it does not relate stress to strain by an energy principle as is conventional. Basically their approach is to quantify the strains in the lung by measuring the displacements at a finite number of points on the surface of the lung. If one assumes a relationship between the displacements at the surface and the internal displacements, the strain can then be calculated. They accomplish this by using tetrahedral elements. The major drawback of the investigation is that there is no relationship developed between the deformations observed and the forces causing them, and hence the observations are difficult to compare with other analyses.

Vawter (1977a, 1978), using a finite element approach based on the incremental approach of Pian and Tong (1971), showed that stresses around airways and blood vessels are significantly altered by the presence of the parenchyma. The material behavior was modeled according to the cubic alveolar model of Fung discussed above, with material constants determined from the biaxial slab test of Vawter et al. (1978). The analysis was general enough to account for the nonisotropic behavior induced by distortion but did not include the effects of surface tension.

The effects of surface tension were considered explicitly in the work of Vawter (1980a, 1980b), Shields and Vawter (1980), and Vawter and Shields (1982) and found to be important. Previous work had implicitly considered the effect of surface tension by using air-filled pressure-volume curves to determine material constants. This can do no more than increase the stiffness of the parenchyma. The above analyses showed that, especially in the apical region of the lung, the regional gas volumes induced by surface tension could not be duplicated by merely increasing the stiffness of the parenchyma.

CONCLUSION

A three-dimensional constitutive relationship for the lung parenchyma which accounts for elastic and interfacial forces and for the nonuniform morphology of the parenchyma, and which explains the experimental observations of tissue behavior, remains an elusive goal. A definitive testing method has not been developed. All tests developed so far either require traumatizing the tissue or utilize unrealistic morphological models of the lung. We believe that tests on intact lungs (or at least lobes), coupled with sophisticated numerical analyses to interpret the observations, have the most potential. Such tests have not been conducted, although preliminary attempts to do so look promising (Shields, 1979).

ACKNOWLEDGMENTS

Many investigators have contributed much to the field of lung mechanics over the past 15 years. Although it is neither possible nor appropriate to recognize all the works in a single chapter, we would like to recommend the following review articles, which should complement the present chapter: F. G. Hoppin and J. Hildebrandt, Mechanical properties of the lung, in J. B. West (ed.), *Bioengineering Aspects of the Lung,* Marcel Dekker, New York, 1977; G. C. Lee and R. C. Tai, Structural mechanics of the lung, in D. N. Ghista (ed.), *Applied Physiological Mechanics,* Harwood, New York, 1979; and T. A. Wilson, Nonuniform lung deformations, *J. Appl. Physiol., 54*(6), 1443–1450, 1983.

REFERENCES

Brown, E. S., Johnson, R. P., and Clements, J. A. (1959). Pulmonary surface tension. *J. Appl. Physiol. 14,* 717–720.

Clements, J. A. (1962). Rat, cat, dog lung surfactant. *Physiologist, 5,* 11.

Flicker, E., and Lee, J-S. (1974). Equilibrium of force of subpleural alveoli: Implications to lung mechanics. *J. Appl. Physiol., 36*(3), 366–74.

Frankus, A., and Lee, G. C. (1974). A theory for distortion studies of lung parenchyma based on alveolar membrane properties. *J. Biomech., 7,* 101.

Fukaya, H., Martin, C. J., Young, A. C., and Katsura, S. (1968). Mechanical properties of alveolar walls. *J. Appl. Physiol., 25*(6), 689–695.

Fung, Y. C. (1974). A theory of elasticity of the lung. *J. Appl. Mech., 41,* 8–14.

Fung, Y. C. (1975). Stress, deformation and atelectasis of the lung. *Circ. Res., 37,* 481–496.

Hajji, M. A. (1978). Indentation of a membrane on an elastic half space. *J. Appl. Mech., 45,* 320–324.

Hajji, M. A., Wilson, T. A., and Lai-Fook, S. J. (1979). Improved measurements of shear modulus and pleural membrane tension of the lung. *J. Appl. Physiol., 47*(1), 175–181.

Hoppin, F. G., Lee, G. C., and Dawson, S. V. (1975). Properties of lung parenchyma in distortion. *J. Appl. Physiol., 39*(5), 742–751.

Karakaplan, A. D., Bieniek, M. P., and Skalak, R. (1980). A mathematical model of lung parenchyma. *J. Biomech. Eng., 102,* 124–136.

Karakaplan, A. D., Bieniek, M. P., and Skalak, R. (1982). Finite element analysis of lung parenchyma. In R. H. Gallagher, B. R. Simon, P. C. Johnson, and J. F. Gross (eds.), *Finite Elements in Biomechanics.* Wiley, New York, 1982.

Krueger, T. W., and Pollack, G. H. (1975). Myocardial sarcomere dynamics during isometric contractions. *J. Physiol. (London), 251,* 627.

Lai-Fook, S. J., Hajji, M. A., and Wilson, T. A. (1980). Deformation of a flexible dish bonded to an elastic half space—Application to the lung. *J. Biomech. Eng., 102,* 234–238.

Lai-Fook, S. J., Wilson, T. A., Hyatt, R. E., and Rodarte, J. R. (1976). Elastic constants of inflated lobes of dog lungs. *J. Appl. Physiol., 40*(4), 508–513.

Lambert, R. K., and Wilson, T. A. (1973). A model for the elastic properties of the lung and their effect on expiratory flow. *J. Appl. Physiol., 34*(1), 34–48.

Lanir, Y., and Fung, Y. C. (1974). Two dimensional mechanical properties of rabbit skin: I. Experimental system. *J. Biomech., 7,* 29–34.

Liu, J-T., and Lee, G. C. (1978). Static finite deformation analysis of the lung. *J. Eng. Mech. ASCE, 104,* 225–239.

Matthews, F. L., and West, J. B. (1972). Finite element displacement analysis of a lung. *J. Biomech., 5,* 591–600.

Mead, J., Takishima, T., and Leith, W. (1970). Stress distribution in lungs: A model of pulmonary elasticity. *J. Appl. Physiol., 28*(5), 596–608.

Pao, Y. C., Chevalier, P. A., Rodarte, J. R., and Harris, L. D. (1978). Finite element analyses of the strain variations in excised lobe of canine lung. *J. Biomech., 11,* 91–100.

Pian, T. H. H., and Tong, P. (1971). Variational formulation of finite displacement analysis. In F. de Veubeke (ed.), *High Speed Computing of Elastic Structures.* University of Liege, Belgium.

Pinto, J. G., and Fung, Y. C. (1973). Mechanical properties of heart muslce in passive state. *J. Biomech., 6,* 597–616.

Scarpelli, E. M. (1965). *The Surfactant System of the Lung,* Lea & Febiger, Philadelphia.

Scarpelli, E. M., and Gabbay, K. H. (1965). Rat lung surfactant. *Science, 148,* 1067.

Shields, W. H. (1979). Feasibility of using intact lung tests to determine mechanical properties of the lung. M.S. thesis, Virginia Polytechnic Institute and State University, Blacksburg, Va.

Shields, W. H., and Vawter, D. L. (1980). The effect of surface tension on regional expansion in the lung. In R. H. Gallagher, B. Simon, P. C. Johnson, and J. F. Gross (eds.), *Finite Elements in Biomechanics.* Wiley, New York, vol. 1, pp. 369–388.

Slama, H., Schodel, W., and Hanson, E. (1973). Goat lung surfactant. *Respir. Physiol., 19,* 233.

Timoshenko, S. P., and Goodier, J. N. (1970). *Theory of Elasticity.* McGraw-Hill, New York, p. 403.

Vawter, D. L. (1977a). Localization of stresses in the lung parenchyma in regions near airways. *1977 Adv. Bioeng.,* 97–98.

Vawter, D. L. (1977b). Stress-free equilibrium volume of the lung. *J. Appl. Physiol., 43*(1), 3–7.

Vawter, D. L. (1978). The influence of lung parenchyma on the boundary conditions for pulmonary hemodynamics. In D. J. Schneck (ed.), *Proceedings, First Mid-Atlantic Conference on Bio-fluid Mechanics.* VPI Press, Blacksburg, Va., pp. 195–203.

Vawter, D. L. (1980a). A finite element model for macroscopic deformation of the lungs. In R. H. Gallagher, B. Simon, P. C. Johnson, and J. F. Gross (eds.), *Finite Elements in Biomechanics.* Wiley, New York, vol. 1, pp. 369–388.

Vawter, D. L. (1980b). A finite element model for macroscopic deformation of the lung. *J. Biomech. Eng., 102,* 1–7.

Vawter, D. L. (1983). Poisson's ratio and incompressibility, *J. Biomech. Eng., 105,* 194–195.

Vawter, D. L., Fung, Y-C., and West, J. B. (1978). Elasticity of excised dog lung parenchyma. *J. Appl. Physiol., 45*(2), 261–269.

Vawter, D. L., Fung, Y-C., and West, J. B. (1979). Constitutive equation of lung tissue elasticity. *J. Biomech. Eng., 101,* 38–45.

Vawter, D. L., Matthews, F. L., and West, J. B. (1975). Effect of shape and size of lung and chest wall on stresses in the lung. *J. Appl. Physiol., 39*(1), 9–17.

Vawter, D. L., and Shields, W. H. (1982). Deformation of the lung: The role of interfacial forces. In R. H. Gallagher, B. Simon, P. C. Johnson, and J. F. Gross (eds.), *Finite Elements in Biomechanics.* Wiley, New York.

Weibel, E. R. (1963). Morphometry of the human lung. Academic Press, New York.

West, J. B., and Matthews, F. L. (1972). Stresses, strains and surface pressures in the lung caused by its weight. *J. Appl. Physiol.*, *32*(3), 332–345.

Wilson, T. A. (1972). A continuum analysis of a two-dimensional mechanical model of lung parenchyma. *J. Appl. Physiol.*, *33*(4), 472–478.

Wilson, T. A. (1980). Surface tension–surface area curves calculated from pressure volume loops. *J. Appl. Physiol.*, *53*(6), 1512–1520.

Wilson, T. A., and Bachofen, H. (1982). A model for mechanical structure of the alveolar duct. *J. Appl. Physiol.*, *52*(4), 1064–1070.

CHAPTER 25

Respiratory Control and Mechanics

Guy S. Longobardo
IBM Corporation
White Plains, New York

Neil S. Cherniack
Kingman P. Strohl
Department of Medicine
Case Western Reserve University
Cleveland, Ohio

Janie M. Fouke
Department of Biomedical Engineering
Case Western Reserve University
Cleveland, Ohio

25.1 THE CONTROL OF RESPIRATION

The respiratory control system adjusts ventilation to maintain the level of blood oxygen and CO_2 in the blood within narrow limits [3, 19, 26].

Figure 25.1 is a schematic representation of the chemically controlled respiratory system, showing the interplay between a controlled system and a controller [19]. The controlled system consists of the respiratory pump, i.e., the lung, the chest wall, the respiratory muscles, and the CO_2 and O_2 contained in chemical combination and physical solution in the blood and body tissues (the body stores of O_2 and CO_2). The controller consists of the bulbopontine respiratory neurons, which produce the regular sequence of inspiration and expiration which characterizes normal breathing; chemoreceptors which respond to changes in the level of P_{CO_2} and P_{O_2} and pH in their immediate locale; and mechanoreceptors in the lung and chest wall. Other receptors in the limbs and cortex may also help adjust ventilation to changes in metabolic rate, and allow ventilation to increase over broad limits during exercise with little or no change in arterial P_{CO_2} [104].

The simple block diagram in Fig. 25.1 belies the complexity of the respiratory system in terms of cross-coupling between components as well as its interactions with other physiological

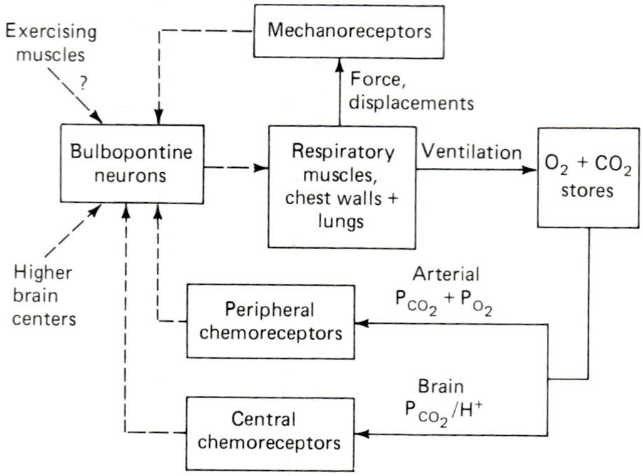

FIG. 25.1 Schematic representation of respiratory control system. Dashed lines indicate neural pathways. Bulbopontine neurons activate respiratory muscles, producing ventilation. Ventilation by determining the amount of O_2 taken into the body and the CO_2 removed affects the amount of the O_2 and CO_2 stored in the body, as well as the levels of P_{CO_2}, P_{O_2}, and H^+ in the regions of operation of chemoreceptors and hence their excitation. Chemoreceptor activity in turn helps modify the activity of the bulbopontine neurons which generate the respiratory rhythm. The activity of the neurons is also influenced by mechanoreceptors, higher brain centers, and possibly by receptors in exercising muscle.

systems such as the cardiovascular system, the temperature regulating system, and renal regulation of blood [H^+]. Also, stimuli, light, and sound can affect breathing levels, allowing breathing to continue even when CO_2 levels are reduced. Further, the respiratory system can be controlled volitionally at least temporarily, for example, during speech.

Much remains to be known about the controller, including the location of central chemoreceptors and the modes of operation of receptors. For example, the control system appears to operate as a proportional controller with a residual steady-state error when CO_2 or hypoxic gas mixtures are inhaled and when acid-base disturbances occur. However, with exercise, it appears to operate as an error-free integral controller or perhaps, as has been suggested by Saunders, a feed-forward controller [96]. In addition, instability of the control system can occur so that ventilation levels cycle and there are periods of recurrent apnea.

Despite these uncertainties, models of the system can be constructed; and even though many of the components are, in the words of Grodins, "gray boxes," useful physiological and clinical insights can be obtained. While it may be impossible to model with total accuracy the entire respiratory cycle, in many cases of interest this may not be necessary.

25.2 OPERATION OF THE CHEMICAL CONTROL SYSTEM: THE CONTROLLER

The *respiratory controller* sets the level of ventilation and the pattern of breathing so that P_{CO_2} and P_{O_2} are maintained at a suitable operating point.

The respiratory rhythm itself appears to depend on a *central pattern generator* which produces inspiration and expiration. This generator seems to be composed of interconnected neurons located in the dorsal and ventral portions of the medulla which excite and inhibit each other. A dorsal group of neurons is located in the vicinity of the nucleus tractus solitarius

(NTS), which has an important influence on phrenic motor activity and hence on diaphragm contraction. A ventral group is located in the nucleus retroambigualis (NRA—nucleus para ambigualis and nucleus retrofacialis) and nucleus ambigualis (NA) [10, 26, 106]. Efferent activity of the cranial nerves which supply upper airway muscles is adjusted by NA activity, and the discharge of inspiratory and expiratory muscles by the other nuclei. These medullary groups of respiratory neurons also receive input from neurons in the rostral pons (pneumotaxic center). In addition, there are pathways which bypass these medullary neuronal networks and allow respiratory muscles to be volitionally controlled by the cortex [88].

The operating point for arterial P_{CO_2} and P_{O_2} depends on the sleep-waking cycle [13, 46, 55, 87]. Arterial P_{CO_2} seems somewhat higher and arterial P_{O_2} somewhat lower in the deeper stages of slow-wave sleep than it is during wakefulness [55, 87]. Peripheral arterial sensors in the carotid bifurcation (the carotid body) and near the aortic arch (the aortic body) detect changes in the level of arterial P_{O_2}. When stimulated, they have both respiratory and circulatory effects and are one of the links by which the ventilatory and vasomotor responses are coordinated [6–8, 72]. In humans the carotid body appears to be mainly responsible for the increase in ventilation observed with hypoxia. The exact mechanism which allows the carotid body to respond to change in P_{O_2} is still under study. It may depend on energy-producing reactions sensitive to hypoxia in these peripheral arterial sensors [3, 6].

Ventilation increases nearly hyperbolically with hypoxia. Further, the response to hypoxia depends on the level of arterial P_{CO_2} [15, 28, 63]; CO_2 increases the ventilatory response to hypoxia, as shown in Fig. 25.2. Most of this interaction between hypoxia and hypercapnia

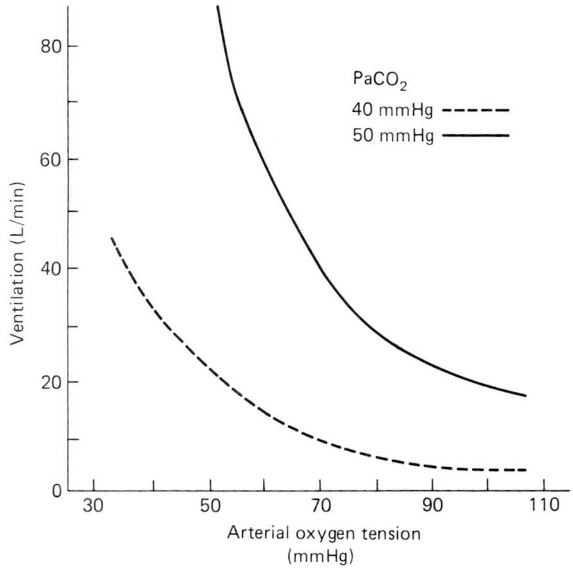

FIG. 25.2 Increasing P_{CO_2} multiplies the ventilatory response to hypoxia.

occurs at the carotid body, which by itself accounts for 10 to 50 percent of the response to hypercapnia even under hyperoxic conditions [3, 101]. An additional multiplication of hypoxic and hypercapnic inputs may take place centrally as well [37, 63]. The carotid body also responds to the swings in arterial P_{CO_2} that occur with inspiration and expiration [6, 36, 66, 90]. Increases in the amplitude of these swings seem to stimulate carotid body activity. Carotid body discharges appear to produce a greater effect in ventilation when the peak of the swing in their activity coincides with the discharge of central inspiratory neurons [41].

The carotid body responds to changes in arterial P_{CO_2} and P_{O_2} within a few seconds, but the total ventilatory response to CO_2 occurs much more slowly and is measured in minutes. This slower component of ventilatory response to CO_2, which makes up 50 to 90 percent of the total CO_2 response, seems to originate in receptors located in the medulla, possibly near its ventral surface, that respond to changes in P_{CO_2} of $[H^+]$ in their immediate environment (probably the interstitial fluid surrounding them) [3, 9, 17, 51, 64, 69, 86]. Ventilation lowers interstitial fluid pH by changing P_{CO_2} levels in the blood perfusing the brain.

The ventilatory response to CO_2 is linear over a wide range of CO_2 tensions (Fig. 25.3). In anesthetized and sleeping animals and humans there may be some threshold level of CO_2

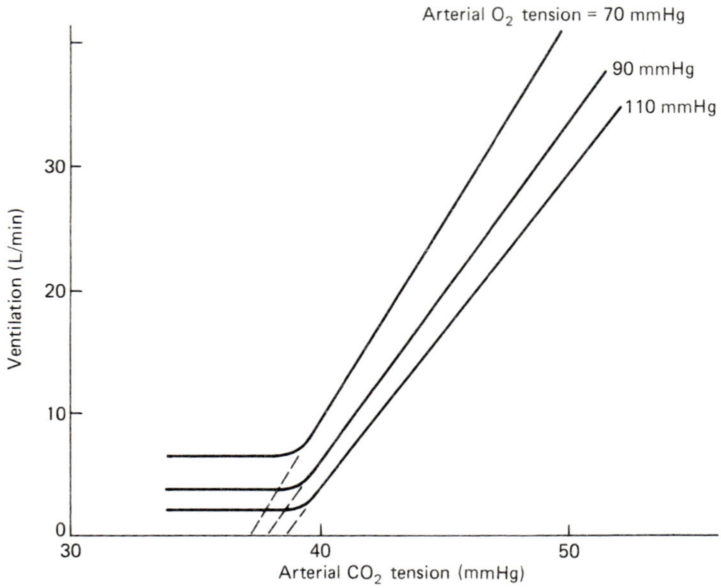

FIG. 25.3 Ventilatory response to CO_2 is linear over a wide range of CO_2 tensions. Hypoxia increases the gain of the CO_2 response. The dashed lines lead to threshold values of CO_2 tension required to initiate breathing in some anesthetized or sleeping humans. (*Adapted from the data of Nielsen and Smith [80].*)

needed to initiate breathing [16, 17, 34, 46, 50, 55, 87]. This threshold P_{CO_2} is decreased by hypoxia and also seems to be reduced by increased activity of the respiratory muscles [15, 40]. For example, voluntary hyperventilation even to extremely low levels of P_{CO_2} fails to produce apnea, while passive hyperventilation can cause breathing to stop in the unconscious subject [40, 46].

Acidosis in the arterial blood increases ventilation at any given level of P_{CO_2}, while alkalosis has the reverse effect. Severe acidosis seems also to increase the slope of the ventilatory CO_2 response line, while severe alkalosis depresses it [28].

Different equations have been devised to describe the effects of chemical stimuli on the controller. While earlier equations (e.g., by Gray [51]) considered the effects of hypoxia and hypercapnia on ventilation to be additive, later equations (e.g., those of Lloyd et al. [67, 68]) represent a more complex multiplicative view of the interplay between these two stimuli.

$$\dot{V} = D(Pa_{CO_2} - B)\left(1 + \frac{A}{Pa_{O_2} - C}\right)$$

where Pa = arterial gas tension, D = the asymptotic limit of the ventilatory response to CO_2 at high levels of Po_2, C = the asymptotic limit of Po_2 at which the ventilatory response to CO_2 approaches infinity, A = a parameter which describes the effect of the sensitivity to hypoxia of the ventilatory response slope to CO_2, and B = a threshold value of Pco_2.

In healthy humans, A is about 10 to 40 mmHg Po_2, B is 34 to 40 mmHg Pco_2, C is approximately 20 to 35 mmHg Pa_{O_2}, and D is about 2 to 6 L/min for each mmHg increase in Pco_2 above B.

In a later version, this equation was refined so that the action of central and peripheral chemoreceptors was considered independently. The central chemoreceptors were considered to respond only to brain Pco_2, while the peripheral chemoreceptors responded both to changes in arterial Po_2 and Pco_2.

Even more complex representations of the effects of chemical stimuli have been developed [107]. In these equations, ventilation depends on both chemical and neural signals.

Most controller equations consider the controller to have no dynamics of its own; i.e., the time needed for transmission of neural signals is assumed to be much shorter than the time required for chemical changes to occur. More recently, it has been proposed that it may take appreciable time for neural networks to reach equilibrium when excited, and this time is of the order of the time required for Pco_2 and Po_2 changes to occur in the body tissues by chemical reactions, diffusion, and perfusion.

Ventilation can be altered by changing tidal volume (the amount of air inspired or expired with each breath) or by changing breathing frequency. In anesthetized animals, changes in breathing frequency seen under conditions of increased chemical drive depend on the activity of pulmonary stretch receptors innervated by the tenth cranial nerve and the vagus, which respond to changes in lung volume [10, 26, 106].

Although average inspiratory flow rate, i.e., tidal volume (V_T) divided by inspiratory time (T_i), under stable thermal conditions, is determined mainly by the level of chemical drive (the level of Pco_2 and Po_2), it has been observed in anesthetized animals that the ratio of inspiratory time (T_i) to expiratory time (T_e) mainly depends on the activity of pulmonary mechanoreceptors (e.g., stretch receptors) and the pontine neurons. Activation of lung stretch receptors during expiration can alter the relationship between T_i and T_e by prolonging T_e independently of changes in T_i [26, 43].

In humans stretch receptor input seems to have a more feeble action, and breathing frequency changes seen with hypoxia and hypercapnia seem to be caused more by change in T_e than in T_i [25, 43]. A mathematical model of the effect of stretch receptor activity on the amplitude and frequency of breathing has been developed by Bradley and coworkers [10].

The stretch receptors are one kind of mechanoreceptors which sense displacements and force changes in the respiratory system. They are strategically located in the smooth muscle layer of the airways of the lung and help the controller adjust to changing mechanical conditions. In general, the information they provide allows the controller to maintain a constant level of ventilation despite altered physical conditions.

The lungs and airways also contain sensory receptors whose natural stimuli, as in other hollow viscera, are irritation of the lining layers and/or changes in distending forces [44, 105].

These receptors are of two types: irritant receptors which ramify among airway epithelial cells and have a distribution similar to the stretch receptors, and J (juxtacapillary) receptors situated in the lung interstitium near alveolar capillaries.

Irritant receptors and J receptors are served by smaller vagal fibers than stretch receptors and, unlike stretch receptors, rapidly decrease their activity when subjected to a sustained stimulus [95, 105].

Irritant receptors are stimulated mechanically by rapid lung inflations and deflations, increases in airflow, and bronchial smooth muscle tone. Irritant receptor stimulation reflexly augments inspiratory motor neuron activity and constricts the airways, and, by interaction with the stretch receptors, promotes rapid, shallow breathing. This pattern of breathing in combination with the airway constriction may limit penetration of dangerous agents into the lung and prevent them from reacting with the gas-exchanging surfaces.

J receptors are also stimulated by edema formation in the lung excited by certain chemicals such as histamine, halothane, and phenyldiguanide. Activation of the J receptors causes laryngeal closure and apnea, followed by a rapid, shallow breathing.

Tendon and spindle receptors in the chest wall signal the respiratory neurons of changes in the force exerted by the respiratory muscles and movement of the chest wall [44, 79].

Tendon organs in the intercostal muscles and the diaphragm monitor the force of muscle contraction and have an inspiratory inhibitory effect.

Muscle spindles, which are abundant in the intercostal muscles but scarce in the diaphragm, are involved in several kinds of inhibitory and excitatory intercostal respiratory reflexes. Spindles also help coordinate breathing during changes in posture and speech and stabilize the rib cage when breathing is impeded by increases in resistance or elastance [76, 79].

Respiration in conscious humans is also under volitional control by higher brain centers as well as automatic (reflex) control. While voluntary control, e.g., breath holding, can upset the usual regulation of P_{CO_2} and P_{O_2} by the respiratory system, it can also be used to preserve homeostasis. For example, the adverse effects of a suffocating environment can be minimized not only by reflexly increasing the level of ventilation but as effectively by escaping from the environment. Both types of responses can occur simultaneously and are physiological.

25.3 THE CONTROLLED SYSTEM

25.3.1 The Gas Stores of the Body

Ventilation alters blood and tissue levels of P_{CO_2} and P_{O_2} by changing the amount of gas stored in the body (the amount of CO_2 and O_2 contained in physical solution and chemical combination) [18]. Only a small amount of O_2 is stored in the body, mainly in the blood, where O_2 is combined with hemoglobin, and in the air (in the functional residual capacity) of the lungs. Hence, large and abrupt changes in arterial P_{O_2} are produced by ventilation changes. If breathing is stopped for 1 min, Pa_{O_2} will fall from about 100 to 40 to 50 mmHg. On the other hand, more CO_2 is stored in the body, mainly in the form of bicarbonate, in the soft tissues, and if the CO_2 were immediately available, P_{CO_2} would rise by only 1 to 2 mmHg each minute even if there were no breathing. Dynamically the CO_2 stores of the body behave as if they were contained in multiple compartments arranged in parallel and interconnected via the circulation. Hence, CO_2 exchange in tissues seems to be limited by perfusion; equilibration of CO_2 between the blood and poorly perfused tissues may be extremely slow, requiring hours. Because of limited perfusion, P_{CO_2} in the blood will increase 6 to 8 mmHg in the first minute of breath holding but still far less than the corresponding fall in P_{O_2} [17, 70].

The dynamics of storage equilibrium affect the rate of respiratory adjustments. High cardiac output increases the speed with which arterial and mixed venous blood gas levels can be restored to usual levels following a disturbance [70]. The dynamics of CO_2 storage in the brain, where the central chemoreceptors are located, are particularly important in determining the dynamics of the response to CO_2 [11, 18]. Active buffering by the brain, as well as active and passive mechanisms which affect $[H^+]$ exchange between the brain and the blood, keep brain and cerebrospinal fluid (CSF) acidity more constant than that of the arterial blood [3, 9]. Also, increases in cerebral blood flow which occur with hypoxia and hypercapnia minimize changes in brain interstitial fluid acidity [3, 9, 32].

25.3.2 Interplay between Controller and Controlled System

Changes in the controller as well as in the stores affect the dynamics of ventilatory control. In general, an increase in the sensitivity to CO_2 (as indicated by greater changes in ventilation with changes in arterial P_{CO_2}) accelerates the speed with which disturbances in blood gas tensions can be corrected [29, 70]. Similarly, disturbances in gas tensions can be corrected rapidly if the resting P_{CO_2} for some reason is higher than its usual value of 40 mmHg [70].

However, the need for rapid responses and for swiftly minimizing deviations in blood gas changes from normal levels must be balanced against the risks of increasing the energy costs of breathing and unstable ventilatory control [13, 16, 19, 70].

For example, increased controller sensitivity allows CO_2 stores to be returned more rapidly to normal levels after a disturbance like breath holding which raises arterial P_{CO_2}. But since the rate of CO_2 elimination from the body is limited by perfusion, increasing controller sensitivity has a decreasing effect on CO_2 removal [70]. On the other hand, increasing controller sensitivity raises ventilation and the use of oxygen by the respiratory muscles (the O_2 cost of breathing); hence, the O_2 cost of removing a given amount of CO_2 may change, reaching some minimum as controller sensitivity is increased from very low to very high values. These energy considerations become very important when the O_2 cost of breathing is very high, as it is in some patients with pulmonary disease.

25.3.3 Models of the Body Stores

A simple model of the body stores similar to that used by Grodins et al. in their 1954 description of the "respiratory chemostat" is schematically shown in Fig. 25.4 [52]. In this model the body is assumed to be composed of two storage sites: a lung CO_2 reservoir, which contains the CO_2 stored in the functional residual capacity of the lung; and a tissue reservoir, in which the CO_2 of the blood and the rest of the body is stored. O_2 storage is not considered.

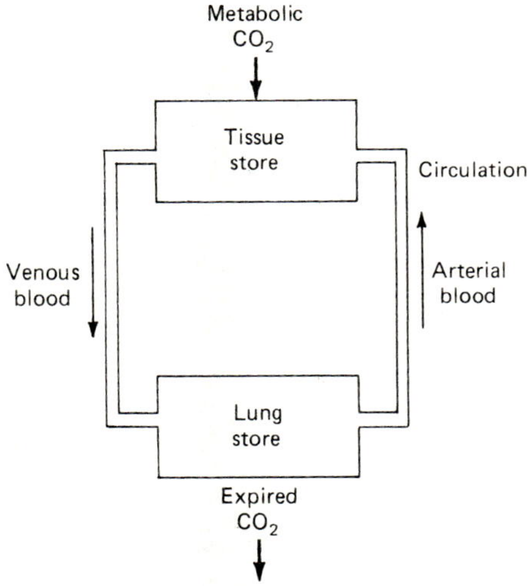

FIG. 25.4 Schematic representation of the Grodins single–body store model for CO_2 [52]. Homogeneous tissues are assumed, with tissue CO_2 tensions equal to blood CO_2 tensions.

For the sake of simplicity the model includes several assumptions in addition to that of a single tissue store. (1) The tissue store is uniform with respect to both CO_2 concentration and tension. Thus the model does not attempt to take into account the distribution of CO_2 with regard to space and time along, and around, capillary beds. (2) The tension and concentration of the CO_2 in the tissues are equal to the tension and concentration of the CO_2 of the venous blood with which the tissue is perfused. Since the assumption is made that there is neither a tension gradient for CO_2 nor a concentration gradient for bicarbonate across the cell membrane, mixed venous CO_2 concentration is assumed to be equal to tissue CO_2 concentration. (3) The in vitro dissociation CO_2 curve of blood applies to both blood and tissue, and a straight line may be used to approximate the curve. (4) Lung tissue per se is a negligible store of CO_2. (5) The lung compartment, like the tissue compartment, is homogeneous with respect to the storage of CO_2.

By applying the principle of conservation of mass across the tissue and lung stores with these assumptions, the equations for this model are written as follows.

For the conservation of CO_2 across the tissue store, the equation* is:

$$\dot{Q}_T(\mathrm{Ca}_{CO_2} - \mathrm{C\bar{v}}_{CO_2}) + \dot{V}_{CO_2} = \frac{d}{dt}(V_{ti}\mathrm{C\bar{v}}_{CO_2}) \tag{25.1}$$

where $\dot{Q}_T$ is the cardiac output, Ca_{CO_2} is the concentration of CO_2 in the arterial blood, $\mathrm{C\bar{v}}_{CO_2}$ is the concentration of CO_2 in the venous blood, $\dot{V}_{CO_2}$ is the metabolic production rate of CO_2, V_{ti} is the tissue volume. The left-hand side of the equation therefore represents the difference between the rates at which CO_2 is brought into and taken away from the tissue store; the right-hand side represents the rate of change in the quantity of CO_2 stored in the tissue. Table 25.1 contains an explanation of symbols.

A similar equation can be written across the lung:

$$\dot{Q}_T(\mathrm{C\bar{v}}_{CO_2} - \mathrm{Ca}_{CO_2}) + \dot{V}_A\mathrm{FI}_{CO_2} - \dot{V}_A\frac{\mathrm{Pa}_{CO_2}}{P_B - 47} = \frac{d}{dt}\left[\frac{(\mathrm{FRC})\mathrm{Pa}_{CO_2}}{P_B - 47}\right] \tag{25.2}$$

where $\dot{Q}_T(\mathrm{C\bar{v}}_{CO_2} - \mathrm{Ca}_{CO_2})$ is the rate at which CO_2 is brought into the lung store by the blood; $\dot{V}_A\mathrm{FI}_{CO_2}$ is the rate at which CO_2 is inspired; $\dot{V}_A\mathrm{Pa}_{CO_2}/(P_B - 47)$ is the rate at which CO_2 is expired; and $(d/dt)[(\mathrm{FRC})\mathrm{Pa}_{CO_2}/(PB - 47)]$ is the rate of change in quantity of CO_2 in the lung store.

During arrested ventilation (apnea) the theoretical time course of arterial CO_2 tension predicted from the model would be

$$\mathrm{Pa}_{CO_2}(t) = \mathrm{Pa}_{CO_2}(0) + C_1(1 - e^{-t/T_1}) + \left[\frac{(P_B - 47)\dot{V}_{CO_2}}{(P_B - 47)V_{ti}\beta_{\mathrm{blood}CO_2} + \mathrm{FRC}}\right]t \tag{25.3}$$

where t is time, $\mathrm{Pa}_{CO_2}(t)$ is the arterial CO_2 tension at any instant of time during arrested ventilation, $\mathrm{Pa}_{CO_2}(0)$ is the arterial CO_2 tension at the time ventilation is stopped, $\beta_{\mathrm{blood}CO_2}$ is the slope of the blood CO_2 dissociation curve, and T_1 is a time constant. $\mathrm{Pa}_{CO_2}(0) + C_1(1 - e^{-t/T_1})$ represents the transient part of the solution of Eqs. (25.1) and (25.2) during apnea, and the remainder of the right-hand side of the equation is the steady-state part of the solution.

The time constant T_1 in Eq. (25.3) can be calculated from

$$T_1 = \frac{V_{ti}\mathrm{FRC}/\dot{Q}_T}{(P_B - 47)V_{ti}\beta_{\mathrm{blood}CO_2} + \mathrm{FRC}} \tag{25.4}$$

and so depends on the tissue volume, functional residual capacity of the lung, cardiac output, and slope of the CO_2 dissociation curve.

The model predicts that the arterial P_{CO_2}, or Pa_{CO_2}, immediately after apnea will finally increase as a straight line after an initial period of curvilinear increase. The smaller the time constant T_1, the sooner the Pa_{CO_2} will change linearly with time. Either an increase in cardiac output or a decrease in the slope of the CO_2 dissociation curve decreases T_1. In humans T_1 is short, about 10 s for a 70-kg person. Also, the greater the tissue volume and CO_2 dissociation slope, the less the rate of linear rise in P_{CO_2}.

This simple model does not adequately reproduce the rapid rise in Pa_{CO_2} that occurs experimentally during apnea, rebreathing, and hyperventilation. For example, during apnea, blood P_{CO_2} rises in a human more rapidly, 5 to 6 mm/min, compared with the model prediction of 2 to 3 mm/min.

The deficiency of this description in matching experiments to determine the CO_2 dissociation slope of the body led to more complicated models, usually including more compartments, namely, muscle, brain, and other tissue stores.

---

* In this equation and in those that follow, it is assumed that all gas volumes and blood gas contents are expressed in the same units of temperature, pressure, and saturation.

TABLE 25.1 Explanation of symbols

Gas concentrations, L/L:		
Ca	=	arterial concentration
$C\bar{v}$	=	mixed venous concentration
$C\bar{v}_i$	=	venous blood concentration leaving *i*th tissue
C_i	=	tissue concentration in *i*th tissue
Gas tensions, mmHg:		
Pa	=	instantaneous arterial tension
P_B	=	barometric pressure
Pv_i	=	venous blood tension at a tissue *i*
P_i	=	tissue tension at a tissue *i*
Pa_{av,dCO_2}	=	average arterial CO_2 tension over previous breath
P_{brain,av,dCO_2}	=	average brain CO_2 tension over previous breath, at the central receptor
Pa_{av,dO_2}	=	average arterial O_2 tension over previous breath
Blood flows, L/min:		
$\dot{Q}_T$	=	cardiac output
$\dot{Q}_i$	=	blood flow to *i*th tissue
$\dot{Q}_b$	=	blood flow to brain
Gas flows, L/min:		
$\dot{V}$	=	instantaneous ventilation rate
$\dot{V}_A$	=	minute ventilation
V_{mag}	=	amplitude of $\dot{V}$
FI	=	inspired concentration
Metabolic production, L/min:		
$\dot{V}_{CO_2}$	=	metabolic production of CO_2
$\dot{V}_{O_2}$	=	metabolic production of O_2
$\dot{V}_{iCO_2}$	=	metabolic production of CO_2 by *i*th tissue
$\dot{V}_{iCO_2}$	=	metabolic consumption of O_2 by *i*th tissue
Volumes, L:		
V_{ti}	=	tissue volume
$V_{ti,i}$	=	tissue volume of *i*th tissue
$V_{blood,i}$	=	blood volume contained in *i*th tissue
Va	=	arterial tissue volume, including blood and lung tissue
FRC	=	functional residual capacity of the lung
Dissociation slopes:		
β_{blood}	=	slope of blood dissociation slope
β_i	=	slope of tissue dissociation curve

25.3.4 Multicompartment Model of CO_2 Stores

Farhi and Rahn [45] suggested that CO_2 was not stored uniformly in the body but rather in multiple tissue compartments with each tissue compartment having a different volume, CO_2 dissociation curve, rate of CO_2 production, and rate of perfusion. This conception of CO_2 storage is analogous to an electrical circuit with a capacitance and a resistance corresponding to each CO_2 storage site. The capacity of a compartment to store CO_2 depended on the slope of its CO_2 dissociation curve and the compartment volume, whereas the level of perfusion of a compartment determined its resistance to the transfer of CO_2.

Figure 25.5 shows a block diagram of the Farhi and Rahn model. The equations that describe the multicompartment model are listed below and are based on the conservation of the mass of CO_2 for each compartment [23, 45].

For each tissue compartment,

$$\dot{Q}_i(Ca_{CO_2} - Cv_{iCO_2}) + \dot{V}_{iCO_2} = \frac{d}{dt}(V_{ti,i}C_{iCO_2} + V_{blood,i}Cv_{iCO_2}) \tag{25.5}$$

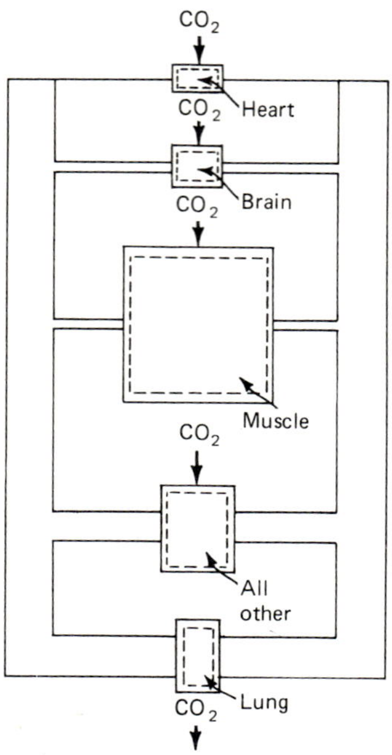

FIG. 25.5 Schematic representation of the Farhi and Rahn [45] multicompartment model of CO_2 stores; one arterial and four tissue CO_2 storage compartments are shown. Entire volume of each tissue compartment (solid line) consists of volume of the tissue named (dotted line) and volume of venous blood equilibrated with it. The arterial compartment is shown to consist of the functional residual capacity (FRC) of the lung and the arterial blood volume, but it also includes the kidney in the Farhi and Rahn model. Tissue compartments have different sizes, representing the differing capacity of each to store CO_2. Channel connecting the compartments (representing the circulation) varies in width since each compartment has a different perfusion level.

where i is the index for a tissue compartment, $\dot{Q}_i$ is the perfusion to that compartment, and Cv_{iCO_2} is the CO_2 concentration of blood leaving the tissue. The tissue compartment is comprised of both the tissue and the blood it contains, that is, V_{ti} and $V_{blood,i}$.

If the tissue CO_2 tension in each compartment is assumed to be equal to the CO_2 tension in the venous blood in that compartment, then

$$Pv_{iCO_2} = P_{iCO_2} \tag{25.6}$$

The CO_2 dissociation curves are linearized so that, for the blood,

$$Cv_{iCO_2} = \beta_{blood,CO_2} Pv_{iCO_2} + B_{blood} \tag{25.7}$$

and, for the tissues,

$$C_{iCO_2} = \beta_i P_{iCO_2} + B_i \tag{25.8}$$

For a tissue compartment, the equation becomes

$$\dot{Q}_i(Ca_{CO_2} - Cv_{iCO_2}) + \dot{V}_{iCO_2}$$

$$= \frac{d}{dt}\left\{\left[V_{ti,i}\left(\frac{\beta_i}{\beta_{blood}}\right) + V_{blood,i}\right] Cv_{iCO_2}\right\} \tag{25.9}$$

The mixed venous CO_2 concentration is a weighted average of the venous blood concentrations leaving each of the different tissues, or

$$C\bar{v}_{CO_2} = \sum \frac{\dot{Q}_i Cv_{iCO_2}}{\dot{Q}_T} \tag{25.10}$$

For the lung and arterial blood compartment,

$$\dot{Q}_T(C\bar{v}_{CO_2} - Ca_{CO_2}) - \frac{\dot{V}aPa_{CO_2}}{P_B - 47}$$

$$= \frac{d}{dt}\left[\frac{(FRC)Pa_{CO_2}}{P_B - 47} + VaCa_{CO_2}\right] \tag{25.11}$$

The assumptions made in computing results from this model are (1) the alveolar and arterial CO_2 tensions are identical, and (2) a change in tissue CO_2 tension is instantaneously reflected by a change in the CO_2 tension in venous blood; i.e., there are no effective diffusion barriers for CO_2 in the body. No assumption is made about the concentration of bicarbonate being the same in a tissue and in the blood equilibrated with it.

Table 25.2 gives, for the multicompartment model, reasonable values for the rate of CO_2 production, the rate of perfusion, the CO_2 dissociation slope, and the volume of each compartment. Farhi and Rahn showed that the model reproduced the values for the slope of the body CO_2 dissociation curve reported in experiments which lasted 30 to 90 min [45]. However, the Farhi and Rahn model was not very successful in duplicating the values for the slopes of the CO_2 dissociation curve reported in experiments in which disturbances in CO_2 balance lasted less than 15 min [45]. The slope of the CO_2 dissociation curve measured in these brief experiments

TABLE 25.2 Important parameters used in multicompartment model of CO_2 stores

Compartment	Volume, % of total	$\dot{V}_{CO_2}$, % of total	$\dot{Q}_i$, % of total	Slope of CO_2 dissociation curve, mL/kg per mmHg P_{CO_2}
FRC of lung	3.3	0	100.0	
Brain	1.93	29.3	14.0	2.9
Heart	0.43	10.9	4.0	2.4
Muscle	40.00	15.1	16.0	4.2
All other	8.55	29.2	51.0	3.0

Source: After Fahri and Rahn [45].

could be reproduced only if the perfusion of the muscle compartment was 70 percent less than the value originally assigned to it.

In Fig. 25.6 the experimental results obtained by Cherniack et al. [23] in two dogs—one studied during hyperventilation and the other studied during apnea—are compared with the results computed from a multicompartment model. The parameters used in this model are like those described by Farhi and Rahn (see Table 25.2), except that they have been modified to describe the CO_2 stores of the dog [99].* It can be seen that both during hyperventilation and apnea the change in arterial and mixed venous P_{CO_2} is greater than that predicted. The rise in PV_{CO_2} during apnea shown in this experiment of 5.7 mmHg/min is similar to that reported by Nahas and Verosky [78] and Halmdahl [59] during apneic oxygenation in dogs and is also similar to the results in humans reported by Fowle and Campbell [48] and by Klocke and Rahn [62].

In the Farhi and Rahn model the compartments in which most of the body CO_2 is produced (e.g., brain and heart) have a small volume and a small storage capacity for CO_2. These compartments with high metabolic rates are separated by the circulation from compartments (e.g., muscle) with a low metabolic rate but a large volume, in which most of the metabolically produced CO_2 of the body is stored. Therefore, a period of time must elapse (the duration of which depends on the relative level of perfusion of the tissue compartments) before the CO_2 can be transferred from the compartments where most of it is produced to the compartments where most of it is stored. Until this transfer of CO_2 is completed, the storage capacity of the body will appear to be less than is true steady-state value. The smaller the storage capacity of the compartments that are storing most of the CO_2, the faster the blood CO_2 tensions will change after a change in ventilation.

Experimental results could be more accurately reproduced if the effective tissue volume, defined as $V_{ti}(\beta_i/\beta_{blood})$, is much lower than defined either because only a portion of body tissues participate in the CO_2 transfer, or because the CO_2 uptake by the tissues, indicated by the slope of the tissue CO_2 dissociation curve, S_i, is less than assumed.

Several alternate possibilities have been suggested, none of which has been proved conclusively. These include, for example, the absence of the enzyme carbonic anhydrase in tissues which regulates the hydration and dehydration of CO_2 according to the equation

$$CO_2 + H_2O \rightleftharpoons H_2CO_3 \rightleftharpoons H^+ + HCO_3^-$$

Since CO_2 must be hydrated before it can be stored as bicarbonate, tissue lacking carbonic anhydrase would have the same low CO_2 dissociation slope during the first few minutes after a ventilation change as does water; effective tissue volume would be low because of a reduction in S_i rather than in V_{ti}. However, as the uncatalyzed reaction for the hydration of CO_2 ap-

* This model also differs from the Farhi and Rahn model in that the kidney is considered to be a venous rather than an arterial store. This change has little effect on the computed results.

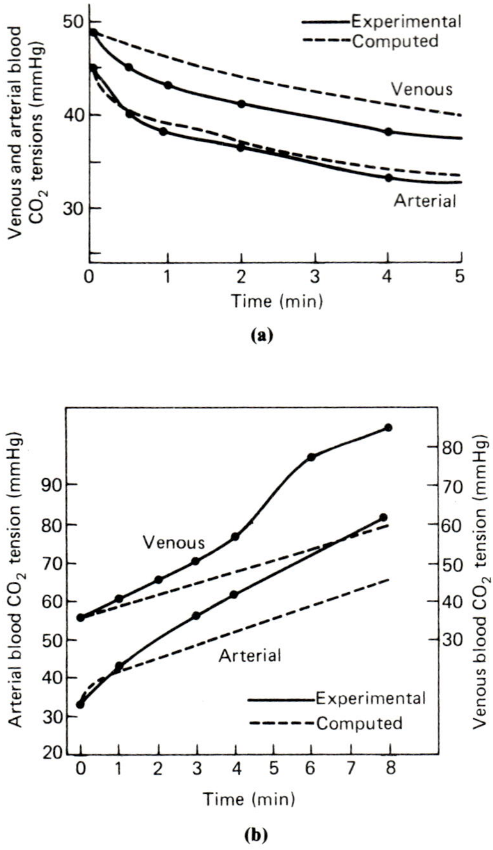

FIG. 25.6 Comparison of computed changes in arterial and mixed venous CO_2 tension during hyperventilation and apnea with those measured by Cherniack et al. [23]. Predicated changes were calculated from the multicompartment model of Farhi and Rahn [45]. (*a*) Hyperventilation; (*b*) apnea.

proached completion, the slope of the CO_2 dissociation curve and the effective tissue volume would increase.

Multicompartment models of CO_2 stores have been devised that consider carbonic anhydrase to be present only in brain and blood, which reproduce short-term changes in P_{CO_2} with reasonable accuracy [18]. Another possibility is that each body tissue is divided into poorly perfused and well-perfused portions, essentially adding more compartments, and speeding up the dynamics by inhibiting participation in the CO_2 transfer process by the poorly perfused tissues [20, 30].

Although the importance of these processes in CO_2 storage is speculative, it is possible that one or more of them affects the rate of CO_2 storage in tissues. The previous discussion also suggests that quantification of the details of intracellular CO_2 storage might significantly increase our understanding of transient changes in CO_2 stores.

25.3.5 Theoretical Analysis of Transient Changes in O_2 Stores

The concept of dividing the body into compartments, as for the CO_2 stores, is also valid for O_2 stores.

Since the primary O_2 store is the blood, one might conceive a simple two-compartment model of O_2 stores: one consisting of a lung and arterial blood compartment, the other composed of a tissue and a venous blood compartment. But, because the CO_2 stores are changing at unequal rates in the various compartments and the level of P_{CO_2} leaving each compartment is not the same, the resulting Bohr effect (and Haldane effect) will be different in each compartment; therefore a more complicated multicompartment model of the O_2 stores is required [22], which parallels the CO_2 model.

For each tissue compartment, an equation for the conversation of mass was written for O_2 as for CO_2 [22]. Thus, for the storage of O_2 in any tissue,

$$\dot{Q}_i(\text{Ca}_{O_2} - \text{Cv}_{iO_2}) - \dot{\text{V}}_{iO_2} = \frac{d}{dt}(V_{ti}\text{C}_{iO_2} + V_{\text{blood},i}\text{Cv}_{iO_2}) \tag{25.12}$$

where $\dot{Q}_i(\text{Ca}_{O_2} - \text{Cv}_{iO_2})$ is the net amount of O_2 brought to the tissue by the blood, $\dot{\text{V}}_{iO_2}$ is the O_2 consumption of the tissue, and $(d/dt)(V_{ti}\text{C}_{iCO_2} + V_{\text{blood},i}\text{Cv}_{iO_2})$ is the change in O_2 stores in the tissue compartment.

The mixed venous concentration of O_2 was obtained using Eq. (25.13), which, as for CO_2, provides a weighted average

$$\text{C}\bar{\text{v}}_{O_2} = \sum \frac{\dot{Q}_i\text{Cv}_{iO_2}}{Q_T} \tag{25.13}$$

Across the lung

$$\dot{Q}_T(\text{C}\bar{\text{v}}_{O_2} - \text{Ca}_{O_2}) + \dot{V}_A\text{FI}_{O_2} - \frac{\dot{V}_A\text{Pa}_{O_2}}{P_B - 47} = \frac{d}{dt}\left[\frac{(\text{FRC})\text{Pa}_{O_2}}{P_B - 47} + V\text{aCa}_{O_2}\right] \tag{25.14}$$

where $\dot{Q}_T(\text{C}\bar{\text{v}}_{O_2} - \text{Ca}_{O_2})$ equals the rate of O_2 delivery to the lungs, $\dot{V}_A\text{Pa}_{O_2}/(P_B - 47)$ equals the rate at which O_2 is expired, $\dot{V}_A\text{FI}_{O_2}$ equals the rate at which O_2 is inspired, $(\text{FRC})\text{Pa}_{O_2}/(P_B - 47)$ equals the O_2 stored in the functional residual capacity of the lung, and $\dot{V}\text{aCa}_{O_2}$ is the amount of O_2 stored in the arterial blood. During asphyxia, both $\dot{V}_A\text{FI}_{O_2}$ and $\dot{V}_A\text{Pa}_{O_2} = 0$.

The following equations, based on the work of Gomez [49], can be used to represent the dissociation curves of blood.

For the relationship between O_2 content and O_2 tension,

$$\text{C}_{O_2} = \text{C}_{\max}\left(\frac{u}{1 + u}\right) + 3 \times 10^{-6}\text{P}_{O_2} \tag{25.15}$$

where $u = 0.925v + 2.8v^2 + 30v^3$; and $v = (0.0004273 + 0.04526)\text{P}_{CO_2}{}^{-0.535}$, and $\text{C}_{\max} = O_2$ capacity of the blood.

For the relation between CO_2 content in blood and CO_2 tension, the following equation, which takes the Bohr and Haldane effects into account, was used:

$$\text{C}_{CO_2} = \left(0.149 - \frac{0.014u}{1 + u}\right)(\text{P}_{CO_2})^{0.35} \tag{25.16}$$

The store of O_2 in physical solution in blood was included in the dissociation curve, but the amount combined with myoglobin in the tissues was neglected because of its small contribution as an O_2 store [4, 47].

The respiratory quotient of each tissue, $\dot{\text{V}}_{CO_2}/\dot{\text{V}}_{O_2}$, was assumed to be the same as the respiratory quotient of the body as a whole. Also, the gradient for O_2 across the lung was neglected and the arterial O_2 tension was considered to be the same as the alveolar O_2 tension, an assumption that may introduce an error of about ±10 mL in calculating the total O_2 stores [103].

The values for the parameters used in calculation from this model are like those in Table 25.2 except that they were scaled for dogs [22]. In the muscle and in all other tissue compartments, the slope of the CO_2 dissociation curve of the tissues was assumed to be the same as that of water.

Since the respiratory quotient of all tissues was assumed to be the same, total body O_2 consumption was assumed to be distributed among the different tissue compartments in the same way as CO_2 production; e.g., if the muscle generated 15.1 percent of the total body $\dot{V}_{CO_2}$, it used 15.1 percent of the total body $\dot{V}_{O_2}$.

Figure 25.7 compares the arterial O_2 and CO_2 tensions computed from this model with those measured in a dog during asphyxia. The measurements made during the control period that were used in the calculations are shown in the legend. Similar experimental results were obtained in eight other animals by Cherniack et al. [22].

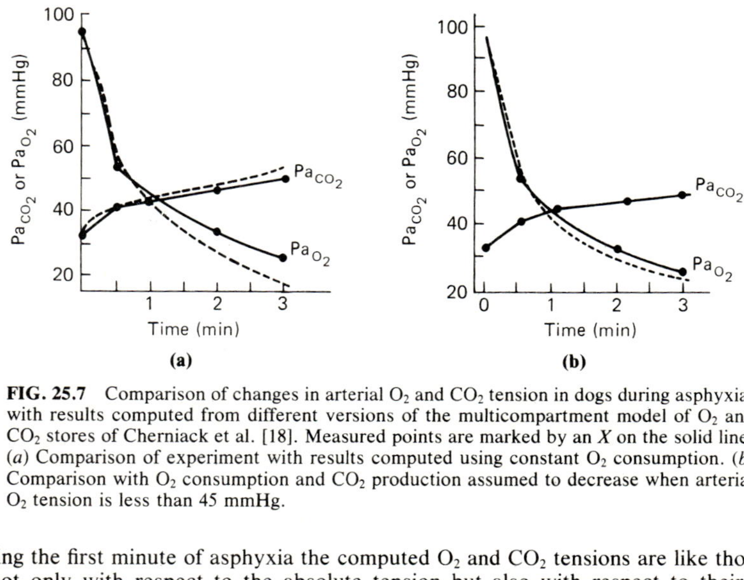

FIG. 25.7 Comparison of changes in arterial O_2 and CO_2 tension in dogs during asphyxia, with results computed from different versions of the multicompartment model of O_2 and CO_2 stores of Cherniack et al. [18]. Measured points are marked by an *X* on the solid line. (*a*) Comparison of experiment with results computed using constant O_2 consumption. (*b*) Comparison with O_2 consumption and CO_2 production assumed to decrease when arterial O_2 tension is less than 45 mmHg.

During the first minute of asphyxia the computed O_2 and CO_2 tensions are like those measured not only with respect to the absolute tension but also with respect to their rate of decrease. After the first minute not only are the calculated absolute tensions incorrect, but the rate of decline of O_2 tension in the second- and third-minute asphyxia is considerably overestimated by the model. The discrepancy noted in the last 2 min of asphyxia could not be explained by the reported affect of asphyxia on the output of the heart, nor by adding to the calculated O_2 stores the amount of O_2 contained in physical solution or O_2 combined with myoglobin. On the other hand, the results might be explicable if, during asphyxia, some mechanism that limits O_2 consumption came into play, such as, for example, oxygen transfer driven solely by oxygen pressure gradient ΔP_{O_2}.

25.3.6 Mathematical Modeling of the Respiratory System

A mathematical model of the respiratory system can be constructed by joining to models of gas stores equations that represent the dynamic effects of changes in P_{CO_2} and P_{O_2} in the arterial blood and brain on ventilation.

Damokosh-Giordano [30] constructed such a model and studied the interrelationship between changes in the controller and changes in the stores on transient ventilation changes. She also examined the effect of the distribution of CO_2 drive between the arterial and cerebral sensors, and found that when CO_2 drive was distributed so that the peripheral chemoreceptor was responsible for only 30 percent of the total drive, one-half of the ventilatory response was achieved in 1.9 min. This is in agreement with the findings of Reynolds, Millhorn, and Holloman, who have conducted detailed studies of the dynamics of the ventilatory response to CO_2 [93]. Responses obtained when the peripheral and central chemoreceptors were assumed to respond to $[H^+]$ rather than to CO_2 showed changes which were slightly slower but essentially similar. Increasing the proportion of the total drive caused by the peripheral receptor also pushed the system toward instability, as manifested by cyclic ventilatory responses (Cheyne-Stokes breathing) in the model after forced hyperventilation. Patients with cerebrovascular disease who have been observed to have more rapid ventilatory responses to CO_2 (suggesting an enhanced peripheral chemoreceptor component) also at times exhibit Cheyne-Stokes respiration [24, 27, 38, 57, 89]. These findings suggest that CO_2 drive may be so distributed between central and peripheral receptors that a balance is achieved between rapidity and stability of ventilatory response.

Damokosh-Giordano [30] reported that doubling the size of the arterial, muscle, or all other tissue compartments produced negligible changes in the ventilatory response during simulations of CO_2 inhalation whether pH or P_{CO_2} was considered to be the direct stimulus to these receptors. Sizable changes in muscle or cerebral blood flow also failed to affect significantly the time course of the ventilation changes whatever the contribution of the peripheral and central chemoreceptors to total ventilatory drive was assumed to be. Subdividing the brain into an intra- and an extracellular fluid compartment separated by a small resistance to CO_2 diffusion (1.0 mmHg difference in the steady state) also had only a slight effect on ventilatory responses. Only large changes in the capacity of the brain to store CO_2, produced by doubling its buffering capacity, significantly slowed ventilatory responses during CO_2 inhalation. This suggested that the ventilation dynamics to CO_2 breathing are insensitive to the construction of plant models. This insensitivity occurs because plant dynamics can have little effect on the time course of arterial P_{CO_2} change and, therefore, on ventilation, when the volume of CO_2 brought into the body by inhalation far exceeds the volume of CO_2 metabolically produced.

In contrast, ventilatory responses to simulated increases in CO_2 production (exercise) are sensitive to the details of CO_2 plant dynamics. Damokosh-Giordano showed that when the CO_2 disturbance to the model originates in a venous compartment, changes in plant parameters (such as the total blood flow to muscle or the distribution of perfusion to the two muscle compartments) have an appreciable effect on ventilatory dynamics. Factors such as total blood flow to muscle become even more important in determining the ventilatory response when the peripheral chemoreceptor accounts for all the drive.

Similarly, the dynamics of CO_2 and O_2 have important effects when continuous swings in ventilation occur, as in Cheyne-Stokes breathing and in sleep apnea, where repetitive episodes of absent ventilation and increased ventilation occur.

In many models of ventilatory control, ventilation is considered continuous in time, as in the Damokosh-Giordano model. That model can be expanded to include the cyclic changes in ventilation rate over a breath and the interaction of O_2 and CO_2 on ventilatory control, as well as the effect of circulation delays to the peripheral and central receptors.

Further, because of the overriding importance of distribution of controller drive, in determining the response of the model to external disturbances a simpler version of body gas stores can be used, if one is not interested in the effects of exercise or in gas transport to different organs.

Accordingly the stores can be reduced to three compartments, a lung or arterial store, and two venous stores, one consisting of the brain, and the second of all other compartments, including muscle and other body organs such as the heart and kidney [30, 71]. The dissociation slope for CO_2 of all the other compartments is taken as approximating that for water, and Eqs. (25.5) and (25.12) apply to the tissue compartments.

Blood flow to brain is taken to vary linearly with Pa_{CO_2} [94] according to

$$\dot{Q}_{\text{brain}} = 0.04 Pa_{CO_2} - 0.93 \tag{25.17}$$

If a "breathing" model is to be used, then the arterial, or lung, store must be described to accommodate the respiratory cycle of inspiration and expiration [71]. Not all inspired air reaches gas-exchanging surfaces; some remains in the conducting airways where gas exchange does not occur. This nonexchanging space is called *dead space* [71]. For the arterial store during inspiration,

$$\frac{d}{dt}\left[\frac{(\mathrm{FRC})\mathrm{Pa}}{P_B - 47} + Va\beta_{\mathrm{blood}}\mathrm{Pa})\right] = \dot{Q}_T\beta_{\mathrm{blood}}(\mathrm{P}\bar{\mathrm{v}} - \mathrm{Pa}) + \dot{V}_A\left[\mathrm{SW}_1\left(\frac{\mathrm{Pa}_{\mathrm{ex,av}}}{P_B - 47}\right) + \mathrm{SW}_2\mathrm{FI}\right] \tag{25.18}$$

where a dead space is assumed, and $\mathrm{Pa}_{\mathrm{ex,av}}$ is the average alveolar gas tension from the previous expiration. $\mathrm{SW}_1 = 1$ and $\mathrm{SW}_2 = 0$ while gas from the previous expiration is being inhaled from the dead space; after that period, ambient air is inspired and $\mathrm{SW}_1 = 0$ and $\mathrm{SW}_2 = 1$.

During expiration,

$$\frac{d}{dt}\left[\frac{(\mathrm{FRC})\mathrm{Pa}}{P_B - 47}\right] = \dot{Q}_T\beta_{\mathrm{blood}}(\mathrm{P}\bar{\mathrm{v}} - \mathrm{Pa}) - \frac{\dot{V}_A\mathrm{Pa}}{P_B - 47} \tag{25.19}$$

For the venous store,

$$\frac{d}{dt}(V_{\mathrm{ti},i}\beta_i\mathrm{P}_i) = \dot{Q}_i\beta_{\mathrm{blood}}(\mathrm{Pa} - \mathrm{Pv}_i) + \dot{V}_i \tag{25.20}$$

where i indicates either the brain or the category of all other compartments, including muscle. The dissociation slope for CO_2 of the category of all other compartments is taken as approximating that for water.

The controller output representing ventilatory demand [71] is represented by

$$\dot{V}_A = 0.3G_1(\mathrm{Pa}_{\mathrm{av,dCO_2}} - G_2) + 0.7G_1(\mathrm{P}_{\mathrm{brain,av,dCO_2}} - G_3) + \left(\frac{45}{\mathrm{Pa}_{\mathrm{av,dO_2}} - 32.44}\right)(\mathrm{Pa}_{\mathrm{av,dCO_2}} - G_2) - G_4 \tag{25.21}$$

CO_2 and O_2 values at controller sites are assumed to be the average arterial and brain CO_2 tensions of the previous breath, delayed by circulation times between the lung and the chemoreceptors. The first and third terms in Eq. (25.21) are the contributions of the peripheral receptors, and the term $45/(\mathrm{Pa}_{\mathrm{av,dO_2}} - 32.44)$ is their oxygen contribution to CO_2 sensitivity. The second term in Eq. (25.21) is the contribution of the central receptor, located in the brain. G_1 is a constant relating ventilation to CO_2 arterial tension. The controller "gain" to CO_2 is represented by the sum of G_1, the CO_2 contribution, and $45/(\mathrm{Pa}_{\mathrm{av,dO_2}} - 32.44)$, the oxygen contribution. G_2, G_3, and G_4 are constants defining the thresholds of the controller characteristics.

The constants in the equation will vary according to the condition of the subject. For example, circulation delays much greater than normal may occur in patients with congestive heart failure; overall gain and threshold may vary from normal, as in chronic obstructive lung disease, or with metabolic acidosis or alkalosis, or as in sleep; and neural or psychological drives may vary according to CO_2 level.

The instantaneous ventilation over a breath is sinusoidal according to

$$\dot{V} = V_{\mathrm{mag}} \sin 2\pi f t \tag{25.22}$$

where V_{mag} is the amplitude of $\dot{V}$, and where f, the breathing frequency, can be calculated from the data of Reynolds et al. [93] as

$$f = 0.4\dot{V}_A + 8.4 \tag{25.23}$$

V_{mag} is calculated as V_A/π since the controller output represents the average ventilation rate over a breath.

The model of respiratory control described in Eqs. (25.18) to (25.23) has been used to describe transient effects of CO_2 inhalation on breathing, and to describe blood and organ $\mathrm{Po_2}$ and $\mathrm{Pco_2}$ changes during controlled ventilation and respiratory abnormalities such as Cheyne-Stokes breathing and sleep apnea [71].

Many other useful models of respiratory control exist and also have been used extensively, for example, models by Yamamoto [107], Grodins et al. [52], Saunders [96], and Fowle and Campbell [48]. The selection of a particular model should be made according to the phenomenon being studied.

Study of the role in respiration of bicarbonate transfer through the brain interstitial and cerebrospinal fluids not only requires a more detailed description of the interconnections among brain neurons but also must obviously include chemical and transport mechanisms more complex than those assumed in the equations presented above. The proliferation of models indicates how much more experimental data are needed.

Models to describe ventilatory response to exercise would, on the surface, appear easy to define; one need only to describe mathematically a "proportional + integral" controller which closes, in the steady state, to zero error. However, the physiological mechanism accounting for integral control is unclear. It is commonly believed that during exercise ventilation increases proportionally to metabolic rate, i.e., CO_2 reduction, so that arterial P_{CO_2} is constant. It is not clear how metabolic rate is sensed. Saunders [96] has proposed that CO_2 production might be sensed by peripheral chemoreceptors from the rise in arterial P_{CO_2} that occurs during expiration. As metabolic rate rises, so does the increase in P_{CO_2} in the expiratory phase of the breathing cycle. Using a model based on "feed-forward" control of metabolism and feedback control of P_{CO_2}, he has been able to reproduce the ventilatory changes seen with CO_2 inhalation and with exercise.

Entirely different models, however, are also able to reproduce ventilatory responses to exercise. In Yamamoto's model the CO_2 receptor is assumed to respond to a matrix of factors which include the metabolism of the brain respiratory neurons, transmembrane concentrations of $[H^+]$ and intracellular $[H^+]$ [107]. All these factors were necessary for the model to successfully simulate both exercise and CO_2 studies.

Poon was able to reproduce ventilation occurring with both exercise and CO_2 inhalation by a model which minimized both ventilation and the deviation of P_{CO_2} from normal values [91].

At the beginning of exercise there is frequently an abrupt increase in ventilation. While this can be roughly simulated by an abrupt increase in cardiac output, it seems to indicate the presence of a neural drive originating in the exercising limbs which increases cardiac output or excites ventilation directly. Only the model proposed by Yamamoto has an explicit neural input [107].

25.3.7 Cheyne-Stokes Breathing and Other Respiratory Rhythm Disorders

Mathematical models have also been used to reproduce the cyclic rise and fall in ventilation with recurrent periods of apnea that characterize Cheyne-Stokes breathing [23, 24, 33, 34, 100]. Often tidal volume and frequency change during the period of hyperpnea in a crescendo-decrescendo manner as shown in Fig. 25.8. The period of the ventilatory oscillation in Cheyne-

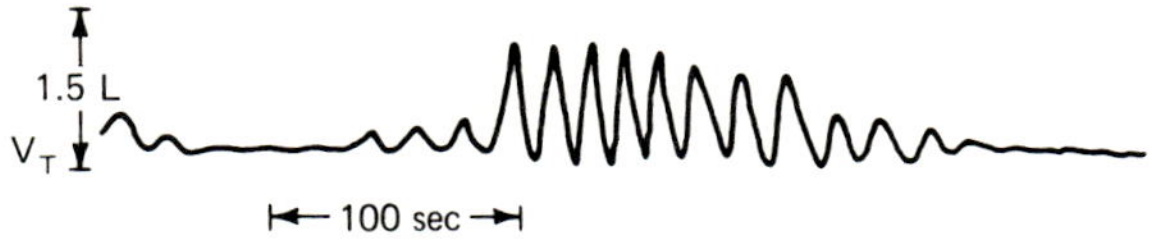

FIG. 25.8 Experimental record of a cycle of Cheyne-Stokes breathing. Note the increase and decrease in tidal volume V_T preceded and followed by periods of apnea.

Stokes breathing is variable but is related to the circulation time measured from the lung to systemic arteries [19, 56]. The longer the circulation time, the more prolonged the cycles. Arterial blood P_{CO_2} is highest during the phase of hyperpnea, while arterial P_{O_2} is at its minimum. Alveolar gas tensions cycle in the opposite way [33].

Cheyne-Stokes breathing has been rather consistently produced in animals by interventions which seemed likely to produce unstable respiratory control [16, 21, 56, 100]. The steadiness of the output in a feedback control system depends on the ability of the control system to be informed adequately of the consequences of its action. If delays in information transfer are sufficiently great, the controller action to correct the effects of a disturbance may result in the output overshooting and undershooting the desired level. These sorts of delays can occur in the respiratory system when circulation time is prolonged between the lungs and the chemoreceptors in the arterial blood and the brain. Guyton and coworkers produced Cheyne-Stokes breathing in anesthetized dogs by inserting loops in the carotid artery which artificially lengthened the circulation time between the lung and the brain to as much as 5 min [56]. Prolonged cycles of hyperventilation and apnea were observed with periods that were proportional to the circulation time through the artificial loop.

Increased controller gains also tend to cause instability in physical control systems. One way to increase the gain of the respiratory controller is to produce hypoxia. Periodic breathing was induced in anesthetized dogs by allowing hypoxia to occur during a period of posthyperventilation apnea [21]. The periodic breathing was characterized by crescendo-descrescendo variations in tidal volume and breathing frequency with periods of apnea as in Cheyne-Stokes breathing.

Persistent periodic breathing was produced in anesthetized cats ventilated with a respirator governed by phrenic nerve output so that feedback loops remained intact [16]. Periodic breathing could be elicited by increasing the gain of the servo-respirator, or in addition by lung deflation and/or respiring the cat with hypoxic gas mixtures. All these procedures increased the effective gain of the respiratory controller. Once periodic breathing appeared, it was sustained. Further increases in servo-respirator gain had little effect on the total length of the cycle but increased the duration of apnea at the expense of the length of the phase of ventilation. Artificial ventilation or cutting the nerves innervating the peripheral chemoreceptors eliminated periodic breathing. Hyperoxia also tended to eliminate periodicity. On the other hand, inhaling CO_2-enriched gas mixtures eliminated periodic breathing.

Recently it has been proposed that the recurrent apneas occurring during sleep in infants and adults result from an instability in respiratory control that affects the activity of both chest wall and upper airway muscles [13]. As a result renewed interest has been generated in the factors that might cause the respiratory control system to become unstable.

25.3.8 Instability in Respiratory Control

Analytical Considerations. A physical system is said to be *unstable* when a bounded disturbance causes an unbounded system response. A simple system such as a mass attached to a spring would, in the absence of friction, behave in this way. The application of a constant force to the mass would cause continuous sinusoidal oscillations in its position. Real systems usually have some amount of frictional resistance or damping which allows the oscillations to decay until the mass finally comes to rest.

The dynamic behavior of a simple spring-mass system can be derived from the equation

$$\text{Sum of forces} = \text{mass} \times \text{accelerations}$$

where the forces applied are a spring, or *elastic,* force Kx proportional to the displacement; a frictional, or *resistive,* force proportional to the rate of change of displacement, $C\,dx/dt$, and a *disturbance,* or externally applied, force F_D. Thus

$$M\frac{d^2x}{dt^2} + C\frac{dx}{dt} + Kx = F_D \tag{25.24}$$

The dynamic behavior in response to a constant applied force is shown in Fig. 25.9, and the response varies according to the damping ratio ζ, defined as

$$\zeta = \sqrt{\frac{C}{KM}} \tag{25.25}$$

Note that there exists a finite steady-state displacement of the system in response to the applied disturbance.

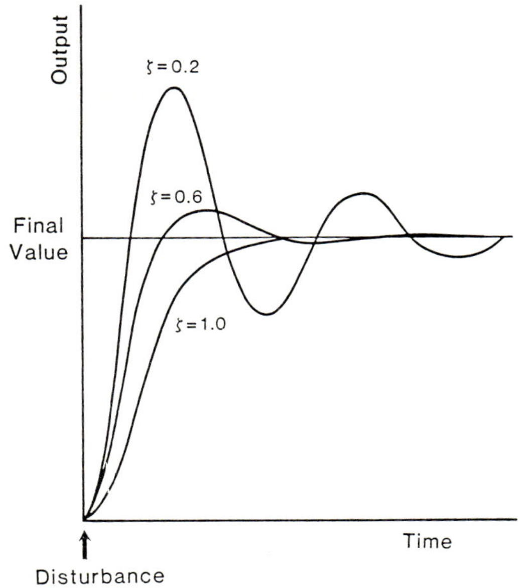

FIG. 25.9 Response of a simple spring-mass system to a constant applied force. The response varies according to the damping ratio ζ.

A controller can be used to minimize the displacements of the mass caused by the disturbance. This usually also has the effect of reducing the damping and has the analogous physical effect of diminishing frictional resistance.

For example, a controller can be used which senses by feedback the displacement of the mass and exerts a restoring force proportional to the displacement of the mass from its initial position or, stated another way, is proportional to the error in displacement.

But accompanying the desirable result of reduced steady-state error is increased oscillation of the mass caused by the lower resistance. In higher-order systems, increasing controller gain can reduce damping so severely that continuous oscillations result from a finite applied disturbance; that is, the system becomes unstable.

Error-free operation can be achieved in physical systems by using an "integral" controller where the controller action is proportional to the integral of the error, or, stated another way, where the rate of change of controller action is proportional to the error. Thus, the controller is in motion as long as there is an error. When controller motion ceases, the system is in equilibrium and the error is zero.

Integral control is rarely used alone in physical systems because it reduces system damping severely, and so in human-made systems is usually combined with proportional control to avoid instabilities. This combination is also called *proportional-plus-reset* control. The term *reset* stems from the idea that the reference point has been reset to a new norm. From the standpoint of the final equilibrium state, the effect is the same as a dynamic change in reference value for a proportional controller. This mode of control could be involved during exercise where there is no, or only a small, deviation at Pa_{CO_2} from normal [54, 104]. A possible explanation for the sort of change of Pa_{CO_2} seen during exercise is that by some mechanism the respiratory controller has lowered the threshold value of the controller characteristic, ventilation versus Pa_{CO_2}, shifting it to the left so that it operates at some new "normal" point. It should be noted that many factors associated with exercise, such as extremity movement and temperature changes, tend to increase ventilation, producing an apparent resetting. Too high a reset rate can reduce system damping to zero and cause continuous oscillations. Under higher reset rates, although the error can be brought to zero, increased oscillation would result from attempts to bring the system back to equilibrium in shorter times.

System damping will also be reduced if the controller does not "know," because of delays, the position of the mass. Such delays increase the tendency for control systems to be unstable, since the controller acting on incorrect information may reinforce the effect of the disturbance rather than suppress it. The controller might apply a restoring force, if the delay is long enough in the direction opposite to what is at that instant required. The controller would be, in effect, reinforcing the excursion from normal. With sufficiently long delays the damping ratio is reduced to zero. In a sense, delays are a more potent stimulus to instability than high controller gain since they can cause instabilities in systems which could not otherwise be made unstable by increasing controller gain.

Like physical systems, the respiratory system and other biological systems have properties analogous to elasticity, friction, delays, and controller gains. Equations describing the action of the respiratory system can also be written and analyzed similarly and described as for spring-mass system.

For example, the relationship between Po_2 and O_2 content described by the oxyhemoglobin dissociation curve is a pressure-volume relationship analogous to elasticity; and the relationship between the changes in ventilation caused by CO_2 inhalation and arterial Pco_2 is analogous to the gain of a proportional controller in a physical system. Delays occur because in the respiratory system [13, 19] information about changes in levels of Pco_2 and Po_2 in the lung are carried by the circulation to chemoreceptors, which requires time.

In linear systems, it is possible to estimate what parameter variations would cause a system to become unstable by observing the effects of sinusoidal inputs of different amplitudes and frequencies on the amplitude of the output and the timing of the output in relation to the input (the phase relationships between the two). But the respiratory control system is not linear. Nonlinearities occur, for example, because of the alinear nature of the oxyhemoglobin dissociation curve, the complex interdependence of ventilation on both Pco_2 and Po_2, and the sharp discontinuities, for example, the appearance of apnea that occurs in the relationship between Pco_2 and Po_2 and ventilation at low levels of chemical drive.

However, the various factors that have been discussed that affected stability in the spring-mass system will tend to have the same action in the respiratory system despite the alinearities. Increases in delays will make breathing more unstable in the respiratory control system. However, the exact effects of delays on breathing patterns and the length of apneic periods are difficult to predict because of the complex interactions that are possible among the various system components.

The Chemoreceptors. The effect of the changes in arterial Po_2 on ventilation depends mainly on the activity of carotid body. While the output of the carotid body is inversely related to Po_2, its gain is not constant. Changes in the level of Po_2 produce a curvilinear increase in both carotid body activity and in ventilation, gain ($\Delta V/\Delta Po_2$) becoming greater as hypoxia becomes more severe [6, 68]. This increase in gain is one of the factors which promotes instability when hypoxia is a major respiratory stimulus.

At constant levels of Pa_{O_2}, the response of the carotid body to changes in Pco_2 is linear. However, hypoxia increases the controller gain to CO_2 ($\Delta V/\Delta Pco_2$) [28, 68, 80, 102].

Although the changing gain of the carotid body with hypoxia increases tendencies to instability when hypoxia is a major respiratory drive, the linear response of the carotid body to steady-state increases in Pco_2 and its ability to respond to the rate of change of Pco_2 should allow respiratory control to be more stable when under the influence of hypercapnia.

Hyperventilation produces apnea in anesthetized and sleeping subjects when arterial Pco_2 falls below some threshold value. The presence of a CO_2 threshold has important consequences for the stability of respiratory control. The apnea produced when hyperventilation lowers Pco_2 below threshold levels allows hypoxia to occur. The hypoxia increases controller gain and decreases the stability to breathing when it eventually resumes. With the onset of spontaneous ventilation, controller gain may be sufficiently high (because hypoxia is not present) to allow ventilation to force Pco_2 once again to subthreshold levels, thus causing a cyclic recurrence of apnea and breathing. All else remaining the same, the higher the Pco_2 threshold, the longer the period of apnea, and the greater the controller gain at the beginning of spontaneous breathing, the more likely it is that instability will occur.

Consciousness seems to decrease the likelihood of hyperventilation producing apnea, although the factors responsible for the preservation of breathing are not clear. Increases in the operating point of the CO_2 controller enhance the effects of ventilation in producing changes in arterial P_{CO_2}, increasing the amount of CO_2 which can be expired by a given volume of ventilation [29]. This in turn enhances tendencies for instability.

The Effect of the Body Gas Stores on Stability of Respiratory Control. Whenever the amount of O_2 brought into the body is different from the O_2 consumed, or the CO_2 removed from the body is different from the CO_2 produced in the tissues, changes occur in the quantity of O_2 and CO_2 contained in the body which affect the level of arterial P_{O_2} and P_{CO_2} [18]. The greater the changes in P_{CO_2} and P_{O_2}, the greater the alteration in the activity of the chemoreceptors and the more likely it will be for unstable breathing to occur. The rate at which changes occur in P_{CO_2} and P_{O_2} will depend on the characteristics of the body gas stores which determine the relations between gas tension and gas content in the store.

The similar volume of oxygen that can be stored in the body either in solution or in chemical concentration is quite small [18]. Changes in ventilation will produce large changes in P_{O_2} and hence in the activity of peripheral chemoreceptors.

By contrast, the body is able to store large amounts of CO_2 both in the lung, in solution in blood, and in the form of bicarbonate in soft tissues and carbonate in bone [18, 23]. If the chemoreceptors were exposed to all the CO_2 available in the body, breathing would be extremely stable since ventilation changes would produce only very little alteration in P_{CO_2}.

Although the peripheral and central CO_2 receptors are not exposed to all of the body CO_2, even the local CO_2 stores to which they are exposed are considerably larger than the corresponding O_2 stores. The larger effective storage capacity for CO_2 and the generally constant gain of the CO_2 controller (except at the P_{CO_2} threshold) under isoxic conditions tend to make breathing stable when CO_2 is the major respiratory stimulus.

The rapid changes that occur in O_2 stores with ventilation changes, coupled with the alinear gain characteristics of the controller with hypoxia, tend to make breathing unstable [16, 34]. These characteristics seem to account for the Cheyne-Stokes breathing seen in newcomers to high altitudes where hypoxia is a particularly important stimulus.

Circulation Effects on Stability. Increases in cardiac output tend to make breathing more stable by decreasing delays in information transfer around the control loop and by promoting equilibration of CO_2 stores [18, 23]. Cardiac output increases with hypoxia and, to a lesser extent, with hypercapnia [18].

Circulation times affect the cycle length of Cheyne-Stokes breathing [33, 56]. As circulation time decreases (time between the periods of hyperventilation), cycle length becomes shorter. Hypoxia-induced speeding of the circulation time is one factor explaining the short ventilation cycles observed during Cheyne-Stokes breathing at high altitudes [65]. On the other hand, when circulation is long, as in congestive heart failure, the breathing cycle length is likewise prolonged.

Interaction of Factors Producing Instability in Respiratory Control. The conditions which produce unstable breathing, the cycle length, and the amplitude of the blood gas fluctuations during the instability will depend on the characteristics of each of the components of the control system and how they are interconnected. In linear control systems, the occurrence of instability and the characteristics of the cyclic output can be predicted, for example, by graphical techniques such as the Bode and Nyquist diagrams [68]. However, the respiratory control system behaves alinearly, and the effects of alterations in system components are better analyzed through the use of mathematical models. Even so, the results obtained from such analyses cannot be considered to be exact since the model cannot mimic exactly the actual respiratory system.

The mathematical model previously proposed can be used to analyze the effects of the interplay of changes in system components on the stability of breathing.

Interaction of Circulation Time and Controller Gain on Stability. Disturbances such as hyperventilation or asphyxia tend to produce instability in breathing in the model. In general, the greater the disturbance the more likely it is that instability will occur; this type of behavior is a characteristic of nonlinear systems.

In the model, 2 min of hyperventilation, which is sufficient to reduce arterial P_{CO_2} by 23 mmHg to $Pa_{CO_2} = 17$, is followed by a brief period of apnea and then by the resumption of regular breathing if the circulation time and controller sensitivity are normal. When the circulation time is increased from 6 to 25 s, instability in breathing occurs with regularly recurring cycles of hyperventilation and apnea. The period of oscillation is 1 min and the apneic interval lasts 36 s. Smaller increases in circulation time produce only one or two apneic periods. With further increases in circulation time above 25 s, the period of the cycle increases by about 15 s for each 6-s increase in delay. As shown in Fig. 25.10, the period of apnea also lengthens as circulation time is increased.

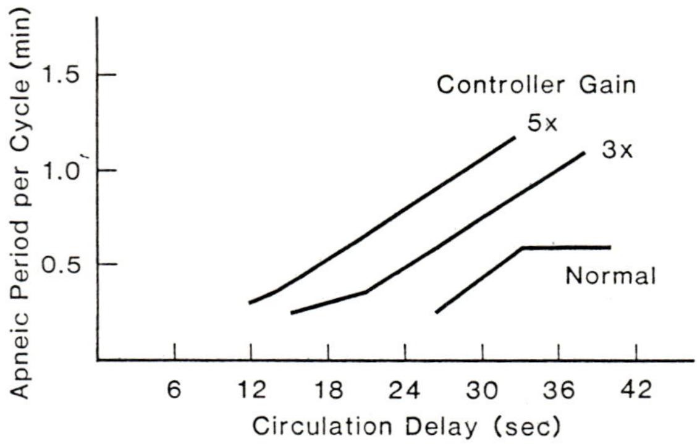

FIG. 25.10 Effect of circulation delays on the period of apnea in Cheyne-Stokes breathing. Results are calculated from the model described in Eqs. (25.17) to (25.23). With increasing controller gain, smaller increases in circulation are needed to produce persistent oscillations in ventilation.

With increasing controller gain, smaller increases in circulation time are needed to produce persistent oscillations in ventilation. For example, if the gain of the controller is tripled, a circulation delay of only 15 s is needed to produce Cheyne-Stokes-like breathing, and when controller gain is multiplied by five times, only a 12-s circulation delay is needed.

Increasing only the hypoxic sensitivity of the controller likewise reduces the increase in circulation time required to produce unstable breathing. With respect to producing instability, a threefold increase in the sensitivity to hypoxia is approximately equivalent to doubling the sensitivity to CO_2.

Controller gain increases also lengthen the apneic period during a cycle, while the overall period of the ventilation cycle remains relatively unaffected. Since the length of the period was little affected by increase in controller gain, the apneic period as a percent of cycle time increases.

When the resting Pa_{CO_2} is increased (as in sleep), the effective gain of the system is higher because of the nonlinear term $\dot{V}Pa_{CO_2}$ [see Eq. (25.19)]. For example, at a high Pa_{CO_2} of 80 mmHg, a circulatory delay of only 12 s was required to produce unstable breathing compared with a 25-s delay at 40 mmHg Pa_{CO_2}, given the same disturbance. The apneic period was 58 s at the higher P_{CO_2} compared to 15 s at 40 mmHg Pa_{CO_2}. Thus the impact of operation at high P_{CO_2} on the length of the apneic period is equivalent to increasing controller gain five times at a resting P_{CO_2} of 40 mmHg. At high levels of P_{CO_2}, instability could be produced even with reduced controller gains.

Reduced oxygen and CO_2 stores have the major effect of reducing the circulation delay which causes onset of irregular breathing. Reduction of both O_2 and CO_2 stores to one-third of normal values each had the effect of an increase of 7 s in circulation delay.

These results suggest that instability results from changes in the relations among components in the system. Moreover they suggest that differences in cycle length, as in the duration of apnea, need not be caused by different mechanisms but instead can occur as a result of alinear behavior in the respiratory system.

25.4 RESPIRATORY SYSTEM MECHANICS

In the preceding sections we have described how the respiratory system responds to maintain constant arterial P_{CO_2} and P_{O_2} when challenged by metabolic or environmental disturbances. The interrelationship between gas tensions in the blood and in other body tissues and the neural activity produced by the respiratory controller has been discussed, but we have not examined in any detail how the changes in neural drive are converted to changes in the size or frequency of breaths. This conversion depends on the forces produced by the respiratory muscles acting on the mechanical properties of the lung and chest wall which resist displacement.

In addition, the respiratory controller must be able to adjust its behavior so that arterial blood gas tensions are maintained within acceptable limits even when the mechanical characteristics of the respiratory system change or the ability of the respiratory muscles to generate force is altered. Such changes occur not only with lung diseases but happen spontaneously as well [14, 74]. Significant changes in the resistance to airflow appear during sleep and whenever mouth breathing is substituted for nasal breathing [87]. Changing from the recumbent to the sitting or standing position affects the alignment of the respiratory muscles and the elastic and resistive properties of opposing breathing [74].

The following sections examine the interaction among controller activity and the physical and mechanical properties of the respiratory system.

As shown in Fig. 25.11, the lung itself can be considered in a simplified way as consisting of a conducting passageway in which virtually no exchange of gas takes place, and alveoli (microscopic air sacs with richly perfused delicate walls) where large amounts of CO_2 can be removed and O_2 added to the circulating blood.

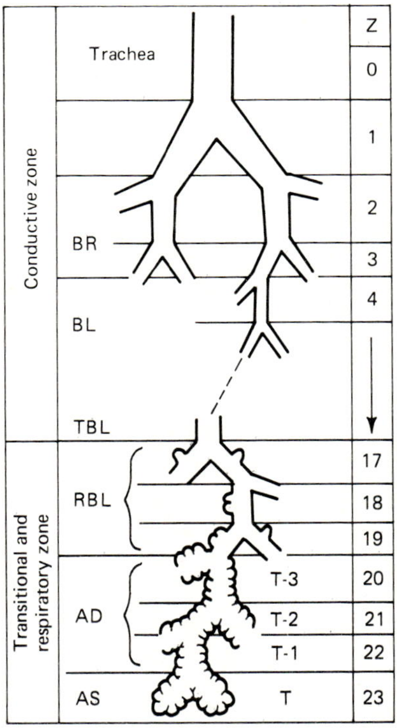

FIG. 25.11 Model of airway branching: BR = bronchus; BL = bronchiole; RBL = respiratory bronchiole; TBL = terminal bronchiole; AD = alveolar duct; AS = alveolar sac. *(Reproduced by permission from E. R. Weibel, Morphometry of the Human Lung, Springer-Verlag, Heidelberg, 1963.)*

25.4.1 Mechanical Considerations during Quiet Breathing

Ventilation is not a continuous process but consists of two stages: inspiration, in which fresh air is sucked into the lungs, and expiration, in which air is expelled from the lungs. The volume of air drawn in and out of the lungs in each breath, or the tidal volume (V_T), multiplied by the frequency of breathing is equal to the minute ventilation rate ($\dot{V}_A$). The volume of air that remains in the conducting airways with each breath and does not reach gas exchanging surface is called the *dead-space volume* (V_{DS}).

The difference between V_T and V_{DS} is the amount of air reaching the alveoli, the alveolar ventilation V_A, so that $V_A = V_T - V_{DS}$.

The level of arterial P_{CO_2} and P_{O_2} in the blood leaving the lungs is proportional to the metabolic rate (the production of CO_2 by the tissues, $\dot{V}_{CO_2}$) and their use of O_2 ($\dot{V}_{O_2}$) and the level of alveolar ventilation according to the equation:

$$Pa_{CO_2} = \frac{\dot{V}_{CO_2}(P_B - 47)}{\dot{V}_A(1 - V_{DS}/V_T)} \tag{25.26}$$

where $\dot{V}_A$ is the minute ventilation.

Although V_{DS} increases with V_T in general, as V_T grows greater so does the level of alveolar ventilation [2, 53]. For the same level of minute ventilation, alveolar ventilation is higher when tidal volume is greater and the frequency of breathing slower than when the opposite occurs (faster breathing rates and smaller tidal volumes).

Air flows from regions of higher pressure to regions of lower pressure. The normal breathing process involves an expansion of the lung during inspiration, which produces subatmospheric pressures in the alveoli and a return to the original resting lung volume (the functional residual capacity) during expiration. Inspiration is almost always an active process caused by the contraction (shortening) of the respiratory muscles, while expiration is passive. During inspiration there is an increase in the total number of molecules in the lung. This is achieved by the bulk, convective flow of gas into the lung. At the same time molecular diffusion helps even out the differences in gas concentration caused by the addition of atmospheric air into the lungs. The relative contributions of convection and diffusion to gas exchange depend on the geometry of the lung. Air is brought into the conducting airways mainly by convection; however, in the periphery of the lung (where the alveoli are located), diffusion becomes more important as a mechanism of O_2 and CO_2 transport. For detailed models of the effects of lung geometry on gas exchange see Engel [42].

The lungs are encased in a chamber formed by the ribs and the respiratory muscles. The diaphragm, the major muscle of inspiration, separates the thoracic cavity from the abdominal cavity. The spaces between the ribs are occupied by the intercostal muscles, which are divided into two groups. The intercostal muscles with an inspiratory action (external intercostals) are located closer to the body surface than are the intercostal muscles which have an expiratory effect (internal intercostals). Contraction of the diaphragm and the external intercostal muscles enlarges the thoracic cavity, creating a greater negative pressure in the space separating the lungs from the chest wall, which causes air to enter the lungs and expand them. This pressure difference acts to overcome elastic and resistive forces and forces opposing the movement of air according to the equation [75]

$$P = EV + R\dot{V} + L\ddot{V} \tag{25.27}$$

where E, R, and L are constants; V = volume; $\dot{V}$ = airflow; and $\ddot{V}$ = acceleration.

The tissue making up the lungs and the thoracic wall has elastic properties which can oppose lung enlargement [60]. Hence, part of the force produced by the inspiratory muscles is used to overcome these elastic forces. Figure 25.12 shows the changes in lung volume (*a*) and thoracic cage volume (*b*) when different static pressures (when there is no airflow) are applied across the respiratory system. Note that the relationships of pressure and volume for the lungs and the thoracic wall are not the same. Also, the pressure-volume characteristics of both the lungs and the thoracic walls are not linear so that, as applied pressure increases, the volume of the lung approaches a limit (called the *total lung capacity*, or TLC). An exponential equation has been developed relating pressure to lung volume, and is of the form $V - V_0(1 - e^{-KP})$.

If there were no thoracic wall, the lungs would be completely collapsed when the pressure difference across the lungs were zero. On the other hand, if the lungs were removed from the thoracic cavity, the thoracic wall would enclose a volume about 60 percent of that achieved at TLC [12]. This is because the structure and configuration of the bony ribs and ligaments which

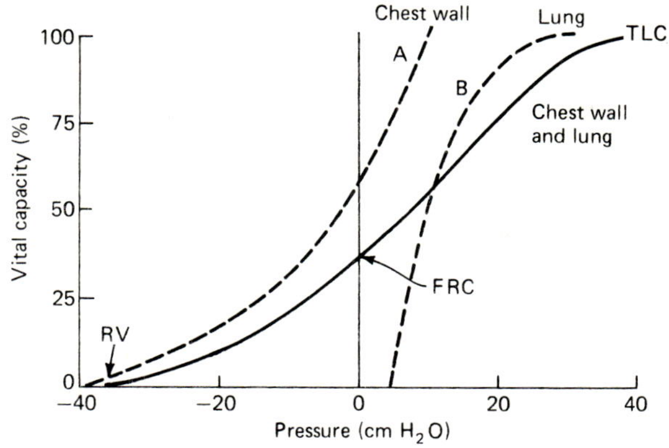

FIG. 25.12 Schematic representation of the pressure-volume relationships of the chest wall and lung combined. The combined curve is obtained by summation of the recoil pressures of the isolated chest wall (*A*) and the isolated lung (*B*).

attach the ribs to the spine prevent collapse. In the intact thorax (with both the lungs and the chest walls in their usual position) the elastic forces of the chest wall are sufficient to prevent lung collapse. A negative pressure of about 5 cm H_2O (fluctuating during the breathing cycle) is maintained in the pleural space, which is sufficient to keep the lungs at volume, called *functional residual capacity* (FRC), which is about 30 percent of TLC.

The relationship of pressure P to volume V is called *elastance,* and the elastance of the lung (E_{L}), chest wall (E_{CW}), and respiratory system (E_{RS}) are defined as follows:

$$\frac{P_{\mathrm{TL}}}{V_{\mathrm{L}}} = E_L$$

$$\frac{P_{\mathrm{TCW}}}{V_{\mathrm{CW}}} = E_{\mathrm{CW}}$$

and

$$\frac{P_{\mathrm{TRS}}}{V_{\mathrm{RS}}} = E_{\mathrm{CW}}$$

where the subscripts TL, TCW, and TRS stand for translung, trans-chest wall, and transrespiratory system, respectively. Disease can affect each of these compliances, separately or in combination. The reciprocals of these relationships are the respective compliances of the lung, C_{L}, chest wall, C_{CW}, and respiratory system, C_{RS}.

In addition to overcoming elastic forces, the forces produced by the inspiratory muscles must overcome frictional or resistive forces. These forces develop when air is flowing into and out of the lungs, and result from the movement of air through the airway, i.e., fluid resistance, and the sliding of tissue elements as the lungs inflate and deflate, i.e., tissue resistance. The mass of the air, lungs, and chest wall are relatively small so that the forces needed to overcome the inertance of the respiratory system are negligible except when breathing is stressed so that flow rates are high.

The conducting airways, although they are represented in Fig. 25.11 as a simple cylindrical tube, actually have a very complex configuration. Air enters the body through the nose and mouth and traverses convoluted chambers in the head. The volume of these chambers and their resistance to airflow can be altered by the activity of muscles in the nose, mouth, and pharynx (the *upper airways*). From the upper airways air passes into the trachea (as shown in Fig.

25.11), which subdivides into multiple smaller channels (called *bronchi* and *bronchioles*) before reaching the alveoli. (This portion of the conducting passages is called the *lower airways.*) Except for the upper part of the trachea the lower airways are within the thoracic cavity. About one-half the resistance to airflow occurs in the geometrically complex upper airways and less than 20 percent of the total resistance in the smaller intrathoracic airways (less than 2 mm in diameter [83, 84].

The walls of the trachea and bronchi are only partially encircled by cartilage. The remainder of their walls and all the walls of the bronchioles are made up of smooth muscle. Hence the walls are not rigid. The intrathoracic airways are subject to the same forces as the alveoli and, like the alveoli, expand and contract with lung inflation and deflation. But the volume changes in the airways are not as great as in the alveoli. Nonetheless, the resistance to airflow is less during inspiration than during expiration. Moreover, it can be shown that resistance is a function of lung volume, decreasing as the lung enlarges, as shown in Fig. 25.13 [35, 74, 94].

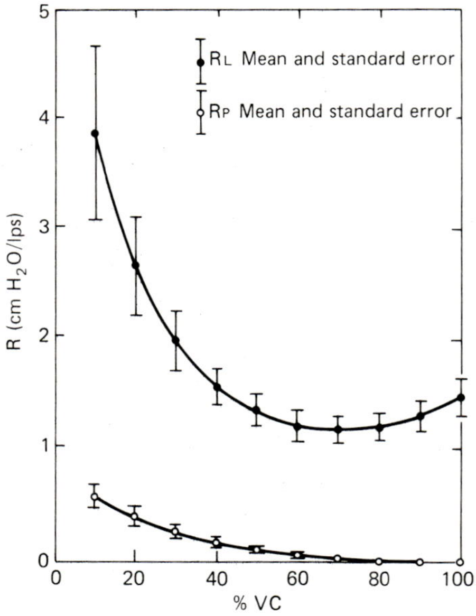

FIG. 25.13 Mean values of measurements of pulmonary flow resistance (R_L) and peripheral resistance (R_P) at different percentages of vital capacity (%VC) in eight living dogs. (*From P. T. Macklem and J. Mead, Resistance of central and peripheral airways measured by a retrograde catheter, J. Applied Physiology, vol. 22, pp. 395–401, 1967.*)

In the larger conducting airways linear velocities are relatively high and flow can become turbulent rather than linear. The cross-sectional area of the airway increases greatly toward the lung periphery so that linear velocities are low.

In contrast to respiration, expiration during quiet, unstimulated breathing is passive. Instead of active muscle contraction, the passive elastic properties of the respiratory system are used to overcome resistive forces.

In fact, it has been shown that the duration of expiration and the airflow rate at the beginning of expiration are greater than would be expected from the resistive and elastic characteristics of the respiratory systems measured during artificial ventilation in subjects that are not breathing spontaneously [1]. This occurs because of a retardation of flow at the start of expiration in spontaneous breathing which is produced by a braking action caused by continued contraction of the inspiratory muscle [110, 111].

The purpose of this postinspiratory activity of the inspiratory muscles is unclear. It may be a mechanism which prolongs the contact of alveolar air and blood which in some way improves gas exchange. A specific group of neurons in the medulla has been identified which seems to be responsible for this postinspiratory activity. When breathing is stimulated, expiratory time is reduced, and thus postinspiratory activity is shortened. In addition muscles which have an expiratory action, such as the internal intercostal muscles, are recruited to provide additional force to push air out of the lungs.

However, expiratory flow does not increase indefinitely as the forces developed by the expiratory muscles grow stronger. Eventually, expiratory flow rates become limited because of airway narrowing caused by compressive expiratory forces and by the development of choke points in the airways as the velocity of flow rises [83, 84, 92].

25.4.2 Models of Lung Mechanics

A simple model of the mechanics of the respiratory system was introduced by Otis et al. [82]. It consists of a single compartment of constant elastance supplied by a single compartment of constant resistance. Elastance and resistance in this model are assumed to be independent of lung volume and airflow rates. In addition, inertial forces are considered to be negligible.

In this model inspiration can be represented as follows:

$$P(t) = EV(t) + R\dot{V}(t) \tag{25.28}$$

where $P(t)$ is the forcing pressure, V is the change in lung volume from FRC, R is resistance, and $\dot{V}$ is the instantaneous flow rate. Since expiration is passively dependent on elastic recoil forces developed during inspiration, $EV(t) = R\dot{V}(t)$ during quiet expiration.

This model is useful, for example, in predicting the changes in volume and flow that will occur in patients whose breathing is controlled artificially (particularly if flow rates are not very high). It might also be used to predict the changes that might occur if breathing were artificially impeded by added resistance and compliances.

A more complex model is needed to relate changes in phrenic nerve activity to changes in tidal volume. Ideally such a model would describe the following transformations:

1. The relationship between nerve activity to the respiratory muscles and chemical and mechanical stimuli.
2. The relationship between nerve activity and the forces produced by the respiratory muscles. This will depend on the length-tension relationship of each of the respiratory muscles and how the force they produce changes with the velocity of their contraction.
3. The relationship between these muscle forces and volume displacement and airflow rates.

The phrenic nerve consists of multiple fibers which produce electrical spikes during inspiration. As inspiration proceeds, the frequency of firing of each fiber increases, and fibers of large diameter (which generate spikes of higher voltage) are recruited. With increasing hypercapnia or hypoxia the activity produced in each inspiratory phase is increased. This activity can be quantitated in several ways. Frequently the signal is rectified so that the negative voltages produced with each spike are made positive. Then the tracing is appropriately filtered and processed so as to give a moving average of the voltage recorded in a unit of time.

It can be shown that this moving average has a predictable time course that can be described by the equation [97, 98]

$$\mathrm{MA} = at^b \tag{25.29}$$

where MA is the moving average of the activity, t is time, and a and b are constants. The constant b usually has a value close to unity, but in different subjects can be greater or less than 1. The term b can also be affected by anesthetics. However, in a given animal anesthetized with a particular anesthetic, b is unaffected by increasing chemical drive while a grows larger. The constant b is also unaffected by the changes in the effective mechanical properties of the respiratory system produced, for example, by agents which cause airway narrowing or by requiring the animal to breathe through devices which increase elastic or resistive forces.

Moreover, the anesthetized animal can be made to inspire from FRC when the airway is completely occluded and a record made of the pressure wave produced (pressures are usually measured through a pinhole opening in the trachea). Since no volume of flow changes can occur when the airway is shut, this pressure wave represents the force produced by the inspiratory muscles when they contract isometrically (without appreciable shortening). It can be shown that this pressure wave, like the moving average with phrenic activity, is a power function of time so that

$$P = at^b \tag{25.30}$$

where P is the inspiratory driving pressure over time t [97, 98]. The constants a and b are similar to those measured in a simultaneous recording of phrenic nerve activity. Thus, in certain very specific circumstances it seems possible to relate nerve activity to the forces produced by the inspiratory muscles contracting isometrically. In addition, it has been shown that there is a linear relationship between the peak of the moving average of phrenic nerve activity and tidal volume produced in a given subject breathing spontaneously [40]. Hence, it has seemed reasonable to use the isometrically developed inspiratory pressure wave to represent the driving pressures developed by the inspiratory muscles when the animal is spontaneously breathing and the airway is not occluded.

However, when this is done and Eq. (25.28) is used to predict volume and flow changes, the model proves to be inaccurate [111, 112]. Volume changes are too great. In addition, when elastic forces are experimentally increased, the decreases in tidal volume predicted by the model are too great. Hence, this simple model has been modified in several different ways.

One approach is to use a value for elastance and resistance that is different from that actually measured [111]. This assumes that "passive elastance" and "passive resistance" measured in animals not spontaneously breathing differ from "effective elastance" and "effective resistance" in the spontaneously breathing animal.

To measure passive elastance the lungs of an animal rendered apneic are inflated with different volumes of air and the pressure developed under conditions of absent airflow are determined. To measure passive resistance the inflated lungs of an apneic animal are allowed to deflate and pressures and flow are measured.

The effective elastance and resistance are determined using the equation [112]

$$P(t) = E_{\mathrm{RS,eff}}V(t) + R_{\mathrm{RS,eff}}\dot{V}(t) \tag{25.31}$$

where the subscript eff stands for *effective*.

The pressure changes obtained during an occluded breath are substituted for P and related to volume and flow changes occurring in the preceding unoccluded breath.

Generally, calculations show that the passive and effective resistances are not very different, but the effective elastance is about twice the passively measured elastance. With this kind of approach the effect of added external elastances and resistances on tidal volume and flow rates can be correctly predicted in anesthetized animals. Also, this kind of analytical approach may be useful in quickly and simply characterizing qualitative changes in respiratory mechanics that occur with treatment of patients with lung disease.

In fact, the discrepancies between predictions made using the simple model described and those observed in spontaneous breathing are caused by several different factors. First the simplified model is probably an inadequate representation of the actual system. Resistance and elastance change alinearly with changes in volume, and nonlinear flow may occur in the upper airways even during quiet breathing. In addition the mechanical properties of the lung are not uniform throughout the lung. Perhaps even more important is the complexity of the relationship between pressures developed by the muscles when they are contracting isometrically and the forces developed by the same muscles during spontaneous breathing when muscle shortening occurs.

25.4.3 Prediction of Driving Pressures

It is well known that the force developed by skeletal muscle depends on its length. The greater the initial length of the muscle up to a maximum, the greater the force it can develop at any fixed level of stimulation. The initial length of the respiratory muscles will depend on functional residual capacity, shortening with a condition such as airway obstruction that can increase the resting and expiratory lung volume.

In addition, the force developed by skeletal muscle (which includes all the respiratory muscles) will vary with the velocity of its contraction. Hill [58] has expressed the relationship between force and contraction velocity by the following equation:

$$(V + b)(P + a) = b(P_0 + a) \tag{25.32}$$

where V is the velocity of shortening, P is the actual pressure developed, and P_0 is the maximum isometric force at any given muscle length.

Recently models have been developed which relate respiratory driving pressures to volume and flow changes using the Hill equation (e.g., Younes et al. [109]. There has as yet been no attempt to formulate a unified model of respiratory control that includes the time course of driving pressure of the respiratory muscles as they are determined by different chemical stimuli, or a realistic model of respiratory system mechanics.

The problem is complicated because the respiratory muscles consist of different groups which can interact in different ways. Expansion of the chest wall seems to depend on the coordinated activity of these different muscles. In addition, the same tidal volume can be achieved, in the face of complete paralysis of the major inspiratory muscle, the diaphragm, by contraction of the external intercostals. Further, under resting conditions the diaphragm is curved and the chest wall development it produces may depend on its radius of curvature.

While the activity of all the respiratory muscles seems to increase with hypoxia and hypercapnia, the degree of increase may not be the same in different muscle groups. It is also clear that different respiratory muscles are affected unequally by spinal and brainstem reflexes as well as by chemical drives [31].

25.4.4 Expiratory Flow during Stimulated Breathing

While inspiratory flow rates increase as increasing negative pressures are developed, expiratory flow reaches a limiting level before maximum expiratory pressures are achieved. This has been explained by narrowing of intrathoracic airways by compressive forces produced during maximum expiratory efforts [83, 84, 92]. Expiratory flow limitation may occur during severe exercise or in other conditions when the drive to breathe becomes intense.

During forced expiration, as air moves downstream from the alveoli toward the airway opening, the molecules of air accelerate and their velocity increases as the total cross-sectional airway diameter decreases. According to the Bernoulli principle, pressure inside the airway falls as velocity grows greater [83, 84, 92]. When the intraluminal airway pressure falls below the pressure surrounding the intrathoracic airways (i.e., pleural pressure), the airways are compressed and their caliber is reduced. This results in a further increase in linear velocity and a further fall in intraluminal pressure. Eventually, the transmural airway pressure becomes sufficiently negative so that the airways eventually collapse at specific locations or choke points.

The speed of propagation of pressure waves along a tube depends on the cross-sectional area of the tube (A) and the elastance of the tube walls (dP/dA), but is inversely related to gas density (ρ), according to the equation:

$$\text{Speed of propagation} = \left(A^3 \frac{dP}{dA} \frac{1}{q\rho}\right)^{1/2}$$

where q is the conversion factor for departure from a blunt flow profile.

In tubes with elastic walls, flow becomes limited at choke points where the flow rate equals the speed of propagation of the pressure waves. The location of the choke point is dependent on lung volume. At large lung volumes, the choke point is situated in the vicinity of the lower trachea.

25.4.5 Effect of Lung Mechanics on Gas Exchange

The millions of alveoli that make up the gas-exchanging regions of the lung do not expand equally or even synchronously. In part this is caused by gravitational forces which allow pressure differences to occur in the pleural cavity. Pleural pressures are more negative in upright humans at the top than at the body of the chest. Hence, at FRC, alveoli at the top of the lung are expanded farther than those at the base. Because elastance increases with increase in volume, equal changes in pleural pressure produce smaller changes in alveoli at the lung bases than in alveoli at the bottom of the lung [77].

Also, the lung and the diameter of the conducting areas and their degree of branching will vary from one lung region to another. Hence, the resistance to airflow is not the same in all areas of the lung. Exactly how the gas in a given breath will be distributed in the lung will depend on airflow rates, their pattern of development, and the length of inspiration. When breathing is stressed, expiration may not be sufficiently long for all lung units to empty completely, further interfering with the distribution of ventilation. For these reasons the composition of the alveolar air during the breathing cycle varies in different parts of the lung.

Because the vessels which carry blood to the alveoli are collapsible and because circulating driving pressures in the lung are low, gravitational forces will also affect the relative distribution of blood flow through the lung. In general, the flow of blood is greater to the more dependent areas of the lung. The blood gas tensions achieved in different regions depend on local ratios of blood flow to ventilation. Inequalities in this ratio in different lung regions will reduce the efficiency of gas exchange and produce a difference between alveolar and arterial gas tensions. Mathematical models that describe these mechanical effects on gas exchange in considerable detail have been published [85]. Here, it is enough to recognize that such mechanical effects on gas exchange occur even in normal lungs.

25.4.6 Optimization of Respiratory Activity

The respiratory muscles consume energy as they contract to produce the forces required for breathing. The amount of O_2 used by the respiratory muscles during quiet breathing is only a small fraction of the total energy requirements of the body (less than 5 percent), but when respiration is stimulated, energy costs of breathing rise and become considerable, especially if the mechanical operation of the respiratory system is impaired [70]. Moreover, because the level of arterial P_{O_2} and P_{CO_2} depend on alveolar ventilation rather than minute ventilation, the energy used to maintain gas tensions at a fixed level will depend on the particular combinations of tidal volume and breathing frequency adopted. It has been proposed that the respiratory controller behaves in an "optimum fashion," choosing a tidal volume and frequency which minimize energy usage [81, 108]. Experiments in humans seem to support the idea that when high levels of alveolar ventilation are needed (as during exercise), the breathing patterns actually used are the ones that would minimize energy consumption. An extension of this idea that has recently been proposed is that breathing is regulated to minimize both the deviation of arterial gas tensions from desired levels and the breathing energy costs. A mathematical model has been developed which was able to reproduce both normal ventilatory responses to inspired CO_2 and the isocapnic response to exercise with resort to additional controller constraints [91]. However, it is unclear what physiological mechanism could inform the controller of the energy costs of breathing.

Another hypothesis is that the respiratory pattern adopted at given levels of alveolar ventilation minimizes the pressures that must be developed during each breath and not the energy costs [73]. Since there are receptors in the airways and force receptors in the muscles, this idea seems to be more reasonable physiologically. The predictions made regarding breathing frequencies that would be adopted at various levels of alveolar ventilation are not very different from models that consider energy cost to be the factor minimized.

Energy costs or pressure deflections may be a consideration in the transient as well as the steady-state behavior of the respiratory controller. Since energy costs rise when ventilation responses to CO_2 are high, it is possible that controller response to chemical stimuli are adjusted to minimize both energy cost and the duration of an arterial P_{CO_2} disturbance. A model based on this idea seemed to explain the changes in ventilatory responses to CO_2 seen in patients with lung diseases [70].

Finally, Yamashiro and Grodins [108] have proposed that the actual time course of airflow (and not just the combination of tidal volume and breathing frequency used) is chosen by the respiratory system to optimize gas exchange and that energy considerations may be a secondary factor.

REFERENCES

1. Agostini, E., and Citerio, G.: Relative decay rate of inspiratory muscle pressure during expiration in man, *Respir. Physiol., 38,* 355–376, 1979.
2. Bates, D. V., Macklem, P. T., and Crystal, R. V.: *Respiratory Function in Disease,* 2d ed., Saunders, Philadelphia, 1971.
3. Berger, A. T., Mitchell, R. R. A., and Severinghaus, J. W.: Regulation of respiration, *N. Engl. J. Med., 297,* 93–97, 1977.
4. Biorck, G.: On myoglobin and its occurrence in man, *Acta Med. Scand., 133* (Suppl. 226), 1–151, 1949.
5. Birchfield, R. I., Sieker, H. O., and Heyman, A.: Alterations in blood gases during natural sleep and narcolepsey, *Neurology, 8,* 107–112, 1958.
6. Biscoe, T. J.: Carotid body: Structure and function, *Physiol. Rev., 51,* 427–495, 1971.
7. Black, A. M. S., McCloskey, D. I., and Torrance, R. W.: The responses of carotid body chemoreceptors in the cat to sudden changes of hypercapnia and hypoxic stimuli, *Respir. Physiol., 13,* 36–40, 1971.
8. Black, A. M. S., and Torrance, R. W.: Respiratory oscillations in chemoreceptor discharge in control of breathing, *Respir. Physiol., 13,* 221–237, 1971.
9. Blesdoe, S. W., and Hornbein, T. F.: Central chemosensors and the regulation of their chemical environment, in T. F. Hornbein (ed.), *Regulation of Breathing,* Marcel Dekker, New York, 1981, pp. 347–406.
10. Bradley, G. W., Euler, C. von, Marttila, I., and Roos, B.: A model of the central and reflex inhibition of inspiration in the cat, *Biol. Cybern., 19,* 105–116, 1975.
11. Bruce, E., Cherniack, N. S., Mitra, J., Weiner, D., and Salamone, J.: Carotid sinus, and medullary chemoreceptor contributions to phrenic and hypoglossal nerve responses during rebreathing, *Physiologist, 23*(4), 182, 1980.
12. Campbell, E. J. M., Howell, J. B. L., and Peckett, E. W.: The pressure-volume relationship of thorax of anesthetized human subjects: A comparison of the effects of expiratory resistance and positive pressure inflation, *J. Physiol. (London), 136,* 563–568, 1957.
13. Cherniack, N. S.: Respiratory dysrhythmias during sleep, *N. Engl. J. Med., 303,* 325–330, 1981.
14. Cherniack, N. S., and Altose, M. D.: Respiratory responses to ventilatory loading, in T. C. Hornbein (ed.), *The Regulation of Breathing,* Marcel Dekker, New York, 1981, pt. II, pp. 905–964.
15. Cherniack, N. S., Edelman, N. H., and Lahiri, S.: Hypoxia and hypercapnia as respiratory stimulants and depressants, *Respir. Physiol., 11,* 113–126, 1970–71.
16. Cherniack, N. S., Euler, C. von, Homma, I., and Kao, F. F.: Experimentally induced Cheyne-Stokes breathing, *Respir. Physiol., 37,* 185–200, 1979a.
17. Cherniack, N. S., Euler, C. von, Homma, I., and Kao, F. F.: Graded changes in central chemoreceptor input by local temperature changes in the ventral surface of the medulla, *J. Physiol. (London), 287,* 191–212, 1979b.
18. Cherniack, N. S., and Longobardo, G. S.: CO_2 and O_2 gas stores of the body, *Physiol. Rev., 50,* 196–243, 1970.

19. Cherniack, N. S., and Longobardo, G. S.: Cheyne-Stokes breathing: An instability in physiologic control, *N. Engl. J. Med., 288,* 952–957, 1973.

20. Cherniack, N. S., Longobardo, G. S., and Fishman, A. P.: The behavior of carbon dioxide stores of the body during unsteady states, in G. Nahas and K. Schaefer (eds.), *Carbon Dioxide and Metabolic Regulations,* Springer-Verlag, New York, 1974.

21. Cherniack, N. S., Longobardo, G. S., Levine, O. R., Mellins, R., and Fishman, A. P.: Periodic breathing in dogs, *J. Appl. Physiol., 21,* 1847–1854, 1968.

22. Cherniack, N. S., Longobardo, G. S., Palermo, F. P., and Heymann, M.: Dynamics of oxygen stores changes following an alteration in ventilation, *J. Appl. Physiol., 24,* 809–816, 1968.

23. Cherniack, N. S., Longobardo, G. S., Staw, I., and Heymann, M.: Dynamics of carbon dioxide stores changes following an alteration in ventilation, *J. Appl. Physiol., 21,* 785–793, 1966.

24. Cheyne, J. A.: A case of apoplexy in which the fleshy part of the heart was converted into fat, *Dublin Hosp. Rep., 2,* 216–223, 1818.

25. Clark, F. C., and Euler, C. von: On the regulation of rate and depth of breathing, *J. Physiol. (London), 222,* 267–295, 1972.

26. Cohen, M. I.: Neurogenesis of respiratory rhythm in the mammal, *Physiol. Rev., 59,* 1105–1173, 1979.

27. Comeace, W. J.: Cheyne-Stokes respiration and auriculoventricular conduction, *N. Engl. J. Med., 219,* 977–982, 1938.

28. Cunningham, D. J. C.: Integrative aspects of the regulation of breathing: A personal view, in *MTP International Review of Science, Physiol. Ser.,* J. G. Widdicombe (ed.), vol. 2: *Respiratory Physiology,* Butterworth, London, 1974, pp. 303–369.

29. Damokosh-Giordana, A., Longobardo, G. S., and Cherniack, N. S.: Transient CO_2 elimination and storage as a function of the ventilatory response to CO_2, *Respir. Physiol., 37,* 219–237, 1979.

30. Damokosh-Giordano, A., Longobardo, G. S., and Cherniack, N. S.: The effect of controlled system (plant) dynamics on ventilatory responses to disturbances in CO_2 balance, in A. S. Iberall and A. C. Guyton (eds.), *Regulation and Control in Physiological Systems,* Instrument Society of America, Pittsburgh, Pa., 1973, pp. 291–299.

31. D'Angelo, E.: Inspiratory muscle activity during rebreathing in intact and vagotomized rabbits, *Respir. Physiol., 47,* 193–218, 1982.

32. Defares, J. G., Derksen, H. E., and Duyff, J. W.: Cerebral blood flow in the regulation of respiration, *Acta Physiol. Pharmacol. Neerlandica, 9,* 327–360, 1960.

33. Douglas, C. G., and Haldane, J. S.: The causes of periodic or Cheyne-Stokes breathing, *J. Physiol. (London), 38,* 401–419, 1909.

34. Dowell, A. R., Backley, E., Cohen, R., Whalen, R. E., and Sieker, H. O.: Cheyne-Stokes respiration, *Arch. Intern. Med., 27,* 712–726, 1971.

35. DuBois, A. B., Botelho, S. Y., and Comroe, J. H.: A new method of measuring airway resistance in man using a body plethysmograph, *J. Clin. Invest., 35,* 327, 1956.

36. Dutton, R. E., Hodson, W. A., Davies, G., and Chernick, V.: Ventilatory adaptation to a step change in P_{CO_2} at the carotid bodies, *J. Appl. Physiol., 23,* 195–202, 1967.

37. Edelman, N. H., Epstein, P. E., Lahiri, S., and Cherniack, N. S.: Ventilatory responses to transient hypoxia and hypercapnia, *Respir. Physiol., 17,* 302–314, 1973.

38. Edelman, N. H., Tuteur, P. G., McHenry, L., and Cherniack, N. S.: Response rates of ventilation and cerebral blood flow to CO_2 in normal goats and abnormal man, *Clin. Res., 19,* 510, 1971.

40. Eldridge, F.: Relationship between phrenic nerve activity and ventilation, *Am. J. Physiol., 212,* 535–543, 1971.

40. Eldridge, F.: Central neural stimulation of respiration in unanesthetized and decerebrate cats, *J. Appl. Physiol., 40,* 23–28, 1976.

41. Eldridge, F. L.: The importance of timing on the respiratory effects of intermittent carotid body chemoreceptor stimulation, *J. Physiol. (London), 222,* 319–333, 1972.

42. Engel, L. A.: Gas mixing within the acinus of the lung, *J. Appl. Physiol.: Respir. Environ. Exercise Physiol., 54*(3), 609–618, 1983.

43. Euler, C. von: The functional organization of the respiratory phase-starching mechanism, *Fed. Proc., 36,* 2375–2360,1977.

44. Euler, C. von: On the role of proprioceptors in perception and execution of motor acts with special references to breathing, in D. Pengelly, A. S. Rebuck, and E. J. M. Campbell (eds.), *Loaded Breathing*, Longmans, London, 1974, pp. 139–149.

45. Farhi, L. E., and Rahn, H.: Dynamics of changes in carbon dioxide stores, *Anesthesiology, 21,* 604–614, 1960.

46. Fink, B. R.: Influence of cerebral activity in wakefulness on the regulation of breathing, *J. Appl. Physiol., 16,* 15–20, 1961.

47. Forster, R. E.: Factors affecting the rate of exchange of O_2 between blood tissues, in F. Dickens and E. Neil (eds.), *Oxygen in the Animal Organism,* Macmillan, New York, 1964, pp. 393–406.

48. Fowle, A. S. E., and Campbell, E. J. M.: The immediate carbon dioxide storage capacity of man, *Clin. Sci., 27,* 41–49, 1964.

49. Gomez, D. M.: Consideration of oxygen hemoglobin equilibrium in the physiological state, *Am. J. Physiol., 200,* 135–142, 1961.

50. Gothe, B., Mantey, P., Goldman, M. D., and Cherniack, N. S.: Effect of quiet sleep in resting and CO_2 stimulated breathing, *J. Appl. Physiol., 50,* 724–730, 1981.

51. Gray, J. S.: The multiple factor theory of the control of respiratory ventilation, *Science, 105,* 466–468, 1947.

52. Grodins, F. S., Gray, J. S.: Schroeder, K. R., Norins, A. L., and Jones, R. W.: Respiratory responses to CO_2 inhalation: A theoretical study of nonlinear biological regulator, *J. Appl. Physiol., 7,* 283–308, 1954.

53. Grodins, F. S., and Yamashiro, S. M.: Respiratory function of the lung and its control, Macmillan, New York, 1978.

54. Grodins, F. S., and Yamashiro, S. M.: What is the pattern of breathing regulated for? in C. von Euler and H. Lagercrantz (eds.), *Central Nervous Control Mechanisms in Breathing: Physiological and Clinical Aspects of Regular, Periodic, and Irregular Breathing in Adults and in the Perinatal Period,* Pergamon Press, New York, 1979, pp. 169–176.

55. Guilleminault, C., Tilkian, A., and Dement, W. C.: The sleep apnea syndromes, *Am. Rev. Med., 27,* 465–484, 1976.

56. Guyton, A. C., Crowell, J. W., and Moore, J. W.: Basic oscillating mechanism of Cheyne-Stokes breathing, *Am. J. Physiol., 187,* 395–398, 1968.

57. Guyton, F. J., Meyer, S., and Takagi, Y.: Cerebral, venous and arterial blood during Cheyne-Stokes respiration, *Am. J. Med., 47,* 534–545, 1969.

58. Hill, A. V.: The design of muscles, *Br. Med. Bull., 12,* 165, 1956.

59. Holmdahl, M. H.: Pulmonary uptake of oxygen, acid-base metabolism, and circulation during prolonged apnea, *Acta Chir. Scand. Suppl., 212,* 1–28, 1956.

60. Hoppin, F. G., Jr., Lee, G. C., and Dawson, S. V.: Properties of lung parenchyma in distortion, *J. Appl. Physiol., 39,* 742–751, 1975.

61. Karp, H. R., Sieker, H. O., and Heyman, A.: Cerebral circulation and function in Cheyne-Stokes respiration, *Am. J. Med., 30,* 861–870, 1971.

62. Klocke, R. J., and Rahn, H.: Breathholding after breathing oxygen, *J. Appl. Physiol., 14,* 689–693, 1959.

63. Lahiri, S., and DeLany, R.: Relationship between carotid chemoreceptor activity and ventilation in the cat, *Respir. Physiol., 24,* 267–286, 1975.

64. Lambertsen, C. J., Semple, S. J. G., Smyth, M. G., and Gelfand, R.: H^+ and P_{CO_2} as chemical factors in respiratory and cerebral circulatory control, *J. Appl. Physiol., 16,* 473–484, 1961.

65. Lange, R. L., and Hecht, H. H.: The mechanism of Cheyne-Stokes respiration, *J. Clin. Invest., 41,* 42–52, 1962.

66. Lewis, G., Ponte, J., and Purves, M. J.: Fluctuations of Pa_{CO_2} with same period as respiration in the cat, *J. Physiol., 298,* 1–11, 1980.

67. Lloyd, B. B.: The interaction between hypoxia and other ventilatory stimuli, *Proceedings of the International Symposium on the Cardiovascular and Respiratory Effects of Hypoxia,* Karger, New York, 1966, pp. 146–165.

68. Lloyd, B. B., Jukes, M. G. M., and Cunningham, D. J. C.: The relation between alveolar oxygen pressure and the respiratory response to carbon dioxide in man, *Q. J. Exp. Physiol., 43,* 214–227, 1958.

69. Loeschcke, H. H.: The respiratory control system: Analysis of steady state solutions for metabolic and respiratory acidosis-alkalosis and increased metabolism, *Pfluegers Arch. Gesamte, Physiol., 341,* 23–42, 1973.

70. Longobardo, G. S., Cherniack, N. S., and Damokosh-Giordano, A.: Possible optimization of respiratory controller sensitivity, *Ann. Biomed. Eng., 8,* 143–158, 1980.

71. Longobardo, G. S., Gothe, B., Goldman, M. D., and Cherniack, N. S.: Sleep apnea considered as a control system instability, *Respir. Physiol., 50,* 311–333, 1982.

72. McDonald, D. M.: Structure-function relations of chemoreceptive nerves in the carotid body, *Am. Rev. Respir. Dis., 115,* 193–207, 1977.

73. Mead, J.: Control of respiratory frequency, *J. Appl. Physiol., 5,* 324–336, 1960.

74. Mead, J.: Responses to loaded breathing, *Bull. Physiopath. Respir., 15* (suppl.), 61–71, 1979.

75. Mead, J., and Agostini, E.: Dynamics of Breathing, in W. O. Fenn and H. Rahn (eds.), *Handbook of Physiology,* sec. 3: *Respiration,* American Physiological Society, Washington, D.C., 1964, vol. 1, pp. 411–427.

76. Milic-Emili, J.: Breathing responses to imposed mechanical loads, in A. P. Fishman, N. S. Cherniak, and J. G. Widdicomb (eds), sec. 3: *The Respiratory System,* of *Handbook of Physiology,* American Physiological Society, Bethesda, Md., 1986, pt. 2.

77. Milic-Emili, J., Henderson, J. A. M., Dolivich, M. B., Trop, D., and Kaneko, K.: Regional distribution of inspired gas in the lung, *J. Appl. Physiol., 21,* 479–759, 1966.

78. Nahas, G. G., and Verosky, M.: The storage of CO_2 during apneic oxygenation, *Ann. N.Y. Acad. Sci., 133,* 134–141, 1966.

79. Newsome-Davis, J., and Sears, T. A.: The proprioceptive control of the intercostal muscles during their voluntary contraction, *J. Physiol., 209,* 711–738, 1970.

80. Nielsen, M., and Smith, H.: Studies on the regulation of respiration in acute hypoxia, *Acta Physiol. Scand., 24,* 293–313, 1951.

81. Otis, A. B.: The work of breathing, in W. O. Fenn and H. Rahn (eds.), *Handbook of Physiology,* sec. 3: *Respiration,* American Physiological Society, Washington, D.C., 1964, vol. 1.

82. Otis, A. B., Fenn, W. O., and Rahn, H.: Mechanics of breathing in man, *J. Appl. Physiol., 2,* 592–607, 1950.

83. Pedley, T. J., Schroter, R. C., and Sudlow, M. F.: Energy losses and pressure drop in models of human airways, *Respir. Physiol., 9,* 371–386, 1970.

84. Pedley, T. J., Schroter, R. C., and Sudlow, M. F.: Flow and pressure drop in systems of repeatedly branching tubes, *J. Fluid Mech., 46,* 365–383, 1971.

85. Pedley, T. J., Schroter, R. C., and Sudlow, M. F.: Gas flow and mixing in the airways, in J. B. West (ed.), *Bioengineering Aspects of the Lung,* Marcel Dekker, New York, 1977.

86. Perkins, J. F., Jr.: Arterial CO_2 and hydrogen ion as independent additive respiratory stimuli: Support for one part of the gray multiple factor theory, in D. J. C. Cunningham and B. B. Lloyd (eds.), *The Regulation of Human Respiration,* Blackwell, Oxford, 1963.

87. Phillipson, E. A.: Control of breathing during sleep: State of the art, *Am. Rev. Respir. Dis., 118,* 909–939, 1978.

88. Plum, F.: Neurological integration of behavioral and metabolic control in breathing, in R. Porter (ed.), *Breathing: Hering Breuer Centenary Symposium,* Churchill, London, 1970, pp. 159–181.

89. Plum, F. H., and Brown, H. W.: The neurologic basis of Cheyne-Stokes respiration, *Am. J. Med., 30,* 849–860, 1961.

90. Ponte, J., and Purves, M. J.: Frequency response of carotid body chemoreceptors in the cat to changes in Pa_{CO_2} and pHa, *J. Appl. Physiol., 37,* 635–647, 1974.

91. Poon, C.: Optimal control of ventilation in hypoxia, hypercapnia and exercise, in B. J. Whipp and P. M. Wilberg (eds.), *Modelling and Control of Breathing,* Elsevier, New York, 1983.

92. Pride, N. B., Permutt, S., Riley, R. L., and Bamberger-Bonea, B.: Determinants of maximum expiratory flow from the lungs, *J. Appl. Physiol., 23,* 646–662, 1961.

93. Reynolds, W. J., Millhorn, H. T., Jr., and Holloman, G. H.: Transient ventilatory responses to graded hypercapnia in man, *J. Appl. Physiol., 35,* 47–54, 1972.

94. Salazar, G., and Knowles, J. H.: An analysis of pressure-volume characteristics of the lungs, *J. Appl. Physiol., 19,* 97–104, 1964.

95. Sampson, S. R., and Vidriek, E. H.: Properties of "irritant receptors" in canine lungs, *Respir. Physiol., 25,* 9–23, 1975.

96. Saunders, K. B.: Oscillations of arterial CO_2 tension in a respiratory model: Some implications for the control of breathing in exercise, *J. Theor. Biol., 84,* 63–197, 1980.

97. Siafakas, N. M., Chang, H. K., Bonora, M., Gautier, H., Milic-Emili, J., and Duron, B.: Time course of phrenic activity and respiratory pressures during airway occlusion in cats, *Environ. Exercise Physiol., 51,* 99–108, 1981.

98. Siafakas, N. M., Peslin, R., Bonora, M., Gautier, H., Duron, B., and Milic-Emili, J.: Phrenic activity, respiratory pressures and volume changes in cats, *J. Appl. Physiol.: Respir. Environ. Exercise Physiol., 51,* 109–121, 1981.

99. Spector, W. S.: *Handbook of Biological Data,* Saunders, Philadelphia, 1956.

100. Stokes, W.: *The Disease of the Heart and Aorta,* Hodges and Smith, Dublin, 1854, pp. 323–324.

101. Swanson, G. D., and Bellville, J. N.: Step changes in end-tidal CO_2: Methods and implications, *J. Appl. Physiol., 39,* 377–385, 1975.

102. Tenney, S. M.: Concepts of threshold and sensitivity of ventilatory control, *Ann. N.Y. Acad. Sci., 109,* 634–648, 1960.

103. Vance, J. W., and Fowler, W. S.: Adjustments of the stores of CO_2 during voluntary hyperventilation, *Dis. Chest, 37,* 304–313, 1960.

104. Wasserman, K., Whipp, B. J., and Davis, J. A.: Respiratory Physiology of Exercise: Metabolism, Gas Exchange, and Ventilatory Control, in J. G. Widdicombe (ed.), *Respiration Physiology,* University Park Press, Baltimore, 1980, vol. III.

105. Widdicombe, J. G., and Fillenz, M.: Receptors of the lungs and airways, in B. Neil (ed.), *Handbook of Sensory Physiology, Enteroceptors,* Springer-Verlag, New York, 1982, pp. 80–112.

106. Wyman, R. T.: Neural generation of the breathing rhythm, *Annu. Rev. Physiol., 39,* 417–488, 1977.

107. Yamamoto, W. S.: Computer simulation of experiments in response to intravenous and inhaled CO_2, *J. Appl. Physiol., 50,* 835–843, 1981.

108. Yamashiro, S. M., and Grodins, F. S.: Optimal regulation of respiratory airflow, *J. Appl. Physiol., 30,* 597–602, 1972.

109. Younes, M., and Riddle, W.: A model for the relation between respiratory neural and mechanical outputs, *J. Appl. Physiol.: Respir. Environ. Exercise Physiol., 51,* 963–978, 1981.

110. Zin, W. A., Pengelly, L. D., and Milic-Emili, J.: Active impedance of the respiratory system in anesthetized cat, *J. Appl. Physiol.: Respir. Environ. Exercise Physiol., 53,* 149–157, 1982.

111. Zin, W. A., Pengelly, L. D., and Milic-Emili, J.: Decay of inspiratory muscle pressure during expiration in anesthetized cats, *J. Appl. Physiol.: Respir. Environ. Exercise Physiol.,* 54, 408–413, 1983.

112. Zin, W. A., Rossi, A., Zocchi, L., and Milic-Emili, J.: Model analysis of tidal volume response to inspiratory elastic loads, *J. Appl. Physiol.: Respir. Environ. Exercise Physiol., 57,* 271–277, 1984.

CHAPTER 26

Respiratory Gas Exchange

Peter D. Wagner
Department of Medicine
University of California, San Diego
La Jolla, California

INTRODUCTION: THE LUNG AS AN ORGAN OF GAS EXCHANGE

This chapter will present a systematic analysis of pulmonary gas exchange, emphasizing fundamental principles and quantitative aspects. At the outset, it should be kept in mind that the primary function of the lung is in fact gas exchange and, more specifically, the uptake of O_2 from the air into blood together with the simultaneous elimination of CO_2 from blood into the air. This is accomplished by delivering to the lungs large quantities of blood returning from the tissues (blood low in O_2 and high in CO_2), and also large quantities of fresh air (high in O_2 and low in CO_2). These constantly turning over volumes of blood and gas come into intimate physical contact, separated only by a thin tissue sheet of average thickness ½ μm. Both O_2 and CO_2 readily diffuse through this sheet between gas and blood phases, sufficiently rapidly that their mass transfer is not subject to limitation of diffusion (at least in healthy subjects). Moment by moment the pulmonary system is never in a steady state (even if tissue metabolism is constant) because both the blood and the gas flows vary periodically. Blood flow in the pulmonary capillaries is pulsatile, due to the pulsatile nature of cardiac activity, with a resting frequency of about 1 Hz, and ventilation is "tidal" (i.e., sequentially inspiratory and expiratory), with a resting frequency of about 0.25 Hz. However, because of the large capacitance of the lungs (due to the large volume of about 3 L), moment-to-moment changes in mass transfer of O_2 and CO_2 due to the time-varying nature of ventilation and blood flow do *not* lead to significant oscillation of O_2 and CO_2 concentrations in the alveolar gas or effluent pulmonary blood [1]. This statement may not be true for the state of very slow, deep breathing or for some specialized research applications, but for most purposes alveolar and end-capillary O_2 and CO_2 concentrations can be regarded as being constant in time.

This conclusion greatly simplifies both qualitative and quantitative aspects of gas exchange, since accounting for pulsatile ventilation and blood flow necessitates solving numerically multiple differential equations. Thus, for the present purposes, all gas concentrations in blood and gas phases are taken to be constant in time. They may be thought of as reflecting average values over several cycles of ventilation and blood flow. There are many studies documenting the adequacy of these simplifications [2–5].

The next consideration in analysis of gas exchange is that of steady-state conditions. This is the question of whether, even when the moment-to-moment minor cyclic changes in O_2 and CO_2 levels are ignored, gas concentrations and fluxes are constant over longer time periods (minutes to hours). This concept involves (1) constancy or variation of tissue metabolism (usually expressed as O_2 uptake, $\dot{V}_{O_2}$, and CO_2 production, $\dot{V}_{CO_2}$), and (2) constancy or variation of several general and pulmonary factors that, as will be seen, affect gas concentrations and fluxes in the lungs. The general factors are cardiac output and alveolar ventilation, and, to a lesser extent, hemoglobin concentration, acid-base balance, and body temperature. The pulmonary factors involve the efficiency of the pulmonary gas-exchange process itself, which depends on the extent to which the distributions of ventilation and blood flow are geographically matched (see below) and also on the extent to which the diffusion process between blood and gas phases is complete.

Regardless of the normalcy of any of the above factors, most analyses of gas exchange assume that all the above are constant in time (over minutes). Once again, to account for actively changing factors requires extremely complex dynamic approaches beyond the expertise of most physiologists and physicians concerned with gas exchange. Philosophically, if gas exchange is changing over minutes sufficiently to void assumptions of steady-state conditions, most workers would not choose to attempt such complex approaches as would be needed because under such conditions other priorities are evident. Thus, clinically and experimentally, one would simply be concerned with rapid deterioration of the patient and would be more interested in therapy to stabilize the patient. The notable exceptions are in research studies (1) of the onset of exercise and (2) involving changes in inspired gas composition imposed specifically to follow the transient pulmonary response. Both these areas are too specialized to be treated in a chapter of this length and scope.

Thus, the discussion in this chapter pertains to both moment to moment and over longer periods of time steady-state conditions with constant flows, fluxes, and gas concentrations. While such an approach is not satisfactory for all applications, it is for the majority; steady-state gas exchange must be understood before variation in time can be taken into consideration.

Another issue which should be discussed at this point is the conceptual framework, or model, of the lungs as a gas-exchanging organ. Morphometric analyses of real lungs [6] demonstrate the complexity, both qualitatively and quantitatively, of both the gas and blood channels that transport gas and blood into, through, and out of the lungs. For example, the airways branch approximately (but not exactly) dichotomously some 23 times [6]. The first 16 of these branching generations are made up of bronchi that essentially do not participate in gas exchange, and hence form what has become known as the *conducting airway dead space* (also known as the *Fowler* or the *series* dead space). On the other hand, generations 17 to 23 contain progressively more densely alveolated tissue (i.e., gas-exchange tissue made up of capillaries and the ½-μm-thick blood-gas barrier mentioned above). During inspiration, air is moved sequentially through this complex system by a combination of convection and diffusion so that the alveoli expand; this process is reversed during expiration. Because of a combination of laminar and turbulent flows, secondary motion at branch points, diffusion occurring constantly, and the fact that precise geometry is impossible to account for (there are some 3×10^8 alveoli as a result of branching), realistic modeling of this process is not feasible. Since diffusive mixing is very rapid, the majority of workers only consider convective elements, especially in light of the time-invariant framework of gas transfer considered above. In addition, the branching parallel system of airways and alveoli naturally leads one to a treatment of the system as a set of discrete gas-exchange units arranged independently in parallel. This concept is also compatible with the arterial and venous branching pattern in the vascular system since, at least over the first 16 or so generations, the blood vessels run and branch with airways.

Thus, the lung has traditionally been considered (in terms of gas exchange, at least) as consisting of a large number of homogeneous "units" arranged in parallel. This concept includes considering each unit as inspiring gas not affected by events in neighboring units; that is, communications between units via collateral ventilatory or circulatory pathways are ignored. While there is considerable anatomical evidence that such channels do exist [7], there is also evidence that their presence does not result in gas-exchange patterns intrinsically different from those of the purely parallel arrangement [8]. In other words, real data are compatible with the parallel model of gas exchange, even when collateral pathways do exist [8].

Assuming that the parallel concept is acceptable, one could still imagine various ways of setting up such a system. Physiologists have traditionally chosen to think of the lung as made up of a collection of discrete units, finite in number, and have usually simplified the lung as an organ consisting of two to five such compartments with specified functional properties. This gross oversimplification can be compensated for either by greatly increasing the number of functional compartments [9] or by considering the lung as a mathematical continuum of function rather than as discrete units [10]. While there is merit in expanding from a small number of compartments, the choice of a large compartment number as opposed to a mathematical continuum is less critical and depends more on the preference of particular investigators and the task at hand.

Since the traditional approach is the compartmental one, based on the anatomical expectation of functionally discrete units, it is worth referencing what little data exist regarding identification of the structural bounds of a gas-exchange unit. Most work in this area has been theoretical, involving calculations of mixing time in peripheral lung regions on the basis of the known geometry and using Fick's law of diffusion [11]. The results indicate that regions distal to generations 17 to 19 (the beginning of the alveolated part of the lung involving airways known as *respiratory bronchioles*) are sufficiently rapidly mixed that they are functionally homogeneous. This is supported by measurements of gas exchange involving graded bead vascular embolism [12]. To a first approximation, there may be some 100,000 such functionally discrete units in the normal mammalian lung [12].

To summarize to this point, the lung is best regarded as consisting of a number of functionally discrete units that exchange gas as if ventilation and blood flow were constant in time rather than pulsatile. This simplification is a compromise between reality and complexity on the one hand and feasibility on the other. Abundant experimental and clinical experiences support the use of this framework. However, on occasion, research requirements demand a more realistic approach to one or more of the assumptions, and while these can be undertaken, it is beyond the scope of this chapter to consider such methods.

26.1 APPROACH TO GAS EXCHANGE: THE LUNG, CIRCULATION, AND TISSUES AS A SYSTEM

Even within the simplified framework above, one must approach gas exchange to account for all the factors that affect O_2 and CO_2 concentrations and fluxes. Figure 26.1 conveniently summarizes these factors and how they involve the lungs, the circulation, and the metabolizing tissues. In an intimately interdependent manner, the peripheral tissues are supplied with a certain cardiac output and a certain O_2 (and CO_2) concentration in the arterial blood. These tissues then extract their metabolic need in terms of O_2 (and add CO_2), giving rise to venous blood depleted of O_2 and augmented by CO_2, which forms the input blood (pulmonary arterial or mixed venous blood) to the lungs. The lungs then allow O_2 uptake and CO_2 elimination from that blood and produce the arterial levels that return to supply the tissues. The process of ventilation brings O_2 from the air to the blood-gas barrier and at the same time transports CO_2 in the opposite direction.

The usual starting point for understanding the relationships implied in Fig. 26.1 is to take any single representative unit of gas exchange in the lung and write down the equations of mass balance for each gas of interest. Solution of such equations shows how ventilation, blood flow, and gas exchange are related. One may then sum the desired variables over all lung units to understand the behavior of the lung as a whole, setting the appropriate boundary conditions (or input) on inspired gas and mixed venous blood. In this regard, it is useful to think of pulmonary gas exchange as a system in which the mixed venous blood and inspired gas form known (or calculable) inputs, which are then processed by the lung to yield expired and arterial gas concentrations that constitute the output.

The remainder of this chapter will, in fact, be just such an analysis to show how gas exchange is conventionally analyzed, and, in the process, how all the variables in Fig. 26.1 are involved in gas exchange.

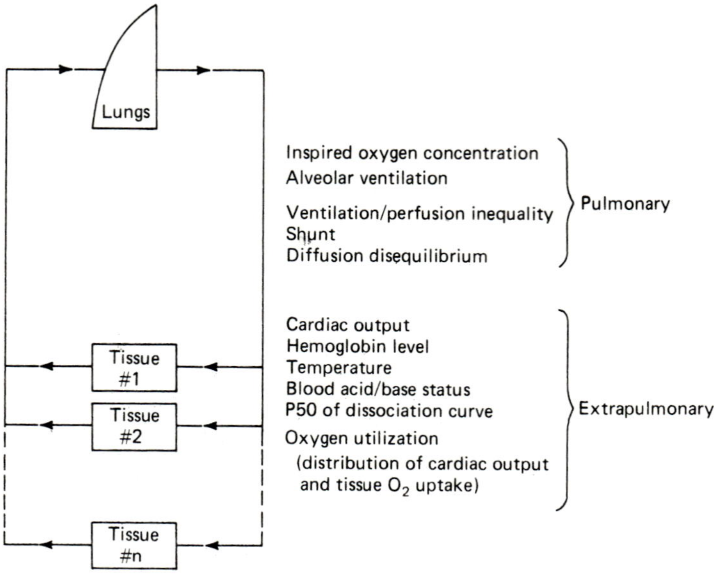

FIG. 26.1 Schematic diagram of the lungs and tissues to illustrate their gas-exchange interdependence, and the multiple factors that affect the concentration of oxygen in arterial blood. Both events taking place within the lungs, together with extrapulmonary factors, can affect the arterial Po_2, as discussed more completely throughout the text. *(From Wagner [31]; reproduced with permission.)*

26.2 GAS EXCHANGE IN A SINGLE LUNG UNIT

We begin with application of mass transfer principles, under the conditions laid out in the introduction, to a single lung unit (Fig. 26.2). For such a unit, we must define inspired alveolar ventilation ($\dot{V}_I$), expired alveolar ventilation ($\dot{V}_A$), and blood flow ($\dot{Q}$), all in liters per minute, and the partial pressures (P) and contents (C) of O_2 and CO_2 in inspired (I) and expired gas (A) and mixed venous ($\bar{v}$) and end-capillary (c′) blood: $P_{I_{O_2}}$, $P_{I_{CO_2}}$, $P_{A_{O_2}}$, $P_{A_{CO_2}}$, $P\bar{v}_{O_2}$, $P\bar{v}_{CO_2}$, Pc'_{O_2}, Pc'_{CO_2}, $C\bar{v}_{O_2}$, $C\bar{v}_{CO_2}$, Cc'_{O_2}, and Cc'_{CO_2}.

Then, by mass balance, the amount of O_2 taken up from ambient air per minute, $\dot{V}_{O_2}$, is

$$\dot{V}_{O_2} = k\left(\frac{\dot{V}_I P_{I_{O_2}}}{P_B - P_{H_2O}} - \frac{\dot{V}_A P_{A_{O_2}}}{P_B - P_{H_2O}}\right) \qquad \text{L/min}$$

where k is constant, allowing $\dot{V}_{O_2}$ to be expressed in standard temperature and pressure, dry (STPD), while $\dot{V}_I$ and $\dot{V}_A$ are traditionally expressed at body temperature and pressure (BTP). Note that

$$k = \frac{273}{273 + T} \times \frac{P_B - P_{H_2O}}{760}$$

where P_B is barometric pressure, from the laws of Boyle and Charles.

For a body temperature T of 37°C, this results in

$$\dot{V}_{O_2} = \frac{1}{863.0}(\dot{V}_I P_{I_{O_2}} - \dot{V}_A P_{A_{O_2}}) \qquad \text{L/min} \tag{26.1}$$

Here, all partial pressures are in torr.

$\dot{V}_A/\dot{Q}$ seen in normal lungs. A log scale of $\dot{V}_A/\dot{Q}$ is used for convenience of representation; experimental methods show that a range of $\dot{V}_A/\dot{Q}$ values can coexist all the way from 100 down to 0.01. Only with a log scale can this range be adequately represented.

26.3 GAS EXCHANGE IN THE LUNG AS A WHOLE

Having analyzed gas exchange in a single lung unit, we can now consider how all the individual lung units will combine to produce gas exchange in the lung as an entire organ. What is different among different units are their values for ventilation, blood flow, and $\dot{V}_A/\dot{Q}$ ratio. In other words, all units see the same inspired and mixed venous concentrations and operate using the same dissociation curves, so that the only basis for variation among lung units is in ventilation and blood flow.

Since gas concentrations depend on the ratio of $\dot{V}_A/\dot{Q}$ rather than the absolute amount of ventilation and blood flow, we speak of a lung as being *homogeneous* from the point of view of gas exchange if in all lung units the $\dot{V}_A/\dot{Q}$ ratio is the same. Conversely, a lung in which the $\dot{V}_A/\dot{Q}$ ratio varies among units is said to have $\dot{V}_A/\dot{Q}$ inequality.

Such inequality virtually always interferes with gas exchange for all gases. Should this occur with *no change* in the inspired gas, mixed venous blood, total alveolar ventilation, or cardiac output, the manifestations of inequality will be (1) a fall in arterial P_{O_2} and rise in arterial P_{CO_2}, (2) a rise in mixed expired alveolar P_{O_2} and fall in mixed expired alveolar P_{CO_2}, and (3) a reduction in $\dot{V}_{O_2}$ and $\dot{V}_{CO_2}$ (i.e., the volume rates of exchange of O_2 and CO_2 in the lung as a whole). How this comes about is shown in Fig. 26.4, where a simple two-compartment example

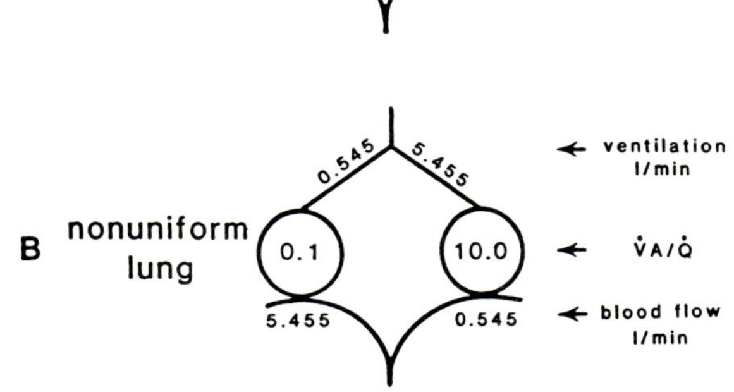

FIG. 26.4 Concept of ventilation-perfusion mismatching. In (A), a two-compartment lung is shown in which the ratio of ventilation to blood flow is 1.0 in each compartment. Such a lung is stated to be uniform from the point of view of gas exchange. However, when ventilation and/or blood flow is unevenly distributed, resulting in unequal ventilation-perfusion ratios, ventilation-perfusion mismatching occurs (B). Note that total ventilation and blood flow are 6 L/min each, both in the uniform and nonuniform lung, and that $\dot{V}_A/\dot{Q}$ mismatching does not depend upon changes in total ventilation or blood flow.

TABLE 26.1 Gas exchange variables in a homogeneous lung

Variable	Unit	Both units of Fig. 26.4*a*
P_{IO_2} (P_{ICO_2})	torr	149
$P\bar{v}_{O_2}$	torr	40
$P\bar{v}_{CO_2}$	torr	45
$C\bar{v}_{O_2}$	torr	15.27
$\dot{V}_A$	L/min	51.85
$\dot{Q}$	L/min	3.0
$\dot{V}_A/\dot{Q}$		1.0
P_{AO_2}	torr	104.1
P_{ACO_2}	torr	39.2
Cc'_{O_2}	mL/100 mL	20.69
Cc'_{CO_2}	mL/100 mL	47.27
Mixed alveolar P_{O_2}	torr	104.1
Mixed alveolar P_{CO_2}	torr	39.2
Arterial O_2 content	mL/100 mL	20.69
Arterial CO_2 content	mL/100 mL	47.27
Arterial P_{O_2}	torr	104.1
Arterial P_{CO_2}	torr	39.2
Mixed alveolar-arterial difference	torr	0
Mixed arterial-alveolar P_{CO_2} difference	torr	0
$\dot{V}_{O_2}$	L/min	0.325
$\dot{V}_{CO_2}$	L/min	0.275

TABLE 26.2 Gas exchange variables in a lung with $\dot{V}_A/\dot{Q}$ inequality

Variable*	Left-hand unit, Fig. 26.4*b*	Right-hand unit, Fig. 26.4*b*
P_{IO_2} (P_{ICO_2} = 0)	149	149
$P\bar{v}_{O_2}$	40	40
$P\bar{v}_{CO_2}$	45	45
$C\bar{v}_{O_2}$	15.27	15.27
$C\bar{v}_{CO_2}$	51.85	51.85
$\dot{V}_A$	0.55	5.45
$\dot{Q}$	5.45	0.55
$\dot{V}_A/\dot{Q}$	0.1	9.9
P_{AO_2}	40.6	141.5
P_{ACO_2}	45.0	16.6
Cc'_{O_2}	15.45	21.3
Cc'_{CO_2}	51.80	33.10
Mixed alveolar P_{O_2}	[(0.55 × 40.6) + 5.45 × 141.5)]/6.0 = 132.3	
Mixed alveolar P_{CO_2}	[(0.55 × 45.0) + (5.45 × 16.6)]/6.0 = 19.2	
Arterial O_2 content	[(5.45 × 15.45) + (0.55 × 21.13)]/6.0 = 15.97	
Arterial CO_2 content	[(5.45 × 51.80) + (0.55 × 33.05)]/6.0 = 50.09	
Arterial P_{O_2}	From O_2 dissociation curve = 43.0	
Arterial P_{CO_2}	From CO_2 dissociation curve = 44.0	
Mixed alveolar-arterial P_{O_2} difference	132.3 − 43 = 89.3	
Mixed arterial-alveolar P_{CO_2} difference	44.0 − 19.2 = 24.8	
$\dot{V}_{O_2}$	0.042	
$\dot{V}_{CO_2}$	0.106	

* Units as for Table 26.1.

of $\dot{V}_A/\dot{Q}$ inequality is depicted. Prior to inequality, Table 26.1 shows this lung exchanged 325 mL O_2/min, 275 mL CO_2/min and the arterial P_{O_2} and P_{CO_2} coming from this lung were 104 and 39 torr, respectively, for the given boundary conditions, ventilation, and blood flow ($P\bar{v}_{O_2} = 40$, $P\bar{v}_{CO_2} = 45$, $P_{I_{O_2}} = 149$, $P_{I_{CO_2}} = 0$ torr, total alveolar ventilation = 6.0 L/min, cardiac output = 6.0 L/min). Suppose ventilation of the left-hand unit in the lower panel was reduced from 3.0 to 0.545 L/min, while at the same time perfusion of the right-hand unit was similarly reduced to 0.545 L/min. Total ventilation and blood flow are unchanged at 6 L/min each, so that the left-hand unit ends up with a low $\dot{V}_A/\dot{Q}$ ratio of 0.545/5.45, or 0.1, while the right-hand unit has a high $\dot{V}_A/\dot{Q}$ ratio of 5.45/0.545, or about 10. These amounts of redistribution are within the observed spectrum in human disease due to airway and vascular obstructive diseases, but the particular two-compartment model is numerically extreme. From Fig. 26.3, a unit with a $\dot{V}_A/\dot{Q}$ of 0.1 has a P_{O_2} of 40.6 and a P_{CO_2} of 45.0 under the above boundary conditions, and a unit with a $\dot{V}_A/\dot{Q}$ of 10 has a P_{O_2} of 141.5 and P_{CO_2} of 16.6. Multiplying the associated difference between end-capillary content and mixed venous content by blood flow yields O_2 and CO_2 volumes exchanged per minute in each unit. The low $\dot{V}_A/\dot{Q}$ unit takes up only 10.0 mL O_2/min and eliminates only 3 mL CO_2/min. The high $\dot{V}_A/\dot{Q}$ unit takes up 32.3 mL O_2/min and eliminates 103.4 mL CO_2/min. Summing the effects of these two units gives a total $\dot{V}_{O_2}$ of only 42.2 mL/min and $\dot{V}_{CO_2}$ of 106.4 mL/min. These calculations are summarized in Table 26.2.

This simple numerical model, while purposefully extreme, shows how maldistribution of ventilation and blood flow has a major influence on gas exchange, resulting in a fall in arterial P_{O_2} and rise in arterial P_{CO_2}, together with a fall in $\dot{V}_{O_2}$ and $\dot{V}_{CO_2}$. The analysis specifically illustrates how the physiologist handles the complicated problem of quantitative gas exchange to account for $\dot{V}_A/\dot{Q}$ mismatching, and what variables are of principal interest.

26.4 COMPENSATORY CHANGES TO RESTORE $\dot{V}_{O_2}$ AND $\dot{V}_{CO_2}$ TO NORMAL

From Table 26.2 it is evident that with no changes in total ventilation or blood flow in boundary conditions, $\dot{V}_A/\dot{Q}$ mismatching leads to a situation incompatible with continued existence. In other words, the lung cannot meet the metabolic demands of the body. Prior to redistribution of ventilation and blood flow, $\dot{V}_{O_2}$ was 325 mL/min, but this fell to only 42 mL/min after mismatch occurred. Similarly, $\dot{V}_{CO_2}$ fell from 275 to 106 mL/min.

In general, one or more of several compensatory mechanisms will take place to restore $\dot{V}_{O_2}$ and $\dot{V}_{CO_2}$ to meet the tissue demands. The first of these is a fall in mixed venous P_{O_2} (and rise in P_{CO_2}) which occurs rapidly and passively as a simple consequence of the tissues still attempting to extract 325 mL O_2/min from the arterial blood reaching them, even though that blood has a very low P_{O_2} (43 torr, Table 26.2).

It turns out that a fall in mixed venous P_{O_2} (and rise in P_{CO_2}) will often allow $\dot{V}_{O_2}$ and $\dot{V}_{CO_2}$ to be returned to normal levels [15], and thus assure continued existence of the organism. However, these changes in mixed venous gas tensions will further reduce arterial P_{O_2} (and raise P_{CO_2}). This can be inferred from Fig. 26.3: the relation between P_{O_2} and $\dot{V}_A/\dot{Q}$ will be lowered (at each $\dot{V}_A/\dot{Q}$ value) if mixed venous P_{O_2} falls. Thus, since P_{O_2} in all units is lower, arterial P_{O_2} will be lower. Consequently, the (passive) compensating tissue responses to $\dot{V}_A/\dot{Q}$ mismatching consist of a *fall* in both arterial and venous P_{O_2} (and rise in P_{CO_2}) and at the same time a restoration of pulmonary O_2 uptake and CO_2 elimination to normal.

Usually, unless $\dot{V}_A/\dot{Q}$ mismatch is quantitatively minimal, there is sufficient fall in arterial P_{O_2} and rise in P_{CO_2} that ventilatory control mechanisms are invoked. Thus, hypoxemia and hypercapnia will separately stimulate various chemoreceptors that result in an increase in rate and/or depth of breathing, raising ventilation. From Eq. (26.1), this will raise P_{O_2} and lower P_{CO_2} in each unit and therefore also in arterial blood. This is a particularly effective compensatory mechanism for lowering P_{CO_2}, but not for raising P_{O_2} (because of the nonlinear shape of the O_2 dissociation curve and the linear shape of the CO_2 dissociation curve). Hence, it is common to see patients with lung disease with a normal arterial P_{CO_2} but considerable hypoxemia. Such a situation reflects the degree of ventilatory compensation, and the normal P_{CO_2} does not mean that CO_2 exchange has not been affected. When ventilatory compensation does occur, it happens rapidly (within seconds to minutes), just as does the mixed venous readjustment described above.

Other compensatory phenomena occur in some situations. For example, cardiac output may increase as a result of sympathetic discharge associated with the event that led to $\dot{V}_A/\dot{Q}$ mismatch in the first place. If this occurs, mixed venous Po_2 will increase (and Pco_2 fall) and (again from Fig. 26.3), this will increase Po_2 in all units and hence in arterial blood over what would have existed in the absence of increase in cardiac output.

The converse is also true: should there be a fall in cardiac output for any reason, arterial Po_2 will fall further than it would from just $\dot{V}_A/\dot{Q}$ mismatch alone. Such a situation is commonly seen in patients with myocardial infarction and pulmonary edema. Here, pulmonary edema causes $\dot{V}_A/\dot{Q}$ mismatch and the infarct leads to heart failure and a reduced cardiac output. Hypoxemia can be quite severe under those conditions.

Longer-term adaptations to the hypoxemia caused by $\dot{V}_A/\dot{Q}$ mismatch occur in an effort to increase O_2 availability to the tissues. The best example of this is polycythemia (increase in hemoglobin and hematocrit), and its effect, by increasing blood O_2 transport capability, is to raise both mixed venous and arterial Po_2.

26.5 INTERPLAY OF FACTORS DETERMINING GAS EXCHANGE

On assimilation of all of the above material, it should be evident how the various factors portrayed in Fig. 26.1 combine and interact to determine O_2 and CO_2 concentrations in arterial and mixed venous blood. Equations (26.1) and (26.2) explicitly demonstrate the influence of ventilation, cardiac output, and inspired gas composition. $\dot{V}_A/\dot{Q}$ mismatching (including shunt, a $\dot{V}_A/\dot{Q}$ of zero) interferes in a complicated manner, as treated above. Application of the Fick principle to the lung as a whole [i.e., Eq. (26.1) applied to the whole lung] shows how tissue O_2 uptake and CO_2 production will influence gas exchange by first affecting mixed venous levels that in turn affect arterial levels. The shape and slope of the O_2 and CO_2 dissociation curves are important, as inferred from the form of Eqs. (26.1) and (26.2), which contain this information implicitly by requiring both partial pressure and concentration of each gas to be known. In turn, the shape and slope of the dissociation curves are influenced by hemoglobin and hematocrit, body temperature, blood acid-base status, and for O_2 the P_{50} of the dissociation curve (the Po_2 at 50 percent saturation of hemoglobin with O_2). The P_{50} is in turn dependent on levels of 2,3-diphosphoglycerate, which can be stimulated by hypoxemia so as to increase P_{50}. Thus all the factors in Fig. 26.1 combine quantitatively to determine how gas exchange takes place. Clearly, to consider all these factors in combination requires digital computation, and the reader is again referred to more complete treatments of this subject [13–17].

26.6 DIFFUSION LIMITATION IN PULMONARY GAS EXCHANGE

The preceding analysis has been carried out on the explicit assumption that within each gas-exchange unit alveolar gas and end-capillary blood reach partial pressure equilibrium. The structural basis of this assumption is reasonable, since the surface area available for diffusion of gases between blood and gas is some 80 m^2 in the normal human lung while the tissue thickness is only of the order of 0.5 μm [18]. There is also functional evidence, based on both theoretical calculations and experimental data, that diffusion equilibration occurs in both the normal and diseased lung at rest and that only rarely is disequilibrium evident in patients when they exercise [19, 20]. It should, however, be pointed out that at high altitudes there is good evidence of diffusion limitation during exercise, even in normal humans [21, 22].

Although diffusion limitation is rarely a problem, it can become one, and thus should be treated separately. The subject has been extensively reviewed [23, 24], and only an outline is given here to acquaint the reader with the approach used in considering the phenomenon.

As with the analysis of $\dot{V}_A/\dot{Q}$ mismatch, certain practical, reasonable, simplifying assumptions are usually made in the analysis of conditions under which diffusion limitation may occur. These include constancy of alveolar gas and mixed venous blood composition, and both uniformity and nonpulsatility of blood flow. Red cell and plasma transit times through the alveolar capillaries are taken to be equal and uniform throughout the lung. The influence of a distribution of transit times has been examined and found to be small [25]. The diffusion pathway from alveolar gas to binding sites in the red cell (Fig. 26.5) is usually simplified into two components. The first represents physical diffusion resistance through alveolar surfactant, epithelial cells, interstitium, endothelial cells, plasma, and the red cell membrane, all lumped into a single variable, while the second accounts for the finite time of chemical reactions of O_2 and CO_2 in blood [26]. Clearly, this is an example of compromise between reality and complexity of modeling a complicated system. Dividing this pathway into two components leads to the definition of a plasma Po_2 and, separately, the Po_2 in the interior of the red cell. In reality there will be some reasonably continuous gradient of Po_2 from alveolar gas to the red cell interior, but the two-compartment approach is seen to be adequate by most workers. In fact, for O_2, the capacitance of the plasma is so small compared with the O_2 transport capability of Hb inside the red cell that considering the system as a *single* resistance by ignoring O_2 in plasma produces insignificantly different results [25]. This is not true for CO_2; its presence in plasma is in about the same amounts as in the interior of the red cell. Thus while CO_2 has to diffuse physically through the same alveolar structures as O_2 (Fig. 26.5), the chemical events in blood are quite different for

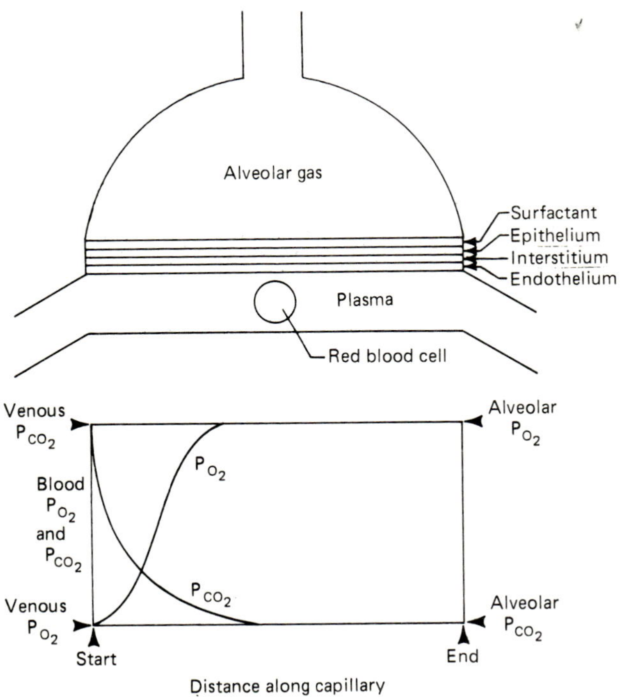

FIG. 26.5 Diagram illustrating components of the diffusion pathway for gases between alveolar gas and the red blood cell. Several identifiable structures must be negotiated in series as shown. Also shown are the rates at which Po_2 and Pco_2 change within a typical red cell as it traverses the pulmonary capillary from left to right.

CO_2. These events include rapid (catalyzed) conversion of red cell bicarbonate into CO_2, which diffuses into the alveolar gas, the fairly slow plasma–red cell bicarbonate-chloride exchange, and the rapid removal of CO_2 from carbamino compounds (principally on hemoglobin) [27]. While all these events can be acceptably modeled by a lumped–reaction rate parameter [25], a more realistic approach is to model the various chemical events separately in plasma and red cell interior (by means of separate differential equations [27]). The former approach is simpler and acceptably accurate for overall computations of CO_2 equilibration, but no separate details of pH or bicarbonate rates of equilibration can be obtained, as can be obtained with the latter method. Thus choice of approach for CO_2 depends on the questions being asked.

The analysis of diffusion equilibration for any gas requires the setting up of differential equations for mass transfer of the gas between alveolar gas and blood. These equations contain boundary conditions and physiological variables just as in the analysis of V_A/Q mismatch. While the equations will be formulated here, their method of solution is generally numerical and beyond the scope of this chapter. The reader is referred to other publications which deal with the numerical methods in more detail [16, 24, 27].

Consider the uptake of O_2 within the framework of Fig. 26.5, where two lumped components of the diffusion pathway exist: that associated with diffusion from gas into the plasma, and that associated with the chemical reaction rate of O_2 with hemoglobin inside the red cell. The former component traditionally is expressed as a conductance term, D_M (where M refers to the combined "membrane" of alveolar wall plus plasma) and the units are mL O_2/min per torr P_{O_2} partial pressure difference between alveolar gas ($P_{A_{O_2}}$) and plasma ($P_{PL_{O_2}}$). The latter is conventionally expressed as a chemical reaction rate (θ) in units of mL O_2/min per torr P_{O_2} difference [between plasma and red cell ($P_{RBC_{CO_2}}$)] per milliliter of blood in the capillaries. If the capillary volume is denoted Vc, then the total conductance of the chemical rate component is θVc. Usually, both D_M and Vc are assumed to be divided uniformly along the length of the capillary. Because alveolar gas is uniform in concentration in a lung unit, the analysis of diffusive processes, which is done by means of Fick's law of diffusion, can ignore the spatial element of that law. Application of that law to O_2 crossing the blood-gas barrier from alveolar gas into blood can then be stated by the following pair of differential equations:

$$\dot{V}_{O_2,t} = D_M(P_{A_{O_2}} - P_{PL_{O_2,t}}) \tag{26.6}$$

and

$$\frac{Vc}{100}\frac{d[HbO_2]}{dt} = \theta Vc(P_{PL_{O_2,t}} - P_{RBC_{O_2,t}}) \tag{26.7}$$

In these equations, $\dot{V}_{O_2,t}$ is instantaneous total rate of transfer of O_2 across the blood-gas barrier into the plasma, mL/min; D_M, θ, Vc, $P_{PL_{O_2,t}}$, and $P_{RBC_{O_2,t}}$ are as defined above, with the subscript t referencing variables changing in time. Alveolar P_{O_2} ($P_{A_{O_2}}$) on the other hand, is taken to be uniform (as discussed earlier) in space as well as in time, within any single lung unit; $[HbO_2]$ is the concentration of O_2 bound to hemoglobin at any instant in mL O_2/100 mL blood, and multiplying by $Vc/100$ puts this side of Eq. (26.7) into the same units as the right-hand side, namely, mL/min for the lung as a whole.

These simple versions of Fick's law explicitly assume "plug flow" in that there is no axial or longitudinal mixing along the capillary, while radial mixing on the other hand, is complete. The added complexity of attempting to allow for axial and radial mixing is not felt appropriate by most workers because of lack of detailed information about rates of such processes in vivo and their probably small effect.

If the capacitance of the plasma for O_2 is ignored (and it normally is only 1.5 percent of that of hemoglobin), one can set $\dot{V}_{O_2,t}$ in Eq. (26.6) equal to $(Vc/100)(d[HbO_2]/dt)$ of Eq. (26.7). Then, dividing Eq. (26.6) by D_M and Eq. (26.7) by θVc and adding the two equations gives

$$\dot{V}_{O_2,t} = \left(\frac{1}{1/D_M + 1/\theta Vc}\right)(P_{A_{O_2}} - P_{RBC_{O_2,t}}) \tag{26.8}$$

The left-hand side is the instantaneous flow of O_2 from alveolar gas to blood, and the right-hand side states that this flux is the product of the total conductance and the instantaneous partial pressure difference between alveolar gas and the red cell.

To make use of this equation, it is useful to replace the left-hand side by concentration terms [as in Eq. (26.7)]:

$$\frac{Vc}{100}\frac{d\mathrm{C{\small RBC}}_{O_2,t}}{dt} = \left(\frac{1}{1/D_M + 1/\theta Vc}\right)(\mathrm{P{\small A}}_{O_2} - \mathrm{P{\small RBC}}_{O_2,t})$$

or

$$\frac{d\mathrm{C{\small RBC}}_{O_2,t}}{dt} = \frac{100}{Vc}\left(\frac{1}{1/D_M + 1/Vc}\right)(\mathrm{P{\small A}}_{O_2} - \mathrm{P{\small RBC}}_{O_2,t}) \tag{26.9}$$

Notice in Eq. (26.9) that CRBC and PRBC are content and Po_2, respectively, for a point on the O_2 dissociation curve, so that knowing either one allows calculation of the other. In using this equation, PA_{O_2}, D_M, θ, and Vc are allocated values, which allows Eq. (26.9) to be integrated. For O_2, since the relation between content and partial pressure is not easily expressed algebraically as an integrable function, this equation is usually numerically integrated over time. We know the initial conditions since at $t = 0$, red cell Po_2 and content are those of mixed venous blood. We can also usually specify the total transit time, which is the ratio of capillary blood volume to cardiac output. For normal values of these two variables (75 mL and 6 L/min, respectively), the transit time is 0.75 s [28]. The most common numerical procedure used is the fourth-order Runga-Kutta integration [16], but any method could be used.

The result of such an integration is the capillary time course showing how rapidly Po_2 rises along the capillary (Fig. 26.5). Most workers agree that > 99 percent equilibration takes only 0.2 to 0.3 s. Thus, equilibration is essentially complete one-third of the distance down the average capillary (Fig. 26.6). This gives considerable reserve, so that the geometry would have to be altered to reduce conductance by a factor of 3 to produce incomplete diffusive equilibration.

During exercise, cardiac output can quadruple from 6 to 24 L/min or more, which might suggest that diffusion limitation will occur in normal subjects during severe exercise. However, owing to higher flows, pulmonary vascular pressures are increased, which increases capillary volume. Thus, transit time is not reduced as much as might be expected at first sight, and available data suggest mean transit time does not fall below about 0.3 to 0.4 s [29], which is still sufficient for equilibration.

It is often stated that while equilibration is not a problem at sea level, diffusion limitation will occur on exercise at high altitudes. The conventional explanation for this correct conclusion is that in Eq. (26.9) the "driving gradient," that is, $PA_{O_2} - PRBC_{O_2,t}$ is reduced because, with altitude, all O_2 partial pressures fall along with the inspired value. However, that is not the reason. The correct reasons are that (1) at high altitudes, because of the fall in Po_2, alveolar and red cell Po_2 are on the steep rather than the flat portion of the O_2 dissociation curve, and (2) the conductance term $1/(1/D_M + 1/\theta Vc)$ has not significantly changed with altitude. On the steep portion of the curve, a given increment in O_2 content leads to a smaller rise in Po_2 and thus for the same conductance a slower rate of equilibration $PRBC_{O_2}$ with PA_{O_2}.

The chemical rate of reaction (θ) depends on O_2 saturation [30]. Its values are still disputed as they pertain to in vivo conditions, but θ is thought to be constant at about 2.5 mL O_2/(mL · mmHg · min) of blood at hemoglobin saturations below about 80 percent [30]. Above this saturation, θ falls essentially linearly toward zero at 100 percent saturation. Thus, in the numerical integration of Eq. (26.9), values of θ at each step in the integration should be appropriate for the current hemoglobin saturation value.

Carbon dioxide can be treated similarly, as stated earlier, by lumping the various chemical reactions into a single value with an appropriate choice of θ for CO_2. Based on a half time of about 0.15 s for the slowest component of these reactions summarized in Ref. 25, θco_2 comes to about 2.0 to 2.5 mL CO_2/(mL · mmHg · min) of blood. While Vc is clearly the same for all gases [Eq. (26.9)], D_M for CO_2 is some 20-fold larger than for O_2 because of the much higher solubility of CO_2 in tissue compared with that of O_2 (this is only slightly offset by the higher molecular weight of CO_2). While at first sight this 20-fold increase in D_M might be thought to cause

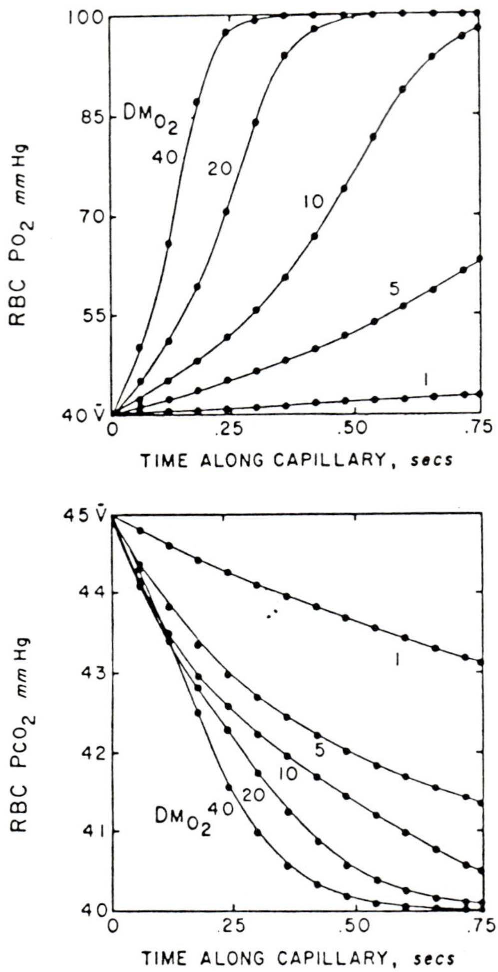

FIG. 26.6 Calculations of the rate of approach to diffusion equilibrium for PO_2 (upper panel) and PCO_2 (lower panel) in a pulmonary capillary allowing 0.75 s contact time. Mixed venous input conditions are those of a normal resting human (PO_2, 40, and PCO_2, 45 torr), and alveolar PO_2 and PCO_2 also correspond to those for a normal resting human (PO_2, 100, and PCO_2, 40 torr). In each panel, five profiles are drawn for different values of membrane conductance. These are referenced to values for oxygen with 40 ml/(min · mmHg) representing a conservative normal estimate. It can be seen that D_{M,O_2} must be reduced by a factor of about 4 before measurable disequilibrium will occur. (*From Wagner and West [25]; reproduced with permission.*)

equilibrium to be reached much sooner for CO_2, it must be remembered that although the conductance is much greater, so is the slope of the CO_2 dissociation curve. In fact, it is about 10 times steeper than the average slope of the O_2 dissociation curve under normal conditions. As explained above, this steeper slope would result in a slower approach to equilibrium. In other words, the increased conductance and increased slope are offsetting factors, and CO_2 reaches equilibrium in the capillary no sooner than O_2 [25, 27]. In fact, because the CO_2 dissociation curve is almost linear, equilibrium is approached initially more rapidly, but later more slowly, for CO_2 than for O_2 (Figs. 26.5 and 26.6).

While altitude has no direct effect on rate of equilibration of CO_2 (unlike O_2), exercise could in theory lead to similar degrees of disequilibrium for the two gases. However, due to the steep slope of the CO_2 dissociation curve, only small increases in arterial P_{CO_2} would result, and these would be rapidly corrected by ventilatory control mechanisms. Diffusion disequilibrium will thus be manifest by hypoxemia and alveolar-arterial P_{O_2} differences, but negligible changes will be seen for CO_2.

Finally, a word should be said about the vulnerability of O_2 and CO_2 to diffusion limitation compared with that of inert gases such as nitrous oxide and halothane as used for anesthesia. In reference to Eq. (26.9), the value of D_M is proportional to *tissue* (not *blood*) solubility (α) of the particular gas, while the rate of approach to equilibrium in blood *for a given conductance* depends on the slope of the concentration–partial pressure relationship in blood (i.e., the effective blood solubility β). For essentially all inert gases, irrespective of the absolute values of α and β, the two variables are similar in magnitude to each other so that their ratio is close to 1. For O_2, $\alpha = 0.003$ mL/100 mL per torr, while β is 0.08 mL/100 mL per torr, evaluated as an average between normal mixed venous and arterial points. For CO_2, $\alpha = 0.067$ mL/100 mL per torr, while β is 0.8 mL/100 mL per torr. Thus, the ratio α/β for O_2 is about 0.04, for CO_2, about 0.08, and for all inert gases, about 1.0. Referring again to Eq. (26.9) with these figures in mind, it is clear that both O_2 and CO_2 will equilibrate an order of magnitude more slowly than any inert gas. Thus, under normal conditions, inert gases will reach equilibrium within the first 0.05 s of contact time. It would be virtually impossible to imagine conditions in which inert gases were seriously subject to diffusion limitation because O_2 uptake would be so low as to be incompatible with life.

SUMMARY

This chapter has examined the conventional physiological approaches to pulmonary gas exchange. The real lung is so complex in time and space that many simplifications and assumptions must be made. However, these assumptions, which have been detailed herein at some length, are probably justified, by all available evidence, except under most unusual conditions. Giving the bioengineer a feel for these assumptions is seen as a major goal of this review.

The phenomenon of gas exchange within small homogeneous lung units was presented on the assumption that diffusion limitation of gas exchange does not occur. It was shown how quantitative gas exchange was a function of (1) the ventilation and blood flow of the lung unit, (2) the slope of the dissociation curve of the gas, and (3) the boundary conditions (i.e., inspired gas and mixed venous blood). Integration over the entire lung was next described and the notion of mismatching of ventilation and blood flow defined. How the lung behaves in the presence of such mismatching and how the body can compensate for the reduced gas transfer that results were explained.

Finally, although rarely a problem in reality, the concept of diffusion limitation of gas exchange was discussed, and the differential equations by which equilibration can be studied were set up.

It is hoped that the basic principles laid out in this chapter will provide a fundamental understanding of the important concepts in gas exchange. As with any subject, a review chapter of this nature cannot deal with all of the more esoteric issues. Therefore, this chapter should be viewed as a focus on the conventional and important principles associated with gas exchange that should first be understood before more controversial and less accepted aspects are studied.

ACKNOWLEDGMENTS

This work was supported by NIH grant HL-17731, and the author thanks Tania Davisson for her assistance in the preparation of this chapter.

REFERENCES

1. DuBois, A. B., Britt, A. G., and Fenn, W. O.: Alveolar CO_2 during the respiratory cycle, *J. Appl. Physiol., 4,* 535–548, 1951–52.
2. Rahn, H., and Fenn, W. O.: *A Graphical Analysis of the Respiratory Gas Exchange,* American Physiological Society, Washington, D.C., 1955.
3. Riley, R. L., and Cournand, A.: 'Ideal' alveolar air and the analysis of ventilation-perfusion relationships in the lungs, *J. Appl. Physiol., 1,* 825–847, 1949.
4. Briscoe, W. A., Cree, E. M., Filler, J., Houssay, H. E. J., and Cournand, A.: Lung volume, alveolar ventilation and perfusion interrelationships in chronic pulmonary emphysema, *J. Appl. Physiol., 15,* 785–795, 1960.
5. Wagner, P. D., Laravuso, R. B., Uhl, R. R., and West, J. B.: Continuous distributions of ventilation-perfusion ratios in normal subjects breathing air and 100% O_2, *J. Clin. Invest., 54,* 54–68, 1974.
6. Weibel, E. R.: *Morphometry of the Human Lung,* Academic Press, New York, 1963.
7. Macklem, P. T.: Airway obstruction and collateral ventilation, *Physiol. Rev., 51,* 368–436, 1971.
8. Wagner, P. D., and Evans, J. W.: Conditions for equivalence of gas exchange in series and parallel models of the lung, *Respir. Physiol., 31,* 117–138, 1977.
9. Evans, J. W., and Wagner, P. D.: Limits on V_A/Q distributions from analysis of experimental inert gas elimination, *J. Appl. Physiol., 42,* 889–898, 1977.
10. Gomez, D. M., Briscoe, W. A., and Cumming, G.: Continuous distribution of specific tidal volume throughout the lung, *J. Appl. Physiol., 19,* 683–692, 1964.
11. Rauwerda, P. E.: Unequal ventilation of different parts of the lung, Ph.D. dissertation, University of Groningen Press, Groningen, 1946.
12. Young, I., Mazzone, R. W., and Wagner, P. D.: Identification of functional lung unit in the dog by graded vascular embolization, *J. Appl. Physiol., 49*(1), 132–141, 1980.
13. Kelman, G. R.: Computer program for the production of O_2-CO_2 diagrams, *Respir. Physiol., 4,* 260–269, 1968.
14. Olszowka, A. J., and Farhi, L. E.: A system of digital computer subroutines for blood gas calculations, *Respir. Physiol., 4,* 270–280, 1968.
15. West, J. B.: Ventilation-perfusion inequality and overall gas exchange in computer models of the lung, *Respir. Physiol., 7,* 88–110, 1969.
16. Olszowka, A. J., and Wagner, P. D.: Numerical analysis in gas exchange, in J. B. West (ed.), *Pulmonary Gas Exchange,* Academic Press, New York, 1980, vol. 1, chap. 8, pp. 263–306.
17. West, J. B., and Wagner, P. D.: Pulmonary gas exchange, in J. B. West (ed.), *Bioengineering Aspects of the Lung,* Marcel Dekker, New York, 1977, chap. 5, pp. 361–457.
18. Weibel, E. R., and Gil, J.: Structure-function relationships at the alveolar level, in J. B. West (ed.), *Bioengineering Aspects of the Lung,* Marcel Dekker, New York, 1977, pp. 1–81.
19. Wagner, P. D., Dantzker, D. R., Dueck, R., Clausen, J. L., and West, J. B.: Ventilation-perfusion inequality in chronic obstructive pulmonary disease, *J. Clin. Invest., 59,* 203–216, 1977.
20. Wagner, P. D.: Ventilation-perfusion inequality and gas exchange during exercise in lung disease, in J. A. Dempsey and C. E. Reed (eds.), *Muscular Exercise and the Lung,* University of Wisconsin Press, Madison, 1977, chap. 24, pp. 345–356.
21. West, J. B., Lahiri, S., Gill, M. B., Milledge, J. S., Pugh, L. G. C. E., and Ward, M. P.: Arterial oxygen saturation during exercise at high altitude, *J. Appl. Physiol., 17,* 617–621, 1962.
22. West, J. B., and Wagner, P. D.: Predicted gas exchange on the summit of Mt. Everest, *Respir. Physiol., 42,* 1–16, 1980.

23. Forster, R. E.: Exchange of gases between alveolar air and pulmonary capillary blood: Pulmonary diffusing capacity, *Physiol. Rev.*, *37*, 391–452, 1957.

24. Wagner, P. D.: Diffusion and chemical reaction in pulmonary gas exchange, *Physiol. Rev.*, *57*(2), 257–312, 1977.

25. Wagner, P. D., and West, J. B.: Effects of diffusion impairment on O_2 and CO_2 time courses in pulmonary capillaries, *J. Appl. Physiol.*, *33*, 426–430, 1972.

26. Roughton, F. J. W., and Forster, R. E.: Relative importance of diffusion and chemical reaction rates in determining rate of exchange of gases in the human lung, with special reference to true diffusing capacity of pulmonary membrane and volume of blood in the lung capillaries, *J. Appl. Physiol.*, *11*, 290–302, 1957.

27. Hill, E. P., Power, G. G., and Longo, L. D.: Mathematical simulation of pulmonary O_2 and CO_2 exchange, *Am. J. Physiol.*, *224*, 904–917, 1973.

28. Roughton, J. W.: The average time spent by the blood in the human lung capillary and its relation to the rates of CO uptake and elimination in man, *Am. J. Physiol.*, *45*, 621–633, 1945.

29. Johnson, R. L., Jr., and Miller, J. M.: Pulmonary capillary blood volume, flow and diffusing capacity during exercise, *J. Appl. Physiol.*, *15*, 893–902, 1960.

30. Staub, N. C., Bishop, J. M., and Forster, R. E.: Importance of diffusion and chemical reactions in O_2 uptake in the lung, *J. Appl. Physiol.*, *17*, 21–27, 1962.

CHAPTER 27

Artificial Lungs and Oxygenation Devices

Peter D. Richardson
Division of Engineering
Brown University
Providence, Rhode Island

INTRODUCTION

The practical history of blood oxygenation has its roots in the nineteenth century. LeGallois suggested that the heart could be replaced by a mechanical pump, and he tried to provide a blood flow in rabbits which had been decapitated. In 1868 Ludovig and Schmidt oxygenated defibrinated blood by shaking it with gas in a hollow vessel. In 1882 von Schroeder made the first bubble oxygenator, using air as the gas, and in 1884 von Frey and Gouber built the first film oxygenator. Landsteiner's description of the ABO blood group system at the turn of the century and the discovery of heparin as an anticoagulant at the time of the First World War paved the way toward ultimate clinical use in which blood transfusion and anticoagulation have important roles. In 1935 Carrel and Lindbergh devised a perfusion system for extracorporeal support of an entire organ. In 1937 Gibbon achieved successful total cardiopulmonary bypass in experimental animals, using a film oxygenator in which blood ran down the side of a vertical rotating cylinder exposed to an oxygen atmosphere; blood was returned to the animal by a piston pump. In 1948 silicone was introduced as a defoamer for blood by Thomas. 1951 saw the first attempts at clinical use in human patients of pump-oxygenator devices by Dogliotti and Constantini, and by Dennis and associates. Of this period Effler [1] has written, "I have never understood why so many physicians and surgeons dedicated much of their early careers to the mechanical aspects of extracorporeal circulation. Certainly a select few had unusual talents and did much for the development of pump oxygenators as we know them today. However, the majority . . . spent much of their substance and contributed little. . . . By the mid 1950s a cardiac surgeon who did not speak authoritatively on one form of oxygenator or another came dangerously close to losing professional face. The compulsion that forced qualified surgeons to become inventors, innovators and gadgeteers reflected a phenomenon that had not been seen before in surgical history."

The phenomenon's transition to a science came quite quickly. One landmark was the book by Galletti and Brecher [1], which organized the circuitry of bypass into a few alternative schemes and discussed their relative merits and applications, and also described different devices; another was the Cincinnati conference on blood oxygenation [2]; and the third was the Copenhagen conference [3, 4] in 1975, the first conference devoted solely to the design and use of blood oxygenators. In the Cincinnati conference volume the final five chapters provide theoretical discussion of gas transport in membrane oxygenators, including the effects of augmented diffusion, and descriptions of designs of two membrane oxygenators (the Kolobow design, which uses membrane sheets coiled spirally, and a prototype capillary membrane design). One of the Copenhagen conference volumes [3] has sections on fundamentals of blood-gas exchange in membrane lungs, blood-material interaction, membrane fabrication and selected artificial lungs, perfusion routes, acute respiratory failure, clinical bypass observations, and lung transplantation, while the other [4] has sections on dimensional considerations in design and construction, aspects of hematology and compatibility, tissue factors, measurements during extracorporeal circulation, and some experimental alternative methods of oxygenation. The specialist literature is extensive.

27.1 METHODS OF EXTRACORPOREAL CIRCULATION USING AN ARTIFICIAL LUNG

Tubes (called *cannulas*) are inserted into blood vessels to provide blood flow access for extracorporeal circulation. Major vessels are either arteries (A) or veins (V), and the drainage and return sites can be V-V, V-A (the most physiological in the sense of being parallel to the natural lungs), A-V, A-A, or mixed. With venous drainage, at least one pump is required, which can be placed upstream or downstream of the artificial lung, depending on the response to blood pressure of the latter. Two pumps can be used, with the oxygenator isolated between them. One-pump or pumpless circuits can be used with arterial drainage. For open-heart surgery it is necessary to bypass the whole cardiac output. The blood flow rate needed is often reduced by cooling the patient with in-line heat exchange. A simple circuit for extracorporeal membrane oxygenation is shown in Fig. 27.1. For treatment of acute respiratory distress only part of the cardiac output needs to be bypassed, and the patient is kept normothermic. Major considerations which determine the pump system and oxygenator placement are the flow resistance to

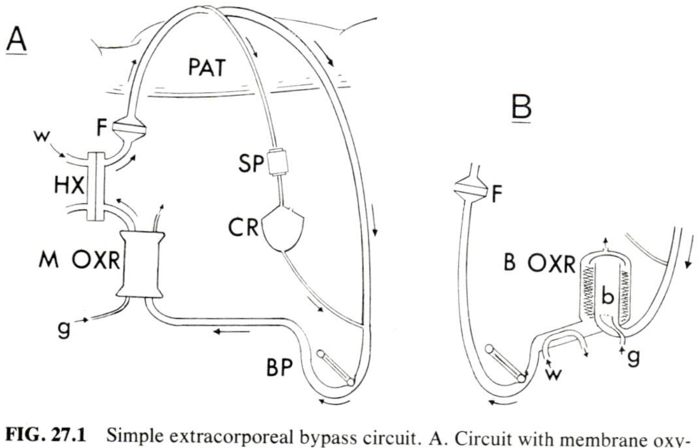

FIG. 27.1 Simple extracorporeal bypass circuit. A. Circuit with membrane oxygenator (M OXR), patient (PAT), blood pump (BP), heat exchanger (HX), filter (F), suction pump (SP), cardiotomy reservoir (CR), gas inlet for oxygenator (g), water inlet for heat exchanger (w). B. Alternate circuit components using bubble oxygenator (B OXR), which is placed on the inlet side of the blood pump; the bubble column (b) of the oxygenator is surrounded by a defoamer column.

blood at the perfusion rate (this resistance can be as much as some hundreds of millimeters of mercury for *some* membrane oxygenators) and the capacity of the oxygenator to hold up blood, which reduces the circulating blood volume in the host. Further considerations are described by Galletti and Brecher [1].

27.2 BLOOD-GAS CHEMISTRY

Blood-gas chemistry is highly nonlinear, in the sense of the oxygen content per unit volume in terms of the partial pressure of oxygen in equilibrium with it. A classic text is the one by Davenport [5]; summaries are found in standard texts on pulmonary physiology [6] and in specialized references [7, 8]; details are not repeated here. Only a few aspects are important enough to be described. The equilibrium thermodynamics for blood are conveniently visualized with the aid of what physiologists call *dissociation curves*. These are shown, for isothermal conditions, for oxygen in Fig. 27.2 and for carbon dioxide in Fig. 27.3. The degree of saturation

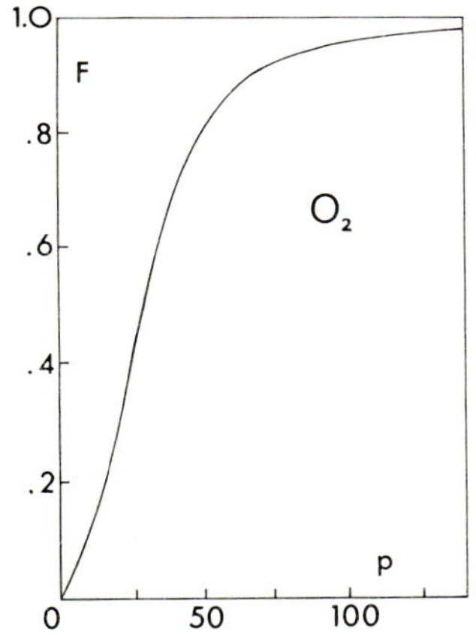

FIG. 27.2 Oxygen saturation curve at pH = 7.4, 38°C. Fraction of hemoglobin saturated, *F*, as a function of oxygen partial pressure P (in mmHg).

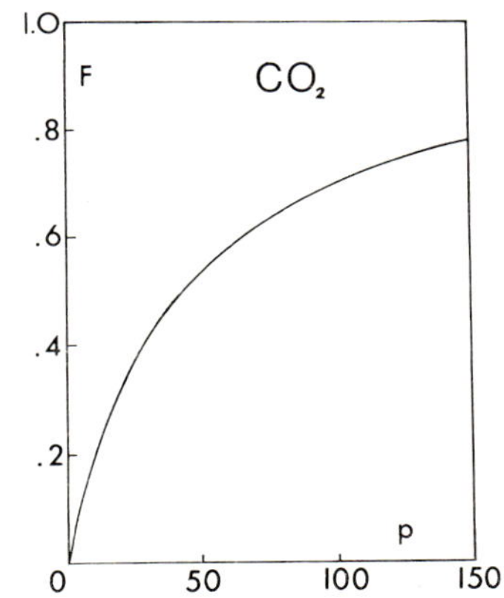

FIG. 27.3 Carbon dioxide saturation of blood components. Fraction *F* as a function of carbon dioxide partial pressure P (in mmHg).

is shown as a function of the partial pressure of the gas with which it is in equilibrium. The nonlinear characteristics of blood, so important physiologically, are seen clearly in these diagrams. It is well appreciated that the reaction velocities are finite, and for some calculations of oxygen exchange with tissues it is useful to consider forward and reverse reaction velocities k_1, k_2 in

$$HbO_2 \underset{k_2}{\overset{k_1}{\rightleftharpoons}} Hb + O_2$$

This is a simplified model in itself [9], and the four-step Adair reaction scheme could be considered a more accurate approach. For purposes of analysis of artificial lungs, it appears satisfactory to assume that local thermodynamic equilibrium exists, although, as with heat exchangers, the thermodynamic state may vary in the fluid across a channel; with blood, the variation of state may include variation of hematocrit for hydrodynamic reasons. In an input-output analysis, such effects are involved only in that they determine the blood-side diffusances (mass transfer coefficients).

A feature significant to the design of artificial lungs is that to oxygenate the blood one can use oxygen at up to 1 atm, and therefore can have a large pressure difference available to drive oxygen transfer. To eliminate carbon dioxide from the blood, one has only the equilibrium pressure of the order of 50 mmHg to drive the carbon dioxide into the oxygen stream. In order to minimize blood trauma and to save on blood priming volume, it is desirable to work with as small an artificial lung as possible, and to use as large gas-exchange pressures as possible. To obtain a satisfactory respiratory gas ratio in a membrane lung (ratio of carbon dioxide removed to oxygen supplied), it is generally necessary to use membranes which transport carbon dioxide more readily than oxygen, and which transport both gases easily and do not transport the plasma. Such differences in gas diffusivities are possible, because diffusion through membranes is not a simple process. It is therefore an important feature of artificial lung mass transfer analysis that the relative carbon dioxide transport should be found as precisely as possible.

27.3 DESIGN AND CONSTRUCTION OF OXYGENATORS

There are a few initial choices available in oxygenator design. Direct contact between blood and oxygenating fluid can be used, or a membrane can be placed between them. In direct-contact designs the only commercial units use blood-gas contact. Some research has been carried out on the use of fluorocarbons to contact blood [3], with the hope of adequate, if indirect, gas exchange accompanied by less trauma to blood in long perfusions than occurs with blood-gas contact. The development of relatively inexpensive membranes has reduced the interest in using fluorocarbons. The direct-contact designs include (predominantly) bubbler designs, which require defoaming of blood before it can be recirculated to a patient, and film designs, which do not have such a requirement. For these, the typical volumetric gas flow rate is one to two times the blood flow rate, and the respiratory quotient is so high that a small percentage of CO_2 must be added to the ventilating gas. Early designs, including the Kay-Cross film oxygenator, in which a spaced row of disks rotated on a shaft and dipped into a slowly advancing pool of blood, were nondisposable and required stripping, cleaning, and resterilization between uses. Later designs are disposable, and use polycarbonate or similar plastic shells to provide enough rigidity to define the blood and gas paths. Many now incorporate heat exchangers for blood.

Membrane oxygenators have flowpath configurations for blood and gas that are broadly similar to conventional heat exchangers. The membranes are generally arranged as small-diameter tubes (hollow fibers) or as sheets. The relative flow directions of blood and gases can be countercurrent (most efficient for the exchange process), crosscurrent (often cheaper to make), or cocurrent. Choice of membrane materials is limited, and the problem of keeping down the membrane's contribution of resistance to CO_2 transport requires that thin membranes be used (around 50 μm). The most common "solid" membrane material used is silicon rubber. The CO_2 resistance problem is much reduced by using a microporous membrane, e.g., microfibrillar polytetrafluoroethylene (PTFE). However, gases can readily bubble into blood through the pores if the blood pressure is lowered enough, and blood can come through if the blood pressure is raised enough. Roughly 2 to 3 m^2 of membrane is used in a simple membrane oxygenator to provide complete gas exchange for an adult human. The earliest (nondisposable) membrane oxygenator, which was also the first used for multiple-day support of an acute respiratory distress patient, was designed by M. Bramson to have 6 m^2 of membrane. The membrane material needs to be tolerable enough to the blood over the period of use [1 to 5 h for heart surgery, a few days for support of a patient with acute respiratory distress syndrome (ARDS)].

For effective gas exchange the blood channels are a few centimeters long (2 to 30 cm) and narrow (100 to 250 μm). To handle typically a 5-L/min blood flow with modest flow resistance, several flow channels are placed in parallel (5 to 50 for sheet membrane designs, many thousands for hollow fiber designs). Some special designs which use particular flow configurations to enhance transport rates (e.g., with extra pulsations applied, or helically coiled tubes) can use wider blood passages (giving lower flow resistance) and smaller membrane areas.

In most oxygenators there are single ports for blood entry and blood exit. With membrane oxygenators these are connected to the many membrane channels or fibers that operate in parallel by blood manifolds. The manifold system must be carefully designed to avoid excessive priming volume, stasis, and thrombogenic potential. Good illustrations of details of various oxygenator designs are given in Nose's book [10].

An additional concern in the choice of materials and of adhesives for a blood oxygenator is that the unit should retain its biotolerability and its mechanical stability following sterilization. With some materials this limits the sterilization procedures which can be used.

27.4 PERFORMANCE

The design objective of an artificial lung is to provide a gas transfer performance comparable to that of a natural lung. The blood leaves an oxygenator without coming to full equilibrium with the gases in the ventilatory spaces. With this characteristic it is possible to consider an artificial lung in a manner very similar to that used for the study of heat exchangers, and it is possible to measure the transfer effectiveness of the transport at any blood and gas flow rates, and for specific inlet conditions for both streams. This concept has been applied successfully in the process of evaluation [11]. Blood oxygenators can experience shunts, just like normal lungs; by plotting the performance in a suitable manner, that is, by examining the transfer effectiveness as a function of blood flow rate and extrapolating to zero flow rate, it is possible to make a numerical estimation of any effective shunt [11]. From an exchange analysis, which is modeled on that for heat exchangers, it is possible to define and determine from results the overall surface-area average diffusance, which is the reciprocal of the average transport resistance [11]. It is worth noting that the mean driving pressure difference for transport should be evaluated appropriately for both the oxygen and carbon dioxide transfers in order to obtain appropriate values for these diffusances [12].

With bubble oxygenators, the more detailed calculation of the expected transport is complicated because bubble motion itself is complicated. The process of transfer of oxygen may be optimum at one bubble size, and transfer of carbon dioxide may be optimum at another; however, in any real device the bubbles move along a column and can coalesce and break up to such an extent that the bubble size varies along the column in any case. Designers commonly rely on empirical results in order to select sizes for the bubbler plates and tubes used in bubble oxygenators. By contrast, there are greater prospects of determining the components of the overall transfer resistance which are associated with the blood flows and gas flows in a membrane oxygenator. The overall transfer resistance across a membrane is made up of the sum of three components: the resistance of the membrane itself, the resistance of the convective transport layer in contact with the membrane on the blood side, and the resistance of the convective transport layer on the gas side. Experimental data from evaluations of membrane oxygenator performance confirm what analyses generally indicate: the transport resistance on the gas-phase side is not very significant, but the convective transport resistance on the blood side is large and can even be the limiting component of the overall resistance. In Ref. 4 the first two chapters are devoted to a discussion of gas transport in blood and the estimation of convective transport resistance in the blood passages of membrane oxygenators for various common geometries which are used. Convective transport calculations for other geometries can be found in the literature [13–17]; it can be helpful to consult also the literature on the heat exchanger [18].

A useful physical perception of the transport resistance to oxygen in blood can be obtained from the approximation of Lightfoot [19]. Lightfoot illustrated the process of gas transfer in a membrane oxygenator by considering in particular the layers of blood close to the membrane. With oxygen at 700 mmHg pressure on the gas side of the membrane, the blood which is in immediate contact with the membrane soon becomes completely saturated, and its partial pressure rises above 150 mmHg. Meanwhile, a layer of venous blood, which may have a saturation of 65 percent or so, flows along the center of the channel. Transport of oxygen toward the unsaturated blood occurs from the membrane through a thin layer of completely saturated blood. At an interface region between the completely saturated blood and the venous blood, there is a thin layer in which the saturation changes from 100 percent to the venous value. Lightfoot was able to demonstrate that this layer of rapid change of oxygen saturation is very thin. Lightfoot described this thin layer as the oxygenation "front." As blood flows along a channel, this front advances into the blood flow from the membrane toward the center of the channel. Because of the position at which blood leaves the channel, it is desirable for the advancing front from the membrane to have come all the way to the center of the channel, or at least so close that the mixed blood leaving the oxygenator channel has a saturation of at least 95 percent. This advancing-front model is sketched in Fig. 27.4.

Most analyses of the transport process have been carried out on the assumption that the channel geometry is a simple one, e.g., a straight tube (hollow fiber) or a parallel-walled channel with flat membrane surfaces. The simple geometry is an approximation, because the membranes which are used are so thin that they are rather readily compliant and can deflect in response to a pressure difference across the membrane. Because the resistance to blood flow along a membrane oxygenator channel can be even as large as 400 mmHg, the distortions of the membrane may be considerable. Under these circumstances it would be more appropriate to describe the blood channel as being a tapered and narrowing channel rather than one with parallel walls. Fortunately, the narrowest part of the channel is at the exit region for the blood, so there is a good prospect in general that the advancing fronts will have the opportunity to meet at the center of the channel. However, with blood flow rates that are high enough, such a prospect becomes less likely, and an incomplete saturation of the blood passing through is possible. An added problem for membrane oxygenators manufactured with flat membranes is that, with real-world manufacturing variabilities, it is possible for the compliance of different parallel channels not to match. Under these circumstances it is possible to find a shunt in a membrane oxygenator. With such a shunt, the channel which holds itself most widely open under given pressure conditions will allow a disproportionate amount of the total flow to pass through it, and at the same time will have the least penetration of the advancing fronts of oxygenation into the flow. A considerable fraction of the total blood flow through the oxygenator can then pass without being oxygenated to saturation. Some manufacturers take care of this problem by assembling the set

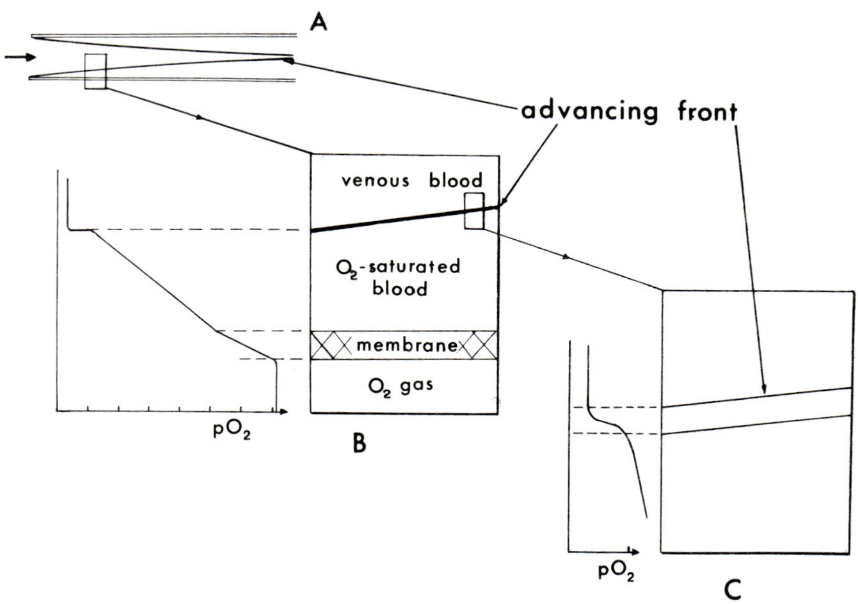

FIG. 27.4 Lightfoot's advancing-front model for oxygenation of a blood stream passing in a channel between membranes. A. Diagram of channel, with blood flowing from the left. B. Enlargement of portion of layer by wall, with corresponding distribution of partial pressure of oxygen, Po_2, shown toward left. C. Further enlargement of advancing front itself, illustrating structure of large change of Po_2. Intervals between marks on Po_2 scales are 100 mmHg.

of parallel blood and gas channels (sometimes described as a stack of blood- and gas-phase envelopes) inside a rigid shell, and they insert a shim bag which can be inflated to compress the stack. Of course, this construction presents another problem in that by this means the membranes could be pushed against each other completely and then occlude flow altogether. This is prevented in general by inserting an open-weave plastic screen into each channel for the blood flow and the gas flow. The spaces between the fibers of the screen allow blood or gas to flow as appropriate, while the span between parallel fibers is narrow enough that the membranes cannot touch. This clearly presents an additional problem in any attempt to calculate the expected transport behavior in the blood and gas fluid streams, because the flow becomes much more complicated in the presence of the screen. However, the situation invites analysis similar to Fung's description for flow in the alveolar capillary network of the natural lung [20].

It has been observed quite early in the process of evaluating the transfer performance of membrane oxygenators that there is a direct relationship between the transport performance measured with blood and that measured with a water stream passing through the blood phase of the oxygenator [21]. It is important to note that the comparison of transport rates must be made at corresponding Reynolds numbers, so that the flow rates associated with water and with blood for such comparisons are different, and that the rates for blood depend upon the hematocrit in particular. This correlation between the convective component of the overall resistance as measured in water and blood flows has been extended subsequently to include many additional oxygenator designs [22]. This observation may help the process of design as well of testing, because it allows designers to make calculations based on simpler representations of transport in water than those required for blood. The effectiveness of the comparison is not altogether surprising to persons who work in convective transport, because the transport processes involved here occur at high Schmidt numbers, for which it is well known that the principal component in determining the convective transport resistance is the region closest to the wall. With both the water and the blood under the advancing front, the principal means of transport is by diffusion of oxygen through a layer of water. The correlation between the convective transport resistance in water and in blood has been demonstrated for oxygenators where the enhancement of the transport is achieved through the generation of secondary flows, such as can be produced by flow in helically coiled tubes [23].

For oxygenators which are to be used entirely extracorporeally, most designers have not felt a need to try to keep the flow resistance low, because with extracorporeal use one or two blood pumps may adequately maintain the flow. In these circumstances the pressures which arise in the oxygenator itself need not be anywhere near physiological pressures. Some existing designs may be used without blood pumps, as demonstrated by Barthelemy et al. [24], who used an artery-to-vein circuit for removal of carbon dioxide while oxygen was being taken up in the natural lungs without respiration. The pressure drop available to drive blood through the natural lungs is very small. Also, the right ventricle which supplies this blood would suffer from conditions which required it to regularly provide a higher pressure. For this reason the development of an implantable artificial lung has always been regarded as a challenge with respect to achieving high enough transport coefficients while maintaining a low resistance to blood flow. Some progress has been made in this direction [25, 26].

27.5 USE OF ARTIFICIAL LUNGS

The use of artificial lungs in extracorporeal circuits requires a number of additional blood-handling components. These other components of the extracorporeal blood circuit include tubing; cannula tips for drainage and return of blood to vessels; connectors; a heat exchanger; pressure gauges, and, sometimes, blood flow meters; blood filters; bubble traps; and malfunction detectors. These latter may include detectors for inadequate drainage of blood into the circuit; detectors for inadvertent cessation of blood flow, e.g., due to improper clamping of some part of the circuit; and detectors for the presence of bubbles, aggregates, or other particles in the blood flow which is being returned to the patient. The number of gaseous microemboli appears to be closely related to the Po_2 in the blood leaving bubble oxygenators [27]. Where the oxygenators are being used during open-heart surgery, it is common for a suction device to be employed to remove blood continuously from the open chest cavity, and this blood is returned to the circulating blood volume after passage through a cardiotomy reservoir in which particles can be trapped and bubbles can rise to a free surface. The total priming volume for the extracorporeal circuit can easily be more than twice that of the oxygenator component of the circuit by itself.

Application of the oxygenator for use during relatively short periods of surgery is now routine, and became so within the span of a few years [28]. Since the mid-1960s, support of patients who have severe acute respiratory failure (ARF) has been attempted through extracorporeal membrane oxygenation (ECMO). Perfusion with the aid of a membrane lung may correct hypoxemia and hypercapnia while the lungs of the patient are relieved of their primary burden of respiratory gas exchange. For more than the first decade of such trials it was the universal practice to maintain ventilation of the natural lungs at the same time that extracorporeal membrane oxygenation was being carried out. The results of clinical trials have been disappointing in this period [29, 30]. In the majority of patients, while life appeared to be somewhat prolonged, there was an inexorable progressive stiffening of the lungs and no possibility of weaning from the ECMO except in a minority of cases. While these results with adult patients were very disappointing, more favorable outcomes were found with some neonates [31, 32]. These results in newborn infants are encouraging and suggest that a trial of ECMO earlier in the course of respiratory failure in adults could be more successful. Pesenti et al. [33], and Gattinoni and his team [34], in collaboration with Kolobow et al. [35], have reported considerable improvement of the survival rate in the treatment of adult respiratory failure, when ECMO has been combined with reduced ventilatory effort applied to the lungs. Additional or alternative removal of CO_2 by dialysis has also been explored [36]. There is speculation that the maintenance of the regular respirator support of such patients may do even more harm than good in the acute phase of this disorder. Reference 37 provides an up-to-date review of the integration of ECMO and respirator support in the intensive care setting.

27.6 PATHOLOGICAL ANATOMY AND PATHOPHYSIOLOGY

Artificial lungs can develop disorders of their own. Some of these mimic disorders of the natural lung. The most common problem in the past has been the development of clots or thrombi in the blood paths. Even when the blood flowing through the organ has been properly anticoagulated with heparin, it has been found that thrombotic episodes can arise [38, 39]. More recently, the additional or alternative use of prostacyclin as an anticoagulant has been explored [40]. Thrombotic material, with its dominant cellular component being platelets, has been found to deposit itself particularly in entrance and exit regions for passages through an artificial lung. This tendency is promoted by any poor manufacturing processes which leave roughnesses on the surface; these seem particularly susceptible to serve as sites from which thrombi grow. In some oxygenators this form of partial or total occlusion can occur in the gas-exchange passages through which the blood also flows. This phenomenon has been observed both during experiments on bypass with animals and in the clinical setting. Over the years manufacturers have progressively improved details of manufacture so that this is now less a problem than in the past. Similarly, techniques of membrane preparation, assembly including the use of adhesives, the provision of support structures, and so on, have developed over the years so that the risk of rupture of a membrane with attendant loss of blood is a smaller risk than it was in the past. Where microporous membranes are used, of course, there is the ever-present risk that blood can be pushed through the pores by an excessive pressure. Very occasionally, adhesives have not been applied well, with the consequence that a dissecting aneurysm can develop in a device. Tests for blood damage caused by oxygenators have been reviewed and assessed [41].

In the gas-exchange passages (sometimes called the *gas phase*) the most common problem encountered is that water vapor which is transmitted through the membrane of a membrane lung may condense. The condensed water, commonly in the form of droplets held within small interstices in the gas phase, may interfere with adequacy of the local ventilation. This problem tends to be greatest when the gas flow rates in the gas phase are low, and when the gas temperature is low, because under these circumstances the gas can readily become saturated with water vapor and allow the condensation to remain. Correspondingly, a remedy for this condition is often to supply a high enough gas flow rate. In many designs it is important that this higher gas flow rate be provided from the outset, because condensed water may be cleared away very slowly once it has formed. It is noteworthy that there is relatively little information about the pathological anatomy of artificial lungs because, in common with other devices and vascular replacements which are made from plastics, the materials of which oxygenators are made do not lend themselves to examination by pathologists using their normal tools. Indeed, the plastic materials normally cause great damage to microtome blades and other tools. Therefore, much of the information concerning pathological anatomy has been developed during oxygenator testing procedures rather than routinely in the course of clinical application.

27.7 COSTS

There is not much information about the overall cost of application of blood oxygenators and membrane lungs in medical care. Although the costs of medical care must come under closer scrutiny as technology progresses, the costs are strongly affected by the local and national systems for health care; one paper has compared costs associated with bubble and membrane oxygenators [42] in the French health care system.

27.8 FUTURE PROSPECTS

New designs continue to be brought to the clinical marketplace [43, 44], but at a slower pace than before. The physiology of bypass with a gas exchanger in the bypass circuit continues to be investigated, and new methodologies of clinical management using bypass are evolving. For the moment the investigative efforts are focused on the means and effects of use more than on new devices, but the need for a wearable lung device remains greater than that for the other most widely used exchange device, the artificial kidney. The next major development may well be a device for a wearable artificial lung.

REFERENCES

1. Galletti, P. M., and Brecher, G. A.: *Heart-Lung Bypass*, Grune & Stratton, New York, 1962.
2. D. Hershey (ed.): *Blood Oxygenation*, Plenum Press, New York, 1970.
3. Dawids, S. G., and Engell, H. C. (eds.): *Physiological and Clinical Aspects of Oxygenator Design*, Elsevier, Amsterdam, 1976.
4. Zapol, W. M., and Qvist, J. (eds.): *Artificial Lungs for Acute Respiratory Failure*, Hemisphere, Washington, 1976.
5. Davenport, H. W.: *The ABC of Acid-Base Chemistry*, 4th ed., University of Chicago Press, Chicago, 1958.
6. West, J. B.: *Respiratory Physiology*, Williams & Wilkins, Baltimore, 1974.
7. Olszowka, A. J., Rahn, H., and Farhi, L. E.: *Blood Gases: Hemoglobin, Base Excess, and Maldistribution; Nomograms for Normal and Abnormal Bloods*, Lea & Feibiger, Philadelphia, 1973.
8. Thews, G., and Hutten, H.: Biophysics of respiratory gas transport, in W. Hoppe et al. (ed.), *Biophysics*, Springer-Verlag, New York, 1983, pp. 503–514.
9. Sheth, B. V., and Hellums, J. D.: Transient oxygen transport in hemoglobin layers under conditions of the microcirculation, *Ann. Biomed. Eng.*, *8*, 183–196, 1980.
10. Nose, Y.: *Manual on Artificial Organs*, vol II: *The Oxygenator*, Mosby, St. Louis, 1973.
11. Galletti, P. M., Richardson, P. D., and Snider, M. T.: Blood oxygenator testing and evaluation: I. Evaluation techniques, Rep. NIH-69-2047-1, 1971; Friedman, L. I., Richardson, P. D., and Galletti,

P. M.: Blood oxygenator testing and evaluation: II. Procedures and Results, Rep. NIH-69-2947-2, National Institutes of Health, Bethesda, Md., 1973.

12. Snider, M. T., Richardson, P. D., Friedman, L. I., and Galletti, P. M.: Carbon dioxide transfer rate in artificial lungs, *J. Appl. Physiol., 36,* 233–239, 1974.
13. Soda, K., and Chow, J. C. F.: Blood oxygenation in tubular membrane oxygenator with turbulent flow, ASME Pap. No. 73-WA/Bio-23, American Society of Mechanical Engineers, New York, 1973.
14. Chow, J. C. F., and Soda, K.: Laminar flow and blood oxygenation in channels with boundary irregularities, *Trans. ASME J. Appl. Mech., 95,* 843–850, 1973.
15. Colton, C. K., and Drake, R. F.: Effect of boundary conditions on oxygen transport to blood flowing in a tube, *Chem. Eng. Symp. Ser., 67,* 88–95, 1971.
16. Chang, H-K., and Mokros, L. F.: Blood-gas transfer in an axial flow annular exchanger, *A.I.Ch.E.J., 17,* 397–401, 1971.
17. Chang, H-K., and Mokros, L. F.: Convective dispersion of blood gases in curved channel exchangers, *A.I.Ch.E.J., 17,* 541–549, 1971.
18. Shah, R. K., and London, A. L.: *Laminar Flow Forced Convection in Ducts,* Academic Press, New York, 1978.
19. Lightfoot, E. N.: Low-order approximations for membrane blood oxygenators, *A.I.Ch.E.J., 14,* 669–670, 1968.
20. Fung, Y. C.: *Biodynamics: Circulation,* Springer-Verlag, New York, 1984, chap. 6, pp. 290–369.
21. Richardson, P. D., Snider, M. T., and Galletti, P. M.: Oxygen transfer to blood in membrane lungs: Correlation with and prediction from in vitro measurements, *Trans. Am. Soc. Artif. Intern. Organs, 17,* 323–329, 1971.
22. Tanishita, K.: Gas transport in artificial lungs: Limitations and improvements, Ph.D. thesis, Brown University, Providence, R.I., 1975.
23. Tanishita, K., Richardson, P. D., and Galletti, P. M.: Tightly wound coils of microporous tubing: Progress with secondary-flow blood oxygenator design, *Trans. Am. Soc. Artif. Intern. Organs, 21,* 216–222, 1975.
24. Barthelemy, R., Galletti, P. M., Trudell, L. A., MacAndrew, J., Richardson, P. D., Puel, P., and Engalbert, A.: Total extracorporeal CO_2 removal in a pumpless artery-to-vein shunt, *Trans. Am. Soc. Artif. Intern. Organs, 28,* 354–358, 1982.
25. Morin, P. J., Gosselin, G., Picard, R., Vincent, M., Guidoin, R., and Nicholl, C. I. H.: Implantable artificial lung, *J. Thorac. Cardiovasc. Surg., 74,* 130–136, 1977.
26. Trudell, L. A., Pierce, E. C., Jr., Teplitz, C., Richardson, P. D., and Galletti, P. M.: A surgical approach to the implantation of an artificial lung, *Trans. Am. Soc. Artif. Intern. Organs, 25,* 462–465, 1979.
27. Pearson, D. T., Holden, M. P., Poslad, S. J., Murray, A., and Waterhouse, P. S.: A clinical evaluation of the gas transfer characteristics and gaseous microemboli production of two bubble oxygenators, *Life Support Sys., 2,* 252–266, 1984.
28. Lillehei, C. W.: A personalized history of extracorporeal circulation, *Trans. Am. Soc. Artif. Intern. Organs, 28,* 5–16, 1982.
29. Zapol, W. M., Snider, M. T., and Schneider, R. C.: Extracorporeal oxygenation for acute respiratory failure, *Anesthesiology, 46,* 272–285, 1977.
30. Hill, J. D., Rodvien, R., Snider, M. T., and Bartlett, R. H.: State of the art address: Clinical extracorporeal membrane oxygenation for acute respiratory insufficiency, *Trans. Am. Soc. Artif. Intern. Organs, 24,* 753–763, 1978.
31. Bartlett, R. H., Gazzaniga, A. B., Wetmore, N. E., Rucker, R., and Huxtable, R. F.: Extracorporeal membrane oxygenation in the treatment of cardiac and respiratory failure in children, *Trans. Am. Soc. Artif. Intern. Organs, 26,* 578–579, 1980.
32. Bartlett, R. H., Andrews, A. F., Toomasian, J. M., et al.: Extracorporeal membrane oxygenation (ECMO) for newborn respiratory failure: 45 cases, *Surgery, 92,* 452–433, 1982.
33. Pesenti, A., Pelizzda, A., Mascheroni, D., Uziel, L., Pirovano, E., Fox, U., Gattinoni, L., and Kolobow, T.: Low frequency positive pressure ventilation with extracorporeal CO_2 removal (LEPPV-$ECCO_2R$) in acute respiratory failure (ARF): Technique, *Trans. Am. Soc. Artif. Intern. Organs, 27,* 263–266, 1981.
34. Gattinoni, L., Pesenti, A., Pelizzola, A., et al.: Reversal of terminal acute respiratory failure by low frequency positive pressure ventilation with extracorporeal removal of CO_2 (LEPPV-$ECCO_2R$), *Trans. Am. Soc. Artif. Intern. Organs, 27,* 289–293, 1981.

35. Kolobow, T., Gattinoni, L., Fumagalli, R., Arosio, P., Pesenti, A., Solca, M., and Chen, V.: Carbon dioxide and the membrane artificial lung: Their roles in the prevention and treatment of respiratory failure, *Trans. Am. Soc. Artif. Intern. Organs, 28,* 20–23, 1982.

36. Gille, J. P., Saunier, C., Baluais, G., and Battistella, P.: Extracorporeal CO_2 removal by hemodialysis, *ASAIO J., 7,* 80–84, 1984.

37. Bartlett, R. H., Whitehouse, W. M., and Turcotte, J. G.: *Life Support Systems in Intensive Care,* Year Book, Chicago 1984.

38. Friedman, L. I., Richardson, P. D., and Galletti, P. M.: Observations of acute thrombogenesis in membrane oxygenators, *Trans. Am. Soc. Artif. Intern. Organs, 17,* 368–375, 1971.

39. Dutton, R. C., and Edmunds, L. H., Jr.: Formation of platelet aggregate emboli in a prototype hollow fiber membrane oxygenator, *J. Biomed. Mater. Res., 8,* 163–183, 1974.

40. Addonizio, V. P., Fisher, C. A., Bowen, J. C., Palatianos, G. C., Colman, R. W., and Edmunds, L. H., Jr.: Prostacyclin in lieu of anticoagulation with heparin for extracorporeal circulation, *Trans. Am. Soc. Artif. Intern. Organs, 27,* 304–307, 1981.

41. Mortensen, J. D.: Evaluation of tests for blood damage produced by oxygenators, *Trans. Am. Soc. Artif. Intern. Organs, 23,* 747–754, 1977.

42. Gille, J-P., Brunet, H., and Andreaux, M.: Utilization d'oxygenateurs à bulle ou à membrane: Étude de couts, *Rev. Eur. Technol. Biomed., 4*(2), 35–38, 1982.

43. Karlson, K. E., Massimino, R., Cooper, G. N., Jr., and Singh, A. K.: Initial clinical experience with a microporous membrane oxygenator utilizing secondary flow in the blood channels, *Trans. Am. Soc. Artif. Intern. Organs, 27,* 267–269, 1981.

44. Dohi, T., Hamada, E., Murakami, T., Nawa, S., Komoto, Y., Teramoto, S., and Kanbayashi, T.: Development and clinical application of a new membrane oxygenator using a microporous polysulfone membrane, *Trans. Am. Soc. Artif. Intern. Organs, 28,* 338–342, 1982.

CHAPTER 28

Neural Conduction

David Landowne
Department of Physiology and Biophysics
University of Miami School of Medicine
Miami, Florida 33101

28.1 PURPOSE OR DESIGN FUNCTION

The information processing which occurs in nerve cells may be divided into two types. Most nerve cells receive input of information from one or more sources and perform an integrative process on this information. The input may be synaptic information coming from other nerve cells or sensory information coming from the environment. After the integrative process many cells produce nerve impulses which propagate down the length of the cell, carrying information to the other end. Nerve cells range in length from a few micrometers to a few meters. The longest cells have sensory endings in the toes and transmit information about the toe to the base of the neck. Contraction in skeletal muscles is initiated by nerve impulses arising in nerve cells within the spinal cord and propagating to the particular muscle cell. The long part of the nerve cell, its *axon,* is specialized for this long-distance transmission of information. This chapter discusses these nerve impulses and the mechanism of their propagation. The next chapter discusses the integration which occurs at sensory receptors.

From a bioengineering perspective, interest in nervous conduction includes the stimulation of nerves to provide input into the nervous system and the detection of nerve impulses, and their relationship to the development of prosthetic mechanical devices. Such interest may ultimately include the construction of functioning nerve cells, although this is clearly beyond present-day technology.

The nerve impulse may be described in terms of its physical properties as follows. For nervous conduction, the nerve axon may be considered a hollow cylinder of nerve membrane which separates two aqueous salt solutions. In life the contents of nerve cells are more complex, containing proteins and cell organelles. Most of this material can be removed and replaced with salt solution (see below) without loss of nerve impulse conduction; this is evidence that the nerve membrane is the functionally interesting portion of the cell. The axoplasm is important in the long-term maintenance of the cell membrane, as well as in supplying the metabolic energy which, indirectly, powers the nerve impulse.

In the resting condition there is a 40- to 90-mV electrical potential difference across the nerve membrane, with the inside negative with respect to the outside. This is thought to be a diffusion potential occurring because of the difference in potassium ion concentration on the two sides of the membrane. There is a high intracellular concentration of potassium, 100 to 300 mM varying with species, and a low extracellular potassium concentration, 2 to 20 mM. If external potassium is experimentally elevated, the resting potential decreases (becomes less negative). With concentrations of external potassium above about 50 mM the resting potential changes about 58 mV for a 10-fold change in concentration. This brings to mind the Nernst equilibrium potential relationship between a diffusion potential and a concentration gradient for a univalent cation:

$$E = \frac{kT}{e} \ln \frac{C_0}{C_i} = (58 \text{ mV}) \log \frac{C_0}{C_i} \qquad \text{at } 20°\text{C} \tag{28.1}$$

where E is the Nernst equilibrium potential, k is Boltzmann's constant, T is the absolute temperature, e is the unit electronic charge, and C_0 and C_i are the concentrations of the ion on both sides of the membrane. The Nernst equilibrium potential is used in the literature on neurophysiology as a theoretical potential at which a particular ion is at thermodynamic equilibrium. When the membrane potential is equal to the Nernst equilibrium potential for potassium, there is no net tendency for potassium ions to move through the membrane. If the membrane potential is different from the potassium equilibrium potential and, in addition, if the membrane has a finite potassium permeability, potassium ions will move through the membrane. The direction of movement will be appropriate to change the membrane potential in the direction of the Nernst potential. Given sufficient time and no other permeant ions, the membrane potential would ultimately become equal to the Nernst potential.

To a first approximation this is believed to be the origin of the resting potential. For high external potassium concentrations the resting potential varies directly as the Nernst equilibrium potential for potassium. For low external potassium concentrations the resting potential is less negative than the theoretical Nernst potential. This is taken as a sign that the resting membrane is permeable to other ions. The theoretical Goldman-Hodgkin-Katz equation, or the constant-field equation [1], has classically been used to describe the membrane potential achieved in the steady state for a membrane which is permeable to several ions. The constant-field equation is widely used, although it is based on an assumption of membrane symmetry which is unlikely to be valid for most membranes (see below). For a membrane permeable to both sodium and potassium ions the constant-field equation predicts

$$V = \frac{kT}{e} \ln \frac{P_{\text{K}}\text{K}_0 + P_{\text{Na}}\text{Na}_0}{P_{\text{K}}\text{K}_i + P_{\text{Na}}\text{Na}_i} \tag{28.2}$$

where P_{K} and P_{Na} are the membrane permeabilities for potassium and sodium. The constant-field equation reduces to the Nernst equation when either ion is absent or impermeable. With two permeable ions the membrane potential is somewhere between the two Nernst potentials; increasing the ratio of the permeabilities $P_{\text{Na}}/P_{\text{K}}$ will move the membrane potential toward the sodium equilibrium potential, whereas lowering this ratio will drive the potential toward the potassium equilibrium potential. By measuring the effects of alterations of ionic concentrations in the external medium, Hodgkin and Katz [1] concluded that at rest $P_{\text{Na}}/P_{\text{K}}$ is about 0.01.

A nerve impulse may be recorded as a propagated brief (about 1 ms) reversal of the membrane potential, from about −70 mV resting potential to about +35 mV at the peak of the action potential (see Fig. 28.1). The propagation velocity varies with axon diameter, the amount of myelination (see below), and the temperature. In humans, conduction velocities range from about 0.5 to 120 m/s. Action potentials are remarkably similar in shape. However, slight differences can be seen between some classes of nerves, and these differences can be used to determine which nerve is firing. For any one nerve, the action potentials are almost exactly the same, and clearly any information-carrying facility comes from the pattern of occurrence of action potentials and not from the size or shape of the individual action potential. The action potential is similar to the familiar monostable multivibrator, or one-shot device, used in constructing digital circuits. The action potential is "all-or-none." If the stimulus is sufficient, a full-blown action potential will occur. The amplitude, duration, and conduction velocity of the action potential are not dependent on the stimulus—exactly analogous to the one-shot device.

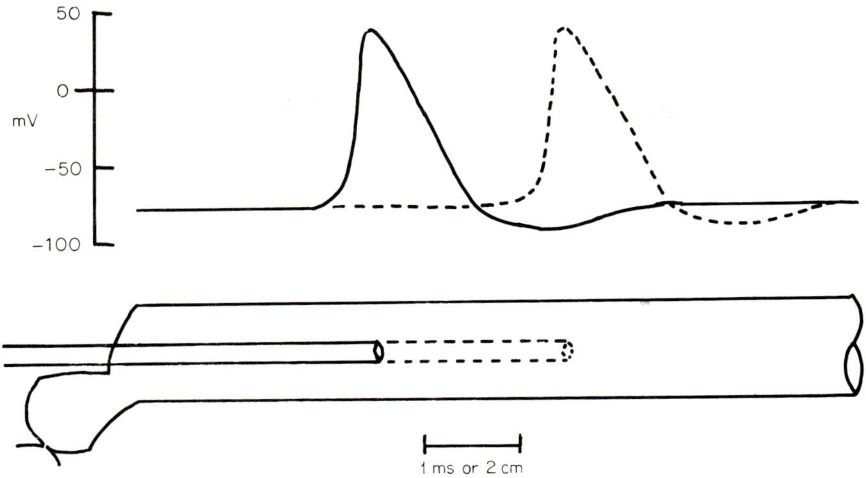

FIG. 28.1 The propagated nerve impulse. Two traces of transmembrane potential versus time, corresponding to the two positions of the intracellular recording electrode, are shown above the giant axon.

Hodgkin and Katz [1] also demonstrated that the amplitude of the action potential was highly dependent on external sodium ions. Using the constant-field equation, they calculated the ratio $P_{Na}/P_K = 20$ at the peak of the action potential. On this basis they postulated that the primary event underlying the nerve impulse was a transient selective increase in permeability to sodium ions. As will be shown below, a change in membrane potential is necessary and sufficient to produce the change in sodium permeability. If the membrane potential is reduced (depolarized) from −70 to about −50 mV, the membrane becomes more permeable to sodium ions. This permits sodium to flow down its electrochemical gradient, bringing the membrane closer to the sodium equilibrium potential. There is much more sodium outside than inside; the sodium equilibrium potential is about +40 mV for most nerve cells. The movement of sodium thus further depolarizes the nerve membrane, which further increases the permeability to sodium. This regenerative response continues until the membrane potential approaches the sodium equilibrium potential. It is this regenerative, or positive-feedback, cycle of depolarization, leading to increased sodium permeability, which in turn leads to further depolarization, that assures the all-or-none property of the action potential.

The propagation of the action potential uses this property. The presence of an action potential in one region of the nerve stimulates the adjacent regions. The electrical coupling between adjacent regions of the membrane is a passive process, as distinguished from the voltage-dependent change in sodium permeability described in the previous paragraph. A simple and completely adequate electrical model is described in the next section. The equations describing the passive properties are often called *cable equations* since they were first worked out for underwater cables by Lord Kelvin. In the analogy the voltage-dependent changes in sodium permeability would represent the "active" repeater stations along a cable. The other electrical properties of the nerve cell are its axoplasmic resistance, analogous to the resistance of the cable core, and membrane resistance and capacitance, analogous to the resistance and capacitance of the insulator of the cable.

There is an additional important property of the nerve cell membrane which will be mentioned here and then ignored. The immediate source of energy for the development and propagation of nerve impulses is the ionic gradients for sodium and potassium across the nerve membrane. The intact nerve membrane has an intrinsic mechanism to maintain these gradients, namely the sodium-potassium pump. This molecular complex actively transports sodium out of the cell and potassium into the cell, both against their electrochemical gradients. In the process adenosine triphosphate (ATP) is degraded into adenosine diphosphate (ADP). The ATP is then resupplied by the cellular metabolism. The pump is clearly not required for the action potential mechanism if other means are arranged to maintain the ionic gradients. The flux of ions via the pump is much smaller (fewer ions per second) than during the nerve impulse. A good analogy is to think of the pump as a trickle charger which charges up the sodium and potassium batteries. The nerve impulse then briefly draws on these batteries to depolarize and repolarize the nerve membrane.

28.2 PASSIVE CONDUCTION—CABLE PROPERTIES

Figure 28.2 shows two equivalent circuit models used to describe the electrical properties of nerve axons. In the first panel the active electrical elements which produce the nerve impulse

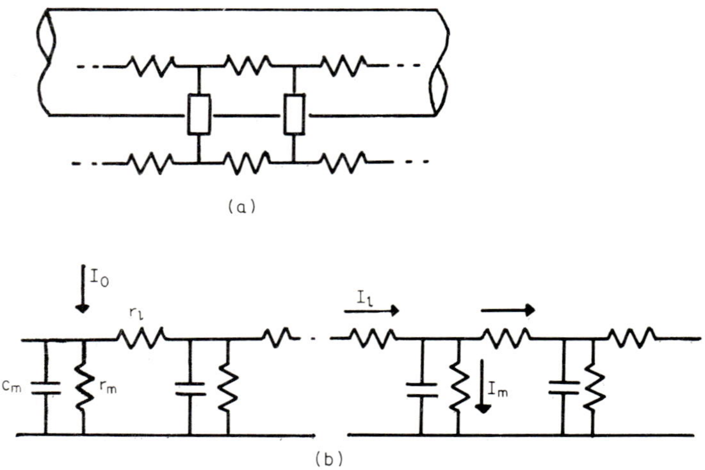

FIG. 28.2 Axon equivalent circuits. (*a*) A section of axon showing internal and external longitudinal resistances and the transmembrane impedance. (*b*) A simplified equivalent circuit of the passive electrical properties used to derive the cable equations for the spread of electricity along an axon.

have been lumped together. These will be described in the next section. The active elements are distributed over the surface of the axon in two dimensions. In this simplified circuit each box represents all the active elements in a narrow ring of membrane perhaps 1 μm in length along the axon. This ring is electrically coupled to adjacent rings by the passive elements. Figure 28.2*b* is a simplification; the active elements have been removed, and the extracellular resistance, which is small compared with the intracellular resistance, is ignored. This is a practical simplification; it is only slightly more complicated to develop the equations with the external resistances. There is a Thévenin equivalence between these circuits.

Let us start with a completely uncharged circuit and inject a steady current I_0 from an external source. After a sufficiently long time the system will come to a new steady state with a potential V_0 at the injection site and $V(x)$ at any position away from the injection site. The bottom line in the schematic is the return path for the current and also the ground potential to which all potentials are referred. Let $I_m(x)$ be the current through a unit length of membrane at a distance x from the injection site. $I_l(x)$ is the longitudinal current through r_l, the longitudinal resistance. By Ohm's law

$$I_l(x) = -\frac{1}{r_l}\frac{dV(x)}{dx} \qquad (28.3)$$

where r_l is the resistance per unit length in the longitudinal direction, in Ω/cm, and $dV(x)/dx$ is the potential gradient in the longitudinal direction. The minus sign means that the direction of the electric current is positive along a falling (negative) voltage gradient.

The longitudinal current decreases as we get farther from the injection site because some of the current goes through the membrane. The membrane current is

$$I_m(x) = \frac{V(x)}{r_m} = -\frac{dI_l(x)}{dx} = \frac{1}{r_l}\frac{d^2V(x)}{dx^2} \qquad (28.4)$$

where r_m is the resistance of a unit length of membrane, in Ω/cm. Thus

$$V(x) = \frac{r_m}{r_l}\frac{d^2V(x)}{dx^2} \qquad (28.5)$$

This second-order differential equation has a solution

$$V(x) = Ae^{-x/\lambda} + Be^{x/\lambda}$$

with

$$\lambda = \sqrt{\frac{r_m}{r_l}}$$

A and B can be determined from the boundary conditions. At $x = \infty$, $V(x) = 0$, so $B = 0$. At $x = 0$, $V = V_0$, so the particular solution is

$$V(x) = V_0e^{-x/\lambda} \qquad (28.6)$$

The voltage falls off exponentially with distance with a characteristic length constant λ. Had we not ignored the external resistance, we would have found a similar equation with $\lambda = \sqrt{r_m/(r_0 + r_i)}$. The length constant of nerves varies with the diameter of the axon with values between a few micrometers and a few millimeters.

The variation with diameter occurs because both r_m and r_i decrease as the diameter increases. r_m decreases linearly with axon diameter because the area of membrane per unit length increases linearly. r_i, the internal axoplasmic resistance, decreases as the square of the diameter, as it is inversely proportional to the cross-sectional area of the axon. Hence the length constant theoretically increases as the square root of the diameter, all other things being equal. In practice some workers use the square root relationship and others prefer to describe their data as a linear function of diameter. There is considerable scatter in the data since it is collected from measurements on many different axons.

The conduction velocity of the nerve impulse increases with the length constant. This is because the axon with a larger length constant "feels" the effects of the action potential farther ahead and thus re-creates the nerve impulse anew at a greater distance. One evolutionary tack to maximize information transmission has been to produce "giant" axons, 50 to 1000 μm in diameter, for specialized signal systems. The giant axon system of the squid, which coordinates its jet-propelled escape response, has been favored in experiments because of its size.

Nature has used another approach to increase the length constant and conduction velocity without the inconvenience of large-diameter fibers. High-speed vertebrate nerves are myelinated. An accessory cell, the Schwann cell, is wrapped around a length of axon and serves as electrical insulation, increasing the resistance between the inside of the axon and the bathing media. There are nodes (of Ranvier) in between adjacent Schwann cells which are not insulated. The active membrane channels are concentrated at the nodes so the nerve impulse is regenerated at each node rather than continuously as in the unmyelinated nerves. Because of this hopping from node to node, myelinated nerve conduction is referred to as *saltatory*. The hops are long compared with the axon diameter but are small compared with the wavelength of the nerve impulse.

The length constant is adequate to describe the properties of the axon in the steady state. When the disturbance is transient, as in the action potential, it is necessary also to consider the effects of membrane capacitance c_m. This confers to the membrane a time constant $\tau = r_m c_m$. If a step of current is injected, the voltage at the site of injection will rise exponentially. That is,

$$V(0) = V_0(1 - e^{-t/\tau}) \tag{28.7}$$

The time constants of nerves are about 1 ms. τ does not vary systematically with diameter, for c_m increases inversely to the decrease in r_m. The effect of myelination is to decrease the capacitance between the intracellular and extracellular fluids, essentially by increasing the distance between the two plates of the capacitor. This also has the effect of speeding up the nerve impulse, as the potential can change more rapidly.

At other positions along the axon the membrane potential rises sigmoidally, with a delay required to charge the membrane capacitance up to the point of measurement. The farther away from the injection site, the longer the delay, the slower the rate of rise, and the lower the final level. The general equation is obtained by adding the capacitive current to Eqs. (28.4) and (28.5). When the membrane potential is changing, the membrane current is given by

$$i_m = \frac{V_m}{r_m} + c_m \frac{\partial V}{\partial t} = \frac{1}{r_l} \frac{\partial^2 V}{\partial x^2} \tag{28.8}$$

This leads to

$$-\lambda^2 \frac{\partial^2 V}{\partial x^2} + \tau \frac{\partial V}{\partial t} + V = 0 \tag{28.9}$$

which has solutions containing error functions of distance and time [2].

28.3 THE MECHANISM OF THE NERVE IMPULSE

Two technical advances allowed Hodgkin, Huxley, and Katz [3] to measure the properties of the active electric circuit elements in Fig. 28.2*a*. Thrusting a fine wire down the inside of a functional segment of axon defeats the cable properties and keeps the whole axon at the same membrane potential. Feedback control of the membrane potential, a process known as *voltage clamping,* allows the study of membrane currents at a constant potential. It has turned out that the active elements are controlled by the membrane potential, so it is entirely appropriate to study them by the voltage-clamp technique.

Presently, squid axons are generally voltage-clamped using a platinized platinum wire, 50 to 100 μm in diameter, as the internal current wire (see Fig. 28.3). This is inserted into the axon from a cut end along with a similar-size KCl bridge. The membrane potential is measured between the internal KCl electrode and another one just outside the axon. The measured membrane potential is then compared with the desired command potential and any difference is fed back into the axon via the internal current wire with the appropriate polarity to reduce the difference between the observed and the desired potential. The current through a central guarded region is measured with a current-voltage converter. With a well-tuned system the membrane potential can follow a step command pulse with a lag of a few microseconds. The limitations are the frequency response of the input stage, which is made with field-effect transistors (FETs), and the slow rate of the driver amplifier. The membrane presents a large capacitive load (1 μF/cm^2) which can throw the feedback loop into uncontrolled oscillations.

Besides squid axons, the voltage-clamp technique has been applied to axons from frogs, lobsters, crabs, worms, rats, and rabbits, as well as to nerve cell bodies of snails, frogs, and a variety of tissue-cultured neurons. Skeletal, smooth, and cardiac muscle cells also have action potentials which have been voltage-clamped. At the time of sperm penetration a variety of eggs from marine animals develop a "fertilization" potential, much like a nerve action potential except much slower. These too have been studied by using the voltage clamp. In all these cells the same general properties are found. They are described in terms of voltage-dependent chan-

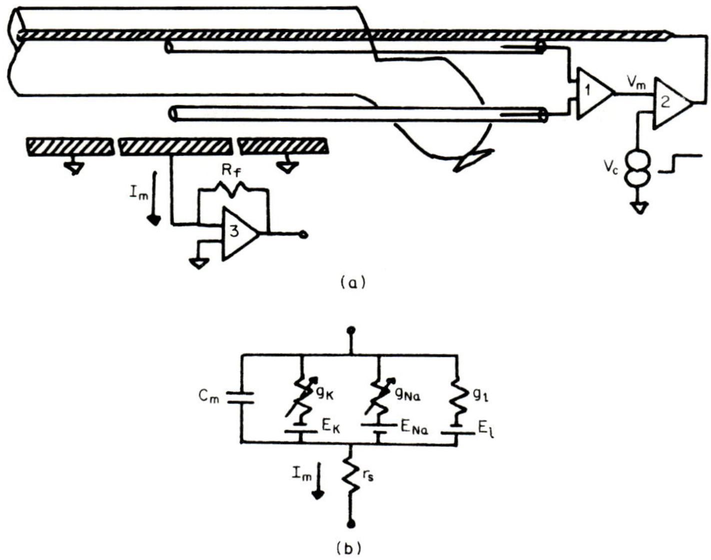

FIG. 28.3 Voltage clamp for a giant axon. (*a*) Amplifier 1 measures the transmembrane potential as the difference between an electrode inside the axon and one outside the axon. Amplifier 2 measures the difference between the actual membrane potential V_m and the desired membrane potential V_c, and sends a current through the axon membrane via an internal axial electrode. The effect of this current is to change the membrane potential such that $V_m = V_c$. Amplifier 3 maintains the central external current electrode at ground. The potential at its output is equal to the current passing through the axon membrane in the central guarded region times the feedback resistor, R_f. (*b*) An electrical equivalent circuit used to describe the axon membrane.

nels—a putative membrane structure which causes a change in membrane conductance toward a specific ion or group of ions. There are variations among the specificity, the voltage sensitivity, and the time course of the conductance changes. A whole garden of channels has been discovered. All of them have been described by using the principles set down by Hodgkin and Huxley [4–7], which will be described in the remainder of this section.

The initial assumption made in analyzing voltage-clamp currents is that the membrane can be represented by a capacitance in parallel with the ion-carrying channels (Fig. 28.3*b*). The total current through the membrane, I_m, can be divided into capacitive (or displacement) current I_c and ionic current I_i, or

$$I_m = I_c + I_i = C\frac{\partial V}{\partial t} + I_i \tag{28.10}$$

This is the simplest equation which can be used. When the command voltage pattern is in steps, $\partial V/\partial t = 0$ when the capacitance is charged, a few microseconds after the start of the step. The current which flows after the capacitive transient is purely ionic and represents the net sum of the current through all the channels.

Evidence that the ionic currents are in parallel with the capacitive current comes from comparing the ionic current-voltage relationship measured with voltage-clamp analysis, where $\partial V/\partial t = 0$, and the current calculated from $-C_m\, \partial V/\partial t$ in free-running membrane action potentials, where $I_m = 0$ [3]. With an axial wire inside the axon, $\partial V/\partial x$ and $\partial^2 V/\partial x^2$ are both zero and Eq. (28.10) is just a special case of Eq. (28.8). By doing the phase plane plot of $\partial V/\partial t$ versus V during the course of a membrane action potential, one obtains a current-voltage relationship

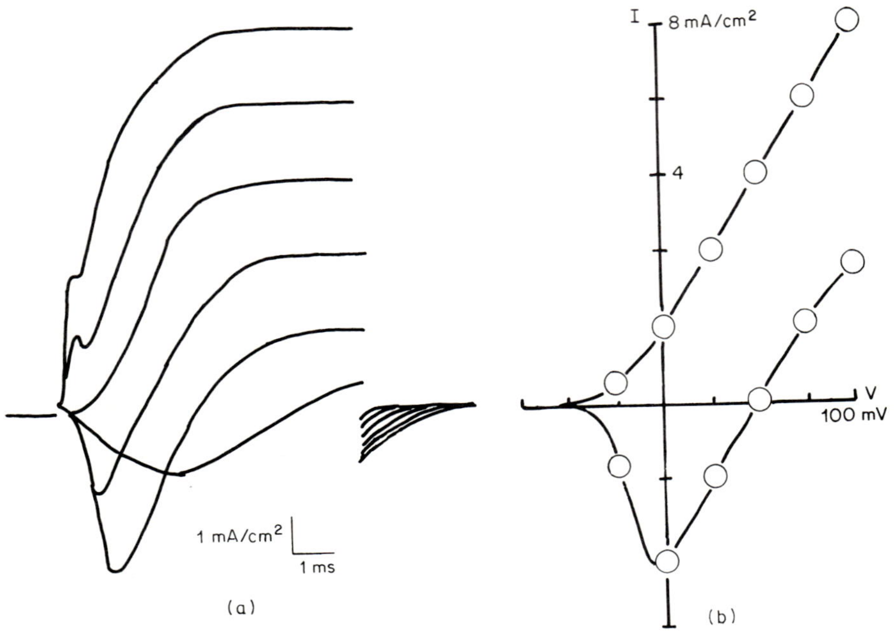

FIG. 28.4 The current-voltage relationship of nerve membrane. (*a*) A family of membrane current traces for six different voltage-clamp pulses. (*b*) The peak early transient current (sodium current) and the steady-state (potassium) current versus the voltage at which they were recorded.

which is very similar to the one obtained with a full voltage-clamp experiment. This phase plane analysis has been used in other cells, notably heart muscle, which have been difficult to voltage-clamp.

Figure 28.4 is a family of voltage-clamp current traces produced with voltage steps from a holding potential of −70 mV to the potentials indicated at the right of each trace. The capacitive transient is not visible in this figure; it occurs during the brief gap in the current record at the beginning and end of the voltage pulse. For pulses between −50 and +50 mV the initial transient ionic current is inward and is followed by a delayed outward current which seems to achieve a steady state. Above +50 mV both the early transient current and the delayed current are outward. Figure 28.4*b* plots the current-voltage relationship at the peak of the early transient and at the end of the voltage pulse.

The early transient current is carried primarily by sodium ions, and the delayed current is carried primarily by potassium ions. Altering the sodium concentration on either side changes the magnitude and/or direction of the early transient current without dramatically altering its time course. The reversal potential, the potential at which the early transient reverses from inward to outward current, varies roughly as would be expected for a Nernst equilibrium potential [Eq. (28.1)] for sodium ions. Hodgkin and Huxley [4] saw that sodium ions were crossing the membrane in both directions under a driving force which is the resultant of the effects of the concentration difference and the electrical potential difference across the membrane.

The next step in the analysis was to separate the ionic currents into components carried by sodium and by potassium ions. Families of voltage-clamp currents were recorded with different sodium concentrations outside the axon. The sodium current, I_{Na}, was taken as the component which changed amplitude, and I_K, the potassium current, was the component which did not change. In addition, Hodgkin and Huxley proposed a leak current, I_l, which was small compared with I_{Na} and I_K, and was purely ohmic and time-invariant. I_l was the small current seen when the membrane potential was made more negative than the resting potential.

Using the concept of a net driving force being the difference between the actual membrane potential V and the reversal potential E_{Na}, Hodgkin and Huxley defined the sodium conductance of the membrane as $g_{Na} = I_{Na}/(V - E_{Na})$, with similar equations for g_K and g_l. The membrane current could then be described as

$$I_m = C_m \frac{\partial V}{\partial t} + g_{Na}(V - E_{Na}) + g_K(V - E_K) + g_l(V - E_l) \qquad (28.11)$$

These membrane conductances are called *chord conductances;* they are the slopes of the chords drawn on the current-voltage curves from the voltage of interest to the equilibrium potential. They are always positive and must be distinguished from the slope conductances $\partial I/\partial V$, which can be either positive or negative. Incidentally, it is the region of negative slope conductance between -40 and $+10$ mV which provides the regenerative powers of the action potential.

There are four regions discernable in the sodium current-voltage relationship (see Figs. 28.4 and 28.7). For potentials more negative than -70 mV there is negligible sodium conductance; the sodium channels are closed. Between -70 and $+10$ mV the sodium conductance increases with the membrane potential as more and more sodium channels are opened. In the initial portion of this voltage region the sodium conductance increases e-fold every 4 mV, which indicates that the process is very voltage-sensitive. From $+10$ to $+70$ mV the sodium current is ohmic; all the sodium channels are open. There is a bend in the current-voltage relationship near $+100$ mV which was not described by Hodgkin and Huxley. It will be discussed along with sodium conductance below. The first three regions—low conductance, transition, and high conductance—can also be seen in the potassium current-voltage relationship.

The value in using ionic membrane conductances defined in this manner becomes apparent when the membrane potential is changed a second time during the current flow. Figure 28.5 shows the sodium current and the sodium conductance when the voltage is returned to the resting potential at the peak of the sodium conductance. Just after the second change in membrane potential there is an increase in the magnitude of the inward sodium current which is due to the increased driving force on sodium ions. The sodium conductance appears to be the same just before and just after the second step. The sodium "tail" current decays away as channels return to the closed condition.

Thus the sodium conductance can be measured by the change in current associated with the second step divided by the magnitude of the second step. If the initial magnitude of the tail current is plotted versus voltage for different second steps, a linear instantaneous current-voltage relationship is obtained. It is just $I_{Na} = g_{Na}(V - E_{Na})$, with the particular value of g_{Na} which was present just before the step. Similar experiments with longer pulses can be used to measure potassium tail currents and the potassium instantaneous current-voltage relationship.

The sodium current is transient; at a constant voltage the sodium conductance increases and then decreases. This decline is called *inactivation,* and several experiments suggest that the sodium conductance mechanism is in a different state when inactivated than when the sodium channels are closed before the voltage is changed.

Starting at a -70-mV holding potential, a step to -60 mV produces a barely noticeable sodium current. However, if the membrane is given a prepulse to -60 mV for about 50 ms and then stepped to a test pulse of $+10$ mV, the amplitude of the resulting sodium current will be less than the current obtained by directly stepping from -70 directly to $+10$ mV. The time course of the sodium current is the same with and without a prepulse. During the time at -60 mV some of the sodium conductance was inactivated. If the experiment is repeated by varying the duration of the prepulse, a time constant of inactivation is obtained—several milliseconds at -60 mV.

The entire prepulse experiment is repeated with different duration and amplitude of prepulses. If the prepulse is to -90 mV, the response to the test pulse is larger than it would be without a prepulse, indicating that some of the channels are already inactivated at -70 mV. The normalized collected results of this type of experiment are shown in Fig. 28.6 as h_∞ and τ_h. h_∞ is

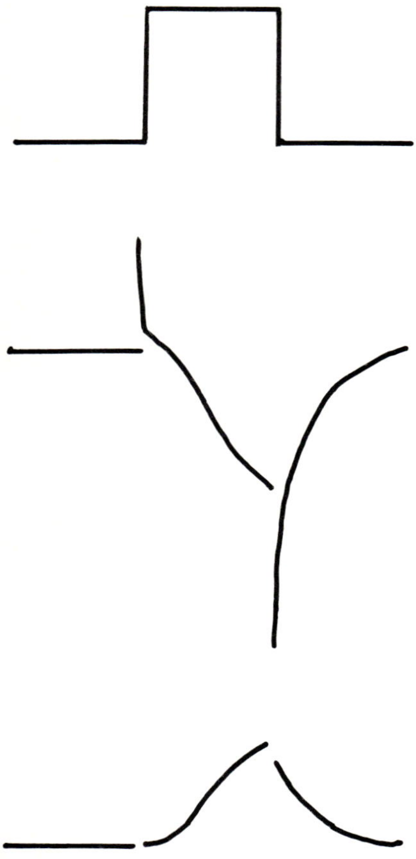

FIG. 28.5 Membrane conductance. Upper trace: membrane potential generated in a voltage-clamp experiment. Middle trace: the resulting membrane current. Lower trace: membrane sodium conductance, $g_{Na} = I_{Na}/(V - E_{Na})$.

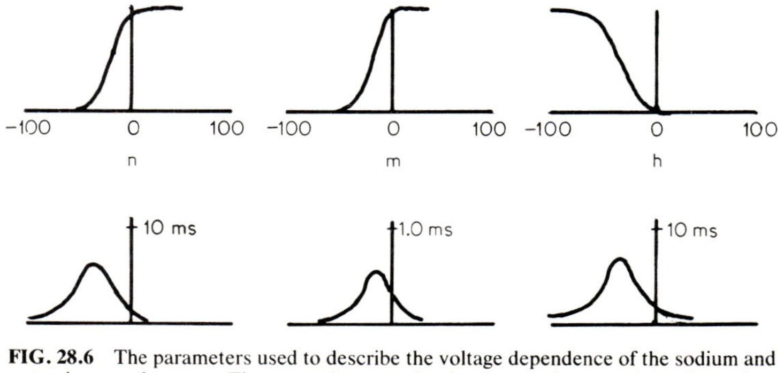

FIG. 28.6 The parameters used to describe the voltage dependence of the sodium and potassium conductance. The upper three graphs plot the steady state of the dimensionless parameters *n*, *m*, and *h* versus membrane potential. The lower graphs show the relaxation time constants for *n*, *m*, and *h* versus membrane potential.

the steady-state value of the fraction of channels available to be opened, that is, the fraction which has not been inactivated. τ_h is the rate at which the conductance inactivates.

One can also measure the rate of removal of sodium inactivation by varying the duration between two pulses. As the time between pulses is made longer, the sodium current of the second pulse becomes larger because the sodium conductance mechanism recovers from the inactivation experienced during the first pulse. Thus there are three methods of measuring the rate of inactivation: by measuring the decline of sodium current in the prepulse experiments, by measuring the recovery of sodium current in two pulse experiments, and by measuring the time course of the decay of sodium current during a single pulse. Hodgkin and Huxley [6] found rough agreement for these three methods, but more recent measurements have suggested differences.

Without structural information about the excitable elements of the membrane there is little hope in calculating the amplitude and time course of the ionic conductances from first principles. Fully cognizant of this, Hodgkin and Huxley [7] provided equations which describe the conductances to show that a mathematical description of the conductances combined with Eq. (28.11) would adequately describe the known electrical properties of the nerve impulse. They also provided a physical basis for the equations but emphasized that it was unlikely to be correct. These equations are used extensively by electrophysiologists today. Even though we know of inadequacies, there is still no generally accepted physical theory, nor has anyone provided a clearer, more intuitively obvious, mathematical description.

When the membrane potential is stepped from −70 mV to a more positive potential, the ionic conductances increase, but there is a definite delay between the voltage change and the conductance change (Fig. 28.5). There is no obvious delay when the pulse is terminated; the tail currents begin to decline as early as measurements can be made. This asymmetry between turn-on and turn-off kinetics was handled by supposing that the conductances were proportional to the third or fourth power of a variable which obeys a first-order equation. This supposition is as if three or four independent particles need be positioned to produce the conductance increase but only one particle need to be removed to permit the conductance decrease. An additional term is required to describe sodium inactivation.

The fully developed Hodgkin and Huxley equations consist of Eq. (28.11) with

$$g_{\mathrm{Na}} = \bar{g}_{\mathrm{Na}} m^3 h \qquad \text{and} \qquad g_{\mathrm{K}} = \bar{g}_{\mathrm{K}} n^4 \tag{28.12}$$

The $\bar{g}$'s are constants with units of conductance per square centimeter, and n, m, and h are dimensionless kinetic variables which can vary between 0 and 1; n, m, and h in turn are defined by

$$\begin{aligned} \frac{dn}{dt} &= \alpha_n(1 - n) - \beta_n n = \frac{n_\infty - n}{\tau_n} \\ \frac{dm}{dt} &= \alpha_m(1 - m) - \beta_m m = \frac{m_\infty - m}{\tau_m} \\ \frac{dh}{dt} &= \alpha_h(1 - h) - \beta_h h = \frac{h_\infty - h}{\tau_h} \end{aligned} \tag{28.13}$$

The α's and β's are first-order rate constants with units of s^{-1}. They are a function of membrane potential and are not time-dependent. The τ's are equal to the reciprocal of the sum of their respective rate constants, and the steady-state values are the forward rate constants times the τ's. That is,

$$\tau_n = \frac{1}{\alpha_n + \beta_n} \qquad \text{and} \qquad n_\infty = \frac{\alpha_n}{\alpha_n + \beta_n}$$

and likewise for m and h. Figure 28.6 shows the voltage dependence of the steady-state values of n, m, and h and their respective time constants. The α's and β's are derived from these data. The voltage dependence of the α's and β's were fitted by Hodgkin and Huxley with

$$\alpha_n = \frac{0.01(v + 10)}{\exp\ [(v + 10)/10 - 1]}$$

$$\beta_n = 0.125\ \exp \frac{v}{80}$$

$$\alpha_m = \frac{0.10(v + 25)}{\exp\ [(v + 25)/10 - 1]}$$

$$\beta_m = 4\ \exp \frac{v}{18} \tag{28.14}$$

$$\alpha_h = 0.07\ \exp \frac{v}{20}$$

$$\beta_h = \frac{1}{\exp\ [(v + 30)/10 + 1]}$$

These rate-constant equations are reproduced with the voltage convention where v represents the change, in millivolts, from the resting potential and with a reversed sign convention. That is, $v = -(V + \text{the resting potential})$. The form of Eqs. (28.14) resembles that of the equations for the movement of a charged particle in a constant field and suggests that the variation of α and β with membrane potential arises from the effect of the electric field on a mobile charge or dipole within the membrane.

Equation (28.11) may now be solved by substituting in Eqs. (28.12) and including Eqs. (28.13) and (28.14). For the case of $I_m = 0$ (the situation with an axial wire) the four simultaneous first-order equations, Eqs. (28.11) and (28.13), may be stepwise numerically integrated. Briefly, for each step the new voltage is estimated from the old voltage and the old $\partial V/\partial t$. The new α's and β's are computed with Eqs. (28.14), and the rate of change of n, m, and h are found with Eqs. (28.13). Then Eqs. (28.12) give the new value of the conductances, and these are substituted into Eq. (28.11) to check the original estimate of the new voltage. If it is sufficiently different, the process is repeated until a consistent new voltage is obtained. With steps of a few microseconds the entire action potential can be computed in a few seconds with a microcomputer.

The equation for a propagated action potential (no axial wire) is more complex because there are longitudinal currents. From Eq. (28.4) or (28.8) we know that

$$I_m = \frac{1}{r_l}\frac{\partial^2 V}{\partial x^2}$$

Because the action potential propagates without change in shape, it obeys the wave equation

$$\frac{\partial^2 V}{\partial x^2} = \frac{1}{\theta^2}\frac{\partial^2 V}{\partial t^2} \tag{28.15}$$

where θ is the conduction velocity. Combining Eqs. (28.11), (28.12), and (28.15), one obtains the equation for a propagated action potential, namely,

$$\frac{1}{\theta^2 r_l}\frac{\partial^2 V}{\partial t^2} = C_m \frac{\partial V}{\partial t} + \bar{g}_{Na} m^3 h(V - E_{Na}) + \bar{g}_K n^4 (V - E_K) + g_l(V - E_l) \tag{28.16}$$

This can be solved by numerical integration.

With Eqs. (28.11) to (28.16) Hodgkin and Huxley were able to compute most of the electrical properties of nerves which had been cataloged. They successfully computed the amplitude, form, and threshold for the membrane action potential. The threshold arises at the balance point when $I_{Na} = I_K + I_l$; any further increase in sodium conductance will cause the system to go on and have a full action potential. They obtained the amplitude, form, and conduction velocity for the propagated action potential. They could calculate the form and amplitude of impedance changes which had been measured during action potentials. The equations successfully described the threshold and response during the refractory period. Following an action potential a nerve is said to be *refractory* because it is more difficult or impossible to elicit a second action potential. This is now explained in terms of sodium inactivation and the relatively slow decline of the potassium conductance. The equations also predict subthreshold responses and other less-well-known electrical responses of axons.

In summary, Hodgkin and Huxley completed the descriptive phase of neural excitation and conduction. All the electrical properties can be explained by these equations, with minor changes and extensions. The current questions in the field revolve around the nature of the reversible alterations in sodium and potassium permeability arising from changes in membrane potential.

28.4 CONDUCTIVITY OF THE SODIUM CHANNEL

The purpose of this section is to describe the open-channel properties, selectivity, and conductance of the sodium channels of axons. The potassium channel and dozens of other known channels will be omitted because of space limitations and because their stories are not as complete as that of the sodium channel. Some of the work done on the kinetics of the sodium channel will be mentioned in the chapter summary.

28.4.1 Tetrodotoxin and the Techniques of Internal Perfusion and Patch Clamping

Tetrodotoxin (TTX) is a 319-molecular-weight drug with a four-ring cage structure which rapidly blocks the early transient currents [8]. The effect is quickly reversible and works with low concentrations of TTX; the dose required to reduce the currents by half is about 3 n*M*. TTX is highly selective, and it apparently has no effect on the delayed potassium currents, the sodium-potassium pump, or any metabolic or other intracellular activities of cells. It is so selective that it is used as a diagnostic tool to detect sodium channels. Occasionally sodium channels are ascribed to a tissue solely because their electrical properties are changed by TTX. It is preferable, however, to demonstrate channels with solution-change experiments.

TTX gives validity to the tacit assumption of Hodgkin and Huxley that the sodium and potassium conductance changes are separate functions and not two aspects of the same function. Before the use of TTX it was possible to suggest there was only one channel which altered its properties. Sodium inactivation and potassium activation have approximately the same time course; it might have been that the selectivity of the one channel was changing. TTX and many other specific drugs provide convincing evidence that there are two channels.

It is now practical to internally perfuse segments of squid axons and maintain the ability to produce action potentials for many hours. Excitability is maintained even after more than 100 times the cell volume of as simple a solution as buffered potassium fluoride has flowed through the axon. This is very convincing evidence that excitability is a property of the cell membrane or of some molecule or molecules that are closely associated and not freely diffusable.

Two methods of initiating internal perfusion are commonly used. They both seem to work equally well, and no systematic study of their differences has been reported. In the "garden roller" technique [9] a small, 6-mm-diameter roller, usually a plastic rod covered with latex tubing, is gently rolled over the axon, squeezing the axoplasm out the cut larger end of the axon. The axon is then reinflated with solution through an inlet cannula tied to the smaller end. An outlet cannula may be inserted if desired and electrodes may be put in from either end. The rolling technique is also commonly used to express samples of axoplasm for physical or chemical analyses.

In the cannula technique [10] the axon is mounted horizontally, usually in a chamber with airgaps and bridges on each end. The inlet cannula is introduced at a cut over one bridge and then guided through the length of the axon and out a cut over the far bridge. The electrode assembly may be inserted inside the inlet cannula and guided into place as the tip of the cannula is withdrawn back through the axon. Often at this stage a proteolytic enzyme is included in the flowing internal perfusion fluid to aid in the washout of axoplasm. When the electrode is in position, the tip of the inlet cannula is withdrawn to near its site of entry, and then the axon is secured to the cannula. It is also possible to include an outlet cannula if necessary.

The technique of patch clamping has been introduced recently and holds great promise. In favorable situations it is possible to record electrical activity from single sodium channels [11]. Basically, a micropipet is placed on the surface of the cell and a small patch of membrane is sucked into the pipet and seals to the glass. The patch may be studied attached to the cell, or it may be pulled off and carried on the pipet through several solution changes. It is possible to control the solution on both sides this way. Single-channel recordings today look like a two-state device, similar to the output of a flip-flop. The current record is at either the open-channel level or the closed-channel level. Some patches have several channels and several different levels in the recordings.

The primary importance of single-channel recordings is that they clearly show there are single channels with the unit of activity being much larger than a single ion crossing the membrane. Thus the macroscopic currents recorded with the conventional voltage-clamp technology represent the average of a large number of smaller unitary events.

28.4.2 Properties of Open Sodium Channels

Hodgkin and Huxley used choline as a sodium substitute in their original experiments. When other monovalent cations are used as substitutes, some support the early transient current and others do not. If lithium replaces sodium in the external solution, the action potential and the voltage-clamp currents are, at first approximation, unchanged. It appears as if the sodium channel will also facilitate the passage of lithium through the membrane. This presents no conceptual difficulty; the axon normally is never required to distinguish between sodium and lithium. However, when perfused axons were voltage-clamped, an outward early transient current was found for large pulses even in the absence of internal sodium ions [12]. This current persists with zero sodium concentration on either side of the membrane, and yet it is blocked by TTX. It represents the outward flow of potassium ions through the sodium channel.

Chandler and Meves [12] suggested the use of the constant-field equation [Eq. (28.2)] to describe the reversal potential for a channel which is permeable to more than one ion. From measurements of the reversal potential with only sodium outside and only potassium inside, they found the ratio P_{Na}/P_K was about 12. When some of the internal potassium was replaced with mixtures of sodium and potassium, there was good agreement with the constant-field equation, but small consistent deviations were seen with high sodium–low potassium internal solutions. Similar experiments were performed with internal rubidium and cesium. The alkali cations were found to have the following permeabilities relative to sodium: Li, 1.1; K, 1/12; Rb, 1/40; and Cs, 1/61.

This list was extended to eight metal and 21 organic cations in frog myelinated nerve [13, 14]. Using space-filling models of these ions, Hille [14] has proposed that the narrowest portion of the channel, the "selectivity filter," has dimensions of about 0.3×0.5 nm, based on the largest ions which will fit through. There also must be one or more charged sites within the channel to exclude anions which could fit through this size hole. Hille discusses why potassium is so poorly permeable in spite of its small size in terms of waters of hydration.

The deviations from the constant-field equation with alterations of the internal solution have been reinvestigated [15]. They can be accommodated by assuming that the P_{Na}/P_K ratio is linearly dependent on the internal potassium ion concentration, decreasing from about 12 in high potassium to a limiting value of about 2 in very low internal potassium. It is the potassium concentration and not the ionic strength which is important; tetramethylammonium ions will not restore the high selectivity. These findings suggested either that the structure of the channel changed with potassium concentration or that there was some problem with the method of analysis. It makes it difficult to use the constant-field equation to evaluate permeability ratios because with only one change of solution it is difficult to know how much of the change in reversal potential is due to the change in concentrations and how much should be ascribed to a change in permeability ratios.

The evaluation of permeabilities from reversal potential measurements is also suspect since it attempts to measure a transport parameter in the case of no net transport. Hille [16] noted that some ions which had comparable permeability estimated from their reversal potential had quite different amplitude of currents at potentials removed from the reversal potential. He constructed a four-barrier Eyring rate theory model to account for his data. Four barriers were required to linearize the current-voltage relationship; only one barrier height (the selectivity filter) was altered to account for the selectivity variations, and one well depth (the binding site) was set by the reduction in current amplitude in a manner similar to the saturation of Michealis-Menton enzyme kinetics. Block of the passage of one ion by the presence of a second ion could also be described by the properties of the binding site. With this view the channel presents a different set of barriers and binding sites to each ion. With two parameters per ion Hille obtained a good fit to the changes in the current-voltage relationship for his experimental changes of the external ionic composition. The binding constant of the sodium channel for sodium ions was higher than his range of solutions. Appropriately the sodium channel doesn't bind sodium very tightly.

Thus Hille replaced the use of the constant-field equation for the description of current-voltage relationships with Eyring rate theory. Begenisich and Cahalan [17, 18] used a three-barrier, two–binding site model to describe their experiments in altering ion concentrations. The effects of internal potassium on the apparent permeability ratio (as estimated by the constant-field equation) were modeled with a higher internal barrier to K than to Na and a lower internal well for K (a tighter binding site). Their barriers and binding sites are asymmetrical; the outer barrier is higher than the inner barrier for both Na and K, and the asymmetry is greater for K. They did not observe saturation of the currents from the outside and thus ascribed no depth to the outer binding site; the inner binding site is deeper.

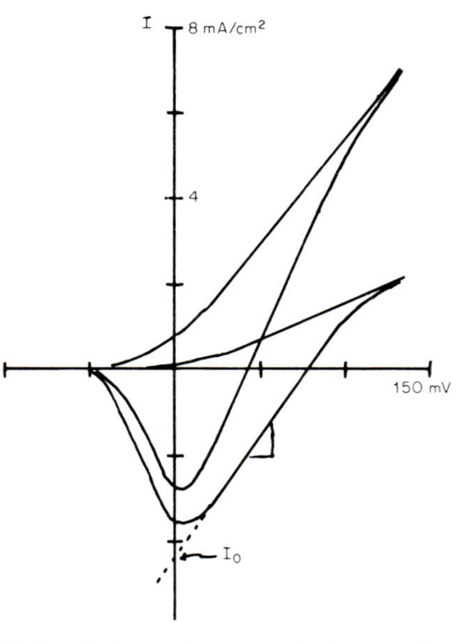

FIG. 28.7 The sodium current-voltage relationship for four different external and internal sodium concentrations. The sodium conductance is obtained from the slope of the ascending limb. The sodium permeability can be calculated from I_0, the current extrapolated to a membrane potential of 0 mV.

In some sense the barrier–binding site models are arbitrary. They do provide a framework for the data which is sufficiently rich to describe almost anything. What is clearly missing is structural data with which to associate these otherwise hypothetical barriers.

Landowne and Scruggs [19] have found a simpler approach to the relationship among current, voltage, and the concentrations on the two sides of the membrane. As can be seen in Fig. 28.7, there is a linear superposition law describing the concentration effects on the current-voltage relationship. That is, the effects of varying internal or external sodium are linearly additive, since the current-voltage relationship with sodium on both sides (460//100) is just the relationship with only sodium outside (460//0) shifted up by the difference between no sodium on either side (0//0) and sodium only on the inside (0//100). The slope of the straight ascending region of the current-voltage relationship, between +20 and +80 mV, is a bilinear function of the two concentrations. That is,

$$g = (ac_0 + bc_i)F$$

where F is Faraday's constant, and a and b are membrane constants, the channel acceptances for sodium ions, analogous to electrical admittances. The short-circuit current, measured at 0 mV, is linearly proportional to the difference between the two concentrations, or

$$I_0 = -P(c_0 - c_i)F$$

where P is the Fickean permeability. When the straight region of the current-voltage relationship is extrapolated back to the current axis, the extrapolated short-circuit current is described by this equation.

Thus the ascending limb of the current-voltage relationship can be described by a simple linear equation

$$I = -P(c_0 - c_i)F + (ac_0 + bc_i)FV \tag{28.17}$$

where a and b are not equal. For sodium going through the sodium channel b/a is about 3; the channel is asymmetrical. As might be expected P is related to a and b. Experimentally $P/(a + b) = 25$ mV, which is the value of kT/e in the Nernst equation [Eq. (28.1)].

SUMMARY

Since 1952, the Hodgkin and Huxley analysis has been extended to include slower electrical phenomena exhibited by the squid axon, and adapted to describe transient electrical phenomena in cells as diverse as aquatic plants and marine eggs. The method works. If the appropriate electrical connections can be made to achieve adequate temporal and spatial control of the membrane potential, the resulting voltage-clamp currents can be resolved into ionic currents by the appropriate solution-change experiments. These can further be resolved into conductance mechanisms or "channels," each with its own selectivity or preference for ions, maximum conductance, and kinetic variables. Sometimes it is necessary to ascribe more than one channel with the same selectivity—e.g., it might be necessary to define g_{K1}, g_{K2}, . . . , each with its own kinetic parameters—to characterize a particular cell. Fortunately, in some cases the different channels will be pharmacologically distinct, with different drugs altering or blocking different conductivity mechanisms, suggesting that different membrane molecules are responsible for the different channels.

The Hodgkin-Huxley equations work very well to describe the electrical properties for a single pair of solutions. They fall short in describing solution changes; in the initial papers this was handled with an arbitrary scaling factor. Use of Eq. (28.17) or the barrier analysis makes the scaling factor unnecessary.

The origin of the kinetics of channels is still obscure. There is controversy in the literature as to whether sodium inactivation follows and is mechanistically linked to sodium activation [20] or is an independent process [21] as originally proposed by Hodgkin and Huxley. There is a large literature concerning treatments that alter the kinetic parameters. Perhaps the most dramatic effect is that proteolytic enzyme treatment of the internal face of the membrane is able to remove sodium inactivation without dramatically altering activation.

One technique used today to study channel kinetics is measurement of nonlinearities in the capacity currents (see Ref. 20). These asymmetric or "gating" currents are thought to represent structural changes within the membrane, reorientation of charged groups or dipoles which leads to the changes in membrane permeability. There is also a recent report of a similar asymmetry in changes in membrane birefringence associated with voltage-clamp pulses [22]. It seems that the movement of impulses along nerves and the movement of ions through channels are fairly well understood today. What remains to be known is how the channel itself moves.

REFERENCES

1. Hodgkin, A. L., and Katz, B.: The effect of sodium ions on the electrical activity of the giant axon of the squid, *J. Physiol.*, *108*, 37–77, 1949.
2. Hodgkin, A. L., and Rushton, W. A. H.: The electrical constants of a crustacean nerve fibre, *Proc. R. Soc. B*, *133*, 444–479, 1946.

3. Hodgkin, A. L., Huxley, A. F., and Katz, B.: Measurement of current-voltage relations in the membrane of the giant axon of Loligo, *J. Physiol., 116,* 424–448, 1952.
4. Hodgkin, A. L., and Huxley, A. F.: Currents carried by sodium and potassium ions through the membrane of the giant axon of Loligo, *J. Physiol., 116,* 449–472, 1952.
5. Hodgkin, A. L., and Huxley, A. F.: The components of membrane conductance in the giant axon of Loligo, *J. Physiol., 116,* 473–496, 1952.
6. Hodgkin, A. L., and Huxley, A. F.: The dual effect of membrane potential on sodium conductance in the giant axon of Loligo, *J. Physiol., 116,* 497–506, 1952.
7. Hodgkin, A. L., and Huxley, A. F.: A quantitative description of membrane current and its application to conduction and excitation in nerve, *J. Physiol., 117,* 500–544, 1952.
8. Narahashi, T.: Chemicals as tools in the study of excitable membranes, *Physiol. Rev., 54,* 813–845, 1974.
9. Baker, P. F., Hodgkin, A. L., and Shaw, T. I.: Replacement of the axoplasm of giant nerve fibres with artificial solutions, *J. Physiol., 164,* 355–374, 1962.
10. Tasaki, I., Watanabe, A., and Takenaka, T.: Resting and action potentials of internally perfused squid giant axon, *Proc. Natl. Acad. Sci. U.S.A., 48,* 117, 1962.
11. Sigworth, F. J., and Neher, E.: Single Na^+ channel currents observed in cultured rat muscle cells, *Nature, 287,* 447–449, 1980.
12. Chandler, W. K., and Meves, H.: Voltage clamp experiments on internally perfused axons, *J. Physiol., 180,* 788–820, 1965.
13. Hille, B.: The permeability of the sodium channel to organic cations in myelinated nerve, *J. Gen. Physiol., 58,* 599–619, 1971.
14. Hille, B.: The permeability of the sodium channel to metal cations in myelinated nerve, *J. Gen. Physiol., 59,* 637–658, 1972.
15. Cahalan, M., and Begenisich, T.: Sodium channel selectivity: Dependence on internal permeant ion concentration, *J. Gen. Physiol., 68,* 111–125, 1976.
16. Hille, B.: Ionic selectivity, saturation and block in sodium channels: A four barrier model, *J. Gen. Physiol., 66,* 535–560, 1975.
17. Begenisich, T. B., and Cahalan, M. D.: Sodium channel permeation in squid axons: I. Reversal potential experiments, *J. Physiol., 307,* 217–242, 1980.
18. Begenisich, T. B., and Cahalan, M. D.: Sodium channel permeation in squid axons: II. Non-independence and current-voltage relations, *J. Physiol., 307,* 243–257, 1980.
19. Landowne, D., and Scruggs, V.: Effects of internal and external sodium on the sodium current-voltage relationship in the squid giant axon, *J. Membrane Biol., 59,* 79–89, 1981.
20. Armstrong, C. M., and Bezanilla, F.: Inactivation of the sodium channel: II. Gating current experiments, *J. Gen. Physiol., 70,* 567–590, 1977.
21. Horn, R., Patlak, J., and Stevens, C. F.: Sodium channels need not open before they inactivate, *Nature, 291,* 426–427, 1981.
22. Landowne, D., and Scruggs, V.: Asymmetric changes in the birefringence of voltage clamped internally perfused squid giant axons, *Biophys. J., 33,* 283a, 1981.

CHAPTER 29

Sensory Receptors

David C. Mountain
College of Engineering
Department of Biomedical Engineering
Boston University
Boston, Massachusetts

INTRODUCTION

Sensory receptor cells are complex physiological systems. Not only do they contain transducers to detect the sensory stimulus, but they often amplify and filter the resulting signal before passing on the information to the central nervous system. In addition, in many receptors these functions are constantly adapting to changes in stimulus conditions.

Sensory receptors are usually classified according to the type of stimulus energy they are most sensitive to. Electroreceptors are cells that have become specialized to detect electric fields. Mechanoreceptors respond to mechanical stimuli such as stretch, vibration, and sound. Photoreceptors are sensitive to light, and chemoreceptors are stimulated by ions or molecules and are involved in the senses of taste and smell.

Despite the great differences in the types of stimuli that various sensory receptors respond to, all these cell types must ultimately convert the stimulus energy into a change in membrane potential so that the information can be relayed to the rest of the nervous system. This potential change is referred to as the *receptor potential*. The change in membrane potential either initiates action potentials in an axon, which is part of the sensory cell, or acts on secondary neurons by way of a synapse. These changes in membrane potential are the results of changes in the ionic currents flowing through the cell membrane. This change in current induced by the stimulus is referred to as the *receptor current*.

The sensory stimuli (except for electroreceptors) cannot themselves alter the ionic currents directly, so they must act through intermediate membrane structures serving as channels for ionic flow or as ion carriers.

These intermediate membrane structures provide the receptor cell with a simple means of amplifying the stimulus energy. The electric potential and the ionic gradients across the cell membrane provide a strong electrochemical driving force which can lead to large standing ionic currents through the membrane. These currents are thought to flow through channels which exist in either an open or a closed state. The state of the channel is due to a conformational change in the molecule or molecular complex which makes up the channel. The sensory stimulus needs only to provide sufficient energy to alter the probability that the channels are in the open state, in order to cause a change in average membrane conductance and hence a significant change in current flow through the cell.

Other membrane channels can play an additional role by amplifying the small voltage changes caused by the sensory stimulus. Voltage-dependent channels which are activated by depolarization of the cell can give rise to large inward currents which further depolarize the cell. The two common ionic species which have the electrochemical gradients necessary to provide large inward currents are Na^+ and Ca^{2+}. Positive-feedback systems of this sort are potentially unstable. It is this instability that leads to action potentials, and both sodium-based and calcium-based action potentials are known to occur.

The large dynamic range of many sensory receptors requires not only amplification mechanisms but also some means of adapting to very large stimulus levels. Most sensory information is contained in the incremental changes of the sensory stimulus rather than its absolute level, so the problem faced by many sensory systems is to be able to respond to incremental changes in stimulus level over a wide range of stimulus intensities.

Sensory receptors must not only resolve changes in stimulus intensity but also be able to respond to them rapidly. However, the requirement for fine temporal resolution implies a wide bandwidth for the sensory system, which in turn would result in an increased noise level as well as decreased sensitivity due to reduced temporal integration of stimulus energy. The cell membrane capacitance, in conjunction with the membrane resistance, acts as a low-pass filter on the receptor potential. The membrane impedance, however, can be extensively modified by the presence of voltage-dependent or ion-dependent membrane channels. The resulting active filter can have a wide range of frequency-response characteristics, depending on the nature of the voltage- and ion-dependent channels.

The importance of active filtering by the receptor cell membrane varies markedly from one receptor type to another. Of the receptor types which will be discussed in this chapter, some exhibit essentially no active filtering while others appear to act as matched filters to a specific stimulus.

It is not possible to review all types of receptors in the space available, so this chapter will concentrate on three receptor types: electroreceptors, hair cells, and photoreceptors. Electroreceptors have been chosen because transduction in the normal sense is not necessary, since an electric field can directly lead to a change in membrane potential. These receptors allow one to study the role of membrane channels in the amplification and filtering of the stimulus signal without the complications of understanding the transduction process. The hair cell has been chosen as an example of an extremely sensitive mechanoreceptor. In addition, it appears that the transduction process is a relatively simple one. Photoreceptors were chosen because they appear to use biochemical means to amplify the response to single photons, and in addition have voltage- and ion-dependent channels which have been extensively studied.

29.1 ELECTRORECEPTORS

Specialized sense organs for electroreception are found in a number of different groups of fish. They are considered part of the acousticolateralis system, which includes the senses of hearing, equilibrium, and the detection of water movement. The receptors of the vertebrate acousticolateralis system are thought to have a common evolutionary origin. The hair cells, which are the mechanoreceptors of the acousticolateralis system, will be discussed in the next section of this chapter. The sensitivity of the electroreception system in marine species is quite impressive. The stingray has been shown to be able to detect voltage gradients as small as 0.05 μV/m (Kalmijn, 1982).

Electroreceptors are separated on the basis of morphology into two classes: tuberous electroreceptors and ampullary electroreceptors. The tuberous electroreceptors are located in small groups which protrude into the epidermis. The ampullary receptors are located at the bottom of canals which are lined with cells which appear to be specialized to prevent leakage of current from the canal. The canal is open to the outer surface of the epidermis and is filled with a conductive gelatinous material. The canal serves to concentrate the voltage gradient which is present over its length at the level of the receptor cells, since the potential difference across the receptor cells is equal to the potential difference between the two ends of the canal. The ampullary receptors are 5 to 10 times more sensitive than the tuberous receptors when compared in a single species (Suga, 1967).

Fish appear to be able to use electroreception in several different ways. There are a number of species of fish which produce weak, periodic discharges from electric organs. Behavioral experiments have shown that these fish can easily discriminate between objects on the basis of their conductivity. It would appear that these fish are able to sense small changes in the electric field created by their electric organs. These fish also use electroreception for communication purposes. Other species may use their electric sense to detect muscle action potentials from prey species.

29.1.1 Tuberous Receptors

The tuberous receptors are of particular interest in the electrolocation process, as they respond best to the frequency of the electric organ discharge (EOD) (Hopkins, 1976). Figure 29.1

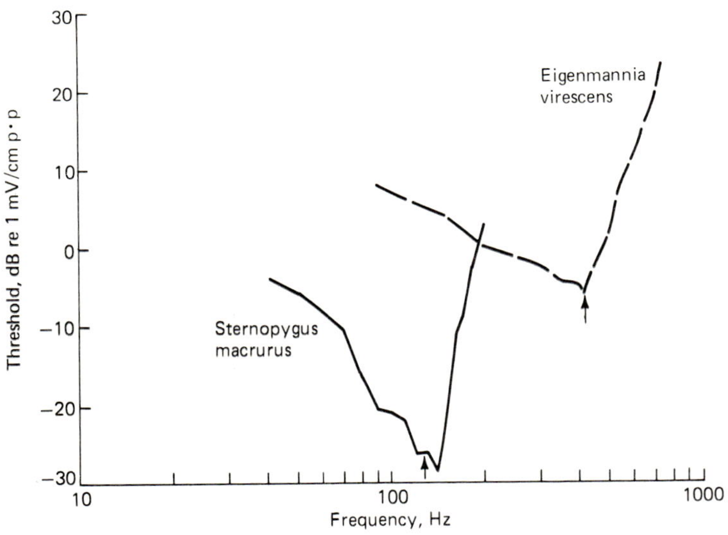

FIG. 29.1 Tuning curves for tuberous electroreceptors from two different species. The arrows indicate the frequency of the EOD for each species. (*Data from Hopkins, 1976.*)

illustrates the frequency tuning observed in single electoreceptor nerve fibers from two different species of gymnotoids. These curves represent the threshold, in millivolts per centimeter, for the firing pattern of the nerve fiber to synchronize the stimulus period. One can think of these threshold-tuning curves as an approximation to the inverse of the transfer function of the electroreceptor. Threshold-tuning curves are often used in sensory physiology instead of transfer functions, because of the inherent nonlinearity present in the transformation between receptor potential and firing rate in sensory nerve fibers. The underlying assumption in this approach is that, if one determines, as a function of frequency, the input level which produces a fixed output level from the system, then the receptor potential is the same for all stimulus frequencies, and the nonlinearity in the receptor potential to action potential transformation will not play a role in the frequency response measurement. This approach of holding the output of the system constant and measuring the input for various frequencies will be valid only for systems comprised of a linear filter followed by a frequency-independent nonlinearity (memoryless nonlinearity). Because of these assumptions, one must be cautious in placing too much emphasis on the quantitative details of these tuning curves. The threshold-tuning curve approach has proved to be an excellent predictor of receptor potential tuning in auditory hair cells where direct measurements of both receptor potentials and nerve fiber firing rates have been made (Russell and Sellick, 1978).

The important point to be noted in Fig. 29.1 is that these receptors show pronounced frequency selectivity, and in each case the most sensitive frequency (best frequency) corresponds to the frequency of the EOD in that species (arrows in Fig. 29.1). The frequency tuning of the tuberous receptors is in contrast to the simple low-pass filter effect one would expect from a linear membrane resistance and capacitance.

Bennett (1967) has suggested a model in which the apical (outward-facing) membrane of the receptor cell is of high resistance, but with substantial capacitance due to a large surface area comprised of microvilli. This membrane is in series with the basal (inward-facing) membrane of the cell from the point of view of currents flowing through the skin of the fish. This basal membrane would have an underdamped impedance due to voltage-dependent conductances. The apical membrane would passively couple the current resulting from the EOD to the basal membrane through its capacitance, and the basal membrane would actively filter the signal. Support for this model has come from experiments in which rectangular currents are applied to the receptor organ. These currents produce a damped oscillatory response in the voltage measured by a microelectrode placed in the skin over the receptor organ (Bennett, 1967). The oscillations are still present after local application of tetrodotoxin, which blocks the afferent nerve action potentials (Bennett, 1971), indicating that the oscillations are not due to the nerve fibers.

The ionic basis for the postulated membrane tuning is unknown. Hopkins (1976) has pointed out that the Hodgkin-Huxley (1952) model of the squid axon can simulate the damped oscillations in membrane potential which are observed experimentally in response to small current steps. The electroreceptor tuning appears to be due to a regenerative process, since drying of the skin leads to sustained oscillation in some receptors (Bennett, 1971). On the other hand, it is unlikely that the voltage-dependent Na^+ channels of the Hodgkin-Huxley model are involved in the electroreceptor tuning, since tetrodotoxin blocks these channels and yet does not affect the oscillatory responses to electric current. It may be that Ca^{2+} supplies the regenerative current, since Ca^{2+} channels are present in a number of receptor cells, including ampullary electroreceptors in elasmobranchii.

29.1.2 Ampullae of Lorenzini

The ampullary organs of marine fish differ morphologically from those of freshwater fish in that the canal between the body surface and the sensory cells is considerably longer in marine fish (Kalmijn, 1974). In addition, several hundred receptor cells converge onto a single afferent nerve fiber; this is a considerably larger convergence than that found in freshwater species (Obara, 1976).

In the marine catfish *Plotosus,* positive current pulses passed into the ampullary pore increased the firing rate of the afferent nerve fibers, and negative current decreased the firing rate (Obara, 1976). If two electrodes are placed inside the ampulla, one for current injection and the other for voltage measurement, it is possible to observe oscillatory and even regenerative responses to depolarizing current steps (Obara, 1976). The regenerative response appears to be due to a Ca^{2+} conductance which is activated by depolarization of the basal (inward-facing) receptor cell membrane.

Unlike the electroreceptors discussed so far, the afferent fibers from the ampullary organs of the skate exhibit a decrease in firing rate when the canal is made positive and an increase in firing rate when it is made negative (Murray, 1962, 1965). This reversed polarity of response means that the electric field is not acting directly on the presynaptic membrane. These receptors are extremely sensitive, with changes in firing rate reported for stimuli as small as 2 μV (Murray, 1967). The great sensitivity of these receptors appears to be the result of two processes. The sensitivity is a consequence of the extreme length of the canals, some of which traverse almost the entire length of the animal's body (Murray, 1962). The walls of the canal are of sufficiently high resistance that the space constant of the canal is long compared with its length so that there is little attenuation of dc potentials (Waltman, 1966). The result of this specialization is that the receptor cells respond to the potential difference between the two ends of the canal, which are widely spaced points in the electric field. In addition to the long canals these organs have receptor cells which are electrically excitable (Obara and Bennett, 1972).

The electrophysiological investigations of this organ have benefited from the ease with which it can be removed and maintained in vitro. Microelectrodes can be placed in the ampulla, and voltage- or current-clamping techniques can be used. When negative current is passed into the ampulla, action potentials can be elicited (Obara and Bennett, 1972).

Clusin and Bennett have investigated the origins of both the action potential (1977a, 1977b) and the oscillations (1979a, 1979b). Their experiments provide evidence for voltage-dependent Ca^{2+} channels in both the apical and basal membranes of the receptor cells. In addition, the apical membrane exhibits a slow, Ca^{2+}-activated current which may be carried by K^+, and the basal membrane exhibits a fast K^+ current which may be activated by depolarization or possibly by Ca^{2+}. Clusin and Bennett (1979b) have proposed a model for the ampullary organ of the skate which is summarized in Fig. 29.2. R_{SH} represents the Thévenin equivalent of current paths through the ampulla exclusive of the receptor cells. The apical and basal membranes of the receptor cells are lumped into the upper and lower three-branch networks, respectively. Capacitive elements were not considered. When a negative potential is applied across the organ, the apical membranes of the receptor cells are depolarized, leading to activation of the apical Ca^{2+} channels (R_{CaA}). The resulting large inward current depolarizes the apical membrane even

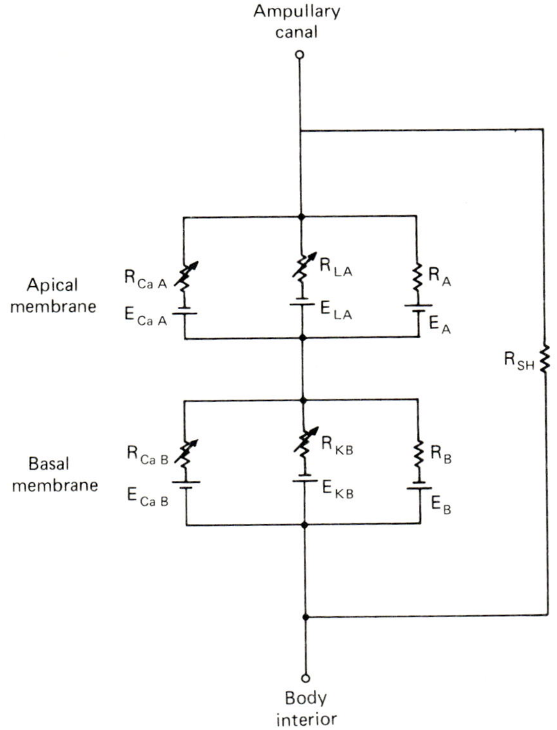

FIG. 29.2 The lumped-parameter circuit model for the ampullary electroreceptor organ of the skate. (*Adapted from Clusin and Bennett, 1979b.*)

more, and depolarizes the basal membrane as well. The basal K^+ channel (R_{KB}) is then activated after a short delay, either by the depolarization of the basal membrane or by the influx of Ca^{2+}. The K^+ current then repolarizes both membranes, inactivating the Ca^{2+} current. It is the delay in activation of the basal K^+ current that leads to the oscillatory potentials observed in the ampulla. The basal Ca^{2+} channels (R_{CaB}) mediate synaptic transmitter release and in addition could serve to amplify the depolarization of the basal membrane.

The apical Ca^{2+}-activated channels (R_{LA}) are too slow to play a role in the oscillatory response. Bennett and Clusin (1979) have suggested that they play a role in adaptation of the electroreceptor to steady voltages. The accumulation of Ca^{2+} inside the cell as a result of a maintained activation of apical Ca^{2+} channels would lead to an increase in the outward current through R_{LA}, and this would repolarize the apical membrane. This process would maintain the incremental sensitivity of the receptor, even in the presence of steady external fields, by maintaining the electroreceptor membrane potential in the most sensitive region.

29.2 HAIR CELLS

Hair cells are mechanoreceptors found in a variety of vertebrate sensory systems. Like the electroreceptors, they are considered part of the acousticolateralis system. Hair cells are found in lateral line, vestibular, and auditory organs. The transduction process responds with great speed to extremely small movement.

The sensory hairs, or stereocilia, protrude from the apical surface of the hair cell. They are in fact modified microvilli with a core of microfilaments made up of the protein actin (Flock and Cheung, 1977). The actin filaments are cross-linked together, which probably increases their rigidity (DeRosier et al., 1980; Tilney et al., 1980). In addition to the stereocilia, all but mammalian auditory hair cells have a single kinocilium. The internal structure of the kinocilium is similar to that of motile cilia (Flock, 1965). The stereocilia are arranged according to height, with the tallest near the kinocilium if it is present. Vertebrate hair cells do not possess axons, and hence they must form synapses with secondary afferent nerve fibers. In addition, afferent fibers synapse on many types of hair cells and can have large effects on their response properties.

29.2.1 Transduction

The transduction process has long been thought to involve a change in conductance in the apical hair-cell membrane induced by displacement of the stereocilia (Davis, 1965). Recent intracellular measurements in hair cells from the frog sacculus have provided direct evidence for this hypothesis (Hudspeth and Corey, 1977). Displacement of the stereocilia toward the kinocilium results in an increase in conductance. The response persists even if the kinocilium is removed (Hudspeth and Jacobs, 1979). The change in conductance is probably due to the fact that the displacement increases the probability that the transduction channels are open. Corey (1980) proposed a model for the transduction channel that results in the following equation relating the probability of a channel being open as a function of displacement:

$$P_t = \frac{1}{1 + e^{(G_2 - Z_2 D)/RT}(1 + e^{(G_1 - Z_1 D)/RT})} \tag{29.1}$$

where P_t is the probability that a transduction channel is open, D is displacement in micrometers, G_1 *is* 0.6 kcal/mol, G_2 is 0.3 kcal/mol, Z_1 is 10 kcal/(mol · μm), Z_2 is 3 kcal/(mol · μm), and $RT = 0.586$ at 22°C. This equation is plotted in Fig. 29.3. The dynamic range of the transduction process spans less than a micrometer and is highly asymmetrical. This asymmetry may be very important in auditory hair cells, as they must be able to respond to frequencies which are very high compared with the time constant of the hair cell. The rectification inherent in the transduction process, coupled with the low-pass filtering of the cell membrane, provides the cell with a means of envelope detection.

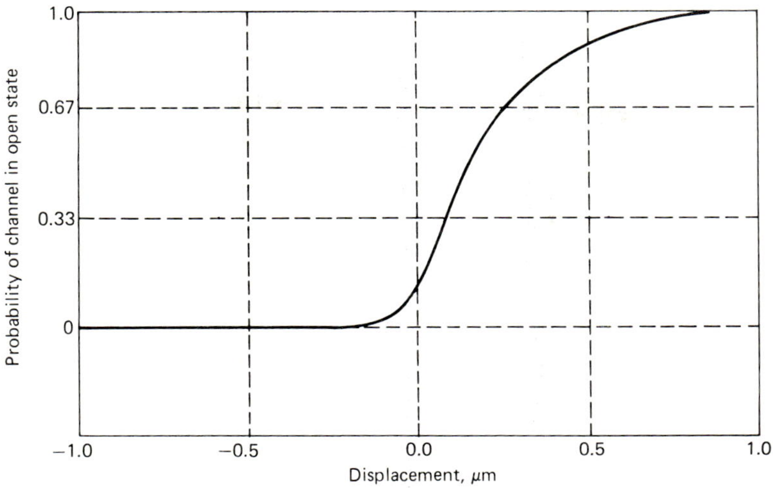

FIG. 29.3 The probability of a hair-cell transducer channel being open as a function of stereocilia displacement as predicted by the model of Corey (1980).

The model from which Eq. (29.1) was derived assumes that the transduction channels can exist in any one of three states. Energy barriers to transitions limit the rate of passage between states. The model can be summarized as follows:

$$A \underset{K_{ba}}{\overset{K_{ab}}{\rightleftharpoons}} B \underset{K_{cb}}{\overset{K_{bc}}{\rightleftharpoons}} C \tag{29.2}$$

where the rate constants (K_{ab}, etc.) are of the form

$$K = \frac{kT}{h} e^{-G/RT} \tag{29.3}$$

where G is the energy difference between the peak of the barrier and the present energy state. In states A and B the channel is closed, and in state C the channel is open. Displacement of the stereocilia results in changes in the rate constants, as a result of changes in the energy of states B and C with respect to state A and changes in the barrier heights. Displacements in the excitatory direction favor transitions toward state C, thereby increasing the probability that the channel is open. Corey chose a three-state model over a two-state model in order to fit the time course of the change in conductance to step displacement.

Corey and Hudspeth (1979b) measured the response latency of the transepithelial receptor potential (microphonic) to step displacements of the otolithic membrane, which moves the hair bundles in the frog sacculus in vitro. They found a response latency of 40 μs at 22°C. At 37°C they reported a value of 13 μs. This extremely fast response is of great importance to auditory receptors, because a number of mammals have high-frequency hearing which extends to 100 kHz or beyond.

Corey and Hudspeth (1979a) have also measured the ionic selectivity of the transduction channel. They found that it was relatively nonspecific, allowing the passage of a variety of cations including Ca^{2+}, K^+, and Na^+. Vertebrate hair cells invariably have their apical surface facing a fluid which is high in K^+ (Smith et al., 1954; Russell and Sellick, 1976; Peterson et al., 1978), so K^+ is probably the ion which carries the majority of the receptor current. On the basis of the size of the ions which pass through the channel, they suggested a diameter of 6.5 μm for the transduction channel.

The current-voltage (IV) relationship for the hair-cell membrane has also been studied in the in vitro preparation of the frog sacculus and is found to be very nonlinear (Corey and Hudspeth,

1979a). The slope resistance is 200 to 300 MΩ at −60 mV, but only about 5 MΩ for potentials more positive than −50 mV. The nonlinear *IV* curve means that the cell can have a relatively large standing current flowing from apex to base which is modulated by small charges in the transduction channels, and yet the cell can still have a high incremental membrane resistance. The high resistance is very important for a sensory cell because it creates large potential changes in response to small charges in the current through the transduction channels.

29.2.2 Auditory Hair Cells

The hair cells in auditory organs are of special interest, as they must respond to high-frequency stimuli. These hair cells are generally situated on a membrane which vibrates in response to sound pressure in the surrounding fluids. The cilia of these cells may be either free or imbedded in an overlying membrane called the *tectorial membrane*. Among the lower vertebrates, the auditory hair cells of the turtle and the alligator lizard have been most extensively studied. Fettiplace and Crawford (1978) have reported that the cochlear hair cells of the turtle exhibit sharp frequency selectivity to sound. Recordings have been obtained from cells with best responses to frequencies ranging between 70 and 670 Hz (Crawford and Fettiplace, 1980). When small steps of current are injected into these cells, the membrane potential exhibits damped oscillation with a frequency identical to the best frequency determined with sound (Fettiplace and Crawford, 1978). The ringing response to current steps could be abolished by hyperpolarization of the cell to −70 to −80 mV (Fettiplace and Crawford, 1978) or by treatment with tetraethylammonium, or TEA (Crawford and Fettiplace, 1981b), suggesting that a K^+ current plays a role in the ringing response. Crawford and Fettiplace (1981a) observed that the steady-state conductance of the cell showed a large increase with depolarization.

The resting potential of the cell has a dramatic effect on the tuning characteristics of the turtle hair cell. Crawford and Fettiplace (1981a) found that hyperpolarizing current caused a dramatic decrease in the resonant frequency and that depolarizing current caused a moderate increase in resonant frequency. The asymmetry with current polarity could be attributed in part to the nonlinear membrane properties mentioned previously. The settling time for the ringing response is decreased for depolarizing current and increased for hyperpolarizing current. From these data they calculated the quality factor (Q) for the tuning by assuming a second-order system. The Q showed a slight increase for small hyperpolarizing current and a dramatic decrease for larger hyperpolarizations. There was also a less dramatic decline in Q for depolarizing currents.

Crawford and Fettiplace (1981a) have proposed a simple linear electric circuit model of the cell membrane to account for the electrical tuning they have observed. The circuit model includes an inductive branch that represents an outward current which is activated by depolarization after a delay. Although this model can generate step responses with appropriate frequency and settling time, it cannot replicate the dependence of these variables on the injected current if only one circuit element is allowed to vary with current. Crawford and Fettiplace (1981a) have pointed out, however, that the presence of a voltage-dependent K^+ channel is probably not sufficient to produce such an underdamped impedance as is observed in the turtle hair cells. It seems likely that additional processes are involved, such as voltage-dependent Ca^{2+} channels, since Ca^{2+} plays an important role in saccular hair cells and in some electroreceptors. A tuned regenerative process would provide amplification as well as filtering of the stimulus.

Weiss et al. (1974) have demonstrated that the auditory hair cells of the alligator lizard basilar papilla also act as tuned filters of the acoustic stimulus. Direct mechanical measurements have been made of the vibration of the membrane upon which the hair cells are located (Weiss et al., 1978). The membrane shows much broader tuning than that observed in the hair cells and in the afferent nerve fibers. In addition, the hair cells have a range of best frequencies which depends on their location on the membrane, yet all regions of the basilar membrane have the same mechanical tuning. In the alligator lizard, there is no evidence at present for the electrical membrane tuning which is seen in the turtle. Recent optical measurements of the movement of the hair bundles in the alligator lizard suggest that the stereocilia act as mechanical resonators (Frishkopf et al., 1982; Holton and Hudspeth, 1982). The hair cells in the low-frequency portion of the organ have longer stereocilia than the cells in the high-frequency region; this would account for the differing resonant frequencies. It should be noted that these hair cells are located in a region of the basilar papilla which has no tectorial membrane over them. This is in contrast to the turtle, which has a tectorial membrane over all the hair cells. If the stereocilia of a hair cell are imbedded in a tectorial membrane, then their movement will be determined by the relative movement of the tectorial and basilar membranes, thus preventing the stereocilia from acting as independent resonators.

The hair cells of the mammalian cochlea also exhibit sharp frequency tuning (Russell and Sellick, 1978). The basilar membrane of mammals differs from that of the alligator lizard in that it exhibits frequency tuning which varies along its length, with the highest resonant frequencies at the basal end and the lowest frequencies at the apical end. The best frequency of the hair cells correlates well with the resonant frequency of the basilar membrane at the same location (Russell and Sellick, 1978, Sellick et al., 1982). Recent measurements of basilar membrane motion suggest that it is responsible for the sharply tuned response of the hair cells (Khanna and Leonard, 1982; Sellick et al., 1982).

In mammals there are two classes of hair cells, referred to as *inner hair cells* and *outer hair cells*. Most of the intracellular recording to date is from the inner hair cells. One striking feature of the receptor potential is that it is a strongly rectified version of the stimulus (Russell and Sellick, 1978). The inner hair cells have a membrane time constant of approximately 0.5 ms (Russell and Sellick, 1978), which means that for high-frequency acoustic stimuli the ac component of the receptor potential will be filtered out, leaving a dc component which follows the envelope of the stimulus. In inner hair cells this dc receptor potential is always in the depolarizing direction (Russell and Sellick, 1978), whereas in the outer hair cells it is hyperpolarizing for lower acoustic frequencies and depolarizing for frequencies near the best frequency (Dallos et al., 1982).

Recent experiments show that stimulation of the efferent nerve fibers which synapse on the outer hair cells alters certain nonlinearities in the mechanics of the cochlea (Mountain, 1980; Siegel and Kim, 1982). Additional experiments in which sinusoidal current was passed through the sensory epithelium suggest that the linear mechanical properties of the cochlea can be altered dynamically (Hubbard and Mountain, 1982; Mountain and Hubbard, 1982). If the hair cells alter their mechanical properties as a function of either voltage or current, then we have an interesting situation in which a receptor cell affects the stimulus it is transducing. Gold (1948) proposed a model in which the hair cells exert a force on the basilar membrane which is proportional to the velocity of the basilar membrane (see Neely and Kim, 1983, and Weiss, 1982, for related models). The force produced by the hair cells would offset the damping caused by the basilar membrane and associated structures vibrating in a viscous fluid. In this model the force is the result of a process mediated by the receptor potential, which Gold assumed was a function of the velocity of the basilar membrane. One difficulty with the Gold model is that hair cells are thought to be displacement transducers rather than velocity transducers. This difficulty can be overcome with a negative-feedback model if the force generated by the hair cells is a low-pass filtered version of displacement (Mountain, 1982; Mountain et al., 1983). For frequencies above the cutoff of the filter the force will be shifted 90° with respect to displacement and hence will be in phase with velocity. This low-pass filter could be due to the filtering of the receptor potential by the cell membrane, if the force is proportional to membrane potential, or could be inherent in the force-generating process, if the force is proportional to transducer current. It seems possible that the hair cells of the mammalian cochlea act as an active filter using electromechanical processes rather than the electrical processes seen in some of the other types of receptor cells.

29.2.3 Adaptation

In this chapter *adaptation* is taken to mean any change in receptor characteristics which takes place in response to stimulus conditions. One of the most interesting forms of adaptation takes place in the hair cells of the frog sacculus. If the cilia are subjected to a static displacement, the entire operating curve of Fig. 29.3 shifts horizontally in the direction of the stimulus but still maintains the same dynamic range (Eatcock et al., 1979). It is as though the mechanical linkage between the cilia and the transducer slips in response to steady displacements. The time course of this adaptation is the result of at least two time constants, one of which is in the range of 10 to 20 ms and the other is in the range of about 100 to 200 ms (Corey, 1980). The mechanism of the adaptation is unknown but appears to be dependent on intracellular Ca^{2+} (Corey, 1980).

Although shifts of the operating curve have not been reported in other types of hair cells, other changes of sensitivity have been observed. In auditory hair cells of the turtle as well as the guinea pig, background sound is known to reduce the sensitivity to a test tone (Crawford and Fettiplace, 1981b; Sellick and Russell, 1979). This phenomenon is usually referred to as *two-tone suppression,* as it was first observed in the response of auditory nerve fibers to stimuli consisting of two tones (Sachs and Kiang, 1968). Crawford and Fettiplace (1981b) have suggested that the suppression is due to a change in the active electrical filter in the turtle hair cell. Sellick and Russell (1979) have argued that in the guinea pig the suppression is related to changes in mechanics.

Changes in the frequency response of hair cells due to two-tone suppression have not been measured directly. Data are available, however, from mammalian auditory nerve fibers which indicate that the Q of the filter is decreased during suppression (Kiang and Moxon, 1974). This result is interesting because it means that background sounds will not only decrease the sensitivity of the receptor but also decrease the settling time of the response to transient stimuli. The trade-off between sensitivity and temporal resolution allows the receptor to follow more faithfully rapidly changing stimuli at high ambient sound levels.

There is an additional way in which the response properties of hair cells can be altered, i.e., by efferent neural control. Intracellular recording techniques have been used in both the turtle (Art et al., 1982) and the guinea pig (Brown et al., 1981) to study directly the effects of efferent stimulation on auditory hair cells. In the turtle, large hyperpolarizing potentials were observed in response to the efferent stimulation. In both the guinea pig and the turtle, the efferent stimulation resulted in a decrease in sensitivity to acoustic stimulation and a decrease in the Q. The effect of efferent stimulation in the turtle may be an alteration of the postulated electrical filter, but the mechanism of the efferent effect in the guinea pig may be somewhat more complicated. The efferent fibers in the mammalian cochlea synapse almost exclusively on the outer hair cells, but Brown et al. (1981) obtained their recording from inner hair cells. The likely explanation is that the efferent stimulation changed the mechanical properties of the outer hair cells, which in turn altered the vibrations of the basilar and tectorial membranes that provide the mechanical stimulus to the inner hair cells. Whether the efferent action on auditory hair cells is due to changes in mechanical or electrical filters, the result is similar to that of two-tone suppression: the sensitivity is decreased and the temporal resolution is increased.

29.3 PHOTORECEPTORS

Photoreceptors and chemoreceptors differ from other types of receptors in that the effective stimulus is quantal in nature. In the case of photoreceptors, the quanta are photons; in the case of chemoreceptors, the quanta are molecules or ions. In photoreceptors, photons trigger conformational changes in photopigment molecules, which in turn initiate a series of biochemical events which are poorly understood. The result is a change in ionic current flow into the receptor cell and a consequent alteration of the membrane potential. In vertebrates the change is a decrease in current flow which leads to a hyperpolarization of the receptor cell (Tomita, 1965, 1970). Like many of the receptor cells discussed earlier in this chapter, the photoreceptors have the ability to amplify the stimulus, but in this case the amplification is biochemical rather than electrical or mechanical in nature. Even though the amplification is biochemical, there is still electrical filtering of the receptor current by voltage-dependent channels in the photoreceptor membrane.

29.3.1 Anatomy

One of the most striking features of photoreceptor cells is the vast quantity of specialized membrane. In vertebrate photoreceptors, invagination of the plasma membrane results in a series of paired membranes which are parallel to the major axis of the cell. The receptor cells are divided into two segments by a severe constriction. The segment with the elaborate membrane structure is referred to as the *outer segment,* and the segment containing the nucleus and the mitochondria is referred to as the *inner segment.* The outer segment is in fact a modified cilium.

The receptor cells of the vertebrate retina are divided into two classes according to their morphology and physiology. Cells which have a cylindrical outer segment are called *rods,* and cells which have a tapered outer segment are called *cones.* Rods are involved in scotopic (dim-light, black-and-white) vision, and cones are involved in photopic (bright-light, color) vision. In rod outer segments the membrane invaginations pinch off to form disks that are physically separate from the plasma membrane. In cones, on the other hand, the invaginated membrane remains attached to the plasma membrane. The photopigment molecules are bound to these parallel membranes. In the case of rod outer segments, 85 percent of the membrane-bound protein is the photopigment rhodopsin (Cavanagh and Wald, 1969; Robinson et al., 1972; Heitzmann, 1972). The parallel membrane layers, with their bound rhodopsin molecules, lead to a high probability of the receptor's capturing a photon traveling along the axis of the cell.

29.3.2 Transduction

The transduction process is the result of a photoisomerization of the retinal chromophore group of rhodopsin. The photon absorption leads to a series of conformational changes in the rhodopsin and finally results in opsin and free retinal (Wald, 1968).

The isomerization of the photopigment must bring about a change in membrane potential, but in rods the pigment is localized in the disks which are internal structures, suggesting that an internal transmitter substance is needed to alter the membrane permeability in response to light. Baylor and Fuortes (1970) proposed such a model for cones as well, in order to explain the relation between light intensity and membrane conductance. The requirement for an internal transmitter gives the photoreceptor an additional possible means of signal amplification, since a single photoisomerization could result in the release or production of many internal transmitter particles, which in turn could interact with ionic channels in the plasma membrane.

When a brief flash of light is presented to a rod, the effect of light is to produce a hyperpolarization lasting several seconds, which is the result of an increase in membrane resistance (Toyoda et al., 1969). Sillman et al. (1969) showed that the light response of rods is very sensitive to the Na^+ and K^+ concentrations of the extracellular fluid. They proposed that light reduces the membrane conductance for Na^+.

The time course of the rod receptor current response to brief flashes or to single photon events can be fitted by a four-stage linear model (Baylor et al., 1974, 1979a, 1979b; Penn and Hagins, 1972). Each stage is a first-order, low-pass filter. These models are usually interpreted as representing a series of sequential chemical reactions, but one should keep in mind that the model is basically a curve-fitting procedure, and there are potentially many other models which could also produce the receptor current response.

Attempts have been made to estimate the number of membrane channels which are affected by the detection of a single photon. The estimates vary from 100 to 500, depending on the technique used.

The identity of the internal transmitter which causes the light-induced closing of the Na^+ channels is not known. Two substances which have been proposed for the internal transmitter are Ca^{2+} and cyclic guanosine monophosphate (cGMP) (for detailed reviews see Hubbell and Bownds, 1979; Pober and Bitensky, 1979; O'Brien, 1982). Increased intracellular Ca^{2+} blocks the Na^+ channels (Yoshikami and Hagins, 1973), whereas increased intracellular cGMP appears to open the channels (Nicol and Miller, 1978).

The difficulty in testing the various hypotheses mentioned above stems from the fact that Ca^{2+} and cGMP appear to interact with each other. It may be that phototransduction uses a chain of internal transmitters which exhibits not only feed forward but also feedback.

29.3.3 Adaptation

The voltage response to small flashes is reduced by the presence of diffuse background light (Fain, 1976; Bastian and Fain, 1979). This change in sensitivity (peak response divided by flash intensity) is called *adaptation* and can also be observed in the receptor current (Baylor et al., 1979a). Adaptation also results in a faster transient response, with both the rise time and the duration becoming shorter in the adapted state. The incremental sensitivity (S_F) to a flash can be related to the background intensity (I_B) by

$$S_F = \frac{S_0}{1 + I_B/I_0} \tag{29.4}$$

where S_0 is the flash sensitivity in the dark and I_0 is the background intensity which is necessary to halve the incremental sensitivity.

Several lines of evidence suggest that a diffusible substance is responsible for the adaptation effect. The adaptation process is very sensitive even to low levels of background light. A background intensity which is sufficient to bleach rhodopsin molecules in only 2 percent of the disks in a rod will reduce the sensitivity of that rod by 50 percent (Bastian and Fain, 1981). For only 2 percent of the disks to have such an impact means that they must alter the response to photons absorbed by other disks. The nature of the diffusible substance involved in adaptation is unknown, as is its site of action.

29.3.4 Voltage-Dependent Channels

While the photoreceptor outer segments are specialized for transduction, the inner segment appears to be specialized for filtering the receptor current.

Bader et al. (1982) have identified five separate currents in rod inner segments. Three of these currents are voltage-dependent, and two are Ca^{2+}-dependent. The current which has the most striking effect on the light response is activated by hyperpolarizing the membrane beyond −30 mV and is blocked by Cs^+ in the extracellular fluid. Although this current builds up with a time constant on the order of 200 ms, it has a major effect on the light response, since the response to even a brief flash can last for several seconds. This channel appears to be permeable to both Na^+ and K^+ ions, resulting in a reversal potential of −32 mV for this current. The reversal potential is near the resting potential for rods, so it has little effect on the low-intensity light response but has a dramatic effect on the high-intensity light response. It is responsible for a rapid decline in the light response at high intensities.

For several reasons, Bader et al. (1982) have not attempted to construct a detailed model of the light response of rods in the intact retina. One of the greatest difficulties is in characterizing the Ca^{2+}-dependent currents, because the intracellular concentration of free Ca^{2+} is not known. They also point out that other authors have found light-induced changes in the extracellular K^+ concentrations which will affect the voltage- and Ca^{2+}-dependent K^+ currents. Modeling of the light response is further complicated by the electrical coupling between photoreceptors, which will be discussed next.

29.3.5 Electrical Coupling

Interactions between photoreceptors were discovered when Baylor et al. (1971) found that the light response of turtle cones was influenced by the illumination of other cones up to 70 μm away. Electrical coupling between adjacent photoreceptors has now been found in both rods and cones in many species and appears to be mediated by gap junctions between receptor cells (see Gold, 1981, for a review).

The network of electrically coupled rods has been modeled by an equivalent electrical circuit in which each rod is modeled as a resistor representing the membrane resistance and four resistors representing the electrical coupling to the interiors of four adjacent rods. This type of model assumes that the membrane capacitance is insignificant and that the extracellular space is isopotential. This type of linear network also neglects the effect of the voltage-dependent membrane conductances discussed earlier. Detwiler et al. (1978) have suggested adding an inductor in parallel with the membrane resistance in each cell to better model the experimental data. Attwell and Wilson (1980) improved on this model by including a current in the membrane model for the rod which was time- and voltage-dependent.

The effect of the voltage-dependent current on the electrical network is to decrease the amplitude and the time to peak of the light response of cells which are distant from the stimulus. The network may act to integrate transient signals spatially over a large area while at the same time allowing signals of longer duration to be processed with greater spacial resolution.

SUMMARY

The discussions in this chapter indicate that sensory receptor cells should not be considered merely passive transducers of the stimulus energy. Instead, many sensory cells amplify and filter the sensory signal. The amplification process can be electrical, biochemical, and perhaps even mechanical in nature. The gain and operating point of the transduction process appear to be able to adapt to the stimulus conditions. Filtering of the sensory signal may be the result of the transduction process itself, or it may be the result of voltage- and ion-dependent channels in the cell membrane. In some cases these membrane channels interact to form highly tuned electrical filters which may be matched to biologically relevant signals. In the mammalian cochlea the major filtering appears to be the result of associated mechanical structures, but the cochlear hair cells may be able to influence the tuning. The filtering process also may adapt to stimulus conditions, often showing a more rapid transient response at high stimulus levels than at low levels. This probably represents a trade-off between the need for temporal integration at low stimulus levels and that for fine temporal resolution of the stimulus signal.

The complexity of these processes in a single receptor cell is so great that detailed mathematical models will probably be needed in most cases to completely describe receptor cell function. The modeling of these processes is being hampered by the difficulty in obtaining the necessary experimental data. This is especially true in cases where the intracellular concentrations and/or activities of ions and molecules play an important role in receptor physiology.

REFERENCES

Art, J. J., Crawford, A. C., Fettiplace, R., and Fuchs, P. A. (1982). Efferent regulation of hair cells in the turtle cochlea. *Proc. R. Soc. B., 216,* 377–384.

Attwell, D., and Wilson, M. (1980). Behavior of the rod network in the tiger salamander retina mediated by membrane properties of individual rods. *J. Physiol. (London), 309,* 287–316.

Bader, C. R., Bertrand, D., and Schwartz, E. A. (1982). Voltage-activated and calcium-activated currents studied in solitary rod inner segments from the salamander retina. *J. Physiol. (London), 331,* 253–284.

Bastian, B. L., and Fain, G. L. (1979). Light adaptation in toad rods: Requirement for an internal messenger which is not calcium. *J. Physiol. (London), 297,* 493–520.

Bastian, B. L., and Fain, G. L. (1981). Calcium and the mechanism of light adaptation in rods. In F. Bonner and A. Kleinzeller (eds.), and W. H. Miller (guest ed.), *Current Topics in Membranes and Transport,* vol. 15: *Molecular Mechanisms of Photoreceptor Transduction.* Academic Press, New York, pp. 341–368.

Baylor, D. A., and Fuortes, M. G. F. (1970). Electrical responses of single cones in the retina of the turtle. *J. Physiol. (London), 207,* 493–520.

Baylor, D. A., Fuortes, M. G. F., and O'Bryan, P. M. (1971). Receptive fields of single cones in the retina of the turtle. *J. Physiol. (London), 214,* 265–294.

Baylor, D. A., Hodgkin, A. L., and Lamb, T. D. (1974). The electrical response of turtle cones to flashes and steps of light. *J. Physiol. (London), 242,* 685–727.

Baylor, D. A., Lamb, T. D., and Yau, K-W. (1979a). The membrane current of single rod outer segments. *J. Physiol. (London), 288,* 589–611.

Baylor, D. A., Lamb, T. D., and Yau, K-W. (1979b). Responses of retinal rods to single photons. *J. Physiol. (London), 288,* 613–634.

Bennett, M. V. L. (1967). Mechanisms of electroreception. In P. Cahn (ed.), *Lateral Line Detectors.* University of Indiana Press, Bloomington, Ind., pp. 313–393.

Bennett, M. V. L. (1971). Electroreception. In W. S. Hoar and D. S. Randall (eds.), *Fish Physiology.* Academic Press, New York, pp. 493–574.

Bennett, M. V. L., and Clusin, W. T. (1979). Transduction at Electroreceptors: Origins of Sensitivity. *J. Gen. Physiol., 73,* 91–116.

Brown, M. C., Nuttal, A. L., and Masta, R. I. (1981). Effects of crossed olivocochlear bundle stimulation on the intracellular dc receptor potential from cochlear inner hair cells. *J. Acoust. Soc. Am., 69,* S52.

Cavanagh, H. D., and Wald, G. (1969). Purification and partial characterization of bovine visual pigment protein. *Fed. Proc., 28,* 344.

Clusin, W. T., and Bennett, M. V. L. (1977a). Calcium-activated conductance in skate electroreceptors: Current clamp experiments. *J. Gen. Physiol., 69,* 121–143.

Clusin, W. T., and Bennett, M. V. L. (1977b). Calcium-activated conductance in skate electroreceptors: Voltage clamp experiments. *J. Gen. Physiol., 69,* 145–182.

Clusin, W. T., and Bennett, M. V. L. (1979a). The oscillatory responses of skate electroreceptors to small voltage stimuli. *J. Gen. Physiol., 73,* 685–702.

Clusin, W. T., and Bennett, M. V. L. (1979b). The ionic basis oscillatory responses of skate electroreceptors. *J. Gen. Physiol., 73,* 703–723.

Corey, D. P. (1980). A biophysical approach to sensory transduction by vertebrate hair cells. Ph.D. thesis, California Institute of Technology, Pasadena, Calif.

Corey, D. P., and Hudspeth, A. J. (1979a). Ionic basis of the receptor potential in a vertebrate hair cell. *Nature (London), 281,* 675–677.

Corey, D. P., and Hudspeth, A. J. (1979b). Response latency of vertebrate hair cells. *Biophys. J., 26,* 499–506.

Crawford, A. C., and Fettiplace, R. (1980). The frequency selectivity of auditory nerve fibres and hair cells in the cochlea of the turtle. *J. Physiol. (London), 306,* 79–125.

Crawford, A. C., and Fettiplace, R. (1981a). An electrical tuning mechanism in turtle cochlear hair cells. *J. Physiol. (London), 312,* 377–422.

Crawford, A. C., and Fettiplace, R. (1981b). Nonlinearities in the responses of turtle hair cells. *J. Physiol. (London), 315,* 317–338.

Dallos, P., Santos-Sacchi, J., and Flock, Å. (1982). Intracellular recordings from cochlear outer hair cells. *Science, 218,* 582–584.

Davis, H. (1965). A model for transducer action in the cochlea. *Cold Spring Harbor Symp. Quant. Biol., 30,* 181–190.

DeRosier, D. J., Tilney, L. G., and Egelman, E. (1980). Actin in the inner ear: The remarkable structure of the stereocilium. *Nature (London), 287,* 291–296.

Detwiler, P. B., Hodgkin, A. L., and McNaughton, P. A. (1978). A surprising property of electrical spread in the network of rods in the turtle's retina. *Nature (London), 274,* 562–565.

Eatcock, R. A., Corey, D. J., and Hudspeth, A. J. (1979). Adaptation in a vertebrate hair cell: Stimulus-induced shift of the operating range. *Soc. Neurosci. Abstr., 5,* 19.

Fain, G. L. (1976). Sensitivity of toad rods: Dependence on wavelength and background illumination. *J. Physiol. (London), 261,* 71–101.

Fettiplace, R., and Crawford, A. C. (1978). The coding of sound pressure and frequency in cochlear hair cells of the terrapin. *Proc. R. Soc. B., 203,* 209–218.

Flock, Å. (1965). Electron microscopic and electrophysiological studies on the lateral-line organ. *Acta Oto-Laryngol. Suppl., 199,* 1–90.

Flock, Å., and Cheung, H. C. (1977). Actin filaments in sensory hairs of inner ear receptor cells. *J. Cell Biol., 75,* 339–343.

Frishkopf, L. S., DeRosier, D. J., and Egelman, E. H. (1982). Motion of basilar papilla and hair cell stereocilia in the excised cochlea of the alligator lizard: Relation to frequency analysis. *Soc. Neurosci. Abstr., 8,* 40.

Gold, G. H. (1981). Photoreceptor coupling: Its mechanism and consequences. In F. Bonner and A. Kleinzeller (eds.), and W. H. Miller (guest ed.), *Current Topics in Membranes and Transport,* vol. 15; *Molecular Mechanisms of Photoreceptor Transduction.* Academic Press, New York, pp. 59–89.

Gold, G. H., and Korenbrot, J. I. (1980). Light-induced calcium release by intact retinal rods. *Proc. Nat. Acad. Sci. U.S.A., 77,* 5557–5561.

Gold, T. (1948). Hearing: II. The physical basis of the action of the cochlea. *Proc. R. Soc. B., 135,* 492–498.

Heitzmann, H. (1972). Rhodopsin is the predominant protein of rod outer segment membranes. *Nature (New Biol.), 235,* 114.

Hodgkin, A. L., and Huxley, A. F. (1952). A quantitative description of membrane current and its application to conduction and excitation in nerve. *J. Physiol. (London), 117,* 500–544.

Holton, T., and Hudspeth, A. J. (1982). Motion of hair-cell stereocilia in the auditory receptor organ of the alligator lizard. *Soc. Neurosci. Abstr., 8,* 40.

Hopkins, C. D. (1976). Stimulus filtering and electroreception: Tuberous electroreceptors in three species of gymnotoid fish. *J. Comp. Physiol., 111,* 171–207.

Hubbard, A. E., and Mountain, D. C. (1982). Injection of AC current into scala media alters the sound pressure at the tympanic membrane: Variations with acoustic stimulus parameters. *J. Acoust. Soc. Am., 71,* S100.

Hubbell, W. L., and Bownds, M. D. (1979). Visual transduction in vertebrate photoreceptors. *Annu. Rev. Neurosci., 2,* 17–34.

Hudspeth, A. J., and Corey, D. P. (1977). Sensitivity, polarity, and conductance change in the response of vertebrate hair cells to controlled mechanical stimuli. *Proc. Nat. Acad. Sci. U.S.A., 74,* 2407–2411.

Hudspeth, A. J., and Jacobs, R. (1979). Stereocilia mediate transduction in vertebrate hair cells. *Proc. Nat. Acad. Sci. U.S.A., 76,* 1506–1509.

Kalmijn, A. J. (1974). The detection of electric fields from inanimate and animate sources other than electric organs. In A. Fessard (ed.), *Handbook of Sensory Physiology.* Springer-Verlag, New York, vol. III, pt. 3, pp. 145–200.

Kalmijn, A. J. (1982). Electric and magnetic field detection in elasmobranch fishes. *Science, 218,* 916–918.

Khanna, S. M., and Leonard, D. G. B. (1982). Basilar membrane tuning in the cat cochlea. *Science, 215,* 305–306.

Kiang, N. Y. S., and Moxon, E. C. (1974). Tails of tuning curves of auditory-nerve fibers. *J. Acoust. Soc. Am., 55,* 620–630.

Mountain, D. C. (1980). Changes in endolymphatic potential and crossed olivocochlear bundle stimulation alter cochlear mechanics. *Science, 210,* 71–72.

Mountain, D. C. (1982). A negative feedback role for outer hair cells in cochlear mechanics. In *Abstracts of the Fifth Midwinter Research Meeting, Association for Research in Otolaryngology.* St. Petersburg Beach, Fla.

Mountain, D. C., and Hubbard, A. E. (1982). Injection of AC current into scala media produces an ear canal emission of cochlear origin. *J. Acoust. Soc. Am., 71,* S100.

Mountain, D. C., Hubbard, A. E., and McMullen, T. A. (1983). Electromechanical processes in the cochlea. In E. de Boer and M. A. Viergever (eds.), *Mechanics of Hearing.* Delft University Press, Delft, The Netherlands.

Murray, R. W. (1962). The response of ampullae of Lorenzini of elasmobranchs to electrical stimulation. *J. Exp. Biol., 39,* 119–128.

Murray, R. W. (1965). Receptor mechanisms in the ampulla of Lorenzini of elasmobranch fishes. *Cold Spring Harbor Symp. Quant. Biol., 30,* 233–243.

Murray, R. W. (1967). The function of the ampullae of Lorenzini of elasmobranchs. In P. Cahn (ed)., *Lateral Line Detectors.* University of Indiana Press, Bloomington, Ind., pp. 277–293.

Neely, S. T., and Kim, D. O. (1983). An active cochlear model showing sharp tuning and high sensitivity. *Hearing Res., 9,* 123–130.

Nicol, G. D., and Miller, W. H. (1978). Cyclic GMP injected into retinal rod outer segments increases latency and amplitude of response to illumination. *Proc. Nat. Acad. Sci. U.S.A., 75,* 5217–5220.

Obara, S. (1976). Mechanism of electroreception in ampullae of Lorenzini of the marine catfish *Plotosus.* In J. P. Reuben, D. P. Pupura, M. V. L. Bennett, and E. R. Kandel (eds.), *Electrobiology of Nerve, Synapse, and Muscle.* Raven Press, New York, pp. 129–147.

Obara, S., and Bennett, M. V. L. (1972). The mode of operation of ampullae of Lorenzini of the skate *Raja. J. Gen. Physiol., 60,* 534–557.

O'Brien, D. F. (1982). The chemistry of vision. *Science, 218,* 961–966.

Penn, R. D., and Hagins, W. A. (1972). Kinetics of the photocurrent of retinal rods. *Biophys. J., 12,* 1073–1094.

Peterson, S. K., Frishkof, L. S., Lechene, C., Oman, C. M., and Weiss, T. F. (1978). Element composition of inner ear lymphs in cats, lizards, and skates determined by electron probe microanalysis of liquid samples. *J. Comp. Physiol., 126,* 1–14.

Pober, J. S., and Bitensky, M. W. (1979). Light-regulated enzymes of vertebrate retinal rods. *Adv. Cyclic Nucleotide Res., 11,* 265–301.

Robinson, W. E., Gordon-Walker, A., and Bownds, D. (1972). Molecular weight of frog rhodopsin. *Nature (New Biol.), 235,* 112—114.

Russell, I. J., and Sellick, P. M. (1976). Measurement of potassium and chloride ion concentrations in the cupulae of the lateral lines of *Xenopus laevis. J. Physiol. (London), 257,* 245–255.

Russell, I. J., and Sellick, P. M. (1978). Intracellular studies of hair cells in the mammalian cochlea. *J. Physiol. (London), 284,* 261–290.

Sachs, M. B., and Kiang, N.Y.-S. (1968). Two-tone inhibition in auditory nerve fibers. *J. Acoust. Soc. Amer., 43,* 1120–1128.

Sellick, P. M., and Russell, I. J. (1979). Two-tone suppression in cochlear hair cells. *Hearing Res., 1,* 227–236.

Sellick, P. M., Patuzzi, R., and Johnstone, B. M. (1982). Measurement of basilar membrane motion in the guinea pig using the Mossbauer technique. *J. Acoust. Soc. Am., 72,* 131–141.

Siegel, J. H., and Kim, D. O. (1982). Efferent neural control of cochlear mechanics? Olivocochlear bundle affects cochlear biomechanical nonlinearity. *Hearing Res., 6,* 171–182.

Sillman, A. J., Ito, H., and Tomita, T. (1969). Studies on the mass receptor potential of the isolated frog retina: II. On the basis of the ionic mechanism. *Vision Res., 9,* 1443–1451.

Smith, C. A., Lowry, O. H., and Wu, M. L. (1954). The electrolytes of the labyrinthine fluids. *Laryngoscope, St. Louis, 64,* 141–153.

Suga, N. (1967). Coding in tuberous and ampullary organs of a gymnoid electric fish. *J. Comp. Neurol., 131,* 437–451.

Tilney, L. G., DeRosier, D. J., and Mulroy, M. J. (1980). The organization of actin filaments in the stereocilia of cochlear hair cells. *J. Cell. Biol., 86,* 244–259.

Tomita, T. (1965). Electrophysiological study of the mechanisms subserving color coding in the fish retina. *Cold Spring Harbor Symp. Quant. Biol., 30,* 559–566.

Tomita, T. (1970). Electrical activity of vertebrate photoreceptors. *Q. Rev. Biophys., 3,* 179–222.

Toyoda, J., Nosaki, H., and Tomita, T. (1969). Light-induced resistance changes in single photoreceptors of *Necturus Gekko. Vision Res., 9,* 453–463.

Wald, G. (1968). The molecular basis of visual excitation. *Nature (London), 219,* 800–807.

Waltman, B. (1966). Electrical properties and fine structure of the ampullary canals of Lorenzini, *Acta Physiologica Scandinavia, 66,* Supplement 264, 1–60.

Weiss, T. F. (1982). Bidirectional transduction in vertebrate hair cells: A mechanism for coupling mechanical and electrical processes. *Hearing Res., 7,* 353–360.

Weiss, T. F., Mulroy, M. J., and Altmann, D. W. (1974). Intracellular responses to acoustic clicks in the inner ear of the alligator lizard. *J. Acoust. Soc. Am., 55,* 606–621.

Weiss, T. F., Peake, W. T., Ling, A., and Holton, T. (1978). Which structures determine frequency selectivity and tonotopic organization of vertebrate cochlear nerve fibers? Evidence from the alligator lizard. In R. F. Naunton and C. Fernandez (eds.), *Evoked Electrical Activity in the Auditory Nervous System.* Academic Press, New York, pp. 91–112.

Yoshikami, S., and Hagins, W. A. (1973). Control of the dark current in vertebrate rods and cones. In H. Langer (ed.), *Biochemistry of Visual Pigments.* Springer-Verlag, New York, pp. 245–255.

CHAPTER 30

Cochlear Mechanics

Charles R. Steele
Applied Mechanics Division
Stanford University
Stanford, California

30.1 GENERAL FEATURES

The human ear is shown in Fig. 30.1. The function of the auditory periphery, the outer, middle, and inner ears, is to transform the pressure fluctuations in the air into the appropriate neural excitation, which is then transmitted to the central nervous system. Yost and Nielson [1] provide a good introductory survey of this system. As more knowledge of its complexities is gained, fewer investigators in the field speak of a simple "theory of hearing." However, some basic features of the periphery can be described.

30.1.1 The Cochlea and Vestibular Apparatus

Independent of the mechanical periphery, nerves can apparently transmit information to the brain, resulting in perception of frequency only in the range of 20 to 300 Hz. For lower frequencies, the information is lost in the spontaneous, random firing of the nerve, while for higher frequencies, the nerve is unable to regain its electrochemical charge quickly enough to fire at an appropriate rate. For this reason, the inner ear consists of two "mechanical" devices: the coiled cochlea, and the utricle, saccule, and semicircular canals of the vestibular apparatus. The latter provides the proper neural stimulation for low frequencies associated with body motion, while the cochlea provides proper neural stimulation for high frequencies.

30.1.2 Neural Tuning

For moderate sound intensities, each of the 40,000 fibers in the cochlear nerve is activated only by a narrow band of frequency [1, 2]. Because of this, the auditory nerve fibers are said to be "tuned" and the cochlea may be considered as a real-time Fourier analyzer: input sound pressure, which is an arbitrary function of time, is decomposed into 40,000 channels of output, each giving a particular frequency component of the input. A comparison of cochear nerve

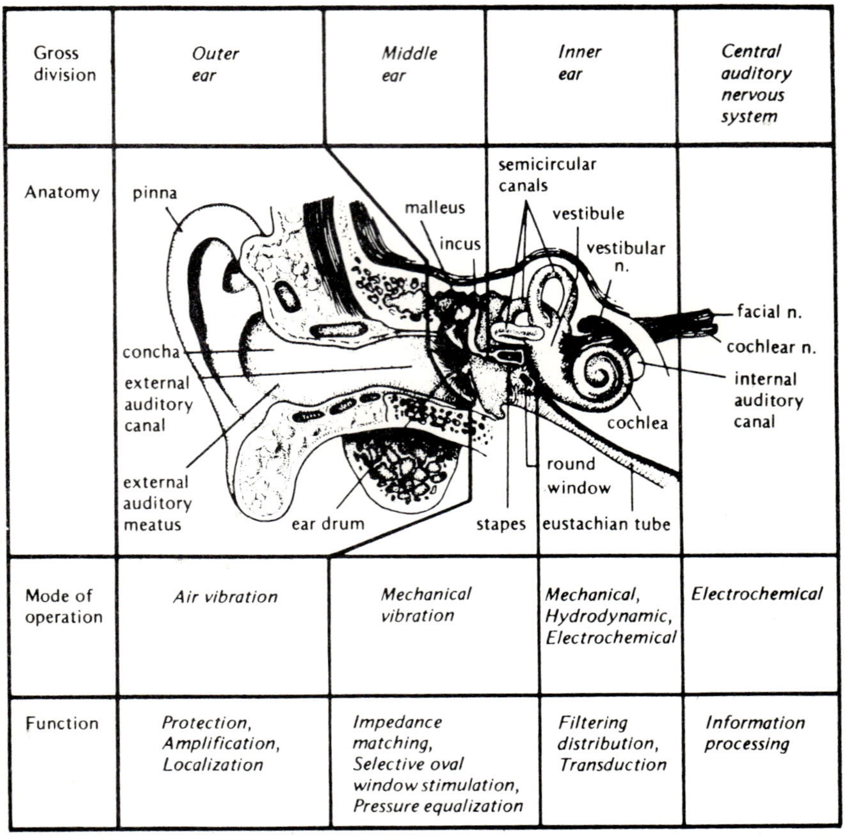

FIG. 30.1 Cross section of human ear, showing divisions into outer, middle, and inner ears. Below are listed the predominant mode of operation of each division and its suggested function. (*From Yost and Nielson [1].*)

behavior with the results of psychoacoustic tests of perception indicates also that the capacity for frequency discrimination is established in the cochlea; higher centers are apparently concerned with more subtle tasks of pattern recognition. The cochlea thus provides a vital first step in information processing.

From the time of the ancient Greeks, through Helmholtz [4] to the present, a discrete set of resonators has been sought to explain the frequency discrimination capability of the ear. Discrete resonators have the characteristic of requiring low damping for sharp tuning but high damping for good tracking of transients. In comparison, cochlear nerve fibers exhibit both sharp tuning and good time response.

30.1.3 Insect Ear Resonators

Discrete resonators are utilized by insects in a variety of ways. Insect ears may be simple antennae or elaborate tympanal membrane constructions, and may be located on the face, abdomen, leg, or wing [5, 6]. The hearing range, as indicated by the two insect entries in Table 30.1, is impressive and seems as broad as the mammalian hearing range. However, the neural

TABLE 30.1 General frequency range of hearing

Species	Frequency, kHz		
	Low	Optimum	High
Sphingid moth	3	20–50	100
Grasshoppers and locusts	0.1	5–10	50
Birds			7
Humans, probably most primates; elephant, camel mole, etc.	0.02	1–4	20
Some primates	....	....	40
Dog	....	....	35
Cat	....	1–10	50
Guinea pig	....	4–12	50
Seals and sea lions	0.3	1–30	60
Rodents	1	5–15, 30–60	100
Dolphin and other toothed whales	0.1	20–70	170
Bats, in general	1	30–100	200
R. ferrumequinum bat		83–86	

Source: Fleischer [8] and Sales and Pye [6].

receptor cells are in close proximity to the mechanical resonator, and the neural response generated seems to be roughly what might be expected from only a moderately damped resonator.

30.1.4 Vertebrate Middle Ear Resonators

The vertebrates retain the resonator (the middle ear in Fig. 30.1) but have developed the elaborate intermediate organ (the cochlea in Fig. 30.1). Even the primitive alligator lizard cochlea, similar in many respects to those generally found in lizards, snakes, and turtles, produces much sharper neural tuning than expected from a damped resonator [7]. Nevertheless, the middle ear resonance is important to establish the general frequency range.

The middle ear of all terrestrial mammals has a form similar to that of a human being (Fig. 30.1), which has two modes of resonance: one of lower frequency with a relatively large displacement of the malleus, and one of higher frequency with a relatively large displacement of the stapes [8]. In lizards and birds, the malleus and incus are absent and the stapes directly connects the eardrum to the cochlea. The resulting single mode of vibration explains their restricted frequency range (Table 30.1). Whales have no pinna or external auditory canal, since these are useless under water, but the eardrum is connected to a massive tympanic bone and the cochlea is embedded in a massive periotic bone. The bones of the middle ear are in an air space similar to that in terrestrial mammals [8]. The relative motion of tympanic and periotic bone in the whale, however, is a third mode of middle ear resonance which provides these creatures with an extended frequency range [8, 9]. The toothed whales, and dolphins and porpoises in particular, are specialists in high frequencies, up to 200 kHz, but they can also perceive low frequencies (20 Hz). Some baleen whales specialize in low frequencies (20 to 30 Hz), for which the resonance of the tympanic bone is essential [8].

The middle and outer ear have nothing to do with the sharpness of neural excitation in the cochlea. Loss of the outer ear seriously impairs perception of directionality, and complete absence of the middle ear causes a hearing loss of about 20 dB in humans. Since such a loss greatly impairs speech perception, surgery in the middle ear has developed to the extent that most middle ear defects can be repaired. The middle ear model of Zwislocki [10] is widely used to calculate the effects of surgical modification.

Even when the middle ear ossicles are completely immobilized (as with otosclerosis), hearing is possible by conduction of the external sound through the bone of the skull to the cochlea. In gnawing animals, echolocating bats, and sea mammals, the entire middle and inner ear complex is suspended from the skull by soft tissue to prevent bone-conducted noise [8].

30.1.5 Cochlear Anatomy

For middle ear vibration, the cochlea supplies primarily dissipation [11], which means that significant energy from the sound field is absorbed by the cochlea. In all vertebrates, the cochlea is designed to function effectively in the same frequency range as the middle ear [9]. Figure 30.2

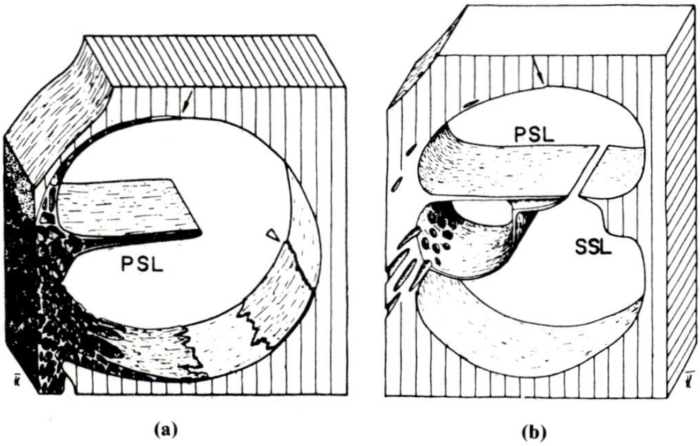

FIG. 30.2 Cross sections of the cochlear bone in the basal turn of (*a*) elephant (*Loxodonta africana*), an extreme form of a low-frequency cochlea, and (*b*) porpoise (*Phocaena phocaena*), an extreme form of a high-frequency cochlea. Great stiffness of the bony support structure is required for the high-frequency cochlea. PSL = primary spiral lamina; SSL = secondary spiral lamina (absent in low-frequency cochleas). (*From Fleischer [9].*)

shows the bone construction in the cochleas of the elephant and porpoise. The elephant is typical of plains, desert, and mountain animals for whom the low frequencies which propagate long distances are important. The porpoise is typical of the echolocating toothed whales and bats, as well as woods animals, who require the more detailed, local information provided by high frequencies. For some reason, humans and the other primates are in the low-frequency category.

The soft tissue is shown in Fig. 30.3. The basilar membrane (BM) is attached to the bony shelf [primary spiral lamina (PSL), on the left in Fig. 30.3] and to the spiral ligament (SL), which is anchored to the bony wall on the right. The basilar membrane and the cells and structure attached to it separate two fluid ducts—the scala media (SM) above and scala tympani (ST) below. Reissner's membrane (RM) separates the scala media from a third fluid duct, the scala vestibuli (SV). The scala vestibuli and scala tympani are connected by an opening at the apex of the cochlea, the helicotrema, and are filled with perilymph. The scala media is filled with endolymph. Perilymph is high in sodium and endolymph is high in potassium. With respect to perilymph, there is a potential of +80 mV in the scala media and −70 mV in the cells on the basilar membrane, an extremely high potential difference to be found in the body [1]. The viscosity and density of both perilymph and endolymph are about those of water.

Other fluid ducts are the spiral sulcus under the tectorial membrane (TM in Fig. 30.3) and the tunnel of Corti (C), which is bounded by the arches of Corti, which support the receptor cells. The three rows of outer hair cells (OHC) are to the right of the arches in Fig. 30.3, and the single row of inner hair cells (IHC) are to the left.

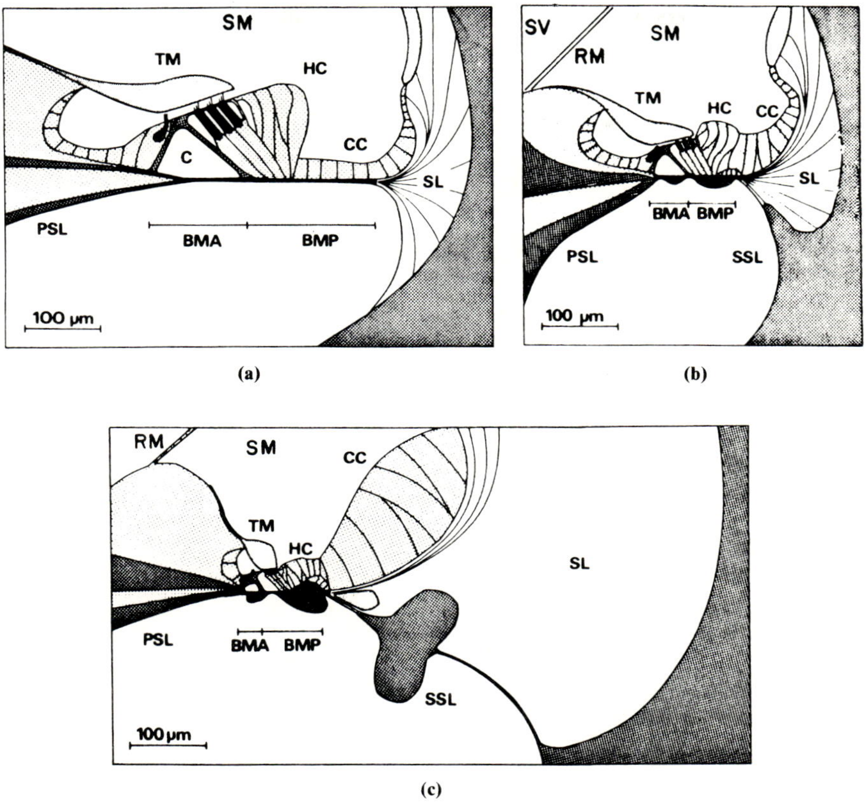

FIG. 30.3 Cross sections of the cochlea in four mammals, emphasizing the hydromechanically important structures. (*a*) Middle turn of the cochlea of a human. (*b*) Generalization of the basal turns of the cochlea of rat, mouse, and bats with a frequency-modulated echolocating signal. (*c*) Specialized basal region of the cochlea of the greater horseshoe bat, *Rhinolophus ferrumequinum*. BMA = basilar membrane, arcuate zone; BMP = basilar membrane, pectinate zone; C = tunnel of Corti; CC = Claudius's cells; HC = Hensen's cells; PSL = primary spiral lamina; SSL = secondary spiral lamina; SL = spiral ligament; TM = tectorial membrane; RM = Reissner's membrane. The BM and RM divide the cross section into three fluid ducts: SV = scala vestibuli; SM = scala media; ST = scala tympani. The receptor cells are shown in black and are just above C; three rows of outer hair cells (OHC) and one row of inner hair cells (IHC). (*From Bruns [14].*)

The human cochlea (Fig. 30.3*a*) is similar to that of the water buffalo [11], except that the tectorial membrane is parallel to the basilar membrane in the latter, and is probably typical of low-frequency cochleas. Figure 30.3*b* is typical of the high-frequency cochlea. The basilar membrane has two pronounced thickenings: the arcuate zone (BMA), under the arches, and the pectinate zone (BMP), extending from the foot of the outer pillar of the arch to the spiral ligament (SL). In the high-frequency cochlea (Fig. 30.3*b*), the bone of the primary spiral lamina (PSL) extends to the inner pillar (e.g., see Table 30.2*b*), while in the low-frequency cochlea (Fig. 30.3*a*), a soft tissue extension of the primary spiral lamina supports the inner pillar. The width of the pectinate zone of the basilar membrane is generally about twice that of the arcuate zone, so in water buffalo (Table 30.2*e*), the distance from the primary spiral lamina bone to the inner pillar is about equal to the total width of both zones of the basilar membrane. The pectinate zone of the basilar membrane is supported on the right by the spiral ligament, which is reinforced by a second bony shelf [secondary spiral lamina (SSL)] that is present in the high-

TABLE 30.2 Dimensions of basilar membrane*

	Distance from base, mm																				
	1	2	3	3.5	4	5	6	8	10	12	14	15	16	17	19	20	23	25	30	33	36
a. Cat [15]																					
Width, BMA + BMP, μm		150				200			230			260				360	430				
Thickness, BMP, μm		13				13			12			9				6	5				
Filament density, $\times 10^3/\mu$m		6				5.2			3.5			1.7				0.6	0.5				
b. Guinea pig [16]																					
Width, PSL to SL, μm	130	150	170	180	180	190			230			290		330							
Width, BMA + BMP, μm	130	150	170	180	180	190			220			250		260							
Thickness, BMP, μm				7	6	5			3			2		1							
c. Mouse (*Mus musculus*) [17]																					
Width, BMA + BMP, μm	100	120	130		150	155	158														
Width, BMP, μm	73	85	95		103	110	105														
Thickness, BMP, μm	15	12	9		6	3	2														
Thickness, BMA, μm	10	9	7		6	5	6														
d. Bat (*R. ferrumequinum*) [13]																					
Width, BMA + BMP, μm	90	103	108		83	90	87	78	82	108	127		150								
Thickness, BMP, μm	31	33	34		30	9	8	7	7	6	5		2								
Thickness, BMA, μm	12	13	13		10	8	7	7	7	6	5		1								
e. Water buffalo (*Anoa bubalis*) [12]																					
Width, PSL to SL, μm									210						350			430	500	550	600
Width, BMA, μm									50						65			73	83	90	105
Thickness, BMA, μm									4.9						4.0			3.3	2.7	2.4	2.0

* See text for explanation of abbreviations.

frequency cochlea (Figs. 30.2*b* and 30.3*b*) but absent in the low-frequency cochlea (Figs. 30.2*a* and 30.3*a*).

Peculiarities can always be found. Figure 30.3*c* shows the cross section in the first 4.5 mm of the greater horseshoe bat (*Rhinolophus ferrumequinum*). The basilar membrane thickenings are immense (Table 30.1*d*) and the secondary spiral lamina is thin but with a heart-shaped knob on the end. This bat uses a pure-tone echolocation signal near 83 kHz, and the cochlea is extraordinarily sensitive to this one frequency [6, 13, 14].

For most mammals the dimensions of the cochlear cross section vary smoothly from the base at the stapes to the apex of the coil, as shown by the basilar membrane measurements in Table 30.2 for the cat [15], guinea pig [15], mouse [17], and water buffalo [12]. The basilar membrane is narrower and thicker at the base than at the apex. Furthermore the support for the basilar membrane is much stiffer in the base. Thus the differences between low- and high-frequency cochleas (Fig. 30.3*a* and *b*) are to some extent found in every cochlea [9]. Even a human being has a small secondary spiral lamina near the base, and even the dolphin loses most of the massive secondary spiral lamina (Fig. 30.2*b*) toward the apex. It is clear from the comparison of the hearing range and the cochlear anatomy that a high stiffness of the partition (PSL, BM, SL, SSL) is associated with high-frequency reception, and a low stiffness is associated with low-frequency reception. There is overwhelming evidence [1, 2] that, in a given cochlea, the stimulation of the receptor cells follows the partition stiffness gradient, with the cells at the stiff basal portion responding to the high frequencies and those in the flexible apical portion responding to low frequencies. Most convincing are recordings made from spiral ganglion cells [19]. These are usually located in the modiolus, to the left in Fig. 30.2*a*, but for the whales are located in the duct formed in the primary spiral lamina (Fig. 30.2*b*).

Intracellular recordings from the inner hair cells [20] show the same sharpness of tuning as the ganglion cells and the auditory nerve fibers. The basilar membrane velocity also exhibits the same sharp tuning [21, 22]. It is clear then, that the mechanical (hydroelastic) response of the cochlea plays an important role in the tuning mechanism. Earlier measurements of basilar membrane vibration in postmortem preparations by stroboscopic illumination and in live preparations by stroboscopic illumination and in live preparations [23] by a radioactive source [24, 25], laser illumination [26], and capacitance probe [27] established the existence of traveling waves and a location of the maximum amplitude, which depends on frequency. Recent measurements [21, 22] show that sharp basilar membrane tuning does exist.

30.1.6 Basilar Membrane Microstructure

Systematic variation in the dimensions of all the tissue of the partition (Fig. 30.3) occurs, as can be seen from the detailed measurements on the water buffalo [12]. Hensen's cells (HC) and Claudius's cells (CC) seem to be extremely soft with no load-carrying capability. Tiedemann [12] suggests that the tectorial membrane is the fundamental resonator, but this is unlikely since the tectorial membrane is huge in some cochleas (pig, cow) and small in others (Fig. 30.3). The significant structure resisting a pressure difference between the scala media and scala tympani seems to be the basilar membrane.

The microstructure [15, 28] shows that the basilar membrane is an amorphous ground substance containing bundles of fibers. These apparently extend from the primary spiral lamina well into the spiral ligament. The diameter of the fibers is 9 to 12 nm, and the fibers may be a protein of the keratin-epidermin-myosin-fibrinogen group [28]. The bundles are in a single layer in the arcuate zone of the basilar membrane, but divide into two layers in the pectinate zone. The thickening of the arcuate zone seen in Fig. 30.3*b* and *c* seems to consist mainly of ground substance. The sandwich construction of the pectinate zone would give this section an increased bending stiffness. The reduction to a single layer at the edges of the pectinate zone, i.e., at the outer pillar and at the entrance to the spiral ligament, means that the pectinate zone has, to a rough approximation, hinged edges. Basilar membrane displacements with significant flexing only at the "hinges" are shown in micrographs [17].

Békésy, in work from 1924 to 1958 [23], made key experiments to determine the mechanical and electrical properties of the cochlea. Only recently has there been an attempt to clarify, quantify, and extend his measurements of the stiffness of the components of the cochlear partition. The need for this is demonstrated by the question of basilar membrane orthotropy. Békésy concluded from the circular impression under a point load that the basilar membrane was isotropic. But Voldrich [29] found that, within 20 min post mortem, the basilar membrane is highly orthotropic, while 24 h post mortem the behavior is as described in Ref. 23. The conclusion is that the basilar membrane consists of stiff radial fibers in a soft ground substance, so that the effective longitudinal stiffness is small in comparison with the radial stiffness. Stiffness is further increased by the fact that in the pectinate zone some fibers leave the outer pillar and join the lower layer, but then cross to the upper layer [15]. These provide sufficient transverse shear stiffness for the sandwich construction to increase the bending stiffness. Thus the microstructure and point-load tests [29] give a consistent model of the pectinate zone as a set of radial beams with negligible longitudinal stiffness. Only in the peculiar *R. ferrumequinum* cochlea (Fig. 30.3*c*) are there pectinate zone fibers in the longitudinal direction.

The first quantitative point-load stiffness measurements since those of Békésy [23] are given by Gummer et al. [30], and intracochlear pressure measurements [11, 31] have been made recently. The in vivo volume compliance, i.e., the volume displacement of the partition per unit length per unit pressure difference between the scala media and the scala tympani, is estimated [31] to be about 20 percent of that given in Ref. 23 for the first turn of the guinea pig (about 0.04 mm^4/N at 2 mm). A similar value is estimated by Lynch et al. [11] for the cat at 6 mm. The point-load values [30] would indicate a volume compliance somewhat larger than that given in Ref. 23, but at least the results from these rather diverse measuring techniques are within a factor of 10 of each other.

30.1.7 Organ of Corti

The structure attached to the basilar membrane has the obviously important task of transferring basilar membrane motion to the receptor cells. The outer hair cells have cilia which are attached to the tectorial membrane and are excited by a shearing force between this membrane and the hair cells. The cilia of the inner hair cells, to the left of the tunnel of Corti (C) in Fig. 30.3*a*, are not attached to the tectorial membrane and are probably excited by a flow of fluid under it. Approximately 90 percent of the afferent nerves synapse with the inner hair cells, so these are the primary receptors [1]. Evidence is accumulating that the outer hair cells may play some active motor function [22, 32].

Understanding of the basic motion of the complex is still lacking. Two possibilities are (1) that the organ of Corti is compliant and has little effect on basilar membrane motion, or (2) that the tectorial membrane restrains the arches and the arcuate zone of the basilar membrane, while the pectinate remains relatively unrestrained.

In support of the second possibility is the sandwich construction of the pectinate zone. The stiffness of the unrestrained basilar membrane in Fig. 30.3*b* would depend primarily on the hinge thickness, which seems about constant (1 to 2 μm) for most cochleas. (One exception is the free-tailed bat, *Molossus ater* [14].) The deliberate thickness change of the pectinate zone (Table 30.2) and careful construction [15] indicate that the quantitative value of bending stiffness is important. The arches also become well developed in the basal region [14]. An advantage of stiff constraint would be that pressure loading of the pectinate zone would be transferred to a shear force on the cilia of the outer hair cells.

In support of the first possibility are the point-load results in the range of normal for a cochlea without outer hair cells (owing to kanamycin treatment) [30]. Of course, the torsional stiffness of the arch-BMA triangle could offer a significant resistance to a point load.

30.2 COCHLEAR MODEL

Many models have been proposed for cochlear function. There is, however, little dispute that the basilar membrane stiffness gradient and fluid flow play a significant role. Much effort has thus been expended on simplified models for the hydroelastic behavior. A typical model is shown in Fig. 30.4; it is similar to experimental models that have been studied [23, 33–36]. The

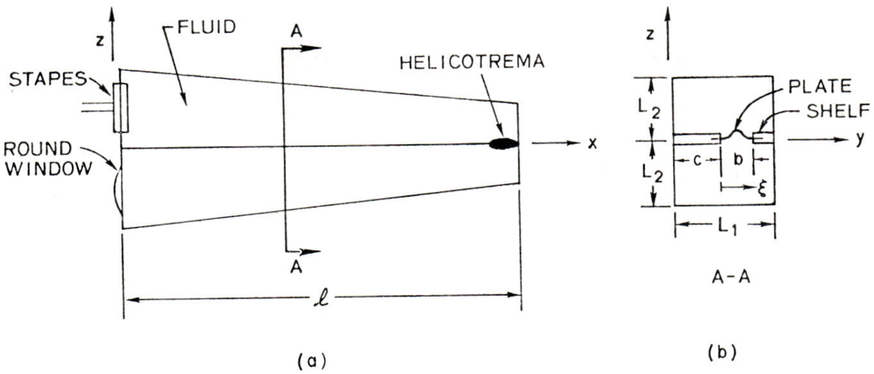

FIG. 30.4 Cochlear model: (*a*) side view, (*b*) cross section. The plate represents the basilar membrane, the upper fluid duct represents the scala vestibuli and scala media, and the lower duct represents the scala tympani. The helicotrema is a hole in the partition through which fluid from one duct can flow to the other. Input sound is simulated by a prescribed displacement of the stapes. The round window is a flexible membrane.

cross section (Fig. 30.4*b*) is symmetric, and the partition consists of a relatively rigid shelf supporting a flexible plate which simulates the basilar membrane. Reissner's membrane (RM), which separates the endolymph of the scala media from the perilymph of the scala vestibuli (Fig. 30.3*b*), seems to have no significant hydroelastic role, since its volume compliance is large in comparison with that of the basilar membrane, and there is little gradient in its dimensions. Thus Reissner's membrane is eliminated and the scalae vestibuli and media are represented by the upper fluid duct in the model (Fig. 30.4).

The striking gross anatomical feature of the mammalian cochlea is the curvature. The exception is in monotremes (egg-laying mammals: platypus, spiny anteater), which have a slightly curved cochlea similar to that of birds and high-frequency capability apparently not much better than that of birds [8]. This most striking feature is omitted from the straight box model in Fig. 30.4. Estimates indicate no significant quantitative effect of the curvature on the response of such a model, particularly for the high-frequency cochlea of the dolphin, which is relatively flat and has relatively large radius of curvature. Curvature can have a large effect on the stiffness of the membranes which are at some angle with respect to the basilar membrane in Fig. 30.3, since they are shells rather than flat plates [37], but the significance for function of the cochlea has not yet been explained.

The calculation of the response of even the straight model (Figure 30.4) is an interesting challenge. The assumption of long wavelength, or, equivalently, that the pressure is constant across the fluid duct cross section or that the y and z components of fluid velocity are negligible in comparison with the x component, yields a familiar, one-dimensional wave equation. Despite behavior demonstrated by experimental models, particularly in Ref. 23, which indicate that the response is not one-dimensional, the approximation has been used by many investigators, as summarized in Ref. 2. The disadvantage of equations which do not reflect the proper physics is that unreasonable adjustment of the physical parameters must be made to obtain reasonable results. A survey covering the more recent work on two- and three-dimensional models is included in Ref. 38.

The cochlear length ratios are unsuitable for direct numerical methods: the length of the fluid duct is 30 times the width, which is 2 to 10 times larger than the width of the plate (the basilar membrane), which is 3 to 10 times the thickness of the viscous boundary layer. Significant wavelengths are from several duct widths to the order of the plate width. However, the slow variation of dimensions with distance along the cochlea makes the problem ideal for asymptotic approximation, in particular the WKB method (Wentzel, Kramers, Brillouin) or, better called by the earlier workers, the CLG method (Carlini, Liouville, Green).

An equivalent method was first used by Ranke [39] for the significant, short-wavelength portion of the response of a two-dimensional model. The method for the calculation of both long and short wavelengths and results for two-dimensional models are in Refs. 40 to 44, with applications to transients in Ref. 43 and nonlinearities in Ref. 44. In Ref. 45, the method is shown to agree well with numerical results in Ref. 46, except for the very low amplitude region past the point of partition resonance, for a two-dimensional model with excessive (for the real cochlea) partition mass. An alternate solution for this region is in Ref. 47, which the more extensive numerical and analytical results of Ref. 48 indicate is actually not better. However, the comparison of three-dimensional calculations with results from the experimental model [34, 35], for which the partition mass is realistic and quantitative measurements of fluid and elastic properties were made, is favorable [49]. The method does yield the physical response to a reasonable degree of accuracy and is very efficient for computer implementation. Calculations for real cochleas are in Refs. 50 to 52, with an alternate derivation of the method in Ref. 52. In light of the evidence for an active mechanism in the normal cochlea [21, 22], which may utilize the electrical potential available, these calculations must be considered as applicable to a "dead" cochlea with inert tissue.

30.2.1 Wave Propagation

The necessary ingredient for the asymptotic analysis of a cochlear model with a small gradient in geometry is the solution for wave propagation in an infinite duct with constant properties. The partition displacement (Fig. 30.4) is zero on the rigid shelves $0 \le y \le c$, $c + b \le y \le L_1$, but is a traveling wave on the flexible plate (basilar membrane) in the region $0 \le \xi = y - c \le b$ described by

$$w_p(x, y, t) = W\eta(\xi)e^{i\theta} \tag{30.1}$$

where

$$\theta = \omega t - \lambda x \tag{30.2}$$

in which ω is the frequency, λ is the wave number, and W is the amplitude, and the shape of the basilar membrane deflection is approximated as shown in Table 30.3.

Fluid Impedance. The solution of Laplace's equation for the motion of an inviscid, incompressible fluid yields the pressure acting on the partition due to the prescribed displacement. The first term of the expansion for the pressure acting on the partition (the basilar membrane) is

$$p(x, y, 0, t) = P\eta(\xi)e^{i\theta} \tag{30.3}$$

in which the pressure amplitude P is related to the displacement amplitude by

$$P = -\rho\omega^2 W h_{eq} \tag{30.4}$$

TABLE 30.3 Shape function

Edges	$\eta(\xi)$	$\bar{\eta}$	$\overline{\eta^2}$	$A_j = \int_0^b \eta(\xi) \cos \frac{j\pi\xi}{L_1} \frac{d\xi}{b}$
Hinged	$\sin \frac{\pi\xi}{b}$	$\frac{2}{\pi}$	$\frac{1}{2}$	$A_j = \frac{1}{\pi}\left[\frac{\cos (j\pi/L_1)(b + c) + \cos (j\pi c/L_1)}{1 - (jb/L_1)^2}\right]$
Clamped	$\sin^2 \frac{\pi\xi}{b}$	$\frac{1}{2}$	$\frac{3}{8}$	$A_j = \begin{cases} \frac{1}{2} & \text{for } j = 0 \\ \frac{L_1}{2j\pi b}\left[\frac{\sin (j\pi/L_1)(b + c) - \sin (j\pi c/L_1)}{1 - (jb/2L_1)^2}\right] & \text{for } j = 1, 2, \ldots \end{cases}$

where ρ is the fluid density, and h_{eq} is a real, positive function, which has the dimension of length, of the geometry and wave number. From the form of Eq. (30.4), the fluid can be seen to behave as a pure mass resisting the basilar membrane displacement, and h_{eq} can be considered as the thickness of an equivalent mass of fluid which does not flow but has only the vertical motion of the basilar membrane:

$$\frac{h_{eq}}{b} = (\overline{\eta^2})^{-1} \sum_{j=0}^{\infty} \frac{A_j^2}{m_j L_1 \tanh m_j L_2} Q_j \tag{30.5}$$

where

$$Q_j = \begin{cases} 1 & \text{for } j = 0 \\ 2 & \text{for } j = 1, 2, 3, \ldots \end{cases}$$

$$m_j^2 = \left(\frac{j\pi}{L_1}\right)^2 + \lambda^2$$

For clamped edges Eq. (30.5) gives

$$\frac{h_{eq}}{b} = \frac{2}{3}\left\{\frac{1}{\lambda L_1 \tanh \lambda L_2} + 2 \sum_{j=1}^{\infty} \frac{(L_1/j\pi b)^2[\sin (j\pi/L_1)(b + c) - \sin (j\pi c/L_1)]^2}{[1 - (jb/2L_1)^2]^2 m_j L_1 \tanh m_j L_2}\right\} \tag{30.6}$$

If the width of the cross section is small in comparison with the wavelength, that is, if $\lambda L_1 \ll 1$, then only the first term in Eq. (30.6) is significant, and we have the two-dimensional model approximation

$$\frac{h_{eq}}{b} \approx \frac{2}{3} \frac{1}{\lambda L_1 \tanh \lambda L_2} \tag{30.7}$$

If the height is also small in comparison with the wavelength ($\lambda L_2 \ll 1$), then Eq. (30.7) reduces to the familiar one-dimensional "long-wavelength" approximation

$$\frac{h_{eq}}{b} \approx \frac{2}{3} \frac{1}{\lambda^2 L_1 L_2} \tag{30.8}$$

Curves of h_{eq}/b calculated from Eq. (30.6) are plotted in Fig. 30.5 for three sections along the model of Helle [35]. The two- and one-dimensional approximations from Eqs. (30.7) and (30.8) are also shown. It is evident that these approximations are substantially in error when the wavelength is short. For short wavelengths, $\lambda L_1 > 10$, Eq. (30.5) is approximated by

$$h_{eq} \approx \frac{1}{\lambda} \tag{30.9}$$

The viscous boundary layer correction to Eq. (30.5) is [49]

$$H_{eq} = h_{eq}\left(1 + \frac{1}{\beta L_2}\right) + \frac{1}{\beta} \tag{30.10}$$

where

$$\beta = \left(\frac{i\rho\omega}{\mu}\right)^{1/2}$$

and μ is the coefficient of viscosity. This adds a dissipative component to the fluid impedance.

For two fluid ducts, the fluid effect is doubled, and the result can be written as

$$\frac{\Delta p}{\Delta A} = 2\rho\omega^2 \frac{H_{eq}\overline{\eta}^2}{b\ \overline{\eta^2}} \tag{30.11}$$

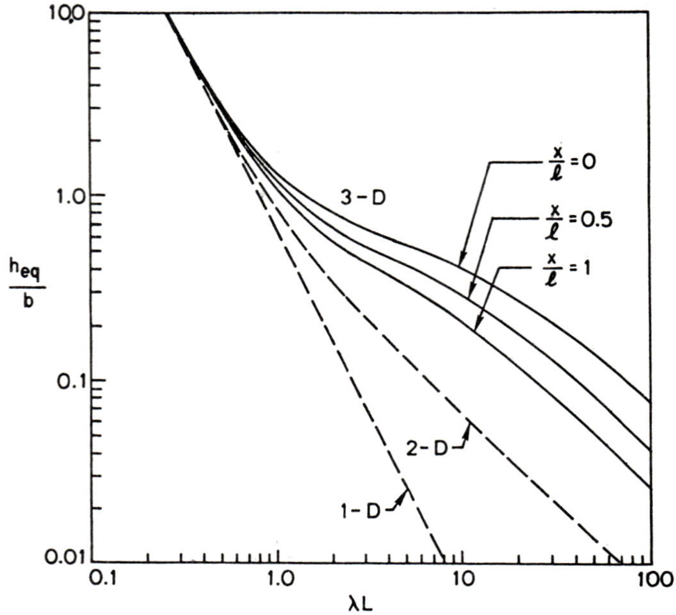

FIG. 30.5 Dependence of effective mass of inviscid fluid measured by h_{eq} on wave number λ; for model with $L_1 = L_2 = L$ and $b/L = 0.14, 0.23$, and 0.39 at beginning, middle, and end, respectively ($x/l = 0, 0.5, 1$). Assumptions of two- and one-dimensional fluid motion are in substantial error for short wavelengths. Typically, λL is around 0.3 near the stapes and around 3.0 at the point of maximum response.

in which Δp is the average pressure difference across the basilar membrane and $\Delta A = W\overline{\eta}b$ is the volume displacement of the basilar membrane unit length (i.e., the area displacement).

Partition Impedance. The volume stiffness of the basilar membrane under a static pressure is

$$\frac{\Delta p}{\Delta A} = C^{-1} = b^{-5}\alpha_1 D \left\{1 + k_x \left[\alpha_2 \left(\frac{\lambda b}{\pi}\right)^2 + \alpha_3 \left(\frac{\lambda b}{\pi}\right)^4\right]\right\} \tag{30.12}$$

where C is the volume compliance, the α_i are in Table 30.4, and k_x is the ratio of the elastic moduli in the longitudinal and transverse directions. For an isotropic plate $k_x = 1$, while for the actual basilar membrane $k_x \approx 0$ [29], in which case C is not dependent of the wave number λ. The bending stiffness depends on the elastic modulus E and thickness h:

TABLE 30.4 Stiffness coefficient factors

Constraint	α_1	α_2	α_3	α_4	α_5	α_6	α_7
Hinged edges	120	2	1	59	48	5.59	4.1
Clamped edges	720	0.5	0.19	138	192	....	1
One edge hinged, one edge clamped	....	....	....	94			
One edge clamped, one edge free	20	....	....	6	3		

$$D = \begin{cases} \dfrac{Eh^3}{12(1-\nu^2)} & \text{for an isotropic plate, } \nu \approx 0.3 \\ \dfrac{Eh^3}{12} & \text{for transverse beams} \\ \dfrac{Eh^2 N_f A_f}{12}\left[1 + \dfrac{h_i}{h} + \left(\dfrac{h_i}{h}\right)^2\right] & \text{for transverse fibers} \end{cases} \tag{30.13}$$

The latter case consists of two layers of fibers in a soft matrix, where h is the total thickness, h_i is the distance between the layers, N_f is the number of fibers per unit of longitudinal length, and A_f is the cross-sectional area of a single fiber.

The total pressure drop across the basilar membrane is

$$\frac{\Delta p}{\Delta A} = C^{-1}(1 + id_E) - \omega^2 \rho_p h \frac{\overline{\eta^2}}{b\overline{\eta}^2} \tag{30.14}$$

in which ρ_p is the density of the basilar membrane and d_E is the ratio of imaginary to real parts of the elastic modulus E, which accounts for material dissipation.

Because of utility in evaluating D, we include the result for the deflection w under a point load P acting on an isotropic, flat plate strip,

$$\frac{P}{w} = \frac{\alpha_4 D}{b^2} \tag{30.15}$$

and the result for a load P distributed on a line of length e parallel to the edges,

$$\frac{P}{ew} = \frac{\alpha_5 D}{b^3} \tag{30.16}$$

For a plate of width b, curved to a quarter circle of radius $2b/\pi$, clamped at one edge and constrained at the other edge from displacement in the tangential direction, and loaded at the edge by normal force P, the result is

$$\alpha_5 = 46.5$$

If both edges of the flat plate are constrained from tangential displacement, the stretching gives the nonlinear corrections

$$\Delta p = b^{-5}\alpha_1 D\, \Delta A \left[1 + \alpha_6 \left(\frac{\Delta A}{bh}\right)^2\right] \tag{30.17}$$

$$P = b^{-3}\alpha_5 Dew \left[1 + \alpha_7 \left(\frac{w}{h}\right)^2\right] \tag{30.18}$$

Some values for Young's modulus, in N/mm^2, are

$$E \approx \begin{cases} 20 \times 10^3 & \text{compact bone} \\ 3 \times 10^3 & \text{keratin} \\ 10^3 & \text{collagen} \\ 4 & \text{rubber, elastin} \\ 3 \times 10^{-3} & \text{Jello} \end{cases} \tag{30.19}$$

From the measurements [23], Reissner's membrane is estimated [37] to have $E \approx 60$, and Hensen's cells to be softer than Jell-O. Values for the basilar membrane fibers of $E = 200$, and for the ground substance of $E \approx 0.2$, may be reasonable. The above value for bone seems to be valid for the primary spiral lamina [43].

Phase Relation. The volume compliance of the fluid [Eq. (30.11)] must be equal to that of the basilar membrane [Eq. (30.14)], which gives the relation between frequency and wave number.

Without the dissipative terms, the relation is

$$f(\lambda, \omega) = \omega^2(2\rho h_{eq} + \rho_p h) \frac{\overline{\eta^2}}{b\overline{\eta}^2} - \frac{1}{C} = 0 \tag{30.20}$$

Since the density of soft tissue is about that of fluid and since the thickness of the basilar membrane is small in comparison with the width, it can be seen from Fig. 30.5 that the mass of the basilar membrane is negligible until the wavelength becomes very short, that is, $\lambda L > 30$. Note that the time-averaged lagrangian density is

$$L = \frac{(\Delta A)^2 f}{4} \tag{30.21}$$

Adding the viscosity of the fluid and of the basilar membrane dispersion relation,

$$F(\lambda, \omega) = f(\lambda_1 \omega) + g(\lambda_1 \omega) = 0 \tag{30.22}$$

where the terms due to viscosity are

$$g(\lambda, \omega) = \frac{2\rho\omega^2}{\beta}\left(\frac{h_{eq}}{L_2} + 1\right)\frac{\overline{\eta^2}}{b\overline{\eta}^2} - \frac{id_E}{G}$$

For small dissipation, the root λ of Eq. (30.22) is approximated by

$$\lambda = \lambda_0 - \frac{g(\lambda_0, \omega)}{f_\lambda(\lambda_0, \omega)} \tag{30.23}$$

where λ_0 is the real root for the undamped problem

$$f(\lambda_0, \omega) = 0$$

Duct with Varying Dimensions. For sufficiently slowly varying dimensions, the WKB first-approximation solution for the model in Fig. 30.4 can be obtained readily. The amplitude and wave number in Eq. (30.1) now vary with position,

$$W = W(x) \tag{30.24}$$

$$\theta = \omega t - \int_0^x \lambda(x, \omega)\, dx = \omega t - \phi - i\xi \tag{30.25}$$

where the contribution to (real) phase is

$$\phi(x, \omega) = \int_0^x \text{Re}\,\lambda\, dx \tag{30.26}$$

and the damping factor is

$$\xi(x, \omega) = \int_0^x \text{Im}\,\lambda\, dx \tag{30.27}$$

The coefficients of Eq. (30.22) depend on position

$$F(\lambda, \omega, x) = 0 \tag{30.28}$$

The procedure for a fixed ω is as follows:

1. Choose a discrete mesh of points x_j.
2. Solve Eq. (30.28) by iteration at each point for the local wave number λ_j.
3. Integrate numerically Eqs. (30.26) and (30.27) for the phase and damping factor.
4. Compute the amplitude function from $W(x) = (bF_\lambda)^{-1/2}$.

The solution for the ratio of basilar membrane displacement w_p to stapes displacement δ_{st} is

$$\frac{w_p}{\delta_{st}} = A_{st}\left(\frac{\lambda}{A}\right)_0 \left[\frac{(bF_\lambda)_0}{bF_\lambda}\right]^{1/2} e^{i(\pi/2-\phi)-\xi} \tag{30.29}$$

where A_{st} is the area of the stapes, A is the cross-sectional area of the duct, and the zero subscript denotes quantities to be evaluated at the end $x = 0$.

30.2.2 Results

The computation requires only about 1 s on a mainframe computer for all details of the three-dimensional fluid and basilar membrane response at a given frequency [49, 52]. The agreement with experimental model results is reasonable [49]. The geometry of the guinea pig cochlea is known, and postmortem values for the volume compliance, which will be designated as C_B, of the partition have been measured [23]. Recent measurements [31] indicate that $C_B/5$ is the in vivo value. Figure 30.6 shows the calculation [Eq. (30.29)] for displacement amplitude and phase with the volume compliance of $C_B/4$, and with the lower fluid duct (scala tympani) full, as normal, and drained, as in the direct measurements of basilar membrane motion [27]. So, with physiological values for basilar membrane compliance, geometry, fluid viscosity (that of water), viscosity ($d_E = 0.05$), and mass (negligible), the model response (Fig. 30.4) is close to that of the real cochlea. The amplitude and phase "plateau" for higher frequencies has yet to be explained.

In Fig. 30.7, the reciprocal of the calculated basilar membrane amplitude with the scala tympani full is compared with neural recordings from the spiral ganglion cells at about the same point. The calculated displacement threshold curve for the basilar membrane corresponds well with the neural recordings from abnormal cochleas in which the outer hair cells have been destroyed by kanamycin treatment. The recordings from normal cochleas have a much sharper tuning at a frequency about a half octave higher, which indicates a "second filter" mechanism between the basilar membrane motion and neural excitation. However, recent evidence [21, 22] shows that, when great care is exercised, the basilar membrane response is similar to that shown by the normal neural curves in Fig. 30.7. An hour after opening the cochlea, the basilar membrane response deteriorates to the calculated curve. The conclusion is that the second filter mechanism does have a significant influence on the "first filter" basilar membrane.

Because of the weaker support of the basilar membrane found in low-frequency cochleas and in the apical portion of high-frequency cochleas (Fig. 30.3*a*), the additional modes of partition response shown in Fig. 30.8 are treated in Ref. 51. Figure 30.9 shows results for a stiff primary spiral lamina corresponding to that of the guinea pig, and for a primary spiral lamina with a reduction to 11 percent of the transverse shear modulus G_y, which corresponds to the geometry of the primary spinal lamina in primates [9]. The flexible shelf causes a more rapid drop in phase but has little effect on the shape of the amplitude curve. The agreement with measurements in the squirrel monkey [25] is good.

30.2.3 Estimate for Localization

The solution for the (inert) cochlear model works rather well; however, the location of the point of maximum response is difficult to judge from Eq. (30.29) without carrying out the numerical details. Steele [41, 42] discusses a cochlear "model 0" of the basilar membrane as one hinged, tapered strip of an infinite, massless plate immersed in an infinite fluid. For a given frequency, short wavelengths occur at points beyond the "transition point" X_{TP}, which can be defined in terms of the volume compliance as

$$2\rho\omega^2 = \frac{\pi^5}{120C(X_{TP})} \tag{30.30}$$

Comparison with Eq. (30.20) shows that X_{TP} occurs at the point at which

$$\frac{h_{eq}}{b} = \frac{1}{\pi} \tag{30.31}$$

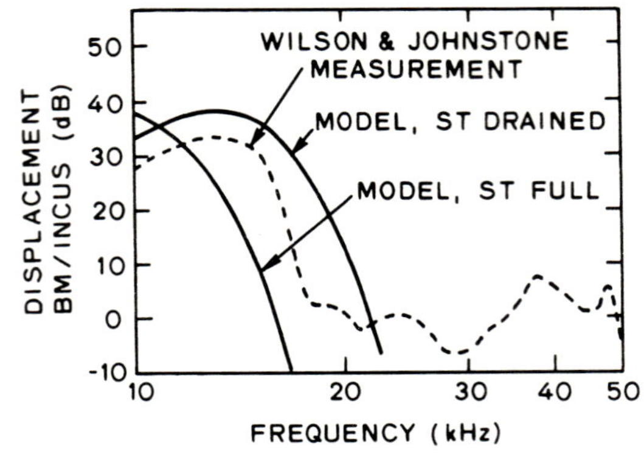

(a)

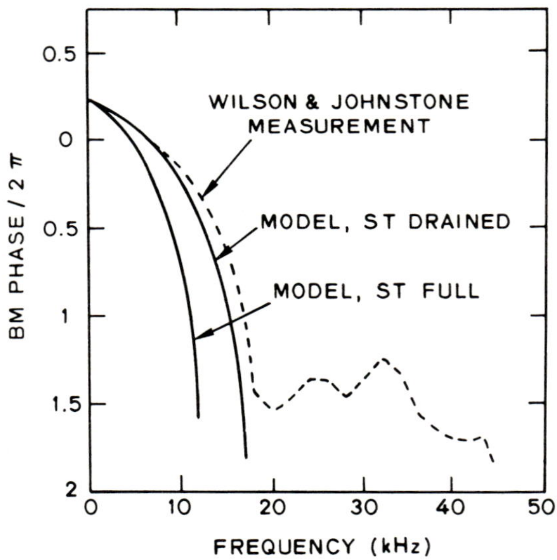

(b)

FIG. 30.6 (*a*) Basilar membrane displacement and (*b*) phase at point 4.1 mm from basal end as measured by Wilson and Johnstone [27] and calculated from model with $C = C_B/4$ with and without fluid in the scala tympani ($k_x = 0$). (*From Steele and Taber [50].*)

Figure 30.5 shows that the fluid inertia is close to the short-wavelength asymptote [Eq. (30.9)]. Since the damping of the traveling wave becomes severe as the wavelength becomes short, the amplitude is roughly 20 dB down from the peak at X_{TP}. In the guinea pig cochlea for $X_{TP} = 4.1$ mm (Figs. 30.6 and 30.7), the compliance $C = C_B/4 = 0.15$ mm^4/N gives the frequency $f = 15$ kHz, which by coincidence is close to the tuning for the neural fibers at that point.

For water, Eq. (30.30) is

$$f = 5.68C^{-1/2} \tag{30.32}$$

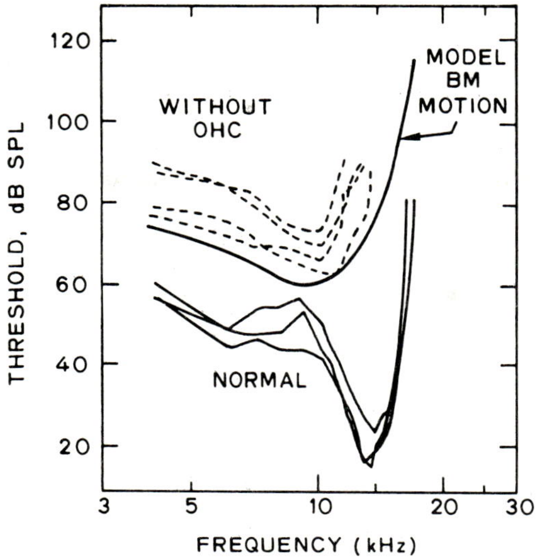

FIG. 30.7 Threshold of ganglion cells at point 4.3 mm from basal end obtained by Robertson and Johnstone [19]. The reciprocal of the calculated basilar membrane response from Fig. 30.6 with the scala tympani full at 4.1 mm is similar to the cell response without outer-hair-cell cilia. (*From Steele and Taber [50].*)

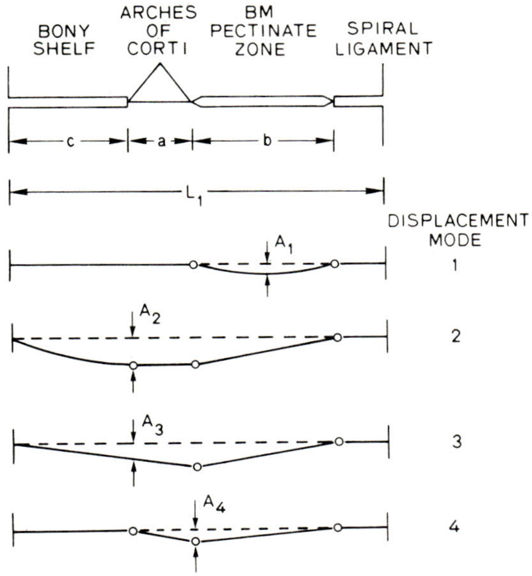

FIG. 30.8 Deflection modes of the cochlear partition. Mode 1 is a half-sine wave deflection of the pectinate zone of the basilar membrane; mode 2 is a quarter-sine wave deflection of the secondary spiral lamina corresponding to the shear displacement due to pressure; mode 3 is the shear displacement due to a concentrated force at the hinge point; and mode 4 is a rotation of the arches with respect to the shelf. (*From Taber and Steele [51].*)

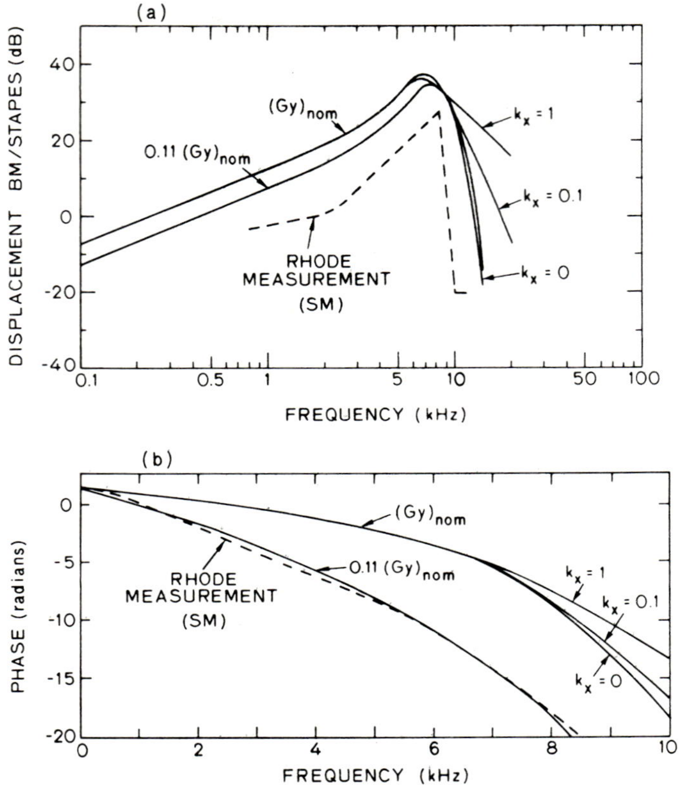

FIG. 30.9 Comparison with measurements of Rhode [25] in the squirrel monkey. Calculations are the nominal GP model (three-mode; 5 mm from the basal end) for relative longitudinal stiffness $k_x = 0, 0.1, 1$, and repeated for $k_x = 0$ with the shear stiffness of the bony shelf reduced to 11 percent of nominal. (*From Taber and Steele [51].*)

where frequency f is in kHz and C is in mm^4/N. For the basilar membrane fibers with $E = 210$ N/mm^2 and $A_f = 10^{-10}$mm^2, and in two layers with $h_i/h = ⅓$ (and $k_x = 0$), Eqs. (30.12) and (30.13) give

$$C = \frac{b^5}{h^2 N_f \gamma}(3.3 \times 10^6) \tag{30.33}$$

where

$$\gamma = \begin{cases} 1 & \text{for hinged edges} \\ 3 & \text{for constrained edges} \\ 6 & \text{for clamped edges} \end{cases}$$

so that

$$f = \frac{h(N_f \gamma)^{1/2}}{b^{5/2}}(3 \times 10^{-3}) \tag{30.34}$$

where b and h are in millimeters and N_f is the number of fibers per millimeter. When used in Eqs. (30.32) and (30.33), the values from Table 30.2 produce the results in Fig. 30.10. Details for

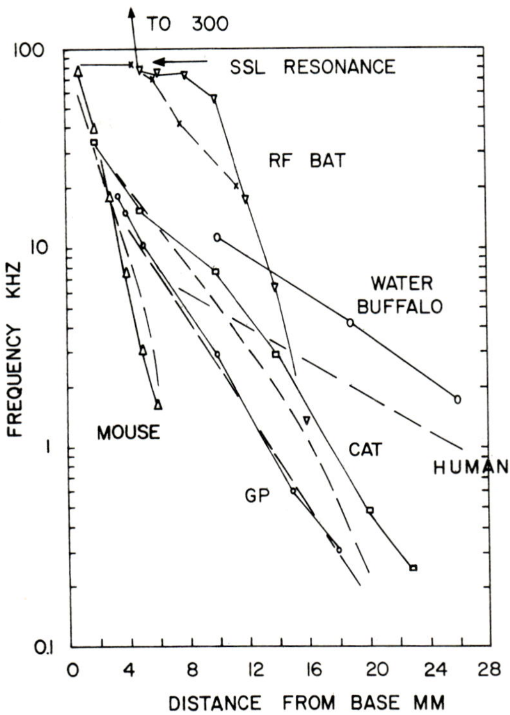

FIG. 30.10 Localization of frequency on the basilar membrane. Solid lines show the calculation of the transition point, assuming that the pectinate zone of the basilar membrane is the only flexible element. Dashed lines indicate results of direct evidence as given in *R. ferrumequinum* bat [13], human [23], cat [53], guinea pig [27], and mouse [18]. Calculations are for fiber $E = 210\ \mathrm{N/mm^2}$ and geometry from Table 30.2. The fiber density distribution for the cat basilar membrane was used for all. For the *R. ferrumequinum* bat, the pectinate zone in the basal 4 mm would flex at 300 kHz; instead, the peculiar secondary spiral lamina resonates at 83 kHz.

the filament density N_f are available only for the cat, so this density distribution was assumed for the other cochleas. When not specifically given, the width of pectinate zone of the basilar membrane was assumed to be twice that of arcuate zone. For the water buffalo, the thickness of pectinate zone was assumed to be the same as that of the arcuate zone, since the micrographs in Ref. 12 do not seem to show any thickening of the pectinate zone for this low-frequency cochlea.

The correlation of Eq. (30.33) with the curves for localization in the cochlea, obtained by a variety of direct measurements [14, 18, 23, 53], is excellent for the cat, guinea pig and mouse, and water buffalo and might be expected to be similar to that of a human. Consistent with the anatomy, hinged edges were assumed for bat, cat, and mouse, while constrained edges were assumed for guinea pig with the less pronounced thinning at the edges of the pectinate zone. Compared with the measurement for C_B in the guinea pig cochlea [23], Eq. (30.33), with $\gamma = 3$, gives about the same measurement for C_B in the apical region and about 25 percent of C_B in the basal region, which is consistent with the measurement given in Ref. 31. The curve in Fig. 30.10

for water buffalo may be about right, since it is for hinged edges (too flexible) and for the cat N_f (probably too stiff).

The tentative conclusions are as follows:

1. Basilar membrane compliance can be computed from measurements of the dimensions and microstructure. The fiber density is significant.
2. Localization of excitation according to frequency depends primarily on compliance of the pectinate zone.

30.3 A COCHLEAR RESONATOR?

The curve for localization in the *R. ferrumequinum* bat is also shown in Fig. 30.10, as computed from Eq. (30.34) by using the dimensions in Table 3.2*d*. Since the cochlea is "normal" (see Fig. 30.3*b*) apical from the 4.5-mm point, the cat N_f was assumed for this region. It is interesting that the calculated frequency in the region $5 < x < 8$ mm is close to 83 kHz, to which this cochlea is extraordinarily sensitive. In the region $0 < x < 4.5$ mm, the cross section is unusual (Fig. 30.3*c*). Because of increase in the thickness of the pectinate zone, Eq. (30.34) indicates a frequency of 300 kHz for this region. The evidence is that the receptor cells in this region are stimulated by 83 to 86 kHz [14]. The conclusion is that in the pectinate zone the basilar membrane is not flexing but is essentially rigid.

Instead, the peculiar secondary spiral lamina (SSL) (Fig. 30.3*c*) seems to act as a simple resonator. For an approximation it may be considered as a plate curved in the form of a quarter circle, with a concentrated mass at the end which is constrained from tangential displacement by the basilar membrane and the spiral ligament. The stiffness from Eq. (30.16) gives the resonant frequency

$$f_{\mathrm{SSL}} = \frac{1}{2\pi}\left(\frac{\alpha_5 D}{\rho A b^2}\right)^{1/2} \tag{30.35}$$

where ρ is the density and A is the cross-sectional area of the knob on the primary spiral lamina. Using $E = 20 \times 10^3$ N/mm^2, $\nu = 0.3$, $h = 0.01$ mm, $b = 0.2$ mm, $\rho = 2$ mg/mm^3, and $A = 0.02$ mm^2, with $\alpha_5 = 46.5$ and $D = Eh^3/12(1 - \nu^2)$, gives

$$f_{\mathrm{SSL}} = 82 \text{ kHz}$$

Despite the oversimplification of this calculation, it shows that it is likely that the resonance of the secondary spiral lamina is the dominating feature in this special cochlea.

CONCLUSION

Research in the mechanics of the cochlea continues to be active and to attract workers with a variety of backgrounds. Several major breakthroughs have been made in recent years. Particularly significant is the ability to measure the response of neurons [19], hair cells [20], and the basilar membrane [21, 22] without causing severe damage in opening the cochlea. A proper model is necessary for understanding cochear mechanics, and the WKB method [50, 51] provides a computationally efficient means of calculating the response of realistic models of the (inert) cochlea. Figure 30.10 indicates that tissue stiffness can be calculated from the microstructure, but direct measurement is important, as shown by Ref. 30 and by work in progress by A. Flock, G. Frommer, and C. Miller.

The high electrical potential in the cochlea is apparently utilized in an active process, but how? The microstructure of the cilia of the outer hair cells [32] may be suitable for active beating. However, a theory of the active process, which is the key to cochlear function, must work equally well for all cochleas—not only for those of the guinea pig and the cat but also for the peculiar resonant region of the *R. ferrumequinum* bat and the alligator lizard.

REFERENCES

1. Yost, W. A., and Nielson, D. W.: *Fundamentals of Hearing—An Introduction,* Holt, Rinehart and Winston, New York, 1977.
2. Dallos, P.: *The Auditory Periphery,* Academic Press, New York, 1973.
3. Zwicker, E.: On psychoacoustical equivalent of tuning curves, in E. Zwicker and E. Terhardt (eds.), *Facts and Models in Hearing,* Springer-Verlag, Berlin, 1974, pp. 132–141.
4. Helmholtz, H.: *Sensations of Tone,* Dover, New York, 1954.
5. Michelsen, A.: Insect ears as mechanical systems, *Am. Sci., 67*(6), 696–706, 1979.
6. Sales, G., and Pye, D.: *Ultrasonic Communication by Animals,* Wiley, New York, 1974.
7. Peake, W. T., and Ling, A.: Basilar membrane motion in the alligator lizard: Its relation to tonotopic organization and frequency selectivity, *J. Acoust. Soc. Am., 67,* 1736–1745, 1980.
8. Fleischer, G.: Evolutionary principles of the mammalian middle ear, *Adv. Anat., Embryol., Cell Biol., 55,* 5–70, 1978.
9. Fleischer, G.: *Studien am Skelett des Gehörorgans der Säugetiere, einschliesslich des Menschen. Saeugetierk. Mitt., 21,* 131–239, 1973.
10. Zwislocki, J.: Analysis of middle ear function, *J. Acoust. Soc. Am., 34*(8), 1514–1523, 1962.
11. Lynch, T. L., Nedzelnitsky, V., and Peake, W. T.: Input impedance of the cochlea in the cat, *J. Acoust. Soc. Am., 72*(1), 108–130, 1982.
12. Tiedemann, H.: A new approach to theory of hearing, *Acta Otolaryngol. Suppl., 277,* 1–50, 1970.
13. Bruns, V.: Peripheral auditory tuning for fine frequency analysis in the CF-FM bat, *Rhinolophus ferrumequinum, J. Comp. Physiol., 106,* 77–97, 1976.
14. Bruns, V.: Functional anatomy as an approach to frequency analysis in the mammalian cochlea, *Verh. Dtsch. Zool. Ges.,* 141–154, 1979.
15. Cabezudo, L. M.: The ultrastructure of the basilar membrane in the cat, *Acta Otolaryngol., 86,* 160–175, 1978.
16. Fernández, C.: Dimensions of the cochlea (guinea pig), *J. Acoust. Soc. Am., 24*(5), 519–523, 1952.
17. Ehret, G., and Frankenreiter, M.: Quantitative analysis of cochlear structures in the house mouse in relation to mechanisms of acoustical information processing, *J. Comp. Physiol., 122,* 65–85, 1977.
18. Ehret, G.: Stiffness gradient along the basilar membrane as a basis for spatial frequency analysis within the cochlea, *J. Acoust. Soc. Am., 64*(6), 1723–1726, 1978.
19. Robertson, D., and Johnstone, B. M.: Aberrant tono topic organization in the inner ear damaged by kanamycin, *J. Acoust. Soc. Am., 57,* 466–469, 1979.
20. Russell, I. J., and Sellick, P. M.: The tuning properties of cochlear hair cells, in E. F. Evans and J. P. Wilson (eds.), *Psychophysics and Physiology of Hearing,* Academic Press, New York, 1977, pp. 71–88.
21. Khanna, S. M., and Leonard, D. G. B.: Basilar membrane tuning in the cat cochlea, *Science, 215,* 305–306, 1982.
22. Sellick, P. M., Patuzzi, R., and Johnstone, B. M.: Measurement of basilar membrane motion in the guinea pig using the Mössbauer technique, *J. Acoust. Soc. Am., 72*(1), 131–141, 1982.
23. Békésy, G.: *Experiments in Hearing,* McGraw-Hill, New York, 1960.
24. Johnstone, B. M., and Boyle, A. J. F.: Basilar membrane vibration examined with the Mössbauer technique, *Science, 158,* 389–390, 1967.
25. Rhode, W. S.: An investigation of post-mortem cochlear mechanics using the Mössbauer effect, in A. R. Moller (ed.), *Basic Mechanism in Hearing,* Academic Press, New York, 1973, pp. 49–68.
26. Kohllöffel, L. U. E.: A study of basilar membrane vibrations: I–III, *Acustica, 27,* 49–89, 1972.
27. Wilson, J. P., and Johnstone, J. R.: Basilar membrane and middle-ear vibration in guinea pig measured by capacitive probe, *J. Acoust. Soc. Am., 57,* 705–723, 1975.

28. Iurato, S.: Functional implications of the nature and submicroscopic structure of the tectorial and basilar membrane, *J. Acoust. Soc. Am., 34,* 1386–1395, 1962.
29. Voldřich, L.: Mechanical properties of basilar membrane, *Acta Otolaryngol., 86,* 331–335, 1978.
30. Gummer, A. W., Johnstone, B. M., and Armstrong, N. J.: Direct measurement of basilar membrane stiffness in the guinea pig, *J. Acoust. Soc. Am., 70*(5), 1298–1309, 1981.
31. Dancer, A., and Franke, R.: Intracochlear sound pressure measurements in guinea pigs, *Hearing Res., 2,* 190–205, 1980.
32. Flock, A.: Physiological properties of sensory hairs on the ear, in E. F. Evans and J. P. Wilson (eds.), *Psychophysics and Physiology of Hearing,* Academic Press, New York, 1977, pp. 15–26.
33. Tonndorf, J.: Shearing motion in scala media of cochlear models, *J. Acoust. Soc. Am., 32,* 228–244, 1960.
34. Cannell, J. K.: Cochlear models, Ph.D. thesis, University of Warwick, United Kingdom, 1969.
35. Helle, R., Beobachtungen an Hydromechanischen Modellen des Innenohres mit Nachbildung von Basilarmembran, Corti-Organ and Deckmembran, Dr.-Ing. dissertation, Technische Universität, München, 1974.
36. Chadwick, R. S., Fourney, M. E., and Neiswander, P.: Modes and waves in a cochlear model, *Hearing Res., 2,* 475–483, 1980.
37. Steele, C. R.: Stiffness of Reissner's membrane, *J. Acoust. Soc. Am., 55,* 1252–1257, 1974.
38. Lighthill, J.: Energy flow in the cochlea, *J. Fluid Mech., 106,* 149–213, 1981.
39. Ranke, O. F.: Theory of operation of the cochlea: A contribution to the hydrodynamics of the cochlea, *J. Acoust. Soc. Am., 22,* 772–777, 1950.
40. Steele, C. R.: A possibility for sub-tectorial membrane fluid motion, in A. R. Moller (ed.), *Basic Mechanisms in Hearing,* Academic Press, New York, 1973, pp. 69–94.
41. Steele, C. R.: Behavior of the basilar membrane with pure tone excitation, *J. Acoust. Soc. Am., 55,* 148–172, 1974.
42. Steele, C. R.: Cochlear mechanics, in W. D. Keidel and W. D. Neff (eds.), *Handbook of Sensory Physiology,* vol. V: *Auditory System,* Springer-Verlag, Berlin, 1976, pt. 3, pp. 443–478.
43. Dotson, R.: Cochlear models with transient excitation, Ph.D. thesis, Stanford University, Stanford, Calif., 1974.
44. Hallauer, W. L.: Nonlinear mechanical behavior of the cochlea, Ph.D. thesis, Stanford University, Stanford, Calif., 1974.
45. Steele, C. R., and Taber, L. A.: Comparison of WKB and finite difference calculations for a two-dimensional cochlear model, *J. Acoust. Soc. Am., 65,* 1001, 1979a.
46. Neely, S. T.: Finite difference solution of a two-dimensional mathematical model of the cochlea, *J. Acoust. Soc. Am., 69*(5), 1386–1393, 1981.
47. Steele, C. R., and Miller, C. E.: An improved WKB calculation for a two-dimensional cochlear model, *J. Acoust. Soc. Am., 68,* 147–148, 1980.
48. Viergerer, M. A.: *Mechanics of the Inner Ear—A Mathematical Approach,* Delft University Press, Delft, Netherlands, 1980.
49. Steele, C. R., and Taber, L. A.: Comparison of WKB and experimental results for three-dimensional cochlear models, *J. Acoust. Soc. Am., 65,* 1007–1018, 1979b.
50. Steele, C. R., and Taber, L. A.: Three-dimensional model calculations for the guinea pig cochlea, *J. Acoust. Soc. Am., 69,* 1107–1111, 1981.
51. Taber, L. A., and Steele, C. R.: Cochlear model including three-dimensional fluid and four modes of partition flexibility, *J. Acoust. Soc. Am., 70*(2), 426–436, 1981.
52. Holmes, M. H.: A mathematical model of the dynamics of the inner ear, *J. Fluid Mech., 116,* 59–75, 1982.
53. Wilson, J. P., and Evans, E. F.: Cochlear frequency map for the cat, in E. F. Evans and J. P. Wilson (eds.), *Psychophysics and Physiology of Hearing,* Academic Press, London, 1977, p. 69.

CHAPTER 31

Vestibular Mechanics

William C. Van Buskirk
Tulane University
New Orleans, Louisiana

J. Wallace Grant
Virginia Polytechnic Institute and State University
Blacksburg, Virginia

INTRODUCTION

The perception of motion is a true sixth sense. It was not included by the ancients in their enumeration of the primary senses because the sense organs for motion are hidden internally—the fact that there are separate sense organs for motion is not at all obvious. The organs for the sensing of motion are located in the nonauditory portion of the labyrinth, or inner ear. They are known collectively as the *vestibular system*.

The principal components of the vestibular system are the semicircular canals and the utricle and saccule. The semicircular canals are the primary sensors of angular motion, and the utricle and saccule contain the otolith organs, which sense both the linear motion of the skull and the orientation of the skull with respect to gravity. In this article we consider the basic mechanics of the semicircular canals and the otolith organs. No attempt will be made to review the transduction of mechanical events into neural signals or the very interesting problems associated with the processing of these signals through the central nervous system. For a review of many of the vestibular problems not addressed in this chapter see Wilson and Melville Jones (1979).

31.1 ANATOMY AND PHYSIOLOGY

Figure 31.1 is a drawing of the inner ear, focusing on the organs of motion in the vestibular apparatus. As seen in the drawing, there are three sets of canals on each side of the head. They are oriented in nearly mutually orthogonal planes so that rotation about any axis may be sensed. As shown, each canal consists of two parts: an outer canal, which is a channel carved in bone, and an inner, membranous canal. The inner canal is filled with a fluid called *endolymph*. The

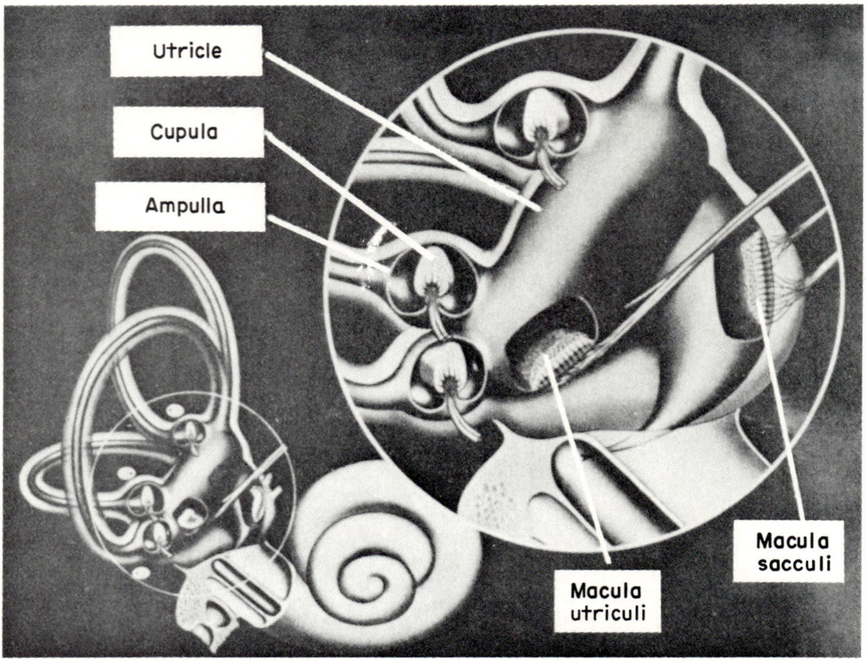

FIG. 31.1 Human vestibular system.

space between the membranous and bony canal is filled with *perilymph,* a fluid different in composition from endolymph.

One end of each semicircular duct is enlarged to form its ampulla. The ampulla nearly fills the cross section of the bony canal and terminates on the utricle. The ampulla contains the cupula, a gelatinous dividing partition with the same density as endolymph. The cupula fills the entire cross section of the ampulla, thus interrupting the otherwise continuous fluid path through the duct, utricle, and ampulla.

The cupula is the transducer for the semicircular canal. It is connected to nervous tissue at its base. Mechanical deflection of the cupula is converted into electrical impulses which transmit the state of angular motion along the vestibular nerve to the central nervous system.

Qualitatively the manner in which a semicircular canal works is as follows. An angular acceleration of the head causes the bony canal and the membranous structure attached to it to accelerate in a similar manner. The inertia of the endolymph, however, causes it to lag behind the motion of the head. Thus, there is a flow of endolymph relative to the duct wall. This flow deflects the cupula, initiating the electrical signal to the brain.

The otolith organs consist of the macula utriculi and macula sacculi of the inner ear. The four organs, two in each ear, appear as thickened portions of the membranous walls of the utricle and saccule of the inner ear. The utricular and saccular membranes are rigidly attached to the bony labyrinth in the temporal bone of the skull, and thus any motion of the skull results in rigid-body motion of the membranous portion of these two structures.

The inertially active mass of these maculae is composed of a rigid, thin bed of heavy calcium carbonate crystals. The calcium carbonate crystals, called *otoconia,* form a conglomerate which is bound together by a weblike membrane structure. The combined structure of web and crystals is called the *otoconial membrane*. Carlstrom et al. (1953) determined that the calcium carbonate crystals are in the calcite form in mammals. Calcite has a specific gravity of 2.71. It is doubtful that the otoconial membrane has a specific gravity this large since it is not a solid mass of calcite, but has a weblike structure binding individual crystals together to form a single rigid bed. The otoconial membrane probably approaches a specific gravity of 2.71 as an upper limit. This is the only tissue or structure within the bony labyrinth whose specific gravity is significantly different from that of the endolymph and perilymph.

The otoconial membrane is rigidly attached to a deformable, thin, gelatinous layer which is composed of the same material as that of the cupula. This gelatinous layer is called the *cupular membrane* and has a specific gravity very nearly equal to that of endolymph. The base of the cupular membrane is attached to the rigid wall of the utricle and saccule. The much denser otoconial membrane is surrounded by endolymph above and by the cupular membrane below, which have specific gravities generally agreed to be near 1. With this type of configuration, any movement of the skull will result in deformation of the cupular membrane due to the inertial effects of the otoconial membrane. In a like manner, gravity will cause the otoconial membrane to move and deform the gelatinous layer. Sensory cells located below the cupular membrane transduce the cupular membrane stresses into trains of nerve action potentials which are carried to the central nervous system.

The utricular macula is relatively flat and lies on the floor of the utricle, oriented such that it is in the approximate plane of the horizontal semicircular canal. The saccular macula lies in a plane orthogonal to that of the utricular macula (see Fig. 31.1). With this orientation and the directional sensitivity of the otoliths, acceleration in three coordinate directions can be sensed.

31.2 FLUID MECHANICS OF THE SEMICIRCULAR CANALS

31.2.1 Governing Equation for Fluid Flow in the Semicircular Canals

In this section, we develop a mathematical model for fluid flow in a single semicircular canal. The membranous semicircular canal duct is approximated by a section of a rigid torus filled with an incompressible newtonian fluid. For the purpose of this analysis, the perilymph is assumed to have no effect on the deflection of the cupula. The governing equation for the flow of fluid in the duct is the classical Navier-Stokes equation

$$\frac{\partial \mathbf{v}}{\partial t} + \mathbf{v} \cdot \nabla \mathbf{v} = -\frac{1}{\rho}\nabla p + \mathbf{b} + \nu \nabla^2 \mathbf{v} \tag{31.1}$$

where $\mathbf{v}$ is the velocity of the fluid with respect to an inertial reference frame, ρ is the density, p is the pressure, $\mathbf{b}$ is the body force, and ν is the kinematic viscosity.

We are interested in the flow of the fluid with respect to the duct. Therefore, we introduce the symbol $\mathbf{u}$, which will represent the velocity of the fluid relative to the duct wall. If $\mathbf{v}_w$ is the velocity of the wall, then

$$\mathbf{v} = \mathbf{u} + \mathbf{v}_w \tag{31.2}$$

Now the velocity of a given point on the duct wall is given by

$$\mathbf{v}_w = \mathbf{v}_c + \boldsymbol{\omega} \times (\mathbf{R} + \mathbf{r}) \tag{31.3}$$

where $\mathbf{v}_c$ is the velocity of the center of curvature of the duct, $\boldsymbol{\omega}$ is the angular velocity of the canal, $\mathbf{R}$ is the position vector of the center of the duct with respect to the center of curvature, and $\mathbf{r}$ is the position vector of the point on the duct wall with respect to the center of the duct (see Fig. 31.2). Since $|\mathbf{r}|/|\mathbf{R}| \ll 1$, we can approximate $\mathbf{v}_w$ by

$$\mathbf{v}_w \approx \mathbf{v}_c + \boldsymbol{\omega} \times \mathbf{R} \tag{31.4}$$

Therefore

$$\mathbf{v} = \mathbf{u} + \mathbf{v}_c + \boldsymbol{\omega} \times \mathbf{R} \tag{31.5}$$

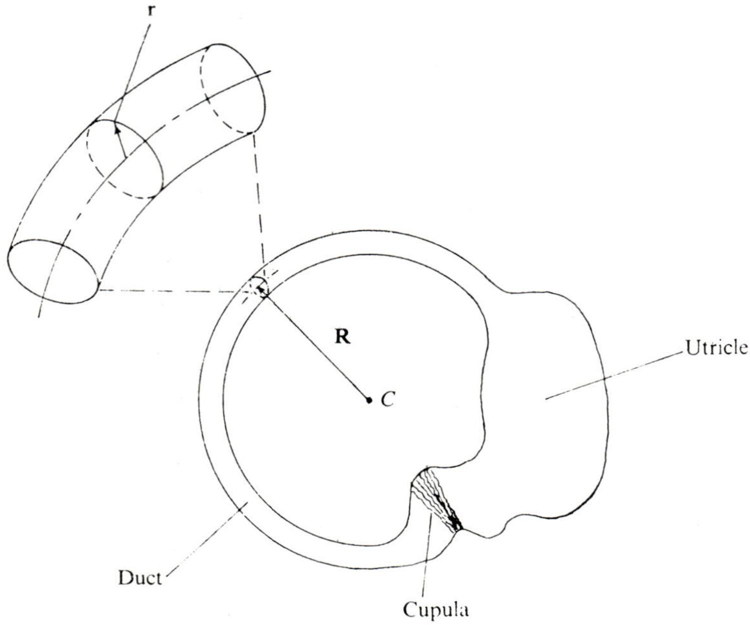

FIG. 31.2 Schematic diagram of a single semicular canal. **R** is a vector from C to the center of the duct, and **r** is a vector from the tip of **R** to the duct wall.

Substituting Eq. (31.5) into Eq. (31.1), we obtain

$$\frac{\partial \mathbf{u}}{\partial t} + \mathbf{u} \cdot \nabla \mathbf{u} + \mathbf{a}_c + \boldsymbol{\alpha} \times \mathbf{R} + \boldsymbol{\omega} \times (\boldsymbol{\omega} \times \mathbf{R}) = -\frac{1}{\rho}\nabla p + \mathbf{b} + \nu \nabla^2 \mathbf{u} \tag{31.6}$$

where $\mathbf{a}_c$ is the acceleration of the center of curvature of the duct and $\boldsymbol{\alpha}$ is the angular acceleration of the canal.

Let the pressure gradient ∇p be split into three parts:

$$\nabla p = \nabla p' + \nabla p'' + \nabla p''' \tag{31.7}$$

where $\nabla p''$ and $\nabla p'''$ are the pressure gradients associated with $\mathbf{a}_c$ and $\mathbf{b}$ (which for this problem is $\mathbf{g}$, the vector acceleration of gravity); that is,

$$\mathbf{a}_c = -\frac{1}{\rho}\nabla p'' \tag{31.8}$$

and

$$\mathbf{b} = \mathbf{g} = \frac{1}{\rho}\nabla p''' \tag{31.9}$$

The body force (gravity) and the acceleration $\mathbf{a}_c$ lead to no flow in the canal.

Canceling the terms indicated by Eqs. (31.8) and (31.9), we obtain

$$\frac{\partial \mathbf{u}}{\partial t} + \mathbf{u} \cdot \nabla \mathbf{u} + \boldsymbol{\alpha} \times \mathbf{R} + \boldsymbol{\omega} \times (\boldsymbol{\omega} \times \mathbf{R}) = -\frac{1}{\rho}\nabla p' + \nu \nabla^2 u \tag{31.10}$$

Since $|\mathbf{r}|/|\mathbf{R}| \ll 1$, we may express this equation in cylindrical coordinates. The axial component is

$$\frac{\partial u}{\partial t} + R\alpha = -\frac{1}{\rho}\frac{\partial p}{\partial z} + \frac{\nu}{r}\frac{\partial}{\partial r}\left(r\frac{\partial u}{\partial r}\right) \tag{31.11}$$

where r is the radial coordinate, z is the axial coordinate, u is the velocity of the fluid in the axial direction relative to the duct wall, R is the radius of curvature, α is the component of $\boldsymbol{\alpha}$ perpendicular to the canal, and $\partial p/\partial z$ is the axial component of $\nabla p'$.

We now consider the possible sources of a pressure gradient. The first source we will examine is the cupula. The cupula, when deflected, exerts a restoring force on the fluid. We model the cupula as a membrane with linear stiffness $K = \Delta p/\Delta V$, where Δp is the pressure difference across the cupula and ΔV is its volumetric displacement. If the angle subtended by the membranous duct is denoted by β, the pressure gradient in the duct produced by the cupula is

$$\frac{\partial p}{\partial z} = \frac{K\,\Delta V}{\beta R} \tag{31.12}$$

Now

$$\Delta V = 2\pi \int_0^t \int_0^a u(r,\ t)r\ dr\ dt \tag{31.13}$$

in which a is the radius of the duct.

A pressure gradient in the duct is also produced by the presence of the utricle. The size of this contribution may be assessed in the following manner. A cylinder of length l being accelerated through inertial space at a rate $\ddot{s}$ will have a pressure difference between its ends of

$$\Delta p = \rho \ddot{s} l \tag{31.14}$$

If the utricle had closed ends, the pressure difference due to rotational motion only could be approximated by

$$\Delta p \approx \rho\gamma R^2\alpha \tag{31.15}$$

where γ is the angle subtended by the utricle. Van Buskirk (1977) has shown that the small acceleration of the fluid in the utricle relative to the walls may be ignored. He has also shown that the pressure drop in the fluid as it moves from the larger utricle into the narrow duct is negligible. Therefore, Eq. (31.15) is an accurate approximation, and the pressure gradient in the duct due to the presence of the utricle is

$$\frac{\partial p}{\partial z} = \frac{\gamma}{\beta}\rho R\alpha \tag{31.16}$$

Again, of course, we have ignored the pressure gradient associated with $\mathbf{a}_c$ and $\mathbf{b}$ [see Eq. (31.8) and following]. Combining this result with the pressure gradient due to the cupula, we have

$$\frac{\partial p}{\partial z} = \frac{\gamma}{\beta}\rho R\alpha + \frac{K\,\Delta V}{\beta R} \tag{31.17}$$

Introducing Eq. (31.13) into Eq. (31.17) and substituting the resulting expression into Eq. (31.11), we obtain the governing equation for the fluid flow in the semicircular canals:

$$\frac{\partial u}{\partial t} + \left(1 + \frac{\gamma}{\beta}\right) R\alpha = -\frac{2\pi K}{\rho\beta R}\int_0^t \int_0^a u(r,\ t)r\ dr\ dt + \frac{\nu}{r}\frac{\partial}{\partial r}\left(r\frac{\partial u}{\partial r}\right) \tag{31.18}$$

Equation (31.18) is nondimensionalized by substituting the following variables:

$$r' = \frac{r}{a} \qquad t' = \frac{t\nu}{a^2} \qquad u' = \frac{u}{R\Omega}$$

where Ω is a characteristic angular velocity of the canal. In terms of these variables, Eq. (31.18) becomes

$$\frac{\partial u'}{\partial t'} + \frac{1 + \gamma/\beta}{\Omega} \alpha(t') = -\varepsilon \int_0^{t'} \int_0^1 u'r' \, dr' \, dt' + \frac{1}{r'} \frac{\partial}{\partial r'} \left(r' \frac{\partial u'}{\partial r'} \right) \tag{31.19}$$

where $\varepsilon = 2K\pi a^6/\rho\beta R\nu^2$.

31.2.2 Solution of the Governing Equation

We now examine the response of the canals to two specific kinds of angular acceleration. We will first obtain an approximate solution for the case of a step input in angular velocity, corresponding to an abrupt angular motion of the head. Since the problem is a linear one, the frequency response to sinusoidal motion of the head can be readily obtained from this first solution.

A step in angular velocity corresponds to an impulse in angular acceleration. Thus, in dimensionless form,

$$\alpha(t) = \Omega\delta(t) \tag{31.20}$$

where $\delta(t)$ corresponds to a unit impulse or Dirac delta function applied at $t = 0$. Substituting Eq. (31.20) into (31.19), we obtain the governing equation for the particular flow we are considering:

$$\frac{\partial u}{\partial t} + \left(1 + \frac{\gamma}{\beta}\right) \delta(t) = \frac{1}{r} \frac{\partial}{\partial r} \left(r \frac{\partial u}{\partial r} \right) - \varepsilon \int_0^t \int_0^1 ur \, dr \, dt \tag{31.21}$$

Note that in Eq. (31.21) we have dropped the primes. From this point on we shall work only with the nondimensionalized form of the governing equation. The boundary and initial conditions for this problem are

$$u(1, t) = 0 \qquad \frac{\partial u(0, t)}{\partial r} = 0 \qquad u(r, 0) = 0 \tag{31.22}$$

Van Buskirk et al. (1976) solved this problem analytically and obtained an approximate solution for the nondimensional volumetric displacement of the fluid given by the expression

$$\phi = \int_0^t \int_0^1 ur \, dr \, dt \tag{31.23}$$

(The actual volumetric displacement is given by $\nabla V = 2R\Omega\pi\alpha^4\phi/\nu$.) Their solution, assuming ε to be very small, is

$$\phi = \sum_{n=1}^{\infty} \frac{2(1 + \gamma/\beta)}{\lambda_n^4} \left[\exp(-\lambda_n^2 t) - \exp\left(-\frac{\varepsilon t}{16}\right) \right] \tag{31.24}$$

in which λ_n represents the roots of the equation $J_0(x) = 0$, where J_0 is the Bessel function of zero order. (The first three roots are $\lambda_1 = 2.405$, $\lambda_2 = 5.520$, and $\lambda_3 = 8.654$.) For $t = \infty$,

$$\phi = \sum_{n=1}^{\infty} \frac{2(1 + \gamma/\beta)}{\lambda_n^4} \tag{31.25}$$

which Peskoff (1977, personal communication) demonstrated reduced to $\phi = (1 + \gamma/\beta)/16$.

Transfer Function. The transfer function relating the nondimensional volumetric displacement ϕ to the angular acceleration α is obtained by finding the Laplace transform of ϕ and dividing by the Laplace transform of α. The result is

$$\frac{\phi}{\alpha}(s) = \sum_{n=1}^{\infty} \frac{2(1 + \gamma/\beta)}{\lambda_n^4} \left[\frac{\varepsilon/16 - \lambda_n^2}{(s + \varepsilon/16)(s + \lambda_n^2)} \right] \tag{31.26}$$

which may be approximated to very high accuracy by the simpler expression

$$\frac{\phi}{\alpha}(s) \approx \frac{-2(1 + \gamma/\beta)/\lambda_1^2}{(s + \varepsilon/16)(s + \lambda_1^2)} \tag{31.27}$$

An even more accurate representation may be obtained by requiring that the final value of the approximation be the same as the final value of the complete series (cf. Grant, 1973). Thus,

$$\frac{\phi}{\alpha}(s) = \frac{-(1 + \gamma/\beta)\lambda_1^2/16}{(s + \varepsilon/16)(s + \lambda_1^2)} \tag{31.28}$$

This may be put into a form

$$\frac{\phi}{\alpha}(s) = \frac{-(1 + \gamma/\beta)\lambda_1^2/16}{(s + \tau_1^{-1})(s + \tau_2^{-1})} \tag{31.29}$$

where $\tau_1 = 16/\varepsilon$ is the nondimensional long time constant, and $\tau_2 = 1/\lambda_1^2$ is the nondimensional short time constant. Converting from dimensionless time back to real time and expressing the transfer function in terms of θ, the mean angular displacement of the endolymph, and ω, the angular velocity of the head, we obtain

$$\frac{\theta}{\omega}(s) = \frac{-(1 + \gamma/\beta)s\lambda_1^2/8}{(s + T_1^{-1})(s + T_2^{-1})} \tag{31.30}$$

where $T_1 = 8\mu\beta R/K\pi a^4$ and $T_2 = a^2/\lambda_1^2\nu$.

Frequency Response. The usefulness of a formulation such as Eq. (31.30) is seen when we display the frequency response of the system in terms of Bode plots. Figure 31.3*a* is a plot of the logarithm of the ratio of the amplitudes of θ and ω as a function of the frequency. The phase lag between θ and ω is plotted in Fig. 31.3*b*. It is clear from these plots that the semicircular canal functions as an angular velocity transducer over the range $2/T_1$ and $1/2T_2$, which includes the range of physiological activity. The important points on Fig. 31.3 are the gain $G = (1 + \gamma/\beta)a^2/8\nu$, the lower cutoff frequency T_1^{-1}, the upper cutoff frequency T_2^{-1}, and the natural frequency $f_n = (T_1T_2)^{-1/2}$.

To make numerical predictions of semicircular canal response it is necessary to introduce numerical values into the parameters involved. The dimensions of the human semicircular canal are $a = 0.15$ mm and $R = 3.2$ mm (Igarashi, 1966). Studies of the physical properties of endolymph (e.g., Steer, 1967) suggest that it has a viscosity and density close to those of water; that is, $\mu = 1.0$ mPa·s and $\rho = 1000$ kg/m^3. Detailed geometrical studies of the labyrinth of the cat (Fernandez and Valentinuzzi, 1968) indicate that $\beta = 1.4\pi$ and $\gamma = 0.42\pi$. We accept these values for humans. No data are available in the literature for the cupula stiffness K of humans. For the purpose of this discussion we adopt the value $K = 3.4$ GPa/m^3. (This is not entirely arbitrary, as it yields results consistent with experiment, as will be seen below.) Using these values, $G = 3.7$ ms, $T_1^{-1} = 0.048$ rad/s $= 7.6$ mHz, $T_2^{-1} = 260$ rad/s $= 41$ Hz, and $f_n = 3.5$ rad/s $= 0.56$ Hz. We can also evaluate the term $\varepsilon = 2K\pi a^6/\rho\beta R\nu^2 = 0.017$. As we see, ε is very small, as was assumed in solving the governing equation for fluid flow in a semicircular canal.

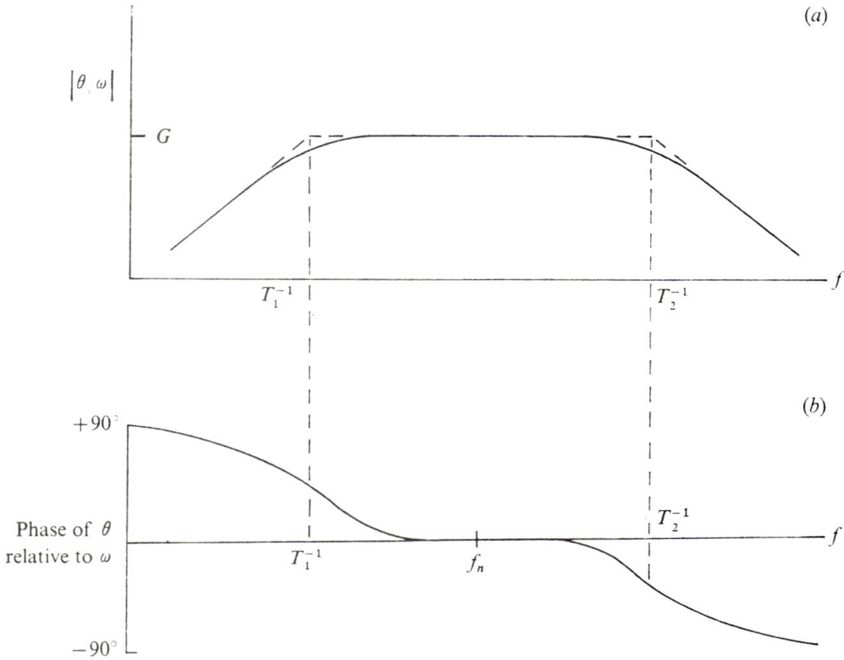

FIG. 31.3 Bode plots of the gain and phase lag between the mean angular displacement of endolymph, θ, and the angular velocity of the head ω, as a function of the frequency f of sinusoidal oscillation.

Impulse Response. The impulse response of the system given by Eq. (31.30) is

$$\phi \approx \frac{(1 + \gamma/\beta)\Omega a^2}{8\nu}(e^{-t/T_2} - e^{-t/T_1}) \tag{31.31}$$

where Ω is the magnitude of the step in angular velocity. The initial displacement of the cupula is governed by the short time constant $T_2 = 3.9$ ms and is shown in Fig. 31.4. The maximum displacement of the endolymph is $3.7 \times 10^{-3}\Omega$ rad, where Ω is given in rad/s. The return phase of the cupula is governed by the long time constant T_1 and is shown in Fig. 31.5. $T_1 = 21$ s, which is approximately what has been observed in experimental studies (Malcolm, 1968).

It is interesting to note the sensation associated with such a stimulation. The time constant T_2 is too short to be felt and the sensation is that of an instantaneous onset of angular velocity. But then, even though the angular velocity remains at a constant level, the sensation of angular velocity decays exponentially. Obviously, one's sensation of angular velocity is linked to the displacement of the cupula.

Deflection of the Cupula. The conventional view of the deflection of the cupula is that it moves like a swinging door rotating about a pivot at its base. Recent evidence suggests that it is far more likely that the cupula behaves like a diaphragm fixed to the ampulla around its perimeter (cf. McLaren and Hillman, 1976, 1977). With this mode of deflection in mind it is possible to determine the average deflection of the cupula, d, using the expression $d = \Delta V/A_c$, where A_c is the cross-sectional area of the cupula. Now $\Delta V = \pi a^2 R\theta$, and anatomical studies (Igarashi, 1966) suggest that $A_c \approx 1.0$ mm^2. Using data given above and Eq. (31.30), we obtain $d/\Omega = 0.83$ μm·s/rad. In the human, most head velocities fall in the range of 200 to 300°/s or 3.5 to 5.2 rad/s (Oman, 1982). For a head velocity of 4 rad/s the average displacement of the cupula would be

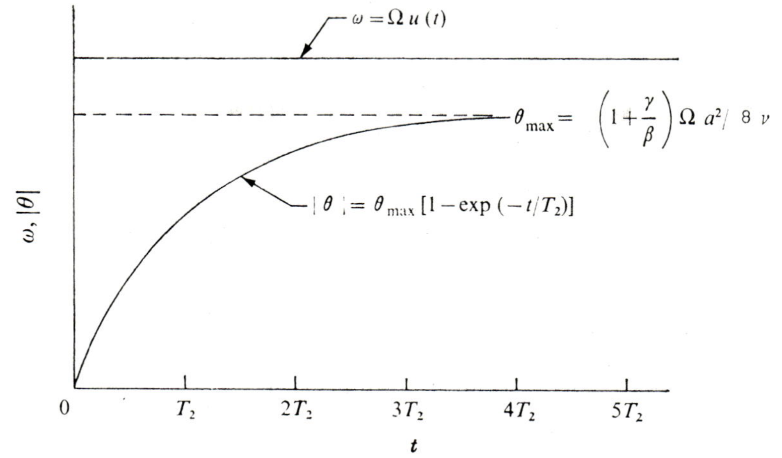

FIG. 31.4 Short-term response of the endolymph to a step input in angular velocity of magnitude Ω.

3.3 μm. The average cupula displacement predicted by this model for the subjective threshold of motion (approximately 35 mrad/s) is 29 nm.

Experimental Observations. Several researchers have attempted to measure experimentally the fluid-dynamic response of the semicircular canals. The first of these were probably van Egmond et al. (1949). Using subjective sensation as an indicator of canal response, they measured the long time constant T_1 and the natural frequency f_n for humans. They found $T_1 \approx 10$ s and $f_n = 0.16$ Hz. Later researchers used a more objective measure of semicircular-canal response, namely nystagmus, a characteristic movement of the eyes associated with semicircu-

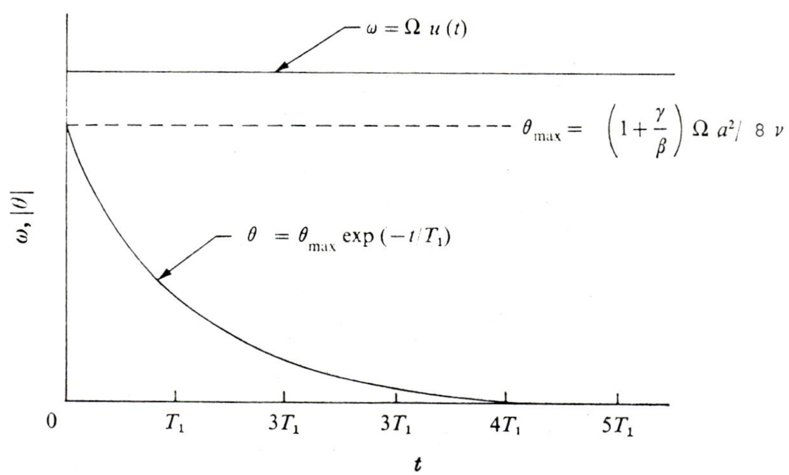

FIG. 31.5 Long-term response of the endolymph to a step input in angular velocity of magnitude Ω. The time scale in this figure is much larger than that in Fig. 31.4.

lar-canal stimulation. Malcolm (1968), using nystagmus and accounting for adaptation of the central nervous system, found a mean value of $T_1 = 21$ s for eight subjects. Niven and Hixson (1961), using sinusoidal oscillation as a stimulus and nystagmus as a measure of canal response, found a mean value of $f_n = 0.21$ Hz for six subjects.

By choosing an appropriate value for K, we have matched our theoretical time constant T_1 to that observed by Malcolm (1968). We should note, however, that our predicted value of f_n is more than twice that observed experimentally. Of course, neural processing could account for that difference, but a definitive answer will have to await further study.

31.3 MECHANICS OF THE OTOLITH ORGANS

31.3.1 Introduction

The otolith organs are the linear motion sensors in the vestibular system. They comprise a second-order system with mass, viscous damping, and spring, and respond to gravity as well as any acceleration of the skull. In mammals this organ consists of the utricular and saccular maculae of the inner ear. These organs have evolved and occupy relative positions to each other so that their combined efforts can respond to accelerations in three coordinate directions of the skull. The otolith organs, like the semicircular canals, are overcritically damped in such a fashion that their displacement is proportional to linear velocity over the frequency range of normal physiological motion.

31.3.2 The Mathematical Model

The otoconial membrane will be considered a rigid flat plate of thickness b and density ρ_0, and having infinite extent. The edge effects of the actual organ, which has finite extent, can be shown to be negligible owing to the small thickness b (15 μm) in relation to its flat plane area (2 mm^2). For the present formulation, only a small elemental surface area dA of the plate will be considered. This plate is in contact with the endolymph fluid above and the cupular membrane below. A two-dimensional cross section of this elemental area is shown in Fig. 31.6, where the cupular membrane is shown in a deformed state of simple shear. The displacement of the otoconial membrane with respect to the skull is δ_c, and V_s is the velocity of the skull.

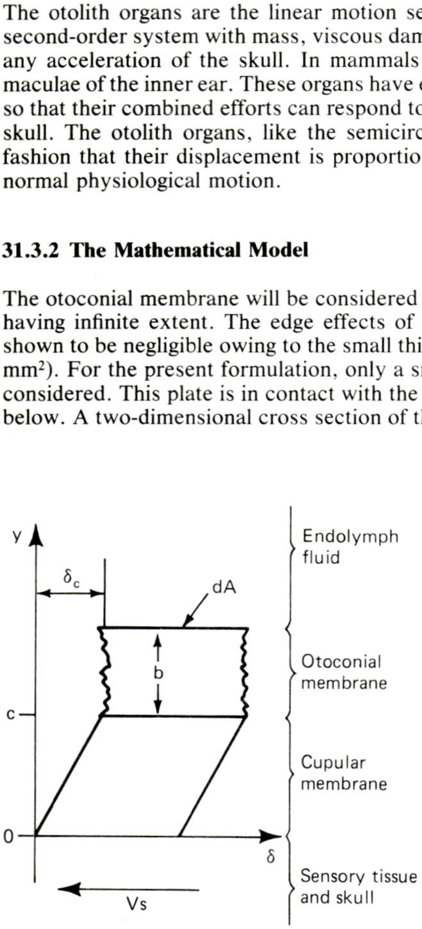

FIG. 31.6 Two-dimensional cross section of an otolith with surface area dA which is perpendicular to the plane of cross section.

The otolith organ structure shown in Fig. 31.6 can be divided into three elemental parts: the endolymph fluid section, the otoconial membrane, or rigid plate, and the deformable cupula membrane. Figure 31.7 shows a free-body diagram of each elemental part with its associated forces. The governing equations of motion can now be developed for each of the three elemental parts (for details of this derivation see Grant, 1984).

Considering the otoconial membrane plate first and applying Newton's second law of motion to the elemental volume shown in Fig. 31.7 yields

$$-\tau_f\, dA - \tau_e\, dA + W_x + B_x = \rho_0(b\, dA)\frac{dV_l}{dt} \tag{31.32}$$

where variables are as defined in the following list.

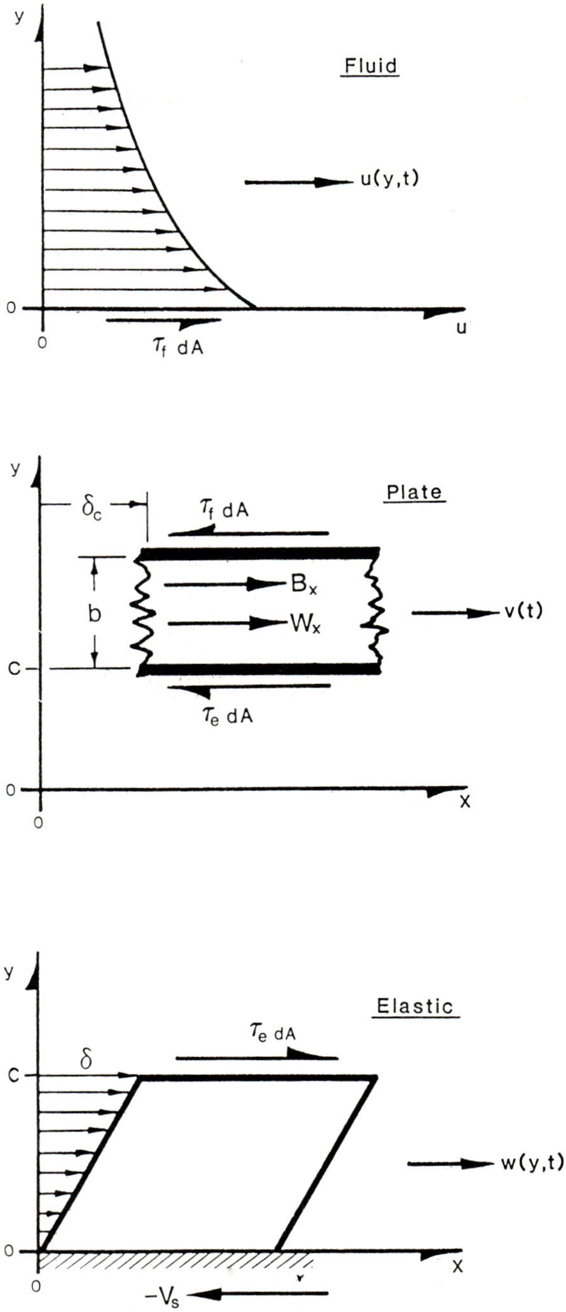

FIG. 31.7 Three elemental parts of an otolith showing free-body diagrams of each.

1. The fluid shear stress can be represented as

$$\tau_f = \mu\left(-\left.\frac{\partial u}{\partial y}\right|_0\right) \tag{31.33}$$

where u is the fluid velocity measured with respect to the skull, the velocity gradient is evaluated at the plate ($y = 0$), and μ is the endolymph viscosity.

2. The weight component in the x direction can be expressed as

$$W_x = \rho_0(b\ dA)g_x \tag{31.34}$$

where g_x is the component of gravity in the x direction.

3. The buoyant force component in the x direction can be represented as

$$B_x = \rho_f\left(\frac{\partial V_s}{\partial t} - g_x\right)(b\ dA) \tag{31.35}$$

where ρ_f is the fluid density and V_s is the skull velocity.

4. The elastic shear stress is developed from linear elasticity theory and can be represented as

$$\tau_e = \frac{1}{2}\left(\frac{E}{1+\nu}\right)\left(\frac{\partial \delta}{\partial y}\right) \tag{31.36}$$

where E is Young's modulus, ν is Poisson's ratio ($\nu = 0.5$, incompressibility), and δ is the deflection of the cupular membrane. Converting the displacement to a velocity integral, Eq. (31.36) becomes

$$\tau_e = \frac{1}{2}\frac{E}{1+\nu}\int_0^t \left.\frac{\partial w}{\partial y}\right|_c dt \tag{31.37}$$

where the subscript c indicates evaluation of the derivative at $y = c$, and w is the cupular material velocity measured with respect to the skull.

5. The inertial velocity V_I is then written as the sum of the relative velocity of the otoconial membrane with respect to the skull, v, and the velocity of the skull measured with respect to an inertial reference frame, V_s, as

$$V_I = v + V_s \tag{31.38}$$

Equation (31.32) then becomes

$$\rho_0 b\frac{\partial y}{\partial t} + (\rho_0 - \rho_f)b\left(\frac{\partial V_s}{\partial t} - g_x\right) = \mu\left.\frac{\partial u}{\partial y}\right|_0 - \frac{1}{2}\frac{E}{1+\nu}\int_0^t \left.\frac{\partial w}{\partial y}\right|_c dt \tag{31.39}$$

The fluid is shown as the upper element of the total system in Fig. 31.7. The linearized Navier-Stokes equation in the x direction describes the fluid motion

$$\rho_f\frac{\partial U_I}{\partial t} = \rho_f g_x - \frac{\partial P}{\partial x} + \mu\frac{\partial^2 U_I}{\partial y^2} \tag{31.40}$$

where U_I is the fluid velocity with respect to an inertial reference frame and

$$U_I = u + V_s \tag{31.41}$$

With an expression for the pressure gradient

$$\frac{\partial P}{\partial x} = \rho_f\left(g_x - \frac{\partial V_s}{\partial t}\right) \tag{31.42}$$

Eq. (31.40) becomes

$$\rho_f \frac{\partial u}{\partial t} = \mu \frac{\partial^2 u}{\partial y^2} \tag{31.43}$$

The boundary conditions for the fluid field equation (31.43) are

$$u(0, t) = v(t) \qquad \text{and} \qquad u(\infty, t) = 0 \tag{31.44}$$

The equation governing cupular membrane motion is developed from the momentum equation. The cupular membrane considered here is shown as the lower element in Fig. 31.7. The momentum equation in the x direction is

$$\rho \frac{\partial W_I}{\partial t} = \rho g_x - \frac{\partial P}{\partial x} + \frac{\partial \tau_e}{\partial y} \tag{31.45}$$

where W_I is the velocity of the cupular membrane measured with respect to an inertial reference frame and

$$W = w + V_s \tag{31.46}$$

The density in Eq. (31.45) is that of the cupular material. For the practical purposes of this formulation, the cupular material density can be considered equal to that of the endolymph fluid. Utilizing Eq. (31.42) for the pressure gradient, Eq. (31.37) for the elastic shear stress, and Eq. (31.46) for the inertial velocity, Eq. (31.45) becomes

$$\rho_f \frac{\partial w}{\partial t} = \frac{1}{2} \frac{E}{1 + \nu} \int_0^t \frac{\partial^2 w}{\partial y^2} \, dt \tag{31.47}$$

The boundary conditions for the field equation are

$$w(c, t) = v(t) \qquad \text{and} \qquad w(0, t) = 0 \tag{31.48}$$

The equation set (31.39), (31.43), and (31.47), with the associated boundary conditions, forms a coupled set of equations which governs the motion of the otoconial membrane in response to skull acceleration and gravity stimuli.

31.3.3 Nondimensionalization

The set of governing equations and boundary conditions can be nondimensionalized in order to determine the relative importance of each term in the equation. The following nondimensional variables, denoted by overbars, will be utilized:

$$\bar{y} = \frac{y}{b} \qquad \bar{t} = \frac{\mu}{\rho_0 b^2} t \qquad \bar{u} = \frac{u}{V} \qquad \bar{v} = \frac{v}{V} \qquad \bar{w} = \frac{w}{V} \qquad \text{and} \qquad \bar{V}_s = \frac{V_s}{V} \tag{31.49}$$

where V is some characteristic velocity in the problem. The governing equations then become

$$R \frac{\partial \bar{u}}{\partial \bar{t}} = \frac{\partial^2 \bar{u}}{\partial \bar{y}^2} \tag{31.50}$$

$$\frac{\partial \bar{v}}{\partial \bar{t}} + (1 - R)\left(\frac{\partial \bar{V}_s}{\partial \bar{t}} - \bar{g}_x\right) = \left.\frac{\partial \bar{u}}{\partial \bar{y}}\right|_0 - \varepsilon \int_0^{\bar{t}} \left.\frac{\partial \bar{w}}{\partial \bar{y}}\right|_1 d\bar{t} \tag{31.51}$$

$$R \frac{\partial \bar{w}}{\partial \bar{t}} = \varepsilon \int_0^{\bar{t}} \frac{\partial^2 \bar{w}}{\partial \bar{y}^2} \, d\bar{t} \tag{31.52}$$

with boundary conditions

$$\bar{u}(0, \bar{t}) = \bar{v} \qquad \bar{w}(1, \bar{t}) = \bar{v}(\bar{t})$$

$$\bar{u}(\infty, \bar{t}) = 0 \qquad \bar{w}(0, \bar{t}) = 0$$

where c is assumed equal to b. The two naturally occurring nondimensional parameters

$$R = \frac{\rho_f}{\rho_0} \tag{31.53}$$

and

$$\varepsilon = \frac{1}{2}\frac{E}{1+\nu}\frac{b^2}{\mu^2}\rho_0 \tag{31.54}$$

were used, and

$$\bar{g}_x = \frac{b^2\rho_0}{\mu V} g_x \tag{31.55}$$

is the nondimensional gravity. Note that when $R = 1$ ($\rho_0 = \rho_f$), there is no disturbance to the system.

The numerical values of ε and R can be determined by using the following physical parameters for humans: $\rho_0 = 2.71$ g/cm^3 (Carlstrom et al., 1953), $\rho_f = 1.00$ g/cm^3 (Steer, 1967), $b = 15$ μm (Igarashi, 1966), $\mu = 0.85$ cP (Steer, 1967), $E = 1.0$ dyn/cm^2 (estimated), and $\nu = 0.5$ (incompressibility). Utilizing these values, $R = 0.37$ and $\varepsilon = 0.028$.

The resulting small values of $\varepsilon = 0.028$ indicates that the elastic term is relatively unimportant in relation to the viscous and inertial terms in Eq. (31.51). Since Eq. (31.51) governs the displacement of the otoconial membrane, it can be concluded that the elastic terms are almost negligible, and otoconial membrane displacement is dominated by inertial and viscous effects.

31.3.4 Solution for and a Step Change in Skull Velocity

Understanding of the system dynamics can be ascertained by the system response to a step change in skull velocity. For this case $g_x = 0$, gravity will not be considered, and elasticity will be assumed to be negligible with $\varepsilon = 0$. For a unit step change in skull velocity of magnitude $-V$ at time $t = 0$, the solution has been determined by Grant et al. (1984) to be

$$\bar{u}(\bar{y}, \bar{t}) = (1 - R)e^{R(\bar{y}+\bar{t})}\,\mathrm{Erfc}\left(\frac{\bar{y}}{2\sqrt{\bar{t}/R}} + \sqrt{R\bar{t}}\right) \tag{31.56}$$

where Erfc is the complementary error function.

The velocity of the otoconial membrane plate can be found by setting $\bar{y} = 0$, and the plate velocity becomes

$$\bar{v}(\bar{t}) = (1 - R)e^{R\bar{t}}\,\mathrm{Erfc}\,\sqrt{R\bar{t}} \tag{31.57}$$

This nondimensional velocity as a function of time is shown in Fig. 31.8. As can be seen, the velocity drops to less than 20 percent of its original value in less than 15 ms. By using the final value theorem, it can be shown that the velocity does go to zero at infinite time. Even though the velocity initially drops rapidly, the plate continues to move because there is no elastic constraint, and the velocity gradient has become small.

The displacement of the otoconial membrane plate as a function of time for the above set of conditions has also been found and is

$$\bar{\delta}(\bar{t}) = \left(\frac{1-R}{R}\right)\left(\sqrt{\frac{4}{\pi}}R\sqrt{\bar{t}} - 1 + e^{R\bar{t}}\,\mathrm{Erfc}\,\sqrt{R\bar{t}}\right) \tag{31.58}$$

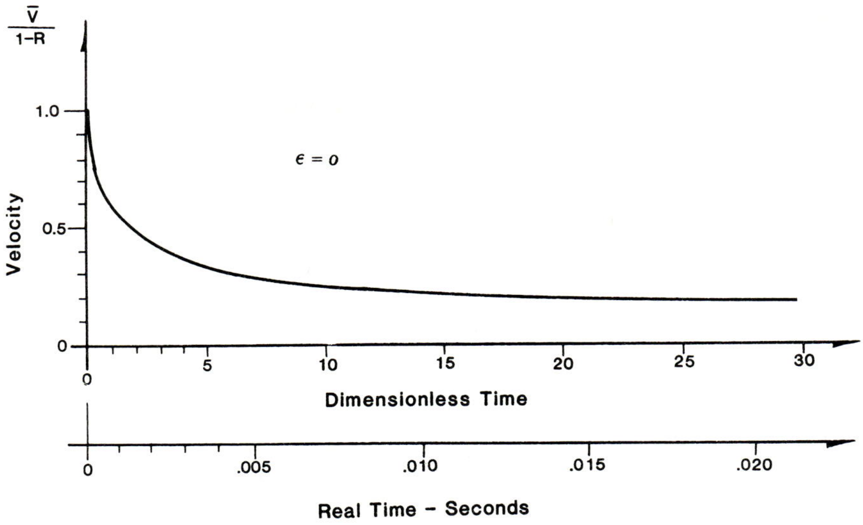

FIG. 31.8 Dimensionless velocity of the otoconial membrane versus real and dimensionless time (ρ_0 = 2.7 g/cm^3, b = 15 μm, and μ = 0.85 cP).

where the dimensionless displacement is

$$\bar{\delta} = \frac{\mu}{V\rho_0 b^2}\,\delta \tag{31.59}$$

The dimensionless displacement as a function of time is shown in Fig. 31.9. Again by using the final value theorem it can be shown that the displacement does go to infinity at infinite time. As can be seen from Fig. 31.9, the displacement is increasing slowly after 15 ms. The elastic term (when $\varepsilon \neq 0$) will limit this displacement and bring the otoconial membrane displacement to some maximum finite value.

A set of solutions to the governing equations for small ε can be obtained by numerical integration of the governing equations. These solutions for various values of ε are also shown in Fig. 31.9. As can be seen, the maximum displacement of the otoconial membrane is limited when the elastic effects ($\varepsilon \neq 0$) are included in the solution. Also, as ε increases, the displacement overshoots, indicating underdamped oscillatory response. The system has been shown to have a highly overdamped response; see de Vries (1950) and Young and Meiry (1968). A reasonable value of ε which will produce overdamped response would be $\varepsilon < 0.02$. When the numerical solutions for nonoscillatory motion are extended beyond those times shown in Fig. 31.9, the displacements tend to remain at their maximum value and then begin an extremely slow return to the equilibrium position.

31.3.5 System Frequency Response

From the previous analysis the upper break-point frequency of the otolith system can be ascertained. The initial response of the otoconial membrane to a step change in skull velocity is a fast displacement to some finite position and a very slow return to the equilibrium position. This type of response, which is similar to that of the canals, characterizes the otolith organs as velocity sensors with otoconial membrane displacement proportional to velocity.

A time constant of T_2 = 3 ms accurately describes the initial displacement of the system. This results in an upper break-point frequency of T_2^{-1} = 330 rad/s = 53 Hz. The lower break-

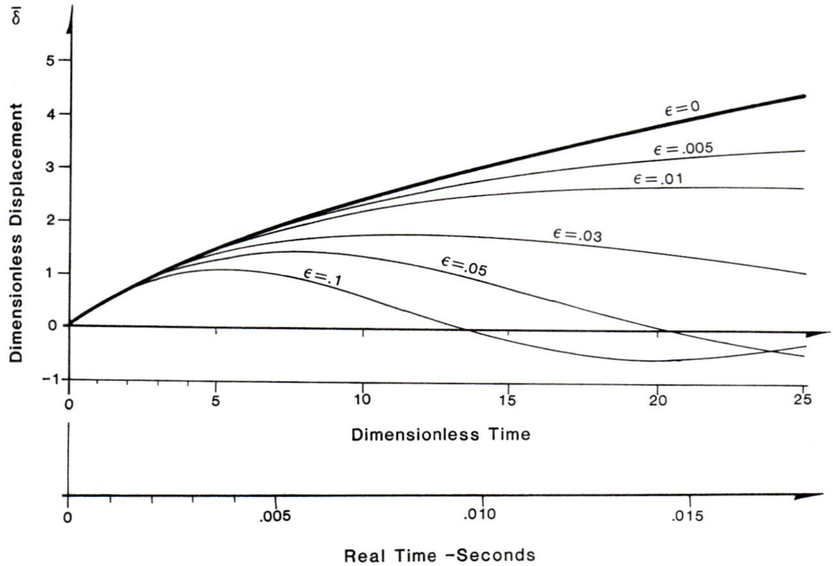

FIG. 31.9 Dimensionless displacement of the otoconial membrane versus real and dimensionless time for various values of ε (ρ_0 = 2.7 g/cm^3, b = 15 μm, and μ = 0.85 cP).

point frequency has been measured by Young (1966) to be T_1^{-1} = 0.1 rad/s = 0.016 Hz. Thus the bandwidth over which the otolith organs are linear is similar to that of the canal system, and again this is the range of normal physiological motion.

SUMMARY

In this chapter we have examined the fluid-dynamic response of the semicircular canals and the mechanics of the otolith organs. We have shown that each of these organs responds like a heavily damped, second-order system. An examination of the frequency response of these two systems indicates that the primary function of each is that of a velocity sensor. Using input from these organs and other sensory modalities, we arrive at not only our sensation of motion but also our sensation of orientation with respect to the gravity field. While the vestibular system works well in transducing ordinary motions, we see from an examination of the impulse response of the semicircular canals that unphysiological motions can lead to the generation of erroneous signals. The otolith organs will also generate false velocity information when subjected to motion with frequency content outside their bandwidth. It has been suggested in the literature that erroneous signals to the central nervous system and signals that conflict with those from other sense organs are possible causes for spatial disorientation and motion sickness (Johnson and Jungkees, 1974).

REFERENCES

Carlstrom, D., Engstrom, H., and Hjorth, S. (1953). Electron microscopic and x-ray diffraction studies of statoconia. *Laryngoscope, 63,* 1052–1057.

de Vries, H. (1950). The mechanics of the labyrinth otoliths. *Acta Oto-Laryngol., 38, 262–273.*

Fernandez, C., and Valentinuzzi, M. (1968). A study on the biophysical characteristics of the cat labyrinth. *Acta Oto-Laryngol., 65, 293–310.*

Grant, J. W., Best, W. A., and Lonegro, R. (1984). Governing equations of motion for the otolith organs and their response to a step change in velocity of the skull. *J. Biomech. Eng., 106,* 302–308.

Grant, J. W. (1973). Biomechanics of the semicircular canals. D. Eng. dissertation, Tulane University, New Orleans, La.

Igarashi, M. (1966). Dimensional study of the vestibular end organ apparatus. *Second Symposium on Role of Vestibular Organs in Space Exploration.* NASA SP-115, Washington, D.C., pp. 47–53.

Johnson, W. J., and Jongkees, L. B. W. (1974). Motion sickness. *Handbook of Sensory Physiology.* Springer-Verlag, New York, vol. VI, pt. 2.

Lowenstein, O. E. (1974). Vestibular system: 1. Basic mechanisms. *Handbook of Sensory Physiology.* Springer-Verlag, New York, vol. VI.

Lowenstein, O., and Roberts, T. D. M. (1950). The equilibrium function of the otolith organs of the thornback ray (*Raja clavata*). *J. Physiol., 110,* 392–415.

Malcolm, R. (1968). A quantitative study of vestibular adaptation in humans. *Fourth Symposium on Role of Vestibular Organs in Space Exploration.* NASA SP-187, Washington, D.C., pp. 369–380.

Mayne, R. (1974). Vestibular system: 2. A systems concept of the vestibular organs. *Handbook of Sensory Physiology.* Springer-Verlag, New York, vol. VI, pt. 2.

McLaren, J. W., and Hillman, D. E. (1976). Configuration of the cupula during endolymph pressure changes. *Neurosci. Abstr., 2,* 1060, Abstr. 1531.

McLaren, J. W., and Hillman, D. E. (1977). Displacement of the semicircular canal cupula during sinusoidal rotation. *Neurosci. Abstr., 3,* 544, Abstr. 1730.

Meiry, J. L. (1965). The vestibular system and human dynamic space orientation. Sc.D. thesis, Massachusetts Institute of Technology, Cambridge, Mass.

Niven, J. I., and Hixson, W. C. (1961). Frequency response of the human semicircular canal: I. Steady-state ocular nystagmus response to high-level sinusoidal angular rotations. *Rep. U.S. Nav. School Aviat. Med.,* No. NSAM-58, Pensacola, Fla.

Oman, C. M. (1982). The influence of duct and utricular morphology on semicircular canal response. (In T. Gaultierotti (ed.), *Vestibular Function and Morphology.* Springer-Verlag, New York.

Peskoff, A. (1977). Personal communication. University of California, Los Angeles.

Spoendlin, H. H. (1965). Ultrastructural studies of the labyrinth in squirrel monkeys. *The First Symposium on The Role of the Vesticular Organ in Space Exploration.* NASA SP-77, Washington, D.C., pp. 7–22.

Steer, R. W. (1967). The influence of angular and linear acceleration and thermal stimulation on the human semicircular canal. Sc.D. thesis, Massachusetts Institute of Technology, Cambridge, Mass.

Van Buskirk, W. C. (1977). The effects of the utricle on flow in the semicircular canals. *Ann. Biomed. Eng. 5,* 1–11.

Van Buskirk, W. C., and Grant, J. W. (1973). Biomechanics of the semicircular canals. *1973 Biomech. Symp.* American Society of Mechanical Engineers, New York, pp. 53–54.

Van Buskirk, W. C., Watts, R. G., and Liu, Y. K. (1976). The fluid mechanics of the semicircular canals. *J. Fluid Mech., 78,* 87–98.

van Egmond, A. A. J., Groen, J. J., and Jongkees, L. B. W. (1949). The mechanics of the semicircular canal. *J. Physiol., 110,* 1–17.

van Holst, E. (1950). Die Arbeitsweise des Statolithenapparates bei Fischchen. *Z. Vgl. Physiol. 32,* 60–120.

Vilstrup, G., and Vilstrup, T. (1952). Does the utricular otolith membrane move on postural change of the head? *Ann. Otol. Rhinol. Laryngol., 61,* 189–197.

Young, L. R. (1974). Role of the vestibular system in posture and movement. In V. B. Mountcastle (ed.), *Medical Physiology.* Mosby, St. Louis, vol. 1, pp. 704–721.

Young, L. R., Lichtenberg, B. K., Arrott, A. P., Crites, T. A., Oman, C. M., and Edelman, E. R. (1981). Occular torsion on earth and in weightlessness. *Ann. N.Y. Acad. Sci.,* **374,** 80–92.

Young, L. R., and Meiry, J. L. (1968). A revised dynamic otolith model. *Aerosp. Med., 39,* 606–608.

Young, L. R., Meiry, J. L., and Li, Y. T. (1966). Control engineering approaches to human dynamic space orientation. *Second Symposium on the Role of the Vestibular Organs in Space Exploration.* NASA SP-115, Washington, D.C., pp. 217–227.

Wilson, Victor J., and Melville Jones, G. (1979). *Mammalian Vestibular Physiology.* Plenum Press, New York.

CHAPTER 32

Computed Tomography and Three-Dimensional Imaging

Lowell D. Harris
Erik L. Ritman
Richard A. Robb
Department of Physiology and Biophysics
Biodynamics Research Unit
Mayo Foundation
Rochester, Minnesota

INTRODUCTION

Computed tomography (CT) combines the fundamental principles of x-ray physics with the disciplines of mathematics, engineering, and computer science to produce images of slices through the body. X-ray CT images are analogous to images obtained by freezing the body, slicing the body into 1- to 10-mm-thick sections perpendicular to the long axis of the body, and forming an x-ray image of the slice by placing the section on a photographic plate and producing a radiograph of the section with an x-ray source situated perpendicular to the plane of the section. Both of the images just described are *tomographic,* from the Greek "*tomo,*" "slice," and *graph,* or "engravings." Unlike the physical slicing process, however, computed tomography is both noninvasive and nondestructive except for the ionizing effects of the radiation and the possible need to inject x-ray contrast medium into a vein.

The significance of imaging slices is the elimination of superposition; i.e., nothing "hides behind," or is obscured by, overlying structures in an image of a thin section. The result is that tomography provides clear, unambiguous views of two-dimensional sections through the body. In conventional radiographic projection imaging, objects such as a tumor in the lungs can hide behind the other structures such as the heart or a rib. Furthermore, because a radiograph is a two-dimensional projection of a three-dimensional structure, depth information is lost; i.e., what is really presented is a silhouette of three-dimensional objects. Conversely, tomographic imaging makes possible the complete three-dimensional "reconstruction" of organs if a "stack" of an adequate number of parallel, sufficiently thin, tomographic images can be obtained. Such a stack of CT images (three-dimensional reconstruction) is referred to as a *volume image.* The ultimate potential of computed tomography is further extended by the realization that many organs not only are three-dimensional but also are continually moving and changing shape. To the extent that adequate temporal as well as spatial and density resolution can be achieved with computed tomography (four-dimensional images), the structure-to-function relationship of the heart, lungs, and circulation can be studied in ways never before possible.

Presented here is a brief introduction to two-, three-, and four-dimensional computed tomography. More detailed descriptions of the mathematics of reconstruction [1–3] and of CT scanner operation [4] are presented elsewhere. The complementary tomographic imaging methods of single-photon-emission tomography (SPECT) [5, 6], positron-emission tomography (PET) [7, 8], nuclear magnetic resonance (NMR) imaging [9, 10], ultrasound sector scanning [11], and ultrasound computer-assisted tomography [UCAT] [12] are only named here. Each imaging modality "probes" the body with a different energy form, yielding unique and useful information about tomographic sections through the body.

32.1 RECONSTRUCTION FROM PROJECTIONS: INTUITIVE DEVELOPMENT

Stated simply, the two-dimensional x-ray reconstruction problem is one of computing a two-dimensional cross-sectional image from a set of one-dimensional projection profiles, obtained by passing an x-ray beam through the body from multiple coplanar angles of view. In short, the procedure is referred to as *reconstruction from projections* [1, 2]. This description is intended to give an intuitive understanding of the reconstruction process by using only enough mathematics to help clarify the basic principles. Excellent reviews of rigorous mathematical descriptions of reconstruction [3], comparison of reconstruction algorithms, and historical development of reconstruction mathematics [4] are given elsewhere.

Before the concepts of reconstruction can be developed, it is necessary to outline how the projection data are obtained and how the projection data are related to the object. The geometry of projection is illustrated in the upper panel of Fig. 32.1. The view is of an infinitely thin cross section through the object of interest. Until otherwise specified, the assumption is that the ray paths of an x-ray beam are parallel. Some commercially available CT scanners still use parallel x-ray beams obtained by translating a highly collimated "pencil" beam source along a path parallel to the detector line. The assumption of parallel ray paths simplifies the present discussion. The extensions to "fan-beam" geometry [13, 14] will be added later. The x-ray detectors, located on the r axis, measure transmitted x-ray beam intensity from which significant information about the object $U(x, y)$ can be derived by means of the relationship between the incident and transmitted x-ray intensities. This relationship for a monoenergetic x-ray beam and objects of uniform density is given by the Lambert-Beer equation:

$$I = I_0 e^{-ul} \tag{32.1}$$

where I_0 and I are the incident and transmitted x-ray intensities, respectively, u is the x-ray absorption coefficient (linear attenuation coefficient), and l is the path length through the object. The linear attenuation coefficient u is closely related to, and will hereafter be referred to as, the density of the object. Although u is dependent on the wavelength (energy) of the x-ray beam [15], the assumption here is that u is independent of photon energy. From measurements of I and I_0, it is possible to estimate the product ul as illustrated with Eq. (32.2):

$$ul = \ln \frac{I_0}{I} = P_{r,\theta} \tag{32.2}$$

where $P_{r,\theta}$ is called the projection *ray sum,* where the relationships between the path l and the orientation variables r and θ are shown in the upper panel of Fig. 32.1.

Of course, in the body, u is not constant but is rather a spatially varying function related to tissue density, represented here mathematically by the function $U(x, y)$. With this additional insight, the next step is to replace the simple ray-sum product ul with a line integral:

$$\int_l U(x, y)\, dl = P_{r,\theta} \tag{32.3}$$

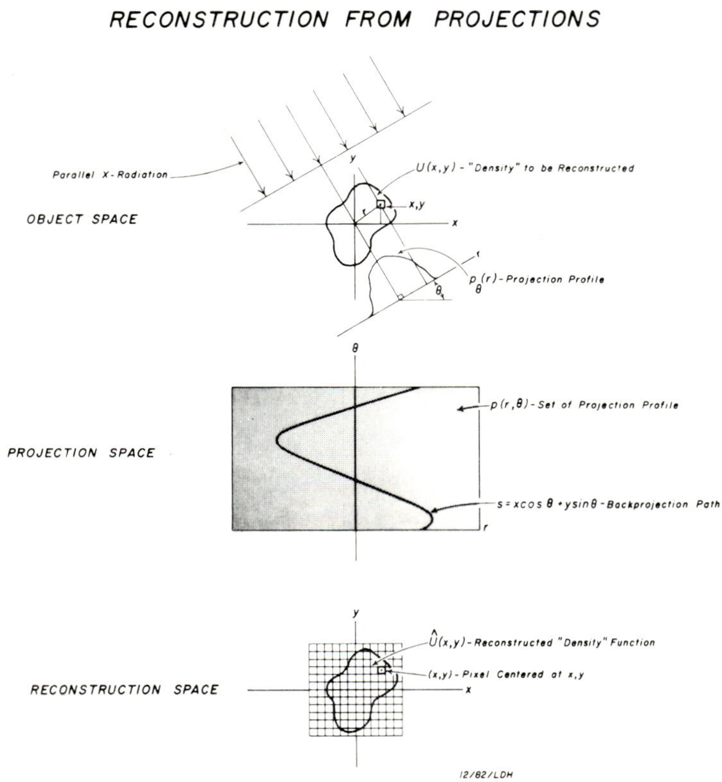

FIG. 32.1 Graphical illustration of reconstruction. Upper panel illustrates process of collecting x-ray projection information assuming parallel x-radiation. Center panel illustrates the back-projection integration path s through theoretical projection space. Lower panel illustrates the discrete nature of reconstruction space. The aim is to compute $\hat{U}(x, y)$, which is an estimate of the density function $U(x, y)$.

The aim of reconstruction is to compute the function $\hat{U}(x, y)$ ("reconstructed" density function), which ideally will be an accurate estimate of the unknown density distribution function $U(x, y)$, from a large number of coplanar ray sums obtained from many angles of view. The set of parallel-path ray sums for a particular detector angle is referred to as a *projection profile* and is written $P_\theta(r)$, as shown in Fig. 32.1. The set of projection profiles from multiple angles of view ($\theta = 0$ to $180°$) is referred to as *projection space,* $P(r, \theta)$, illustrated in the center panel of Fig. 32.1.

$P(r, \theta)$ exists as a continuous function only mathematically. In reality, projection space is a finite set of discrete ray sums, $P_{r,\theta}$, defined at a finite number of detector angles (θ) and for a finite number of detector positions (r). Moreover, the individual detectors have finite dimensions and therefore measure transmitted x-ray intensity along a small segment of the r axis, i.e., bands through the object. Finally it must be realized that to facilitate computation in a digital computer, the reconstruction space "object" is divided into many small picture elements (upper and lower panel in Fig. 32.1).

One method for computing $\hat{U}(x, y)$ (yielding rather poor results) is simple back projection [16], which is illustrated in Fig. 32.2. Conceptually, back projection is accomplished by "smearing" the projection profile back across the reconstruction space picture elements. In this exam-

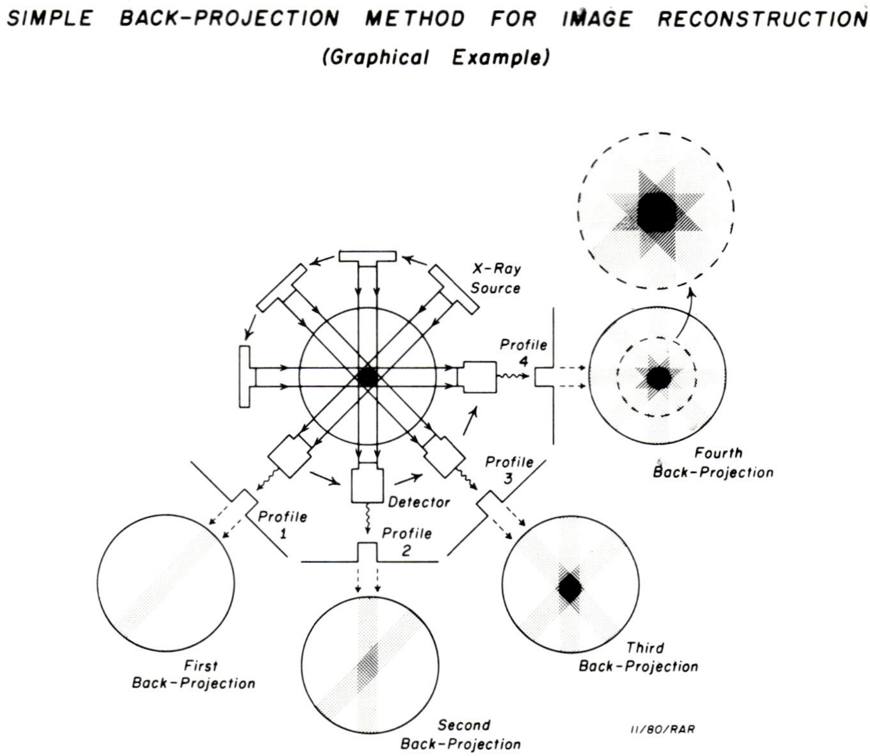

FIG. 32.2 Graphical example of simple back-projection method for image reconstruction. For each angle of view, the profile values are projected back across the reconstruction space by adding them to the values of the pixels in the reconstruction grid. For a single point, the image is thus reconstructed by successive strips. Where values in successive profiles are nonzero, the back-projected values add together at locations of intersection. As more and more profiles recorded at successive angular increments are back-projected, the central intersection of the strips defines the reconstructed region. However, because of the finite width of the beam and the finite number of views used, a starlike pattern rather than a round point is reconstructed. As the number of back-projected strips approaches infinity, the density distribution along the spikes of the star varies as $1/r$, where r is the distance from the center of the star (i.e., the true point). (*From Robb [4]; reproduced with permission.*)

ple, which is a reconstruction of a single dark circular region, back projection of the first profile produces a band across the image. The back projection of the second profile produces a band corresponding to the new angle, resulting in reinforcement at the intersection of the two bands. As more projection profiles from additional views are back-projected, the region corresponding to the original dense (dark) central region is reinforced, resulting in a reconstruction of that region. In addition, however, a star-shaped artifact appears to emanate from the dark region. Assume for the moment that the dark circular region is an infinitely small, high-density spike (impulse); then the reconstruction of the impulse from an infinite number of equally spaced views around 180° will be a point with an associated artifact. It can be shown [17] that the artifact will be a smooth continuous function with value $1/r$, where r is the distance from the impulse point. Therefore, simple back projection results in a reconstruction where every point in the image has been convolved with the $1/r$ function.

To see how the $1/r$ artifact can be removed (deconvolved) from the back-projected image, it is helpful to look at back projection in a different way. Rather than considering this process as one of smearing the projection profile across the image, one may think of back projection as a process of sampling all the projection profiles at the location which corresponds to where the point projected onto the detector strip. The contribution from all the projection profiles for each point (pixel in the discrete case) are summed to form the back-projected value at that point.

Back projection then is a summation, which in the continuous projection space is described as integration along a specific path, as illustrated in the center panel of Fig. 32.1. The path of integration is defined by the geometrical relationship between x and y in the object space and r and θ in the projection space:

$$r = x \cos \theta + y \sin \theta \tag{32.4}$$

This relationship defines a unique path s through projection space for each point in $U(x, y)$. The back-projection line integral can be written

$$\int_x P(r, \theta)\, d\theta \tag{32.5}$$

By substituting Eq. (32.4) into (32.5) and setting the limits of integration, Eq. (32.5) can be rewritten

$$\int_0^{\pi} P(x \cos \theta + y \sin \theta, \theta)\, d\theta \tag{32.6}$$

where the path s traces out a sine wave through the projection space. Note that the variable of integration is θ and not s. Evaluation of a line integral along the path s with respect to θ is a legitimate line integration. The result is that the area under $P(r, \theta)$ along s is projected onto the plane defined by the θ and P axes [18]. This step is necessary to compensate for the fact that the integration path length is a function of how far the point x, y is from the center of rotation.

To remove the $1/r$ artifact the back projection previously described is combined with the process of convolution (deconvolution). The basic idea is to convolve the projection profile with a filter which removes the $1/r$ artifact before it is added in back projection and then back-project the filtered projection profiles. The filtering is given by

$$P'_\theta(r) = P_\theta(r) * f \tag{32.7}$$

where f is the filter function, $P_\theta(r)$ is the one-dimensional projection profile, $*$ represents the convolution operator, and $P'_\theta(r)$ is the one-dimensional filtered projection profile. Because this system is linear and stationary, it is possible to filter, i.e., to remove the $1/r$ artifact before it is introduced during back projection. The filtered back-projection integral can finally be written

$$\hat{u}(x, y) = \int_0^{\pi} P'(x \cos \theta + y \sin \theta, \theta)\, d\theta \tag{32.8}$$

Although the derivation of the filter function f [19, 20] will not be discussed here, a few intuitive points are illustrated by use of the discrete representation of two often-used filters shown in Fig. 32.3. Note that the filter is symmetrical and contains a negative component on either side of the center point. It is this negative portion which removes the $1/r$ artifact.

The discrete representation of the continuous integration given by Eq. (32.8) is

$$\hat{u}(x, y) = \frac{\pi}{k} \sum_{n=0}^{k-1} P_n^* \left(x \cos \frac{n\pi}{k} + y \sin \frac{n\pi}{k} \right) \tag{32.9}$$

where k is the number of projection profiles around 180°. The entire convolution and back-projection processes are implemented numerically by means of interpolation to compensate for the fact that the discrete sampling location of the back-projection summation and the projection space are usually not coincident.

Other methods have been developed to remove the $1/r$ artifact, such as two-dimensional deconvolution of the simple back-projection image, a process referred to as *rho filtering*. Addi-

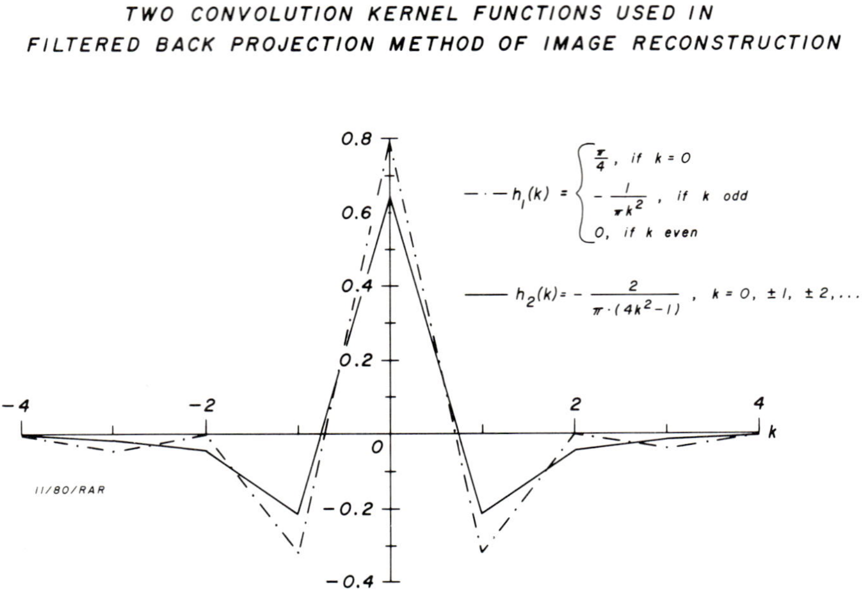

FIG. 32.3 Two convolution filter functions used in the filtered back-projection method of image reconstruction. The width of $h_1(k)$, often called the *Ram-Lak filter,* and the width of $h_2(k)$, often called the *Shepp-Logan filter,* are nearly the same for the central lobe, so that image resolution for noiseless data will be similar with either filter. However, the side lobes of $h_2(k)$ are more damped, producing greater attenuation of higher frequencies than $h_1(k)$. (*From Robb [4]; reproduced with permission.*)

tional methods involve taking the Fourier transform of a simple back projection, removing the $1/r$ artifact by multiplication of the Fourier transform of the filter function and taking the inverse Fourier transform [21, 22]. Iterative methods which invert the system of equations generated directly by using a discrete representation of Eq. (32.3) have also been used successfully [23, 24].

The extension of filtered back projection to systems using a divergent (fan x-ray) beam is done quite simply [13, 14]: (1) the back-projection path must be calculated by assuming a fan beam scanning geometry; (2) the projection profile must also be preweighted to compensate for the diverging x-ray beam, before filtering; and (3) the limits of integration must be extended to go from 0 to $\pi + \alpha$, where α is the fan angle. The same filter function can be utilized.

32.2 THE CT SCANNER

The CT scanner is a machine which uses the physical and mathematical principles outlined in the previous section to produce digital images of tomographic sections [4]. Functionally the scanner is made up of three interrelated subsystems: (1) the scanning subsystem (gantry, x-ray source, detectors, patient table, etc.), (2) the computing subsystem, and (3) the display console. In commercial CT equipment, the scanning subsystem generates the requisite digital projection data for a single tomographic image, passing these data onto the computer system, which computes the digital tomographic image, which in turn is passed on to the display system, which produces a visible image on film or television (TV) monitor. The fabrication of a highly complex CT scanner involves literally thousands of design decisions which in major or minor ways affect the performance of the machine. Since a full discussion of the history of or rationale behind different approaches is not feasible here, we will describe how scanner performance can be evaluated and how major design considerations affect performance.

CT scanner performance is generally evaluated by three standard objective measures: temporal, spatial, and density resolution [25, 26]. For most commercial scanners, temporal resolution, the time required to scan (collect) all the projection data, is specified by the manufacturer. This time interval, which may be variable for different scanning modes, should not be confused with the time interval between the onset of the scan and display of the CT image, an interval which includes reconstruction and postprocessing times. Ideally the magnitude of the scan interval should be set by biological factors. For example, scanning the heart clearly calls for higher temporal resolution than scanning the head, with scanning of the abdomen somewhere in between. If the object changes shape or density during the scan interval, an inconsistent projection set (i.e., the object appears to be different at different angles of view) is produced, which may result in artifacts in the reconstructed image [27]. In general, the ideal scanner would have infinite temporal resolution, i.e., a scan time of zero, but practical factors such as mechanical motion and x-ray tube heat loading make this impossible.

The second measure of scanner performance, spatial resolution, has improved dramatically in the past few years. It is generally measured by using specially designed test objects (phantoms) and given in terms of line or hole pairs per centimeter [25, 26]. The phantoms generally are made up of sets of rows of holes drilled in a material such as acrylic fiber. The phantom is physically positioned in the scanner so that the slice is perpendicular to the axis of the holes. The holes are arranged in rows of increasing size, with the centers of the holes in each row separated by a distance equal to twice the hole diameter. The holes are filled with a material which differs in density from the surrounding material and then scanned. Spatial resolution of the total system is estimated by visually observing the smallest set of holes which appear as discrete entities in the CT scans. More quantitative, full-width, half-maximum measures can also be made from the images.

Spatial resolution is affected by many factors, such as the number of rays per projection, number of projections, angle between the projections, range of angles for which projections are obtained, size of the detectors, detector efficiency, reconstruction algorithms, pixel size, and x-ray dose. In general, spacial resolution is measured only in the plane of the tomographic image by use of a phantom which is uniform in the direction perpendicular to the scanning plane. Since the body seldom cooperates to the extent of being axially uniform, i.e., uniform over a 10-mm thickness, these indices of resolution can be misleading.

Density resolution is a measure usually expressed as the minimum density difference which can be "detected" [25] from the CT image. Like spatial resolution, density resolution is measured by specially constructed phantoms with regions of varying density so that minimum detectable density differences can be directly observed. Density resolution, like spatial resolution, is affected by such factors as the x-ray dose, detector efficiency, analog system noise, analog-to-digital conversion noise, and reconstruction algorithm. Not surprisingly, the temporal, spatial, and density resolution of any scanner are not independent of one another. For example, a reduction in scan time achieved by accepting fewer angles of view inevitably results in a loss of both spatial and density resolution.

The scanning subsystem is made up of the gantry, the x-ray source, and the detector system. The geometric configuration of the scanning subsystems for several existing CT scanners is illustrated in Fig. 32.4. The two bottom panels show the configuration of most modern scanners, although all four geometries have advantages and are still in use. The geometry in the lower left is the so-called third generation, which is a fan beam system in which the detector array and x-ray source rotate in tandem as a fixed unit about the patient. The main advantage of the third-generation geometry over the first- and second-generation geometries (upper left- and right-hand panels, respectively, in Fig. 32.4) is improved temporal resolution due to minimal mechanical motion. Scan times with first-generation scanners are 3–5 min; second generation, 20 s to 1 min; and third generation, 2 to 10 s. The substitution of a fan beam for a highly collimated pencil beam results in increased x-ray scatter and hence noise in the projection data, resulting in poorer spatial and density resolution. If time and x-ray dose were not factors, the first-generation geometry would still be superior.

Shown in the lower right-hand panel of Fig. 32.4 is the detector-source geometry of the fourth-generation scanners, which have a large number (600 to 1200) of individual detectors of a

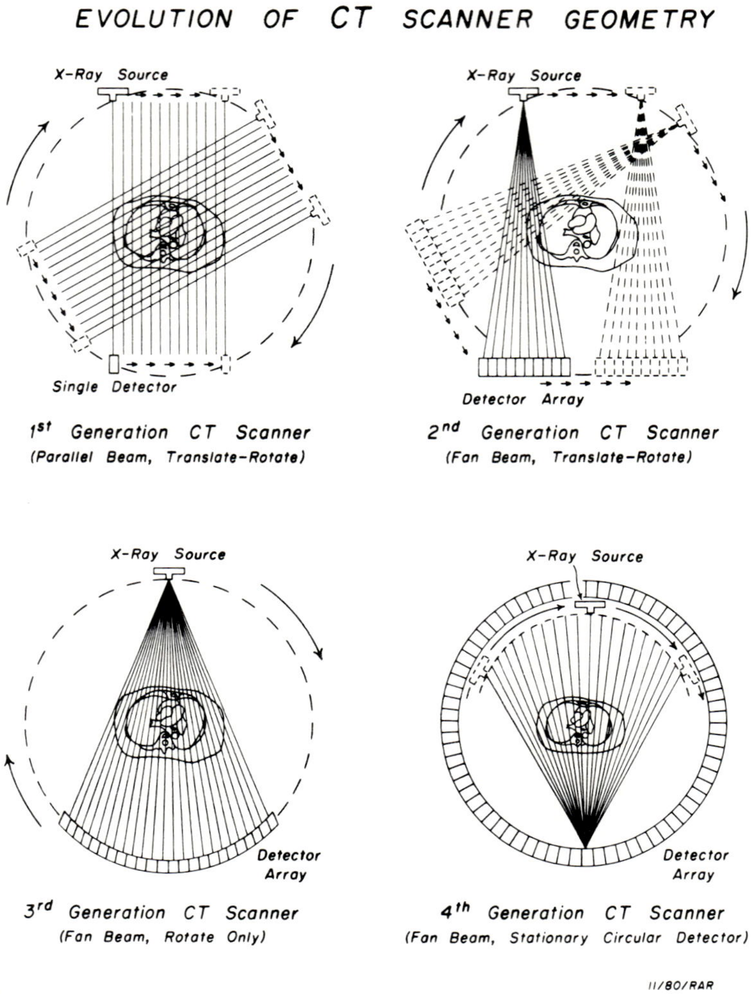

FIG. 32.4 Diagram of evolution of CT scanner geometries. First-generation CT scanners use a pencil beam defined by a well-collimated x-ray source and single detector. The source and detector are first translated in tandem in a rectilinear pattern, with successive beam measurements parallel to each other. Then the source and detector are rotated through a small angle and the rectilinear scan repeated. This translate-rotate procedure continues for 180°. Second-generation scanners use a small linear array of detectors instead of a single detector in order to speed up data acquisition with fan beam geometry. However, translation and rotation of source and detectors are still required. Third-generation systems use a larger array of detectors (usually circular) so that the entire body is subtended by the x-ray fan beams, and only rotation of the source and detector array is required to obtain the readings for image reconstruction. Fourth-generation systems use a fixed circular array of many detectors and a rotating x-ray source inside the detector ring. In this geometry, each position of the detector defines the apex of the fan beam as x-rays are recorded from the various source positions. The geometric considerations for image reconstruction are the same, however, as for third-generation systems, where the source position defines the apex of the fan beam. (*From Robb [4]; reproduced with permission.*)

fixed ring. The x-ray source moves around the patient on a circle inside the detector ring. The fan beam projection is formed by time-based sampling of the output of each detector as the tube moves through the fan angle. Since the x-ray beam is fan-shaped, the outputs of several detectors are sampled simultaneously so that all projections are obtained in a single rotation of the source. An important advantage of the fourth-generation geometry is x-ray detector calibration. As the source moves from left to right in Fig. 32.4, each of the exposed detectors "sees" a raw x-ray beam from either side of the patient. This information is used to calibrate the detector output relative to all other detectors and to itself. Conversely, the central set of detectors in the third-generation geometry are never exposed to the raw x-ray beam so long as an object is in the scanner. Therefore, the detectors must be stable for long periods of time in the third-generation system since drift from calibrations will result in a characteristic "ring" artifact. On the other hand, the fourth-generation system requires many more detectors, especially if it is to have both high-dose efficiency and high density and spatial resolution. To achieve high spatial resolution, the azimuthal dimension of each detector must be small (around 1 mm). For example, if the diameter of the detector ring is 900 mm (about 3 ft), the circumference of the detector ring is 2827.4 mm, then the system would require 2827 detectors! Reducing the number of detectors but keeping the azimuthal detector dimension constant, i.e., collimation behind the patient, results in dose inefficiency because approximately one-half the x-ray flux must pass through the body but cannot be utilized to form the image.

The computer subsystem must reconstruct from 65,536 (256×256) to 262,144 (512×512) picture elements from 5×10^5 to 3×10^6 measurements of x-ray intensity. With fan beam convolution reconstruction, the number of additions and multiplications required to perform the reconstruction is approximately $4M(m^2 + N)$, where M is the number of profiles, m is the number of ray sums per profile, and N is the number of pixels in the resulting image [28]. For a low-end case of 65,536 pixels, 180 profiles, and 256 ray sums, the figure is 9.4×10^7. A more typical number is 7.5×10^8 operations, requiring on a minicomputer approximately 8 min for reconstruction of the first case and around 50 min for the second. As a result, all commercially available machines utilize high-speed arithmetic array processors to achieve reconstruction times which are generally less than 1 min.

The display system converts the digital image output of the reconstruction process to a visible image by use of a frame buffer. The frame buffer is a peripheral device attached to the computer system which usually has enough memory to contain a single image, i.e., up to 512×512 pixels with 12 bits per pixel. The output of the frame buffer is a video signal which is input into a monitor. The display system controls are incorporated into a console which allows the viewer to interact with the image data. Since each pixel is reconstructed to 12 bits (or, potentially, 4096 discrete gray levels) but the observer can appreciate no more than 8 bits (256 levels), the frame buffer digital-to-analog conversion systems generally display no more than 8 bits. To allow the user to select quickly a desired range of pixel magnitudes (Fig. 32.5), there is a 12-bit-in to 8-bit-out high-speed look-up table incorporated into the display system. The map is controlled with two potentiometers to set the desired window level and range.

32.2.1 Results

The utility of computed tomography can in large measure be judged by its overwhelming acceptance as a clinical diagnostic and research tool. Computed tomography is used to study virtually every organ system of the body in situations where clear, unambiguous views of tomographic sections are useful. The nature of CT images and the diagnostic power of computed tomography are best illustrated with a few examples. Figures 32.6 and 32.7 show representative CT scans of the head. Figure 32.6 shows two pictures of the same image data through the orbits and represents a thin, 1.5-mm transverse section. The thickness of the section is controlled by axial collimation of the x-ray beam so that projection information is collected from the specified section through the body. Slice thickness is important, since the thicker the slice the lower the noise ("speckle" in the image) if the x-ray exposure is held constant [29, 30]. It is important to remember, however, that pixels really have volume (*voxels* for volume picture element). For example, typical CT-image pixel dimensions in the transaxial plane are $0.75 \times$

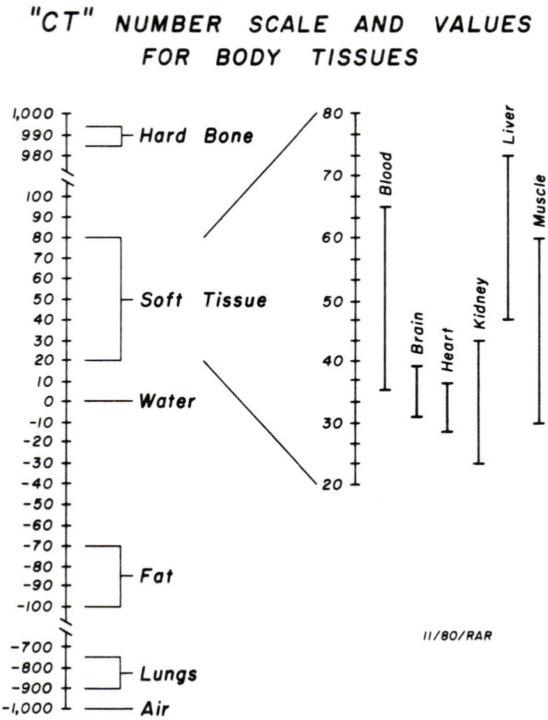

FIG. 32.5 Typical CT number scale and values for various body tissues. Values are based on measurements with effective x-ray energy of 72 keV. Range of values for a given tissue reflects both measurement errors and actual attenuation differences among samples. Data include both in vitro measurements made by assembling monoenergetic sources and in vivo measurements made with actual CT scanners. *(From Robb [4]; reproduced with permission.)*

0.75 mm, but the section may be 10 mm thick. As a result, the pixel actually represents mean "density" in a parallelepiped of dimensions 0.75 × 0.75 × 10 mm.

Figure 32.6 also illustrates the tremendous range of CT numbers obtained with commercial CT scanners. In this case, the reconstruction spans 12 bits, or 2048, levels from −1000 (in Hounsfield units) for air, to +1000 for dense bone. Before display of the image in Fig. 32.6*a*, the full range (−1000 to +1000) was scaled to 8 bits (0 to 255 levels) and the resulting image transferred to the memory of a frame buffer, resulting in a television image which was in turn photographed. In order to better visualize the soft tissues, the image was windowed before display (Fig. 32.6*b*). In this picture, a limited range of 105 CT numbers was displayed over the full 8-bit range of the frame buffer. The displayed CT range is referred to as the *window width* and the magnitude of the center of the window range as the *window level.* In this display, all pixels with magnitude greater than or equal to the upper bound of the window are white (display level 255), and all pixels with magnitude less than or equal to the lower bound are black (display level 0).

Figure 32.7 shows four routine diagnostic CT scans of a patient's head. The use of contrast materials to enhance the visibility of certain structures such as a tumor is illustrated in Fig. 32.7*b*. The visibility of the tumor is enhanced by the infusion of x-ray contrast medium into the

A

B

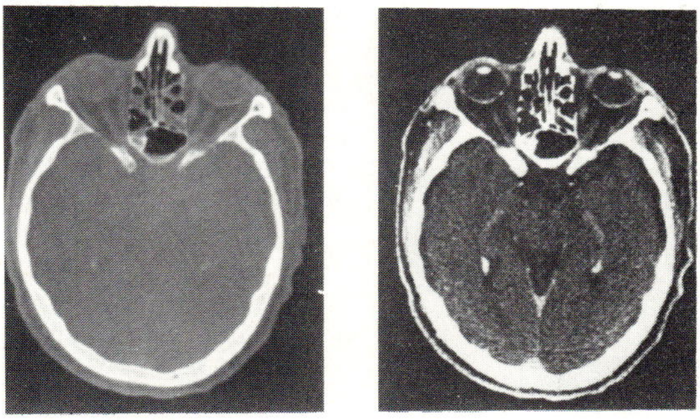

FIG. 32.6 Effect of display of limited range of CT numbers. CT scan of the human orbits. Images are of a 1.5-mm-thick section illustrating the effect of "windowing." The panel on the left shows the entire range of pixel values, 1000 to +1000, in Hounsfield units, compressed into 256 shades of gray. For the left-hand panel a limited range of 105 CT numbers are displayed, showing the soft tissue.

A

B

C

D

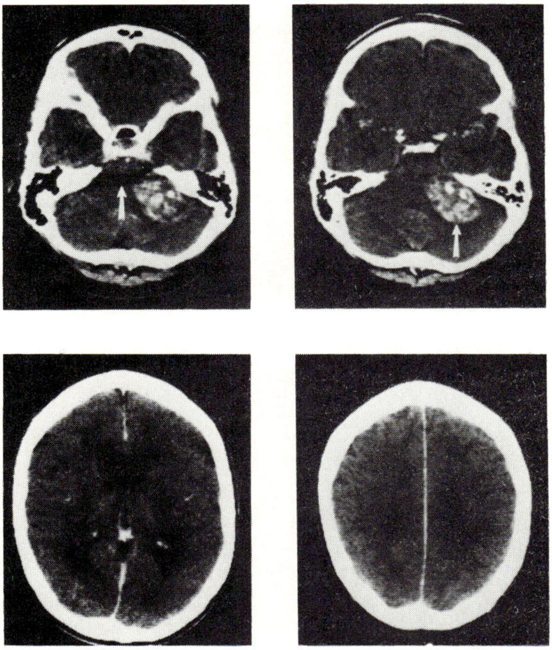

FIG. 32.7 Four parallel CT scans of the human head (5-mm sections—windowed). The images have been windowed to display a range of 105 CT numbers. Before the scan, an iodinated contrast agent was injected into the blood to preferentially enhance the visibility of a tumor [arrow in *B*]. Shown also is a reconstruction artifact which is a dark band [arrows in *A*] caused by the very dense petrous bone. (*Courtesy of Dr. R. A. Hattery, Department of Radiology, Mayo Clinic, Rochester, Minn.*)

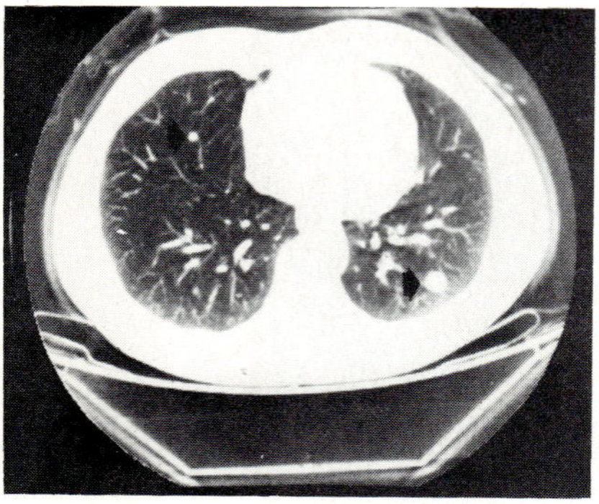

FIG. 32.8 CT image of patient's chest at basal level of lungs displayed to visualize a small range of densities at full range. The heart, spine, and chest wall are above the density levels windowed, but the low-contrast blood vessels in the lung are easily seen. A 6-mm-diameter nodule (metastatic carcinoma) is visible (black arrow) in the anterior portion of the right lung, and a 13-mm-diameter nodule is seen (lower arrow) in the posterior portion of the left lung. (*Courtesy of Dr. J. R. Muhm, Department of Radiology, Mayo Clinic, Rochester, Minn.; reproduced with permission from Robb [4].*)

blood. This increases the "density" of the blood, and because the tumor has a higher ratio of blood to tissue than normal tissue, the tumor becomes preferentially enhanced.

Figure 32.8 shows a CT scan through the thorax, showing a cancerous nodule in the right lung (black arrow). In this case the display window has been set to visualize the density range in the lung itself. As a result, the higher-density structures of the ribs and heart are all combined as the white regions.

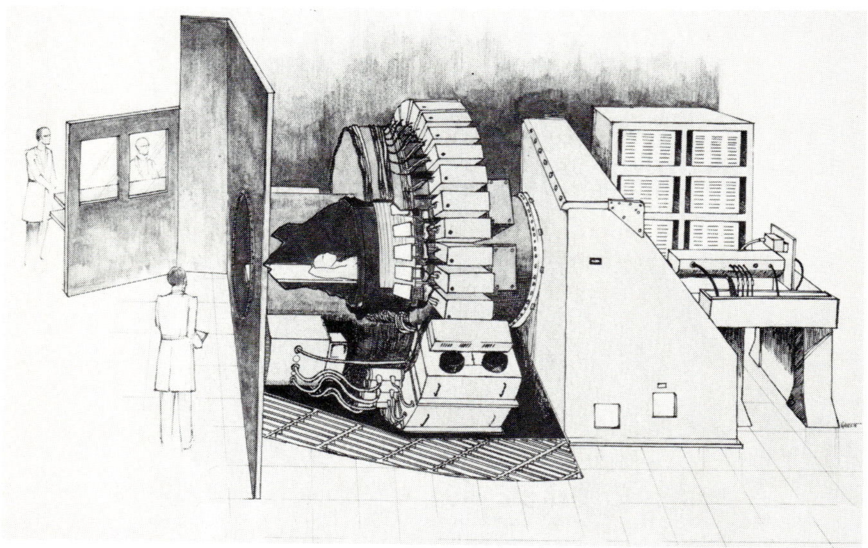

FIG. 32.9 Drawing of the dynamic spatial reconstructor (DSR). The system has multiple x-ray tubes and directly opposing imaging systems. An entire 21.5 diameter by 21.5 high cylindrical volume is scanned. The signals from the multiple video cameras are digitized and used to reconstruct 120 parallel CT images of the scanned volume (computer system not shown). The entire gantry system rotates at 15 r/min for additional scanning flexibility. [*Reproduced with permission from R. S. Schwartz and E. L. Ritman, Computed tomography in cardiac imaging, Pract. Cardiol. (submitted for publication).*]

32.3 FOUR-DIMENSIONAL COMPUTED TOMOGRAPHY: DYNAMIC VOLUME SCANNING

The extension of the principles of computed tomography to the study of moving organs of the body, especially the heart, lungs, and circulation, is the objective of an ongoing project of the Biodynamics Research Unit at the Mayo Foundation [31, 32], which has led to the development of the dynamic spatial reconstructor (DSR) illustrated in Fig. 32.9. The DSR can be described as a repetitive volume scanner with high temporal resolution. Although based on the principles of computed tomography, the DSR scan is inherently a volume scanner, meaning that a complete volume is scanned synchronously. The scanning volume is a cylinder with transverse diameter of either 21.5 or 39 cm and a head-to-foot (axial) height of 21.5 cm. Instead of one-dimensional projection profiles, the DSR generates two-dimensional projection images which in turn are utilized to compute 120 parallel tomographic images of slices 1.8 mm thick. The necessary projection data (14 views) to reconstruct a single volume image is obtained in an 11-ms time interval. This, approximately 1/100 s, scanning time results in stop-action volume images of the entire heart and associated major intrathoracic blood vessels. Motion of organ systems can be followed because the scanning of the volume can be repeated every 16.7 ms for up to 20 s, resulting in 1200 time-sequential volume images.

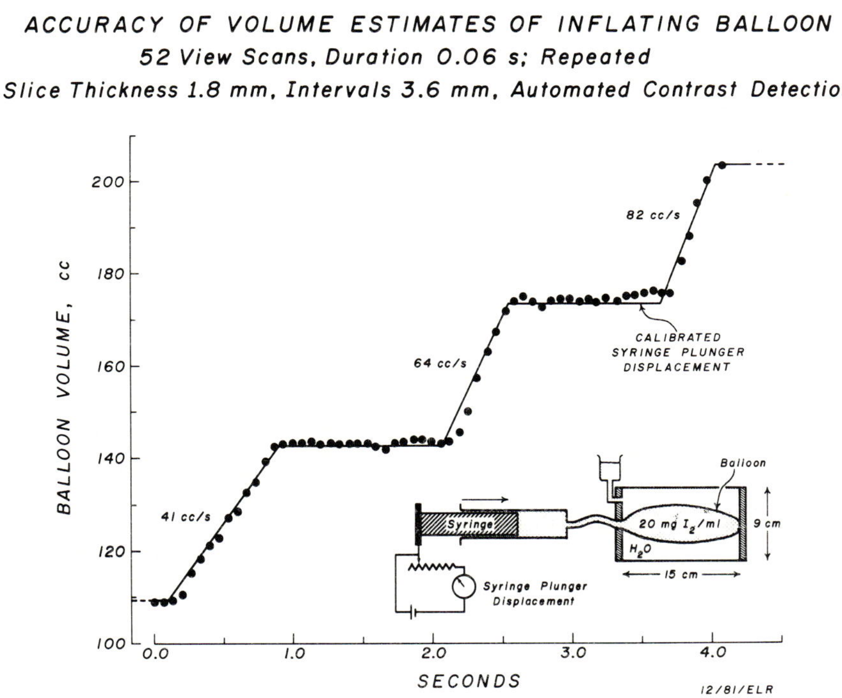

FIG. 32.10 Roentgen contrast agent (20 mg iodine per milliliter) was injected into a partially filled balloon at rates indicated by the changing slope of the solid line. The injection sequence included three phases, each of increasing rate of injection. The water in the cylinder surrounding the balloon was passively vented into a thistle tube, as illustrated in the inset. Each datum point represents the volume computed from reconstructed images of 44 parallel cross sections, each 1.8 mm thick, with midpoints separated by 1.8-mm intervals and generated from four sequential scans so as to generate 52-view, 0.061-s reconstructions. (*From Behrenbeck et al. [33]; reproduced with permission.*)

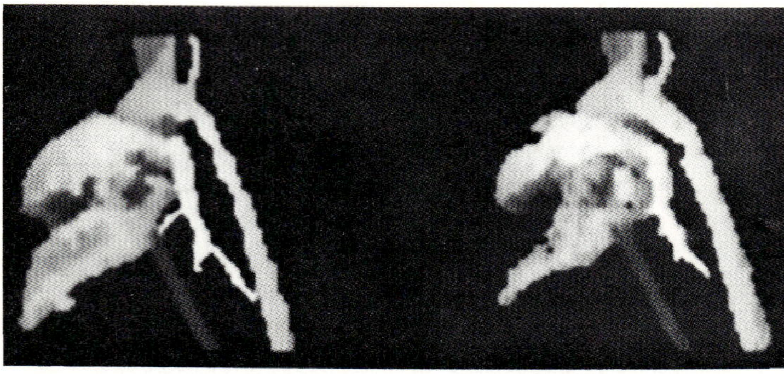

FIG. 32.11 Stop-action volume image capabilities of the DSR illustrated with computer-generated shaded surface display of the contrast-enhanced blood in the heart at end diastole (right) and end systole (left). The surface detection and display were done with the cubrille approach of Herman and Liu [36]. [*Reproduced with permission from E. A. Hoffman et al., Accurate quantitative tomographic imaging of the shape and dimensions of the cardiac chambers, Circulation (submitted for publication).*]

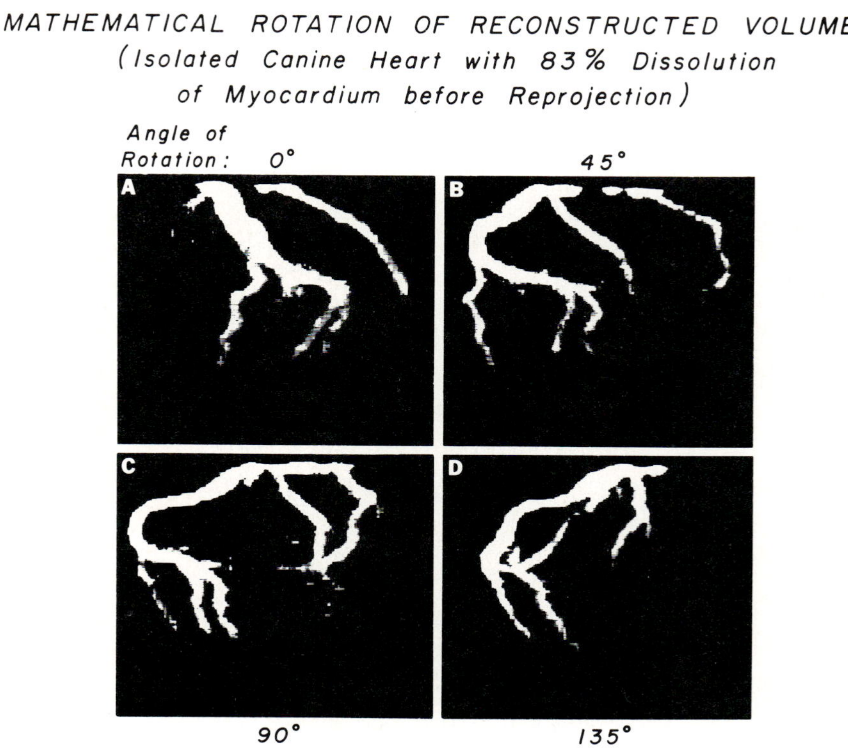

FIG. 32.12 Volume image display capabilities of the DSR illustrated by utilizing numerical projections [34] of a three-dimensional reconstruction of a dog heart with contrast-enhanced coronary arteries. Projections are formed from four angles of view to depict the three-dimensional nature of the volume image. (*From Harris et al. [34]; reproduced with permission.*)

Detailed description of the scanning, reconstruction, and performance characteristics of the DSR are given elsewhere [33]. The DSR project involves ongoing development of imaging systems and reconstruction algorithms, as well as methods for the multidimensional display and analysis of DSR image data [34, 35]. The system is now being utilized to investigate biological questions in experimental animals and humans.

The high-temporal-resolution repetitive scanning capabilities of the DSR are illustrated by the graph in Fig. 32.10. These measures involve the use of a partially inflated balloon filled with iodinated contrast agent (20 mg iodine per milliliter). The balloon was attached to a system to inject fluid into the balloon at selectable constant injection rates and then suspended in a vented water bath, as shown in the lower right-hand corner of Fig. 32.10. The closed dots represent volume calculations obtained by using the DSR images and the solid line represents true balloon volume determined by syringe travel. Volume estimation is within 3 percent for injection rates up to 82 mL/s.

The capability of the DSR to obtain stop-action synchronous volume image information is illustrated in Fig. 32.11. Shown are two shaded-surface displays of the contrast agent in the heart chambers at end diastole (left) and end systole (right). These images were obtained from two 0.06-s volume images (120 synchronous parallel tomographic images) scanned at end systole and end diastole. The section images are submitted to a threshold detection algorithm to identify the endocardial surface, resulting in a "description" of the surface which in turn is submitted to a shading algorithm [36] to compute the images shown. As illustrated in the figure, surface display is utilized to visualize the shape and the change in shape of complex surfaces.

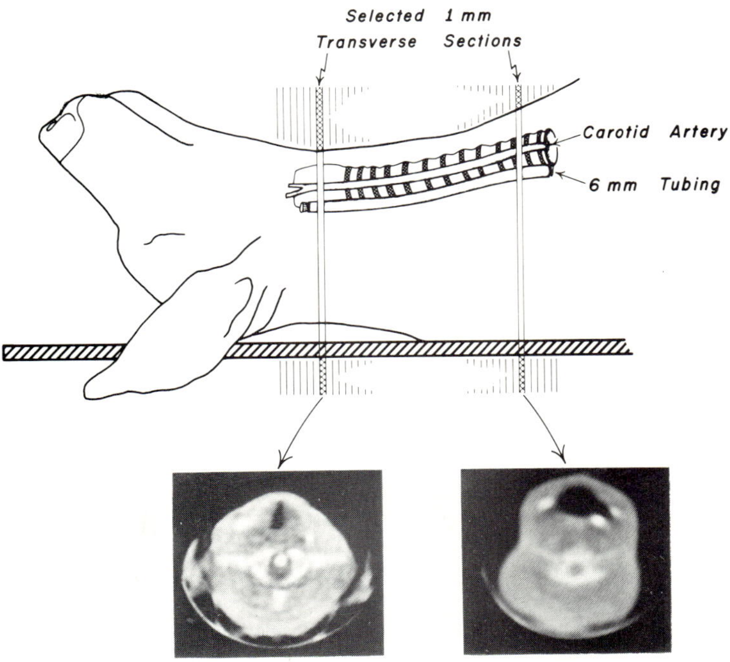

FIG. 32.13 Use of the DSR four-dimensional scanning to compute blood flow. Shown are two cross-sectional images of the neck of the dog after an injection of an iodinated contrast material into the carotid artery. Analysis of the downstream and upstream images through time yields specific flow measures. [*Reproduced with permission from L. J. Sinak and E. L. Ritman, Dynamic spatial reconstructor, in C. B. Higgins (ed.), CT of the Heart and Great Vessels:—Experimental Evaluation and Clinical Application, Futera Publishing Co. Mount Kisco, New York, 1983, pp. 61–73.*]

The volume imaging nature of the DSR system is illustrated in Fig. 32.12. These photographs are direct views of a volume image of the contrast-filled coronary arteries of a dog heart. The reconstruction is made up of 96 parallel CT scans from the base to the apex of the heart. This volume image is directly displayed by a technique called *numerical projection* [34] whereby the volume image is mathematically projected onto a plane in a process analogous to forming a radiograph. Numerical projection facilitates the direct visualization of volume images from multiple angles of view, as illustrated in the four panels of Fig. 32.12. Direct displays complement surface displays by allowing investigators to edit or explore volume image sets. Currently, developments are under way to produce direct three-dimensional displays of volume images utilizing the principle of a vibrating mirror [37, 38]. Mirror display not only will facilitate three-dimensional display of volume images, but will allow the operator to interactively dissect, i.e., cut, peel, or open, the image.

The use of the repetitive high-temporal-resolution scanning of the DSR to measure blood flow is illustrated in Fig. 32.13. A sequence of volume images (four-dimensional images) of a dog's neck were obtained during the injection of x-ray contrast agent into the carotid artery. The amount of contrast agent at many points along the artery as a function of time is estimated by analysis of the region of the artery. Two representative images at a single point in time are shown in Fig. 32.13.

SUMMARY

The objective of computed tomography is the computation of images of two-dimensional sections through the body. From a physical viewpoint, the great utility of tomographic images is the removal of superposition and thereby the production of clear, unambiguous views into the body. In addition, computed tomography provides accurate measurements of organ shape and dimensions in two and even three dimensions if the scanning tool can produce the requisite volume image data. Moreover, the extension to the noninvasive study of moving organs through use of four-dimensional images is now becoming possible so that organ function can be directly studied in ways never before possible.

ACKNOWLEDGMENTS

The authors express appreciation to their colleagues and technical staff for discussions and development of many of the concepts presented here. A special thanks to Darlene Kasten, Elaine Quarve, and Steven Richardson for their assistance in the preparation of this manuscript.

REFERENCES

1. Gordon, R., and Herman, G. T.: Three-dimensional reconstruction from projections: A review of algorithms, *Int. Rev. Cytol., 38,* 111–151, 1974.
2. Brooks, R. D., and DiChiro, G.: Principles of computer assisted tomography (CAT) in radiographic and radioisotopic imaging, *Phys. Med. Biol., 21,* 689–732, 1976.
3. Rowland, S. W.: Computer implementation of image reconstruction formulas, in G. T. Herman (ed.), *Topics in Applied Physics,* vol. 32: *Image Reconstruction from Projections,* Springer-Verlag, New York, 1979, pp. 9–79.
4. Robb, R. A.: X-ray computed tomography: An engineering synthesis of multiscientific principles, *CRC Crit. Rev. Biomed. Eng., 7*(4), 265–334, 1982.
5. Ter-Pogossian, M. M.: Basic principles of computed axial tomography, *Semin. Nucl. Med., 7,* 109–127, 1977.
6. Phelps, M. E.: Emission computed tomography, *Semin. Nucl. Med., 7,* 337–365, 1977.

7. Brownell, G. L., Correia, I. A., and Zamenhof, R. G.: Positron instrumentation, in J. H. Lawrence and T. F. Budinger (eds.), *Recent Advances in Nuclear Medicine,* Grune & Stratton, New York, 1978, vol. 5, pp. 1–49.
8. Budinger, T. F., Gallberg, G. T., and Huseman, R. H.: Emission computed tomography in topics in applied physics, in G. T. Herman (ed.), *Image Reconstruction from Projections,* Springer-Verlag, New York, 1979, pp. 147–246.
9. Kaufman, L., Crooks, L., and Margulis, A. (eds.): *Nuclear Magnetic Resonance Imaging in Medicine,* Igaku-Shoin, New York, Tokyo, 1981.
10. Hinshaw, W. S., and Lent, A. H.: An introduction to NMR imaging: From the Bloch equation to the imaging equation, *Proc. IEEE, 71,* 338–350, March 1983.
11. James, E. M., and Charboneau, J. W.: Diagnostic ultrasound—Coming of age, *Mayo Clin. Proc., 57,* 198–200, 1982.
12. Greenleaf, J. F., and Bahn, R. C.: Clinical imaging with transmissive ultrasonic computerized tomography, *IEEE Trans. Biomed. Eng., BME-28*(2), 177–185, February 1981.
13. Lakshminarayanan, A. V.: Reconstruction from divergent ray, Tech. Rep. 92, Department of Computer Science, State University of New York, Buffalo, 1975.
14. Herman, G. T., and Naparstek, A.: Fast image reconstruction based on radon inversion formula appropriate for rapidly collected data, *SIAM J. Appl. Math., 33,* 511, 1977.
15. Bursch, J., Johs, R., and Heintzen, P. H.: Validity of Lambert-Beer's law in roentgen densitometry of contrast material using continuous radiation in roentgen-, cine-, and videodensitometry, Georg Thieme Verlag, Stuttgart, 1971, p. 81.
16. Crowther, R. A., DeRosier, D. J., and Klug, A.: The reconstruction of a three-dimensional structure from projections and its application to electron microscopy, *Proc. R. Soc. London Ser. A, 317,* 319–340, 1970.
17. Smith, P. R., Peters, T. M., and Bates, R. H. T.: Image reconstruction from finite numbers of projections, *J. Phys. Appl. Math. Nucl. Gen., 6,* 361–382, 1973.
18. Wylie, C. R., Jr. (ed.): *Advanced Engineering Mathematics,* 2d ed., McGraw-Hill, New York, 1960, p. 493.
19. Ramachandran, G. N., and Lakshminarayanan, A. V.: Three-dimensional reconstruction from radiographs and electron micrographs: Application of convolutions instead of Fourier transform, *Proc. Nat. Acad. Sci. U.S.A., 68,* 2236–2240, 1971.
20. Shepp, L. A., and Logan, E. C.: The Fourier reconstruction of a head section, *IEEE Trans. Nucl. Sci., 21,* 21, 1974.
21. Rowley, P. D.: *J. Opt. Soc. Am., 59,* 1496, 1969.
22. DeRosier, D. J., and Klug, A.: The reconstruction of three-dimensional structures from electron micrographs, *Nature (London), 217,* 130–136, 1968.
23. Gordon, R., Bender, R., and Herman, G. T.: Algebraic reconstruction techniques (ART) for three-dimensional electron microscopy and x-ray photography, *J. Theor. Biol., 29,* 471–481, 1970.
24. Gilbert, P. F. C.: Iterative methods for the three-dimensional reconstruction of an object from projections, *J. Theor. Biol., 36,* 105–117, 1972.
25. McCullough, E. C.: Specifying and evaluating the performance of computed tomography (CT) scanners, *Med. Phys., 7,* 291–298, 1980.
26. Goodenough, D. J., Weaver, K. G., and Davis, D. O.: Development of a phantom for evaluation and assurance of image quality in CT scanning, *Opt. Eng., 16,* 52, 1977.
27. Alfidi, R. J., MacIntyre, W. J., and Haaga, J. R.: The effects of biological motion on CT resolution, *Am. J. Roentgenol., 127,* 11–15, 1976.
28. Shepp, L. A., and Kruskal, J. B.: Computerized tomography: A new medical x-ray technology, *Am. Math. Mo., 85,* 420, 1978.
29. Brooks, R. A., and DiChiro, G.: Statistical limitations in x-ray reconstructive tomography, *Med. Phys., 3,* 237–240, 1976.
30. Chesler, D. A., Riederer, S. J., and Pelc, N. J.: Noise due to photon counting statistics in computed x-ray tomography, *J. Comput. Assisted Tomog., 1,* 64–74, 1977.
31. Ritman, E. L., Kinsey, J. H., Robb, R. A., Gilbert, B. K., Harris, L. D., and Wood, E. H.: Three-dimensional imaging of the heart, lungs, and circulation, *Science, 210,* 273–280, 1979.

32. Kinsey, J. H., Robb, R. A., Ritman, E. L., and Wood, E. H.: The DSR: A high temporal resolution volumetric roentgenographic CT scanner, *Herz, 5*(3), 177–188, June 1980.

33. Behrenbeck, T., Kinsey, J. H., Harris, L. D., Robb, R. A., and Ritman, E. L.: Three-dimensional spatial, density, and temporal resolution of the dynamic spatial reconstructor, *J. Comput. Assisted Tomog., 6*(6), 1138–1147, 1982.

34. Harris, L. D., Robb, R. A., Yuen, T. S., and Ritman, E. L.: Display and visualization of three-dimensional reconstructed anatomic morphology: Experience with the heart, lungs, and coronary vasculature of dogs, *J. Comput. Assisted Tomog., 3*(4), 439–446, 1979.

35. Harris, L. D.: Identification of the optimal orientation of oblique sections through multiple parallel CT images, *J. Comput. Assisted Tomog., 5*(6), 881–887, 1981.

36. Herman, G. T., and Liu, H. K.: Display of three-dimensional information in computed tomography, *J. Comput. Assisted Tomog., 1,* 155–160, 1977.

37. Traub, A. C.: A new three-dimensional display technique, Rep. No. M684, Mitre Corporation, Bedford, Mass., 1968.

38. Baxter, B., Hitchner, L. E., and Anderson, R. E.: Applications of a three-dimensional display in diagnostic imaging, *J. Comput. Assisted Tomog., 6,* 1000–1005, 1982.

CHAPTER 33

Blood Pressure and Flow Measurements

M. Intaglietta
Department of Applied Mechanics and Bioengineering
University of California, San Diego
La Jolla, California

33.1 BLOOD PRESSURE MEASUREMENTS

The measurement of blood pressure was first reported by Stephen Hales, [1], who in 1773 measured mean arterial pressure by inserting an open-ended tube in the artery of the neck of an unanesthetized horse. The measurement was made by noting the height to which blood rose in the tube relative to the point of insertion. Many techniques have been developed since, including direct methods which provide instantaneous data, indirect methods which provide information on the time-dependent features of the arterial pressure pulse, and microscopic methods that measure directly and indirectly the pressure in vessels as small as the mammalian capillaries.

Pressure is defined as force applied per unit area. Physiological pressures are usually given in millimeters of mercury. Centimeters of water are also used, particularly in microvascular studies. For the purpose of modeling and performing hydrodynamic calculations it is more practical to express pressures with the units of the centimeter-gram-second (cgs) system, i.e., bars or dynes per square centimeter. Cobbold [2] reports the following useful relationships:

$$1 \text{ mmHg} = 1 \text{ torr} = 12.9 \text{ mm blood} = 13.1 \text{ mm saline} = 1330 \text{ dyn/cm}^2$$

The pressure of a fluid at a point is the same in all directions only in the absence of flow. When the fluid is in motion, the pressure measured by a transducer includes a factor determined by the kinetic energy of the fluid and the orientation of the sensing element relative to the direction of the flow at the point of measurement. This factor is significant when pressure is transmitted to a transducer through a catheter, where the open tip is in the moving fluid. The pressure correction necessary in the presence of relative motion between fluid and sensor is given by Bernoulli's equation as

$$\Delta P = \frac{1}{2} \rho v^2 \qquad \text{dyn/cm}^2 \qquad (33.1)$$

where ρ is the density of the fluid in g/cm^3, and v, in cm/s, is the relative velocity between fluid and sensor perpendicular to the sensing surface.

Arterial blood pressure waveforms are periodic, where the maximum pressure in a cycle is called the *systolic pressure*, and the minimum is the *diastolic pressure*. The mean pressure is the time-weighted average over one cardiac cycle. The range of amplitude of blood pressure in humans is variable; instrumentation that accurately measures pressure in the range of 0 to 300 mmHg is considered adequate for most applications. The fundamental frequency in humans is in the range of 60 to 120 beats per minute. Harmonic analysis of the pressure waveform shows that the amplitude of the fifth harmonic is of the order of 10 percent of the fundamental, and the fifteenth harmonic shows an amplitude of 1 percent of the fundamental. According to Geddes [3], the required frequency response for measurements in humans should be from direct current (dc) to 30 Hz.

33.1.1 Direct Pressure Measurements: Macroscopic Methods

The determination of blood pressure where the sensors come in direct contact with the blood, or where the fluid in the region of interest is coupled hydraulically to the pressure sensor, is termed a *direct measurement*. This includes contacting blood by inserting a cannula, catheter, needle, or pressure transducer into the blood vessels or heart. Although there is no clear demarcation between micro- and macromethods, blood vessels at the level of a 1-mm lumen constitute the lower limit for macromethods.

Modern developments in the transduction of pressure to electrical signals provide accurate representation of the hydraulic events; however, the transducers, primarily because of cost and the requirement of sterilization for human applications, can seldom be brought in direct contact with the pressure source at the point in the vascular system under study. The connection is usually made by means of fluid-filled catheters, which, owing to their flexibility, have compliance, i.e., volumetric changes directly related to the pressure to which they are subjected that is significant when compared with that of the transducer. The compliance of the connection to the transducer affects the time-dependent characteristics of the response of the pressure measuring system.

The *compliance* is defined as the change in volume per unit pressure change. The total overall system compliance is the composite of the catheter, the fluid, and the transducer. We should note that although the fluid (saline, water) is usually assumed to have negligible compliance, this factor becomes significant when we attempt to measure pressure in very small volumes such as the capillaries. The compliance of the catheter-transducer system is in part localized at the transducer, and the balance is distributed throughout. Consequently, the transmission of pressure pulses is described by the same techniques used for electrical transmission lines, as shown by Hansen [4, 5] and Fry [6]. According to this electrical analog the transmission line is a filter that damps the amplitude of signals above a characteristic cutoff frequency, and if the impedance at the end of the line is not properly matched, there will be pressure reflection phenomena that distort the waveform to be measured, as shown by Latimer [7, 8]. Geddes [3] gives the following expressions for the natural frequency f and the damping coefficient β in the catheter-transducer system:

$$f = k_1 D (cL\rho)^{-1/2} \tag{33.2}$$

$$\beta = k_2 \eta \left(\frac{Lc}{\rho}\right)^{1/2} D^{-3} \tag{33.3}$$

where k_1 and k_2 are constants that depend upon the system of units; η is the viscosity of the fluid in the catheter, in poise (P); L is the length of the catheter, in cm; D is the internal diameter of the catheter, in cm; C is the overall compliance of the system, i.e., the sum of catheter, fluid, and transducer compliances, in cm^5/dyn; and ρ is the density of the fluid in g/cm^3.

Equation (33.2) indicates that the natural frequency of the system is greatest for short, large-diameter, and low-compliance catheters. Currently, stiff (very low compliance) transducers are available; however, this development is somewhat negated by the inherent catheter compliance, since compliances are additive. Conversely, damping increases as the inner diameter of the catheter decreases; therefore a balance must be struck between these competing factors. Various methods have been proposed for dealing with this problem, such as an increase in damping through either series or parallel hydraulic connections, and compensation through electrical networks [9–12].

A major problem with liquid-filled catheter systems is the presence of air bubbles which notably increase the overall compliance of the system in an unpredictable way. Shinozaki et al. [13] have carried out comprehensive tests of currently available tubing of various materials, as well as catheters, to determine their relative capability of transmitting physiological waveforms with a minimum of distortion. They also report the propensity of various materials and connectors to trap bubbles. An overall test of thin, fluid-filled catheters for the measurement of pressure and its derivatives has been reported by Lomholt et al. [14], and Gardner [15] summarized the frequency response and damping requirements for accurate pressure measurements made with catheter systems.

Pressure Transducers. Electrical output pressure transducers sense the pressure to be measured by converting hydraulic pressure into the mechanical displacement of a thin-walled elastic member such as a plate, shell, or tube, which provides a surface for the pressure to act upon. The unbalance in pressure between the opposite sides of this element cause its deflection, which is transduced into an electrical output proportional to displacement. Accurate pressure transduction at the macroscopic level is virtually universally accomplished by the deflection of diaphragms, which consist of thin circular plates rigidly clamped at the edges. The theory that describes the displacement of the surface of a diaphragm as a function of the pressure difference between its two sides and its natural frequency is developed in the classic text by Timoshenko and Woinowsky-Krieger [16]. The significant result is that for small displacements and thin diaphragms the motion of the center is proportional to the fourth power of the radius. Conversely, the natural frequency is directly related to the thickness of the plate. As a rule the thickness required to ensure that the pressure encountered in physiological applications will not exceed the elastic limits of the diaphragm's material is such that the natural frequency is well above 500 Hz. A detailed analysis of the time-dependent aspects of manometric theory is provided by Milnor [17], who develops the equations and principles presented by McDonald [18].

The transduction of the diaphragm's displacement into an electrical signal is accomplished by strain gauges, variable inductive and capacitive circuits, and optical methods. Bonded and unbonded strain gauges are most commonly used. Bonded strain gauges are permanently attached to the diaphragm and yield a compact, light-weight transducer. Since the deflection of the diaphragm includes both compression and elongation of its surface, the gauges can be placed in a full Wheatstone bridge configuration and therefore temperature fluctuations have a relatively small effect. Bonded strain gauges usually consist of a conducting wire that is cemented to the surface to be measured. The same effect can be obtained by applying a film of insulating material on the surface and vacuum-depositing a film of metal in the desired location and configuration. Semiconductor strain gauges have also been utilized, and have the advantage of a strain gauge factor (i.e., relative output versus relative deformation) that is of the order of 50 times that of metal; however their temperature sensitivity is proportionately greater.

A number of specialized texts [3, 19–21] report the characteristics, design considerations, and methods of use of the current devices. The popular Statham P23 series are pressure transducers of the unbonded type. In such a transducer the deflection of the center of the diaphragm is transmitted by a rigid link to strain gauges which consist of wires stretched between pegs, some on the sliding element that is rigidly attached to the membrane and the others fixed to the transducer case. The wires form a Wheatstone bridge and are arranged so that a change in pressure increases the strain in two wires and decreases the strain in the other two. Since the wires are substantially stiffer than the diaphragm, virtually all the force developed at the membrane is transmitted to the wires. In addition, once the sensing element is removed from the diaphragm, there is no need for miniaturized assemblies, and the transducer can be built with features that ensure high sensitivity and stability. From the viewpoint of sensitivity the output of this type of transducer is of the order of 1 to 10 μV/(V·mmHg), and volume displacements (compliances) are in the range of 0.01 to 0.001 mm^3/100 mm Hg.

The miniaturization inherent to semiconductor devices have set the stage for the development of catheter tip transducers, which in physiological applications allow measurement of the pressure directly where it develops, circumventing the need for transmitting the hydraulic signal through a fluid-filled line. This kind of a device was described as early as 1898 by Graubman [3], who used two electrodes to measure the resistance of the electrolyte in a small, deformable capsule. As the surrounding pressure increased, the spacing of the electrodes decreased, thus providing the pressure-dependent change of electrical resistance.

A commercially produced device was first described by Allard [22], who transmitted the motion of a diaphragm mounted at the end of an 8F catheter to a metallic core within a coil, thus changing its inductance and the frequency of an FM oscillator.

Statham Instruments introduced the first strain-gauge catheter tip pressure transducer, based on designs first reported by Warnick and Drake [23]. This device consists of a thin elastic cylinder that fits into a 6.5F catheter. The cylinder is wrapped with strain-gauge wire which measures the strain in the cylinder. Since the cylinder is deformed by the pressure applied to the inside of the catheter, it can remain open throughout and be used to obtain blood samples. The high sensitivity of silicon strain gauges has been repeatedly utilized to miniaturize the diaphragm-gauge assembly in such a fashion that the whole transducer can be mounted with the membrane directly exposed to the point of measurement. Two types of sensing techniques are used. In one, a silicon cantilever beam detects the motion of the diaphragm. In a different application a silicon disk which includes the strain gauges is used as the diaphragm. The primary drawbacks of these approaches are the comparatively high temperature sensitivity and fragility.

Fiber optics have successfully been implemented as a means of transmitting a pressure-dependent signal by nonhydraulic means. The method uses two fiber-optic bundles, which terminate at the inner surface of a diaphragm placed at a small distance from the end of the fibers [24]. One group of fibers brings light to the diaphragm, and the second bundle collects the light reflected from the diaphragm, which is a function of the spacing between fiber ends and diaphragm and therefore of the outside pressure. An important advantage of this system is the lack of any electrical connection within the catheter, which makes it notably safer than any of the other systems.

33.1.2 Indirect Pressure Measurements: Macroscopic Methods

Indirect, noninvasive pressure measurements in humans were first developed by Korotkoff [25], who, in 1905, described the method of sphygmomanometry, which consists in balancing the pressure in an air-filled cuff against the blood pressure of the vessels. In a routine measurement the cuff is wrapped around an arm, over the brachial artery, and is inflated to a pressure that exceeds systolic blood pressure, thus collapsing the artery and stopping blood flow. Air is then slowly released, until arterial pressure exceeds cuff pressure and a sound related to the reestablishment of blood flow (Korotkoff sound) is heard by means of a stethoscope. It is assumed that the cuff pressure at which blood flow just begins is systolic pressure and that at which it becomes continuous is diastolic pressure.

Korotkoff sounds can be detected by a stethoscope or any electronic sound pickup that can be acoustically coupled to the artery under study. Geddes [3] shows that the nature of the sound produced by the establishment of flow through the partially constricted artery varies as a function of blood pressure, and that, as a rule, systolic and diastolic pressures will be underestimated and overestimated, respectively, even when careful consideration is given to the relative size of the cuff and arm, rate of deflation, and nature of the sound.

Automated methods for implementing the sequence of inflation, deflation, and detection of the sound associated with establishment of pulsatile flow have been developed by utilizing a variety of detection techniques. Lynwood [26] provides a summary of the early attempts. The principal variation between systems lies in the methods for detecting the onset of blood flow. Such methods include the use of a microphone, the measurement of the volume change downstream from the point of occlusion [27], the measurement of the pulse-wave propagation velocity as a function of the degree of occlusion of the artery [28], optical methods which determine the color changes in the digits [29], and ultrasonic techniques which measure the Doppler shift of a carrier frequency (2 MHz) caused by the flow velocity of blood in the brachial artery. This latter method was first described by Ware and Laenger [30], and has been implemented as a commercial instrument by Roche Medical Electronics [31].

The cuff method has been applied also to measurements in the digits, where it can be used over longer periods without discomfort, and can be used in the sleeping patient. Several commercially available systems have been described [9, 26, 32]. Comparative studies between direct and indirect methods have been reported by Rubin et al. [33], Bruner et al. [34], and Bruner [35].

33.1.3 Pressure Measurements in the Microcirculation

Pressure measurements in the microcirculation (i.e., vessels in the 5- to 100-μm-diameter range) are associated with mechanical signal levels that are only marginally capable of actuating passive transducers, in terms of both the amount of fluid that can be transferred from the vessel to make up for the system's compliance and the available mechanical energy. The limitations determined by the small dimensions of these hydraulic conduits can be estimated by noting the order of magnitude of the compliance of water, 6.1×10^{-8} $cm^3/(cm^3 \cdot mmHg)$, and the volume flow through a 1-μm-tip-diameter glass micropipet, 3×10^{-11} $cm^3/(s \cdot mmHg)$ [36].

Micropressures undergo periodic fluctuations synchronous with the action of the heart, vasomotor activity at the level of arterioles, and the respiratory cycle. From the viewpoint of instrumentation, the presence of pulsatile signals in the record is taken as evidence that the measuring system is functioning properly.

Two types of indirect methods have been implemented: the first is based on the principle of the capillary tonometer reported by Roy and Brown [37] and consists of inflating a balloon against the vessel under observation. The balloon pressure is increased until it causes the vessel to collapse, thus stopping the flow. Inaccuracies appear as a consequence of the contributions of vessel wall structure and tissue properties which cannot be fully accounted for. Furthermore, in regions where the microvasculature has little collateral circulation the pressure readings provide values in between the arterial supply and venous exits, and if there is no collateral circulation, arterial pressure is measured throughout.

A method for measuring pressure indirectly was devised by Intaglietta and Zweifach [38], based on the methods for measuring exchange devised by Landis [39]. A single capillary is occluded by compressing it with a microneedle until the flow of blood stops. Any subsequent, sustained motion of the red cells trapped in the capillary must be due to a gain or loss of fluid to the surroundings, the assumption being made that a red cell will move with the same velocity as the fluid that it displaces. With the stoppage of blood flow by mechanical occlusion, the pressure on the arterial side of the occlusion will be in equilibrium with the pressure of the nearest supply vessel with an active blood flow. Similarly, on the venous side of the occlusion, the pressure reflects that of the venule into which the capillary empties. These phenomena are interpreted by means of the so-called Starling hypothesis, which states that fluid exchange between a capillary and surrounding tissue is driven by the difference in hydrostatic pressure between the intra- and extraluminal compartments as well as the corresponding differences in colloidal osmotic pressure. Since all the parameters, with the exception of capillary pressure, can be either measured systemically or assumed to remain constant during a measurement, the actual net pressure (i.e., the pressure which causes filtration) can be deduced through a simple algorithm.

A method based on a passive transducer for microvascular pressure measurements was first developed by Rappaport et al. [40], who devised an electromanometer for measuring pressures from glass microcannulas with 20-μm-tip openings, with an overall frequency response of the order of 20 Hz. Similar results were obtained by Levasseur et al. [41], who utilized a Statham P23Gb pressure transducer with a dome whose net internal volume is 6 percent of that of the standard dome, and which can be coupled directly to the microcannulas. They utilized approximately 15 cm of PE-10 polyethylene tubing whose tip was shaped to an 18-μm beveled opening. The tubing was directly connected to the pressure dome, and the reported overall damped natural frequency for the system was 45 Hz.

The limits of direct pressure measurements have been systematically explored as a function of microcannula tip diameter by Wunderlich and Schermann [42] and Intaglietta [43]. These studies show that the response time falls in proportion to the fourth power of the tip diameter, indicating that viscous, Poiseuille-type flow is the major determinant of the hydraulic resistance. Extrapolating this result to the 1- to 2-μm-tip cannulas required, the 95 percent response time to pressure step inputs for conventional pressure transducers is of the order of 10 min. On the basis of these considerations it is generally accepted that pressure detection with passive systems is limited to microcannulas with 15-μm tips and above if it is desirable to detect pulsatile pressure variations.

The comparison of intravascular pressure to an external source was pioneered by Landis [44]. He inserted dye-filled micropipets into microvessels, manually manipulating the pipet pressure until the dye-plasma interface was observed to remain stationary in the tip of the micropipet. Eichna and Bordley [45] and Levick and Michel [46] refined the method and utilized it in humans. Wiederhielm et al. [47] extended the null balance method of pressure measurement by replacing the manual control of the external reservoir pressure with a servomechanism. In this system a glass microcannula is filled with an aqueous solution containing 2 M NaCl whose electrical resistivity is lower than that of blood plasma. The electrical resistance measured between the shank of the micropipet and the tissue is primarily determined by the position of the interface between the plasma and pipet solution along the length of the micropipet tip. This resistance is measured in an ac bridge and maintained at a constant value by controlling the pressure in the shank of the micropipet. The pressure required to maintain the resistance constant is assumed to be the pressure in the blood plasma. The validity of this assumption has been experimentally confirmed by measuring pressures in test chambers simultaneously with macroscopic and resistance servo-nulling systems.

The technique has been further developed by Intaglietta and Tompkins [48] and Intaglietta [49], who introduced the negative capacitance amplifier as a means of canceling the pipet's impedance. Fein [50] and Fox and Wiederhielm [51] reported a method for obtaining measurements from micropipet tips smaller than 1 μm. Muheim [52] has described the fabrication of micropipets with a defined tip configuration. An instrument is now commercially available (IPM Inc., San Diego, California). Klinger [53] developed the mathematical analysis of the micropipet capacitive impedance and related this to the ion distribution in the lumen of the pipet [54]. Mahler et al. [55] used the servo system for the first time for the measurement of pressure in the capillaries of the human nail fold.

33.2 BLOOD FLOW MEASUREMENTS

The measurement of blood flow begins with the work of Harvey [56], who estimated cardiac output from determinations of heart rate and anatomical dissection of the ventricles. In principle this method could be implemented by multiplying the measured volume of blood ejected from the ventricle during a single contraction of the heart by the heart rate. At present, this is not practical, and direct measurement of cardiac output in humans is only available during procedures involving thoracic surgery, where blood flow can be measured in the exposed vessels in immediate proximity to the heart.

Blood flow is not measured by the conventional industrial instruments such as the rotameter, because such techniques require access to the inside of the conduit and expose blood to sharp edges which have the potential for causing the formation of blood clots. As a consequence the measurement of blood flow has evolved into a highly specialized technology based on physical principles whose application comprises various degrees of "invasiveness."

33.2.1 Dilution Techniques

Cardiac output is routinely measured by means of dilution techniques, based on the Fick principle [57], in which a known concentration of a tracer material is injected into the vasculature prior to the heart and its concentration is measured at any point downstream. The so-called Fick principle is an expression of the conservation of mass; therefore, if the concentration of a marker (indicator) is known at the input of an organ C_i, and is measured at the output C_o, and the rate of uptake or utilization $\dot{m}$ of the substance in the organ is known, then the flow rate Q is given by

$$Q = \frac{\dot{m}}{C_i - C_o} \tag{33.4}$$

For example, we can determine the oxygen uptake ($\dot{m}$) through respiration, by measuring the rate of air flow with a spirometer and utilizing a gas analyzer to establish the difference in oxygen concentration between inspired and expired gas; $C_i - C_o$ is measured as the difference between the arterial and venous O_2 contents of blood across the lung, and Q (cardiac output) can then be calculated according to Eq. (33.4).

The same principle can be implemented by the rapid injection of an optical dye (indocyanine green), whose peak absorption is at 805 nm, a radioactive tracer, or a bolus of saline at a lower temperature than blood, and then measuring either the concentration or the temperature somewhere in the arterial circulation. This approach was developed by Stewart [58, 59] and would be quite precise if it were not for the effect of recirculation, which has to be corrected for by an approximate method such as that introduced by Kinsman, Moore, and Hamilton [60] and subsequently discussed in greater generality by Zierler [61] and Hamilton [62]. The influence of recirculation can be minimized by neglecting the tail of the concentration curve, but this underestimates the area and, consequently, overestimates the flux. If the concentration is taken over too long an interval, some of the indicator is making its second trip past the sampling station or, in an extreme case, some of it may travel past the sampling station several times.

Thermodilution [63] is a widely used technique which avoids the recirculation problem, since the cold saline injected in amounts of the order of 10 mL is rewarmed rapidly, rendering recirculation artifacts negligible. The procedure is usually implemented by means of the Swan-Ganz catheter [64], which has four lumens, two of which are used for thermodilution procedures, i.e., injection of cold saline and temperature measurement downstream from the point of injection. It should be noted that this method uses only one venous puncture. Recirculation is also minimized through the use of xenon, which is injected in the blood in the vena cava, and measurement of its concentration in the pulmonary artery by means of a catheter. Xenon is subsequently exchanged with the atmosphere in the lung, thus diminishing the amount of circulating indicator.

Dienes [65] developed a complete analysis of the problem of recirculation based on the formalism of Laplace transforms, and concluded that the flow rate in a recirculating system is the largest eigenvalue of an integral equation, and developed a program that provides the solution.

33.2.2 Electromagnetic Flowmeters

The electromagnetic flowmeter is based on the phenomenon by which the motion of a conductor through a magnetic field generates a voltage perpendicular to the magnetic field and the direction of motion, and proportional to the velocity. If the magnetic field is uniform and perpendicular to the direction of flow, then the voltage V, in volts, measured across the outer diameter of a circular flow conduit is given by

$$V = Hd\bar{v} \times 10^{8} \tag{33.5}$$

where H is the magnetic field in gauss (G), d is the diameter in cm, and $\bar{v}$ is the average velocity in the cross section, in cm/s. An important feature is that the measured voltage is not dependent on the flow profile, as shown by Shercliff [66]. Furthermore, asymmetries of the flow have little effect as a consequence of the low velocity of the fluid proximal to the vessel wall electrode and the comparatively high conductivity of plasma.

The first implementation of the principle of electromagnetic flowmeters was reported by Fabre [67]. Kolin [68] and Wetterer [69] devised the first practical instruments. A primary consideration in the design of the system is the method for generating a uniform magnetic field across the vessel. Permanent magnets, or dc fields, are not practical because of the residual polarization in any type of electrode system which is necessary for detecting the voltage generated by the flow. Since typical signals range from 10 to 500 μV, the problems of noise and stability of electronics are severe. Kolin [70] introduced the sine wave system where the alternating field and concomitant voltage signal provided a notable increase in the stability and signal-to-noise ratio. Denison et al. [71] replaced the sine wave excitation with a square wave, and Yanof et al. [72] further improved this approach by introducing the trapezoidal excitation method.

Two basic types of flow probes have been constructed: the perivascular probe [73] and the intravascular catheter tip configuration [74]. Perivascular probes have been built and operated with lumen diameters as small as 0.75 mm. A variety of catheter tip flow probes have been designed and subsequently implemented [75, 79].

An important parameter in the use of electromagnetic flowmeters is that of static and dynamic calibration, since Eq. (33.5) is not strictly applicable as a consequence of shunting of the signal due to the conductivity of the medium surrounding the fluid in motion and to the electrical resistance of blood. Calibration is usually made relative to a standard flow [76]. Milnor [17] provides an updated summary of commercially available instruments.

33.2.3 Ultrasonic Methods

The invasive nature of electromagnetic flow measurements circumscribes their application to exposed vessels and experimental conditions, and therefore the method is not suitable for use in humans. The use of ultrasound circumvents these problems and provides a noninvasive method for detecting blood flow velocity and, in some instances, blood flow. Two techniques are utilized: (1) the measurement of transit time of ultrasound in the flowing blood and (2) the Doppler technique, where the frequency of an acoustic signal relative to a stationary detector varies as a function of the velocity of medium in which it travels. The principal developers of these techniques are Franklin and coworkers [77, 78]. Reneman [79] provides an exhaustive review of the early developments in this field.

In the transit time method (pulsed ultrasonic flowmeters) two piezoelectric crystals are mounted in cuff, facing each other tangentially. Each can emit as well as receive (transduce) sound, and alternate in duty cycle in such a fashion that the circuitry will detect the difference between upstream and downstream transit time. This type of instrument produces signals from each portion of the velocity profile in the blood vessel, and therefore the output must be carefully analyzed and calibrated [80]. Typical instruments measure differences in the arrival time in order of nanoseconds and have frequency responses from 0 to 50 Hz.

In the Doppler velocimeter [77] a beam of ultrasound is directed toward the moving blood, where the red blood cells act as moving reflectors. Cobbold [2] gives the following expression for the Doppler frequency change Δf:

$$\Delta f = f_2 \left(1 - \frac{c - v \cos\theta}{c + v \cos\phi}\right) \tag{33.6}$$

where f_s is the frequency of the ultrasound generator, c is the velocity of sound in the medium, v is the velocity of the red blood cells, θ is the angle of the ultrasound beam relative to the direction of flow, and ϕ is the angle of the detector relative to the flow. Since $c \gg v$, Δf is proportional to the velocity v, and the direction of flow can be detected by the use of additional circuitry. As in the previous case, the ultrasonic beam is reflected from red blood cells that have a distribution of velocities according to the flow profile in the vessel, and the signal output has a characteristic spectrum. Various methods have been implemented for the detection of the flow direction, including the use of an additional oscillator [81], a single sideband receiver using a quadrature phase detector [82], and a two-sideband method [83], to cite a few.

Blood flow measurement can be accomplished with ultrasound techniques by reconstructing the velocity profile. Such an approach is possible with the pulsed Doppler system, which allows one to discriminate between the echoes reflected from different positions in the bloodstream. The same result is also accomplished with multichannel systems operating simultaneously [84, 85].

The use of catheter tip instruments has been first reported by Stone et al. [86], Benchimol et al. [87], and Kalmanson et al. [88]. The most important characteristic of the ultrasound methods is that they can be used transcutaneously, providing flow velocity data on the carotid artery and the arteries and veins of the extremities. As a result, this instrument has become a standard part of the diagnostic vascular laboratory [89]. The noninvasive method is not necessarily restricted to surface vessels: Light [90] described a technique for obtaining measurements from the aortic arch, and Reid et al. [91] described the measurement of flow in the renal artery by the noninvasive Doppler method.

33.2.4 Thermal Methods

The dependence of the rate of heat transfer on flow velocity has been utilized in two different types of instruments. Rein [92] pioneered this methodology through the development of a procedure for heating one portion of a blood vessel and measuring the rise in the temperature of blood downstream. This device was subsequently implemented as a blood vessel cuff, where the temperature was sensed by thermocouples. The development of intravascular catheters has provided a better configuration for this methodology, in that the blood can be heated directly and the sensor can be placed in the vicinity of the heater [93]. In this type of instrument the heater is maintained at a temperature which is 5°C above that of normal blood. Since the flowing blood cools this element, the supply of a current which keeps the temperature constant provides a direct indication of the blood flow around the heater. The response of this system has been reported to be as high as 150 Hz by Grahn et al. [94]. A similar method is the so-called hot wire anemometer, where the resistance of a heated wire is used as a temperature sensor. The flow of fluid across the wire carries away thermal energy, decreasing its temperature; therefore, the current needed to maintain the resistance constant is directly proportional to the flow rate in the vicinity of the resistive element. Blood flow velocity has been measured by this technique utilizing a variety of resistive heaters including platinum and gold [95].

33.2.5 Methods for Measurement in the Microcirculation

Blood flow measurements in the microcirculation have developed into a specialized technique that uses distinct physical principles and methods which are not common to other types of flow measurement. The small size of the vessels excludes volumetric determinations and the use of electromagnetic flow measuring systems. Attention has been directed to the optical properties of flowing blood as a means of making microvascular flow measurements. One of the first attempts was that of Hugues [96], who devised a streak-image method to estimate velocity. At high magnification, blood acquires a granular appearance due to the presence of blood cells. The magnified image of flowing blood gives the appearance of streak lines parallel to the centerline of the blood vessel. When this image is projected onto photographic film which is being moved in a direction at right angles to that of the blood vessel, the photographic image of the streaks will have an orientation determined by both the corpuscular velocity and the film velocity; thus, when the latter is known, the former can be calculated.

A modification of this method was used by Monro [97], who developed a rotating prism device from which visual estimates of red cell velocity can be made. In this modification, the speed of the prism (and thus the angle of the streak lines) is manually adjusted.

In capillaries it is possible to compare the position of a red cell from one frame to another of either a cinefilm or a video tape recording, and the velocity of cells can be calculated in terms of their displacement as a function of time. This approach was used in conjunction with high-speed cinematographic methods by Bloch [98].

Branemark and Jonsson [99] designed a system in which the velocity of the red blood cells was compared by eye with that of a moving illuminated spot. This technique, termed the *flying spot system,* was utilized in the measurement of flow velocity in capillaries.

Asano et al. [100] first proposed the use of the changes in the optical light transmission pattern of red blood cells, as they move across a photodetector, as a means of measuring their transit time, and Muller [101] implemented such a technique for capillaries. Wayland and Johnson [102] developed a method for the study of transit time (time delay) of the photometric pattern generated by individual red cells or groups of red cells at two different photometric areas placed along the centerline of the blood vessel (hence the name *dual-slit method*). In their method the image of the vessel under study was projected through the microscope optics onto a screen, so that it could be viewed directly by reflection. The screen was penetrated by two slits optically coupled via a lucite rod to a photomultiplier. The delay between upstream and downstream signals was measured either directly from chart recordings of the photometric signals or by computing the delay to maximum cross correlation between signals. Intaglietta et al. [103] further developed this method by implementing on-line cross-correlation computations.

The use of the television camera as a photodetector for red blood cell flow velocity in capillaries was introduced by Intaglietta et al. [104]. This method is implemented by isolating the area within the televised scene, where the red cells are passing, and analyzing the video signal within this area. Implementing this procedure simultaneously at two stations along the capillary axis provides the necessary information to determine transit times. The isolation of video image areas for average light-level (or densitometric) analysis of the video signal within the given area is accomplished with a commercially available dual-channel video photometric analyzer (IPM Inc., San Diego, California). This instrument isolates from the composite video signal two temporal windows, which correspond to square elements of the image area. The measurement of transit time is accomplished by a special-purpose cross correlator specifically designed for velocity measurements, on the basis of the observation that the correlogram for red blood cell velocity measurements is approximately symmetrical about the maximum cross correlation [105]. The conversion of flow velocity to volumetric flow requires that special consideration be given to exact location of the source of photometric signals. This problem has been analyzed by Baker and Wayland [106]. Bollinger and coworkers [107] and Fagrell et al. [108] have developed this technique into a method for study of the human skin.

The methods described are based on temporal cross-correlation analysis implemented to measure delay between signals which have a common origin. A different approach has been used by several investigators [109–111] who selected a spatial correlation technique. In this approach the optical trace along a microvessel taken at two subsequent times is cross-correlated; therefore the displacement to maximum cross correlation is directly proportional to velocity [112].

Grunwald and Sinclair [113], LeCong and Loveberg [114], and Einav et al. [115] described a method based on optical laser velocimetry. Slaaf and coworkers [116] developed a technique where the image of individual red blood cells or groups of cells is projected onto an optical grating in such a fashion that the signal detected by a photosensor aimed at the grating is modulated at a frequency proportional to the velocity of red blood cells. A three-stage prism grating is utilized, and the direction of the flow is determined by the phase relationship between the signals from each of the gratings.

33.2.6 Whole-Body Methods

In this category are grouped dynamic methods where the whole body is immersed in a field and the data on flow is derived by the interaction of the motion of blood with the given field. Within this category of measurement falls the magnetorheographic method reported by Okai et al. [117], who used a permanent magnet placed on the outside of the thigh of a dog and placed two needles at either side of the femoral artery. This concept was also utilized by Kolin [118], who introduced a significant variation in the general scheme for electromagnetic measurements by placing the whole animal within the magnetic field and recording the voltage generated by introducing the electrodes into the vessel under study by means of a catheter. Bowman and Kudravcev [119] described a blood flowmeter based on the phenomenon of nuclear magnetic resonance.

Blood flow measurements have been obtained from various arteries noninvasively by utilizing the proton resonance frequencies 21.4 and 75 MHz. In these studies the subjects are placed within a superconducting solenoid electromagnet. Blood flowing in arteries is first magnetized by the steady field due to the electromagnet, and is demagnetized by an externally applied magnetic field which varies at radio-frequency rates. The flow is detected by a crossed-coil detector sensitive to the resonant motion of the hydrogen nucleus. The technique is primarily useful for the detection of pulsatile flow and is being developed to yield quantitative data by Batocletti et al. [120] and Halbach et al. [121].

Blood flow in various parts of the body can also be determined by measuring the transit time of boli of radiopaque material as described by Rutishouser [122] and implemented by video techniques by Silverman et al. [123]. Within this general area a method has been reported by which blood flow is measured by computer tomography [124].

REFERENCES

1. Hales, S.: Containing haemastaticks, in *Statical Essays,* no. 22, *History of Medicine Series,* Library of New York Academy of Medicine, Hafner, New York, 1733, reprinted 1964.
2. Cobbold, R. C. S.: *Transducers for Biomedical Measurements: Principles and Applications,* Wiley, New York, 1974.
3. Geddes, L. A.: *The Direct and Indirect Measurement of Blood Pressure,* Year Book, Chicago, 1970.
4. Hansen, A. T.: The theory for elastic liquid-containing membrane manometers: Special part, *Acta Physiol. Scand., 19,* 333–343, 1950.
5. Hansen, A. T., and Warberg, E.: A theory for elastic liquid-containing membrane manometers: General part, *Acta Physiol. Scand., 19,* 306–332, 1950.
6. Fry, D. L.: Physiologic recording by modern instruments with particular reference to pressure recording, *Physiol. Rev., 40,* 753–788, 1960.
7. Latimer, K. E.: The transmission of sound waves in liquid-filled catheter tubes used for intravascular blood-pressure recording, *Med. Biol. Eng., 6,* 29–40, 1968.
8. Latimer, K. E., and Latimer, R. D.: Measurements of pressure-wave transmission in liquid-filled tubes used for intravascular blood-pressure recording, *Med. Biol. Eng., 7,* 143–167, 1969.
9. Crul, J. F.: Measurement of arterial pressure, *Acta Anesth. Scand., 6,* suppl. 11, 135–169, 1962.
10. Sutterer, W. F., and Wood, E. H.: Strain-gauge manometers: Application to recording of intravascular and intracardiac pressures, in O. Glasser (ed.), *Medical Physics,* Year Book, Chicago, 1960, vol. III, pp. 641–651.
11. Noble, F. W., and Barnett, G. O.: An electric circuit for improving the dynamic response of the conventional cardiac catheter system, *Med. Electron. Biol. Eng., 1,* 537–543, 1963.
12. Intaglietta, M., Pawula, R. F., and Tompkins, W. R.: Pressure measurements in the mammalian microvasculature, *Microvasc. Res., 2,* 212–220, 1970.
13. Shinozaki, T., Deane, R. S., Ch(Rand), B., and Mazuzan, J. G.: The dynamic response of liquid-filled catheter systems for direct measurement of blood pressure, *Anesthesiology, 53*(6), 498–504, 1980.
14. Lomholt, N., Bille-Brake, N. E., Rask Andersen, H. Djuinhus, J. C., and Bredgaard, M.: Thin fluid-filled catheters for more accurate determinations of pressures and their time derivatives, *Acta Chir. Scand Suppl., 502,* 27–32, 1980.
15. Gardner, R. M.: Direct pressure measurements, dynamic response requirements, *Anesthesiology, 54,* 227–236, 1981.
16. Timoshenko, S., and Woinowsky-Krieger, S.: *Theory of Plates and Shells,* McGraw-Hill, New York, 1959.
17. Milnor, W. R.: *Hemodynamics,* Williams & Wilkins, Baltimore, 1982.
18. McDonald, D. A.: *Blood Flow in Arteries,* Edward Arnold, London, 1974.
19. Yanof, H. M.: *Biomedical Electronics,* Davis, Philadelphia, 1965.
20. Gabe, I. T.: Pressure measurement in experimental physiology, in D. H. Bergel (ed.), *Cardiovascular Fluid Dynamics,* Academic Press, London, 1972, vol. I.
21. Webster, J. G. (ed.): *Medical Instrumentation,* Houghton Mifflin, Boston, 1978.
22. Allard, E. M.: Sound and pressure signals obtained from a single intracardiac transducer, *IRE Trans. Bio-Med. Electron., BME-9,* 74–77, April 1962.
23. Warnick, A., and Drake, E. H.: A new intra-cardiac pressure measuring system for infants and adults, *IRE Nat. Conv. Rec.,* pt. 9, 68–73, 1958.
24. Lindstrom, L. H.: Miniaturized pressure transducer intended for intravascular use, *IEEE Trans. Bio-Med. Eng., BME-17,* 207–219, July 1970.
25. Korotkoff, N. S.: On the subject of methods of measuring blood pressure. *Bull. Imp. Military Med. Acad., St. Petersburg, 11,* 365–372, 1905.
26. Lywood, D. W.: Blood pressure, in *Manual of Psychophysiological Methods,* North-Holland, Amsterdam, 1967.
27. Lange, K.: A recording sphygmotonograph: A machine for the continuous recording of systolic and diastolic arterial pressure in man, *Am. Int. Med., 18,* 367–383, 1943.
28. de Dobbeleer, G.: Procède nouveau de mésure de la tension arterielle maxima et minima, *Acta Anesthesiol. Belg., 14,* 21–36, 1963.

29. Yamakoshi, K. I., Shimazu, H., and Togawa, T.: Indirect measurement of instantaneous arterial blood pressure in the human finger by the vascular unloading technique, *IEEE Trans. Biomed. Eng.*, *27*(3), 150–155, 1980.

30. Ware, R. W., and Laenger, C. J.: Indirect blood pressure measurement by Doppler ultrasonic kinetoarteriography, *Proc. 20th Annu. Conf. Eng. Med. Biol.*, Boston, 1967, vol. IX, pap. 27.3.

31. Raber, S.: Blood pressure instruments: Non-invasive arterial measurements, in R. B. Spooner (ed.), *Hospital Instrumentation, Care and Servicing*. Instrument Society of America, Pittsburgh, Pa., 1977.

32. Ball, G. R., Pallet, J., and Shillingford, J. P.: An automatic finger blood pressure recorder, *Lancet, 2,* 1178–1180, 1961.

33. Rubin, P., McLean, K., and Reid, J.: A comparative study of automated blood pressure recorders, *Post Grad. Med. J., 56,* 815–817, 1980.

34. Bruner, J. M., Krenis, L. J., Kermsman, J. M., and Sherman, A. P.: Comparison of direct and indirect measurement of arterial blood pressure, *Med. Instrum., 15,* 182–188, 1981.

35. Bruner, J. M. R.: *Handbook of Blood Pressure Monitoring,* PSG Publishing, Littleton, Mass., 1978.

36. Intaglietta, M.: The measurement of pressure and flow in the microcirculation: Application to physiological and clinical problems, *Med. Prog. Tech., 4,* 55–60, 1976.

37. Roy, C. S., and Brown, J. G.: The blood pressure and its variation in the arterioles, capillaries and smaller veins, *J. Physiol., 2,* 323–359, 1980.

38. Intaglietta, M., and Zweifach, B. W.: Indirect method for the measurement of pressure in blood capillaries, *Circ. Res., 19,* 199–205, 1966.

39. Landis, E. M.: Micro-injection studies of capillary permeability: II. The relation between capillary pressure and the rate at which fluid passes through the walls of single capillaries, *Am. J. Physiol., 82,* 217–238, 1927.

40. Rappaport, M. B., Bloch, E. H., and Irwin, J. W.: A manometer for measuring dynamic pressures in the microvascular system, *J. Appl. Physiol., 14,* 651–655, 1959.

41. Levasseur, J. E., Funk, F. C., and Patterson, J. L., Jr.: Physiological pressure transducers for microhemocirculatory studies, *J. Appl. Physiol., 27,* 422–425, 1969.

42. Wunderlich, P., and Schermann, J.: Continuous recording of hydrostatic pressure in renal tubules and blood capillaries by use of a new pressure transducer, *Pfluegers Arch. Ges. Physiol., 313,* 89–94, 1973.

43. Intaglietta, M.: Pressure measurements in the microcirculation with active and passive pressure transducers, *Microvasc. Res., 5,* 317–323, 1973.

44. Landis, E. M.: Micro-injection studies of capillary blood pressure in the human skin, *Heart, 15,* 209–228, 1930.

45. Eichna, L. W., and Bordley, J. III: Capillary blood pressure in man: Comparison of direct and indirect methods of measurement, *J. Clin. Invest., 18,* 695–704, 1939.

46. Levick, J. R., and Michel, C. C.: Capillary pressures in the fingers and toes, *J. Physiol. (London), 274,* 97–109, 1978.

47. Wiederhielm, C. A., Woodbury, J. W., Kirk, S., and Rushmer, R. F.: Pulsatile pressure in the microcirculation of the frog's mesentery, *Am. J. Physiol., 207,* 173–176, 1964.

48. Intaglietta, M., and Tompkins, W. R.: Micropressure measurement with 1 micron and smaller cannulae, *Microvasc. Res., 3,* 211–214, 1971.

49. Intaglietta, M.: Microvascular pressure measurements by cannulation: Independency of concentration gradient and deviations in micro-servonulling, *Microvasc. Res., 3,* 396–399, 1971.

50. Fein, H.: Microdimensional pressure measurements in electrolytes, *J. Appl. Physiol., 32,* 560–564, 1972.

51. Fox, J. R., and Wiederhielm, C. A.: Characteristics of the servo-controlled micropipet pressure system, *Microvasc. Res., 5,* 324–335, 1973.

52. Muheim, M. H.: Fabrication of glass micropipettes with well defined tip configuration by hydrofluoric acid etching, *Pfluegers Arch., 372,* 101–102, 1977.

53. Klinger, D. R.: Capacitive modulation of micropipette impedance, *IEEE Trans., BME-24,* 129–134, 1977.

54. Firth, D. R., and DeFelice, L. J.: Electrical resistance and volume flow in glass micropipettes, *Can. J. Physiol. Pharmacol., 49,* 324–335, 1971.

55. Mahler, F., Muheim, M. H., Intaglietta, M., Bollinger, F., and Anliker, M.: Blood pressure fluctuations in human nailfold capillaries, *Am. J. Physiol., 326,* H888–H893, 1979.

56. Harvey, W.: Movement of the heart and blood in animals, in K. J. Franklin, *De Motu Cordis,* Blackwell, Oxford, 1957.

57. Fick, A.: Über die Anderung der Elasticitat des Muskels wahrend der Zuckung, *Pfluegers Arch., 4,* 301–312, 1870.

58. Stewart, G. N.: Researches on the circulation time and on the influences which affect it, *J. Physiol., 22,* 159–183, 1897.

59. Stewart, G. N.: The pulmonary circulation time, the quantity of blood in the lungs and the output of the heart, *Am. J. Physiol., 58,* 20–44, 1921.

60. Kinsman, J. M., Moore, J. W., and Hamilton, W. F.: Studies on the circulation: I. Injection method: Physical and mathematical considerations, *Am. J. Physiol., 89,* 331–339, 1929.

61. Zierler, K. L.: Theoretical basis of indicator-dilution methods for measuring flow and volume, in E. H. Wood (ed.), *Symposium on Use of Indicator-Dilution Technics,* American Heart Association, Washington, D.C., 1962.

62. Hamilton, H. F.: Measurement of cardiac output, in H. F. Hamilton, (ed.), *Handbook of Physiology,* vol. 1: *Circulation,* Williams & Wilkins, Baltimore, 1962.

63. Hosie, K. F.: Thermal dilution techniques, *Circ. Res., 10,* 491–504, 1962.

64. Ganz, W., Donoso, R., Marcus, H. S., Forrester, J. S., and Swan, H. J. C.: A new technique for measurement of cardiac output by thermodilution in man, *Am. J. Cardiol., 27,* 392–396, 1971.

65. Dienes, J. K.: The mathematics of recirculation and the measurement of cardiac output, *Math. Biosci., 32,* 141–153, 1976.

66. Shercliff, J. A.: *The Theory of Electromagnetic Measurement,* Cambridge University Press, Cambridge, 1962.

67. Fabre, P.: Utilisation des forces electromotrices d'induction pour l'enregistrement des variations de vitesse des liquides conducteures: Un novel hemodromographe sans palette dans le sang, *C. R. Acad. Sci. (Paris), 194,* 1097–1103, 1932.

68. Kolin, A.: An electromagnetic flowmeter: Principle of the method and its application to blood flow measurements, *Proc. Soc. Exp. Biol. Med., 35,* 53–56, 1936.

69. Wetterer, E.: Eine neue Methode zur Registrierung der Bluststromungsge schwindigkeit am uneroffneten Gefass, *Z. Biol., 98,* 26–36, 1937.

70. Kolin, A.: Electromagnetic rheometry and its application to blood flow measurements, *Am. J. Physiol., 122,* 797–804, 1938.

71. Denison, A. B., Spencer, M. P., and Green, H. D.: A square-wave electromagnetic flowmeter for application to intact blood vessels, *Circ. Res., 3,* 39–46, 1955.

72. Yanof, H. M., Salz, P., and Rosen, A. L.: Improvements in trapezoidal-wave electromagnetic flowmeter, *J. Appl. Physiol., 18,* 230–232, 1963.

73. Khouri, E. M., and Gregg, D. E.: Miniature electromagnetic flowmeter applicable to coronary arteries, *J. Appl. Physiol., 18,* 224–227, 1963.

74. Mills, C. J.: A catheter-tip electromagnetic velocity probe, *Phys. Med. Biol., 11,* 323–324, 1966.

75. Millar, H. D., and Baker, L. E.: A stable ultraminiature catheter tip pressure transducer, *Med. Biol. Eng. Comput., 11,* 86–89, 1973.

76. Mills, C. H.: Measurement of pulsatile flow and flow velocity, in D. H. Bergel (ed.), *Cardiovascular Fluid Dynamics,* Academic Press, New York, 1972, pp. 51–90.

77. Franklin, D. L., Schlegel, W., and Rushmer, R. F.: Blood flow measured by Doppler frequency shift back-scattered sound, *Science, 134,* 564–565, 1961.

78. Franklin, D. L., Baker, D. W., and Rushmer, R. E.: Ultrasonic transit-time flowmeter, *IRE Trans. Biomed. Eng., BME-9,* 44–49, 1962.

79. Reneman, R. S. (ed.): *Cardiovascular Applications of Ultrasound,* North-Holland, Amsterdam, 1974.

80. Gessner, U.: The performance of the ultrasonic flowmeter in complex velocity profiles, *IEEE Trans. Biomed. Eng., BME-16,* 139–142, 1969.

81. Mackay, R. S.: Non-invasive cardiac output measurement, *Microvasc. Res., 4,* 438–452, 1972.

82. McLeod, F. D.: A directional Doppler flowmeter, *Digest, 7th Int. Conf. Med. Biol. Eng.,* Stockholm, 1967, pp. 213–217.

83. Sainz, A., Roberts, V. C., Pinardi, G., and Lidenaur, S. M.: Blood velocity and acceleration measurements by Doppler shift ultrasound, in D. E. M. Taylor and J. W. Lamond (eds.), *Non-invasive Clinical Measurements,* University Park Press, Baltimore, 1977.

84. Baker, D. W.: Pulsed ultrasonic Doppler blood-flow sensing, *IEEE Trans. Sonics Ultrasonics, SU-17,* 170–185, 1970.

85. Peronneau, P., Higlais, J., Pellet, M., and Leger, F.: Velocimètre sanguin par éffet Doppler à émission ultrasonore pulsée: Description de l'appareil, *Result. Onde Electr., 50,* 369–372, 1970.

86. Stone, H. L., Stegall, H. F., Bishop, V. S., and Laenger, C.: Continuous measurement of blood flow velocity with an intravascular Doppler flowmeter, *Dig. Tech. Pap., 7th Int. Conf. Med. Biol. Eng.,* Stockholm, 1967, pp. 215–219.

87. Benchimol, A., Stegall, H. F., Maroko, P. R., Garitlan, J. L., and Brenner, L.: Aortic flow velocity in man during cardiac arrhythmias measured with the Doppler catheter-flowmeter system, *Am. Heart J., 78,* 649–659, 1969.

88. Kalmanson, D., Toutain, G., Novikoff, N., and Derai, C.: Retrograde catheterization of left heart cavities in dogs by means of an orientable directional Doppler catheter-tip flowmeter: A preliminary report, *Cardiovasc. Res., 6,* 309–318, 1972.

89. Fronek, A.: Analysis of recent developments in blood flow measurement, in Y. C. Fung, N. Perrone, and M. Anliker (eds.), *Biomechanics: Its Foundation and Objectives,* Prentice-Hall, Englewood Cliffs, N.J., 1972.

90. Light, L. H.: Transcutaneous observation of blood velocity in the ascending aorta in man, in F. De Freitas (ed.), *Ballistocardiography and Cardiovascular Therapy, Bibl. Cardiol.,* no. 26, Karger, Basel, 1970.

91. Reid, M. H., Mackay, R. S., and Lantz, B. H.: Review—Non-invasive blood flow measurements by Doppler ultrasound with application to renal artery flow determination, *Invest. Radiol., 15,* 323–331, 1980.

92. Rein, H.: Die Thermo-Stromuhr: Ein Verfahren zur fortlaufenden Messung der mittleren absoluten Durchflussmengen in uneroffneten Gefassen in Situ, *Z. Biol., 87,* 394–418, 1928.

93. Mellander, S., and Rushmer, R. F.: Venous blood flow recorded with an isothermal flowmeter, *Acta Physiol. Scand., 48,* 13–19, 1960.

94. Grahn, A. R., Paul, M. H., and Wessel, H. U.: A new directional-sensitive probe for catheter-tip thermal velocity measurements, *J. Appl. Physiol., 27,* 407–412, 1969.

95. Seed, W. A., and Wood, N. B.: Development and evaluation of a hot film velocity probe for cardiovascular studies, *Cardiovasc. Res., 4,* 253–263, 1970.

96. Hugues, J.: Contribution à l'études des factures vasculairs et sanguins dans l'hemostase spontanée, *Arch Int. Physiol., 61,* 565–711, 1952.

97. Monro, P. A. G.: A method for measuring the velocity of moving particles under the microscope, in E. Barrer and V. E. Cosselett (eds.), *Advances in Optical and Electron Microscopy,* Academic Press, New York, 1966.

98. Bloch, E. H.: High speed cinephotography of the microvascular system, in A. L. Copley (ed.), *Proc. 1st Int. Hemorheol. Conf.,* Pergamon Press, London, 1966, pp. 655–658.

99. Branemark, P. I., and Jonsson, I.: Determination of the velocity of corpuscles in blood capillaries: A flying spot device, *Biorheology, 1,* 143–146, 1963.

100. Asano, M., Yoshida, K., and Tatai, K.: Blood flow rate in the microcirculation as measured by photoelectric microscopy, *Bull. Inst. Publ. Health, 13,* 201–207, 1964.

101. Muller, H.: Erythrocyte transit time technique, *Methods Med. Res., 11,* 207–211, 1966.

102. Wayland, H., and Johnson, P. C.: Erythrocyte velocity measurement in microvessels by two-slit photometric method, *J. Appl. Physiol., 22,* 333–337, 1967.

103. Intaglietta, M., and Tompkins, W. R.: System for measurement of velocity of microscopic particles in liquids, *IEEE Trans. Biomed. Eng., BME-18,* 376–377, 1971.

104. Intaglietta, M., Silverman, N. R., and Tompkins, W. R.: Capillary flow velocity measurements in vivo and in situ by television methods, *Microvasc. Res., 10,* 165–179, 1975.

105. Tompkins, W. R., Monti, R., and Intaglietta, M.: Velocity measurement by self-tracking correlator, *Rev. Sci. Instrum., 45,* 647–649, 1974.

106. Baker, M., and Wayland, H.: On-line volume flow rate and velocity profile measurements for blood in microvessels, *Microvasc. Res., 7,* 131–143, 1974.

107. Bollinger, A., Butti, P., Barras, J. P., Traschler, H., and Siegenthaler, W.: Red blood cell velocity in nailfold capillaries of man measured by television microscope technique, *Microvasc. Res., 7,* 61–72, 1974.

108. Fagrell, B., Fronek, A., Intaglietta, M.: A microscope-television system for studying flow velocity in human skin capillaries, *Am. J. Physiol.*, *233*(2), H318–H321, 1977.

109. Anliker, M., Casty, M., Friedli, P., Kubli, R., and Keller, H.: Noninvasive measurement of blood flow, in N. H. C. Hwang and N. A. Norman (eds.), *Cardiovascular Flow Dynamics and Measurements*, University Park Press, Baltimore, 1977, pp. 43–88.

110. Fu, S. E., and Lee, J. S.: A video system for measuring the blood flow velocity in microvessels, *IEEE Trans. Biomed. Eng.*, *BME-25*, 295–297, 1978.

111. Goodman, A. H., Guyton, A. C., Drake, R., and Laflin, J. H.: A television method for measuring capillary red cell velocities, *J. Appl. Physiol.*, *37*, 126–130, 1974.

112. Tyml, K., and Sherebrin, M. H.: A method for on-line measurement of red cell velocity in microvessels using computerized frame-by-frame analysis of television images, *Microvasc. Res.*, *20*, 1–8, 1980.

113. Grunwald, J. E., and Sinclair, S. H.: Laser Doppler measurements of relative blood velocity in the human optic-nerve head, *Invest. Ophthalmol.*, *22*(2), 241–248, 1982.

114. LeCong, P., and Loveberg, R. H.: Analysis of a dual beam laser velocimeter applied to microcirculation studies, *Rev. Sci. Instrum.*, *51*, 565–574, 1980.

115. Einav, S., Berman, H. J., Fuhro, R. L., DiGiovanni, P. R., Fridman, J. D., and Fine, S.: Measurement of blood flow in vivo by laser Doppler anemometry through a microscope, *Biorheology*, *12*, 203–205, 1975.

116. Slaaf, D. W., Toop, P. S., Rood, M., Tangelder, G. T., Teurens, T. J. M., Alewymse, R., Reneman, R. S., and Arts, T.: A bidirectional optical (BDO) three-stage prism grating system for on-line measurement of red blood cell velocities in microvessels, *Microvasc. Res.*, *22*, 110–112, 1981.

117. Okai, O., Togawa, T., and Oshima, M.: Magnetorheography: Nonbleeding measurement of blood flow, *J. Appl. Physiol.*, *30*, 564–566, 1971.

118. Kolin, A.: New approaches to electromagnetic measurements of blood flow and vasomotion without recourse to surgery, in *The arterial system: Dynamics, Control Theory, and Regulation*, Springer-Verlag, Berlin, 1978, pp. 122–150.

119. Bowman, R. L., and Kudravcev, V.: Blood flowmeters utilizing nuclear magnetic resonance, *IRE Trans. Med. Electron.*, *ME-6*, 267–269, 1959.

120. Batocletti, J. H., Halbach, R. G., Selles-Cunha, S. X., and Sences, A., Jr.: The NMR blood flowmeter—Theory and history, *Med. Phys.*, *8*, 435–443, 1981.

121. Halbach, R. E., Battocletti, J. H., Salles Cunha, S. X., and Sences, A., Jr.: The NMR flowmeter—Design, *Med. Phys.*, *8*, 444–451, 1981.

122. Rutishouser, W.: Application of roentgendensitometry to blood flow measurements in models, animals and in intact conscious man, in P. H. Heintzen (ed.), *Roentgen- Cine- and Video densitometry*, Springer-Verlag, Stuttgart, 1977.

123. Silverman, N. R., Intaglietta, M., and Tompkins, W. R.: Renal cortical transit times: A simple parameter for evaluation of renal perfusion, *Radiology*, *150*, 25–31, 1972.

124. Tonge, K. A., Wright, C. H., Mathew, J., Leach, R. D., and Saunders, J. E.: Flow rate determination using computed tomography, *Br. J. Radiol.*, *53*, 946–949, 1980.

CHAPTER 34

Electrical Safety in Medicine

Alvin Wald
Department of Anesthesiology
College of Physicians and Surgeons
Columbia University
New York, New York

INTRODUCTION

The disabilities of patients and the special activities within health care facilities mandate special considerations of electrical safety above those required in public facilities. First, electricity often powers equipment that is used to support impaired physiological systems and sustain life. Ventilators, heart-lung machines, and kidney dialysis units must function reliably. Second, patients are in an unfamiliar environment and are often physically incapacitated or mentally impaired. Patients may be unaware or unresponsive to normally occurring warnings, either from their own body or from equipment alarms, and may be unable to call for help. In addition to this lack of responsiveness, there are general and specific medical conditions which lower the patient's threshold to electrical hazards. A situation that a healthy person in the home could readily cope with might be disastrous for a hospitalized individual.

This section will discuss three interrelated major categories of electrical safety needs for medical applications: (1) physiological hazards of current, (2) adequacy and safety of electrical service, and (3) safety of electromedical appliances.

34.1 PHYSIOLOGICAL HAZARDS OF CURRENT

The physiological hazards of current can be considered in two broad categories: macroshock and microshock. Macroshock involves current generally larger than 1 mA which passes through the intact body. Microshock involves low current, in the microampere range, and applied directly to the heart. Many of the studies defining macroshock were carried out on human volunteers in the 1930s to 1950s, prior to the present restrictive regulations on human experimentation. Other data have come from animal studies and investigations of accidental electrocutions. Concern with microshock hazards came about after the development of medical techniques which involve direct access to the heart. There are some data available from human studies, but the majority of data on microshock have come from animal experiments. Since there can never be an exact correlation between animal experimentation and human conditions, a certain amount of controversy and uncertainty remains in the interpretation of results and conclusions from the studies.

34.1.1 Electric Current

Voltage in conjunction with resistance at the point of contact determines the magnitude of the current, which, in turn, determines the physiological effect. For intact dry skin, the upper layer of the epidermis has a resistivity of 100,000 to 300,000 Ω/cm^2 [1]. This value varies greatly among individuals, with different conditions, and at various body sites. Moisture, from either external sources or perspiration, can lower skin resistance to less than 1 percent of its dry value. Minimum resistance across intact skin is a nominal 1000 Ω. Once the skin is bypassed, the internal body tissue, which acts as a volume conductor, determines resistance. Internal resistance is usually taken to be 500 Ω or less. In the typical accident involving electricity, current flows between the two arms or from an upper to a lower limb. The most serious consequences occur when current passes through the chest, affecting the heart and respiratory muscles, or through the brain, affecting the central nervous system. For current flow along the body's vertical axis, up to 10 percent will pass through the heart; along the horizontal axis, 3 percent [2]. Microampere currents directly applied to the myocardium can cause either ventricular fibrillation or a sustained arrhythmia which results in the loss of ventricular contraction (pump failure). Other types of injury which can be caused by electricity are trauma due to sustained muscle contraction, and burns at areas of high current density.

34.1.2 Cellular Effects

For a resting cell, ions on both sides of the cell membrane maintain the interior in a polarized state of approximately 100 mV negative with respect to the extracellular fluid. If this potential is reduced to approximately 60 mV, the cell membrane becomes unstable and depolarizes, causing the cell to become slightly positive. Under ordinary conditions, the cell potential then reverses itself, repolarizing and returning to its resting state. This sequence of depolarization and repolarization is the action potential. Typically, a localized action potential propagates along an entire nerve membrane. At the myoneural junction, acetylcholine released at the nerve ending stimulates the muscle membrane to depolarize. The action potential then propagates throughout the entire muscle, causing it to contract. A muscle can be made to contract by external electrical stimulation, either directly or through its enervating nerve. If electric current is supplied to a muscle for less than 10 ms, whether or not depolarization occurs depends on the total energy supplied and the relationship between the electric field and tissue ions. For longer time periods, once the externally supplied electron flow exceeds a threshold, depolarization will propagate.

The myocardium, made up of specialized muscle cells, reacts similarly to other muscle tissue. If electric current contracts the heart during the period when it is most vulnerable, at ventricular repolarization, or the T-wave portion of the electrocardiogram, multiple sites of depolarization having different refractory responses can develop. The myocardium will not contract synchronously and the heart will cease to pump blood, i.e., ventricular fibrillation. Outside of the vulnerable interval, an impulse of current of less than 10 ms may not allow sufficient time for repolarization to occur, and there will be only a single muscular contraction. This impulse effect is the basis for the action of pacemakers and defibrillators and for their classification in terms of total energy delivered.

34.1.3 Microshock Current Limits

Microshock hazards have become significant for medical techniques that require a direct electrically conductive connection between the heart and an external device. With the heart no longer protected by skin resistance, which limits the magnitude of current flow, or by the whole-body volume conductor effect, which reduces current density, microampere current levels become significant. Clinical situations that place a patient at risk to microshock occur when there is a catheter filled with conductive fluid or an electrically conductive lead wire in direct contact with the heart muscle or within one of its chambers. These situations include:

1. Cardiac pacing with an externalized wire
2. Intracardiac electrocardiogram monitoring
3. Intracardiac pressure measurement
4. Injection of radiopaque substances for x-ray visualization

Some of the conflict over an acceptable, safe microshock current limit has been caused by different experimental conditions. Also, failure of the heart as a pump due to a disruption of rhythm occurs at a lower current than does sustained ventricular fibrillation [3–5]. After electrical stimulation is discontinued, the heart in pump failure will resume normal activity, while in true fibrillation the heart will remain in a desynchronous state. If the current is not discontinued in the former case, or if the heart is not defibrillated in the latter, the results will be the same—death due to the absence of cardiac output. Although there is a clear distinction between normal cardiac rhythm and ventricular fibrillation, other abnormal rhythms may also be induced by external electrical stimulation.

The factors which determine the safe limit for current applied to the heart fall into two broad categories: intrinsic physiological conditions and external electrical factors.

Physiological Conditions. Physiological factors which may alter the threshold to fibrillation include:

1. Status of the heart muscle and valves
2. Heart size
3. Coronary circulation
4. Cardiac electrical conduction system
5. Body temperature
6. Fluid and electrolyte balance
7. Tissue oxygenation
8. Acid-base balance
9. Metabolic state
10. Hormonal homeostasis
11. Drugs and anesthetic agents

Nonphysiological Factors. Other factors which can affect the fibrillation threshold for current applied directly to the heart are:

1. Electrode position
2. Electrode size and current density
3. Time factors

Electrode position: Using a saline-filled catheter with an active electrode in the heart and return electrode on a rear leg, Roy [3] found the most sensitive area in dogs' hearts to be at the apex of the right ventricle for internal placement, and at approximately the midpoint of the right ventricle, 10 to 15 mm above the apex, for external placement on the pericardium. For any particular size of electrode, threshold fibrillatory current varied fivefold with position and orientation of the contact electrode. The effect of electrode position can be summarized by the work of Ledsome et al. [5], as shown in Table 34.1. The lowest current which caused fibrillation was in the order of 20 μA. From these and similar studies, the leakage current limit for medical

TABLE 34.1 Canine ventricular arrhythmia, pump failure, and fibrillation threshold currents of 60 Hz and 3 s*

	Catheter position	
	Random (n = 54)	Most sensitive (n = 50)
Ventricular arrhythmia	65 ± 62	17 ± 2.5
Pump failure	122 ± 102	21 ± 4.4
Fibrillation	335 ± 395	108 ± 36

Source: Ledsome et al. [5].
* Values in microamperes, mean ± SD.

appliances that can have leads in direct contact with the heart has been set at 10 μA by standards developed in this country [6–8].

Electrode size and current density: Roy [4] combined results from 10 studies and showed how fibrillatory threshold level is inversely proportional to current density. Current was applied to the exposed heart for 3 s or longer. For the smallest-area electrode used, 2 mm^2, the average minimum current density for fibrillation was 80 μA/mm^2. With increasing area the current density fell, approaching a lower limit of 5 μA/mm^2 for areas of 200 mm^2 and larger.

Time factors: For fibrillation to occur, electric current must be present during the T-wave portion of the electrocardiogram, which corresponds to the relative refractory period of the myocardium. Aside from this limitation, the fibrillatory threshold decreases with increasing duration of current over the range of 8.3 ms (½ of 60-Hz wave) to 5 s for intact animals [9], and from 16.6 ms to 2.1 s with direct stimulation of the heart [10].

Fibrillatory Window. Other work [10, 11] has shown that there is a variable upper limit of current above which fibrillation will not occur. That is, there is a current window in which fibrillation can occur. The longer the duration of the applied current, the wider the window. For relatively short pulses, the window is three to four times the threshold limit. Above the upper threshold, which is 2 to 3 A through the intact chest, the heart will remain contracted for as long as current is applied. Here, a sufficiently large number of cardiac muscle cells are simultaneously depolarized so that fibrillation is not initiated. When the stimulating current ceases, the cells repolarize and the heart resumes its normal rhythm. The recognition of this phenomenon gave rise to the original ac defibrillator.

34.1.4 Macroshock Current Limits

Many studies have been carried out to determine the limits of the effects of electric current on humans. Most of these studies were done with young, healthy volunteers. Hospitalized patients fall into a different category. Aged, debilitated, and unable to respond to stimuli, these patients may be more susceptible. In addition, the patient may be immobilized, with open wounds, and in contact with various body or therapeutic fluids which can provide low-resistance pathways for stray currents. The current limits discussed below should be considered in terms of these limitations.

Perception Current. *Perception current* is the minimum, or threshold, current perceived by a subject. Mucous membranes are the most sensitive tissue, but for practical purposes perception current is measured at the fingertips or in the grasped hand. Fingertips are more sensitive, threshold values being one-third of those for grasping [12].

With a conductor grasped in the hand and 60-Hz stimulation, there is a normal distribution of perception currents, with a mean value of 1.1 mA for men, as shown in Fig. 34.1. Because of their smaller muscle mass, predicted values for women are taken as two-thirds of those for men (as found when measuring let-go currents; see below).

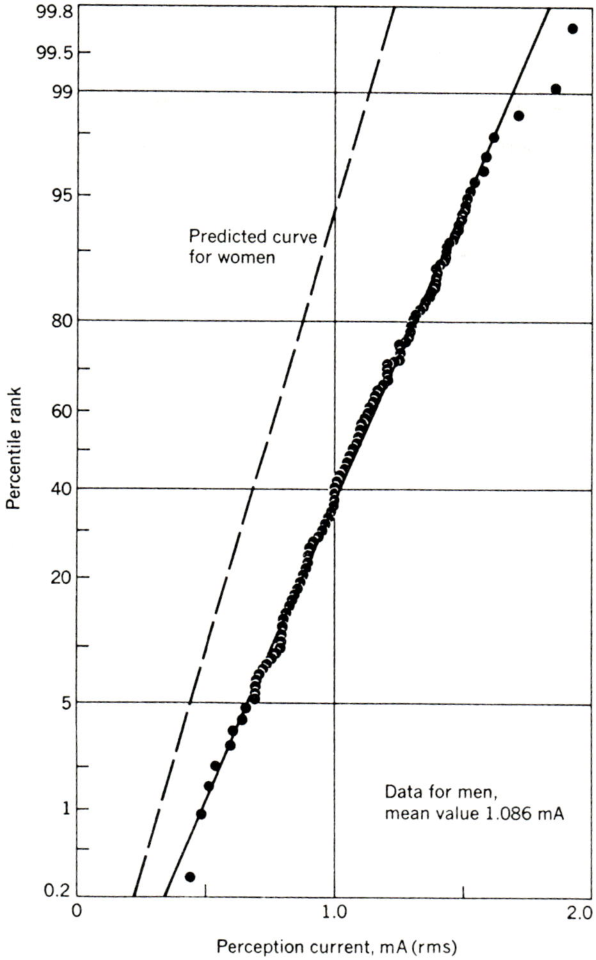

FIG. 34.1 Perception current versus percentile, measured for men and predicted for women. (*Reprinted from Dalziel [1], p. 42. Copyright © 1972, IEEE.*)

Frequency and perception current: Perception current threshold depends on the stimulating frequency [12, 13]. Perception current is a minimum and remains relatively constant between 60 and 200 Hz, and monotonically increases up to 200,000 Hz, as seen in Fig. 34.2. At 1000 Hz, mean perception current is 2 mA, twice the minimum value. Above 100,000 Hz, the sensation felt is that of heat rather than that of muscle stimulation. At very low frequencies (5 to 20 Hz), muscle contractions may follow the current alterations, though threshold data have not been obtained. DC stimulation on the skin is perceived as warmth, in contrast to 60 Hz, which produces a tingling sensation. On the tongue, both dc stimulation and 60 Hz are perceived as taste, where threshold currents range between 0.6 and 783 (median 16) μA for dc stimulation, and between 4 and 315 (median 30) μA for 60 Hz. The mean value is 45 μA for both dc stimulation and 60 Hz, indicating nonnormal distributions.

In itself, the perception of current is not hazardous. When unexpectedly encountering a

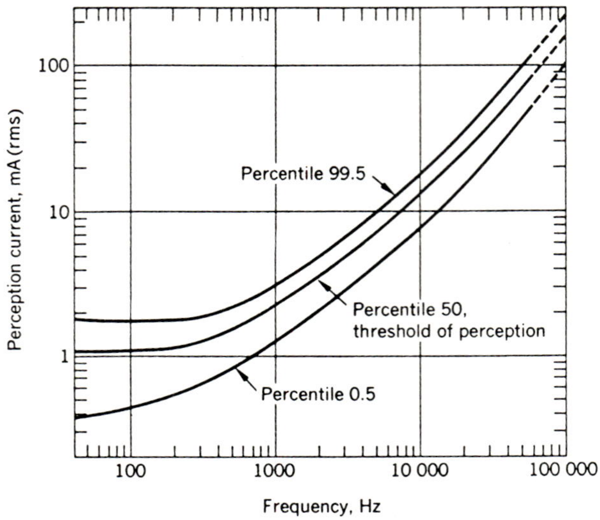

FIG. 34.2 Perception current versus frequency for men at different percentiles. (*Reprinted from Dalziel [1], p. 42. Copyright © 1972, IEEE.*)

shock, however, a person is likely to react in a startled manner, possibly causing an accident. Domestic and industrial appliances have a leakage current limit of 500 μA [14], somewhat below the estimated mean grasping perception threshold for women. Current limits for medical devices will be considered below.

Let-Go Current. *Let-go current* is that current at which a subject is unable to release an energized conductor grasped in the hand [15]. Current stimulates all the forearm muscles, causing tetanic contraction. Since the flexor muscles are stronger than the extensor muscles, the subject cannot release the grip. The location of the return electrode, the size of the stimulating conductor, and moisture conditions at the sites of electrode contact all have little effect on the results. Values for let-go currents also follow a normal distribution, as shown in Fig. 34.3. The mean value for women, 10.5 mA, is two-thirds that for men, 16 mA.

For men subjected to currents above 18 mA, psychological motivation plays an important role. The highest let-go current reported was 26 mA "on a student in physiology who boasted that he was as good as any engineering student. Although he made no complaints, it is more than likely that he had a sore arm for at least a week" [16]. At currents of 18 to 22 mA and higher, respiration ceased during the time the current was applied, and there were some instances of temporary paralysis of a digit. The subjects all recovered without any adverse effects. This type of experiment is unlikely to be repeated in these days of strict control with regard to human experimentation.

Frequency and let-go current: Let-go current is also frequency-dependent [15, 17]. Between 10 and 200 Hz there is a relative minimum, with a mean of about 16 mA (for men). At both higher and lower frequencies, let-go current increases as shown in Fig. 34.4. Both at 5 and 1000 Hz, mean let-go current is approximately 24 mA.

Direct current is perceived as internal heating in the hands and arms, with little muscular contraction as is experienced with alternating currents. The limiting factor for tolerance to direct current is not the pain of muscular contraction which prevents the let-go response, but rather a subject's limit of endurance to the burning sensation that is produced. Mean let-go current of 74 mA for direct current is equal to that for a frequency of 10,000 Hz.

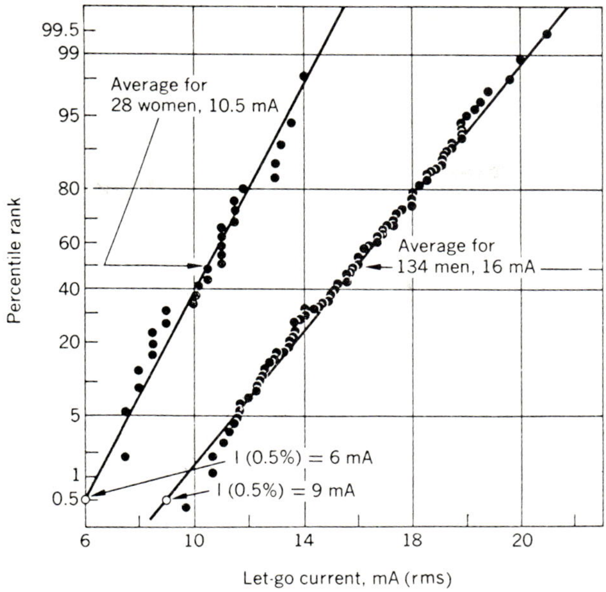

FIG. 34.3 Let-go current versus percentile for men and women. (*Reprinted from Dalziel [1], p. 43. Copyright © 1972, IEEE.*)

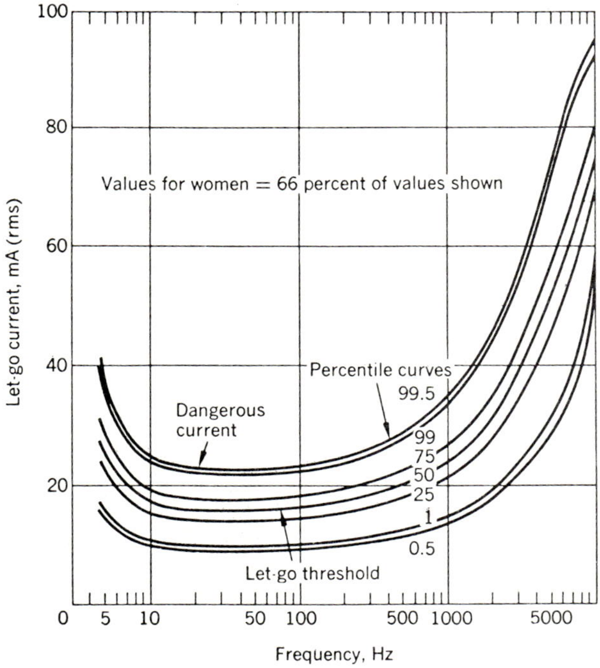

FIG. 34.4 Let-go current versus frequency for men at different percentiles. (*Reprinted from Dalziel [1], p. 44. Copyright © 1972, IEEE.*)

Fibrillatory Current. Experiments using higher currents, which cause fibrillation, can be carried out only on animals, with the results extrapolated to humans. Other information can be obtained by analyzing accidental electrocutions. For a 60-Hz current passing between the limbs and through the trunk, Dalziel and Lee [16, 18] combined results from a number of animal experiments and derived a generalized expression for fibrillatory current passing between the extremities and through the chest. Probability of fibrillation or nonfibrillation and body weight were combined into a single parameter, K, to give current I as a function of stimulus time T from 0.083 to 5 s.

$$I = \frac{K}{\sqrt{T}} \quad \text{mA (rms)} \qquad \text{for } 5 > T > 0.083$$

Specific values of K for adults and children are given in Table 34.2. For 95 percent probability of fibrillation with a 1-s shock, a 50-kg adult requires 185 mA, and an 18-kg child, 69 mA; the same probability for nonfibrillation is 116 mA for adults and 52 mA for children.

TABLE 34.2 Fibrillatory parameter K at 60 Hz, 95% probability level*

	50-kg adult	18-kg child
Nonfibrillating	116	52
Fibrillating	185	69

Source: Dalziel and Lee [16, 18].
* $I = K/\sqrt{T}$ mA rms; $5 > T > 0.083$ s.

Frequency and fibrillation current: The amount of current necessary to cause fibrillation in the intact animal also depends on frequency. Geddes and Baker [19] experimented with dogs, using alternating current from 10 to 10,000 Hz. There was a relative minimum, in the order of 250 mA, up to about 200 Hz, and then a substantial rise in current necessary for fibrillation at higher frequencies. At 1000 Hz, approximately 1500 mA was required.

Another series of experiments on sheep by Ferris et al. [11] used direct current and frequencies of 25 and 60 Hz. The minimum current to fibrillate at 60 Hz was in the order of 250 mA. At zero frequency (dc) 1320 mA was needed.

Very High Currents. Currents of 6 A or more can cause respiratory muscle contracture, with cessation of breathing for as long as the current is applied. This magnitude of current can also halt breathing by affecting the central nervous system's respiratory and cardiac control centers. Large currents can also cause severe deep tissue burns, especially if the contact area is small.

34.1.5 Summary of Effects of Current

The effects of current on the body may be summarized as follows:

1. In the frequency range of 10 to 200 Hz, the body is most susceptible to the effects of current.
2. To the exposed healthy canine heart, currents as low as 20 mA (60 Hz rms) can cause fibrillation. Factors such as electrode size and location and stimulus time can raise this threshold by an order of magnitude. On the other hand, this threshold may be lowered in a nonideal physiological state.
3. The range of perception current through the intact skin follows a normal distribution in the general population. The mean value for men for 60 Hz is 1.1 mA, and is approximately constant over the range 10 to 200 Hz. Perception current threshold increases at higher frequencies. Mucous membranes are the most sensitive tissue.

4. Let-go current also follows a normal distribution, with a mean of 16 mA for men. The threshold increases at frequencies lower than 10 and greater than 200 Hz.
5. Threshold values of perception and let-go currents for women are, on the average, two-thirds those for men.
6. Fibrillation current threshold through the intact chest depends on body size and time of shock application. A 95 percent probability of fibrillation with a 1-s shock to a 50-kg adult requires 185 mA. At frequencies outside the range of maximum susceptibility, fibrillatory current thresholds are higher.
7. Current greater then 2 A through the intact chest may not result in fibrillation but can cause temporary cardiac arrest and paralysis. Current greater than 6A can cause cardiac arrest, paralysis, and deep burns.
8. Table 34.3 summarizes results presented in the text.

TABLE 34.3 Current thresholds* (minimum mean values)

Current	DC	60 Hz rms	1000 Hz rms
Fibrillation (direct)	. . . .	0.020	
Density (2 mm^2)	. . . .	0.080/mm	
Perception (mean)	. . . .	1.1	2
Tongue (median)	0.016	0.030	
Let-go (men)	74	16	24
Fibrillation (animal)	1320	250	1500
Calculated (1 s, 50 kg, $p = 0.95$)	. . . .	185	

* Current in milliamperes.

34.2 SAFETY AND ADEQUACY OF ELECTRICAL SERVICES

Health care facilities require safe and adequate electric power. Standard engineering and architectural design principles should be employed, and local building and health codes followed. Advancing technology ineluctably increases the need for electric power throughout the entire facility. New developments, such as in x-ray technology and nuclear medicine, inconceivable 10 years ago, now must be serviced in an often inadequately based structure. Hospital buildings are often used for 50 years and more, and no one can predict the next generation of medical technological advancement. At the other end of the scale, the number of electrical outlets at any location is traditionally underestimated. Here, too, prediction of local needs must take into account future developments.

34.2.1 Codes and Standards

Various published standards give guidance for safe and adequate electrical services in medical facilities. Standards are written by various governmental and private organizations. Government standards are mandatory and have the force of law. Privately written standards are voluntary but, if adopted into governmental codes and regulations, take on the aspects of mandatory standards. When a voluntary standard is adopted by legislation, the particular version, by date, of the standard is specified. If the standard is then revised, becoming either more or less restrictive, the original version is still in effect unless altered by legislation. Different standards by different, or even the same, organizations may be in conflict with each other. State and local codes and regulations may be more or less stringent than some voluntary standards. The interactions among various standards writing organizations and the effect of these standards on health care facilities must be constantly assessed [20].

One example will be given here to illustrate how a user of standards can be confounded. Specifications for electric power in anesthetizing locations were addressed by the National Fire Protection Association (NFPA) [21]. Their published standard (1978) called for the use of isolated power systems. State and local jurisdictions may have incorporated this standard into their own hospital and building codes. The latest version of this standard was published in 1984 [8, Chap. 3]. The new standard no longer requires an isolated power supply in anesthetizing locations that are designated as nonflammable. On the other hand, the *National Electrical Code 1984* [22], published by the NPFA, does require isolated power supplies for all anesthetizing locations, both flammable and nonflammable. Whether or not isolated power is required thus depends on the specific jurisdictional code under which a hospital is constructed, renovated, or operated. In other aspects of hospital electrical safety, requirements in the two standards are identical or cross-referenced. It is hoped that in the next cycle of revisions, this and other code conflicts can be resolved. In general, an attempt was made to confine the *National Electrical Code 1984* [22] to design and construction aspects of the physical plant, while the *Standard for Health Care Facilities* [8] is primarily concerned with operating and maintenance considerations.

In addition to voluntary standards formulated by a consensus of interested individuals and organizations, different state and county or municipal public authorities that have jurisdiction in health care or construction matters may have their own building, construction, health, and fire codes. Local professional engineers, architects, and building contractors are most knowledgeable in these matters and should always be consulted when new construction or renovations are being planned.

National Electrical Code 1984. For details of proper electrical power installation and grounding methods, the NFPA's *National Electrical Code 1984* [22] is the definitive guide. This code is revised every 3 years in order to incorporate new developments and improve clarity where needed. It has been designated an American National Standard by the American National Standards Institute (ANSI), a voluntary organization for coordination of standards. Various jurisdictional authorities, both governmental and nongovernmental, may require compliance with this code as well as other NFPA standards.

Specific chapters of the *National Electrical Code* consider wiring design and protection, wiring methods and materials, equipment for general use, special occupancies, special equipment, special conditions, and communications. The chapter on special occupancies, Sec. 517, "Health Care Facilities," is of particular relevance. Electrical system design engineers, architects, and building contractors, as well as code enforcers and insurance inspectors, must be familiar with the provisions of this code.

Standard for Health Care Facilities. Through the 1970s, the NFPA published a series of individual standards relating to various aspects of hospital safety. There was often confusion as these standards, revised at different times and by different committees, came into conflict with each other. In 1982, 12 individual NFPA standards were combined into a single volume, with the latest version published in 1984 [8]. The chapters in this standard relating to electrical safety, along with their original NFPA designations, are given below:

Chapter 3: "Use of Inhalation Anesthetics (Flammable and Nonflammable)" (NFPA 56A)

Chapter 8: "Essential Electrical Systems for Health Care Facilities" (NFPA 76A)

Chapter 9: "Safe Use of Electricity in Patient Care Areas of Hospitals" (NFPA 76B)

Appendix E: "Safe Use of High-Frequency Electricity in Health Care Facilities" (NFPA 76C)

34.2.2 Essential Electrical Systems

Hospitals must be protected against failure of electric power generation. Alternate power source requirements for emergency use are specified in Chap. 8 of Ref. 8. Briefly, the alternate power source generator must be located on the premises and be driven by a prime mover (i.e., internal combustion engine or gas turbine). The generator must function automatically within 10 s of failure of the primary power source, and supply power to the emergency system consisting of a life-safety branch and a critical branch. The life-safety branch includes power for fire and gas alarms, illumination for directions and means of exit, emergency communications, and illumination and electric receptacles at the site of the alternate power source. The critical branch provides power to selected receptacles and for illumination in patient areas and auxiliary locations necessary to provide safe hospital operation.

Auxiliary power must also be supplied to a system which allows continued operation of hospital equipment. On this system, equipment is energized after an appropriate delay that allows the emergency system to function first. Equipment may be connected either automatically or manually after the delay. Typical items on the system include major hospital apparatus such as elevators, automatic doors, heating and ventilating units, and other selected equipment.

34.2.3 Patient Care Areas

The NFPA divides patient care areas into general categories plus some auxiliary areas (Art. 517 in Ref. 22; Chap. 9 in Ref. 8). It is the responsibility of the authorities having jurisdiction in the hospital to designate to which category an area belongs. Requirements for electrical service and safety are specified in terms of these areas.

1. *General care area.* Patients in this area may come into contact with ordinary electrical appliances, including electrical beds, lamps, and television sets, as well as hospital equipment such as monitors, examination lights, and peripheral intravenous lines. These areas may include patient bedrooms and examination and treatment rooms.
2. *Critical care area.* Patients in this area are subject to invasive procedures, such as cardiac catheterization and angiography, and are connected to line-operated electromedical devices.
3. *Wet location.* This is a patient area normally subject to conditions such as routine dousing of the work area, or where there is standing water on the floor. A hydrotherapy area is a typical example of a wet location. Wet locations require special electrical safety precautions—either a ground-fault current interrupter, which removes power if current in the ground conductor exceeds 6 mA, or if it is not safe to interrupt the power, an isolated power system.
4. *Pediatric and psychiatric locations.* Because of the special precautions required for patients in these areas, electrical receptacles must be tamper-proof, so that there can be no improper access to energized contacts.
5. *Inhalation anesthetizing locations.* These are separately classified areas within a hospital, in three categories; special requirements for these areas are specified in Art. 517 of Ref. 22 and in Chap. 3 of Ref. 8.
   - *a.* A *flammable location,* where only flammable anesthetics are used, is classified as a *hazardous location.*
   - *b.* A *mixed location,* where both flammable and nonflammable anesthetics may be used, is also classified as a *hazardous location.*
   - *c.* A *nonflammable location,* where only nonflammable anesthetics may be used, is classified as an *other-than-hazardous location.*

Flammable anesthetizing locations require a number of special construction, operational, and maintenance procedures. These special requirements include isolated power, conductive floors, specially constructed or located electrical receptacles, and wiring which meets installation specifications for a class I, division I hazardous area. Extra precautions must be taken against static electricity, and the use of electromedical appliances is severely limited. The cost of a flammable anesthetizing location, in most cases, is simply not warranted. In modern anesthesia practice, there is no surgical procedure which cannot be better managed with the use of nonflammable agents. Life in an operating room is sufficiently tense without the added burden of potential explosion and fire. From any point of view, mixed locations are even more troublesome. All the special safety requirements for flammable locations must be maintained, while some requirements may be relaxed, at times, when nonflammable agents are used in that same location. The danger here is, of course, oversight, forgetfulness, or carelessness. Changing from nonflammable to flammable agents in the middle of a procedure has been a noted cause of accidents in a mixed location.

34.2.4 Electrical Power Distribution

An electrical distribution panel is used to supply power to the receptacles in a particular hospital area. These panels often terminate three-phase wye 208/120 V or delta 240/120 V circuits from a more central high-voltage distribution point or the building service entry point.

The distribution panel, or panelboard, contains fuses, or, more likely, modern circuit breakers (overcurrent protection devices) for individual branch circuits. (Branch circuits are the part of the electrical distribution system between the overcurrent protection device and the wall outlet or receptacle.) For standardization and safety, the wiring insulation and receptacle contacts, as well as power cords on appliances, are color-coded as listed in Table 34.4. All wiring is

TABLE 34.4 Color coding for electric power distribution

	Insulation		
Wire	United States	Europe (EE 22)	Contacts
Hot (active)	Black	Brown	Gold
Neutral	White	Blue	Silver
Ground (earth)	Green/yellow stripe	Green/yellow stripe	Green

installed in grounded and bonded metal raceways. These grounded raceways offer an alternate redundant ground-circuit pathway.

Three-Wire Branch Circuits. Perhaps the most important improvement in electrical distribution for health care facilities, as well as for industry and home, is the use of three-wire branch circuits. This type of wiring has been required by the *National Electrical Code* for all new construction since 1962. This requirement, along with the corresponding ground wire of electrical appliances, has greatly reduced hazards due to faulty electrical equipment. A concomitant development has been Underwriters' Laboratory listing of the "hospital-grade" open-blade type of plug with U-shaped ground prong and corresponding sockets [23]. These high-grade electrically and mechanically sturdy components have improved safety in the hospital environment as well as significantly reduced maintenance and replacement requirements. Some nonflammable anesthetizing locations may use twist-lock plugs and receptacles. This configuration, however, often leads to incompatibility with other hospital equipment or to the use of adaptor cords, which is heartily disapproved.

The third, or ground, wire must be a minimum equivalent size of no. 10 AWG. It must be insulated to guard against corrosion and arcing at stress points under fault conditions. The ground wire returns appliance chassis leakage current to the electrical distribution system. More important, in the event of failure to an internal component, or a live wire coming in contact with an external metal part, the ground wire provides a low-impedance path for current flow away from the patient. For example, with a chassis leakage current of 100 μA, a patient-to-ground resistance of 500 Ω, and a chassis-to-ground resistance of 1 Ω, if the patient comes in contact with the chassis, leakage current flowing through the patient will be $1/500 \times 100$, or 0.2 μA, which is well below the level of microshock hazard. In the case of equipment fault, high-resistance ground, or direct connection between patient and electrified chassis, a higher current could flow, but the patient still is protected against macroshock by the shunting action of the ground wire.

Patient Care Areas. Some of the requirements for electrical power in patient care areas are discussed below. Each bed must be supplied with safe and adequate electric power. At present, there exists some conflict between the minimum requirements of NFPA 70 [22] and NFPA 99 [8]:

1. NFPA 70 (Art. 517 F) states that the location of each bed in a general patient care area shall be supplied with at least two branch circuits, at least one of which shall originate from a normal (sic) system panelboard. For critical care areas, of the two required branch circuits, one must be from the emergency system. All branch circuits from either system must originate from only one respective panelboard. Each bed in a general care area is required to have a minimum of four electrical receptacles, and each bed in a critical care area must have at least six.
2. NFPA 99 (Chap. 9, Sec. 9-2.3.1.1) states that every patient care area shall be served by not more than one normal branch-circuit distribution panel and/or one emergency branch-circuit distribution panel. The minimum number of receptacles specified is four (total 20 A), and the number of receptacles shall be determined by the intended use of the area.

Of course, these conflicting requirements may be reconciled by having each patient bed location served from one normal and one emergency branch distribution panel and by providing an adequate number of receptacles.

34.2.5 Grounding

The grounding system, an array of interconnected conductors, is used to protect the patient against fault (macroshock) and leakage (microshock) currents. Under ideal conditions, there are neither fault nor leakage currents, and the grounding conductors are everywhere at the same potential. However, due to nonideal insulation there always is some leakage current in the electrical distribution system itself, and in any electrical appliances connected to it. Also, equipment can malfunction, leading to the generation of a large fault current. The grounding system must provide isolation and limit potential differences between conducting surfaces everywhere in the patient vicinity, under both normal and fault conditions.

General Grounding Specifications. General grounding specifications and performance guidelines are specified in NFPA 70 (Art. 250) [22]. Specific requirements for hospitals are covered in NFPA 70 (Art. 517) and NFPA 99 (Chap. 9) [8]. In addition to the electrical system, fixed exposed conductive surfaces in a patient vicinity that are likely to become energized must also be grounded.

Ground system performance is specified in terms of both voltage and impedance measurements. NFPA 70 (Art. 517 F) specifies differences in ground-circuit potential to be measured between any two conductive surfaces in the patient vicinity. NFPA 99 (Chap. 9) specifies the voltage and impedance measurements that are to be made with respect to a reference point. This reference point can be (1) a reference grounding point, i.e., an equipment grounding bus or extension of one, that is a convenient collection point for grounding and bonding wires used in that patient area; (2) a grounding point, in or near the room being tested, that is electrically remote from the receptacles (i.e., cold water pipe); or (3) the grounding contact of a receptacle that is on a branch circuit different from the one being tested.

The specifications for measuring voltage are equivalent, though stated differently, in the two standards. NFPA 70 calls for measurement of potential differences for frequencies of 1000 Hz or less across a 1000-Ω resistor. NFPA 99 calls for measurement by an instrument having an input resistance of 1000 Ω ± 10%, at frequencies of 1000 Hz or less. Table 34.5 lists the allowable limits for the grounding system voltage and impedance.

TABLE 34.5 Grounding system: Voltage and impedance limits*

	Voltage, mV rms	Impedance, Ω
New construction†	20	0.1
Quiet ground	....	0.2
Existing construction:		
General care and wet areas	500	0.2
Critical care areas†	40	0.2

* NFPA 99 [8, Chap. 7] allows a ±20 percent accuracy for these measurements.

† Including anesthetizing locations.

Equipotential Grounding. Equipotential grounding uses various techniques to minimize the potential differences generated by stray currents. These techniques restrict leakage and fault currents of the electrical system and of electrical appliances to a very short-length, well-defined path, thereby limiting their effects. Features of equipotential grounding include the following:

1. The primary requirement is that at each patient bed location there be only one *reference grounding point*. All electrical activity for that patient is referred to that point, which is connected directly to the building ground, usually the structural steel or cold water pipe system.
2. All fixed exposed conductive surfaces in the room which are in the patient vicinity and are likely to become energized are connected to one or more *room bonding points* or directly to the reference grounding point. All room bonding points are connected to the reference grounding point.
3. Although not required, there may be a *patient equipment grounding point* (buss) with jacks that accept redundant ground wires from patient equipment. This patient equipment grounding point is connected directly to the reference grounding point or to the grounding conductor of a nearby electrical receptacle.
4. Power receptacles for a particular patient bed are grouped together, to reduce ground path length, and are grounded to the reference grounding point. Power for a patient location is supplied from single normal and emergency power panels, with grounding terminal bars bonded together. In practice, the reference grounding point is often located in the distribution panel box.
5. Ground wire connections are made with an insulated (green) continuous copper conductor not smaller than no. 10 AWG.

34.2.6 Isolated Power

An isolated power system has both power conductors isolated from ground, typically accomplished with the ungrounded secondary winding of an isolation transformer. Thus, even if an energized conductor makes direct contact to ground, there will be no fault current. Originally, isolated power systems were used specifically in flammable anesthetizing locations to protect against ignition sources caused by fault currents. A secondary benefit was to protect personnel from macroshock hazards of defective equipment. Even though flammable anesthetics are now rarely used, the requirements for isolated power in inhalation anesthetizing locations are still included in NFPA 70 (but not in NFPA 99). In order to clearly identify isolated power wiring, the insulation is color-coded as listed in Table 34.6.

There is also some belief that patients in critical care areas, in particular where there is an indwelling cardiac catheter or lead wire, are protected against microshock hazards by virtue of

TABLE 34.6 Color coding for isolated power system insulation

Isolated conductor no. 1	Orange
Isolated conductor no. 2	Brown
Grounding conductor	Green
Third wire of three-phase system	Yellow

an isolated power supply. Some state and local codes may require isolated power in cardiac care units as well as in operating rooms. The protection afforded by isolated power systems in critical care areas is less against microshock than against shocks due to faulty or defective equipment. A good preventive maintenance and inspection program can help protect against macroshock hazard.

34.2.7 Line-Isolation Monitor

Along with an isolation transformer, a line-isolation monitor is required to measure leakage current from each isolated conductor to ground and to sound an alarm in event of excess leakage or fault current. A single fault, by itself, is not a hazard but rather only a potential hazard, since the patient is not in a direct pathway between the two power conductors. A second fault, however, may provide this pathway and hence be a direct hazard to the patient or personnel. There are two types of line-isolation monitors. The older style is essentially a differential transformer that measures the difference in current between the two lines, which is then assumed to flow to ground. Modern line-isolation monitors are high-impedance devices that alternate connection between each line and ground, measuring actual leakage current. Because insulation of isolated power supply wiring tends to degrade with time, continual testing and maintenance are required.

Current between a conductor and ground in isolated power systems is known as *hazard current*. The line-isolation monitor should alarm for a total hazard current of 5 mA. There should be no alarm for a fault hazard current of 0.7 mA (NFPA 70, Art. 517 [22]) or 3.7 mA (NFPA 99, Chap. 3 [8]), another example of conflict between standards.

34.2.8 Ground-Fault Interrupter Circuit

Ground-fault interrupter circuits (GFICs) are devices which respond to excess current in the ground wire of a branch circuit. If this current exceeds a specified limit, power is automatically disconnected at the distribution panel.

Modern GFICs do not actually measure ground-circuit current, but instead use a differential transformer to monitor the difference in current between hot and neutral wires. This current difference is assumed to flow in the ground wire, and if it exceeds the specified limit, the circuit is interrupted. The response time of a GFIC is similar to that of an ordinary fuse or circuit breaker. The higher the current, the faster the response. For a nominal current limit of 6 mA, typical trip time is 100 ms; for a current of 50 mA, trip time may be in the order of 15 ms.

The current activation point of the GFIC is not low enough to protect against the microshock hazards of leakage current, but can prevent large fault currents from creating significant potential differences in a grounding system. GFICs are not ordinarily used in patient care areas, since interruption of power might prove more hazardous than the relatively low ground-circuit current detected. Their principal application is in wet locations, where there is the possibility of large ground-circuit currents due to moisture-induced leakage pathways and insulation degradation, but where interruption of power can be tolerated. The major advantage of a GFIC over an isolation transformer is the considerably lower cost.

34.3 SAFETY OF ELECTROMEDICAL APPLIANCES

There are many types of electromedical appliances used within a hospital for patient treatment and diagnosis. These devices must meet special requirements for safety as well as efficacy. *Efficacy* means that the appliance has a beneficial application, performs as specified, and has no deleterious or inherently dangerous features which outweigh its potential beneficial use. *Safety,* on the other hand, refers to the intrinsic design, construction, and operation of the device.

34.3.1 Federal Regulations

For the most part, the federal government's Food and Drug Administration regulates the safety and efficacy of medical devices under the Food, Drug, and Cosmetic Act Amendment of 1976, known also as the Medical Device Amendments of 1976 (Public Law 94-295). This law divides medical devices into three categories: (1) class I, "General Controls," covers good manufacturing techniques and proper labeling; (2) class II, "Performance Standards," includes general controls plus specific device standards for safety and efficacy; (3) class III, "Premarket Approval," requires formal application and approval before the device is made available on a commercial basis. Devices are placed in class III when sufficient information is not available for class I or II classification, if the device is used to support or sustain human life or to prevent impairment of human health, or if the device presents a potential unreasonable risk of illness or injury [24].

Radiation-emitting devices fall under the Radiation Control for Health and Safety Act of 1968 (Public Law 90-602). Under this law, federal standards have been developed for x-ray systems, ultrasonic devices, lasers, and sunlamps.

34.3.2 Classification

Electrical apparatus for patient use can be divided into four categories [7]:

1. Apparatus with isolated patient connections
2. Apparatus with nonisolated patient connections
3. Apparatus likely to come into contact with a patient
4. Apparatus with no patient contact

In addition, equipment can be classified as either permanently installed (fixed), cord-operated, or battery-powered (rechargeable or nonrechargeable batteries). Devices which deliver some type of energy to the patient (electrical, thermal, mechanical) for therapeutic or diagnostic purposes may also be separately classified.

34.3.3 Leakage Current

Leakage current is any current, including capacitively coupled current, not intended to be applied to a patient but which may be conveyed from exposed metal parts of an appliance to ground or to other accessible parts of an appliance [8]. The term *risk current* is also used in this context [7].

Leakage Current Measurements. Leakage current is measured either as direct current, or in rms values for sinusoidal waveforms for frequencies to 1 kHz. For higher frequencies, a weighting factor equal to the frequency in kilohertz, up to a maximum of 100, reduces the true rms value. With a nominal 1000-Ω resistor as the load across which leakage current is measured, the proper weighting function is obtained by using the circuit shown in Figure 34.5. Measurement with a true rms reading high-impedance voltmeter gives a sensitivity of 1 mV/μA over the specified frequency range. Leakage current limits (Tables 34.7 and 34.8) are based on physiolog-

	R_1 (ohms)	R_2 (ohms)	C (microfarads)
UL 544*[6]	1000	10.2	0.15
ANSI/AAMI SCL[7]	1000 ± 1%	10 ± 1%	0.15 ± 5%

*500 Ω for appliance with low source impedance

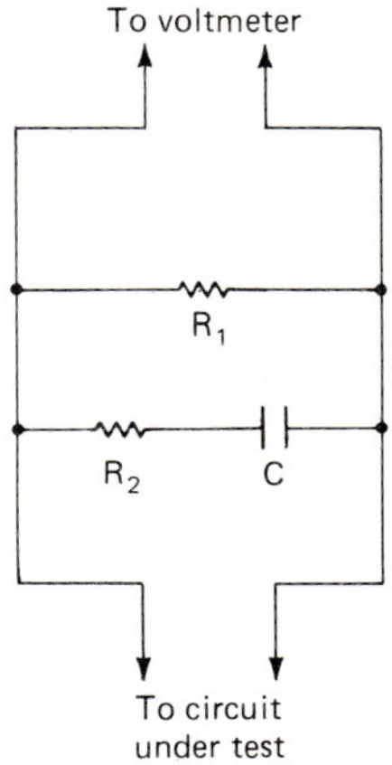

FIG. 34.5 Circuit for leakage current measurements. Frequency response, as specified by three different standards, are essentially the same. The NFPA-recommended circuit [8] is slightly more complex, including an additional 10,000-Ω series decoupling resistance.

ical effects, engineering capabilities, and economic considerations. Equipment should be tested for leakage currents at regular intervals, depending on the type of service for which it is intended.

Permanently Wired Equipment. Permanently wired equipment typically is installed at the time of original construction or during major renovation of a hospital. It is set in a fixed location, never moved about, and permanently wired to the power mains. Examples are scanners, radiation therapy equipment, and x-ray devices.

Permanently wired apparatus is tested for leakage current when first installed. With all ground connections temporarily opened, the leakage current between frame and ground is to be no greater than 5 mA. Apparatus which may come into contact with a patient must have its

TABLE 34.7 Chassis leakage current limits*

Source	Patient equipment	Nonpatient equipment	Permanently wired
UL 544 [6]	100 (140 dc)	500 (700 dc)	5000
ANSI/AAMI SCL† [7]	100	500	5000
NFPA 99 [8]	100‡	5000	5000

* Current is milliamperes rms (weighted), unless noted otherwise.

† Limits apply for double-insulated apparatus with functional or supplementary insulation effectively bypassed by test. Otherwise limits are one-half those given.

‡ Exceptions are allowed; up to 250 mAs if an appropriate inspection and maintenance program is supplemented to ensure proper grounding; up to 500 mA for special equipment with a documented maintenance schedule (nominal interval of 3 months).

TABLE 34.8 Lead leakage current limits*

			Sink	
Source	Nonisolated	Isolated	Input†	Patient end‡
UL 544 [6]	50 (70 dc)	10 (14 dc)	10	20
ANSI/AAMI SCL [7]	50	10	10	20
NFPA 99 [8]	50	10	10	20

* Current is microamperes rms (weighted), unless otherwise noted.
† At input to instrument with isolated input.
‡ At patient end of cable connected to instrument.

conductive exposed elements tested to specifications of the patient care category for which it is used.

Cord-Connected Equipment. Most electrical appliances used in health care facilities fall into the category of cord-connected line-powered equipment. A power cord attachment, either fixed or detachable, is used to connect the appliance to a power receptacle. Separate leakage current limits are specified for cord-connected equipment chassis and leads.

Chassis Leakage Current. Chassis leakage current is current that flows in the ground wire of the electrical power cord. This current is due to leakage of electrical and electronic components within the instrument itself, and leakage from the hot and neutral power wires to ground. In itself, this current is not a hazard; rather, it is more correctly classified as a potential hazard. It becomes a true hazard only if the ground wire is broken or otherwise compromised. Then the current which normally would flow in this wire directly to ground can find a different, possibly more dangerous path. A patient or hospital worker in contact with an exposed conductive element of the chassis could then become part of an alternate pathway to ground.

The following conditions of measurement of chassis leakage current are listed in various standards.

1. Leakage current is measured from exposed metal of the equipment chassis to ground.
2. Measurements are made with electric power turned on and off, and with operating controls adjusted for maximum leakage.
3. Measurements are made with normal and reversed polarity of electric power.
4. If an appliance enclosure has no exposed conducting surface, one is simulated with a 10×20 cm bare metal foil in intimate contact with an exposed insulated surface. If there is no accessible insulated surface this size, the foil used is the same as the surface, to a maximum of 200 cm^2.

The three different standards which address chassis leakage current for cord-connected equipment all give essentially the same limits, except for different nomenclature and special considerations, as listed in Table 34.7.

Equipment with Patient Leads. Many types of electromedical apparatus have leads that come in contact with the patient. This type of equipment is classified as either ordinary (nonisolated) or isolated input. Ordinary equipment is meant to be used with the general patient population. Equipment with isolated input is designed to be used safely with patients having an indwelling cardiac lead or catheter, and hence at risk to microshock.

Methods of lead leakage current measurement are specified in the various safety standards, with some of the basic conditions given below:

1. All controls are set to produce maximum leakage.
2. Electric power is turned off and on.

3. Measurements are made with normal and reverse polarity of electric power.
4. The ground wire is intact and open at the equipment plug.
5. For equipment with more than one patient lead, leakage current is measured with all leads connected together, between pairs of leads, and between one lead and all other leads.
6. Sink current for equipment with isolated input is measured by applying line voltage, through the measurement circuit (Fig. 34.5), to the patient leads.
7. Therapeutic apparatus, or apparatus designed to deliver energy to the patient, is tested when powered and operational, but not while delivering energy.
8. Appliances with integral battery chargers that are capable of being operated while being charged must meet all applicable requirements for lead and chassis leakage currents.
9. Battery-powered appliances operating on nonrechargeable batteries must meet all applicable requirements for lead leakage current.

The three standards which address lead leakage current all give essentially the same limits, except for special considerations, as listed in Table 34.8.

34.3.4 Grounding

Some of the major grounding requirements for electrical appliances (equipment) used in health care facilities for patient care as specified in the various recognized standards are as follows:

1. A three-wire electrical attachment cord must be used between appliance and electrical receptacle. The third wire must be a separately insulated (green) ground conductor with a maximum resistance of 0.15 Ω between appliance chassis and power cord ground pin.
2. Two-wire cords may be permitted only with double-insulated appliances. If there is no ground wire, and none is required, the plug must not have a ground pin.
3. All exposed metal or other electrically conductive parts which are connected to the chassis or other internal components or circuits of an appliance must be grounded. (Exceptions are permitted for small metal parts not likely to become energized, such as nameplates and screws.)
4. Double-insulated appliances must have both functional and supplemental insulation.
5. In order to reduce electromagnetic interference problems, in any patient vicinity a jack or terminal strip may serve as an equipment grounding point for the joining of redundant grounds of electrical appliances and other equipment such as metal tables and stands.
6. In flammable inhalation anesthetizing locations, additional grounding precautions must be taken to reduce potential differences between appliances and to prevent buildup of static electricity.

34.4 GENERAL SAFETY

The following guidelines apply to general safety in any medical establishment:

1. The most important rule for electrical safety is constant vigilance coupled with a questioning attitude.
2. There should be regularly scheduled inspections of all equipment and electrical services, with complete records kept.
3. All users of equipment should be instructed in proper operation and be alert for signs of potential hazards.
4. Specific categories of appliances or equipment and individual types of devices may be covered by standards of performance, efficacy, labeling, and safety.

REFERENCES

1. Dalziel, C. F.: Electric shock hazard, *IEEE Spectrum, 9*(2), 41–50, 1972.
2. Kouwenhoven, W. B., Hooker, D. R., and Langworthy, O. R.: The current flowing through the heart under conditions of electric shock, *Am. J. Physiol., 100,* 344–350, 1932.
3. Roy, O. Z.: 60 Hz ventricular fibrillation and rhythm thresholds and the nonpacing intracardiac catheter, *Med. Biol. Eng., 13,* 228–233, 1975.
4. Roy, O. Z.: Summary of cardiac fibrillation thresholds for 60 Hz currents and voltages applied directly to the heart, *Med. Biol. Eng. Comput., 18,* 657–659, 1980.
5. Ledsome, J. R., Graystone, P., Coppin, J., and Leung, T.: Ventricular fibrillation thresholds in the presence of leakage currents, *Cardiovasc. Res., 8,* 769–774, 1974.
6. Underwriters Laboratories: *Standard for Medical and Dental Equipment,* 2d ed., UL 544, Northbrook, Ill., 1976, rev. 1982.
7. Association for the Advancement of Medical Instrumentation: *Safe Current Limits for Electromedical Apparatus,* AANSI/AAMI SCL, Arlington, Va., 1978.
8. National Fire Protection Association: *Standard for Health Care Facilities,* NFPA 99-1984, Quincy, Mass., 1984.
9. Kouwenhoven, W. B., Chestnut, R. W., Knickerbocker, G. G., Milnor, W. R., and Sass, D. J.: A-C shocks of varying parameters affecting the heart, *Trans. Am. Inst. Electr. Eng., 78,* 163–169, 1959.
10. Roy, O. Z., Park, G. C., and Scott, J. R.: Intracardiac catheter fibrillation thresholds as a function of the duration of 60 Hz current and electrode area, *IEEE Trans. Biomed. Eng., 24,* 430–435, 1977.
11. Ferris, L. P., King, B. G., Spence, P. W., and Williams, H. B.: Effect of electric shock on the heart, *Electr. Eng., 55,* 498–515, 1936.
12. Dalziel, C. F.: The threshold of perception currents, *AIEE Trans,* pt. III: *Power Apparatus and Systems, 73,* 990–996, 1954.
13. Dalziel, C. F., and Mansfield, T. H.: Effect of frequency on perception currents, *Trans. Am. Inst. Electr. Eng., 69,* 1162–1168, 1950.
14. American National Standards Institute: *Leakage Current for Appliances,* ANSI C 101.1, New York, 1973.
15. Dalziel, C. F., Logen, J. B., and Thurston, J. L.: Electric shock, *Trans. Am. Inst. Electr. Eng., 60,* 1073–1079, 1941.
16. Dalziel, C. F., and Lee, W. R.: Lethal electric currents, *IEEE Spectrum, 6*(2), 44–50, 1969.
17. Dalziel, C. F., Ogden, E., and Abbott, C. E.: Effect of frequency on let-go currents, *AIEE Trans. Electr. Eng., 62,* 745–750, 1943.
18. Dalziel, C. F., and Lee, W. R.: Reevaluation of lethal electric currents, *IEEE Trans. Ind. Gen. Appl., IGA-4,* 467–476, Discussion, 676–677, 1968.
19. Geddes, L. A., and Baker, L. E.: Response to passage of electrical current through the body, *J. Assoc. Adv. Med. Instrum., 5,* 13–18, 1971.
20. Wald, A.: A standards primer, *J. Clin. Eng., 8,* 135–140, 1983.
21. National Fire Protection Association: *Standard for the Use of Inhalation Anesthetics (Flammable and Nonflammable),* NFPA 56A 1978, Quincy, Mass., 1978.
22. National Fire Protection Association: *National Electrical Code 1984 (NEC),* ANSI/NFPA 70, Quincy, Mass., 1983.
23. Underwriters Laboratories: *Standard for Attachment Plugs and Receptacles,* 10th ed., UL 498, Northbrook, Ill., 1981, rev. 1982.
24. Weitzman, S. A.: A lawyer's perspective, *IEEE Eng. Med. Biol. Mag., 3*(1), 17–20, 1984.

CHAPTER 35

Stereology

Odile Mathieu-Costello
Department of Medicine
University of California, San Diego
La Jolla, California

INTRODUCTION

The development of model concepts for the understanding of function at different levels of biological organization includes the evaluation of the structural properties of connected components. In this context, quantitative morphological analysis is important to sort out structural characteristics relevant to the phenomenon under study among unimportant detail and nonsignificant variations (Weibel, 1982). Some of the important aspects of the quantitative formulation of the structural properties of a system are the following:

1. Structural information averaged for a given compartment and simplified to its most pertinent attributes can be obtained and directly included in the mathematical formulation of the dynamic properties of a system.
2. Basic assumptions on structural design, necessary for the modeling of complex interrelationships between connected compartments, can be tested.
3. Objective correlation of structural information with physiological and biochemical data can be made.

Among the quantitative methods available to analyze system components, stereological methodology relates to the quantitative evaluation of structures in microscopy. The word *stereology* was coined in 1961 by a small group of scientists gathered by Hans Elias to discuss problems in interpreting and quantifying structures on histological sections. As defined by Ewald Weibel (1979), stereology is "a body of mathematical methods relating three-dimensional parameters defining the structure to two-dimensional measurements obtainable on sections of the structure." In this chapter, basic principles and practical aspects of stereology are summarized. Our purpose is (1) to compile the methods presently available to quantify structures from the images of their sections: which parameters can be measured, which equations apply, in which cases, and (2) to briefly review practical advice for stereological work: methodological errors and their correction, specimen preparation and sampling designs, optimization of measurement efficiency. References and monographs on the subject (DeHoff and Rhines, 1968;

Underwood, 1970; Weibel, 1979 and 1980; Elias and Hyde, 1983) should be consulted for more detail.

35.1 TERMINOLOGY AND NOMENCLATURE

For clarity and to facilitate reading of papers from different sources, it is important to use a standardized and well-defined system of stereological nomenclature, symbols, and terminology. Over the past decade, effort has been made in this direction by the International Society for Stereology, although complete agreement on basic stereological nomenclature has still not been reached (Underwood, 1981). Common trends and some degree of concordance exist, however, in frequently used terms and sets of symbols. It is recommended to clearly define the system of nomenclature adopted in a particular study and, as far as possible, to choose terms and symbols among those commonly in use.

The definition of terms (Table 35.1) and symbols (Table 35.2) used in this text is taken from Weibel (1979, 1980). Double symbols are used for relative dimensions: the first letter refers to the component parameter; the second, written as a subscript, refers to the parameter of the reference system. As examples, volume, surface, and numerical densities (see Table 35.1 for definitions) of structures in a containing space are written as V_V, S_V, and N_V, respectively. To clearly identify the components to which each parameter relates, Weibel (1979, Chap. 1) uses lowercase letters attached to the parameter symbol in parentheses: the parameter component is

TABLE 35.1 Definitions of stereological terms

Anisotropic: Having different properties in different directions of space.

Containing space: Structural space to which parameters describing internal components are related.

Convex solid: Solid for which any straight line segment connecting any of two points lies totally within the solid.

Curvature: Deviation of curved surface (line) from plane (straight line), measured by reciprocal of radius of the largest tangent circle.

Density: Quantity per unit reference entity (volume, surface, or length).

Frequency distribution: Distribution of frequencies of occurrence of an event over the range of all possible events.

Intercept: Segment or chord of test line contained within object and extending between two points on the object's surface.

Intersection: Intersection between test line and object surface or profile boundary.

Isotropic: Having the same properties in all directions of space.

Isotropic uniform random: All orientations and all positions in space equally likely.

Object, particle: Discrete element of structure.

Phase: Aggregate of all parts of structure which are identical in nature (e.g., mitochondrial phase).

Profile: Plane figure resulting from sectioning a three-dimensional object with a plane.

Random sample: Sample obtained by random process (e.g., by using random numbers), i.e., without regard to the properties of structure.

Random section: Section of structure obtained by random sampling; i.e., all components and all orientations have equal chance of being cut.

Reference system: Quantity to which feature characteristics are related (containing space, test system, etc.).

Section: Plane intersecting a structure; can be a thin slice of finite thickness.

Slice: Section of finite thickness.

Test system: Set of geometric elements (points, lines, areas) used to obtain stereological estimates.

Trace, boundary: Line resulting from sectioning a two-dimensional structure with a plane, i.e., outline (or perimeter) of a profile.

Uniform random sample: All points in space equally likely to be included in the sample.

Source: Compiled from Weibel (1980).

TABLE 35.2 List of symbols

Symbol	Quantity	Dimension
	Structure in space	
V	Volume	3
S	Surface	2
J	Length	1
K	Mean curvature	1
G	Total (gaussian) curvature	0
N	Number (of objects)	0
V_V	Volume density	0
S_V	Surface density	−1
J_V	Length density	−2
N_V	Numerical density	−3
	Single particle	
D	Diameter	1
R	Radius	1
H	Caliper diameter	1
L	Length	1
λ	Ratio D/L	0
	Profiles on section	
A	Area	2
B	Boundary length	1
Q	Transection of linear structure	0
T	Tangent to boundary	0
N	Number	0
A_A	Areal density	0
B_A	Boundary length density	−1
N_A	Numerical density	−2
Q_A	Transection density	−2
T_A	Tangent density	−2
	Test sets	
V_T	Test volume	3
A_T	Test area	2
L_T	Test line length	1
P_T	Test point number	0
	Test sets on profiles	
L_L	Intercept length density	0
I_L	Intersection density	−1
P_P	Point density	0
	Section micrograph	
t	Section thickness	1
M	Magnification	0

written first, and the reference system second, after a comma. For example, the volume density (V_V) of mitochondria (m) in hepatocytes (h) is written as

$$\frac{V(m)}{V(h)} = V_V(m, h) \tag{35.1}$$

This notation is used in this text. The parameter X ($= V_V, S_V, N_V, \ldots$) of a structure a in a containing space c is written as $X(a, c)$.

35.2 BASIC STEREOLOGICAL PARAMETERS

Stereological procedures are based on geometrical probability. They exploit the fact that there are quantitative relationships between parameters of structures in space and specific attributes of their profiles on sections. The entity accessible to measurements is the section which is used to visualize, by microscopy, the internal arrangement of the material under investigation. Mathematical equations are used which relate fundamental parameters such as volume, surface, length, curvature, and numerical densities of structures to simple profile, point, and/or intersection counts on sections.

Among the basic parameters described below, the choice of the relevant and most informative descriptor(s) of a system is an important step in the design of a stereological study (see Sec. 35.3.1). It obviously depends on the structure and/or the problem under investigation. For example, the relative amount of different tissue compartments is best estimated by their volume density (e.g., cell types, extracellular spaces), surface density (exchange interfaces, cellular membranes), length density (tubular networks), or numerical density (particles, e.g., ribosomes). The surface-to-volume ratio is a useful parameter to estimate particle size variations between structures and/or different experimental conditions. Curvature densities are related to surface pressure (integral of mean curvature; see Weibel, 1980, Chap. 3) and topological properties of complex structures (total or gaussian curvature; see DeHoff, 1968). When estimating the numerical (or size) distribution of objects, remember that the reconstruction of particle number (or sizes) from measurements of profiles on sections depends on the model adopted. Branching, or single branched, mitochondria have been described in various organisms and cellular types (e.g., Keyhani, 1980; Rambourg and Segretain, 1980); in such cases, the reconstruction of numerical (or size) density of mitochondria from profile counts (or sizes) on sections is obviously meaningless.

Depending on properties of the structures (three-dimensional orientation and/or spacial distribution; structure size as compared with section thickness), different equations and/or sampling designs apply for the collection of measurements on sections. Solutions to the evaluation of basic stereological parameters of large structures (measurements considered to be performed on true planes of sectioning, i.e., sections of zero thickness), homogeneously distributed in the sample, are given below. Correction of errors due to section thickness, and sampling designs for nonhomogeneously distributed structures, are summarized in Sec. 35.3, along with other practical aspects of stereological analysis.

35.2.1 Isotropic Structures

Volume Density V_V. There is an equality of the volume density $V_V(a, c)$ of a phase (or objects) a in a containing specimen c and the following parameters measurable on sections [Fig. 35.1a to c; for notation, see Eq. (35.1)]:

1. The areal density $A_A(a, c)$ of profiles of a per unit area of section of c (Delesse, 1847)
2. The linear density $L_L(a, c)$ of intercepts of profiles of a per unit length of a test line superimposed on the section of c (Rosiwal, 1898)

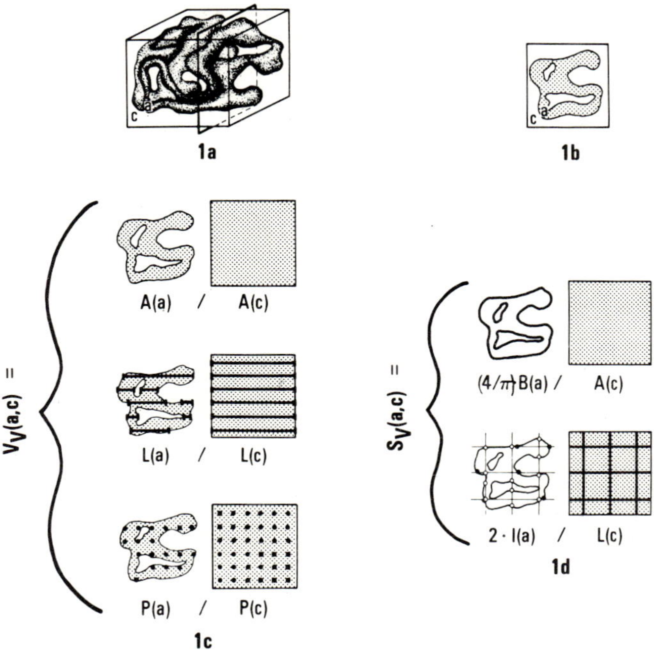

FIG. 35.1 Model for estimating volume and surface densities. (*a*) Model structure *a* contained in a cube *c*. (*b*) Plane of sectioning through the cube illustrated in (*a*). (*c*) Estimation of volume density by areal fraction, intercept length density, and point counting. (*d*) Estimation of surface density by perimeter length density and intersection density; intersections with horizontal and vertical lines are separated for clarity of illustration.

3. The density of points $P_P(a, c)$ overlaying the profiles of *a* in a test point system superimposed on sections of *c* (Glagoleff, 1933; Chalkley, 1943):

$$V_V(a, c) = A_A(a, c) = L_L(a, c) = P_P(a, c) \tag{35.2}$$

Surface Density S_V. The density of the surface, $S(a)$, of an isotropic structure *a* in a containing volume $V(c)$ is proportional to the following attributes of its profiles on sections (Fig. 35.1*d*):

1. The boundary length $B(a)$ of profiles of *a* per unit area of section of *c*

2. The number of intersections $I(a)$ of the boundary of profiles of *a* per unit length of test line superimposed on sections of *c* (Tomkeieff, 1945; Saltykov, 1946):

$$S_V(a, c) = \frac{4}{\pi} B_A(a, c) = 2I_L(a, c) \tag{35.3}$$

Length Density J_V. The length density $J_V(a, c)$ of an isotropic structure *a* in a containing specimen *c* is proportional to the number of transections of the structure, $Q(a)$, per unit area of section of *c* (Saltykov, 1946; Smith and Guttman, 1953; Hennig, 1963; Fig. 35.2):

$$J_V(a, c) = 2Q_A(a, c) \tag{35.4}$$

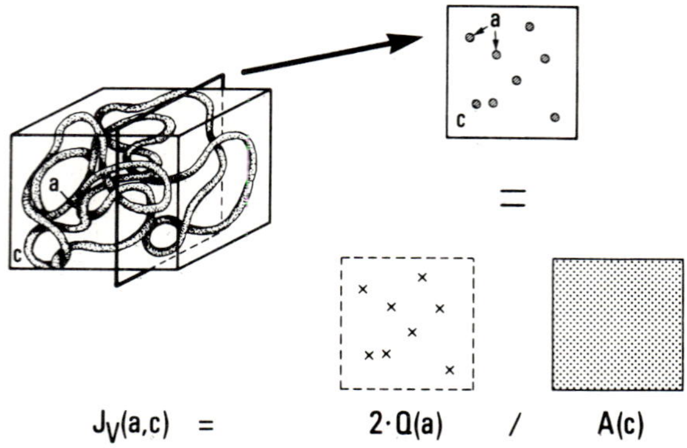

FIG. 35.2 Model for estimating length density of structure a in cube c.

Mean Curvature Density K_V; Total (Gaussian) Curvature Density G_V. The mean curvature K and the total (gaussian) curvature G of a surface are defined locally at a given point by

$$K = \frac{1}{2}\left(\frac{1}{r_1} + \frac{1}{r_2}\right) \tag{35.5}$$

$$G = \frac{1}{r_1}\frac{1}{r_2} \tag{35.6}$$

where r_1 and r_2 are the radii of the largest and smallest circles, respectively, which can be fitted to the surface, in normal sections passing through the point in all possible directions.

The *mean curvature density* $K_V(a, c)$ of a surface a in a containing volume $V(c)$ is the integral of the mean curvatures of all points of a in c. The stereological estimate of $K_V(a, c)$ can be obtained by sweeping a line across the sections, and counting the number of "positive," $T_+(a)$, and "negative," $T_-(a)$, tangents formed with convex and concave elements of the trace of a, respectively (Cahn, 1967; DeHoff, 1967; Fig. 35.3):

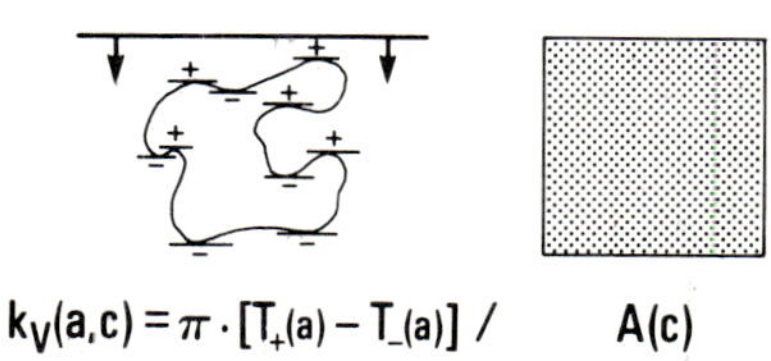

FIG. 35.3 Estimation of mean curvature density of model structure a in cube c (Fig. 35.1a). On the section of the structure (Fig. 35.1b), positive (+) and negative (−) tangents are generated by sweeping a test line across the section; the internal boundary has been omitted for clarity of illustration.

$$K_V(a, c) = \pi T_A(a, c) \tag{35.7}$$

where $T_A(a, c)$ is the net tangent count [$= T_+(a) - T_-(a)$] per unit area of section of c.

For convex particles, two positive tangents, $T_+(a)$, are formed with each of the $N(a)$ profiles of a found per unit area of section of c; then

$$K_V(a, c) = 2\pi N_A(a, c) \tag{35.8}$$

The *gaussian curvature density* $G_V(a, c)$ of the surface a of simple particles, not necessarily convex, in a containing volume $V(c)$ is proportional to the number of objects $N(a)$ per unit volume of c:

$$G_V(a, c) = 4\pi N_V(a, c) \tag{35.9}$$

In case of particles containing $N'(a)$ holes (Miles and Davy, 1978):

$$G_V(a, c) = 4\pi \frac{N(a) + N'(a)}{V(c)} \tag{35.10}$$

***Numerical Density* N_V.** The numerical density of particles a in a containing specimen c can be estimated from the numerical density of profiles, $N(a)$, per unit area of section of c, and the mean tangent diameter, $\overline{H}(a)$, of the particles (DeHoff and Rhines, 1961):

$$N_V(a, c) = \frac{N_A(a, c)}{\overline{H}(a)} \qquad (35.11)$$

Hilliard (1967) gave equations to calculate $\overline{H}$ for particles of various shape from characteristic dimension(s)—i.e., radius, in the simple case of spheres; eccentricity and radius for spheroids; length and radius for cylinders; edge lengths for polyhedrons. DeHoff (1964) published plots from which $\overline{H}$ can be read for cylinders and spheroids of various axial ratios. In many cases, however, it is difficult, if not impossible, to obtain a reliable estimate of $\overline{H}$ from measurements on sections. Alternative methods, outlined below, have been proposed to estimate N_V from N_A and other stereological measurements for particles of simple shape and/or when some assumptions on particle shape can be made.

1. *Method of Weibel and Gomez (1962) for constant-shape particles:* The numerical density $N_V(a, c)$ of constant-shape particles a in a containing specimen c can be estimated by the equation

$$N_V(a, c) = \frac{K}{\beta}\frac{[N_A(a, c)]^{3/2}}{[V_V(a, c)]^{1/2}} \qquad (35.12)$$

where $N_A(a, c)$ is the numerical density of profiles of a per unit area of section of c; $V_V(a, c)$ is the volume density of a and can be estimated via Eq. (35.2); and β is a dimensionless shape coefficient which relates the volume of the particles, $v(a)$, to their mean cross-sectional area $\overline{A}(a)$:

$$v(a) = \beta[\overline{A}(a)]^{3/2} \qquad (35.13)$$

For spheres, $\beta = \sqrt{6/\pi}$. For cylinders of diameter-to-length ratio λ (Baudhuin, 1968; Lindberg and Vorwerk, 1970),

$$\beta = \frac{\pi(\lambda + 2)^{3/2}}{4\pi^{1/2}\lambda} \qquad (35.14)$$

For rotational ellipsoids of axial ratio ε (Baudhuin, 1968),

$$\beta = \frac{1}{\varepsilon}\left(\frac{3}{4\pi}\right)^{1/2}\left[1 + \frac{\varepsilon^2}{(1 - \varepsilon^2)^{1/2}} \operatorname{arcsinh}\left(\frac{1}{\varepsilon^2} - 1\right)^{1/2}\right]^{3/2} \qquad (35.15)$$

K is a size distribution coefficient which is proportional to the ratio of the third moment (M_3) to the first moment (M_1) of the size distribution:

$$K = \frac{M_3^{1/2}}{M_1^{3/2}} \geq 1 \qquad (35.16)$$

For particles of equal size, $K = 1$; when the standard deviation of the size distribution of the particles is less than 20 percent of their mean diameter, K varies between 1 and 1.05, and can be neglected (Weibel, 1979, Chap. 2).

2. *Method of Loud et al. (1965) for circular cylinders:* The surface-to-volume ratio $S_V(a) = S(a)/V(a)$ is a useful parameter to determine the shape of particles from the analysis of their profiles on sections (Nebel et al., 1960). The numerical density $N_A(a, c)$ of cylindrical particles a in a containing space c is related to the volume density $V_V(a, c)$ of the particles, and to their surface-to-volume ratio $S_V(a)$, by the equation (Loud et al., 1965):

$$N_V(a, c) = \frac{V_V(a, c)\lambda[S_V(a)]^3}{2\pi(\lambda + 2)^3} \qquad (35.17)$$

where λ is the ratio of diameter to length of a.

Combining Eqs. (35.12) and (35.17) gives, after rearrangement,

$$\frac{[V_V(a, c)]^{3/2}[S_V(a)]^3}{[N_A(a, c)]^{3/2}} = \frac{2\pi(\lambda + 2)^3}{\beta\lambda} \tag{35.18}$$

The coefficient β depends on λ only [Eq. (35.15)]:

$$\frac{2\pi(\lambda + 2)^3}{\beta\lambda} = F(\lambda) \tag{35.19}$$

A graph from which λ can be determined for different values of the left side of Eq. (35.18) has been worked out (Loud et al., 1965). The procedure to estimate the numerical density $N_V(a, c)$ of cylinders a in a containing space c is the following:

The parameters $V_V(a, c)$, $S_V(a)$, and $N_A(a, c)$ are estimated from measurements on sections

From the numerical value of the left side of Eq. (35.18), the diameter-to-length ratio λ of the cylinders is determined graphically

With λ known, the diameter $D(a)$, the length $L(a)$, the volume $v(a)$, and the numerical density $N_V(a, c)$ of the particles are estimated via the equations

$$D(a) = \frac{2(\lambda + 2)}{S_V(a)} \tag{35.20}$$

$$L(a) = \frac{D(a)}{\lambda} \tag{35.21}$$

$$v(a) = \frac{\pi[D(a)]^3}{4\lambda} \tag{35.22}$$

$$N_V(a, c) = \frac{V_V(a, c)}{v(a)} \tag{35.23}$$

Baudhuin (1968) worked out graphs from which the axial ratio of prolate (or oblate) ellipsoids can be estimated from the numerical values of $V_V(a, c)$, $S_V(a)$, and $N_A(a, c)$ measured on sections.

3. *Method of Dehoff (1964) for constant-size particles:* The numerical density $N_V(a, c)$ of constant-size particles a in a containing specimen c is related to the numerical density $N_A(a, c)$ of their profiles on sections:

$$N_V(a, c) = \frac{N_A^2(a, c)}{2I_L(a, c)} \frac{\gamma_2}{\gamma_1^2} \tag{35.24}$$

and

$$N_V(a, c) = N_A(a, c) \frac{2I_L(a, c)}{P_P(a, c)} \frac{\gamma_1\gamma_2}{\gamma_3} \tag{35.25}$$

where $I_L(a, c)$ is the number of intersections, $I(a)$, of the boundary of profiles of a per unit length of a test line superimposed on sections of c; $P_P(a, c)$ is the density of points overlaying profiles of a in a test point system superimposed on sections of c; and γ_1, γ_2, and γ_3 are shape coefficients relating mean caliper diameter $\overline{H}(a)$, volume $v(a)$, and surface $s(a)$, respectively, to a single dimension x of the particles:

$$\overline{H}(a) = \gamma_1 x \tag{35.26}$$

$$s(a) = \gamma_2 x^2 \tag{35.27}$$

$$v(a) = \gamma_3 x^3 \tag{35.28}$$

For spheres, the values of γ_1, γ_2, and γ_3 are 2, 4π, and $4\pi/3$, respectively; the shape coefficients for particles of various shape are given by Weibel (1979, Chap. 2).

Particle Size Distribution and Shape Factors. Methods exist to estimate the size distribution of particles in a containing volume from the size distribution of their profiles on sections. Different sizes of profiles are found which can result from the interception of the same size of particles at different levels with respect to their center (Fig. 35.4). One needs to determine the

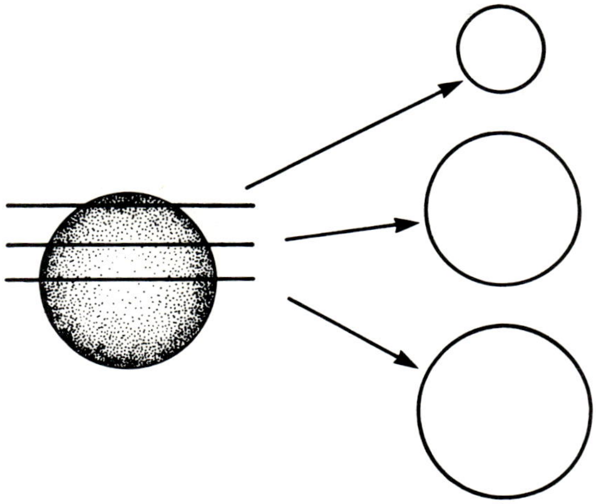

FIG. 35.4 Different sizes of profiles are generated by sectioning a sphere at different levels.

number of particles of different size which have contributed to the number of profiles of each size found on sections.

Briefly, the procedure to construct the size distribution of spherical particles from measurements of the size of their profiles on sections is the following:

1. The profile size distribution is broken into classes.
2. The profiles of the largest size, x, are necessarily sections of particles from the largest-size class, X. The probability $\mathrm{Pr}_{i,j}$ of a plane intersecting a sphere of diameter j to produce profiles of diameter i is

$$\mathrm{Pr}_{i,j} = \frac{1}{j}\left[\sqrt{j^2 - (i-1)^2} - \sqrt{j^2 - i^2}\right] \qquad (35.29)$$

   From the number of profiles of size x counted on sections, one can calculate, via Eq. (35.29), (1) the number of particles of size X existing in the distribution and (2) the number of profiles of other sizes produced by the section of these particles.
3. The number of profiles produced in each class by the section of particles of size X calculated above is subtracted from the profile size distribution.
4. In the residual profile size distribution obtained, the largest-size class of profiles is $x - 1$; by repeating the procedure above (steps 2 and 3), one successively calculates, and subtracts from the profile size distribution, the number of profiles produced in each class in intercepting (1) particles of size $X - 1$, (2) particles of size $X - 2$, etc., until the particles of the smallest size have been analyzed.

The procedure outlined above requires several intermediate calculations and is cumbersome to apply. Several methods have been proposed to calculate directly, by means of a set of coefficients, the size distribution frequency of spherical particles from the size distribution of the following attributes of their profiles on sections: diameter (Wicksell, 1925; Saltykov, 1958), sectional area (Saltykov, 1958), and chord intercept length (Spektor, 1950). A detailed description of these methods is given by Underwood (1970, Chap. 5).

For particles of shape other than spherical, the problem is complicated by the fact that profile distributions obtained on sections are mixtures of figures of different shape (e.g., circles and ellipses in the case of ellipsoidal particles). Wicksell (1926) devised a method to calculate the frequency distribution of the eccentricity of prolate (or oblate) ellipsoids from the eccentricity distribution of their profiles on sections. The method of DeHoff (1962) allows reconstruction of the size distribution of constant-shape ellipsoidal particles (prolate or oblate) from the size distribution of their profiles. A mathematical model for the solution of the general problem of the size-shape distribution of spheroids from measurements of their profiles on sections is given by Cruz-Orive (1976). Length-frequency distributions of intercepts have been analytically derived for cubes (Coleman, 1969; Itoh, 1970) and general cylinders (Kellerer, 1971). Computer simulation was used to determine distributions of intercept length (Itoh, 1970; Warren and Naumovich, 1977) and sectional areas (Paul, 1981) of convex particles such as ellipsoids and various families of polyhedrons.

Shape factors: The mathematical expression of shape and topological characteristics of convex solids is analyzed in detail by Underwood (1970, Chap. 7). The relative frequencies of *n*-polygons on sections can be used to characterize the shape of the sectioned polyhedrons. Computer-simulated random sectioning was used to determine frequencies for various families of polyhedrons (Durand et al., 1977; Warren and Durand, 1981; Voss, 1982).

Different aspects of the shape of convex objects (angularity, degree of symmetry, etc.) have been related to the shape of distribution curves of sectional areas (Paul, 1981). Empirical equations have also been given which relate dimensionless shape parameters such as $4\pi A(a)/B(a)^2$ [$A(a)$ and $B(a)$ = area and boundary length, respectively, of profiles of a measured in sections] to three-dimensional parameters such as volume-to-surface ratio for a variety of particle shapes (Warren and Durand, 1981).

35.2.2 Anisotropic Structures

The volume density $V_V(a, c)$ of an anisotropic phase (or objects) a in a containing specimen c can be estimated via Eq. (35.2) using profile measurements obtained on an arbitrarily oriented section. The stereological estimate of V_V is not affected by the sectioning angle.

In contrast, the surface density $S_V(a, c)$ and the length density $J_V(a, c)$ of the same structures both depend on the angle of sectioning and on the degree of orientation of a. For anisotropic structures, the estimators S_V and J_V given by Eqs. (35.3) and (35.4), respectively, are still unbiased for S_V and J_V, *provided* that profile measurements are performed on isotropic uniform random sections (see "Section Sampling" in Sec. 35.3.3). If sections are taken in the customary way (e.g., perpendicular or parallel) with respect to the axis of orientation, then the estimators S_V and J_V given by Eqs. (35.3) and (35.4) are biased.

For straight structures, strictly parallel to each other (complete anisotropy),

$$S_V(a, c) = B_A(a, c) \tag{35.30}$$

and

$$J_V(a, c) = Q_A(a, c) \tag{35.31}$$

where $B_A(a, c)$ is the boundary length of profiles of a per unit area of section of c, and $Q_A(a, c)$ is the number of transections of a per unit area of section of c. Both parameters, $B_A(a, c)$ and $Q_A(a, c)$, are measured on sections taken *perpendicular* to the axis of orientation of the structure.

The accurate estimation of the surface and length densities of *only partially* oriented structures (a commonly encountered case in the analysis of biological structures) is a difficult problem. In order to estimate the surface density S_V and the length density J_V of partially oriented structures from the analysis of their profiles on sections, it is necessary to adopt an orientational distribution model for the unit segments of the structure in space. An in-depth description of the problem and possible solutions are given by Weibel (1980, Chap. 10). We have proposed a method for quantifying the degree of anisotropy and estimating the length density of skeletal muscle capillaries (partially oriented structures) by means of sections which are parallel and perpendicular to the muscle fiber axis (Mathieu, Hoppeler, et al., 1981).

The principle of our method is the following:

- Under the Fisher axial distribution model for partial anisotropy, the length density $J_V(a, c)$ of a structure a in a containing specimen c may be estimated via

$$J_V(a, c) = \gamma(k; \alpha)^{-1} Q_A(a, c) \tag{35.32}$$

where the coefficient γ depends on the three-dimensional concentration parameter k of the unit segment's orientation and on the sectioning angle α relative to the preferred orientation (Weibel, 1980).

- On two sets of sections taken perpendicular ($\alpha = 0$) and parallel ($\alpha = \pi/2$) to the axis of anisotropy, the respective transectional densities $Q_{A1}(a, c)$ and $Q_{A2}(a, c)$ are related to one three-dimensional quantity, $J_V(a, c)$, via

$$J_V(a, c) = \gamma(k; 0)^{-1} Q_{A1}(a, c) = \gamma(k; \pi/2)^{-1} Q_{A2}(a, c) \tag{35.33}$$

so that

$$\frac{Q_{A1}(a, c)}{Q_{A2}(a, c)} = \frac{\gamma(k; \pi/2)^{-1}}{\gamma(k; 0)^{-1}} \tag{35.34}$$

Since $\gamma(k; \pi/2)^{-1}$ and $\gamma(k; 0)^{-1}$ are monotonic functions of k, there is only one value of the ratio $\gamma(k; \pi/2)^{-1}/\gamma(k; 0)^{-1}$ for each single value of k (see Mathieu et al., 1983, and Table 35.3).

- Therefore, the method proceeds as follows:
  1. The ratio $Q_{A1}(a, c)/Q_{A2}(a, c)$, obtained from measurements on sections, estimates the ratio $\gamma(k; \pi/2)^{-1}/\gamma(k; 0)^{-1}$.
  2. Knowing the latter ratio, one obtains the value of the coefficients k, $(k; 0)^{-1}$, and $(k; \pi/2)^{-1}$ from Table 35.3.
  3. The parameter $J_V(a, c)$ can then be estimated via either of the two equations (35.33).

TABLE 35.3 Coefficients $\gamma(k; \alpha)^{-1}$ for transverse ($\alpha = 0$) and longitudinal ($\alpha = \pi/2$) sections, and their ratios, for different values of k, based on the Fisher axial distribution (see Sec. 35.2.2)

k	$\gamma(k; 0)^{-1}$	$\gamma(k; \pi/2)^{-1}$	$\dfrac{\gamma(k; \pi/2)^{-1}}{\gamma(k; 0)^{-1}}$
0.00	2.00	2.00	1.00
0.20	1.87	2.07	1.11
0.40	1.76	2.16	1.23
0.60	1.65	2.25	1.36
0.80	1.56	2.36	1.51
1.00	1.48	2.48	1.68
2.00	1.23	3.23	2.63
4.00	1.08	4.79	4.42
6.00	1.05	5.98	5.71
8.00	1.03	6.96	6.73
10.00	1.03	7.82	7.61
20.00	1.01	11.14	10.99
.....	.....	.....	.....
∞	1	∞	∞

The *spherical normal* (or *Fisher axial distribution*) model adopted for the orientational distribution function of the capillary segments in the material under study was found to fit the relevant data satisfactorily (Mathieu et al., 1983).

35.3 PRACTICAL ASPECTS OF STEREOLOGICAL ANALYSIS

35.3.1 Planning of a Study

General Design. At the beginning of this chapter we emphasized the importance of quantitative morphological analysis for an objective evaluation and appropriate modeling of system arrangement and modulation. At the same time it should be kept in mind that the performance of such studies is generally very costly. Since they are based on statistical and probabilistic concepts, stereological methods require a large number of data to be collected and treated. The existence of significant regional variations may require a large number of sections to be examined in order to obtain data representative of the whole material under study.

The following points need to be kept in mind in the design of stereological studies:

1. How relevant is the information to be obtained to the phenomenon under study? Is quantitative evaluation imperative (e.g., to test subtle differences, for further mathematical formulation, to correlate data with physiological or biochemical measurements) or does it resume to simply "demonstrate" a change already visible by visual microscopical examination? Is the procedure chosen appropriate, or could the less costly evaluation of another parameter provide comparable insight? How relevant to the problem under study is a global parameter, averaged to a whole tissue or organ, versus regional patterns?

2. Is each step of the study adequately planned? Stereological measurements are performed on sections, i.e., on *portions* of material which have been *manipulated* through several preparative steps. Whatever the precision and/or the accuracy of profile measurements on micrographs may be, artifacts and/or bias introduced at any step in the preparation or in the selection of the material examined makes further effort useless and stereological estimation erroneous.

Reference System(s). Stereological estimates are relative parameters. *Densities* (volume, surface, curvature, numerical, etc.; see Sec. 35.2) of structures in a containing specimen or with respect to another structure are evaluated from ratios of specific attributes of their profiles on sections. The numerical values of such parameters evidently depend on the phase and/or objects chosen as reference. The more meaningful quantities to which to refer stereological parameters in a particular study need to be carefully defined and clearly specified. Depending on the parameter analyzed and the reference system chosen, different levels of magnification and a multiple-stage sampling procedure, or *cascade sampling* (see Weibel, 1979, Chap. 3; Cruz-Orive and Weibel, 1981), may be required. The possibility that the reference quantities do not remain constant under given experimental conditions needs to be kept in mind in the interpretation of results (see Sec. 35.3.5).

Sampling. As emphasized previously, stereological principles are based on statistical and probabilistic concepts. Unbiased estimates of the parameters are obtained *provided* that measurements are performed on sections (or portions of sections) taken randomly, and are representative of the material under study.

Preparational procedures for microscopy include a series of steps in which successive restrictions of the size of the material evaluated occur:

- Tissue fixation and embedding: the standard size of tissue samples processed for light microscopy (LM) is about 1 cm^3. In order to obtain adequate preservation and impregnation of material prepared for electron microscopy (EM), tissues need to be cut to much smaller blocks ($\sim$1 mm^3).

- Sectioning: 1- to 10-μm sections are used for light microscopy, and the thickness of thin sections for electron microscopy is about 70 nm; only one section per block is usually used for morphometry (see "Section Sampling" in Sec. 35.3.3).
- Microscopical examination: depending on the size of the structures under examination, and the reference space(s) chosen, it may be necessary to use different levels of magnification. At each level, obviously, the effective size of the field examined is inversely proportional to the magnification used.

In other words, depending on the resolution and magnification required (LM or EM) and the size of the tissue under study, extremely small portions of the material can actually be examined by microscopy and reasonably measured for stereology. The procedure used to "select" material at each preparative step, from tissue removal to collection of counts on micrographs, is of paramount importance for the quality of final stereological estimates.

Basically, the following sampling procedures may be used:

- *Random sampling:* Each subsample is obtained randomly by using random numbers.
- *Systematic sampling:* Subsample units are regularly dispersed in the material under study, by following a predetermined pattern (see "Tissue Sampling" in Sec. 35.3.2).
- *Stratified random sampling:* Combines the two preceding methods; the material is systematically sliced in regions, or strata, from each of which subsamples are taken randomly.

Theoretical and practical aspects of sampling for stereology are described in detail by Weibel (1979, 1980). Sampling techniques as they apply to each step of the preparation of the material are outlined below and briefly evaluated. Overall optimization of sampling design, for reduced cost and increased efficiency, is summarized in the subsection "Standard Error of Estimates and Sample Size" in Sec. 35.3.5.

35.3.2 Tissue Preparation

Tissue Sampling. Depending on the size of the organ (or tissue) under study, the following sampling procedures can be designed (Cruz-Orive and Weibel, 1981):

1. *One-stage systematic sampling:* The organ (or tissue) c is sliced into n sections which are cut in a predetermined direction u according to the following procedure:

- A random number x is chosen from zero to $d = h/n$, where d is the distance between consecutive sections and h is the caliper length of c along u.
- The first sectioning plane, X_1, is laid at a distance x from 0, and perpendicular to u; the subsequent planes ($X_2, X_3, \ldots, X_n$) are laid parallel to X_1, at respective distances $x + d$, $x + 2d, \ldots, x + (n - 1)d$ from 0. Each slice is then processed and embedded separately.

2. *Two-stage systematic sampling:* When each slice is too large to be processed completely, the following procedure can be used:

- The preceding design (paragraph 1) is followed.
- On each slice obtained, a regular lattice with a preestablished pattern (e.g., chessboard lattice with black and white squares) is overlaid at random; evenly dispersed portions of the slice (chosen with respect to the lattice) are subcut and processed.

3. *One-stage, independent, area-weighted periodic sampling:* The procedure is as follows:

- The preceding design (paragraph 1) is followed; the slices obtained are laid closely, with no overlapping, on a flat area.
- Two uniform random points are generated (1) on the flat area above and (2) on a regular lattice.

- The two uniform random points are brought to coincide; the intersection of the slices with the quadrats of the lattice form an *area-weighted periodic* section (Fig. 35.5*a*) which is then sampled. When the tissue samples need to be very small with respect to the size of the slices, a grid may be used, and small pieces of material overlapped by points of the grid sampled (Fig. 35.5*b*).

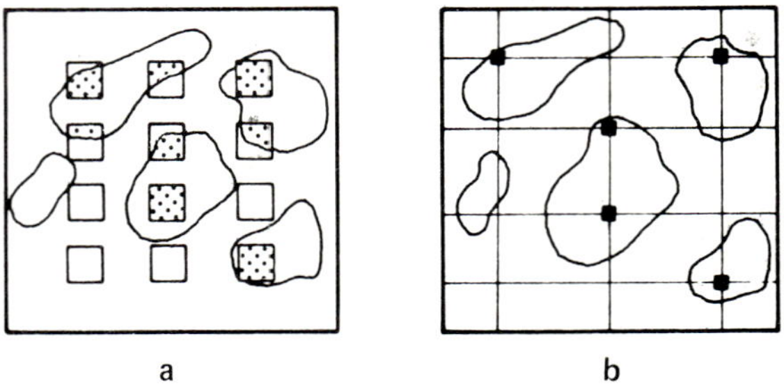

FIG. 35.5 (*a*) Area-weighted periodic (AWP) section (stippled). (*b*) AWP fields selected from a section. (*After Cruz-Orive and Weibel, 1981.*)

In procedure 3 above, the number of slices required to estimate the volume density $V_V(a, c)$ of a structure a in c, with a relative error E at a probability $1 - \alpha$, is (Cruz-Orive and Weibel, 1981):

$$[n] = (z_{1-\alpha/2})^2 n_0 \left(\frac{K_0}{E}\right)^2 \tag{35.35}$$

where K_0 is the coefficient of variation of the estimate of $V_V(a, c)$ found from a pilot sample of n_0 sections; and $z_{1-\alpha/2}$ is the $100(1 - \alpha/2)$ percentile of the standard normal distribution. After n has been estimated [Eq. (35.35)], the procedure is repeated, replacing $z_{1-\alpha/2}$ by $t_{1-\alpha/2;[n]-1}$ {= $100(1 - \alpha/2)$ percentile of the t-distribution with $[n] - 1$ degrees of freedom}, until n converges to a fixed value. In procedures 1 and 2 above, the first-stage sample is systematic; consequently variance estimators of $V_V(a, c)$ are not available from the data (Cruz-Orive and Weibel, 1981).

In the application of the sampling procedures above, care must be taken that possible periodicities in the structures do not coincide with slices positioning (schemes 1 and 2) and/or with the pattern of the lattices overlaid on the tissue (schemes 2 and 3). The evaluation of ordered systems may also call for different sampling considerations; in addition to estimating "global" parameters averaged to the whole material under study, it may be of interest to evaluate regional gradients and/or to follow their fate under various conditions. This is achieved by subdividing the organ or tissue into subregions which are then considered separately, i.e., in each of which adequate tissue sampling is performed in order to obtain representative and evenly distributed sampling units (schemes 1 to 3 above, depending on the size of the subregions considered). Such analysis may be very demanding. In the calculation of overall parameters, the weight of each subregion needs to be taken into consideration.

A practical example of the stratified random sampling procedure as applied to lung tissue in small and large mammals is given by Muller et al. (1981).

Specimen Fixation and Processing. Specimen preparation is a critical step where artifacts and/or dimensional changes of tissue need to be carefully evaluated. Depending on the tissue under study, prefixation treatment such as anesthesia may induce structural changes. The effect of anesthetics and/or other forms of euthanasia on cellular ultrastructure has not been systematically investigated. It is known, however, that barbiturates affect synaptic ultrastructure (Jones and Devon, 1978) and microtubular assembly (Ventilla and Brown, 1976), as well as lysosomal enzyme activity and distribution (Al-Azzawi and Stoward, 1970). Anesthetics also induce catecholamine release and affect vascular smooth muscle (Altura et al., 1980). Injection of a vasodilator prior to vascular perfusion of fixation was found to induce ultrastructural changes in liver (Frenzel et al., 1976).

Fixation and paraffin embedding for light microscopy are known to induce considerable changes in tissue and cellular dimensions. For electron microscopy, carefully controlled procedures are used to preserve cell and tissue fine structure through steps which generally include fixation in glutaraldehyde (GA) solution in buffer, rinse in buffer, postfixation in osmium tetroxide in buffer, dehydration in a series of alcohol or acetone, impregnation in resin mixtures, and embedding. A number of studies have been done to evaluate artifacts and/or dimensional changes occurring in various tissues during the preparative steps above and/or during alternative preparation procedures for electron microscopy (for review, see Hayat, 1970; Glauert, 1974; Hayat, 1981).

The following considerations need to be kept in mind when preparing tissue for subsequent morphometric analysis:

1. *Organ weight:* The importance of analyzing stereological parameters in absolute terms is emphasized in the subsection "Interpretation of Results" in Sec. 35.3.5. In many studies, it is imperative to record organ weight prior to tissue sampling; this is easily done by liquid displacement (Scherle, 1970) in osmotically adjusted fixative or buffer solution.

2. *Vascular perfusion* of fixative in situ, followed by immersion of tissue blocks in the fixative, is a method of choice (1) to immediately expose whole organs or tissues to the fixative, and thereby reduce postmortem changes during tissue removal and sampling, and (2) to preserve the in vivo three-dimensional arrangement of cell and tissue structures. In situ vascular perfusion is necessary for adequate preservation of a number of tissues, including skeletal and cardiac muscle sarcomere arrangement and ultrastructure, kidney nephron fine structure, and distended preparations of vessel walls (for a review, see Hayat, 1981). It is also required for stereological analysis of anisotropic structures (e.g., skeletal muscle capillaries) as they are in situ rather than artifactually convoluted by immersion fixation (see Mathieu et al., 1983).

3. The pH and *osmolarity of fixative solutions* influence the dimensions of tissue and cellular structures. It has been a matter of discussion for the past decade whether, in glutaraldehyde fixation, the osmolarity of the fixative vehicle rather than the total solution should be made isotonic. Glutaraldehyde per se was found to contribute little or nothing to the osmotic pressure of GA fixatives (Bone and Denton, 1971; Pentilla et al., 1974; Brunk et al., 1981; Arborgh et al., 1976). Other authors have reported that the osmotic contribution of GA in the fixative cannot be neglected (Fahimi and Drochmans, 1965; Arnold et al., 1971; Iqbal and Weakley, 1974; Barnard, 1976; Mathieu et al., 1978; Lee et al., 1982). In the study of lung tissue, we derived a formula which predicts the rate of dimensional changes in the thickness of the air-blood tissue barrier as GA and total fixative osmolarity are varied (Mathieu et al., 1978).

4. There is *no "universal" fixative solution for electron microscopy.* Depending on the tissue under study, different solution osmolarities need to be used. Specific fixation methods for a number of tissues and organisms are given by Hayat (1981). Comparing the effect of different fixative solutions on cell and tissue dimensions, we found that isotonic solutions (300 to 360 mosmol) adequately preserve lung tissue (Mathieu et al., 1978); a 1100-mosmol GA fixative was required for the preservation of skeletal muscle ultrastructure (Hoppeler et al., 1981); a 610-mosmol GA fixative led to a generally well fixed appearance of perfused arterial wall structures (Mathieu-Costello and Fronek, 1985).

5. For reasons not fully understood, vascular perfusion fixation usually requires higher fixative osmolarity than immersion fixation (Hayat, 1981; Bachofen et al., 1982).

6. Finding the best fixation procedure for a given tissue in a particular study is a matter of empirical trial. Pilot experiments need to be run to analyze the effect of different fixative osmolarities on tissue and cellular preservation. In most cases, in situ "true" dimensions are not known; fixation quality is evaluated from cellular matrix density, preservation of cell organelles (e.g., mitochondrial ultrastructure), absence of plasma membrane disruption, etc. (see Hayat, 1981).

7. Aldehyde solutions do not destroy the osmotic activity of cellular membranes (for review, see Hayat, 1970; Glauert, 1974). Tissues fixed with GA fixatives are sensitive to changes in the molarity of the buffer rinse (Fahimi and Drochmans, 1965; Valdivia, 1971; Korneliussen, 1972). Postfixation in an isotonic solution of osmium tetroxide may reverse shrinkage induced by a highly hypertonic primary fixative (Iqbal and Weakley, 1974).

8. The effect of various buffers on cell structures depends on the tissue and the fixative(s). Results have indicated that the tonicity rather than the type of buffer system is of importance for the preservation of cellular dimensions (Maunsbach, 1966; Ericsson et al., 1965). On the other hand, artifacts and/or differences in the morphological appearance of cellular components (for review, see Maunsbach, 1966; see also Gil and Weibel, 1968; Skaer and Whytock, 1977; Simson, 1977; Lee et al., 1982) may also occur when different buffers are used in different fixatives.

9. Tissue swelling and possible cell membrane disruptions during the first steps of dehydration are avoided by starting the procedure in 70% alcohol (Weibel and Knight, 1964). Different degrees of shrinkage are produced in various tissues by the last steps of dehydration, resin impregnation, and polymerization (for review, see Hayat, 1981). A slow and continuous change of concentrations during dehydration and resin infiltration considerably reduced cellular shrinkage, in contrast to a conventional, step-by-step procedure (Rostgaard and Tranum-Jensen, 1980).

10. Evidently, the use of rigorously identical solutions and preparative steps for all tissue samples in different sets of experiments is a prerequisite for meaningful comparative morphometry. A point needs to be considered, however. The same fixative solution may induce different structural changes in the same tissue type, depending on changes in membrane properties in different experimental or pathological conditions. Rested and exhausted skeletal muscle were found to be differently affected by various fixation procedures (Bowers et al., 1974; Gale, 1974). We found that normal and atherosclerosis-susceptible portions of the arterial wall (dorsal and ventral segments, respectively, in the same animals, fixed by vascular perfusion) respond differently to the same fixative (Mathieu and Costello, 1983). Dimensional changes in tissue may be only secondary events induced by the preparation procedure, whereas changes may have occurred elsewhere, e.g., in cell and/or mitochondrial membrane properties. In many studies, it is of importance to run pilot fixation tests on both control and experimental tissue (1) to determine which fixative or fixatives need to be used for an adequate preservation of all the specimens, and (2) to thereby identify which tissue changes possibly relate to fixation artifacts rather than the experimental condition under study.

35.3.3 Tissue Sectioning

Section Sampling. In order to save effort in measuring duplicated profiles of the same structures, only one section per block (or, if necessary, sections sufficiently apart from each other) are used for morphometry.

For electron microscopy, specimen portions (chosen randomly during tissue sampling; see Sec. 35.3.2) are subcut into smaller blocks. The position and orientation of the tissue in these blocks are generally random. Therefore, it can be assumed that an arbitrary section cut from such blocks is an area-weighted random section (see below) which yields unbiased stereological estimates of the parameters under study.

In other cases (e.g., section of whole organ examined in toto by light microscopy; EM analysis of oriented structures), it may be necessary to randomize the sectioning procedure. Theoretical and practical aspects of section sampling are discussed by Miles and Davy (1976), Miles (1978), and Weibel (1979, 1980). Briefly, the direction of sectioning is defined in a three-dimensional coordinate system by two angles determined by random numbers. The position of the sectioning plane within the block is then defined by using one of the following procedures:

- A *uniform random section* is obtained by using a random number to define the position of the sectioning plane from the origin of the coordinate system (note that with this procedure the edges and the center of the block have an equal chance to be sampled).

- The position of an *area-weighted section* is defined by using a set of three random numbers and considering the block as intersected by a cubic lattice (in this procedure, the probability of a given portion of the block to be sampled is proportional to its size).

The practical application of the second procedure above is difficult. However, it is the only method to obtain unbiased samples from tissue blocks which are not regular cubes (Mayhew and Cruz-Orive, 1974).

Section Preparation

Artifacts: Compression of tissue and consequent distortion of cell components occur during sectioning; the degree of compression varies with embedding media, tissue properties, and cutting conditions (Hayat, 1970; Peachey, 1958; Loud et al., 1965). When uniform tissue compression occurs, dimensionless parameters such as volume densities are not affected. Other parameters of the same structures (e.g., surface density, length density) are overestimated. A compression factor f_c may be measured as follows (Weibel, 1979, Chap. 4):

$$f_c = \frac{l_c}{l_0} \tag{35.36}$$

where l_0 is the axis of the block sectioning face, in the direction perpendicular to the cutting edge; and l_c is the same length as measured on the section.

The compressed area of the section, A_c, is

$$A_c = A_0 \frac{l_c}{l_0} \tag{35.37}$$

where A_0 is the area of the block sectioning face.

Folding of specific tissue compartments (e.g., "empty" alveolar spaces in lung parenchyma; elastic lamellae in large arteries) may occur during sectioning. For electron microscopy, thin section flattening is performed by solvent vapor treatment (Satir and Peachey, 1958), after sections have been floated off the knife edge; flattening of sections to be used for morphometry needs to be rigorously standardized.

Section thickness: The thickness of paraffin sections for light microscopy can be read directly on the micrometer scale of the fine focus knob of the microscope, after focusing successively on the upper and lower surfaces of the sections.

Several methods have been proposed to estimate the average thickness of ultrathin sections (for review, see Weibel, 1979, Chap. 4). Approximate values (within 10 to 30 nm) are obtained from interference colors as light is reflected from sections floating off the knife edge (Peachey, 1958). Accurate measurements of section thickness (for calculating correction factors for stereological estimates) are obtained by interference microscopy (Gillis and Wibo, 1971; Pluta, 1971), by microdensitometry (Silverman et al., 1969), or by comparing differences in electron current between sections and standard (Casley-Smith and Crocker, 1975). Another method, proposed by Small (1968), consists of selecting "minimal folds" on sections, i.e., folds which are two section thicknesses wide (Fig. 35.6). Such folds are recognized from their parallel outline on some distances and, at high magnification, from a fine line running along the middle of the fold, indicating the region where the two section thicknesses are opposed with no overlap (see Weibel, 1979, Chap. 4); section thickness is easily measured as half the width of minimal folds.

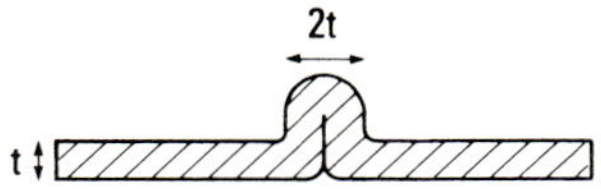

FIG. 35.6 Illustration of minimal fold.

Bias due to section thickness: Stereological principles (Sec. 35.2) assume that measurements are performed on true sectioning planes, i.e., on sections of zero thickness. Sections used for light and/or electron microscopy have a finite thickness; their *entire* content is projected and measured. Consequently, the following errors on stereological measurements may occur.

With the *Holmes effect* (Holmes, 1927), the boundaries of the projected image of opaque objects delineate the widest projections of the slice of objects present in the section (Fig. 35.7). Consequently, opaque structures are overestimated (and translucent phases underestimated), whereas profile shapes and numbers are affected by the overlapping of the projected images. The amount of resulting bias on stereological parameters depends on the size of the structures (as compared with section thickness) and on their shape; the error is partly compensated by loss of small profiles due to insufficient contrast.

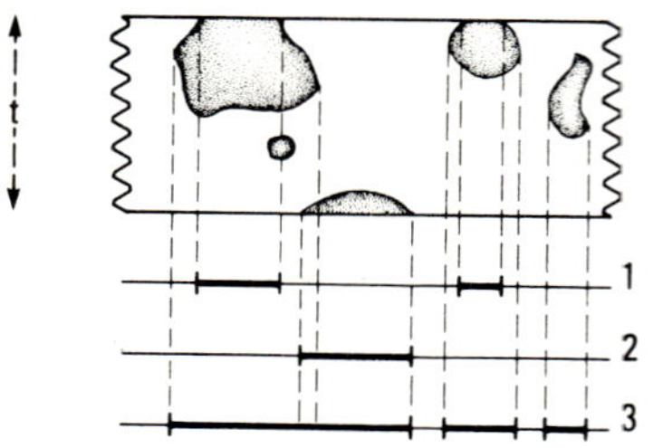

FIG. 35.7 Effect of section thickness t on the projection image (3) of opaque structures. Compare with observations of planes of sectioning [e.g., upper and lower surfaces of the section, (1) and (2), respectively].

The correction factor f_H on the volume density of spheres a of average diameter $D(a)$ is (Holmes, 1927; Hennig, 1957):

$$f_H = 1 + \frac{3t}{D(a)} \tag{35.38}$$

where t is the section thickness.

Section thickness correction factors have been formulated for basic stereological parameters of discrete convex particles (Cahn and Nutting, 1959) and random aggregates of convex elements (Miles, 1976). From these analyses, Weibel and Paumgartner (1978) derived correction factors for V_V and S_V estimates on discrete particles and/or interpenetrating structures of different shape; they worked out graphs from which correction factors for spheres, disks, and tubules may be read as a function of section thickness. In liver cells measured on 40-nm-thick sections it was found that (Weibel and Paumgartner, 1978):

- Correction for section thickness should be made when $t >$ 1/10 to 1/5 of the smallest dimension of the structure under study.
- For a given structure, relative bias for the volume density estimate is generally larger than that for surface density.
- Overestimation of volume density of cellular organelles varies largely, depending on the shape of the structures; it was 37, 46, and 29 percent for smooth endoplasmic reticulum tubules, mitochondrial cristae disks, and rough endoplasmic reticulum cisternae, respectively.

Gunderson (1979) gave a useful method to correct for overprojection and profile "fuzziness" in estimations of volume and surface densities of tubular structures on thick sections. Under some assumptions regarding the size and the shape of the structures, more efficient and less biased estimates of volume and surface densities of tubular structures a in a containing specimen c are obtained *indirectly*, as a function of the length density $J_V(a, c)$ of the structures, i.e., via the following equations (Gundersen, 1979):

$$J_V(a, c) = 2N_A(a, c)\frac{1}{[1 + \bar{D}(a)/\bar{L}(a)][\frac{1}{2} + \pi/4 + 2\bar{t}/\bar{D}(a)]} \tag{35.39}$$

$$S_V(a, c) = \pi\bar{D}(a)J_V(a, c) \tag{35.40}$$

$$V_V(a, c) = \frac{\pi}{4}\bar{D}^2(a)J_V(a, c) \tag{35.41}$$

where $N_A(a, c)$ is the number of profiles of a found per unit area of section of c; $\bar{D}(a)$ and $\bar{L}(a)$

are the mean diameter and length of the tubules, respectively; and $\bar{t}$ is the mean section thickness, averaged over the section area.

The advantage of this method is that profile counts [estimation of $N_A(a, c)$, Eqs. (39.39) to (39.41)] are not affected by overprojection and/or profile fuzziness, whereas estimates of $V_V(a, c)$ and $S_V(a, c)$ via Eqs. (35.2) and (35.3) depend directly on the measurement of profiles size on sections. The disadvantage is that the estimate of $J_V(a, c)$ is extremely sensitive to anisotropy.

The equations for numerical density of particles, given in Sec. 35.2.1, apply when measurements are performed on sections of zero thickness, i.e., in practice, when the diameter $D(a)$ of particles a is much larger than the section thickness t. When $D(a) < t$, the profiles examined on projected images of the sections represent two kinds of objects: (1) particles entirely contained in the slice and (2) other particles which are partially truncated by the upper or lower surfaces of the section. The numerical density $N_V(a, c)$ of small spherical particles a in a containing specimen c is evaluated from the numerical density $N_A(a, c)$ of their profiles on thick sections as follows (Abercrombie, 1946):

$$N_V(a, c) = \frac{N_A(a, c)}{t + D(a)} \tag{35.42}$$

where $D(a)$ is the particle diameter and t is the section thickness. When $t \ll D(a)$, Eq. (35.42) reduces to Eq. (35.11).

Two methods are available to correct for the error on $N_A(a, c)$, Eq. (35.42), due to lack of recognition of small cap sections:

1. Calculating the height $h(a)$ of the smallest visible cap from the radius $r(a)$ of the smallest visible profile of a as (Floderus, 1944)

$$h(a) = \frac{D(a)}{2} - \sqrt{\left[\frac{D(a)}{2}\right]^2 - [r(a)]^2} \tag{35.43}$$

Then

$$N_V(a, c) = \frac{N_A(a, c)}{t + D(a) - 2h(a)} \tag{35.44}$$

2. Or estimating the maximum equatorial perimeter of the particles, $b_e(a)$, via (Aherne, 1967)

$$b_e(a) = \frac{\pi}{2} \bar{I}_e(a) d \tag{35.45}$$

where $\bar{I}_e(a)$ is the mean number of intersections with a grid of parallel lines spaced at distance d, per maximum equatorial perimeter. The value of $\bar{I}_e(a)$ is approximated by counting the number of intersections $I(a)$ with all the profiles of a, $N(a)$:

$$\bar{I}_e \approx \frac{I(a)}{N(a)} \tag{35.46}$$

Then

$$N_V(a, c) = N_A(a, c) \frac{2}{\bar{I}_e(a) d \cdot 2t} \tag{35.47}$$

Although this method is not free from bias, in cases of contrast deficiency and/or large size variations between particles (Freedman, 1974), the approximation of $\bar{I}_e(a)$ via Eq. (35.46) may be reasonable on thick sections where many profiles actually display the maximum perimeter (Weibel, 1980, Chap. 5).

An interesting method was given by Ebbeson and Tang (1965): the measurement of particle size [Eqs. (35.42) to (35.47)] is avoided by using two sections of different thickness, $t_1 \neq t_2$, and by counting the respective numerical densities of profiles of a per unit section of c, $N_{A1}(a, c)$ and $N_{A2}(a, c)$:

$$N_V(a, c) = \frac{N_{A1}(a, c) - N_{A2}(a, c)}{t_1 - t_2} \qquad (35.48)$$

35.3.4 Microscopy

Field Sampling. When the whole section cannot be pictured at the required magnification, subsampled fields (quadrats) need to be selected. Several methods can be used to select quadrats randomly.

Systematic sampling (systematic quadrats, or SQ, that is, even spacing of quadrats over the section) may be achieved in light microscopy by using an automatic stage drive system (Weibel, 1970). In electron microscopy, pictures may be taken in predetermined portions (e.g., upper left-hand corner) of every (or every other) window of the mesh grid supporting the section (Weibel et al., 1966).

It is known that after systematic sampling, the standard error of stereological estimates is smaller than when pictures are selected by simple random sampling [due to clustering and/or field overlapping in the latter method (Hilliard and Cahn, 1961; Hennig, 1967; Cochran, 1977)]. In systematic sampling, care must be taken, however, that systematic arrangements or periodicity of the structures does not coincide with the regular spacing of the pictured fields. Possible bias of systematic sampling, as compared with area-weighting procedures, has been questioned (Miles and Davy, 1977). Consequently, the following methods were successively developed.

In *independent area-weighted sampling* (area-weighted quadrats, or AWQ; Muller et al., 1981), (1) the specimen stage of the microscope is digitized into XY rectangular coordinates by an electronic device, and a point is determined by two random numbers defining X and Y, respectively; (2) from another random number, a point is defined on a regular lattice of nine points marked on the screen; (3) the preparation is translated to make the two points coincide. This procedure is very time-consuming. Sampling seven quadrats on each of four sections of lung required approximately 10 H (Muller et al., 1981); consequently, AWQ was replaced by the following procedure.

The *systematic area-weighted sampling* (SAWQ) procedure was developed as an attempt to improve the accuracy obtained by systematic quadrats, at a reduced cost compared with that of AWQ (Muller et al., 1981). A fixed rule is used to position the section (e.g., a predetermined portion of the window of the mesh grid supporting the section; see description of systematic sampling). Then steps 2 and 3 in the AWQ procedure are executed.

In EM sampling of lung tissue (final magnification of ~10,000×), SQ and SAWQ methods yielded about equal precision on $S_V(a, s)$, $S_V(c, s)$, and $V_V(c, s)$ ratio estimators [a = alveoli, c = capillaries, s = septum; for nomenclature, see Eq. (35.1)]. However, the SAWQ procedure was about 10 percent more costly than the SQ procedure; systematic sampling, or SQ, is considered as an adequate quadrat-sampling procedure in most practical situations (Muller et al., 1981).

Micrograph Preparation

Picture recording: Using LM and/or EM picture prints for morphometry is very expensive and should be avoided. In many studies, LM measurements can be done directly on the microscope (see "Instrumentation" in Sec. 35.3.5 below). For EM morphometry, a very efficient method is to record micrographs on 35-mm film, and to use contact positives (Weibel, 1969) to perform measurements directly on projection systems (see Sec. 35.3.5).

EM artifacts: Depending on the embedding medium, 20 to 50 percent reduction of section thickness is produced by exposure to the electron beam (Cosslett, 1961; Willis, 1973; Bennett, 1974). Section thinning affects (1) the three-dimensional arrangement of structures in the preparation, and (2) correction factors for section thickness (see "Section Preparation" in Sec. 35.3.3), when portions of tissue and plastic (or other medium, e.g., standards embedded in resin) used for measuring section thickness are not affected homogeneously.

Image distortion due to lens aberration occurs in electron microscopy, especially at low magnification. The amount of distortion can be evaluated from micrographs of the grating replica used to calculate magnification (see below; Loud et al., 1965). On the double print of the image of the replica, at $\pi/2$ rotation, the distortion factor is evaluated from the center square size (i.e., real magnification) as compared with the periphery (distorted image). Image distortion has the opposite effect that tissue compression has on stereological estimates; with Epon embedding, both effects were found to compensate (Loud et al., 1965).

Magnification measurement: For electron microscopy, magnification is best recorded by taking micrographs of a carbon grating replica (E. F. Fullam, Inc., Schenectady, New York) on each filmstrip used for morphometry (above); the magnification factor after projection in any measurement device (see "Instrumentation" in Sec. 35.3.5 below) is thereby obtained directly.

Slight fluctuations in magnification are not uncommon when different sections are examined at a given EM magnification setting (because of variations in intermediate lens current and other factors). Cruz-Orive (1982) recently pointed out the bias introduced on stereological estimates by such differences, when averaged magnification factors are used, and raw data collected on different sections with a fixed-size test system are pooled [see Eq. (35.59)]; it is advisable to routinely handle raw data points $P_{i,j}$ and intersections $I_{i,j}$, both collected on sections examined at magnification $M_{i,j}$, as (Cruz-Orive, 1982)

$$\frac{a_i}{P_i} M_{ij}^{-2} P_{ij} \quad \text{and} \quad \frac{a_i}{l_i} M_{ij}^{-1} I_{ij} \tag{35.49}$$

where P_i, l_i, and a_i are characteristics of the test system used (Fig. 35.8). In the collection of data by a small desktop computer (see Sec. 35.3.5), this procedure introduces negligible computing time and effort. It should be kept in mind, however, that differences in the position of the sections with respect to the calibration grid also affect the accuracy of the magnification factor $M_{i,j}$ (see Cruz-Orive, 1982).

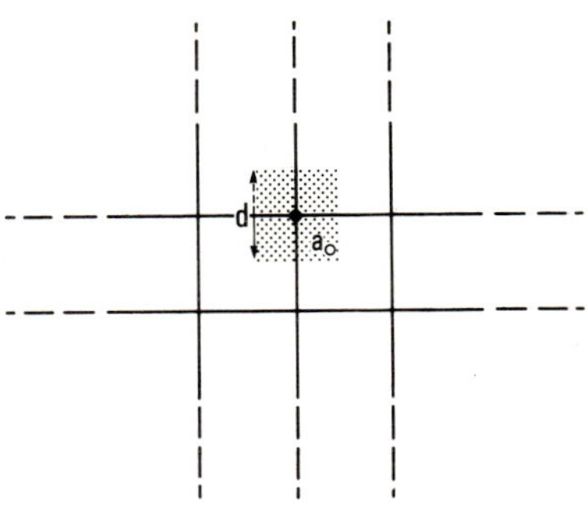

FIG. 35.8 Characterization of a coherent square lattice; $p_0 = 1$; $l_0 = 2d$; $a_0 = d^2$.

Effect of resolution: Different estimates of surface and volume densities are obtained when the same structures are examined at different magnification (Keller et al., 1976). Such differences could explain different results obtained in the same material in different laboratories (see review by Paumgartner et al., 1981). The different results are due to resolution, by which increasing amounts of detail are seen as structures are examined at higher magnification. This effect has been called the "coast of Britain" effect (Weibel, 1979), by analogy with the well-known differences obtained in cartography in the relative length of area outlines when different scales are used. Weibel (1979, Chap. 4) analyzed theoretically the effect of resolution on stereological estimates, by use of what was called a *fractal dimensions model* (Mandelbrot, 1977). On this basis, Paumgartner et al. (1981) analyzed resolution effects on measurements of mitochondrial and endoplasmic reticulum (ER) membranes in hepatocytes. From magnification 18,000 to 130,000× threefold increases were found in the estimates of surface and volume densities of ER, and in the estimate of the surface density of inner mitochondrial membranes, while the estimate of the surface density of the outer mitochondrial membranes increased by only 20 percent. No further differences were found above magnification 130,000×, called *critical* magnification. Resolution correction factors were given to estimate "true" stereological values at the critical magnification, from measurements performed at any magnification (Paumgartner et al., 1981).

35.3.5 Morphometric Measurements

Instrumentation. Stereological methods require a large number of data to be collected and treated. In pilot studies, measurements of profile attributes may be performed by hand on a few micrographs superimposed with a test lattice drawn on an acetate sheet. When a larger number of pictures are evaluated, several areas of the counting procedure can be optimized.

Projection systems: In light microscopy measurements can be done directly on the microscope by using an eyepiece graticule, by using a microscope where the binocular tube is replaced by a projection head with a built-in graticule slide (Wild Heerbrugg, Inc.), or by using a system adapted from a camera lucida (Elias and Boltz, 1976). Schwartz and Elias (1970) developed a device which classifies particles by size and which can be used in connection with a camera lucida system.

In electron microscopy, it is very efficient to record images on 35-mm filmstrips (see "Micrograph Preparation" in Sec. 35.3.4) and to use projection units as developed by Weibel et al. (1966) or microfilm readers (e.g., Jena DL-2 Macro Viewer Projector, Polysciences, Inc.) fitted with a test system, to perform measurements directly on positive films.

Data acquisition devices: Points, intersections, and/or profile counts on micrographs may be recorded via a variety of practical procedures, as outlined below.

1. *Simple system.* Small mechanical counters (as used in hematology) may be used to collect limited data for pilot studies.

2. *Computer-assisted point counting.* When a large number of data need to be handled, small computers may be used very efficiently for point counting and data processing. Different systems have been developed as computer hardware has improved (Ratz et al., 1974; Hoppeler et al., 1980; Briarty and Fisher, 1981). The advent of small desktop calculators (e.g., Hewlett Packard 9815) and desktop computers (Hewlett Packard 85; Apple, IBM) has made it possible to optimize basic steps of stereological work (Hoppeler et al., 1980). One of the main advantages of such systems is that function keys of the calculator, or any letter key of the computer, are used for point counting directly. Thereby, different counts on different structures are collected rapidly with minimum effort during one single manual scanning of a picture. For comparison, the pen of semiautomatic image digitizers, often used for point counting, may require as many picture scannings as parameters are recorded. Primary data (picture counts) are stored on magnetic tape and automatically recalled for calculation of parameters and statistical treatment; the system can be easily interfaced to larger computer units and/or peripheral devices (plotter, line prints, tracing device, etc.) when necessary. It is possible to perform statistical tests on the precision achieved (relative standard error of the estimate; see "Standard Error of Estimates and Sample Size" below) at any time during the counting procedure.

3. *Semiautomatic tracing devices.* Image digitizers (i.e., computer-assisted tracing devices; Dunn et al., 1975; Redman, 1977; Wolff, 1977) are now available which permit analysis of micrographs with a much larger density of test points than feasible by manual point counting. A pen is used to trace manually, on a tracing board, the contour of profiles to be measured; several parameters (e.g., areal fraction, trace length, profile numbers, etc.) may be obtained.

4. *Automatic image analyzers* (for review, see Hougardy, 1976; Weibel, 1979, Chap. 7). Automatic image analyzers are computers and television-based systems for collecting and processing morphometric data automatically. They are still of limited use for most biological studies, mainly because of limitations in image detection and pattern recognition. Automatic image analysis has been used successfully for cell counting and diagnosis in clinical pathology (e.g., see Bartels and Wied, 1981; Voss et al., 1981) and in a variety of studies such as morphometry of bone tissue (Green et al., 1981) and pattern analysis in the lung (Keller and Burri, 1981). The recent introduction of new technology, in particular, "framestores" image digital memory systems (for review, see Smith, 1982), could considerably improve the capability of computer systems for automatically analyzing tissue and cell ultrastructure.

Choice between different systems for data acquisition: Point counting using systematic lattices is superior to lineal integration or planimetry [Eq. (35.2)] for estimating areal fractions (Hilliard and Cahn, 1961; Weibel, 1979, Chap. 4). With the advent of semiautomatic tracing devices, the question arose about the advantage of evaluating each individual picture with more precision (see above). We compared the relative efficiency of different methods: semiautomatic computer image analysis with MOP AM03 (Kontron A.G, Zurich), automatic image analysis with Quantimet 720 (Cambridge Instruments), and point counting with different densities of test points (P_T per micrograph: 4, 25, 64, 100, 196, 400), by comparing the reduction of the relative standard error (RSE) of the estimates per unit time of measurement. Volume density ($V_V \approx$ 20%) and surface density ($S_V \approx 0.34$ cm^{-1}) of profiles were determined in synthetic models (10 micrographs each; coefficient of variation between micrographs: 10, 30, and 70%), and in a biological sample (skeletal muscle tissue). The most *efficient* methods were Quantimet or point counting with only 100 points per micrograph. The slowest method (increasing order) were (1) point counting with 196-point grid; (2) point counting with 400-point grid; and (3) MOP, for RSEs *comparable* to those obtained with Quantimet or point counting with only 100 points per micrograph (Mathieu, Cruz-Orive, et al., 1981).

These results demonstrated that one can considerably optimize the efficiency of stereological measurements without losing the precision of the estimates. The standard error of an estimate is mainly determined by the coefficient of variation between sampling units (micrographs in our study), so that measuring each sample unit with very high precision does not necessarily improve the precision of the estimates. From a statistical point of view, when dealing with a collection of sample units (e.g., sections), it is much more efficient to use the available time to measure less precisely more pictures from a large number of sections than to achieve a very high precision on a few subsampled fields (Mathieu, Cruz-Orive, et al., 1981).

Test Systems

Choice of test system: A variety of test systems can be used for morphometry (see Weibel, 1979, App. 3). Systematic point counting using regular arrangements of points is superior to random point test systems for reducing the standard error of stereological estimates (Hilliard and Cahn, 1961). Square lattices where intersections between lines are used as test points and lines are used to count intersections are often used. In a coherent test system (Weibel, 1969) total reference measurements (points P_T, length L_T, area A_T) are related to each other. Thereby, reference lengths and areas (Sec. 35.2) are estimated directly by

$$L(c) = 2P(c)d \tag{35.50}$$

$$A(c) = P(c)d^2 \tag{35.51}$$

where d is the spacing between two points.

Statistically, stereological estimations are optimized when test points are independent from each other, i.e., when no more than one point can fall on the same profile (Hilliard and Cahn, 1961). Consequently, the minimum spacing distance d between two points should be defined so that

$$d^2 > \bar{A}(a) \tag{35.52}$$

where $\bar{A}(a)$ is the mean profile area.

In addition, the precision on stereological parameters depends on the coefficient of variation between sampling units, so that measuring each picture with high precision, e.g., by using a relatively fine grid, is not necessary (see "Instrumentation" in Sec. 35.3.5; Mathieu, Cruz-Orive, et al., 1981). For estimating the volume density of mitochondria in skeletal muscle, the use of a coarser grid [144 test points per picture, instead of 576, both numbers satisfying Eq. (35.52)] resulted in parameter estimates with the same final precision but in less time (Hoppeler et al., 1981).

The total number of points P_T which need to be applied per representative sample (i.e., on the n quadrats subsampled), in order to estimate volume and surface densities of homogeneous particle aggregates with given standard errors, SE(V_V) and SE(S_V), respectively, depends on the value of the parameters V_V and S_V:

For volume densities (Hennig, 1956; Weibel, 1963),

$$P_T = \frac{1 - V_V}{V_V} \frac{1}{\mathrm{SE}^2(V_V)} \tag{35.53}$$

For surface densities of convex particles of low volume and surface density (Hilliard, 1976),

$$L_T = \frac{4S_V}{\mathrm{SE}^2(S_V)} \tag{35.54}$$

Nomograms relating the expected standard error of V_V (or S_V) estimates, from V_V and P_T (or S_V and L_T), are given by Weibel (1979, Chap. 4). From Eqs. (35.53) and (35.54), it can be seen that different cell structures may require different test grids, depending on their volume and/or surface density. In such cases, point (or intersection) counts may be simultaneously recorded by using double lattice systems (Weibel, 1969). In a test system (total number of points P_T spaced by distance d), every x^{th} (e.g., second or third) line is drawn heavier, forming a superimposed test system of P'_T points spaced by a distance xd. All the points (i.e., the finer test system) are used to evaluate the sparse structures, while the coarser system (P'_T points) is used to measure the more abundant structures. In such test systems (Weibel, 1969),

$$P_T = x^2 P'_T \tag{35.55}$$

Using test systems: Test systems may be used as follows:

- For estimating volume densities. Point identification may be ambiguous because of the relative thickness of test lines (e.g., in double lattice systems; see above). Systematic rules (for example, the intersection between the upper and the right edges of lines = one point; see Weibel 1979, Chap. 4) need to be applied to localize test points precisely.
- For estimating volume and surface density of the same structure. Square lattices contain an excessive length of lines for a given density of points (Weibel, 1969). The critical density of points of square lattices to achieve a given precision on stereological estimates was found to be smaller for surface density than for volume density (Mathieu, Cruz-Orive, et al., 1981). Systems such as multipurpose test systems (Weibel et al., 1966), where the length of test lines for a given point density is reduced, should be used.
- For counting profile number. When profiles intersect the edge of the quadrats, it may be erroneous to apply rules such as systematically counting profiles intersecting two predetermined edges of the test system (and "rejecting" the profiles intersecting the other edges), since the same structures may hit different corners of consecutive quadrats. To avoid this bias, Gundersen (1977) proposed the "forbidden line rule" (Fig. 35.9). Note that this rule applies for counts of profile numbers and *not* for counts of points and/or intersections with profiles.

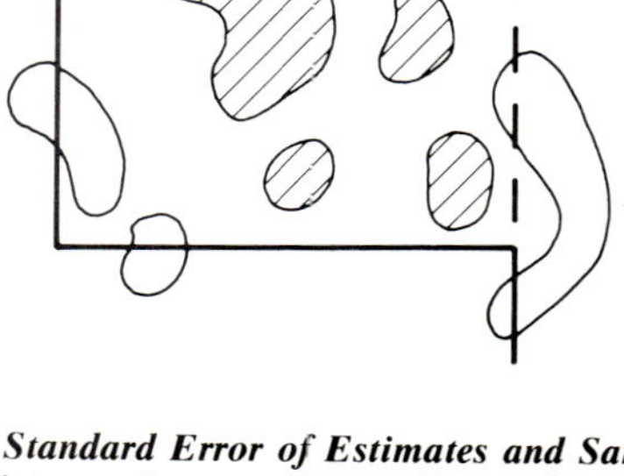

FIG. 35.9 Test frame with forbidden line (solid) for counting profile numbers; cross-hatched line indicates profiles counted. (*After Gundersen, 1977.*)

Standard Error of Estimates and Sample Size. Stereological estimates are ratios of points, intersections, and/or profile numbers (Sec. 35.2). The standard error, SE, of a ratio,

$$X = \frac{\sum_{i=1}^{n} x_i}{\sum_{i=1}^{n} y_i}$$

estimated over n sampling units (e.g., sections), is calculated as follows (Cochran, 1977):

$$\mathrm{SE}(X) = \left(\sum_{i=1}^{n} \frac{y_i}{n}\right)^{-1} \left[\sum_{i=1}^{n} \frac{(x_i - Xy_i)^2}{n(n-1)}\right]^{1/2} \tag{35.56}$$

that is, from raw data computation:

$$\mathrm{SE}(X) = \left[\frac{1}{n-1}\left(\frac{\sum_{i=1}^{n} x_i^2/y_i}{\sum_{i=1}^{n} y_i} - X^2\right)\right]^{1/2} \tag{35.57}$$

[*Biometrics, 36,* 595–605, 1980, eq. (16)].

Sample size at each level of the sampling procedure outlined above (i.e., tissue, sections, micrographs, and test grid selection) needs to be carefully determined in order to obtain parameter estimates with an acceptable precision (for example, RSE ~ 5 to 10%) at reasonable cost. Optimization of sample sizes at different levels of cascade sampling is discussed by Weibel (1979, Chap. 3), Gundersen and Osterby (1981), and Cruz-Orive and Weibel (1981). The following points need to be taken into account in statistically evaluating and/or allocating sample sizes in the design of stereological studies:

1. Obviously, the ratio of two means, $\bar{x}$ and $\bar{y}$, estimated as

$$\frac{1}{n}\sum_{i=1}^{n} x_i \qquad \text{and} \qquad \frac{1}{n}\sum_{i=1}^{n} y_i$$

respectively, is not equal to the mean of the ratios:

$$\frac{\bar{x}}{\bar{y}} \neq \frac{1}{n}\sum_{i=1}^{n} \frac{x_i}{y_i} \tag{35.58}$$

In other words, means of ratio estimated (e.g., from different sections) can be used only when reference quantities are equal (Mayhew and Cruz-Orive, 1974) or when weighted sampling is used (Cruz-Orive and Weibel, 1981). When reference areas are not equal, all primary data need to be summed up, and then ratio estimates must be calculated. For example, in a sample of p sections subsampled each in m quadrats (see "Field Sampling" in Sec. 35.3.4), where test points superimposed on nuclei (x) and on cytoplasm (y) are collected, nucleo-cytoplasmic ratio X is estimated as

$$X = \frac{\sum_{i=1}^{p}\sum_{j=1}^{m} x_{ij}}{\sum_{i=1}^{p}\sum_{j=1}^{m} y_{ij}} \tag{35.59}$$

In the estimation of X, overall variation (1) between sections and (2) between pictures is evaluated by calculating SE(X) via Eq. (35.56) or (35.57), with $n = p$ and $n = pm$, respectively.

2. Cumulative graphs of standard error of estimates can be used to evaluate the number of quadrats (or sections) required to achieve a given precision in the estimation of the parameters

(Chalkley, 1943). In the design of cascade sampling procedures (Sec. 35.3.1) it should be remembered that the variation between sampling units at the last level of the sampling procedure (animals in the comparison between animal groups, tissue sites in overall estimation of organs) is the main contribution to the overall variance of final estimates (Cruz-Orive and Weibel, 1981; Gundersen and Osterby, 1981). In addition, the standard error of an estimate X is mainly determined by the coefficient of variation, CV(X), between sampling units. Provided CV(X) $\neq 0$, the lowest relative standard error of the estimate, on a finite number of sampling units, cannot be reduced arbitrarily; it is CV(X)/$\sqrt{n}$ (Mathieu, Cruz-Orive, et al., 1981). When the coefficient of variation between sections (or sampling sites) is large, measuring each section (or sampling site) with a very high precision is not necessary.

Interpretation of Results

Relative and absolute parameters: As stressed above, stereological parameters are *relative* parameters. The analysis of parameters in absolute terms may be of paramount importance when reference quantities do not remain constant under given experimental or pathological conditions. In the morphometric study of the oxidative metabolism of skeletal muscles in mammals (Mathieu, Krauer, et al., 1981), comparing *absolute* mitochondrial contents in different muscles across the wide range of animal sizes analyzed was imperative, since reference volumes for mitochondrial densities (i.e. muscle volumes) scaled differently against body mass in the different muscles under study.

Regional variation: Stereological methods are statistical in nature. *Average estimates,* representative of the material under study, are obtained provided that measurements are performed on random sections (or portions of sections). Depending on the tissue or organ under study, it may be more meaningful to evaluate regional patterns rather than global parameters averaged to the whole tissue or organ. In such cases, the most meaningful designs for tissue sampling (Sec. 35.3.2) and parameter calculation (Sec. 35.3.5) need to be carefully evaluated. As an example, regional gradients exist in the distribution of oxidative enzymes not only within muscles but also within muscle groups (Armstrong, 1980). As pointed out by Hoppeler et al. (1981), it may be advantageous, depending on the problem under study, to refer mitochondrial content to functionally relevant muscle groups (or compartments) rather than anatomically defined muscles.

REFERENCES

Abercrombie, M. (1946). Estimation of nuclear population from microtomic sections. *Anat. Rec., 94,* 239.

Aherne, W. (1967). Methods of counting discrete tissue components in microscopical sections. *J. R. Microsc. Soc., 87,* 493.

Al-Azzawi, H. T., and Stoward, P. J. (1970). The effects of various forms of euthanasia on the histochemically demonstrable activities of lysosomal enzymes. *Proc. R. Microsc. Soc., 5,* 13.

Altura, B. M., Altura, B. T., Carella, A., Turlapaty, P., and Weinberg, J. (1980). Vascular smooth muscle and general anesthetics. *Fed. Proc., 39,* 1584.

Arbough, B., Bell, P., Brunk, V., and Collins, V. P. (1976). The osmotic effect of glutaraldehyde during fixation: A transmission electron microscopy, scanning electron microscopy and cytochemical study. *J. Ultrastruct. Res., 56,* 339.

Armstrong, R. B. (1980). Properties and distributions of the fiber types in the locomotory muscles of mammals. In K. Schmidt-Nielsen (ed.), *Comparative Physiology: Primitive Mammals.* Cambridge University Press, Cambridge, p. 243.

Arnold, J. D., Berger, A. E., and Allison, O. L. (1971). Some problems of fixation of selected biological samples of S.E.M. examination. In *Scanning Electron Microscopy: Proc. 4th Annu. Scanning Electron Microsc. Symp.* IIT Research Institute, Chicago, pt. I, p. 249.

Bachofen, H., Ammann, A., Wangensteen, D., and Weibel, E. R. (1982). Perfusion fixation of lungs for structure-function analysis: Credits and limitation. *J. Appl. Physiol., 53,* 528.

Barnard, T. (1976). An empirical relationship for the formulation of glutaraldehyde-based fixatives, based on measurements of cell volume change. *J. Ultrastruct. Res., 54,* 478.

Bartels, P. H., and Wied, G. L. (1981). Automated image analysis in clinical pathology. *Am. J. Clin. Pathol. Suppl., 75,* 489.

Baudhuin, P. (1968). L'Analyse morphologique quantitative de fractions subcellulaires. Thèse d'agregation, Université Catholique, Louvain.

Bennett, P. M. (1974). Decrease in section thickness on exposure to the electron beam; the use of tilted sections in estimating the amount of shrinkage. *J. Cell Sci., 15,* 693.

Bone, Q., and Denton, E. J. (1971). The osmotic effects of electron microscope fixatives. *J. Cell Biol., 49,* 571.

Bowers, W. D., Jr., Hubbard, R. W., Smoake, J. A., Daum, R. C., and Nilson, E. (1974). Effects of exercise on the ultrastructure of skeletal muscle. *Am. J. Physiol., 227,* 313.

Briarty, L. G., and Fischer, P. J. (1981). A general-purpose microcomputer program for stereological data collection and processing. *J. Microsc., 124,* 219.

Brunk, U., Collins, V. P., and Arro, E. (1981). The fixation, dehydration, drying and coating of cultured cells for SEM. *J. Microsc., 123,* 121.

Cahn, J. W. (1967). The significance of average mean curvature and its determination by quantitative metallography. *Trans. AIME, 239,* 610.

Cahn, J. W., and Nutting, J. (1959). Transmission quantitative metallography. *Trans. AIME, 215,* 526.

Casley-Smith, J. R., and Crocker, K. W. J. (1975). Estimation of section thickness, etc., by quantitative electron microscopy. *J. Microsc., 103,* 351.

Chalkley, H. W. (1943). Methods for quantitative morphological analysis of tissue. *J. Nat. Cancer Inst., 4,* 47.

Cochran, W. C. (1977). *Sampling Techniques* (3d ed.). Wiley, New York, Sec. 2.11.

Coleman, R. (1969). Random paths through convex bodies. *J. Appl. Prob., 6,* 430.

Cosslett, A. (1961). The effect of the electron beam on thin sections. In *Proc. Eur. Reg. Conf. Electron Microsc.,* Delft, vol. 2, p. 678.

Cruz-Orive, L. M. (1976). Particle size shape distributions: The general spheroid problem: I. Mathematical model. *J. Microsc., 107,* 235.

Cruz-Orive, L. M. (1982). The use of quadrats and test systems in stereology, including magnification corrections. *J. Microsc., 125,* 89.

Cruz-Orive, L. M., and Weibel, E. R. (1981). Sampling designs for stereology. *J. Microsc., 122,* 235.

DeHoff, R. T. (1962). The determination of the size distribution of ellipsoidal particles from measurements made on random plane section. *Trans. AIME, 224,* 474.

DeHoff, R. T. (1964). The determination of the geometric properties of aggregates of constant-size particles from counting measurements made on random plane section. *Trans. AIME, 230,* 764.

DeHoff, R. T. (1967). The quantitative estimation of mean surface curvature. *Trans. AIME, 239,* 617.

DeHoff, R. T. (1968). Curvature and topological properties of interconnected phases. In R. T. DeHoff and F. N. Rhines (eds.), *Quantitative Microscopy.* McGraw-Hill, New York, p. 291.

DeHoff, R. T., and Rhines, F. N. (1961). Determination of the number of particles per unit volume from measurements made on random plane sections: The general cylinder and the ellipsoid. *Trans. AIME, 221,* 975.

DeHoff, R. T., and Rhines, F. N. (1968). *Quantitative Microscopy.* McGraw-Hill, New York.

Delesse, M. A. (1847). Procédé mécanique pour déterminer la composition des roches. *C. R. Acad. Sci. (Paris), 25:* 544, 1847.

Dunn, R. F., O'Leary, D. P., and Kumley, W. E. (1975). Quantitative analysis of micrographs by computer graphics. *J. Microsc., 105,* 205.

Durand, M. C., Barbery, G., Hucher, J., and Grolier, J. (1977). Stéréologie de quelques polyèdres simples. In *Proc. 2d Eur. Symp. Stereol.,* p. 288.

Ebbesson, S. O. E., and Tang, D. (1965). A method for estimating the number of cells in histological sections. *J. R. Microsc. Soc., 84,* 449.

Elias, H., and Hyde, D. M. (1983). *A Guide to Practical Stereology.* Karger, Basel.

Elias, H., and Boltz, E. (1976). Simple devices for stereology and morphometry. In E. E. Underwood,

R. deWit, and G. A. Moore (eds.), *Proc. 4th Int. Congr. Stereol.* National Bureau of Standards, Pub. No. 431, Washington, p. 431.

Ericsson, J. L. E., Saladino, A. F., and Trump, B. F. (1965). Electron microscopic observations of the influence of different fixatives on the appearance of cellular ultrastructure. *Z. Zellforsch. Mikroskop. Anat., 66,* 161.

Fahimi, M. D., and Drochmans, P. (1965). Essais de standardisation de la fixation au glutaraldehyde: II. Influence des concentrations en aldehyde et de l'osmolalité. *J. Microsc., 4,* 737.

Floderus, S. (1944). Untersuchungen uber den Bau der menschlichen Hypophyse mit besonderer Beruksichtigung der quantitativen mikromorphologischen Verhaltnisse. *Acta Pathol. Microbiol. Scand. Suppl., 53,* 1.

Freedman, L. S. (1974). A note on Aherne's method of counting tissue components in relatively thick sections. *J. Microsc., 100,* 219.

Frenzel, H., Kremer, B., Richter, I-E., and Hucker, H. (1976). Der Einfluss des Perfusionsdruckes bei der Perfusionsfixation auf die feinstruktur der Lebersinusoide: Transmissions-und rasterelektronenmikroskopische Untersuchung. *Res. Exp. Med., 168,* 229.

Gale, J. B. (1974). Mitochondrial swelling associated with exercise and method of fixation. *Med. Sci. Sport, 6,* 182.

Gil, J., and Weibel, E. R. (1968). The role of buffers in lung fixation with glutaraldehyde and osmium tetroxide. *J. Ultrastruct. Res., 25,* 331.

Gillis, J. M., and Wibo, M. (1971). Accurate measurement of the thickness of ultrathin sections by interference microscopy. *J. Cell. Biol., 49,* 947.

Glagoleff, A. A. (1933). On the geometrical methods of quantitative mineralogic analysis of rocks. *Trans. Inst. Econ. Min. (Moscow), 59,* 1.

Glauert, A. M. (1974). *Practical Methods in Electron Microscopy.* North-Holland, Amsterdam, Elsevier, New York, vol. 3, pt. 1.

Green, D., Howells, G. R., and Thorne, M. C. (1981). Morphometric studies in mouse bone using a computer-based image-analysis system. *J. Microsc., 122,* 49.

Gundersen, H. J. G. (1977). Notes on the estimation of the numerical density of arbitrary profiles: The edge effect. *J. Microsc., 111,* 219.

Gundersen, H. J. G. (1979). Estimation of tubule or cylinder L_V, S_V and V_V on thick sections. *J. Microsc., 117,* 333.

Gundersen, H. J. G., and Osterby, R. (1981). Optimizing sampling efficiency of stereological studies in biology: or "Do more less well." *J. Microsc., 121,* 65.

Hayat, M. A. (1970). *Principles and Techniques of Electron Microscopy.* Van Nostrand Reinhold, New York, vol. 1.

Hayat, M. A. (1981). *Fixation for Electron Microscopy.* Academic Press, New York.

Hennig, A. (1956). Diskussion der Fehler bei der Volumenbestimmung mikroskopisch kleinen Korper oder Hohlraume aus den Schnitt projektionen. *Z. Wiss. Mikrosk., 63,* 67.

Hennig, A. (1957). Zur Geometrie von Schnitten. *Z. Wiss. Mikrosk., 63,* 362.

Hennig, A. (1963). Length of a three-dimensional linear tract. In H. Haug (ed.), *Proc. 1st Int. Congr. Stereol.* Congressprint, Wien, p. 44.

Hennig, A. (1967). Fehlerbetrachtungen zur volumenbestimmung aus der Integration ebener Schnitte. In E. R. Weibel and H. Elias (eds.), *Quantitative Methods in Morphology.* Springer-Verlag, Heidelberg, p. 99.

Hilliard, J. E. (1967). The calculation of the mean caliper diameter of a body for use in the analysis of the number of particles per unit volume. In H. Elias (ed.), *Stereol. Proc. 2d Int. Congr. Stereol.* Springer-Verlag, New York, p. 211.

Hilliard, J. E. (1976). Assessment of sampling error in stereological analysis. In E. E. Underwood, R. deWit, and G. A. Moore (eds.), *Proc. 4th Int. Congr. Stereol.,* National Bureau of Standards, Pub. No. 431, Washington, p. 59.

Hilliard, J. E., and J. W. Cahn (1961). An evaluation of procedures in quantitative metallography for volume-fraction analysis. *Trans. AIME, 221,* 344.

Holmes, A. (1927). *Petrographic Methods and Calculations.* Murby, London.

Hoppeler, H., Mathieu, O., Bretz, R., Krauer, R., and Weibel, E. R. (1980). The use of small computer systems for stereology. *Mikroskopie (Wien), 37* (suppl.), 408.

Hoppeler, H., Mathieu, O., Krauer, R., Claassen, H., Armstrong, R. B., and Weibel, E. R. (1981). Design of the mammalian respiratory system: VI. Distribution of mitochondria and capillaries in various muscles. *Respir. Physiol., 44,* 87.

Hougardy, H. P. (1976). Automatic image analyzing instruments today. In E. E. Underwood, R. deWit, and G. A. Moore (eds.), *Proc. 4th Int. Congr. Stereol.* National Bureau of Standards, Pub. No. 431, Washington, p. 141.

Iqbal, S. J., and Weakley, B. S. (1974). The effects of different preparative procedures on the ultrastructure of the hamster ovary: I. Effects of various fixative solutions on ovarian oocytes and their granulosa cells. *Histochemistry, 38,* 95.

Itoh, H. (1970). An analytical expression of the intercept length distribution of cubic particles. *Metallography, 3,* 407.

Jones, D. G., and Devon, R. M. (1978). An ultrastructural study into the effects of Pentobarbitone on synaptic organization. *Brain Res., 147,* 47.

Keller, H. J., and Burri, P. H. (1981). Automated pattern analysis of growing rat lung. *J. Microsc., 121,* 119.

Keller, H. J., Friedli, H. P., Gehr, P., Bachofen, M., and Weibel, E. R. (1976). The effects of optical resolution on the estimation of stereological parameters. In E. E. Underwood, R. deWit, and G. A. Moore (eds.), *Proc. 4th Int. Congr. Stereol.* National Bureau of Standards, Pub. No. 431, Washington, p. 409.

Kellerer, A. M. (1971). Considerations of the random transversal of convex bodies and solution for general cylinders. *Radiat. Res., 47,* 359.

Keyhani, E. (1980). Observations on the mitochondrial reticulum in the yeast *Candida utilis* as revealed by freeze-fracture electron microscopy. *J. Cell Sci., 46,* 289.

Korneliussen, H. (1972). Elongated profiles of synaptic vesicles in motor end-plates: Morphological effects of fixative variations. *J. Neurocytol., 1,* 279.

Lee, R. M. K. W., McKenzie, R., Kobayashi, K., Garfield, R. E., Forrest, J. B., and Daniel, E. E. (1982). Effects of glutaraldehyde fixative osmolarities on smooth muscle cell volume, and osmotic reactivity of the cells after fixation. *J. Microsc., 125,* 77.

Lindberg, L. G., and Vorwerk, P. (1970). On calculating volumes of transsected bodies from two dimensional micrographs. *Lab. Invest., 23,* 315.

Loud, A. V., Barany, W. C., and Pack, B. A. (1965). Quantitative evaluation of cytoplasmic structures in electron micrographs. *Lab. Invest., 14,* 996.

Mandelbrot, B. B. (1977). *Fractals: Form, Chance, and Dimension.* Freeman, San Francisco.

Mardia, K. V. (1972). *Statistics of Directional Data.* Academic Press, London.

Mathieu, O., Claassen, H., and Weibel, E. R. (1978). Differential effect of glutaraldehyde and buffer osmolarity on cell dimensions: A study on lung tissue. *J. Ultrastruct. Res., 63,* 20.

Mathieu, O., and Costello, M. L. (1983). Differential effect of glutaraldehyde fixative osmolarity on the ultrastructure of normal and atherosclerotic pigeon aorta. *Acta Stereol., 2,* suppl. I., 277.

Mathieu, O., Cruz-Orive, L. M., Hoppeler, H., and Weibel, E. R. (1981). Measuring error and sampling variation in stereology: Comparison of the efficiency of various methods for planar image analysis. *J. Microsc., 121,* 75.

Mathieu, O., Cruz-Orive, L. M., Hoppeler, H., and Weibel, E. R. (1983). Estimating length density and quantifying anisotropy in skeletal muscle capillaries. *J. Microsc., 131,* 131.

Mathieu, O., Hoppeler, H., Cruz-Orive, L. M., and Weibel, E. R. (1981). Stereological estimation of the length density of anisotropic structures: Capillaries in skeletal muscle. In *Proc. 3d Eur. Symp. Stereol.,* Ljubljana, *Stereol. Jugosl., 3,* suppl. 1, p. 643.

Mathieu, O., Krauer, R., Hoppeler, H., Gehr, P., Lindstedt, S. L., Alexander, R. M., Taylor, C. R., and Weibel, E. R. (1981). Design of the mammalian respiratory system: VII. Scaling mitochondrial volume in skeletal muscle to body mass. *Respir. Physiol., 44,* 113.

Mathieu-Costello, O., and Fronek, K. (1985). Morphometry of the amount of smooth muscle cells in the media of various rabbit arteries. *J. Ultrastruct. Res., 91,* 1.

Maunsbach, A. B. (1966). The influence of different fixatives and fixation methods on the ultrastructure of rat kidney proximal tubule cells: II. Effects of varying osmolality, ionic strength, buffer systems and fixative concentration of glutaraldehyde solutions. *J. Ultrastruct. Res., 15,* 283.

Mayhew, T. M., and Cruz-Orive, L. M. (1974). Caveat on the use of the Delesse principle of areal analysis for estimating component volume densities. *J. Microsc., 102,* 195.

Miles, R. E. (1976). Estimating aggregate and overall characteristics from thick sections by transmission microscopy. *J. Microsc., 107,* 227.

Miles, R. E. (1978). The sampling, by quadrats, of planar aggregates. *J. Microsc., 113,* 257.

Miles, R. E., and Davy, P. (1976). Precise and general conditions for the validity of a comprehensive set of stereological fundamental formulae. *J. Microsc., 107,* 211.

Miles, R. E., and Davy, P. J. (1977). On the choice of quadrats in stereology. *J. Microsc., 110,* 27.

Miles, R. E., and Davy, P. (1978). Particle number or density can be stereologically estimated by wedge sections. *J. Microsc., 113,* 45.

Muller, A. E., Cruz-Orive, L. M., Gehr, P., and Weibel, E. R. (1981). Comparison of two subsampling methods for electron microscopic morphometry. *J. Microsc., 123,* 35.

Nebel, B. R., Tyler, S. A., and Murphy, C. J. (1960). A note on the statistical determination of shape of chromatin elements in human spermatid and *Tradescantia* microsporocyte. *Biophys. Biochem. Cytol., 7,* 377.

Paul, J. L. (1981). Distribution curves of sectional areas through some families of convex particles. *J. Microsc., 122,* 165.

Paumgartner, D., Losa, G., and Weibel, E. R. (1981). Resolution effect on the stereological estimation of surface and volume and its interpretation in terms of fractal dimensions. *J. Microsc., 121,* 51.

Peachey, L. D. (1958). Thin sections: I. A study of section thickness and physical distortion produced during microtomy. *Biophys. Biochem. Cytol., 4,* 233.

Pentilla, A., Kalimo, H., and Trump, B. F. (1974). Influence of glutaraldehyde and/or osmium tetroxide on cell volume, ion content, mechanical stability, and membrane permeability of Ehrlich ascites tumor cell. *J. Cell Biol., 63,* 197.

Pluta, M. (1971). On the accuracy of microinterferometric measurements of optical path difference by means of the half-shade method. *J. Microsc., 93,* 83.

Rambourg, A., and Segretain, D. (1980). Three-dimensional electron microscopy of mitochondria and endoplasmic reticulum in the red muscle fiber of the rat diaphragm. *Anat. Rec., 197,* 33.

Ratz, H. U., Gnagi, H. R., and Weibel, E. R. (1974). An on-line computer system for point counting stereology. *J. Microsc., 101,* 267.

Redman, G. (1977). Halbautomatische Bildanalyse mit dem MOP und dem MOP/AM 01. *Mikroskopie (Wien), 33,* 102.

Rosiwal, A. (1898). Uber Geometrische Gesteinsanalysen. *Verch. K. K. Geol. Reichsanstalt, 6,* 143.

Rostgaard, J., and Tranum-Jensen, J. (1980). A procedure for minimizing cellular shrinkage in electron microscope preparation: A quantitative study on frog gall bladder. *J. Microsc., 119,* 213.

Satir, P. G., and Peachey, L. D. (1946). Thin sections: II. A simple method for reducing compression artifacts. *J. Biophys. Biochem. Cytol., 4,* 345.

Saltykov, S. A. (1946). *Zavod. Lab., 12,* 816.

Saltykov, S. A. (1958). *Stereometric Metallography* (2d ed.). State Publishing House for Metals and Sciences, Moscow.

Scherle, W. (1970). A simple method for volumetry of organs in quantitative stereology. *Mikroskopie (Wien), 26,* 57.

Schwartz, D. E., and Elias, H. (1970). An optico electric particle size classifier. *J. Microsc., 91,* 57.

Silverman, L., Schreiner, B., and Glick, D. (1969). Measurement of thickness within sections by quantitative electron microscopy. *J. Cell Biol., 40,* 768.

Simson, J. A. V. (1977). The influence of fixation on the carbohydrate cytochemistry of rat salivary gland secretory granules. *Histochem. J., 9,* 645.

Skaer, R. J., and Whytock, S. (1977). Chromatin-like artifacts from nuclear sap. *J. Cell Sci., 26,* 301.

Small, J. V. (1968). Measurements of section thickness. In D. S. Bocciatelli (ed.), *Proc. 4th Eur. Congr. Electron Microsc.* Tipographia Poliglotta Vaticana, Rome, vol. 1, p. 609.

Smith, C. S., and Guttman, L., (1953). Measurement of internal boundaries in three dimensional structures by random sectioning. *Trans. AIME, 197,* 81.

Smith, K. C. A. (1982). On-line digital computer techniques in electron microscopy: General introduction. *J. Microsc., 127,* 3.

Spektor, A. G. (1950). Distribution analysis of spherical particles in nontransparent structures. *Zavod. Lab., 16,* 173.

Tomkeieff, S. I. (1945). Linear intercepts, areas and volumes. *Nature, 155,* 24.

Underwood, E. E. (1970). *Quantitative Stereology.* Addison-Wesley, Reading, Mass.

Underwood, E. E. (1981). A standardized system of notation for stereologists. In: *Proc. 3d Eur. Symp. Stereol. Ljubljana Stereol. Jugosl., 3,* suppl. 1, 715.

Valdivia, O. (1971). Methods of fixation and the morphology of synaptic vesicles. *J. Comp. Neurol., 142,* 257.

Ventilla, M., and Brown, W. J. (1976). The effect of barbiturates on microtubular assembly. *J. Ultrastruct. Res., 54,* 325.

Voss, K. (1982). Frequencies of *n*-polygons in planar sections of polyhedrons. *J. Microsc., 128,* 111.

Voss, K., Simon, H., and Wenzelides, K. (1981). Hardware and software of a system for automatic microscopical picture analysis and its application in medicine. In *Proc. 3d Europ. Symp. Stereol. Ljubljana Stereol. Jugosl., 3,* suppl. 1, 139.

Warren, R., and Durand, M. C. (1981). The stereology of particle shape using computer simulation. In *Proc. 3d Europ. Symp. Stereol., Ljubljana Stereol. Jugosl., 3,* suppl. 1, 145.

Warren, R., and Naumovich, N. (1977). Relative frequencies of random intercepts through convex bodies. *J. Microsc., 110,* 113.

Weibel, E. R. (1963). *Morphometry of the Human Lung.* Springer-Verlag, Berlin, Academic Press, New York, p. 21.

Weibel, E. R. (1969). Stereological principles for morphometry in electron microscopic cytology. *Int. Rev. Cytol., 26,* 235.

Weibel, E. R. (1970). An automatic sampling stage microscope for stereology. *J. Microsc., 91,* 1.

Weibel, E. R. (1979). *Stereological Methods.* Academic Press, New York, vol. 1.

Weibel, E. R. (1980). *Stereological Methods.* Academic Press, New York, vol. 2.

Weibel, E. R. (1982). A bridge between morphology and physiology. *Acta Stereol., 1,* 23.

Weibel, E. R., and Gomez, D. M. (1962). A principle for counting tissue structures on random sections. *J. Appl. Physiol., 17,* 343.

Weibel, E. R., Kistler, G. S., and Scherle, W. F. (1966). Practical stereological methods for morphometric cytology. *J. Cell Biol., 30,* 23.

Weibel, E. R., and Knight, B. W. (1964). A morphometric study on the thickness of the pulmonary air-blood barrier. *J. Cell Biol., 21,* 367.

Weibel, E. R., and Paumgartner, D. (1978). Integrated stereological and biochemical studies on hepatocytic membranes: II. Correction of section thickness effect on volume and surface density estimates. *J. Cell Biol., 77,* 584.

Wicksell, S. D. (1925). The corpuscle problem: I. A mathematical study of a biometric problem. *Biometrica, 17,* 84.

Wicksell, S. D. (1926). The corpuscle problem: II. Case of ellipsoidal corpuscles. *Biometrica, 18,* 152.

Willis, R. A. (1973). Optimization of stereospecific and high angle tilting procedures for biological thin sections. *J. Microsc., 98,* 379.

Wolff, B. (1977). A new easily constructed semi-automatic morphometer. *Mikroskopie (Wien), 33,* 209.

CHAPTER 36

Instrumentation for Pulmonary Function Tests

Russell Pimmel
Department of Electrical Engineering
University of Missouri
Columbia, Missouri

INTRODUCTION

This chapter discusses pulmonary function testing, with emphasis on the protocols and instruments used. It begins with a general discussion of pulmonary function testing and the problems in interpreting the resulting data, and follows with a description of the procedures, instrumentation, and interpretations used in three standard pulmonary function tests performed routinely in most clinical pulmonary function laboratories. These include the measurement of forced expiratory volume and flow rate parameters, absolute lung volumes, and diffusing capacity. In the final section, several more specialized, developmental, or research-oriented tests are discussed briefly. The measurement of arterial blood pH and oxygen and carbon dioxide levels, perhaps the most effective measures of gas exchange, has not been included because of space limitations and because of the decision to focus on physiologically based function tests.

The presentation assumes that the reader is familiar with basic pulmonary physiology as presented in a standard textbook such as that by West [1]. It also assumes some understanding of the operation of certain biomedical instruments including differential pressure transducers, position-sensing potentiometers, heated thermal elements, and linear variable differential transformers. Some background in the effect of respiratory diseases on pulmonary function may be beneficial; such information is available in several standard textbooks (e.g., [2]).

36.1 PULMONARY FUNCTION TESTING

In general, pulmonary function tests are used in three ways: (1) to evaluate the current clinical status of an individual from a single set of measurements, (2) to characterize changes in an individual's pulmonary function during the course of an illness or in response to a pharmacological or chemical challenge, or (3) to evaluate the effect of various factors on pulmonary function in a clinical or epidemiological study of a selected population.

36.1.1 Evaluating the Current Clinical Status of an Individual

Frequently, the current clinical status of an individual must be evaluated from a single set of pulmonary function data, for example, at the beginning of an illness or during a well checkup, screening test, or preemployment physical examination. Ideally pulmonary function tests should detect any abnormality, define the severity of the abnormality, identify the region of the respiratory system that is abnormal, and define the specific pathological basis of the abnormality. In practice, existing techniques are moderately successful in detecting and classifying the severity of abnormal function, but they fail almost completely in defining the region or pathological changes involved.

In order to detect an abnormality and define its severity, it is necessary to define normal values for the measured pulmonary function parameters for the specific subject being evaluated. Since there is no theoretical basis for defining these normal values, an empirical approach is commonly used. This empirical approach is complicated because pulmonary function parameters depend strongly on body size. In addition, differences in sex and race, as well as the natural aging process of the lung, affect this dependence. Thus defining normal values empirically requires that data from a large number of subjects thought to be normal must be collected and analyzed to define regression equations for predicting normal values based on the height, age, and sex of the individual subject under evaluation. Established regression equations are summarized in most pulmonary function testing manuals [3, 4].

Even when large-population studies are done carefully, their usefulness still is limited by the fairly large residual variation that is not explained by height, age, and sex. As a result, standard errors about the regression lines are fairly broad, resulting in large normal ranges. This means that an individual's measurements must differ substantially, that is by 20 percent or more, from his or her predicted values before they are considered abnormal [5]. Thus, individuals whose baseline function is above average must undergo large changes before their measured parameters become abnormal; conversely, subjects whose baseline function is well below average may appear to be abnormal even when healthy.

Defining the region and pathological changes involved in an abnormality is difficult because most standard measurements of pulmonary function are affected by the condition of both airways and parenchyma. Radically different changes in the airways or parenchyma can affect a pulmonary function parameter in the same way [5]. For example, maximum expiratory flow at a given point in the vital capacity can be reduced by obstruction of the airway due to hypersecretion, hypertrophy, bronchospasm, or debris; by reduction of lung recoil due to emphysema; by an increase in lung recoil due to interstitial disease; or by a neuromuscular problem.

Localizing abnormalities is further complicated because most disease processes are heterogeneous and affect only a portion of the lung. This heterogeneity is seen in parenchymal diseases, which frequency have a patchy appearance on morphological examination. Another particularly important example of heterogeneous changes occurs in small-airway disease, in which the peripheral or small airways become obstructed while the central or large airways remain relatively normal. Since the peripheral airways contribute only a small fraction, nominally 20 percent, of the total resistance to airflow, marked small-airway obstruction can occur without being detected.

Although standard pulmonary function tests cannot localize abnormalities or define pathological changes, they can be used to classify the abnormality as having an obstructive or a restrictive pattern, which is useful because these two patterns are indicative of two groups of diseases [2]. The obstructive pattern, which is characterized by a reduction in forced expiratory flow and frequently by an increase in residual volume (RV), functional residual capacity (FRC), and total lung capacity (TLC), is seen in chronic bronchitis, emphysema, and asthma. The restrictive pattern, which is characterized by a reduction in RV, FRC, and TLC and by a secondary reduction in forced expiratory flow, is seen in interstitial fibrosis and other parenchymal diseases, with pneumothorax or pleural effusions, and with chest wall and neuromuscular diseases. With few exceptions, pulmonary function measurements do not provide any more specific diagnosis, and so it must be made by means of other information, such as radiographs and medical history.

36.1.2 Characterizing Changes in an Individual's Pulmonary Function

Pulmonary function tests are used to follow a patient's progress during the course of an illness and to evaluate efficacy of the selected therapy. They also are used to define an individual's response to pharmacological or chemical challenge. Important examples of pharmacological challenge are defining an individual's response to bronchodilators in assessing reversibility of airway obstruction and to bronchoconstrictors in assessing hyperreactivity of the airways. Chemical challenges include the response to environmental pollutants such as ozone or to occupational irritants such as toluene diisocyanate.

Ideally, in characterizing changes in an individual's pulmonary function, the tests should detect the change, define its magnitude, identify the region of the respiratory system that has changed, and define the specific pathophysiological basis of the change. These goals are analogous to those listed earlier for evaluating the current clinical status of an individual, and most of the problems identified in that discussion, especially those concerning specific location and pathophysiology, apply here.

Pulmonary function tests generally are better at defining changes in an individual than in characterizing the normality of results of a single test in an individual because standard errors of repeated measurements on an individual are smaller than standard errors about the regression lines used to predict normal values. Typically, repeated measurements on an individual have a standard error of 5 to 10 percent of the mean values [6], while the corresponding standard error for the predicted values is 10 to 20 percent [5]. Thus, a change of 15 percent in an individual's function is considered real, and this percentage is used clinically, for example, in evaluating the response to bronchodilators using forced expiratory volume and flow rate parameters [3]. This good reproducibility within an individual makes serial pulmonary function tests useful for following a patient through the course of an illness and for evaluating the efficacy of therapeutic interventions, especially because the tests are noninvasive and yield objective measures of lung function.

36.1.3 Characterizing Population Differences

In clinical and epidemiological studies pulmonary function tests are used to characterize (1) differences in two populations or (2) the change in one population after some intervention or spontaneous event. Comparing pulmonary function in two patient groups treated with different drugs or two groups representative of people living in high- and low-pollution areas are examples of the first type of study. Comparing pulmonary function in a group before and after treatment with a drug or before and after a controlled exposure to a lung irritant are examples of the second type of study.

In characterizing the differences between two populations or the change in some population after an intervention, pulmonary function tests should detect the difference, define its magnitude, identify the region of the respiratory system that is different, and define the specific pathophysiological basis for the difference. Again, these goals are analogous to those listed in the previous two sections, and the problem identified there, especially those concerned with location and pathophysiology, apply here. One advantage of comparing different groups, or the same group at different times, is that statistical analysis can be used to detect subtle differences that are undetectable on an individual basis.

36.2 FORCED EXPIRATORY VOLUMES AND FLOW RATE PARAMETERS—SPIROMETRY

This section describes the procedures, instrumentation, and interpretations used in the simplest and most common pulmonary function test, the measurement of forced expiratory volumes and flow rate parameters. Frequently this test is referred to as *forced expiratory spirometry* or, simply, *spirometry*.

In measuring forced expiratory volume and flow rate parameters, the subject inhales to total lung capacity and forcibly expires to residual volume. The test is based on the well-established finding that, at a given lung volume, the forced expiratory flow rate is independent of muscular effort so long as a certain minimum effort is achieved, and thus it depends only on the mechanical properties of the lung. As a result, the forced expiratory maneuver provides a convenient measure of pulmonary function because only expiratory volume or flow must be measured.

Signals from forced expirations are displayed in two ways: volume versus time, which is referred to as a *spirogram,* or flow versus volume, which is referred to as a *flow-volume loop*. Several numerical parameters are calculated from these displays, including forced vital capacity (FVC), forced expiratory volume in 1 s (FEV_1), average flow rate (forced expiratory flow, or FEF) over a given range of the vital capacity, with the 25 to 75 percent range ($FEF_{25-75\%}$) being the most common, and instantaneous flow rates with given fractions of the vital capacity remaining, with 25, 50, and 75 percent ($FEF_{25\%}$, $FEF_{50\%}$, and $FEF_{75\%}$) being most common. Several of these parameters are defined graphically in Fig. 36.1.

36.2.1 Procedures

The procedures for making forced expiratory volume and flow rate measurements are described in many pulmonary function testing manuals [3, 4] and in the "Epidemiology Standardization Report" sponsored by the National Heart, Lung, and Blood Institute and the American Thoracic Society [7]. Generally these manuals recommend the performance of three or so acceptable maneuvers in which FVC and FEV_1 are within a certain tolerance of each other, for example, 5 percent [7]. To compensate for the distortion in the flow and volume signals at the very beginning of the maneuver, back extrapolation is used to define its effective start. Numerical definitions of the end of the maneuver have been reported, for example, a volume change less than 25 mL for a 0.5-s interval or a flow rate that is less than 0.05 L/s for 0.5 s [7].

Values of various volume and flow rate parameters may be calculated manually by using graphical analysis of the spirogram or flow-volume curve, or they may be calculated automatically by computer analysis of the digitized volume or flow signal. Calculated values for the various parameters frequently are normalized by and reported as a percent of the subject's predicted values. The most commonly used parameters are FVC, FEV_1, $FEF_{25-75\%}$ (previously called maximum midexpiratory flow, or MMEF), and the ratio FEV_1/FVC.

The temperature in the spirometer must be recorded so that measured volumes can be converted to body temperature. Calibration of the volume measurement should be verified daily with a large calibrated syringe. In an operational laboratory, technicians frequently use their own forced expiratory measurements to verify the system's calibration.

36.2.2 Instrumentation

Instrumentation for measuring forced expiratory volume and flow rate parameters ranges from the simple direct-writing, water-sealed spirometer to complicated computerized systems in which either expired volume or flow, or both, are transduced and processed electronically.

In an electronic spirometer, volume may be detected directly by using a water-sealed, dry-rolling-sealed, or wedge spirometer with a position-sensing transducer, such as a potentiometer or a linear variable differential transformer, to detect displacement, thereby providing a signal proportional to spirometer volume. The corresponding flow signal is obtained either by electronic differentiation of the transduced volume signal or by numerical differentiation in a digital computer. Alternatively, flow may be detected directly with a conventional pneumotachometer or with a heated thermal element [8]. Pneumotachometers contain either a set of capillary tubes or a fine mesh screen designed so that the pressure drop is a linear function of flow. Heated thermal elements are maintained at a temperature above that of the body; flow across the element tends to cool them so that the amount of current required to maintain this temperature provides a measure of the flow. When flow is detected, the expired volume signal is obtained by electronic or numerical integration of the flow signal.

Flow-sensing devices have a better frequency response than volume-sensing devices, and

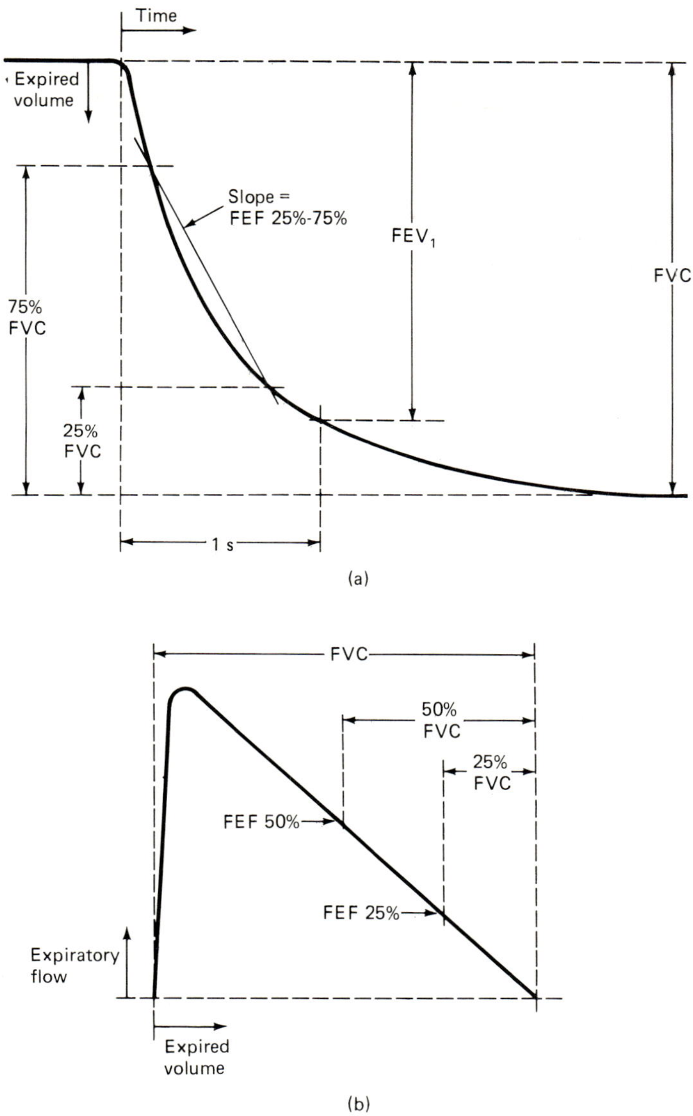

FIG. 36.1 (*a*) A forced expiratory spirogram and (*b*) a flow-volume loop, showing standard numerical parameters.

they are easier to clean and sterilize; however, they are more complex and more difficult to calibrate because the calibration depends on the composition, temperature, and humidity of the gas, which usually differ between inspiration and expiration. This problem makes it particularly difficult to balance the integrators used to derive a volume signal from the transduced flow signal.

Specifications for the equipment for measuring forced expiratory volume and flow are given in many pulmonary function testing manuals [3, 4] and in the report of the Snowbird Conference

[7, 9]. The spirometer must be capable of accumulating volume for at least 10 s and have a minimum range of 7 L with an accuracy of ±3% of the reading or ±0.05 L, whichever is greater. This accuracy must hold for flow rates ranging from 0.2 to 12 L/s. Resistance to flow should be less than 1.5 $cmH_2O/(L \cdot s)$. There must be a graphical display with a resolution of at least 0.5 s/cm on the time axis and 1 L/cm on the volume axis.

Gardner et al. [10] found that 8 of 17 commercial spirometers did not meet the Snowbird Conference specifications. They noted that the noncounterweighted water-sealed, dry-rolling-sealed, and wedge spirometers generally were satisfactory, while counterweighted water-sealed spirometers and flow-sensing systems were unsatisfactory.

Displaying the flow-volume loop in real time requires an *xy* recorder with a very high slew rate. For example, with a slew rate of 30 in/s the display should not be more than 4 or 5 in high. In many systems the flow-volume loop is displayed on an oscilloscope, or it is generated from stored digitized data so that slew rate is no longer a problem.

In the automated system where the volume signal is digitized, a sampling rate of 15 samples per second and an analog-to-digital converter that provides a volume resolution of 20 mL are recommended [7]. For systems sampling expiratory flow, considerably higher sampling rates are required.

36.2.3 Interpretation

Measured forced expiratory flow parameters are interpreted by using normal values derived from established equations. As discussed earlier, the range of normal values is 20 percent for FVC and FEV_1, 30 percent for $FEF_{75-25\%}$, and even larger for the instantaneous flow rate parameters. Because these ranges are so broad, forced expiratory flow parameters are not considered to be sensitive indicators of early lung disease.

Both obstructive and restrictive disease result in a reduced FVC, FEV_1, and $FEF_{25-75\%}$. With obstructive disease the ratio FEV_1/FVC will be reduced, with a value of 0.7 being considered the lower normal bound [3]. With restrictive disease this ratio generally will be normal or high. Although the ratio FEV_1/FVC provides some distinction between obstructive and restrictive disease, accurate classification requires lung volume measurements along with forced expiratory flow measurements.

Subjective analysis of the configuration of the spirogram or flow-volume loop can provide additional information that may be useful in distinguishing obstructive and restrictive patterns, as shown in Fig. 36-2. With obstructive disease the spirogram will be a prolonged curve with a gradual slope; with restrictive disease, it will be a brief curve with a steep slope. The flow-volume loop is generally concave upward with obstructive disease and concave downward with restrictive disease. Several groups have described quantitative parameters characterizing the configuration of the spirogram and flow-volume loop, but none has been accepted widely.

Standard forced expiratory volume and flow rate parameters provide little indications of heterogeneity of disease. Theoretically, instantaneous flow rates at low lung volumes, for example $FEF_{75\%}$, should be sensitive to small-airway function, but the large variability in this measurement makes interpretation difficult. The degree of concavity in the flow-volume loop provides a subjective measure of heterogeneity, but quantitative measures of this subjective feature are not in general use.

36.3 LUNG VOLUMES

This section describes the procedures, instruments, and interpretations used with a second routine pulmonary function test, the measurement of absolute lung volumes. The three physiologically defined lung volumes depend on the mechanical properties of the lung and chest wall, on the strength of the respiratory muscles, and on the output from the respiratory center in the central nervous system (CNS); their measurement provides information about respiratory function.

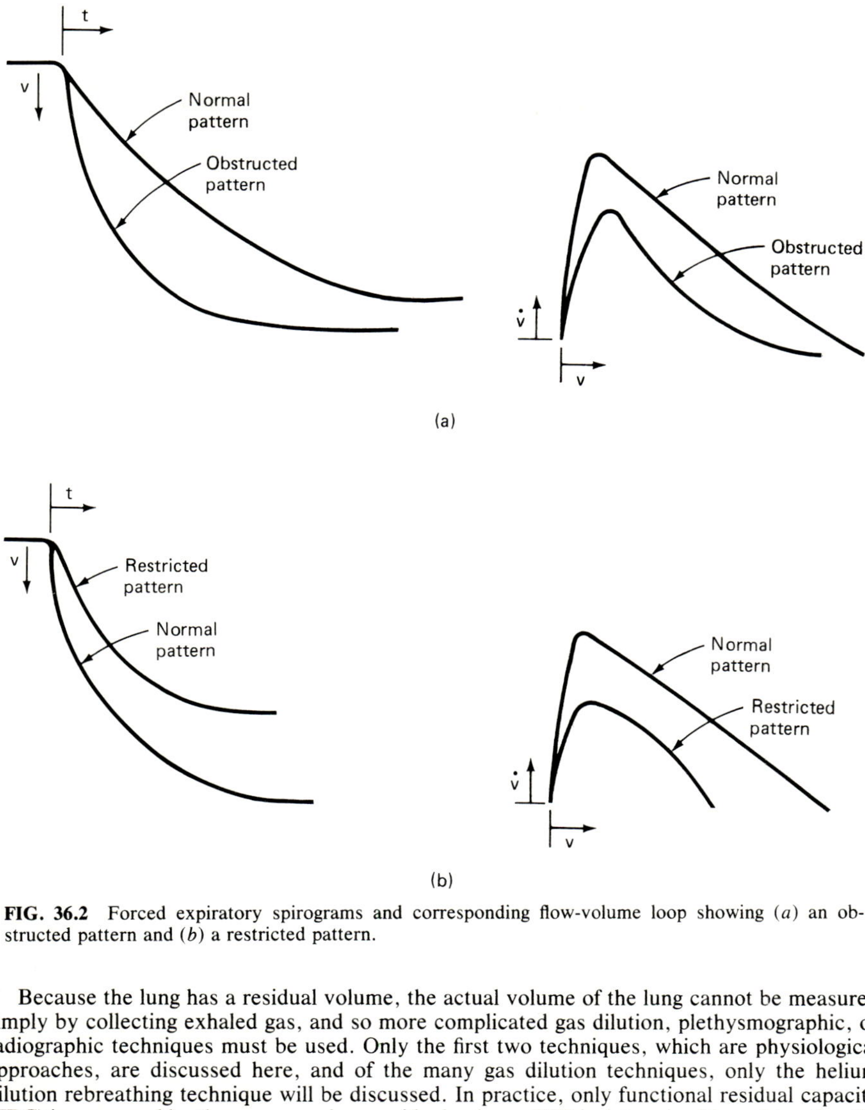

FIG. 36.2 Forced expiratory spirograms and corresponding flow-volume loop showing (*a*) an obstructed pattern and (*b*) a restricted pattern.

Because the lung has a residual volume, the actual volume of the lung cannot be measured simply by collecting exhaled gas, and so more complicated gas dilution, plethysmographic, or radiographic techniques must be used. Only the first two techniques, which are physiological approaches, are discussed here, and of the many gas dilution techniques, only the helium dilution rebreathing technique will be discussed. In practice, only functional residual capacity (FRC) is measured by these approaches; residual volume (RV) is determined by subtracting the maximum volume that is exhaled from FRC, while total lung capacity (TLC) is determined by adding the maximum volume that is inhaled from FRC.

36.3.1 Procedures

Helium Dilution Technique. Techniques for measuring lung volumes by helium dilution are described in several pulmonary function testing manuals [3, 4] and in the "Epidemiology Standardization Report" [7]. Briefly, the subject is connected to a spirometer containing a known volume of gas with a measured initial helium concentration, typically 10 percent. As rebreathing proceeds the helium concentration is diluted by the gas in the lung. Because helium is very insoluble, little is absorbed by the lung tissue or blood flowing through the lung. Rebreathing continues for 7 min or until the helium concentration changes less than 0.02 percent during a 30-s interval [7]. During the rebreathing, O_2 is added to replace that consumed by the subject, and CO_2 is absorbed by circulating the gas in the spirometer through a CO_2 absorber by means of a small blower. Also the subject should perform several slow maximum inspirations to facilitate gas mixing in all lung units. At the end of the rebreathing period, the final helium concentration is measured and volume changes during a slow maximum inspiration and expiration are recorded for calculating RV and TLC.

FRC is calculated from the initial and final helium concentrations (He_I and He_F) and from the initial volume in the spirometer (V_s) by means of the following equation:

$$\text{FRC} = V_s \frac{\text{He}_I - \text{He}_F}{\text{He}_I} \tag{36.1}$$

If the dead space of the mouthpiece is less than 50 mL, then no correction is made for this volume. Generally, no correction is made for the small amount of helium absorbed by the lung tissue and blood; however, Ruppel [4] suggests subtracting 100 mL from the measured FRC to correct for this error. However, when the FRC is small, for example, in children or in patients with severe lung disease, these corrections may be important. RV and TLC are calculated by subtracting and adding volume changes recorded during the slow maximum expiration and inspiration, respectively. Measured values of RV, FRC, and TLC are compared with predicted values calculated from standard regression equations, and measured values are frequently expressed as a percentage of the predicted value. Often the RV/TLC ratio is calculated and reported.

Plethysmographic Technique. Procedures for measuring FRC by plethysmography are described in pulmonary function testing manuals [3], in the "Epidemiology Standardization Report" [7], and in a special report from the National Heart and Lung Institute [11]. Briefly, the subjects sits in a rigid, sealed box called a *constant-volume plethysmograph*. The subject breathes into the box through a pneumotachometer or other flow-sensing element. At the end of a normal expiration the airway is occluded, usually by means of a solenoid valve, and the subject then pants against the closed airway. Pressure changes at the mouth (ΔP_{mouth}) and in the box (ΔP_{box}) are measured by appropriate transducers.

If constant temperature is assumed, changes in box pressure as the chest moves in and out are proportional to volume changes in the gas surrounding the subject in the box. Because the change in this gas volume must equal the change in the gas volume in the lung, the measured changes in the box pressure reflect changes in lung volume as well. Furthermore, because there theoretically is no gas flow in the lungs, changes in mouth pressure equal changes in alveolar pressure. Thus, the two transducers provide signals proportional to the changes in lung volume and alveolar pressure as the subject pants. Again with a constant-temperature assumption, the relationship shown in Eq. (36.2) can be derived and used to calculate FRC:

$$\text{FRC} = K(P - 47)\left(\frac{\Delta P_{\text{box}}}{\Delta P_{\text{mouth}}}\right) \tag{36.2}$$

In this relationship K contains all calibration factors, and $P - 47$ is barometric pressure minus the saturated water vapor pressure.

Usually the two transduced signals, that is, box pressure and mouth pressure, are displayed on an oscilloscope using the *xy* mode so that the slope of the resulting line represents $\Delta P_{\text{box}}/\Delta P_{\text{mouth}}$. With appropriate calibration, FRC can be read directly from the scope display. In practice, the two signals usually generate a narrow loop with irregular patterns at both ends so that defining the slope is somewhat subjective. Usually only the center portion of the display is used, while the two end portions are ignored. In computerized systems, the two transduced signals are digitized and subjected to linear regression analysis to define the slope.

With the plethysmographic technique, RV and TLC cannot be obtained during the FRC measurement since a sealed, rigid plethysmograph cannot measure accurately the large volume changes that occur with maximum inspiration or expiration. However, once FRC is defined plethysmographically, RV and TLV can be obtained on a simple spirometer, as noted in the discussion of the helium dilution technique.

Just as with the FRC measured by helium dilution, values obtained plethsmographically are compared with predicted values obtained from standard regression equations.

36.3.2 Instrumentation

Helium Dilution Technique. Measurement of FRC by helium dilution requires a continuous sampling helium analyzer and a 6- to 9-L spirometer with a recirculating path containing a CO_2 absorber and a blower with a nominal capacity of 30 L/min. The spirometer should meet all the specifications identified earlier in the discussion of forced expiratory spirometric measurements.

Helium analyzers usually measure helium concentration by monitoring changes in thermal conductivity caused by its presence [8]. In one approach four heated thermal elements are connected in a bridge configuration; two elements are placed in a stream of a standard helium-free gas, while the other two are placed in a stream of the test gas. Helium in the test gas alters the thermal conductivity between the two heated thermal elements in the test gas stream, thereby altering the resistance of these two elements. This change in resistance unbalances the bridge so that its output voltage provides a signal that is related to the concentration of helium in the test gas. Usually the analyzer contains electronic circuits to linearize and scale the transduced signal so that the output signal or the displayed value provides a direct measure of helium concentration.

For measuring FRC a helium analyzer should have a range that is consistent with the initial concentration, typically 10 or 15 percent. Its response should be both linear and accurate within 1 percent of the full-scale reading, and the response to a step change in helium concentration should reach 90 percent of the final value in less than 10 s [7]. The gas-sampling rate for the analyzer should be extremely small unless it has a provision for returning sampled gas back to the system after analysis. Detailed techniques for calibrating helium analyzers are presented elsewhere [3].

The overall FRC measuring system should be calibrated daily by using a 3-L syringe. The syringe is filled with room air and connected to the helium dilution system. Helium is equilibrated by repeatedly withdrawing and reinjecting 0.5 L to simulate rebreathing with a normal tidal volume. The syringe volume is calculated from the spirometer volume, and the initial and final helium concentrations are calculated by means of Eq. (36.1); the calculated volume should be within 0.05 L of the original syringe volume.

Plethysmographic Technique. The plethysmograph must be rigid and tightly sealed, with the net leak having a time constant greater than 1.5 s when a bolus of air is injected. This ensures that the plethysmograph is sealed at panting frequencies of 1 Hz or so. The mouth pressure transducer and amplifier should have a range of ± 20 cmH_2O, while the box pressure transducer and amplifier should be more sensitive with a range of ± 2 cmH_2O. Both should be linear and accurate within 3 percent. Their frequency response characteristics should be flat without any phase distortion up to the panting frequency, typically 1 or 2 Hz. The performance of the overall system can be verified by using a rigid container with a volume of approximately 4 L connected to an oscillating pump with a stroke volume of 50 to 100 mL and a frequency of oscillation at the panting rate [3]. The measured volume should be within 5 percent of the true gas volume of the container.

An oscillating pump also is useful for calibrating transduced pressure changes in the box in terms of actual volume changes, and many commercial plethysmographs have a built-in pump for this purpose. Many also have a solenoid valve that opens the box quickly, returning its pressure to atmospheric pressure. This quick venting is useful in eliminating the effect of a slow drift in pressure due to temperature changes just before a measurement is made.

36.3.3 Interpretation

Predicted values derived from standard regression equations are used when interpreting measured values of RV, FRC, and TLC. As with forced expiratory volume and flow rate parameters, there is a wide range of normal values, and so any measurement within 25 percent of the appropriate predicted value usually is considered normal. A decreased RV, FRC, and TLC indicate restrictive lung disease, while an increased RV, FRC, and TLC indicate hyperinflation, which is usually, but not always, a sign of obstructive disease [2, 3]. TLC, when expressed as a percentage of its predicted value, has been used to classify the severity of restrictive disease as mild, moderate, or severe by using thresholds of 80, 65, and 50 percent of predicted values, respectively [3].

With obstructive disease, lung volumes measured by helium dilution must be interpreted very carefully because the obstructed lung frequently contains either completely or partially occluded units with which helium does not completely equilibrate. Since the plethysmographic technique does not require ventilation to all units, it is the preferred technique for studying patients with obstructive disease. In fact, the ratio of FRCs measured by the two approaches provides a measure of trapped gas and therefore airway obstruction [3].

36.4 DIFFUSING CAPACITY

This section describes the procedures, instrumentation, and interpretations used with a third routine pulmonary function test, the measurement of diffusing capacity. This parameter indicates how effectively gas is transported from the alveolar compartment to the red blood cells in the pulmonary capillaries. Diffusing capacity is defined as the ratio of the rate of gas exchange to the difference between the partial pressure of the tracer gas in the alveolar compartment and in the capillary blood. Usually carbon monoxide is used as a tracer gas in measuring diffusing capacity because at low concentrations it is absorbed rapidly by hemoglobin so that its partial pressure in capillary blood is near zero. Thus only alveolar partial pressure is important in the transfer of this gas. To calculate the diffusing capacity, the rate of carbon monoxide absorption and the alveolar partial pressure must be measured. The test may be made by using a single inspiration and expiration or after a rebreathing period. Because the single-breath approach is much simpler and more convenient, it is used more frequently in clinical laboratories.

36.4.1 Procedure

The procedure for measuring carbon monoxide diffusing capacity by the single-breath method is described in many pulmonary function testing manuals [2, 3] and in the "Epidemiology Standardization Report" [7]. Briefly, the procedure is as follows. The subject inspires rapidly from RV to TLC from a bag containing measured amounts of carbon monoxide and helium, typically 0.3 and 10 percent, respectively. The bag is sealed inside a box which is connected to a spirometer so that the inspired volume can be measured. After an approximately 10-s breath-holding period, the subject exhales rapidly. Initially, the exhaled gas is directed to the box-spirometer system. After 1 L is expired to allow the dead space to clear, a sample of expired gas, approximately 1 L, is collected in a sample bag for analysis. In children and in patients with reduced vital capacities, it may be necessary to collect the sample after only 0.5 L is exhaled. Sampled gas is considered to be representative of alveolar gas, and its carbon monoxide and helium concentrations are measured. The actual duration of the breath-holding period, defined from midinspiration to the time when the alveolar sample is collected, also is measured from the spirometer tracing.

Usually a four-way valve is used to facilitate diffusing capacity measurements. As shown in Fig. 36.3, the common port is connected to the subject's mouthpiece. One of the selectable ports is open to room air, and the subject breathes through this port while becoming accustomed to the system. The second selectable port is connected to the bag containing the two tracer gases which the subject inspires. The third selectable port is connected to the box-spirometer

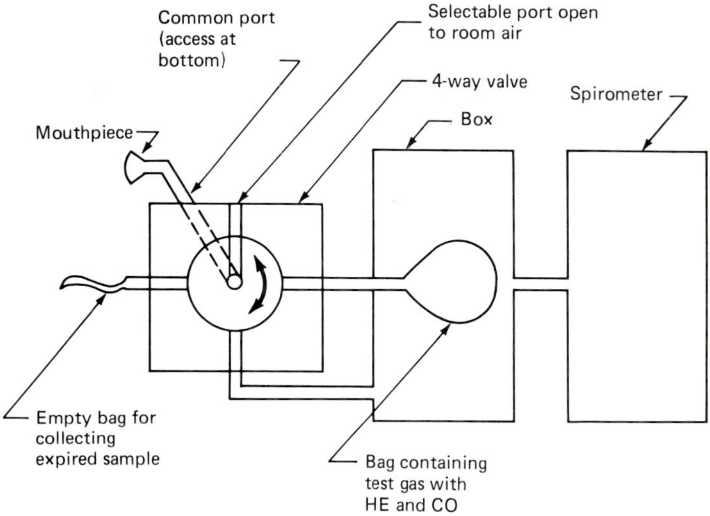

FIG. 36.3 Valve and bag-in-box connections for measuring carbon monoxide diffusing capacity.

system so that the initial exhaled volume can be measured, and the fourth selectable port is connected to the sample bag.

Initial inspired and expired alveolar concentrations of helium (He_I and He_E) and carbon monoxide (CO_I and CO_E) are measured along with the inspired volume (V_I), the duration of the breath-holding period (T), and the barometric pressure (P). Diffusing capacity ($D_{\text{L-CO}}$) is calculated from these measurements by the following equation:

$$D_{\text{L-CO}} = \frac{63 V_I \text{He}_I \ln (\text{He}_E\text{CO}_I/\text{He}_I\text{CO}_E)}{\text{He}_E T(P - 47)} \tag{36.3}$$

36.4.2 Instrumentation

The equipment used in measuring diffusing capacity consists of a spirometer and appropriate valves arranged in a bag-in-box configuration and helium and carbon monoxide analyzers. The spirometer must meet the specifications given earlier. In addition, the resistance to airflow in the inspiratory path and in the two expiratory paths must be less than 0.2 $cmH_2O/(L\cdot s)$ [7].

The operation and specifications for the helium analyzer were discussed above. Carbon monoxide analyzers usually use absorption spectroscopy to detect the concentration of this gas [8]. In this scheme infrared energy is transmitted through two cells, with one containing a standard gas and one containing the test gas. Energy emerging from these two cells is absorbed by a detector with two chambers separated by a diaphragm which is part of a differential pressure transducer. Carbon monoxide in the test gas absorbs the infrared energy, thus reducing the amount of energy incident on the corresponding chamber in the detector. This results in a temperature difference between the two chambers, which causes a pressure difference deforming the diaphragm to produce an electrical signal related to carbon monoxide concentration. Frequently, a condenser microphone serves as the pressure transducer.

The carbon monoxide analyzer should have a range consistent with the concentration in the inspired gas, typically 0.3 percent. Its response should be both linear and accurate within 1 percent of the full-scale reading. Its response to a step change in carbon monoxide should reach

90 percent of its final value in less than 10 s. Detailed techniques for calibrating carbon monoxide analyzers are presented elsewhere [3].

The calibration of the overall system can be verified with a calibrated syringe by filling the syringe from the bag containing the test gas, and then emptying it into the spirometer and the sample bag. After corrections have been made for the dead space of the syringe and mouthpiece, which can be measured by filling it with water, the two measured gas concentrations should be within 2 percent of those in the inspired test gas [7].

36.4.3 Interpretation

Predicted values derived from standard regression equations are used when interpreting measured values of $D_{L\text{-}CO}$. If the measured value is greater than 140 percent of the predicted value, it is considered to be increased, and if it is less than 80 percent of the predicted value, it is considered to be reduced [3]. If the hemoglobin level in the blood is abnormal, then the measured $D_{L\text{-}CO}$ must be corrected by dividing it by 0.0697 times the actual hemoglobin level in grams per 100 mL of blood [3].

36.5 OTHER PULMONARY FUNCTION TESTS

The previous three sections described procedures, instrumentation, and interpretations used with three pulmonary function tests performed routinely in most clinical pulmonary function laboratories. This section discusses several more specialized or developmental approaches. Many of these approaches were developed to detect lung disease at an early stage or to provide quantitative measures of heterogeneous changes. Because of space limitations, many approaches have been omitted, and the presentation of those included is only an overview.

36.5.1 Forced Expired Volume and Flow with a Helium-Oxygen Mixture

When a normal subject performs a maximum forced expiration with a gas that is less dense than air, for example, a mixture containing 80% helium and 20% oxygen, higher flow rates are achieved at high lung volumes. However, as residual volume is approached, differences in flow rates with air and the less-dense gas diminish and eventually disappear. In some patients with obstructive lung disease, a similar density dependence in maximum flow is observed, while in others, no density dependence is observed. The first response is associated with large-airway obstruction and the second with small-airway obstruction.

The degree of density dependence in maximum expiratory flow can be quantified by the ratio of the maximum flow rates with air and the less dense gas at specific lung volumes, for example, the change in $FEF_{50\%}$. In addition, the volume where the flow rates with the two gases become equal, referred to as the *volume of isoflow,* also has been used.

Theoretical explanations for differences in maximum flow rates with air and a less dense gas are not completely satisfactory, but the approach is still considered to have some potential for distinguishing large- and small-airway obstruction. However, additional developmental and theoretical work needs to be completed before this approach becomes accepted. An important practical problem is matching the two forced expirations obtained with the two gas mixtures when the vital capacities differ. When large differences in vital capacities occur, and 5 percent is not unusual, matching the maneuvers at total lung capacity will produce substantially different values for the flow ratios, and especially for the volume of isoflow from the values obtained by matching at residual volume.

36.5.2 Configuration of Forced Expiratory Spirogram

Many experts believe that the configuration of the forced expiratory spirogram and of the maximum expiratory flow-volume curve contains information about the nature of the airway obstruction. There have been several attempts to quantify the configuration of these curves by means of various parameters, including moments and moment-derived parameters [12], slope-ratio parameters [13], and time-constant parameters [14]. These approaches generally use data from the entire vital capacity rather than the few specific points used in computing conventional forced expiratory spirometric parameters. Also parameters defined in these approaches are reported to be sensitive to heterogeneities in the emptying pattern, so that they have the potential for localizing abnormalities or changes in pulmonary function. At the present time, all these configurational analyses must be considered to be in a developmental state because none has been widely tested.

36.5.3 Airway Resistance by Plethysmography

Airway resistance is the ratio of the pressure drop across the airways to the flow through the airways. This flow can be measured directly at the mouth, for example, by using a pneumotachometer. However, measuring the pressure drop across the airway requires a measure of alveolar pressure which can only be obtained indirectly, for example, by using a constant-volume plethysmograph identical to that described earlier for lung volume measurements [11].

Airway resistance by plethysmography is determined from two successive measurements. First the subject pants into the box while flow and box pressure changes are monitored and used to define the ratio of the two, that is, $\Delta P_{box}/\Delta \dot{V}$. After several panting cycles, the shutter is closed, and the changes in box pressure and mouth pressure are monitored and used to calculate the ratio $\Delta P_{mouth}/\Delta P_{box}$. Because the airway is occluded, mouth pressure is assumed to equal alveolar pressure so that this second ratio calibrates the box pressure measurement used in the first ratio, $\Delta P_{box}/\Delta \dot{V}$, in terms of alveolar pressure. Thus airway resistance is found by multiplying the two ratios to obtain $\Delta P_{mouth}/\Delta \dot{V}$, which approximates $\Delta P_{alveolar}/\Delta \dot{V}$, the airway resistance. In practice, the two ratios usually are found by displaying the appropriate signals on an oscilloscope used in an *xy* mode, as noted earlier in the discussion of FRC measurements. In computerized versions of these measurements, the computer monitors the various signals, closes the shutter, and calculates airway resistance, usually by means of linear regression analysis of digitized data to define the two slopes.

Because airway resistance is related theoretically to the length and diameter of the airways and to the physical properties of the gas in the airways, this parameter provides a more mechanistic and direct characterization of mechanical function than either forced expiratory flow and volume parameters or RV, FRC, and TLC measurements. Moreover, measuring airway resistance does not require a maximum inspiration, and so it is free of the complex bronchodilating and bronchoconstricting reflexes associated with that maneuver. It also has the potential advantage of providing explicitly a self-normalized parameter, that is, specific airway conductance, which is obtained by dividing the reciprocal of airway resistance by the lung volume at which resistance was measured. Among normal subjects, specific airway conductance generally is independent of size and lung volume so that a single normal value can be defined, thereby eliminating the need for empirical regression equations.

36.5.4 Dynamic Lung Resistance and Compliance

In order to characterize the dynamic resistance and compliance of the lung, including the airways and lung tissue, the flow at the airway opening, the change in volume, and the pressure surrounding the lung, that is, pleural pressure, must be measured. Flow can be detected by using a standard pneumotachometer, and a volume signal can be obtained by electronically integrating the flow signal. Pleural pressure measurements are more difficult since noninvasive methods are nonexistent, and so esophageal pressure generally is measured and used as an approximation of pleural pressure [15]. Esophageal pressure is measured by placing a balloon-covered catheter in the esophagus and connecting the catheter to a pressure transducer; optimal balloon characteristics, placement, and volume have been defined empirically.

Dynamic pulmonary compliance is obtained from the flow, volume, and pressure waveforms by defining the two points in the cycle where flow is zero and then computing the ratio of the change in volume between the two points to the corresponding change in pressure. Dynamic pulmonary resistance is obtained by defining midinspiration and midexpiration, and then computing the ratio of the change in pressure between these two points to the corresponding change in flow. Resistance and compliance values from 5 to 10 consecutive breaths are averaged to reduce estimate variability.

With a slightly different approach it is possible to compute estimates of lung resistance at any point in the respiratory cycle by calculating the ratio of the magnitude of the flow at that point to the magnitude of the viscous component of pressure. The latter is obtained by subtracting the elastic component of pressure from the total pressure; the elastic component is obtained by dividing the inspired volume by the compliance estimate derived from the two points of zero flow, as noted above.

Dynamic lung resistance and compliance provide separate measures of the mechanical properties of the airways and the parenchyma. Moreover, since these parameters are related theoretically to airway geometry and more fundamental physical properties of the gas and tissue, they provide more direct measures of mechanical function than forced expiratory spirometric parameters and FRC, TLC, and RV measurements. However, the measurement of dynamic lung resistance and compliance requires placement of an esophageal balloon, an inconvenient and sometimes uncomfortable intervention. Also they require fairly sophisticated equipment, namely, a pneumotachometer, two pressure transducers and associated amplifiers, an electronic integrator, and a strip-chart recorder. The calculations are tedious; however, they can be computerized. Finally, the measurements provide no indication of heterogeneities or separate measures of large- and small-airway function, although a much more involved version of this approach has been used to define the frequency dependence of dynamic compliance, which is considered to be a sensitive measure of small-airway disease.

36.5.5 Forced Excitation Respiratory Resistance and Impedance

While high-frequency fluctuation is applied at the airway opening, the resulting pressure across the total respiratory system and the associated flow can be measured at the airway opening. The relationship between the pressure and flow depends on the mechanical properties of the entire respiratory system, including the airways, lung tissue, and chest wall. Thus parameters derived from these two signals can be used to characterize mechanical function [16]. When sinusoidal excitation is used, these pressure and flow signals can be displayed on a strip-chart recorder and visually separated from the superimposed spontaneous breathing component. From the tracing, peak-to-peak pressure and peak-to-peak flow can be determined and used to define the magnitude of the total respiratory impedance. The time interval between the peaks of the sinusoidal components of the pressure and flow signals can be used to calculate the phase angle of this impedance. The real part of this complex impedance, that is, total respiratory resistance, is calculated by multiplying the magnitude of the impedance by the cosine of its phase angle.

In addition to graphical analysis of the chart recording, special signal-processing units and general-purpose computers have been designed to perform these calculations. Nonsinusoidal excitation has been used with Fourier analyzers or general-purpose computers performing spectral analysis before calculations of impedance or resistance are made.

The forced excitation approach requires no special vital capacity or panting maneuvers and little subject contact with the equipment. Even with single-frequency oscillations the approach provides a direct measure of the respiratory resistance, which is, theoretically, related to the geometry of the airway and the viscous properties of the gas and lung tissue. When multiple frequencies are used, the frequency dependence of resistance, which can be quantified by arbitrary parameters such as the ratio or difference in low- and high-frequency resistance measurements, provides an indirect measure of mechanical heterogeneities. Algorithms involving more detailed analysis of this frequency dependence may provide estimates of various mechanical parameters, including central and peripheral resistance, but these algorithms are still developmental.

36.5.6 Single-Breath Nitrogen Washout Measurements

As lung volume approaches residual volume, various units in the lung begin closing. In normal subjects, the range of volume over which these units close is fairly narrow, but in abnormal lungs where heterogeneous changes have occurred, this range is broadened because abnormal units close at a higher lung volume than normal units. On inspiration from RV, heterogeneous disease also spreads the range of lung volumes over which closed units open.

This feature of heterogeneous lung disease can be detected by a single-breath nitrogen washout maneuver where the subject slowly inspires 100% oxygen from RV to TLC and then slowly exhales to RV. On inspiration, the early opening units are filled with dead-space gas while the later opening units receive 100% oxygen so that there is a gradation in nitrogen concentration among lung units, with normal units having a higher nitrogen concentration than abnormal units. During early expiration, all units contribute to the exhaled gas, but as the abnormal units begin closing, a large fraction is supplied by the normal units and so nitrogen concentration in the exhaled gas abruptly increases. This abrupt increase defines the start of the so-called phase IV of the single-breath nitrogen washout. It can be detected by plotting nitrogen concentration measured at the mouth versus expired volume, and then noting the beginning of the phase IV rise. The difference between RV and the volume where phase IV begins, after normalization by the vital capacity, is called the *closing volume. Closing capacity,* the ratio of the volume where phase IV begins to the TLC, also is used to quantify this maneuver, but it requires an independent measurement of RV [17].

Measuring closing volume is a fairly simple procedure requiring standard equipment, including an electronic spirometer, nitrogen analyzer, and *xy* recorder. As noted above, measuring closing capacity requires a separate measurement of RV, which can be accomplished by helium dilution or by plethysmography. Closing volume may provide a sensitive, although indirect, measure of heterogeneous changes in lung function. Interpreting changes in these parameters is difficult because both parameters are increased by a reduction in lung recoil or by an increase in small-airway obstruction.

REFERENCES

1. West, J. B.: *Respiratory Physiology—The Essentials,* 2d ed., Williams & Wilkins, Baltimore, 1979.
2. West, J. B.: *Pulmonary Pathophysiology—The Essentials,* Williams & Wilkins, Baltimore, 1977.
3. Kanner, R. E., and Morris, A. H. (eds.): *Clinical Pulmonary Function Testing,* Intermountain Thoracic Society, Salt Lake City, Utah, 1975.
4. Ruppel, G.: *Manual of Pulmonary Function Testing,* 2d ed., Mosby, St. Louis, 1979.
5. Mead, J.: Problems in interpreting common tests of pulmonary mechanical function, in P. T. Macklem and S. Permutt (eds.), *The Lung in Transition Between Health and Disease,* Marcel Dekker, New York, 1979.
6. Becklake, M. R., and Permutt, S.: Evaluation of tests of lung function for screening for early detection of chronic obstructive lung disease, in P. T. Macklem and S. Permutt (eds.), *The Lung in Transition Between Health and Disease,* Marcel Dekker, New York, 1979.
7. Ferris, B. G.: Epidemiology standardization report, pt. 2, *Am. Rev. Respir. Dis., 118,* 55–88, 1978.
8. Primiano, F. P., Jr.: Measurements of the respiratory system, in J. G. Webster (ed.), *Medical Instrumentation—Application and Design,* Houghton Mifflin, New York, 1978.
9. Gardner, R. M.: Report on the snowbird workshop on standardization of spirometry, *ATS News, 3*(4), 5, 1977.
10. Gardner, R. M., Hankinson, J. L., and Est, B. J.: Testing spirometers—ATS standards, *ATS News, 3*(3), 24, 1977.
11. Leith, D. E., and Mead, J.: *Principles of Body Plethysmography,* National Heart, Lung, and Blood Institute, Washington, D.C., 1974.
12. Permutt, S., and Menkes, H. A.: Spirometry—Analysis of forced expiration within the time domain, in P. T. Macklim and S. Permutt (eds.), *The Lung in the Transition Between Health and Disease,* Marcel Dekker, New York, 1977.
13. Mead, J.: Analysis of the configuration of maximum expiratory flow-volume curves, *J. Appl. Physiol.: Respir. Environ. Exercise Physiol., 44,* 607–613. 1977.
14. Pimmel, R. L., Miller, T. K. III, Fouke, J. M., and Eyles, J. G.: Time-constant histograms from the forced expired volume signal, *J. Appl. Physiol.: Respir. Environ. Exercise Physiol., 51,* 1581–1593, 1981.

15. Macklem, P. T.: *Procedures for Standardized Measurements of Lung Mechanics,* National Heart, Lung, and Blood Institute, Washington, D.C., 1974.

16. Pimmel, R. L., and Fullton, J. M.: Characterizing respiratory mechanics with forced excitation techniques, *Ann. Biomed. Eng., 9,* 475–488, 1981.

17. Engel, L. A., and Farhi, L. E.: Airway closure, closing volume, and changes in inert gas mixing, in P. T. Macklem and S. Permutt (eds.), *The Lung in the Transition Between Health and Disease,* Marcel Dekker, New York, 1977.

CHAPTER 37

Theory and Design of Implantable Cardiac Pacemakers

Wilson Greatbatch
State University of New York at Buffalo
School of Engineering
Buffalo, New York

William M. Chardack
State University of New York at Buffalo
School of Medicine
Buffalo, New York

INTRODUCTION

The concept of pacemaking, which dates back to the early 1930s [1], concerns the delivery of repetitive electrical stimuli to cardiac muscle in the immediate vicinity of the electrode. The remainder of the myocardium is then activated by a propagated action potential. Originally intended for resuscitation of the arrested heart, the concept was applied in 1952 to the treatment of heart block by electric shocks delivered to the chest wall [2]. *Heart block* refers to a block of conduction between auxiliary (atrial) and principal chambers of the heart (ventricles). The former contract at a normal rate and the latter at a slower rate (the idioventricular rate): 30 to 40 beats per minute or lower. Some adaptation to the slow rate occurs through enlargement and hypertrophy of the heart, each stroke propelling a larger volume of blood. As the limit of compensation is reached, perfusion of organs becomes inadequate.

Syncopal attacks (faintness to loss of consciousness) are the typical manifestation of heart block, caused by a sudden reduction of cardiac output and cerebral blood flow. The mechanism may be bradycardia, asystole, ventricular tachycardia, or fibrillation. Each can be lethal. The prognosis of the disease, untreated, is ominous; 50 percent of the patients die within the first year after the diagnosis has been made.

37.1 USE OF PACEMAKERS IN THE TREATMENT OF HEART DISEASE

Chest wall stimulation, because of its painful side effects, was unsatisfactory treatment for heart block on a long-term basis. Some progress was made by wires sutured to the myocardium, brought out through the chest wall (or by transvenously placed electrodes), and connected to external pulse generators. The currents required to pace through such installations are below the threshold of sensation, but any electrical conductor traversing the skin had limitations because of infection, accidental disconnection, and the inconvenience of an externally worn device. A pacemaker powered by a cell rechargeable from an external coil was implanted in a human being in 1958 but failed after a few days [3]. The first successful long-term use of an implanted system carrying its own power supply was reported in 1960 [4] and became the treatment of choice for chronic heart block.

Implantable cardiac pacemakers consist of a pulse generator (power source and electronic circuit), lead, and electrodes. Electrical impulses are transmitted through the lead, an insulated electrical conductor. The electrode is the contact between the implanted system and cardiac tissue. Lead systems may be bipolar (both cathode and anode being on or in the heart) or unipolar (the cathode being myocardial or endocardial and the anode being on the metal housing of the pulse generator) (see Fig. 37.3).

The normal heart is excited into a coordinated contraction by an electrical event (the propagated action potential) traveling along the conduction pathways of the heart (Fig. 37.1). The

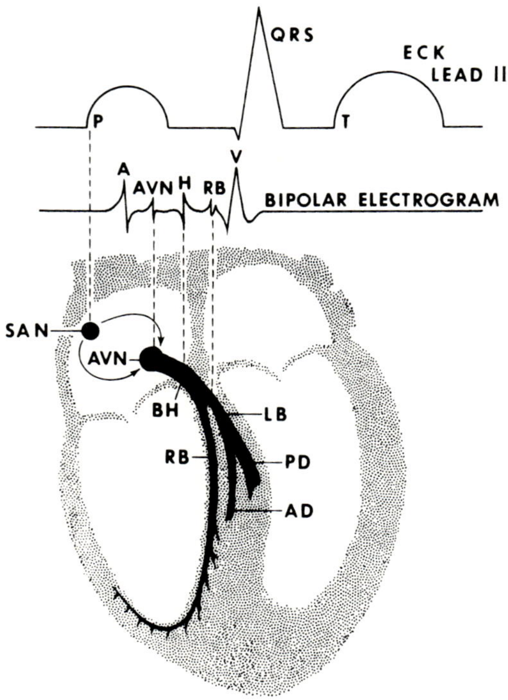

FIG. 37.1 Diagram showing progression of conduction in the heart and the relationship between anatomic structures and the events observed in a limb electrocardiogram and in an electrogram recorded from a transvenously inserted bipolar electrode positioned close to the conduction bundle (see text). SAN = sinoatrial node, AVN = atrioventricular node, BH = bundle of His, RB = right bundle branch, LB = left bundle branch (PD = posterior division; AD = anterior division). [*Reproduced with permission from William M. Chardack, Andrew A. Gage, and Anthony J. Federico, "Cardiac Pacemakers," in David Sabiston (ed.), Christopher-Davis Textbook of Surgery, Saunders, Philadelphia, 1981, chap. 21.*]

natural pacemaker of the heart regulating its rate is located in the sinoatrial (SA) node in the left atrium (Fig. 37.1). The node receives nerve stimuli and responds by adjusting the rate. From the SA node, the impulse spreads through the atrial wall to the atrioventricular (AV) node. Atrial contraction contributes to the filling of the ventricles, the principal pumping chambers of the heart. Activation of the latter takes place by progression of the stimulus from the AV node along tracts of specialized tissue, the bundle of His, which then divides into the left and right branches and several subdivisions (Fig. 37.1). The propagated potential then reaches the peripheral ramifications of the system (Purkinje fibers), activating the bulk of muscle. Following each contraction, cardiac tissue is refractory until repolarization is complete.

Progression of electrical activity of the heart can be recognized on the surface electrocardiogram (Fig. 37.1), which represents the summation of the potentials generated by individual cells. Details of conduction can be obtained by electrodes introduced through a peripheral vein and positioned close to the bundle of His (Fig. 37.1). Placement into the right atrium and ventricle permits the retrieval of electrical events associated with the activation of these chambers (the intracardiac electrogram) (Fig. 37.2). All three electrical manifestations are important in the diagnosis of disease as well as in the design and use of pacemaker instrumentation.

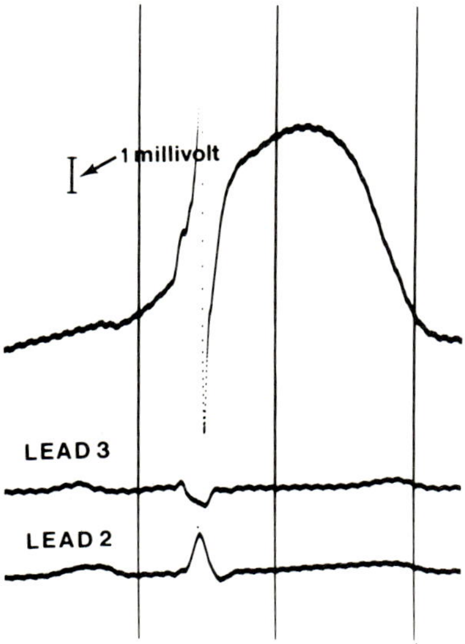

FIG. 37.2 *Upper trace:* Unipolar right ventricular electrogram obtained at initial insertion of the electrode. As the electrode matures (weeks), peak-to-peak amplitude decreases somewhat. Relatively greater decreases may occur in the rate of change *(DV/DT)*. *Lower tracings:* Lead 3 and lead 2 of a surface electrocardiogram.

Pulse generators which have a sensing function must respond to a wide range of right atrial and ventricular electrograms. Most manufacturers use a square sine wave of 4 ms duration from the ventricle and of 10 ms duration from the atrium. The amplitude (sensing threshold) used as a standard falls into an approximate range from 0.5 to 4 mV. In programmable devices, this sensitivity to cardiac input signals is adjustable over this range; in nonprogrammable pulse generators, a midrange value is selected empirically. [*Reproduced with permission from S. Furman, "Physiologic Basis of Cardiac Pacing," in Philip Varriale and Emil A. Naclerio (eds.), Cardiac Pacing—A Concise Guide to Clinical Practice, Lea & Febiger, Philadelphia, 1979.*]

Since the advent of implantable pacemakers, advances led to the development of:

- Flex-resistant leads.
- Long-lived power sources of small size.
- Sophisticated pulse generators based on modern microelectronics. They operate in synchrony with spontaneous cardiac activity.
- Their functional parameters can be programmed, noninvasively, and assessed by telemetry.

Implantable pacemakers are now used to treat bradyarrhythmias and tachyarrhythmias other than heart block. Their utilization has increased considerably. About 100,000 implantations are performed per year (1982) in the United States, and 1 person in 800 (in 1984) has an implantable pacemaker [5]. In the western world, about 120 models are commercially available from over 20 manufacturers.

The surgical installation of an implantable pacemaker system has also changed. In the early era of pacemaking, the lead electrodes were placed on the surface of the heart, which required opening of the chest under general anesthesia. At present, most installations are performed with transvenously introduced endocardiac leads, requiring only local anesthesia. The lead or leads are threaded through a peripheral vein into one or both chambers of the right heart. Typical surgical approaches are illustrated in Fig. 37.3. The principal drawback of endocardiac leads

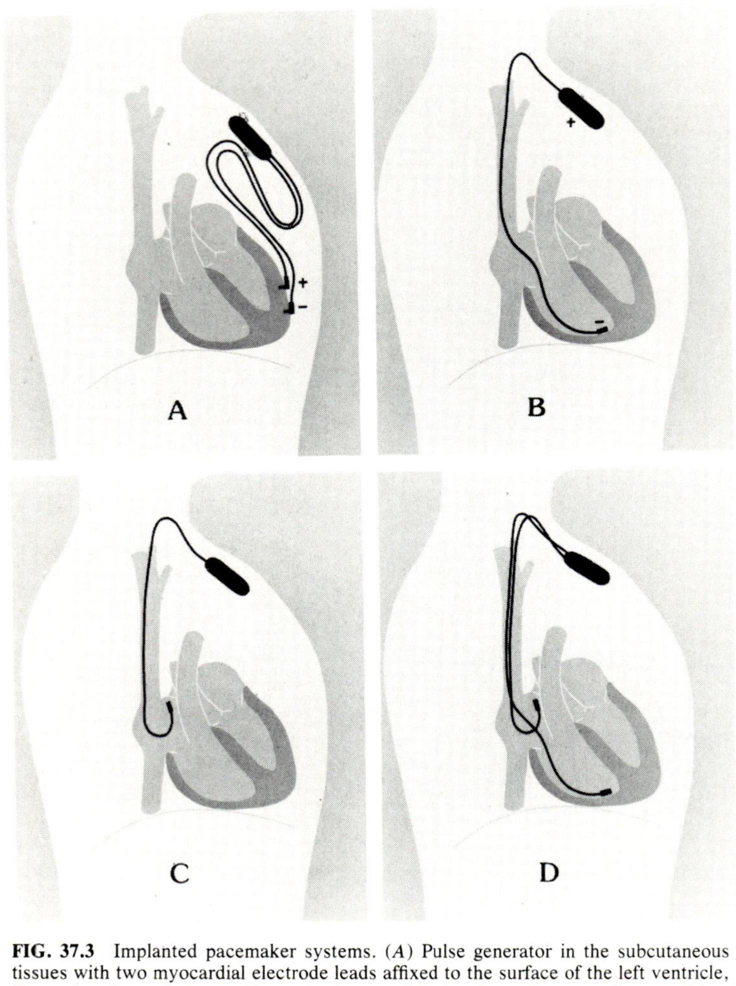

FIG. 37.3 Implanted pacemaker systems. (*A*) Pulse generator in the subcutaneous tissues with two myocardial electrode leads affixed to the surface of the left ventricle, illustrating the typical operative procedure used during the early years of pacemaking. (*B*) Unipolar endocardial system with the electrode (cathode) positioned in the apex of the right ventricle, representative of a VVI system used in the majority of patients treated. (*C*) Unipolar atrial lead electrode hooked into the atrial appendage. (*D*) Atrial and ventricular endocardial leads for sensing and stimulating (both are cathodes) with atrioventricular systems (for example, a DDD system). (See Table 37.1.) [*Reproduced with permission from William M. Chardack, Andrew A. Gage, and Anthony J. Federico, "Cardiac Pacemakers," in David Sabiston (ed.), Christopher-Davis Textbook of Surgery, Saunders, Philadelphia, 1981, chap. 21.*]

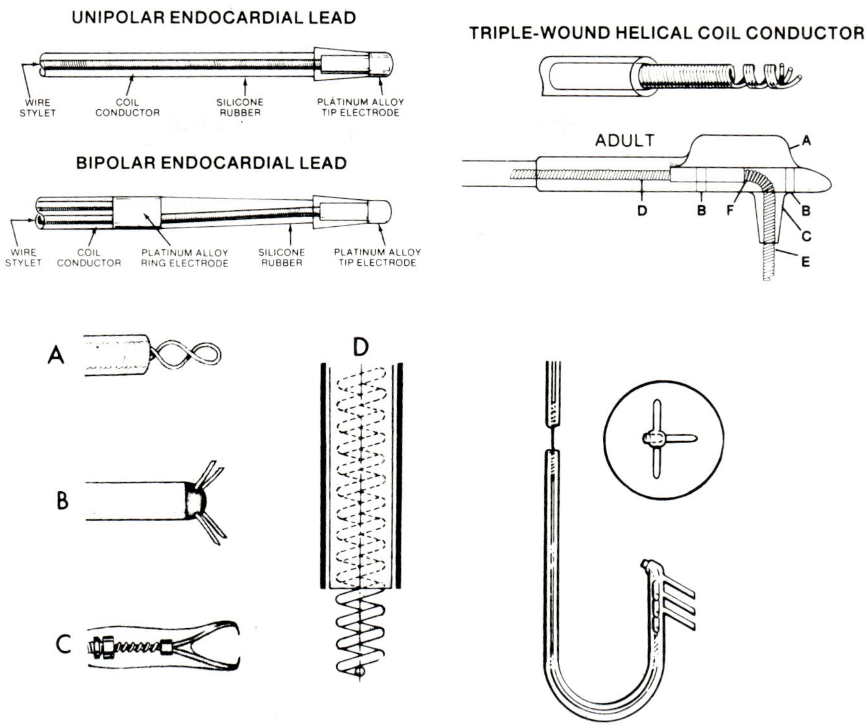

FIG. 37.4 *Top left:* Construction of coiled uni- and bipolar endocardial electrode leads. *Top right:* Triple-wound helical coil conductor (above) and diagram (below) of adult-sized myocardial electrode lead. A = silicone rubber head and sleeve; B = suture holds; C = silicon shield; D = helical coil conductor; E = coil electrode; F = solid reinforced platinum-iridium core. (*Courtesy of Cardiac Pacemakers, Inc., Minneapolis, Minnesota.*)

Bottom left: Diagram of fixation designs for endocardial electrodes. (*A*) Helix (passive). (*B*) Prongs (active). (*C*) Grasping (active). (*D*) Retractable screw (active). *Bottom right:* Tined atrial lead for retention in atrium. Stylet keeps lead straight during introduction. When stylet is withdrawn, lead resumes J shape and hooks into atrial appendage. [*Courtesy of Medtronic, Inc., Minneapolis, Minn. From Paul Citron, "Clinical Experience with a New Transvenous Atrial Lead," Chest, 73, 193, 1978. Reproduced with permission from William M. Chardack, Andrew A. Gage, and Anthony J. Federico, "Cardiac Pacemakers," in David Sabiston (ed.), Christopher-Davis Textbook of Surgery, Saunders, Philadelphia, 1981, chap. 21.*]

was dislodgement. The incidence of this complication has been reduced by the design of tines, hooks, and other means of fixation; a variety of these are illustrated in Fig. 37.4.

37.2 ELECTRIC CURRENT REQUIRED TO STIMULATE

The electrical requirements for acute and chronic stimulation are now well known. A given current density must be achieved at the electrode surface to elicit a propagated action potential. After placement of the electrode, scar tissue develops around it. The periphery of that layer becomes the interface with excitable cardiac tissue. It is considered desirable (at the time of installation) to achieve a margin of approximately 100 percent with respect to the current requirements that are expected to exist at some time after the initial implant.

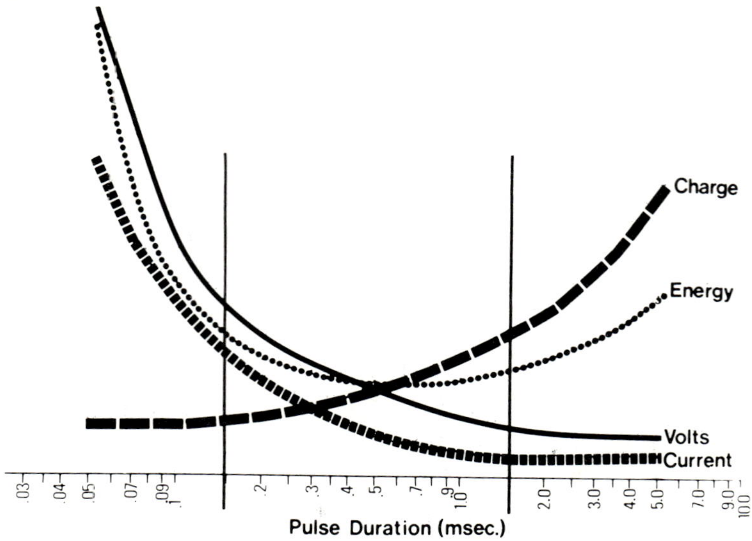

FIG. 37.5 Stimulation thresholds expressed in voltage, current, energy, and charge. Vertical intercepts (center section) indicate range of values of interest for the design of pulse generators. Energy thresholds would indicate a slight advantage at 0.5 ms duration. Charge is an important determinant since, in practice, the available voltage (and therefore energy) is fixed by the characteristics of the battery. [*Reproduced with permission from S. Furman, "Physiologic Basis of Cardiac Pacing," in Philip Varriale and Emil A. Naclerio (eds.), Cardiac Pacing—A Concise Guide to Clinical Practice, Lea & Febiger, Philadelphia, 1979.*]

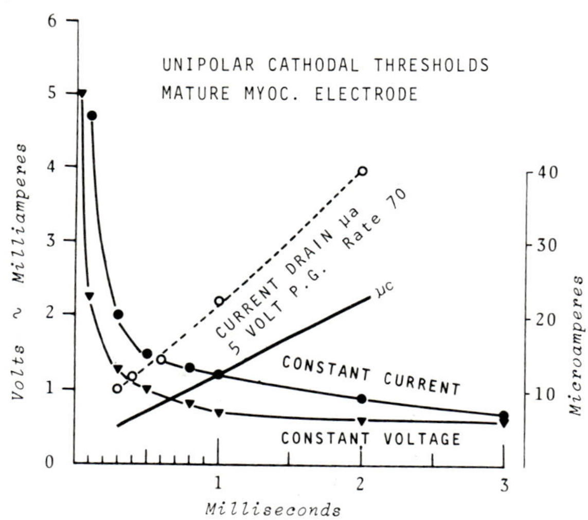

FIG. 37.6 Typical strength-duration curve obtained from a mature epicardial electrode 3 years after implantation. Thresholds were determined by rectangular current and voltage pulses of variable pulse durations. Current and voltage decrease as pulse duration increases. When threshold is expressed in microcolumbs, charge increases with pulse duration, as does current drain (microamperes) (measured from an implanted pulse generator which had an adjustable pulse duration and an output voltage of about 5 V). It is more efficient to pace at shorter pulse widths. [*Reproduced with permission from William M. Chardack, Andrew A. Gage, and Anthony J. Federico, "Cardiac Pacemakers," in David Sabiston (ed.), Christopher-Davis Textbook of Surgery, Saunders, Philadelphia, 1981, chap. 21.*]

The energy required to stimulate is related to a number of factors, including the dimension and geometry of the electrodes, the surrounding scar tissue, the polarization due to electrode material, the ohmic resistance of the lead, electrode, and tissue, and the pulse duration. The latter is the most relevant factor. Charge (current × time) is a function of pulse duration, which determines current consumption and battery life. These relationships are illustrated in Figs. 37.5 and 37.6. Over the years, much work has gone into optimizing electrode size, material, and other factors. At present, available battery capacity has made these efforts less important, and programmability of pulse generators permits the adjustment of electrical output to the minimum adequate for a given patient, thus optimizing battery life.

37.3 PULSE GENERATORS

Pulse generators fall into two categories: asynchronous devices which operate at a fixed rate unrelated to cardiac activity, and noncompetitive pulse generators, which sense cardiac electrical potentials, operate in synchrony with them, and deliver, or withhold, stimuli as required. In 1974, a three-position code was adopted to classify pulse generators. To deal with the increasing complexity of present-day instrumentation, two additional positions have been proposed (Table 37.1) [6]. Representative examples of pulse generators and their functions are shown in Figs. 37.7 and 37.8.

37.3.1 Asynchronous Pulse Generators

Asynchronous fixed-rate pulse generators were the most widely used devices in the early 1960s, when the indication for pacemaker therapy was complete and fixed heart block. Stimuli are delivered to the ventricles (VOO); in some cases delivery to the atrium (AOO) may be indicated.

Fixed-rate pacemakers can lead to competitive rhythms in the presence of spontaneous ventricular extrasystoles, or to the resumption of normal conduction. Asynchronous devices, of great benefit during the early years of pacemaking, are now obsolete.

37.3.2 Noncompetitive Pulse Generators

Noncompetitive pulse generators sense spontaneous cardiac activity (the intracardiac electrogram) and through the same electrode deliver stimuli as required (Fig. 37.2).

Several devices are available. The *ventricular-inhibited demand pacemaker* (VVI) (Figs. 37.7 and 37.8) has been the most commonly used noncompetitive device since the mid-1960s. Its circuitry responds to ventricular QRS potentials that appear at the electrode. The amplified R

TABLE 37.1 Classification of pulse generators by letters used in abbreviations

Position and category				
I. Chamber(s) paced	II. Chamber(s) sensed	III. Mode of response(s)	IV. Programmable functions	V. Special tachyarrhythmia functions
V = Ventricle	V = Ventricle	T = Triggered	P = Programmable (rate and/or output)	B = Bursts
A = Atrium	A = Atrium	I = Inhibited	M = Multiprogrammable	N = Normal rate competition
D = Double	D = Double	D = Double* O = None		S = Scanning
	O = None	R = Reverse	O = None	E = External

Source: Parsonnet et al. [6].
* Atrial-triggered and ventricular-inhibited.

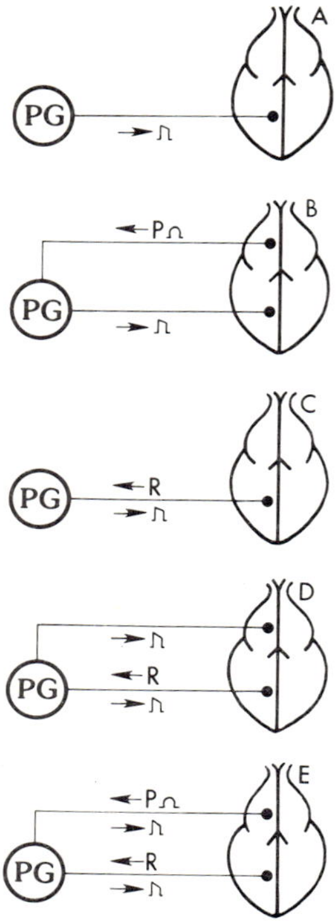

FIG. 37.7 Diagram of the modus operandi of five pulse generators. From top to bottom: PG = pulse generator; P, R, and arrow above line (which represents lead electrode) denote sensed cardiac activity. Arrow below line indicates stimulus delivered. A, VDD; B, VAT; C, VVI; D, DVI; E, fully automatic, DDD. [*Reproduced with permission from William M. Chardack, Andrew A. Gage, and Anthony J. Federico, "Cardiac Pacemakers," in David Sabiston (ed.), Christopher-Davis Textbook of Surgery, Saunders, Philadelphia, 1981, chap. 21.*]

wave resets the timing cycle of the pulse generator. With the device set to 70 pulses per minute, any ventricular depolarization, ectopic or conducted, that appears before the end of the 857-ms interval (RR interval corresponding to a rate of 70 pulses per minute) will suppress the output for another cycle. As long as spontaneous ventricular activity occurs at a beat-to-beat rate exceeding 70 beats per minute, the VVI pacemaker remains quiescent and no stimuli appear on the electrocardiogram (Fig. 37.8). Following sensing of a QRS potential, or after emission of a stimulus, the pulse generator is refractory to another input signal. This refractory period must be long enough to prevent sensing of T waves, and short enough to detect early ectopic spontaneous depolarizations.

During periods of normal conduction, the output of the device is suppressed and the integrity of the system (on standby) must be verifiable. This is accomplished by placing a magnet over the implanted pulse generator to convert the system into an asynchronous mode of stimulation. Stimuli falling outside of the refractory period should then produce a competitive rhythm, or completely capture cardiac activity if the magnet rate is set to a high frequency (90 to 100 beats per minute).

The bifocal atrioventricular sequential pulse generator (DVI) (Fig. 37.7*d*) is a ventricular-suppressed demand device that sequentially activates the atrium and the ventricle with an AV delay of approximately 120 to 150 ms. Two electrodes, atrial and ventricular, are required. Current drain is greater than with single-ventricle pacing.

P-wave synchronous pulse generators (VAT) (Fig. 37.7*b*) were introduced in the early 1960s. Restoring normal, "physiologic" AV conduction has always aroused interest. The device senses atrial potentials which trigger the pulse generator to deliver a stimulus to the ventricle after an appropriate PR delay. VAT pulse generators require protection against abnormal atrial electrical activity. To prevent atrial-triggered tachycardias, a blocking circuit with a 2 : 1 ratio comes into play when atrial rate exceeds a preset limit (between 125 and 150 beats per

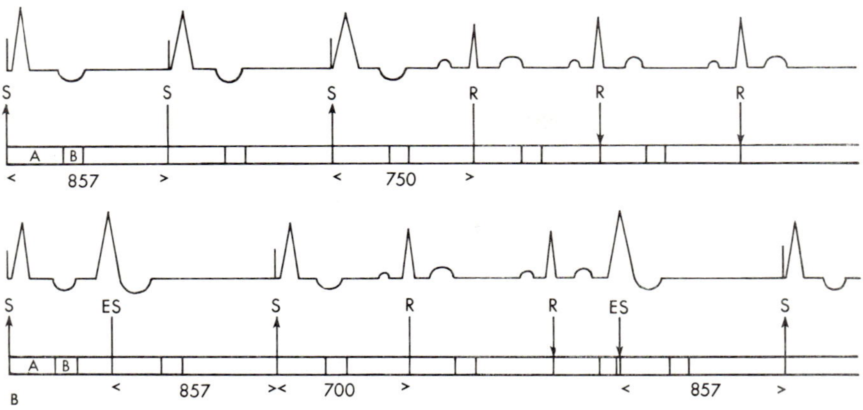

FIG. 37.8 Diagram of function of ventricular-inhibited demand pacemaker (VVI). Upper row shows electrocardiogram, and the bar below illustrates the timing function of the pulse generator circuit. The first three events are three pacemade depolarizations occurring at the basic escape interval of the pulse generator, i.e., at 857 ms, corresponding to a rate of 70 beats per minute. A plus B represents the refractory period of the pulse generator cycle. B is the "noise sampling" period. During this brief period, the pulse generator circuit searches for spurious electrical signals that might originate from interference from external electrical fields. After the third emitted stimulus (S), a conducted beat occurs 750 ms after the last stimulus (S). The R wave of that event is sensed by the pulse generator (inverted arrow) and resets the timing circuit. The last two events also occur at intervals shorter than 850 ms, and thus the pulse generator remains inhibited. The lower row shows the effects of premature extrasystoles (ES). The first complex at the left is a pacemade depolarization. A premature extrasystole occurs (ES), and a new timing cycle of 857 ms is started. Since no spontaneous activity occurs during the cycle, the pulse generator escapes and emits a stimulus (S) after 857 ms. Then, two conducted depolarizations occur (R). The pulse generator remains inhibited and is inhibited again by a premature extrasystole (ES), which starts another timing cycle of 857 ms. Because of a compensatory pause following the extrasystole, no spontaneous activity occurs and the pulse generator escapes after 857 ms and delivers the last stimulus (S). [*Reproduced with permission from William M. Chardack, Andrew A. Gage, and Anthony J. Federico, "Cardiac Pacemakers," in David Sabiston (ed.), Christopher-Davis Textbook of Surgery, Saunders, Philadelphia, 1981, chap. 21.*]

minute). A higher blocking ratio comes into play if the rate increases farther. Other means have been devised to deal with abnormally high atrial rates. Some devices will not exceed a maximum rate and will revert to an asynchronous mode of pacing; others periodically will omit a stimulus after a number of impulses have been delivered at a high rate.

In case of failure of the atrial electrode, or if atrial potentials become inadequate or abnormally slow, the VAT device reverts to an asynchronous mode at a preset rate of approximately 70 beats per minute. If normal conduction occurs, the atrial-triggered stimulus falls into the QRS interval (i.e., into the refractory period.) This is a harmless reaction, but it distorts the electrocardiogram and results in an unnecessary expenditure of battery current.

A more recent version of the P-wave synchronous system combines the atrial-triggered mode with inhibition of the pulse generator by sensing of ventricular depolarizations.

Recent advances in microelectronics have made possible the construction of *"fully automatic" pulse generators* (DDD) (see Figs. 37.7*e* and 37.9 and Table 37.2). They require atrial and ventricular electrodes, sense activity in both chambers, and respond by delivering or withholding stimuli into the atrium or ventricle or both. The design objectives of the device are as follows:

- To remain silent in the presence of normal sinus rhythm
- To stimulate the atrium for correction of sinus bradycardia in the presence of normal AV conduction

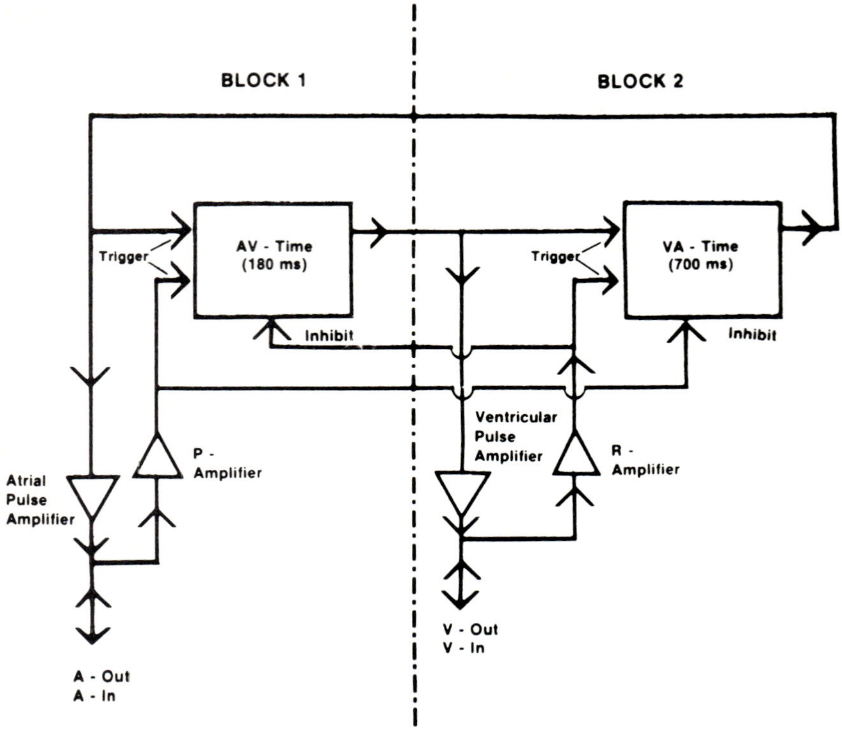

FIG. 37.9 Simplified block diagram of an early DDD phase generator. The behavior of this pulse generator is controlled by interaction between the two sides of the circuits. In block 1, the atrial sensing circuits detect natural or stimulated atrial contractions and trigger a ventricular stimulating pulse if ventricular activity is not sensed within a preset time (180 ms). In block 2, ventricular sensing circuits detect natural or stimulated ventricular contractions and trigger an atrial pulse if no atrial activity is detected after a second preset interval (700 ms) has elapsed. [*Reproduced with permission from Hermann D. Funke, M.D., "Cardiac Pacing with Universal DDD Pulse Generator," in S. S. Barrold and J. Mugica, M.D. (eds.), Third Decade of Cardiac Pacing, Futura, Mount Kisco, N.Y., 1982, chap. 2, p. 191.*]

TABLE 37.2 Approximate range of specifications of currently available pacemakers

Parameter	Specification*
Rate	40–170 beats per minute, in steps or continuously adjustable
Amplitude	Approximately 5 V; 10 V available in special models
Pulse width	0.1–2 ms
Atrial sensitivity	0.75–3 mV
Ventricular sensitivity	1.0–5 mV
Atrial and ventricular refractory period (values still under consideration)	150–300 ms
Programmable pacing modes	Usually DDD (DVI, VVI)

* These are approximate values only and are given only for the purpose of indicating a general range of magnitude.

- To sense the atrial signal (when atrial function is normal) and to deliver appropriately timed stimuli to the ventricle when AV conduction is blocked
- To sequentially stimulate atrium and ventricle for correction of atrial bradycardia in the presence of AV block
- To sense premature ventricular extrasystoles, and to avoid competition by putting the pulse generator into the ventricular- or atrial-inhibited mode

These systems require protection against undesirable high atrial rates. The retrograde transmission of ventricular stimuli back into the atrium has led to difficulties such as pacemaker-induced ventricular tachyarrhythmias. Some of these problems have been dealt with by adjustment of the refractory periods following sensed or emitted stimuli of the PR interval and of the frequency of the device, as well as of the pacing mode. This area requires further clinical evaluation.

Physiologically Controlled Pacemakers. The generic form of rate-responsive pacemakers is the "synchronous" VAT pacemaker first implanted in 1963 [7]. Since the atrium cannot be used in the presence of atrial or sinus node dysfunction, a number of rate-responsive systems have been devised which are based upon implanted sensors monitoring various physiologic parameters. These modalities have recently been reviewed in depth [8]. Two approaches to rate-responsive pacing have reached the stage of restricted experimental clinical use.

One method is based upon monitoring the QT interval as it varies with exercise. Its use in 15 patients was recently reported in Europe [9].

The other system is rate-responsive to body activity, which is monitored by a piezoelectric motion transducer within the pacemaker, identifying exercise by detecting head-foot acceleration [10].

Other approaches have been explored. A pacemaker system using an implanted sensor of the pH of the blood has been used [11], but electrochemical problems with the implanted sensor have been experienced. A number of approaches have been used that are based upon monitoring the respiratory rate by a variety of sensing systems, and a more recent development uses colorimetric detection by a phototransistor of the oxygen saturation of the blood [12]. Neuroelectric potentials on the carotid bodies (related to blood pressure) are the basis for yet another approach, but the placement of an electrode on this small and fragile nerve presents obvious difficulties. The use of the body temperature as a rate-control parameter was first used a decade ago [13] and has been used subsequently in several systems. The exact indications, the usefulness, and the preferred modality of nonatrial rate-responsive pacing remain to be determined.

Although a discussion of physiologic pacing is beyond the scope of this review, there are some problems that must still be resolved. Protection against high atrial rates has been mentioned, but because AV synchrony is of greatest importance at high atrial rates, protection against the latter is in conflict with the design objective of the device. The mean age of patients requiring a pacemaker is 70 years, and it is questionable whether high rates are necessary in the age group of 70 and above because physical work performance is frequently limited by other factors before the cardiovascular system is stressed. Pacemaking with much simpler devices (VVI demand pacemaker) has produced excellent clinical results. Over two decades, the survival rate of patients has virtually equaled that of a normal population of similar age and sex. Work performance, although limited, has been adequate. The incremental benefits, if any, of a vastly more complex device must be evaluated against the increase in cost. In a few patients, the alternation between periods of normal sinus rhythm and ventricular pacing has produced symptoms (reduced cardiac output and pressure during ventricular pacing). Usually these complications can be anticipated before operation.

37.3.3 Programmability of Pulse Generators

Noninvasive programmability of functional parameters of implanted pacing devices can now be implemented with all types of pulse generators. This programmability is achieved by coded magnetic, electromagnetic, or radio-frequency pulses delivered from an externally applied programmer (Fig. 37.10). There is general agreement about the desirability of adjusting amplitude and rate, and these objectives had indeed been reached by more primitive means in the early days of pacemaking.

Programmability of other parameters, such as sensitivity to intracardiac potentials, the refractory period, AV delay, and pacing mode, is a feature in a number of commercially

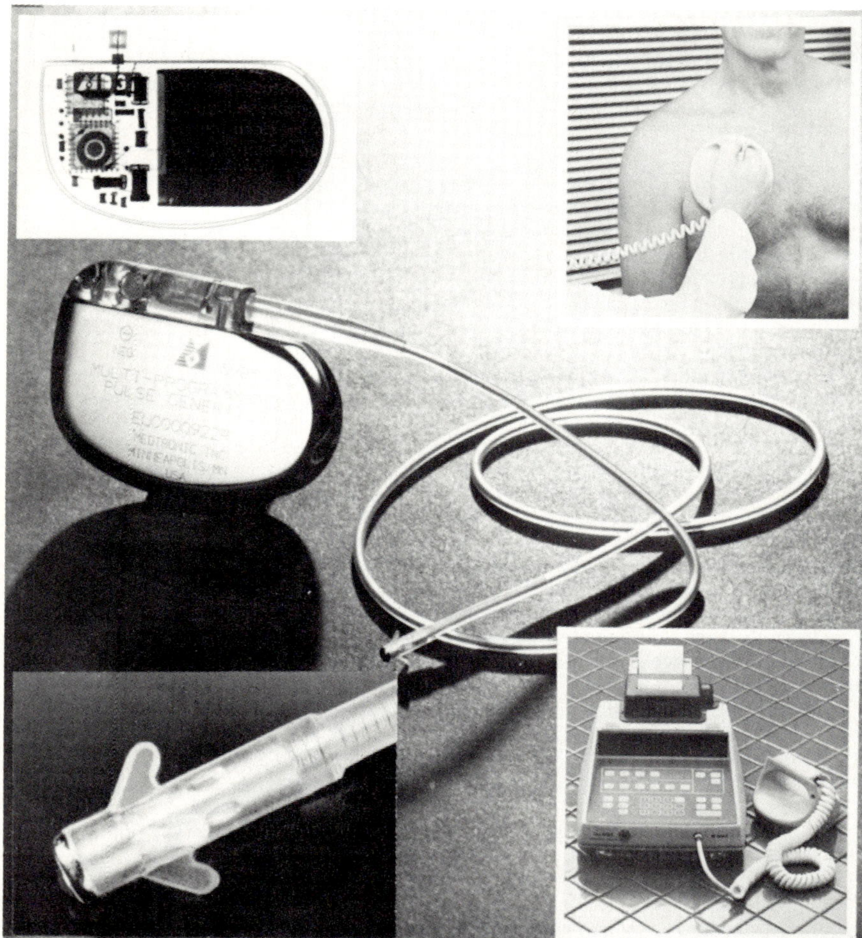

FIG. 37.10 Programmable implanted system, operating through transmission of coded radio-frequency pulses. *Center:* Pulse generator (Spectrax, manufactured by Medtronic, Inc., Minneapolis, Minn.) and unipolar "tined" lead electrode (enlarged in photograph, lower left). Tines are designed for endocardial fixation. *Lower right:* Programmer. Adjustable parameters are pulse duration, rate amplitude, R-wave sensitivity, and so forth (see text). *Upper right:* Application of programmer. *Upper left:* X-ray showing construction of pulse generator lithium cell on right, hermetically sealed. Hybrid circuitry on left: major integrated circuits are hermetically sealed. Pulse generator housing (titanium) is also hermetically sealed. (Similar programmable pulse generators are commercially available from Cardiac Pacemakers, Inc., Minneapolis, Minn.; Intermedics, Inc., Freeport, Tex.; Pacesetter Systems, Inc.; and others.) (*Courtesy of Medtronic, Inc., Minneapolis, Minn.*)

available devices. It is also possible to use telemetry from the implanted device to transmit data such as current drain from the battery and the sensed endocardiac electrogram.

37.3.4 Pacemakers with Antiarrhythmic Capabilities

Ventricular pacing at high rates has been used for many years to suppress premature ventricular extrasystoles, runs of repetitive extrasystoles, and episodes of prolonged ventricular trachycardia. In recent years, interest has focused on a number of supraventricular and ventricular tachycardias and their treatment by implanted pacemakers delivering timed stimuli either singly or in bursts to the atrium or ventricle. Some devices must be activated by the patient or physician; others scan cardiac electrical activity and operate automatically. This type of electrophysiologic intervention is still under evaluation.

37.3.5 Pulse Generator Design

The first generation of pulse generators was based on the advent of the mercury-zinc (Rubin-Mallory) cell and the transistor (Fig. 37.11). The former used a highly corrosive electrolyte, and hydrogen was a by-product of the electrode chemical reaction. Power source and circuitry had

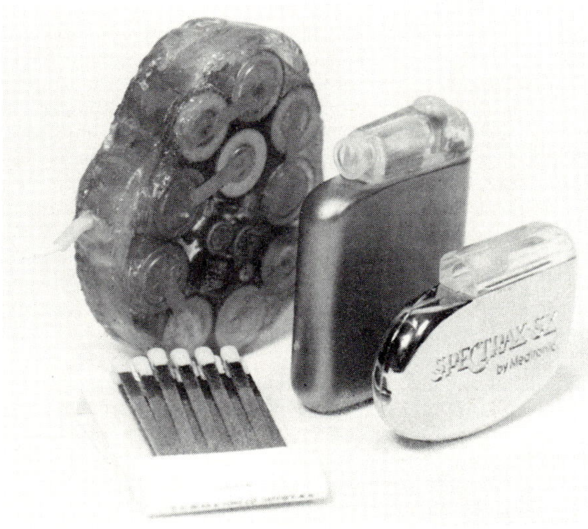

FIG. 37.11 *Left:* Pulse generator used in first successful long-term correction of heart block in a human being (1960). It was developed and constructed by our group [2]. Mercury cells and circuitry in center are cast in epoxy. Outer silicone rubber capsule has been removed. *Middle:* Multiprogrammable pulse generator of recent manufacture (Pacesetter Systems, Inc.) with telemetry capability, including transmission of current drain on lithium battery (weight, 56 g). *Right:* Another example of a multiprogrammable pulse generator (Medtronic, Inc.) (weight, 45 g). Note decrease in size and weight over 20 years while pulse generator longevity and sophistication have increased (see text). Further reduction in size at the expense of battery capacity is not warranted. [*Reproduced with permission from William M. Chardack, Andrew A. Gage, and Anthony J. Federico, "Cardiac Pacemakers," in David Sabiston (ed.), Christopher-Davis Textbook of Surgery, Saunders, Philadelphia, 1981, chap. 21.*]

to be encapsulated in epoxy (with or without an outer silicone rubber layer) which was permeable to gas, including water vapor. The system had to operate in body water. Since permeability did not extend to salt ions and organic molecules, the environment was that of distilled water, noncorrosive and nonconductive. However, vulnerability of the package stemmed from salt ion contamination during manufacture. In spite of these disadvantages (including a power source that could not be hermetically sealed), some pulse generators reached a 5-year life span.

The circuit of the first successfully implanted pulse generator was a two-transistor blocking oscillator [14]. With the advent of the demand pacemaker (VVI), the number of transistors increased to seven. The addition of refractory and interference rejection circuits raised this number to several dozen. Programmability and telemetric function required integrated circuits, at first consisting of the addition of separate transistor chips and culminating later in monolithic circuits with all components and wires diffused onto a ceramic substrate. A recent report described a pulse generator consisting of a complete microprocessing chip capable of storing data on the patient, power source, electrode impedance, and circuitry. This complex information could be asked for and transmitted by telemetry. Current technology can provide a monolithic chip for the pulse generator which has the memory and computing capability of a standard desktop computer. The remaining problem is to match these capabilities in a cost-effective way to the requirements of clinical application.

37.3.5 Clinical Aspects of Pacemaking

Originally, pacemakers were developed for the treatment of symptomatic fixed and complete, acquired heart block. Today diagnosis is made and treatment is administered at a much earlier phase of the disease (for instance, intermittent block and precursor conditions). Heart block accounts for approximately 40 percent of implantations, and an equal percentage is performed for a group of entities categorized as *sinus disease*. These include sinus arrest, sinus slowing, sinoatrial block, and the brady-tachycardia syndrome. Heart block and sinus disease affect the same age group (mean age, 70 years). Approximately 15 to 20 percent of implantations are performed for congenital heart block, conduction disturbances complicating cardiac operations, and a number of other rare entities.

In advanced heart block, pacing prevents syncopal attacks and relieves congestive failure, as well as cerebral and renal impairment due to low cardiac output. Improvement may be striking in many patients but less so in others, depending upon the degree of coexisting cardiac disease. Symptomatic relief is obtained by pacing in bradyarrhythmias and tachyarrhythmias not due to block.

Data covering more than two decades show that pacing prolongs survival rates, and the life expectancy of patients with pacemakers approaches that of the normal population. The lifetime of the pulse generator has been improved and now exceeds the average life expectancy of patients with pacemakers. Pulse generator replacements are performed under local anesthesia (as is the initial procedure), and the risk of the operation is minimal compared with that associated with other surgical interventions on the cardiovascular system.

37.4 POWER SOURCES

37.4.1 The Mercury-Zinc Cell

The first fully implantable pacemakers (1960 to 1970) were based on the Rubin-Mallory zinc oxide–mercuric battery. This cell was modified for medical application by employing a slightly less active electrolyte system. Magnesium dioxide was added to inhibit internal cell discharge in storage. The electrolyte was sodium hydroxide, which is highly corrosive, and the separator consisted of layers of webril, an organic material. Zinc oxide tended to migrate through the cell, forming internal conductive pathways which drained the cell. The electrochemical reaction produced hydrogen gas which had to be vented through a neoprene seal, precluding hermeticity not only of the battery but of the entire pulse generator structure.

Initially, a 5-year battery lifetime was predicted on the basis of data supplied by the manufacturer. These data had not been obtained under conditions prevailing in the body (temperature of 98°F and humidity of 100 percent) since there had been no conceivable reason for the manufacturer to conduct such tests. Because of the high current output of early pacemakers and the unforeseen internal losses in the mercury cell, the battery became the limiting factor, reducing the lifetime of the pulse generator to approximately 2 years. Today the mercury cell is obsolete, but it should not be forgotten that it made possible the construction of implanted pacemaker systems which have saved and prolonged countless lives, requiring only a periodic and relatively innocuous replacement of the pulse generator performed under local anesthesia.

37.4.2 Biological Batteries

The body contains many of the chemicals necessary for the electrochemical generation of electricity. Zinc has been mostly used as an anode in biogalvanic cells, and fuel cells have been constructed. However, no biological battery system has been developed to the point of clinical use.

37.4.3 Rechargeable Batteries

Rechargeable systems employing nickel-cadmium batteries have been implanted. The concept of a small battery, periodically rechargeable from an external source, is attractive, but two requirements must be met. First, the lifetime of the battery (with recharging) must exceed the lifetime of a primary cell of comparable size without recharging. Second, it must be acceptable to physician and patient to give the latter control of the recharging process. Because of these problems, the rechargeable system has achieved only very limited acceptance.

37.4.4 Nuclear Batteries

Although several radioactive materials have been considered as a power source for an implanted pulse generator, only plutonium 238 has proved to be practical. It has a half-life of 87 years and degrades in an exponential and totally predictable fashion, generating heat which is then used to generate electricity through thermocouples. The first implantation of such a system was performed in 1970. Second-generation nuclear powered pacemakers are comparable in size to conventional power sources and have an estimated system lifetime exceeding 60 years. About 2500 isotopic pulse generators have been implanted in the last 12 years (about 1500 in the United States). Retrieval and follow-up of these systems have been virtually complete, and not a single failure caused by the power source has come to light. Under the most stringent controls, the nuclear-powered pulse generator has had the best performance record [15] and meets all the requirements for a lifetime pulse generator. Yet, because of the cumbersome regulatory restrictions imposed on the physician and the patient and the generally prevailing antinuclear sentiment of public opinion, comparatively few of these devices have been used.

37.4.5 Lithium Batteries

Introduced in 1971 [16], the lithium battery has added a new dimension to the longevity and performance of the implanted pacemaker system. The battery, now used in virtually all pacemakers, is the lithium-iodine cell. Lithium is the most active of the alkali metals and is relatively easily handled. The various systems differ from each other by the material used for the cathode.

Lithium-Iodine Batteries. The chemical reactions involved in the lithium-iodine battery are as follows:

$$\mathrm{Li} \rightarrow \mathrm{Li}^+ + e^-$$

$$\mathrm{I}_2 + 2e^- \rightarrow 2\mathrm{I}^-$$

The combined reaction is

$$2Li + I_2 \rightarrow 2\ LiI$$

The cathode material is a complex of iodine and polyvinyl-2-pyridine (PVP-2), neither of which is a conductor of electricity. When mixed together and held at 300°C for 3 days, they react into a black viscous paste which is electronically conductive. At the beginning of life, the lithium iodide electrolyte impedance rises more or less linearly during the first phase of discharge and is much greater in magnitude than the cathode impedance. In the second phase of discharge, the cathode becomes starved of iodine and develops a higher impedance than the lithium-iodine electrolyte. This produces the shoulder of the discharge curve, after which the loaded voltage falls more rapidly with discharge. If the load is removed for a reasonable period (3 days) during phase I or phase II of discharge, the open-circuit voltage (OCV) of the cell will return to within millivolts of its original 2.8-V level. During the third phase, the concentration of available iodine in the cathode has dropped so low that the Nernst equation is no longer satisfied for the 2.8-V level and the OCV then begins to drop. This third phase occurs well beyond the normal design end of life of the cell.

The three principal problems associated with zinc-mercury batteries are nonexistent with the lithium-iodine cell. (1) There is no gas generation. (2) Because of this, the lithium-iodine cell can be hermetically sealed in a metal can with a glass-metal feed-through. No body moisture can get in and no battery effluents can get out. A true hermetic seal will not pass more than 30×10^{-9} cm^3 of helium per second, after an overnight bomb at 60 lb/in^2 (gage) of helium. No organic material will meet this test. Thus, a true hermetic seal can be achieved only with glass, metal, or ceramic materials. (3) The lithium-iodine cell has no fabricated separator. When lithium comes into contact with the iodine complex, a monomolecular layer of lithium iodide spontaneously forms. Should it be pierced or shattered for any reason, new lithium iodide will form to replace it. The separator is thus self-healing. These three factors permit a pacemaker cell which has proved to be over 10 times more reliable than the older zinc-mercury battery.

The Lithium–Silver Chromate Cell. The lithium–silver chromate cell uses a cathode of silver chromate and a polypropylene separator, and has an OCV of 3.35 V. The liquid electrolyte is a lithium perchlorate solution in an organic solvent. The reaction is

$$2Li + Ag_2CrO_4 \rightarrow Li_2CrO_4 + 2Ag$$

Two cells are usually used in parallel to achieve the desired capacity. The system has been used by European manufacturers.

The Lithium–Copper Sulfide Cell. The OCV is about 2V, and thus three cells are usually used in series. This cell too requires a fabricated separator. The reactions are as follows. For the anode,

$$Li \rightarrow Li^+ + e^-$$

The cathode discharges in two steps:

$$2CuS + 2e^- \rightarrow Cu^+ + S^{2-}$$

$$Cu_2S + 2e^- \rightarrow 2Cu + S^{2-}$$

The second reaction creates a lower plateau voltage at 1.75 V during the last portion of life, providing an end-of-life warning indicator.

The overall cell reactions are as follows:

$$2CuS + 2Li \rightarrow Cu_2S + Li_2S \qquad V = 2.12\ V$$

$$Cu_2S + 2Li \rightarrow 2CU + Li_2S \qquad V = 1.75\ V$$

37.5 DESIGN OF ELECTRODES

The first electrodes used were multistranded stainless steel wires. Within the first year of clinical use, electrode disruption from metal fatigue becomes apparent. In 1961 [17], the helical coil spring myocardial electrode (90% platinum, 10% iridium) was introduced, and the helical design is still utilized in most current leads (Fig. 37.4). Some are multifilar in construction, reducing electrical resistance.

Alloys presently used in lead construction are flex- and corrosion-resistant. Commonly employed alloys are Elgiloy (with cobalt as a major constituent) and MP-35 (which does not contain iron). The coil lead terminates at the electrode, which is made of platinum or one of its alloys. Porous materials are also used to increase surface area and adhesion of the electrode to cardiac tissue. The platinum terminal can be alloyed up to 30 percent with iridium, rhodium, or palladium without appreciable sacrifice of coupling properties. Nonnoble metals cannot be used as small surface anodes because of corrosion.

Electrode leads are insulated in polyethylene, silicone rubber (Silastic), or polyurethane sleeves. The sleeve is heat-sealed to both ends of the lead. One terminates at the electrode, the other at the connector. The connector–pulse generator junction is a pressure seal, and although the seal is not hermetic, the potential current leakage path is insignificant. Current leakage through rupture of the sleeve (accidental, from manufacturing defects, or biodegradation) can lead to undesirable stimuli of adjacent muscle tissue and battery depletion. The latter problem recently has been experienced with newly introduced polyurethane leads.

The electrochemistry of the electrodes is important because electricity flows through metals by means of electrons and by ions through fluids in body tissues. This implies a chemical reaction at the interface between electrode and body. The reaction differs at the cathode and the anode and takes place in a boundary layer of a few micrometers in thickness.

37.6 ELECTRICAL HAZARDS OF PACEMAKING

37.6.1 Hazards Arising from the Pacemaker Stimulus

Since the advent of pacing, there has been concern about inducing ventricular fibrillation (uncoordinated and mechanically ineffective contraction of cardiac muscle fibers) by the pacemaking stimulus itself. This concern does not exist so long as the cardiac rhythm is dominated by the artificial pacemaker, because all stimuli are delivered in late diastole beyond the "vulnerable period" which coincides with the apex of the T wave. Delivery of pacemaker stimuli into the vulnerable period can occur during competitive rhythms and be caused by various malfunctions of a pulse generator. Experimental data show that, in a normal myocardium, ventricular fibrillation can be induced by stimuli delivered into the vulnerable period, but the currents to do so exceed the output current amplitudes of pacemakers by one to two orders of magnitude. There has been no difference in the number of sudden unexplained deaths, conceivably due to induced fibrillation, in patients with competitive devices as compared with a similar patient population with noncompetitive devices. One might conclude that this discussion is irrelevant since today's devices are noncompetitive; however, some external as well as implantable units designed to control tachyarrhythmias are intentionally competitive.

In summary, electrical parasystole may be hazardous when myocardial vulnerability is increased. Relevant examples are temporary pacing in the course of myocardial infarction, after cardiac surgery, during the delivery of stimuli for diagnostic purposes, and with implanted pulse generators devised to deliver intentionally competitive stimuli to interrupt tachyarrhythmias.

37.6.2 Leakage Currents from Line-Powered Instruments

Pacing electrodes are low-resistance current pathways to the myocardium, and currents of the order of 100 μA can induce ventricular fibrillation. This danger exists when pacing leads are installed or exteriorized during pulse generator replacements. Only battery-driven devices should be used under such circumstances, and all terminals and connections must be protected against contact with potential sources of line current.

37.6.3 Hazards Arising from External Electric and Electromagnetic Fields

There is considerable electrical and electromagnetic "noise" in the present-day environment, and the effects of these fields on implantable pulse generators have been studied extensively. They vary with the source of interference and the type of pulse generator. Ventricular-programmed pacemakers, which must respond to QRS potentials of a few millivolts, are more sensitive to external fields; interference with their output has been observed when patients are exposed to radiated electromagnetic energy from sources such as radio and television transmitters, microwave ovens, diathermy, automobile ignition, electric razors, and fluorescent lights. Because of the antenna configuration of unipolar leads, they are more vulnerable than bipolar leads. Ventricular-programmed pacemakers also can be interfered with by conducted electricity, such as 60-Hz leakage currents (which may be below the level of sensory perception) from household appliances and machinery. Effective protection of modern ventricular-programmed pulse generators against radiated fields is obtained through shielding of any metal housing, by electronic filters, and by digital access codes.

In the presence of strong alternating conduction currents or magnetic fields, the demand pulse generator will revert to fixed-rate operation. In a few instances, extremely low conducted currents and magnetic fields have been observed to cause reversion either to an asynchronous mode or to a low stimulation rate or, in some cases, to complete inhibition. Consequently, pacemaker wearers should be discouraged from operating appliances such as magnetic tape erasers, tool demagnetizers, large transformer-driven devices, arc welding apparatus, and television transmitters. If they must operate such equipment, they should do so initially in the presence of another person capable of removing them to some distance from the device. These electromagnetic interferences affect the pulse generator only temporarily, and normal function will resume as soon as contact with the source is broken. In the past, instances of permanent damage have been exceptional and have resulted mostly from the application of defibrillating shocks.

37.6.4 Hazards of the Hospital Environment

Paradoxically, hazards to patients with pacemakers exist in the medical environment. Notable examples are the use of 60-Hz devices on externalized electrodes or the use of electrosurgical and diathermy apparatus on patients with implanted pulse generators. Diathermy should never be used on body areas overlying an implanted pulse generator. If the latter is shielded by the metal, severe localized heating of tissue adjacent to the metal capsule can occur. The device itself may be permanently damaged by the heating effects of diathermy. In the use of electrosurgical apparatus, the administration of electroconvulsive therapy, and the use of certain dental instrumentation, interference with the pulse generator must be guarded against. Electrosurgical current should be applied in bursts separated by several seconds, and the pulse should be monitored. Many successful operations on patients with pacemakers have been performed with these precautions. Cautery and cutting currents should not be used within 2 in of implanted pulse generators.

37.6.5 Effects of Ionizing Radiation

The effects of ionizing radiation on implantable pacemakers have changed with the advent of newer designs. Early devices using discrete components and large-area junction semiconductors were immune to the levels of radiation employed by diagnostic and therapeutic radiation. Newer pacemaker designs use field-effect semiconductor chips (metal oxide semiconductor field-effect transistors, or MOSFETs) having a very small area. Recent reports [18] deal with several mechanisms which can cause temporary malfunction and permanent damage to CMOS (complementary metal oxide semiconductor) chips at radiation levels as low as 1000 rad. Radiation therapy commonly exceeds those levels by a factor of 4 to 6.

37.6.6 Interference from Muscle Potentials

Muscle activity in the body generates electrical potentials which fall within the passband of the sensing amplifier of the demand pulse generator. Unipolar electrode configurations are more sensitive (at least by one order of magnitude) to such interference signals than bipolar configurations. The recent literature [19] contains an increasing number of reports dealing with suppression of pulse generator activity by electrical potentials arising in the body, and the sole use of bipolar lead configurations are advocated by some groups.

It is fair to state that with the very rapid developments that are now occurring in pacemaker-related electronic technology, stringent and continuous testing of all new designs has become an absolute necessity and obligation. This process of evaluation must be an ongoing one, and results of tests performed on earlier generations of devices cannot be relied upon. A typical example is the behavior of implanted pacemakers when exposed to nuclear magnetic resonance scanners.

37.7 END-OF-BATTERY-LIFE SIGNALS

The indication of approaching end of life of a battery is far more reliable with the newer lithium power sources than it was with the older zinc-mercury systems. Lithium-iodine batteries exhibit a predictable increase in resistance with depletion which can be converted into a rate change of a few beats per minute. Lithium–silver chromate and lithium–copper sulfide batteries both exhibit a voltage plateau later in life which can similarly be converted into a rate change. Pulse generators of current manufacture provide an end-of-life signal several months ahead of battery exhaustion. This signal may be a decrease of the rate, an increase of the pulse width, or both.

In older systems, the "runaway pacemaker" was a possibility from failure of any of the rate-determining components or from fluid leakage into their environment. Modern designs incorporate some sort of rate-limiting features. Most use a crystal clock with a mechanically oscillating quartz crystal that can physically oscillate at only one frequency. Also, a software function is often included to drop beats if needed, thus slowing the rate. Such a circuit might come into play in a VAT or a DDD pacemaker which senses an atrial tachycardia.

SUMMARY AND CONCLUSIONS

Most of the technical problems of implantable pacemakers have been solved. The lithium cell provides a power source that can be hermetically sealed and allows for the construction of a second hermetic enclosure of the electronic circuitry. The capacity of the cell is adequate to fulfill the requirements of approximately 90 percent of patients whose mean age is 70 years. Radioisotope-powered pacemakers, although rarely used, have a projected pulse generator life of at least 60 years.

The flex resistance, as well as methods of fixation, of the lead electrodes has been greatly improved, although the limits of their useful life have not been completely established.

The availability of modern microcircuitry has provided the technical basis for an almost unlimited combination of function, adjustability, and monitoring of implantable pulse generators. The problems to be solved relate to matching available technological capabilities in a cost-effective way to the requirements of clinical application.

REFERENCES

1. Hyman, A. S.: Resuscitation of the stopped heart by intracardiac therapy: II. Experimental use of an artificial pacemaker, *Arch. Intern. Med., 50,* 283, 1932.
2. Zoll, P. M.: Resuscitation of the heart in ventricular standstill by external electric stimulation, *N. Engl. J. Med., 274,* 768, 1952.
3. Lagergren, H.: How it happened: My recollection of early pacing, *PACE, 1,* 140, 1978.
4. Chardack, W. M., Gage, A. A., and Greatbatch, W.: A transistorized, self-contained, implantable pacemaker for the long-term correction of complete heart block, *Surgery, 48,* 643, 1960.
5. Parsonnet, V.: The proliferation of cardiac pacing: Medical, technical, and socioeconomic dilemmas, *Circulation, 65*(5), 841–845, 1982.
6. Parsonnet, V., Furman, S., and Smyth, N. P. D.: A revised code for pacemaker identification, *PACE, 4*(4), 400, 1981.
7. Nathan, D., Center, S., Wu, C., and Keller, W.: An implantable synchronous pacemaker for the long-term correction of complete heart block, *Circulation, 27,* 682, 1963.
8. Geddes, L., Fearnot, N., and Smith, H.: The exercise-responsive pacemaker, *IEEE/BMES Trans., BME-31*(12), 733–760, 1984.
9. Donaldson, R., and Rickards, A.: Rate responsive pacemaker using the evoked QT principle, *PACE, 6*(6), 1344, 1983.
10. Anderson, K., and Brumwell, D.: Rate-variable pacemaker, U.S. Patent 4,428,378, Jan. 31, 1984.
11. Camilli, L.: A new pacemaker autoregulating the rate of pacing in relation to metabolic needs, *5th International Symposium Excerpta Medica,* Amsterdam, 1977, p. 414.
12. Wirtzfeld, A., Goedel-Meinen, L., Beck, T., et al.: Central venous oxygen saturation for the control of automatic rate-responsive pacing, *PACE, 5,* 829, 1982.
13. Fischell, R.: Fixed-rate rechargeable pacemaker, U.S. Patent 3,867,950, Feb. 25, 1975.
14. Greatbatch, W., and Chardack, W.: A transistorized implantable pacemaker for the long-term correction of atrioventricular block, *Trans NEREM Conf.,* Boston, 1959, vol. 1, p. 8.
15. Bilitch, M., Hauser, R. G., Goldman, B. S., Furman, S., and Parsonnet, V.: Performance of cardiac pacemaker pulse generators, *PACE, 5,* 788, 1982.
16. Greatbatch, W., Lee, J. H., Mathias, W., Eldridge, M., Moser, J. R., and Schneider, A. A.: The solid-state lithium battery: A new improved chemical power source for implantable cardiac pacemakers, *IEEE Trans. Biomed. Eng., BME18*(5), 317–324, 1971.
17. Chardack, W. M., Gage, A. A., and Greatbatch, W.: Correction of complete heart block by a self-contained and subcutaneously implanted pacemaker, *J. Thorac. Cardiovasc. Surg., 42,* 814, 1961.
18. *Symposium on Radiation Effects on Implanted Pacemakers, PACE, 5*(2), 146–160, 1982.
19. *Symposium on Electromagnetic Interference of Muscular Origin, PACE, 5*(1), 4–9, 1982.

CHAPTER 38

Circulatory Assist Devices

Henry M. Spotnitz, M.D.
Department of Surgery
Columbia University College of Physicians and Surgeons
New York, New York

INTRODUCTION

Successful clinical application of cardiopulmonary bypass by Gibbon in 1953 [1] and the development of intraaortic balloon pumping (IABP) in the 1960s [2–4] represent important milestones in the brief history of circulatory assist devices. The clinical efficacy of extracorporeal circulation and IABP are well established. Continuing engineering development, as well as laboratory and clinical experience, has brought us to the threshold of successful chronic circulatory assistance in humans.

This discussion will outline forms of circulatory assistance currently in use and under development. A number of relevant physiologic, medical, and technological issues are discussed. For further technical detail, the reader is referred to National Institutes of Health (NIH) progress reports and related reviews [5–7].

In 1983, a dedicated team at the University of Utah successfully implanted a total artificial heart in a human [8]. Though the patient was tethered to a large external compressor and control mechanism, and though his clinical course was stormy and controversial, successful circulatory support with a total artificial heart was demonstrated for 112 days. This served as a dramatic incentive to those who work in this area, and has stimulated renewed dedication by investigators which will lead to further progress.

While development of the total artificial heart progresses, the climate for circulatory assistance is changing. Sudden death and cardiogenic shock due to myocardial infarction are becoming increasingly amenable to treatment as a result of recent advances. These advances include rapid triage, transport, and stabilization of patients critically ill with heart disease [9]. Coronary thrombolysis [10] and control of cardiac arrhythmias [11] are also important factors in this evolution. Moreover, balloon angioplasty has proved effective in relieving coronary stenosis [12], and laser therapy [13] appears feasible. These methods may prevent death and severe, chronic heart failure in patients who survive myocardial infarction. The necessity for emergency cardiac surgery may also decrease with continued development of these methods. In addition, the incidence of death from complications of atherosclerosis and hypertension is decreasing in the United States [13]. Still, the problem is one of enormous proportions, with sudden death affecting more than 600,000 persons annually in the United States alone [14].

Advances in the materials and methods of cardiac surgery have also been impressive. The mortality rate for patients subjected to a variety of surgical procedures now approaches 1 to 2 percent [15, 16]. The incidence of acute postoperative power failure requiring circulatory assistance has been 2 to 6 percent over the past decade [17]. The use of IABP seems to be decreasing with improved surgical methods. These developments may well reduce the number of candidates for circulatory assistance in the United States. Furthermore, current techniques for partial or total cardiac bypass are extraordinarily expensive and demanding of both personnel and facilities. Pressures to reduce medical expenditures are also increasing. These pressures will tend to work against the widespread application of circulatory assist devices unless these devices can be made comparable to other high-technology medical procedures in cost, reliability, and eventual productivity of the patient.

While these issues make the scope of eventual clinical application uncertain, it is also clear that improved circulatory assistance is sorely needed in many areas. The population of young patients with terminal heart failure due to irreversible cardiomyopathy may be larger than can be supported by cardiac transplantation unless methods of organ donation are improved. Current scarcity of hearts for transplantation is already becoming critical. Many patients unsuitable for transplantation because of age, pulmonary hypertension, or other factors might be successfully treated by mechanical cardiac replacement. Furthermore, despite recent progress, patients continue to die of heart failure after both myocardial infarction and cardiac surgery. Circumstances surrounding these events frequently suggest that survival would be possible with improved methods for circulatory support. Improved diagnostic methods are also required to define which patients are likely to be salvaged by temporary support and which are not.

It appears likely that the role of a circulatory assistance in clinical medicine will continue to grow. The directions that this growth will follow will be influenced not only by evolutionary bioengineering but also by improved understanding of the most appropriate and cost-effective role for our precious medical resources.

38.1 INTRAAORTIC BALLOON PUMPING FOR LEFT VENTRICULAR POWER FAILURE

Counterpulsation devices, of which IABP is the most widely utilized [17], are relatively simple and operate by inducing synchronized pressure waves into the arterial circulation. Their success requires a depressed but functional left ventricle whose performance can be improved by increasing coronary perfusion and decreasing systolic afterload. IABP employs no intravascular valves and therefore is not effective for volume pumping, extremes of left ventricular failure, or during ventricular fibrillation.

The intraaortic balloon pump is an evolutionary development of a modification of counterpulsation proposed by Moulopoulos et al. [18]. Originally conceived as a technique for therapy of postinfarction cardiogenic shock [19], IABP has proved to have wider utility as a temporary support for patients with acute heart failure after cardiac surgery [20] or for temporary control of unstable angina in patients awaiting surgery [17]. While IABP can frequently stabilize patients in cardiogenic shock, long-term survival rates in treatment of this condition have been disappointing [20]. Results can be improved somewhat if IABP is combined with emergency revascularization surgery [21]. This has led to the view, as yet not refuted by clinical experience, that cardiogenic shock can be palliated but cannot be cured by circulatory support once myocardial necrosis becomes extensive. According to this view, when application of IABP is successful, reversible ischemia and not myocardial necrosis is being treated. As yet, clinical medicine has not provided adequate tools for judging when ischemic heart failure is due to severe ischemic injury and when it is due to irreversible myocardial necrosis. Indeed, it is not completely certain when ischemic damage becomes irreversible [22].

Clinically, the physiologic benefits of IABP appear to be due to improved contractility of the ischemic left ventricle. IABP is believed to increase contractility by increasing diastolic coronary perfusion pressure and coronary blood flow [23]. In laboratory studies, counterpulsation decreases steady-state oxygen requirements by decreasing systolic wall stress [24], but in clinical ischemia left ventricular systolic pressure often increases with institution of IABP, suggesting increased contractility and increased myocardial oxygen consumption as ischemia is relieved. IABP also tends to suppress ventricular arrhythmias, an effect which is probably due both to relief of ischemia and to withdrawal of sympathetic tone. The latter effect has been demonstrated experimentally [25] and is attributed to increased frequency of carotid sinus nerve discharge during counterpulsation [26]. The acute improvement in left ventricular function produced by IABP initially does not persist if IABP is abruptly terminated. Subsequent natural reparative processes are necessary to produce hemodynamic stability, which allows IABP to be weaned. The period of IABP required after cardiac surgery is generally 1 to 7 days, but longer intervals have been reported [27].

The most frequent complication of IABP has been limb ischemia due to arterial obstruction by the balloon catheter. This has been reported in 10 to 20 percent of treated patients and, in the extreme, has led to amputation and even (rarely) to death [27]. It appears that the incidence of such complications has been increased by the development of balloons designed for percutaneous insertion, which require an external sheath larger than the balloon catheter [27]. This disadvantage is offset by the simplicity and the increased speed of insertion of the percutaneous balloon, which can be lifesaving [23].

The balloon catheter requires a large lumen for mechanical efficiency, particularly at rapid heart rates. Thus, circulation to the lower leg in patients with small arteries and/or peripheral vascular disease may be severely compromised by balloon insertion. This has led some authors to suggest that the percutaneous technique (which requires a large indwelling sheath throughout balloon pumping) should be utilized for patients with large arteries, while surgical insertion techniques should be utilized for patients with smaller arteries, including women. Intensive engineering efforts are being devoted to decreasing catheter diameter. The use of helium as a driving gas decreases viscosity, allowing effective IABP with reduced lumen size or at high heart rates. The hazard of embolism with helium as opposed to carbon dioxide is theoretically increased [28]. Improved materials and design also allow smaller catheter lumens and reduced catheter wall thickness. Recently, tiny balloons designed for use in young children have been successfully tested [29]. Computerized, self-regulating control systems, using the electrocardiogram (ECG) and aortic pressure, have been developed [30]. For the present, IABP remains a clinically proven technique which can only become more generally useful and less hazardous as equipment designs and methods are refined. In appropriately selected patients, it is unquestionably lifesaving and can substantially reduce the risks associated with cardiac surgery [23].

38.2 PULSATILE ASSIST DEVICES FOR USE DURING CARDIOPULMONARY BYPASS

Standard roller pumps for cardiopulmonary bypass are essentially constant-flow devices which do not produce physiologic variation in pressure or flow. Analogues of the intraaortic balloon pump have been used since 1976 to produce pulsatile perfusion [31]. A polyurethane balloon within a rigid housing is inserted in series between the roller pump of the heart-lung machine and the arterial perfusion cannula. A console similar to those for IABP introduces and exhausts gas from the housing surrounding the balloon. The pulsations produced resemble normal arterial pressure waves and can be synchronized with ventricular contraction. This permits pulsatile flow on total bypass. During partial bypass or following bypass, the device can be utilized for counterpulsation, as long as anticoagulation is maintained.

Another technique which has been developed is a variable-speed roller pump, capable of repetitive cycles of pump speed to produce pulsatile pressure waves. A third technique involves a standard roller pump compressing tubing of variable diameter.

Two controversial questions surround the need for intraoperative pulsation devices [32]: (1) Are they physiologically advantageous? (2) Are they worth the cost?

Several investigators have demonstrated that pulsatile bypass improves urine output and ameliorates the increase in peripheral vascular resistance which usually develops during cardiopulmonary bypass without pulsatile flow [33]. Others claim that similar results can be obtained by high-flow perfusion techniques without pulsatile bypass [34]. Analyses of effects of pulsatile bypass on cardiac performance, myocardial injury, and mortality following coronary bypass surgery have shown no clear benefit [35]. However, the availability of counterpulsation when unanticipated ischemic failure occurs during weaning from cardiopulmonary bypass can be advantageous. Pulsation via the arterial perfusion line appears hemodynamically inferior to true IABP, but critical analysis of waveform effects on arterial pressure with IABP and arterial line pulsation has not been reported. In addition, pulsation through the arterial line must be discontinued when heparinization is reversed. Pulsatile perfusion would be more attractive if costs of implementation were reduced. For this reason, pulsations based on variable-speed roller pumps deserve further attention. The principal disadvantage of current systems is their inability to function effectively in the absence of bypass flow.

38.3 PULMONARY ARTERY COUNTERPULSATION FOR RIGHT-SIDED HEART FAILURE

The success of IABP in left ventricular power failure has stimulated speculation about utility of pulmonary artery counterpulsation in right ventricular failure. Experimental studies have demonstrated the potential efficacy of this approach, particularly when pulmonary hypertension is associated with subendocardial ischemia of the right ventricle [36, 37]. Recently, techniques for pulmonary artery counterpulsation during cardiac surgery have been demonstrated in patients with acute right ventricular failure; however, these methods are cumbersome and require thoracotomy for initiation and removal [38]. The hazards of bleeding associated with surgical manipulation of the pulmonary artery can become extreme with severe pulmonary hypertension. The relatively small volume of blood contained in the pulmonary artery compared with that in the aorta decreases maximum theoretical pulmonary balloon stroke volume. (Specially fabricated balloons may partially overcome this problem.) Since pulmonary artery balloon pumping is technically more difficult to implement than IABP, and since issues of anticoagulation and embolism may be less critical in the pulmonary than in the systemic circulation, prolonged volume pump support may be of greater potential utility than pulmonary artery counterpulsation for acute right ventricular failure.

A disadvantage of counterpulsation from the pulmonary artery is the absence of diastolic augmentation of coronary artery perfusion pressure, an important effect of IABP. Since increased coronary artery perfusion could reduce right ventricular ischemia, aortic counterpulsation might also favorably affect right ventricular failure due to pressure overloads [37].

38.4 VOLUME PUMPS

38.4.1 Temporary Left Ventricular Assistance

When IABP fails to reverse left ventricular power failure, a pump capable of volume work (as opposed to the pressure work provided by IABP) is required. Such systems have usually been employed for perioperative left ventricular power failure. Design requirements are less stringent for this form of assistance than for long-term blood pumps. Flows of 3 to 5 L/min are generally adequate, and recovery of left ventricular function usually occurs within 3 to 7 days if the injury is reversible [39–44]. The simplest "device" of this kind is a roller pump, with cannulation from the left atrium [39] or left ventricular apex [45] to the aorta or femoral artery. A left atrium–aortic bypass system employing Silastic cannulas which can be permanently obturated has been used successfully and is commercially available and FDA-approved [39]. The principal difficulty with a roller pump system is the need for anticoagulation and constant attendance by a perfusionist to maintain the system and to avoid excessive negative pressure during low-flow conditions. IABP is usually added separately if counterpulsation or pulsatile flow is desired.

More complex pumps also utilized successfully for perioperative left ventricular power failure include magnetically driven centrifugal pumps, extracorporeal [40] and paracorporeal [41] pneumatic pumps, and implantable pneumatic pumps [5]. Potential advantages of these systems include pulsatile flow, simplified operation [46], minimal need for anticoagulants, and decreased risk of air and particulate embolism.

The application and evolution of these systems have provided experience useful in developing blood pumps intended for chronic application to humans. The success of total left heart bypass in restoring adequate left ventricular function in isolated left ventricular failure refractory to IABP and pharmacologic support is approximately 50 percent in most series reflecting substantial experience [5, 39–44]. The survival rate when biventricular support is required is much lower and may not exceed 10 percent [44]. Problems include bleeding, sepsis, embolism, multiorgan failure, pump dysfunction, and irreversible heart failure. When implantable pumps are used for perioperative heart failure, only a single, small pneumatic drive line must leave the patient's body, whereas two substantial cannulas are needed for extracorporeal assistance. Problems related to power supply and control are simplified by maintaining the relevant systems outside the patient's body.

The use of short-term devices is predicated on the observation that there are forms of left ventricular injury which are reversible even though pharmacologic or IABP support is insufficient to maintain life. Experimental evidence suggests that myocardial infarct size can be decreased by total left heart bypass [47], but the theory that patients with fresh infarctions can be successfully treated with left ventricular bypass [48] has yet to be proved. At present, the most suitable population for volume pump support of the left ventricle consists of surgical patients who have suffered severe global ischemic injury without frank myocardial necrosis, though survival has been reported even following appearance of a major intraoperative myocardial infarction [41].

38.4.2 Support or Replacement of the Left Ventricle

Chronic bypass or replacement of the left ventricle requires resolution of most of the problems inherent in total cardiac replacement. The power output of the left ventricle greatly exceeds that of the right; control requirements and sensitivity to embolism are also most critical in the systemic circulation. The probability that valvular regurgitation will lead to hemolysis is also greater on the left side, because pressure gradients across the mitral and aortic valves are greater than those across the tricuspid and pulmonic valves during retrograde flow. A formidable list of bioengineering problems are being resolved in developing pumps for chronic left ventricular bypass. For purposes of discussion, these will be divided into factors affecting the pump itself, its drive system and power supply, the control system, and the "compliance chamber."

The pump must be capable of 10-L/min flow at 120 mmHg mean aortic pressure and 15 mmHg left atrial pressure, according to requirements of the NIH-sponsored artificial heart program (Figs. 38.1 and 38.2). The pump should also respond to rising or falling left atrial pressure with parallel changes in cardiac output and stroke volume, in a manner resembling the Starling ventricular function curve of the normal left ventricle (Figs. 38.3 and 38.4) [49–51]. In addition to providing capability for physiologic response to exercise, the Starling response provides an autoregulatory capacity which will compensate for temporary inequality of stroke volume of the left and right ventricles. The pump must also be extremely durable, capable of a wide range of pressure and volume work, and must possess a long fatigue life. It must be compatible with the dimensions and environment of the thoracic cage or abdomen and must not cause excessive damage to blood components.

The architecture of the pump involves inlet and outlet valves, a blood pathway, and a pumping chamber. Several current designs are based on a seamless pumping chamber containing a flexible diaphragm for force transmission. A wide variety of configurations have been proposed and tested [5–7]. The chamber is usually constructed of segmented polyurethane [52–54]. Occlusive contact of blood pumping surfaces is now avoided, having been recognized as a cause of excessive hemolysis in early designs. Hemolysis in implantation studies in animals has intensified efforts to minimize factors contributing to this problem [55]. These efforts have focused on the valves employed as well as all blood contact surfaces within the pump. Approaches to the design of the blood contact surfaces have included coating those surfaces with metal beads [56] or Dacron filaments [57] to encourage a stable endothelial layer. Direct seeding

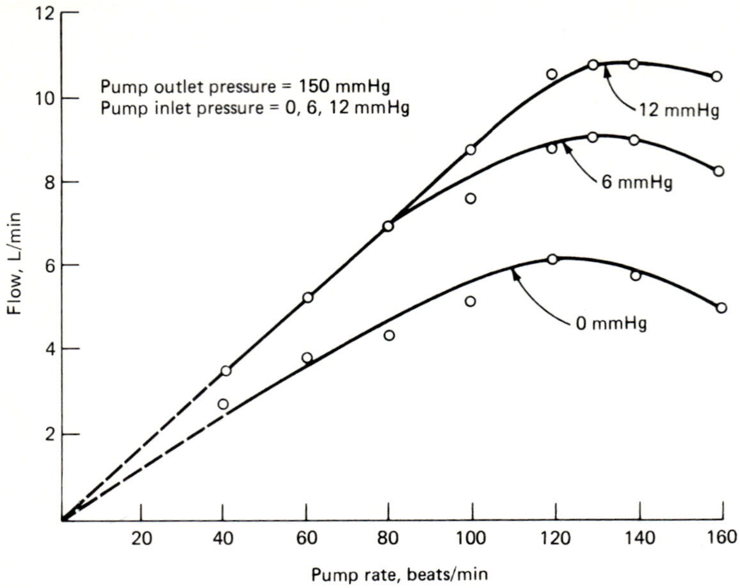

FIG. 38.1 Representative performance curves for E-type artificial left ventricular assist device (ALVAD) demonstrate relation between pump rate and pump flow with outflow resistance of 150 mmHg. Level of performance increases as inlet pressure increases from 0 to 6 to 12 mmHg. Peak flow of 10 L/min is achieved at a pump rate of 140 cycles per minute. (*From Norman et al. [60].*)

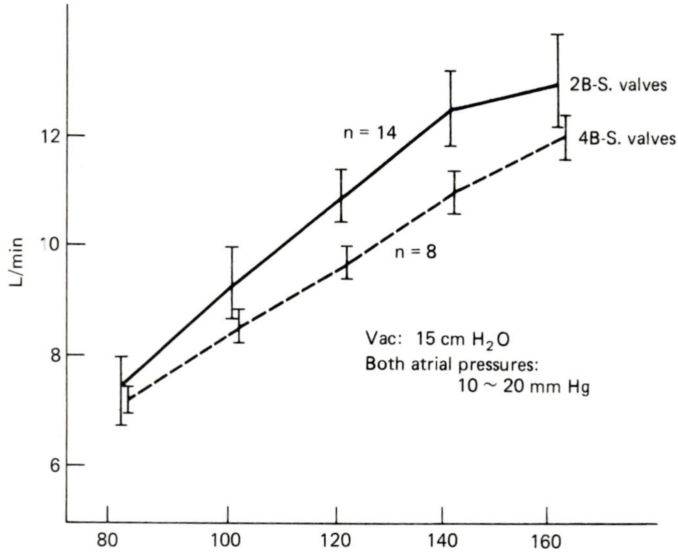

FIG. 38.2 As in Fig. 38.1, these data describe in situ performance of the Jarvik-7 artificial heart. The solid line defines the mean relation between pump rate and flow in five calf implants with prosthetic mitral and tricuspid valves but with maintenance of the animal's intrinsic aortic and pulmonic valves. The broken line represents data from a single calf in which all four cardiac valves were replaced with prostheses. Flow rates greater than 10 L/min are readily achieved. (*From Mochizuki et al. [59].*)

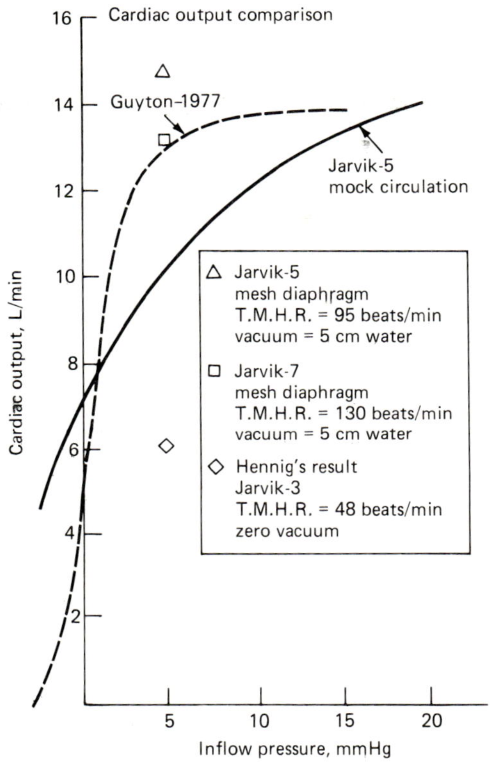

FIG. 38.3 Physiologic relationship between cardiac output and filling pressure (dashed line) is compared with measured performance of artificial hearts in mock circulation studies. A steep relationship between filling pressure and output provides autoregulation of excessive flow to either chamber of the heart and allows a physiologic increase in flow in response to increased venous return with exercise. (*From Lioi et al. [50].*)

of prosthetic surfaces with cultured endothelial cells has been actively explored [58], as has "biolizing" the surface with proteins or conditioning it with hydrogels [54]. The current preference for a smooth polyurethane surface is related to a tendency for stimulated endothelial linings (pseudoneointima) to proliferate, resulting in a pannus which obstructs blood flow or calcification which interferes with pump function [55, 59].

The structure of the diaphragm which transmits force to the blood is becoming increasingly complex, with multilayered construction utilized to improve fatigue life and function (Fig. 38.5) [50] or to minimize permeability to fluids [60]. The design of the blood flowpath has also proved unexpectedly critical, with reports of significant obstruction to flow intrinsic to basic designs (Fig. 38.6) [50, 61], as well as progressive obstruction due to connective tissue deposition [59]. High stress levels within mechanical ventricles have resulted in premature fractures in mechanical valves [8] and have led to insertion of tissue valves in Dacron conduits which separate the valves from the pump housing [41, 60]. Both the cross-sectional area for blood flow and the regurgitant fraction in artificial valves influence overall pump function [50], and this has renewed interest in artificial valve design [62] and in preservation of native valves whenever possible.

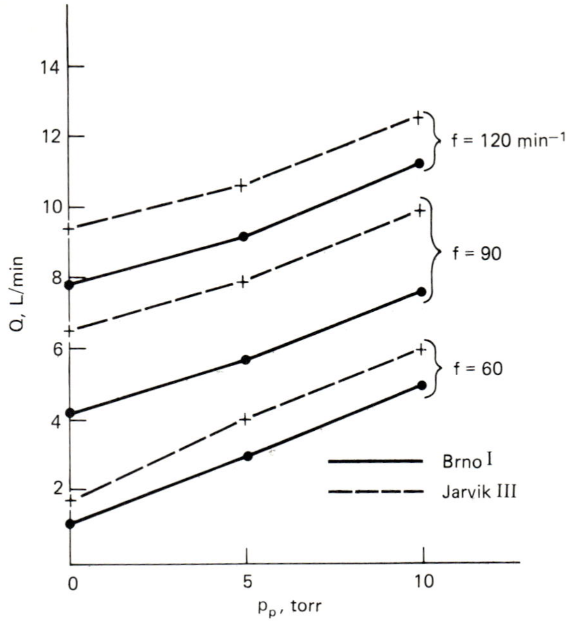

FIG. 38.4 The performances of two circulatory assist devices, Jarvik-3 and BRNO-I [51], are compared at pump rates of 60, 90, and 120 cycles per minute. Increasing filling pressure (horizontal axis) is associated with increasing pump flow at any given pump rate.

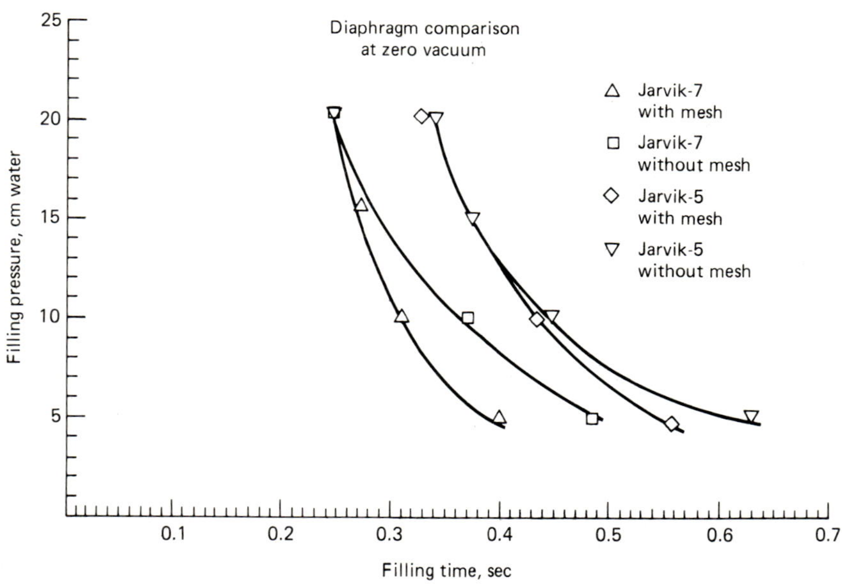

FIG. 38.5 Effect of design of pump diaphragm on performance. Filling time (horizontal axis) is related to filling pressure (vertical axis). The addition of a Dacron mesh layer to a multilayered composite diaphragm results in more rapid filling at any given inlet pressure in both the Jarvik-5 and Jarvik-7 designs. (*From Lioi et al. [50].*)

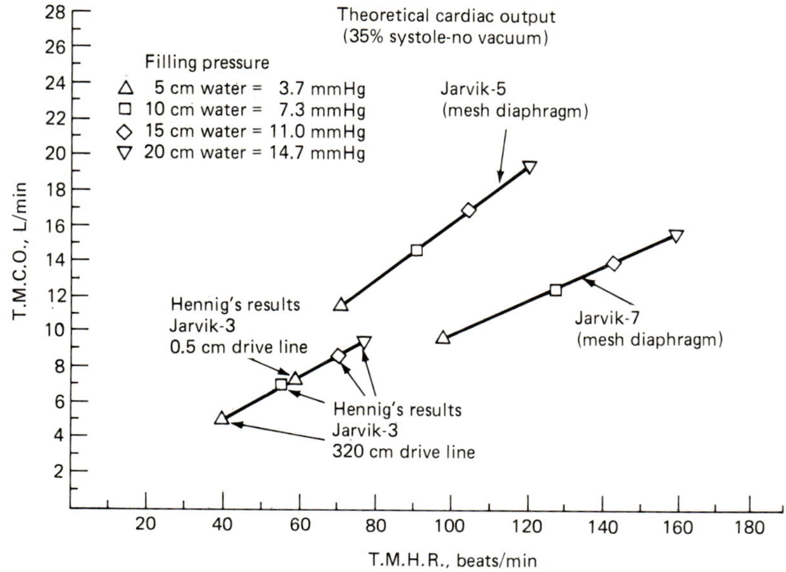

FIG. 38.6 Effect of pump design on performance. Reduced output of the Jarvik-3 device may be related to restrictive design of inlet portion of pump. (*From Lioi et al. [50].*)

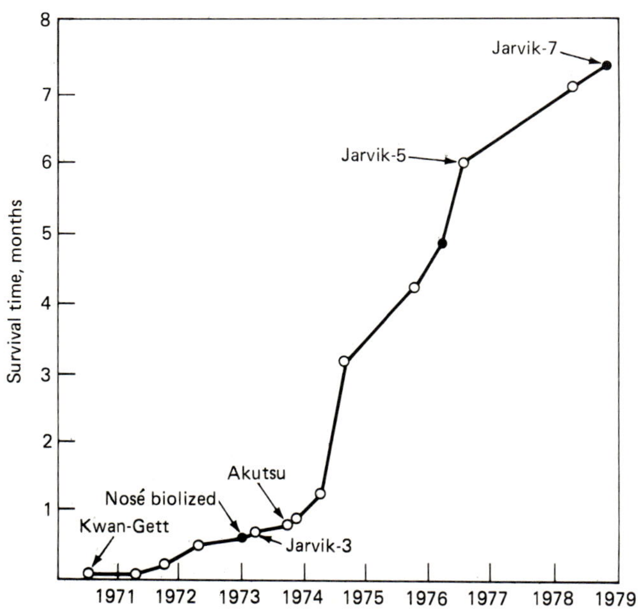

FIG. 38.7 Survival of experimental animals with totally implanted artificial hearts has improved markedly in recent years, reflecting improved bioengineering, surgery, and pharmacology. (*From Jarvik [68].*)

Extensive testing of the pump components is obviously essential before human trials can be conducted, and millions of cycles have been observed in mock circulatory systems [5–7, 60, 63] and in experimental animals (Fig. 38.7) [55, 59].

Current clinical trials of the Jarvik total artificial heart are based on exteriorization of the drive, power, and control components, similar to designs for short-term blood pumps [8]. This approach has expedited development of the artificial heart and has highlighted difficulties with bleeding and systemic embolism as opposite extremes of a delicate balance of anticoagulation. While exteriorizing as many components as possible has a number of obvious advantages in early development, the artificial heart should ultimately allow an active lifestyle and partly pay for itself in restored productivity of the recipient. Improved portability is clearly essential. Whether the advantages of a tethered system with external components that can be easily changed will be preferable to a totally epithelialized system which is less accessible but also less likely to become infected remains to be seen.

The pivotal issue in regard to portability is how power will be transmitted to the drive system, estimated to require at least 10 W [6]. Among the principal alternatives are a radioisotope-based thermal source, which might be combined with a Stirling cycle engine [64]. Heat

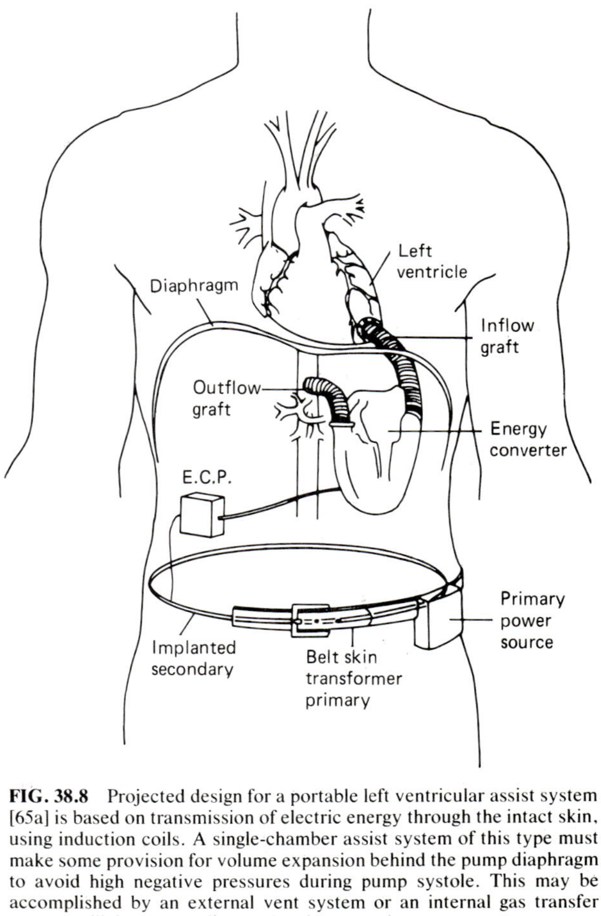

FIG. 38.8 Projected design for a portable left ventricular assist system [65a] is based on transmission of electric energy through the intact skin, using induction coils. A single-chamber assist system of this type must make some provision for volume expansion behind the pump diaphragm to avoid high negative pressures during pump systole. This may be accomplished by an external vent system or an internal gas transfer system utilizing a compliance chamber (not shown).

dissipation, space constraints, expense, and shielding from radiation injury would be significant problems. Transcutaneous placement of electrical or pneumatic drive lines is a second alternative, which allows external placement of the control and drive systems. Essential to this approach would be a permanent, infection-resistant coupling [6–8]. A third approach would involve percutaneous transmission of electric power using subcutaneous and external induction coils [6, 7, 60, 63]. The latter system could be combined with a rechargeable internal battery pack that would allow for 30 min away from the external power source [60]. Batteries in an externally worn vest would allow mobility for daily activities (Figs. 38.8 and 38.9) [60].

Considerable effort has been devoted to drive systems capable of converting electric energy to mechanical work with high efficiency (50 to 60 percent). These systems have been based on brushless dc motors [60] and solenoids [63], as well as on other approaches [5–7, 64].

A variety of control systems, as well as varied algorithms for pump control, have been devised. Monitoring of pressure within the drive lines of pneumatic systems can be used to detect end systole and end diastole on the basis of the rate of pressure change [50]. Minipotentials in electric drive lines can similarly be utilized as an rpm counter to monitor cardiac cycles and track the phase of function [60]. A permanent magnet attached to the diaphragm within the pumping chamber can be used to track diaphragm position by the Hall effect [60]. Electromagnetic or Doppler flowmeters might also be employed. These inputs would be combined with permanent pressure-monitoring devices within the atria, pump, and arteries to monitor function. The ECG would also be monitored. The input data could be used to adjust pump stroke volume, cycle length, and relative duration of systole and diastole. Depending on the state of the native circulation, pump algorithms would allow totally autonomous function, ECG synchronization, or counterpulsation [60, 65]. Control systems have not yet been sufficiently miniaturized to be implanted, but this is a planned objective [60, 63, 65a].

Compliance chambers are required to deal with the volume replacement problem when blood is rapidly ejected from a completely internalized univentricular pump. The concept involves a gas-filled chamber (e.g., with sulfur hexafluoride) connected to allow reversible flow of

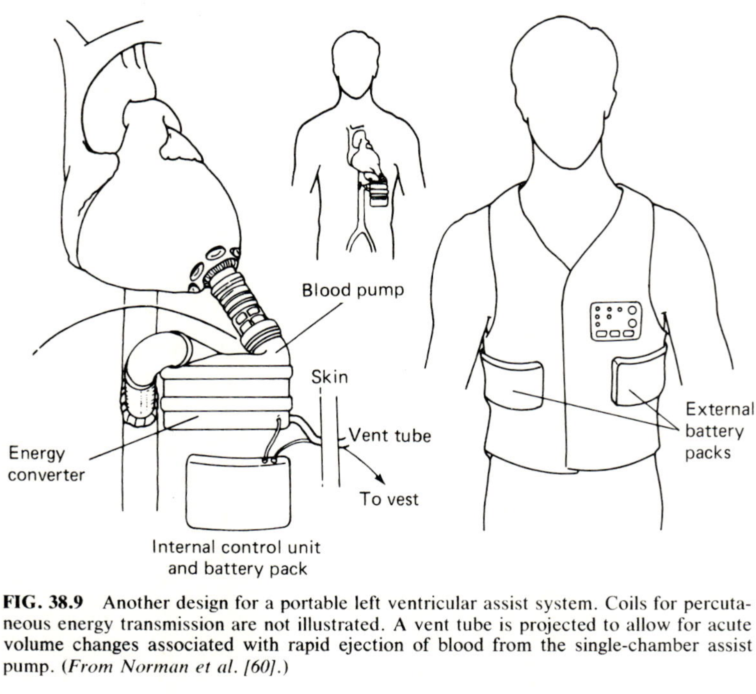

FIG. 38.9 Another design for a portable left ventricular assist system. Coils for percutaneous energy transmission are not illustrated. A vent tube is projected to allow for acute volume changes associated with rapid ejection of blood from the single-chamber assist pump. (*From Norman et al. [60].*)

gas into (during systole) and out of (during diastole) the housing behind the pumping surface, as required by the phase of the cardiac cycle [60, 66]. Practical problems encountered include gas leakage and encasement of the compliance chamber itself by a fibrous sac which restricts function. Alternative solutions include a permanent vent to the atmosphere or an alternating biventricular device in which blood ejected by one chamber is replaced by blood filling the second.

It appears in 1985 that a totally implanted pump for ventricular replacement can be achieved, though many engineering problems remain to be solved. The critical issues now are whether costs can be limited, recipient lifestyle and productivity facilitated, and pump durability and reliability extended. Resolving these problems will be necessary before the vast expenditures needed for frequent application of such devices become politically acceptable.

38.4.3 Support of the Right Ventricle

In the absence of pulmonary hypertension, the right ventricle is not thought to be essential for life. Indeed, for certain forms of congenital heart disease (e.g., tricuspid atresia), right atrial function alone is frequently adequate for perfusion of lungs possessing normal flow resistance [67]. Right ventricular failure which is life-threatening therefore requires both right ventricular dysfunction and primary or secondary pulmonary hypertension. Such a combination of events is most likely to occur with chronic mitral stenosis, congenital heart disease, or severe left ventricular failure. As mentioned above, physical dimensions make the use of intravascular balloon pumps less attractive for the pulmonary artery, though specially molded balloons may optimize the use of the irregular space available. Furthermore, small emboli are better tolerated by the pulmonary than by the systemic circulation, since the effects of even small cerebrovascular accidents can be crippling. Thus, the use of volume pumps for right heart bypass appears extremely attractive, though experience is quite limited [40, 44]. The simplest cannulation technique for right ventricular assistance withdraws blood from the femoral vein and returns it to the femoral artery, but an oxygenator is needed if flows are high. In the presence of hypoxia due to simultaneous respiratory and cardiac failure, extracorporeal membrane oxygenation has frequently been lifesaving in infants and children. This method is discussed elsewhere in this handbook.

Hazards in open-chest methods of cannulation for right heart bypass relate to mechanical fragility of the right ventricle and pulmonary artery in chronic pulmonary hypertension. The safest method for open-chest cannulation therefore involves a right atrial drainage cannula and a pulmonary return cannula inserted via right atrium and thence across the tricuspid and pulmonic valves. Repeat thoracotomy is required for removal. The possibility of a completely transvenous method for right or left ventricular bypass without oxygenators, thoracotomy, or anticoagulation is a tantalizing, but thus far elusive, goal.

38.5 BIVENTRICULAR SUPPORT (TOTAL ARTIFICIAL HEART)

Many designs for the total artificial heart resemble duplicated left ventricular assist pumps (see Sec. 38.4.1). Animal implants of designs intended for use in humans have been successful for periods of months (Fig. 38.7), and performance has been encouraging [5–8, 55, 59, 68], but some problems remain. Though space limitations within the body are even more restrictive for biventricular pumping than for univentricular pumping, there are some benefits as well. In one design, two assist pumps are hydraulically coupled by a powerful, reversible dc motor driving a turbine pump (Figs. 38.10 and 38.11) [68]. Fluid is withdrawn from the housing of the right ventricular pump and simultaneously injected into the housing of the left ventricular pump to produce simultaneous right ventricular diastole and left ventricular systole. The direction of the motor and of fluid transfer is reversed to complete the cycle. Compliance chambers are eliminated and the drive system simplified by the basic design (Fig. 38.11). However, recent experi-

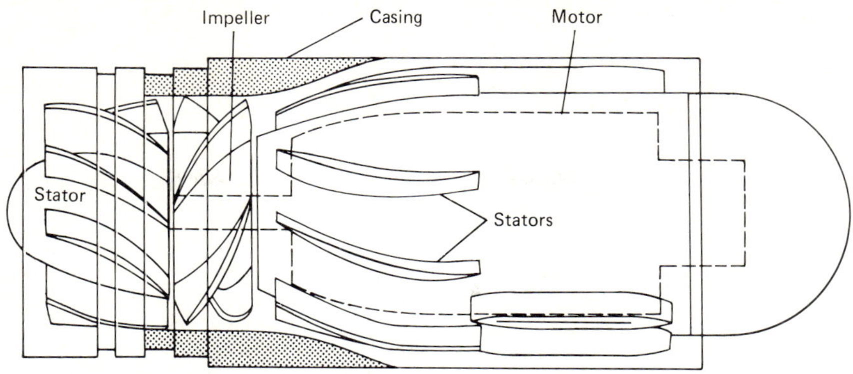

FIG. 38.10 Miniaturized, reversible, brushless dc motor and turbine projected for use as the driver in simplified design for total artificial heart illustrated in Fig. 38.11. (*From Jarvik [68].*)

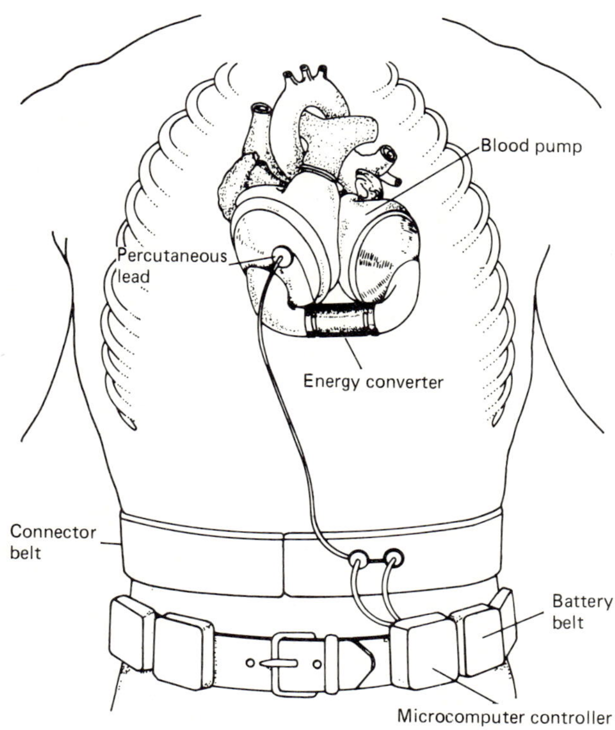

FIG. 38.11 Simplified design for total artificial heart is based on fluid coupling of the space behind the pump diaphragms. Systolic ejection of one chamber is hydraulically linked with diastolic filling of the other chamber. The system is driven by sequentially altering the direction of fluid flow in the hydraulic system, using the turbine drive shown in Fig. 38.10 [68]. While this design appears simple, experience suggests that elaborate volume transfer systems may still be required to compensate for small differences in flow in the systemic and pulmonary circulations [69].

ence indicates that modifications will be needed to compensate for variation of relative flow requirements in the pulmonary and systemic circulations [69].

One pressing clinical requirement for total, but temporary cardiac support is severe biventricular failure following open-heart surgery. The use of biventricular support presumes that the cardiac injury is reversible. This problem can be handled even by roller pumps. However, the superior function, durability, blood compatibility, and reduced thrombogenicity of more sophisticated designs have encouraged their use, where feasible, for temporary support. Unfortunately, the development of right ventricular failure during left ventricular support generally indicates that the underlying cardiac injury is also progressive and that recovery will not occur. As a result, the temptation to perform emergency cardiac transplantation in such circumstances is strong [70]. On the other hand, present experience is rather limited, and the possibility that 30-day support may allow cardiac recovery has not been fully explored.

Problems with donor supply in cardiac transplant programs are also creating pressure for temporary biventricular support. Some physicians feel that terminal heart failure in young patients awaiting a cardiac transplant should be treated by temporary insertion of even an incompletely tested total artificial heart in preference to death. Life support requirements in such circumstances are generally less than 2 weeks. Cardiac transplantation following mechanical circulatory assistance is presumed to be associated with increased risk of infection [71].

Consideration of these factors suggests a need for development of a new class of devices specifically intended for use in patients who require temporary support prior to transplantation. This would include young patients with end-stage cardiomyopathy, cardiogenic shock due to massive primary myocardial infarction, or catastrophic intraoperative cardiac injury. The device would need to function for 2 to 4 weeks until a donor heart could be found on an emergent basis. It should be totally internalized to eliminate the possibility of entry of organisms along drive lines or blood conduits, and requirements for flow, control, and durability would be relatively simple, because the patient would be kept at bed rest while awaiting transplantation. The device would differ somewhat from present 30-day blood pumps in that the patient's own ventricles would be removed, and avoidance of infection would be a primary concern. The availability of such a device would fill a specific need in current clinical medicine, and would alleviate the current moral, scientific, and political dilemma in which inappropriately tested and designed devices may be utilized to relieve an emergent problem.

38.6 BIOLOGICAL ASSIST DEVICES

The use of striated (skeletal or cardiac) muscle would appear to be an attractive alternative to electrical or pneumatic devices for force generation in circulatory assistance. Potential advantages include the ability to utilize circulating substrates as a power supply, some ability for self-repair and self-maintenance, and built-in responsiveness to volume overloading and circulating catecholamines. In testing of limited duration, both force of contraction and potential stroke volume with skeletal muscle appear adequate for left ventricular replacement [72]. Uses of skeletal muscle which have been explored experimentally include counterpulsation and left ventricular bypass. The principal impediment to generalized application of these experimental observations has been the rapid deterioration of developed force when applied to chronic work loads. The only successful clinical application of striated muscle to hemodynamic work in clinical medicine has been through myocardial transplantation in cardiac allografts [71]. Though cardiac transplantation has been extremely successful in recent years, the potential for this technique is sharply limited by scarcity of suitable donors.

Heterotopic transplants, with donor hearts transplanted in parallel with the recipient heart [73], have been successful. Advantages of this method are that the recipient heart can act as an assist while the donor heart recovers from any perioperative injury. This arrangement also may allow survival of the patient in the presence of severe pulmonary hypertension, which is not tolerated by the normal right ventricle of the donor heart and therefore precludes standard transplants. On the other hand, the hypertrophied right ventricle of the recipient's heart may tolerate abnormal work loads well in combination with the normal left ventricle of a donor heart. In addition, if rejection occurs, the donor heart can act as an assist until another suitable donor becomes available. There have been instances in which donor heart assistance has been accompanied by some recovery of recipient heart function [74]. Drawbacks of this parallel arrangement include thrombus formation and embolism due to low flow through the recipient's left ventricle. Further experimental and clinical experience with this procedure will help determine the most suitable candidates, thus lowering the incidence of complications.

In 1985, the possibility of cross-species transplantation for permanent or temporary cardiac replacement (until a human donor can be found) is a subject of increasingly intense investigation.

ACKNOWLEDGMENTS

I. John Uygur provided strong support for the preparation of this manuscript; his efforts are gratefully acknowledged. The author also thanks Lucia Paronetto, Mary Ann Kral, and Dorothy Lowry for their editorial assistance.

REFERENCES

1. Gibbon, J. H., Jr.: Application of a mechanical heart and lung apparatus to cardiac surgery, *Minn. Med., 37,* 171, 1954.
2. Kantrowitz, A., and Kantrowitz, A.: Experimental augmentation of coronary flow by retardation of the arterial pressure pulse, *Surgery, 34,* 678–687, 1953.
3. Claus, R. H., Birtwell, W. C., Albertal, G., Sunzer, S., Taylor, W. J., Fosberg, A. M., and Harken, D. E.: Assisted circulation: I. The arterial counterpulsator, *J. Thorac. Cardiovasc. Surg., 41,* 447–458, 1961.
4. Kantrowitz, A., Tjonneland, S., Freed, P. S., Phillips, S. J., Butner, A. N., and Sherman, A. L.: Pumping in cardiogenic shock, *JAMA, 203,* 135, 1968.
5. Norman, J. C.: Mechanical ventricular assistance: A review, *Artif. Organs, 5,* 103–117, 1981.
6. Department of Health and Human Services: Devices and technology branch contractors meeting program, Division of Heart and Vascular Diseases, NHLBI, National Institutes of Health, Bethesda, Md., 1982.
7. Department of Health and Human Services: Devices and technology branch contractors meeting program, Division of Heart and Vascular Diseases, NHLBI, National Institutes of Health, Bethesda, Md., 1984.
8. DeVries, W. C., Anderson, J. L., Joyce, L. D., Anderson, F. L., Hammond, E. H., Jarvik, R. K., and Kolff, W. J.: Clinical use of the total artificial heart, *N. Engl. J. Med., 310,* 273–278, 1984.
9. Stuls, K. R., Brown, D. D., Schug, V. I., and Bean, J. A.: Prehospital defibrillation performed by emergency medical technicians in rural communities, *N. Engl. J. Med., 310,* 219–223, 1984.
10. Rentrop, P., Blanks, H., Karsch, K. R., Kaiser, M. D., Kostering, H., and Leitz, K.: Intracoronary thrombolysis in acute myocardial infarction and unstable angina pectoris, *Circulation, 63,* 307, 1981.
11. Singh, B. N.: Prevention of myocardial reinfarction and sudden death in survivors of acute myocardial infarction: Role of prophylactic B-adrenoreceptor blockade, *Am. Heart J., 107,* 189–200, 1984.
12. Hamby, R. I., and Katz, S.: Percutaneous transluminal coronary angioplasty: Its potential impact on surgery for coronary artery disease, *Am. J. Cardiol., 45,* 1161, 1980.
13. Lee, G., Ikeda, R. M., Stobbe, D., Ogata, C., Theis, D., Hussein, H., and Mason, D. T.: Laser irradiation of human atherosclerotic obstructive disease: Simultaneous visualization and vaporization achieved by dual fiberoptic catheter, *Am. Heart J., 105,* 163, 1983.
14. National Center for Health Statistics: *Final Mortality Statistics, 1978,* DDHS Pub. No. (PHS) 80-1120, no. 6, suppl. 2, 1980.
15. Loop, F. D., Cosgrove, D. M., Lytle, B. W., Thurer, R. L., Simpfendorfer, C., Taylor, P. C., and Proudfit, W. L.: An eleven year evolution of coronary arterial surgery (1967–1978), *Ann. Surg., 190,* 444, 1979.

16. Kirklin, J. W., and Kouchoukos, N. T.: Aortic valve replacement without myocardial revascularization (editorial), *Circulation, 63,* 252, 1981.

17. Mundth, E. D.: Assisted circulation, in D. C. Sabiston, Jr., and F. C. Spencer (eds.), *Gibbon's Surgery of the Chest,* Saunders, Philadelphia, 1983, pp. 1490–1514.

18. Moulopoulos, S. D., Topaz, S., and Kolff, W. J.: Diastolic balloon pump assistance and early surgery in cardiogenic shock, *Am. Heart J., 63,* 669, 1962.

19. McEnany, M. T., Kay, H. R., Buckley, M. J., Dagget, W. M., Erdmann, A. J., Mundth, E. D., Rao, R. S., DeToeuf, J., and Austen, W. G.: Clinical experience with intraaortic balloon support in 728 patients, *Circulation Suppl., 58,* 124, 1978.

20. O'Rourke, M. R., Sammel, N. C., and Chang, V. P., Arterial counterpulsation in severe refractory heart failure complicating acute myocardial infarction, *Br. Heart J., 41,* 508, 1979.

21. Mundth, E. D., Buckley, M. J., Dagget, W. M., McEnany, M. T., Leinbach, R. C., Gold, H. K., and Austen, W. G.: Intraaortic balloon pump assistance and early surgery in cardiogenic shock, *Adv. Cardiol., 15,* 159, 1975.

22. Tranum-Jensen, J., Janse, M. J., Fiolet, J. W. T., Krieger, J. G., d'Alnoncourt, C. N., and Durrer, D.: Tissue osmolarity, cell swelling, and reperfusion in acute regional myocardial ischemia in the isolated porcine heart, *Circ. Res., 49,* 364–381, 1981.

23. Bregman, D.: Intraaortic balloon counterpulsation, in M. I. Ionescu (ed.), *Techniques in Extracorporeal Circulation,* 2d ed., Butterworth, London, 1981, pp. 561–600.

24. Spotnitz, H. M., Covell, J., Ross, J., and Braunwald, E.: Left ventricular mechanics and oxygen consumption during arterial counterpulsation, *Am. J. Physiol., 217,* 1352–1358, 1969.

25. Spotnitz, H. M., Beach, P. M., Bregman, D., Truccone, N. J., Parodi, E. N., Manger, W. M., and Malm, J. R.: Oxygen conservation by propanolol and intraaortic balloon pumping in the normal canine heart, *J. Surg. Res., 22,* 453–462, 1977.

26. Brown, B. G., Goldfarb, D., and Gott, V. L.: A vasomotor reflex contribution to systolic pressure reduction during diastolic augmentation, *Trans. Am. Soc. Artif. Intern. Organs, 12,* 63–67, 1966.

27. Pennington, D. G., Schwartz, M., Codd, J. E., Merjavy, J. P., and Kaiser, G. C.: Intraaortic balloon pumping in cardiac surgical patients: A nine year experience, *Ann. Thorac. Surg., 36,* 125–131, 1983.

28. Kayser, K. L., Dudley, J. W., and Shore, R. T.: Comparison of driving gases for IABP's, *Med. Instrum., 15N*(1), 51–54, 1981.

29. Veasy, G. L., Orsmond, G. S., Ruttenberg, H. D., and Boucek, M. M.: Pediatric use of intraaortic balloon pumping (IABP) (abstract), *Circulation Suppl., 68*(III), 182, 1983.

30. Phillipe, E., Clark, J. W., Lande, A., and Ellis, J. R.: Microprocessor control of IABP, *Ann. Biol. Med. Eng., 8N3,* 209–224, 1980.

31. Bregman, D., Bailin, M., Bowman, F. O., Parodi, E. N., Haubert, S. M., Edie, R. N., Spotnitz, H. M., Reemstma, K., and Malm, J. R.: A pulsatile assist device (PAD) for use during cardiopulmonary bypass, *Ann. Thorac. Surg.,* 24, 574–581, 1977.

32. Edmunds, H. L.: Pulseless cardiopulmonary bypass, *J. Thorac. Cardiovasc. Surg.,* 84, 800–804, 1982.

33. Philbin, D. M., Levine, F. H., Emerson, C. W., Coggins, C. H., Buckley, M. J., and Austen, W. G.: Plasma vasopressin levels and urinary flow during cardiopulmonary bypass in patients with valvular heart disease: Effect of pulsatile flow, *J. Thorac. Cardiovasc. Surg., 78,* 779–783, 1979.

34. Frater, R. W. M., Wakayama, S., Oka, Y., Becker, R. M., Desai, P., Oyama, T., and Blaufox, M. D., Pulsatile cardiopulmonary bypass: Failure to influence hemodynamics or hormones, *Circulation Suppl.,* 62(I), 19–25, 1980.

35. Levine, F. H., Phillips, H. R., Carter, J. E., Philbin, D. M., Osbakken, M. D., Boucher, C. A., Pohost, G. M., Okada, R. D., and Buckley, M. J.: The effect of pulsatile perfusion of left ventricular function after aortocoronary bypass grafting, *Circulation Suppl.,* 64(II), 40–44, 1981.

36. Spotnitz, H. M., Berman, M. A., Reis, R. L., and Epstein, S. E.: The effects of synchronized counterpulsation of the pulmonary artery on right ventricular hemodynamics, *J. Thorac. Cardiovasc. Surg., 61,* 167, 1971.

37. Spotnitz, H. M., Berman, M. A., and Epstein, S. E.: Pathophysiology and experimental treatment of acute pulmonary embolism, *Am. Heart J., 82,* 511–520, 1971.

38. Miller, D. C., Moreno-Cabral, R. J., Stinson, E. B., Shinn, J. A., and Shumway, N. E.: Pulmonary artery balloon counterpulsation for acute right ventricular failure, *J. Thorac. Cardiovasc. Surg., 80,* 760–763, 1980.

39. Koffsky, R. M., Litwak, R. S., Mitchell, B. L., and Jurado, B. A.: A simple heart assist device for use after intracardiac surgery: Development, deployment, and clinical experience, *Artif. Organs, 2,* 257–262, 1978.

40. Pierce, W. S., Parr, G. V. S., Myers, J. L., Pae, W. E., Jr., Bull, A. P., and Waldhausen, J. A.: Ventricular assist pumping in patients with cardiogenic shock after cardiac operations, *N. Engl. J. Med., 305,* 1606–1610, 1981.

41. Bernhard, W. F., Berger, R. L., Stetz, J. P., Carr, J. G., Colo, N. A., McCormick, J. R., and Fishbein, M. C.: Temporary left ventricular bypass: Factors affecting patient survival, *Circulation Suppl., 60*(I), 131–141, 1979.

42. Schoen, F. J., Bernhard, W. F., Khuri, S. F., Koster, J. K., Van Devanter, S. J., and Weintraub, R. M.: Pathologic findings in postcardiotomy patients managed with a temporary left ventricular assist pump, *Am. J. Surg., 143,* 508–514, 1982.

43. Pennock, J. L., Pierce, W. S., Wisman, C. B., Bull, A. P., and Waldhausen, J. A.: Complications and survival following ventricular assist pumping for cardiogenic shock (abstract), *Am. Surg. Assoc. Abstr.,* 63, 1983.

44. Pennington, D. G., Merjavy, J. P., Schwartz, M. T., Codd, J. E., Barner, H. B., Kaiser, G. C., and Willman, V. T.: The importance of biventricular failure in patients with postoperative ventricular failure (abstract), *STS Abstr.,* 100, 1984.

45. Peters, J. L., McRae, J. C., Fukumasu, H., Fumio, I., and Kolff, W. J.: Transapical left ventricular bypass: A method for partial or total circulatory support, *Artif. Organs, 2,* 263–267, 1978.

46. Clark, J. W., Phillipe, E., Peterson, R., Lande, A., and Ellis, J. R.: Automatic control of a series-parallel mechanical circulatory assist system in severe uni- or biventricular failure, *Ann. Biomed. Eng., 8n,* 57–74, 1980.

47. Pennock, J. L., Pae, W. E., Pierce, W. S., and Waldhausen, J. A.: Reduction of myocardial infarct size: Comparison between left atrial and left ventricular bypass, *Circulation, 59,* 275–279, 1979.

48. Pae, W. E., Jr., and Pierce, W. S.: Temporary left ventricular assistance in acute myocardial infarction and cardiogenic shock, *Chest, 79,* 692–694, 1981.

49. Sarnoff, S. J., and Berglund, E.: Ventricular function: I. Starling's law of the heart studied by means of simultaneous right and left ventricular function curves, *Circulation, 9,* 706, 1954.

50. Lioi, A. P., Nielsen, S. D., and Olsen, D. B.: Are the Jarvik artificial ventricles limited by inflow resistance?, *Artif. Organs, 5,* 118–124, 1981.

51. Hanzelka, P., Krcma, V., Svoboda, P., Trbusek, V., Urbanek, P., Vasku, J., Urbanek, E., and Dostal, M.: BRNO-I: An implantable diaphragm-type total artificial heart: Technical aspects of design, *Artif. Organs, 4,* 65–67, 1980.

52. Boretos, J. W., and Pierce, W. S.: Segmented polyurethane: A polyether polymer, *J. Biomed. Mater. Res., 2,* 121, 1968.

53. Szycher, M., Poirier, V. L., and Dempsey, D.: Polyurethane elastomers: Their crucial role in artificial hearts, *Elastomerics, 115,* 11–15, 1983.

54. Berry, G. C., and Dror, M.: Modification of polyurethanes by interpenetrating polymer network formation with hydrogels, *Am. Chem. Soc. Div. Org. Coat. Plast. Chem. Prepr., 38,* 465–470, 1978.

55. Kolff, J., Hershgold, E. J., Hadfield, C., Olsen, D. B., Lawson, J., and Kolff, W. J.: The improving hematologic picture in long-term surviving calves with total artificial hearts, *Artif. Organs, 3,* 97–103, 1979.

56. Harasaki, H., Snow, J., Gerrity, R., Whalen, R., Ozawa, K., and Nose, Y.: Powdered metal surfaces for blood pumps, *Trans. Am. Soc. Artif. Intern. Organs, 25,* 225, 1979.

57. Bernard, W. F., LaFarge, C. G., Robinson, T., Yun, I., and Shirahaga, K.: An improved blood pump interface for left ventricular bypass, *Ann. Surg., 168,* 750, 1968.

58. Pierce, W. S.: Polymers in biomedical devices: Materials for artificial heart and circulatory assist devices, in *Johnson & Johnson Symp. Proc.: Polymers in Medicine and Surgery,* Plenum Press, New York, 1975, pp. 263–286.

59. Mochizuki, T., Lawson, J. H., Olsen, D. B., Fukumasu, H., Daitoh, N., Jarvik, R., Kessler, T. R., Pons, A. B., Hastings, L., Razzeca, K. J., Nielsen, S. D., and Kolff, W. J.: A seven month survival of a calf with an artificial heart designed for human use, *Artif. Organs, 5,* 125–131, 1981.

60. Norman, J. C., McGee, M. G., Fuqua, J. M., Igo, S. R., Turner, S. A., Sterling, R., Urrutia, C. O., Frazier, O. H., Clay, W. C., and Chambers, J. A.: Development and evaluation of a long-term implantable, electrically actuated left ventricular assist system: THI/Gould LVAS, *Artif. Organs, 7,* 64–73, 1983.

61. Hennig, E., Mohnhaupt, A., and Bucherl, E. S.: The influence of the filling pressure on the output of pneumatically driven blood pumps, *Proc. Eur. Soc. Artif. Organs, 3,* 19, 1976.

62. Kiraly, R., Yozu, R., Hillegass, D., Harasaki, H., Murabayashi, S., Snow, J., and Nose, Y.: Hexsyn trileaflet valve: Application to temporary blood pumps, *Artif. Organs, 6,* 190–197, 1982.

63. Portner, P. M., Oyer, P. E., Miller, P. J., Jassawalla, J. S., Ream, A. K., Corbin, S. D., and Skytte, K. W.: Evolution of the solenoid-actuated left ventricular assist system: Integration with a pusher-plate pump for intra-abdominal implantation in the calf, *Artif. Organs, 2,* 402–412, 1978.

64. Sandquist, G. M., Smith, L. M., and Kolff, W. J.: Plutonium-238 as a heat source for the artificial heart, *Intersoc. Energy Convers. Eng. Conf., IEEE Publ.,* 1473–1480, 1975.

65. Kovacs, S. G., Johnson, T. D., Weber, D., and Yarnoz, M. D.: Magnetically actuated heart assist pump, *IEEE Trans. Magn., 16,* 1068–1070, 1980.

65a. Oyer, P. E., Stinson, E. B., Portner, P. M., Ream, A. K., and Shumway, N. E.: Development of a totally implantable, electrically actuated left ventricular assist system, *Am. J. Surg., 140,* 17–25, 1980.

66. Wisman, C. B., Rosenberg, G., Weis, W. J., Landis, D. L., Donachy, J. H., Snyder, A. J., Richenbacher, W. E., and Pierce, W. S.: Development and successful application of an intrathoracic compliance chamber for the implantable electric motor driven ventricular-assist pump, *Surg. Forum, 34,* 253–255, 1983.

67. Sanders, S. P., Wright, G. B., Keane, J. F., Norwood, W. J., and Castaneda, A. R.: Clinical and hemodynamic results of the Fontan procedure for tricuspid atresia, *Am. J. Cardiol., 49,* 1733, 1982.

68. Jarvik, R. K.: The total artificial heart, *Sci. Am., 244,* 74–80, 1981.

69. Rosenberg, G., Snyder, A. J., Landis, D. L., Geselowitz, D. B., Donachy, J. H., and Pierce, W. S.: An electric motor–driven total artificial heart: Seven months survival in the calf, *Trans. Am. Soc. Artif. Intern. Organs, 30,* 69–74, 1984.

70. Norman, J. C., Brook, M. I., Cooley, D. A., et al.: Total support of the circulation of patient with post cardiotomy stone heart syndrome by a partial artificial heart (ALVAD) for 5 days followed by heart and kidney transplantation, *Lancet, 1,* 1125–1127, 1978.

71. Reemstma, K., Drusin, R., Edie, R., Bregman, D., Dobelle, W., and Hardy, M.: Cardiac transplantation for patients requiring mechanical circulatory support, *N. Engl. J. Med., 298,* 670–671, 1978.

72. Spotnitz, H. M., Merker, C., and Malm, J. R.: Applied physiology of the canine rectus abdominus: Force-length curves correlated with functional characteristics of a rectus powered "ventricle": Potential for cardiac assistance, *Trans. Am. Soc. Artif. Intern. Organs, 20,* 747–755, 1974.

73. Notvisky, D., Cooper, D. K. C., and Barnard, C. N.: The surgical technique of heterotopic heart transplantation, *Ann. Thorac. Surg., 36,* n4, 476–482, 1983.

74. Cooper, D. K. C.: Orthotopic and heterotopic transplantation of the heart: The Cape Town experience, *Ann. R. Coll. Surg. Eng., 66,* 228–234, 1984.

CHAPTER 39

The Artificial Kidney

Allen Zelman
Department of Biomedical Engineering
Rensselaer Polytechnic Institute
Troy, New York

39.1 THE BIOLOGICAL KIDNEY

The biological kidney performs numerous regulatory functions in addition to manufacturing important biochemicals. Primarily, the kidneys function to (1) remove nitrogeneous metabolic waste products, (2) regulate the volume of body water, (3) maintain acid-base and electrolyte composition, (4) assist in the regulation of blood pressure, and (5) assist in red blood cell production (erythropoiesis).

The biological kidney regulates the chemical composition of the plasma by filtering a plasma fraction through the glomerulus into the tubule. The wall of the tubule uses both active and passive transport processes to concentrate the filtrate into a product called *urine*. The tubule wall has the means for molecular identification of specific solutes and uses chemical energy to evoke selective reabsorption of these substances from the filtrate to the plasma and selective secretion from the plasma to the filtrate; this process is well characterized, but not well understood from a molecular point of view, and is called *active transport*. A great number of solutes have been identified as being actively transported by the tubule wall, including naturally occurring substances (creatinine, guanidine, thiamine, choline, etc.) and foreign substances [EDTA (ethylenediaminetetraacetic acid), phenol red, penicillin, etc.]. The active transport systems were probably evolved to secrete yet *unknown substances* which, because of high toxicity, must be maintained in very low concentrations in the body fluids.

Passive transport is often coupled to active transport; for example, the active reabsorption of Na^+ from the filtrate in the tubule causes a voltage to draw Cl^- also from the filtrate. Then the removal of NaCl from the filtrate causes osmosis and the reabsorption of water. Thus, the active transport of Na^+ can evoke the passive transport of Cl^- and H_2O. The active secretion of H^+ may evoke the passive reabsorption of Na^+ by ion exchange in order to maintain electroneutrality. Typically substances do not "diffuse" by concentration gradients from the plasma to the tubule lumen because the urine is concentrated relative to the plasma. However, the tubule wall is specialized in that its permeability is regulated to allow only limited back diffusion of solutes or water to the plasma. The daily normal excretion of the kidney is given in Table 39.1, and the normal solute concentration levels in the plasma are maintained within the limits given in Table 39.2.

TABLE 39.1 Daily normal excretion of biological kidneys or a single dialysis for a 70-kg person*

Substance	Excretion: Normal kidneys (24 h), g	Excretion: Uremic patient (4–5 h on 1-m² dialyzer), g
Sodium	4.2	10
Potassium	2.4	5
Phosphorus (inorganic)	0.84	0.75
Urea	24.5	20
Creatinine	1.6	3
Uric acid	0.14	1
Water	1400	1500

* The dialyzer values will vary greatly depending on patient's diet, level of exercise, and metabolic rate, but these values are reasonable for the average patient.

TABLE 39.2 Solute concentration levels in plasma

Substance	Normal	Uremic* range
Sodium, meq/L	140 (133–144)	138 (135–150)
Potassium, meq/L	4.4 (3.5–5.5)	5 (4–6)
Phosphorus, inorganic, mg/dL	3.5 (2.7–4.3)	4 (3–5)
Blood urea nitrogen,† mg/dL	12.5 (7–18)	67 (50–150)
Creatinine, mg/dL	1.1 (0.7–1.5)	13 (5–16)
Uric acid, mg/dL	4.5 (3.0–6.1)	5 (4–6)

* Patient undergoing hemodialysis 12 to 15 h/week.

† Divide values by 0.466 to determine urea concentrations in the absence of abnormal ammonia levels.

Each kidney contains approximately 1 million tubules, and as many as 90 percent of these tubules will fail before symptoms of uremia will appear. The term *uremia* literally means urine in the blood. Symptoms include nervous manifestations (apathy, drowsiness, muscular weakness, headache, vertigo, and neuralgic pains), alimentary tract disturbances (anorexia, nausea, ammoniacal breath odor), respiratory alterations (dyspnea of hypertensive heart failure, metabolic acidosis, and pulmonary edema), circulatory manifestations (alterations in rate and rhythm of the heart, myocardial contractability, altered coagulation, and anemia), and skeletal manifestations (osteomalacia and rickets). In addition, there will be an abnormal electrolyte composition and osmolarity of the plasma and excess body water.

If the kidneys fail altogether, death will ensue within a few days or weeks. However, many of the functions of the biological kidney can be assumed by the artificial kidney, and the life and well-being of kidney-failure patients may be extended by a great many years. Many patients have lived longer than 10 years on an artificial kidney; the average life span is about 6 years and is increasing every decade [1]. Today some 200,000 patients worldwide are maintained by hemodialysis.

39.2 GENERAL DESCRIPTION OF THE ARTIFICIAL KIDNEY

The *artificial kidney* is a device which can partially simulate the biological kidney in three functions. It can (1) remove nitrogenous metabolic waste products, primarily small-molecular-weight solutes such as urea, uric acid, and creatinine; (2) remove excess body water; and (3) partially reestablish appropriate plasma acid-base and electrolyte composition and concentrations. Upon failure of the biological kidney, the patient's blood must be treated on the artificial kidney for 4 to 5 h per treatment, 3 days a week; the treatment is called *hemodialysis*.

The artificial kidney is composed of two primary components: a *hemodialyzer* in which all mass transport takes place, and an *automatic dialysate delivery system* which continuously supplies an aqueous solution called a *dialysate* to the hemodialyzer. The dialysate carries away the waste products removed by the hemodialyzer. The relationship between patient, hemodialyzer, and automatic dialysate delivery system is shown in Figure 39.1. The term *hemodialyzer* is usually shortened to simply *dialyzer*.

Figure 39.2 shows a schematic diagram for a dialyzer, indicating notation for blood flows Q_B, dialysate flows Q_D, and hydrostatic pressures P. Blood enters the device, separated from the dialysate by a hydrophilic membrane which is permeable to water and solutes up to a few thousand daltons.

The transfer processes for nonelectrolyte, electrolyte, and water which occur across the membrane between the blood and dialysate are in principle enormously complex but can be simplified for this preliminary discussion. Nonelectrolytes migrate primarily as a result of their transmembrane concentration differences across the membrane; this process is often referred to as *diffusion*. The only nonelectrolyte in the dialysate is dextrose, which migrates into the blood, whereas all other nonelectrolytes migrate from the blood into the dialysate, which is discarded. Electrolytes do not migrate across the membranes per se; ions do. As a consequence, there is an

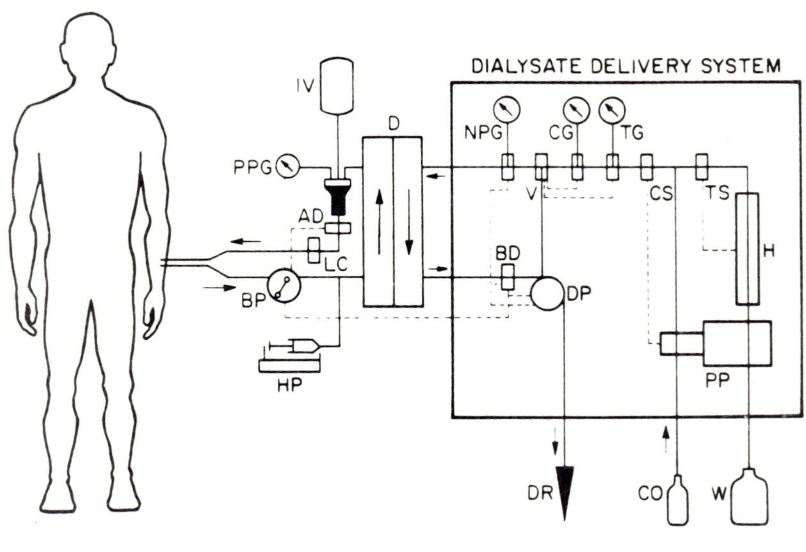

AD AIR DETECTOR
BD BLOOD DETECTOR
BP BLOOD PUMP
CG CONDUCTIVITY METER
CO CONCENTRATE
CS CONDUCTIVITY FEEDBACK SENSOR
D DIALYZER
DP DIALYSATE PUMP
DR DRAIN
H HEATER
HP HEPARIN INFUSION PUMP
IV INTRAVENOUS FLUIDS
LC LINE CLAMP
NPG NEGATIVE PRESSURE GAUGE
PP PROPORTIONING PUMP
PPG POSITIVE PRESSURE GAUGE
TG TEMPERATURE GAUGE
TS TEMPERATURE FEEDBACK SENSOR
V BYPASS VALVE
W WATER

FIG. 39.1 Hemodialysis.

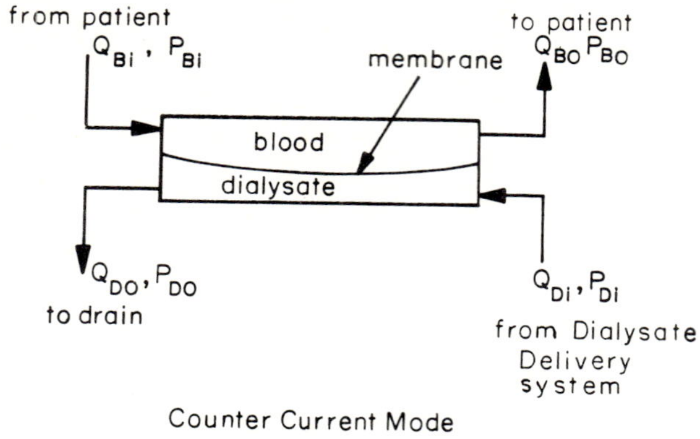

FIG. 39.2 Countercurrent mode.

enormously complex ion exchange–diffusion process which occurs with the following results: H^+, Na^+, K^+, Ca^{2+}, Mg^{2+}, HCO_3^-, $H_2PO_4^-$, and Cl^- leave the plasma, enter the dialysate, and are discarded; acetate leaves the dialysate and enters the plasma, where it is metabolized to HCO_3^-. Some patients who have little capacity to metabolize acetate to bicarbonate must be dialyzed using a dialysate containing HCO_3^-; this enormously complicates dialysate delivery because Ca^{2+} tends to form insoluble calcium carbonate in the presence of bicarbonate and the dialysate pH must be carefully controlled to prevent this chemical reaction. As a result of these ion exchange processes, a voltage must occur across the membrane; however, this transmembrane voltage is not thought to be of clinical significance and it has not been seriously investigated, nor has it been taken into account for characterization of mass transport during hemodialysis. The osmolality of the plasma of dialysis patients is between 300 and 320 mosmol and that of the dialysate is about 265 mosmol, and water tends to leave the dialysate and enter the plasma by osmosis. However, the effective osmotic pressure difference is small and due solely to plasma proteins (~21 mmHg) [see Eq. (39.2)]. Since dialysate patients cannot urinate, they must always lose water during dialysis. To ensure the loss of water from the patient, the hydrostatic pressure of the dialysate is lowered below 1 atm so that the transmembrane hydrostatic pressure is sufficiently great to overcome osmosis and force water from the patient.

The complexity of regulating these transfer processes led, in the early years of hemodialysis, to a problem called *disequilibrium syndrome*. When the nitrogenous waste products are removed too quickly, the osmolality of the plasma is lowered more quickly than that of the cerebrospinal fluid; hence water will enter the brain by osmosis, inducing a hydrostatic pressure which stresses the brain and spinal column. Such a condition produces severe clinical symptoms, including headaches and seizures, and may cause death. The clinical practice of hemodialysis requires the rate of solute removal to be sufficiently rapid to minimize the time the patient must spend connected to the machine, but sufficiently slow to prevent the disequilibrium syndrome; thus the schedule of dialysis must be tailored to the needs of each patient.

Essentially by trial and error, dialysate formulas, blood and dialysate flows, hydrostatic pressures, and dialyzer designs have been adopted to minimize the disequilibrium syndrome and maximize solute and water removal in order to minimize the patient's time on the artificial kidney. A commonly used dialysate formula is given in Table 39.3. The blood flow is usually set from 200 to 300 mL/min, with the lower value being more common. The dialysate flow should be as high as economically practical and is generally set at about 500 mL/min for single-pass dialysate flows.

The rate at which water is removed from the patient, i.e, ultrafiltration, Q_v, is regulated by measuring $P_{B,o}$ (hydrostatic pressure of outgoing blood) and adjusting $P_{D,i}$ (hydrostatic pressure of incoming dialysate); only $P_{B,o}$ and $P_{D,i}$ are measured during dialysis. For clinical purposes,

$$Q_{v,\text{clin}} = A_m L_p (P_{B,o} - P_{D,i}) = A_m L_p (P_{B,o} + |P_{D,i}|) \tag{39.1}$$

TABLE 39.3 Composition of a typical dialysate

	Unit		
	g/L	meq/L	mmol/L
$CaCl_2 \cdot 2H_2O$	0.184	2.50	1.25
$MgCl_2 \cdot 6H_2O$	0.153	1.50	0.75
KCl	0.149	2.00	2.00
NaCl	5.844	100.00	100.00
$NaC_2H_3O_2$	2.707	33.00	33.00
Dextrose	2.000	....	11.11

where A_m is the membrane area in m^2, L_p is the hydraulic permeability in mL/(h · m^2 · mmHg), and $A_m L_p$ is the *ultrafiltration index* for the dialyzer. A more accurate equation for ultrafiltration is

$$Q_v = A_m L_p(\overline{\Delta P_m} - \pi_p) \tag{39.2}$$

where $\overline{\Delta P_m}$ is called the *mean transmembrane hydrostatic pressure drop*, or TMP, and is defined by

$$\text{TMP} = \overline{\Delta P_m} = \frac{P_{B,i} + P_{B,o}}{2} - \frac{P_{D,i} + P_{D,o}}{2} \tag{39.3}$$

and π_p is the effective transmembrane osmotic pressure drop. For all practical purposes, π_p is due entirely to plasma proteins, and for reclining patients is 21 mmHg. Equation (39.1) is a reasonable approximation, for $P_{B,i} - P_{B,o} \approx 20$ mmHg and $P_{D,i} - P_{D,o} \approx 50$ mmHg for most dialyzers. Then

$$Q_v = A_m L_p(P_{B,o} + |P_{D,i}| + 14 \text{ mmHg}) \tag{39.4}$$

By comparing Eqs. (39.1) and (39.4) one sees that the clinical value for controlling Q_v is in error only by about 14 mmHg. Most dialyses proceed with $(P_{B,o} - P_{D,i}) \approx 250$ to 400 mmHg; hence Q_v can be predicted to within ±6 percent during dialysis. When Q_v is too rapid, the patient's blood pressure tends to drop, so this parameter is measured several times during dialysis; also the patient is usually weighed once or twice during dialysis.

Dialyzers have been designed with three possible blood flow–dialysate flow patterns. These are depicted in Fig. 39.3. The countercurrent mode is the most efficient mode; it maximizes both transmembrane solute concentration differences and hydrostatic pressures. The cocurrent

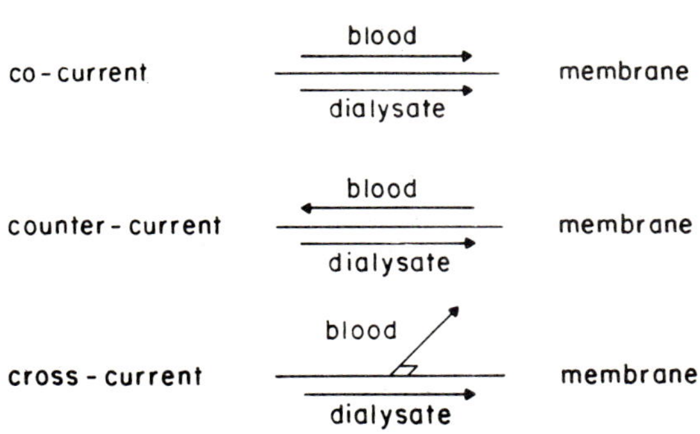

FIG. 39.3 Patterns of flow in dialyzers.

mode is the least efficient because it minimizes solute removal and the hydrostatic pressure differences. The efficiency of the cross-current mode lies between the extremes of cocurrent and countercurrent modes. In standard operation the dialysate flow is approximately 2.5 times that of the blood flow, so the solute removal is not substantially different in these three designs, i.e., ±15 percent difference; however, the water removal *can* be substantially different, as we shall see.

39.2.1 Types of Dialyzers

Flat-Plate Dialyzer. Figure 39.4 is a schematic diagram of a flat-plate dialyzer. This dialyzer is composed of multiple layers of two sheets of membrane and plastic spacers. The blood flows

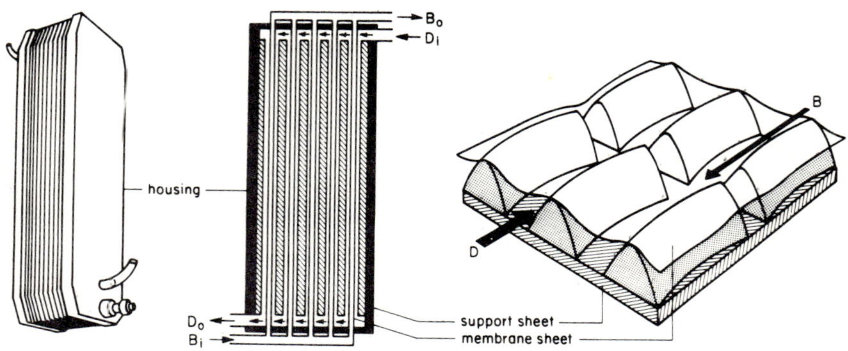

FIG. 39.4 Flat-plate dialyzer.

between the two sheets of membrane, whereas the dialysate flows on both sides of the blood-membrane channel between the membrane and spacer. The plastic spacer has a pattern molded onto its surface to enhance dialysate and blood mixing. Dialysate is pumped in a single pass, i.e., once through the dialyzer, and then drains at ~500 mL/min. Typically these dialyzers have about the same low flow resistance to blood and dialysate, i.e., ~0.1 mmHg · min/mL. They can be sterilized by gamma radiation or ethylene oxide; there is no need to prerinse these dialyzers since all dialyzers are initially primed with sterile saline and dialysate to remove air, which is sufficient to rinse out the high concentrate of glycerol used to keep the membrane pliable. Presently manufactured flat-plate dialyzers generally have moderate compliance with an increase in volume of no more than 0.5 mL/mmHg TMP. Because of their low flow resistance and compliance, these dialyzers can be operated in the cocurrent mode even with "high-flux" membranes. The compliant membrane system will act to minimize the TMP, thus allowing simple operation of the dialyzer from zero to high ultrafiltration rates with considerable control and accuracy of ultrafiltration rate [2].

Coil Dialyzer. In the coil dialyzer (Fig. 39.5), the blood flows through a flattened tubing which is wound, together with a plastic mesh, around a cylindrical core. Blood pressure (supplied by a blood pump) forces the membrane against the mesh, forming a pattern in the blood path; thus the mesh acts to enhance blood and dialysate mixing. The dialysate delivery is generally single-pass recirculation; fresh dialysate is fed into and removed from a small chamber at ~500 mL/min while another pump circulates the dialysate between the dialyzer and the chamber at ~1500 mL/min. The high flow through the dialyzer tends to enhance mass transfer by mitigating boundary layer effects, whereas the recirculating dialysate tends to accumulate nitrogenous wastes, which lowers mass transport. As a consequence, the coil dialyzer does not perform more effectively than other designs. Coil dialyzers have been designed with a high resistance to blood flow and require a significant hydrostatic pressure drop between the blood ports. Conse-

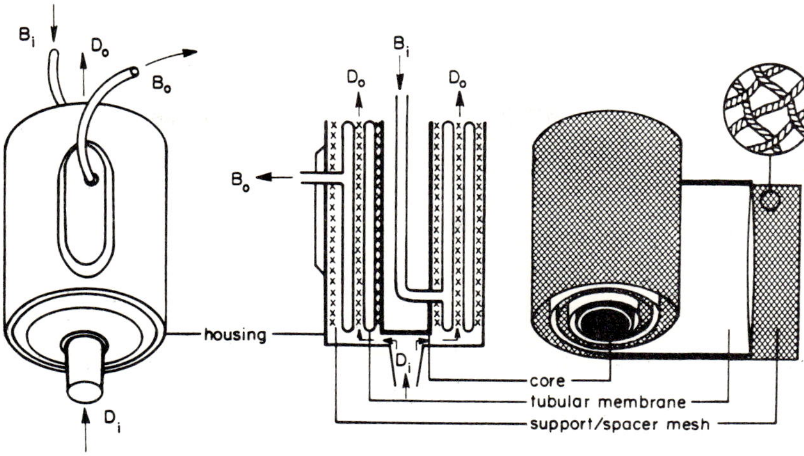

FIG. 39.5 Coil dialyzer.

quently, they tend to have a high minimum ultrafiltration rate, and patients often require sterile saline during dialysis to prevent dehydration. These dialyzers can be sterilized by gamma rays or ethylene oxide and are easily rinsed and set up. They do have a high compliance (~1 mL/mmHg TMP), which means that as a TMP is increased the prime volume increases also. This feature increases the blood layer thickness and lowers the mass transfer rate. However, ultrafiltration increases when TMP is increased, which in turn tends to increase mass transfer, and these two factors tend to offset each other. Because of these design features, the coil dialyzer is more difficult to control and predict clearance, but its simplicity and low cost have provided continued popularity.

Hollow-Fiber Dialyzer. The hollow-fiber dialyzer is the most recent development in dialyzer design (see Fig. 39.6). Blood flows through the hollow fibers. A typical unit may have 5092 hollow fibers with a bore size of 127 μm radius and 25 cm long to make a 1-m²-surface-area dialyzer. There is generally some clotting of fibers as dialysis proceeds, so the number of fibers is often increased by 20 to 50 percent to ensure adequate dialysis. The hollow fibers are glued together by an epoxy on each end; the ends are then cut perfectly flat with a microtome. A

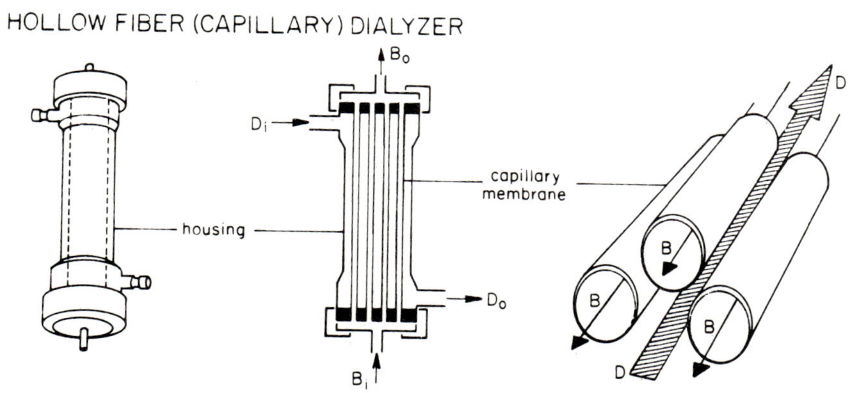

FIG. 39.6 Hollow-fiber (capillary) dialyzer.

uniformly distributed flow of dialysate through the bundle is essential for adequate mass transfer, and considerable ingenuity has gone into this design feature. The hollow-fiber dialyzer has nearly zero compliance. In addition, the hydraulic resistance of both the blood and dialysate paths are fairly low. It is the smallest of all dialyzers and has the lowest prime volume. These many positive features have made this style very popular. In the past the dialyzer had to be stored in formaldehyde for sterile shipment, which was a nuisance; this problem is being overcome by use of ethylene oxide or heat sterilization.

39.2.2 The Dialysate Delivery System

The hemodialyzer operates with a 37°C, balanced electrolyte, nonsterile solution, the dialysate, flushing the side of the membrane opposite the blood. The dialysate delivery system (DDS) is a machine which continuously manufactures a biologically compatible dialysate from tap water and preprepared concentrate (see Fig. 39.1). Tap water is often pretreated with activated charcoal to remove chlorine and chloramines, and demineralized to ensure low mineral content. Water entering the DDS is continuously mixed with a concentrate (34 parts water to 1 part concentrate) to form the dialysate. The dialysate is heated to 37 ± 1°C and monitored by sensors for temperature and conductivity (see Sec. 39.5 for calculations). Proper temperature is maintained by a proportional controller and a stainless steel heat exchanger. Conductivity is measured, but usually not controlled.

A positive TMP is used to remove water from the patient. In practice it has been found safer to reduce the dialysate pressure rather than raise the blood pressure. A negative dialysate pressure (0 to −500 mmHg) will produce air bubbles which could be trapped inside the dialyzer and decrease the surface area for mass transfer. A system has been developed to deaerate the dialysate before it enters the negative pressure portion of the system. The dialysate then enters the dialyzer with a very low partial pressure of oxygen, which causes oxygen to leave the blood and enter the dialysate. The venous return line from the dialyzer becomes a dark red, indicating deoxygenation. A negative pressure system (separate from the deaeration system) can maintain an adjustable degree of negative pressure. The negative pressure, $P_{D,i}$, alone is adjusted to control the rate of ultrafiltration from the patient. The dialyzer and dialysate delivery system must act in concert so that the blood pressure in the dialyzer is greater than the dialysate pressure everywhere along the membrane. In this way the patient is protected if the membrane should rupture because sterile blood will be forced into the nonsterile dialysate rather than vice versa. Since water and salt solutions are relatively inexpensive, the solution leaving the dialyzer is discarded. Regeneration of the dialysate is technically feasible, and one such system is commercially available but at considerable additional cost.

Each dialysate delivery system is equipped with the following safeguards. (1) A light detector is used to detect air bubbles in the blood lines by changes in opacity of the flowing blood. When bubbles are detected in the venous line, a clamp is closed, preventing blood from entering the patient, and the blood pump is turned off. Approximately 50 cm^3 of air entering the patient at one time could cause death. (2) A spectrophotometric hemoglobin detector which detects the presence of hemoglobin in the dialysate also activates the line clamp and shuts off the blood pump. (3) If the temperature or conductivity of the dialysate vary outside preset limits (usually set by the dialysis staff), the dialyzer is bypassed and the inappropriate dialysate is diverted to the drain. As a result of the high reliability of machine fail-safe systems and the high level of competence among dialysis staff, hemodialysis is now one of the safest medical procedures.

39.3 BRIEF HISTORICAL ACCOUNT OF THE ARTIFICIAL KIDNEY

A considerable amount of human effort has been devoted to the development of the artificial kidney. The first successful dialysis on a live animal was reported in 1913 by Abel, Rowntree, and Turner [3]. They manufactured a "celloidin" membrane, used arterial and venous cannulas, and described the major groups of electrolytes and nonelectrolytes which are dialyzable.

Two developments were necessary for practical application of the dialysis procedure, an appropriate membrane and a biologically compatible anticoagulant. In the 1930s the commercial production of cellophane membranes and a standard heparin solution made dialysis of humans possible.

In 1944 Kolff and Berk [4] at Kempen, Netherlands, reported the first dialysis on a human being using the rotary drum dialyzer which they designed. By 1960 B. H. Scribner initiated the first program for the long-term use of hemodialysis in the treatment of chronic end-stage uremia. Quinton, Dillard, and Scribner [5] developed the Silastic-Teflon cannula system. This is an extracorporeal arteriovenous shunt which separates for insertion of the hemodialyzer into the patient's blood circuit for the dialysis treatment. In 1965 Cole et al. [6] reported the first use of an on-line dialysate proportioning system. A major advance in blood access came in 1966 when Brescia et al. [7] reported a subcutaneous arteriovenous (AV) fistula. Now blood access could be obtained without risk of infection by insertion of needles through skin during each dialysis. Eschbach et al. [8] introduced in 1966 a fail-safe dialysis system with monitors for dialysate temperature, conductivity, blood and dialysate pressures, and blood leaks; these devices were incorporated into a single patient station with a small dialysate delivery system. In 1971 Gordon et al. introduced a system for regenerating dialysate by ion exchange, sorption, and enzymatic decomposition of urea [9].

There are several techniques and developments that at present are primarily research novelties but may prove to be valuable treatments. Henderson et al. [10] introduced a hemofiltration technique which utilizes a highly porous membrane and a TMP high enough to remove 15 to 80 L of ultrafiltrate per dialysis; this fluid is replaced by sterile dialysate. This process removes larger quantities of higher-molecular-weight solutes than standard hemodialysis. The technique of *hemoperfusion,* the process of flushing the blood directly over a biologically compatible material which sorbs solutes, was introduced in 1964 by Yatzidis [11], and has been refined by coating the sorbents to prevent destruction of formed blood elements and particle emboli [12, 13]. In 1976 Lai et al. [14] introduced the first "smart" dialysate regeneration system incorporating feedback control for the dialysate concentration of Na^+, H^+, and HCO_3^-.

When hemodialysis is not technically possible, one resorts to peritoneal dialysis. This method injects fluid into the peritoneal cavity of the patient and removes waste products by diffusion through the vascular walls of the stomach and intestine. A major change in the character of the artificial kidney came with the recent introduction of *continuous ambulatory peritoneal dialysis* (CAPD) by Popovitch, Montcrief, and Nolph [15]. They determined that the process of "adding 2 L of fluid to the peritoneal cavity, allowing 4 h for equilibration, then draining, five times a day 6 days a week," was sufficient dialysis to maintain an anephric patient. This form of peritoneal dialysis has gained some 10 percent of the dialysis market in the United States; this figure is expected to increase to 25 percent.

39.4 CHARACTERIZATION OF MASS TRANSPORT DURING HEMODIALYSIS

39.4.1 Introduction

When chemical substances migrate from one region to another, there has been a traditional but arbitrary differentiation between migration associated with diffusion and migration associated with convection. *Diffusional transport* is depicted as a random motion of each individual molecule while the component as a whole slowly migrates, or "drifts," from one region to another. The diffusion, or drift, velocity of the component is restricted by molecular frictional interactions between the migrating species and all the other chemical components in the system. *Convective transport* is depicted as the nonrandom migration of a volume element while the molecules within the volume element continue their random motion insensitive to the net displacement of the volume element. The convective velocity is restricted by molecular frictional interaction between one volume element and an adjacent volume element. Inasmuch as volumes of mass are composed of molecules and the molecules themselves are what interact with one another, it should be obvious that the designation of volume-volume interaction in a convective process and molecule-molecule interaction in a diffusion process is an arbitrary distinction and may or may not be a practical resolution of the transport phenomena, depending upon the circumstances. The transport coefficient used to depict convective transport is called *viscosity,* whereas the transport coefficient used to depict diffusion is called *diffusivity.*

It should be noted that *all* two-chemical-component systems are in reality three-component systems: *container, solvent,* and *solute.* The container could be pipe, trough, or capillary. The Fick and Navier-Stokes equations neglect the material of the container through which the transport occurs. This is a valid approach so long as the molecular interactions between the container and solvent-solute system are not sufficiently strong to invoke a space dependence on the viscosity or diffusivity.

Now consider the container (pipe, trough, or capillary through which transport takes place) to become very small. Allow it to become as small as 0.5 to 2 nm in diameter, i.e., the typical pore size for a membrane. Now the container no longer acts as an inert part of the system, for the molecular interactions between it and the solute and solvent become strong. Consequently, what previously was classified as a two-component system now is a bona fide three-component system with the container now called a *membrane.* When the dimensions of the membrane pore size become inordinately small, the physical concepts of diffusivity and viscosity become ambiguous.

At this time *no* set of mass transport equations has been generally accepted as the most correct means of characterizing membrane transport phenomena. In addition, conducting a search of the literature is difficult because of the many approaches used to catalog data.

The problems associated with proper characterization of membrane transport phenomena are further compounded when the membrane is placed in a hemodialyzer. These problems are easy to describe but difficult to resolve:

1. Because of volume flow through the membrane and tortuosity of flow parallel to the membrane, concentration polarization and boundary layer effects are enormously difficult to attack rigorously.
2. The channel geometry and membrane are not mass-produced in a perfectly uniform manner. As a consequence, every type of dialyzer, as well as every individual dialyzer, will have somewhat different transport characteristics. Typical variations between different batches of similar dialyzers have mass transport characteristics within only ± 10 to 20 percent.
3. The transport characteristics of the membranes for dialysis may be altered during shipment unless carriers take precautions to ensure small variations in humidity and temperature.
4. During dialysis, proteins adhere to the membrane surface, clotting occurs within the dialyzer, and blood pressure can, in some dialyzers, force the membrane against the membrane support blocking area exposed to the dialysate.

With such a multitude of problems it is remarkable that mass transport during hemodialysis can be accurately quantitated at all. Three developments have made accurate characterization of hemodialysis possible:

1. Introduction of simple, yet accurate, phenomenological equations for water and solute transport during dialysis.
2. Substantiation that only plasma proteins contribute significantly to osmosis during hemodialysis; i.e., proteins bound to the membrane do not contribute to osmosis.
3. Introduction of a wholly intuitive, but accurate, equation for solute clearance as a function of blood flow and ultrafiltration rate at constant dialysate flow rates.

The literature contains dozens of different chemically formulated membranes proposed for hemodialysis. Of these only four have withstood the rigors of clinical trials: cellulose, cellulose acetate, polyacrylonitrile, and polycarbonate. The clinical attributes of these materials continue to be debated and practitioners prescribe dialyzers more on the basis of cost and convenience than on uncertain scientific evidence demonstrating the merits of one material over another. The cellulose membrane, Cuprophan PT-150, produced by the West German firm Enka-Glanstoff, has become the standard to which all other membranes are evaluated because of its uniform quality, clinical effectiveness, and low cost. This membrane is fabricated into sheets, rolls, and hollow fibers. All membranes with solute and water permeabilities approximately equal to PT-150 are considered *standard* membranes, whereas membranes with vitamin B_{12} permeabilities and water permeabilities three or more times greater than PT-150 are called *high-flux* membranes. The mechanical and permeability parameters of these membranes are given in Ref. 16.

39.4.2 Ultrafiltration during Hemodialysis

Staverman derived a formula relating the ultrafiltration rate Q_v to the mean transmembrane hydrostatic pressure $\overline{\Delta P_m}$ and the individual solute transmembrane osmotic pressure differences, $\Delta\pi_i$:

$$Q_v = A_m L_p(\overline{\Delta P_m} - \Sigma\, \sigma_i\, \Delta\pi_i) \tag{39.5}$$

where A_m is the membrane area in m^2; L_p is the hydraulic permeability in $mL/(h \cdot m^2 \cdot mmHg)$; σ_i is the Staverman reflection coefficient, which is dimensionless; and $\Delta\pi_i$ is the transmembrane osmotic pressure difference of each solute i, in mmHg. Q_v is positive out of the blood and Δ is taken as blood side minus dialysate side. This equation has gained considerable popularity because of its simplicity and accuracy over a wide range of experimental conditions. So long as the membrane is nonionic (as are all membranes used for hemodialysis) and properly supported to prevent significant tension or compression of the membrane by the applied hydrostatic pressure, L_p and σ_i are essentially independent of $\overline{\Delta P_m}$ and $\Delta\pi_i$. One can show that the value of σ_i may be between $-\infty$ and $+1$. When $\sigma_i = 1$, the solute is impermeable, and for all hemodialysis membranes σ_i has never been shown to be less than 0. Equation (39.5) can be greatly simplified because $\sigma_1\, \Delta\pi_i \ll \overline{\Delta P_m}$ except for protein, which is for all practical purposes impermeable and has $\sigma_p = 1$. Thus for hemodialysis Eq. (39.5) reduces to

$$Q_v = A_m L_p(\overline{\Delta P_m} - \pi_p) \tag{39.6}$$

where π_p is the osmotic pressure of the plasma proteins, ~21 mmHg for a reclining patient and ~25 mmHg for a standing patient. For dialyzers using Cuprophan PT-150 membranes, L_p is a constant during hemodialysis and $L_p \approx 3\ mL/(h \cdot m^2 \cdot mmHg)$ [17]. However, when very porous (high-flux) membranes are used, that is, $L_p > 6\ mL/(h \cdot m^2 \cdot mmHg)$, then L_p decreases linearly with time during dialysis according to the equation [17]

$$L_p(t) = L_p^0(1 - \alpha t) \tag{39.7}$$

where $$L_p^0 = L_p \quad \text{at } t = 0$$

and the rate constant, α, depends on the particular membrane and/or dialyzer, but has been shown to be about 6 to 7 percent per hour for all so-called high-flux dialyzers. Table 39.4 gives

TABLE 39.4 In vivo ultrafiltration parameters of high-flux dialyzers: $Q_v = A_m L_p^0(1 - \alpha t)(\overline{\Delta P_m} - \pi_p)$*

High-flux dialyzer	L_p^0, mL/(h · m² · mmHg) ± %	α, h⁻¹ ± %	Q_v, mL/h	NDP†	N†
Gambro HF, 1.36 m² (cocurrent)	8.09 ± 1.1	0.0616 ± 4.0	±72	575	8
RP-610, 1.0 m² (cocurrent)	17.4 ± 2.0	0.0707 ± 6.0	±101	480	8
CDAK, 1.8 m² (cocurrent)	6.17 ± 2.5	0.0657 ± 8.0	±88	295	4
CDAK, 1.8 m² (countercurrent)	7.28 ± 4.5	0.0597 ± 12	±86	209	2

* For all dialyzers $\pi_p = 21 \pm 6$ mmHg; range, 12–29 mmHg. The symbol ± indicates standard deviation; actual error estimated at ±10 to 15%. Input dialysate flow, 500 ± 5 mL/min; input blood flow, 225 mL/min; $0 \le Q_v \le 10$ mL/min.

† NDP = number of data points; N = number of dialyses.

values of L_p^0 and α for a few high-flux dialyzers. Extensive data for standard dialyzers can be found in Ref. 16. Although there is a protein deposit on the membrane during hemodialysis which could affect the osmotic pressure across the membrane, by direct measurement of Q_v, ΔP_m, and π_p, Eqs. (39.6) and (39.7) have been shown to accurately characterize hemodialysis under all clinical conditions. That is, the protein deposition on the membrane may affect L_p but does not contribute to π_p.

39.4.3 Solute Transfer during Hemodialysis

According to the theory of nonequilibrium thermodynamics, in the transfer of n independent solutes, there may be up to $(n)(n + 1)/2$ independent transport coefficients; e.g., 10 transported species may need as many as 55 transport coefficients to characterize the process.

The complexity of the mass transfer processes during hemodialysis has necessitated the introduction of a simple, yet useful, parameter for characterization of all solute removal, including ions. This parameter is *clearance C*. If in 1 min all the solute removed by the dialyzer were taken from a single volume of blood, this volume per minute would be the clearance. Thus the clearance is related to solute removal by

$$Q_s = c_{B,i}C \tag{39.8}$$

where Q_s is the rate of solute i removed by the dialyzer, in mg/min; and $c_{B,i}$ is the concentration of that solute in the blood entering the dialyzer, in mg/100 mL. It is obvious from Fig. 39.2 that

$$Q_s = c_{B,i}C = Q_{B,i}c_{B,i} - Q_{B,o}c_{B,o} = Q_{D,o}c_{D,o} - Q_{D,i}c_{D,i} \tag{39.9}$$

The clearance of solute is expressed solely in terms of blood flows, dialysate flows, and concentrations of solute. The clearance C of any solute is then given by

$$C = \frac{Q_{B,i}c_{B,i} - Q_{B,o}c_{B,o}}{c_{B,i}} = \frac{Q_{D,o}c_{D,o} - Q_{D,i}c_{D,i}}{c_{B,i}} \tag{39.10}$$

The relationship

$$Q_{B,i} - Q_{B,o} = Q_v \tag{39.11}$$

permits modification of Eq. (39.10) into a great many forms. The most easily measured clinical parameters are $Q_{B,i}$ (measured by setting an accurate blood pump), Q_v (measured directly [17, 18] or by measuring ΔP_m and calculating Q_v), $Q_{D,o}$ (measured directly), and the solute concentrations, so the clearance is generally calculated from the equations

$$C = \left(\frac{c_{B,i} - c_{B,o}}{c_{B,i}}\right)(Q_{B,i} - Q_v) + Q_v = \frac{Q_{D,o}(c_{D,o} - c_{D,i}) + Q_v c_{D,i}}{c_{B,i}} \tag{39.12}$$

The clearance is a nonphysical parameter indicating neither diffusive nor convective mass transport. The clearance for a single solute may have a wide range of values and is a strong function of $Q_{B,i}$, Q_v, $Q_{D,i}$, $C_{D,i}$, and temperature, as well as dialyzer design and type of membrane. Fortuitously, the nonelectrolyte and electrolyte clearances are essentially independent of $c_{B,i}$, all other solute concentrations, and time. This concentration and time independence are probably due to the facts that both plasma and dialysate composition must be maintained within narrow bounds, i.e., within a few millimoles per liter, for each solute and that the gelatinous protein buildup on the membrane is not a significant diffusion barrier.

As $c_{D,i}$ is varied from 0 to $c_{B,i}$, the clearance varies from a maximum to a minimum. This concentration dependence of clearance can be avoided by introduction of another transport coefficient, called the *dialysance D*, which is defined by

$$Q_s = D(c_{B,i} - c_{D,i}) \tag{39.13}$$

and

$$D = \frac{Q_{B,i}c_{B,i} - Q_{B,o}c_{B,o}}{c_{B,i} - c_{D,i}} = \frac{Q_{D,o}c_{D,o} - Q_{D,i}c_{D,i}}{c_{B,i} - c_{D,i}} \tag{39.14}$$

For all its simplicity, the dialysance is for all "practical" purposes independent of $c_{D,i}$; thus time-dependent processes for solutes in the dialysate can be easily modeled. The numerical value of D is generally found by performing a clearance experiment with $c_{D,i} = 0$; then $D = C$.

The values of the clearance for most dialyzers have been adequately catalogued [16, 19]. A composite of clearance values for standard dialyzers is given in Table 39.5, and current values for high-flux dialyzers are given in Table 39.6.

TABLE 39.5 Typical values of clearance for standard 1-m² dialyzers*

Substance	Molecular weight	Approximate clearance, mL/mm	Approximate variation, mL/mm
Urea	60	125	±15
Creatinine	113	100	±15
Uric acid	168	80	±10
Vitamin B_{12}	1355	25	±6

* $Q_{B,i} \approx 200$ mL/min, $Q_{D,i} \approx 560$ mL/min, $Q_v \leq 10$ mL/min, $c_{D,i} = 0$, and $T = 37°C$.

TABLE 39.6 In vivo clearance (mL/min)* for high-flux dialyzers

Dialyzer	Urea	Creatinine	Inorganic phosphorus	Uric acid	NDP†	N†
Gambro, HF, 1.36 m² (cocurrent)	125 ± 6	112 ± 6	103 ± 10	104 ± 9	10	2
RP-610, 1.0 m² (cocurrent)	126 ± 4	110 ± 4	96 ± 5	88 ± 6	15	3
CDAK, 1.8 m² (cocurrent)	146 ± 4	134 ± 5	123 ± 15	123 ± 9	7	2
CDAK, 1.8 m² (countercurrent)	173 ± 6	148 ± 2	128 ± 7	134 ± 5	5	1

* The symbol ± indicates standard deviation; actual error estimated at ±10–15%. Input dialysate flow, 500 ± 5 mL/min; input blood flow, 225 mL/min; $0 \leq Q_v \leq 10$ mL/min.

† NDP = number of data points; N = number of dialyses.

To achieve a reasonable description of the mass transport characteristics of hemodialyzers, one resorts to a graphical representation called a *nomogram*. Nomograms can be very useful clinically because at a glance one can see how a particular dialyzer is supposed to perform. However, traditional methods for characterization of hemodialysis use parameters that are not independent; thus, nomograms presently in use are not as useful as they should be. In particular, one would like to know clearance as a function of $Q_{B,i}$ and Q_v. Again a strictly intuitive approach proved to be adequate. Zelman and Gisser proposed the simple formula

$$C_i = \alpha_0 + \alpha_1 Q_v + \alpha_2 \frac{1}{Q_{B,i}} \tag{39.15}$$

Equation (39.15) permits a reasonably unambiguous separation of ultrafiltration rate and blood flow rate in the characterization of in vitro dialyzer clearance at a specified constant $Q_{D,i}$. Figures 39.7 and 39.8 depict the in vitro clearance of a standard dialyzer, the Vivacell, and a high-flux dialyzer, the RP-6. One notices that ultrafiltration has almost no effect on the clearance of urea which has a molecular weight of 60, while Q_v significantly increases the clearance of vitamin B_{12} with a molecular weight of 1355.

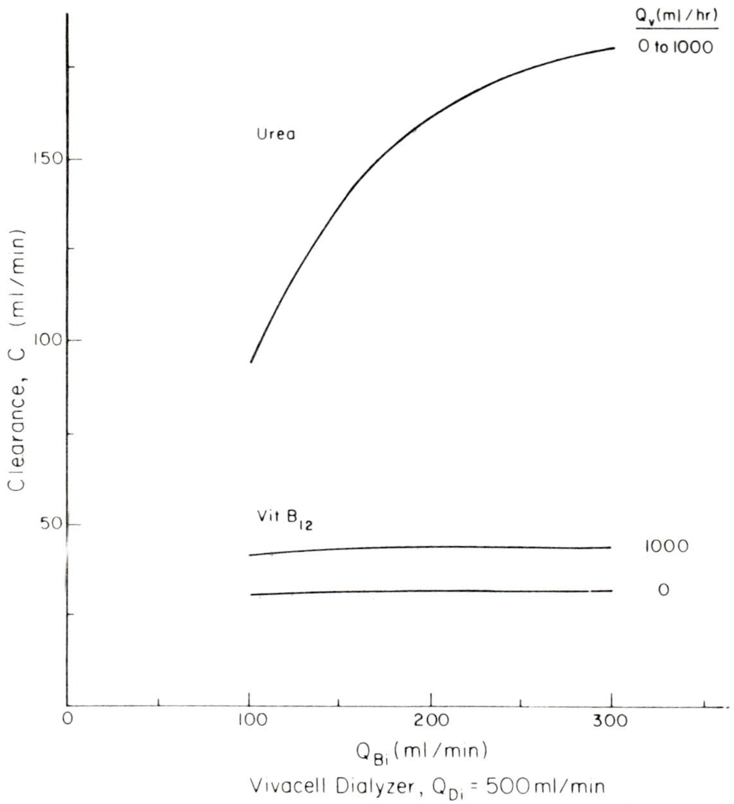

FIG. 39.7 Clearance of Vivacell dialyzer.

39.4.4 Mass Transport Coefficients

The mass transport coefficient K is defined by the equation

$$\delta Q_s = K(C_B - C_D)\, dA_m \tag{39.16}$$

where δQ_s is the flux of solute per unit time through the differential area element dA_m. Equation (39.16) has been integrated over the surface of the membrane for countercurrent dialysis, with the assumption that $Q_v = 0$, with the result

$$Q_s = KA_m \frac{C_{B,o} - C_{D,i} - (C_{B,i} - C_{D,o})}{\ln (C_{B,o} - C_{D,i})/(C_{B,i} - C_{D,o})} \tag{39.17}$$

or for cocurrent dialyzers as

$$Q_s = KA_m \frac{C_{B,o} - C_{D,o} - (C_{B,i} - C_{D,i})}{\ln (C_{B,o} - C_{D,o})/(C_{B,i} - C_{D,i})} \tag{39.18}$$

Explicit derivations of Eqs. (39.17) and (39.18) can be found in Ref. 16. Because boundary layers are ignored, K is found to be a function of blood and dialysate flow rate; and because ultrafiltration was ignored in the derivation, K is also a function of Q_v. However, K is much

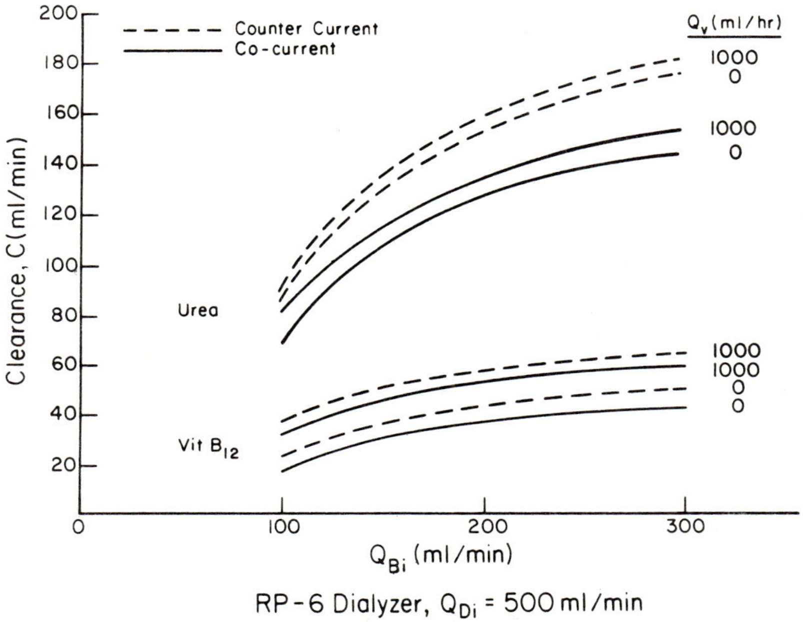

FIG. 39.8 Clearance of RP-6 dialyzer.

more of a constant than are either clearance or dialysance, as depicted in Figs. 39.9 and 39.10. Once again these nonograms were constructed by fitting several data to the intuitive curve

$$K = \alpha_0 + \alpha_1 Q_v + \alpha_2 \frac{1}{Q_{B,i}} \tag{39.19}$$

The mass transport coefficient has never been very popular in the characterization of hemodialysis. Even though they are much less dependent on Q_v and $Q_{B,i}$ than clearance, transcendental functions such as Eqs. (39.17) and (39.18) offer essentially no analytical advantages over the clearance function.

39.5 CALCULATIONS

The following sections give examples of calculations useful in a clinical hemodialysis setting.

39.5.1 Dialysate Conductivity

Both peritoneal and hemodialysate delivery systems have conductivity meters which are used for quality control of the dialysate. The proper use and calibration of these gauges require knowledge of conductivity g as a function of electrolyte and nonelectrolyte concentration and temperature. The effect of nonelectrolyte (glucose) concentration and temperature on conductivity are approximately linear, but conductivity is a nonlinear function of electrolyte concentration. An empirical equation relating these variables was developed by Zelman and is given by

$$g = (g_{25°C} - 3.39 \times 10^{-4} c_g)[1 + 0.0187(T - 25°C)] \tag{39.20}$$

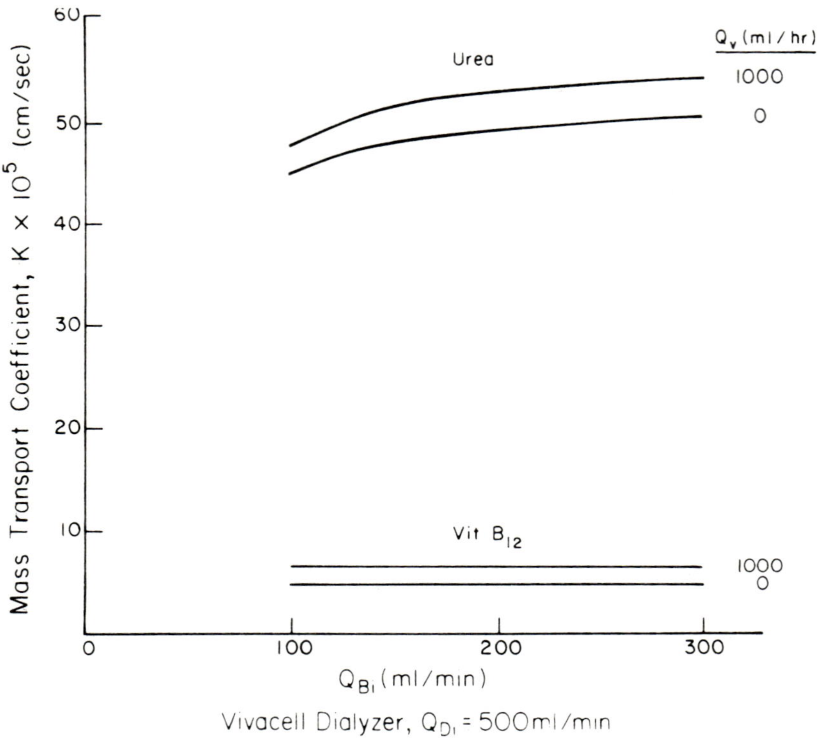

FIG. 39.9 Mass transport coefficient of Vivacell dialyzer.

with

$$g_{25°C} = \sum_{i=1}^{N} \Lambda_{25°i}(Z_{+i}\nu_{+i})c_{S,i} \tag{39.21}$$

where g is the conductivity of the solution, in $(\Omega \cdot \text{cm})^{-1}$; $g_{25°C}$ is the conductivity of the electrolytes alone at the same molar concentrations but at 25°C; c_g is the concentration of glucose in g/100 mL; T is the dialysate temperature, °C; $\Lambda_{25°i}$ is the equivalent conductivity, in $\text{cm}^2/\Omega \cdot \text{eq}$ of species i at 25°C and must be determined at the same ionic strength of the solution; N is the number of independent electrolytes; Z_{+i} and ν_{+i} are the positive valence and stoichiometric coefficient, respectively; and $c_{S,i}$ is the molar concentration, in mol/cm³, of the ith electrolyte. The quantity $\Lambda_{25°i}$ of each species should be found for each independent solute at the ionic strength of the total solution. For example, a typical dialysate composition is given by Table 39.3. The ionic strength, I, is defined as

$$I = \frac{1}{2} \sum Z_i^2 c_{\text{ion},i} \tag{39.22}$$

where $C_{\text{ion},i}$ is the molar concentration of the ith ion. For Table 39.3,

$$I = \tfrac{1}{2}\{(+2)^2[Ca^{2+}] + (+2)^2[Mg^{2+}] + (+1)^2[Na^+] + (+1)^2[K^+] + (-1)^2[Cl^-] + (-1)^2[C_2H_3O_2^-]\}$$

$$= \tfrac{1}{2}[4(1.25) + 4(0.75) + 133 + 2 + 102 + 33] = 139 \text{ mmol/L}$$

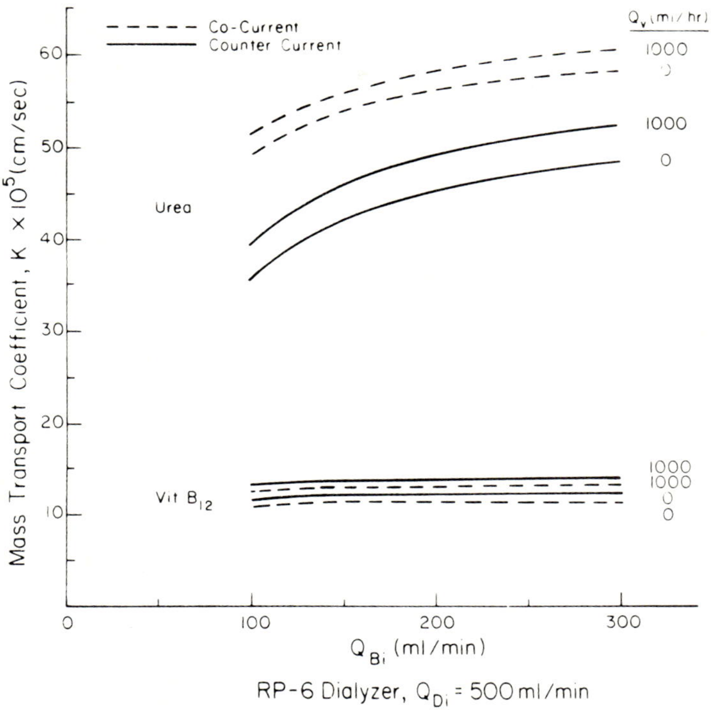

FIG. 39.10 Mass transport coefficient of RP-6 dialyzer.

Now determine which molar concentration of each solute would produce an ionic strength of 139 mmol/L and look up the equivalent conductance of each electrolyte under that concentration. For mono-monovalent salts (KCl, NaCl, $NaC_2H_3O_2$)

$$I = C_{S,i}$$

For dimonovalent salts ($MgCl_2$, $CaCl_2$)

$$I = 3C_{S,i}$$

So find Λ_i of NaCl, KCl, and $NaC_2H_3O_2$ under 139 mmol/L = 0.139 mol/L and find Λ_i of $MgCl_2$ and $CaCl_2$ under ⅓(139 mmol/L) = 0.0463 mol/L. The equivalent conductance of most electrolytes can be fitted to

$$\Lambda_{25°i} = \beta_0 + \beta_1 c^{1/2} + \beta_2 c + \beta_3 c^{3/2} + \beta_4 c^2 \tag{39.23}$$

where the values of the β's can be obtained from Table 39.7. The equivalent conductance for the dialysate concentration given in Table 39.3 is (from Table 39.7)

$$\Lambda_{25°\mathrm{NaCl}} = 104.46 \ \mathrm{cm^2/(\Omega \cdot eq)}$$

$$\Lambda_{25°\mathrm{Cl}} = 126.74 \ \mathrm{cm^2/(\Omega \cdot eq)}$$

$$\Lambda_{25°\mathrm{NaC_2H_3O_2}} = 70.56 \ \mathrm{cm^2/(\Omega \cdot eq)}$$

$$\Lambda_{25°\mathrm{MgCl}} = 103.70 \ \mathrm{cm^2/(\Omega \cdot eq)}$$

$$\Lambda_{25°\mathrm{CaCl}} = 109.10 \ \mathrm{cm^2/(\Omega \cdot eq)}$$

TABLE 39.7 Curve fit for equivalent conductance*

Substance	λ_0, $cm^2/(\Omega \cdot eq)$	λ_1, $cm^2/(\Omega \cdot eq \cdot M^{1/2})$	λ_2, $cm^2/(\Omega \cdot eq \cdot M)$	λ_3, $cm^2/(\Omega \cdot eq \cdot M^{3/2})$	λ_4, $cm^2/(\Omega \cdot eq \cdot M^2)$	R_λ, $cm^2/(\Omega \cdot eq)$	R_λ^f, %	R^2 (dimensionless)
NaCl	126.45	−89.058	103.11	−51.253	−23.755	±0.0069	±0.754	1.0000
KCl	149.85	−95.872	104.16	−5.2950	−81.330	±0.0376	±1.706	1.0000
$CaCl_2$	135.84	−183.19	297.47	9.3260	−555.05	±0.0890	±2.831	0.99999
$MgCl_2$	129.40	−177.06	292.61	12.929	−597.95	±0.0309	±1.900	1.0000
$NaHCO_3$	89.341†	−35.661	−319.69	1320.5	−1496.7	±0.3617	±7.675	0.99881
$NaC_2H_3O_2$	91.00	−82.060	106.91	−105.25	38.761	±0.0192	±1.616	1.0000

Source: Original data from (1) *International Critical Tables of Numerical Data: Physics, Chemistry, and Technology*, McGraw-Hill, New York, 1933; (2) R. A. Robinson and R. H. Stokes, *Electrolyte Solutions*, Butterworth, London, 1959.

* R_λ is the standard error of curve fit; R_λ^f is the fractional standard error; R^2 is the correlation coefficient; M indicates mol/L.

† This value predicted from curve fit.

Note that 1 meq/L = 10^{-6} eq/cm^3. The electrolyte conductivity at 25°C is given by

$$g_{25°C} = [104.46(100_{NaCl}) + 126.74(2_{KCl}) + 70.56(33_{NaC_2H_3O_2})$$
$$+ 103.70(2)(0.75_{MgCl_2}) + 109.10(2)(1.25_{CaCl_2})] \times 10^{-6}(\Omega \cdot \text{cm})^{-1}$$
$$= 1.346 \times 10^{-2}(\Omega \cdot \text{cm})^{-1}$$

At 37°C and with glucose, the conductivity of the dialysate would be

$$g_{37°C} = [1.346 \times 10^{-2} - 3.39 \times 10^{-4}(0.2)][1 + 0.0187(37° - 25°)]$$
$$= 1.640 \times 10^{-2}(\Omega \cdot \text{cm})^{-1}$$

Equation 39.20 is a reasonable estimate of solution conductivity and is accurate to within ±2 percent up to 45 g/L of glucose so long as the electrolyte concentration and temperature range remain suitable for hemodialysis, that is, 35 ± 5°C.

39.5.2 Time-Dependent Water Loss by the Patient

The water loss, in grams, is numerically equal to the volume loss, in milliliters, of the patient. The empirical equation relating volume loss, $V_p(t) - V_p^0$, of the patient to the transport parameters of dialysis is given by

$$\frac{dV_p(t)}{dt} = -Q_v = -A_m L_p^0(1 - \alpha t)(\overline{\Delta P_m} - \pi_p) \tag{39.24}$$

where $V_p(t)$ is the volume of the patient at time t. For all practical applications A_m, L_p^0, α, and π_p are independent of time, and some values are given in Table 39.4. If $\overline{\Delta P_m}$ is held constant, as can be accomplished for example by the Gambro dialysate delivery system, then Eq. (39.24) can be integrated,

$$V_p(t) - V_p^0 = A_m L_p^0(\overline{\Delta P_m} - \pi_p) \int_0^t (1 - \alpha t)\, dt \tag{39.25}$$

and

$$V_p(t) - V_p^0 = A_m L_p^0(\overline{\Delta P_m} - \pi_p) \left(t - \frac{\alpha t^2}{2}\right) \tag{39.26}$$

For the 1-m^2 Gambro Lundia dialyzer using Cuprophan PT-150 membranes ($\alpha = 0$) and L_p = 3 mL/(h · mm^2 · mmHg) and 150 mmHg for 4 h, the patient's weight would decrease according to

$$V_p(4) - V_p^0 = -1548 \text{ mL} = -1.55 \text{ kg} = -3.41 \text{ lb}$$

For the 1-m^2 RP-610 at 150 mmHg and using values from Table 39.4, one obtains

$$V_p(4) - V_p^0 = 7710 \text{ mL} = 7.71 \text{ kg} = 17.0 \text{ lb}$$

The large rate of water loss by the RP-610 even at moderate TMPs mandates using this dialyzer with special precautions such as an ultrafiltration control device or by running the dialyzer in the cocurrent mode. The ultrafiltration dialyzers using Cuprophan PT-150 membranes may require TMPs of 300 to 400 mmHg to remove sufficient water from severely hydrated patients.

39.5.3 Time-Dependent Solute Loss by the Patient

The mass transfer through the dialyzer can be used in a simple way to predict the concentration of only a few solutes because most solutes are compartmentalized in anatomical structures. If the rate of solute removal by the dialyzer is sufficiently fast, then the rate of transfer from the patient's organs into the plasma will be the rate-limiting step for dialysis. However urea and Na^+ have sufficiently fast intrabody transfer rates that one may consider the patient's water which is in equilibrium with the urea or Na^+ as a well-mixed, dialyzable volume. Urea is in equilibrium with essentially the total body water or 598 mL/kg of body weight for men and 510 mL/kg for women. The Na^+ space consists primarily of the extracellular volume, 230 mL/kg, plus plasma, (5 L) (1 − Hct), where Hct means hematocrit or volume fraction of blood occupied by the red blood cells. A 70-kg person having a hematocrit of 30% (dialysis patients have low hematocrits) will have a urea volume space of 41.9 L and a Na^+ volume space of 19.6 L. Because of biological variability these values are only reasonable order-of-magnitude approximations. See Ref. 20 for these data.

Consider a standard dialyzer with $\alpha = 0$. The solute removal of urea for a single-pass dialysate delivery system can be calculated from

$$\frac{d(c_p V_p)}{dt} = -c_p C \tag{39.27}$$

where $c_p V_p$ is the total urea in the urea space. Then

$$c_p \frac{dV_p}{dt} + V_p \frac{dc_p}{dt} = -c_p C \tag{39.28}$$

Combining Eqs. (39.24), (39.25), and (39.28) and integrating, one finds

$$c_p(t) = c_p(0) \left(\frac{V_p^0 - Q_v t}{V_p^0} \right)^{1+C/Q_v} \tag{39.29}$$

For a typical Q_v of 10 mL/min and urea clearance values from Table 39.5, for a 70-kg person one can calculate that after 4 h of dialysis the concentration of urea in the patient will drop from 144 to 64.9 mg/100 mL [from 67 to 30.2 mg/100 mL blood urea nitrogen (BUN)].

For Na^+ the dialysance of a standard 1-m² dialyzer is about 140 mL/min. Since Na^+ is in the dialysate, one uses dialysance for time-dependent Na^+ problems. The governing equation is

$$\frac{d(c_p V_p)}{dt} = -D(c_p - c_D^0) \tag{39.30}$$

where from Table 39.3 $c_D = c_D^0$, a constant at 133 meq/L. Again assuming $\alpha = 0$, one can integrate Eq. (39.30) to obtain

$$c_p(t) = \frac{D}{Q_v + D} c_D^0 + \frac{c_p(0)(Q_v + D) - Dc_D^0}{Q_v + D} \left(\frac{V_p^0 - Q_v t}{V_p^0} \right)^{1+D/Q_v}$$

For a typical Q_v of 10 mL/min and a 70-kg person, one can calculate that after 4 h of dialysis the concentration of Na^+ in the patient may drop from 150 to 141 meq/L.

39.5.4 Error Analysis of Clearance

The most common error in the literature on dialysis is the reporting of the error of measured clearance values. The error in C, δC, can be estimated from

$$\delta C = \left(\frac{\delta c_{B,o}}{c_{B,i}} + \frac{c_{B,o}\delta c_{B,i}}{c_{B,i}^2} \right) (Q_{B,i} - Q_v) + \left(1 - \frac{c_{B,o}}{c_{B,i}} \right) (\delta Q_{B,i} + \delta Q_v) + \delta Q_v \tag{39.31}$$

The following are very typical values for clinical laboratories and dialysis equipment.

$$Q_{B,i} = 200 \pm 1 \text{ mL/min} \qquad Q_v = 10 \pm 1 \text{ mL/min}$$

$$c_{B,i} = 100 \pm 2 \text{ mg/100 mL} \qquad c_{B,o} = 34 \pm 2 \text{ mg/100 mL}$$

Then

$$\delta C = 7.5 \text{ mL/min} \qquad \text{or} \qquad C = 125 \pm 6\%$$

A 6 percent error is really a "best estimate," and often one finds clearance values calculated from the blood side and dialysate side differing by 40 percent. Values in the literature nearly always reflect standard deviation rather than experimental error.

CONCLUSION

The bioengineer has contributed to the development of hemodialysis to the point where it is one of the most reliable and safe instrumented medical practices. Strangely, there remains enormous uncertainty concerning the role hemodialysis plays in detoxifying the body. Today no "uremic toxin" has been chemically identified. Until such biochemical knowledge is brought forth, the role of the engineer is no longer of primary importance in the development of new treatment techniques. The biochemists must now lead the way before hemodialysis can take another giant "engineering" step forward.

REFERENCES

1. Giordano, C., and Friedman, E. A. (eds.): *Uremia—Pathobiology of Patients Treated for 10 Years or More,* Wichtig Editore, Milano, Italy, 1981.
2. Kablitz, C., Stephen, R. L., Zelman, A., Harrow, J., Deeter, B., and Kolff, W. J.: Pressure control of ultrafiltration rate with high-flux dialysis membranes, *Artif. Organs, 5,* 162–167, 1981.
3. Abel, J. J., Rowntree, L. G., and Turner, B. B.: On the removal of diffusable substances from the circulating blood of living animals by dialysis, *J. Pharmacol. Exp. Ther., 5,* 275, 1913–1914.
4. Kolff, W. J., and Berk, H. T. J.: The artificial kidney: A dialyzer with great area, *Acta Med. Scand., 117,* 121, 1944.
5. Quinton, W. E., Dillard, D., and Shribner, B. H.: Cannulation of blood vessels for prolonged hemodialysis, *Trans. Am. Soc. Artif. Intern. Organs, 6,* 104, 1960.
6. Cole, J. J., Fritzen, J. R., Vizzo, J. E., van Paasschen, W. H., and Grinsrud, L.: One year's experience with a central dialysate supply system in a hospital, *Trans. Am. Soc. Artif. Intern. Organs, 11,* 22, 1965.
7. Brescia, M. J., Cimino, J. E., Appel, K., and Hurwich, B. J.: Chronic hemodialysis using venipuncture and a surgically created arteriovenous fistula, *N. Engl. J. Med., 275*(20), 1089, 1966.
8. Eschbach, J. W., Wilson, W. E., Jr., Peoples, R. W., Wakefield, A. W., Babb, A. L., and Schribner, B. H.: Unattended overnight home hemodialysis, *Trans. Am. Soc. Artif. Intern. Organs, 12,* 346, 1966.
9. Gordon, A., Bettes, O., Greenbaum, M., Marantz, L., Gral, T., and Maxwell, M.: Clinical maintenance hemodialysis with a sorbent based, low-volume dialysate regeneration system, *Trans. Am. Soc. Artif. Intern. Organs, 17,* 253, 1971.
10. Henderson, L. W., Besarab, A., Michaels, A., and Bluemle, L. W., Jr.: Blood purification by ultrafiltration and fluid replacement (diafiltration), *Trans. Am. Soc. Artif. Intern. Organs, 13,* 216, 1967.
11. Yatzidis, H.: A convenient haemoperfusion micro-apparatus over charcoal for the treatment of endogenous and exogenous intoxification: Its use in an artificial kidney, *Proc. F.D.T.A., 1,* 83, 1964.
12. Andrade, J. B., van Wangene, R., Chen, C., Ghavamian, M., Volder, J., Kirkham, R., and Kolff, W. J.: Coated absorbants for direct blood perfusion: II, *Trans. Am. Soc. Artif. Intern. Organs, 18,* 473, 1972.
13. Chang, T. M. S., Migchelsen, M., Coffy, J. F., and Stark, A.: Serum middle molecules levels in uremia during long-term intermittent hemoperfusions with the ACAC coated charcoal microcapsule artificial kidney, *Trans. Am. Soc. Artif. Intern. Organs, 20,* 364, 1974.
14. Lai, F., Tankersley, R., Scott, M., Wayt, H., and Zelman, A.: Third generation artificial kidney: Automatic concentration and pH control, *Trans. Am. Soc. Artif. Intern. Organs, 21,* 345, 1975.
15. Popovich, R. P., Moncrief, J. W., and Nolph, K. D.: Continuous ambulatory peritoneal dialysis, *J. Int. Soc. Artif. Organs, 2,* 84, 1978.

16. Klein, E. (ed.): Evaluation of hemodialyzers and dialysis membranes, DHHS Pub. No. (NIH)77-1294, 1977.

17. Stephen, R. L., Zelman, A., Kablitz, C., Gisser, D. G., Harrow, J. P., Strait, G., Deeter, B., Duffy, D. P., and Kolff, W. J.: Evaluation of artificial kidney, in C. Giordano and E. A. Friedman (eds.), *Uremia-Pathobiology of Patients Treated for 10 Years or More,* Wichtig Editore, Milano, Italy, 1981, pp. 221–230.

18. Keshaviah, P. R., Constantini, E. G., Leuhmann, A., and Shapiro, F. L.: Dialyzer ultrafiltration coefficients: Comparison between in vitro and in vivo values, *Artif. Organs, 6,* 23, 1981.

19. Hoenich, N. A., Frost, T. H., and Kerr, D. N. S.: Dialyzers, in W. Drukker, F. M. Parsons, and J. F. Maher-Hagen (eds.), *Replacement of Renal Function by Dialysis,* Martinus Nijhoff, the Hague, 1978, pp. 80–124.

20. Altman, P. L., and Dittmer, D. S. (eds.): *Biology Data Book,* Federation of American Societies for Experimental Biology, Bethesda, Md., 1974.

CHAPTER 40

Bioengineering of Total Joint Replacement

Jack L. Lewis
William D. Lew
Departments of Civil Engineering and Orthopedic Surgery
Northwestern University
Evanston, Illinois

INTRODUCTION

The replacement of diseased joints by artificial materials has become a common procedure in orthopedic surgery. The success of the procedure in restoring function to patients and its widespread use, coupled with active input by engineers in the development of the implants, has made this an important bioengineering subject. The goal of this chapter is to introduce the reader to total joint replacement, to describe the rationale and current design approaches to joint implants, and to identify research issues.

The primary rationale for joint replacement is degenerative joint disease leading to erosion and wear of articular cartilage and consequent pain. Joint instability due to ligamentous laxity is a further causative factor, but even this necessitates joint replacement only after the excess motion due to the instability has led to cartilage erosion. Joint replacement is mainly a treatment of cartilage wear, and is thus called a *surface replacement arthroplasty*. The essence of joint replacement is the replacement of diseased bearing surfaces of articulating joints with artificial bearing materials. The replacement materials must maintain their position relative to the bone and not wear out during the life of the patient.

Joint replacement is relatively new. Although using artificial materials to replace diseased joint surfaces has been tried several times over the past decades, widespread use of the procedure did not begin until Charnley introduced polymethyl methacrylate (PMMA) as an aid to fixation of implant components to bone in the early 1960s [1]. This procedure greatly reduced problems encountered with earlier use of metal joint implants fixed by wedging the implant into a surgically prepared cavity in the bone.

Hips were the first joints replaced in large numbers. Total knee replacement began several years later. From 1970 onward, joint replacement proceeded at a rapid rate, with most major joints being replaced (hip, knee, ankle, finger, wrist, shoulder, elbow). As shown in Fig. 40.1,

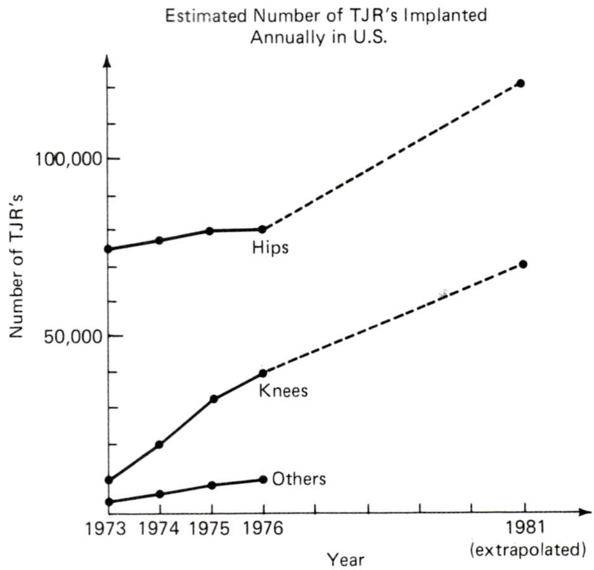

FIG. 40.1 Estimated number of total joint replacements implanted annually in the United States.

the number of total joints implanted has grown tremendously [2]. In 1976, it was estimated that 130,000 total joints were implanted in the United States [2]. Somewhat more, on the order of 200,000, were probably implanted in 1982.

Charnley also introduced high-density polyethylene (HDPE) as a bearing material, and although metal-on-metal total hips were popular for awhile, nearly all total joints today employ metal articulating against HDPE, with the components fixed to bone with PMMA, although other methods for fixation are now being explored. Figure 40.2 shows typical components for the hip and knee. Component geometry, metal type, and method for using HDPE vary considerably between designs.

In spite of this widespread usage of total joints and apparent success, there is still concern with design, surgical procedures, and definition of suitable patient populations. On the order of 5 percent of the implants currently installed fail each year. Also there is concern that the factors leading to failure are time-dependent and that the failure rate may increase with time. Total joint replacement is still a relatively new procedure, with follow-ups of generally less than 10 years for implants often required to remain functional for 20 to 30 years. The longer-term success of total joint replacement thus is unknown. This factor, coupled with changing designs and surgical procedures, leaves considerable doubt as to long-term performance ability of joint implants. Therefore, there is considerable ongoing effort to improve their reliability and longevity.

The nature of problems occurring in patients receiving joint replacement has been studied extensively, either by follow-up of patients after surgery [3] or by study of failed implants [4]. The major problems identified are as follows:

1. Loosening of components from bone with consequent pain and immobility
2. Dislocation or other mechanical instability
3. Infection
4. Restricted range of motion due to component impingement or ligamentous constraint

FIG. 40.2 Typical components for total hip and knee replacements.

These problems are, in part, a function of surgeon-controlled factors such as component orientation, soft tissue handling, aseptic environment, and PMMA handling, and have been greatly improved upon as more experience has been gained with the surgical procedures. These factors are generally not related to design of the components. The major problems of loosening from unknown causes and joint instability still remain. These are the primary problems currently of interest to the bioengineer. Wear has not been a significant problem to date, but with the improvement in reliability and use of joint replacement in younger patients, wear will become an issue of more importance.

Since the cause of the loosening is unknown, it is probably not the case that all loosening is due to mechanical factors. Some may be due to undetected iatrogenic factors or infection. However, apparent loss of fixation to the bone is the most frequent clinical correlate with failure; this must be the major concern in understanding functioning and failure in the total joint system. It is not clear what is actually failing. Rigid bonding to the bone is obviously failing, but this would not necessarily lead to clinical failure, since some total joints are put in without cement. These are intended to be loose, and most work well [5]. Although it is not certain, it appears that if firm fixation can be maintained with cement, clinical failure will not occur. The failure process for a nonbonded implant is poorly understood.

Thus, careful review of past performance of total joints gives a fairly clear picture of the problem areas. The goal of current research on total joints and their design refinement is to reduce the incidence of these problems. This would not only make more reliable devices for the current population of patients but would also make total joints available for younger, more active patients for whom they are not now commonly given because of fear of eventual failure.

Current joint implant components are constantly undergoing redesign evolution. This process involves an extensive research activity in both engineering and medicine, as well as engineers working with surgeons directly to design devices based upon the best available information. In the following sections, the bioengineering elements going into this process will be discussed. It must be kept in mind, however, that actual component design is highly dependent on surgical expertise and that engineering knowledge is used only in a collaborative context. The paucity of current knowledge on interaction of an implant with the body, as well as the imprecision of design techniques, must also be appreciated. In this regard, the engineering results can serve as design guides and as a means of identifying research topics.

Because the current materials appear to perform adequately, and since there is reluctance to introduce new materials because of questions of biocompatibility, the emphasis of most recent engineering work on total joints has been on the mechanics of the implants as opposed to questions about materials. And because the problems in need of solution are, first, fixation and, second, stability, structural analysis of the implant system has been the main focus of research and design activity; joint kinematics has been a related, but less well developed area of activity. Relevant work in these two areas will be described briefly in the following sections. The aim of the presentation is to show what has been done and where research questions remain. The discussion will focus on the two most commonly replaced joints, the hip and knee, for two reasons: (1) most work has been done on these joints because of their overwhelming importance, and (2) the engineering techniques and issues pertaining to these joints are relevant for all joints. The results of experience with and research on the hip and knee joints can guide work with the other joints.

40.1 GENERAL DESIGN CONSIDERATIONS

Fixation of the implants to bone has been a major consideration in design modification and in the development of improved surgical techniques. PMMA was originally introduced as a space-filling grouting agent rather than an adhesive cement, but lasting adhesion soon became the guiding criterion and firm fixation by mechanical interlock became the goal. Clinical loosening can be seen, from one point of view, as a mechanical failure of the implant structural system. This view leads logically to a structural analysis of the implant system. Firm fixation by rigid adhesive bonding may not be necessary, however, as pointed out previously. An alternative scheme for fixation involves the use of fibrous tissue between implant and bone, since this tissue is biologically stable and allows only minimal relative motion. Both these modes of fixation, as well as combinations of the two, will be explored.

Structural analysis of the implant system, with or without a compliant fibrous tissue between implant and bone, is the primary analytic tool used to study total joints. Kinematic joint models are a second tool, since kinematic constraint requirements can influence surface forces on the implant, which in turn can effect structural failure. Approaches to structural and kinematic analyses will be described in the following sections.

40.2 TOTAL HIP REPLACEMENT

40.2.1 Design Requirements

The hip, shown schematically in Fig. 40.3, has been the joint most frequently replaced. The main area of concern is loosening of the components, primarily the femoral component, but potential failure of the acetabular component is worrisome as well. Dislocation is of concern, but surgical technique and current component designs have minimized problems associated with dislocation [6–8].

Areas of concern to the bioengineer in the design of total hip components are (1) configuration of the femoral head and neck, (2) fixation of the femoral stem, (3) fixation of the acetabular component, and (4) wear of the articulating surfaces.

Configuration of the head-neck region influences the forces on the implant, as can be seen by the simple model of the hip system in Fig. 40.4. Shortening the neck changes the forces on the head, influencing both stem stresses and wear. There is controversy as to the desirability of an abnormally short neck, but most designs have attempted to reproduce the normal anatomy. These have worked well insofar as range of motion and minimal dislocations are concerned, so this does not seem to be an active design area. Refer to Refs. 9 and 10 for further discussion on this issue.

Fixation of the femoral component has received considerable attention. Structural analysis of this component has been widely used.

Failure of cemented implants has been of primary concern; therefore, the system of interest has been composed of cortical bone, cancellous bone, PMMA, and metal on polyethylene.

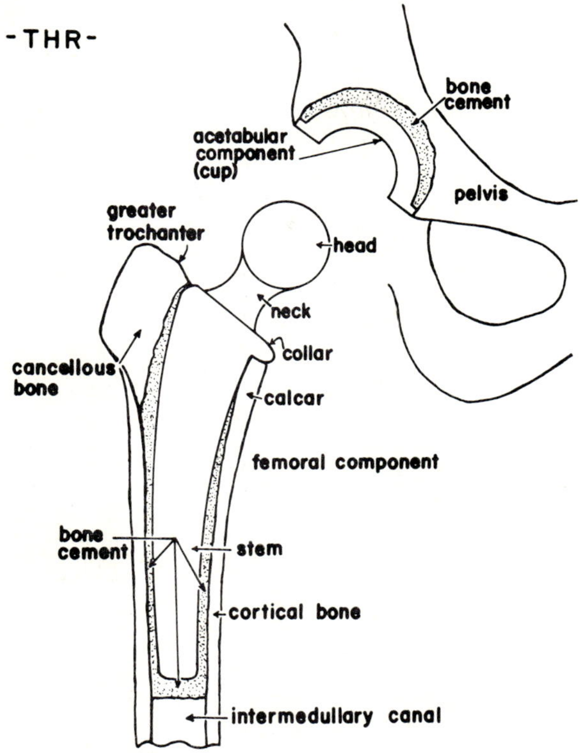

FIG. 40.3 Schematic diagram of the components of a total hip replacement.

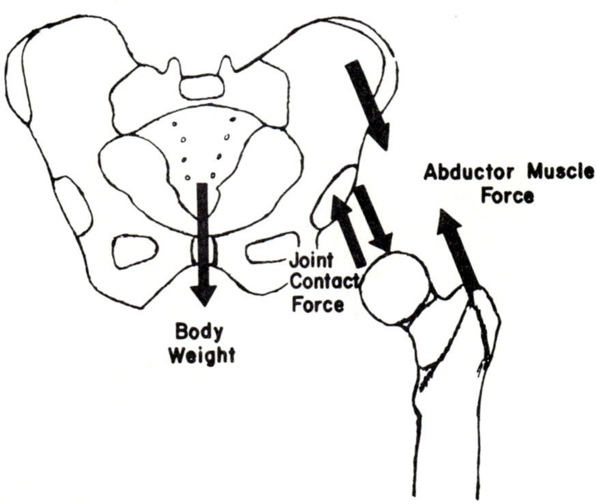

FIG. 40.4 Simple system of forces acting at the hip.

Since the site of initial failure has not been identified, stresses in each of these materials have been examined to be used as guides in explaining failure and predicting how design changes would change system stresses.

40.2.2 Structural Analysis of Total Hip Femoral Component

The femoral component of a total hip consists of a slender, tapered metal stem with a metal sphere replacing the natural femoral head (see Fig. 40.3). The stem is fixed to the bone by PMMA, which is intended to fill all spaces between the stem and bone. Cancellous bone is located in the proximal femur, extending to a level about halfway down the stem. The cancellous bone is nonhomogeneous and anisotropic in stiffness and strength [11, 12]. The cortical bone is similar, but is more uniform and shows a relatively smaller anisotropy. Approximate elastic moduli for the system materials are given in Table 40.1. Only single values are given for bone. More anisotropic elastic constants are given in Refs. 13 and 14 for cancellous and cortical bone, respectively. Spatial distribution of properties are given in Refs. 15 and 16.

Analysis of system stresses has revolved around several specific questions. Among these are (1) stem length, (2) stem cross-sectional shape, (3) stem material (cobalt-chromium alloy versus titanium), and (4) effect of a metal collar which transfers loads to the cortical calcar bone at the most proximal location. There has also been speculation on where loosening failure might initiate, with PMMA, metal, metal-PMMA interface, and PMMA-bone interfaces as the possible elements in failure.

Various structural models have been used to approximate the femoral implant system [17–20]. Figure 40.5 shows several of these models, in increasing order of detail. The straight composite beam model [17] and the Huiskes model [18] are relatively easy to construct for computerized solution and are useful for parametric study, but it is not clear how well the models simulate the actual system.

The realistic, three-dimensional finite element models [19, 20] more nearly simulate the actual implant, but they are time-consuming to construct, and the increased cost of large amounts of computer time make parametric studies more difficult. Current understanding of the implant stress distribution has used results from all the models.

The question of accuracy of the models has been addressed in some depth, primarily by comparison with experiments. The beam-type models are intended to be approximations, so a high level of accuracy is not expected. The three-dimensional finite element models are intended to be accurate simulations. Their success has been checked in two separate studies, both using strain gauges on a cadaver femur and implant, with this femur modeled as accurately as was practical by a three-dimensional finite element model. All surfaces were assumed bonded in the model. Rohlmann et al. [21] modeled the complete femur by 1950 eight-noded elements. The model, shown in Fig. 40.6, was of the same geometry as an experimental femur in which 34 strain-gauge rosettes were used to measure strains over the surface of the femur. Materials in the model were assumed isotropic, but variable stiffness properties were assigned values de-

TABLE 40.1 Material properties for structural analysis of system components

Material	Modulus, MPa	Poisson's ratio	Failure stress, MPa	Fatigue stress, MPa
Titanium alloy (Ti)	110,000	0.30		
Cobalt-chromium alloy (cast Co–Cr)	210,000	0.30		
PMMA	2,070	0.35	30 (tension) 92 (compression)	14
Cortical bone	15,500	0.28	150 (compression)	
Cancellous bone I	330	0.32		
Cancellous bone II	600	0.32		
PMMA-bone interface			3.7 (tension)	0.6
PMMA-metal interface			7 (tension)	2.0

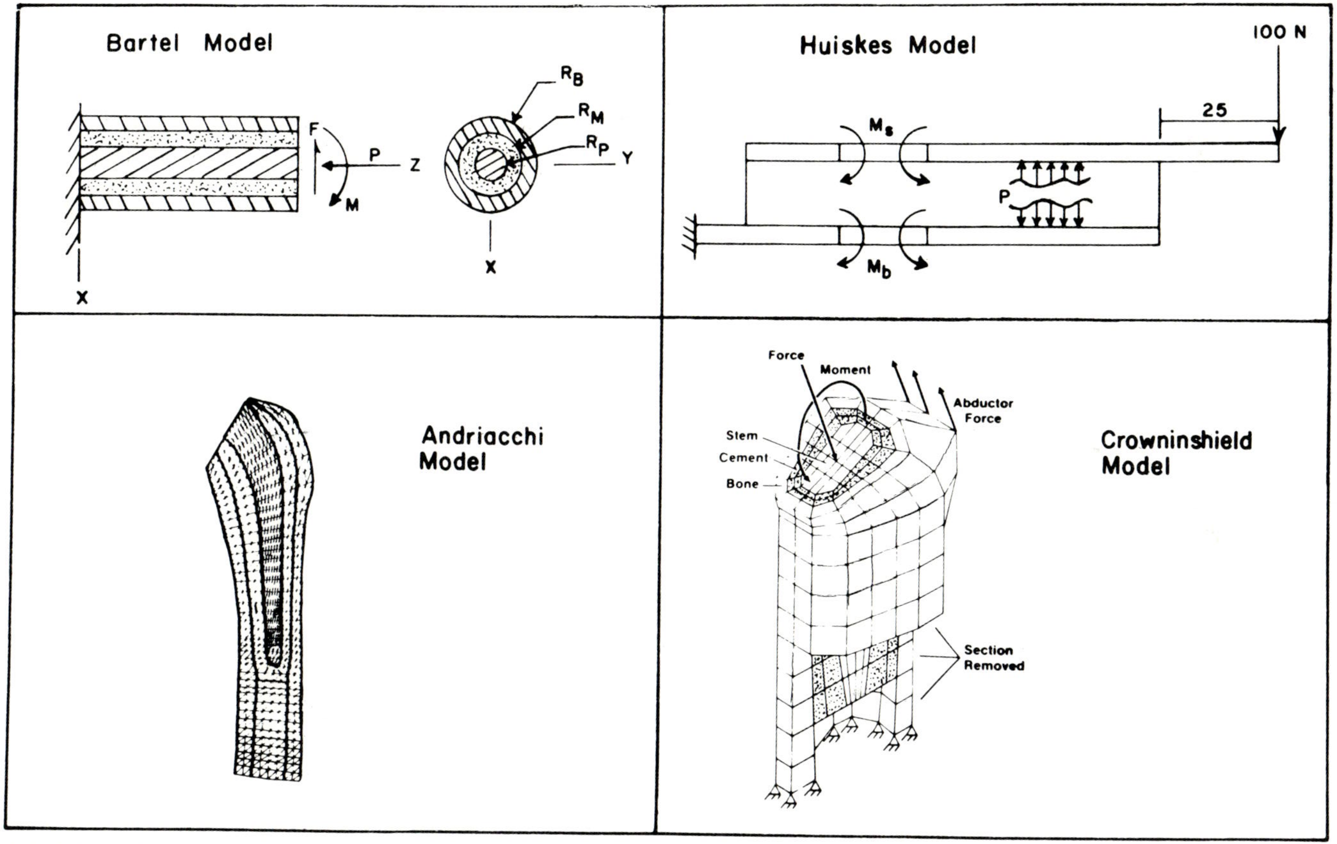

FIG. 40.5 Typical structural models of total hip replacements: Straight composite beam model by Bartel [17], the Huiskes model [18], a two-dimensional model by Andriacchi, and a three-dimensional model by Crowninshield [19]. (*Reprinted with permission from the Journal of Biomechanics, Pergamon Press, New York.*)

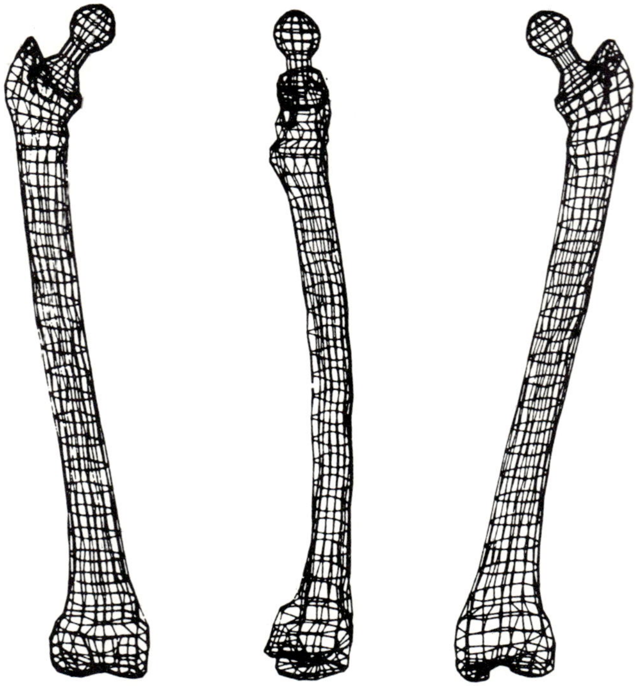

FIG. 40.6 Three-dimensional finite element model of the proximal femur with hip endoprosthesis. (*From Rohlmann et al. [21].*)

pending on x-ray density. Measured head and trochanter loads were applied. Results from experimental and finite element models of maximum and minimum principal strains differed by 12.7 percent on average, with a standard deviation of 16.4 percent. Tarr et al. [22] performed a similar comparison between a three-dimensional finite element model and a strain-gauge experiment with the femur. They found agreement within 15 percent on average for bone strains. Stem strains were less accurate, of the order of 35 percent on average. The stem mesh was only two elements thick, probably causing some of the error. However, the results suggest that predicted bone stresses can be quite accurate, although internal stresses can be in greater error. Differences between results from finite element models and experiments will arise because of inexact geometry, approximations in material properties, and imprecision in measuring experimental loads. Even with these differences, however, accuracy of the three-dimensional models is quite good. As will be shown, larger errors are caused by unknown surface load values and unknown interface conditions in the living human.

With the accuracy of the three-dimensional finite element models estimated, the models can be used to analyze the femoral implant system as well as to establish limitations of the more approximate models. A model similar to that of Tarr et al. [22] will be used to demonstrate dominant features of the femoral component system. Consider the model in Fig. 40.7. A head load acting at an angle of 18° to the shaft axis and a trochanter force will be assumed. The design features examined were calcar collar, stem material, stem length, and head-force location. In addition to addressing specific design questions, the results for the various combinations demonstrated some fundamental features of the stress distribution for this system. Figure 40.8 shows results for various design parameter combinations, for the maximum compressive stress down the medial wall of the femur. Several conclusions are evident from these curves. First, the effect of the collar is limited to the most proximal region. Eliminating the collar has no effect on system stresses at the middle and distal stem. Conversely, changes at the stem tip have no

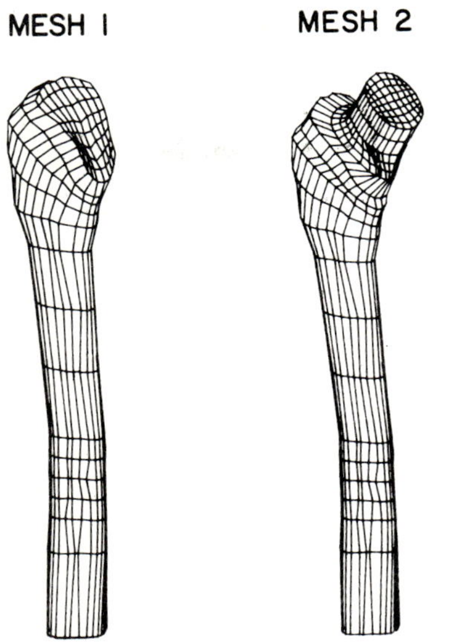

FIG. 40.7 Three-dimensional finite element models of the proximal femur. Mesh 1 is a model of the normal femur, while Mesh 2 is a femur with prosthesis implanted.

influence on the proximal region. Thus, for example, lengthening the stem will have no effect on proximal stresses. This phenomenon demonstrates the basic feature of the stem stress system: that the proximal portion is a complex three-dimensional stress state generated by bending moments, shear, and compressive forces applied at the proximal surface. The middle and distal regions, however, are dominated by bending moments. Stresses in the midportion can be described fairly accurately by composite beam theory. The distal region requires a three-dimensional theory, because of geometric discontinuity, and is subject only to pure bending. The unsuspected feature of this stress system is that the stem is not being pushed down the shaft in shear. All shear loads are taken up in the proximal region.

This stress state was first hinted at in the early work of Bartel [23], and then further articulated by Huiskes [24]. It allows a simple interpretation of a complex stress state. For example, this interpretation indicates that the effect of lengthening the stem is to put the tip in a new bending moment region; that is, stresses around the tip will change only if the bending moment changes or if tip geometry changes. Huiskes [24] labels the three regions the proximal, middle, and distal regions. The rela-

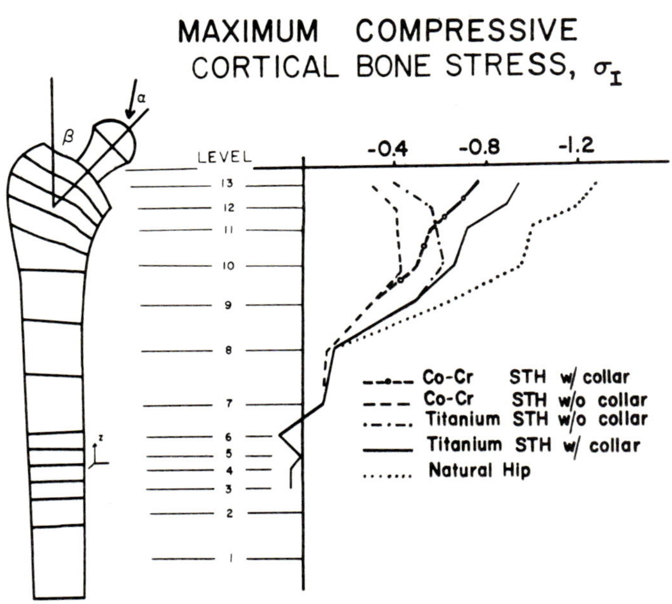

FIG. 40.8 Maximum longitudinal compressive stresses in the cortical bone at levels along the medial femur.

tive sizes of each region vary depending on the relative cross-sectional sizes of bone and stem and on the stem material. Depending on the values of these parameters, the middle region can disappear, and the proximal and distal regions will interact.

Use of the three-dimensional finite element models has clarified some issues related to clinical behavior of the implant, but it has not necessarily settled these issues. For example, the role of a calcar collar has been a much-debated topic, with some supporting it [27] and others against it [25]. It is now fairly clear, from both the finite element models [26] and comparable experiments with strain gauges on femurs [27], that the metal stem unloads the proximal bone. This is an important point because bone that is loaded appreciably less than normal will resorb, possibly contributing to loosening. The degree of unloading depends on stem stiffness, which in turn depends on the stem materials and the cross-sectional configuration. A stiff cobalt-chromium stem will unload the calcar cortical bone with or without a collar. As the stem is made less stiff, the calcar loading increases, and this loading is always greater with a collar. In fact, the most proximal bone never reaches more than about 30 percent of normal without a collar [25, 26], even with the more compliant titanium. With a collar, stresses up to 70 percent of normal have been predicted and measured experimentally [22]. Thus, there is general agreement that a collar is necessary to get calcar bone loading, and that calcar loading is possible. The counterargument is that effective contact cannot be achieved consistently during surgery, and so the collar will ultimately be ineffective [25]. It is further claimed that proximal load transfer with a cemented compliant stem will endanger the most proximal PMMA, leading to system failure. The validity of these claims is not established, and the controversy continues.

A second clinical question concerns the relative merit of the more compliant titanium versus a cobalt-chromium combination as a stem material. The significant difference between these two, for the same-sized stem, is that titanium will load the proximal region more and will have lower cement stresses distally [22]. Whether proximal or distal system stresses are more important depends on magnitudes of the forces acting on the femur; these are not known with sufficient accuracy to resolve the question. The finite element results tell what happens to system stresses when going from a titanium to a cobalt-chromium stem, but they cannot provide answers to the main clinical question of which material generates less PMMA and interface stresses, because of other unknowns in the problem.

As discussed earlier, the finite element models appear to be quite accurate as simulations of the initial fixed state. It is known that as the body repairs the initial damage, it responds to the presence of the implant, with the interface between bone and PMMA changing its mechanical state [28]. The details of this change and the nature of the final state are only now being investigated. It is likely, however, that the final state is not like the initial well-bonded interface state, and the finite element models will no longer strictly apply. Thus, there is the question of what relevance the well-bonded models have on eventual clinical performance. There is no answer to this question at this time. The models are used partially because they are the most appealing approach available, even with their recognized limitations. There are also other uses, however. First, debate on implant performance and design has been based on the assumed mechanical state of the implant system, and, for the most part, the assumption has been of a bonded interface. The stress analyses have addressed questions based on these assumptions, and thus have helped clear the air and better focus the questions even if they have not provided definitive answers. The finite element models are also used in developing a better understanding of the true implant mechanical state. By assuming a mechanical state, making predictions on behavior, and then comparing those predictions with in vivo performance, these models suggest how to alter initial assumptions according to differences between predictions and performance. Iteration of this procedure can lead to eventual understanding of how the real system behaves. Thus, the structural analysis models currently in use assume features of the system suspected to be in error. The results can be useful in furthering understanding of the system, but the models are not exact and cannot be used alone to dictate design of the implants.

Because of this imprecision in design, there are many designs for the femoral component, and most apparently are reasonably successful. The designs are not very different, however. Better comparative evaluation of the designs will have to await long-term clinical results and better understanding of body forces and the in vivo mechanical state.

40.2.3 Component Wear

Another factor of concern in total hip replacement is wear. During the early use of total hip replacements, there was considerable research activity related to wear [29, 30], since it was assumed to be a problem. However, it has not been a significant problem in devices remaining in patients up to 10 years postoperatively, so there is little current research activity in this area. Improved reliability will probably increase the attention given to wear as a problem of interest, with younger patients receiving the implants.

40.2.4 Other Components of Hip Replacement

The acetabular (pelvic) component has been studied in a manner similar to the femoral component, but less extensively. Both two-dimensional plane strain and axisymmetric geometry with nonaxisymmetric loading have been analyzed [31, 32]. These analyses have addressed the question of material used in cemented components, in particular, the question of whether a metal backing should be used to reinforce the HDPE bearing surface cup. The consensus is that a metal-backed cup should have lower bone stresses and cement stresses than a component of HDPE alone. The basis of this conclusion will be shown in Sec. 40.3, which deals with the tibial component of total knee replacements.

Other types of total hip replacement have been proposed and studied, in particular, the surface replacement total hip. In design, a thin metal shell is placed over the spherical surface of the femoral head, avoiding placement of the metal stem down the femoral shaft. Clinical experience with these components has been disappointing [33, 34], and they are currently in disfavor. Stress analyses of the system have been performed [35], and although some plausible reasons for failure were suggested, no conclusive results are available and work continues to decide whether a design of this type is feasible.

40.3 TOTAL KNEE REPLACEMENT

Total knee replacement, shown schematically in Fig. 40.9, differs from conventional total hip replacement in two fundamental ways from a bioengineering viewpoint. First, the tibial component of the total knee is fixed in the cancellous bone of the proximal tibia. The femoral component of the total hip, on the other hand, is fixed to the cortical bone of the femoral shaft. Second, ligaments about the knee influence the joint function and component design to a much greater degree than for the hip. These two factors have led to separately identifiable areas of investigation and design questions.

As shown in Fig. 40.2, total knee replacements were introduced after total hip replacements, and consequently are still going through a design evolution phase. Total knee surgery is technically more difficult than surgery at the hip, which partially explains the higher failure rate of the early total knee implants, reported at 5 to 15 percent at 2 years [36, 37]. The major problem clinically has again been component loosening, almost exclusively of the tibial component. Component impingement and motion constraint or laxity due to ligament interaction [38] have also been related to clinical failure. Because of this experience, the design questions studied have been (1) fixation of the tibial component to bone, and (2) geometry of the articulating surfaces.

40.3.1 Component Fixation and Structural Analysis

Nearly all total knee implants are fixed to bone with PMMA, and, as with the total hip, loss of fixation has been viewed as a mechanical failure. Consequently, stress analysis of the tibial component fixed with PMMA has been performed to examine different designs.

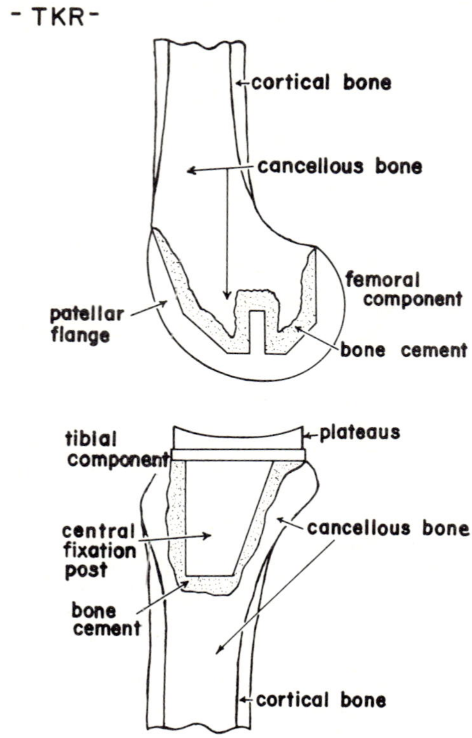

FIG. 40.9 Schematic diagram of the component of a total knee replacement.

Approaches have been similar to those for total hips, although the simpler models are less used because of the more complex geometry of the knee joint. Early models were two-dimensional plane-strain models [39, 40]. Axisymmetric models with nonaxisymmetric loading were next in level of complexity [41], and finally complete three-dimensional models have been developed [42, 43]. These computer modeling efforts have been complemented by experimental work, although the models and experiments have not been for the same conditions, and so comparison is limited. As with the models for the total hip femoral component, models for knee implants have been motivated and directed by clinical questions and controversies, including (1) metal versus polyethylene components, (2) connected versus separate medial and lateral components, and (3) size and number of fixation posts. As with the hip, it is not clear what initiates failure, but frequent radiolucent lines at the bone-PMMA interface [44], plus intact implant-PMMA structures removed at revision surgery [38], suggest that the bone-PMMA interface is the major failure site. The interface, cancellous bone, and PMMA are possible sites of initiation of the failure.

The relatively rigid platelike component sitting on the compliant bone behaves like a beam on an elastic foundation, as shown in Fig. 40.10. A characteristic of this structure is that loading near the edge causes beam tilting and tensile stresses at the opposite end. Since the bone-PMMA interface is thought to be especially weak in tension as opposed to shear [45], this weakness may cause failure of the interface. Another feature of this system where the beam is polyethylene-backed with PMMA is that central loading will cause "winging up" and tension again at the bone-PMMA interface, as shown in Fig. 40.11. Also, since the PMMA is stiffer than the polyethylene or bone, the PMMA carries most of the load and is subjected to significant

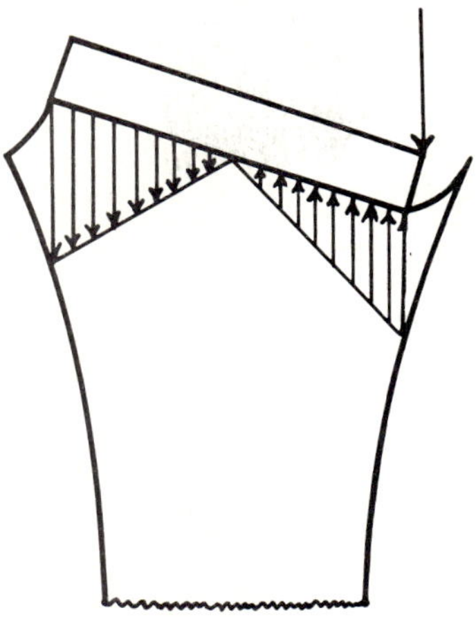

FIG. 40.10 Schematic of a beam on an elastic foundation, showing the resulting tensile and compressive stresses due to an applied load near end of the beam.

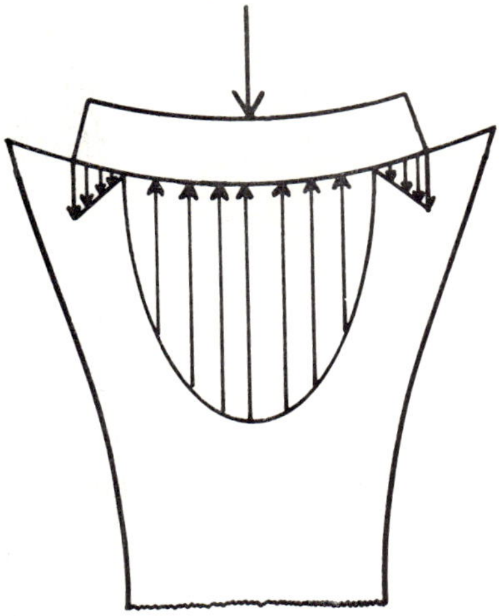

FIG. 40.11 Schematic diagram of the winging-up effect of a plate on an elastic foundation.

bending stresses for which it is not well suited. These effects are prominent in the behavior of the system, but their magnitude and importance are modulated by deviations of the real tibial structure from the idealized geometries, particularly by the cortical bone and nonuniformity of cancellous bone properties. The general phenomena are also influenced by posts extending into the cancellous bone.

The influence of these variables was explored by Askew and Lewis [39], who used plane-strain finite element models to examine the effect of the cortical shell, the cancellous bone nonhomogeneity, and the cancellous bone anisotropy. Their model is shown in Fig. 40.12. Their results [39] show that the cortical shell and cancellous bone nonhomogeneity are important variables in the model.

The plane models can examine only a few of the significant problems. The component geometries have no planes or axes of symmetry for simpler modeling. Consequently, complete three-dimensional models are necessary to address the questions of interest. Two studies using three-dimensional models have been reported [42, 43]. The authors' work will be used for convenience to illustrate results. The model used by Lewis et al. [42] is shown in Fig. 40.13. Variable-stiffness isotropic cancellous bone was assumed. Stress throughout the system was computed. As discussed earlier, the current view of implant failure is that it is primarily mechanical failure. Again, the limitations of this assumption are recognized, but the biological responses and their interaction with loading are not sufficiently well understood to guide design. The mechanical approach has been successful and is the one used here. The basic approach is to compare designs on the basis of stress or some other mechanical failure criteria. Failure in each of several possible structural elements is considered, and simple failure criteria for each of these are assumed. These possible criteria were considered: (1) compressive failure of cancellous bone, (2) tensile failure of the bone-PMMA interface, and (3) tensile failure of the PMMA.

Results of the analysis for each of the several criteria are shown in Fig. 40.14. From these, the following conclusions are evident: (1) Metal components generally give lower bone-cement interface tensile stresses and lower PMMA tensile stresses. The latter feature is due to the metal taking most of the bending loads. (2) Separated medial and lateral components have higher bone compressive stresses and bone-cement interface tensile stresses. (3) Fixation post number and geometry can drastically influence system stresses. These features are incorporated in most

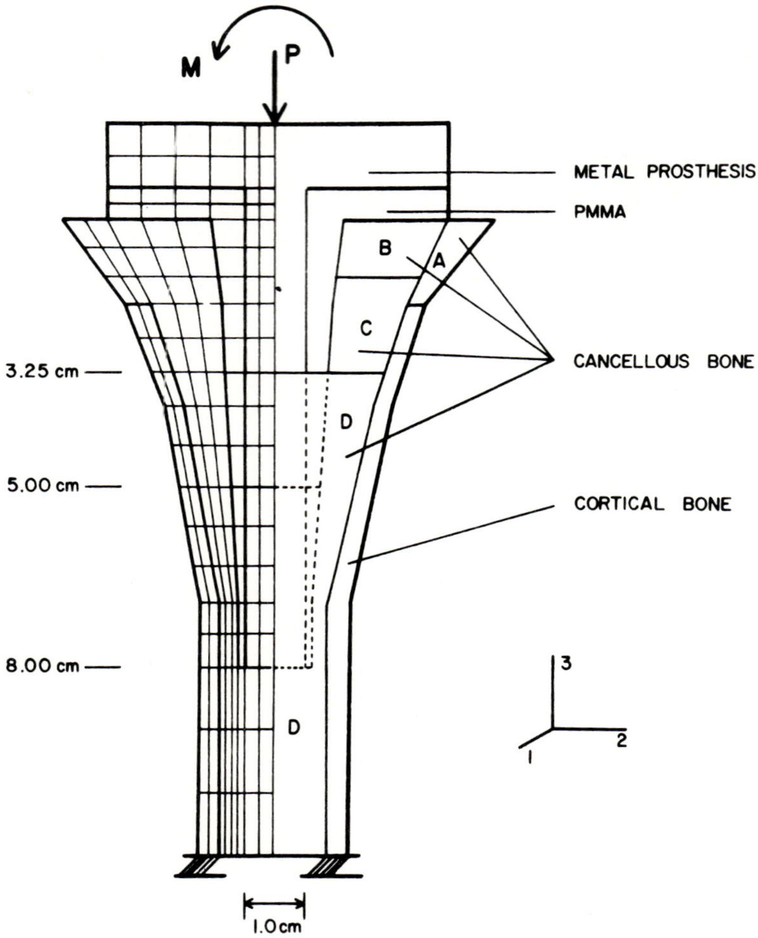

FIG. 40.12 Two-dimensional finite element model of the proximal tibia with prosthesis, indicating the geometry, the loading, the mesh density, and the material properties. (*Reprinted with permission from the Transactions of the ASME, Journal of Biomechanical Engineering, American Society of Mechanical Engineers, New York.*)

currently used total knee designs. Early results look promising in that there are fewer early failures than with the older designs.

Structural analysis of more constrained knee implants, such as hinge designs or semiconstrained devices, has not been performed, and significant clinical studies have not been reported. Consequently, although the rationale for design of these devices is clear, i.e., to provide stability not given by the missing ligaments, the success of such devices and optimal guidelines for design are not yet known.

40.3.2 Kinematic Requirements for Implants

In addition to fixation to bone, the other important features of total knee performance concern the range of motion allowed by the implant and the forces on the implant and surrounding

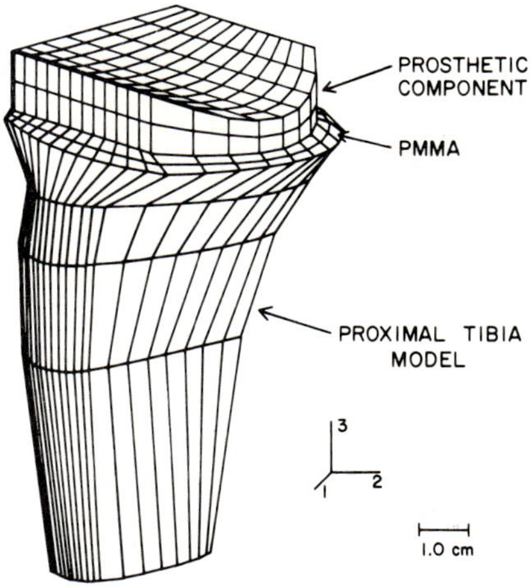

FIG. 40.13 Three-dimensional finite element model of the proximal end of the tibia with prosthesis. (*Reprinted with permission from the Journal of Bone and Joint Surgery, Boston.*)

tissue. Both these factors depend upon the geometry of the articulating surfaces, the manner in which the mating articular surfaces are restrained relative to each other, and the location or orientation of implant components on the bones.

Impingement problems occurred with some of the early knee implant designs. For the design shown in Fig. 40.15, called the *Geometric design*, the anterior lips of the tibial component would often impinge with the bone of the anterior femur in extension (Fig. 40.16). In their retrieval study of removed components, Hori et al. [38] reported that 83 percent of the loosened tibial components showed evidence of this type of bony impingement, while most of the components of this design which failed because of other reasons did not have such signs of impingement. This suggests that impingement, such as in this example, leads to excessively high implant-interface stresses due to abnormally high tilting forces which in turn can lead to loosening.

A different type of problem related to implant kinematics is also illustrated by the early Geometric knee implant design. The Geometric components are highly conforming; that is, flexion essentially occurs about a single axis of rotation, unlike the natural knee. When the natural knee is flexed from extension, the ligaments assume their proper length patterns so that high loads are not generated. If the joint kinematics are abnormal, such as they are for a fixed axis of rotation, these length patterns are not maintained and the ligaments may develop abnormally high loads at the wrong time within the range of motion or may be abnormally lax. Thus, it is important that the articulating surface geometry be such that knee motion is compatible with ligament function.

This issue is complicated by the variety of ligaments absent in patients due to the arthritic disease process. The anterior cruciate ligament is often absent; the posterior cruciate ligament is less often absent. The collateral ligaments and other capsular structures are usually present, but may be abnormally contracted and consequently surgically altered. The state of ligaments can thus be quite variable among patients.

Whatever the alterations from the natural joint condition are, the implant should (1) allow the necessary range of motion without component impingement, dislocation, or subluxation; (2) not cause abnormal ligament tightening or laxity; (3) generate proper joint contact forces.

Ligament mechanics of the normal knee must be understood as a basis for looking at implant-ligament compatibility. Using a variety of techniques, many investigations of normal knee ligament mechanics have been reported [47–51]. Using ligament force-measuring buckle transducers in in vitro specimens, Lewis et al. [50] have shown that, for normal knees, the ligaments are not significantly loaded during unforced flexion and extension. Under applied loads, the medial collateral ligament is highly loaded in hyperextension, as well as in external-internal rotation of the tibia and valgus throughout the flexion range. The lateral collateral ligament is highly loaded in hyperextension, as well as in varus and external rotation, through-

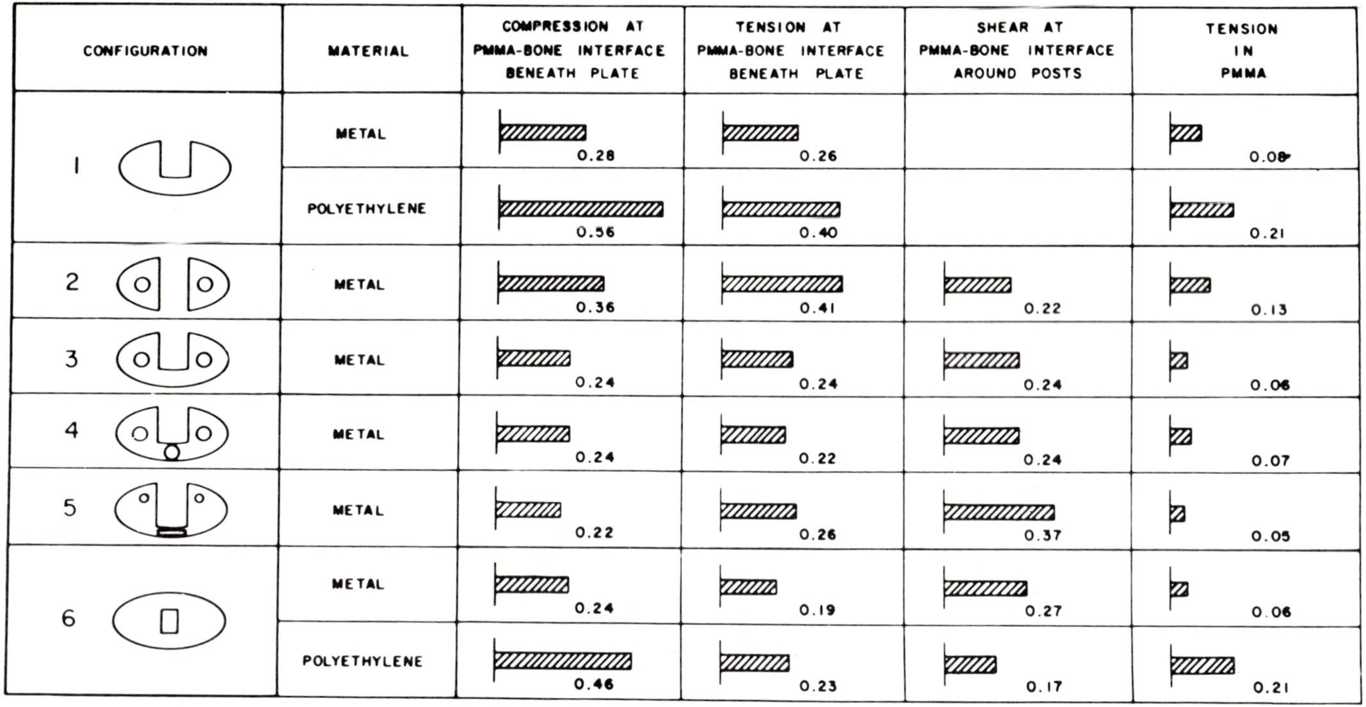

FIG. 40.14 Maximum stress indices; averages of maximum stress fractions over all load types, for different tibial component fixation configurations. (*Reprinted with permission from the Journal of Bone and Joint Surgery, Boston.*)

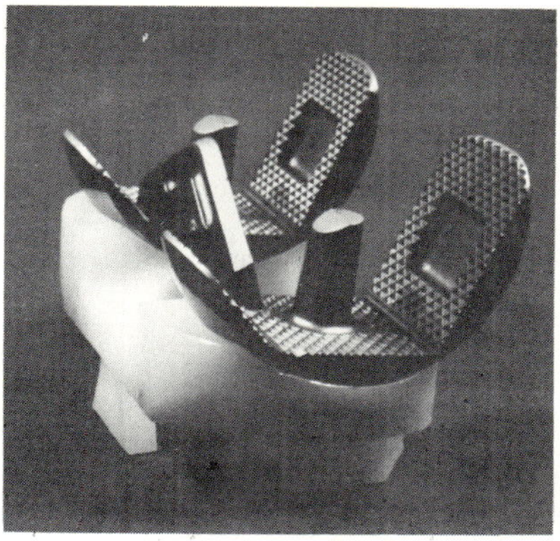

FIG. 40.15 Geometric total knee replacement design (trademark of Howmedica, Inc., Rutherford, N.J.).

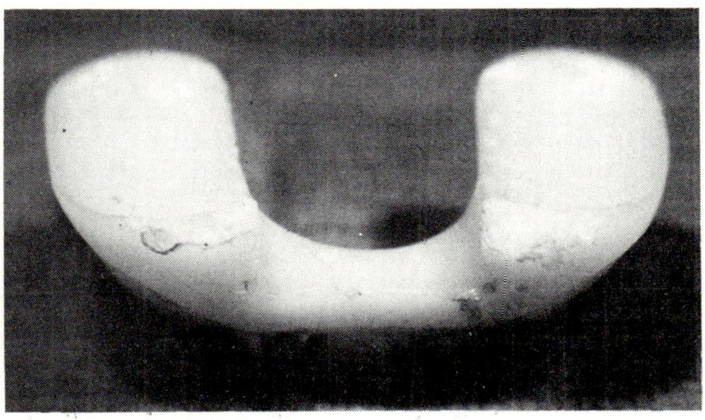

FIG. 40.16 Geometric tibial component with deformed anterior lips due to impingement with distal femur in extension.

out the flexion range. The posterior cruciate ligament is highly loaded at external loads near 90° flexion (posterior-directed force, varus-valgus, internal-external rotation). The anterior cruciate ligament is highly loaded in hyperextension, as well as at external loads near extension (anterior-directed force, varus, and internal rotation). The ligaments with their original anatomy are required to interact with the geometry and placement of the total knee implant. It should be noted that in spite of the large amounts of biomedical research done on ligaments, in vivo ligament mechanics remains poorly understood.

Rigid-body mathematical models can be used to predict the locations of ligamentous attachments (and subsequent length patterns) during knee joint motion, as a function of implant geometry parameters and placement. Lewis and Lew have reported such a model [51], in which total knee replacement motions (and resulting ligament length patterns) during knee joint motion are described in terms of implant geometry parameters and placement (Fig. 40-17). These motions and resulting length patterns can be simulated for the articular surface geometry of any implant by mathematically generating rotations about one or more axes, as well as displacements of the articular curvature of one component over the other (Fig. 40.18). By using the mathematical model to simulate clinical laxity tests, the maximum unloaded lengths (or resting length) for each of the ligamentous fibers can be estimated. Note that the model does not include stiffness characteristics of the joint. As an implant motion is simulated, ligaments are allowed to reach their loaded length (and continue to increase in length), without restraining further motion

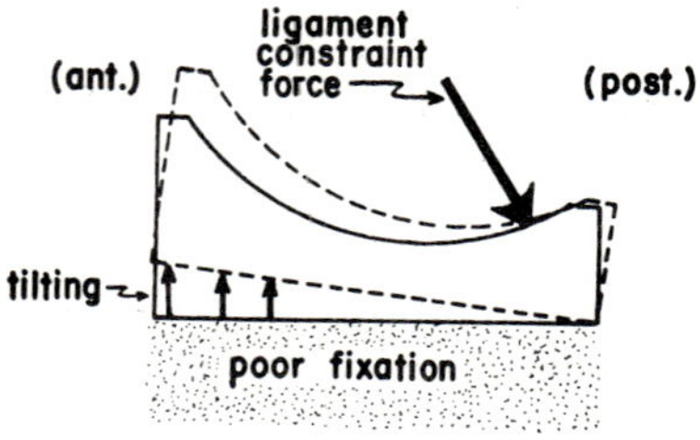

FIG. 40.17 Abnormally high posterior cruciate ligament constraint force in the presence of poor fixation can potentially contribute to geometric tibial component tilting and loosening.

or disturbing the distribution of forces in the other ligaments. In reality, this does not happen, however; the value of the model is that it predicts whether ligaments will become tight or lax within a range of motion, and where in the range this will occur.

For example, the mathematical model was applied to predict the cruciate and collateral length patterns during flexion of knees containing Geometric and Kinematic knee implants [46]. These two were chosen because they can retain both cruciate ligaments and represent extremes of component conformity; that is, the Geometric design is nonanatomical in the sense that its components are highly conforming with a fixed single axis of rotation, while the Kinematic design is less conforming and allows anterior-posterior rolling and sliding to accompany flexion and extension, as occurs in the natural joints (see Fig. 40.2). Figure 40.19 shows predicted length patterns of cruciate and collateral ligaments during high-conforming Geometric and low-conforming Kinematic flexion. Note that in the Kinematic case, an arbitrarily chosen 1 cm of posterior displacement (rolling and sliding) accompanies flexion, similar to that occurring in the

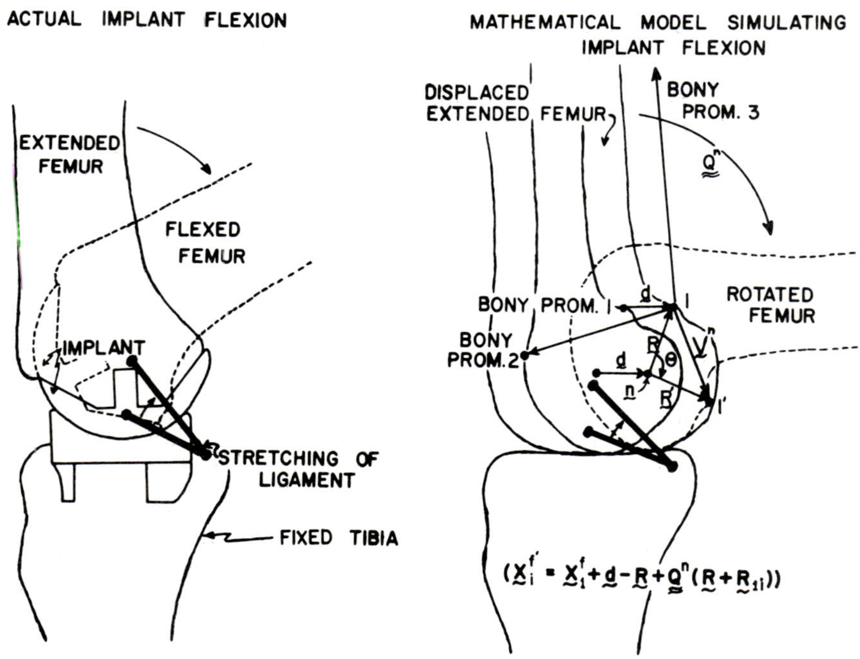

FIG. 40.18 Mathematical model of the knee with total knee implant, which predicts ligament lengths during joint motion as a function of implant geometry and placement, thereby simulating the actual joint function.

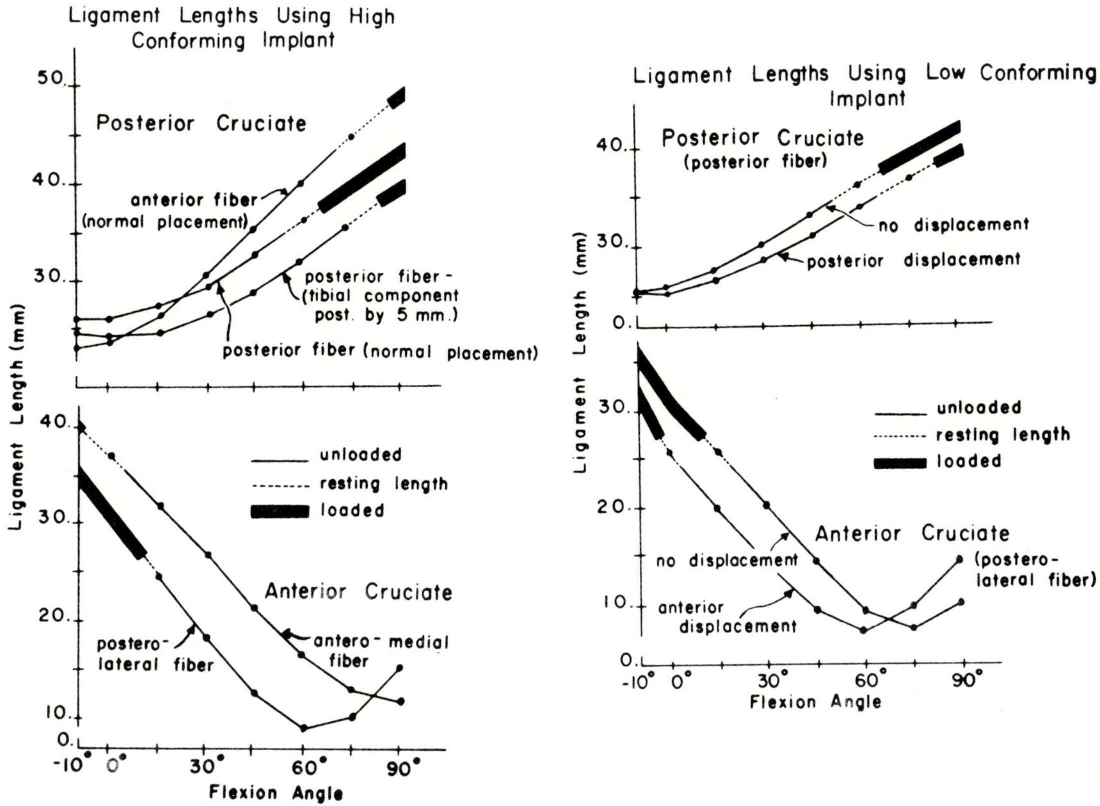

FIG. 40.19 Length patterns of the cruciate ligaments predicted by a mathematical model for flexion of a knee with high-conforming Geometric components and low-conforming Kinematic components (trademarks of Howmedica, Inc., Rutherford, N.J.). (*Reprinted with permission from the Journal of Bone and Joint Surgery, Boston.*)

natural joint. The distance between the medial collateral ligament attachment points remains approximately the same throughout flexion of both implant types, while the lateral collateral ligament increases in length with flexion, with the ligament becoming loaded earlier in the flexion range with the high-conforming Geometric design. The anterior cruciate ligament lengthens when approaching extension for both implant cases, as the ligament reaches its resting length earlier with the high-conforming design than with the low-conforming design. The distance between posterior cruciate ligament attachment points increases with flexion, with the ligament reaching its resting length much earlier in the flexion range with the high-conforming design than with the low-conforming design.

Several points are evident from these results. First, the predictions show that the anterior cruciate ligament becomes loaded near extension, while the posterior cruciate ligament is loaded in later flexion. Secondly, the posterior cruciate gets loaded very early in the flexion range with a high-conforming implant design, while it gets loaded much later in the flexion range with the anatomically shaped Kinematic design, where posterior displacement is allowed by the low-conforming components. When the model is used to generate a hypothetical Kinematic flexion without posterior displacement, the posterior cruciate length pattern differs little from the Geometric length pattern (Fig. 40.19).

When the natural knee is flexed from extension, the ligaments assume their proper length patterns so that high loads are not generated when no load is applied to the joint. If the joint kinematics are abnormal, such as they are for a fixed axis of rotation, these length patterns are not maintained, and the ligament may develop abnormally high loads at the wrong time, within the range of motion, or be abnormally lax. With the single-axis Geometric components, the posterior cruciate ligament would often tighten abnormally during the normal range of flexion, preventing unforced flexion. When flexion was forced beyond this point, large constraint forces were generated on the component, forces which could lead to loss of component fixation (Fig. 40.17). Thus, it is important that the articulating surface geometry be such that knee motion is compatible with ligament function. Thus, the implant design parameter which allows ligament length patterns to approximate the natural knee is the degree of conformity of the mating articular surfaces with resulting potential for anterior-posterior displacement to accompany flexion and extension. The high-conforming Geometric design does not allow such displacements to occur, so that the posterior cruciate ligament gets loaded too early within the range of flexion, producing the potential for loosening to occur.

Current knee replacement designs are mostly of the condylar (similar to the Kinematic) type, that is, semi-constrained, low-conforming components. Femoral components are either duocurvature or multicurvature in nature, so as to ensure ligament compatibility with rotational motions and anterior-posterior displacements, as well as a properly located contact point allowing an adequate moment arm of the quadriceps muscle. Some of the condylar implants retain both cruciate ligaments, while others require the routine excision of the anterior and/or posterior cruciate ligaments, which is accompanied by a slight increase in component conformity.

Using low-conforming tibial components which require the routine excision of the anterior and/or posterior cruciate ligaments creates another question regarding ligament-implant compatibility: What effect does the presence or absence of these cruciate ligaments have on the magnitude and direction of the joint contact forces (in the absence of obvious ligament constraint forces as previously described)? For given functional load states, would the joint contact force and collateral ligament forces have to increase in magnitude, and would the contact force approach an "edge-loading" condition, in order to maintain equilibrium in the absence of an anterior cruciate ligament? Such a question has important implications in terms of loosening of the tibial components. Conclusions regarding this issue are not definitive, since any analysis depends upon the load levels carried by an in vivo anterior cruciate ligament during actual functional activities. Such information is not known at this time, and is an important topic for future research.

40.4 OTHER JOINTS

The design issues for other joint types are very similar to those of the hip and knee, i.e., fixation, ligamentous compatibility, and wear. There are some obvious special considerations, however, and the reader is referred to other sources for detailed information [52].

40.5 FUTURE RESEARCH

The problems associated with total joint replacement are well defined; their solutions are not so clear. In order to improve the design of joint implants, research is needed in several areas: (1) forces acting on the implant, (2) biological state and response and in vivo mechanical state at the bone-cement, or implant, interface, (3) mechanical state at the metal- or polyethylene-cement interface, (4) wear.

40.5.1 Forces on the Implant

Structural analysis requires knowledge of surface forces. These are known approximately for several well-defined positions, but how forces vary during the activities of daily living and what the forces are during other functional states are unknown. Consequently, design loads are poorly defined. For many design questions this is not important because the conclusions are the same for many different loads. However, other conclusions do depend on design loads, so knowledge of these forces is essential. This problem is a part of the larger problem of lack of effective methodology for determining forces in musculoskeletal structures, i.e., muscles, ligaments, tendons, and bones. There is still no agreement on how to do this; it is an area of active research. Consequently, the forces acting on the implants are only approximately known and there will not likely be better-known values in the near future until the larger problem is solved. The consequence for design and evaluation is significant, but several important questions can be resolved without knowing the forces precisely. However, the limited knowledge of joint forces must be kept in mind and the research in this area continued.

40.5.2 Biological and Mechanical States at the Bone-Cement (Implant) Interface

The mechanical state at the implant interface is assumed to be rigidly bonded, and this assumption is supported by good agreement between three-dimensional finite element models and "strain-gauge-on-bone" experiments. This is probably a reasonable simulation of the initial implantation state. However, it is well known that major changes occur at the bone-cement interface over time in vivo. With implants in humans, a line 0.5 to 2 mm thick is seen on x-rays beginning a few months after surgery. This so-called radiolucent line is seen in nearly all implants to some extent [3, 44]. In animal experiments and in human autopsy material, a fibrous liner is always observed histologically [e.g., 55, 56]. Hori and Lewis [57] measured the compressive stiffness of liner material taken from an uncemented implant after 2 months in an animal knee joint. Histologically, this material was identical to tissue found around cemented implants. The unconfined compressive Young's modulus of the material was $E = 2 \times 10^6$ MPa. This material was then assumed to be around a tibial component of a knee implant in a layer 0.5 mm thick on an interface, with harmonic interdigitation of amplitude 0.28 cm and length 0.72 cm. A two-dimensional plane-strain model was used (Fig. 40.20). Representative results [57] show the interface stress with and without the fibrous layer for various assumed values of Poisson's ratio. It is evident that the layer has a significant effect on the system stresses. This model was a first attempt at approximating the tissue effect and has many limitations. Nevertheless, it shows the potential dramatic effect and emphasizes the importance of better quantifying the phenomenon. The uncemented implant can be considered one end of the spectrum of fixation, being the least fixed. As fibrous liner thickness decreases, fixation would be expected to become more nearly rigid. Complete ingrowth into a porous surface would probably be the most rigidly fixed, since nearly direct bony apposition has often been observed in histological sections [58, 59]. There is thus a need to know the properties of this interfacial tissue, but there is also a need to understand the relation between interface stress or motion and tissue generation. Different designs of implants generate different amounts of fibrous tissue. We need to determine the effect of this interface state on the implant system stress distribution, and to understand the relationship between mechanical input variables such as stress and displacement and tissue formation.

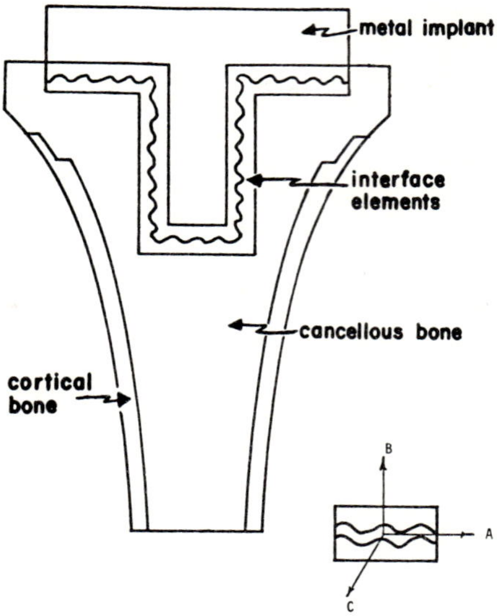

FIG. 40.20 Schematic view of tibial model with transversely isotropic interface elements around the metal prosthesis.

40.5.3 Mechanical State at the Metal-Cement Interface

The three-dimensional finite element models predict the metal-cement interface to be loaded near failure. Whether this interface is indeed failing in vivo needs to be established. If the interface does fail, system stresses will obviously be altered, as has been demonstrated [24]. Various methods for improving the strength of this interface have been explored, such as prebonding a thin layer of PMMA [60], or coating with a porous surface [61]. The long-term viability of these interfaces must be determined.

40.5.4 Wear

Wear of implants is not a major current problem, but may become one as implants are required to last longer. The long-term wear rates and fate of wear debris as well as design methodologies need to be established.

CONCLUSION

Total joint replacement is working well clinically, and although some improvements are needed, the basic concept appears to be sound. The current approaches will probably remain the basis for evolution of future new designs. The newer fixation modes, such as cementless ingrowth of bone into a porous surface, or fixation by a fibrous layer without cement, are still quite experimental and it is not clear how widespread their use will become. Either approach poses unique problems, such as extensive bone remodeling with bony ingrowth, and research issues will likely be redefined. In any case, total joint replacement will remain a successful orthopedic treatment for severe joint disease and an active area for research and development in bioengineering.

REFERENCES

1. Charnley, J.: A biomechanical analysis of the use of cement to anchor the femoral head prosthesis, *J. Bone Jt. Surg., 47B,* 354–363, 1965.
2. Hori, R. Y., Lewis, J. L., Zimmerman, J. R., and Compere, C. L.: The number of total joint replacements in the United States, *Clin. Orthop. Rel. Res., 132,* 46–52, 1978.
3. Stauffer, R. N.: Ten-year follow-up study of total hip replacement: With particular reference to roentgenographic loosening of the components, *J. Bone Jt. Surg., 64A,* 983, 1982.
4. Hori, R. Y., Lewis, J. L., Wixson, R. L., and Kramer, G. M.: An analysis of 130 removed total hips and knees, *Proceedings of NBS/NIH/FDA Conference on Implant Retrieval: Material and Biological Analysis,* NBS Pub. SP-601, Washington, D.C., 1980.
5. Blaha, J. D., Insler, H. P., Freeman, M. A. R., Revell, P. A., and Todd, R. C.: The fixation of a proximal tibial polyethylene prosthesis without cement, *J. Bone Jt. Surg., 64B,* 326, 1982.
6. Lewinnek, G. E., Lewis, J. L., Tarr, R., Compere, C. L., and Zimmerman, J. R.: Dislocations after total hip replacement arthroplasties, *J. Bone Jt. Surg., 60A,* 217–220, 1978.
7. Ali Khan, M. A., Brakenbury, P. H., and Reynolds, I. S. R.: Dislocation following total hip replacement, *J. Bone Jt. Surg., 63B,* 217–218, 1981.
8. Fraser, G. A., and Wroblewski, B. M.: Revision of the Charnley low-friction arthroplasty for recurrent or irreducible dislocation, *J. Bone Jt. Surg., 63B,* 552–555, 1981.
9. Tarr, R. R., Clarke, I. C., Gruen, T. A., Sarmiento, A., Espiritu, E., Hull, D. B., McGuire, P., Sewhoy, A. L., and McKellop, H. A.: Total hip femoral component design, *Orthop. Rev., 11,* 23–35, 1982.
10. Charnley, J.: *Low Friction Arthroplasty of the Hip: Theory and Practice,* Springer-Verlag, New York, 1979.
11. Behrens, J. C., Walker, P. S., and Shoji, H.: Variation in strength and structure of cancellous bone at the knee, *J. Biomech., 7,* 201–207, 1974.
12. Goldstein, S., Wilson, D. L., Sonstegard, D. A., and Matthews, L. S.: An experimental determination of numerical values for elastic modulus and ultimate strength of human trabecular bone as a function of location, *Orthop. Trans., 4,* 92, 1980.
13. Williams, J. L., and Lewis, J. L.: Properties and an anisotropic model of cancellous bone from the proximal tibial epiphysis, *J. Biomech. Eng., 104,* 50–56, 1982.
14. Van Buskirk, W. C., Cowin, S. C., and Ward, R. N.: Ultrasonic measurement of orthotropic elastic constants of bovine femoral bone, *J. Biomech. Eng., 103,* 67–71, 1981.
15. Brown, T. D., and Ferguson, A. B., Jr.: Mechanical property distributions in the cancellous bone of the human proximal femur, *Acta Orthop. Scand., 51,* 429, 1980.
16. Reilly, D. T., and Burstein, A. H.: The mechanical properties of cortical bone, *J. Bone Jt. Surg., 56A,* 1001–1022, 1974.
17. Bartel, D. L., and Ulsoy, G. A.: The effects of stem length and stem material on stresses in bone-prosthesis systems, *Trans. 21st Orthop. Res. Soc., 2,* 25, 1975.
18. Huiskes, R.: Stress analyses of intramedullary fixation systems: Some fundamental aspects of human joint replacement, *Acta. Ortho. Scand. Suppl., 185,* 109–200, 1980.
19. Crowninshield, R. D., Brand, R. A., Johnston, R. C., and Milroy, J. C.: An analysis of femoral component stem design in total hip arthroplasty, *J. Bone Jt. Surg., 62-A,* 68–78, 1980.
20. Lewis, J. L., Askew, M. J., Wixson, R. L., Kramer, G. M., and Tarr, R. R.: The influence of prosthesis stem stiffness and calcar collar on stresses in the proximal femur with a cemented femoral component, *J. Bone Jt. Surg.* (in press).
21. Rohlmann, A., Mössner, U., Bergmann, G., and Kölbel, R.: Finite-element analysis and experimental investigation in a femur with hip endoprosthesis, *J. Biomech., 16,* 727–742, 1983.
22. Tarr, T., Lewis, J. L., Ghassemi, F., Sarmiento, A., Clarke, I., and Weingarten, V.: Anatomic three-dimensional finite element model of the proximal femur with total hip prosthesis, *Proc. Int. Conf. Finite Elements Biomech., 2,* 487–510, 1980.
23. Bartel, D. L., and Desormeaux, S. G.: Femoral stem performance, in *Proceedings of the Symposium on Retrieval and Analysis of Orthopaedic Implants,* National Bureau of Standards Special Pub. 472, Washington, D.C., 1977, pp. 51–59.
24. Huiskes, R.: Some fundamental aspects of human joint replacement, *Acta Orthop. Scand. Suppl., 185,* 109–199, 1980.

25. Crowninshield, R. D., Brand, R. A., Johnston, R. C., and Pedersen, D. R.: An analysis of collar function and the use of titanium in femoral prostheses, *Clin. Orthop., 158,* 270–277, 1981.

26. Hampton, S. J., Andriacchi, T. P., and Galante, J. O.: Three-dimensional stress analysis of the femoral stem of a total hip prosthesis, *J. Biomech., 13,* 443–448, 1980.

27. Oh, I., and Harris, W. H.: Proximal strain distribution in the loaded femur: An in-vitro comparison of the distributions in the intact femur and after insertion of different hip replacement femoral components, *J. Bone Jt. Surg., 60-A,* 75–85, 1978.

28. Radin, E. L., Rubin, C. T., Thrasher, E. L., Lanyon, L. E., Crugnola, A. M., Schiller, A. S., Paul, I. L., and Rose, R. S.: Changes in the bone-cement interface after total hip replacement, *J. Bone Jt. Surg., 64A,* 1188–2000, 1982.

29. Weightman, B., Paul, J. L., Rose, R. M., Simon, S. R., and Radin, E. L.: A comparative study of total hip replacement prostheses, *J. Biomech., 6,* 299, 1973.

30. Clark, I.: Polyethylene wear in prosthetic joints—A new interpretation, *Proceedings of the Conference on the Evaluation of Artificial Joints,* University of Leeds, 1976.

31. Vasu, R., Carter, D. R., and Harris, W. H.: Stress distributions in the acetabular region: I. Before and after total joint replacement, *J. Biomech., 15,* 155–164, 1982.

32. Pedersen, D. R., et al.: An anisymmetric model of acetabular components in total hip arthroplasty, *J. Biomech., 15,* 305–315, 1982.

33. Capello, W. N., Trancik, T. M., Misamore, G., and Eaton, R.: Analysis of revision surgery and resurfacing hip arthroplasty, *Clin. Orthop., 170,* 50, 1982.

34. Amstutz, H. C., Graff-Radford, A., Gruen, T. A., and Clark, I. C.: Tharies surface replacements: A review of the first 100 cases, *Clin. Orthop., 134,* 87, 1978.

35. Shybut, G. T., Askew, M. J., Hori, R. Y., and Stulberg, S. D.: Theoretical and experimental studies of femoral stresses following surface replacement hip arthroplasty, in the *Proceedings of the 8th Annual Open Scientific Meeting of the Hip Society,* Mosby, St. Louis, 1980, chap. 9.

36. Bryan, R. S., and Peterson, L. F. A.: Polycentric total knee arthroplasty: A prognostic assessment, *Clin. Orthop., 145,* 23–28, 1979.

37. Coventry, M. B., Upshaw, J. E., Riley, L. H., Finerman, G. A. M., and Turner, R. H.: Geometric total knee arthroplasty: II. Patient data and complications, *Clin. Orthop., 94,* 177–186, 1973.

38. Hori, R. Y., Lewis, J. L., Wixson, R. L., and Kramer, G. M.: An analysis of 130 removed total hips and knees, in A. Weinstein, D. Gibbons, S. Brown, and W. Ruff (eds.), *Proceedings of the Conference on Implant Retrieval, Material and Biological Analyses,* National Bureau of Standards, Gaithersburg, Md., 1980, pp. 509–588.

39. Askew, M. J., and Lewis, J. L.: Analysis of model variables and fixation post length effects on stresses around a prosthesis in the proximal tibia, *J. Biomech. Eng., 103,* 239–245, 1981.

40. Vicknin, H. H., Hayes, W. C., and Lotke, P. A.: Parametric finite element studies of tibial component fixation in the total condylar knee prostheses, *Trans. 25th Orthop. Res. Soc.,* San Francisco, Calif, 1979, p. 99.

41. Murase, K., Crowninshield, R. D., Pedersen, D. R., and Chang, T. S.: An analysis of tibial component design in total knee arthroplasty, *J. Biomech., 16,* 13–22, 1983.

42. Lewis, J. L., Askew, M. J., and Jaycox, D. P.: A comparative evaluation of tibial component designs of total knee prostheses, *J. Bone Jt. Surg., 64A,* 129–135, 1982.

43. Bartel, D. L., Santavicca, E. A., and Burstein, A. H.: The effects of pegs and trays on stresses associated with loosening of knee prostheses, *Trans. Orthop. Res. Soc., 5,* 17, 1980.

44. Insall, J. N., Ranawat, C. S., Aglietti, P., and Shine, J.: A comparison of four models of total knee replacement prostheses, *J. Bone Jt. Surg., 58A,* 754–765, 1979.

45. Kolbel, R., Bergmann, G., and Boenick, U.: Mechanical properties of the cement/bone bond, in M. Schaldach and M. Hehman (eds.), *Engineering in Medicine,* Springer-Verlag, New York, 1976, vol. 2, pp. 347–357.

46. Lew, W. D., and Lewis, J. L.: The effect of knee-prosthesis geometry on cruciate ligament mechanics during flexion, *J. Bone Jt. Surg., 64A,* 734–739, 1982.

47. Butler, D. L., Noyes, F. R., and Grood, E. S.: Ligamentous restraints to anterior-posterior drawer in the human knee, *J. Bone Jt. Surg., 62-A,* 259–270, 1980.

48. Piziali, R. L., Rastegar, J., Nagel, D. A., and Schurman, D. J.: The contribution of the cruciate ligaments to the load-displacement characteristics of the human knee joint, *J. Biomech. Eng., 102,* 277–283, 1980.

49. Wang, C. J., Walker, P. S., and Wolf, B.: The effects of flexion and rotation on the length patterns of the ligaments of the knee, *J. Biomech., 6,* 587–596, 1973.

50. Lewis, J. L., Lew, W. D., Shybut, G. T., Jasty, M., and Hill, J. A.: Biomechanical function of knee ligaments, *Proceedings of Symposium on the Knee in Sports Medicine,* Denver, 1982.

51. Lewis, J. L., and Lew, W. D.: A Method for locating an optimal "fixed" axis of rotation for the human knee joint, *J. Biomech. Eng., 100,* 187–193, 1978.

52. Allan Inglis (ed.): *Symposium on Total Joint Replacement of the Upper Extremity,* Mosby, St. Louis, 1982.

53. Miller, J. E., Burke, D. L., Bobyn, D., Tremblay, G. R., and Kelebay, L. C.: Experimental studies on the pathogenesis of the fibrous membrane which develops at the interface between implants and bone, *Trans. 25th Orthop. Res. Soc.,* San Francisco, Calif., 1979, p. 302.

54. Charnley, J., Follacci, F. M., and Hammond, B. T.: The long-term reaction of bone to self-curing acrylic cement, *J. Bone Jt. Surg., 50-B,* 822–829, 1968.

55. Vernon-Roberts, B., and Freeman, M. A. R.: Morphological and analytical studies of the tissues adjacent to joint prostheses: Investigations into the causes of loosening of prostheses, in M. Schaldach and D. Hohmann (eds.), *Engineering in Medicine,* vol. 2: *Advances in Artificial Hip and Knee Joint Technology,* Springer-Verlag, Berlin, 1976, pp. 148–186.

56. Charnley, J.: The reaction of bone to self-curing acrylic cement: A long-term histological study in man, *J. Bone Jt. Surg., 52-B,* 340–353, 1970.

57. Hori, R. Y., and Lewis, J. L.: Mechanical properties of the fibrous tissue found at the bone-cement interface following total joint replacement, *J. Biomed. Mater. Res., 16,* 921, 1982.

58. Hedley, A. K., Clarke, I. C., Kozinn, S. C., Coster, I., Gruen, T., and Amstutz, H. C.: Porous ingrowth fixation of the femoral component in a canine surface replacement of the hip, *Clin. Orthop., 163,* 300, 1982.

59. Pilliar, R. M.: Powder metal-made orthopaedic implants with porous surface for fixation by tissue ingrowth, *Clin. Orthop., 176,* 42–51, 1983.

60. Raab, S., Ahmed, A. M., Pfeiffer, D., Miller, J., and Burke, D. L.: The influence of surface preparation on the strength of bone-cement/implant interface, *Trans. 25th Orthop. Res. Society,* San Francisco, Calif., 1979, p. 65.

61. Clemow, A. J. T., Weinstein, A. M., Klawitter, J. J., Koeneman, J., and Anderson, J.: Interface mechanics of porous titanium implants, *J. Biomed. Mater. Res., 15,* 73–82, 1981.

CHAPTER 41

Biomechanics of the Human Spine and Trunk

Albert B. Schultz
Department of Mechanical Engineering and Applied Mechanics
University of Michigan
Ann Arbor, Michigan

INTRODUCTION

There are good reasons to study the mechanics of the musculoskeletal system that the human spine and trunk constitute. It has been estimated (National Health Survey, 1973) that 8 million people in the United States suffer from back impairments. Mechanical factors are involved in both the cause and treatment of many of these impairments. Considerable progress has been made over recent years in understanding the relevant mechanics, although much still remains unknown. This chapter will review what is presently known about the mechanical functioning of the healthy human spine. The bulk of that knowledge pertains to the low-back region.

41.1 MAJOR MECHANICAL COMPONENTS OF THE SPINE AND TRUNK AND THEIR PROPERTIES

41.1.1 Bones

The bones of the spine and the bones of the rib cage are the two major sets of bones that comprise the trunk skeleton. The spine (Fig. 41.1) consists usually of 33 bones called *vertebrae*. In the adult, the 9 lower vertebrae are fused together to form the sacrum and coccyx, which are parts of the pelvis; and the 24 higher vertebrae are stacked longitudinally to form, with intervening soft tissues, a flexible column. These 24 mobile vertebrae are classified, proceeding upward, into 5 lumbar (low back), 12 thoracic (chest), and 7 cervical (neck). These are designated L5 to L1, T12 to T1, and C7 to C1.

The 12 pairs of ribs articulate posteriorly with T1 to T12, and anteriorly all but the lowest two pairs articulate through cartilaginous elements with the sternum to form the rib cage. The

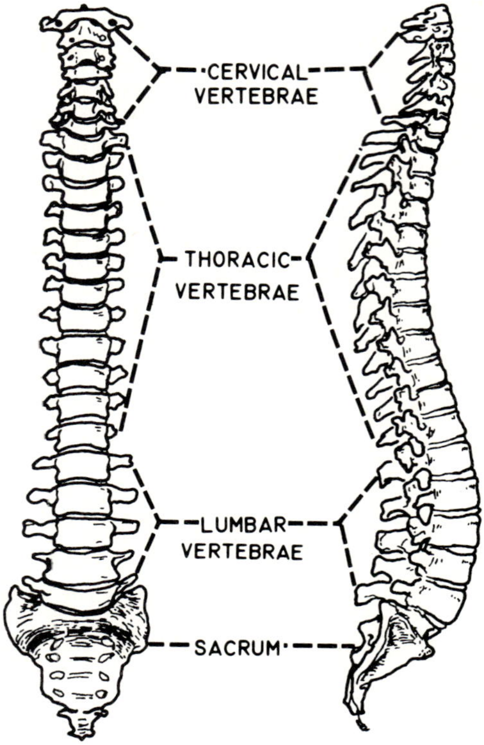

FIG. 41.1 Schematic diagram of the spinal column. On the left, the column is shown as viewed from the front, and on the right, as viewed from the right side. Normally, the vertebrae fall in a vertical line when viewed from the front in upright standing, but when viewed from the side, sagittal plane curves are always present.

skull articulates with C1, while two clavicles and two scapulae form the shoulder girdles to which the arms are attached.

A vertebra consists of an anterior body and a set of posterior elements through which the spinal cord passes. Adjacent vertebral bodies join through an intervertebral disk, and the posterior elements of adjacent vertebrae join through two articular facet joints. Lanier (1939), among others, provides data concerning mean vertebral body heights and diameters. For example, the L3 vertebral body typically has a height of 2.8 cm, and transverse and sagittal diameters of 5.4 and 3.5 cm, respectively.

Each rib consists of a curved and twisted bony portion which articulates with its thoracic vertebra through a series of soft tissue connections, and an anterior cartilaginous portion which joins with that of other ribs and the sternum. Rib and rib articulation geometries have been quantified (Roberts and Chen, 1970; Schultz, Benson, and Hirsch, 1974a, 1974b).

41.1.2 Soft Tissues

One-quarter to one-third of the length of the adult spinal column is occupied by the intervertebral disks. Each disk contains an elastic annular wall which surrounds a central nucleus. The nucleus is filled with a gelatinous material, and occupies 30 to 50 percent of the disk cross-sectional area (Perey, 1957; Eie, 1966; Farfan et al., 1970). These structures are capped above and below by cartilaginous end plates.

In contrast to the disks, whose structures are unique to the spine, there are many other soft tissues which have structures commonly found throughout the body. They include ligaments, synovial joint capsules, muscles, tendons, fascia, and skin. These tissues interconnect the trunk skeletal components, and connect the head and arms to the trunk. A description of them can be found in standard anatomical texts. Descriptions of the organs, or viscera, contained within the

TABLE 41.1 Mean mechanical properties of 42 lumbar spinal motion segments*

Parameter	Posterior elements intact	Posterior elements removed
Displacement, cm, due to 145-N force		
Compression†	0.018	0.019
Posterior shear	0.085 (0.082)	0.124 (0.021)
Anterior shear	0.121 (0.038)	0.142 (0.046)
Lateral shear	0.100 (0.039)	0.111 (0.050)
Displacement, cm, due to 400-N force		
Compression	0.051 (0.024)	0.050 (0.026)
Rotation (degrees) due to 10.6 N · m moment		
Flexion	5.51 (1.00)	5.93 (0.67)
Extension	2.99 (0.02)	
Lateral bending	4.90 (0.79)	4.68‡ (0.77)
Torsion	1.50 (0.67)	2.28‡ (0.77)
Initial pressure, kPa, in unloaded state		
	11 (4)	8 (3)
Pressure increase, kPa, due to 400-N force		
Compression	251 (75)	282 (81)
Additional pressure increase, kPa, due to 145-N force		
Compression†	86	99
Posterior shear	10 (85)	88 (95)
Anterior shear	38 (55)	120 (54)
Lateral shear	45 (40)	98 (26)
Additional pressure increase, kPa, due to 10.6 N · m moment		
Flexion	267 (122)	321 (143)
Extension	49 (202)	
Lateral bending	256 (134)	248‡ (39)
Torsion	32 (37)	161 (65)

Source: Data from Schultz, Warwick, et al. (1979) and Berkson et al. (1979).

* Parentheses indicate standard deviation.

† Interpolated values.

‡ Three specimens only.

thoracic and abdominal cavities are also available from anatomical texts. The viscera serve as significant mechanical components of the trunk musculoskeletal structure when their pressurized contents are used to transmit in part loads applied to the trunk structures (for example, see Bartelink, 1957; Morris et al., 1961).

41.1.3 Component Mechanical Properties

Fairly comprehensive data on the mechanical properties of individual components of the spine and rib cage are available. These have almost all been measured, for good reasons, in cadaver material. While the use of cadaver material can be questioned, available evidence seems to indicate that mechanical properties determined in carefully handled, fresh cadaver material do reflect, at least for short-term behavior, those that exist under conditions in vivo.

Representative earlier reports of the mechanical properties of whole vertebrae, vertebral bone, intervertebral disks, and spinal ligaments are cited by Schultz (1974) and in the comprehensive text by White and Panjabi (1978).

The basic mechanical unit of the spinal column is the motion segment. A motion segment consists of two adjacent vertebrae and their intervening disk and ligamentous tissues. Recent reports of moment-rotation and force-deformation characteristics of whole motion segments under small loads include those of Markolf and Morris (1974), Liu et al. (1975), Panjabi et al. (1977), Lin et al. (1978), Schultz, Warwick, et al. (1979), Berkson et al. (1979), and Andersson and Schultz (1979a).

Typical stiffness data (Table 41.1) indicate that in response to a 100-N force, a lumbar motion segment will displace about 0.1 mm in compression or about 0.5 mm in shear if the accompanying bending is not prevented. In response to a 10-N · m moment, a lumbar motion segment will tilt on the order of 5° in flexion, extension, or lateral bending, and twist about 2° in torsion. The papers cited also provide data on intradiskal pressure increases under load, and on the effect on mechanical properties of articular facet joint destruction or of the injection of additional fluid into the nucleus.

The mechanical properties of lumbar motion segments vary considerably from specimen to specimen. Class differences in mechanical behaviors because of differences in age, sex, disk level, and even degree of visually observed disk degeneration are not pronounced. The individual variations that exist in the behavior of motion segments are so large that they often overwhelm these class differences in mean behavior (Fig. 41.2). Because of the variability, it is often of little use to strive for great precision in analyses of trunk mechanics when data on mechanical properties critically affect their outcomes. A semiquantitative understanding of the underlying mechanics is usually adequate.

The maximum loads that lumbar motion segments can sustain without mechanical damage or clinical pathological changes are not well known, nor is the mechanical role of the articular facet joints understood adequately in quantitative terms.

The mechanical properties of ribs and rib articulations have been measured (Schultz et al., 1974a, 1974b). Unidirectional loading applied to a spine motion segment or to an articulated rib typically results in multidirectional displacements and rotations.

41.2 EXPERIMENTAL DETERMINATIONS OF OVERALL MECHANICAL BEHAVIOR

41.2.1 Data on Trunk Dimensions and Mass Distributions

Data on the gross dimensions of the trunk, such as heights and diameters, and on the weights and mass-center locations of its several parts in various populations, are available from the literature on body anthropometry (for example, see Clauser et al., 1969; National Aeronautics and Space Administration, 1978).

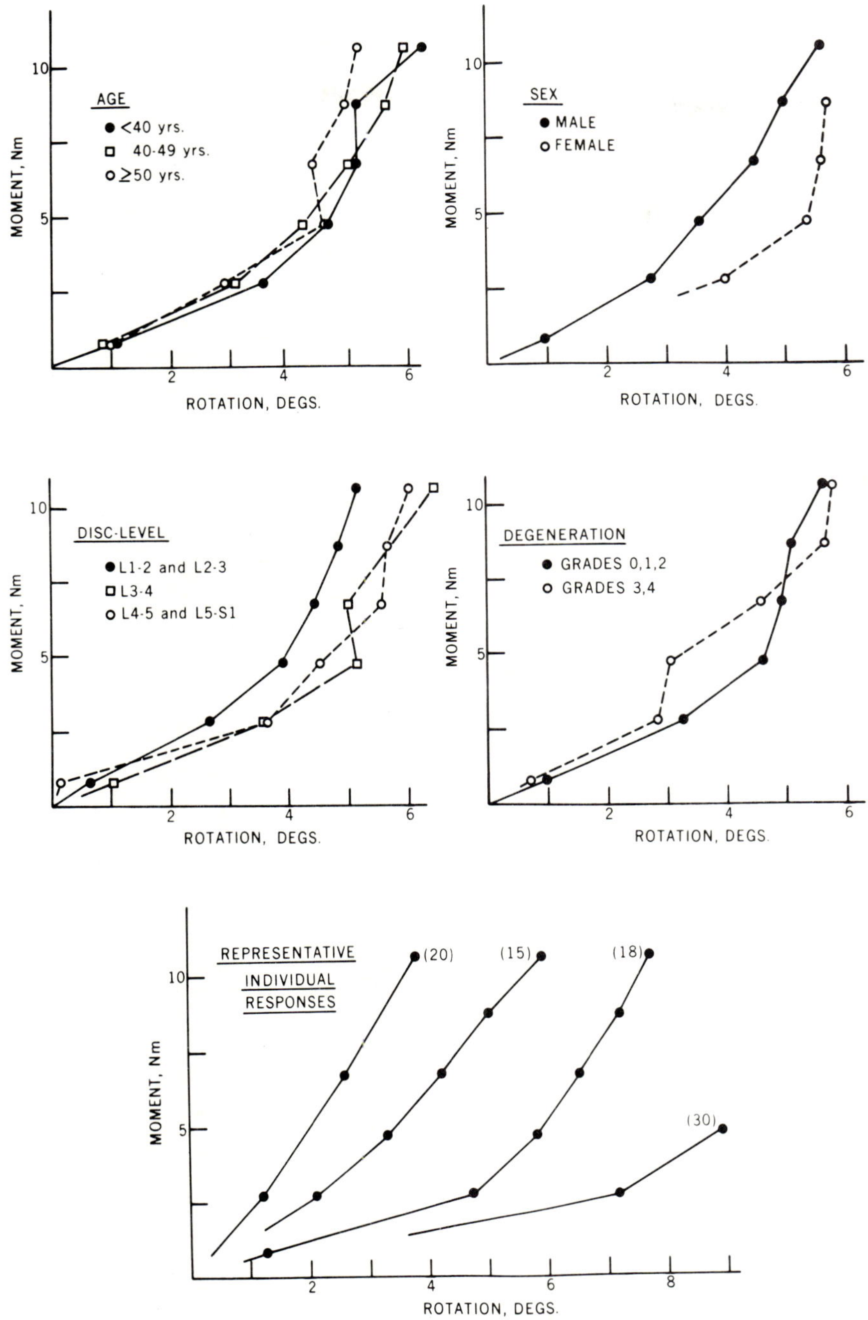

FIG. 41.2 Mean moment-rotation responses in lateral bending and their class differences in 42 cadaver lumbar motion segments. Response variation is also shown. (*From Nachemson et al., 1979.*)

41.2.2 Vertebral Column Configurations and Motions

Few data are available that quantify vertebra by vertebra the configurations the spine assumes in everyday activities across a normal population. Even for relatively simple configurations this is the case. In upright standing, the vertebrae lie in a nearly straight vertical line when viewed from the front, and only the configuration as viewed from the side is needed. Lateral roentgenographs might provide those data, but these are usually taken with the arms raised so as not to obscure the field of view. This action alters the configuration. Moreover, small cyclic motions due to respiration and postural sway continually occur.

Simple configurational measurements can be made in cadaver material (Farfan et al., 1972), but cadaver spines are usually flattened and often not subject to the same body-segment weights or soft tissue constraining forces that are present in vivo. More complex configurations, such as those involving bending and twisting of the trunk, are difficult to describe in detail because of their three-dimensional nature. Few configurational data seem available for these situations. As a result, when detailed configurational data are needed, they usually must be estimated.

Vertebral column motions have been studied both in vivo and in cadaver material. Flexion (forward), extension (backward), and lateral (side) bending are essentially planar motions, and so are relatively easy to measure in vivo from roentgenographs. A number of such studies have been reported (Bakke, 1931; Tanz, 1953; Moll and Wright, 1980). Even if a single motion segment is capable of only a few degrees of rotation in a given mode, there are enough segments in the column so that a large total motion capability results. For example, Bakke found total spine motion capabilities to be on the order of 80° in bending forward or to the side, and 140° in bending backward. Attempts to correlate changes in motion ranges with various spinal abnormalities have succeeded only sometimes.

Longitudinal axis rotation, often referred to simply as *rotation*, is more difficult to measure in vivo, as are combinations of sagittal and frontal plane motions. Few studies of such motions have been reported. Several careful attempts have been made to evaluate complex three-dimensional configurational changes through biplanar radiography (for example, see Brown et al., 1976). In these schemes, displacements are calculated from roentgenographic observations of changes in the coordinates of a set of points on each vertebra by using rigid-body kinematics theory. The radiographic process is such that the actual displacement of a point may be significantly different from its radiographically apparent displacement (Benson et al., 1976). Thus, there still seems to be no completely satisfactory scheme to measure longitudinal axis rotations or other complex motions in a clinical setting, despite their possible diagnostic importance.

41.2.3 Voluntary Trunk Strengths

A knowledge of voluntary trunk strengths is useful in obtaining a basic understanding of trunk mechanics in the setting of industrial work standards and in clinical diagnoses of neuromuscular system abnormalities, for example.

Lifting strengths have often been measured, usually under isometric conditions. That is, the position of the body is fixed during the test, and motion is not allowed. A typical study (Singh and Ashton, 1970) reports lifting strengths of approximately 1500 N in adult males.

Studies of lifting strength usually enable only crude estimates of trunk muscular strengths to be made. Fairly complete knowledge of the body configuration is required to calculate trunk reactions from the measured lifting forces, and these data have seldom been reported. When leg lifts are performed, with the back straight and the knees bent, considerably higher lifting strengths have been reported than in back lifts with legs straight and back bent. Whether the power for the lift is generated in the back muscles or in the leg muscles seems immaterial. In either case the soft tissues of the back must be capable of at least passive transmission of the necessary forces. Greater strengths are found in leg lifts probably because of more favorable body configurations rather than because of intrinsic differences in the methods of accomplishing the lift. When body configuration is controlled, leg-lift and back-lift strengths are found to be comparable (Andersson and Schultz, 1979b).

More useful information is provided by isometric tests of the ability of the trunk muscles to develop flexion, extension, and lateral bending moments (Troup and Chapman, 1969; Poulsen

TABLE 41.2 Mean maximum voluntary strengths of the lumbar trunk muscles, in N · m*

Subjects	In attempted flexion	In attempted extension	In attempted lateral bending
27 healthy adult males†	149	210	147
30 healthy adult females†	87	117	79
19 healthy girls, age 12‡	33	42	38

* Strengths were measured in 5-s quasi-static exertions, in upright standing positions. Loads were applied to a chest harness placed just under the arms. The maximum voluntary moments reported were those developed about the L5 level in the adults and about the L3 level in the children.

† Data from McNeill et al. (1980).

‡ Data from Portillo et al. (1982).

and Jorgensen, 1971; McNeill et al., 1980). Typical data (Table 41.2) show that the trunk muscles can produce bending moments on the lumbar spine that in adults are on the order of 150 N · m. This can be contrasted with the bending stiffnesses of lumbar spine motion segments, which are on the order of 1 N · m per degree for small amounts of bending. For the many common activities that involve nearly upright trunk postures, the bending resistance of the spine motion segments is very small relative to the capabilities of the trunk muscles to bend them. So, in many situations the bending resistance of the spine itself can be considered to be negligible.

Few voluntary trunk strength data concern longitudinal twisting, and almost no data are available for isometric strength tests carried out in positions other than upright standing. The effects on isometric muscle strength of age (Asmussen and Heeboll-Nielsen, 1961), body motions (Asmussen et al., 1965), repetitive loading (Jorgensen and Poulsen, 1974), training (Ashton and Singh, 1973), fatigue (Ortengren et al., 1975), and suggestion, noise, and drugs (Ikai and Steinhaus, 1961) have been reported. The latter report concluded that in voluntarily executed maximal efforts, psychological rather than physiological factors determine the limits of performance.

41.2.4 Myoelectric Measurements

It is difficult to measure directly in vivo the forces developed internal to the trunk in various activities. Some indirect techniques are available. These include myoelectric, intradiskal pressure, and trunk cavity pressure measurements.

Muscles contract when electrically stimulated by motor nerves. This motor nerve activity can be sensed by appropriate needle, wire, or skin-surface electrodes. These myoelectric measurements provide data on muscle contraction force levels, but, until recently, most of these measurements have been interpreted only qualitatively. Recent studies in which myoelectric measurements were interpreted quantitatively are described in Sec. 41.3.3.

A comprehensive review of myoelectric measurements relating to the trunk musculature (Ortengren and Andersson, 1977) cites studies that show, for example, that in relaxed standing, approximately 75 percent of subjects have a low level of myoelectric activity in their abdominal wall muscles, while the remainder show similar small activities in their back muscles. In trunk forward flexion there is often pronounced muscular activity when flexion is begun, but once a bent position is achieved, this activity quiets. This results probably because the ligaments of the spine become taut in flexed positions and begin to sustain the required forces passively.

41.2.5 Intradiskal Pressure Measurements

Except in severely desiccated (degenerated) intervertebral disks, the gelatinous contents of the disk nucleus exhibit hydrostatic behavior (Nachemson, 1960). In response to applied loads intradiskal pressures increase, and typically are 1.3 to 1.5 times the applied compression per unit disk cross-sectional area. Intradiskal pressure measurements are thus indicative of compression load in vitro and, presumably, in vivo. Disk pressure increase measurements have been employed extensively to estimate compressive loads on the spine (Nachemson and Elfstrom, 1970; Nachemson, 1981; Schultz et al., 1982a). These measurements show that the L3–L4 pressure is often on the order of 1 MPa. That disk often experiences compression loads in vivo of at least three to four times the superincumbent body weight, and sometimes much more. These large longitudinal loads arise whenever the trunk muscles contract strongly.

41.2.6 Trunk Cavity Pressure Measurements

In some physical activities the pressure in the thoracic, gastric, and rectal cavities may increase significantly over resting levels. A number of studies have examined the role of cavity pressures in trunk mechanics (Bartelink, 1957; Morris et al., 1961; Eie and Wehn, 1962; Schultz et al., 1982a). In very strenuous activities gastric pressures up to 150 to 200 mmHg (20 to 27 kPa) have been found.

Interpretation of these pressure measurements has remained controversial, centering around questions of to what extent and over what time periods internal viscera are able to serve as a structural compression support for the flexed trunk.

41.2.7 Behavior of the Rib Cage

Overall rib cage dimensions (Clauser et al., 1969) and rib cage motions in respiration (Jordanoglu, 1969) have been reported. The change in sagittal and frontal thoracic cage diameters that result from lateral squeezing forces up to 150 N have been measured in vivo (Agostoni et al., 1966). Volunteers were reported to tolerate loads of 300 to 400 N applied to the sternum (Patrick et al., 1965), but their chest deformations were not given. Responses to static chest loading in embalmed cadavers and to both static and dynamic chest loading in fresh cadavers have been measured (Patrick et al., 1965; Nahum et al., 1970).

41.3 MODEL ANALYSES OF OVERALL MECHANICAL BEHAVIOR AND THEIR VALIDATION

As stated previously, it is often difficult to make experimental measurements of trunk mechanical behavior, particularly in vivo. Complex behaviors are frequently studied more readily by mathematical model analysis on the basis of experimental data concerning individual trunk constituents and limited other experimental measurements.

The mathematical models described in the literature on human trunk biomechanics seem to fall into three classes: rigid-body models, lumped-parameter deformable-element models, and distributed-parameter deformable-element models. In the last class the models reported all seem to use finite element techniques.

Analyses of the mechanical behaviors of the trunk or of its components using such models have ranged from the simple to the complex. Representative analyses of several types will be described, and sample findings cited.

41.3.1 Estimates of Trunk Strengths from Muscle Cross-Sectional Areas

Trunk strengths can be estimated using simple ideas. Several studies (e.g., see Ikai and Fukunaga, 1968) have provided estimates of the maximum contractile strength of human arm or leg muscles. That strength usually lies in the range of 400 to 1000 kPa. Trunk muscle cross-sectional areas and centroidal locations can be estimated from scaled drawings of cross-sectional anatomy (Eycleshymer and Schoemaker, 1911) or by direct measurement in cadavers (Rab and Chao, 1975). If it is assumed that all voluntary muscle possesses the same intrinsic strength, estimates can be made of the maximum moment-developing capabilities of the trunk muscles. Schultz, Haderspeck, et al. (1983) provide representative data so obtained for the lumbar trunk muscles.

41.3.2 Determination of Forces Internal to the Trunk: Rigid-Body Models

The human body seldom can be considered rigid in the lay sense of that term, but it often can usefully be considered rigid for the purposes of biomechanical analysis. As long as the deformations of a structure are not of interest, a rigid-body analysis of it can provide considerable quantitative information on the relationships between the external forces and moments to which it is subjected and the internal forces and moments its structural elements provide in response. Mostly, rigid-body models have been used to analyze physical activities performed quasi-statically (isometrically). Rigid-body model analyses are usually indeterminate; the number of unknowns exceeds the number of equations of motion in dynamic problems, or of equilibrium in static problems.

Internal force estimates can be obtained by rendering the problem determinate through assumptions. Typical assumptions are that certain forces, such as those in muscle antagonists, are zero; or that groups of muscles and other soft tissues act in concert, and so may be replaced by a single force equivalent. A number of investigations have followed this plan of action. These ideas work well in situations that are mechanically simple. Two-dimensional sagittally symmetric analyses of quasi-static lifting activities (Morris et al., 1961; Chaffin and Baker, 1970; Schultz et al., 1982b) and some three-dimensional analyses (Andersson, et al., 1980) have been carried out using these schemes.

In more complex situations, more sophisticated analysis schemes are needed; the following description is adopted from Schultz and Andersson (1981). For example, to determine internal forces at the L3 level in such situations, a transverse cutting plane is imagined to be passed at this level. The forces in the structures cut by that plane are conveniently determined in two steps. First, the net reaction is computed, and then a set of internal forces that can provide that net reaction is estimated.

The net reaction consists of the three net force components and three net moment components that must be supplied by the part of the body below the imaginary cutting plane to act on the part of the body above the plane, in order to keep the upper body in equilibrium. The net reaction is determined from the six requirements for equilibrium that result from Newton's laws.

To calculate the net reaction, a coordinate system is established at the center of the vertebral body cut by the imaginary plane. Suppose that the x axis is positive to the right, the y axis positive anteriorly, and the z axis positive superiorly. The x and y axes lie in the transverse cutting plane, with their orientations referred to the orientation of the vertebral body; the z axis is perpendicular to the cutting plane (see Fig. 41.3).

The external loads acting on the upper body are of two types: loads applied in order to perform the task and loads from body-segment weights. The applied loads are the forces needed to maneuver (lift, push, etc.) any object being handled; they are usually applied to the hands and can be measured directly. The weight loads act at the centers of mass of the upper body segments. The weights of different body segments and mass-center locations can be found in the literature. If the applied and body weight loads are known, the net reaction is obtained from the equations of equilibrium, as illustrated in Fig. 41.3.

The net reaction is not affected directly by anything occurring in the parts of the body below the cutting plane, nor is it affected by the material properties of the hard and soft tissues of the body. Anatomical variables affect the net reaction only insofar as they influence mass distributions and moment arms. The major difficulty in the determination of a net reaction is to gather the needed configurational data. The process of gathering precise three-dimensional coordinate data for many points can be tedious. Approximate data will frequently suffice; z-coordinate data are not needed for loads that act vertically, nor are x-coordinate data needed to analyze sagittally symmetric activities.

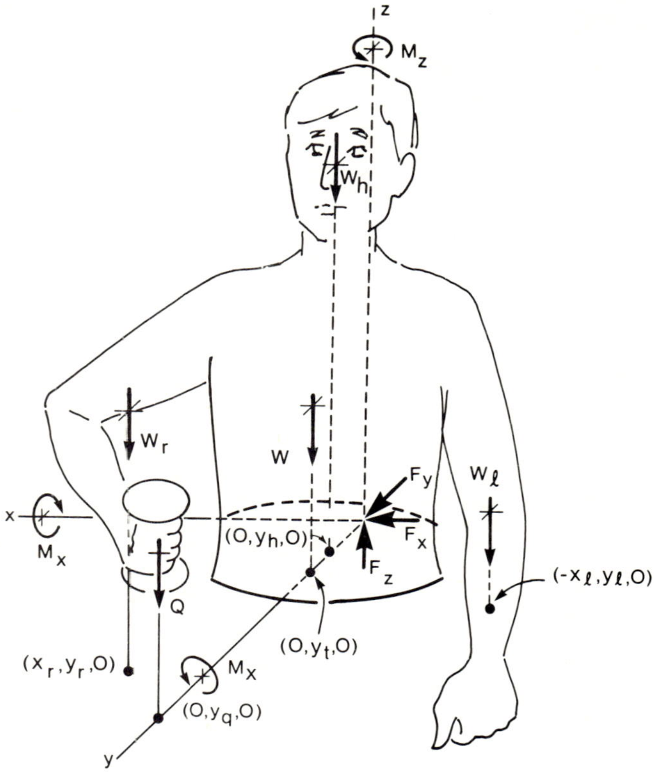

FIG. 41.3 Holding a weight in the right hand: external forces and net reaction components. The net reaction is computed by considering both (1) equilibrium of forces in the x direction ($F_x = 0$), y direction ($F_y = 0$), and z direction ($F_z = Q + W_h + W_l + W_r + W_t$), and (2) equilibrium of moments about the x axis ($M_x = y_q Q + y_h W_h + y_l W_l + y_r W_r + y_t W_t$), y axis, ($M_y = x_r W_r - x_l W_l$), and z axis ($M_z = 0$). In this example, there are no forces in the x or y direction, and no force has a moment about the z axis. In addition, none of W_h, W_q, or W_t has a moment about the y axis. However, it is a straightforward task to generalize the net-reaction computational scheme illustrated here to more complicated situations. (*From Schultz and Andersson, 1981.*)

The net reaction needed to equilibrate the upper part of the body is supplied in the lumbar trunk by muscle contractions, connective tissue tensions, intraabdominal pressure, and resistances supplied by the spinal motion segments. These are forces internal to the trunk. The problem is to determine the motion-segment resistances, or, in other words, the loads on the spine. But in the course of estimating these loads, the other internal forces will need to be estimated as well.

On the basis of the mechanical property and trunk strength data cited earlier, the lumbar motion segments can be assumed not to supply moments of significant resistance in ordinary activities, but they can resist significant shear and compression forces.

A cross-sectional model of the trunk, as shown in Fig. 41.4, is now considered. This model can be used to estimate trunk internal forces in a variety of circumstances. It incorporates 10 muscle equivalents to represent most of the major muscles groups spanning the lumbar region. The five on the right side are the right latissimus dorsi equivalent L_r, assumed to act at $(x_l, -y_l)$

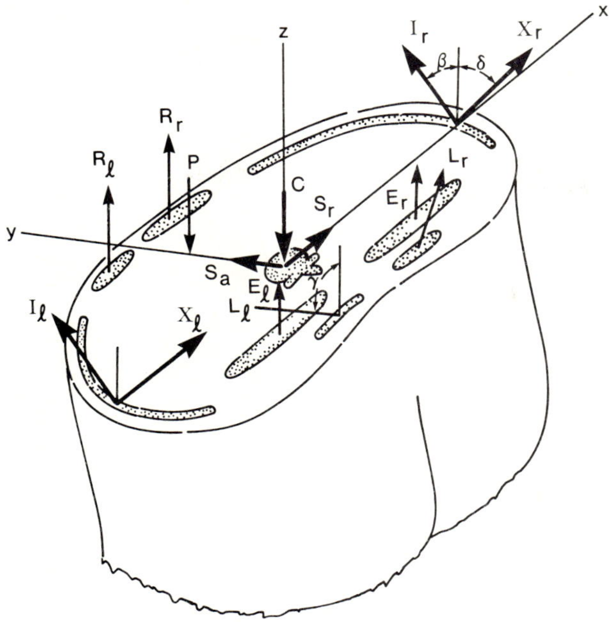

FIG. 41.4 A general model for internal force estimation. Determination of the internal forces is governed by the six equations of equilibrium:

$$F_x = (L_l - L_r) \sin \gamma + S_r$$
$$F_y = (I_l + I_r) \sin \beta - (X_l + X_r) \sin \delta + S_a$$
$$F_x = C + P - (E_l + E_r) - (R_l + R_r) - (I_l + I_r) \cos \beta - (L_l + L_r) \cos \gamma - (X_l + X_r) \cos \delta$$
$$M_x = y_e(E_l + E_r) - y_r(R_l + R_r) + y_p + y_l(L_l + L_r) \cos \gamma - y_0[(I_l + I_r) \cos \beta + (X_l + X_r) \cos \delta]$$
$$M_y = x_e(E_r - E_l) + x_r(R_r - R_l) + x_l(L_r - L_l) \cos \gamma + x_0[(I_r + I_l) \cos \beta + (X_r - X_l) \cos \delta]$$
$$M_z = y_l(L_r - L_l) \sin \gamma + x_0[(I_r + I_l) \sin \beta - (X_r - X_l) \sin \delta]$$

at an angle of γ with the z axis; the right erector spinae equivalent E_r at $(x_e, -y_e)$; the right external oblique X_r at (x_0, y_0), at an angle of δ with the z axis; the right internal oblique I_r, also at (x_0, y_0), at angle β; and the right rectus abdominus R_r, at (x_r, y_r). Five corresponding muscle equivalents are symmetrically placed on the left side. An intraabdominal pressure force P is assumed to act at $(0, y_p)$. The spinal motion-segment resistances consist of a compression C, a right lateral shear S_r, and an anterior shear S_a. The three motion-segment resistances are assumed to act at the coordinate system origin.

The requirement that these 14 internal forces provide the six components of the net reaction that are needed for equilibrium is expressed by the six equations presented in Fig. 41.4. The intraabdominal pressure resultant can be determined from experimental measurements. The other 13 internal forces are unknown, and only six equations are available to find them, so use of this model leads to a statically indeterminate problem. This indeterminate problem can be solved through use of optimization techniques, as proposed, for example, by Seireg and Arvikar (1973). Linear programming (Dantzig, 1968) is probably the simplest of these techniques.

The major question that arises when such procedures are used for solution is the choice of the so-called objective function, or the quantity to be optimized. Suppose, for illustration, that the compression on the lumbar vertebra is to be minimized, and use the equation for z-direction

TABLE 41.3 Typical forces, in newtons, on the L3 motion segment predicted using the 10-muscle equivalent model of Fig. 41.4 and a minimum-intensity objective function

Task performed	Compression	Lateral shear	Anterior-posterior shear
Stand relaxed	500	0	0
Resist:			
44-N · m flexion	1460	0	150
40-N · m extension	970	0	−170
31-N · m lateral bend	1150	110	−10
51-N · m twist	1100	50	−50

Source: Data from Schultz et al. (1982a).

force equilibrium to do so. The equations for x and y force equilibrium are used to find S_r and S_a after the solution is obtained. The three equations of moment equilibrium, and the 10 requirements that the muscle contraction intensities not exceed a reasonable level (1000 kPa), are used as constraints.

Use of the minimum compression objective function works well (that is, leads to predictions that correlate well with validation measurements, as explained subsequently) in some situations. An optimization scheme that works well in a wider range of situations is to minimize approximately the largest muscle contraction intensity needed to perform a task. To do this, the maximum allowed contraction intensity is set to, say, 100 kPa and a minimum compression solution is attempted. If no solution to this linear programming problem exists at this intensity, increase that allowed intensity by 100 kPa and make another solution trial. This intensity incremention process is then repeated until a solution is found. Tables 41.3 and 41.4 illustrate typical results obtained through the use of this internal force prediction scheme.

Internal force analyses demonstrate that the loads imposed on the spine and the lumbar trunk muscles by physical task performances are determined much more by the moments of the external loads about the spine than by the external loads themselves. For example, it is much more stressful to hold a weight with the arms extended anteriorly than to hold the same weight close to the chest, because the moment about the spine of the arm weights and the weight held is much larger when the arms are extended.

TABLE 41.4 Typical mean trunk muscle contraction forces, in newtons, required for the tasks listed in Table 41.3*

Task performed	E_L	E_R	I_L	I_R	$\overline{X}_L$	$\overline{X}_R$	Required intensity, N/cm²
Stand relaxed	60	60	0	0	0	0	6
Resist:							
44-N · m flexion	510	510	0	0	0	0	23
40-N · m extension	0	0	160	160	140	140	23
31-N · m lateral bend	310	120	50	50	120	40	23
51-N · m twist	80	260	290	0	0	220	31

Source: Data from Schultz et al. (1982a).

* Forces calculated using minimum-intensity objective function. Rectus abdominis and latissimus dorsi contraction forces were small, and are not shown. E = erector spinae, I = internal abdominal oblique, $\overline{X}$ = external abdominal oblique. Subscripts indicate left or right side.

41.3.3 Validation of Internal Force Estimates

The foregoing explanations show that while there is little ambiguity about the nature of the net reductions, major assumptions are used in the estimation of the internal forces. These assumptions need to be validated by experimental measurements.

Direct measurement of trunk internal forces in vivo is impractical, but, as pointed out earlier, there are at least two means to determine internal forces indirectly in the laboratory: by measurement of intradiskal pressures and by measurement of myoelectric activities. Evidence showing that intradiskal pressure measurements are indicative of compression loads on the spine and that myoelectric activity in the back muscles relates monotonically to muscle contraction forces was cited earlier.

Using these two measurement techniques, a series of studies has amply validated the spinal load calculation schemes presented here in a number of circumstances. Good agreement between computed muscle tensions and measured muscle myoelectric activities over a range of tasks is reported by Andersson et al. (1980) and by Schultz et al. (1982a, 1982c, 1983). Schultz et al. (1982c) also report good agreement between predicted spinal compressions and spinal compressions determined from intradiskal pressure measurements. In addition to providing validation data, the reports cited provide a body of quantitative data on the loads imposed on the various lumbar trunk structures by a wide range of isometric physical activities.

41.3.4 Analyses of Dynamic Problems

Performance of a physical activity is sometimes called *dynamic* if it is accompanied by body motion of any kind. However, many activities involving body motion can be analyzed without significant error as if they were static. Dynamic considerations are important mechanically only when a motion involves significant linear or angular accelerations; that is, only when the inertial forces and moments produced are of magnitudes that are significant when compared with the forces and moments needed for equilibrium. If the inertial forces and moments are small when compared with the forces and moments that would be required for static equilibrium, an activity involving body motion can safely be analyzed as a quasi-static activity.

In truly dynamic activities in which significant inertial forces and moments do arise, these forces and moments can be accounted for easily enough. First, the inertial forces and moments need to be computed. This can be done through measurement or estimation of the appropriate linear and angular accelerations and use of anthropometric data on body-segment masses, centers of mass, and moments of inertia. Once the inertial forces and moments have been determined, they are applied at the body-segment centers of mass as if they were additional external forces. The remaining procedures already outlined are then executed without alteration. In other words, the inertial forces and moments directly enter the procedures used to calculate the net reaction, but not those used to estimate the internal forces, under almost all circumstances of practical interest.

41.3.5 Lumped-Parameter Deformable-Element Models

Rigid-body models are not suitable for analyses of structures when predictions are needed of what deformations will be experienced in response to the loads applied to these structures. Models that incorporate deformable elements are required for this type of analysis. Such models are mathematically more complicated than rigid-body models, and their use requires a knowledge of the mechanical properties of all incorporated deformable elements. Trunk models of this kind (Panjabi and White, 1971; Belytschko et al., 1973) have been used to study the progression and correction of idiopathic scoliosis, and spinal injuries that result from emergency ejection of pilots from aircraft, for example.

These models typically employ the direct stiffness method of structural analysis, and so involve a large set of simultaneous algebraic equations. Each mobile vertebra or calcified rib segment, for example, is idealized as a rigid body. If N bodies are so modeled in three dimen-

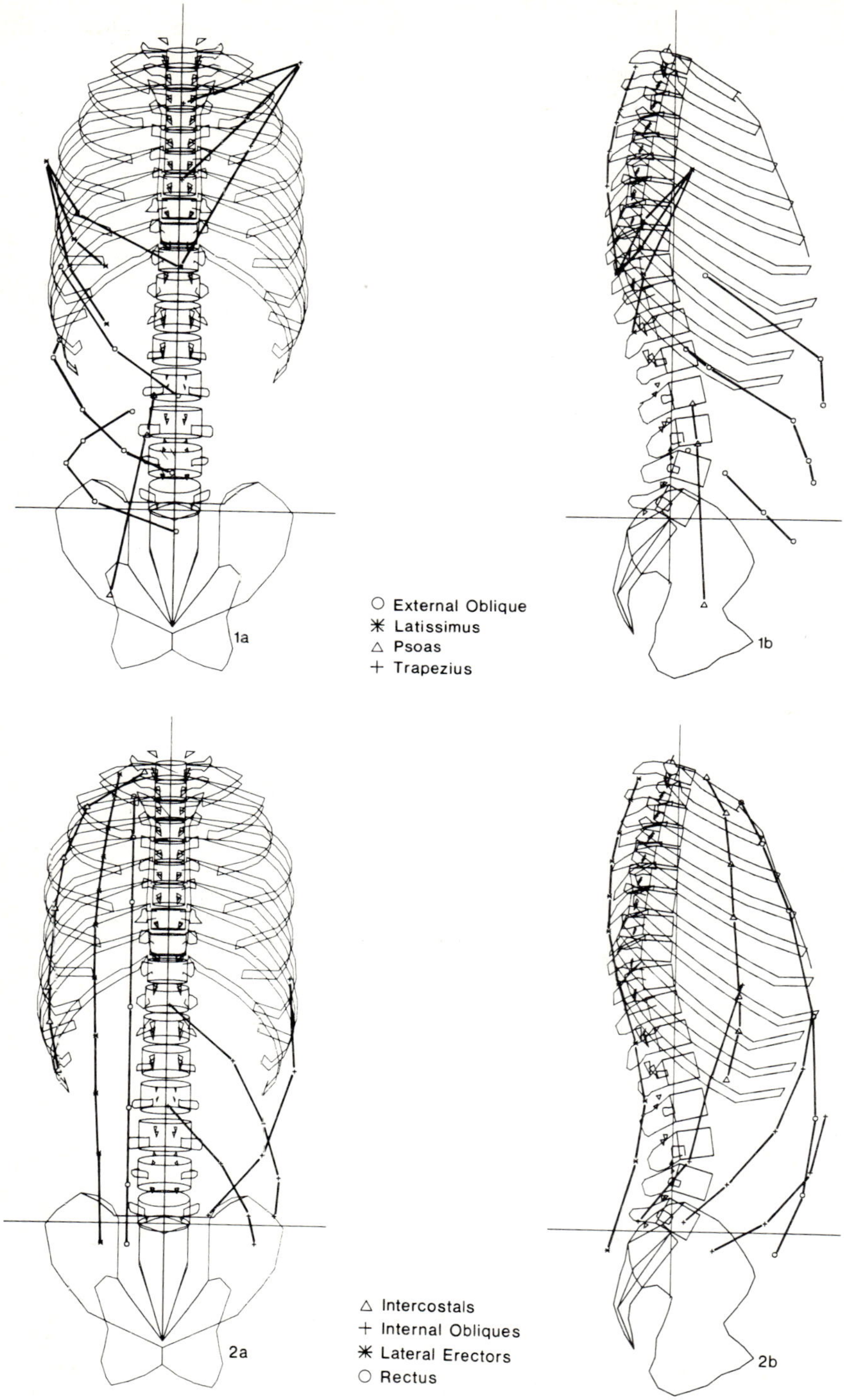

FIG. 41.5 Placement of selected trunk muscles in a lumped-parameter deformable-element model of the trunk. The model will predict what deformations result in response to known contraction forces in each muscle slip. (*From Takashima et al., 1979.*)

TABLE 41.5 Muscles incorporated into the deformable-element model of Fig. 41.5*

Muscle	Superior-inferior placement	Cross-sectional area per side, cm^2
Abdominal wall muscles:		
Rectus	H	3.3
	L	3.9
Internal oblique	H	3.9
	M	3.2
	L	3.8
External oblique	H	2.6
	M	3.2
	L	3.8
Transverse	H	3.4
	M	3.4
	L	3.4
Chest wall muscles:		
Pectoralis major	H	7.4
	L	6.5
Pectoralis minor		1.6
Serratus, anterior superior	H	3.7
	M	7.9
	L	4.9
Intercostals, lateral	H	8.6
	M	8.4
	L	6.3
Superficial back muscles:		
Trapezius	H	2.4
	M	2.4
	L	1.9
Latissimus dorsi	H	1.9
	H-M	1.9
	M-L	1.9
	L	2.4
Deep back muscles:		
Psoas		11.6
Quadratus		2.6
Lateral erectors:		
Cervical		6.9
Thoracic	H	6.0
	L	8.7
Lumbar		18.6
Midline erectors	H	5.6 (unilateral)
	L	13.2 (unilateral)

Source: Data from Takashima et al. (1979).

* H = high, M = middle, L = low. With the exception of the two midline erectors, each entry represents two model muscles symmetrically located on the right and left sides.

sions, there are 6*N* degrees of freedom, so that the model is governed by 6*N* simultaneous equations.

The rigid bodies are interconnected by deformable elements representing ligaments, intervertebral disks, and so on. The force-deformation or moment-rotation (that is, stiffness) relation for each deformable element is needed as input. Using the data on stiffness and model geometry, a global stiffness matrix $[k]$ is assembled that specifies the 6*N* simultaneous equations in the form $\{f\} = [k]\{d\}$. The usual situation is that the external forces and moments f are given, and the equations are solved for the resulting displacements and rotations $\{d\}$. After the equations are solved, the internal forces in every deformable element can be computed. It is the particulars of the geometry and the deformable-element properties that make the 6*N* equations representative of the human trunk.

Figure 41.5 illustrates a model of this kind that possesses 108 degrees of freedom. The model was constructed to examine the trunk deformations that might result from the contraction of each of the major trunk muscles. Table 41.5 provides a list of those muscles and gives typical data on their cross-sectional areas in different regions of the trunk. Table 41.6 shows some of

TABLE 41.6 Effects of body-segment weight application on the lumped-parameter, deformable-element spinal model of Fig. 41.5

Kyphosis increase, T5–T12, degrees:	
Nonrighted	17.5
Righted	6.2
Lordosis increase, T12–L5, degrees:	
Nonrighted	−11.2
Righted	12.2
Righting forces, N:	
Midline spinal erector slips:	
Low (T10–S1)	258
High (T1–T10)	93
Largest muscle contraction intensity, N/cm^2	20
Forces internal to the L5–S1 motion segment,* N	
Nonrighted:	
Disk compression	532
Disk anterior shear	110
Disk flexion moment, N · m	14.6
Articular facet spring forces	698
Righted:	
Disk compression	513
Disk anterior shear	171
Disk flexion moment, N · m	14.8
Articular facet spring forces	990

* These model-derived estimates need to be interpreted with caution. See Takashima et al. (1979).

the configurational and internal force data predicted when this model spine must support the weights of the head, neck, arms, and superior trunk segments, both when it responds passively to those loads and when the motor control systems act to restore an upright posture following weight application. Detailed information on models of this kind and the predictions obtained from them are given by Takashima et al. (1979).

41.3.6 Finite Element Models

Finite element techniques are widely used in engineering for full-field analyses of stresses and deformations in structures. The equations expressing the equilibrium, compatibility, and constitutive relations that the structures must satisfy frequently are too complicated to allow exact solution. Finite element techniques permit the efficient attainment of approximate solutions to those equations, and so make analyses of stress and deformation in these structures practical.

Finite element techniques have been used to study several constituent structures of the human trunk. Chen (1978) used them to study the behavior of the rib cage, and Hakim and King (1979) used them to analyze stress distributions in a vertebra and the effect of hemilaminectomy on those stress distributions.

For studies of the biomechanics of low-back disorders, the structure of most interest is the lumbar intervertebral disk. Finite element analyses of intervertebral disk behavior are useful to examine relationships among motion-segment gross displacements and rotations, disk bulge, intradiskal pressure, and stresses and strains in the annular wall and the vertebral body. The relationships between these quantities and the forces and moments to which the segments are subjected, their gross morphologies, and the mechanical properties of their constituent mate-

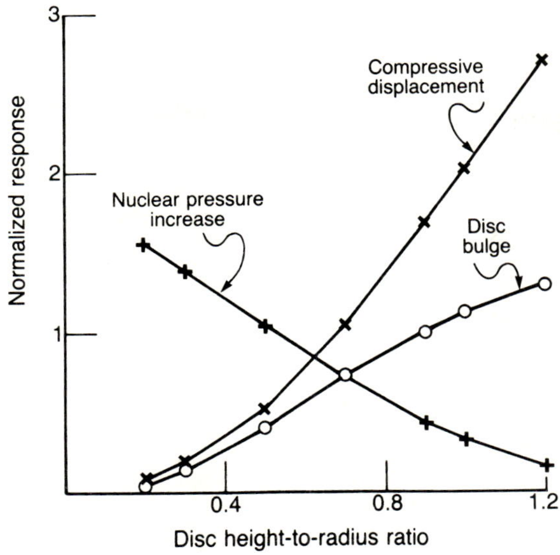

FIG. 41.6 Responses of an anisotropic, axisymmetric finite element model intravertebral disk to compressive load, as a function of disk geometry. Annular fiber angles are 40° to the transverse plane, and the modulus E_2, transverse to the annular fibers, is 9.5 MPa. Material constants were chosen to match experimental observations of behavior. Pressure increase is normalized to $F/\pi r^2$, and bulge and displacement are normalized to $F/\pi r E_2$, where F is applied load and r is disk radius. (*See Spilker and Schultz, 1983, for details.*)

rials, including possible effects on these properties of disk degeneration and nuclear hydration, can also be examined. It would be quite difficult to determine these relationships exclusively through a program of experimental measurements; structural analyses are certainly called for. Figure 41.6 illustrates results from an anisotropic, axisymmetric, finite element model analysis of the effect of intervertebral disk geometry on the compressive displacement, disk transverse bulge, and intradiskal pressure increase in response to a compressive load, as reported by Spilker and Schultz (1983).

CONCLUSION

The human spine and trunk are the sites of many kinds of diseases and injuries of the musculoskeletal system. The costs of these disorders are overwhelming. Mechanical factors are often centrally involved (see, for example, Schultz 1982). Biomechanical studies have begun to show in quantitative terms how those structures function mechanically, at least in healthy people. Without doubt, that growing understanding will be utilized in coming years to make significant advances in the diagnosis, treatment, and even the prevention of diseases and injuries of the spine and trunk.

REFERENCES

Agostoni, E., Mognoni, G., Torri, G., and Miserocchi, G. (1966). Forces deforming the rib cage. *Respir. Physiol., 2,* 105–117.

Andersson, G. B. J., Ortengren, R., and Schultz, A. (1980). Analysis and measurement of the loads on the lumbar spine during work at a table. *J. Biomech., 13,* 513–520.

Andersson, G., and Schultz, A. (1979a). Effects of fluid injection on mechanical properties of intervertebral discs. *J. Biomech., 12,* 453–458.

Andersson, G. B. J., and Schultz, A. B. (1979b). Transmission of moments across the elbow joint and the lumbar spine. *J. Biomech., 12,* 747–755.

Ashton, T., and Singh, M. (1973). The effect of training on maximal isometric back-lift strength and mean peak voltage of the erector spinae. *Biomechanics.* University Park Press, Baltimore, vol. 4.

Asmussen, E., Hansen, O., and Lammert, O. (1965). The relation between isometric and dynamic muscle strength in man. *Commun. Dan. Nat. Assoc. Infantile Paralysis, 20,* 3–12.

Asmussen, E., and Heeboll-Nielson, K. (1961). Isometric strength of adult men and women. *Commun. Dan. Nat. Assoc. Infantile Paralysis, 11,* 1–14.

Bakke, S. (1931). Roentgeonologische Beobachtunen uber die Bewegungen der Wirbelsaule. *Acta Radiol., 13.*

Bartelink, D. (1957). The role of abdominal pressure in relieving the pressure on the lumbar intervertebral discs. *J. Bone Jt. Surg., 39B,* 718–725.

Belytschko, T., Andriacchi, T., Schultz, A., and Galante, J. (1973). Analog studies of forces in the human spine: Computational techniques. *J. Biomech., 6,* 361–371.

Benson, D., Schultz, A., and DeWald, R. (1976). Roentgenographic evaluation of vertebral rotation. *J. Bone Jt. Surg., 58A,* 1125–1129.

Berkson, M. H., Nachemson, A., and Schultz, A. B. (1979). Mechanical properties of human lumbar spine motion segments: II. Responses in compression and shear; influence of gross morphology. *J. Biomech. Eng., 101,* 53–57.

Brown, R., Burstein, A., Nash, C., and Schock, C. (1976). Spinal analysis using a three-dimensional radiographic technique. *J. Biomech., 9,* 355–365.

Chaffin, D., and Baker, W. (1970). A biomechanical model for analysis of symmetric sagittal plane lifting. *AIEE Trans., II,* 16–27.

Chen, P. (1978). Finite element dynamic structural model of the human thorax for chest impact response and injury studies. *Aviat. Space Environ. Med., 49,* 143–149.

Clauser, C., McConville, J., and Young, J. (1969). Weight volume and center of mass of segments of the human body. AMRL-TR-69-70, Wright-Patterson Air Force Base, Dayton, Ohio.

Cossette, J., Farfan, H., Robertson, G., and Wells, R. (1971). The instantaneous center of rotation of the third lumbar intervertebral joint. *J. Biomech., 4,* 149–153.

Dantzig, G. B. (1968). *Linear Programming.* McGraw-Hill, New York.

Eie, N. (1966). Load experiments on autopsy specimens. *J. Oslo City Hosp., 16,* 75–98.

Eie, N., and Wehn, P. (1962). Measurement of the intra-abdominal pressure in relation to weight bearing of the lumbo-sacral spine. *J. Oslo City Hosp., 12,* 205–217.

Eycleshymer, A. C., and Schoemaker, D. M. (1911). *A Cross-Section Anatomy.* Appleton-Century-Crofts, New York.

Farfan, H., Cossette, J., Robertson, G., Wells, R., and Kraus, H. (1970). The effects of torsion on the lumbar intervertebral joint: The role of torsion in the production of disc degeneration. *J. Bone Jt. Surg., 52A,* 468–497.

Farfan, H., Huberdeau, R., and Dubow, H. (1972). Lumbar intervertebral disc degeneration. *J. Bone Jt. Surg., 54A,* 492–510.

Hakim, N., and King, A. (1979). A three-dimensional finite element dynamic response analysis of vertebra with experimental verification. *J. Biomech., 12,* 277–292.

Ikai, M., and Fukunaga, J. (1968). Calculation of muscle strength per unit cross-sectional area of human muscle by means of ultrasonic measurement. *Int. A. Angew Physiol., 26,* 26–32.

Ikai, M., and Steinhaus, A. (1961). Some factors modifying the expression of human strength. *J. Appl. Physiol., 16,* 157–163.

Jordanoglou, J. (1969). Rib movement in health, kyphoscoliosis, and ankylosing spondylitis. *Thorax, 24m,* 407–413.

Jorgensen, K., and Poulsen, E. (1974). Physiological problems in representative lifting with special reference to tolerance limits to the maximum lifting frequency. *Ergonomics, 17,* 31–29.

Lanier, R. (1939). The presacral vertebrae of American white and Negro males. *Am. J. Phys. Anthropol., 25,* 341–420.

Lin, H., Liu, Y., and Adams, K. (1978). Mechanical response of the lumbar intervertebral joint under physiological (complex) loading. *J. Bone Jt. Surg., 60A,* 41–55.

Liu, Y., Ray, G., and Hirsch, C. (1975). The resistance of the lumbar spine to direct shear. *Orthop. Clin. North Am., 6,* 33–49.

Markolf, K., and Morris, J. (1974). The structural components of the intervertebral disc: A study of their contributions to the ability of the disc to withstand compressive forces. *J. Bone Jt. Surg., 56A,* 675–687.

McNeill, T., Warwick, D., Andersson, G., and Schultz, A. (1980). Trunk strengths in attempted flexion, extension and lateral bending in healthy subjects and patients with low-back disorders. *Spine, 5,* 529–538.

Moll, J., and Wright, V. (1980). Measurement of spinal movement. In M. Jayson (ed.), *The Lumbar Spine and Back Pain.* Grune & Stratton, New York, pp. 157–183.

Morris, J. M., Lucas, D. B., and Bresler, R. (1981). Role of the trunk in stability of the spine. *J. Bone Jt. Surg., 43A,* 327–351.

Nachemson, A. (1960). Lumbar intradiscal pressure. *Acta. Orthop. Scand., 43,* 104.

Nachemson, A. L. (1981). Disc pressure measurements. *Spine, 6,* 93–97.

Nachemson, A., and Elfstrom, G. (1970). Intravital dynamic pressure measurements in lumbar discs. *Scand. J. Rehab., 1,* 5–40.

Nachemson, A., Schultz, A., and Berkson, M. (1979). Mechanical properties of human lumbar spine motion segments: Influence of age, sex, disc level, and degeneration. *Spine, 4,* 1–8.

Nahum, A., Gadd, C., Schneider, D., and Kroell, C. (1970). Deflections of the human thorax under sternal impact. *Int. Auto. Safety Conf.* Detroit and Brussels, pp. 797–807.

National Aeronautics and Space Administration. (1978). *Anthropometric Source Book.* Ref. Pub. 1024.

National Health Survey. (1973). Prevalence of selected imparments. U.S. Department of Health, Education, and Welfare Ser. 10, no. 87.

Ortengren, R., and Andersson, G. B. J. (1977). Electromyographic studies of trunk muscles, with special reference to the lumbar spine. *Spine, 2,* 44–52.

Ortengren, R., Andersson, G., Broman, H., Magnusson, R., and Petersen, I. (1975). Vocational electromyography: Studies of localized muscle fatigue at the assembly line. *Ergonomics, 18,* 157–174.

Panjabi, M., Krag, M., White, A., and Southwick, W. (1977). Effects of preload on load displacement curves of the lumbar spine. *Orthop. Clin. North Am., 8,* 181–193.

Panjabi, M., and White, A. (1971). A mathematical approach for three-dimensional analysis of the mechanics of the spine. *J. Biomech., 4,* 203–211.

Patrick, L., Kroell, C., and Mertz, H. (1965). Forces on the human body in simulated crashes. *Proc. 9th Stapp Car Crash Field Demon. Conf.,* University of Minnesota, Minneapolis, pp. 237–259.

Perey, O. (1957). Fracture of the vertebral end-plate in the lumbar spine. *Acta Orthop. Scand., 25.*

Portillo, D., Sinkora, G., McNeill, T., Spencer, D., and Schultz, A. (1982). Trunk strengths in structurally normal girls and girls with idiopathic scoliosis. *Spine, 7,* 193–196.

Poulsen, E., and Jorgensen, K. (1971). Back muscle strength, lifting and stooped working postures. *Ergonomics, 2,* 133–137.

Rab, G., and Chao, E. (1975). Muscle force analysis of the lumbar spine. *J. Bone Jt. Surg., 57A,* 582.

Roberts, S., and Chen, P. (1970). Elastostatic analysis of the human thoracic skeleton. *J. Biomech., 3,* 527–545.

Schultz, A. (1974). Mechanics of the human spine. *Appl. Mech. Rev., 27,* 1487–1497.

Schultz, A. B. (1982). Mechanical factors in the etiology of idiopathic low back disorders. *Symposium on Idiopathic Low Back Pain.* Mosby, St. Louis, pp. 201–219.

Schultz, A., and Andersson, G. (1981). Analysis of loads on the lumbar spine. *Spine, 6,* 76–82.

Schultz, A., Andersson, G., Haderspeck, K., Ortengren, R., Nordin, M., and Bjork, R. (1982a). Analysis and measurement of lumbar trunk loads in tasks involving bends and twists. *J. Biomech., 15,* 669–675.

Schultz, A. B., Andersson, G. B. J., Ortengren, R., Bjork, R., and Nordin, M. (1982b). Analysis and quantitative myoelectric measurements of loads on the lumbar spine when holding weights in standing postures. *Spine, 7,* 390–397.

Schultz, A. B., Andersson, G. B. J., Ortengren, R., Nachemson, A., and Haderspeck, K. (1982c). Loads on the lumbar spine: Validation of a biomechanical analysis by measurements of intradiscal pressures and myoelectric signals. *J. Bone Jt. Surg., 64A,* 713–720.

Schultz, A., Benson, D., and Hirsch, C. (1974a). Force-deformation properties of human ribs. *J. Biomech., 7,* 303–309.

Schultz, A., Benson, D., and Hirsch, C. (1974b). Force-deformation properties of human costo-sternal and costo-vertebral articulations. *J. Biomech., 7,* 311–318.

Schultz, A., Haderspeck, K., Warwick, D., and Portillo, D. (1983). Use of lumbar trunk muscles in idometric performance of mechanically-complex standing tasks. *J. Orthop. Res.* (in press).

Schultz, A. B., Warwick, D. N., Berkson, M. H., and Nachemson, A. L. (1979). Mechanical properties of human spine motion segments: I. Responses in flexion-extension, lateral bending, and torsion. *ASME J. Biomed. Eng., 101,* 46–52.

Seireg, A., and Arvikar, R. J. (1973). A mathematical model for evaluation of forces in lower extremities of musculo-skeletal system. *J. Biomech., 6,* 313–326.

Singh, M., and Ashton, T. E. J. (1970). Study of back-list strength with electrogoniometric analysis of hip angle. *Res. Q., 41,* 562–568.

Spilker, R. L., and Schultz, A. B. (1983). Effects of geometry and material properties on a composite model intervertebral disc. (To be published.)

Takashima, S. T., Singh, S. P., Haderspeck, K. A., and Schultz, A. B. (1979). A model for semi-quantitative studies of muscle actions. *J. Biomech., 12,* 929–939.

Tanz, S. (1953). Motion of the lumbar spine. *Am. J. Roentgenol., 69,* 399, 412.

Troup, J., and Chapman, A. (1969). The strength of the flexor and extensor muscles of the trunk. *J. Biomech., 2,* 46–62.

White, A. A., and Panjabi, M. M. (1978). *Clinical Biomechanics of the Spine.* Lippincott, New York.

Index

ABOUT THE EDITORS

Richard Skalak, Ph.D., is the Director of the Bioengineering Institute and the James Kip Finch Professor of Engineering Mechanics, Department of Civil Engineering and Engineering Mechanics, both at Columbia University, where he also received his degree. His numerous and various research projects have focused on solid and fluid mechanics problems in medical and biological fields. Dr. Skalak is Chairman of the U.S. National Committee for Biomechanics, President of both the Biomedical Engineering Society and the Society of Engineering Science, and the Editor of the *Journal of Biomechanical Engineering*.

Shu Chien, M.D., Ph.D., is Professor of Physiology and Cellular Biophysics and the Director of Division of Circulatory Physiology and Biophysics, both at Columbia University. His primary research interest focuses on the mechanisms of circulatory regulation in health and disease. His investigations have encompassed the whole spectrum of biological systems, and feature the use of an interdisciplinary approach involving biological experimentation and engineering modeling. Dr. Chien is the recipient of many research awards, including the Fåhraeus Medal and the Landis Award. He is Chairman of the Steering Committee of the American Society of Biorheology, a Councillor of the American Physiological Society, a Past President of the Microcirculatory Society, and an editor of *Clinical Hemorheology*.